Volcanism, Impacts, and Mass Extinctions: Causes and Effects

edited by

Gerta Keller
Department of Geosciences
Princeton University
Princeton, New Jersey 08544, USA

Andrew C. Kerr
School of Earth and Ocean Sciences
Cardiff University
Park Place
Cardiff CF10 3AT, Wales, UK

THE GEOLOGICAL SOCIETY OF AMERICA®

Special Paper 505

3300 Penrose Place, P.O. Box 9140 ▪ Boulder, Colorado 80301-9140, USA

2014

Published by The Geological Society of America, Inc.
3300 Penrose Place, P.O. Box 9140, Boulder, Colorado 80301-9140, USA
www.geosociety.org

Printed in U.S.A.

GSA Books Science Editors: Kent Condie and F. Edwin Harvey

Library of Congress Cataloging-in-Publication Data

Volcanism, impacts, and mass extinctions : causes and effects / edited by Gerta Keller and Andrew C. Kerr.
pages cm. — (Special paper ; 505)
Includes bibliographical references.
Summary: "Comprises articles stemming from the March 2013 international conference at London's Natural History Museum. Researchers across geological, geophysical, and biological disciplines present key results from research concerning the causes of mass extinction events"— Provided by publisher.
ISBN 978-0-8137-2505-5 (pbk.)
1. Mass extinctions. 2. Volcanism and climate. 3. Volcanism—Effect of environment on. I. Keller, Gerta, author. II. Kerr, Andrew C., 1968- author.
QE721.2.E97V65 2014
576.8'4—dc23

2014015429

Cover: (Front [top, left] to back [bottom, right] wrap) Illustration of a *Ceratosaurus*; all non-avian dinosaurs went extinct by the end of the Cretaceous. Image credit: DM7/Shutterstock. (Front, upper right) Photo of an active lava lake in the Overlook crater within the Halema'uma'u Crater and eruption center of the Kīlauea volcano, Hawai'i (22 October 2012). Photo credit: David Dow, U.S. Geological Survey, Hawaiian Volcano Observatory. It is illustrative of the massive flood basalt eruptions that are now believed to have caused five of the six major mass extinctions in Earth's history. Increasingly, scientists believe that the likely killing mechanism in all or most mass extinctions was the massive and rapid lava eruptions and near-continuous emissions of sulfur dioxide, carbon dioxide, and chlorine gases leading to rapid climate changes, acid rain on land, and ocean acidification that prevented calcium carbonate secretion by marine calcifiers leading to rapid extinctions. (Front, bottom) This artist's rendering of Earth from space is representative of the Chicxulub meteorite entering Earth's atmosphere, which then crashed into the Yucatan Peninsula (Mexico) ca. 66 Ma. Image credit: solarseven/Shutterstock. For over three decades, this meteorite has been believed to be the sole cause of the end-Cretaceous mass extinction and possibly of other mass extinctions. Decades of intensive research have produced no evidence of a major impact associated with any other mass extinction besides the end-Cretaceous, which also experienced one of the largest catastrophic volcanic eruptions.

10 9 8 7 6 5 4 3 2 1

Contents

Foreword

The nature and causes of mass extinctions in the geological past have been intensely debated for the past three decades. Central to this debate are the questions of whether one or several bolide impacts, the eruption of large igneous provinces, or a combination of the two were the primary mechanisms driving the environmental and habitat changes that are universally regarded as the proximate causes for four of the five major Phanerozoic mass extinction events. Considerable research efforts have been directed toward understanding the environmental changes associated with all major mass extinctions. But no topic has received as much attention as bolide impacts in general and in particular the end-Cretaceous mass extinction and the Chicxulub impact on Yucatan, Mexico, as the presumed cause. Between 1982 and 2010 the Geological Society of America alone has published at least 11 Special Papers (volumes 190, 247, 302, 307, 339, 347, 356, 384, 437, 458, and 465) devoted to impacts. Large igneous province volcanism, the other catastrophe related to mass extinctions, has scarcely been mentioned and just two Special Papers have been published on large igneous provinces at large (volumes 352 and 430).

Recent years have seen a revolution in our understanding of both the interplanetary environment and large igneous province eruptions and their environmental effects, such that the widely accepted simple impact-kill scenario no longer seems adequate for the end-Cretaceous or any other mass extinction events. Important new data and observations have been published that bear on the interpretation of these events from the fields of paleontology, stratigraphy, sedimentology, geochronology, geochemistry, mineralogy, volcanology, geophysics (notably paleomagnetism), astronomy, and astrophysics.

For example, planetary and space scientists have questioned single large impacts (e.g., Chicxulub) as the primary cause of mass extinctions because such impactors originating from the asteroid belt are random and rare. The breakup of giant comets thrown into Earth-crossing orbits may be more likely as there is a strong correlation between comet bombardment and mass extinctions. Just how asteroids and comets might affect the terrestrial environment is still under debate.

In Earth sciences, critical advances have been made in radiometric age dating, which closely ties the Viluy, Siberian, Central Atlantic magmatic province, and Deccan volcanism to the end-Devonian, end-Permian, end-Triassic, and end-Cretaceous mass extinctions, respectively. In India, massively large volcanic pulses occurring in quick succession that may result in runaway effects that could cause mass extinctions have been identified during a short time interval below the Cretaceous-Tertiary (K-T) boundary. Also in India, the mass extinction in planktic foraminifera, which suffered near total extinctions at the end of the Cretaceous, has been documented in intertrappean sediments between these massive volcanic eruptions, confirming the deadly nature of large igneous province volcanism.

Consequently, the body of scientific evidence from both large igneous province volcanism and planetary/space science has matured to the point that it requires a new critical review of hypotheses concerning the causes of mass extinctions as well as eco-evolutionary events before and after. To this end, an international, multi-disciplinary conference was held at The Natural History Museum in London from 27–29 March 2013, bringing together researchers from across the geological, geophysical, and biological disciplines to assess the state of research into the causes of mass extinction events and evaluate the respective roles of volcanism, bolide impacts, sea-level fluctuations, and associated climate and environmental changes in major episodes of species extinction.

Keller, G., and Kerr, A.C., 2014, Foreword, *in* Keller, G., and Kerr, A.C., eds., Volcanism, Impacts, and Mass Extinctions: Causes and Effects: Geological Society of America Special Paper 505, p. v–ix, doi:10.1130/2014.2505(000). For permission to copy, contact editing@geosociety.org.

The conference program was designed to cover many aspects of the five major mass extinction events and their hypothesized causes ranging from impacts and large igneous provinces to other environmental causes. This multi-disciplinary conference succeeded in bringing together 150 researchers, including seasoned experts as well as younger researchers and students in geology, geophysics, geochemistry, volcanology, sedimentology, paleontology, and astronomy. A new, collaborative, multi-disciplinary approach to resolving outstanding problems was explored through listening and learning from each other and by spirited, constructive discussion. The data and concepts presented and discussed at the meeting, and published in this Special Paper, have value well beyond the geosciences, particularly with regard to understanding modern environmental crises.

The main conclusions of this conference were:

1) Large igneous province volcanism, along with associated climate and environmental changes, is likely to have played a significant role in at least four of the five major mass extinctions in Earth history: the end-Cretaceous, end-Triassic, end-Permian (comprising two distinct extinction events), and end-Devonian (also two extinction events). However, the exact causal mechanisms by which 50–90% of the species preserved in the fossil record went extinct at each event remains to be determined. Better age control for individual lava units and extinction events is critical to establish the relationship between the causes and effects.

2) There was overwhelming agreement that a single large asteroid impact (Chicxulub) could not have been the sole cause of the end-Cretaceous mass extinction, but rather was a contributing factor. The long-term biological, environmental, and climatic changes before, at, and after the bolide impact horizon call for a multi-causal scenario, certainly involving volcanism and possibly multiple impacts or comet showers.

3) Participants gained an improved understanding of how large igneous province eruptions affect the biosphere. This included data and conclusions derived from atmospheric chemistry, geochronology of eruptions, associated mechanisms of climatic changes, and the direct effects on species-level extinctions.

4) Mass extinction patterns can tell us much about the age, tempo, and nature of the catastrophe and the type of environments that were most affected. We know most about the end-Cretaceous mass extinction. Similarly, detailed records are still needed for the other mass extinctions. If common patterns, or critical differences, between all mass extinctions are found, these can yield critical evidence for or against specific extinction mechanisms.

5) Ultimately, the effects of volcanism, impacts, sea-level and climate changes (warming and cooling), ocean acidification, ocean anoxia, and atmospheric changes have to be considered in any extinction scenario in order to understand the causes and consequences of mass extinctions. Moreover, these data hold the keys to help us understand, and cope with, the looming environmental and extinction crises in the modern world.

The 24 contributions in this Special Paper embody many of the salient conclusions of the meeting, and the papers are grouped into three broad themes: Records of mass extinctions, volcanism and mass extinctions, and impacts and mass extinctions. A brief summary of each paper is given below.

MASS EXTINCTIONS, VOLCANISM, AND IMPACTS

Norman MacLeod reviews the history of research into extinction events and discusses the nature and implications of the uncertainties inherent in mass extinction data, including the spatial and temporal resolutions of recorded data, variations in data collections, fluctuations in sediment accumulation rates, and presence of hiatuses. All of these factors render the temporal and geographic fossil records incomplete in many local stratigraphic sections and cores. He concludes that testing hypotheses of global extinction causality requires acknowledgment of the uncertainties inherent in extinction data.

David Bond and Paul Wignall review the temporal link between mass extinctions and large igneous provinces and suggest that the environmental damage can be controlled by a range of factors including continental configuration, the latitude, volume, rate, and duration of eruption, its style and setting (continental versus oceanic), the preexisting climate state, and the resilience of the extant biota to change. They conclude that the most significant causal factor in mass extinction events correlated with large igneous provinces is rapid gas (CO_2 and SO_2) release from single eruptive events and not the total volume of lava erupted in the province.

Gerta Keller reviews the evidence that directly links Deccan volcanism to the Cretaceous-Tertiary boundary mass extinction as recorded in planktic foraminifera, and evaluates the age of the Chicxulub impact based on the stratigraphically oldest impact spherule layer in Texas, northeast Mexico, and the impact crater on Yucatan. She demonstrates the coincidence between these two catastrophes with the rapid global warming in C29r near the onset of the main Deccan phase 2 (80% of total 3500-m-thick lava pile) about 100–150 k.y. before the K-T boundary, and correlates the mass extinction directly with the onset of the world's longest lava megaflows at the end of the Maastrichtian.

Jahnavi Punekar and others assess the environmental and biological effects of Deccan volcanism in a global survey of carbon and oxygen isotopes and blooms of opportunistic planktic foraminifera. They demonstrate that the global high-stress environment before the K-T boundary is coeval with the main Deccan volcanic phase 2 (80% of total lava pile) in C29r, which ended with the mass extinction. They further document that the last Deccan phase 3 (14% of total lava pile) correlates with the delayed Danian recovery in C29n. They conclude that Deccan volcanism can account for both the mass extinction and delayed marine recovery.

Robert Spicer and Margaret Collinson review the plant fossil record and floral changes during the Maastrichtian and across the K-T boundary with special emphasis on North America and New Zealand. They note that the floral turnover and extinction patterns varied considerably with rapid climate changes. They conclude that the extinctions cannot be solely ascribed to the Chicxulub impact, but appear to be a continuation of environmental instability, possibly driven by large-scale volcanism that began well before the arrival of the meteorite.

Iain Gilmour and others investigate the sediments that infilled the Boltysh meteorite impact crater in Ukraine based on carbon isotopes and palynological data. They determined a major negative carbon isotope excursion in the early Danian about 200 k.y. after the K-T boundary, which they correlate with the global Dan-C2 excursion of the same age. They conclude that this hyperthermal event had a similar short duration as the Paleocene-Eocene Thermal maximum.

Sergei Vikulin and others present plant fossil leaf data from the Paleocene oil shales of the Boltysh crater (Ukraine) and assess the implications for reconstruction of post-impact vegetation and paleoecology. The presence of the Mesozoic fern *Weichselia reticulata* suggests survival in the refugial geothermal ecosystem of Boltysh into the early Paleogene, and as such represents the youngest record of this fern. They conclude that the local survival of this fern is likely due the unique combination of edaphic environmental factors in the Boltysh hydrothermal area.

Bandana Samant and Dhananjay Mohabey report results of palynological studies on the infratrappean and multiple intertrappean sedimentary beds within the Deccan Traps. They observe a distinct floral turnover in the sediments with the initiation of volcanic activity, and suggest that this floral change, and striking decline in the diversity of dinosaurs, is strongly linked to the initiation of the main phase 2 of Deccan volcanism. They conclude that Deccan volcanism, during its peak period of eruption, played a significant role in the global mass extinction at the K-T boundary.

Guntupalli Prasad and Ashok Sahni review the vertebrate fossil record from the infra- and intertrappean beds within the Deccan volcanic province. They note a decline in dinosaur diversity as a result of the early phase of volcanism and that the freshwater/terrestrial vertebrate faunas were least affected. This survival is attributed to the episodic nature of Deccan volcanism and its periods of quiescence. They conclude that any impact on life due to Deccan volcanism occurred during the main phase 2 when 80% of the total volume erupted.

David Archibald reviews the dinosaur record and notes that over the past 10 m.y. of the Cretaceous in the western interior of North America, nonavian dinosaur species dropped by almost 50%. These losses are attributed in part to the last major regressive cycle of the Cretaceous that reduced the low coastal plains inhabited by nonavian dinosaurs. He concludes that although Deccan volcanism was likely the major driver of climatic change that resulted in decreased dinosaur diversity, the cause for their final extinction remains unresolved; however, the Chicxulub impact undoubtedly played a major role.

Andrew Caruthers and others compare paleontological and geochemical data from North America with published correlative data in Europe to assess the nature of the Early Jurassic Pliensbachian–Toarcian anoxic event. Recently, the timing and geographic extent of this event have been questioned with more emphasis placed on regional conditions in the Tethys rather than volcanogenic outgassing of CO_2 during Karoo-Ferrar large igneous province eruptions causing global warming and an extinction event. However, the authors' data indicate that this environmental crisis was concurrent with Karoo magmatism and strongly suggest a cause-effect relationship.

József Pálfy and Ádám Kocsis review Central Atlantic magmatic province volcanism as a possible cause for the environmental and biotic changes leading to the end-Triassic mass extinction. Significantly, doubts about the temporal coincidence are removed by more precise radiometric dates and new biostratigraphic data that confirm the brevity of these changes. They conclude that Central Atlantic magmatic province volcanism, particularly rapid climate change due to CO_2 and SO_2 emissions, is a viable driver of Earth system changes across the Triassic-Jurassic boundary.

Grzegorz Pieńkowski and others distinguish eight climatic events in the Triassic–Jurassic continental sediments of Poland and assess their effect on the end-Triassic mass extinction. The first climatic change predated the Central Atlantic magmatic province flood basalts by 100–200 k.y. and was followed by five further climatic fluctuations up to the Triassic-Jurassic boundary. Periodic atmospheric loading by CO_2, CH_4, or by SO_2, sulfate aerosols, and toxic compounds is inferred to have caused this series of rapid climatic reversals and resulting extinctions. The appearance of highly diversified tetrapod ichnofauna in the earliest Jurassic strata indicates a rapid recovery and refill of ecological niches by dinosaurs.

Howard Armstrong and David Harper review the end-Ordovician mass extinction, which is unusual in that it is associated with glaciation. They note that there is limited understanding of how glaciation directly causes mass extinctions. They further argue that extinctions occurred at the onset and termination of glaciation and were due to climatically induced changes in sea level, ocean redox, and sea-surface temperatures, which resulted in a reduction in marine niche space. This in turn led to increased competition and selection pressures, resulting in extinctions where the carrying capacities of particular ecological niches were exceeded.

VOLCANISM AND ENVIRONMENTAL EFFECTS

Vincent Courtillot and Frédéric Fluteau review the time scales of continental flood basalt volcanism with a focus on dramatically short, large volcanic pulses occurring in quick succession, which may result in runaway effects that could cause mass extinctions. Flow-by-flow magnetic stratigraphy has identified many single eruptive events with a volume larger than 1000 km^3 and some in excess of 10,000 km^3 emplaced in possibly less than a decade. The authors hypothesize that the temporal sequence of lava extrusions determines the severity of extinctions. These findings have significant implications for the role of flood basalts in all major mass extinction events.

Steve Self and others summarize current knowledge of continental flood basalt (CFB) eruptions and emplacement characteristics to define a set of eruption source parameters of magmatic volatiles that could be used as inputs for Earth system modeling studies. They caution that the limited knowledge of the number and length of non-eruptive phases (hiatuses) during the continental flood basalt is a key unknown parameter that is critical for better constraining the severity and duration of environmental effects caused by continental flood basalt eruptions.

Mickaël Mussard and others evaluate climate changes caused by continental flood basalts (e.g., CO_2, SO_2 release) based on a biogeochemical model coupled to a climate model. They find that CO_2-related warming is enhanced when large-scale SO_2 injections are added. Their results show that the succession of drastic cooling events induced by sulfate aerosols decreases the efficiency of silicate weathering and destabilizes the carbon cycle during the full time span of trap emplacement. They conclude that SO_2-related climate feedback may have enhanced the long-term warming due to CO_2 emissions.

Eric Font and others investigate the low magnetic susceptibility (MS) interval just below the K-T boundary at Bidart, France. Results show a decrease in detrital magnetite and presence of rare akaganeite, which they hypothesize is a product formed in aerosols derived from the reaction of a volcanic plume with water and oxygen in the high atmosphere. Modelling the consequences of acidic rains on a continental regolith shows nearly complete magnetite dissolution over about 30 ka. They conclude that the age and duration of the MS interval falls within Deccan phase 2 and highlights the importance of volcanically related environmental changes leading up to the K-T boundary mass extinction.

IMPACTS AND ENVIRONMENTAL EFFECTS

Adrian Jones reviews impact flux and mass extinction events and suggests that impacts are much more deadly than conventional volcanism due to the rapid transfer of energy. Based on impact flux rates, he notes there could have been sufficient large impactors to cause one or more of the "big five" mass extinctions in the past 300 Ma. He concludes that the best contender to date is the combination of the Chicxulub impact and Deccan volcanism (though not triggered by the impact as they are not antipodal) for the end-Cretaceous mass extinction, whereas the possible contributions of multiple impacts are still uncertain.

William Napier argues that the asteroid belt is an inadequate source of large terrestrial impactors (100–300 km across) and that these are far more likely to derive from comets orbiting between Jupiter and Neptune. The breakup of these comets results in prodigious dust influx into the stratosphere, and repeated fireball storms that may destroy stratospheric ozone and enhance incident UV. He concludes that these storms may be major contributors to biological trauma with the potential to create mass extinctions by way of prolonged stress. He observes a strong correlation between comet bombardment episodes and mass extinctions of marine genera.

Duncan Steel reviews the discovery over the past two decades of a large number of substantial objects in the outer solar system. He calculates the impact probability with Earth for all known asteroids, and discusses why the old concept of single, random, large asteroid impacts causing mass extinctions is deficient. He further assesses how often giant comets might be thrown directly into Earth-crossing orbits and implications for terrestrial catastrophism. Significantly, he notes that there is a wide disparity of ideas among planetary and space scientists as to how asteroids and comets might affect the terrestrial environment.

Dallas Abbott and others (Chapter 22) use the occurrence of unusual, or out-of-season, dust storms and dissolved ion data as proxies for dust to propose a calendar-year chronology for a portion of the GISP2 ice core during the early sixth century A.D. Their new time scale is consistent with a small volcanic input to the A.D. 536–537 climate downturn. The time range of Ni-rich fragments and cosmic spherules matches historical observations of "dancing stars" starting in the summer of A.D. 533 and lasting until A.D. 539 or 540. These dancing stars have been previously attributed to cosmogenic dust loading of the Earth's atmosphere.

Dallas Abbott and others (Chapter 23) propose that Sn- and Ni-rich particles and cosmic spherules found in the GISP2 ice core indicate increased atmospheric dust loading between A.D. 533 and 540. The Sn enrichments suggest a cometary source for the dust and could explain the climate downturn in A.D. 533. They conclude that significant global dimming during A.D. 536 and 537 cannot be explained solely by a combination of cometary dust and volcanic ash. A low latitude explosion in the ocean (A.D. 535–536) is postulated to explain both additional dust and the presence of tropical marine microfossils. They suggest that these results imply that mass extinction events may also have multiple volcanic and cosmogenic drivers.

Steve Miller assesses the role the media play in the mass extinction debates and, in particular, the public impact of meteor impact scenarios. Significantly, he notes that the original paper by Alvarez et al. on the death of the dinosaurs attracted limited media attention at the time (1980). Miller's analysis of media coverage from 1980 to 2008 reveals that the period from 1991 to 1995 was critical in terms of changing public perceptions. He argues that the collision of Comet Shoemaker-Levy 9 with Jupiter played a fundamental role in propelling the impact scenario for the death of the dinosaurs into the public eye.

Acknowledgments

In addition to the hosting and organizational support of the conference by the Natural History Museum (in particular Norm MacLeod, the local organizer) and the Mineralogical Society of Great Britain and Ireland, financial donations to offset conference expenses were gratefully received from the Solid Earth Composition and Evolution Working Group of the International Mineralogical Association, the Volcanic and Magmatic Studies Group (VMSG) of the UK, the Society for Sedimentary Geology (SEPM), USA, and Earth Sciences Institute, ISTE, GEOPOLIS, University of Lausanne, Switzerland. The Department of Geosciences, Princeton University, USA, provided website and secretarial support.

Gerta Keller
Andrew C. Kerr

The Geological Society of America
Special Paper 505
2014

The geological extinction record: History, data, biases, and testing

Norman MacLeod*
The Natural History Museum, Cromwell Road, London SW7 5BD, UK

ABSTRACT

The geological record represents the only source of data available for documenting long-term historical patterns of extinction intensity and extinction susceptibility. Such data are critical for testing hypotheses of extinction causality in the modern world as well as in deep time. The study of extinction is relatively new. Prior to 1800, extinctions were not accepted as a feature of the natural environment. Even after extinctions were recognized to have occurred in Earth's geological past, they were deemed to have played a minor role in mediating evolutionary processes until the 1950s. Global extinction events are now recognized as having been a recurring feature of the history of life and to have played an important role in promoting biotic diversification. Interpretation of the geological extinction record is rendered complex as a result of several biasing factors that have to do with the spatial and temporal resolutions at which the data used to study extinctions have been recorded: fluctuations in sediment accumulation rates, the presence of hiatuses in the stratigraphic sections/cores from which fossils are collected, and variation in the volumes of sediments that can be searched for fossils of different ages. The action of these factors conspires to render the temporal and geographic records of fossil occurrences incomplete in many local stratigraphic sections and cores. In some cases, these stratigraphic and sampling uncertainties can be quantified and taken into account in interpretations of that record. However, their effects can never be eliminated entirely. Testing hypotheses of global extinction causality requires acknowledgment of the uncertainties inherent in extinction data, the search for unique predictions of historical patterns of variation or associations that can, in principle, be preserved in the fossil record and tied logically to the operation of specific causal processes, and to adoption of an explicitly comparative approach that establishes the presence of multiple instances of the predicted cause-effect couplets within a well-documented chronostratigraphic context.

*N.MacLeod@nhm.ac.uk

MacLeod, N., 2014, The geological extinction record: History, data, biases, and testing, *in* Keller, G., and Kerr, A.C., eds., Volcanism, Impacts, and Mass Extinctions: Causes and Effects: Geological Society of America Special Paper 505, p. 1–28, doi:10.1130/2014.2505(01). For permission to copy, contact editing@geosociety.org.

INTRODUCTION

For the past 30 years, a significant proportion of the scientific community has been intensely interested in the idea of extinctions, especially the mass extinction of the dinosaurs at or close to the boundary between the Cretaceous (K) and Paleogene (Pg) intervals of Earth history. This interest predates the current concern with the so-called sixth (mass) extinction, a potential event that may occur in the future and that takes its name from the "Big Five" ancient mass extinction events of the fossil record (see following). The reasons for this interest are many and varied, but they share a common source. The concept of extinction elicits a deep emotional reaction in most people—including scientists—to no small extent because we all share an intuitive concern about the transformations being wrought in the contemporary environment. When we see changes taking place in landscapes, animals, and plants at local, regional, and global scales, we cannot help but experience a sense of foreboding that comes from drawing obvious parallels between the status of own species and the fates of other, far more ancient, species that "ruled the Earth" in the distant past. In this review, I (1) recount the history of extinction studies, (2) describe the types of paleontological data used to study ancient extinction events, (3) discuss the uncertainties and biases that affect the interpretation of these data along with some of the methods available to quantify them, (4) describe three examples of data collected from the end-Cretaceous extinction interval to illustrate points (2), (3), and (4), and (5) offer a perspective on the manner in which hypotheses that seek to address the causes of ancient extinction events can be tested in the paleontological record.

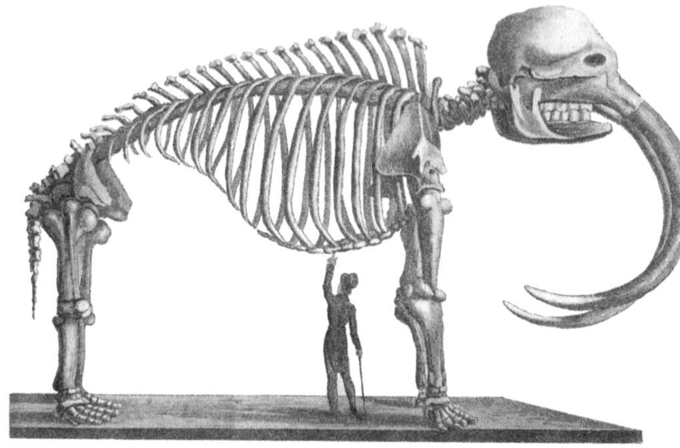

Figure 1. Edouard de Montule's 1816 painting of Charles Wilson Peale's reconstruction of the *American incognitum*. Specimens like Peale's supplied critical evidence for Cuvier's reinterpretation of the fossil as an extinct probiscidean—one of the first animals to be recognized as extinct by the scientific community. Note the tusks in Peale's original reconstruction were oriented upside down, a mistake that was later corrected when additional specimens came to light. Image used with permission from the Rare Book Room, Buffalo & Erie County Public Library.

History of Extinction Studies: 1700–1950

Prior to the late 1700s, the western social and scientific consensus held that extinction was impossible. Since all creatures had been created by God, and since God's creation was perfect, none of the species He fashioned could disappear. Nonetheless, as the new science of paleontology began bringing its first major discoveries to light, an increasingly large number of researchers began to wonder whether the large vertebrate fossils of strange creatures that were then being unearthed by Enlightenment explorers were indeed the remains of extinct species.

The most famous of these fossils was recovered from a site in the territory that would later become the American state of Ohio. In 1739, French soldiers under the command of Baron Charles le Moyne de Lougueuil (1656–1729) recovered a tusk, femur, and three curious molar teeth from an unknown locality, possibly the site that would later become known as Big Bone Lick (Hedeen, 2008). Lougueuil sent these specimens to the Cabinet du Roi (Royal Cabinet of Curiosities) in Paris. Subsequent study of this material by the naturalists Jean-Étienne Guettard (1715–1786) in 1756 and Louis-Jean-Marie Daubenton (1716–1799) in 1762 led them to conclude that the tusk and femur belonged to a modern elephant, the "Siberian Mammoth," but that the molars belonged to a giant hippopotamus.

Additional piecemeal specimens recovered from the Ohio country after the 1739 find made their way to London and to Philadelphia, where they were studied by naturalists from Britain and the United States with inconclusive results. These finds did manage, however, to throw doubt on Daubenton's hippopotamus interpretation because the tusks, long bones, and teeth clearly belonged to the same animal. Nevertheless, the molars of the "*American incognitum*"—as the animal came to be known—were unlike those of any living elephant.

After inspecting the London specimens, the British anatomist William Hunter interpreted them as belonging to a giant carnivore and was the first to speculate that these remains might be from an extinct species. Then, in 1799, discovery of an *American incognitum* femur from Quaternary deposits in the Hudson River Valley led to excavations organized by Charles Wilson Peale (1741–1827). By 1801, these excavations had resulted in the recovery of an almost complete skeleton. Peale reconstructed the skeleton with help from the American anatomist Caspar Wistar (1761–1818), cast the bones that were missing, and displayed the mounted skeleton in public in December of that year (Fig. 1). It is not an exaggeration to say that pandemonium over Peale's skeleton quickly swept the United States and later the world.

Based on drawings of the mounted American specimen, the French anatomist Georges Cuvier (1769–1832; Fig. 2) resolved the *American incognitum* controversy in 1806 by demonstrating that both the Siberian mammoth and the "animal de l'Ohio" were elephants, but of different species. Cuvier argued that these species were both distinct from modern elephants, of which there

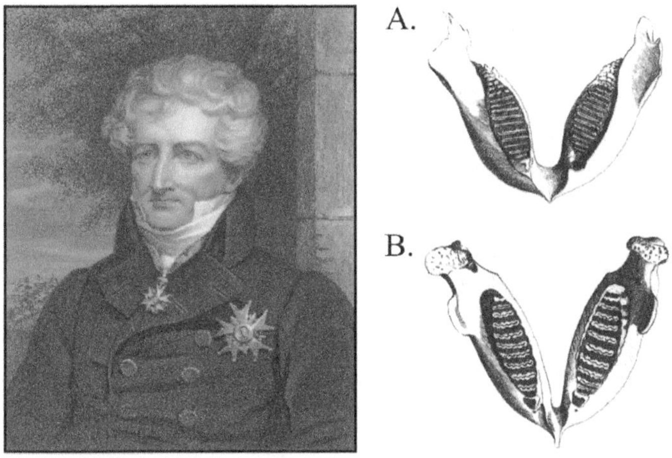

Figure 2. James Thompson's ca. 1800 engraving of Baron Georges Cuvier (1769–1832), describer of the first extinct species and originator of the concept of mass extinction. Right: Illustration of the mandible and molar teeth of a mammoth (A) and an Indian elephant (B) that appeared in Cuvier's 1799 monograph on living and fossil elephants. Sources: Wikimedia Commons.

were two species.[1] Cuvier went on to suggest that the Ohio elephant, which he described as a "mastodon" on account of the rounded, conical shape of its molar cusps, was too large to have been overlooked by naturalists on various voyages and expeditions of discovery in the 1700s and so probably represented an extinct species. Such was Cuvier's stature in the scientific community that this pronouncement is widely regarded as the first convincing evidence for the existence of an extinct species.

Georges Cuvier also formulated the concept of mass extinction in the early 1800s after noting the juxtaposition of fossils of unusual horse-like and dog-like creatures with seashells in several localities within the Paris Basin. Along with his collaborator Alexandre Brongniart (1770–1847), Cuvier interpreted these associations as indicating an environmental catastrophe had taken place in Earth's distant past, the scale and intensity of which were unprecedented in human history. This event affected both terrestrial and shallow marine environments together, sweeping up organisms that lived in separate habitats indiscriminately and depositing them in assemblages that could not be considered natural. Cuvier referred to such events as "revolutions," drawing on the most obvious political parallel of his own day: the French Revolution. Originally in a 1810 report on his Paris Basin stratigraphic investigations, and from 1812 onwards in his monograph on fossil fish (see Rudwick, 1997), Cuvier expanded and developed this concept of revolutionary extinctions that resulted in whole groups of organisms being annihilated suddenly such that only a few relict survivors were left.

The first diagram of the biodiversity, or richness of life, through geological time was published by John Phillips (1800–1874) in 1860. This diagram (Fig. 3), which was based on data

[1]To that point the African and Indian elephants had been grouped into a single species.

Figure 3. John Phillips' 1860 diagram of Phanerozoic biodiversity (left). A photograph of John Phillips taken in 1860 (right) (Source: Wikimedia Commons).

collected primarily from UK stratigraphic sections, shows life's history as having a broadly ascending pattern that can be divided into three phases by apparent diversity radiations, each of which is separated by distinct intervals in which biodiversity declined, one at the close of the Paleozoic and one at the close of the Mesozoic time intervals. Phillips also proposed a smaller and less well-defined drop in richness as characterizing the middle Paleozoic. Though Phillips' biodiversity diagram was used in his own day to advocate acceptance of the Paleozoic, Mesozoic, and Cænozoic (= Cenozoic) as useful subdivisions of geological time, it took more than a century for paleontologists to become seriously interested in the study of the extinction patterns evident therein. This delay reflects the fact that many scientists of Phillips' time regarded the data at their disposal as being insufficient to portray the major outlines of life's diversity history correctly.

In contrast to Cuvier, Charles Darwin (1809–1882; Fig. 4) viewed extinction as a gradual process that comes about as a result of natural selection. In particular, Darwin opposed the idea that large numbers of extinctions were caused directly by physical processes in the form of natural catastrophes. This was in keeping with the theory of gradualism in natural processes to which many of Darwin's philosophical and scientific colleagues subscribed (e.g., the geologist Charles Lyell, 1797–1875) and that many of those who opposed Darwin's theory of natural selection rejected (e.g., Sir Richard Owen, 1804–1892). As a result, the topic of extinction does not figure greatly in Darwin's seminal book on evolution, *On the Origin of Species by Means of Natural Selection, or the Preservation of Favored Races in the Struggle for Life* (1859), or, indeed, in any of Darwin's writings.

Owing to the popularity of Darwin's arguments in favor of evolution as a natural process, and his theory of natural selection as the means by which evolution comes about, his views on species extinction went on to become scientific orthodoxy throughout the late 1800s and early 1900s. By the 1940s and 1950s, however, the scientific community's view of extinction had drifted from

A. B.

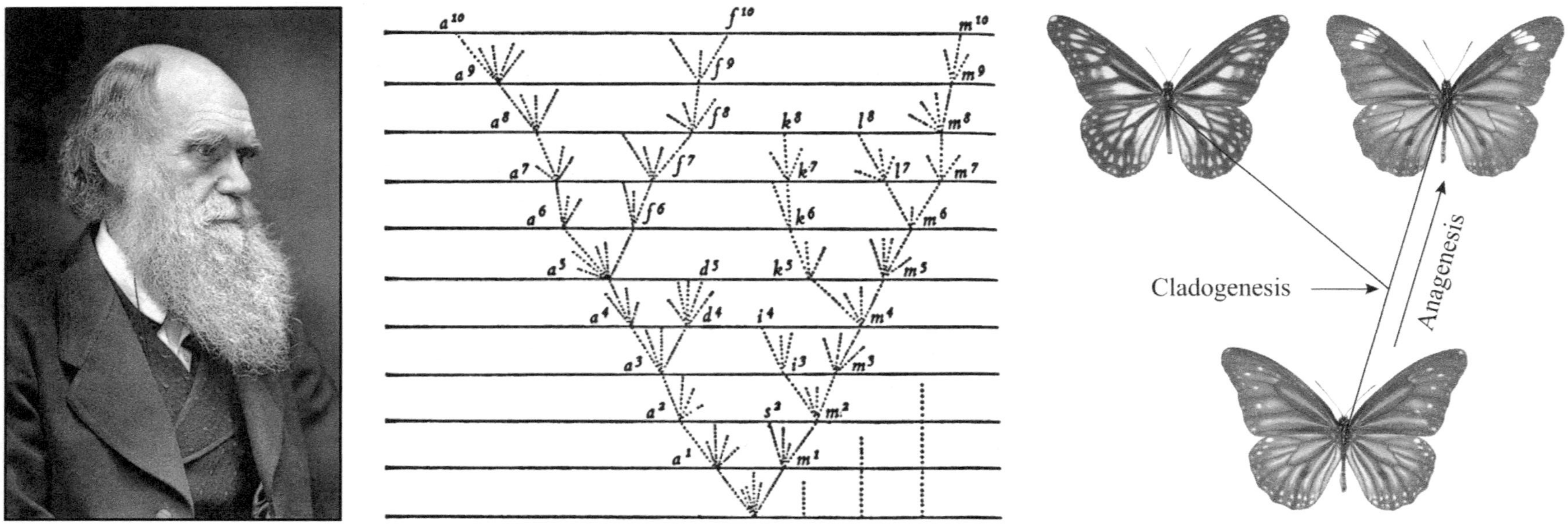

Figure 4. Charles Darwin (1809–1882), naturalist originator of the theory of natural selection (left). Image of Darwin used with permission; © The Trustees of the Natural History Museum, London. (A) Darwin's graphical depiction of the patterns of morphological change predicted as a result of his theory of natural selection. (B) An illustration of two different modes of evolutionary change: anagenesis, which involves the long-term change of an entire species in response to environmental change with no disruption to gene flow, and cladogenesis, which involves a short-term branching event from one or more ancestral populations of a species (usually accompanied by a disruption to gene flow) after which the descendent species sets off on an independent evolutionary trajectory.

regarding extinction as a by-product of the struggle for existence to being quite a passive process in which the ecological roles vacated by extinct species might lay fallow for (literally) tens of millions of years before natural selection was able to fashion a new occupant for the former role. For example, in his book *Tempo and Mode of Evolution* (1944), the influential vertebrate paleontologist George Gaylord Simpson (1902–1984) pointed to the fact that, according to the fossil record, millions of years separated vacation of the marine, air-breathing predator role once occupied by ichthyosaurs from its reoccupation by toothed whales. Similarly, tens of millions of years passed between the demise of the dinosaurs as the dominant large terrestrial vertebrates and the appearance of a similarly diverse assemblage of large mammals. On the whole Simpson (1944, 1953) and others of his time regarded extinction as playing even less important a role in evolution than that envisioned by Darwin. One could—and many across the broad spectrum of western culture did—lament the passing of such magnificent creatures as dinosaurs, pterodactyls, ichthyosaurs, plesiosaurus, saber-toothed cats, mammoths, and mastodons from the ecological scene, but informed scientific opinion throughout the 1930s, 1940s, and 1950s held that extinctions were incidental by-products of normal natural selection and that the biosphere seemed to take its time replacing groups that became extinct.

The modern view of extinction is decidedly different. Beginning with the publications of Otto Schindewolf (1955, 1962) in Germany and Norman Newell (1963, 1967) in the United States during the 1960s, and continuing throughout the 1980s, catastrophism—the idea that Earth has been repeatedly subject to massive environmental convulsions—was rehabilitated as a demonstrable feature of the natural environment. Previous generations of earth scientists, including Darwin and Simpson, had rejected nineteenth-century catastrophist theory because of its appeal to vague and mysterious processes that were, by definition, unavailable for scientific study.[2] However, by the 1960s, paleontological evidence for the coordinated loss of major animal groups over relatively short intervals of geological time had become so overwhelming that Lyell's and Darwin's suggestions that these patterns could be accounted for by imperfections in the fossil record were no longer credible. Similarly, by the 1970s, the physical evidence for massive volcanic eruptions (much larger than had ever occurred in human history), astonishing changes in sea level, and even whole continents drifting across the surface of Earth, all having occurred within surprisingly short geological time intervals, became too compelling to ignore in terms of their implications for the history of life.

History of Extinction Studies: 1950–2014

Otto Schindewolf (1896–1971) was a highly respected, though decidedly iconoclastic, figure in mid-twentieth-century paleontology. Drawing from his training in German typological taxonomy, Schindewolf emphasized the consistency of the type form each species displays irrespective of the character of static and/or developmental variation that distinguishes individuals from one another. This observation led him to postulate a saltational theory of macroevolution in which major transitions in form are accomplished suddenly and driven by factors internal to the organism. Schindewolf proposed this model of morphological and genetic fixity in opposition to Darwin's model of environmentally mediated natural selection as being the only force directing evolutionary change (Schindewolf, 1950). In this same publication, Schindewolf cited the sudden appearance of many morphologically unique species in the fossil record as evidence for his saltational view of macroevolution and the equally sudden disappearance of many well-established species at the close of the Paleozoic and Mesozoic as the result of overspecialization in the face of sudden and externally forced changes in Earth's environment. Later, Schindewolf (1954, 1963) argued that the lack of any unique changes in the character of the end-Permian and end-Cretaceous sediments associated with these mass extinction events in Earth history implicated extraterrestrial forces of great intensity but short duration, and that left no fossilizable trace (e.g., cosmic radiation from a nearby supernova), as the ultimate cause of these extinction events.

In contrast to Schindewolf's anti-Darwinian stance, Norman Newell (1919–2005) was considered a more conservative member of both the U.S. and global paleontological establishments. Also, rather than relying on anecdote, rhetoric, speculation, and appeals to authority as Schindewolf had, Newell supported his arguments on the nature of extinction through data derived from the biostratigraphic ranges of taxonomic families. These formed the basis for his innovative empirical and quantitative approach to the study of biodiversity changes through time (Newell, 1963, 1967)—an approach that still forms an important part of extinction research programs to this day.

Newell chose the family level of taxonomic resolution for his studies for purely pragmatic reasons. Even in the 1960s, published works describing fossil species' distributions in space and time were too numerous and scattered among too many journals, monographs, books, and technical reports issued worldwide for any one person to be able to assemble them into a single data set. Periodically, specialists in various fossil groups issued summaries of the literature for their group from which patterns of change in the number of fossil families and genera over time could be derived. Newell used these data as proxies for species-level patterns. Also, Newell was concerned about the comparability of species, and even genus, concepts that were then being employed by taxonomists as the characteristics used to define lower levels of taxonomic organization varied widely. The feeling at the time was that the family level of taxonomic resolution was (1) reasonably comparable across different fossil animal groups, (2) resulted in patterns of change over time that varied markedly between groups, and (3) involved a sufficiently large number of groups to comprise an accurate statistical summary of the history of biodiversity changes for the biosphere as a whole.

[2]Catastrophism was also rejected for political (association with the French revolution) and religious (association with biblical fundamentalism) reasons.

The level of stratigraphic, or temporal, resolution of these data is also important to appreciate. As with taxonomic resolution, the biostratigraphic literature detailing fossil species' distribution in time was, by 1960, already too large and too scattered to be able to be gathered together and summarized by any one person. The same specialist-written summaries Newell used for his family-level taxonomic data provided him with a means whereby he could infer the temporal distribution of these family-level groups. However, here he ran into the same comparability problem. The biostratigraphic units used to describe these distributions varied from group to group, and the manner in which these biozone-level stratigraphies of different animal groups were correlated with each other was by no means clear. The most consistently used, common, chronostratigraphic time unit employed to map these biostratigraphic units onto the geological time scale was that of the chronostratigraphical series or stage. Accordingly, Newell referenced his biodiversity summaries to the stage-level time scale.

When Newell plotted stratigraphic range data for 2500 animal families collected into stage-level time bins, it was immediately clear that, while the biodiversity history for each major organismal group was unique, cross-group commonalities were also apparent. Since only a third of this total included extant representatives, Newell concluded that the fate of most families—and so most species—was to become extinct. In addition, and as had been noted by Phillips previously, Newell recognized that, on average, the total number of families comprising the fossil record increased over time (Fig. 5A). This meant that the rate at which new families—and so new species—appeared in the fossil record was more than sufficient to compensate for those that disappeared in successive extinction events. However, an even more intriguing observation from the perspective of extinction studies was the fact that certain times in Earth history were characterized by far higher rates of family—and so species—disappearances than appearances and that these intermittent reductions in biodiversity often occurred during the same time intervals in markedly different groups. This suggested a broad synchroneity of response to some perturbing factor, or set of factors, the effect(s) of which crossed both taxonomic and ecological boundaries. When Newell summed all his data together and expressed them in terms of percent extinction intensity (Fig. 5B), six time intervals stood out from the background of historical variation as intervals of extraordinarily high rates of family-level extinctions: the end-Cambrian, end-Ordovician, end-Devonian, end-Permian, end-Triassic, and end-Cretaceous.

Newell (1963) referred to these intervals of increased extinction intensity as "mass extinctions," and he was the first to note some of the differences between these events that have since passed into common (scientific) knowledge. For example, Newell correctly identified the end-Permian extinction as the largest of these "big six" extinction events, as well as noting that this event involved the loss of both terrestrial and marine families. He also noted that a decline in many marine groups was discernible more than 30 m.y. before the end-Cretaceous event and that benthic marine organisms (e.g., benthic foraminifera), fishes, nautiloid cephalopods, mammals, turtles, and crocodiles survived its effects with only minor evolutionary modifications.

After reviewing the broad range of then-current explanations for mass extinctions, Newell came down on the side of sea-level fluctuations as the most likely general cause of mass extinction, noting that changes in sea level would not only have profound effects on both marine and terrestrial ecosystems, but would also be associated with substantial climate change. The failure of substantial extinctions to be associated with the Pleistocene glacial advances had been—and still continues to be—cited as proof that sea-level change per se cannot be invoked as a cause for

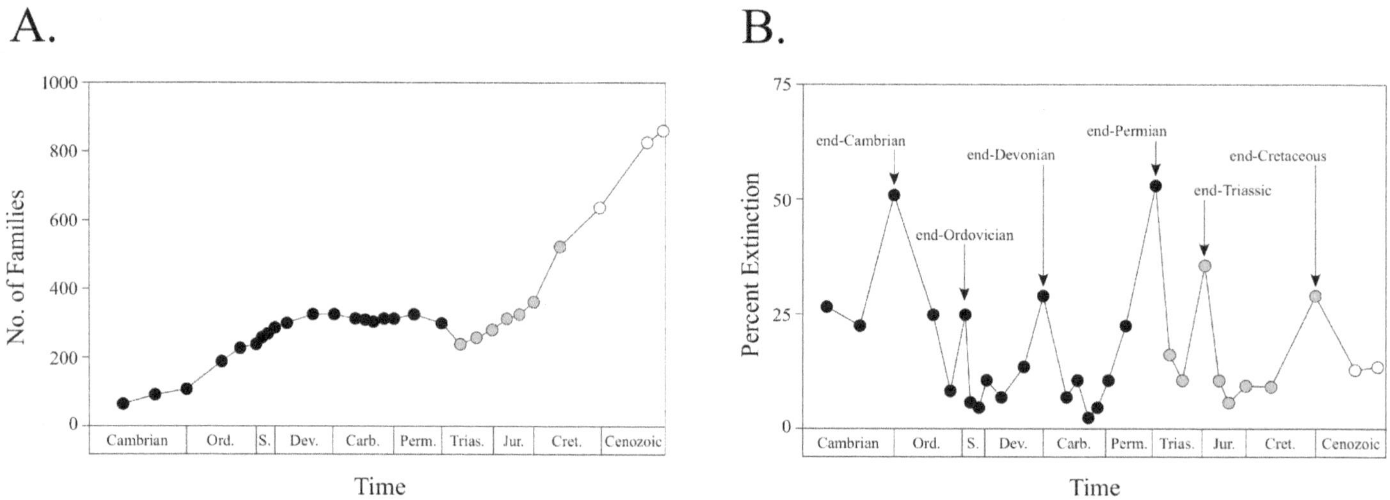

Figure 5. Norman Newell's original diagram of the changing pattern of stage-level biodiversity (A) and extinction intensities (B) over the course of the Phanerozoic. The original six mass extinction events are labeled in B. Figure redrawn from Newell (1963). Black—Paleozoic data; gray—Mesozoic data; white—Cenozoic data.

mass extinction. Newell noted (correctly) that sea level already stood at quite a low level historically during the Pleistocene, thus preventing the continental platforms from being drained during these glacial advances and subsequently reflooded during the glacial retreats. These short-lived Pleistocene variations on a low sea-level theme had a different character from the far larger and longer-duration regressions that characterized the ends of the Cambrian, Ordovician, Permian, Triassic, and Cretaceous periods, when extensive epicontinental seas disappeared for time intervals of significance to long-term evolutionary-ecological processes. More importantly though, Newell took the repeated association between evidence for operation of the putative causal factor and direct paleontological evidence for increased extinction intensity as his main criteria for evaluating mass extinction causal hypotheses. He recognized explicitly (1) the necessity of using extinction selectivity predictions to test alternative causal models, (2) the importance of information gained through study of the ways in which modern organisms cope with environmental changes as an aid to interpretation of the fossil record, and (3) the value of keeping an open mind, and using the multiple working hypotheses approach (Chamberlain, 1897) to understanding these enigmatic episodes of Earth history.

Newell's and Schindewolf's work through the interval from 1954 to 1967 represent a critical juncture in extinction studies insofar as both signal the re-emergence of catastrophism—the rejection of which formed a key tenet of Darwinian gradualism—on the basis of the data of the fossil record. Up to this point data that documented an apparent association between mass extinction in the fossil record and physical disruption of the sedimentation record, first described by Cuvier and Brongniart, had been "explained away" by geologists such as James Hutton (1726–1797), Charles Lyell, Charles Darwin, and their intellectual descendants (e.g., Ernst Mayr, George Gaylord Smith) as the result of imperfections in the fossil record: erosional unconformities, nonpreservation, imperfect knowledge of all fossil species that lived during particular intervals, lack of appropriately detailed geological investigation, incorrect correlation, etc. Under this explanation, the apparent association between patterns of profound biotic turnover and disruptions to physical sedimentation processes was regarded as one of passive linkage, with whatever factor(s) responsible for the physical disruption(s) being unrelated to the processes of organic evolution. Hutton, Lyell, Darwin, Ernst Mayr (1905–2005), and many others believed implicitly that, if these imperfections could be corrected, the true record of organic evolution would be revealed as a stately, progressive unfolding of morphological, ecological, behavioral, and developmental changes. Oddly among this group, Simpson (1944, 1953) was also the foremost advocate of his time for the idea that saltational change was an important pattern of evolution, a point of view he felt carried profound implications for understanding the nature of evolutionary processes. However, even Simpson did not believe that extinctions were driven primarily by physical events imposed on organismal populations by the external environment.

Schindewolf and Newell's primary conceptual contributions were to rejuvenate and restate Cuvier's original interpretation of mass extinctions that saw this juxtaposition as indicating active linkage—that whatever was disrupting patterns of physical sedimentation was also disrupting the biotic processes responsible for species survival. The difference between Cuvier's time and the time of Schindewolf and Newell was that Cuvier had only one good, but local, example of this juxtaposition and was working at a time when the science of paleontology was relatively new. Schindewolf and Newell saw extinctions not just in terms of the juxtaposition of marine and terrestrial fossils in local sections confined to a single region, but in biostratigraphic data drawn from many sections globally, most of which had been amassed in the interval between 1800 and 1950. Moreover, they saw this association occurring repeatedly at the ends of major chronostratigraphic periods and at times (then) thought to be relatively quiescent in terms of major intervals of crustal reorganization (e.g., mountain building events; see Newell, 1963).

Several taxonomic compendia were produced in the wake of Newell's publications (e.g., Harland et al., 1967; Van Valen, 1973). These were used to explore various long-term trends in the fossil record. However, no single compiler of such data was more assiduous than the late John J. (Jack) Sepkoski Jr. (1948–1999). In 1981, Sepkoski published a summary of the stratigraphic ranges of 2800 marine invertebrate families, the data from which he assembled into a diagram that constituted an update of Phillips' 1860 summary (Fig. 6A). While the level of stratigraphic resolution, number of taxa employed, level of documentation, and overall quantitative treatment of the data had improved greatly in the almost 150 years since Phillips' original work, the same general patterns are present in Sepkoski's 1981 diagram, including diversity reductions at the end of the Paleozoic and Mesozoic, fluctuations in the middle Paleozoic, and the overall increase in biodiversity throughout the Phanerozoic. Sepkoski continued to collect data from the paleontological literature after 1981, revising the family-level curve and releasing a series of genus-level data sets from the middle 1980s (see Fig. 6B). In both the expanded family and genus-level data sets, the general patterns in his original 1981 family-level Phanerozoic biodiversity curve were retained.

Sepkoski performed a *Q*-mode factor analysis (see MacLeod, 2005a, 2006) on both the family and genus time-series data sets that identified suites of taxa that exhibited broadly similar diversification patterns. For his 1981 analysis, Sepkoski chose to redescribe his family-level compendia as an interaction between three factors (= multitaxon suites), which together accounted for over 90% of the biodiversity variation recorded in his data. When the diversification histories of these data were plotted across the geological time scale, distinct sets of diversification histories were observed. Based on these results, Sepkoski defined three broad "evolutionary faunas" (Table 1), which he interpreted as exhibiting an evolutionary-ecological coherence that reflected patterns of either passive or active competition/displacement over evolutionary time.

The net extinction record is part of the diversification pattern shown for the Sepkoski family and generic data sets in Figure 5,

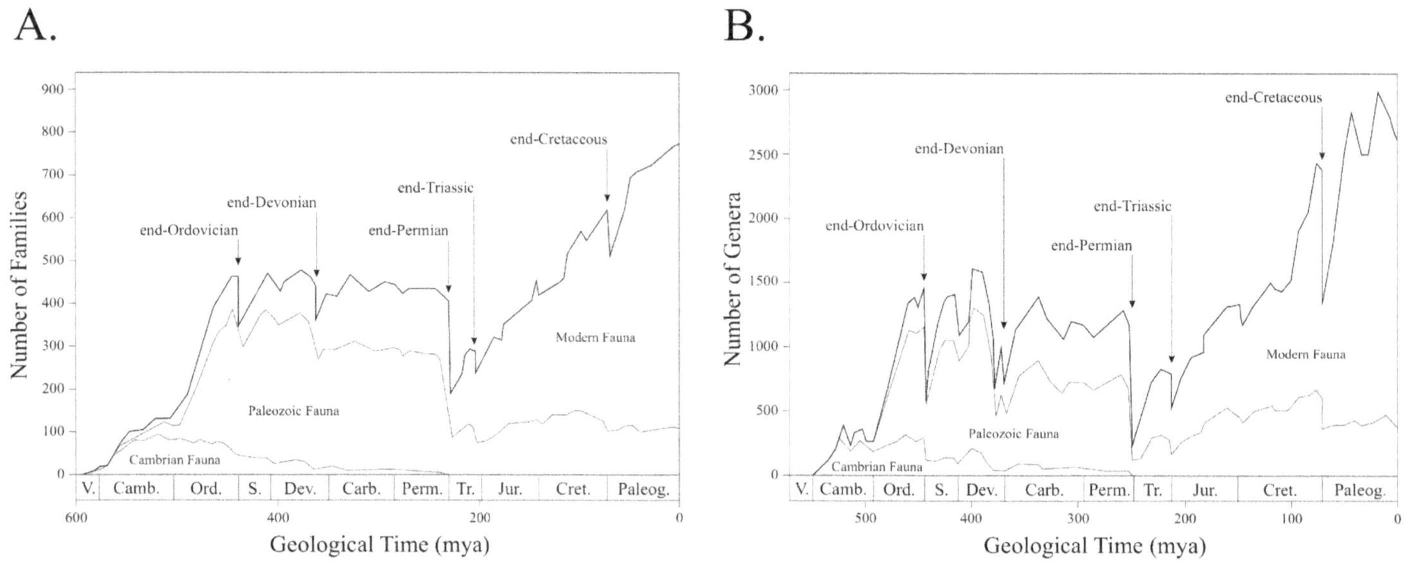

Figure 6. Jack Sepkoski's (1948–1999) update of John Phillips' Phanerozoic biodiversity curves for family-level (A) and genus-level (B) data. These diagrams also show historical patterns of variation for Sepkoski's (1981) three evolutionary faunas (see text and Table 1 for discussion). Figure redrawn from MacLeod (2013).

but owing to the complex interplay between the evolutionary processes of origination (= speciation) and extinction, the detailed features of the extinction record are difficult to pick out of these biodiversity diagrams and assess accurately. Nevertheless, given the Sepkoski data tabulations it is an easy matter to tease these two aspects of biodiversity control apart.

Traditionally, four extinction/origination indices or metrics have been used to describe such data. The forms of these indices, and the issues associated with each, are the same for origination and extinction, though the following discussion will focus on the extinction record exclusively.

$$\text{Total extinction} = \text{No. of taxa becoming extinct during interval}$$

$$\text{Proportional (or percent) extinction} = \frac{\text{Total extinction}}{\text{No. of taxa existing during interval}}$$

$$\text{Time-normalized extinction} = \frac{\text{Total extinction}}{\text{Duration of time interval}}$$

$$\text{Per-taxon extinction} = \frac{\text{Proportional extinction}}{\text{Duration of time interval}}$$

The total extinction index is the simplest to calculate, but it is biased by a number of factors. The proportional extinction index accounts for the number of taxa at risk of extinction in any time interval, but not of the size of the time interval over which the ratio is taken. The time-normalized index takes the duration of the interval over which extinction-related process operate into consideration, but not the number of taxa at risk of extinction. The per-taxon index is a probabilistic measure of extinction intensity that takes into account the number of taxa at risk and the

TABLE 1. SEPKOSKI'S THREE EVOLUTIONARY FAUNAS (SEE ALSO FIG. 6)

Evolutionary fauna	Composition	Diversification pattern
Cambrian fauna	Trilobite arthropods Inarticulate brachiopods Monoplacophoran molluscs Polychaete annelids	Originated in the Tommotian; dominated the Cambrian; declined through the Ordovician; maintained a very low diversity from the Late Paleozoic through the Holocene.
Paleozoic fauna	Articulate brachiopods Crinoid and stelleroid echinoderms Stenolaemate bryozoans Anthrozoan corals Ostracode arthropods Cephalopod molluscs	Originated in the Cambrian; diversified in the Ordovician; dominated remainder of the late Paleozoic; suffered a precipitous decline at the end of the Permian; staged a moderate resurgence in the Mesozoic; began a second decline in the Late Cretaceous; declined rapidly through the Paleogene, but more moderately through Neogene to Holocene.
Modern fauna	Demosponge poriferans Gastropod and bivalve molluscs Echinoid echinoderms Gymnolaemate bryozoans Chondricthyan and osteichthyan vertebrates	Originated in the Ordovician; diversified moderately through the late Paleozoic; suffered a precipitous decline at the end of the Permian; diversified rapidly through the Mesozoic and Cenozoic; underwent moderate to precipitous declines at the end of the Triassic and end of the Cretaceous.

duration of the time interval, but is not referenced to any underlying statistical distribution that would help assess the level of significance of this index's value. Sometime later, Foote (2000) proposed a fifth extinction rate index that, when used properly, can take phylogenetic branching into account.

$$q = -\ln \frac{N_{bt}/N_b}{\text{Duration of time interval}}$$

where N_{bt} is the total number of taxa that range through interval, and N_b is total number of taxa entering interval.

None of these indices is perfect. All capture important aspects of extinction data. It is noteworthy that each of these indices was formulated at a time when the best global measures of extinction intensity relied on stage-level temporal resolutions. More modern approaches to stratigraphic analysis are now available that offer much higher temporal resolutions (see following). The indices listed here are not optimal ones to use to summarize the results of these higher-resolution investigations. While each of these metrics can be (and has been) used for particular investigations, by far the most common metric used to represent and analyze extinction patterns is the proportional (or percent) extinction index. Figure 7 summarizes the stage-binned time series of total extinction intensities for both the Sepkoski familial and generic data sets.

THE "BIG FIVE" MASS EXTINCTIONS

In 1982, David Raup and Jack Sepkoski published a summary of Sepkoski's family-level extinction data for stratigraphic stages (Fig. 7A). At the time of its publication, this was the most comprehensive summary of shifts in Phanerozoic extinction intensity available since Newell's (1967) biodiversity charts. As such, this summary, and the data that stood behind it, provided researchers with an opportunity to conduct comparative forensic analyses of extinction magnitudes, patterns, and trends.

With respect to the question of extinction intensity, Raup and Sepkoski, like Newell before them, realized that, whereas most stratigraphic stages exhibited relatively low levels of extinction intensity (< 8%), a small subset exhibited much higher extinction frequencies. Raup and Sepkoski (1982) used a regression-based statistical test to identify stages that exhibited extinction intensities higher than expected given the scale of variation exhibited by the majority of the data (extinction rates of less than eight families per million years). Based on the results of this analysis, only four stages were identified as exhibiting extinction intensities significantly higher ($\alpha = 0.95$) than would be expected under a null statistical model: the end-Ordovician, end-Permian, end-Triassic, and end-Cretaceous. The values for the Late Devonian peaks were not found to be significant statistically in this analysis. Nevertheless, this Late Devonian interval was admitted to their "mass extinction" category on the basis of historical precedent and, quixotically, under the dual argument that this peak represented a prolonged extinction event (the magnitude of which was low because it lasted for several stages) and had been associated with an implied bolide impact by one researcher (McLaren, 1970). Based on Raup and Sepkoski's (1982) analyses and interpretations, these five time intervals have come to be regarded as the "big five" mass extinctions.

Subsequent analysis of Sepkoski's genus-level data for Phanerozoic stages (Fig. 7B) identified these same time intervals as containing extinction-intensity peaks, but within a more

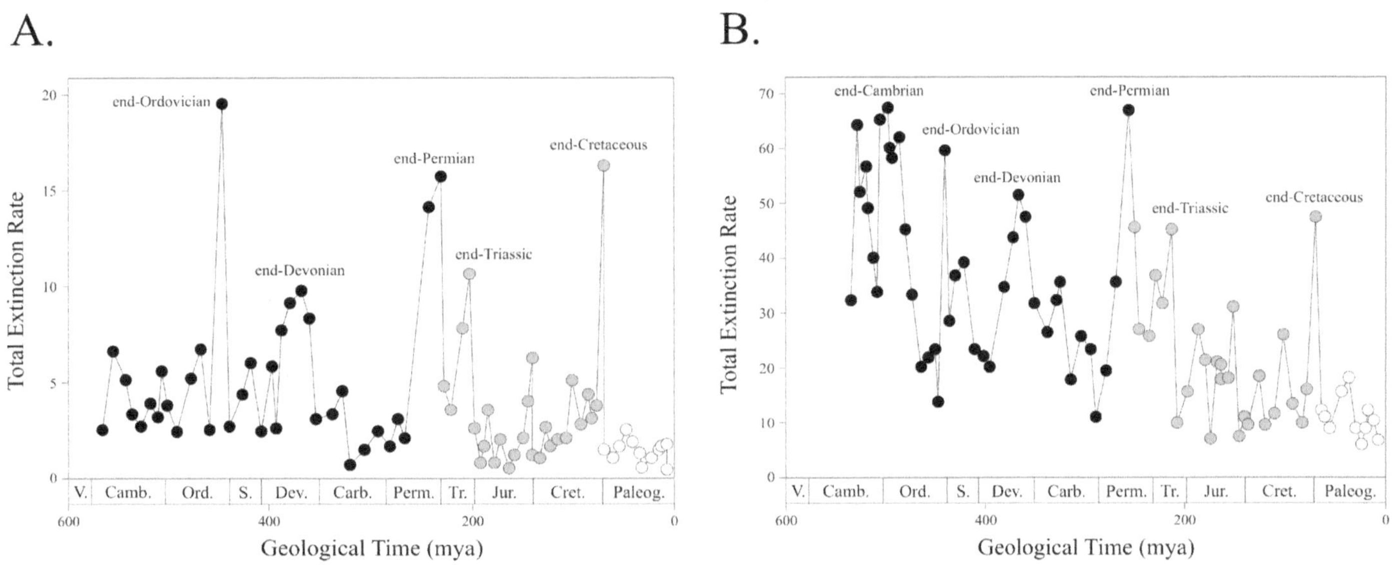

Figure 7. Jack Sepkoski's update of Norman Newell's Phanerozoic extinction-intensity curves for family-level (A) and genus-level (B) data. Note: these data represent all extinctions recorded to have occurred anywhere in a stratigraphic stage as a single point placed at the upper stage boundary. See text for discussion. Figure redrawn from MacLeod (2013). Black—Paleozoic data; gray—Mesozoic data; white—Cenozoic data.

complex and highly variable system in which the peaks do not stand out from the background variation as clearly as they did in the family-level data set. In addition, this genus-level analysis reinstated Newell's original end-Cambrian interval as a time of extraordinary extinction intensity. However, the end-Cambrian biomere extinction events have, virtually without exception, been regarded as events of a different sort from the Raup-Sepkoski "big five" mass extinction events, and all Cambrian data have been, and continue to be, routinely excluded from discussions of mass extinctions and analyses of mass extinction data (e.g., Bambach et al., 2004).

A natural question to ask of these data is: How many species were eliminated during the "big five" mass extinction events? It is impossible to answer this question directly because the original Sepkoski databases were compiled by counting numbers of families and genera, not species. Nevertheless, it is possible to obtain estimates of species-level extinction intensities by conducting simulation experiments. In 1979, Raup developed his "field of bullets" model for this purpose.

If we construct a mathematical model of the manner in which species are assigned to genera and genera to families (e.g., 10 species per genus, 10 genera per family), and then eliminate a proportion of the simulated species randomly—as if they were struck down by a random field of bullets—it is possible to determine how many genera, families, orders, etc., would be eliminated through the extinction of differing proportions of species. In order to obtain accurate estimates, the model needs to be run many times and its initial conditions varied to take into account different species subdivision schemes (e.g., number of monospecific genera and/or families in the simulated data set). Raup's (1979) model assumed that all species are equally susceptible or resistant to extinction, an assumption he acknowledged as unrealistic. Regardless, Raup summarized his data as a set of rarefaction curves that mathematically relate observed proportions of species loss to estimated proportions of loss of higher taxa (Fig. 8). To the extent that Raup's field of bullets simulation is an accurate description of the character of extinction data and extinction processes (see following discussion), these simulated curves could be used to address the reverse problem of estimating species loss from the observed level of extinction in higher taxonomic categories.

Raup originally used his field of bullets approach to estimate how many species-level extinctions were implied by Sepkoski's original tabulations of the size of the end-Permian event. Based on his initial simulations, Raup estimated this loss as between 88% and 96% of marine species during the last stratigraphic stage of the Permian. This was an astonishingly large number, much higher than anyone expected at the time. The accuracy of this estimate—and all other estimates obtained using the field of bullets model—remains uncertain, as no empirical data on the actual distribution of species among higher taxonomic categories and/or the extinction susceptibilities of different taxa exist (for an empirical test of Raup's model, see Poag, 1997). However, in a scientifically and sociologically interesting aside, Raup (1991) noted that, despite his taking great pains to explain the nature of his simulation strategy and the caveats associated with his estimates in the 1979 article, subsequent authors have almost always cited only the upper species loss estimate, often portraying this as the only figure available. More recent maximum and minimum field of bullets estimates for the magnitude of the other "big five" extinction events based on a more recent version of the Sepkoski database are provided in Table 2.

For extinction studies, a core point of scientific interest (and contention) with regard to the Sepkoski extinction data hinges on the extent to which they can be used to understand aspects of mass extinction causality. The stage-level data shown in Figure 6B can be rearranged in rank order so that they form a distribution of extinction intensities rather than illustrating the history of variations in extinction intensity. The resulting graph (Fig. 9A) indicates that, when pooled into stage-level intervals, the distribution of extinction intensities is continuous. This continuity may account for the difficulty previous investigators have had in developing a truly objective and consistent definition of mass extinction. As noted earlier herein, Raup and Sepkoski (1982) first identified their "big five" mass extinction events on the basis of a (quasi)statistical analysis of extinction magnitudes (see Fig. 3A), but even in their analysis, the distribution of deviations from the regression trend line was continuous. Some events that fell slightly above their arbitrary 95% confidence interval about the regression trend (e.g., the Tithonian peak) were not identified as mass extinction events, while others that fell below the cutoff

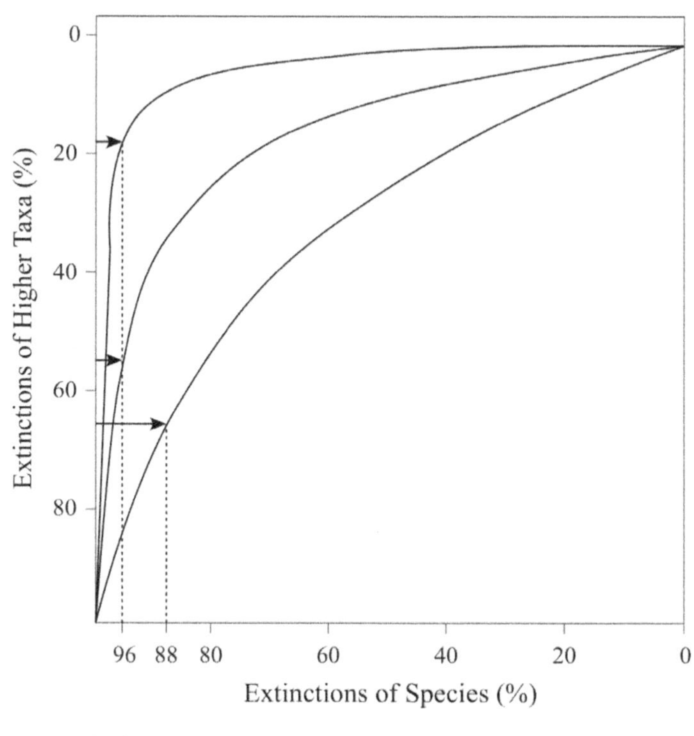

Figure 8. Graph relating observed percent extinction of taxonomic genera, families, and orders to percent extinction of species estimated under the "field of bullets" model. Figure redrawn from Raup (1991).

TABLE 2. FIELD OF BULLETS ESTIMATES OF LOSS FOR WELL-SKELETONIZED TAXA SPECIES DURING THE "BIG FIVE" EXTINCTION EVENTS (DATA FROM JABLONSKI, 1995)

Extinction	System/stage	Family data		Genus data	
		Observed (%)	Est. species loss (%)	Observed (%)	Est. species loss (%)
End-Ordovician	Ashgillian	24.1–27.9	77–91	55.6–64.4	82–88
Devonian	Frasnian	20.3–23.7	70–88	53.7–60.3	79–87
End-Permian	Changhsingian	48.7–53.3	93–97	78.2–85.8	93–97
End-Triassic	Norian	19.8–24.2	70–88	48.6–57.4	76–84
End-Cretaceous	Maastrichtian	14.5–17.5	57–83	42.9–51.1	71–81

significance value (e.g., the Late Devonian extinction peaks) were. More importantly, the ontological acceptance of mass extinction as a discrete category of extinction events has prompted many researchers to assume that these events have a character, and perhaps are caused by a set of processes, that differ(s) from the character and cause(s) of other, smaller extinction events. Such a set of distinguishing features may exist (e.g., distinctions in the taxonomic, ecological, or geographic patterns of extinction susceptibility and survivorship; see Jablonski, 1986).

Recently, Bambach et al. (2004) reanalyzed these stage-level data and, in the context of their analysis, proposed that elimination of the Cambrian stages from consideration as "mass extinctions" could be justified on the grounds that doing so shifted the "big five" Raup and Sepkoski events to the high end of the extinction-intensity spectrum. Leaving aside the artificiality of this post hoc justification, as can be seen in Figure 9B, ad hoc removal of the Cambrian values from the Sepkoski (2002) data does not result in these "big five" events occupying the first five positions along the extinction-intensity spectrum. With regard to the issue of the distribution's shape, in the partial data set from which all Cambrian data have been excluded (Fig. 9B), the end-Permian event does stand out as distinct in terms of its magnitude. However, the Tremadoc event (59.3% extinction)—which has never been listed as a mass extinction—occupies a position above those of the end-Devonian and end-Cretaceous events. In addition, the difference in magnitude between the Visean and Arenig events (both 45% extinction) and the end-Cretaceous event (47.3% extinction) is far less than the difference between the end-Triassic (60.3% extinction) event and either the end-Devonian (52.1% extinction) or end-Cretaceous events. Based on the distribution of extinction magnitudes for geological stages, no consistent criterion or set of criteria has yet been devised that can identify the "big five" mass extinction events as unique in any way unless various stage-level data are discarded and/or ignored for ad hoc reasons.

Two important implications derive from the current failure to identify a set of unique characteristics that clearly separates mass extinction events from other extinction events. First, the same cause or set of causes may be responsible for generating all extinction events irrespective of their magnitude. Under this scenario, smaller extinction events are those in which the set of extinction-causing processes operate at relatively weak levels. To the extent that the extinction events visible in the fossil record are the result of multiple causes, the smaller-scale end of the extinction-intensity spectrum may also represent time intervals in which a single cause (e.g., sea-level change, climate change,

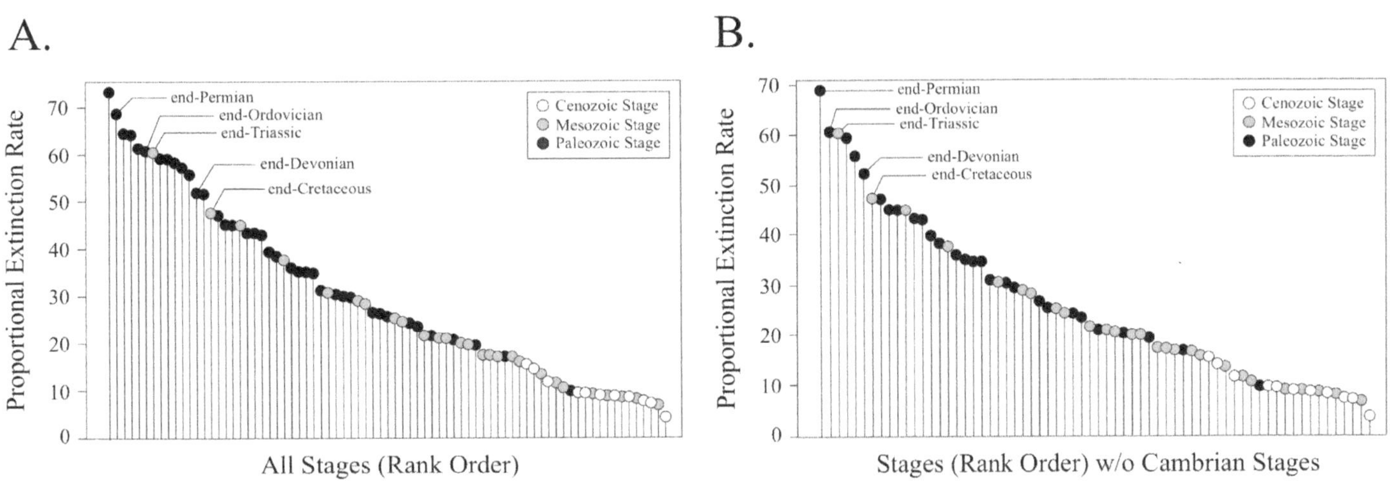

Figure 9. The Sepkoski (2002) stage-level Phanerozoic extinction data for marine invertebrate genera rank ordered by percent extinction intensity. (A) Data from all Phanerozoic stages. (B) Data from non-Cambrian Phanerozoic stages. See text for discussion. Black—Paleozoic data; gray—Mesozoic data; white—Cenozoic data.

volcanism, small- or medium-scale bolide impact) predominated. Similarly, extinction events at the larger end of the extinction-intensity spectrum may represent time periods characterized by a coincidental alignment of two or more major causal factors, both of which operated at high levels. While this scenario remains one of a number of plausible explanations for the geometry of extinction magnitudes, analysis of these data suggests that truly unique sets of processes are likely not associated with either end of the extinction-intensity spectrum.

Second, extinction-intensity peaks are defined as much by the character of extinction intensities in the time intervals preceding and succeeding them as by the event itself. In other words, there is an irreducibly historical aspect to the structure of the extinction-intensity record. Any attempt to understand or account for this record needs to take this historical factor into account. Extinction intervals must be seen—and studied—in a holistic manner that takes account of the patterns, victims, survivors, and causes of low extinction intensities in time intervals preceding and following a local intensity peak, as well as the patterns, victims, survivors, and causes confined to the time interval containing the intensity peak.

"Background" Extinctions

In order to distinguish the lesser, stage-level extinction events that (historically) surround their "big five" mass extinctions from the "big five" events themselves, Raup and Sepkoski (1982) coined the term "background" extinctions as the set of stages, and stage-level data, for which extinction-intensity values were less than approximately eight families per million years. If, as has been assumed by many (but see earlier discussion), the "big five" mass extinctions represent unusual events that require an unusual explanation, by inference the background extinctions represent taxonomic diversity loss events caused by more usual evolutionary-ecological processes, presumably the normal Darwinian processes of competition and natural selection. In terms of raw extinction numbers though, the background extinction category is by far the more important. Estimates suggest that over 95% of all species extinctions that have occurred in the history of life have taken place during background extinction intervals (Raup, 1986, 1991). This difference alone makes the background extinction category both interesting and important. More than this, as a set of observations, the background extinction data are interesting and important because they exhibit an unusual and somewhat unexpected long-term structure of their own: the background extinction-intensity gradient.

This gradient is usually portrayed as a linear ramp with a negative slope through time (see Fig. 10A). Most commonly, this declining pattern of extinction-intensity magnitudes is described using trend-line (linear regression) analysis, sometimes of the entire extinction-intensity data set, more often with the "big five" mass extinctions removed. While such analyses indicate that

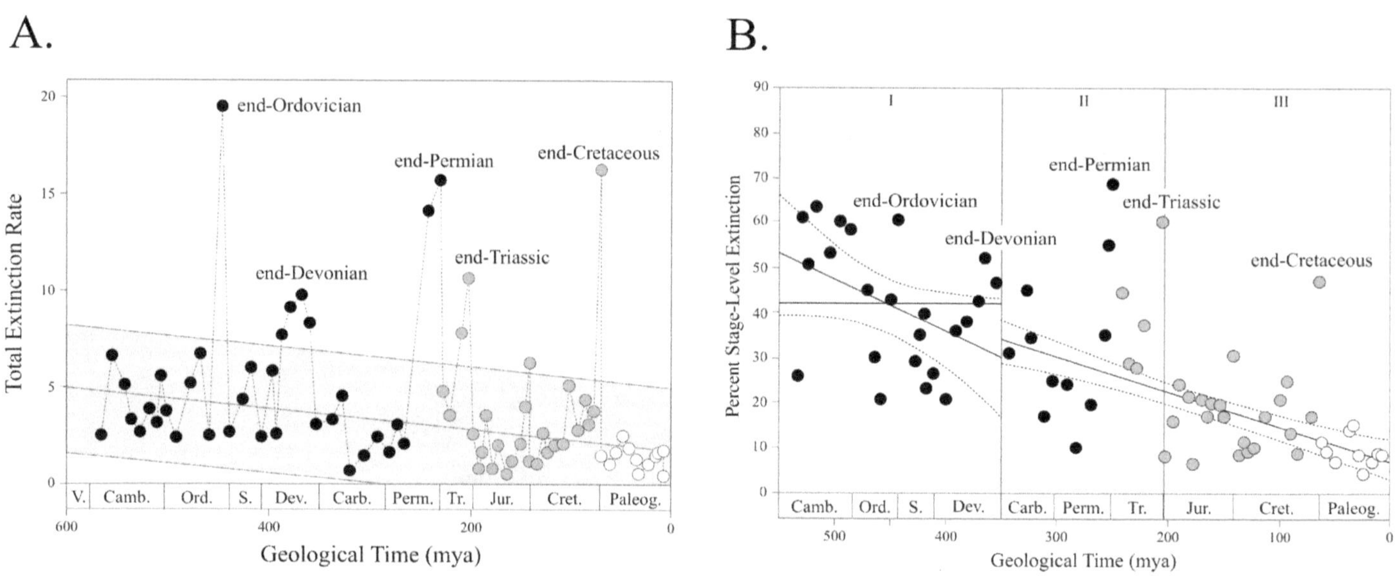

Figure 10. Stage-level Phanerozoic background (= low-intensity) extinction-intensity data for marine invertebrate genera. (A) Raup and Sepkoski's (1982) analysis of family-level extinction data with a 95% confidence interval (shaded region) drawn about the trend line that quantifies the declining stage-level extinction magnitudes across the Phanerozoic. Raup and Sepkoski's identification of the "big five" mass extinction (= high-intensity) events was based, in part, on this analysis. Figure redrawn from Raup and Sepkoski (1982). (B) Genus-level extinction-intensity data, parsed by stratigraphic stage, with the extinction history subdivided into three phases: an early phase (I) in which extinction intensities were highly volatile but over which there was no statistically significant trend, a middle phase (II) in which a statistically significant background extinction-intensity gradient was present, but over which variation remained high, and a late phase (III) in which the background gradient was present and variation about this trend was damped. Figure redrawn from MacLeod (2004). Black—Paleozoic data; gray—Mesozoic data; white—Cenozoic data.

these data do exhibit a statistically significant negative slope, the variation of extinction-intensity values about the linear trend line is highly structured, such that the ramp can be subdivided into segments. There is, at best, a very slight and statistically insignificant trend to the background extinction-intensity data older than ca. 350 Ma (Fig. 10B) in both the family and genus-level data. Following this early Paleozoic interval, a consistent declining trend in extinction-intensity data is present, but the variation of points about the trend line is stepped, with high variation characterizing the interval from 200 to 350 Ma and markedly lower levels of variation characterizing the interval from 0 to 200 Ma.

Various hypotheses have been advanced to account for this unexpected feature of life's extinction history. These hypotheses can be grouped into two broad categories: intrinsic factors that reflect some aspect of adaptation and/or secular increase in environmental complexity (see Martin, 1996; MacLeod, 2003, 2004), and extrinsic factors that reflect the operation of sampling biases in the extinction data (see Pease, 1992).

Intrinsic factors include:

1. A trend toward the adaptation of species to marginal environments, with improved longevity of genera and families over time conferred by the inclusion of species adapted to these habitats being inherently resistant to extinction pressures caused by environmental changes;
2. Evolutionary/ecological responses of the global biota to secular changes in the abiotic environment that, over time, have operated to (a) increase marine and atmospheric circulation rates, (b) increase rates of photosynthesis among primary producers, and decrease rates of nutrient delivery to biotic systems by locking up nutrients in natural sinks, all of which result in the development of extinction resistance by diversification via the promotion of competition between, and adaption within, species; and
3. Macroevolutionary effects of the invasion of new habitats by ecologically critical groups through the development of key adaptations.

Extrinsic factors include:

1. A secular trend of increase in the number of species per lineage over time, which confers extinction resistance on younger lineages as a result of the existence of greater numbers of species with broad geographic distributions;
2. An apparent trend of increase in the number of species per lineage over time due to greater familiarity with modern species and their close relatives (e.g., resulting in a less finely divided taxonomy), which confers an apparent extinction resistance to younger lineages; and
3. An apparent trend of increase in the number of species per lineage over time due to greater outcrop areas of younger ages that are available for paleontologists to search, which confer an apparent extinction resistance on younger lineages as a result of an apparently broader geographic distribution.

Most likely, all these factors, and possibly others yet to be discovered, have played a role in the development and maintenance of the background extinction gradient over geological time. Nonetheless, some aspects of the structure of the data (e.g., abrupt changes in patterns of variation about the trend line) cannot be accounted for under models that consist purely of extrinsic factors.

Following on from Jack Sepkoski's successful—and career-long—development and analysis of his taxonomic compendia (the final installment of which appeared in 2002 and is available for download at http://strata.geology.wisc.edu/jack/), subsequent efforts to refine and further develop his database have focused on three initiatives. In 1993, Mike Benton published a reference volume, *The Fossil Record 2*, which over the course of 45 chapters written by 90 internationally recognized taxonomic specialists offered family-level taxonomic data for essentially all groups of fossils within a consistent chronostratigraphic framework (Benton, 1993). This compendium benefited from being assembled by a group of experienced and knowledgeable practitioners who could ensure consistency and correct the data presented for various types of bias (see later herein). The original Sepkoski family- and genus-level compendia were assembled by a single person who was not a taxonomic specialist in any group, but who relied on the informal advice of a large group of taxonomic specialists offered over a long period of time to resolve ambiguities and ensure quality-control standards. Irrespective of Sepkoski's best efforts in this regard, Patterson and Smith (1987; see also Smith and Patterson, 1988) raised a series of troubling issues over the quality of the data contained in the Sepkoski compendia in their analysis of extinction periodicity. The *Fossil Record 2* project was conceived, in part, to address these concerns.

Shortly after this compendium's publication, Benton made all the published family-level data available as a series of MS-Excel spreadsheets that could be downloaded free-of-charge from *The Fossil Record 2* Web site (http://www.fossilrecord.net/fossilrecord/index.html). As of this writing, these data are temporarily unavailable. However, *The Fossil Record 2* Web site currently contains mention of plans to issue an update of *The Fossil Record 2* family-level database to reflect taxonomic changes and new discoveries made since 1993, and then to issue a genus-level compendium at some point in the future.

In a parallel effort, the Paleobiology Database Project (PBDB, http://paleodb.org/) seeks to provide researchers and the public with information about the entire fossil record in the form of a searchable online database. This project encourages paleontological professional and postgraduate students to upload data on fossil collections they have made, which, after a degree of vetting by a board of specialists, are added to the database. This database has expanded continuously since 2000 and, as of this writing, summarizes the information contributed by 387 individuals from over 51,000 published references, 295,000 taxa, 156,000 fossil collections, and 1,000,000 fossil occurrences. While the Paleobiology Database Project contains more records than *The Fossil Record 2* database and is in the process of being extended to the species level, it does not impose the same quality-control standards and cannot claim to represent as consistent and nuanced (through the experience of world-class specialists) a treatment.

Similarly, the most recent global database of relevance to regional and global extinction studies is the Geobiodiversity Database (GBDB, http://www.geobiodiversity.com; see also Fan et al., 2013). The GBDB is designed to facilitate public access to a broad range of rock and fossil data worldwide. As of this writing the GBDB database included information from over 81,000 published references, 94,000 identified taxa, 53,000 fossil collections, and 275,000 fossil occurrences from 12,000 localities from three countries, with plans to increase its rate of record entry to 100,000 records per year by 2015. It should also be noted that, unlike the PBDB, the GBDB includes built-in export/analysis software for conducting biodiversity, quantitative stratigraphy, and paleobiogeographic investigations. In 2012, the Geobiodiversity Database was selected as the reference database of the International Commission on Stratigraphy (Fan et al., 2013). However, like the PBDB, the GBDB relies on an informal quality-control system founded on users who have been given "authorizer" status, so they may, if they so choose, verify and improve the GBDB database records. Otherwise, the data that populate the GBDB are taken largely at face value from published sources.

In the most recent synoptic analysis of the Sepkoski (2002) data set, Bambach et al. (2004) concluded that, so long as all Cambrian stage-level extinctions are ignored, the remaining extinction record can be used to subdivide the Phanerozoic into six stratigraphically coherent intervals of alternating low and high extinction intensity, each of which, other than the last (= Holocene), terminates with a "mass depletion event." This term was preferred by these authors based on their analysis of the contribution of variations in the rate of genus appearances over time to the extinction record. When origination data were taken into consideration, only three of the Raup and Sepkoski "big five" mass extinction events—the end-Ordovician, end-Permian, and end-Cretaceous—were found to result from elevated levels of extinction. These analyses also indicated that the end-Devonian and end-Triassic events were due primarily to a failure of species originations during these time intervals as opposed to elevated levels of species extinction (though this might be as much a comment on taxonomic convention and practice as a genuine feature of evolutionary patterns). In addition, Bambach and his coauthors concluded that, since each of their three genuine mass extinction events differ from one another in terms of both pattern and diversity dynamics, the hope that all might result from the same cause(s) seemed unlikely.

Biases in the Fossil Record

Before the specifics of individual extinction records can be evaluated, an assessment of the set of biases acknowledged to effect extinction patterns in the fossil record is necessary. Unlike the blanket attribution of the coincident disappearance of large numbers of species to "imperfections of the stratigraphic record," paleontologists understand the sources of bias in the fossil record in much more detail now than they did in Cuvier's and Darwin's day. In many cases, procedures to identify, and in some cases, compensate for, these biases have been developed, at least in terms of quantifying their potential effects.

Temporal Patterning in Extinction Data

It is standard practice for biostratigraphic range charts to represent the chronostratigraphic ranges of fossil groups as solid lines joining individual occurrence horizons that denote intervals of time along a temporal axis (Fig. 11). Because we personally experience time as a continuum, it is tempting to regard these axes as representing time as a continuous variable. In fact, these charts often represent time as a discontinuous variable—akin to the binning convention used to construct histograms—with the vertical range line usually being drawn through the entire interval (or time bin), irrespective of whether the actual time of extinction is known. Of course, this uncertainty is often compounded when (observed) depth scales, rather than (inferred) time scales, are used as the standard for stratigraphic comparison and especially when time scales are plotted against biozones (which can be observed but have no formal temporal significance) and/or chronozones (which have temporal significance but cannot be observed).

In the more recent primary technical literature, many authors have paid more attention to the manner in which their stratigraphic data are described and represented in the diagrams and tables that accompany their research reports. Naturally, the aforementioned criticisms do not apply to data published in this more refined manner. Nevertheless, the majority of stratigraphic range charts published over the past 75 years do contain these problems, especially secondary and tertiary reference works such as the *Treatise on Invertebrate Paleontology*, which has served as a primary reference source for the Sepkoski, PBDB, and GBDB compendia. For diagrams drawn in this manner, a stratigraphic range that terminates within a chronostratigraphic interval (e.g., chronozone, stage/age) should be interpreted as simply recording the taxon's inferred presence in the interval, not the time of the taxon's extinction event relative to other taxa in the section/core or relative to the interval's boundaries unless the pattern of samples is shown on a diagram or can be accessed in supporting data. Faulty understanding of the description and graphing conventions used in much of the older, primary biostratigraphic literature has led a number of researchers to the erroneous conclusion that all extinction events occur at stage/age boundaries and/or that there is a pronounced tendency for extinctions to occur together in time (e.g., Raup, 1992). This is often not the case (see Fig. 11).

Similarly, use of higher taxonomic categorizations (e.g., families, genera) as proxies for species in extinction studies often leads to underappreciated distortions of the fossil record. As noted earlier, the taxonomic categories of family and genus are often employed because many paleontologists regard them as more stable and comparable than species-level data across the broad scope of life's diversity. However, owing to their composite character, the presence of a family at one point in time may represent a rather large number of species, whereas at another point in time (especially if the latter is close to the group's extinction horizon), the actual number of species represented by familial presence may be much smaller. As a reviewer of a previous version of this article pointed out, examples of database compilations exist in which range-through genus richness exceeds or equals range-through species richness! This problem is often the result of disjunct genus ranges, i.e., genera that range through intervals in which no species were collected. In such cases, either the assignment of species to genera is faulty, or, more likely, the geographic scope is inadequate to capture all the species in a genus.

Pseudoextinction is another problem encountered commonly, but acknowledged rarely, in extinction studies. In an evolving lineage, two types of morphological transformations can occur. The first (anagenesis) results in the progressive transformation of an entire species from one morphological condition or state into another. The other (cladogenesis) results when a single species is split into two or more daughter species, one of which may continue to exhibit the morphological condition of the ancestral population(s) (see Fig. 4). While the extinction of a lineage by the physical death of the last individual within the population must be regarded as a true extinction event, the physical death of the last individual within an anagenetically transforming population that happens to exhibit an ancient or atavistic morphological condition is not the same sort of event. Although the phenomenon of pseudoextinction need not be distinguished from true extinction in most routine biostratigraphic studies (by taxonomic convention, the pseudoextinction event does take place at a particular time), the difference between pseudoextinction and true extinction is profound in the context of extinction studies. Mammal systematist/paleobiologist David Archibald has estimated that as many as 25% of the extinctions recorded in three different early Puercan (= lowermost Danian) mammal lineages are pseudoextinctions. Patterson and Smith (1987) estimated that even higher numbers of pseudoextinctions were recorded in the echinoid and fish sections of the Sepkoski genus-level taxonomic compendium.

In addition, the distribution of fossil occurrences in a section or core can be deceiving in terms of the simple documentation of a fossil group's last occurrence. The last observed

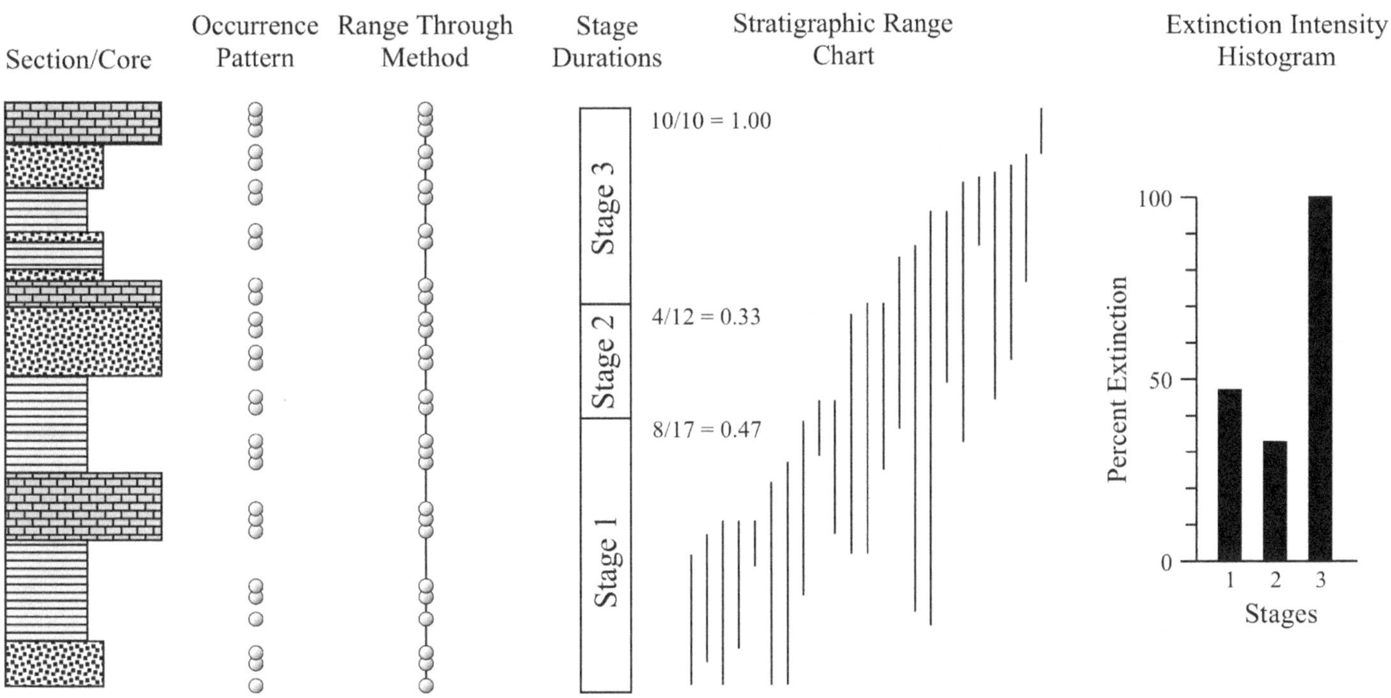

Figure 11. Illustration of the manner in which fossil occurrences in a section or core are summarized into extinction data. Stratigraphic ranges are estimated from the spatial distribution of fossil occurrences in a stratigraphic section or core via the straightforward and eminently sensible range-through method in which the lowest (oldest) and highest (youngest) occurrences define the time interval over which the taxon is regarded to have existed. Collections of such stratigraphic ranges are then graphed together to form stratigraphic range charts. These ranges are usually plotted against stratigraphic stages to facilitate comparisons with other range charts globally. Finally, extinctions occurring anywhere within a stratigraphic state are counted and combined to form an index of extinction intensity (here percent extinction) for the stage. These extinction-intensity data are usually summarized in the form of a histogram or scatterplot. Note that (1) a wide variety of possible extinction geometries will yield precisely the same extinction-intensity value, and (2) there is no way to recover the true spatial (or temporal) pattern of extinctions from examination of the extinction-intensity index data alone.

occurrence of each species is only an estimate of its true extinction in time and space. Given what is known about the spatial distribution of Holocene species, about the vagaries of fossil preservation, and about sampling theory, a complex set of factors will usually conspire to distort the observed extinction record for each species (and, by extrapolation, for each genus, family, etc.) to a greater or lesser extent. Phil Signor and Jere Lipps (1982) pointed out that this phenomenon could result in an abrupt or coincident extinction pattern being smeared out by chance sampling artifacts so as to make it appear to be gradual or progressive (see Fig. 12)., a possibility that Raup (1991) termed the "Signor-Lipps effect" (although the stratigrapher Alan Shaw offered an essentially identical description of the same phenomenon as early as 1964). As Signor and Lipps (1982) noted, observed patterns of species disappearances in local sections or cores are inherently ambiguous. According to the Signor-Lipps model, it is erroneous to conclude that the observation of a progressive extinction pattern necessarily constitutes evidence favoring identification of an abrupt or progressive extinction process. Rather, ancillary forms of evidence are necessary to ensure that the observed patterns of extinction in the fossil record reflect local, regional, and/or global patterns of species extinction in time and space accurately.

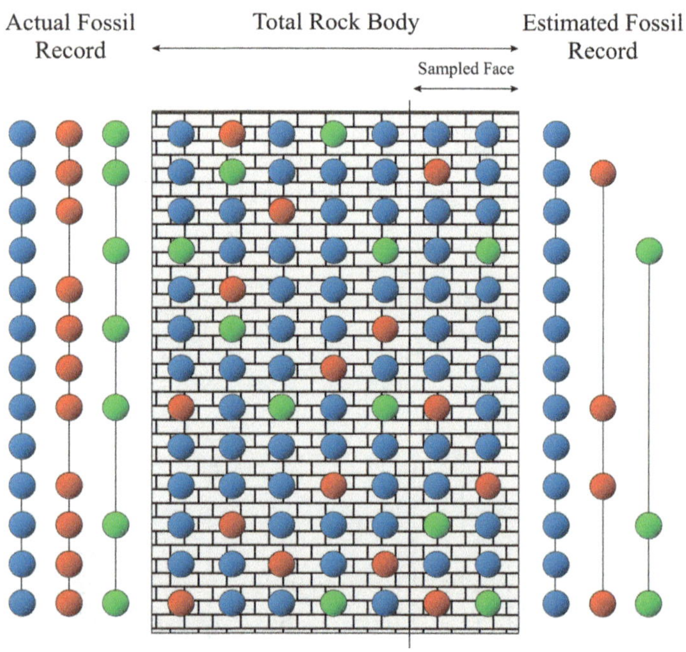

Figure 12. A hypothetical example of the Signor-Lipps effect. The total fossil record of any species is provided by the sum total of all occurrences of the fossil within a body of rock. Because a relatively small portion of that total rock body is ever available for sampling, the known fossil record represents an estimate of the total fossil record. Owing to this distinction between the total and known fossil records, and because of additional biases that might arise as a result of changes in the relative abundance of fossil species over time, changes in facies within single stratigraphic successions, and sampling patterns, estimates of the pattern of last-appearance data (e.g., coincident vs. progressive) are usually biased toward the recovery of progressive patterns. In this example, note that, despite all three species ranging through the entire section, ranges based on the subset of occurrences in the two right-hand columns yield a sequential extinction pattern. This phenomenon has been used to argue that progressive extinction patterns imply an abrupt extinction mechanism. Such arguments are invalid—as was pointed out by Signor and Lipps (1982)—because the progressive bias applies to genuinely progressive as well as genuinely abrupt true extinction patterns. The best one can do is not to assume the apparent pattern of biostratigraphic last-appearance data provides a reliable estimate of true extinction times or rates.

There is no way to correct for the Signor-Lipps effect because it represents a fundamental uncertainty in the specification of both local and global extinction horizons,[3] though given a sufficiently rich fossil record, simulation experiments can be performed that might suggest that a progressive apparent extinction pattern would be unlikely to have been produced by random preservation/sampling processes. This is not the same thing as correcting (or normalizing) a data set for this source of bias. However, by using the distribution of gaps between known occurrences within a species' or higher taxon's stratigraphic range, and making use of several simplifying assumptions, it is possible to estimate a confidence interval within which the true local extinction event can be expected to have occurred and above which it is appropriate to regard the species or group as being truly extinct, at least locally (Strauss and Sadler, 1989; Marshall, 1990, 1994, 1997; see Fig. 13).

Stratigraphic confidence interval (SCI) analysis is inherently probabilistic and so does not enable an investigator to retrospectively pinpoint a taxon's "true" extinction level. For groups of taxa, various scenario-based extinction geometries can be evaluated using extensions of the confidence-interval method. The strength of this approach is that different extinction geometries can be evaluated for their probability of being consistent with an apparent extinction record. However, below a certain level of stratigraphic resolution, SCI analysis will always identify a spectrum of geometries—ranging from randomized progressive extinction patterns to a strictly simultaneous geometry—as being essentially equiprobable (Marshal, 1995).

Last, but by no means least, simple taxonomic uncertainty conspires to complicate data reported in the paleontological literature. Certain species or higher groups that appear to become extinct at particular stratigraphic levels are often discovered subsequently at higher stratigraphic levels (= later in time) at the same locality or in a different region altogether. Sometimes, the morphological condition of these newly discovered specimens at the higher stratigraphic level is similar to that of older specimens, and sometimes it is not. If the second appearance is regarded as a continuation of the species, it is referred to as a Lazarus taxon and should not be regarded as having undergone anything other than a local extinction at the lower horizon. However, if these younger specimens differ in some consistent and characteristic way from their older forms, such that their placement within the

[3]It is important to note that the same uncertainty arguments apply to the apparent pattern of taxon appearances in the fossil record.

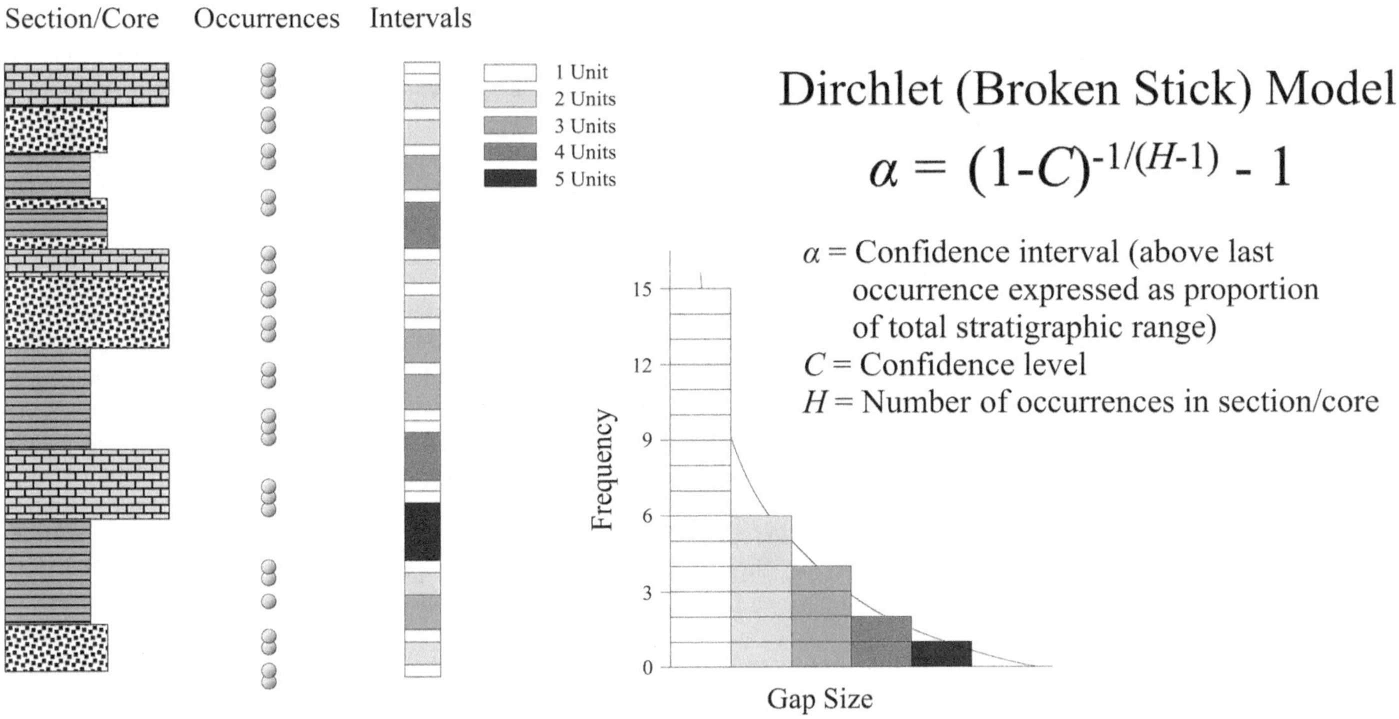

Figure 13. Illustration of the theory behind the calculation of stratigraphic confidence intervals. Assuming continuous scrutiny of a fossiliferous section or core for fossils of a certain species, group, or type, the recorded occurrence pattern is almost always discontinuous, with gaps of various sizes separating the recorded occurrences. These gaps form a frequency distribution that can be modeled using various mathematical expressions, here the Dirchlet or "broken stick" distribution. Once quantified in this manner, the expected gap frequency distribution can be used to estimate the interval above (or below) the last observed occurrence in which the true extinction horizon is likely to be found. Note: the true extinction horizon is only estimated to lie somewhere within the confidence interval and should not be regarded as being placed at any particular level within the interval in question.

previously existing lineage is questionable, the younger specimen's "reappearance" should be regarded as suspect, belonging to an "Elvis" taxon, in which case the status of the group's extinction is uncertain. Of course, specimens that are identical in form to older representatives species but are found in younger (postextinction) sediments must either be specimens that have been eroded from older deposits and redeposited at later times or representatives of populations that survived the extinction event. In these cases, unless unequivocal evidence for reworking of all anomalously placed specimens can be presented, the survivorship hypothesis cannot be discounted.

Geographic Patterning in Extinction Data

As alluded to earlier, the distinction between extirpation and true extinction always needs to be kept in mind when interpreting extinction data. If a species' geographic range is limited (e.g., insular species confined to physical or ecological islands), these two concepts merge. In all other cases, though, the disappearance of a fossil species from any single locality is best thought of as an extirpation event rather than an extinction event.

This distinction is especially problematic when interpreting paleontological data insofar as it is often impossible to know whether a set of species' disappearance patterns from a local stratigraphic succession represents their youngest occurrence globally (a true extinction pattern). The record of Late Cretaceous dinosaurs is perhaps the best known—and one of the most dramatic examples—of this phenomenon. Campanian dinosaurs are known from a number of different continents. However, Upper Maastrichtian dinosaurs are well known only from the Hell Creek and Lance Formations that crop out in Montana, Wyoming, North Dakota, and South Dakota. The fossil record of dinosaurs, along with other terrestrial vertebrates and invertebrates, is quite well known in this region. However, the extent to which the pattern of Late Cretaceous extinctions observed here is an accurate representation of global extinction patterns is wholly uncertain. Unless new Upper Maastrichtian dinosaur-bearing localities are found in the few geologically unexplored regions of Earth's surface that remain, this situation is unlikely to change.

In the modern world, we can be reasonably sure that isolated, small populations of large terrestrial quadrupeds (e.g., mammoths, mastodons) could not have escaped the attention of scientifically literate observers. Accordingly, it is logical to regard these species as being extinct. It was just this difference between extirpation and extinction that was first used by naturalists as varied as Carl Linnaeus, Thomas Jefferson, and Charles Lyell to deny the possibility of extinction by arguing that species that had apparently disappeared from certain areas of the globe may still be found alive in remote regions. Yet, as the Age of Discovery proceeded, it was this same distinction that led Georges Cuvier and others to respond that this "survival in unexplored regions" argument was no longer tenable for large species in light of the observational evidence collected from an increasingly large proportion of Earth's surface. Because of this "discovery factor," much uncertainty regarding the extinction status of Holocene species of smaller size and more cryptic habits remains. In such cases, the apparent extinction record should be treated as a record of extirpation rather than true extinction. As a result, the difference in the geographic range of the data available for widely distributed species and for those species for which geographic distribution is unknown is a matter of direct relevance both to estimates of the state of individual species' extinction risk and to the assessment of the magnitude of the effects that physical changes associated with species extinctions will have on the structure of the biosphere.

With respect to the former factor, it is well known that species that exhibit limited geographic ranges are far more likely to undergo extinction than species for which their geographic scope spans entire continents and/or oceans. For example, among the 802 recognized species extinctions that have occurred since A.D. 1600, over 75% involved species characterized by small populations indigenous to islands. Significant proportions of the continental extinctions have also occurred among species indigenous to relatively small areas within larger continents (e.g., South Africa, Germany, India).

The reason why species indigenous to spatially limited areas are more susceptible to extinction is well understood. All populations undergo fluctuations in abundance in response to changes in their local physical (e.g., temperature, rainfall for terrestrial species, salinity for marine species) and/or biotic (food items, predators, competitors) conditions. If populations are distributed over only a small area, local or regionally adverse conditions can operate on all populations simultaneously and with the same intensity of effect. Under this scenario, the coordinated extirpation of a small number of local individuals will result in the extinction of the species as a result of, essentially, the same set of factors. In species with broader geographic distributions, it is far less likely that adverse conditions will operate at the same level of intensity across the entire species' range. It is also likely that, for species for which wide geographic ranges are subdivided into a number of local populations with limited interpopulation gene flow, natural background fluctuations in environmental parameters across the combined range of all populations will operate to encourage the adaptation of local populations to local conditions, thereby increasing the extinction resilience of the species as a whole. Species characterized by larger numbers of individuals, subdivided into larger and more spatially heterogeneous populations, also increase extinction resilience as a result of the greater likelihood of favorable adaptations arising in a local population and for those mutations to spread to other conspecific populations via emigration. In modern populations, uncertainty regarding the number and location of remnant populations of species can result in a species being listed as "at risk," "endangered," or even "extinct in the wild" based on assessments of difficulties experienced by a local population when viable populations of the same species exist elsewhere.

With respect to the latter factor, several interesting paleontological studies have concluded that, during "background" (= lesser intensity) extinction events, species (and taxa) with wide geographic distributions and/or dispersal capabilities exhibit a greater likelihood of survival than more indigenous species and/or those with limited dispersal capabilities. Nonetheless, during "mass" (= higher intensity) extinction events, both types of species exhibit statistically indistinguishable survivorship likelihoods. In perhaps the best known of these studies, Jablonski (1986) examined extinction and survivorship patterns in bivalves and gastropods across the Cretaceous-Paleogene boundary. Based on data collected from a variety of sources, he showed that planktotrophic larval development (= broad larval dispersal capability), broad geographic range, and high species richness values were all positively associated with extinction survivorship during background extinction intervals in the Late Cretaceous, early Cenozoic, and late Cenozoic, but these same factors were not associated significantly with survivorship across the Cretaceous-Paleogene boundary. This result supported a general, but to that time untested, belief that geographic range and taxonomic partitioning could confer a degree of resilience in the face of moderate environmentally driven extinction pressure. The interesting, and somewhat unexpected, finding was that this long-suspected association appeared to break down during major extinction events.

On the basis of these results, Jablonski proposed the existence of two macroevolutionary realms, a lower level in which the normal processes of Darwinian selection and adaptation operate and another, higher-level realm within which entire species or species groups are sorted randomly by group-level selection processes without regard to the normal processes of natural selection that operate on individuals within a population. This remains an intriguing idea, but one for which empirical evidence from extinction events other than (possibly) the end-Cretaceous event is lacking at present. Regardless, Jablonski's study underscores the importance of geographic range as factor that may be of use in distinguishing between, and understanding the significance of, different types of extinction patterns.

Case Studies from the End-Cretaceous Extinction Event

The temporal and geographic patterns inherent in data pertaining to most extinction events are well exemplified by comparing high-resolution invertebrate, vertebrate, and microfossil stratigraphic records across the best known of all mass extinction events, the end-Cretaceous event, at three different stratigraphic sections and in three different depositional environments.

Ammonoid Extinction Record in Western Europe

Ammonoids are an ancient group of cephalopod molluscs that dominated many marine ecosystems throughout from the Late Paleozoic through to the end of the Mesozoic. Throughout their long evolutionary history, ammonoid diversity waxed and waned, dropping to remarkably low levels during the end-Devonian, end-Permian, and end-Triassic extinction events prior to their last recorded occurrence at or near the Cretaceous-Paleogene boundary. In between these first-order diversity reduction events, ammonoid family and genus richness values were also volatile, often exhibiting marked reductions during stages that record intermediate-level extinction intensities on the Sepkoski curves (e.g., Pliensbachian, Aptian). Accordingly, ammonoids have long been held to represent a particularly good example of a stratigraphically and ecologically important fossil group for which their evolutionary history has been affected strongly by the factors that cause mass extinctions.

The ammonoid extinction event has been studied in the greatest detail in western Europe, especially at the Zumaya section and environs, located on the coast of the Bay of Biscay, by W. J. (Jim) Kennedy and Peter Ward (Ward and Kennedy, 1993). A summary of these data is reproduced in Figure 14A. These collections record a progressive extinction event beginning well below the local Cretaceous-Paleogene boundary, with the last ammonoid fossil being recorded approximately 10 cm below this boundary. Marshall and Ward (1996) claimed that the Zumaya section is chronostratigraphically complete with "no missing microfossil or macrofossil zones" (see also Smit and Romein, 1985). However, the observation of biostratigraphic

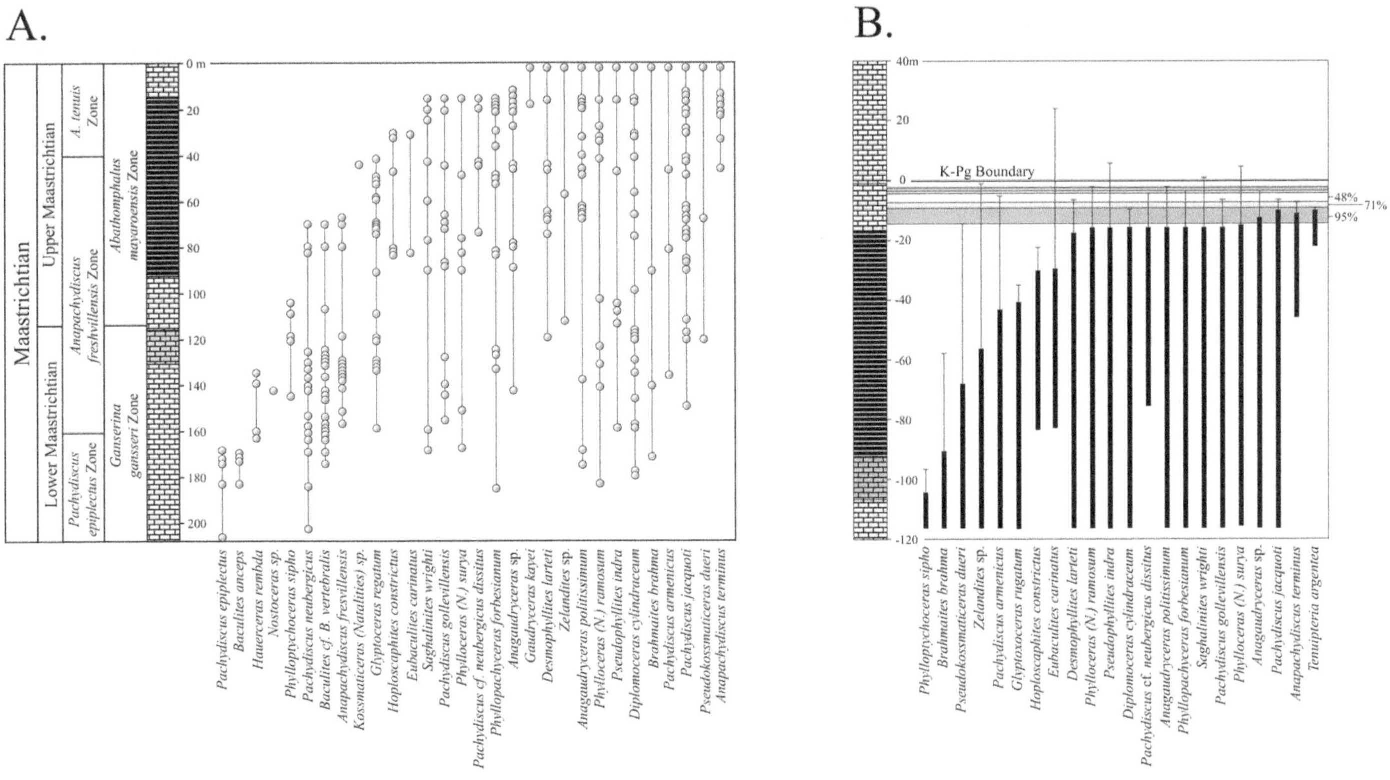

Figure 14. Marshall and Ward's summary of the end-Cretaceous ammonite extinction pattern in northern Spain and southern France projected onto the Cretaceous-Paleogene (K-Pg) boundary section from Zumaya, northern Spain. (A) Overall extinction pattern in the Maastrichtian portion of the section. (B) Extinction pattern for the ammonoid "candidates for extinction" (Marshall and Ward, 1996, p. 1362) with 50% confidence intervals on the individual stratigraphic ranges of the "most likely" candidates for a coincident extinction at the K-Pg boundary. Results of application of the Springer (1990) test for a coincident extinction horizon indicate that the observed biostratigraphic record is nonuniquely consistent with the hypothesis that all species for which the confidence interval includes the horizon located 8 m below the local K-Pg boundary (lower limit of a 6.5 m sampling gap) were extant at that time. However, this result does not exclude the possibility that some or most of these taxa did not, in fact, reach this horizon, much less become extinct at the K-Pg boundary itself. Subsequent rarefaction analysis results support, at the very least, an ammonoid abundance drop in the sampling gap, a result that receives additional support from the Hendaye section in southern France (Marshall and Ward, 1996). On the basis of these stratigraphic data, Marshall and Ward (1996) concluded that a nonunique confidence-interval–based case could be made for the coincident extinction of five ammonite species (a mere 17% of the total Spanish and French Upper Maastrichtian ammonoid fauna) at the K-Pg boundary. However, even this result may be an overestimate, because very large numbers of other, noncoincident extinction geometries, including the survivorship of some ammonoid species into the Paleogene, are also consistent with these data. Both figures redrawn from Marshall and Ward (1996).

zones only demonstrates that there are no hiatuses large enough to have removed an entire ammonite biozone. An analysis of the Zumaya microfossil data (MacLeod and Keller, 1991) demonstrated that the lowermost Danian planktonic foraminiferal zone (P0) is missing at Zumaya, along with the lowermost portion of the succeeding zone (P1a). In addition, absence of the Ir anomaly, which marks virtually all temporally complete Cretaceous-Paleogene boundaries worldwide and serves as the formal definition of the Cretaceous-Paleogene boundary, along with the absence of the characteristic boundary clay layer containing shocked minerals, suggests that an as-yet-undetermined, uppermost Cretaceous (Maastrichtian) chronostratigraphic interval is also missing from the Zumaya section.

Irrespective of these chronostratigraphic complexities, it is clear that the vast majority of ammonoid species (81%) disappear from the fossil record in an interval well below (>20 cm) the Cretaceous-Paleogene boundary, an interval that in total represents at least 10,000 yr (Sadler, 1981; MacLeod and Keller, 1991). Use of stratigraphic confidence intervals to represent uncertainty in these data further complicates the interpretation of these data. Marshall and Ward (1996) applied SCI analysis to these data and concluded that they were most consistent with the recognition of a major extinction event occurring prior to the Cretaceous-Paleogene interval—within an interval in which there is no sedimentological evidence for major disruption in sedimentation patterns.

Unfortunately, Marshall and Ward's (1996) SCI analysis was compromised by their use of stratigraphic depth as the scale over which to calculate the confidence intervals. This approach is suboptimal insofar as it presumes implicitly that no hiatuses, gaps in sediment accumulation, or fluctuations in rates of sediment accumulation are present within their data. This presumption it tied logically to the fact that the confidence interval calculation is referenced to the total stratigraphic range of the taxon (see Fig. 13). If an unrecognized hiatus exists in the sampled part of the section or core, the total stratigraphic range will be underestimated, and so the duration of the confidence interval will be underestimated. Moreover, if a stratigraphic hiatus terminates the taxon's range, the stratigraphic confidence interval may fall within the boundaries of the missing time interval, in which case it would be inappropriate to extend the interval (artificially) into sediments of an age that is younger than the confidence interval itself. For example, if a species' existence was represented by (say) 20 occurrences in a stratigraphic section that spanned the Upper Maastrichtian (estimated duration of 1 m.y.), the 95% stratigraphic confidence interval would be estimated as 0.17 of the observed range or 170,000 yr (see Fig. 13). However, if this species' range was terminated by a stratigraphic hiatus at the Cretaceous-Paleogene boundary, it would not be appropriate to draw the confidence interval at 0.17 of the thickness of the observed range if the sediments immediately overlying the Maastrichtian interval are (say) Eocene in age. This discrepancy does not mean the SCI method is wrong, only that care needs to be taken in its application and the interpretation of its results.

The best that can be said for these data is that, if taken at face value, they demonstrate that ammonoids became extinct in a progressive manner in western Europe over a considerable time interval at the end of the Maastrichtian. Some (an as-yet-undetermined number) of the ammonoid species may have become extinct simultaneously at the Cretaceous-Paleogene boundary (see Marshall and Ward, 1996). However, the observed record is not compatible with this interpretation uniquely. It is also consistent with the inference that some ammonoids may have survived into the Lower Danian (see Machalski and Heinberg 2006). Taking the uncertainties inherent in the use of paleontological data into consideration, it must also be admitted that, with regard to these data, it cannot be determined with accuracy whether (1) any coincident extinctions of ammonoid species occurred over this interval, (2) any ammonoid extinctions coincided with the chronostratigraphic Cretaceous-Paleogene boundary, or (3) any ammonoid species persisted into the lowermost Danian. These uncertainties can only be resolved by developing high-resolution regional and global chronostratigraphic models for Cretaceous-Paleogene boundary sections and cores (e.g., MacLeod and Keller, 1991) and subjecting those data to confidence interval analysis, being careful to migrate all errors associated with the former into the latter.

Nonavian Dinosaur Extinction at Hell Creek, Montana

Dinosaurs, the perennial poster children of extinction research, symbolize in iconic form not only all terrestrial victims of extinction processes but humanity's deep-seated fear of its own mortality as a species (see Mitchell, 1998). Taxonomic discussions of the dinosaur record are complicated by the fact that, from an evolutionary point of view, dinosaurs are not extinct. The clade Dinosauria includes two prominent subgroups: Order Ornithischia and Order Saurischia, both characterized by unique arrangements of the bones of the sacral skeleton and presence in many Ornithischia of a predentary bone. Order Saurischia is further divided into two suborders, the Sauropodomorpha and the Theropda, into which most contemporary taxonomists place birds, based originally on similarities in the arrangement of avian pubic bones, but in the context of more recent discoveries based on shared possession of feathers, nesting behaviors, etc. This placement of an extant clade within the Dinosauria makes the discussion of dinosaur extinction difficult because representatives of this group are one of the most common animals observed today; a nomenclatural complexity usually resolved by referring to extinct members of the Dinosauria by the admittedly awkward phrase "nonavian dinosaurs."

Dinosaurs originated in the Late Triassic from archosaurian ancestors. Their stratigraphic record continues to be revised as new discoveries are made. However, as of this writing, the clade appears to have diversified steadily throughout the Jurassic to an Early Cretaceous peak, suffered a decline in the Albian-Aptian interval and the late Early Cretaceous, and then rebounded in the Late Cretaceous to achieve maximum family and generic richness in the Campanian. From this point, the stratigraphic record shows falling numbers of dinosaur family and generic counts to the disappearance of all nonavian dinosaurs in the Maastrichtian Stage at or close to the Cretaceous-Paleogene boundary.

Figure 15 summarizes the stratigraphic distribution of the entire vertebrate fauna—including all dinosaurs—from the Hell Creek Formation, which crops out in the western U.S. states of Montana, Wyoming, North Dakota, and South Dakota. Data collected from these localities represent the sum total of our stratigraphic knowledge of all dinosaur species at the very end of the nonavian dinosaurs' evolutionary history. The stratigraphy of the Hell Creek Formation is complex and open to a wide variety of interpretations. What is known is that these deposits were formed by river systems meandering across the alluvial plain of the Laramide mountains located to the west of the outcrop belt as these rivers flowed into the Pierre Seaway, which occupied the central portion of the North American craton at this time. During the latest Maastrichtian, this seaway was drained by an approximately 200 m fall in eustatic sea level, moving the shoreline south to the region now occupied by the Gulf Coast states (Texas, Louisiana, Mississippi) and so creating a regional erosional surface in the North American interior. High-resolution stratigraphy shows the fluvial sediments of the Hell Creek Formation to be conformably overlain by a carbonaceous claystone with root traces, the characteristic Cretaceous-Paleogene boundary clay complex, which contains shocked minerals and an Ir anomaly, and a lignite layer currently assigned to the Paleocene Tullock Formation.

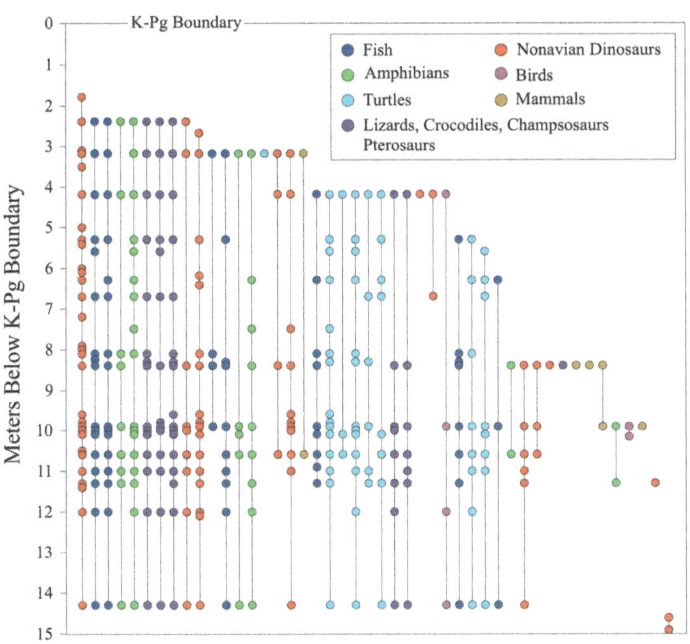

Figure 15. The terrestrial vertebrate faunal extinction record in the Hell Creek Formation. Note the progressive pattern of taxon last occurrences and the fact that all extinctions are recorded as having taken place well below the local Cretaceous-Paleogene (K-Pg) boundary, as marked by a well-developed boundary clay interval and an Ir anomaly. Figure redrawn from Archibald (2011).

As can be seen from Figure 15, the stratigraphic record of vertebrate species occurrences records a stepped (owing to sampling issues), but progressive, extinction pattern with prominent extinction horizons occurring at 9.9 m, 8.4 m, 4.2 m, and 2.4 m below the local Cretaceous-Paleogene boundary. The stratigraphically highest occurrence of any confirmed dinosaur remains (assignable to *Triceratops*) was recorded at 1.77 m below the Cretaceous-Paleogene boundary. No stratigraphic confidence interval analysis for these data has been published to date. Nevertheless, all the uncertainties and complications described earlier herein for the western European ammonite analysis pertain to these data as well. What can be said with some measure of confidence (again) is that, if taken at face value, these data demonstrate that vertebrates (including all nonavian dinosaurs) declined in a progressive manner in the Hell Creek sections over a considerable time interval at the end of the Maastrichtian. Taking the inherent uncertainties and biases in the stratigraphic record into consideration, it cannot be determined from these data whether (1) any coincident extinction of vertebrate (or dinosaur) species occurred within this interval, (2) any vertebrate (or dinosaur) extinctions coincided with the chronostratigraphic Cretaceous-Paleogene boundary, or (3) any nonavian dinosaur species persisted into the lowermost Paleogene.

Global Planktonic Foraminifer Extinction

While planktonic foraminifera cannot claim to exert the same hold on the popular imagination as dinosaurs, or even ammonites, few groups can claim superior practical utility for high-resolution biostratigraphic, paleobiogeographic, and paleoecological analysis. This group first appears in the fossil record in the Late Jurassic and remains extant in modern oceans today. Through this long evolutionary history, species numbers have waxed and waned. Nevertheless, substantial turnovers in species numbers and characteristic species morphologies occurred at the end of the Cretaceous and during the late Eocene–Oligocene transition (Cifelli, 1969; Cermeñño et al., 2013).

Existence of a planktonic foraminiferal mass extinction at the Cretaceous-Paleogene boundary had been accepted by most micropaleontologists (see Luterbacher and Premoli-Silva, 1964; Cifelli, 1969; Olsson, 1970) for decades prior to the 1980 publication of the Alvarez bolide impact scenario (Alvarez et al., 1980). Before 1980, this extinction had been attributed to a variety of causes, including collapse of the thermal structure of the global marine water column (Cifelli, 1969), climate change induced by Deccan volcanism (McLean, 1978, 1985a, 1985b), and food-chain collapse induced by changes in marine surface-water salinity (Gartner and McGurik, 1979). In terms of taxonomic scope, estimates have varied, but most regarded well over 50% of the standing Upper Maastrichtian species stock to have disappeared coincident with the Cretaceous-Paleogene boundary (e.g., Harland et al., 1967; Van Valen, 1973).

Following publication of the 1980 Alvarez et al. bolide impact article in *Science* magazine, a number of paleontologists set out to reexamine the Cretaceous-Paleogene fossil record for evidence of a precise association between dramatic declines in standing species richness and either direct or correlative associations with the physical evidence of bolide impact. The first planktonic foraminiferal results to be published were Jan Smit's studies of the southern Spanish sections of Agost and Caravaca (Smit, 1977a, 1977b, 1979, 1982), at the Gubbio section (Smit and ten Kate, 1982), and later at the Cretaceous-Paleogene stratotype section (El Kef, Tunisia; Smit, 1982). In all these successions, Smit reported an association among Cretaceous species disappearances, the Cretaceous-Paleogene boundary, and physical evidence of bolide impact (e.g., Ir anomalies). The only exception to this pattern that he and a few others noted was the morphologically enigmatic species *Guembelitria cretacea*, though, in the earliest publications, Smit did allude to a few, possibly reworked specimens and/or species in the earliest overlying Danian sediments and mentioned the possibility of a short-lived transitional interval dominated by Cretaceous survivor species (Smit, 1977b). However, this survivor fauna was later dropped from all subsequent reports of data from, and discussions of, the earliest Danian foraminiferal record by Smit and colleagues (e.g., Smit, 1982). Later confirmatory work showed the Gubbio section to be biostratigraphically incomplete (Herm et al., 1981; Smit, 1982; Keller, 1988, 1989). Both the Spanish and Tunisian sections were also shown to record a rich, diverse, and abundant fauna of Cretaceous species above the Cretaceous-Paleogene boundary, in what are indisputably Danian sediments (e.g., Keller, 1988, 1989; Canudo et al., 1991; Keller and Benjamini, 1991; Luciani, 1997; Arenillas et al., 2004; Molina et al., 2005, 2006).

Keller and her colleagues pointed to the excellent preservational state of the specimens comprising these anomalous Cretaceous occurrences, structured patterns of abundance variation in some cases throughout an extended stratigraphic interval, their morphotypic similarity to the emerging Danian planktonic foraminiferal fauna (suggesting an overall ecological similarity), the highly structured pattern of biogeographic variation characteristic of these occurrences (MacLeod and Keller, 1994), and patterns of isotopic variation (Barrera and Keller, 1990) all as independent lines evidence suggesting that at least some Cretaceous taxa survived the end-Cretaceous extinction event, whatever its cause. On the basis of comparative data, this group of putative end-Cretaceous "survivors" appeared to dominate the lowermost Danian planktonic foraminiferal fauna, where they were progressively replaced by fully Danian species, first in the tropics and later in high-latitude boreal refugia (MacLeod and Keller, 1994). Subsequently, Keller and colleagues documented similar patterns of apparent "Cretaceous" species survivorship in other classic Cretaceous-Paleogene boundary sections in Texas (Keller, 1989), Denmark (Schmitz et al., 1992), at other Spanish localities (Canudo et al., 1991), in Israel (Keller and Benjamini, 1991), and in the Antarctic (Keller, 1993).

In an attempt to resolve the enigma of the Paleogene occurrences of "Cretaceous" planktonic foraminiferal species, Paul (2005, p. 300) suggested a simple, straightforward method for placing this determination of mass extinction survivorship on an analytic footing.

> "If a Cretaceous species genuinely survived the mass extinction, one would expect it to be present in Tertiary sediments in most of the sections from which it is recorded. There is no such expectation for reworked specimens, which one would expect to occur much more randomly in Tertiary sediments.
>
> To establish this requires a number of sections in which the species is recorded. If in the majority of these sections the species is last recorded in the Cretaceous, it is more likely to be a Cretaceous species which genuinely became extinct at or below the Cretaceous-Tertiary boundary. If, on the other hand, the species is recorded in the Tertiary in most sections, it is more likely to be a genuine survivor."

Paul (2005) applied his analytic approach to sections and cores from 16 Cretaceous-Paleogene boundary sections and cores localities, represented by 25 successions, and concluded that as many as 28 (21.5%) of "Cretaceous" species survived into the lowermost Danian biozone P0, with 12 (11%) of these surviving into the next succeeding biozone, P1a. These results are not dissimilar from the probable survivorship figures cited by Keller and her colleagues (e.g., Keller, 1989; MacLeod and Keller, 1991; MacLeod, 1995a, 1995b), but they are decidedly inconsistent with the stance on the "Cretaceous" planktonic foraminiferal survivorship issue taken by other planktonic foraminiferal biostratigraphers (e.g., Huber, 1991, 1996; Huber et al., 1994; Smit, 1996; Liu and Olsson, 1992; Arenillas et al., 2004; Molina et al., 2005, 2006).

MacLeod (2011) corrected several taxonomic inconsistencies in Paul's preliminary study to objectively identify the suite of "Cretaceous" species based on Paul's (2005) test for survivorship. Results indicate that of the 73 planktonic foraminifer species recovered from these sections/cores for which stratigraphic ranges are known to reach as high as the Cretaceous-Paleogene boundary, 22 (30.1%) pass Paul's survivorship test. Moreover, when this test is applied to the stratigraphic record of "Cretaceous" species found routinely in succeeding Paleogene biozones, 14 species (19.1%) are identified as zone P1a survivors (~30,000+ yr into the Paleogene) and two appear to have survived for approximately 2 m.y. into the Cenozoic. These results were then used to develop a common, synthetic model for the chronostratigraphy of 30 Cretaceous-Paleogene sections and cores using Kemple, Sadler, and Strauss' constrained optimization (CONOP) method for inferring the correct sequence of biostratigraphic datums in time (see also Kemple et al., 1995; MacLeod and Sadler, 1995). Results of this analysis are summarized in Figure 16.

As in the case of the ammonite and terrestrial vertebrate data sets discussed earlier, planktonic foraminifera exhibit a pattern of progressive extinction across the Cretaceous-Paleogene horizon, with significant numbers of "Cretaceous" species occurring

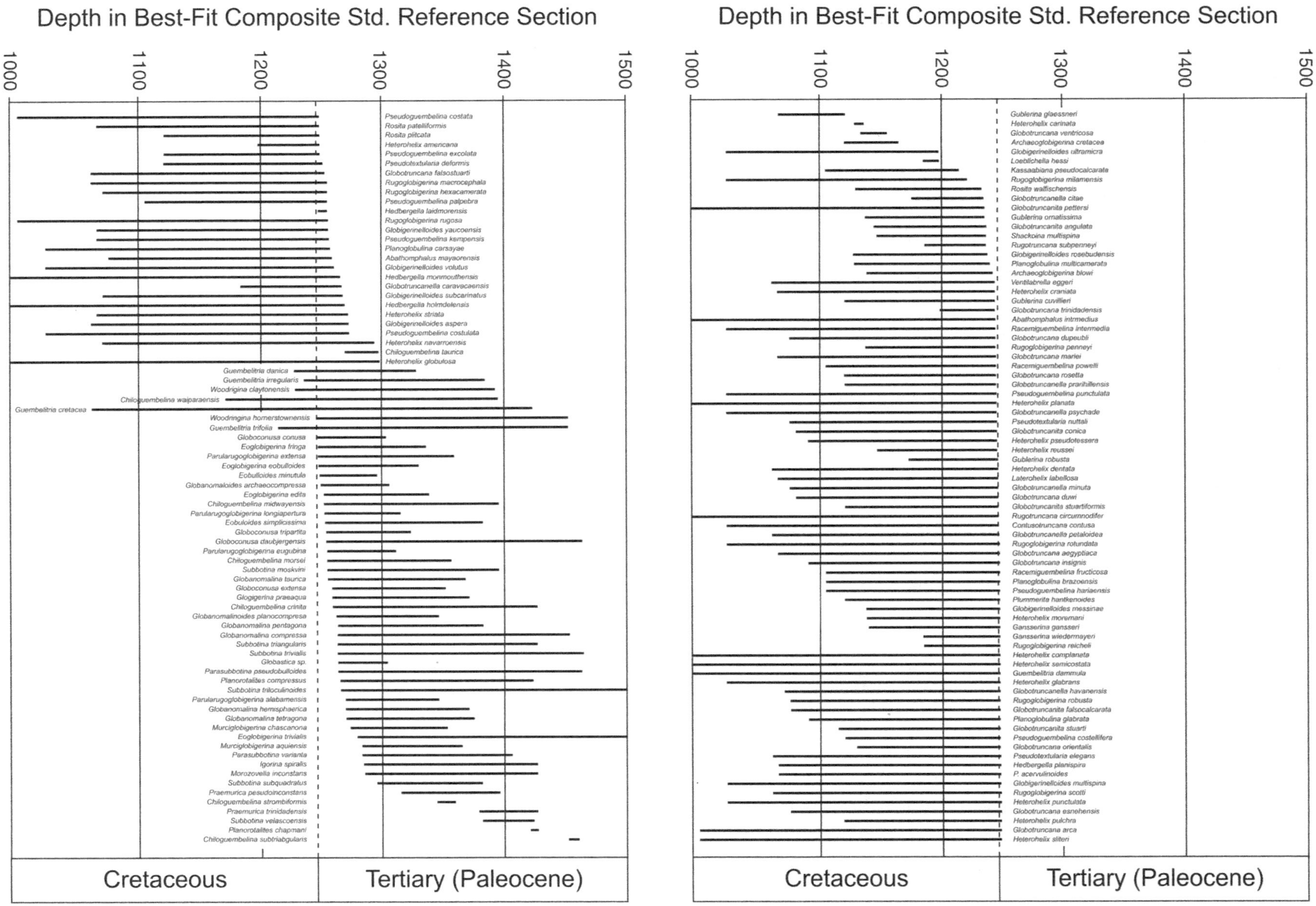

Figure 16. Inferred global pattern of planktonic foraminifer extinctions and originations in the interval across the Cretaceous-Paleogene (K-Pg) boundary (dashed line) estimated by combining data from 24 K-Pg boundary sections and cores into a single chronostratigraphic model and taking care to include only those data for which demonstrable evidence of reworking is lacking. See MacLeod (2011) for more detail.

below (37), at (8), and above (65) the boundary. If all species found above the Cretaceous-Paleogene boundary in any section are included as survivors, the survivorship ratio rises to 87%. If only the 22 species that pass the very conservative Paul (2005) test are used, this ratio drops back to 30%. The actual value probably lies between these two extremes.

The true chronostratigraphic ranges of these taxa remain uncertain. Some proportion probably did vanish coincident with the boundary. Others probably became extinct prior to the boundary horizon, and others are probably part of the survivor fauna. Stratigraphic confidence interval analysis could be applied to these data. However, for the reasons stated earlier herein, such results will not resolve the issue of whether this progressive extinction pattern can be interpreted as a catastrophic extinction pattern.

TESTING EXTINCTION CAUSE HYPOTHESES

Establishing the correct temporal sequence of biotic events in the distant past is a challenging and necessary step toward testing hypotheses relating to the cause(s) of extinction events, but it is only the first step. There are at least two further steps that need to be negotiated successfully before speculations about extinction causes can be fairly said to have entered the realm of scientific hypotheses. The first of these is that the various scenarios advanced as causes of ancient extinctions must make unique predictions about the patterns that should be found in the fossil record if (and when) these mechanisms are in operation. This is quite a stringent requirement. To date, none of the advocates of various mass extinction causal mechanisms has done a very good job in making predictions that could be used to test and (possibly) confirm their favored model. Most simply specify that large numbers of extinctions will be recorded in the fossil record at the times over which their mechanisms operated in Earth's past (but see Finnegan et al. [2012] for an example of the application of this approach to investigation of the end-Ordovician extinction event).

The problem with this simplistic approach to extinction cause hypothesis testing is that (1) if more than a single mechanism is operating at any one time in Earth history, it will be impossible to know which extinction to assign to which individual mechanism and/or to the synergistic effects of them operating in concert, and (2) owing to the Signor-Lipps effect (1982), along with the effects of hiatuses (MacLeod et al., 1997; MacLeod, 2003), fluctuations in sediment accumulation rates (Sadler, 1981), the availability of outcrops that can be searched for fossils (Smith and McGowan, 2007), and the incomplete biogeographic records of most fossil groups (e.g., dinosaurs; see Archibald, 2011), the fossil record may not record the actual time of any species' extinction accurately. Moreover, since the majority of the paleontological data on which extinction studies are based are known only to the level of the stratigraphic stage, strict correspondence between the operation of a putative extinction causal mechanism and the data of the fossil record at resolutions below this rather coarse level will be very difficult to achieve for any but a few groups (e.g., planktonic foraminifera) or a few localities (e.g., the Zumaya and Hell Creek investigations summarized herein). It is possible that some causal mechanisms may imply the creation of novel ecological, geographic, morphological, or other patterns in fossil taxa in terms of differential extinction susceptibility. The so-called Lilliput effect (Urbanek, 1993; Harries and Knorr, 2009) is one example of a potential morphological pattern that has been produced by a number of major and minor mass extinction events. However, to date, none of the research into of the various causal mechanisms has reached the stage at which predictions of species sorting patterns unique to that mechanism have been identified. Indeed, since all of the ultimate extinction causal mechanisms (sea-level regression, anoxia, large igneous province [LIP] volcanic eruption, bolide impact) exert pressure on organismal populations worldwide, not through their direct effects but by changing the rates and intensities at which other, secondary mechanisms operate (e.g., cloud cover, rainfall, mean annual temperature, injection of particulate material into the stratosphere, atmospheric composition), the unique aspects of singular mechanisms may be irreducibly masked when seen from the vantage point of the fossil record.

Even if it proves ultimately impossible to identify the signatures of extinction causal mechanisms in biotic data directly, it still might be possible to gather sufficient circumstantial evidence to demonstrate a statistically significant association between the operation of particular causal mechanisms and biosphere responses of a particular type. Doing this will require another step change in extinction studies in the form of a series of cause-effect comparisons across extinction events of various magnitudes. The logic of this hypothesis test was summarized eloquently by David Raup (1991, p. 151).

> "There is no way of assessing cause and effect [in historical data] except to look for patterns of coincidence—and this requires multiple examinations of each cause-and-effect pair. If all extinction events are different the deciphering of any one of them will be next to impossible."

Raup's observation is a timely restatement of the classic scientific experimental method in which laboratory conditions are arranged such that all variables save one are held constant, and the experiment is run multiple times in the hope that a common response pattern emerges. Of course, the data of paleontology rarely allow controlled laboratory experiments to be undertaken, especially on extinct species. Nevertheless, each of the ultimate extinction causal mechanisms has operated multiple times over the course of Earth history, and those times of operation are known with certainty to the level of the stratigraphic stage. In principle, it is a simple matter to perform a statistical test of the level of association between the time series of an extinction causal process and time series of fluctuations in extinction intensities over time. In practice, this is proving fiendishly difficult owing (again) to biasing factors in the stratigraphic record.

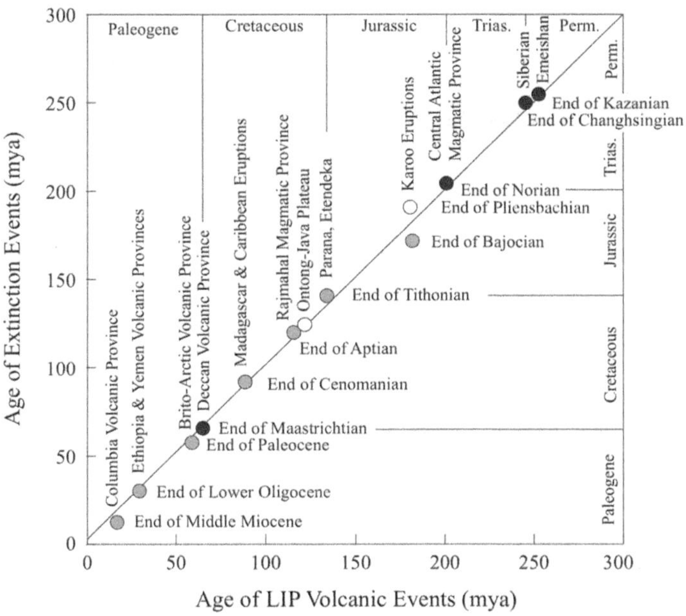

Figure 17. Comparison of the estimated ages of large igneous province (LIP) volcanic eruption events with the estimated ages of intermediate (gray) and large (black) extinction events. Redrawn from Courtillot (1999).

A low-temporal-resolution analysis of this sort was carried out by Rampino and Stothers (1988) for the association between large igneous province volcanic events and peaks in extinction intensity over the last 250 m.y. (Fig. 17). This association is highly significant statistically and forms a convincing demonstration of a cause-effect link between these two event classes. MacLeod (2004) developed a more robust statistical test for the association between the time series of various causal mechanisms and the eight largest extinction-intensity peaks of the last 250 m.y. based on a Monte Carlo simulation model. Unlike the statistics associated with Rampino and Stothers' (1988) regression-based approach to time-series analysis, this test does not involve assumptions about the underlying distribution of the data or independence of observation. It also allows precise levels of association significance to be calculated. Results of this stage-level test for the time series of four putative causal processes are shown in Table 3. Of these potential mechanisms, both large igneous province volcanism and eustatic sea-level regression exhibit a statistically significant stage-level association with the time series of extinction-intensity peaks. Large bolide impacts (= crater size > 50 km) exhibited a high (0.09), but nonsignificant association.

This test should only be judged as preliminary, since the level of temporal resolution used was (necessarily) coarse, and since neither the intensities of the causal mechanisms nor sizes of the extinction peaks were taken into consideration. More sophisticated analyses will be required to explore the character of these more detailed associations further. Nevertheless, these results do illustrate the strategy of adopting an explicitly comparative approach to the investigation of extinction causality. This model has also been extended to test multiple-cause hypotheses (MacLeod, 2005b).

As an enticement, and an encouragement, to continue this line of investigation, I can do no better than to point out that, when the records of extinction intensity, bolide impact, large igneous province volcanism, marine anoxia events, and eustatic sea-level change are plotted against one another (Fig. 18), the three largest extinction events of the last 250 m.y. are associated uniquely with large igneous province volcanic eruptions and lowstands in global sea level. In one case (end-Cretaceous), a large extinction-intensity peak is associated with an anomalously large bolide impact. The most important implication of these associations may be that many natural factors can raise global extinction levels when they operate singly or in groups at low intensities. High levels of extinction in the global record may require the operation of multiple factors at high intensities. This is a hypothesis that awaits testing, and, perhaps most importantly, one that can be tested.

ACKNOWLEDGMENTS

I hereby acknowledge and thank the many colleagues, collaborators, students, reviewers, and intellectual sparring partners with whom I have had the pleasure of discussing and debating many of the topics covered in this article over the years. Special thanks go to Peter Sadler and Charles Marshall, whose comments, criticisms, objections, and encouragements on an initial draft resulted in numerous improvements. None of the aforementioned, however, are responsible in any way for any remaining errors in fact or interpretation.

TABLE 3. ESTIMATED PROBABILITIES OF A CAUSATIVE STAGE-LEVEL ASSOCIATION BETWEEN MAJOR- AND INTERMEDIATE-LEVEL PEAKS IN THE TIME SERIES OF EXTINCTION INTENSITIES AND THE TIME SERIES OF A VARIETY OF PHYSICAL EVENTS THAT WOULD BE EXPECTED TO DISRUPT THE GLOBAL ENVIRONMENT

Putative extinction cause	No. of occurrences	No. of stage-level associations	Median probability
Bolide impact (>10 km)	20	6	0.172
Major bolide impact (>50 km)	6	3	0.088
Major eustatic sea-level regression	16	6	0.040
Major eustatic sea-level lowstand	7	3	0.135
Major marine anoxia interval	10	2	0.609
Major LIP volcanic eruption	9	6	0.000

LIP—large igneous province.

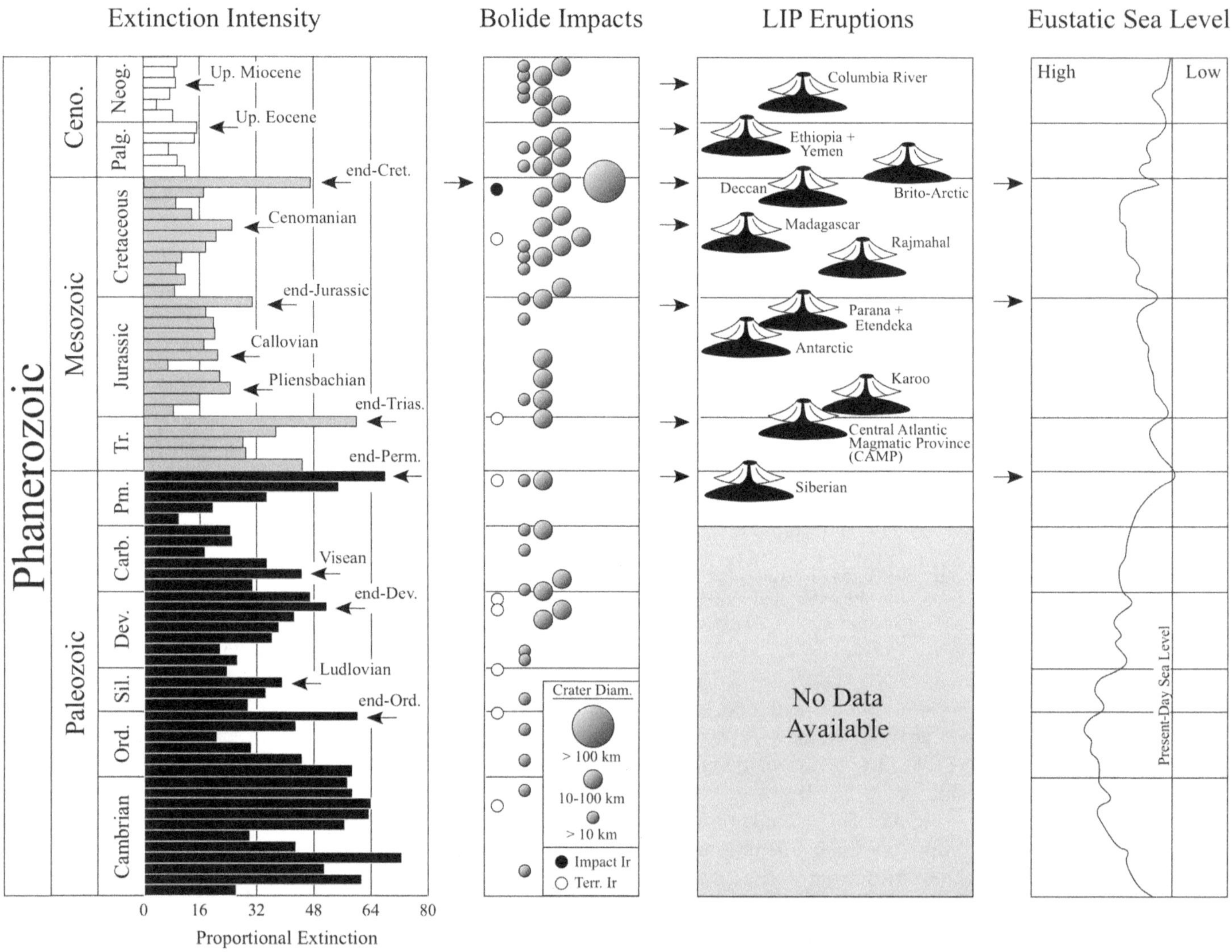

Figure 18. Graphical comparison of the Phanerozoic time series of stage-level proportional extinction intensities (far left), bolide impact events (middle left), large igneous province (LIP) eruptions (middle right), and fluctuations in eustatic sea level (far right).

REFERENCES CITED

Alvarez, L.W., Alvarez, W., Asaro, F., and Michel, H.V., 1980, Extraterrestrial cause for the Cretaceous-Tertiary extinction: Science, v. 208, no. 4448, p. 1095–1108, doi:10.1126/science.208.4448.1095.

Archibald, J.D., 2011, Extinction and Radiation: How the Fall of Dinosaurs Led to the Rise of Mammals: Baltimore, Maryland, Johns Hopkins University Press, 108 p.

Arenillas, I., Arz, J.A., and Molina, E., 2004, A new high-resolution planktic foraminiferal zonation and subzonation for the lower Danian: Lethaia, v. 37, p. 79–95, doi:10.1080/00241160310005097.

Bambach, R.K., Knoll, A.K., and Wang, S., 2004, Origination, extinction, and mass depletions of marine diversity: Paleobiology, v. 30, p. 522–542, doi:10.1666/0094-8373(2004)030<0522:OEAMDO>2.0.CO;2.

Barrera, E., and Keller, G., 1990, Foraminiferal stable isotope evidence for gradual decrease of marine productivity and Cretaceous species survivorship in the earliest Danian: Paleoceanography, v. 5, p. 867–890, doi:10.1029/PA005i006p00867.

Benton, M.J., 1993, The Fossil Record 2: London, Chapman & Hall, 845 p.

Canudo, J.I., Keller, G., and Molina, E., 1991, Cretaceous/Tertiary boundary extinction pattern and faunal turnover at Agost and Caravaca, S.E. Spain: Marine Micropaleontology, v. 17, p. 319–341, doi:10.1016/0377-8398(91)90019-3.

Cermeño, P., Castro-Bugallo, A., and Vallina, S. M., 2013, Diversification patterns of planktic foraminifera in the fossil record: Marine Micropaleontology, v. 104, p. 38–43, doi:10.1016/j.marmicro.2013.09.003.

Chamberlain, T.C., 1897, The method of multiple working hypotheses: Journal of Geology, v. 5, p. 837–848.

Cifelli, R.S., 1969, Radiation of Cenozoic planktonic foraminifera: Systematic Zoology, v. 18, p. 154–168, doi:10.2307/2412601.

Courtillot, V., 1999, Evolutionary Catastrophes: The Science of Mass Extinction: Cambridge, UK, Cambridge University Press, 173 p.

Darwin, C., 1859, On the Origin of Species by Means of Natural Selection, or the Preservation of Favoured Races in the Struggle for Life: London, John Murray, 513 p.

Fan, Junxuan, Chen, Qing, Hou, Xudong, Miller, A.I., Melchin, M.J., Shen, Shuzhong, Wu, Shuangye, Goldman, D., Mitchell, C.E., Yang, Qun, Zhang, Yuandong, Zhan, Renbin, Wang, Jun, Leng, Qin, Zhang, Hua, and Zhang, Linna, 2013, Geobiodiversity database: A comprehensive section-based integration of stratigraphic and paleontological data: Newsletters on Stratigraphy, v. 46, no. 2013, p. 111–136.

Finnegan, S., Heim, N.A., Peters, S.E., and Fischer, W.W., 2012, Climate change and the selective signature of the Late Ordovician mass extinction: Proceedings of the National Academy of Sciences of the United States of America, v. 109, no. 18, p. 6829–6834, doi:10.1073/pnas.1117039109.

Foote, M., 2000, Origination and extinction components of taxonomic diversity: Paleozoic and post-Paleozoic dynamics: Paleobiology, v. 26, no. 4, p. 578–605, doi:10.1666/0094-8373(2000)026<0578:OAECOT>2.0.CO;2.

Gartner, S., and McGurik, J.P., 1979, Terminal Cretaceous extinction scenario for a catastrophe: Science, v. 206, p. 1272–1276, doi:10.1126/science.206.4424.1272.

Harland, W.B., Holland, C.H., House, M.R., Hughes, Norman, F., and Reynolds, A.B., 1967, The Fossil Record: A Symposium with Documentation: Geological Society of London Special Publication 2, 827 p.

Harries, P.J., and Knorr, P.O., 2009, What does the 'Lilliput effect' mean?: Palaeogeography, Palaeoclimatology, Palaeoecology, v. 284, p. 4–10, doi:10.1016/j.palaeo.2009.08.021.

Hedeen, S., 2008, Big Bone Lick: The Cradle of American Paleontology: Lexington, Kentucky, University of Kentucky Press, 202 p.

Herm, D., Hillebrandt, A.V., and Perch-Nielsen, K., 1981, Die Kriede/Tertiar-Grenze im Lattenbirge (Nordliche Kalkalpen) in mikropalaontologischer Sicht: Geologica Bavarica, v. 82, p. 319–344.

Huber, B.T., 1991, Maestrichtian planktonic foraminiferal biostratigraphy and the Cretaceous/Tertiary boundary at ODP Hole 738C (Kerguelen Plateau): Proceedings of the Ocean Drilling Project, Scientific Results, v. 119, p. 451-465.

Huber, B.T., 1996, Evidence for planktonic foraminifer reworking versus survivorship across the Cretaceous-Tertiary boundary at high latitudes, *in* Ryder, G., Fastovsky, D., and Gartner, S., eds., The Cretaceous-Tertiary Event and Other Catastrophes in Earth History: Geological Society of America Special Paper 307, p. 319–334.

Huber, B.T., and Boersma, A., 1994, Cretaceous origination of *Zeauvigerina* and its relationship to Paleocene biserial planktonic foraminifera: Journal of Foraminiferal Research, v. 24, p. 268–287, doi:10.2113/gsjfr.24.4.268.

Huber, B.T., Liu, C., Olsson, R.K., Berggren, W.A., Keller, G., and MacLeod, N., 1994, MicroForum: Comment and response on "The Cretaceous-Tertiary transition in the Antarctic Ocean and its global implications", by G. Keller: Marine Micropaleontology, v. 24, no. 1994, p. 91–118.

Jablonski, D., 1986, Background and mass extinctions: The alternation of macroevolutionary regimes: Science, v. 231, no. 4734, p. 129–133, doi:10.1126/science.231.4734.129.

Jablonski, D., 1995, Extinctions in the fossil record, *in* Lawton, J.H., and May, R.M., eds., Extinction Rates: Oxford, UK, Oxford University Press, p. 25–44.

Keller, G., 1988, Extinction, survivorship and evolution of planktic foraminifera across the Cretaceous/Tertiary boundary at El Kef Tunisia: Marine Micropaleontology, v. 13, p. 239–263, doi:10.1016/0377-8398(88)90005-9.

Keller, G., 1989, Extended Cretaceous/Tertiary boundary extinctions and delayed population change in planktonic foraminiferal faunas from Brazos River, Texas: Paleoceanography, v. 4, p. 287–332, doi:10.1029/PA004i003p00287.

Keller, G., 1993, The Cretaceous-Tertiary boundary transition in the Antarctic Ocean and its global implications: Marine Micropaleontology, v. 21, p. 1–45, doi:10.1016/0377-8398(93)90010-U.

Keller, G., and Benjamini, C., 1991, Paleoenvironment of the eastern Tethys in the early Paleocene: Palaios, v. 6, p. 439–464, doi:10.2307/3514984.

Kemple, W.G., Sadler, P.M., and Strauss, D.J., 1995, Extending graphic correlation to many dimensions: Stratigraphic correlation as constrained optimization, *in* Mann, K.O., and Lane, H.R., eds., Graphic Correlation: Tulsa, Oklahoma, Society for Sedimentary Geology (SEPM) Special Publication 53, p. 65–82.

Liu, C., and Olsson, R.K., 1992, Criteria for recognizing survivor microfossil species during mass extinction, planktonic foraminifera at K/T boundary as example: Geological Society of America Abstracts with Programs, v. 24, no. 7, p. A29.

Luciani, V., 1997, Planktonic foraminiferal turnover across the Cretaceous-Tertiary boundary in the Vajont Valley (southern Alps, northern Italy): Cretaceous Research, v. 18, p. 799–821, doi:10.1006/cres.1997.0088.

Luterbacher, H., and Premoli Silva, I., 1964, Biostratigrafia del limite Cretaceo-Terziario nell'Appennino centrale: Rivista Italiana di Paleontologia e Stratigrafia, v. 70, p. 67–88.

Machalski, M., and Heinberg, C., 2006, Evidence for ammonite survival into the Danian (Paleogene) from the Cerithium Limestone at Stevns Klint, Denmark: Bulletin of the Geological Society of Denmark, v. 52, no. 2, p. 97–111.

MacLeod, N., 1995a, Graphic correlation of high latitude Cretaceous-Tertiary boundary sequences at Nye Kløv (Denmark), ODP Site 690 (Weddell Sea), and ODP Site 738 (Kerguelen Plateau): Comparison with the El Kef (Tunisia) boundary stratotype: Modern Geology, v. 19, p. 109–147.

MacLeod, N., 1995b, Graphic correlation of new Cretaceous/Tertiary (K/T) boundary sections, *in* Mann, K.O., and Lane, H.R., eds., Graphic Correlation: Tulsa, Oklahoma, Society for Sedimentary Geology Special Publication 53, p. 215–233.

MacLeod, N., 2003, The causes of Phanerozoic extinctions, *in* Rothschild, L., and Lister, A., eds., Evolution on Planet Earth: London, Academic Press, p. 253–277.

MacLeod, N., 2004, Identifying Phanerozoic extinction controls: Statistical considerations and preliminary results, *in* Beaudoin, A.B., and Head, M.J., eds., The Palynology and Micropaleontology of Boundaries: Geological Society of London Special Publication 230, p. 11–33.

MacLeod, N., 2005a, Factor analysis: Palaeontological Association Newsletter, v. 60, p. 38–51.

MacLeod, N., 2005b, Mass extinction causality: Statistical assessment of multiple-cause scenarios: Russian Journal of Geology and Geophysics, v. 9, p. 979–987.

MacLeod, N., 2006, Minding your Rs and Qs: Palaeontological Association Newsletter, v. 61, p. 42–60.

MacLeod, N., 2011, Cretaceous-Tertiary planktonic foraminiferal biostratigraphy and survivorship (revisited), *in* López Oliva, J.G., ed., Epístolas Geológicas: Linnares, Neuvo Léon, Mexico, Universidad Autónoma de Nuevo León, p. 185–249.

MacLeod, N., 2013, The Great Extinctions: What Causes Them and How They Shape Life: London, The Natural History Museum, 208 p.

MacLeod, N., and Keller, G., 1991, How complete are Cretaceous/Tertiary boundary sections? A chronostratigraphic estimate based on graphic correlation?: Geological Society of America Bulletin, v. 103, p. 1439–1457, doi:10.1130/0016-7606(1991)103<1439:HCACTB>2.3.CO;2.

MacLeod, N., and Keller, G., 1994, Comparative biogeographic analysis of planktic foraminiferal survivorship across the Cretaceous/Tertiary (K/T) boundary: Paleobiology, v. 20, p. 143–177.

MacLeod, N., and Sadler, P., 1995, Estimating the line of correlation, *in* Mann, K., and Lane, H.R., eds., Graphic Correlation and the Composite Standard: Tulsa, Oklahoma, Society for Sedimentary Geology Special Publication 53, p. 51–64.

MacLeod, N., Rawson, P. F., Forey, P. L., Banner, F. T., BouDagher-Fadel, M. K., Bown, P. R., Burnett, J. A., Chambers, P., Culver, S., Evans, S. E., Jeffrey, C., Kaminski, M. A., Lord, A. R., Milner, A. C., Milner, A. R., Morris, N., Owen, E., Rosen, B. R., Smith, A. B., Taylor, P. D., Urquhart, E., and Young, J.R., 1997, The Cretaceous-Tertiary biotic transition: Journal of the Geological Society of London, v. 154, p. 265–292, doi:10.1144/gsjgs.154.2.0265.

Marshall, C.R., 1990, Confidence intervals on stratigraphic ranges: Paleobiology, v. 16, no. 1, p. 1–10.

Marshall, C.R., 1994, Confidence intervals on stratigraphic ranges: Partial relaxation of the assumption of randomly distributed fossil horizons: Paleobiology, v. 20, p. 459–469.

Marshall, C.R., 1995, Distinguishing between sudden and gradual extinctions in the fossil record: Predicting the position of the Cretaceous-Tertiary iridium anomaly using the ammonite fossil record on Seymour Island, Antarctica: Geology, v. 23, no. 8, p. 731–734.

Marshall, C.R., 1997, Confidence intervals on stratigraphic ranges with nonrandom distributions of fossil horizons: Paleobiology, v. 32, no. 2, p. 165–177.

Marshall, C.R., and Ward, P.D., 1996, Sudden and gradual molluscan extinctions in the latest Cretaceous of western European Tethys: Science, v. 274, no. 5291, p. 1360–1363, doi:10.1126/science.274.5291.1360.

Martin, R.E., 1996, Secular increase in nutrient levels through the Phanerozoic: Implications for productivity, biomass, and diversity of the marine biosphere: Palaios, v. 11, p. 209–219, doi:10.2307/3515230.

McLaren, D.J., 1970, Time, life, and boundaries: Journal of Paleontology, v. 44, p. 801–815.

McLean, D.M., 1978, A terminal Mesozoic "greenhouse": Lessons from the past: Science, v. 201, p. 401–406, doi:10.1126/science.201.4354.401.

McLean, D.M., 1985a, Deccan Traps mantle degassing in the terminal Cretaceous marine extinctions: Cretaceous Research, v. 6, p. 235–259, doi:10.1016/0195-6671(85)90048-5.

McLean, D.M., 1985b, Mantle degassing unification of the trans-K–T geobiological record, *in* Hecht, M.K., Wallace, B., and Prance, G.T., eds., Evolutionary Biology, Volume 9: New York, Plenum Press, p. 287–313.

Mitchell, W.J.T., 1998, The Last Dinosaur Book: The Life and Times of a Cultural Icon: Chicago, Illinois, University of Chicago Press, 321 p.

Molina, E., Alegret, L., Arenillas, I., and Arz, J.A., 2005, The Cretaceous/Paleogene boundary at the Agost section revisited: Paleoenvironmental reconstruction and mass extinction pattern: Journal of Iberian Geology, v. 31, p. 135–148.

Molina, E., Alegret, L., Arenillas, I., Arz, J. A., Gallala, N., Hardenbol, J., von Salis, K., Steurbaut, E., Vandenberghe, N., and Zaghbib-Turki, D., 2006, The global boundary stratotype section and point for the base of the Danian Stage (Paleocene, Paleogene, Tertiary, Cenozoic) at El Kef, Tunisia—Original definition and revision: Episodes, v. 29, no. 4, p. 263–273.

Newell, N.D., 1963, Crises in the history of life: Scientific American, v. 208, no. 2, p. 76–93, doi:10.1038/scientificamerican0263-76.

Newell, N.D., 1967, Revolutions in the history of life, *in* Albritton, C.C., Hubbert, M.K., Wilson, L.G., and Newell, N.D., eds., Uniformity and Simplicity: A Symposium on the Principle of the Uniformity of Nature: Geological Society of America Special Paper 89, p. 63–91.

Olsson, R.K., 1970, Planktonic foraminifera from base of Tertiary, Miller's Ferry, Alabama: Journal of Paleontology, v. 44, p. 598–604.

Patterson, C., and Smith, A.B., 1987, Is the periodicity of extinctions a taxonomic artefact?: Nature, v. 330, p. 248–251, doi:10.1038/330248a0.

Paul, C.R.C., 2005, Interpreting bioevents: What exactly did happen to planktonic foraminifera across the Cretaceous-Tertiary boundary: Palaeogeography, Palaeoclimatology, Palaeoecology, v. 224, p. 291–310, doi:10.1016/j.palaeo.2005.03.038.

Pease, C.M., 1992, On the declining extinction and origination rates of fossil taxa: Paleobiology, v. 18, no. 1, p. 89–92.

Poag, C.W., 1997, Roadblocks on the kill curve: Testing the Raup hypothesis: Palaios, v. 12, no. 6, p. 582–590, doi:10.2307/3515413.

Rampino, M.R., and Stothers, R.B., 1988, Flood basalt volcanism during the past 250 million years: Science, v. 241, p. 663–668, doi:10.1126/science.241.4866.663.

Raup, D.M., 1979, Size of the Permo-Triassic bottleneck and its evolutionary implications: Science, v. 206, no. 4415, p. 217–218, doi:10.1126/science.206.4415.217.

Raup, D. M., 1986, The Nemesis Affair: A Story of the Death of the Dinosaurs and the Ways of Science: New York, W.W. Norton & Co., 220 p.

Raup, D.M., 1991, Extinction: Bad Genes or Bad Luck: New York, W.W. Norton and Co., 210 p.

Raup, D.M., 1992, Large-body impact and extinction in the Phanerozoic: Paleobiology, v. 18, no. 1, p. 80–88.

Raup, D.M., and Sepkoski, J.J., Jr., 1982, Mass extinctions in the marine fossil record: Science, v. 215, p. 1501–1503, doi:10.1126/science.215.4539.1501.

Rudwick, M.J.S., 1997, Georges Cuvier, Fossil Bones, and Geological Catastrophes: New Translations & Interpretations of the Primary Texts: Chicago, Illinois, University of Chicago Press, 318 p.

Sadler, P.M., 1981, Sediment accumulation rates and the completeness of stratigraphic sections: The Journal of Geology, v. 89, p. 569–584, doi:10.1086/628623.

Schindewolf, O.H., 1950, Der Zeitfaktor in Geologie und Paläontologie: Stuttgart, Germany, E. Schweizerbart, 114 p.

Schindewolf, O.H., 1954, Über die möglichen Uraschen der grossen erdgeschictlichen Faunenschnitte: Neues Jahrbuch für Geologie und Paläontologie, Monatshefte, v. 1954, p. 457–465.

Schindewolf, O.H., 1955, Die Entfaltung des Lebens im Rahmen der geologischen Zeit: Sondersband: Studium Generale, v. 8, p. 489–497.

Schindewolf, O.H., 1962, Neokatastophismus?: Deutsch Geologische Gesellschaft Zeitschrift Jahrgang, v. 114, p. 430–445.

Schmitz, B., Keller, G., and Stenvall, O., 1992, Stable isotope and foraminiferal changes across the Cretaceous-Tertiary boundary at Stvens Klint, Denmark: Arguments for long-term oceanic instability before and after bolide-impact event: Palaeoceanography, Palaeoclimatology: Palaeogeography, v. 96, p. 233–260, doi:10.1016/0031-0182(92)90104-D.

Sepkoski, J.J., Jr., 1981, A factor analytic description of the Phanerozoic marine fossil record: Paleobiology, v. 7, no. 1, p. 36–53.

Sepkoski, J.J., Jr., 2002, A compendium of fossil marine animals, *in* Jablonski, D., and Foote, M., eds., Bulletins of American Paleontology 363: Ithaca, New York, Paleontological Research Institution, p. 563.

Shaw, A., 1964, Time in Stratigraphy: New York, McGraw-Hill, 365 p.

Signor, P.W., III, and Lipps, J.H., 1982, Sampling bias, gradual extinction patterns and catastrophes in the fossil record, *in* Silver, L.T., and Schultz, P.H., eds., Geological Implications of Impacts of Large Asteroids and Comets on the Earth: Geological Society of America Special Paper 190, p. 291–296.

Simpson, G.G., 1944, Tempo and Mode in Evolution: New York, Columbia University Press, 237 p.

Simpson, G.G., 1953, The Major Features of Evolution: New York, Columbia University Press, 237 p.

Smit, J., 1977a, Discovery of a planktonic foraminifera association between the *Abathomphalus mayaroensis* zone and the '*Globigerina*' *eugubina* zone at the Cretaceous/Tertiary boundary in the Barranco del Gredero (Caravaca, SE Spain): A preliminary report I: Koninklijke Nederlandese Akademie van Wetenschappen Proceedings, ser. B, v. 80, p. 280–289.

Smit, J., 1977b, Discovery of a planktonic foraminifera association between the *Abathomphalus mayaroensis* zone and the '*Globigerina*' *eugubina* zone at the Cretaceous/Tertiary boundary in the Barranco del Gredero (Caravaca, SE Spain): A preliminary report II: Koninklijke Nederlandese Akademie van Wetenschappen Proceedings, ser. B, v. 80, p. 290–301.

Smit, J., 1979, The Cretaceous/Tertiary transition in the Barranco del Gredero, Spain, *in* Christensen, W.K., and Birkelund, T., eds., Cretaceous-Tertiary Boundary Events Symposium, Proceedings II: Copenhagen, Denmark, University of Copenhagen, p. 156–163.

Smit, J., 1982, Extinction and evolution of planktonic foraminifera after a major impact at the Cretaceous/Tertiary boundary, *in* Silver, L.T., and Schultz, P.H., eds., Geological Implications of Impacts of Large Asteroids and Comets on the Earth: Geological Society of America Special Paper 190, p. 329–352.

Smit, J., 1996, The K/T boundary Chicxulub impact event: A review, *in* Drobne, K., Gorican, S., and Kotnik, B., eds., The Role of Impact Processes in the Geological and Biological Evolution of Planet Earth: Ljubljana, Slovenia, Scientific Research Centre, SAZU, p. 83–84.

Smit, J., and Romein, A.J.T., 1985, A sequence of events across the Cretaceous-Tertiary boundary: Earth and Planetary Science Letters, v. 74, p. 155–170, doi:10.1016/0012-821X(85)90019-6.

Smit, J., and ten Kate, W.G.H.Z., 1982, Trace element patterns at the Cretaceous-Tertiary boundary—Consequence of a large impact: Cretaceous Research, v. 3, p. 307–332, doi:10.1016/0195-6671(82)90031-3.

Smith, A.B., and McGowan, A.J., 2007, The shape of the Phanerozoic marine palaeodiversity curve: How much can be predicted from the sedimentary rock record of western Europe: Palaeontology, v. 50, no. 4, p. 765–774.

Smith, A.B., and Patterson, C., 1988, The influence of taxonomic method on the perception of patterns of evolution: Evolutionary Biology, v. 23, p. 127–216, doi:10.1007/978-1-4613-1043-3_5.

Springer, M.S., 1990, The effect of random range truncations on patterns of evolution in the fossil record: Paleobiology, v. 16, p. 512–520.

Strauss, D., and Sadler, P.M., 1989, Classical confidence intervals and Bayesian probability estimates for ends of local taxon ranges: Mathematical Geology, v. 21, p. 411–427, doi:10.1007/BF00897326.

Urbanek, A., 1993, Biotic crises in the history of Upper Silurian graptoloids: A paleobiological model: Historical Biology, v. 7, p. 29–50, doi:10.1080/10292389309380442.

Van Valen, L., 1973, A new evolutionary law: Evolutionary Theory, v. 1, p. 1–30.

Ward, P.D., and Kennedy, W.J., 1993, Maastrichtian Ammonites from the Biscay Region (France, Spain): Paleontological Society Memoir 34, 58 p.

Manuscript Accepted by the Society 31 January 2014

Printed in the USA

The Geological Society of America
Special Paper 505
2014

Large igneous provinces and mass extinctions: An update

David P.G. Bond*
Department of Geography, Environment and Earth Science, University of Hull, Hull HU6 7RX, UK, and
Norwegian Polar Institute, Fram Centre, 9296 Tromsø, Norway

Paul B. Wignall
School of Earth and Environment, University of Leeds, Leeds LS2 9JT, UK

ABSTRACT

The temporal link between mass extinctions and large igneous provinces is well known. Here, we examine this link by focusing on the potential climatic effects of large igneous province eruptions during several extinction crises that show the best correlation with mass volcanism: the Frasnian-Famennian (Late Devonian), Capitanian (Middle Permian), end-Permian, end-Triassic, and Toarcian (Early Jurassic) extinctions. It is clear that there is no direct correlation between total volume of lava and extinction magnitude because there is always sufficient recovery time between individual eruptions to negate any cumulative effect of successive flood basalt eruptions. Instead, the environmental and climatic damage must be attributed to single-pulse gas effusions. It is notable that the best-constrained examples of death-by-volcanism record the main extinction pulse at the onset of (often explosive) volcanism (e.g., the Capitanian, end-Permian, and end-Triassic examples), suggesting that the rapid injection of vast quantities of volcanic gas (CO_2 and SO_2) is the trigger for a truly major biotic catastrophe. Warming and marine anoxia feature in many extinction scenarios, indicating that the ability of a large igneous province to induce these proximal killers (from CO_2 emissions and thermogenic greenhouse gases) is the single most important factor governing its lethality. Intriguingly, many voluminous large igneous province eruptions, especially those of the Cretaceous oceanic plateaus, are not associated with significant extinction losses. This suggests that the link between the two phenomena may be controlled by a range of factors, including continental configuration, the latitude, volume, rate, and duration of eruption, its style and setting (continental vs. oceanic), the preexisting climate state, and the resilience of the extant biota to change.

*d.bond@hull.ac.uk

Bond, D.P.G., and Wignall, P.B., 2014, Large igneous provinces and mass extinctions: An update, *in* Keller, G., and Kerr, A.C., eds., Volcanism, Impacts, and Mass Extinctions: Causes and Effects: Geological Society of America Special Paper 505, p. 29–55, doi:10.1130/2014.2505(02). For permission to copy, contact editing@geosociety.org.

INTRODUCTION

The possibility that volcanism is capable of driving mass extinctions has long been posited (e.g., Kennett and Watkins, 1970; Vogt, 1972), and large igneous provinces—the most voluminous manifestation of volcanism on Earth—have been the most frequently preferred culprits. Improvements of radio-isotopic dating in the past two decades have considerably strengthened this notion because four of the "Big 5" extinction events of the Phanerozoic, as well as every minor crisis since the Permian, are shown to coincide with large igneous province eruptions (Fig. 1; Courtillot, 1999; Wignall, 2001). Only the first of the "Big 5," the end-Ordovician crisis, remains uncorrelated to a large igneous province culprit, although one has been mooted for this interval (Courtillot and Olson, 2007; Lefebvre et al., 2010).

The volcanism-extinction link is now well documented (e.g., Rampino and Stothers, 1988; Wignall, 2001, 2007; Courtillot and Renne, 2003; Kravchinsky, 2012), but temporal coincidence does not prove a causal link, although the frequency of the association is sufficiently high to imply this. Perhaps the strongest criticism comes from the observation that many large igneous province eruptions, especially those of the oceanic plateaus of the Cretaceous, are not associated with significant extinction losses (Figs. 1 and 2). Were post-Jurassic biotas more resistant to the deleterious effects of massive volcanism, or was the lack of lethality a consequence of other factors?

The current challenge for earth scientists is to better understand the variable environmental effects of large igneous province eruptions and identify the causal mechanism(s) whereby they sometimes cause catastrophic extinctions. This paper reviews our knowledge of the potential climatic effects of volcanism for several crises that show the best correlation with mass volcanism: the Frasnian-Famennian Viluy Traps; the Capitanian Emeishan flood basalts; the end-Permian Siberian Traps; the end-Triassic Central Atlantic magmatic province; and the Toarcian Karoo and Ferrar Traps. The celebrated extinction at the end of the Cretaceous is associated with both the mighty Deccan Traps and the Chicxulub impact crater, making for a rather special and contentious case study. This event is treated in detail elsewhere in this volume and is not expanded on here.

At this stage, it is pertinent to define what is meant by both "large igneous province" and "mass extinction." The former term was introduced by Coffin and Eldholm (1991), but we follow the succinct revised definition of large igneous provinces of Bryan and Ernst (2008). The latter authors define large igneous provinces as magmatic provinces with >0.1 × 10^6 km^2 extent, volumes >0.1 × 10^6 km^3, and a maximum eruption duration of <50 m.y. They are characterized by igneous pulse(s) of short duration (~1–5 m.y.), during which a large proportion (>75%) of the total

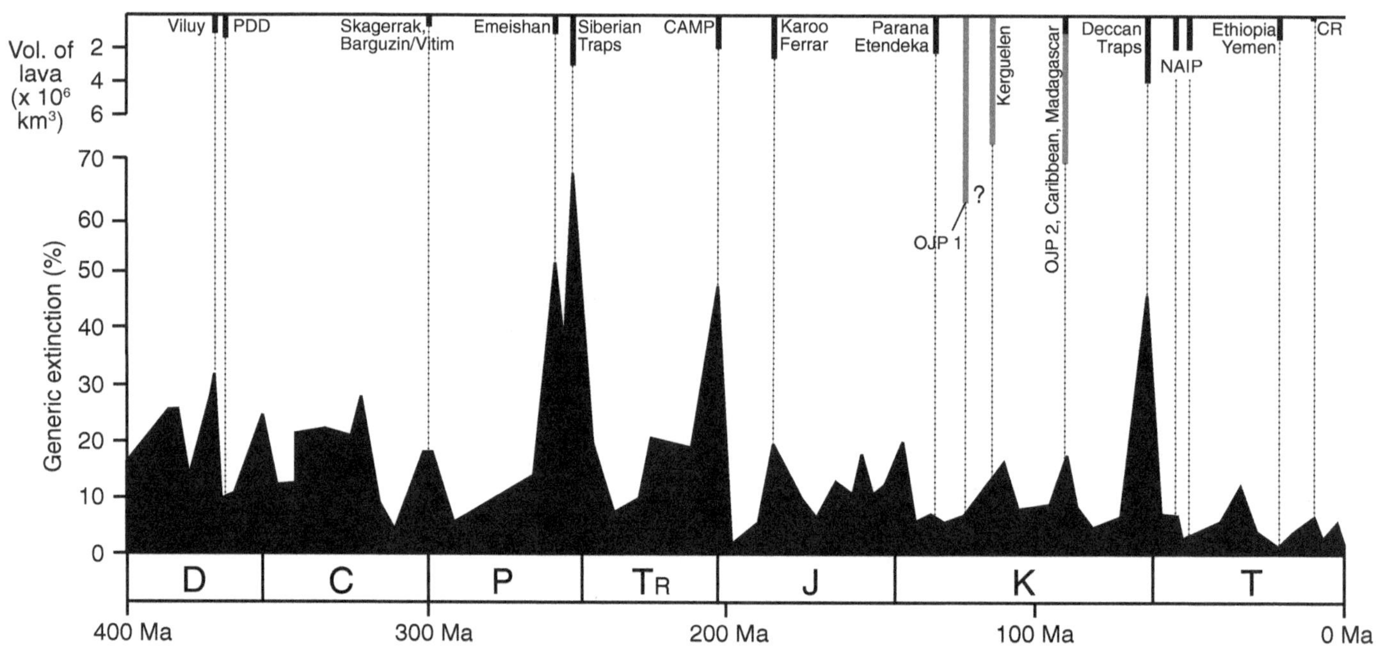

Figure 1. Generic extinction magnitude through the past 400 m.y. (based on Sepkoski, 1996, 2002) vs. the age and estimated original volume of large igneous provinces (volume estimates based on Courtillot and Renne, 2003; Kravchinsky, 2012). PDD—Pripyat-Dnieper-Donets rift; CAMP—Central Atlantic magmatic province; OJP 1/OJP 2—Ontong Java Plateau phases 1 and 2; NAIP—North Atlantic igneous province; CR—Columbia River Basalt Group. Continental flood basalt–volcanic rifted margin provinces (after Bryan and Ernst, 2008) are shown as black bars, while oceanic plateaus are shown as gray bars. Note the apparent correlation between mass extinction events (peaks in generic extinction) and large igneous province emplacement, but no major extinctions associated with the three volumetrically largest episodes of volcanism in the Cretaceous. This figure does not show the possible range of ages for the Viluy Traps (ca. 377–364 Ma). Figure is adapted from Bond et al. (2010a). D—Devonian; C—Carboniferous; P—Permian; Tr—Triassic; J—Jurassic; K—Cretaceous; T—Tertiary.

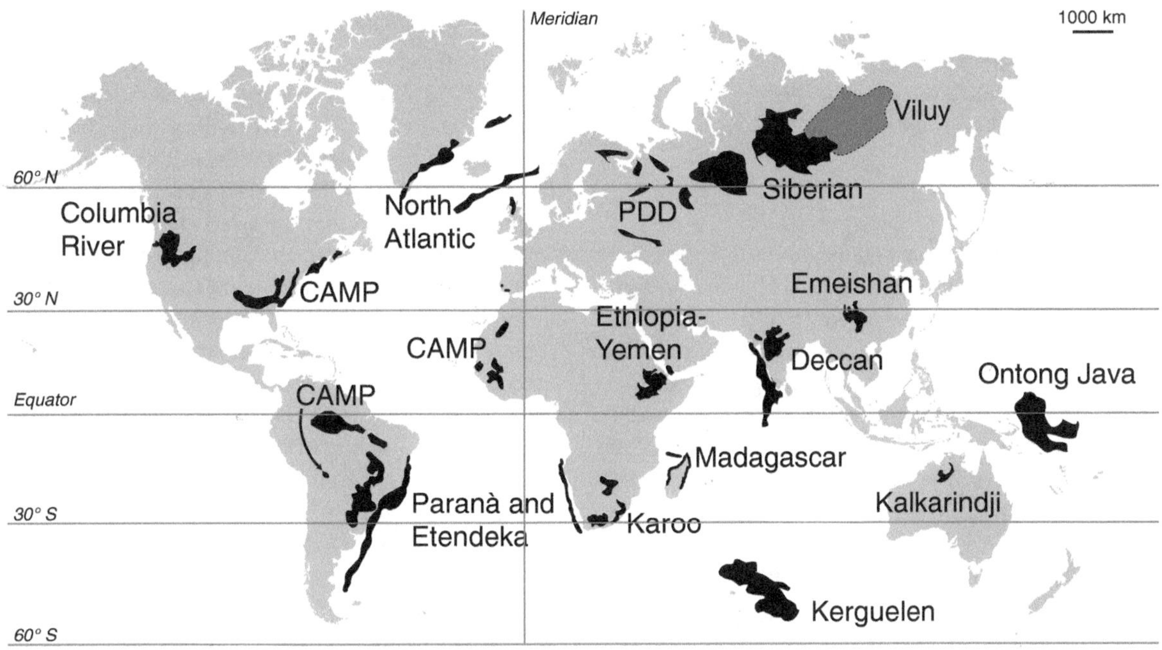

Figure 2. Map of major large igneous provinces mentioned in the text (adapted from Coffin and Eldholm, 1994; Bryan and Ernst, 2008; Marzoli et al., 2011). The Ferrar Traps, located in Antarctica, are not shown on this map. Postulated area of the Viluy Traps is very uncertain. The apparently vast distribution of the Siberian Traps reflects the suggested westward extension of volcanics beneath Jurassic and Cretaceous basin fill, based on Early Triassic basalts found in boreholes (Westphal et al., 1998; Reichow et al., 2005; Saunders et al., 2005).

igneous volume is emplaced (Bryan and Ernst, 2008). The majority of large igneous provinces fall into two subdivisions, continental flood basalt provinces and oceanic plateaus, with a few being a combination of continental flood basalts and volcanic rifted margins. A mass extinction records a globally widespread and rapid loss of species from numerous environments, and the "Big 5" extinctions of the Phanerozoic are those in which >50% of species disappeared from the fossil record (sensu Raup and Sepkoski, 1982; Sepkoski, 1996). The Capitanian event should probably also be included in a revised "Big 6." It has been proposed that the Frasnian-Famennian and end-Triassic "mass extinctions" be demoted to "mass depletions," since those events were likely a diversity decrease due to failure to originate, rather than elevated species loss (e.g., Bambach et al., 2004). Indeed, different treatments of taxonomic databases have rendered some interesting results (Table 1). Nevertheless, for now, the Frasnian-Famennian and end-Triassic events maintain their place at the top table of mass extinctions.

EFFECTS OF VOLCANISM ON CLIMATE

Is it realistic to suggest that large igneous province volcanism can affect global climate to the extent that extinction is inevitable, even in regions distant from the site of eruptions? Here,

TABLE 1.TAXONOMIC-SEVERITY RANKING OF THE 11 LARGEST PHANEROZOIC CRISES SINCE THE ORDOVICIAN*

Rank	Event	%[1]	Event	%[2]	Event	%[3]	Ecological severity ranking[†]
1	End-Permian	–58	End-Permian	–57	End-Permian	–83	End-Permian
2	End-Ordovician	–49	End-Ordovician	–43	End-Triassic	–73	End-Cretaceous
3	Capitanian	–47	Capitanian	–36	End-Ordovician	–52	End-Triassic
4	End-Triassic	–40	End-Cretaceous	–34	End-Devonian	–50	Frasnian-Famennian
5	End-Cretaceous	–39	End-Triassic	–33	End-Cret., Late Dev. (=)	–40	Capitanian
6	Frasnian-Famennian	–35	Frasnian-Famennian	–22	N.A.	N.A.	Serpukhovian
7	Givetian	–30	Serpukhovian	–13	Serpukhovian	–39	End-Devonian, End-Ordovician (=)
8	End-Devonian	–28	Givetian	–10	Givetian	–36	N.A.
9	Eifelian	–24	End-Devonian, Ludford. (=)	–7	Eifelian	–32	Givetian
10	Serpuk., Ludford. (=)	–23	N.A.	N.A.	Capitanian	–25	Eifelian, Ludfordian (=)
11	N.A.	N.A.	Eifelian	–6	Ludfordian	–9	N.A.

Abbreviations: N.A.—not applicable; Serpuk.—Serpukhovian; Ludford.—Ludfordian; Dev.—Devonian; Cret.—Cretaceous.

*Ranked by the percentage marine genera extinction magnitude in the analyses of [1]Sepkoski (1996), [2]Bambach et al. (2004), and [3]McGhee et al. (2013).

[†]Ecological severity ranking (McGhee et al., 2013) is a measure of the "ecological impact" of a crisis (see McGhee et al., 2004). The Ludfordian, Eifelian, Givetian, and Serpukhovian are relatively minor intra-Silurian, Devonian, and Carboniferous bioevents.

we review the major climate-changing products of volcanism and provide a brief overview of recent developments in modeling work that provide clues to the "missing link" between large igneous province eruptions and mass extinctions.

Products of Volcanism

Volcanic eruptions inject gases and ash into the troposphere and stratosphere, where their residence time, behavior, and dispersal vary greatly. Apart from water vapor (H_2O), carbon dioxide (CO_2) and sulfur dioxide (SO_2) are volumetrically the most important volcanic gases. Both are greenhouse gases, but their warming effects operate over very different time scales: Only CO_2 causes significant warming over geological time. While SO_2 causes localized short-term warming over periods of days to weeks, its major effect is that of cooling because it forms sunlight-blocking aerosols. Chlorine and fluorine are other important products of volcanism, contributing to ozone depletion and acid rain (e.g., Sigurdsson, 1990; Thordarson and Self, 1993, 2003).

Sulfur Dioxide

Basaltic lavas, such as those associated with large igneous provinces, are especially rich in sulfur dioxide (Palais and Sigurdsson, 1989; Sigurdsson, 1990), and its atmospheric conversion to aerosols is one of the key drivers of cooling during eruptions (e.g., Robock, 2000). This effect can last for 2–3 yr after an eruption, but rarely longer, because rain out occurs quite rapidly. For SO_2 to be effective in causing cooling, it must be injected into the stratosphere, where it is rapidly dispersed around the hemisphere (crossing the equator is a slower process). This has been observed for a number of modest (by large igneous province standards) volcanic eruptions over the past few centuries that have caused significant global cooling (Briffa et al., 1998; de Silva and Zielinski, 1998). A recent example is provided by the Mount Pinatubo eruption of 1991, which injected 20 megatons (Mt) of SO_2 more than 30 km into the stratosphere. The result was a global temperatures decrease approaching 0.5 °C for 3 yr (although this cooling was probably exacerbated by the contemporaneous Mount Hudson eruption in Chile).

One of the largest historical eruptions occurred in 1783–1784 from the Laki fissure in Iceland when an ~15 km^3 volume of basaltic magma was extruded, releasing ~122 Mt of SO_2, 15 Mt of HF, and 7 Mt of HCl (Thordarson and Self, 1993, 2003). Laki's eruption columns extended vertically up to 13 km, injecting sulfate aerosols into the upper troposphere and lower stratosphere, where they reacted with atmospheric moisture to produce ~200 Mt of H_2SO_4. This aerosol-rich fog hung over the Northern Hemisphere for 5 mo (Thordarson and Self, 2003), leading to short-term cooling (Williams-Jones and Rymer, 2000), and harmful acid rain in both Europe and North America. Additionally, HCl and HF emissions damaged terrestrial life in Iceland and mainland Europe, as a low-level fluorine-rich haze stunted plant growth and acidified soils (Frogner Kockum et al., 2006).

The temporal link between the Laki eruptions and cooling is apparently compelling; Iceland was 5 °C colder than normal (Wood, 1992), and the eastern United States recorded its coldest ever winter in 1783–1784, with average temperatures 4.8 °C below the 225 yr average (Scarth, 2001). However, the causal link between the two has been questioned by D'Arrigo et al. (2011), who argued that the 1783–1784 winter saw a combined negative phase of the North Atlantic Oscillation and an El Niño–Southern Oscillation warm event (a pattern repeated during the unusually harsh Northern Hemisphere winter of 2009–2010). Thus, Laki may not have been the sole contributor to cooling during 1783–1784. Another link between volcanism and cooling was seen during one of the largest episodes of volcanic stratospheric H_2SO_4 loading of the past 500 yr from Huaynaputina, in Peru, which erupted in 1600 coincident with the coldest global average temperatures of the time (de Silva and Zielinski, 1998).

With a volume of 15 km^3, Laki was tiny in comparison to the scale of eruptions encountered during large igneous province emplacement. However, it is unclear if the cooling effects of larger eruptions can be simply scaled up from the known effects of observed historical eruptions. Sulfate aerosol formation during large eruptions may be limited by the amount of water vapor in the atmosphere, with the result that not all of the injected SO_2 is converted into aerosols immediately (Pinto et al., 1989). Larger eruptions could in theory cause more prolonged cooling than small eruptions, but the cooling effect need not necessarily be any more intense because of the limiting role of water vapor availability.

The past few years have seen major advances in the use of global aerosol microphysics models, partially stemming from the geoengineering debate on injecting SO_2 into the stratosphere to mitigate global warming (e.g., English et al., 2012). These models have supported the assertion that there is an upper limit to the radiative forcing that can be obtained with sulfate aerosols due to the particle size distribution and growth effect (Heckendorn et al., 2009; Timmreck et al., 2009, 2010; Niemeier et al., 2011; Hommel and Graf, 2011; English et al., 2012). For a given sulfate load, the scattering of shortwave radiation is modulated by particle size, and as aerosol particle size increases, scattering of radiation decreases (Rasch et al., 2008; Timmreck et al., 2009).

There is a positive correlation between mass of sulfur released during historic volcanic eruptions and Northern Hemisphere temperature decline (Fig. 3; Devine et al., 1984; Sigurdsson, 1990), but it does not appear to be a linear one (Blake, 2003). Thus, eruptions between A.D. 1400 and 1994 with <2 km^3 magma volume generated insignificant cooling, while those with >4 km^3 volumes induced cooling of 0.35 °C on average (Blake, 2003). Indeed, by far the largest eruption of the late Quaternary, at Toba, Sumatra (73,000 yr ago), injected ~4400 Mt of sulfate aerosols into the stratosphere (Rampino and Self, 1992; Zielinski et al., 1996), but contemporaneous cooling is not clearly evident from the $\delta^{18}O$ record (Rampino and Self, 1992). Even the time of year that eruption occurs may affect the impact of aerosols. Frölicher et al. (2013) postulated that the El Niño–Southern Oscillation controls atmospheric response to volcanic gases, implying that past continental configuration and ocean circulation may have been a factor in the response of climate to large igneous province volcanism.

Schmidt et al. (2010, 2012a, 2012b) modeled the importance of aerosol-induced indirect radiative forcing of climate from Laki-style eruptions. While direct radiative forcing derives from the scattering of incoming solar radiation in the stratosphere, indirect forcing is caused by the effect of aerosols on the radiative properties of clouds in the troposphere, usually restricted to the region of volcanism. Schmidt et al.'s simulations showed that tropospheric volcanic aerosols are an important contributor to cloud condensation nuclei, and that the additional indirect climate forcing caused by increased cloud formation can be substantial (the cloud albedo effect). Intriguingly, Schmidt et al. (2012a) found that the global annual mean cloud albedo effect was about twice as strong under pre-industrial conditions (-1.06 W m^{-2}) as under present-day conditions (-0.56 Wm^{-2}), suggesting that there may be an upper limit to aerosol-induced indirect radiative forcing. Nevertheless, it is possible that indirect radiative forcing was an additional contributor to volcanic darkness and cooling in past episodes of large igneous province volcanism and/or extinction, albeit localized to the eruption site.

The preceding section concentrated on relatively small, recent injections of SO_2 into the atmosphere. Modeling work has emphasized the complexity of the behavior of SO_2 in the atmosphere, and we should take great care in extrapolating these models to large-scale large igneous province volcanism. While it is apparent that the relationship between the size of the gas flux and cooling or radiative forcing is not linear, it is possible that volcanism on a large igneous province scale would induce changes

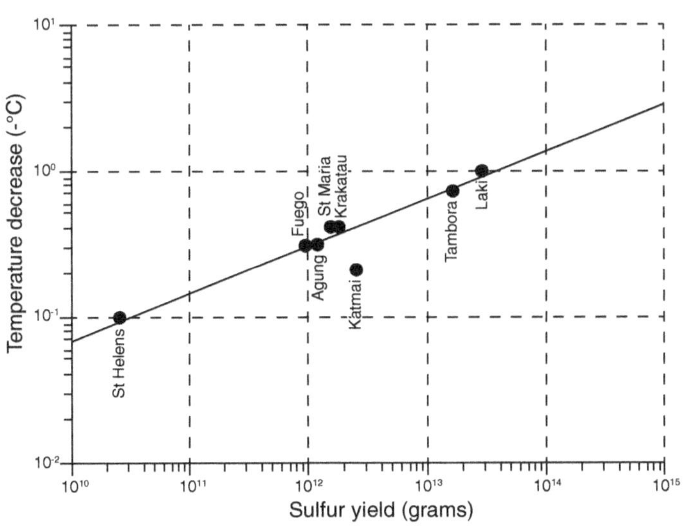

Figure 3. Correlation between volcanic sulfur yield to the atmosphere and the observed Northern Hemisphere temperature decrease for several historical eruptions. Sulfur yield is based on petrologic estimate (Devine et al., 1984; Palais and Sigurdsson, 1989). Figure is after Sigurdsson (1990).

to Earth's climate that are far greater than anything associated with historical eruptions. Thus, Self et al. (2005) proposed that each huge eruption (10^2–10^3 km^3) associated with continental flood basalt volcanism could last for a decade or more and could inject ~1 Gt of SO_2 per year. Potentially, decadal-length fluxes of gas could cause longer-term cooling, especially if the recurrence interval between flows was brief (<100 yr).

Carbon Dioxide

Carbon dioxide is the other principal volcanogenic gas, alongside SO_2, and its greenhouse effects are another potential consequence of large igneous province eruptions, especially given its longer atmospheric residence time. However, the key debate is whether volcanic CO_2 degassing is sufficiently voluminous to have significant effects.

The estimated annual volcanic CO_2 flux of 130–440 Mt (Gerlach, 2011) is injected into an atmosphere of 3000 Gt of CO_2. This is two orders of magnitude smaller than the current anthropogenic CO_2 flux (Sabine et al., 2004). In comparison, Self et al. (2005) calculated that a single large igneous province eruption with a volume of 1000 km^3 (with a top-end estimate of 0.5 wt% CO_2 in the magma and 100% degassing) would inject around 13 Gt of CO_2 into the atmosphere, i.e., less than 1/200th the mass of the atmospheric reservoir. In isolation, a single eruption is unlikely to cause a climatically significant increase in atmospheric CO_2. However, the atmospheric lifetime of CO_2 is at least several hundred years, far longer than that of SO_2, and about a fifth of CO_2 remains in the atmosphere for several thousands of years (Archer, 2005). The cumulative effects of repeated, closely spaced flood basalt eruptions could potentially promote global warming. The total CO_2 release from a large igneous province such as the Siberian Traps has been estimated at 30,000 Gt CO_2, i.e., 10 times the mass in today's atmosphere (Courtillot and Renne, 2003), based on an original volume of ~3×10^6 km^3 for the Siberian Traps.

Underlying the discussions of large igneous province CO_2 release is the assumption that plumes of undepleted mantle are the source of large igneous province magma and their volatiles. However, recent modeling work and study of mantle inclusions in the Siberian Traps led Sobolev et al. (2011) to suggest that recycled ocean crust may form a substantial proportion (15%) of plumes. The consequences for volatile release are considerable because this material would yield substantial volumes during partial melting. Driven off ahead of the ascending magma, the initial volatile release is therefore calculated to release 170,000 Gt of CO_2 and about a tenth this amount of HCl (Sobolev et al., 2011). If this gas was released in a few, closely spaced blasts, the environmental effects would be devastating.

Another potentially important effect of CO_2 release is increased acidity of the oceans. Huybers and Langmuir (2009) modeled the effect of increased atmospheric CO_2 on carbonate saturation and showed that a rapid injection of ~3000 Gt of CO_2 into the ocean, accompanied by a 4 °C ocean warming and 100 ppm increase in atmospheric CO_2 concentration, would cause the carbonate saturation horizon to shoal by ~1 km (Fig. 4). If correct, Huybers and Langmuir's (2009) modeling implies that the largest subaerial large igneous province eruptions were able to generate significant ocean acidification, as has been proposed for the end-Permian mass extinction (Payne et al., 2007). However, these effects would only be severe in cooler, higher-latitude oceans and in deeper waters.

Other Gases and Ash

Water vapor (H_2O) is volumetrically the largest volcanic gas, and its immediate effect is to act as a positive radiative forcing agent in the stratosphere that may in part counteract the cooling effects of sulfate aerosols. Joshi and Shine (2003) applied a general circulation model to the Pinatubo eruption scenario and calculated a global average radiative forcing due to H_2O of +0.1 W m^{-2}. Radiative forcing due to Pinatubo's volcanic aerosols peaked at −3.5 W m^{-2} (Myhre et al., 2001), and so it seems probable that the effect of water vapor is to partly counteract the effect of aerosols.

Hydrogen chloride and hydrogen fluoride (HCl and HF), volcanic ash in the form of silicate particulate matter, and methane (CH_4) are other important products of volcanism. Some of these have been documented as causing environmental harm during recent eruptions, leading to their effects being invoked in mass extinction scenarios (without much acclaim). In particular, HCl causes damage downwind from volcanoes because HCl and H_2O condense readily on ash particles and fall as acidic rain. This mechanism affected crops and livestock in Iceland and mainland Europe during the 1783–1784 Laki eruptions, which released 7 Mt of HCl (Thordarson and Self, 1993, 2003). However, this rainfall also ensures the swift removal of volcanic HCl from the atmosphere, and it is unlikely that HCl could cause long-term damage on a global scale. HCl also destroys ozone as reactive chlorine atoms are released through interaction with sulfate aerosols, but due to the rapid scrubbing of HCl from the troposphere by rain, there is limited opportunity for stratospheric ozone destruction.

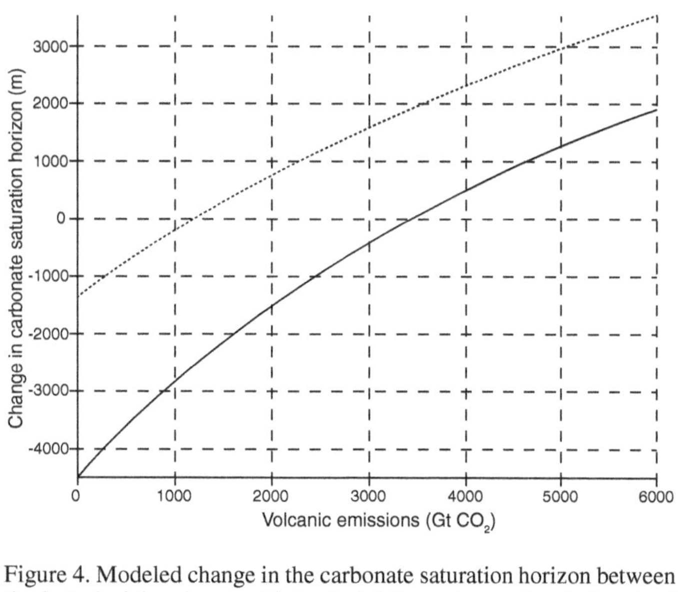

Figure 4. Modeled change in the carbonate saturation horizon between the last glacial and present interglacial for various volcanic inputs of CO_2 into the ocean-atmosphere system. One scenario considers only changes in CO_2 and temperature (dotted line), and the other also includes organic storage of carbon on the continents (solid line). Figure is after Huybers and Langmuir (2009). Negative values on the *y* axis will not affect the carbonate saturation horizon.

Hydrogen fluoride also promotes acid rain downwind of the eruption site, but its primary deleterious effect is that of poisoning. The gas combines with fine ash particles, poisoning animal and plant life wherever it falls in volume. The 15 Mt HF released during the Laki eruptions caused widespread crop failure as well as bone damage and excess mortality in livestock and humans. Like HCl, though, HF is rapidly removed from the atmosphere, and it is unlikely to be of great relevance in global mass extinction scenarios.

Early attempts to understand the climatic effects of volcanism by Humphreys (1940) assumed that volcanic ash was responsible for major cooling by backscattering incident solar radiation, an effect that we now know actually derives from sulfate aerosols. In fact, ash particles have only a very minor, localized effect on climate (lasting days), because they remain in the atmosphere for such a brief duration. Even the 1991 Pinatubo eruption, which saw a huge umbrella plume extending 35 km vertically, deposited almost all of its 5.2 km^3 ash over the South China Sea in just a few days (Wiesner et al., 2004). Clearly, with such short atmospheric residence times, ash is not a great threat to global biodiversity. Ash layers are, however, extremely useful to paleontologists, stratigraphers, and geochronologists, because they can provide a datable record of volcanism (the "smoking gun") in stratigraphic sequences distant from the eruption site that contain the record of mass extinctions (e.g., Wignall et al., 2009b).

Thermogenic Greenhouse Gases

In recent years, an additional source of large igneous province–related gas flux has been identified. In order to explain the abrupt warming associated with the Paleocene-Eocene thermal maximum, Svensen et al. (2004) suggested that baking of organic-rich sediments by high-level intrusives could generate great volumes of greenhouse gases, perhaps thousands of gigatons of CH_4 (rapidly oxidized to CO_2 in the atmosphere). Such thermogenic gases will have the light isotopic signature of their source sedimentary organic matter and so can account for the negative $\delta^{13}C$ excursions often encountered during large igneous province eruption intervals. Tangible evidence for gas release comes in the form of thousands of hydrothermal vent complexes (small diatremes) developed above igneous intrusions of large igneous provinces (Svensen et al., 2004, 2007). If such structures were simultaneously active, then it is possible their gas effusion rates were considerable. Svensen's hypothesis has gained considerable favor as a cause of the rapid global warming often observed during extinction crises (Heydari et al., 2008; Retallack and Jahren, 2008; Ganino and Arndt, 2009).

What Makes a Large Igneous Province a Global Killer?

One of the great enigmas of the geological record is the great variability of the consequential effects of large igneous province eruptions (Wignall, 2001). Prior to the Cretaceous, most large igneous province eruptions can be convincingly linked with biotic crises, but the effects of younger provinces seem much more variable. Understanding the cause of this variability may lie in factors like continental configuration, latitude of eruption, scale and duration of eruptions, their style and setting (continental vs. oceanic), and the pre-eruption climate state.

Volume, Rate, and Style of Eruption

Large igneous provinces are mostly constructed of individual sheet flows, primarily of tholeiitic basalt, that range in volume from several hundred to several thousand cubic kilometers. Other volumetrically important extrusive components of continental flood basalt provinces are mafic volcaniclastic deposits (e.g., Ukstins Peate and Bryan, 2008) and typically late-stage silicic ignimbrites (Bryan, 2007). Estimating the original volume of large igneous provinces is notoriously difficult, but the best-known continental flood basalt provinces emplaced between 0.2×10^6 km^3 (Columbia River flood basalts) and $>3 \times 10^6$ km^3 of lava (e.g., Siberian and Deccan Traps). There is no direct correlation between total volume of lava and extinction magnitude (Wignall, 2001; Courtillot and Renne, 2003; Fig. 1). This conclusion is not surprising, given the observations that there is likely always sufficient recovery time between individual eruptions to negate any cumulative effect of successive flood basalt eruptions. Instead, the environmental and climatic damage must be attributed to single-pulse gas effusions. Radioisotopic dating has constrained the total duration of many continental flood basalt province eruptions to 1–2 m.y. (Courtillot and Renne, 2003; Bryan et al., 2010; Kravchinsky, 2012), but the magnitude and recurrence time of individual eruptions are poorly understood.

The relatively small Columbia River flood basalts represent the most recently emplaced large igneous province, and its eruptive history is amongst the best known. It consists of stacked, pahoehoe lava flow fields ranging from 1 to 2000 km^3 in volume (Tolan et al., 1989) that were the products of individual, potentially sustained, eruptions (Self et al., 1997). Each flow consists of multiple, sheet-like lobes that are 20–30 m thick (Bryan et al., 2010). Each flow field is thought to have originated from eruptions lasting perhaps years or decades and with eruption rates of 4000 m^3 s^{-1} (Self et al., 1996, 1997, 1998). The stratigraphy of the province is well constrained and has shown that, although basaltic eruptions range in age from 17 Ma to 6 Ma, >90% of the total volume was erupted between 16.6 Ma and 15.3 Ma (Tolan et al., 1989; Camp et al., 2003; Hooper et al., 2007). Peak magmatism occurred during the initial eruptions and lasted for a much shorter duration than 1.3 m.y., possibly just tens of thousands of years (Bryan et al., 2002; Bryan and Ernst, 2008). A similar eruptive history has been proposed for some of the larger continental flood basalt provinces such as the Central Atlantic magmatic province (Knight et al., 2004) and the Deccan Traps (Chenet et al., 2007), which were probably emplaced during a series of very short-lived (hundreds of years or less) eruptions. If peak magmatism occupies only a short time interval, then it follows that gas fluxes peaked during the onset of eruptions—an observation that has additional pertinence when examining the relationship between large igneous province timing and extinction (see following).

Role of Eruption Site

Tropical eruptions generate heating in the tropical stratosphere that creates anomalous temperature and density gradients between the equator and the poles, causing a strengthening of zonal winds and a stronger stratospheric polar vortex (Driscoll et al., 2012). This strengthened vortex is associated with positive North Atlantic Oscillation and Arctic Oscillation, which generate high pressure at midlatitudes and low pressure at the poles (Black, 2002; Kolstad and Charlton-Perez, 2011). This mechanism ensures that aerosols from tropical eruptions, such as Pinatubo, quickly spread over the entire globe. In contrast, aerosols derived from high-latitude eruptions remain in the hemisphere in which they were injected, unable to escape the polar vortex (Oman et al., 2005). Although this modeling is based on historical-scale eruptions with a modern continental configuration, it suggests that eruption site is a key factor in the ability of aerosols to affect global climate and that only tropical large igneous province eruption could cause global mass extinctions through the cooling/darkness effect of sulfate aerosols. The troposphere is thinner at higher latitudes, and polar volcanic eruptions are more likely to inject aerosols into the stratosphere.

Of the five extinction-associated large igneous provinces discussed in this paper, the Emeishan Traps (paleolatitude of 10°S–10°N) and Central Atlantic magmatic province (paleolatitude of ~20°S–20°N) straddled the equator. The Karoo and Ferrar Traps (45°S) erupted in southern midlatitudes, while the Viluy Traps (~50°N–60°N) and Siberian Traps (>60°N) were emplaced in northern Boreal latitudes. So, clearly, eruption site alone is not crucial for a large igneous province–extinction link.

FRASNIAN-FAMENNIAN (LATE DEVONIAN) MASS EXTINCTION

Extinction Record

The Frasnian-Famennian mass extinction (at the stage boundary, dated to 372.2 Ma according to Gradstein et al., 2012) has long been considered one of the "Big 5" extinction events (Table 1). Many marine groups suffered marked extinctions, and spectacularly large Devonian reef ecosystems never fully recovered (Kiessling et al., 2000; Copper, 2002). The close temporal association between the Frasnian marine extinctions and the development of two discrete, geographically widespread anoxic

"Kellwasser events" supports an anoxia-extinction causal link in offshore level-bottom communities (Buggisch, 1972; House, 1985; Buggisch, 1991; Bond et al., 2004). The same cause has recently been implicated in the demise of shallow-water reefs (Bond et al., 2013). Both anoxic events correspond to pulses of sea-level rise that together form transgressive-regressive (T-R) cycle IId of the Late Devonian eustatic curve (Johnson et al., 1985; Bond and Wignall, 2008). A significant warming, derived from oxygen isotope data from conodont apatite (Joachimski et al., 2009), saw low-latitude sea-surface temperatures rise from 23 °C in the Middle Devonian to 30–32 °C by the end-Frasnian. This apparently simple climatic history is complicated by two Kellwasser-related cooling phases (Joachimski and Buggisch, 2002; Balter et al., 2008), perhaps induced by atmospheric CO_2 drawdown during black shale deposition, and it has been suggested that the biotic crisis has its origins in a "destabilised greenhouse" (Racki, 1998, p. 192). Could eustatic sea-level rise, climate changes including ~9 °C global warming, and widespread marine anoxia have their origins in contemporaneous large igneous province volcanism?

Late Devonian Volcanics

The Viluy Traps, which cover most of the northeast margin of the Siberian Platform, have been known for four decades (Masaitis et al., 1975; Gaiduk, 1988), but only relatively recently has their age become well constrained (Kravchinsky et al., 2002; Kiselev et al., 2006, 2007; Kuzmin et al., 2010; Kravchinsky, 2012). The lavas partially fill the Viluy rift, a graben 800 km long by 450 km wide (Kiselev et al., 2006), that is the western branch of a Devonian triple-junction rift (Kuzmin et al., 2010). Volcanism began in the Frasnian and continued into the Mississippian, emplacing a volcanic pile up to 7 km thick (Kuzmin et al., 2010). Most of the lava has been eroded or lies buried beneath the younger Siberian Traps to the west, making volume estimates difficult, but values of ~1×10^6 km^3 have been suggested (Courtillot et al., 2010; Kuzmin et al., 2010). Until recently, radioisotopic dates from the volcanics were rather sparse and gave a broad range of ages suggesting that magmatism lasted from 380 to 340 Ma (K-Ar ages in Shpount and Oleinikov, 1987; Ar-Ar ages in Kiselev et al., 2006). Courtillot et al. (2010) constrained the major pulse of Viluy volcanism between 370 and 360 Ma (K/Ar and Ar-Ar ages), concluding that the $^{40}Ar/^{39}Ar$ plateau ages of 370.0 ± 0.7 Ma (conventional calibration) or 373.4 ± 0.7 Ma (recalculated per Renne et al., 2010, 2011) are the most reliable ages for Viluy Traps volcanism. In accord with paleomagmatic data, the entire large igneous province could have been emplaced around 370 Ma, very close to the Frasnian-Famennian boundary (Kravchinsky et al., 2002). The latest time scale (Gradstein et al., 2012) places this stage boundary at 372.2 Ma (based on a cubic spine fit on 17 selected radioisotopic dates from the Devonian and lowermost Carboniferous), actually much closer to Courtillot et al.'s (2010) age determination for Viluy volcanism than previously thought. (In 2010, the Frasnian-Famennian boundary was placed at 376 Ma.) The most recently obtained K-Ar and Ar-Ar dates indicate multiphase emplacement of the Viluy Traps between 376.7 ± 1.7 Ma and 364.4 ± 1.7 Ma (Ricci et al., 2013), clearly strengthening its temporal link with the Frasnian-Famennian extinction (and also implying a causal link with the lesser crisis at the Devonian-Carboniferous boundary).

Other large traps are well known from the Late Devonian. The 2000 km Pripyat-Dnieper-Donets rift system, in the southwest of the Russian Platform, is estimated to have hosted a minimum of 1.5×10^6 km^3 of magma, making the Pripyat-Dnieper-Donets large igneous province potentially even larger than the Viluy Traps (Kravchinsky, 2012). Conversely, Courtillot and Renne (2003) dismissed the Pripyat-Dnieper-Donets system as both volumetrically minor, with <10,000 km^3 magma, and too young (367–364 Ma; Famennian Stage, based on forward modeling rather than radioisotopic dating; Kusznir et al., 1996; Wilson and Lyashkevich, 1996) to be implicated in the Frasnian-Famennian extinction. Extensive, but volumetrically relatively minor, alkaline magmatism on the Kola Peninsula in Russia has been dated to 380–375 Ma (Wu et al., 2013), i.e., slightly older than the Frasnian-Famennian boundary.

The Late Devonian was clearly a time of voluminous trap magmatism on the Russian and Siberian Platforms that could feasibly have injected enough CO_2 into the atmosphere to generate the estimated late Frasnian +9 °C sea-surface temperature shift (Joachimski et al., 2009), which in turn may have led to the development of widespread marine anoxia and mass extinction (Racki, 1998; Bond et al., 2004; Bond and Wignall, 2008). Ongoing improvements in radioisotopic dating of the Viluy Traps and Pripyat-Dnieper-Donets, as well as the marine extinction level and key developments in terrestrial ecosystems, are continuously improving the temporal link between Late Devonian volcanism and extinction(s).

CAPITANIAN (MIDDLE PERMIAN) MASS EXTINCTION

Extinction Record

The volumetrically relatively small Emeishan flood basalts and Capitanian (Capitanian Stage of the Guadalupian Series) mass extinction provide the clearest temporal link between these two phenomena, because the record of both coexists in the same sections in southwest China. It was in that region that a Capitanian mass extinction was first discovered in the record of fusulinacean foraminifera (Jin et al., 1994; Stanley and Yang, 1994), and it was subsequently ranked the third most severe biodiversity crisis of the Phanerozoic (Sepkoski, 1996; Bambach et al., 2004). These early studies were based on literature reviews that used a coarse temporal resolution of the available data with age ranges only resolved at the stage level. Consequently, the extinction was attributed to the end of the Guadalupian. Recent studies of fossil range truncations of diverse shallow-marine taxa in conodont-dated sections have shown that extinction occurred earlier (in the *Jinogondolella altudaensis–J. prexuanhanensis* conodont zones) in South China, making this a mid-Capitanian crisis (Fig. 5; Shen and Shi, 2009; Wignall et al., 2009a, 2009b; Bond et al., 2010a, 2010b).

The southwest China sections are probably unique, globally, in that they preserve not only the record of a mass extinction, but also of the smoking gun in the form of the Emeishan large igneous province. Thus, the provinces of Guizhou, Yunnan, and Sichuan preserve the remnants of an extensive Middle Permian carbonate platform composed of the 700-m-thick Maokou Formation (Yang et al., 1999, 2004; Wang and Sugiyama, 2000; Lai et al., 2008; Wignall et al., 2009a, 2009b; Bond et al., 2010a, 2010b). In western Guizhou and Yunnan, volcanics of the Emeishan large igneous province interdigitate with and overlie the Maokou Formation, enabling the fossil record both before and during the eruptions to be examined. This reveals that the mass extinction occurred at the onset of explosive Emeishan volcanism, and it coincides with the beginning of a major (–6‰) $\delta^{13}C_{carbonate}$ excursion (Fig. 5; Wignall et al., 2009a; Bond et al., 2010b). The Capitanian extinction record is best known from South China, but the event was clearly of global scale. Middle to Late Permian foraminiferal ranges and geographic distributions show provincialism throughout the Middle Permian, and all realms suffered severe losses (Bond and Wignall, 2009).

The plant fossil record in South China reveals a significant 24% species loss, suggesting that the Capitanian crisis also occurred on land. An intra-Capitanian extinction of 56% of plant species in the North China block sequences may also have coincided with these losses (Fig. 5; Bond et al., 2010a). Dating of these terrestrial sequences is notoriously difficult, but correlation using the paleomagnetic record (for examples, see Jin et al., 1998; Ali et al., 2002, 2005) provides two alternatives: Either turnover amongst plant species was contemporaneous with the marine extinction, or plant losses postdate the marine extinction and occurred during the waning phase of Emeishan volcanism. The Middle Permian also saw a major tetrapod crisis with the loss of the dominant dinocephalians (Lucas, 1998, 2009). Current understanding of the dinocephalian losses suggests that this event occurred during the preceding stage (the Wordian), but future improvements in both sampling and dating of this tetrapod crisis may reveal a synchronicity of plant, animal, and marine invertebrate extinctions.

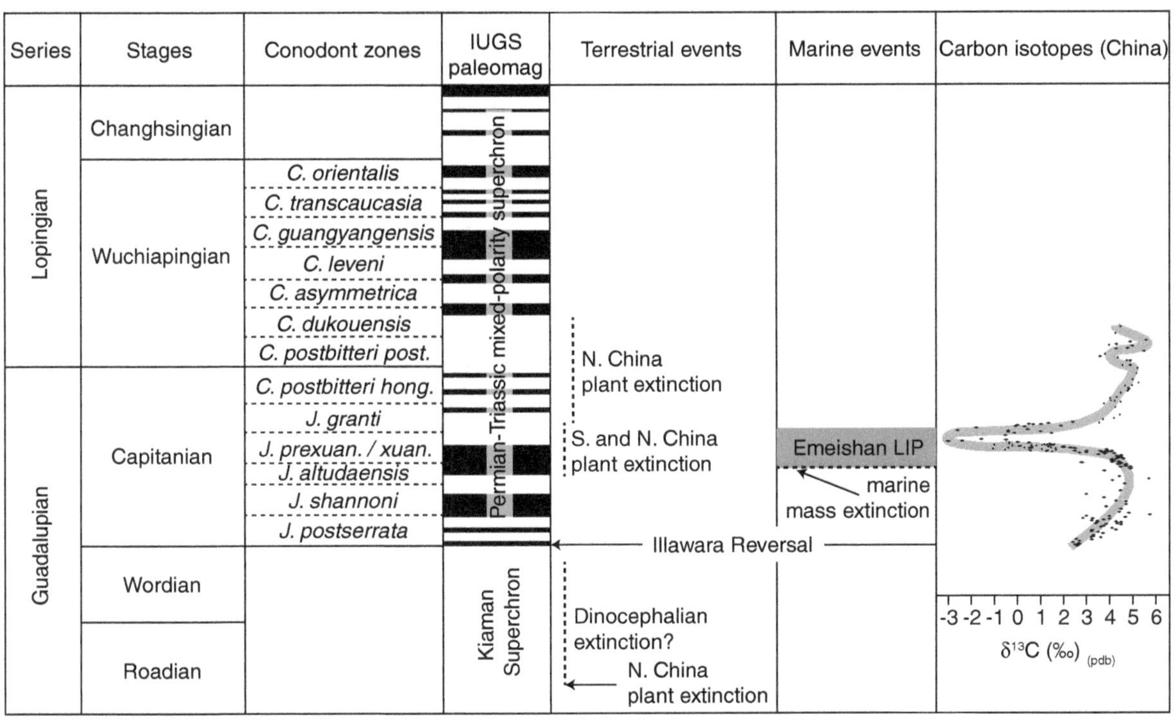

Figure 5. Summary of key Middle Permian events from China with carbon isotope curve (based on Wang et al., 2004; Wignall et al., 2009a; Bond et al., 2010b). Note the intra-Capitanian extinction level, contrasting assignments to an "end-Guadalupian" event in the literature. The International Union of Geological Sciences (IUGS) paleomagnetostratigraphic "bar code," in which periods of normal and reversed polarity are shown as black and white intervals, respectively, is currently the most reliable method of correlating between marine and terrestrial sequences. Placement of the Illawara reversal is after Isozaki (2009). LIP—large igneous province; pdb—Peedee belemnite. Abbreviations for conodont zones are as follows: *J.—Jinogondolella*; *C.—Clarkina*; *prexuan./xuan.—prexuanhanensis/xuanhanensis*; *hong.—hongshuensis*; *post.—postbitteri*. Figure is modified from Bond et al. (2010a).

Capitanian Volcanics

The Emeishan Basalt Formation is widely distributed in the provinces of Sichuan, Yunnan, and Guizhou in southwest China, and it represents one of the most recently identified large igneous provinces (Chung and Jahn, 1995, who ascribed a Permian-Triassic boundary age; Chung et al., 1998). In comparison to other large igneous provinces, the Emeishan traps are relatively small: Estimates of their areal extent range from 0.25–0.3 × 10^6 km^2 (Ali et al., 2005) to ~0.5 × 10^6 km^2 (Zhou et al., 2002). The thicknesses of the basalt piles range from several hundred meters in western Guizhou (Liu and Xu, 1994) to 5 km in its westernmost outcrops in northern Yunnan (Sun et al., 2010). Most estimates place the original volume at ~0.5 × 10^6 km^3 (Ali et al., 2005), but some have quoted a volume >1 × 10^6 km^3 (Courtillot and Renne, 2003). Extensive erosion and deformation ensure that estimations of their original volume are difficult.

The Emeishan Basalt was erupted onto the Maokou Formation platform carbonates, and there has been intensive debate as to the nature of this contact. He et al. (2003, 2004) favored an emplacement model that features mantle plume updoming prior to eruption, and thus they suggested a karstic surface with up to 200 m of relief developing prior to infill by lava flows. This observation and model have been challenged by several authors who contend that the initial eruptions occurred into a shallow sea (Ukstins Peate and Bryan, 2008; Ali et al., 2010; Sun et al., 2010). In some parts of the province, volcanism was preceded by a phase of partial platform carbonate collapse at the start of the *Jinogondolella altudaensis* zone (Sun et al., 2010). The age of onset of volcanism is thus tightly constrained to the mid-Capitanian *J. altudaensis–J. xuanhanensis* zonal interval. In contrast, radioisotopic age dating suffers from large errors. The compilation of Liu and Zhu (2009) shows the oldest and youngest U-Pb dates from Emeishan samples spanning 267–251 Ma, which was recently more tightly constrained to 260–257 Ma (Shellnutt et al., 2012), consistent with a Capitanian to Wuchiapingian age (see Gradstein et al., 2012).

Elsewhere, the Panjal Volcanics of northwest India include basaltic lavas and lesser volumes of rhyodacitic tuffs emplaced between supposedly Lower and Middle Permian marginal marine strata and Late Permian pelagic strata (Nakazawa et al., 1975), and thus they might be considered contemporaneous with the Capitanian extinction level. Recently, however, Shellnutt et al. (2011) reported a zircon U-Pb laser-ablation inductively coupled plasma–mass spectrometry (ICP-MS) date from rhyolite in the lower-middle part of the volcanic sequence that yielded a mean $^{206}U/^{238}Pb$ age of 289 ± 3 Ma. This suggests that the Panjal Traps are considerably older than previously interpreted and are therefore unrelated to either the Capitanian or end-Permian extinctions.

The clear temporal link between Capitanian marine extinctions and shallow-water, explosive Emeishan volcanism implies that these eruptions triggered the crisis. The extinction appears to have been particularly selective toward shallow-water photosynthetic taxa (e.g., calcareous algae) or those that likely harbored photosymbionts (e.g., fusulinacean foraminifera, alatoconchiid bivalves), suggesting a role for volcanically induced darkness in the extinction scenario. Many of the extinction victims had poor physiological buffering (e.g., calcareous sponges and corals), implying a possibility of acidification-driven extinction (McGhee et al., 2013) and by implication significant fluxes of CO_2 to the atmosphere. However, Ganino and Arndt's (2009) suggestion that the Emeishan province released thermogenic CO_2 into the atmosphere does not explain the observed negative isotope excursion, because the basalts intruded into platform carbonates would have supplied isotopically heavy carbon.

In the best-dated Chinese sections, the main carbon isotope excursion is found to slightly postdate the extinction, which occurs at the end of an interval of exceptionally heavy $\delta^{13}C$ values that has been called the "Kamura event" (Isozaki, 2007a, 2007b), a supposedly long-lasting major global cooling episode. The small mismatch between the extinction level and the main carbon isotope excursion perhaps accords with an extinction scenario involving abrupt cooling and death-by-photosynthetic shutdown at the onset of volcanism, although there is little direct evidence for cooling. It is perhaps significant that Emeishan volcanism occurred at the equator, and thus had the potential to inject SO_2 into the stratosphere in both the Northern and Southern Hemispheres, causing short-term cooling prior to longer-term warming from volcanic CO_2 emissions.

END-PERMIAN MASS EXTINCTION

Extinction Record

The end-Permian mass extinction needs little introduction. With >90% marine species loss (e.g., Erwin, 1994) and widespread devastation on land (e.g., Retallack, 1995; Smith and Ward, 2001), it was the greatest crisis in Earth's history. With its temporal connection to the Siberian Traps (e.g., Renne et al., 1995), the end-Permian event has become the quintessential example of volcanically induced mass extinction. Almost all animals and plants in almost all environmental settings were affected.

The marine crisis has generally been attributed to a single pulse of extinction of relatively short duration (<20,000 yr) that affected marine and terrestrial ecosystems simultaneously in the latest Permian. Shen et al. (2011) dated the marine extinctions to just before 252.28 Ma with precise U-Pb dating of sections in South China. In biostratigraphic terms, this level is within the Late Permian *Neogondolella yini* zone (top of bed 24e at the Meishan global stratotype section), a level that was thought to account for >90% species loss (Jin et al., 2000). However, recently, more extensive collecting in numerous marine sections throughout South China has revealed that this first extinction pulse was followed by a 200 k.y. period of recovery and modest diversification before a second extinction phase in the *Isarcicella staeschi* conodont zone of the earliest Triassic (Song et al., 2012).

The severity of the end-Permian extinction on land has been questioned by Gastaldo et al. (2009) and Xiong and Wang (2011), the latter suggesting that land plant microfossil genera actually increased across the Permian-Triassic boundary, before suffering a major extinction in the early Early Triassic (possibly contemporaneous with the second pulse of marine extinctions identified by Song et al., 2012). Nevertheless, the majority of evidence suggests that terrestrial and marine communities were devastated at around the same time in the Permian-Triassic boundary interval (e.g., Shen et al., 2011; Song et al., 2012). The Karoo Basin and Greenland sections record a catastrophic event for terrestrial vegetation and vertebrates in high southern and northern latitudes (Twitchett et al., 2001; Ward et al., 2005). Thus, the extinction mechanism must include kill factors that simultaneously affect marine and terrestrial environments globally. A major negative carbon isotope shift, in the region of 5‰–8‰, accompanies the marine extinction (Holser et al., 1991; Holser and Magaritz, 1995; Wignall et al., 1998), pointing to rapid and large-scale destabilization of the carbon cycle.

Contemporaneous environmental changes at the time of the extinctions include the widespread development of anoxic conditions in a broad range of water depths (Wignall et al., 2010; Bond and Wignall, 2010), a rapid increase of sea-surface temperatures (Joachimski et al., 2012; Sun et al., 2012), and the widespread development of microbial carbonate deposition in equatorial waters (Pruss et al., 2006). The demise of terrestrial plant communities saw the cessation of coal deposition (Retallack et al., 1996) and a change in fluvial environments to braided-river–dominated settings (Ward et al., 2000).

Siberian Traps

The Siberian Traps outcrops cover 1.5×10^6 km^3 of the Tunguska Basin of northwest Siberia (Fig. 6). Magmas were erupted through the Tunguska sedimentary sequence, consisting of thick deposits of Cambrian evaporites (Meyerhoff, 1980) that include a volume of ~585,000 km^3 of rock salt (Zharkov, 1984). Flood basalts account for a relatively modest 3.4×10^5 km^3 of the outcrop area. Their original extent was undoubtedly greater because a major western extension of the volcanics occurs beneath the Jurassic and Cretaceous layers of the West Siberian Basin (Figs. 1 and 6; Westphal et al., 1998; Reichow et al., 2005; Saunders et al., 2005). Estimating the original volume of the Siberian Traps and West Siberian rift system is problematic, but upper-end figures of 3–4 × 10^6 km^3 (Courtillot et al., 1999; Fedorenko et al., 2000; Kuzmin et al., 2010) indicate that it is one of the largest known continental flood basalt provinces.

Age of Trap Volcanism and Extinction

The eruption of the Siberian Traps was first put forward as the trigger mechanism for end-Permian extinction over two decades ago (e.g., Renne and Basu, 1991; Campbell et al., 1992), and the temporal connection has greatly strengthened subsequently (Renne et al., 1995; Bowring et al., 1998; Kamo et al., 2003; Mundil et al., 2004; Shen et al., 2011). The age of an ash layer, Bed 25, 16 cm below the conodont-defined boundary at the Meishan stratotype in South China, has been variously dated to 251.2 ± 3.4 Ma (sensitive high-resolution ion microprobe [SHRIMP] U-Pb date from Claoué-Long et al., 1991), 249.91 ± 1.52 Ma (^{40}Ar-^{39}Ar date from Renne et al., 1995), 250.04 ± 1.13 Ma (^{40}Ar-^{39}Ar date of the same level at Shangsi from Renne et al., 1995), 251.4 ± 0.3 Ma (zircon U-Pb date from Bowring et al., 1998), 252.4 ± 1.8 Ma (SHRIMP U-Pb date from Metcalfe, 1999), and 253 Ma (zircon U-Pb date from Mundil et al., 2001). Most recently, Shen et al. (2011) dated the Bed 25 ash as 252.28 ± 0.08 Ma (U-Pb date) and an ash band from 12 cm above the boundary as 252.10 ± 0.06 Ma, and by interpolation placed the boundary itself to 252.17 ± 0.06 Ma.

Volcanism in western Siberia began with the formation of the predominantly basaltic Tuffaceous Series (Sadovnikov and Orlova, 1993, 1998). These underlie the flood basalts in most of the province, and they dominate the entire succession in the south of the region. Zircons are rare in this series, making the dating of

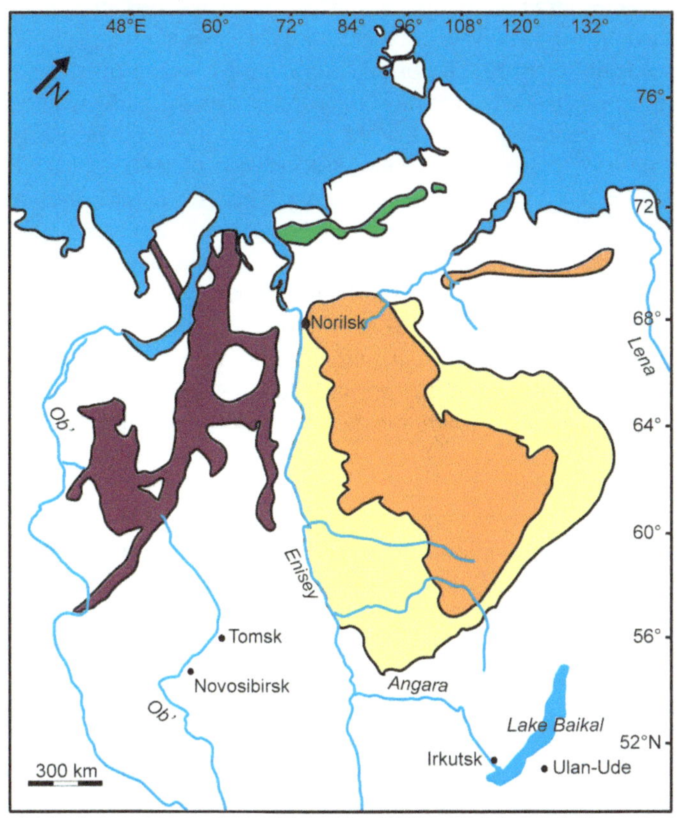

Figure 6. Map of Late Permian and Early Triassic magmatism of the Siberian Platform and West Siberian plain. Orange—extrusive volcanic rock exposure; yellow—intrusive volcanic rock exposure; purple—West Siberian Rift basalts, tuffs, and tuffites, where borehole samples have revealed basalts of the same age and chemistry as the main Siberian Traps basalts; green—Early Triassic Taimyr Traps; blue—water courses (names in italics). Major towns and cities are shown in regular text. Figure is modified from Kuzmin et al. (2010).

the onset of volcanism difficult, although the presence of conchostracans within intertuff sediments allows biostratigraphic correlation (Kozur, 1998; Kozur and Weems, 2011). Thus, Kozur and Weems (2011) suggested a middle Changhsingian to latest Permian age for the Tuffaceous Series, with the overlying flood basalt eruptions ensuing just below the Permian-Triassic boundary. Recent radioisotopic dates from within the flood basalt stratigraphy support synchronicity with the boundary in South China. Kamo et al. (1996) obtained a U-Pb age of 251.2 ± 0.3 Ma from the lower third of the lava pile in the Noril'sk area. Reichow et al. (2009) obtained an ^{40}Ar-^{39}Ar date of 250.3 ± 1.1 Ma for the final stages of extrusive volcanism at Noril'sk. Although the two techniques are not directly comparable, Renne et al.'s (2010, 2011) Ar-Ar recalibration would render Reichow et al.'s (2009) dates ~1% older and therefore close to, but slightly older than, Kamo et al.'s (1996) U-Pb dates. It seems likely that the entire Noril'sk lava pile erupted rather rapidly (also indicated by Venkatesan et al.'s [1997] much younger, but closely spaced ages for the base and top of the Noril'sk succession).

The Noril'sk succession is only a minor component of the Siberian Traps, and thus dates from this region may not agree with ages from other regions of the large igneous province. In the Maymecha-Kotuy region, the lava flows appear to be a chemically distinct suite of alkali-ultrabasics, which overlap in age within errors, but are most likely older than the Noril'sk succession according to a ^{40}Ar-^{39}Ar age of 253.0 ± 2.6 Ma (Basu et al., 1995). However, Kamo et al. (2003) obtained dates of 251.7 ± 0.4 Ma from the basal (Arydzhangsky) flow, and of 251.1 ± 0.3 Ma from the Delkansky flow, near the top of the pile. These ages suggest that the majority of volcanism may have been synchronous in Noril'sk and Maymecha-Kotuy, and the entire lava pile might have erupted in substantially less than a million years, but see also the recent compilation of Ivanov et al. (2013), who suggested a much more prolonged duration of magmatism. Again, further high-resolution dating is urgently required.

Style of Volcanism

The relatively high proportion of basaltic pyroclastics (>10% of the total volcanic pile in Noril'sk, and a significant contribution in Maymecha-Kotuy and Putorana to the south) and the relatively small volume of individual lava flows are unusual aspects of the Siberian Traps. Individual flows around Noril'sk are small compared to those of some other continental flood basalt provinces, rarely exceeding a few tens of meters thickness and a few tens of kilometers in extent (Sharma, 1997).

If the volumes of Siberian flood basalts are unusually small, then, perhaps more significant, their volatile content may have been unusually high. Black et al. (2012) measured S, Cl, and F in melt inclusions from 10 Siberian Traps samples and discovered anomalously high concentrations. S peaked at 0.51 wt% in Maymechinsky lava flows, and Cl and F peaked at 0.78 wt% and 1.95 wt%, respectively, in an Ust-Ilimsk dolerite sill. Total magmatic degassing from the Siberian Traps alone (not including a likely substantial input from contact metamorphism of sedimentary rocks in the Tunguska Basin) is estimated as ~6300–7800 Gt sulfur, ~3400–8700 Gt chlorine, and ~7100–13,600 Gt fluorine (Black et al., 2012). Even these concentrations may be an underestimate of volatile release: Sobolev et al.'s (2011) estimates noted previously suggest even greater volumes.

Extinction Mechanisms

As one might expect for the greatest mass extinction, wide varieties of kill mechanisms have been put forward, including marine anoxia, volcanic winter, hypercapnia, ocean acidification, global warming, increased sediment flux to the oceans, ozone destruction, extreme atmospheric oxygen depletion, and poisoning by toxic trace metals (Fig. 7; see reviews of Benton and Twitchett, 2003; Racki and Wignall, 2005; Wignall, 2007). With the exception of some claims for bolide impact (Becker et al., 2001; Kaiho et al., 2001), all these mechanisms ultimately view the Siberian Traps as the origin of these environmental woes.

Of the competing theories, anoxia has become a popular kill mechanism in the marine scenario due to the abundance of evidence for oxygen depletion in boundary sections in regions as diverse as the Paleotethys (Wignall and Twitchett, 1996), marginal seas (Australia; Grice et al., 2005; Bond and Wignall, 2010), Panthalassa (Isozaki, 1994, 1997), the western United States (Wignall and Hallam, 1992; Woods and Bottjer, 2000), and the Boreal realm (Wignall et al., 1998; Dustira et al., 2013). The causal chain reaction has long related the warming of the oceans to the onset of anoxia (Wignall and Twitchett, 1996), although the role of increased nutrient supply is also important (Fig. 7). H_2S degassing from anoxic surface waters onto land has also been proposed as a terrestrial kill mechanism (Kump et al., 2005).

While oxygen restriction probably contributed to the first pulse of extinction, it was not the sole killer. The open-ocean record of radiolaria reveals that the main losses occurred before the onset of anoxia (Wignall and Newton, 2003; Wignall et al., 2010). The taxonomic selectivity of the first pulse of extinction, which preferentially affected warm, shallow-water habitats and particularly reef taxa, could accord with either a rapid phase of warming or an ocean acidification-driven crisis (e.g., Payne et al., 2007; Montenegro et al., 2011; Hinojosa et al., 2012; Heydari et al., 2013). The anoxia-extinction link is much clearer for the second pulse of extinction, in the Early Triassic, because it saw the replacement of benthos-dominated communities with nekton-dominated ones, the loss of all deep-water benthos, and the relative success of dysoxia-tolerant bivalves (Wignall and Hallam, 1992; Bond and Wignall, 2010; Song et al., 2012).

The Permian-Triassic boundary coincides with a well-known negative carbon isotope excursion of ~5‰–7‰ from bulk marine carbonates (e.g., Baud et al., 1989; Holser and Magaritz, 1995; and many subsequent studies) and marine organic material (e.g., Wignall et al., 1998; Riccardi et al., 2007; Grasby and Beauchamp, 2008). Although apparently abrupt in condensed sections, the isotopic shift is seen to be gradual (lasting

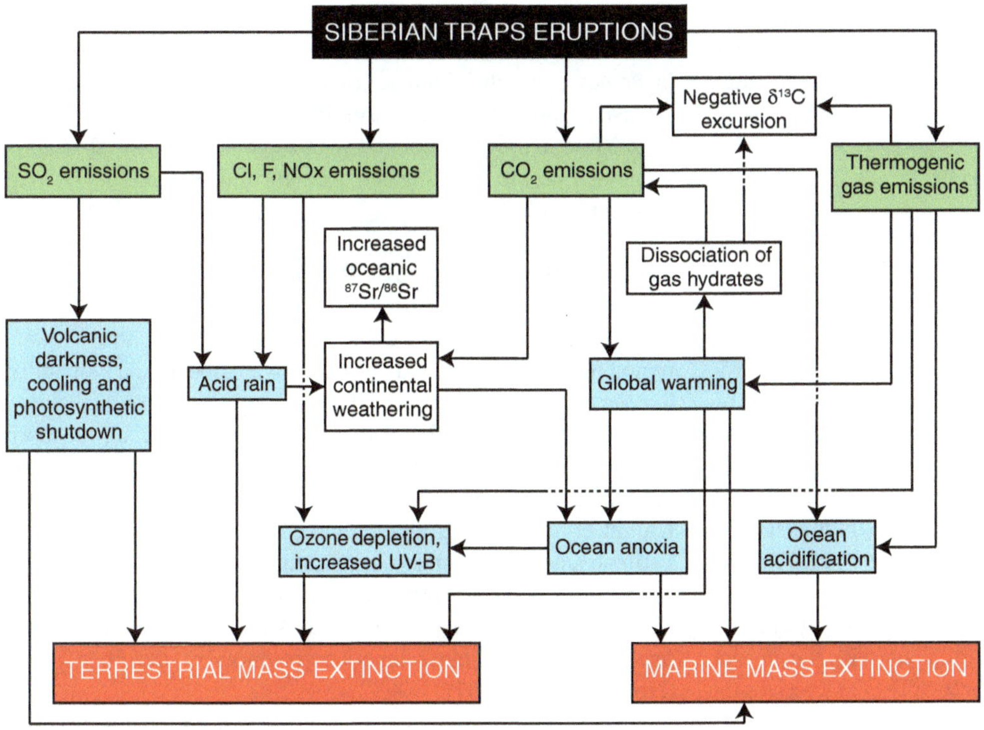

Figure 7. Flow chart summarizing proposed cause-and-effect relationships during the end-Permian extinction and Early Triassic protracted recovery period, modified from earlier versions by Wignall (2001, 2007) and Algeo et al. (2011). Green boxes—direct products of volcanism; blue boxes—kill mechanisms. The link between oceanic anoxia and ozone depletion was proposed by Kump et al. (2005) and derives from surface-water H_2S degassing. Thermogenic gas has been suggested to cause ozone depletion by injecting CH_3Cl from coal combustion (Beerling et al., 2007). UV-B—ultraviolet-B radiation.

0.5 m.y.) in more expanded sections (Korte et al., 2010). This suggests that short-lived events such as destabilization of methane hydrate reservoirs or catastrophic bolide impact are unlikely to have caused this shift. Instead, the onset and peak of the isotopic shift are seen to coincide with the start-up and peak effusion of the Siberian Traps (Korte et al., 2010). However, mantle-derived CO_2 has a $\delta^{13}C$ value of –5‰, which is only a little lighter than the ocean-atmosphere system of the latest Permian. Potentially, the carbon release from Siberian volcanism may have been substantially derived from recycled ocean crust incorporated into a mantle plume (Sobolev et al., 2011). Analyses of olivine grains separated from crushed lavas suggest that the $\delta^{13}C$ values of the C may have been substantially lighter (–12‰), and the volume of volatiles release substantially greater than that inferred by simply scaling up measurements from modern basaltic eruptions. If this is the case, then the $\delta^{13}C$ record could simply be read as a monitor of Siberian eruptions.

The ramifications of the Sobolev et al. (2011) proposal have yet to be incorporated into Earth system models for the end-Permian crisis. Instead, researchers frequently propose that the volcanic C flux was only a component of total C release. It may have acted as a stimulus: Modest global warming may have triggered destabilization of methane hydrates stored in permafrost soils and continental shelves (Racki and Wignall, 2005). Alternatively, Retallack and Jahren (2008) suggested that the $\delta^{13}C$ excursion could be attributed to injection of nearly 1 trillion tonnes of carbon in the form of thermogenic CO_2 release from coal beds beneath the Siberian Traps (Payne and Kump, 2007; Retallack and Jahren, 2008; Svensen et al., 2009). Support from such a major coal-burning episode comes from the discovery of coal fly ash (a product of combustion) in boundary beds of the Sverdrup Basin, Arctic Canada (Grasby et al., 2011). Ogden and Sleep (2012) tested the coal-volcanism scenario by modeling the effects of a massive mafic sill intruding, heating, and mixing with thick coal seams. The hot coal-basalt mixture is expected to extrude at numerous surface locations, combusting on contact with air, and injecting vast quantities of basaltic dust, CO_2, and methane into the atmosphere. All of the carbon in 1000 km^3 of coal would need to be liberated in order to generate the observed isotope excursion—a scenario not beyond the realms of possibility.

Volcanism and coal burning also contribute gases to the atmosphere, such as Cl, F, and CH_3Cl from coal combustion, that suppress ozone formation (Black et al., 2012). The mooted presence of abundant H_2S in the atmosphere (degassed from anoxic surface waters; Kump et al., 2005; Kaiho et al., 2006) would also destroy ozone by suppressing OH and H radicals involved in its formation. A further effect of H_2S may have been to prolong the residence time of methane in the atmosphere and thus indirectly reduce ozone generation rates (Lamarque et al., 2007). Ozone destruction and the consequent increase of ultraviolet-B radiation have become a popular terrestrial-extinction cause (e.g., Visscher et al., 2004; Kump et al., 2005; Sephton et al., 2005; Collinson et al., 2006; Beerling et al., 2007). This theory neatly ties Siberian Traps volcanism and thermogenic gas as the ultimate cause of extinctions on land and in the oceans, but modeling work has yet to produce the required level of ozone damage. Beerling et al.'s (2007) model of the input of large volumes of CH_3Cl from coal combustion at the eruption site indicated that the only significant damage to the ozone layer would occur at high latitudes, rather weakening the argument for ozone destruction as a cause of global mass extinction.

END-TRIASSIC MASS EXTINCTION

Extinction Record

In terms of severity of ecological disturbance, the end-Triassic crisis is behind only the end-Permian and end-Cretaceous events (Table 1; McGhee et al., 2013). The terminal Triassic Rhaetian Stage was notable for the extinction of reef communities that included the scleractinian corals' most severe crisis (Flügel, 2002; Flügel and Kiessling, 2002; Martindale et al., 2012). Bivalves and ammonoids were also prominent victims, although their record suggests a gradual decline in the latest Triassic before a major extinction a little below the ammonoid-defined Triassic-Jurassic boundary (Hallam, 2002).

The plant record also records a severe crisis, although it is unclear if it was precisely contemporaneous with the marine extinctions. Thus, a major floral turnover locally characterizes the (palynologically defined) Triassic-Jurassic boundary in the Newark Basin of the United States (Fowell and Olsen, 1993). Plant macrofossils from East Greenland reveal even greater species losses of 85% (McElwain et al., 1999, 2007), but, curiously, there are no associated losses in the palynological record (Raunsgaard Pedersen and Lund, 1980; Koppelhus, 1996). In parts of Europe, the end-Triassic sporomorph crisis is only marked by the loss of *Ovalipollis*, which was followed by diversification amongst microfloras (Hesselbo et al., 2004). The terrestrial record therefore suggests that globally, plants experienced disturbance and a major turnover, rather than catastrophic extinction losses.

The Triassic-Jurassic transition was marked by numerous paleoenvironmental changes. Stomatal density indices in fossil leaves suggest a fourfold increase in atmospheric CO_2 and global warming of 3–4 °C (McElwain et al., 1999) or possibly as much as 6 °C (Beerling and Berner, 2002; Huynh and Poulsen, 2005), presumably driven by volcanic CO_2 release in the same manner as the Siberian Traps–driven warming. A potential consequence of this enormous rise in atmospheric CO_2 is ocean acidification: a popular kill mechanism for the end-Triassic event that is consistent with the selective loss of shelly marine forms with little physiological buffering and especially of the corals (Hautmann, 2004; Hautmann et al., 2008a, 2008b; Greene et al., 2012; Martindale et al., 2012).

Destabilization of the carbon cycle is once again recorded by an initial sharp but brief negative ^{13}C shift, followed by recovery and then a more prolonged negative ^{13}C excursion, the onset of which marks the formal Triassic-Jurassic boundary. In total, there is an ~6‰ negative shift in the record of both organic and inorganic carbon (e.g., Pálfy et al., 2001; Hesselbo et al., 2002, 2004; Jenkyns et al., 2002; Korte et al., 2009; Ruhl et al., 2009, 2011). In the British Isles, the main excursion immediately postdates the main marine extinctions and occurs at a level of intensively deformed strata (often referred to as "seismite"). The seismite layer has been attributed to a single earthquake of exceptional intensity caused by a meteorite impact (Simms, 2003, 2007) or by seismic activity associated with the onset of Central Atlantic magmatic province volcanism (Hallam and Wignall, 2004; Wignall and Bond, 2008).

Central Atlantic Magmatic Province

The rifting of Pangea and breakup of the central Atlantic region were accompanied by extensive flood basalt volcanism that emplaced the Central Atlantic magmatic province across four continents (Fig. 2). The estimated original area is poorly constrained but may have approached 10×10^6 km^2, although most of the extrusives have been eroded away (McHone, 2003; Knight et al., 2004; Nomade et al., 2007; Greene et al., 2012). Estimates of the original volume range from ~2×10^6 km^3 (Holbrook and Kelemen, 1993; Marzoli et al., 1999) to ~4×10^6 km^3 (Olsen, 1999), making the Central Atlantic magmatic province one of the largest large igneous provinces of the Phanerozoic (Nomade et al., 2007).

The $^{40}Ar/^{39}Ar$ dates of Central Atlantic magmatic province basalts reveal that intrusive magmatism started around 202 Ma and was followed soon after by extrusive activity that peaked at 199 Ma (Nomade et al., 2007). Marzoli et al. (2011) recalibrated various Ar-Ar dates using the method described by Renne et al. (2010, 2011) and showed peak activity around 201 Ma, an age in accordance with recent U-Pb dates. Thus, Schoene et al.'s (2010) U-Pb dating places the earliest extrusive activity at 201.38 Ma, an age that provides the current definition of the Triassic-Jurassic boundary (Gradstein et al., 2012). Blackburn et al. (2013) applied U-Pb dating to Central Atlantic magmatic province lavas in several different regions, establishing ages of activity from ca. 201.5 Ma to 201.00 Ma. Early flows within Rhaetian-dated sediments in Morocco and Canada confirm that Central Atlantic magmatic province eruptions began in the latest Triassic (Marzoli et al., 2004; Cirilli et al., 2009; Deenen et al., 2010) and were therefore coincident with latest Triassic extinctions (Ruhl et al., 2010, 2011; Deenen et al., 2010), or at least within one precession cycle (20 k.y.; Deenen et al., 2010). In the United States, the oldest lavas postdate the palynological turnover (Fowell and Olsen, 1993) by a similar duration (Blackburn et al., 2013) and are of basal Jurassic Hettangian age. Thus, in the Newark Basin, the lowest lava flow is 30 m above the palynologically defined Triassic-Jurassic boundary. Some have suggested that pre-eruptive dike emplacement affected the global carbon cycle before the first lavas were extruded (e.g., Ruhl and Kürschner [2011], who discovered a small negative C excursion that slightly preceded the "initial" excursion of Hesselbo et al. [2002]). This is a convenient explanation for the slight diachroneity in the records of volcanism and extinction in the northeastern United States. In fact, the lava flows in the Newark and Hartford Basins probably do not record the full temporal extent of eruptions because the oldest flows in the Culpeper Basin of Virginia are of Rhaetian age (Kozur and Weems, 2011), just like their equivalents in Morocco and Canada.

The main phase of Central Atlantic magmatic province volcanism appears to have been of short duration. This has been estimated as ~1.6–2 m.y. (Marzoli et al., 1999, 2004; Nomade et al., 2007). An even shorter duration is estimated as 580 ± 100 k.y. for the Newark Basin (Olsen et al., 1997, 2003) and 840 ± 60 k.y. for the Hartford Basin (Schaller et al., 2012), although neither basin records the full eruption history. Knight et al. (2004) suggested that the entire volume of the Central Atlantic magmatic province lava pile in Morocco was erupted within 3–5 pulses, each potentially as brief as 450 yr. Such claims have major implications for the rate of volatile release. McHone (2003) calculated that total emissions of CO_2, S, F, and Cl ranged from 1110 to 5190 Gt (based on whole-rock analyses). Beerling and Berner (2002) calculated that as much as 17,500 Gt of carbon was released as CO_2, with methane hydrate degassing adding a further 12,000 Gt of carbon as CH_4 (based on magmatic CO_2 contents from modern Hawaiian basalts; Berner and Beerling, 2007), although thermogenic methane release has also been invoked (Ganino and Arndt, 2009; van de Schootbrugge et al., 2009). Such volumes can potentially explain the contemporaneous $\delta^{13}C$ excursions (Pálfy et al., 2001; Ruhl et al., 2011). However, the rate of gas effusion is a key factor when interpreting the isotopic excursions, and if the duration of peak eruptions was as brief as suggested by Knight et al. (2004), with intense eruption pulses lasting around 400 yr and most volcanism occurring in less than 20 k.y., then the Central Atlantic magmatic province represents a potent trigger for environmental disturbances that include global warming and ocean acidification.

EARLY JURASSIC (TOARCIAN) MASS EXTINCTION

The extinctions in the Toarcian Stage of the Early Jurassic are well understood thanks to the excellent northwest European record (e.g., Jenkyns, 1985, 1988; Hesselbo et al., 2000; McArthur et al., 2000; van de Schootbrugge et al., 2005; Wignall et al., 2005; Cohen et al., 2007; Sandoval et al., 2012). Estimates of 5% family-level global diversity loss indicate a relatively minor crisis, although significant losses were experienced by shallow-marine mollusks (Little and Benton, 1995). The extinction was originally thought to have been localized to northwest Europe (Hallam, 1986, 1996), but studies in South America (Aberhan and Fürsich, 1996), North America (Caruthers et al., 2011; Caruthers and Smith, 2012), Tibet (Wignall et al., 2006), and Japan (Hori, 1993; Wignall et al., 2010; Gröcke et al., 2011), as well as deeper-water facies of western Tethys (Vörö, 1993), have revealed the global nature of the crisis. Elevated extinction rates through five ammonite zones spanning the Pliensbachian-Toarcian boundary (Little and Benton, 1995) culminated with peak losses during an interval of widespread anoxia (the Toarcian oceanic anoxic event [OAE]) in the *Falciferum* zone (Little, 1996), suggesting a role for marine anoxia.

Pálfy et al. (2002) ascribed a U-Pb age of ca. 183 Ma to volcanic layers interbedded within strata of the contemporaneous North American ammonite zone—an age assignment that confirms a close temporal link between the Toarcian extinctions and the Karoo and Ferrar continental flood basalt province (Figs. 1 and 8). The Early Jurassic saw major basaltic volcanism that emplaced >2.5 × 10^6 km^3 of sills, dikes, and flows in southern Africa (Karoo Traps) and Antarctica (Ferrar Traps; Encarnación et al., 1996). Outcrops of the province cover an area of ~3.1 × 10^6 km^2 (Eldholm and Coffin, 2000; Courtillot and Renne, 2003). The province is well dated (Encarnación et al., 1996; Duncan et al., 1997; Jones et al., 2001; Le Gall et al., 2002; Jourdan et al., 2004, 2005, 2007a, 2007b; Riley et al., 2004, 2006), mainly from work in southern Africa. Jourdan et al. (2007b, 2008) provided a succinct summary of the main geochronologic findings: (1) The entire province was active for a duration of 10–12 m.y. (184–172 Ma), but the main volume of the basaltic sequence was emplaced over 3–4.5 m.y. around 180 Ma (Ar-Ar dating) and was therefore coincident with the main extinction pulse (Pálfy et al. [2002] used U-Pb dating to show that peak magmatism lasted only 2 m.y. and was clustered around 183 Ma; note that recalibration of Jourdan et al.'s [2007b, 2008] Ar-Ar dates as per Renne et al. [2010, 2011] renders them very close in age to those quoted by Pálfy et al. [2002]); (2) brief (1 m.y. or less), chemically distinct events such as the Okavango dike swarm (179.2 ± 0.4 Ma), the 800-m-thick southern Botswana lava pile (178.6 ± 0.5 Ma), and the 1.9-km-thick Lesotho lava pile (181.6 ± 0.7 Ma) have been identified; (3) a huge sill complex crops out in the main Karoo sedimentary basin, but it lacks precise dating; (4) basaltic magmatism was followed by late-stage silicic magmatism between 178 and 174 Ma; and (5) magmatic activity ended with the intrusion of the mid-ocean-ridge basalt (MORB)–like Rooi Rand dikes at 174–172 Ma.

The Toarcian marine losses in Europe coincided with global warming (Bailey et al., 2003; McElwain et al., 2005), black shale deposition, and carbon cycle perturbations. An ~2‰–3.5‰ negative $\delta^{13}C$ marine carbonate excursion was identified from the early *Falciferum* zone by Hesselbo et al. (2000), who also reported a –7‰ shift in the $\delta^{13}C$ record of marine organic matter and wood. The negative excursion is followed immediately by a positive $\delta^{13}C_{carb}$ shift of equal magnitude (Jenkyns, 1988), which is generally regarded to reflect the enhanced burial of organic matter in anoxic seas (Wignall et al., 2005). Superimposed on this negative-positive $\delta^{13}C$ isotope history, there are three 2‰–3‰ negative shifts. The lower two of these are purported to have lasted no more than 20 k.y. each (Fig. 8; Kemp et al., 2005).

The ultimate origin of the carbon isotope excursions may lie in volcanogenic CO_2 release from the Karoo and Ferrar Traps, but the magnitude of the excursion suggests that at least some component came from methane hydrate dissociation as a consequence of volcanically induced global warming (Hesselbo et al., 2000, 2007; Kemp et al., 2005). With the advent of the thermogenic gas hypothesis (Svensen et al., 2004), the addition of methane from magmatically heated coal and organic-rich shale beds

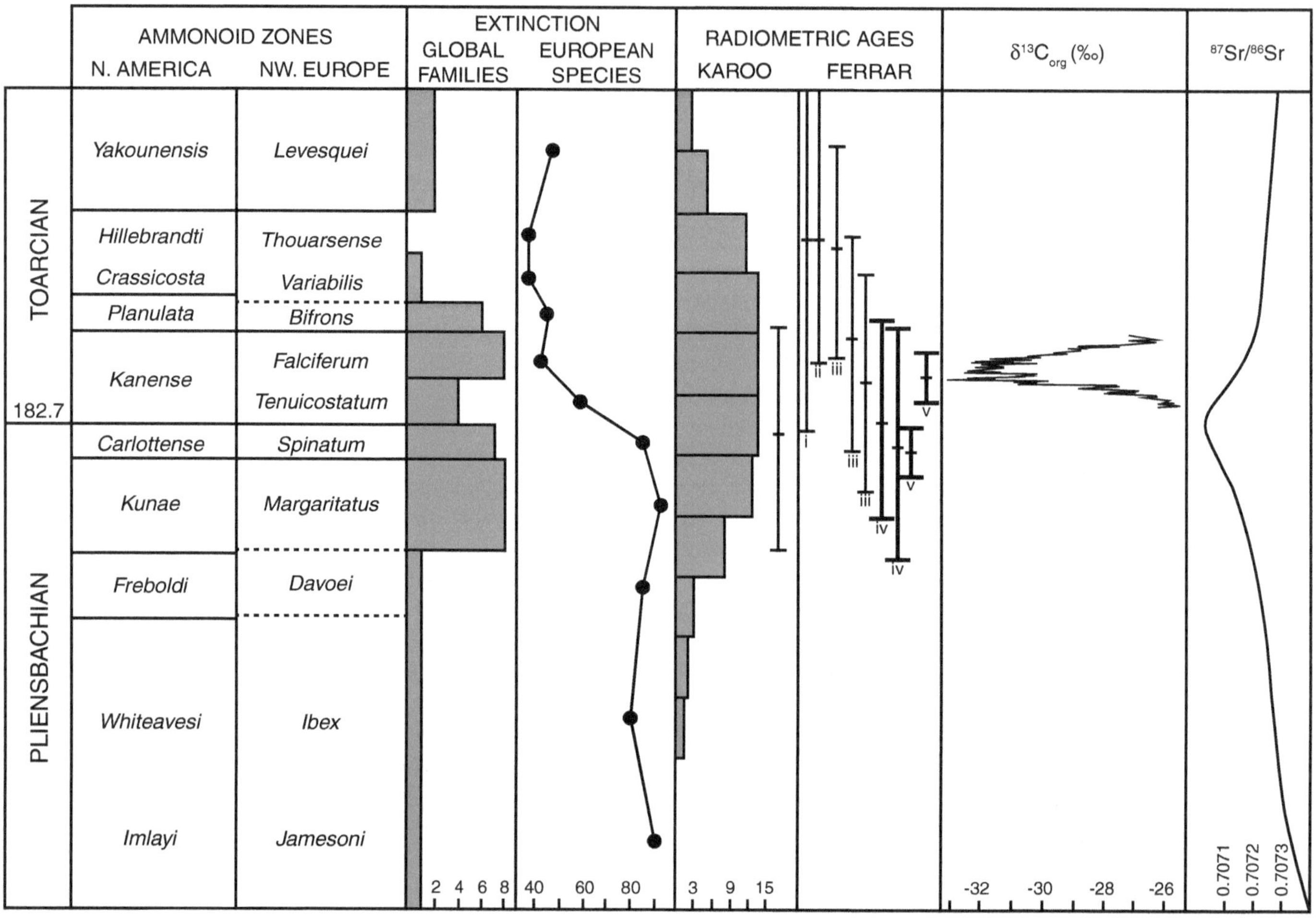

Figure 8. Pálfy et al.'s (2002) correlation of the Toarcian marine extinction, Karoo and Ferrar flood basalt volcanism, and carbon and strontium isotope stratigraphy within the ammonoid zonal framework. The histogram of number of global family extinctions by zone is based on Little and Benton (1995). The "European Species" column shows cumulative species diversity per zone, expressed in number of species of bivalves, ammonoids, rhynchonellid brachiopods, crinoids, foraminifera, and ostracods from Britain (Hallam, 1996). Radioisotopic ages from the Karoo Group were recalculated by Pálfy et al. (2002) from published sources. The histogram in this column represents the age spectrum of 28 $^{40}Ar/^{39}Ar$ dates from Duncan et al. (1997); the vertical bar is the error bar of the Karoo U-Pb age (Encarnación et al., 1996). Radioisotopic ages for the Ferrar Traps are based on $^{40}Ar/^{39}Ar$ (thin lines) and U-Pb (heavy lines) and were recalculated by Pálfy et al. (2002) from published sources. Ages and error bars for Ferrar dates are from left to right: (i) composite of 11 $^{40}Ar/^{39}Ar$ ages by Heimann et al. (1994); (ii) composite of two $^{40}Ar/^{39}Ar$ ages by Foland et al. (1993); (iii) three $^{40}Ar/^{39}Ar$ ages by Duncan et al. (1997); (iv) two U-Pb ages by Encarnación et al. (1996); and (v) two U-Pb ages by Minor and Mukasa (1997). The carbon isotope profile is from Kemp et al. (2005), rescaled for this figure. The seawater $^{87}Sr/^{86}Sr$ curve is simplified from Jones et al. (1994). Figure is modified from Pálfy et al. (2002). The Pliensbachian-Toarcian boundary is placed at 182.7 Ma in the *Geologic Time Scale 2012* (Gradstein et al., 2012).

in the Karoo Basin has also been implicated in the isotope excursion (McElwain et al., 2005; Svensen et al., 2007). This model has been contested by Gröcke et al. (2009), who noted that the low vitrinite and liptinite contents of Karoo and other Gondwanan coals are at odds with an ability to release abundant methane. The Karoo and Ferrar Traps are marked by comparatively narrow contact aureoles surrounding intrusions, further limiting thermogenic gas release, some of which might have been captured as coalbed CH_4 or condensed as pyrolytic carbon (Gröcke et al., 2009). Instead, Wignall et al. (2006) noted that some of the rapid $\delta^{13}C$ excursions could be a local signature of recycling of isotopically light carbon from the lower water column. However, similarly large shifts have been reported in the $\delta^{13}C_{carb}$ record of the Paris Basin in France (Hermoso et al., 2009) and that of $\delta^{13}C_{org}$ from Haida Gwaii (Queen Charlotte Islands), Canada (Caruthers et al., 2011), suggesting that it is a global signal.

The association of the Karoo and Ferrar Traps with ocean anoxia and the Toarcian extinctions provides another clear case study of the large igneous province–extinction link. In this case, however, one of the largest continental flood basalts of the Phanerozoic is associated with only a minor ecological crisis, which returns us to the key theme of this review: Why do some large igneous provinces apparently trigger major extinctions, while other do not?

The Karoo magma erupted in contrasting styles: Initial brief and temporally distinct magmatic pulses emplaced a large proportion of the total volume of the province; these were followed by more persistent, but diffuse magmatism that lasted for several million years and also contributed a large component of total volume. Some of the thickest lava piles, such as that in Lesotho, were erupted in a brief interval (<0.8 m.y.; Jourdan et al., 2007b), but these individually contributed only a modest proportion of the total volcanogenic volatile flux. A relatively slow rate of volatile input to the atmosphere may have induced only gradual climate change, leaving enough time for the ecosystem to recover between each individual volcanic event (Jourdan et al., 2008). Modeling suggests that even adding thermogenic and/or methane hydrate emissions into the equation would have a limited impact on climate if individual pulses are spread over a protracted period (Beerling and Brentnall, 2007).

These subdued effects of Karoo-Ferrar volcanism are comparable to the Ontong Java and Kerguelen Plateaus eruptions. These two large oceanic plateaus emplaced unparalleled volumes of lava (the former has an estimated volume of 100 × 10^6 km^3; Gladczenko et al., 1997), and yet neither is associated with mass extinction. Potentially, most of the CO_2 released by these giant submarine eruptions would have been taken up by the oceanic sink, limiting the effects of global warming. However, at least some of the Ontong Java Plateau erupted close to sea level (Mahoney et al., 2001), and so it is difficult to envisage how at least some CO_2 could not have reached the atmosphere.

DISCUSSION AND CONCLUSIONS

Temporal Link

The temporal link between large igneous province eruptions and mass extinctions is well established and has been refined in recent years thanks to much effort and improvements in radioisotopic dating techniques. All but one of the "Big 5" extinctions are associated with large igneous provinces, the exception being the Late Ordovician crises. This suggests that the largest extinctions require a volcanic trigger, and yet the most voluminous examples from the Cretaceous are only associated with minor or no biotic crises. Of the 16 large igneous province episodes summarized in Table 2 (see also Fig. 1), only around half are temporally constrained to intervals of mass extinction—and that includes the tenuous deep time examples of the Volyn and Kalkarindji flood basalts. Previous assertions of links between the Paraná and Etendeka Traps and an end-Jurassic extinction, the North Atlantic igneous province and an end-Paleocene extinction, and the Columbia River Flood Basalts and an end–early Miocene extinction (Courtillot and Renne, 2003) do not stand up to scrutiny, because those events were not "mass extinctions."

Several of the post-Jurassic large igneous provinces triggered similar environmental responses to those associated with earlier large igneous province eruptions, chiefly global warming and marine anoxia (Table 2). The Deccan Traps apparently warmed global climates prior to the terminal Cretaceous bolide impact (see review of Keller et al., this volume). The massive Ontong Java phase 1 volcanism straddles the Barremian-Aptian OAE (Selli event). The Caribbean-Colombian and Madagascar Plateaus have been implicated in the Cenomanian-Turonian OAE (Bonarelli event) and minor extinction (although these probably predated volcanism). The North Atlantic igneous province is associated with extreme warming of the Paleocene-Eocene thermal maximum, but contemporaneous global extinction losses were minor. The most intriguing aspect of these large igneous provinces is their failure to trigger more severe biotic events, which raises several questions: Were post-Jurassic biotas more resilient to change? Have ecosystems adapted following repeated crises during the late Paleozoic and early Mesozoic or has Earth become more efficient at dealing with excess CO_2 fluxes and other volcanic phenomena? Or is the causal link between

TABLE 2. LARGE IGNEOUS PROVINCES AND ASSOCIATED EXTINCTION EVENTS AND ENVIRONMENTAL CHANGES

Large igneous province	Age (peak activity, Ma)	Coeval extinction	Global warming	C isotope excursion (‰)	Marine anoxia	Ocean acidification
Volyn CFB	580–545	Ediacaran? ca. 545–542 Ma				
Kalkarindji CFB	510–503	Early Cambrian? ca. 510 Ma	?	−4	✓	
Viluy Traps	380–340 (ca. 370)	Frasnian-Famennian? ca. 372 Ma	✓	+2 to +4	✓	
PDD Rift	367–364					
Emeishan Traps	260–257	Capitanian	?	−6		
Siberian Traps	253–250	End-Permian	✓	−5 to −7	✓	?
CAMP	202–199	End-Triassic	✓	−6 (^{carb} and ^{org})	?	?
Karoo and Ferrar Traps	184–172 (183–180)	Toarcian	✓	−7 then +7 (^{org})	✓	
Paraná and Etendeka Traps	134–132					
Ontong Java Plateau (phase 1)	122				✓	
Rajmahal Traps/Kerguelen Plateau	119–109					
Madagascar/Caribbean-Columbia/ Ontong Java Plateau (ph. 2)	90–87	End-Cenomanian? ca. 94 Ma	✓	+2	✓	
Deccan Traps	67–65	End-Cretaceous	✓	−2		
NAIP (ph. 1, ph. 2)	61, 56–54		✓ (PETM)	−3	✓	
Ethiopia and Yemen Traps	31–29					
Columbia River Flood Basalts	17–6 (16.6–15.3)					

Note: Carbon isotope excursions are based on the $\delta^{13}C$ record of marine carbonates, unless specified as "org". CFB—continental flood basalt province; PDD—Pripyat-Dnieper-Donets; CAMP—Central Atlantic magmatic province; NAIP—North Atlantic igneous province; PETM—Paleocene-Eocene thermal maximum. ✓—environmental changes with strong evidence; ?—environmental changes with tentative evidence.

volcanism and mass extinction broken, and their apparent temporal coincidence just that, a coincidence?

Key factors in large igneous province lethality may be the style, rate, and site of eruptions, and thus their ability to rapidly inject large fluxes of climate-changing volatiles into the stratosphere. The potential of thermogenic greenhouse emissions to add to the volatile budget has come to light in the past decade, and this represents a plausible explanation for the negative carbon isotope shifts that accompany several extinctions. The state of climate at the onset of volcanism, and the resilience of the extant biota may also be important. The key challenge for earth scientists now is to model, analyze, and understand *how* the products of volcanism can cause extinction, within an ever-improving age framework.

Triggers and Killers

This review has focused on five extinction events—including four consecutive crises in the 80 m.y. interval between the Middle Permian and Early Jurassic—that are clearly linked to large igneous province volcanism. The temporal relationship between the Frasnian-Famennian biodiversity crisis and the Viluy Traps has yet to be resolved but is becoming more convincing. In contrast, the Capitanian, end-Permian, end-Triassic, and Toarcian crises were all associated with a swathe of environmental changes consistent with the effects of volcanism. Of these five events, three (maybe four) are associated with marine anoxia, four (maybe five) are associated with warming, and four are associated with major negative carbon isotope excursions (Table 2). Ocean acidification has been invoked in two of these five extinctions, with the selectivity of extinctions touted as key evidence (e.g., Payne et al., 2007; Foster, 2008; Kiessling and Simpson, 2011), although a reliable proxy for past pH changes has yet to be devised.

Short-term global cooling and dimming are other obvious consequences of explosive volcanism, but the duration of such events, measured in years, is almost impossible to resolve in the geological record. The best evidence for cooling comes from extinction selectivity, and none of the crises documented here yields strong evidence for this kill mechanism, except perhaps the Capitanian crisis. Increasingly high-resolution extinction studies may yet reveal a role for volcanogenic cooling in the early stages of ecosystem perturbation.

Concluding Remarks

(1) The temporal link between large igneous provinces and mass extinction events has been recognized for nearly three decades. Recent advances in radioisotopic dating techniques have refined and confirmed the link for at least half of the major extinctions of the Phanerozoic. Almost all of Earth's most severe biotic crises, including four of the "Big 5" extinctions, are associated with large igneous provinces, implying that large-scale volcanism is a prerequisite for a really big extinction.

(2) Modeling, observations of recent eruptions, and detailed volcanological, geochemical, and paleontological studies have yielded clues to the causal mechanisms that link volcanism and extinctions. The relative importance of many factors in this extinction-volcanism nexus, including the continental configuration at the time of the eruption, the latitude and altitude of eruption, the volume, rate, and duration of eruption, its style, the preexisting climate state, and the resilience of the extant biota to change, have all to be evaluated and understood. However, the total volume of magma is clearly not an important factor, because some of the smallest large igneous provinces are associated with extinctions, whereas some of the largest are not.

(3) Of the five events examined here, four are clearly associated with global warming and other proximal killers such as marine anoxia. Three of these extinction and warming episodes are accompanied by large negative carbon isotope excursions, supporting a volcanogenic origin. In all three cases, the scale of the excursion is too big to be attributed to volcanic CO_2 and CH_4 alone (unless large igneous province CO_2 emissions were considerably lighter than mantle values). Methane hydrate release during warming episodes, and more recently thermogenic gas release, is commonly invoked.

(4) Volcanically induced warming is clearly a powerful driver of marine anoxia: All post-Paleozoic oceanic anoxic events coincide with intervals of large igneous province formation. These warming-anoxia events are associated with major extinctions (e.g., the end-Permian), minor extinction events (e.g., the Toarcian, and the Cenomanian-Turonian Bonarelli event), and very minor extinctions (e.g., the late Paleocene, the Aptian-Albian Selli event).

(5) Volcanic darkness and cooling have been invoked in several extinction scenarios, in particular the Capitanian, when shallow, warm-water, photosynthetic taxa suffered preferential losses. Both of these climatic effects are the anticipated result of the injection of large amounts of SO_2 into the stratosphere. Perhaps significantly, of all the large igneous provinces discussed in this paper, only the Emeishan Traps and Central Atlantic magmatic province were located at equatorial latitudes where they were potentially able to inject sulfate aerosols into both hemispheres and cause a global cooling event. Again, the site of eruption is seen to be important: Modeling and observations of recent events have shown that aerosols from mid- to high-latitude eruptions fail to cross the equator.

ACKNOWLEDGMENTS

Work on this review paper began during Bond's Marie Curie Intra European Fellowship at the Norwegian Polar Institute, Tromsø. The paper was completed during Bond's Natural

Environment Research Council Advanced Research Fellowship at the University of Hull. Funding from the Research Executive Agency (project FP7-PEOPLE-2011-IEF-300455), and from the Natural Environment Research Council (grant NE/J01799X/1) is gratefully acknowledged. We thank the University of Hull for kindly providing financial support for the open-access publication of this paper. Research data underpinning this article derive from the recent literature in the fields of volcanism and mass extinctions; articles are available from the publishers and original authors. Finally, we thank Grzegorz Racki and Andrea Marzoli for their helpful reviews of this manuscript.

REFERENCES CITED

Aberhan, M., and Fürsich, F., 1996, Diversity analysis of Lower Jurassic bivalves of the Andean Basin and the Pliensbachian-Toarcian mass extinction: Lethaia, v. 29, p. 181–195, doi:10.1111/j.1502-3931.1996.tb01874.x.

Algeo, T.J., Chen Zhongqiang, Fraiser, M.L., and Twitchett, R.J., 2011, Terrestrial-marine teleconnections in the collapse and rebuilding of Early Triassic marine ecosystems: Palaeogeography, Palaeoclimatology, Palaeoecology, v. 308, p. 1–11, doi:10.1016/j.palaeo.2011.01.011.

Ali, J.R., Thompson, G.M., Song, Xieyan, and Wang, Yunliang, 2002, Emeishan Basalts (SW China) and the "end-Guadalupian" crisis: Magnetobiostratigraphic constraints: Journal of the Geological Society of London, v. 159, p. 21–29, doi:10.1144/0016-764901086.

Ali, J.R., Thompson, G.M., Zhou, Mei-Fu, and Song, Xieyan, 2005, Emeishan large igneous province, SW China: Lithos, v. 79, p. 475–489, doi:10.1016/j.lithos.2004.09.013.

Ali, J.R., Fitton, J.G., and Herzberg, C., 2010, Emeishan large igneous province (SW China) and the mantle plume up-doming hypothesis: Journal of the Geological Society of London, v. 167, p. 953–959, doi:10.1144/0016-76492009-129.

Archer, D., 2005, Fate of fossil fuel CO_2 in geologic time: Journal of Geophysical Research–Oceans (1978–2012), v. 110, C09S05, doi:10.1029/2004JC002625.

Bailey, T.R., Rosenthal, Y., McArthur, J.M., van de Schootbrugge, B., and Thirlwall, M.F., 2003, Paleoceanographic changes of the late Pliensbachian–early Toarcian interval: A possible link to the genesis of an oceanic anoxic event: Earth and Planetary Science Letters, v. 212, p. 307–320, doi:10.1016/S0012-821X(03)00278-4.

Balter, V., Renaud, S., Girard, C., and Joachimski, M.M., 2008, Record of climate-driven morphological changes in 376 Ma Devonian fossils: Geology, v. 36, p. 907–910, doi:10.1130/G24989A.1.

Bambach, R.K., Knoll, A.H., and Wang, S.C., 2004, Origination, extinction, and mass depletions of marine diversity: Paleobiology, v. 30, p. 522–542, doi:10.1666/0094-8373(2004)030<0522:OEAMDO>2.0.CO;2.

Basu, A.R., Poreda, R.J., Renne, P.R., Teichmann, F., Vasiliev, Y.R., Sobolev, N.V., and Turrin, B.D., 1995, High-He-3 plume origin and temporal-spatial evolution of the Siberian flood basalts: Science, v. 269, p. 822–825, doi:10.1126/science.269.5225.822.

Baud, A., Magaritz, M., and Holser, W.T., 1989, Permian-Triassic of the Tethys: Carbon isotope studies: Geologische Rundschau, v. 78, p. 649–677, doi:10.1007/BF01776196.

Becker, L., Poreda, R.J., Hunt, A.G., Bunch, T.E., and Rampino, M., 2001, Impact event at the Permian-Triassic boundary: Evidence from extraterrestrial noble gases in fullerenes: Science, v. 291, p. 1530–1533, doi:10.1126/science.1057243.

Beerling, D.J., and Berner, R.A., 2002, Biogeochemical constraints on the Triassic-Jurassic boundary carbon cycle event: Global Biogeochemical Cycles, v. 16, p. 10-1–10-13, doi:10.1029/2001GB001637.

Beerling, D.J., and Brentnall, S., 2007, Numerical evaluation of mechanisms driving Early Jurassic changes in global carbon cycling: Geology, v. 35, p. 247–250, doi:10.1130/G23416A.1.

Beerling, D.J., Harfoot, M., Lomax, B.H., and Pyle, J.A., 2007, The stability of the stratospheric ozone layer during the end-Permian eruption of the Siberian Traps: Royal Society of London Philosophical Transactions, ser. A, v. 365, p. 1843–1866, doi:10.1098/rsta.2007.2046.

Benton, M.J., and Twitchett, R.J., 2003, How to kill (almost) all life: The end-Permian extinction event: Trends in Ecology & Evolution, v. 18, p. 358–365, doi:10.1016/S0169-5347(03)00093-4.

Berner, R.A., and Beerling, D.J., 2007, Volcanic degassing necessary to produce a $CaCO_3$ undersaturated ocean at the Triassic-Jurassic boundary: Palaeogeography, Palaeoclimatology, Palaeoecology, v. 244, p. 368–373, doi:10.1016/j.palaeo.2006.06.039.

Black, B.A., Elkins-Tanton, L.T., Rowe, M.C., and Ukstins Peate, I., 2012, Magnitude and consequences of volatile release from the Siberian Traps: Earth and Planetary Science Letters, v. 317–318, p. 363–373, doi:10.1016/j.epsl.2011.12.001.

Black, R.X., 2002, Stratospheric forcing of surface climate in the Arctic Oscillation: Journal of Climatology, v. 15, p. 268–277, doi:10.1175/1520-0442(2002)015<0268:SFOSCI>2.0.CO;2.

Blackburn, T.J., Olsen, P.E., Bowring, S.A., McLean, N.M., Kent, D.V., Puffer, J., McHone, G., Rasbury, E.T., and Et-Touhami, M., 2013, Zircon U-Pb geochronology links the end-Triassic extinction with the Central Atlantic Magmatic Province: Science, v. 340, p. 941–945.

Blake, S., 2003, Correlations between eruption magnitude, SO_2 yield and surface cooling, *in* Oppenheimer, C., Pyle, D.M., and Barclay, J., eds., Volcanic Degassing: Geological Society of London Special Publication 213, p. 371–380.

Bond, D.P.G., and Wignall, P.B., 2008, The role of sea-level change and marine anoxia in the Frasnian–Famennian (Late Devonian) mass extinction: Palaeogeography, Palaeoclimatology, Palaeoecology, v. 263, p. 107–118, doi:10.1016/j.palaeo.2008.02.015.

Bond, D.P.G., and Wignall, P.B., 2009, Latitudinal selectivity of foraminifer extinctions during the late Guadalupian crisis: Paleobiology, v. 35, p. 465–483, doi:10.1666/0094-8373-35.4.465.

Bond, D.P.G., and Wignall, P.B., 2010, Pyrite framboid study of marine Permian-Triassic boundary sections: A complex anoxic event and its relationship to contemporaneous mass extinction: Geological Society of America Bulletin, v. 122, p. 1265–1279, doi:10.1130/B30042.1.

Bond, D.P.G., Wignall, P.B., and Racki, G., 2004, Extent and duration of marine anoxia during the Frasnian-Famennian (Late Devonian) mass extinction in Poland, Germany, Austria, and France: Geological Magazine, v. 141, p. 173–193, doi:10.1017/S0016756804008866.

Bond, D.P.G., Hilton, J., Wignall, P.B., Ali, J.R., Stevens, L.G., Sun Yadong, and Lai Xulong, 2010a, The Middle Permian (Capitanian) mass extinction on land and in the oceans: Earth-Science Reviews, v. 102, p. 100–116, doi:10.1016/j.earscirev.2010.07.004.

Bond, D.P.G., Wignall, P.B., Wang Wei, Védrine, S., Jiang Haishui, Lai Xulong, Sun Yadong, Newton, R.J., Cope, H., and Izon, G., 2010b, The mid-Capitanian (Middle Permian) mass extinction and carbon isotope record of South China: Palaeogeography, Palaeoclimatology, Palaeoecology, v. 292, p. 282–294, doi:10.1016/j.palaeo.2010.03.056.

Bond, D.P.G., Zatoń, M., Wignall, P.B., and Marynowski, L., 2013, Evidence for shallow-water 'Upper Kellwasser' anoxia in the Frasnian-Famennian reefs of Alberta, Canada: Lethaia, v. 46, p. 355–368, doi:10.1111/let.12014.

Bowring, S.A., Erwin, D.H., Jin Yugan, Martin, M.W., Davidek, K., and Wang Wei, 1998, U/Pb zircon geochronology and tempo of the end-Permian mass extinction: Science, v. 280, p. 1039–1045, doi:10.1126/science.280.5366.1039.

Briffa, K.R., Schweingruber, F.H., Jones, P.D., Osborn, T.J., Shiyatov, S.G., and Vaganov, E.A., 1998, Reduced sensitivity of recent tree-growth to temperature at high northern latitudes: Nature, v. 391, p. 678–682, doi:10.1038/35596.

Bryan, S.E., 2007, Silicic large igneous provinces: Episodes, v. 30, p. 20–31.

Bryan, S.E., and Ernst, R.E., 2008, Revised definition of large igneous provinces (LIPs): Earth-Science Reviews, v. 86, p. 175–202, doi:10.1016/j.earscirev.2007.08.008.

Bryan, S.E., Riley, T.R., Jerram, D.A., Leat, P.T., and Stephens, C.J., 2002, Silicic volcanism: An under-valued component of large igneous provinces and volcanic rifted margins, *in* Menzies, M.A., Klemperer, S.L., Ebinger, C.J., and Baker, J., eds., Magmatic Rifted Margins: Geological Society of America Special Paper 362, p. 99–120.

Bryan, S.E., Ukstins Peate, I., Peate, D.W., Self, S., Jerram, D.A., Mawby, M.R., Marsh, J.S., and Miller, J.A., 2010, The largest volcanic eruptions on Earth: Earth-Science Reviews, v. 102, p. 207–229, doi:10.1016/j.earscirev.2010.07.001.

Buggisch, W., 1972, Zur Geologie und Geochemie der Kellwasserkalke und ihrer begleitenden Sedimente (Unteres Oberdevon) (Geology and geochemistry of the lowermost upper Devonian Kellwasser Limestone and its associated sediments): Abhandlungen des Hessischen Landesamtes für Bodenforschung, v. 62, p. 1–67.

Buggisch, W., 1991, The global Frasnian-Famennian 'Kellwasser event': Geologische Rundschau, v. 80, p. 49–72, doi:10.1007/BF01828767.

Camp, V.E., Ross, M.E., and Hanson, W.E., 2003, Genesis of flood basalts and Basin and Range volcanic rocks from Steens Mountain to the Malheur River Gorge, Oregon: Geological Society of America Bulletin, v. 115, p. 105–128, doi:10.1130/0016-7606(2003)115<0105:GOFBAB>2.0.CO;2.

Campbell, I.H., Czamanske, G.K., Fedorenko, V.A., Hill, R.A., and Stepanov, V., 1992, Synchronism of the Siberian Traps and the Permian-Triassic boundary: Science, v. 258, p. 1760–1763, doi:10.1126/science.258.5089.1760.

Caruthers, A.H., and Smith, P.L., 2012, Pliensbachian ammonoids from the Talkeetna Mountains (Peninsular terrane) of southern Alaska: Revue de Paléobiologie, volume spécial 11, p. 365–378.

Caruthers, A.H., Gröcke, D.R., and Smith, P.L., 2011, The significance of an Early Jurassic (Toarcian) carbon-isotope excursion in Haida Gwaii (Queen Charlotte Islands), British Columbia, Canada: Earth and Planetary Science Letters, v. 307, p. 19–26, doi:10.1016/j.epsl.2011.04.013.

Chenet, A.L., Quidelleur, X., Fluteau, F., Courtillot, V., and Bajpai, S., 2007, 40K-40Ar dating of the Main Deccan large igneous province: Further evidence of KTB age and short duration: Earth and Planetary Science Letters, v. 263, p. 1–15, doi:10.1016/j.epsl.2007.07.011.

Chung Sunlin, and Jahn Borming, 1995, Plume-lithosphere interaction in generation of the Emeishan flood basalts at the Permian-Triassic boundary: Geology, v. 23, p. 889–892, doi:10.1130/0091-7613(1995)023<0889:PLIIGO>2.3.CO;2.

Chung Sunlin, Jahn Borming, Wu Genyao, Lo Chinghua, and Cong Bolin, 1998, The Emeishan flood basalt in SW China; a mantle plume initiation model and its connection with continental breakup and mass extinction at the Permian-Triassic boundary, *in* Flower, M.F.J., Chung Sunlin, Lo Chinghua, and Lee Tungyi, eds., Mantle Dynamics and Plate Interactions in East Asia: American Geophysical Union Geodynamics Monograph 27, p. 47–58.

Cirilli, S., Marzoli, A., Tanner, L., Bertrand, H., Buratti, N., Jourdan, F., Bellieni, G., Kontak, D., and Renne, P.R., 2009, Latest Triassic onset of the Central Atlantic magmatic province (CAMP) volcanism in the Fundy Basin (Nova Scotia): New stratigraphic constraints: Earth and Planetary Science Letters, v. 286, p. 514–525, doi:10.1016/j.epsl.2009.07.021.

Claoué-Long, J.C., Zhang Zichao, Ma Guogan, and Du Shaohua, 1991, The age of the Permian-Triassic boundary: Earth and Planetary Science Letters, v. 105, p. 182–190, doi:10.1016/0012-821X(91)90129-6.

Coffin, M.F., and Eldholm, O., 1991, Large Igneous Provinces, JOI/USSAC Workshop Report: Austin, Texas, University of Texas at Austin Institute for Geophysics Technical Report 114, 79 p.

Coffin, M.F., and Eldholm, O., 1994, Large igneous provinces: Crustal structure, dimensions, and external consequences: Reviews of Geophysics, v. 32, p. 1–36, doi:10.1029/93RG02508.

Cohen, A.S., Coe, A.L., and Kemp, D.B., 2007, The late Palaeocene–early Eocene and Toarcian (Early Jurassic) carbon isotope excursions: A comparison of their time scales, associated environmental changes, causes and consequences: Journal of the Geological Society of London, v. 164, p. 1093–1108, doi:10.1144/0016-76492006-123.

Collinson, J.W., Hammer, W.R., Askin, R.A., and Elliot, D.H., 2006, Permian-Triassic boundary in the central Transantarctic Mountains, Antarctica: Geological Society of America Bulletin, v. 118, p. 747–763, doi:10.1130/B25739.1.

Copper, P., 2002, Reef development at the Frasnian-Famennian mass extinction boundary: Palaeogeography, Palaeoclimatology, Palaeoecology, v. 181, p. 27–65, doi:10.1016/S0031-0182(01)00472-2.

Courtillot, V., 1999, Evolutionary Catastrophes: The Science of Mass Extinction: Cambridge, UK, Cambridge University Press, 173 p.

Courtillot, V., and Olson, P., 2007, Mantle plumes link magnetic superchrons to Phanerozoic mass depletion events: Earth and Planetary Science Letters, v. 260, p. 495–504, doi:10.1016/j.epsl.2007.06.003.

Courtillot, V.E., and Renne, P.R., 2003, On the ages of flood basalt events: Comptes Rendus Geoscience, v. 335, p. 113–140, doi:10.1016/S1631-0713(03)00006-3.

Courtillot, V., Jaupart, C., Manighetti, I., Tapponnier, P., and Besse, J., 1999, On causal links between flood basalts and continental breakup: Earth and Planetary Science Letters, v. 166, p. 177–195, doi:10.1016/S0012-821X(98)00282-9.

Courtillot, V., Kravchinsky, V.A., Quidelleur, X., Renne, P.R., and Gladkochub, D.P., 2010, Preliminary dating of the Viluy Traps (eastern Siberia): Eruption at the time of Late Devonian extinction events?: Earth and Planetary Science Letters, v. 300, p. 239–245, doi:10.1016/j.epsl.2010.09.045.

D'Arrigo, R., Seager, R., Smerdon, J.E., Legrande, A.N., and Cook, E.R., 2011, The anomalous winter of 1783–1784: Was the Laki eruption or an analog of the 2009–2010 winter to blame?: Geophysical Research Letters, v. 38, p. L05706.

Deenen, M.H.L., Ruhl, M., Bonis, N.R., Krijgsman, W., Kuerschner, W.M., Reitsma, M., and van Bergen, M.J., 2010, A new chronology for the end-Triassic mass extinction: Earth and Planetary Science Letters, v. 291, p. 113–125, doi:10.1016/j.epsl.2010.01.003.

de Silva, S.L., and Zielinski, G.A., 1998, Global influence of the AD 1600 eruption of Huaynaputina, Peru: Nature, v. 393, p. 455–458, doi:10.1038/30948.

Devine, J.D., Sigurdsson, H., Davis, A.N., and Self, S., 1984, Estimates of sulphur and chlorine yield to the atmosphere from volcanic eruptions and potential climatic effects: Journal of Geophysical Research, v. 89, p. 6309–6325, doi:10.1029/JB089iB07p06309.

Driscoll, S., Bozzo, A., Gray, L.J., Robock, A., and Stenchikov, G., 2012, Coupled Model Intercomparison Project 5 (CMIP5) simulations of climate following volcanic eruptions: Journal of Geophysical Research, v. 117, D17105, doi:10.1029/2012JD017607.

Duncan, R.A., Hooper, P.R., Rehacek, J., Marsh, J.S., and Duncan, A.R., 1997, The timing and duration of the Karoo igneous event, southern Gondwana: Journal of Geophysical Research, v. 102, p. 18,127–18,138, doi:10.1029/97JB00972.

Dustira, A.M., Wignall, P.B., Joachimski, M., Blomeier, D., Hartkopf-Fröder, C., and Bond, D.P.G., 2013, Gradual onset of anoxia across the Permian-Triassic boundary in Svalbard, Norway: Palaeogeography, Palaeoclimatology, Palaeoecology, v. 374, p. 303–313, doi:10.1016/j.palaeo.2013.02.004.

Eldholm, O., and Coffin, M.F., 2000, Large igneous provinces and plate tectonics, *in* Richards, M.A., Gordon, R.G., and van der Hilst, R.D., eds., The History and Dynamics of Global Plate Motions: American Geophysical Union Geophysical Monograph 121, p. 309–326.

Encarnación, J., Fleming, T.H., Elliot, D.H., and Eales, H., 1996, Synchronous emplacement of Ferrar and Karoo dolerites and the early breakup of Gondwana: Geology, v. 24, p. 535–538, doi:10.1130/0091-7613(1996)024<0535:SEOFAK>2.3.CO;2.

English, J.M., Toon, O.B., and Mills, M.J., 2012, Microphysical simulations of sulfur burdens from stratospheric sulfur geoengineering: Atmospheric Chemistry and Physics, v. 12, p. 4775–4793, doi:10.5194/acp-12-4775-2012.

Erwin, D.H., 1994, The Permo-Triassic extinction: Nature, v. 367, p. 231–236, doi:10.1038/367231a0.

Fedorenko, V., Czamanske, G., Zen'ko, T., Budahn, J., and Siems, D., 2000, Field and geochemical studies of the melilite-bearing Arydzhangsky Suite, and an overall perspective on the Siberian alkaline-ultramafic flood-volcanic rocks: International Geology Review, v. 42, p. 769–804, doi:10.1080/00206810009465111.

Flügel, E., 2002, Triassic reef patterns, *in* Flügel, E., Kiessling, W., and Golonka, J., eds., Phanerozoic Reef Patterns: Society for Sedimentary Geology (SEPM) Special Publication 72, p. 391–463.

Flügel, E., and Kiessling, W., 2002, Patterns of Phanerozoic reef crises, *in* Flügel, E., Kiessling, W., and Golonka, J., eds., Phanerozoic Reef Patterns: Society for Sedimentary Geology (SEPM) Special Publication 72, p. 691–733.

Foland, K.A., Fleming, T.H., Heimann, A., and Elliot, D.H., 1993, Potassium-argon dating of fine-grained basalts with massive Ar-loss: Application of the $^{40}Ar/^{39}Ar$ technique to plagioclase and glass from the Kirkpatrick Basalt, Antarctica: Chemical Geology, v. 107, p. 173–190, doi:10.1016/0009-2541(93)90109-V.

Foster, G.L., 2008, Seawater pH, $p\mathrm{CO}_2$ and $[\mathrm{CO}^{2-}_3]$ variations in the Caribbean Sea over the last 130 kyr: A boron isotope and B/Ca study of planktic foraminifera: Earth and Planetary Science Letters, v. 271, p. 254–266, doi:10.1016/j.epsl.2008.04.015.

Fowell, S.J., and Olsen, P.E., 1993, Time calibration of Triassic Jurassic microfloral turnover, eastern North America: Tectonophysics, v. 222, p. 361–369, doi:10.1016/0040-1951(93)90359-R.

Frogner Kockum, P., Herbert, R., and Gislason, S., 2006, A diverse ecosystem response to volcanic aerosols: Chemical Geology, v. 231, p. 57–66, doi:10.1016/j.chemgeo.2005.12.008.

Frölicher, T.L., Joos, F., Raible, C.C., and Sarmiento, J.L., 2013, Atmospheric CO2 response to volcanic eruptions: The role of ENSO, season, and variability: Global Biogeochemical Cycles, v. 27, p. 239–251, doi:10.1002/gbc.20028.

Gaiduk, V.V., 1988, The Middle Paleozoic Vilyui Rift System: Yakutsk, YaF SO AN SSSR, 128 p. (in Russian).

Ganino, C., and Arndt, N.T., 2009, Climate changes caused by degassing of sediments during the emplacement of large igneous provinces: Geology, v. 37, p. 323–326, doi:10.1130/G25325A.1.

Gastaldo, R.A., Neveling, J., Kittinger Clark, C., and Newbury, S.S., 2009, The terrestrial Permian-Triassic boundary event bed is a nonevent: Geology, v. 37, p. 199–202, doi:10.1130/G25255A.1.

Gerlach, T., 2011, Volcanic versus anthropogenic carbon dioxide: Eos (Transactions, American Geophysical Union), v. 92, p. 201–202, doi:10.1029/2011EO240001.

Gladczenko, T.P., Coffin, M.F., and Eldholm, O., 1997, Crustal structure of the Ontong Java Plateau: Modeling of new gravity and existing seismic data: Journal of Geophysical Research–Solid Earth (1978–2012), v. 102, p. 22,711–22,729.

Gradstein, F.M., Ogg, J.G., Schmitz, M., and Ogg, G., 2012, The Geologic Time Scale 2012: Amsterdam, Elsevier, 1176 p.

Grasby, S.E., and Beauchamp, B., 2008, Intrabasin variability of the carbon-isotope record across the Permian-Triassic transition, Sverdrup Basin, Arctic Canada: Chemical Geology, v. 253, p. 141–150, doi:10.1016/j.chemgeo.2008.05.005.

Grasby, S.E., Sanei, H., and Beauchamp, B., 2011, Catastrophic dispersion of coal fly ash into oceans during the latest Permian extinction: Nature Geoscience, v. 4, p. 104–107, doi:10.1038/ngeo1069.

Greene, S.E., Martindale, R.C., Ritterbush, K.A., Bottjer, D.J., Corsetti, F.A., and Berelson, W.M., 2012, Recognising ocean acidification in deep time: An evaluation of the evidence for acidification across the Triassic-Jurassic boundary: Earth-Science Reviews, v. 113, p. 72–93, doi:10.1016/j.earscirev.2012.03.009.

Grice, K., Cao Changqun., Love, G.D., Böttcher, M.E., Twitchett, R.J., Grosjean, E., Summons, R.E., Turgeon, S.C., Dunning, W., and Jin Yugan, 2005, Photic zone euxinia during the Permian-Triassic superanoxic event: Science, v. 307, p. 706–709, doi:10.1126/science.1104323.

Gröcke, D.R., Rimmer, S.M., Yoksoulian, L.E., Cairncross, B., Tsikos, H., and van Hunen, J., 2009, No evidence for thermogenic methane release in coal from the Karoo-Ferrar large igneous province: Earth and Planetary Science Letters, v. 277, p. 204–212, doi:10.1016/j.epsl.2008.10.022.

Gröcke, D.R., Hori, R.S., Trabucho-Alexandre, J., Kemp, D.B., and Schwark, L., 2011, An open marine record of the Toarcian oceanic anoxic event: Solid Earth Discussions, v. 3, p. 385–410, doi:10.5194/sed-3-385-2011.

Hallam, A., 1986, The Pliensbachian and Tithonian extinction events: Nature, v. 319, p. 765–768, doi:10.1038/319765a0.

Hallam, A., 1996, Major bio-events in the Triassic and Jurassic, *in* Walliser, O.H., ed., Global Events and Event Stratigraphy in the Phanerozoic: Berlin, Springer-Verlag, p. 265–283.

Hallam, A., 2002, How catastrophic was the end-Triassic mass extinction?: Lethaia, v. 35, p. 147–157, doi:10.1080/002411602320184006.

Hallam, A., and Wignall, P.B., 2004, Discussion on sea-level change and facies development across potential Triassic-Jurassic boundary horizons, SW Britain: Journal of the Geological Society of London, v. 161, p. 1053–1056, doi:10.1144/0016-764904-069.

Hautmann, M., 2004, Effect of end-Triassic CO2 maximum on carbonate sedimentation and marine mass extinction: Facies, v. 50, p. 257–261, doi:10.1007/s10347-004-0020-y.

Hautmann, M., Benton, M.J., and Tomašových, A., 2008a, Catastrophic ocean acidification at the Triassic-Jurassic boundary: Neues Jahrbuch für Geologie und Palaontologie, Abhandlungen, v. 249, p. 119–127, doi:10.1127/0077-7749/2008/0249-0119.

Hautmann, M., Stiller, F., Cai Huawei, and Sha Jingeng, 2008b, Extinction-recovery pattern of bottom-level faunas across the Triassic-Jurassic boundary in Tibet: Implications for potential killing mechanisms: Palaios, v. 23, p. 711–718, doi:10.2110/palo.2008.p08-005r.

He Bin, Xu Yigang, Chung Sunling, Xiao Long, and Wang Yamei, 2003, Sedimentary evidence for a rapid, kilometer-scale crustal doming prior to the eruption of the Emeishan flood basalts: Earth and Planetary Science Letters, v. 213, p. 391–405, doi:10.1016/S0012-821X(03)00323-6.

He Bin, Wang Yamei, and Jiang, Xiaowei, 2004, Palaeokarst landforms on top of limestone of the Maokou Formation in the west of the Upper Yangtze Platform, China, and its geological significance: Chinese Geology, v. 31, p. 46–50.

Heckendorn, P., Weisenstein, D., Fueglistaler, S., Luo, B.P., Rozanov, E., Schraner, M., Thomason, L.W., and Peter, T., 2009, The impact of geoengineering aerosols on stratospheric temperature and ozone: Environmental Research Letters, v. 4, p. 045108, doi:10.1088/1748-9326/4/4/045108.

Heimann, A., Fleming, T.H., Elliot, D.H., and Foland, K.A., 1994, A short interval of Jurassic continental flood basalt volcanism in Antarctica as demonstrated by ^{40}Ar/^{39}Ar geochronology: Earth and Planetary Science Letters, v. 121, p. 19–41, doi:10.1016/0012-821X(94)90029-9.

Hermoso, M., Minoletti, F., Le Callonnec, L., Jenkyns, H.C., Hesselbo, S.P., Rickaby, R.E.M., Renard, M., de Rafélis, M., and Emmanuel, L., 2009, Global and local forcing of early Toarcian seawater chemistry: A comparative study of different paleoceanographic settings (Paris and Lusitanian Basins): Paleoceanography, v. 24, p. PA4208, doi:10.1029/2009PA001764.

Hesselbo, S.P., Gröcke, D.R., Jenkyns, H.C., Bjerrum, C.J., Farrimond, P., Morgans Bell, H.S., and Green, O.R., 2000, Massive dissociation of gas hydrate during a Jurassic oceanic anoxic event: Nature, v. 406, p. 392–395, doi:10.1038/35019044.

Hesselbo, S.P., Robinson, S.A., Surlyk, F., and Piasecki, S., 2002, Terrestrial and marine extinction at the Triassic-Jurassic boundary synchronized with major carbon-cycle perturbation: A link to initiation of massive volcanism?: Geology, v. 30, p. 251–254, doi:10.1130/0091-7613(2002)030<0251:TAMEAT>2.0.CO;2.

Hesselbo, S.P., Robinson, S.A., and Surlyk, F., 2004, Sea-level change and facies development across potential Triassic-Jurassic boundary horizons, south west Britain: Journal of the Geological Society of London, v. 161, p. 365–379, doi:10.1144/0016-764903-033.

Hesselbo, S.P., Jenkyns, H.C., Duarte, L.V., and Oliveira, L.C.V., 2007, Carbon-isotope record of the Early Jurassic (Toarcian) oceanic anoxic event from fossil wood and marine carbonate (Lusitanian Basin, Portugal): Earth and Planetary Science Letters, v. 253, p. 455–470, doi:10.1016/j.epsl.2006.11.009.

Heydari, E., Arzani, N., and Hassanzadeh, J., 2008, The invisible serial killer: Application to Permian-Triassic boundary mass extinction: Palaeogeography, Palaeoclimatology, Palaeoecology, v. 264, p. 147–162, doi:10.1016/j.palaeo.2008.04.013.

Heydari, E., Arzani, N., Safaei, M., and Hassanzadeh, J., 2013, Ocean's response to a changing climate: Clues from variations in carbonate mineralogy across the Permian-Triassic boundary of the Shareza Section, Iran: Global and Planetary Change, v. 105, p. 79–90, doi:10.1016/j.gloplacha.2012.12.013.

Hinojosa, J.L., Brown, S.T., Chen Jun, DePaolo, D.J., Paytan, A., Shen Shuzhong, and Payne, J.L., 2012, Evidence for end-Permian ocean acidification from calcium isotopes in biogenic apatite: Geology, v. 40, p. 743–746, doi:10.1130/G33048.1.

Holbrook, W.S., and Kelemen, P.B., 1993, Large igneous province on the US Atlantic margin and implications for magmatism during continental breakup: Nature, v. 364, p. 433–436, doi:10.1038/364433a0.

Holser, W.T., and Magaritz, M., 1995, The Late Permian carbon isotope anomaly in the Bellerophon Basin, Carnic and Dolomite Alps: Jahrbuch der Geologischen Bundesanstalt, v. 128, p. 75–82.

Holser, W.T., Schönlaub, H.P., Boeckelmann, K., Magaritz, M., and Orth, C.J., 1991, The Permian-Triassic of the Gartnerkofel-1 core (Carnic Alps, Austria): Synthesis and conclusions: Abhandlungen der Geologischen Bundesanstalt, v. 45, p. 213–235.

Hommel, R., and Graf, H.-F., 2011, Modelling the size distribution of geoengineered stratospheric aerosols: Atmospheric Science Letters, v. 12, p. 168–175, doi:10.1002/asl.285.

Hooper, P.R., Camp, V.E., Reidel, S.P., and Ross, M.E., 2007, The Columbia River Basalts and their relationship to the Yellowstone hotspot and Basin and Range extension, *in* Foulger, G.R., and Jurdy, J.M., eds., Plumes, Plates, and Planetary Processes: Geological Society of America Special Paper 430, p. 635–668.

Hori, S.R., 1993, Toarcian oceanic event in deep-sea sediments: Geological Survey of Japan Bulletin, v. 44, p. 555–570.

House, M.R., 1985, Correlation of mid-Palaeozoic ammonoid evolutionary events with global sedimentary perturbations: Nature, v. 313, p. 17–22, doi:10.1038/313017a0.

Humphreys, W.J., 1940, Physics of the Air: New York, McGraw Hill, 665 p.

Huybers, P., and Langmuir, C., 2009, Feedback between deglaciation, volcanism, and atmospheric CO_2: Earth and Planetary Science Letters, v. 286, p. 479–491, doi:10.1016/j.epsl.2009.07.014.

Huynh, T.T., and Poulsen, C.J., 2005, Rising atmospheric CO_2 as a possible trigger for the end-Triassic mass extinction: Palaeogeography, Palaeoclimatology, Palaeoecology, v. 217, p. 223–242, doi:10.1016/j.palaeo.2004.12.004.

Isozaki, Y., 1994, Superanoxia across the Permo-Triassic boundary: Recorded in accreted deep-sea pelagic chert in Japan, *in* Embry, A.F., Beauchamp, B., and Glass, D.J., eds., Pangea: Global Environments and Resources: Canadian Society of Petroleum Geologists Memoir 17, p. 805–812.

Isozaki, Y., 1997, Permo-Triassic boundary superanoxia and stratified superocean: Records from lost deep sea: Science, v. 276, p. 235–238, doi:10.1126/science.276.5310.235.

Isozaki, Y., 2009, Illawarra reversal: The fingerprint of a superplume that triggered Pangean breakup and the end-Guadalupian (Permian) mass extinction: Gondwana Research, v. 15, p. 421–432, doi:10.1016/j.gr.2008.12.007.

Isozaki, Y., Kawahata, H., and Ota, A., 2007a, A unique carbon isotope record across the Guadalupian–Lopingian (Middle–Upper Permian) boundary in mid-oceanic paleoatoll carbonates: The high-productivity "Kamura event" and its collapse in Panthalassa: Global and Planetary Change, v. 55, p. 21–38, doi:10.1016/j.gloplacha.2006.06.006.

Isozaki, Y., Kawahata, H., and Minoshima, K., 2007b, The Capitanian (Permian) Kamura cooling event: The beginning of the Paleozoic-Mesozoic transition: Palaeoworld, v. 16, p. 16–30, doi:10.1016/j.palwor.2007.05.011.

Ivanov, A.V., He, Huayiu, Yan, Liekun, Ryabov, V.V., Shevko, A.Y., Palesskii, S.V., and Nikolaeva, I.V., 2013, Siberian Traps large igneous province: Evidence for two flood basalt pulses around the Permo-Triassic boundary and in the Middle Triassic, and contemporaneous granitic magmatism: Earth-Science Reviews, v. 122, p. 58–76, doi:10.1016/j.earscirev.2013.04.001.

Jenkyns, H.C., 1985, The early Toarcian and Cenomanian-Turonian anoxic events in Europe: Comparisons and contrasts: International Journal of Earth Sciences, v. 74, p. 505–518.

Jenkyns, H.C., 1988, The early Toarcian (Jurassic) event: Stratigraphy, sedimentary and geochemical evidence: American Journal of Science, v. 288, p. 101–151, doi:10.2475/ajs.288.2.101.

Jenkyns, H.C., Jones, C.E., Gröcke, D.R., Hesselbo, S.P., and Parkinson, D.N., 2002, Chemostratigraphy of the Jurassic system: Applications, limitations and implications for paleoceanography: Journal of the Geological Society of London, v. 159, p. 351–378, doi:10.1144/0016-764901-130.

Jin Yugan, J., Yichun, Z., and Qinghua, S., 1994, Two phases of the end-Permian mass extinction, *in* Embry, A.F., Beauchamp, B., and Glass, D.J., eds., Pangea: Global Environments and Resources: Canadian Society of Petroleum Geologists Memoir 17, p. 813–822.

Jin Yugan, Mei Shilong, Wang Wei, Wang Xiangdong, Shen Shuzhong, Shang Qinghua, and Chen Zhongqiang, 1998, On the Lopingian series of the Permian System: Palaeoworld, v. 9, p. 1–18.

Jin Yugan, Wang Yue, Wang Wei, Shang Qinghua, Cao Changqun, and Erwin, D.H., 2000, Pattern of marine mass extinction near the Permian-Triassic boundary in South China: Science, v. 289, p. 432–436, doi:10.1126/science.289.5478.432.

Joachimski, M.M., and Buggisch, W., 2002, Conodont apatite $\delta^{18}O$ signatures indicate climatic cooling as a trigger of the Late Devonian mass extinction: Geology, v. 30, p. 711–714, doi:10.1130/0091-7613(2002)030<0711:CAOSIC>2.0.CO;2.

Joachimski, M.M., Breisig, S., Buggisch, W., Talent, J.A., Mawson, R., Gereke, M., Morrow, J.R., Day, J., and Weddige, K., 2009, Devonian climate and reef evolution: Insights from oxygen isotopes in apatite: Earth and Planetary Science Letters, v. 284, p. 599–609, doi:10.1016/j.epsl.2009.05.028.

Joachimski, M.M., Lai Xulong, Shen Shuzhong, Jiang Haishui, Luo Genming, Chen Bo, Chen Jun, and Sun Yadong, 2012, Climate warming in the latest Permian and the Permian-Triassic mass extinction: Geology, v. 40, p. 195–198, doi:10.1130/G32707.1.

Johnson, J.G., Klapper, G., and Sandberg, C.A., 1985, Devonian eustatic fluctuations in Euramerica: Geological Society of America Bulletin, v. 96, p. 567–587, doi:10.1130/0016-7606(1985)96<567:DEFIE>2.0.CO;2.

Jones, C.E., Jenkyns, H.C., and Hesselbo, S.P., 1994, Strontium isotopes in Early Jurassic seawater: Geochimica et Cosmochimica Acta, v. 58, p. 1285–1301, doi:10.1016/0016-7037(94)90382-4.

Jones, D.L., Duncan, R.A., Briden, J.C., Randall, D.E., and MacNiocaill, C., 2001, Age of the Batoka basalts, northern Zimbabwe, and the duration of Karoo large igneous province magmatism: Geochemistry Geophysics Geosystems, v. 2, p. 1022, doi:10.1029/2000GC000110.

Joshi, M.M., and Shine, K.P., 2003, A GCM study of volcanic eruptions as a cause of increased stratospheric water vapor: Journal of Climate, v. 16, p. 3525–3534, doi:10.1175/1520-0442(2003)016<3525:AGSOVE>2.0.CO;2.

Jourdan, F., Féraud, G., Bertrand, H., Kampunzu, A.B., Tshoso, G., Le Gall, B., Tiercelin, J.J., and Capiez, P., 2004, The Karoo triple junction questioned: Evidence from 40Ar/39Ar Jurassic and Proterozoic ages and geochemistry of the Okavango dike swarm (Botswana): Earth and Planetary Science Letters, v. 222, p. 989–1006, doi:10.1016/j.epsl.2004.03.017.

Jourdan, F., Féraud, G., Bertrand, H., Kampunzu, A.B., Tshoso, G., Watkeys, M.K., and Le Gall, B., 2005, The Karoo large igneous province: Brevity, origin, and relation with mass extinction questioned by new $^{40}Ar/^{39}Ar$ age data: Geology, v. 33, p. 745–748, doi:10.1130/G21632.1.

Jourdan, F., Féraud, G., Bertrand, H., and Watkeys, M.K., 2007a, From flood basalts to the inception of oceanization: Example from the $^{40}Ar/^{39}Ar$ high-resolution picture of the Karoo large igneous province: Geochemistry Geophysics Geosystems, v. 8, p. Q02002, doi:10.1029/2006GC001392.

Jourdan, F., Féraud, G., Bertrand, H., Watkeys, M.K., and Renne, P.R., 2007b, Distinct brief major events in the Karoo large igneous province clarified by new 40Ar/39Ar ages on the Lesotho basalts: Lithos, v. 98, p. 195–209, doi:10.1016/j.lithos.2007.03.002.

Jourdan, F., Féraud, G., Bertrand, H., Watkeys, M.K., and Renne, P.R., 2008, The 40Ar/39Ar ages of the sill complex of the Karoo large igneous province: Implications for the Pliensbachian-Toarcian climate change: Geochemistry Geophysics Geosystems, v. 9, p. Q06009, doi:10.1029/2008GC001994.

Kaiho, K., Kajiwara, Y., Nakano, T., Miura, Y., Kawahata, H., Tazaki, K., Ueshima, M., Chen Zhongqiang, and Shi Guangrui, 2001, End-Permian catastrophe by a bolide impact: Evidence of a gigantic release of sulfur from the mantle: Geology, v. 29, p. 815–818, doi:10.1130/0091-7613(2001)029<0815:EPCBAB>2.0.CO;2.

Kaiho, K., Chen Zhongqiang, Kawahata, H., Kajiwara, Y., and Sato, H., 2006, Close-up of the end-Permian mass extinction horizon recorded in the Meishan section, South China: Sedimentary, elemental, and biotic characterization and a negative shift of sulfate sulfur isotope ratio: Palaeogeography, Palaeoclimatology, Palaeoecology, v. 239, p. 396–405, doi:10.1016/j.palaeo.2006.02.011.

Kamo, S.L., Czamanske, G.K., and Krogh, T.E., 1996, A minimum U-Pb age for Siberian flood-basalt volcanism: Geochimica et Cosmochimica Acta, v. 60, p. 3505–3511, doi:10.1016/0016-7037(96)00173-1.

Kamo, S.L., Czamanske, G.K., Amelin, Y., Fedorenko, V.A., Davis, D.W., and Trofimov, V.R., 2003, Rapid eruption of Siberian flood-volcanic rocks and evidence for coincidence with the Permian-Triassic boundary and mass extinction at 251 Ma: Earth and Planetary Science Letters, v. 214, p. 75–91, doi:10.1016/S0012-821X(03)00347-9.

Keller, G., 2014, this volume, Deccan volcanism, the Chicxulub impact, and the end-Cretaceous mass extinction: Coincidence? Cause and effect?, *in* Keller, G., and Kerr, A.C., eds., Volcanism, Impacts, and Mass Extinctions: Causes and Effects: Geological Society of America Special Paper 505, doi:10.1130/2014.2505(03).

Kemp, D.B., Coe, A.L., Cohen, A.S., and Schwark, L., 2005, Astronomical pacing of methane release in the Early Jurassic period: Nature, v. 437, p. 396–399, doi:10.1038/nature04037.

Kennett, J.P., and Watkins, N.D., 1970, Geomagnetic polarity change, volcanic maxima and faunal extinction in the South Pacific: Nature, v. 227, p. 930–934, doi:10.1038/227930a0.

Kiessling, W., and Simpson, C., 2011, On the potential for ocean acidification to be a general cause of ancient reef crises: Global Change Biology, v. 17, p. 56–67.

Kiessling, W., Flügel, E., and Golonka, J., 2000, Fluctuations in the carbonate production of Phanerozoic reefs, *in* Insalaco, E., Skelton, P.W., and Palmer, T.J., eds., Carbonate Platform Systems: Components and Interactions: Geological Society of London Special Publication 178, p. 191–215, doi:10.1144/GSL.SP.2000.178.01.13.

Kiselev, A.I., Yarmolyuk, V.V., Egorov, K.N., Chernyushov, R.A., and Nikiphorov, A.V., 2006, Middle Paleozoic basite magmatism of the north-west part of the Viluy rift: Composition, sources, geodynamics: Petrology, v. 6, p. 660–682.

Kiselev, A.I., Yarmolyuk, V.V., Nikiforov, A.V., and Egorov, K.N., 2007, Composition and sources of Devonian volcanism in the Vilyui rift: Doklady Earth Sciences, v. 415, p. 744–750, doi:10.1134/S1028334X07050194.

Knight, K.B., Nomade, S., Renne, P.R., Marzoli, A., Betrand, H., and Youbi, N., 2004, The Central Atlantic magmatic province at the Triassic-Jurassic boundary: Paleomagnetic and $^{40}Ar/^{39}Ar$ evidence from Morocco for brief, episodic volcanism: Earth and Planetary Science Letters, v. 228, p. 143–160, doi:10.1016/j.epsl.2004.09.022.

Kolstad, E.W., and Charlton-Perez, A.J., 2011, Observed and simulated precursors of stratospheric polar vortex anomalies in the Northern Hemisphere: Climate Dynamics, v. 37, p. 1443–1456, doi:10.1007/s00382-010-0919-7.

Koppelhus, E.B., 1996, Palynology of the Lacustrine Kap Stewart Formation, Jameson Land, East Greenland: Copenhagen, Danmarks og Groenlands Geologiske Undersoegelse, 32 p.

Korte, C., Hesselbo, S.P., Jenkyns, H.C., Rickaby, R.E., and Spötl, C., 2009, Palaeoenvironmental significance of carbon- and oxygen-isotope stratigraphy of marine Triassic-Jurassic boundary sections in SW Britain: Journal of the Geological Society of London, v. 166, p. 431–445, doi:10.1144/0016-76492007-177.

Korte, C., Pande, P., Kalia, P., Kozur, H.W., Joachimski, M.M., and Oberhänsli, H., 2010, Massive volcanism at the Permian-Triassic boundary and its impact on the isotopic composition of the ocean and atmosphere: Journal of Asian Earth Sciences, v. 37, p. 293–311, doi:10.1016/j.jseaes.2009.08.012.

Kozur, H.W., 1998, Some aspects of the Permian-Triassic boundary (PTB) and of the possible causes for the biotic crisis around this boundary: Palaeogeography, Palaeoclimatology, Palaeoecology, v. 143, p. 227–272, doi:10.1016/S0031-0182(98)00113-8.

Kozur, H.W., and Weems, R.E., 2011, Detailed correlation and age of continental late Changhsingian and earliest Triassic beds: Implications for the role of the Siberian Trap in the Permian-Triassic biotic crisis: Palaeogeography, Palaeoclimatology, Palaeoecology, v. 308, p. 22–40, doi:10.1016/j.palaeo.2011.02.020.

Kravchinsky, V.A., 2012, Paleozoic large igneous provinces of northern Eurasia: Correlation with mass extinction events: Global and Planetary Change, v. 86, p. 31–36, doi:10.1016/j.gloplacha.2012.01.007.

Kravchinsky, V.A., Konstantinov, K.M., Courtillot, V., Valet, J.-P., Savrasov, J.I., Cherniy, S.D., Mishenin, S.G., and Parasotka, B.S., 2002, Palaeomagnetism of East Siberian traps and kimberlites: Two new poles and palaeogeographic reconstructions at about 360 and 250 Ma: Geophysical Journal International, v. 148, p. 1–33, doi:10.1046/j.0956-540x.2001.01548.x.

Kump, L.R., Pavlov, A., and Arthur, M.A., 2005, Massive release of hydrogen sulfide to the surface ocean and atmosphere during intervals of oceanic anoxia: Geology, v. 33, p. 397–400, doi:10.1130/G21295.1.

Kusznir, N.J., Kokhuto, A., and Stephenson, R.A., 1996, Synrift evolution of the Pripyat Trough: Constraints from structural and stratigraphic modeling: Tectonophysics, v. 268, p. 221–236, doi:10.1016/S0040-1951(96)00231-4.

Kuzmin, M.I., Yarmolyuk, V.V., and Kravchinsky, V.A., 2010, Phanerozoic hot spot traces and paleogeographic reconstructions of the Siberian continent based on interaction with the African large low shear velocity province: Earth-Science Reviews, v. 102, p. 29–59, doi:10.1016/j.earscirev.2010.06.004.

Lai Xulong, Wang Wei, Wignall, P.B., Bond, D.P.G., Jiang Haishui, Ali, J.R., John, E.H., and Sun Yadong, 2008, Palaeoenvironmental change during the end-Guadalupian (Permian) mass extinction in Sichuan, China: Palaeogeography, Palaeoclimatology, Palaeoecology, v. 269, p. 78–93, doi:10.1016/j.palaeo.2008.08.005.

Lamarque, J.F., Kiehl, J.T., and Orlando, J.J., 2007, Role of hydrogen sulfide in a Permian–Triassic boundary ozone collapse: Geophysical Research Letters, v. 34, L02801, doi:10.1029/2006GL028384.

Lefebvre, V., Servais, T., François, L., and Averbuch, O., 2010, Did a Katian large igneous province trigger the Late Ordovician glaciation? A hypothesis tested with a carbon cycle model: Palaeogeography, Palaeoclimatology, Palaeoecology, v. 296, p. 310–319, doi:10.1016/j.palaeo.2010.04.010.

Le Gall, B., Tshoso, G., Jourdan, F., Féraud, G., Bertrand, H., Tiercelin, J.J., Kampunzu, A.B., Modisi, M.P., Dyment, J., and Maia, M., 2002, $^{40}Ar/^{39}Ar$ geochronology and structural data from the giant Okavango and related mafic dyke swarms, Karoo igneous province, Botswana: Earth and Planetary Science Letters, v. 202, p. 595–606, doi:10.1016/S0012-821X(02)00763-X.

Little, C.T.S., 1996, The Pliensbachian-Toarcian (Lower Jurassic) extinction event, *in* Ryder, G., Fastovsky, D.E., and Gartner, S., eds., The Cretaceous-Tertiary Event and Other Catastrophes in Earth History: Geological Society of America Special Paper 307, p. 505–512.

Little, C.T.S., and Benton, M.J., 1995, Early Jurassic mass extinction: A global long-term event: Geology, v. 23, p. 495–498, doi:10.1130/0091-7613(1995)023<0495:EJMEAG>2.3.CO;2.

Liu Baojun, and Xu, X. (chief eds.), 1994, Atlas of the Lithofacies and Paleogeography of South China (Sinian–Triassic): Beijing, Science Press, 192 p.

Liu Chengying, and Zhu Rixiang, 2009, Geodynamic significances of the Emeishan Basalts: Earth Science Frontiers, v. 16, p. 52–69, doi:10.1016/S1872-5791(08)60082-2.

Lucas, S.G., 1998, Global Triassic tetrapod biostratigraphy and biochronology: Palaeogeography, Palaeoclimatology, Palaeoecology, v. 143, p. 347–384, doi:10.1016/S0031-0182(98)00117-5.

Lucas, S.G., 2009, Timing and magnitude of tetrapod extinctions across the Permo-Triassic boundary: Journal of Asian Earth Sciences, v. 36, p. 491–502, doi:10.1016/j.jseaes.2008.11.016.

Mahoney, J.J., Fitton, J.G., Wallace, P.J., Antretter, M.J., Banerjee, N.R., Bergen, J.A., Cairns, G., Castillo, P.R., Chambers, L.M., Chazey, W.J., III, Coffin, M.F., Godard, M.M., Hall, S.A., Honnorez, J., Ingle, S.P., Kroenke, L.W., MacLeod, K.G., Naruse, H., Neal, C.R., Ogg, J.G., Riisager, P., Sano, T., Sikora, P.J., van der Werff, W., White, R.V., and Zhao Xixi, 2001, Proceedings of the Ocean Drilling Program Initial Reports 192 (online only): College Station, Texas, Ocean Drilling Program, doi:10.2973/odp.proc.ir.192.2001.

Martindale, R.C., Berelson, W.M., Corsetti, F.A., Bottjer, D.J., and West, A.J., 2012, Constraining carbonate chemistry at a potential ocean acidification event (the Triassic-Jurassic boundary) using the presence of corals and coral reefs in the fossil record: Palaeogeography, Palaeoclimatology, Palaeoecology, v. 350–352, p. 114–123, doi:10.1016/j.palaeo.2012.06.020.

Marzoli, A., Renne, P.R., Piccirillo, E.M., Ernesto, M., Bellieni, G., and De Min, A., 1999, Extensive 200-million-year-old continental flood basalts of the Central Atlantic magmatic province: Science, v. 284, p. 616–618, doi:10.1126/science.284.5414.616.

Marzoli, A., Bertrand, H., Knight, K.B., Cirilli, S., Buratti, N., Vérati, C., Nomade, S., Renne, P.R., Youbi, N., Martini, R., Allenbach, K., Neuwerth, R., Rapaille, C., Zaninetti, L., and Bellieni, G., 2004, Synchrony of the Central Atlantic magmatic province and the Triassic-Jurassic boundary climatic and biotic crisis: Geology, v. 32, p. 973–976, doi:10.1130/G20652.1.

Marzoli, A., Jourdan, F., Puffer, J.H., Cuppone, T., Tanner, L.H., Weems, R.E., Bertrand, H., Cirilli, S., Bellieni, G., and De Min, A., 2011, Timing and duration of the Central Atlantic magmatic province in the Newark and Culpeper Basins, eastern USA: Lithos, v. 122, p. 175–188, doi:10.1016/j.lithos.2010.12.013.

Masaitis, V.L., Mikhailov, M.V., and Selivanovskaya, T.V., 1975, Volcanism and Tectonics of the Patom-Vilyui Aulacogen: Moscow, Nedra, 182 p.

McArthur, J.M., Donovan, D.T., Thirlwall, M.F., Fouke, B.W., and Mattey, D., 2000, Strontium isotope profile of the early Toarcian (Jurassic) oceanic anoxic event, the duration of ammonite biozones, and belemnite palaeotemperatures: Earth and Planetary Science Letters, v. 179, p. 269–285, doi:10.1016/S0012-821X(00)00111-4.

McElwain, J.C., Beerling, D.J., and Woodward, F.I., 1999, Fossil plants and global warming at the Triassic-Jurassic boundary: Science, v. 285, p. 1386–1390, doi:10.1126/science.285.5432.1386.

McElwain, J.C., Wade-Murphy, J., and Hesselbo, S.P., 2005, Changes in carbon dioxide during an oceanic anoxic event linked to intrusion into Gondwana coals: Nature, v. 435, p. 479–482, doi:10.1038/nature03618.

McElwain, J.C., Popa, M.E., Hesselbo, S.P., Haworth, M., and Surlyk, F., 2007, Macroecological responses of terrestrial vegetation to climatic and atmospheric change across the Triassic/Jurassic boundary in East Greenland: Paleobiology, v. 33, p. 547–573, doi:10.1666/06026.1.

McGhee, G.R., Sheehan, P.M., Bottjer, D.J., and Droser, M.L., 2004, Ecological ranking of Phanerozoic biodiversity crises: Ecological and taxonomic severities are decoupled: Palaeogeography, Palaeoclimatology, Palaeoecology, v. 211, p. 289–297, doi:10.1016/j.palaeo.2004.05.010.

McGhee, G.R., Clapham, M.E., Sheehan, P.M., Bottjer, D.J., and Droser, M.L., 2013, A new ecological-severity ranking of major Phanerozoic biodiversity crises: Palaeogeography, Palaeoclimatology, Palaeoecology, v. 370, p. 260–270, doi:10.1016/j.palaeo.2012.12.019.

McHone, J.G., 2003, Volatile emissions from Central Atlantic magmatic province basalts: Mass assumptions and environmental consequences, *in* Hames, W., Mchone, J.G., Renne, P., and Ruppel, C., eds., The Central Atlantic Magmatic Province: Insights from Fragments of Pangea: American Geophysical Union Geophysical Monograph 136, p. 241–254.

Metcalfe, I., 1999, Gondwana dispersion and Asian accretion: An overview, *in* Metcalfe, I., ed., Final Results Volume for IGCP Project 321: Rotterdam, Netherlands, A.A. Balkema, p. 9–28.

Meyerhoff, A.A., 1980, Geology and petroleum fields in Proterozoic and Lower Cambrian strata, Lena-Tunguska Petroleum Province, eastern Siberia, USSR, *in* Halbouty, M.T., ed., Giant Oil and Gas Fields of the Decade 1968–1978: American Association of Petroleum Geologists Memoir 30, p. 225–252.

Minor, D.R., and Mukasa, S.B., 1997, Zircon U/Pb and hornblende ^{40}Ar/^{39}Ar ages for the Dufek layered mafic intrusion, Antarctica: Implications for the age of the Ferrar large igneous province: Geochimica et Cosmochimica Acta, v. 61, p. 2497–2504, doi:10.1016/S0016-7037(97)00098-7.

Montenegro, A., Spence, P., Meissner, K.J., Eby, M., Melchin, M.J., and Johnston, S.T., 2011, Climate simulations of the Permian-Triassic boundary: Ocean acidification and the extinction event: Paleoceanography, v. 26, p. PA3207, doi:10.1029/2010PA002058.

Mundil, R.L., Metcalfe, I., Ludwig, K., Paull, R., Oberli, F., and Nicoll, R., 2001, Timing of the Permian-Triassic biotic crisis: Implications from new zircon U/Pb age data: Earth and Planetary Science Letters, v. 187, p. 131–145, doi:10.1016/S0012-821X(01)00274-6.

Mundil, R., Ludwig, K.R., Metcalfe, I., and Renne, P.R., 2004, Age and timing of the Permian mass extinctions: U/Pb dating of closed-system zircons: Science, v. 305, p. 1760–1763, doi:10.1126/science.1101012.

Myhre, G., Myhre, A., and Stordal, F., 2001, Historical evolution of radiative forcing of climate: Atmospheric Environment, v. 35, p. 2361–2373, doi:10.1016/S1352-2310(00)00531-8.

Nakazawa, K., Ishu, K., Kato, M., Okimura, Y., Nakamura, K., and Haralambous, D., 1975, Upper Permian fossils from island of Salamis, Greece: Memoirs of the Faculty of Science, Kyoto University, Series of Geology and Mineralogy, v. 41, p. 21–44.

Niemeier, U., Schmidt, H., and Timmreck, C., 2011, The dependency of geoengineered sulfate aerosol on the emission strategy: Atmospheric Science Letters, v. 12, p. 189–194, doi:10.1002/asl.304.

Nomade, S., Knight, K.B., Beutel, E., Renne, P.R., Verati, C., Féraud, G., Marzoli, A., Youbi, N., and Bertrand, H., 2007, Chronology of the Central Atlantic magmatic province: Implications for the Central Atlantic rifting processes and the Triassic-Jurassic biotic crisis: Palaeogeography, Palaeoclimatology, Palaeoecology, v. 244, p. 326–344, doi:10.1016/j.palaeo.2006.06.034.

Ogden, D.E., and Sleep, N.H., 2012, Explosive eruption of coal and basalt and the end-Permian mass extinction: Proceedings of the National Academy of Sciences of the United States of America, v. 109, p. 59–62, doi:10.1073/pnas.1118675109.

Olsen, P.E., 1999, Giant lava flows, mass extinctions, and mantle plumes: Science, v. 284, p. 604–605, doi:10.1126/science.284.5414.604.

Olsen, P.E., Kent, D.V., and Fowell, S.J., 1997, Causal? association of the Triassic-Jurassic mass extinction and Pangean flood basalt—A matter of timing: Eos (Transactions, American Geophysical Union), v. 78, p. F721.

Olsen, P.E., Kent, D.V., Et-Touhami, M., and Puffer, J., 2003, Cyclo-, magneto-, and bio-stratigraphic constraints on the duration of the CAMP event and its relationship to the Triassic-Jurassic boundary, *in* Hames, W., Mchone, J.G., Renne, P., and Ruppel, C., eds., The Central Atlantic Magmatic Province: Insights from Fragments of Pangea: American Geophysical Union Geophysical Monograph 136, p. 7–32.

Oman, L., Robock, A., Stenchikov, G., Schmidt, G.A., and Ruedy, R., 2005, Climatic response to high-latitude volcanic eruptions: Journal of Geophysical Research–Atmospheres (1984–2012), v. 110, p. D13103, doi:10.1029/2004JD005487.

Palais, J.M., and Sigurdsson, H., 1989, Petrologic evidence of volatile emissions from major historic and pre-historic volcanic eruptions, *in* Berger, A., Dickinson, R.E., and Kidson, J.W., eds., Understanding Climate Change: American Geophysical Union Geophysical Monograph 52, p. 31–53.

Pálfy, J., Demény, A., Haas, J., Hetényi, M., Orchard, M., and Vetö, I., 2001, Carbon isotope anomaly and other geochemical changes at the Triassic-Jurassic boundary from a marine section in Hungary: Geology, v. 29, p. 1047–1050, doi:10.1130/0091-7613(2001)029<1047:CIAAOG>2.0.CO;2.

Pálfy, J., Smith, P.L., and Mortensen, J.K., 2002, Dating the end-Triassic and Early Jurassic mass extinctions, correlative large igneous provinces, and isotopic events, *in* Koeberl, C., and MacLeod, K.G., eds., Catastrophic Events and Mass Extinctions: Impacts and Beyond: Geological Society of America Special Paper 356, p. 523–532.

Payne, J.L., and Kump, L.R., 2007, Evidence for recurrent Early Triassic massive volcanism from quantitative interpretation of carbon isotope fluctuations: Earth and Planetary Science Letters, v. 256, p. 264–277, doi:10.1016/j.epsl.2007.01.034.

Payne, J.L., Lehrmann, D.J., Follett, D., Seibel, M., Kump, L.R., Riccardi, A., Altiner, D., Sano, H., and Wei Jiayong, 2007, Erosional truncation of uppermost Permian shallow-marine carbonates and implications for Permian-Triassic boundary events: Geological Society of America Bulletin, v. 119, p. 771–784, doi:10.1130/B26091.1.

Pinto, J.P., Turco, R.P., and Toon, O.B., 1989, Self-limiting physical and chemical effects in volcanic eruption clouds: Journal of Geophysical Research, v. 94, no. D8, p. 11,165–11,174, doi:10.1029/JD094iD08p11165.

Pruss, S.B., Bottjer, D.J., Corsetti, F.A., and Baud, A., 2006, A global marine sedimentary response to the end-Permian mass extinction: Examples from southern Turkey and the western United States: Earth-Science Reviews, v. 78, p. 193–206, doi:10.1016/j.earscirev.2006.05.002.

Racki, G., 1998, Frasnian-Famennian biotic crisis: Undervalued tectonic control?: Palaeogeography Palaeoclimatology Palaeoecology, v. 141, p. 177–198.

Racki, G., and Wignall, P.B., 2005, Late Permian double-phased mass extinction and volcanism: An oceanographic perspective: Developments in Palaeontology and Stratigraphy, v. 20, p. 263–297, doi:10.1016/S0920-5446(05)80010-X.

Rampino, M.R., and Self, S., 1992, Volcanic winter and accelerated glaciation following the Toba super-eruption: Nature, v. 359, p. 50–52, doi:10.1038/359050a0.

Rampino, M.R., and Stothers, R.B., 1988, Flood basalt volcanism during the past 250 million years: Science, v. 241, p. 663–668, doi:10.1126/science.241.4866.663.

Rasch, P.J., Crutzen, P.J., and Coleman, D.B., 2008, Exploring the geoengineering of climate using stratospheric sulfate aerosols: The role of particle size: Geophysical Research Letters, v. 35, L02809, doi:10.1029/2007GL032179.

Raunsgaard Pedersen, K., and Lund, J.J., 1980, Palynology of the plant-bearing Rhaetian to Hettangian Kap Stewart Formation, Scoresby Sund, East Greenland: Review of Palaeobotany and Palynology, v. 31, p. 1–69, doi:10.1016/0034-6667(80)90022-6.

Raup, D.M., and Sepkoski, J.J., Jr., 1982, Mass extinctions in the marine fossil record: Science, v. 215, p. 1501–1503, doi:10.1126/science.215.4539.1501.

Reichow, M.K., Saunders, A.D., White, R.V., Al'Mukhamedov, A.I., and Medvedev, A.Y., 2005, Geochemistry and petrogenesis of basalts from the West Siberian Basin: An extension of the Permo-Triassic Siberian Traps, Russia: Lithos, v. 79, p. 425–452, doi:10.1016/j.lithos.2004.09.011.

Reichow, M.K., Pringle, M.S., Al'Mukhamedov, A.I., Allen, M.B., Andreichev, V.L., Buslov, M.M., Davies, C.E., Fedoseev, G.S., Fitton, J.G., Inger, S., Medvedev, A.Ya., Mitchell, C., Puchkov, V.N., Safonova, I.Yu., Scott, R.A., and Saunders, A.D., 2009, The timing and extent of the eruption of the Siberian Traps large igneous province: Implications for the end-Permian environmental crisis: Earth and Planetary Science Letters, v. 277, p. 9–20, doi:10.1016/j.epsl.2008.09.030.

Renne, P.R., and Basu, A.R., 1991, Rapid eruption of the Siberian Traps flood basalts at the Permo-Triassic boundary: Science, v. 253, p. 176–179, doi:10.1126/science.253.5016.176.

Renne, P.R., Black, M.T., Zichao Zhang, Richards, M.A., and Basu, A.R., 1995, Synchrony and causal relations between Permian-Triassic boundary crises and Siberian flood volcanism: Science, v. 269, p. 1413–1416, doi:10.1126/science.269.5229.1413.

Renne, P.R., Mundil, R., Balco, G., Min, K., and Ludwig, K.R., 2010, Joint determination of ^{40}K decay constants and ^{40}Ar*/^{40}K for the Fish Canyon sanidine standard, and improved accuracy for ^{40}Ar/^{39}Ar geochronology: Geochimica et Cosmochimica Acta, v. 74, p. 5349, doi:10.1016/j.gca.2010.06.017.

Renne, P.R., Balco, G., Ludwig, K.R., Mundil, R., and Min, K., 2011, Response to the comment by W.H. Schwarz et al. on "Joint determination of ^{40}K decay constants and ^{40}Ar*/^{40}K for the Fish Canyon sanidine standard, and improved accuracy for ^{40}Ar/^{39}Ar geochronology" by P.R. Renne et al. (2010): Geochimica et Cosmochimica Acta, v. 75, p. 5097, doi:10.1016/j.gca.2011.06.021.

Retallack, G.J., 1995, Permian-Triassic life crisis on land: Science, v. 267, p. 77–80, doi:10.1126/science.267.5194.77.

Retallack, G.J., and Jahren, A.H., 2008, Methane release from igneous intrusion of coal during Late Permian extinction events: The Journal of Geology, v. 116, p. 1–20, doi:10.1086/524120.

Retallack, G.J., Veevers, J., and Morante, R., 1996, Global coal gap between Permian-Triassic extinctions and Middle Triassic recovery of peat forming plants: Geological Society of America Bulletin, v. 108, p. 195–207, doi:10.1130/0016-7606(1996)108<0195:GCGBPT>2.3.CO;2.

Riccardi, A., Kump, L.R., Arthur, M.A., and D'Hondt, S., 2007, Carbon isotopic evidence for chemocline upward excursions during the end-Permian event: Palaeogeography, Palaeoclimatology, Palaeoecology, v. 248, p. 73–81, doi:10.1016/j.palaeo.2006.11.010.

Ricci, J., Quidelleur, X., Pavlov, V., Orlov, S., Shatsillo, A., and Courtillot, V., 2013, New ^{40}Ar/^{39}Ar and K-Ar ages of the Viluy Traps (eastern Siberia): Further evidence for a relationship with the Frasnian-Famennian mass extinction: Palaeogeography, Palaeoclimatology, Palaeoecology, v. 386, p. 531–540, doi:10.1016/j.palaeo.2013.06.020.

Riley, T.R., Millar, I.L., Watkeys, M.K., Curtis, M.L., Leat, P.T., Klausen, M.B., and Fanning, C.M., 2004, U-Pb zircon (SHRIMP) ages for the Lebombo rhyolites, South Africa: Refining the duration of Karoo volcanism: Journal of the Geological Society of London, v. 161, p. 547–550, doi:10.1144/0016-764903-181.

Riley, T.R., Curtis, M.L., Leat, P.T., Watkeys, M.K., Duncan, R.A., Millar, I.L., and Owens, W.H., 2006, Overlap of Karoo and Ferrar magma types in KwaZulu-Natal, South Africa: Journal of Petrology, v. 47, p. 541–566, doi:10.1093/petrology/egi085.

Robock, A., 2000, Volcanic eruptions and climate: Reviews of Geophysics, v. 38, p. 191–219, doi:10.1029/1998RG000054.

Ruhl, M., and Kürschner, W.M., 2011, Multiple phases of carbon cycle disturbance from large igneous province formation at the Triassic-Jurassic transition: Geology, v. 39, p. 431–434, doi:10.1130/G31680.1.

Ruhl, M., Kürschner, W.M., and Krystyn, L., 2009, Triassic-Jurassic organic carbon isotope stratigraphy of key sections in the western Tethys realm (Austria): Earth and Planetary Science Letters, v. 281, p. 169–187, doi:10.1016/j.epsl.2009.02.020.

Ruhl, M., Deenen, M.H.L., Abels, H.A., Bonis, N.R., Krijgsman, W., and Kürschner, W.M., 2010, Astronomical constraints on the duration of the Early Jurassic Hettangian Stage and recovery rates following the end-Triassic mass extinction (St. Audrie's Bay/East Quantoxhead, UK): Earth and Planetary Science Letters, v. 295, p. 262–276, doi:10.1016/j.epsl.2010.04.008.

Ruhl, M., Bonis, N.R., Reichart, G.J., Damsté, J.S.S., and Kürschner, W.M., 2011, Atmospheric carbon injection linked to end-Triassic mass extinction: Science, v. 333, p. 430–434, doi:10.1126/science.1204255.

Sabine, C.L., Feely, R.A., Gruber, N., Key, R.M., Lee, K., Bullister, J.L., Wanninkhof, R., Wong, C.S., Wallace, D.W.R., Tilbrook, B., Millero, F.J., Peng Tsunghung, Kozyr, A., Ono, T., and Rios, A.F., 2004, The oceanic sink for anthropogenic CO2: Science, v. 305, p. 367–371, doi:10.1126/science.1097403.

Sadovnikov, G.N., and Orlova, E.F., 1993, The lower boundary and biostratigraphy of the nonmarine Triassic in Siberia, *in* Lucas, S.G., and Morales, M., eds., The Nonmarine Triassic: 3: Albuquerque, New Mexico, New Mexico Museum of Natural History, p. 421–422.

Sadovnikov, G.N., and Orlova, E.F., 1998, Lower boundary of nonmarine Triassic of Siberia: Hallesches Jahrbuch für Geowissenschaften, v. 5, p. 156–157.

Sandoval, J., Henriques, M.H., Chandler, R.B., and Ureta, S., 2012, Latest Toarcian–earliest Bajocian (Jurassic) Grammoceratinae (Hildoceratidae, Ammonitina) of the western Tethys: Their palaeobiogeographic and phylogenetic significance: Geobios, v. 45, p. 109–119, doi:10.1016/j.geobios.2011.11.001.

Saunders, A.D., England, R.W., Reichow, M.K., and White, R.V., 2005, A mantle plume origin for the Siberian Traps: Uplift and extension in the West Siberian Basin, Russia: Lithos, v. 79, p. 407–424, doi:10.1016/j.lithos.2004.09.010.

Scarth, A., 2001, Vulcan's Fury: Man Against the Volcano: New Haven, Connecticut, Yale University Press, 299 p.

Schaller, M.F., Wright, J.D., Kent, D.V., and Olsen, P.E., 2012, Rapid emplacement of the Central Atlantic magmatic province as a net sink for CO_2: Earth and Planetary Science Letters, v. 323, p. 27–39, doi:10.1016/j.epsl.2011.12.028.

Schmidt, A., Carslaw, K.S., Mann, G.W., Wilson, M., Breider, T.J., Pickering, S.J., and Thordarson, T., 2010, The impact of the 1783–1784 AD Laki eruption on global aerosol formation processes and cloud condensation nuclei: Atmospheric Chemistry and Physics, v. 10, p. 6025–6041, doi:10.5194/acp-10-6025-2010.

Schmidt, A., Carslaw, K.S., Mann, G.W., Rap, A., Pringle, K.J., Spracklen, D.V., Wilson, M., and Forster, P.M., 2012a, Importance of tropospheric volcanic aerosol for indirect radiative forcing of climate: Atmospheric Chemistry and Physics, v. 12, p. 7321–7339, doi:10.5194/acp-12-7321-2012.

Schmidt, A., Thordarson, T., Oman, L.D., Robock, A., and Self, S., 2012b, Climatic impact of the long-lasting 1783 Laki eruption: Inapplicability of mass-independent sulfur isotopic composition measurements: Journal of Geophysical Research–Atmospheres (1984–2012), v. 117, D23116, doi:10.1029/2012JD018414.

Schoene, B., Guex, J., Bartolini, A., Schaltegger, U., and Blackburn, T.J., 2010, Correlating the end-Triassic mass extinction and flood basalt volcanism at the 100 ka level: Geology, v. 38, p. 387–390, doi:10.1130/G30683.1.

Self, S., Thordarson, T., Keszthelyi, L., Walker, G.P.L., Hon, K., Murphy, M.T., Long, P., and Finnemore, S., 1996, A new model for the emplacement of Columbia River basalts as large, inflated pahoehoe lava flow fields: Geophysical Research Letters, v. 23, p. 2689–2692, doi:10.1029/96GL02450.

Self, S., Thordarson, T., and Keszthelyi, L., 1997, Emplacement of continental flood basalt lava flows, *in* Mahoney, J.J., and Coffin, M.F., eds., Large Igneous Provinces: Continental, Oceanic, and Planetary Flood Volcanism: American Geophysical Union Geophysical Monograph 100, p. 381–410.

Self, S., Keszthelyi, L., and Thordarson, T., 1998, The importance of pahoehoe: Annual Review of Earth and Planetary Sciences, v. 26, p. 81–110, doi:10.1146/annurev.earth.26.1.81.

Self, S., Thordarson, T., and Widdowson, M., 2005, Gas fluxes from flood basalt eruptions: Elements, v. 1, p. 283–287, doi:10.2113/gselements.1.5.283.

Sephton, M.A., Looy, C.V., Brinkhuis, H., Wignall, P.B., De Leeuw, J.W., and Visscher, H., 2005, Catastrophic soil erosion during the end-Permian biotic crisis: Geology, v. 33, p. 941–944, doi:10.1130/G21784.1.

Sepkoski, J.J., Jr., 1996, Patterns of Phanerozoic extinction: A perspective from global databases, *in* Walliser, O.H., ed., Global Events and Event Stratigraphy in the Phanerozoic: Berlin, Springer-Verlag, p. 35–51.

Sepkoski, J.J., Jr., 2002, A compendium of fossil marine animal genera: Bulletins of American Paleontology, v. 363, p. 1–560.

Sharma, M., 1997, Siberian Traps, *in* Mahoney, J.J., and Coffin, M.F., eds., Large Igneous Provinces: Continental, Oceanic, and Planetary Flood Volcanism: American Geophysical Union Geophysical Monograph 100, p. 273–295.

Shellnutt, J.G., Jahn Borming, and Zhou Meifu, 2011, Crustally-derived granites in the Panzhihua region, SW China: Implications for felsic magmatism in the Emeishan large igneous province: Lithos, v. 123, p. 145–157, doi:10.1016/j.lithos.2010.10.016.

Shellnutt, J.G., Denyszyn, S.W., and Mundil, R., 2012, Precise age determination of mafic and felsic intrusive rocks from the Permian Emeishan large igneous province (SW China): Gondwana Research, v. 22, p. 118–126, doi:10.1016/j.gr.2011.10.009.

Shen Shuzhong, and Shi Guangrui, 2009, Latest Guadalupian brachiopods from the Guadalupian/Lopingian boundary GSSP section at Penglaitan in Laibin, Guangxi, South China, and implications for the timing of the pre-Lopingian crisis: Palaeoworld, v. 18, p. 152–161, doi:10.1016/j.palwor.2009.04.010.

Shen Shuzhong, Crowley, J.L., Wang Yue, Bowring, S.A., Erwin, D.H., Sadler, P.M., Cao Changqun, Rothman, D.H., Henderson, C.M., Ramezani, J., Zhang Hua, Shen Yanan, Wang Xiangdong, Wang Wei, Mu Lin, Li Wenzhong, Tang Yuegang, Liu Xiaolei, Liu Lujun, Zeng Yong, Jiang Yaofa, and Jin Yugan, 2011, Calibrating the end-Permian mass extinction: Science, v. 334, p. 1367–1372, doi:10.1126/science.1213454.

Shpount, B.R., and Oleinikov, B.V., 1987, A comparison of mafic dyke swarms from the Siberian and Russian platforms, *in* Halls, H.C., and Fahrig W.G., eds., Mafic Dyke Swarms: Geological Association of Canada Special Paper 34, p. 379–383.

Sigurdsson, H., 1990, Evidence for volcanic loading of the atmosphere and climatic response: Palaeogeography, Palaeoclimatology, Palaeoecology, v. 89, p. 277–289, doi:10.1016/0031-0182(90)90069-J.

Simms, M.J., 2003, Uniquely extensive seismite from the latest Triassic of the United Kingdom: Evidence for bolide impact?: Geology, v. 31, p. 557–560, doi:10.1130/0091-7613(2003)031<0557:UESFTL>2.0.CO;2.

Simms, M.J., 2007, Uniquely extensive soft-sediment deformation in the Rhaetian of the UK: Evidence for earthquake or impact?: Palaeogeography, Palaeoclimatology, Palaeoecology, v. 244, p. 407–423, doi:10.1016/j.palaeo.2006.06.037.

Smith, R.M., and Ward, P.D., 2001, Pattern of vertebrate extinctions across an event bed at the Permian-Triassic boundary in the Karoo Basin of South Africa: Geology, v. 29, p. 1147–1150, doi:10.1130/0091-7613(2001)029<1147:POVEAA>2.0.CO;2.

Sobolev, S.V., Sobolev, A.V., Kuzmin, D.V., Krivolutskaya, N.A., Petrunin, A.G., Arndt, N.T., Radko, V.A., and Vasiliev, Y.R., 2011, Linking mantle plumes, large igneous provinces and environmental catastrophes: Nature, v. 477, p. 312–316, doi:10.1038/nature10385.

Song Haijun, Wignall, P.B., Tong Jinnan, and Yin Hongfu, 2012, Two pulses of extinction during the Permian-Triassic crisis: Nature Geoscience, v. 6, p. 52–56.

Stanley, S.M., and Yang Xiangning, 1994, A double mass extinction at the end of the Palaeozoic Era: Science, v. 266, p. 1340–1344, doi:10.1126/science.266.5189.1340.

Sun Yadong, Lai Xulong, Wignall, P.B., Widdowson, M., Ali, J.R., Jiang Haishui, Wang Wei, Yan Chunbo, Bond, D.P.G., and Védrine, S., 2010, Dating the onset and nature of the Middle Permian Emeishan large igneous province eruptions in SW China using conodont biostratigraphy and its bearing on mantle plume uplift models: Lithos, v. 119, p. 20–33, doi:10.1016/j.lithos.2010.05.012.

Sun Yadong, Joachimski, M.M., Wignall, P.B., Yan Chunbo, Chen Yanlong, Jiang Haishui, Wang Lina, and Lai Xulong, 2012, Lethally hot temperatures during the Early Triassic greenhouse: Science, v. 338, p. 366–370, doi:10.1126/science.1224126.

Svensen, H., Planke, S., Malthe-Sørenssen, A., Jamtveit, B., Myklebust, R., Eidem, T.R., and Rey, S.S., 2004, Release of methane from a volcanic basin as a mechanism for initial Eocene global warming: Nature, v. 429, p. 542–545, doi:10.1038/nature02566.

Svensen, H., Planke, S., Chevallier, L., Malthe-Sørenssen, A., Corfu, F., and Jamtveit, B., 2007, Hydrothermal venting of greenhouse gases triggering Early Jurassic global warming: Earth and Planetary Science Letters, v. 256, p. 554–566, doi:10.1016/j.epsl.2007.02.013.

Svensen, H., Planke, S., Polozov, A.G., Schmidbauer, N., Corfu, F., Podladchikov, Y.Y., and Jamtveit, B., 2009, Siberian gas venting and the end-Permian environmental crisis: Earth and Planetary Science Letters, v. 277, p. 490–500, doi:10.1016/j.epsl.2008.11.015.

Thordarson, T., and Self, S., 1993, The Laki (Skaftár Fires) and Grímsvötn eruptions in 1783–1785: Bulletin of Volcanology, v. 55, p. 233–263, doi:10.1007/BF00624353.

Thordarson, T., and Self, S., 2003, Atmospheric and environmental effects of the 1783–1784 Laki eruption: A review and reassessment: Journal of Geophysical Research–Atmospheres (1984–2012), v. 108, p. AAC 7-1–AAC 7-29.

Timmreck, C., Lorenz, S.J., Crowley, T.J., Kinne, S., Raddatz, T.J., Thomas, M.A., and Jungclaus, J.H., 2009, Limited temperature response to the very large AD 1258 volcanic eruption: Geophysical Research Letters, v. 36, L21708, doi:10.1029/2009GL040083.

Timmreck, C., Graf, H.-F., Lorenz, S.J., Niemeier, U., Zanchettin, D., Matei, D., Jungclaus, J.H., and Crowley, T.J., 2010, Aerosol size confines climate response to volcanic super-eruptions: Geophysical Research Letters, v. 37, L24705, doi:10.1029/2010GL045464.

Tolan, T.L., Reidel, S.P., Beeson, M.H., Anderson, J.L., Fecht, K.R., and Swanson, D.A., 1989, Revisions to the estimates of the areal extent and volume of the Columbia River Basalt Group, *in* Reidel, S.P., and Hooper, P.R., eds., Volcanism and Tectonism of the Columbia River Flood Basalt Province: Geological Society of America Special Paper 239, p. 1–20.

Twitchett, R.J., Looy, C.V., Morante, R., Visscher, H., and Wignall, P.B., 2001, Rapid and synchronous collapse of marine and terrestrial ecosystems during the end-Permian biotic crisis: Geology, v. 29, p. 351–354, doi:10.1130/0091-7613(2001)029<0351:RASCOM>2.0.CO;2.

Ukstins Peate, I., and Bryan, S.E., 2008, Re-evaluating plume-induced uplift in the Emeishan large igneous province: Nature Geoscience, v. 1, p. 625–629, doi:10.1038/ngeo281.

van de Schootbrugge, B., Bailey, T.R., Rosenthal, Y., Katz, M.E., Wright, J.D., Miller, K.G., Feist-Burkhardt, S., and Falkowski, P.G., 2005, Early Jurassic climate change and the radiation of organic-walled phytoplankton in the Tethys Ocean: Paleobiology, v. 31, p. 73–97, doi:10.1666/0094-8373(2005)031<0073:EJCCAT>2.0.CO;2.

van de Schootbrugge, B., Quan, T.M., Lindström, S., Püttmann, W., Heunisch, C., Pross, J., Fiebig, J., Petschick, R., Röhling, H.-G., Richoz, S., Rosenthal, Y., and Falkowski, P.G., 2009, Floral changes across the Triassic/Jurassic boundary linked to flood basalt volcanism: Nature Geoscience, v. 2, p. 589–594, doi:10.1038/ngeo577.

Venkatesan, T.R., Kumar, A., Gopalau, K., and Al'Mukhamedov, A.I., 1997, 40Ar-39Ar age of Siberian basaltic volcanism: Chemical Geology, v. 138, p. 303–310, doi:10.1016/S0009-2541(97)00006-5.

Visscher, H., Looy, C.V., Collinson, M.E., Brinkhuis, H., Van Konijnenburg-Van Cittert, J.H., Kürschner, W.M., and Sephton, M.A., 2004, Environmental mutagenesis during the end-Permian ecological crisis: Proceedings of the National Academy of Sciences of the United States of America, v. 101, p. 12,952–12,956, doi:10.1073/pnas.0404472101.

Vogt, P.R., 1972, Evidence for global synchronism in mantle plume convection and possible significance for geology: Nature, v. 240, p. 338–342, doi:10.1038/240338a0.

Vörö, A., 1993, Jurassic brachiopods from the Bakony Mts. (Hungary): Global and local effects on changing diversity, *in* Pálfy, J., and Vörö, A., eds., Mesozoic Brachiopods of Alpine Europe: Budapest, Hungarian Geological Society, p. 179–187.

Wang Wei, Cao Changqun, and Wang Yue, 2004, The carbon isotope excursion on GSSP candidate section of the Lopingian-Guadalupian boundary: Earth and Planetary Science Letters, v. 220, p. 57–67, doi:10.1016/S0012-821X(04)00033-0.

Wang Xiangdong, and Sugiyama, T., 2000, Diversity and extinction patterns of Permian coral faunas of China: Lethaia, v. 33, p. 285–294, doi:10.1080/002411600750053853.

Ward, P.D., Montgomery, D.R., and Smith, R., 2000, Altered river morphology in South Africa related to the Permian-Triassic extinction: Science, v. 289, p. 1740–1743, doi:10.1126/science.289.5485.1740.

Ward, P.D., Botha, J., Buick, R., De Kock, M.O., Erwin, D.H., Garrison, G.H., Kirschvink, J.L., and Smith, R., 2005, Abrupt and gradual extinction among Late Permian land vertebrates in the Karoo Basin, South Africa: Science, v. 307, p. 709–714, doi:10.1126/science.1107068.

Westphal, M., Gurevitch, E.L., Samsonov, B.V., Feinberg, H., and Pozzi, J.P., 1998, Magnetostratigraphy of the Lower Triassic volcanics from deep drill SG6 in western Siberia: Evidence for long-lasting Permo–Triassic volcanic activity: Geophysical Journal International, v. 134, p. 254–266, doi:10.1046/j.1365-246x.1998.00557.x.

Wiesner, M.G., Wetzel, A., Catane, S.G., Listanco, E.L., and Mirabueno, H.T., 2004, Grain size, areal thickness distribution and controls on sedimentation of the 1991 Mount Pinatubo tephra layer in the South China Sea: Bulletin of Volcanology, v. 66, p. 226–242, doi:10.1007/s00445-003-0306-x.

Wignall, P.B., 2001, Large igneous provinces and mass extinctions: Earth-Science Reviews, v. 53, p. 1–33, doi:10.1016/S0012-8252(00)00037-4.

Wignall, P.B., 2007, The end-Permian mass extinction—How bad did it get?: Geobiology, v. 5, p. 303–309, doi:10.1111/j.1472-4669.2007.00130.x.

Wignall, P.B., and Bond, D.P.G., 2008, The end-Triassic and Early Jurassic mass extinction records in the British Isles: Proceedings of the Geologists' Association, v. 119, p. 73–84.

Wignall, P.B., and Hallam, A., 1992, Anoxia as a cause of the Permian/Triassic extinction: Facies evidence from northern Italy and the western United States: Palaeogeography, Palaeoclimatology, Palaeoecology, v. 93, p. 21–46, doi:10.1016/0031-0182(92)90182-5.

Wignall, P.B., and Newton, R., 2003, Contrasting deep-water records from the Upper Permian and Lower Triassic of South Tibet and British Columbia: Evidence for a diachronous mass extinction: Palaios, v. 18, p. 153–167, doi:10.1669/0883-1351(2003)18<153:CDRFTU>2.0.CO;2.

Wignall, P.B., and Twitchett, R.J., 1996, Ocean anoxia and the end-Permian mass extinction: Science, v. 272, p. 1155–1158, doi:10.1126/science.272.5265.1155.

Wignall, P.B., Morante, R., and Newton, R., 1998, The Permo-Triassic transition in Spitsbergen: $\delta^{13}C_{org}$ chemostratigraphy, Fe and S geochemistry, facies, fauna and trace fossils: Geological Magazine, v. 135, p. 47–62, doi:10.1017/S0016756897008121.

Wignall, P.B., Newton, R.J., and Little, C.T.S., 2005, The timing of paleoenvironmental change and cause-and-effect relationships during the Early Jurassic mass extinction in Europe: American Journal of Science, v. 305, p. 1014–1032, doi:10.2475/ajs.305.10.1014.

Wignall, P.B., Hallam, A., Newton, R.J., Sha, Jingeng, Reeves, E., Mattioli, E., and Crowley, S., 2006, An eastern Tethyan (Tibetan) record of the Early Jurassic (Toarcian) mass extinction event: Geobiology, v. 4, no. 3, p. 179–190, doi:10.1111/j.1472-4669.2006.00081.x.

Wignall, P.B., Sun Yadong, Bond, D.P.G., Izon, G., Newton, R.J., Védrine, S., Widdowson, M., Ali, J.R., Lai Xulong, Jiang Haishui, Cope, H., and Bottrell, S.H., 2009a, Volcanism, mass extinction and carbon isotope fluctuations in the Middle Permian of China: Science, v. 324, p. 1179–1182, doi:10.1126/science.1171956.

Wignall, P.B., Védrine, S., Bond, D.P.G., Wang Wei, Lai Xulong, Ali, J.R., and Jiang Haishui, 2009b, Facies analysis and sea-level change at the Guadalupian-Lopingian global stratotype (Laibin, South China), and its bearing on the end-Guadalupian mass extinction: Journal of the Geological Society of London, v. 166, p. 655–666, doi:10.1144/0016-76492008-118.

Wignall, P.B., Bond, D.P.G., Kuwahara, K., Kakuwa, K., Newton, R.J., and Poulton, S.W., 2010, An 80 million year oceanic redox history from Permian to Jurassic pelagic sediments of the Mino-Tamba terrane, SW Japan, and the origin of four mass extinctions: Global and Planetary Change, v. 71, p. 109–123, doi:10.1016/j.gloplacha.2010.01.022.

Williams-Jones, G., and Rymer, H., 2000, Hazards of volcanic gasses, *in* Sigurdsson, H., ed., Encyclopedia of Volcanoes: San Diego, Academic Press, p. 997–1004.

Wilson, M., and Lyashkevich, Z.M., 1996, Magmatism and the geodynamics of rifting of the Pripyat-Dnieper-Donets rift, East European Platform: Tectonophysics, v. 268, p. 65–81, doi:10.1016/S0040-1951(96)00234-X.

Wood, C.A., 1992, Climatic effects of the 1783 Laki eruption, *in* Harrington, C.R., ed., The Year Without a Summer: World Climate in 1816: Ottawa, Canadian Museum of Nature, p. 58–77.

Woods, A.D., and Bottjer, D.J., 2000, Distribution of ammonoids in the Lower Triassic Union Wash Formation (eastern California): Evidence for paleoceanographic conditions during recovery from the end-Permian mass extinction: Palaios, v. 15, p. 535–545, doi:10.1669/0883-1351(2000)015<0535:DOAITL>2.0.CO;2.

Wu Fuyuan, Arzamastsev, A.A., Mitchell, R.H., Lia Qiuli, Suna Jing, Yanga Yueheng, and Wang Rucheng, 2013, Emplacement age and Sr-Nd isotopic compositions of the Afrikanda alkaline ultramafic complex, Kola Peninsula, Russia: Chemical Geology, v. 353, p. 210–229, doi:10.1016/j.chemgeo.2012.09.027.

Xiong Conghui, and Wang Qi, 2011, Permian-Triassic land-plant diversity in South China: Was there a mass extinction at the Permian/Triassic boundary?: Paleobiology, v. 37, p. 157–167, doi:10.1666/09029.1.

Yang Xiangning, Zhou Jianping, Liu Jiarun, and Shi Guijun, 1999, Evolutionary pattern of fusulinacean foraminifer in Maokouan: Middle Permian: Science in China, ser. D, v. 42, p. 456–464.

Yang Xiangning, Liu Jiarun, and Shi Guijun, 2004, Extinction processes and patterns of Middle Permian Fusulinaceans in southwest China: Lethaia, v. 37, p. 139–147, doi:10.1080/00241160410005114.

Zharkov, M.A., 1984, Paleozoic Salt Bearing Formations of the World: Berlin, Springer-Verlag, 427 p.

Zhou Meifu, Malpas, J., Song Xieyan, Robinson, P.T., Sun Min, Kennedy, A.K., Lesher, C.M., and Keays, R.R., 2002, A temporal link between Emeishan large igneous province (SW China) and the end-Guadalupian mass extinction: Earth and Planetary Science Letters, v. 196, p. 113–122, doi:10.1016/S0012-821X(01)00608-2.

Zielinski, G.A., Mayewski, P.A., Meeker, L.D., Whitlow, S., Twickler, M.S., and Taylor, K., 1996, Potential atmospheric impact of the Toba mega-eruption ~71,000 years ago: Geophysical Research Letters, v. 23, p. 837–840, doi:10.1029/96GL00706.

Manuscript Accepted by the Society 31 January 2014

Printed in the USA

The Geological Society of America
Special Paper 505
2014

Deccan volcanism, the Chicxulub impact, and the end-Cretaceous mass extinction: Coincidence? Cause and effect?

Gerta Keller*
Geosciences Department, Princeton University, Princeton, New Jersey 08544, USA

ABSTRACT

The recent discovery of the direct link between Deccan volcanism and the end-Cretaceous mass extinction also links volcanism to the late Maastrichtian rapid global warming, high environmental stress, and the delayed recovery in the early Danian. In comparison, three decades of research on the Chicxulub impact have failed to account for long-term climatic and environmental changes or prove a coincidence with the mass extinction. A review of Deccan volcanism and the best age estimate for the Chicxulub impact provides a new perspective on the causes for the end-Cretaceous mass extinction and supports an integrated Deccan-Chicxulub scenario. This scenario takes into consideration climate warming and cooling, sea-level changes, erosion, weathering, ocean acidification, high-stress environments with opportunistic species blooms, the mass extinction, and delayed postextinction recovery.

The crisis began in C29r (upper CF2 to lower CF1) with rapid global warming of 4 °C in the oceans and 8 °C on land, commonly attributed to Deccan phase 2 eruptions. The Chicxulub impact occurred during this warm event (about 100–150 k.y. before the mass extinction) based on the stratigraphically oldest impact spherule layer in NE Mexico, Texas, and Yucatan crater core Yaxcopoil-1. It likely exacerbated climate warming and may have intensified Deccan eruptions. The reworked spherule layers at the base of the sandstone complex in NE Mexico and Texas were deposited in the upper half of CF1, ~50–80 k.y. before the Cretaceous-Tertiary (K-T) boundary. This sandstone complex, commonly interpreted as impact tsunami deposits of K-T boundary age, was deposited during climate cooling, low sea level, and intensified currents, leading to erosion of nearshore areas (including Chicxulub impact spherules), transport, and redeposition via submarine channels into deeper waters. Renewed climate warming during the last ~50 k.y. of the Maastrichtian correlates with at least four rapid, massive volcanic eruptions known as the longest lava flows on Earth that ended with the mass extinction, probably due to runaway effects. The kill mechanism was likely ocean acidification resulting in the carbonate crisis commonly considered to be the primary cause for four of the five Phanerozoic mass extinctions.

*gkeller@princeton.edu

Keller, G., 2014, Deccan volcanism, the Chicxulub impact, and the end-Cretaceous mass extinction: Coincidence? Cause and effect?, *in* Keller, G., and Kerr, A.C., eds., Volcanism, Impacts, and Mass Extinctions: Causes and Effects: Geological Society of America Special Paper 505, p. 57–89, doi:10.1130/2014.2505(03). For permission to copy, contact editing@geosociety.org.

INTRODUCTION

Two catastrophes coincided with the end-Cretaceous mass extinction: the Chicxulub impact on Yucatan and Deccan volcanism in India. For the past 30 yr, many scientists have claimed the Chicxulub impact to be the sole cause for the mass extinction (Schulte et al., 2010), though not without controversy (Archibald et al., 2010; Courtillot and Fluteau, 2010; Keller et al., 2010). Over the past 5 yr, research advances in Deccan volcanism have revealed its central role in this mass extinction, calling for reevaluation of Chicxulub and its presumed catastrophic effects. How each of these catastrophes contributed to or caused the mass extinction, either by itself or in combination, and whether they were coincident in time are still under investigation. Intense research of large igneous province volcanism identified four of the five big mass extinctions in Earth's history as coincident with large, rapid volcanic eruptions, revealing these catastrophes as much more deadly than previously assumed (Bond and Wignall, this volume; Courtillot and Fluteau, this volume). Only the K-T boundary mass extinction (also known as the Cretaceous-Paleogene boundary) coincides with both a large impact and one of the largest volcanic episodes in Earth's history. This study reviews recent advances in Deccan volcanism tying this catastrophe directly to the K-T boundary mass extinction, summarizes the best age estimate for the Chicxulub impact, and places these two catastrophes within their stratigraphic context relative to the mass extinction. This is a first step toward identifying the respective roles of impact and volcanism in the K-T boundary mass extinction.

Deccan volcanic eruptions once covered most of India, with an estimated 1.5 million km^2 and 1.2 million km^3 lava extruded, which even today, with about two thirds eroded or buried, still covers an area the size of France or Texas (Fig. 1). Flow after flow of volcanic eruptions piled up in horizontally layered sequences reaching several thousand meters, which today still form mountains up to 3500 m (Fig. 1; Bodas et al., 1988; Khadri et al., 1988; Chenet et al., 2007). Some massive eruptions reached over 1500 km across India and out to the Bay of Bengal via intracanyon transport (Baksi et al., 1994), forming the longest lava flows known on Earth (Self et al., 2008; Jay and Widdowson, 2008).

Until the early 1980s, Deccan eruptions were believed to have occurred over several million years (68–63 Ma, magnetic anomalies C31R to C28R), spanning most of the Maastrichtian to Early Danian. Subsequent studies determined that the bulk of eruptions clustered within less than 1 m.y. (C29r) spanning the K-T boundary, although the estimated uncertainty in the $^{40}Ar/^{39}Ar$ ages is 1%–2.5% (Courtillot et al., 1986, 1988; Vandamme et al., 1991; Hofmann et al., 2000; Sheth et al., 2001; Pande et al., 2004; Kuiper et al., 2008). More recently, detailed paleomagnetic studies combined with $^{40}K/^{40}Ar$ and $^{40}Ar/^{39}Ar$ dating of Deccan lava flows revealed three main volcanic phases of relatively short durations in the early late Maastrichtian, in C29R, and in the early Danian C29N (Chenet et al., 2007, 2008, 2009).

A potential cause-and-effect relationship between Deccan volcanism and the K-T boundary mass extinction was advocated by the early 1980s (McLean, 1978, 1985; Courtillot et al., 1986, 1988; Rice, 1990; Venkatesan et al., 1993; Jaiprakash et al., 1993; Raju et al., 1995). The evidence implicating Deccan volcanism in the K-T boundary mass extinction rested primarily on the sheer size of this volcanic province, the C29R age, and the association of three other mass extinctions with major volcanic provinces

Figure 1. (A) Map of India with current distribution of the Deccan Traps, including the longest lava flows to the Krishna-Godavari Basin, and localities analyzed. KTB—K-T boundary. (B) Photo of Deccan Traps in the mountains of Mahabaleshwar.

(Permian-Triassic, Triassic-Jurassic, and end-Devonian; Wignall, 2001; Courtillot and Renne, 2003; Bond and Wignall, this volume). This suggested that large igneous provinces could have been the general cause of mass extinctions.

Acceptance of Deccan volcanism as a catastrophe that could have led to the extinction of the dinosaurs and many other groups has lagged for three main reasons: (1) the popularity of the Chicxulub impact as the cause for the mass extinction, (2) the belief that Deccan volcanism occurred over at least 1 m.y., leaving sufficient time for recovery between eruptions, and (3) the absence of data that directly linked the mass extinction with Deccan volcanism in India. Research discoveries over the past several years have largely laid these concerns to rest. Critical factors in this development have been the paleomagnetic and $^{40}K/^{40}Ar$ and $^{40}Ar/^{39}Ar$ dating of Deccan lava flows demonstrating three main phases of eruptions (Chenet et al., 2007, 2008, 2009; Jay and Widdowson, 2008; Jay et al., 2009), and the discovery that the main phase 2 eruptions are directly linked to the end-Cretaceous mass extinction (Keller et al., 2008, 2009a, 2009b, 2009c, 2011a, 2012).

This paper reviews and summarizes the evidence accumulated over the past few years that links Deccan volcanism to the K-T boundary mass extinction as recorded in planktic foraminiferal populations in India. These unicellular calcareous shelled organisms suffered the most severe mass extinctions among marine organisms worldwide. In addition, the link between Deccan volcanism and the Chicxulub impact is explored, and the age of the Chicxulub impact is estimated based on the stratigraphically oldest impact ejecta (e.g., impact spherules, impact breccia). Key points addressed include: (1) the direct link between Deccan volcanism and the K-T boundary mass extinction in India, (2) the nature and timing of the mass extinction associated with volcanism, (3) the geochemical markers of climate and environmental changes related to volcanism, (4) the global biotic effects of Deccan volcanism as recognized in species dwarfing and blooms of the disaster opportunist *Guembelitria cretacea*, (5) the age of the Chicxulub impact based on the stratigraphic position of the oldest (primary) impact ejecta layer, (6) the coincidence between Deccan volcanism and the Chicxulub impact relative to the K-T boundary and mass extinction, and (7) the climate changes associated with phase 2 Deccan volcanism and the Chicxulub impact.

STRATIGRAPHY OF DECCAN VOLCANISM AND CHICXULUB IMPACT

Radiometric Dating

Age determinations of the Deccan Traps are based on paleomagnetism and $^{40}K/^{40}Ar$ and $^{40}Ar/^{39}Ar$ dating in west-central India, where basalts form nearly horizontally deposited beds (dip ~1°) for the entire 3500-m-thick lava pile that extend from 16°N to 25°N and from 70°E to 79°E (Fig. 1; Chenet et al., 2007). It is now generally accepted that Deccan volcanism occurred mainly in three major phases. The initial Deccan eruption phase 1 is dated early late Maastrichtian (ca. 67.4 Ma, base C30N) and accounts for ~6% of the 3500-m-thick total Deccan lava pile (Fig. 2). For the next ~1.5 m.y. (C30N into lower C29R), volcanic activity seems to have been absent or inconsequential. The main Deccan phase 2 eruptions occurred over a relatively short time interval in C29R and account for ~80% of the total lava pile (Chenet et al., 2007, 2008, 2009). Biostratigraphic data demonstrate that this volcanic phase ended with the mass extinction, as reviewed herein (Keller et al., 2008, 2011a, 2012). Over the next 200–300 k.y. of the early Paleocene (Danian, C29R above the K-T boundary), volcanic activity was inconsequential or absent. The last Deccan eruption, phase 3, began near the base of C29N and accounts for ~14% of the total Deccan lava pile (Fig. 2). Associated with phase 2 and phase 3 are the largest and longest lava megaflows known on Earth, spanning 1500 km across India and out into the Bay of Bengal (Fig. 1; Self et al., 2008; Keller et al., 2008, 2011a, 2012).

Although the three main phases of Deccan volcanism are well documented, the dating techniques based on $^{40}K/^{40}Ar$ and $^{40}Ar/^{39}Ar$ are imprecise due to uncertainties in the standards used and radioactive decay rates, which reduce absolute age accuracy up to 2.5% (Kuiper et al., 2008). Nevertheless, a recent study by Renne et al. (2013) claimed that the Chicxulub and the K-T boundary mass extinction were synchronous within 32,000 yr based on $^{40}Ar/^{39}Ar$ ages from bentonite layers from Hell Creek, Montana, and 16 impact spherules from Haiti. To date, only U-Pb isotope dilution–thermal ionization mass spectrometry (ID-TIMS) zircon

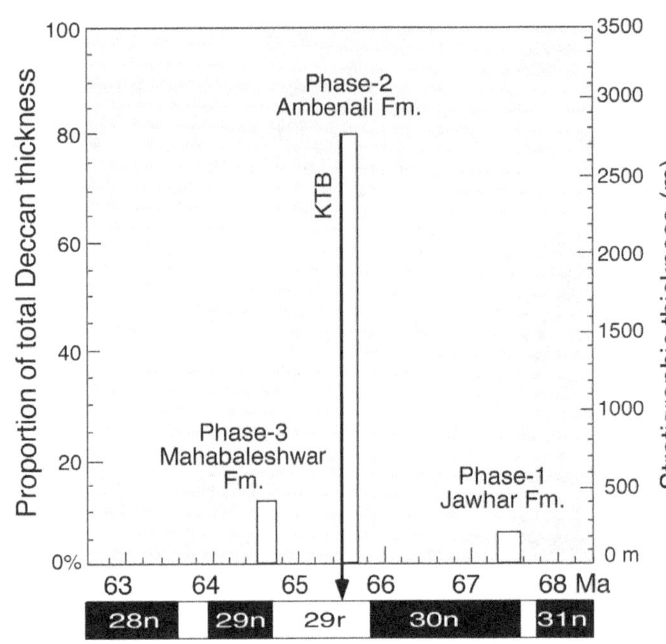

Figure 2. Relative thickness of Deccan lava flows in each of the three phases of volcanic eruptions calculated as percent of total Deccan Trap thickness. Ages are based on paleomagnetic time scale (modified from Chenet et al., 2007, 2008).

geochronology is capable of achieving better than ±60–100 k.y. accuracy based on single-zircon crystal analyses with analytical precision of ±0.02%. Dating of the Deccan Traps material based on U-Pb ages has been unsuccessful to date due to the rarity or absence of zircon crystals in the lava flows, and U-Pb age determinations for the K-T boundary are lacking. Thus, the establishment of better age control for Deccan volcanism and the Chicxulub impact based on radiometric dating has remained elusive.

Relative Dating—Biostratigraphy

Even the best absolute dating methods may not distinguish the order of closely spaced events within the narrow time interval of the K-T boundary mass extinction, Chicxulub impact, and Deccan volcanism, all of which may have occurred within an interval of less than 160 k.y. (planktic foraminiferal zone CF1). Identifying the relative order of such closely spaced events remains the domain of relative dating based on lithostratigraphy (relative order of sediment layers from old to young), biostratigraphy, and stable isotope stratigraphy (Keller, 2008).

Biostratigraphy focuses on correlating rock strata based on shared fossil species or species assemblages without regard to lithology and is arguably the most reliable relative age dating technique in stratigraphy. It makes use of the fossil record and its unique evolutionary events, such as species originations and extinctions, unique population turnovers, and individual species blooms, and ties these to the paleomagnetic and radiometric records. Relative age dating of marine sediments globally, and specifically sequences with Chicxulub impact ejecta and Deccan volcanism, is based on the high-resolution planktic foraminiferal zonal scheme developed by Keller et al. (1995, 2002a) for the Danian and by Li and Keller (1998a, 1998b) for the Maastrichtian (Fig. 3). This zonation was developed based on planktic foraminifera from the El Kef stratotype and Elles sections in Tunisia and sections from the eastern Tethys, Madagascar, Spain, and South Atlantic.

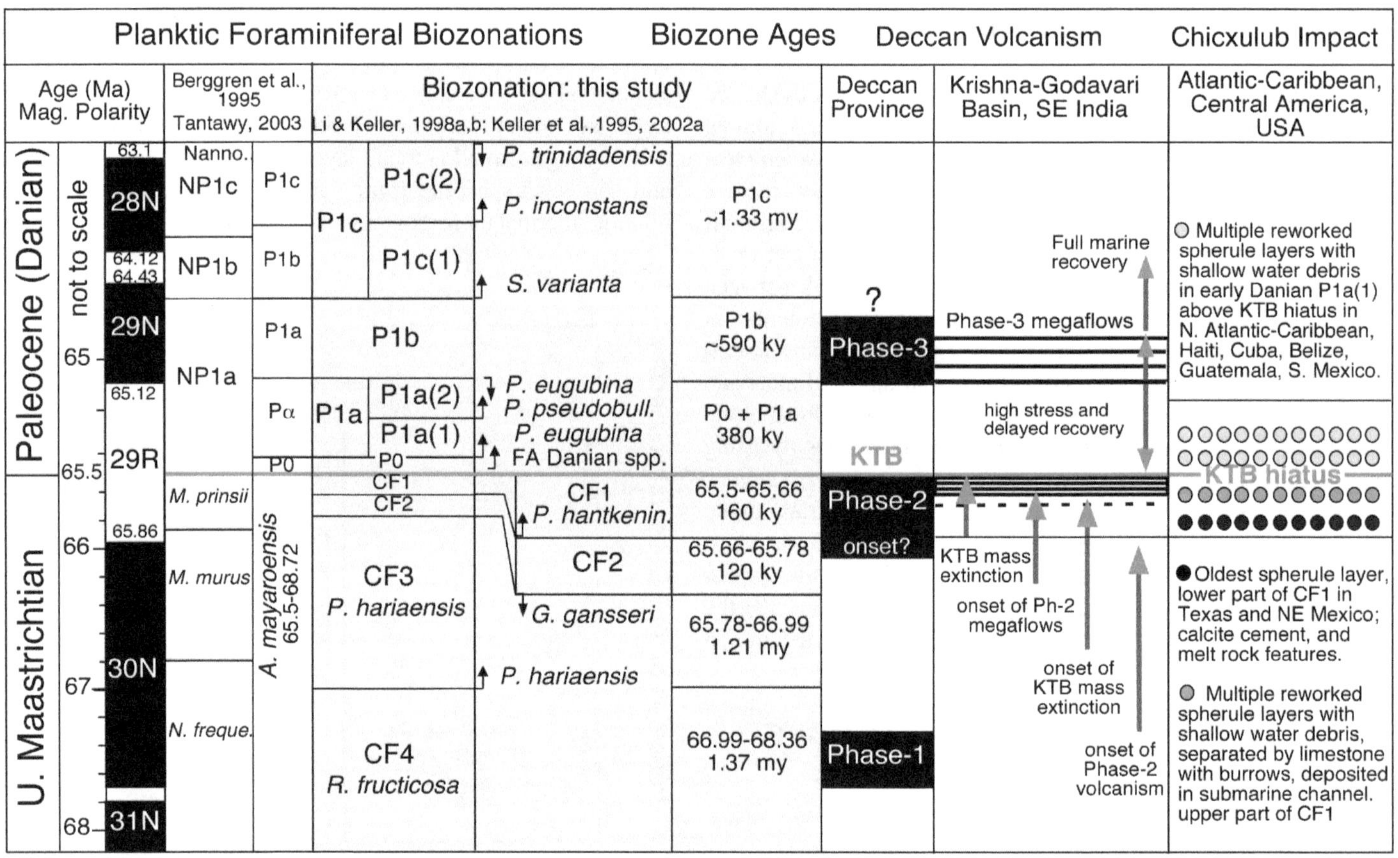

Figure 3. Late Maastrichtian and early Paleocene (Danian) planktic foraminiferal zonal schemes and ages of biozones calculated based on the time scale of Gradstein et al. (2004) with the K-T boundary (KTB) at 65.5 Ma. The stratigraphic position of Deccan megaflows and the onset of the K-T boundary mass extinction in the Krishna-Godavari Basin are based on this study (modified from Keller et al., 2011a). Genus abbreviations: *M. prinsii—Micula prinsii*; *M. murus—Micula murus*; *N. freque.—Nephrolithus frequens*; *A. mayaroensis—Abathomphalus mayaroensis*; *G. gansseri—Gansserina gansseri*; *P. hariaensis—Pseudoguembelina hariaensis*; *P. hantkenin.—Plummerita hantkeninoides*; *P. eugubina—Parvularugoglobigerina eugubina*; *P. pseudobull.—Parasubbotina pseudobulloides*; *P. inconstans—Praemurica inconstans*; *P. trinidadensis—Praemurica trinidadensis*; *R. fructicosa—Racemiguembelina fructicosa*; *S. varianta—Subbotina varianta*.

K-T Boundary Identifying Criteria

From the stratigraphic and paleontologic point of view, the most critical event in Deccan history is identifying the position of the mass extinction in the Deccan lava pile and determining whether there is a cause-and-effect relationship. Similarly, determining the age of the Chicxulub impact relative to the mass extinction is the most important stratigraphic problem of the K-T boundary event. To date, the Chicxulub impact is *assumed* to have caused the mass extinction, and this *belief*, rather than stratigraphic data, has led to a redefinition of the K-T boundary based on Chicxulub impact ejecta (discussed herein).

The K-T boundary is one of the most easily identified boundary events in geological history (Keller, 2011). A set of five K-T boundary–identifying criteria, originally proposed by the International Commission on Stratigraphy (ICS) and the International Union of Geological Sciences (IUGS) working group during 1980-1990s, has proven globally applicable and independently verifiable: (1) the mass extinction of Cretaceous planktic foraminifera, (2) evolution of the first Danian species, (3) K-T boundary clay and red layer, (4) Ir anomaly, and (5) $\delta^{13}C$ shift (Cowie et al., 1989; Keller et al., 1995; Remane et al., 1999).

Planktic foraminifera are the only marine microfossil group that suffered near-total and relatively abrupt mass extinction across the K-T boundary, leaving just one long-term survivor—*Guembelitria cretacea*. The evolution of new species from this survivor followed almost immediately. These unique bioevents, the mass extinction and evolution of new planktic foraminifera, have remained the most reliable K-T boundary–defining criteria worldwide. All other K-T boundary markers, such as the $\delta^{13}C$ shift, clay layer, red layer, and Ir anomaly, including impact glass spherules, are useful supporting evidence but not unique signals in the geological record due to reworking and redeposition. Therefore, impact signals cannot define the K-T boundary in the absence of unique biomarkers (Keller, 2011). In the Deccan Traps, the K-T boundary has now been identified in several localities in central and southeastern India based on the standard defining criteria (mass extinction, evolution of first Danian species; Keller et al., 2011a, 2012) and in one locality (Meghalaya) based on the defining criteria as well as all supporting criteria (Gertsch et al., 2011a).

The same defining and supporting criteria have been successfully applied to hundreds of K-T boundary localities worldwide, including over 100 sequences in the North Atlantic, Caribbean, Central America, and the United States, many of which contain Chicxulub impact ejecta, particularly melt-rock glass spherules. In these sequences, the stratigraphic position of the impact ejecta relative to the K-T boundary provides critical information on the timing of this impact. However, the presence of multiple impact glass spherule layers in many sequences and their stratigraphic position below, at, and above the K-T boundary indicate that impact ejecta can be easily eroded, transported, and redeposited after primary deposition (Fig. 3; Keller, 2008). Stratigraphy's challenge is to decode these multiple depositional events and identify the primary (oldest) impact ejecta layer that pinpoints the time of the Chicxulub impact.

Impact signals are not of significant concern in India. No impact glass spherules have been found, and the only major Ir anomaly is known from the Meghalaya section (Bhandari et al., 1993, 1994; Gertsch et al., 2011a). A minor Ir anomaly reported from the Anjar section in Kutch and originally identified as K-T boundary age (Bhandari et al., 1996; Courtillot et al., 2000) is of volcanic origin and late Maastrichtian age (e.g., Bajpai and Prasad, 2000; Sant et al., 2003). Chatterjee et al. (2006) suggested a crater (Shiva) off Mumbai and impact ejecta at various localities in India, though none has been confirmed.

Chicxulub Defined as K-T Boundary Age

In recent years, impact markers, such as Ir anomalies and impact glass spherules, have frequently been used as the sole markers for the K-T boundary based on the *belief* that the Chicxulub impact caused the mass extinction (e.g., Olsson et al., 1997; Norris et al., 1999, 2000; Arenillas et al., 2006; Molina et al., 2006; MacLeod et al., 2007; Schulte et al., 2010). This practice resulted in redefinition of the K-T boundary based solely on impact markers, particularly impact spherules and Ir anomaly (Molina et al., 2006), which resulted in circular reasoning and delayed recognition of the true age of the Chicxulub impact (Keller, 2008, 2011; Keller et al., 2013). Impact spherules are frequently reworked in late Maastrichtian and early Danian sediments; therefore, designating a layer close to the K-T boundary as the age of Chicxulub is arbitrary.

Iridium anomalies are easily remobilized and reconcentrated at redox layers (e.g., Colodner et al., 1992; Miller et al., 2010; Gertsch et al., 2011a, 2011b). Therefore, small (<1–1.5 ppb) Ir anomalies are unreliable impact markers. To date, out of 345 K-T boundary sections worldwide, just 85 have Ir enrichments, and only five localities have large Ir anomalies: Denmark, Tunisia (El Kef and Elles), Meghalaya, India, and New Zealand (Schmitz et al., 1992; Bhandari et al., 1994; Gertsch et al., 2011a; Premovic et al., 2012). These large Ir anomalies are concentrated at redox boundaries (red clay layers) and reflect significant Ir influx from nearby marine and continental sites (Premovic et al., 2012). No Ir anomaly has ever been detected in Chicxulub impact ejecta. The small (<1 ppb) Ir enrichments commonly reported (e.g., Norris et al., 1999, 2000; Arenillas et al., 2006; Schulte et al., 2010) are above a K-T boundary hiatus and above reworked impact spherules in early Danian sediments (Keller et al., 2013).

DECCAN VOLCANISM LINKED TO MASS EXTINCTION

Although numerous studies over the past 30 yr have advocated Deccan volcanism as the potential cause for the K-T boundary mass extinction (e.g., McLean, 1985; Courtillot et al., 1986, 1988; Jaiprakesh et al., 1993), no direct link could be established in these continental flood basalts. Ideally, a direct link consists of basalt flows separated by marine sediments with microfossils recording the mass extinction. Only in extreme outlying areas of the Deccan volcanic province, such as the Krishna-Godavari (K-G) Basin of southeastern India, are marine sediments present together with Deccan Trap basalts (Fig. 4). For example, in the Rajahmundry area, numerous quarries expose Deccan Traps once deposited along the paleoshoreline (Keller et al., 2008). Further seaward in the Krishna-Godavari Basin, the Deccan Traps are buried 2500–3500 m below the surface. In central India (Jhilmili, Chhindwara), an intertrappean bed representing a marine incursion was discovered (Keller et al., 2009a, 2009b). The best K-T boundary sequence in India is known from Meghalaya, NE India, ~800 km from the Deccan volcanic province. This section is critically important because it records the stress conditions related to Deccan volcanism and ties these with the global record (Garg et al., 2006; Gertsch et al., 2011a).

Um Sohryngkew, Meghalaya

The Um Sohryngkew section of Meghalaya is located along the river that bears its name and runs along the border with Bangladesh (Fig. 1). This section has one of the most complete K-T boundary records worldwide, containing all K-T boundary defining and supporting criteria, including the mass extinction, evolution of first Danian species, $\delta^{13}C$ shift, the boundary clay, and a thin red layer with an Ir anomaly of 12 ppb (Fig. 5; Gertsch et al., 2011a). This large Ir anomaly and other platinum group element anomalies are attributed to an extraterrestrial impact, condensed sedimentation, redox fluctuations, and volcanism (Pandey, 1990; Bhandari et al., 1993, 1994; Garg et al., 2006; Gertsch et al., 2011a).

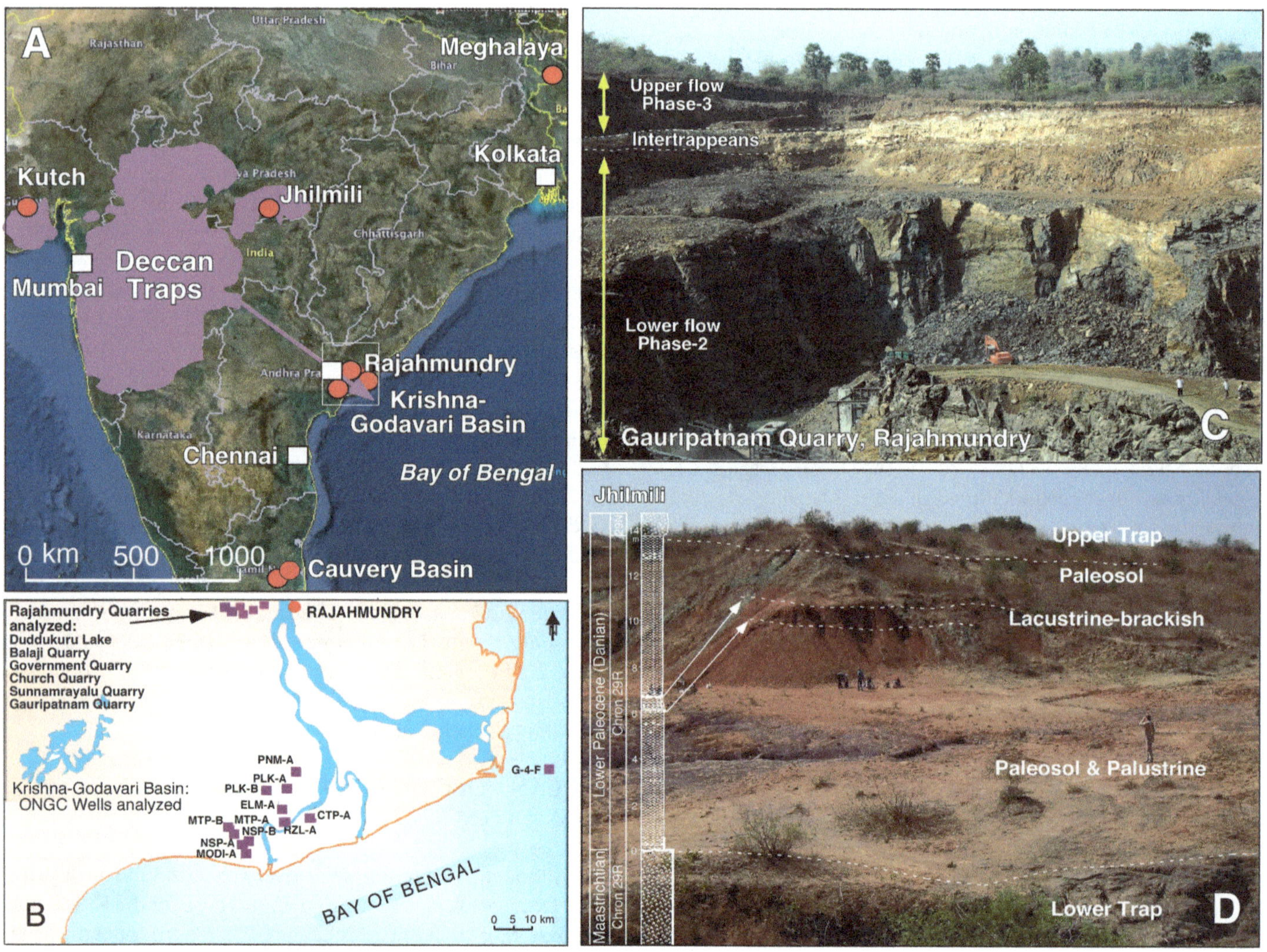

Figure 4. Location map and photos of outcrops analyzed at Jhilmili and the Rajahmundry area, and locations of deep wells analyzed in the Krishna-Godavari Basin and outcrop in Jhilmili, central India. (A) Location map; (B) outcrop and bore well locations in the Krishna-Godavari Basin; (C) Gauripatnam Quarry with intertrappean sediments of earliest Danian age in Rajahmundry; (D) outcrop in Jhilmili, central India, with intertrappean sediments of earliest Danian zone P1a age.

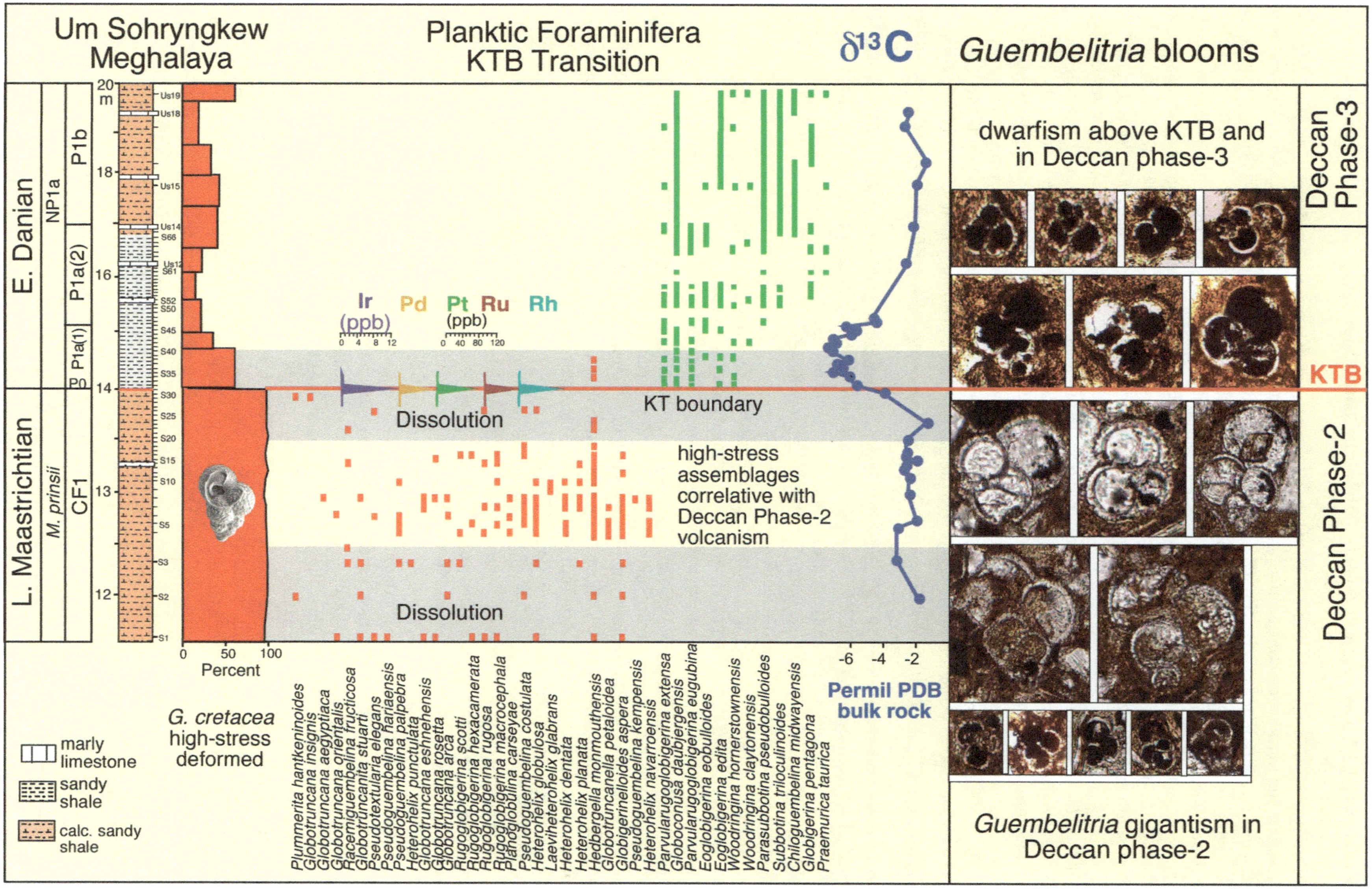

Figure 5. The K-T boundary (KTB) transition at Um Sohryngkew, Meghalaya, one of the most complete boundary sequences worldwide, shows superstress conditions in planktic foraminifera of the latest Maastrichtian (zones CF1–2, *Micula prinsii*) as evident in *Guembelitria* blooms, dwarfing and gigantism, sporadic occurrences of other species, and effects of ocean acidification. Size changes in *Guembelitria cretacea* in zone CF1 (*M. prinsii*) correlative with volcanic phase 2 (magnification is the same for all specimens): small dwarf size (bottom row) near the base of the CF1 interval; gigantism (rows 2 and 3) through most of CF1; normal size of species (row 4) above the K-T boundary mass extinction; return to small dwarfed morphotypes in zone P1b (row 5) correlative with Deccan phase 3. PDB—Peedee belemnite. Figure is modified from Gertsch et al. (2011a).

The latest Maastrichtian zone CF1 below the K-T boundary has low planktic foraminiferal diversity (total 28 species), but less than half are generally present in any one sample due to rare and sporadic occurrences and variable carbonate dissolution (Fig. 5). Species assemblages are dominated by *Guembelitria cretacea*, which account for more than 95% of the total planktic foraminiferal population. The low species diversity, rare and sporadic occurrences, and dominance of *Guembelitria cretacea* are characteristics of high-stress environments, whereas the carbonate dissolution suggests lower pH due to ocean acidification.

Above the K-T boundary, the $\delta^{13}C$ shift and earliest Danian species indicate zone P1a (subzones P1a[1] and P1a[2]; Fig. 3). In P1a(1), *Guembelitria* specimens decrease in abundance correlative with minimum $\delta^{13}C$ values and increase again in zone P1b after the extinction of *Parvularugoglobigerina eugubina*. This pattern and the sequence of species evolution in the early Danian follow global patterns (Punekar et al., this volume).

The most unusual aspects in the Um Sohryngkew K-T boundary transition are the extreme abundance (>95%) and very large size of *Guembelitria cretacea* in zone CF1. This species is normally a very small triserial morphotype, frequently common to abundant in the 36–63 μm size fraction and almost always <100 μm in size. However, in this section, very large morphotypes frequently exceeding 150 μm dominate (~85%). A size comparison is illustrated in Figure 5 based on the same magnification for all specimens. The bottom row shows normal small species of zone CF1, compared with rows 2 and 3 illustrating the predominantly large morphotyes of the assemblage. Above the K-T boundary, the normal small morphotypes dominate with few large ones. A survey of zone CF1 *Guembelitria cretacea* populations in the eastern Tethys reveals similar dominance, but large morphotypes are rare (Punekar et al., this volume).

Guembelitria blooms are indicative of high-stress environments and are most commonly recorded from the aftermath of the K-T boundary mass extinction (Pardo and Keller, 2008; Keller and Abramovich, 2009). Their dominance in zone CF1 below the K-T boundary records extreme high-stress conditions correlative with the main phase 2 of Deccan volcanism, whereas their high abundance in the early Danian zone P1b is correlative with phase 3 (Figs. 2 and 5; Keller et al., 2012). In Meghalaya, the *Guembelitria* dominance in the latest Maastrichtian is associated with geochemical and mineralogical data that reveal humid climate conditions and periodic acid rains (carbonate dissolution) related to pulsed Deccan eruptions, and strong continental weathering and runoff resulting in mesotrophic waters (Gertsch et al., 2011a). These superstress conditions likely led to the demise of nearly all planktic foraminifera and blooms of the disaster opportunist *Guembelitria cretacea*. The Meghalaya section thus provides strong evidence for a close correlation between the global mass extinction record and Deccan volcanism, but a direct link within the Deccan Traps remained elusive until the discoveries in central and southeastern India.

Jhilmili, Central India

A surprising discovery of early Danian zone P1a planktic foraminifera was made through routine ostracod analysis of a 14-m-thick predominantly terrestrial intertrappean bed at Jhilmili, Chhindwara District, Madhya Pradesh (Figs. 4D and 6; Keller et al., 2009a, 2009b). These intertrappean sediments were believed to be terrestrial in origin and late Maastrichtian in age. The presence of rare early Danian freshwater and brackish marine ostracods (Khosla et al., 2009; Sharma and Khosla, 2009; Bajpai et al., 2013) and zone P1a planktic foraminifera in a narrow (0.60 m) interval between lower (phase 2) and upper (phase 3) basalt flows confirmed the link between the main phase 2 Deccan volcanism and the mass extinction. However, this discovery also revealed the presence of a major seaway depositing marine plankton in central India. The seaway likely followed the Narmada-Tapti rift, extending from the west 800 km across India (Keller et al., 2009a). Such a seaway already existed during the late Cenomanian marine transgression (Badve and Ghare, 1977) and was hypothesized to have existed during the late Maastrichtian to explain the high concentration of dinosaur nesting sites and other remains along the Narmada-Tapti rift zone (Sahni, 1983; Keller et al., 2009c).

Rajahmundry Quarries, SE India (Krishna-Godavari Basin)

The world's longest lava megaflows are known to have reached the Krishna-Godavari Basin during the latest Maastrichtian and again in the early Danian, ~300 k.y. after the mass extinction (Self et al., 2008; Keller et al., 2008). In the Rajahmundry area, Deccan basalts are quarried, exposing one thick intertrappean bed (Figs. 4C and 4D). The basalt megaflows below and above the intertrappean bed are generally known as lower and upper traps, which correspond to phase 2 and phase 3 of Deccan volcanism, respectively (Fig. 2). Paleomagnetic data identify the lower traps and the intertrappean sediments as C29R and the upper traps as C29N (Subbarao and Pathak, 1993). Age determinations to date are based on $^{40}K/^{40}Ar$ and $^{40}Ar/^{39}Ar$ dates of plagioclase separates (Knight et al., 2003, 2005; Chenet et al., 2007, 2008), which have large errors (1%–2.5%) that permit no determination of the K-T boundary position.

Early studies of intertrappean sediments in Rajahmundry quarries yielded a Danian age based on ostracods (Bhandari, 1995; Khosla and Nagori, 2002). A study of planktic foraminifera in four Rajahmundry basalt quarries recovered an earliest Danian zone P1a assemblage near the base of the intertrappean bed (Figs. 7 and 8), revealing the first direct link between the K-T boundary mass extinction and Deccan volcanism (Keller et al., 2008; Malarkodi et al., 2010; subsequently confirmed at Jhilmili, see previous section "Jhilmili, Central India"; Fig. 6). Intertrappean sediments in all quarries examined show similar biostratigraphy and depositional environments, although the thicknesses of individual units are variable, and some may not be present in

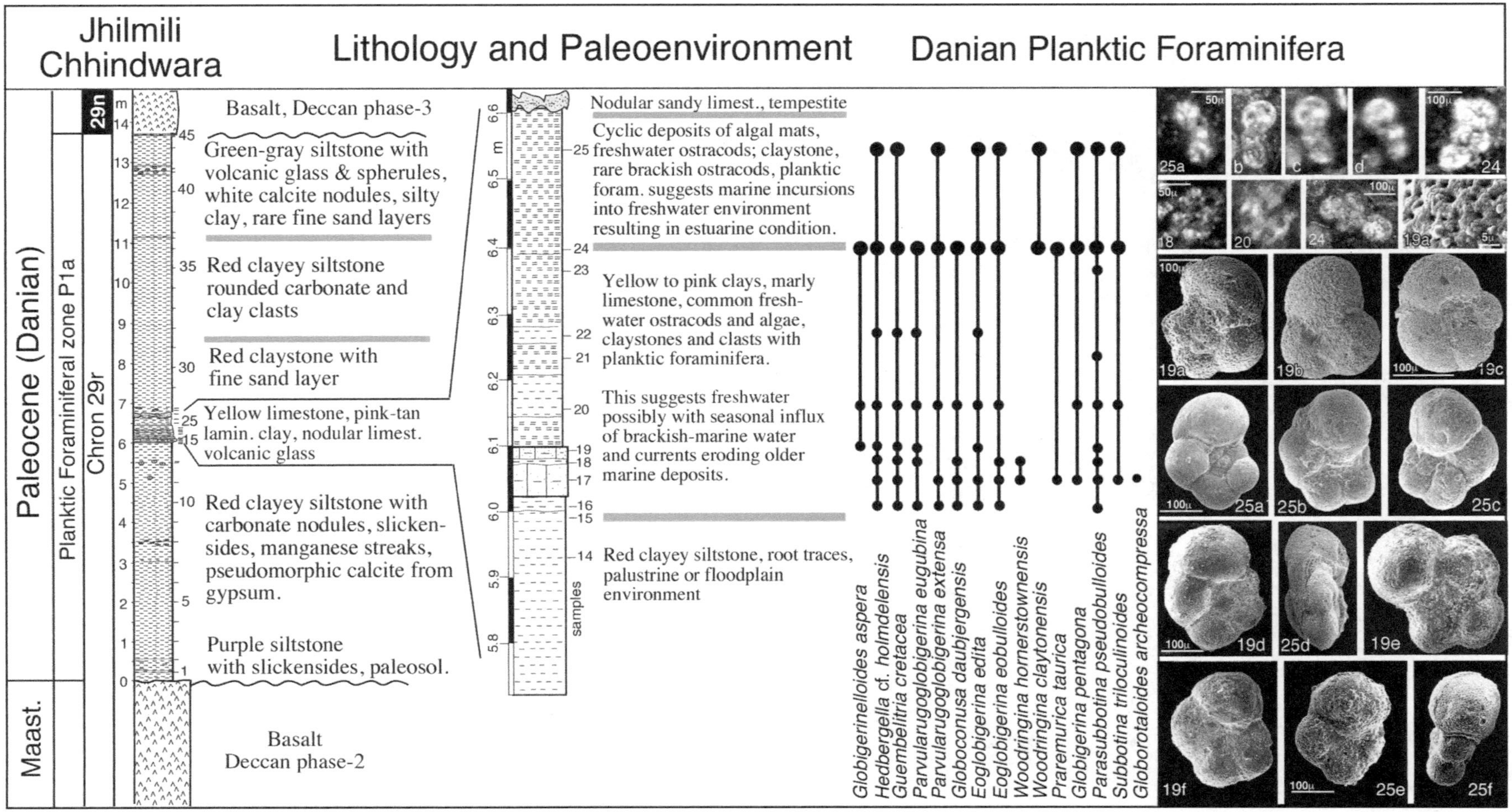

Figure 6. Early Danian intertrappean bed at Jhilmili, Chhindwara District, Madhya Pradesh, shows temporary marine incursion, with earliest Danian zone P1a planktic foraminifera indicating a seaway existed into central India. Larger circles mark samples with common foraminifera. Small planktic foraminiferal species are illustrated from thin section micrographs. Scale bar is the same for each row, unless otherwise indicated; numbers refer to samples. Top row images: 24, 25a–d—*Parasubbotina pseudobulloides*; 18, 20—*Parvularugoglobigerina eugubina*; 19a—scanning electron microscope images of cancellate spinose wall texture with calcite overgrowth of chambers in: 19a, 19b, 19c—*Subbotina triloculinoides*; 25a–c, 19d—*Parasubbotina pseudobulloides*; 19e, 25d—*Hedbergella* cf. *holmdelensis*; 19f—*Globigerina pentagona*; 25e—*Subbotina triloculinoides*; 25f—*Globigerinelloides aspera*. Figure is modified from Keller et al. (2009b).

Figure 7. Photos of two Deccan basalt quarries (Government and Church Quarries) in the Rajahmundry area showing thick intertrappean beds with blowup of 1-m-thick unit 3b at the Church Quarry with earliest Danian zone P1a planktic foraminifera directly overlying the main phase 2 of Deccan volcanism.

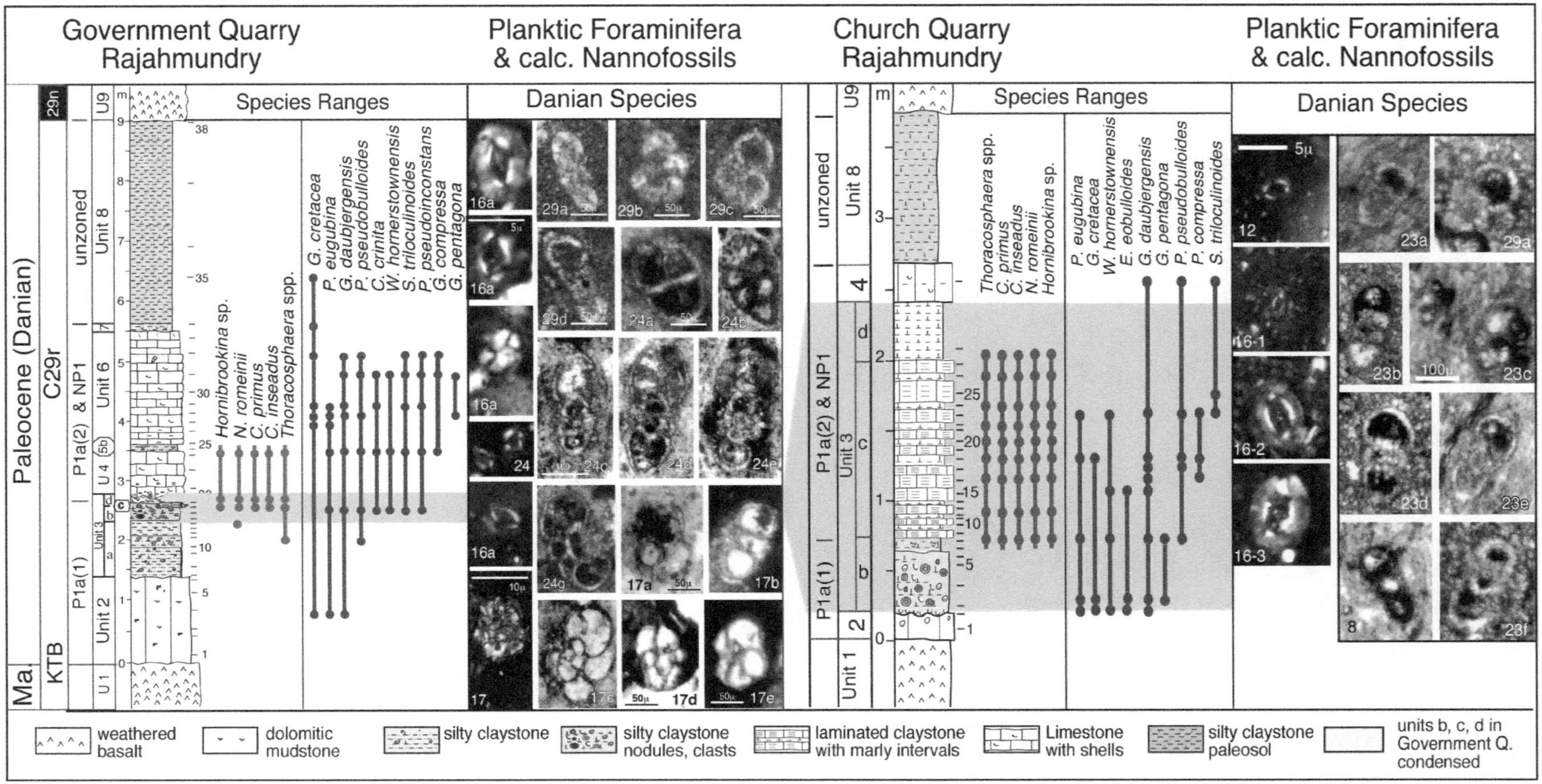

Figure 8. Government and Church Quarry biostratigraphy and species ranges of planktic foraminifera and calcareous nannofossils illustrated by photomicrographs. KTB—K-T boundary. Government Quarry (numbers keyed to samples in lithologic column): Calcareous nannofossils (scale bar = 5 µm for the first five species): 16a-1—*Cruciplacolithus* cf. *C. inseadus*; 16a-2—*Cruciplacolithus primus*; 16a-3—*Octolithus multiplus*; 24—*Neobiscutum romeinii*; 16a-4—*Hornibrookina* sp.; 17—*Thoracosphaera* spp. fragment. Planktic foraminifera: 29a, b, 17d, e—*Parvularugoglobigerina eugubina*; 29c, 24a, 17a—*Globoconusa daubjergensis*; 24b—*Chiloguembelina* sp.; 29d—*Globanomalina compressa*; 24c, d, g, 17c—*Parasubbotina pseudobulloides*; 24e—*Subbotina triloculinoides*. Church Quarry species (numbers keyed to samples in lithologic column): Calcareous nannofossils (scale bar = 5 µm): 12—*Neobiscutum romeinii*; 16-1—*Cruciplacolithus*? cf. *C. primus* (small morphotype); 16-2—*Cruciplacolithus* cf. *C. inseadus*; 16-3—*C. primus*. Planktic foraminifera (scale bar = 100 µm): 8—*Parvularugoglobigerina eugubina*?; 23a, 29a—*Globoconusa daubjergensis*; 23b, c, e, f—*Parasubbotina pseudobulloides*; 23d—*Subbotina triloculinoides*. Figure is modified from Keller et al. (2008).

all outcrops, as shown for the Government and Church Quarries (Figs. 7 and 8).

Sediments above the lower trap (phase 2) basalt begin with dolomitic mudstone with clasts containing small, earliest Danian planktic foraminifera indicative of zone P0 or base of zone P1a (Figs. 3 and 8). Above this interval, sediments vary from silty claystone with rare shells and foraminifera to limestones with common shells, calcareous nannofossils (early Danian NP1), and foraminifera indicative of zone P1a (Fig. 8). The lower and upper parts (units 3a and 6) of this interval are well represented in the Government Quarry, whereas the middle part (units 3b, 3c, and 3d) is well developed in the Church Quarry. The upper intertrappean sediments consist of paleosol directly underlying the upper trap phase 3 basalt.

Sediments and planktic foraminifera thus indicate a sea-level transgression at the base of the Danian, followed by fluctuating estuarine to inner neritic environments. The return to terrestrial deposition occurred prior to phase 3 eruptions (Fig. 8; Keller et al., 2008). Biostratigraphic data from the Rajahmundry quarries represent the first direct evidence of the earliest Danian evolution in planktic foraminifera immediately above the lower trap phase 2 megaflows. This indicates that the mass extinction must have coincided with the longest megaflows at the end of phase 2 Deccan volcanism, but this hypothesis cannot be tested in the Rajahmundry area because the prevailing sandy estuarine environment prior to Deccan phase 2 megaflows lacks planktic foraminifera. The best chance for a marine record is in deep wells of the Krishna-Godavari Basin seaward from Rajahmundry.

Deep Wells of the Krishna-Godavari Basin

Earlier studies of the Krishna-Godavari Basin, ~75 km east of Rajahmundry toward the Bay of Bengal, reported marine microfossils spanning the K-T boundary in deep wells (2500–3500 m below surface) drilled by India's Oil and Natural Gas Corporation (ONGC; Figs. 4A and 4B). Although the precise position of the K-T boundary event could not be determined by these studies, the information presented was promising. In most wells, up to eight lava flows were observed correlative with Deccan phase 3 and phase 2 Deccan Traps of Rajahmundry (Fig. 9). Most basalt layers are separated by relatively thin beds of sediments, and one thick sediment layer separates lower and upper basalt flows with reported Danian planktic foraminifera (Govindan, 1981; Jaiprakash et al., 1993; Raju et al., 1994, 1995), dinoflagellates (Mehrotra and Sargeant, 1987), calcareous nannoplankton (Saxena and Misra, 1994), and palynomorphs (Prasad and Pundeer, 2002).

In collaboration with ONGC scientists, we set out to study 12 wells in the Krishna-Godavari Basin (Fig. 4B) to test two hypotheses: (1) Deccan volcanism may have been the main cause of the K-T boundary mass extinction, possibly as a result of the giant phase 2 megaflow eruptions and their environmental consequences (the Meghalaya section provides support for this hypothesis although no direct link to the Deccan Traps; Fig. 5; see "Um Sohryngkew, Meghalaya" section); and (2) the long delayed post–K-T boundary recovery was the result of Deccan phase 3 volcanism in C29N. Results were published in Keller et al. (2011a, 2012), and key results are summarized next.

Lithology, Depositional Setting, and Megaflows

In the ONGC deep wells from the Krishna-Godavari Basin, sediment deposition occurred in a middle shelf environment that deepened eastward (~100–150 m; Jaiprakash et al., 1993; Raju et al., 1994; Keller et al., 2011a). The K-T boundary transition separates two phases of Deccan volcanism at current well depths between 2500 m and 3500 m below the surface. The deep burial is due to the high sediment input and rapid subsidence after volcanism ended (e.g., Raju et al., 1995, 1996; Misra, 2005; Raju, 2008). In most wells, four phase 2 and three to four phase 3 megaflows are present and can be correlated across the basin based on lithology, biostratigraphy, and well log data (Fig. 9). These basalt megaflows are generally 5 m to 15 m thick, except for two wells, where phase 3 megaflows are 60 m thick (Fig. 9; CTP-A and RZL-A). Intertrappean sediments vary between 5 m and 15 m but occasionally reach 50 m (wells CTP-A, RZL-A). These anomalously thick basalt megaflows and intertrappean beds are likely due to infilling of local topographic lows. In contrast, near the paleoshoreline (Rajahmundry area), megaflows are stacked with no intertrappean sediments, except between Deccan phase 2 and phase 3 (Fig. 4C; Keller et al., 2008). Self et al. (2008) suggested that the overall pattern of the basalt megaflows is consistent with sheet flows of very large-volume pahoehoe flow fields.

Lithostratigraphy

Resistivity and gamma values show pronounced signals for the megaflows in all Krishna-Godavari Basin wells (Fig. 9), which are also correlated based on biostratigraphy (Fig. 3; Keller et al., 2011a). Resistivity for normal sandstones, siltstones, and shale varies between 1 and 6 Ωm, unless they are hydrocarbon bearing. In contrast, resistivity values for the basalts vary anywhere from 50 Ωm to >200 Ωm, depending on the amount of sediments incorporated as a result of erosion at the base and top of megaflows. In all Krishna-Godavari Basin wells, e-logs show high- to very high-resistivity peaks against the basalt flows, whereas some basalt beds are less distinct due to mixed sediments. Gamma logs show significantly lower values in basalt flows relative to the intertrappean clastic sediments (Fig. 9). The cores and drill cuttings from these megaflows are dark gray to green-gray in color, very hard, and compact.

Biostratigraphy

Planktic foraminiferal assemblages of the Krishna-Godavari Basin average between 20 and 30 species, which is typical diversity in middle neritic environments (Keller and Abramovich, 2009). Variations in species richness are due to preservation (dissolution effects) and availability of sample size, with cored intervals yielding the most reliable data (Fig. 10). Faunal assemblages below phase 2 megaflows are typical of zones CF1–CF3. In the absence of index species *Plummerita hantkeninoides* and

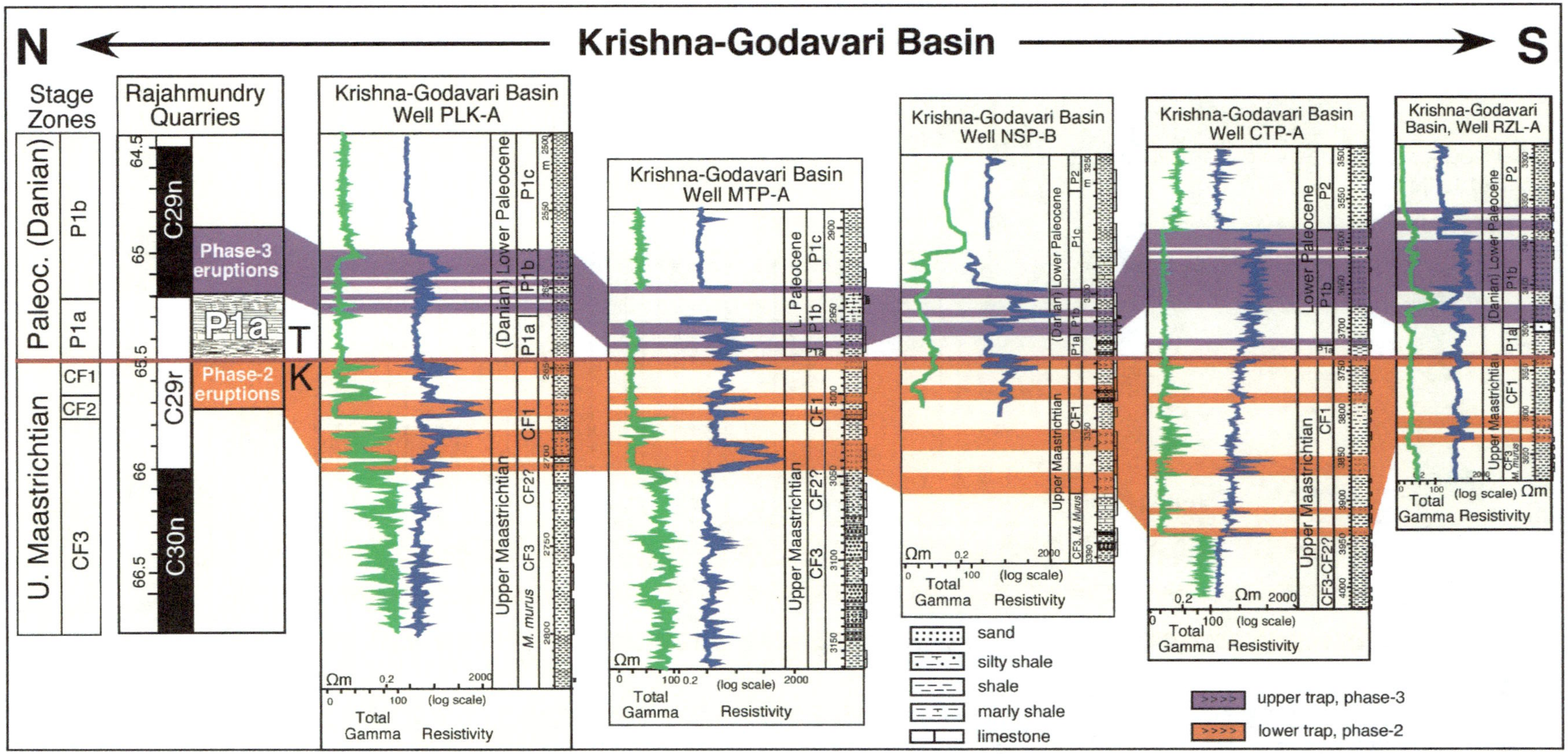

Figure 9. Correlation of Krishna-Godavari Basin wells based on biostratigraphy, Deccan lava megaflows, and well-log data (gamma and resistivity). Note that Deccan phase 2 and phase 3 each contains three to four basalt beds that represent Earth's longest megaflows. All megaflows are separated by relatively thin sandy (5–15 m) intertrappean beds. These megaflows correlate with the lower and upper traps in Rajahmundry quarries, where no intertrappean beds are present, except between phase 2 and phase 3 volcanic episodes (modified from Keller et al., 2011a). KTB—K-T boundary.

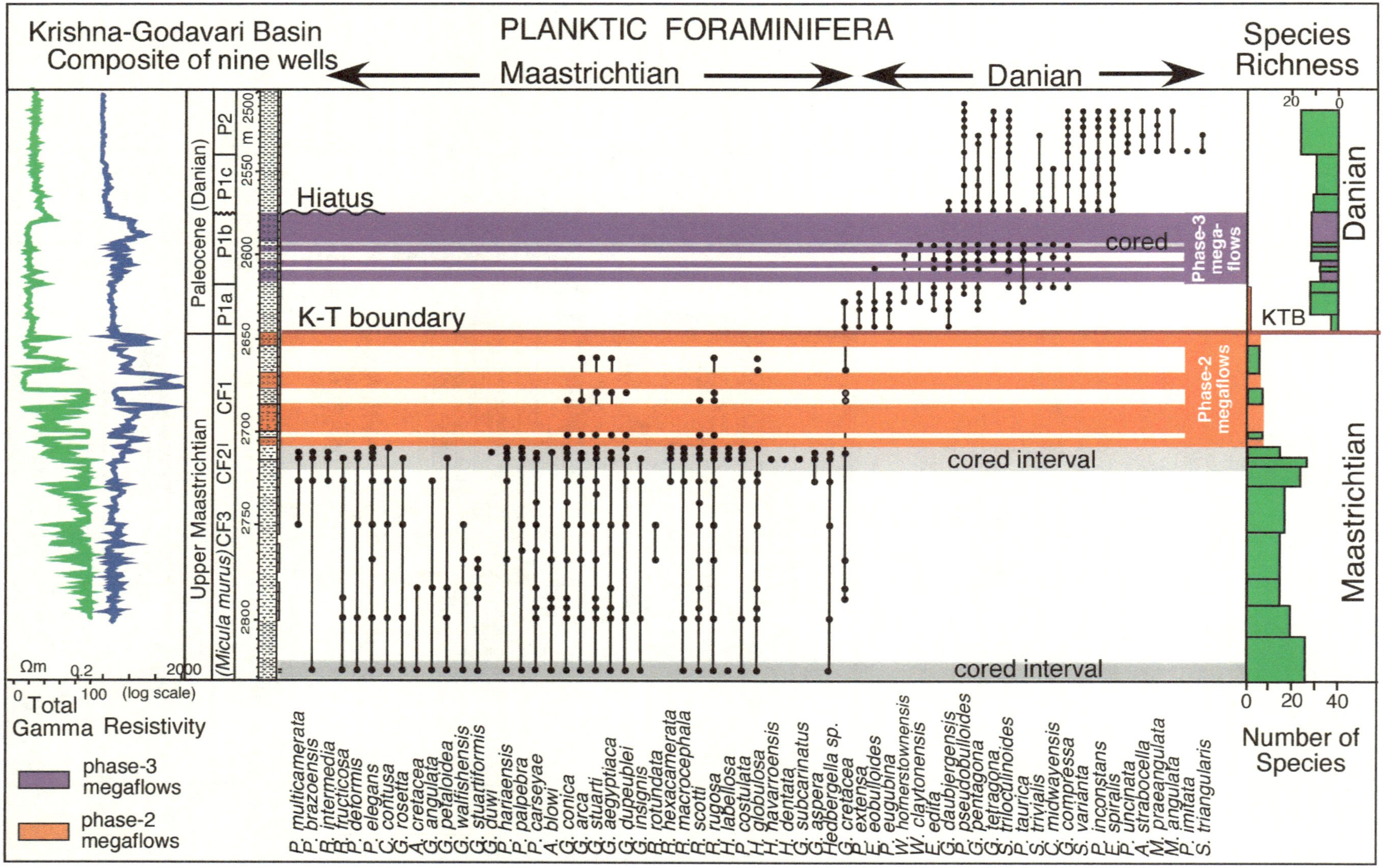

Figure 10. Composite species ranges of nine wells in the Krishna-Godavari Basin plotted against biostratigraphy and phase 2 and phase 3 lava flows of the PLK-A well. Gray shadings mark cored intervals that record the most reliable species richness data. Note the mass extinction began with 50% species extinctions before the first megaflow eruptions, followed by another 50% drop after the first megaflow, and was complete by the last megaflow at the K-T boundary. Figure is modified from Keller et al. (2011a).

Gansserina gansseri, these zones cannot be differentiated. However, zones CF1 and CF2 are nearly equivalent to the *Micula prinsii* nannofossil zone, and CF3 is nearly equivalent to the *Micula murus* zone, both of which have been identified in wells within and below phase 2 megaflows, respectively (Saxena and Misra, 1994, 1995; von Salis and Saxena, 1998).

Paleomagnetic and radiometric analyses place phase 2 megaflows in C29R below the K-T boundary (e.g., Knight et al., 2003, 2005; Chenet et al., 2007, 2008) correlative with zones CF1–CF2 and the *Micula prinsii* zone. Based on this correlation, Deccan phase 2 eruptions may span zones CF1–CF2, but the megaflows and intertrappean sediments of the Krishna-Godavari Basin were most likely deposited near the end of phase 2 and more specifically in zone CF1, which spans the last 160 k.y. of the Maastrichtian (Fig. 3). Based on magnetic analysis of individual lava flows, Chenet et al. (2007, 2008, 2009) and Courtillot and Fluteau (this volume) suggest that Deccan phase 2 may have occurred over as little as 100 k.y., with active eruptions over perhaps as little as 10 k.y. Possible support for such short-term deposition comes from Font et al. (2011), who identified low magnetic susceptibility in a short interval below the K-T boundary at Bidart, France, which they attributed to HCl-rich volcanic gas and aerosols derived from Deccan plumes. Based on sediment accumulation rates, this short interval marks the top 30,000 yr of zone CF1 at Bidart, where most planktic foraminifera are dissolved due to ocean acidification likely related to Deccan phase 2 (Font et al., this volume). If this interpretation is correct, then the most devastating effects of Deccan phase 2 volcanism were concentrated over a very short time interval at the end of the Maastrichtian.

Intertrappean sediments between phase 2 and phase 3 megaflows contain the earliest Danian zone P1a assemblage, similar to Meghalaya, Jhilmili, and Rajahmundry quarries (Figs. 5, 6, 8, and 10; Keller et al., 2008; Malarkodi et al., 2010). This demonstrates that the mass extinction of Cretaceous species was coeval with the underlying Deccan phase 2, suggesting a cause-effect relationship. Evolution of Danian species in the intertrappean sediments above phase 2 follows the same pattern as observed globally, with small, simple morphologies that reflect adaption for survival in highly stressed environments (e.g., Luciani, 2002; Coccioni and Luciani, 2006; Pardo and Keller, 2008). The fact that these early Danian assemblages mirror the global evolutionary pattern demonstrates that environmental conditions after the main Deccan phase 2 remained stressed in India and globally.

Phase 3 intertrappean assemblages are similar to zone P1a, except for the disappearance of the zone P1a index species *P. eugubina*, which indicates deposition in zone P1b (Fig. 3). Species remained dwarfed, and abundances of early zone P1a species decreased (e.g., *Parvularugoglobigerina extensa*, *Eoglobigerina edita*, *Woodringina claytonensis*, *Woodringina hornerstownensis*, *Eoglobigerina eobulloides*). This reflects continued high-stress conditions, as also observed globally (Pardo and Keller, 2008). After phase 3 volcanism, planktic foraminifera developed larger morphologies (>150 μm) and increased diversity globally (Fig. 10). Thus, the long delay in the global ecosystem recovery in the early Danian was likely due to climate change, increased weathering of Deccan basalts, and terrestrial runoff prolonging environmental stress conditions through P1a, followed by intense stress conditions associated with Deccan phase 3 (see Punekar et al., this volume).

Mass Extinction—Volcanic Phase 2

Results from nine deep wells analyzed in the Krishna-Godavari Basin provide strong evidence that the end-Cretaceous mass extinction was directly related to the main phase 2 of Deccan volcanism (Fig. 10). More specifically, we can conclude that this mass extinction began shortly before the eruption of the first of four megaflows that mark the world's largest and longest lava flows. A 4 m cored interval just below the first megaflow shows normal diversity near the base, followed by a rapid and permanent decrease in diversity of 50% toward the top of the core (Fig. 11; Keller et al., 2011a). Diversity and extinctions in intertrappean sediments, summarized based on nine wells (Fig. 10), show that another 50% species disappeared after the first megaflow, leaving just seven survivors. No recovery occurred between the next three megaflows, and the mass extinction of Cretaceous species was complete by the last megaflow at the K-T boundary.

This pattern of extinctions seems unequivocal in the rapid demise of planktic foraminifera. What is less clear is whether even the seven "survivors" were actually survivor species or simply reworked shells. Five of the seven species are thick-walled and robust globotruncanids, which globally tend to disappear early because of their environmental sensitivities. Their presence in the intertrappean may be an artifact of erosion and transport. Also uncertain is how rapid the four megaflow eruptions followed each other. Based on flow-by-flow magnetic stratigraphy, Courtillot and Fluteau (this volume) suggest that single eruptive Deccan events could have been emplaced in less than a decade (Chenet et al., 2009). The thickness of intertrappean sediments between the four megaflows generally ranges from 5 m to 15 m (Fig. 10), although occasionally wells with thicker deposits have been observed in the Krishna-Godavari Basin possibly due to topographic variations (Keller et al., 2011a). These intertrappean sediments are generally sandy and could have been deposited by either normal sedimentation over relatively short time intervals, or current transport from erosion of nearby topographic highs. In either case, deposition of the four megaflows and intertrappean sediments that mark the mass extinction just below the K-T boundary record a time interval possibly as short as a few thousand to tens of thousands of years.

Strong carbonate dissolution effects are observed in foraminifera from the intertrappean and infratrappean sediments of the Krishna-Godavari wells, as also noted in the Meghalaya section (Fig. 5). This indicates ocean acidification due to volcanic CO_2 and SO_2 resulting in decreased pH of ocean waters. Similar strong dissolution effects attributed to ocean acidification as a result of Deccan volcanism are also observed over a short interval (~30,000 yr) below the K-T boundary at Bidart, southern

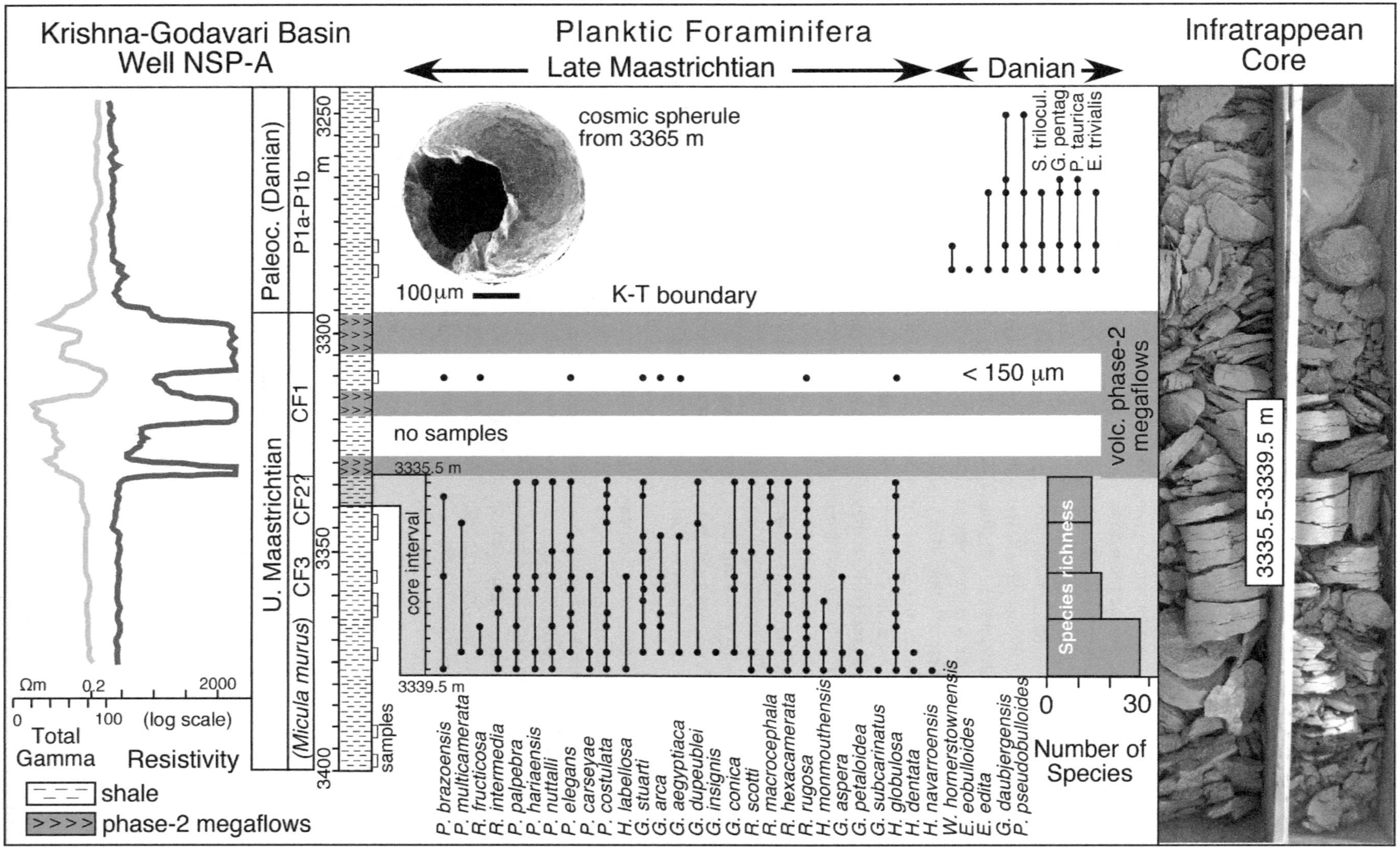

Figure 11. Onset of K-T boundary mass extinction in a 4 m core of India's Oil and Natural Gas Corporation (ONGC) well NSP-A. This core interval shows that 50% of the species disappeared prior to the first megaflow in the Krishna-Godavari Basin, which suggests active phase 2 volcanism in the main Deccan province prior to the megaflows that reached Rajahmundry and ended with the mass extinction of nearly all planktic foraminifera. A cosmic spherule was found 25.5 m below the cored interval, though its significance is unclear (from Keller et al., 2011a).

France (Font et al., 2011). Dissolution effects below the K-T boundary mass extinction are well known but have garnered little attention to date because their cause remained a mystery. A systematic survey remains to be done to evaluate this potential link to Deccan volcanism.

How rapid was this mass extinction? In deep-sea sections globally, the mass extinction occurred over a very short time interval, which in condensed sections appears instantaneous. This is perhaps the main reason an instantaneous extraterrestrial cause for the mass extinction is the preferred interpretation by many scientists. Large igneous eruptions such as the Deccan Traps, Siberian Traps, and the Central Atlantic magmatic province are commonly believed to have occurred over one or more million years. This perception is erroneous, and better dating techniques (U-Pb) for Central Atlantic magmatic province volcanism reveal a rapid mass extinction associated with the first major eruption pulse lasting just a few tens of thousands of years (Blackburn et al., 2013). Other indicators for voluminous short-duration eruptions come from weathering rates, red bole layers, absence of intertrappean sediments in the Deccan Traps, and flow-by-flow magnetic stratigraphy (Chenet et al., 2007, 2008; Courtillot and Fluteau, this vol). Absolute dating of Deccan basalt flows by U-Pb methods similar to that conducted for the Central Atlantic magmatic province still remains to be done.

Climatic and Environmental Effects of Deccan Volcanism

Upper Maastrichtian to early Danian paleoclimatic effects of Deccan volcanism can be assessed based on the chemical index of alteration (CIA), which is commonly used to estimate the intensity of weathering related to climatic conditions (Nesbitt and Young, 1982, 1989). Late Maastrichtian CIA values of zone CF3 are relatively high (mean 69), indicating dry seasonal conditions (Fig. 12). In the 4 m of infratrappean sediments below the first phase 2 megaflows, CIA values gradually decrease (mean 62). This suggests that significant volcanic activity had initiated well before the first megaflows in the Krishna-Godavari Basin. During this time, climate turned increasingly arid, correlative with the onset of the mass extinction in planktic foraminifera.

Minimum CIA values (mean 53) are reached in the intertrappean sediments of phase 2 (zone CF1) and in the earliest Danian intertrappean sediments (zone P1a; Fig. 12), which indicates arid conditions. In the uppermost interval of phase 3 (zone P1b), CIA values (mean 76) are similar to those measured in sediments below phase 1 volcanism of the early late Maastrichtian (Keller et al., 2012). This suggests a return to more humid conditions, possibly linked to reduced or intermittent volcanic activity.

Similar arid to semiarid conditions with seasonal wet and dry cycles have been observed in smectite-enriched paleosols of the early Danian intertrappean sediments between volcanic phase 2 and phase 3 at Rajahmundry and Jhilmili (Keller et al., 2008, 2009a), in intertrappean sediments at Anjar, Kutch, close to the Deccan volcanic province (Khadkikar et al., 1999), and in Meghalaya, 800 km from the Deccan volcanic province (Gertsch et al., 2011a). Localized arid conditions surrounding the Deccan volcanic province are interpreted as "mock aridity" resulting from volcanically induced xeric conditions and extreme geochemical alkalinity in a regionally more humid climate (Harris and Van Couvering, 1995; Khadkikar et al., 1999; Gertsch et al., 2011a).

Na/K and K/(Fe + Mg) ratios reveal the balance between detrital and volcanogenic input, which may also reflect climate conditions (Gertsch et al., 2011a; Keller et al., 2012). In the late Maastrichtian zones CF3 and CF2–CF1 (intervals A and B), below phase 2 megaflows (Fig. 12), sediments show steady Na/K ratios close to mean shale values, whereas K/(Fe + Mg) ratios are closer to mean Deccan basalt values. In the latest Maastrichtian zone CF1 and early Danian zone P1a (intervals C and D), these ratios reflect a dominant basaltic source consistent with deposition of the main phase 2 megaflows in zone CF1. In the early Danian phase 3 megaflows (interval E), Na/K decouples from K(Fe + Mg) ratios, which may reflect increasing humidity, as suggested by the CIA index, and increased weathering of Fe-Mg–enriched minerals.

Additional information can be gained from weathering of basalts. During the early late Maastrichtian (zone CF3, interval A), the Ca/Na and Mg/Na ratios are stable and close to mean shale values, typical for rivers draining silicate rocks, such as granite, gneiss, and shale (Gertsch et al., 2011a). However, just below the onset of phase 2 megaflows (interval B), both proxies reach values normally recorded in basaltic river waters (Dessert et al., 2003). These variations are interpreted to result from chemical weathering of Ca- and Mg-rich volcanic rocks (Fig. 12). This suggests that the Krishna-Godavari Basin was part of the drainage of the Deccan volcanic province and that significant basalt outcrops were already present prior to phase 2 megaflows. Surprisingly, both ratios decrease considerably in the CF1 intertrappean sediments of phase 2, especially in the lower part, suggesting reduced weathering of basalt linked to the increased aridity (mock aridity) evident in the CIA proxy (Fig. 12). During phase 3, increased ratios reflect a return to more humid conditions and enhanced weathering.

HIGH-STRESS ENVIRONMENTS: *GUEMBELITRIA* BLOOMS

Guembelitria blooms are well-known indicators of high-stress environments first documented in the aftermath of the K-T boundary mass extinction, notably in the Tunisian sections of El Kef and Elles and subsequently worldwide (review in Keller and Pardo, 2004). More recently, *Guembelitria* blooms were also documented in the latest Maastrichtian zone CF1 marking high-stress conditions in the ~160,000 k.y. preceding the K-T boundary mass extinction (review in Pardo and Keller, 2008), during the early late Maastrichtian C30n (67.4 Ma), and early Danian C29n (review in Punekar et al., this volume). These three bloom intervals correlate with the three main phases of Deccan volcanism and climate warming (Fig. 13).

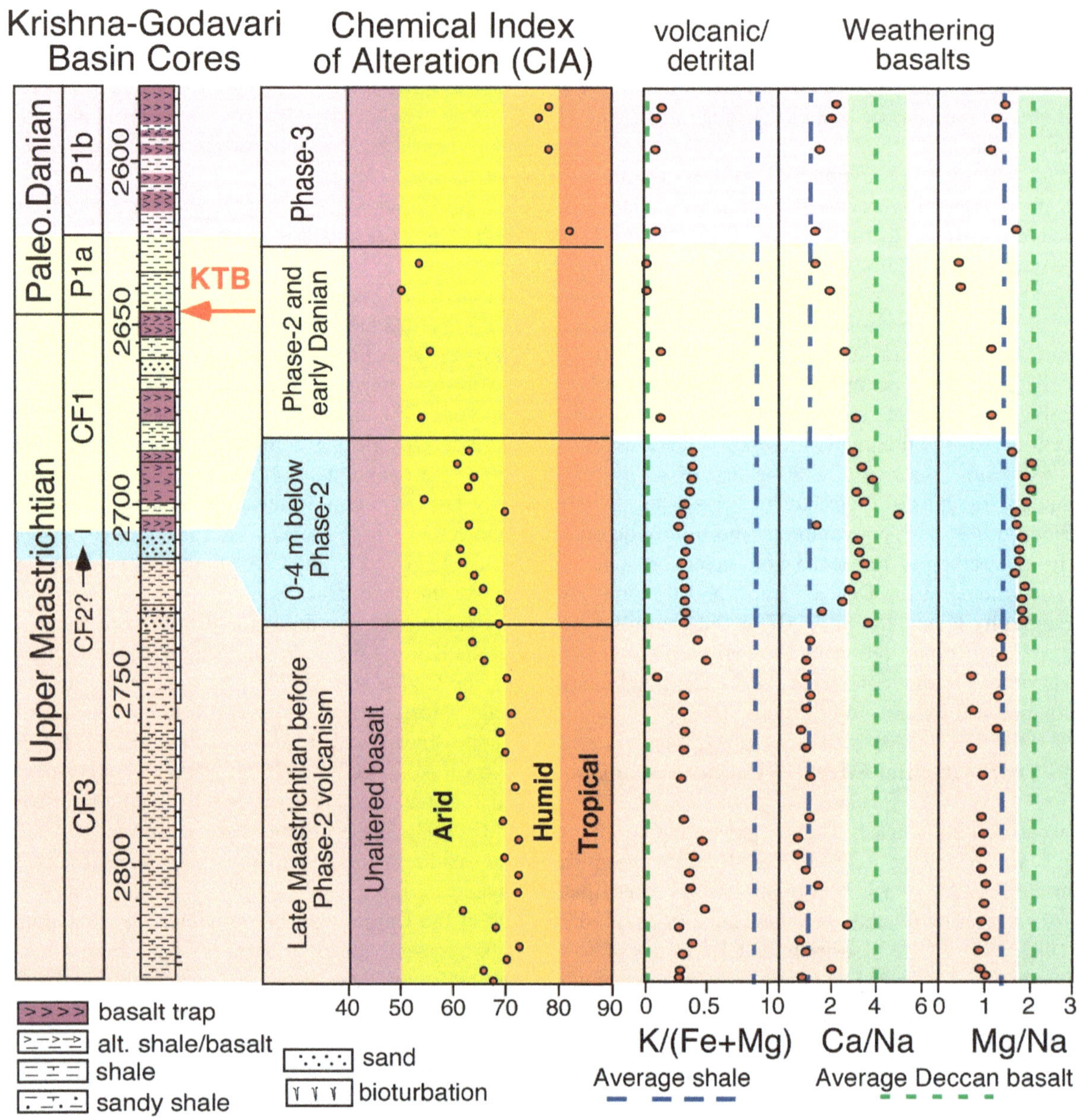

Figure 12. Geochemical proxies of volcanism (weathering of basalt, volcanism vs. detritism) and the chemical index of alteration (CIA) based on major-element geochemistry. Note the change in the weathering index and CIA from humid to arid with the onset of Deccan main phase 2 in the late Maastrichtian and the return to normal humid tropical conditions after the last Deccan phase 3 in the early Danian. Blue dashed line—mean average shale (Wedepohl, 1971); green dashed line—Deccan mean average basalt (Crocket and Paul, 2004); green shaded field—range of Deccan basalt river values (Dessert et al., 2003); KTB—K-T boundary. Figure is modified from Keller et al. (2012).

Phase 2 Volcanism

In Meghalaya, India, *Guembelitria* blooms reach 95% of the foraminiferal population coincident with Deccan phase 2, which indicates high environmental stress as a direct result of Deccan volcanism (Fig. 5). In addition, the giant test size of *Guembelitria* suggests superstress conditions surrounding the Deccan volcanic province. The large *Guembelitria* test size appears to be indirectly related to Deccan volcanism–induced climate warming and associated effects. On the basis of mineralogical, geochemical, and trace-element analyses of the Meghalaya (Um Sohryngkew) section, Gertsch et al. (2011a) argued that intense physical and chemical weathering of the continental areas was likely due to acid rains as a result of high SO_2 emissions during the latest Maastrichtian when Deccan phase 2 volcanic activity reached its maximum (Self et al., 2006). During this time, high terrestrial influx of silt and nutrients led to eutrophication and muddy waters, which resulted in reduced penetration of sunlight

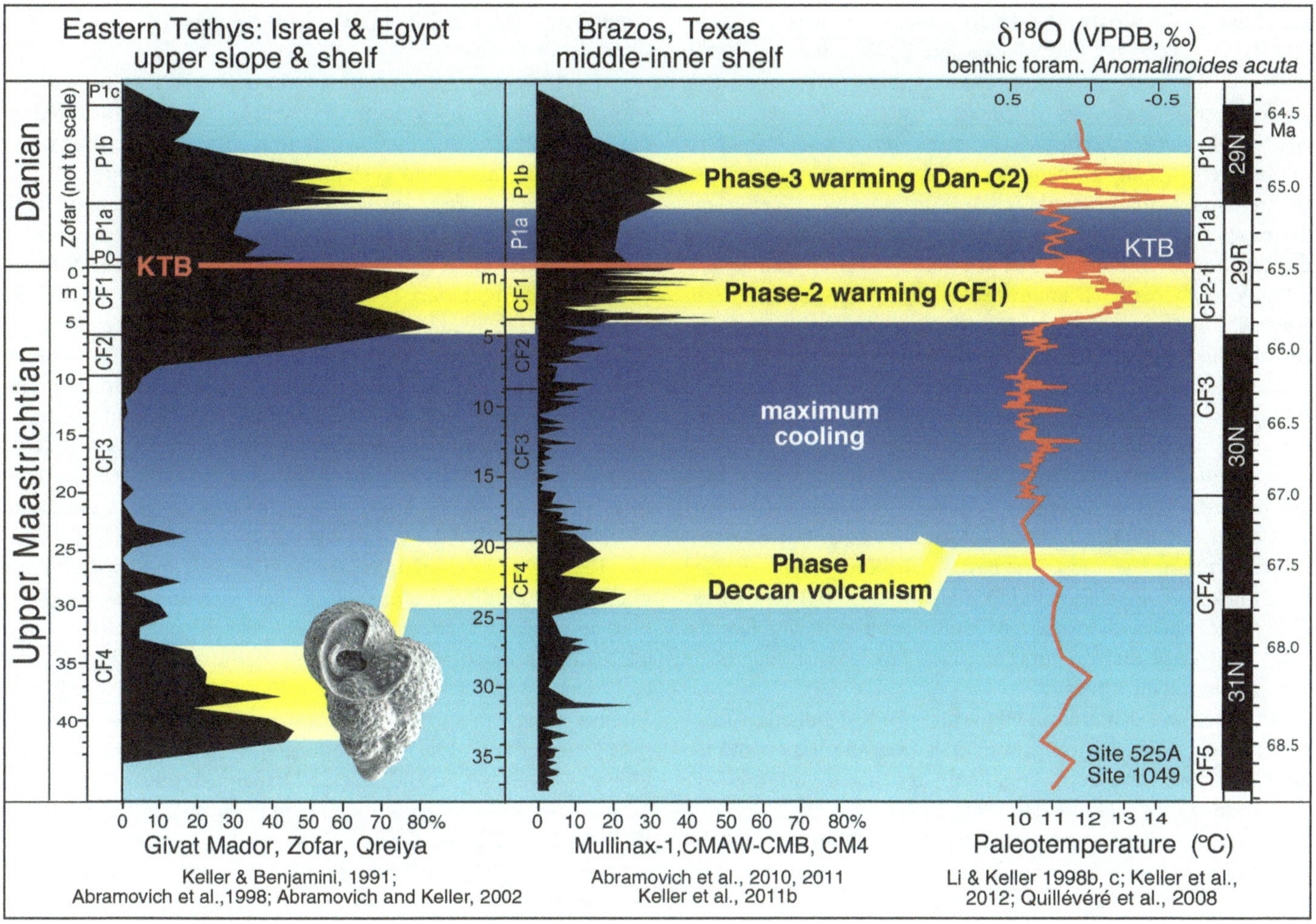

Figure 13. *Guembelitria cretacea* blooms recorded from the eastern Tethys and Texas correlative with the three phases of Deccan volcanism. Deccan phase 2 and phase 3 are associated with times of rapid warming in zones CF1 and P1b. Insufficient data are available to evaluate the environmental effects of the initial volcanic phase 1. KTB—K-T boundary; VPDB—Vienna Peedee belemnite. Stable isotope data are from Li and Keller (1998c) and Quillévéré et al. (2008). Faunal data are from Keller and Benjamini (1991), Keller et al. (2011b), and Abramovich et al. (2011).

in the shallow Um Sohryngkew area. Similar volcanism-induced climate change and resultant high physical and chemical erosion leading to high ecosystem stress were inferred for the Permian-Triassic boundary mass extinction (Algeo and Twitchett, 2010).

At the K-T boundary in Meghalaya, *Guembelitria cretacea* may have adapted to these high-stress conditions by inflating chambers, resulting in higher buoyancy to live near the water-atmosphere interface. Acidic ocean waters inhibited $CaCO_3$ production, particularly of thicker-shelled larger species living in subsurface water, which favored blooms of the thin-shelled *Guembelitria*, the only species dwelling in the uppermost surface waters (Abramovich et al., 2010). This may explain the near absence of all other species, which account for only 2%–5% of the total foraminiferal population during this time (Fig. 5). *Guembelitria* blooms, though not the morphologic gigantism, have been observed throughout the Tethys, and Atlantic and Indian Oceans (e.g., Pardo and Keller, 2008; Keller and Abramovich, 2009). *Guembelitria* blooms are thus excellent indicators of high-stress environments particularly related to Deccan volcanism.

Phase 3 Volcanism—Delayed Biotic Recovery

Guembelitria blooms prevailed in the aftermath of the K-T boundary mass extinction worldwide, reflecting the continued high-stress environment in the early Danian zones P0 and P1a, but they decreased in the upper part (P1a[2]), suggesting incipient recovery (Pardo and Keller, 2008). Throughout this time, species diversity is very low (<12 species), characterized by simple species morphology and dwarfed size (<63 μm; Keller and Abramovich, 2009). However, in zone P1b, correlative with phase 3 volcanism, large *Guembelitria* blooms reappear, marking the return to higher-stress conditions.

The last phase 3 of Deccan volcanism began near the base of C29N, with the largest eruptions resulting in three to four megaflows in the Krishna-Godavari Basin (Knight et al., 2003, 2005; Baksi, 2005; Jay and Widdowson, 2008; Jay et al., 2009). Intertrappean sediments contain typical early Danian zone P1b foraminiferal assemblages dominated by *Guembelitria cretacea*, *Globoconusa daubjergensis*, and biserial species, which are observed globally (Punekar et al., this volume), and species sizes remained small (generally <100 µm; review in Keller and Abramovich, 2009). In India, no major differences are observed between faunal assemblages of the three intertrappean layers, and only a few earlier Danian species disappeared (e.g., *Parvularugoglobigerina eugubina*, *Parvularugoglobigerina longiapertura*). This suggests that high-stress environmental conditions persisted but that sufficient time elapsed between eruptions to permit environmental recovery, thus avoiding the extinctions evident in volcanic phase 2. Alternatively, these early Danian species, which evolved during high-stress conditions, may have been primed for survival, unlike the highly specialized species of the late Maastrichtian. Similar to volcanic phase 2, *Guembelitria* blooms of phase 3 are observed through the Tethys (Fig. 13; see also Punekar et al., this volume), indicating global high-stress environments but with less severe conditions than during Deccan phase 2. The age and duration of phase 3 volcanism cannot be estimated based on current data.

Full recovery in marine plankton after the K-T boundary mass extinction occurs in zone P1c, well after Deccan phase 3 and the Dan-C2 global warm event commonly associated with this volcanic event (Fig. 13; Quillévére et al., 2008), when diversity, species size, and species complexity increased in India and globally (Keller and Abramovich, 2009; Keller et al., 2011a; Punekar et al., this volume). The coincidence of planktic foraminiferal recovery, and by extension ecosystem recovery, with the last phase of Deccan volcanism suggests that the long delayed recovery after the mass extinction was likely due to Deccan volcanism.

CHICXULUB IMPACT—LINKED TO MASS EXTINCTION?

The Chicxulub impact on Yucatan is commonly assumed to have caused the K-T boundary mass extinction. This interpretation was sanctified by the redefinition of the K-T boundary as impact ejecta associated with the mass extinction (Gradstein et al., 2004; Molina et al., 2006; Arenillas et al., 2006; Schulte et al., 2010). The problem with defining Chicxulub impact ejecta as K-T boundary in age is not just circular reasoning, but that it has actually prevented establishing the true age of this impact (Keller, 2011). Many researchers simply ignored the stratigraphic position of impact ejecta (impact glass spherules, Ir enrichments), unless positioned at or near the K-T boundary, which is considered the ultimate proof of Chicxulub's K-T boundary age. The many localities with older and younger impact spherule layers are ignored or dismissed with ad hoc explanations lacking evidence (Schulte et al., 2010). The age of the Chicxulub impact is the main basis for the decades of controversy. A brief review of the Chicxulub impact spherule evidence reveals this age problem.

North Atlantic, Caribbean, Central America

The biostratigraphic position of impact spherule layers of all known K-T boundary sequences is well known and illustrated in Figure 14. The age distribution pattern is complex, revealing multiple spherule layers and systematic reworking. Throughout the North Atlantic, Caribbean, Haiti, Cuba, Belize, Guatemala, and southern Mexico, multiple impact spherule layers are reworked in early Danian zone P1a sediments above an unconformity that spans the K-T boundary (Fig. 3; reviews in Keller et al., 2003a, 2003b). In a few localities of the North Atlantic and Caribbean, a single thin (2 cm to 15 cm) spherule layer overlies Maastrichtian sediments (e.g., Bass River, New Jersey, Ocean Drilling Program [ODP] Sites 999B, 1001B, 1049A–C, and 1259B) and has been hailed as the ultimate proof that the Chicxulub impact is K-T boundary in age (Olsson et al., 1997; Norris et al., 1999, 2000; MacLeod et al., 2007). This is the main basis for the redefinition of Chicxulub as K-T boundary age (Schulte et al., 2010).

Recent high-resolution stable isotope and quantitative planktic foraminiferal analyses of these sections reveal a K-T boundary hiatus that spans from the late Maastrichtian zones CF1–CF2 and sometimes also CF3 (NJ Bass River) through the early Danian zones P0 and most of P1a(1) (Keller et al., 2013). Moreover, impact spherules are reworked in early Danian sediments (upper zone Pla[1]) as indicated by abundant *Parvarugoglobigerina eugubina* and diverse Danian assemblages of up to 12 species, most of which do not evolve until ~100 k.y. after the K-T boundary. Despite the abundance of *P. eugubina* in the spherule layer of ODP Site 1049, previous investigators claimed that this species first appeared immediately above the spherule layer, indicating spherule deposition at the K-T boundary (Olsson et al., 1997; Norris et al., 1999, 2000). Despite the absence of zone CF1 (ODP Site 1049, New Jersey Bass River core), they claimed that the K-T boundary is complete, attributing the absence of the index species *Plummerita hantkeninoides* to environmental exclusion, although this species has been reported from Spain as well as ODP Site 1259 (Pardo et al., 1996; MacLeod et al., 2007). Recently, Schulte et al. (2010) hailed these NW Atlantic localities as proof that Chicxulub is K-T boundary in age. Given the fragmented record of the North Atlantic and Caribbean, as a result of the ancient Gulf Stream and complex tectonic activity (Fig. 14; Keller et al., 1993, 2013; Watkins and Self-Trail, 2005), this area is unlikely to yield complete sediment records across the K-T boundary. Such records are most likely obtained from NE Mexico and Texas.

NE Mexico and Texas

Impact Tsunami or Long-Term Deposition?

Throughout NE Mexico and along the Brazos River, Texas, dozens of outcrops reveal multiple impact spherule layers in sediments of zone CF1 that spans the last 160 k.y. of the Maastrichtian. These spherule layers are generally at the base of submarine channels and contain mixed shallow-water debris from nearshore

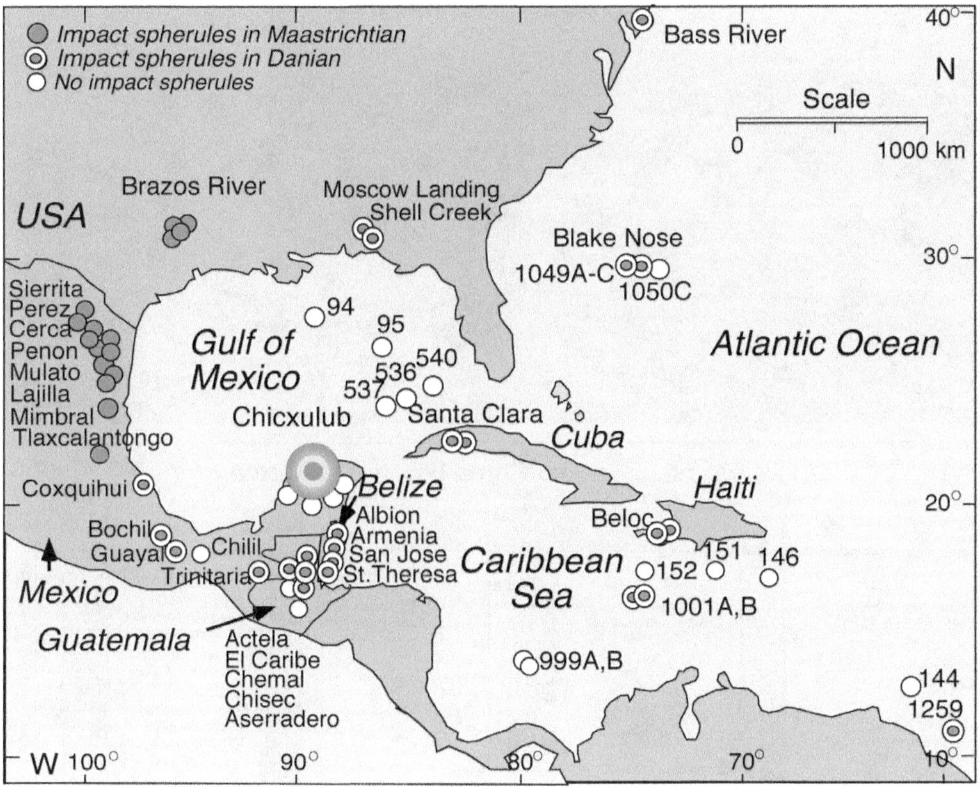

Figure 14. K-T boundary localities with impact spherules surrounding the Chicxulub impact crater on Yucatan with stratigraphic positions of impact spherules. Note that impact spherule layers in NE Mexico and Texas are in zone CF1, which spans the last 160 k.y. of the Maastrichtian. In all other localities, impact spherule layers are reworked in early Danian sediments of zone P1a(1) (lower part of *Parvularugoglobigerina eugubina* zone) and frequently overlie a hiatus at the K-T boundary.

areas (plant remains, shallow-water benthic foraminifera, glauconite) transported via submarine channels to depths between 500 and 1000 m in NE Mexico (Fig. 15; Adatte et al., 1996; Keller et al., 1997, 2002b, 2003a; Alegret et al., 2001; Schulte et al., 2003). Along the Brazos River, deposition occurred in a shallow middle to inner neritic environment (Keller et al., 2011b; Hart et al., 2011). In NE Mexico localities, additional impact spherule layers are present in marls and marly limestone several meters below the channel deposits (Fig. 15).

Clues to the depositional nature of these submarine channel deposits come from sedimentary features. In NE Mexico, two spherule layers at the base of the channel are separated by a 20 cm thick sandy limestone layer (e.g., El Mimbral, El Peñon, La Lajilla) and by sandstone layers in the Brazos sections (Keller et al., 2011b; Adatte et al., 2011). The limestone layer at El Peñon contains rare truncated J-shaped burrows infilled with spherules (Figs. 16A and 16B). Similar spherule-filled truncated burrows are observed near the base of the overlying sandstone unit (Keller et al., 1997; Ekdale and Stinnesbeck, 1998). The top unit of this channel deposit consists of alternating fine sand and laminated shale layers with multiple horizons of bioturbation by *Chondrites*, *Planolites*, *Ophiomorpha*, and *Zoophycos* (Ekdale and Stinnesbeck, 1998), and similar burrowed horizons are reported from the Brazos sections (Gale, 2006). Two discrete zeolite layers have been traced in the sandstone units across NE Mexico (Adatte et al., 1996). In addition, there are the thick impact spherule layers 4–9 m below the submarine channel deposits. All of these sedimentary features require long-term deposition.

However, deposition via impact-generated tsunami waves is the most common interpretation of these deposits because of the presence of impact spherules and their link to Chicxulub (e.g., Bourgeois et al., 1988; Smit et al., 1992, 1996; Heymann et al., 1998; Arenillas et al., 2006; Hart et al., 2011; Schulte et al., 2003, 2006, 2010). In this scenario, the sandy limestone between spherule layers is interpreted as large-scale tectonic disturbance (none has been documented), the J-shaped spherule-filled burrows are fluid-escape structures, the bioturbated horizons do not exist or are the result of downward burrowing from the K-T boundary, and the spherule layers below the K-T boundary were caused by tsunami and earthquake disturbance or liquefaction. They conclude, "A range of sedimentary structures and lack of evidence for ocean floor colonization within the clastic unit in northeastern Mexico indicate rapid deposition" (Schulte et al., 2010, p. 1215).

A more parsimonious interpretation compatible with the evidence is long-term deposition via normal sedimentation processes during a low sea level and subsequent rise, as documented from sections in NE Mexico and Texas (Adatte et al., 1996, 2011). In this scenario, impact spherules were eroded in nearshore environments at repeated intervals (explains shallow-water debris and multiple spherule layers) and transported via submarine channels into deeper water. More massive sandstone above the spherules may represent normal gravity flows or slumps along the slope of the Gulf of Mexico. The fine sand and laminated shale alternation

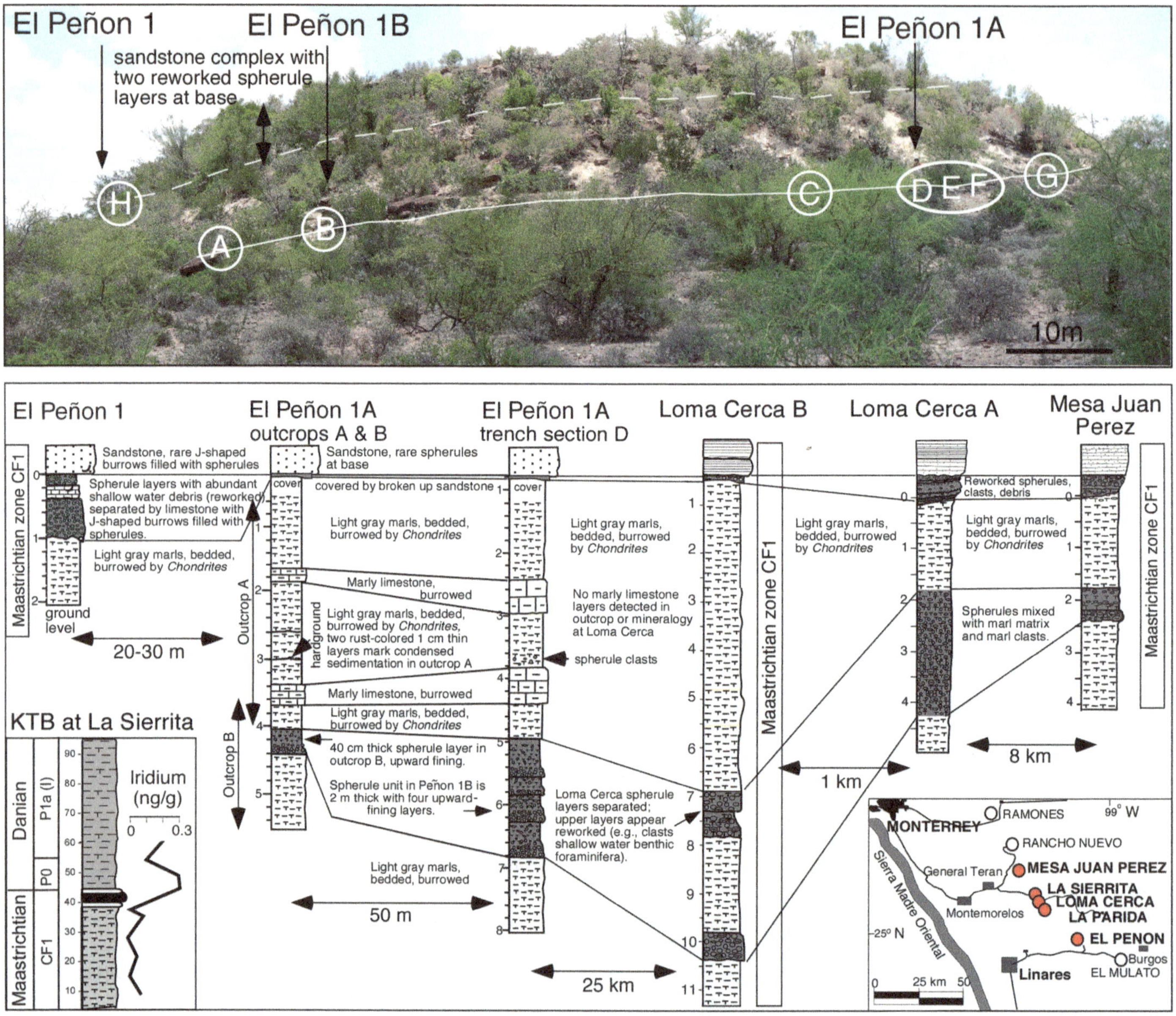

Figure 15. Photo showing sequence of El Peñon outcrops and associated stratigraphic logs. Dashed line in photo traces the sandstone complex with reworked impact spherules at the base of a submarine channel; solid line 4–5 m below traces the impact spherule layer near the base of zone CF1, with an estimated age 100–150 k.y. below the K-T boundary. El Peñon spherule layers can be correlated to Loma Cerca and Mesa Juan Perez sections at 25 km and 35 km to the north, respectively. Variable erosion in submarine channels below the reworked spherule unit at the top accounts for the reduced marl layers at Loma Cerca A and Mesa Juan Perez. The K-T boundary (KTB) is exposed at La Sierrita and contains no evidence of Chicxulub impact spherules (modified from Keller et al., 2009d).

represents the rising sea level and return to normal sedimentation along with colonization of the seafloor by invertebrates (e.g., Adatte et al., 1996, 2011; Ekdale and Stinnesbeck, 1998; Keller et al., 1997, 2007, 2011b; Gale, 2006).

Oldest Impact Spherule Layer

The age of the Chicxulub impact is best evaluated based on the stratigraphic position of the oldest impact glass spherule layer (Keller, 2008). At Loma Cerca, the stratigraphically oldest spherule layer was discovered 9 m below the channel deposits (Fig. 15; Keller et al., 2002b) and was interpreted by Schulte et al. (2003) as slump or tectonic disturbance. However, the only evidence of disturbance is a small (<2 m) overturned gravity flow within a spherule layer sandwiched between more resistant marl layers along a slope at Loma Cerca.

During a Princeton undergraduate student field trip in 2002, we tested the hypothesis that a stratigraphically older spherule layer should be present below the reworked spherule layers of

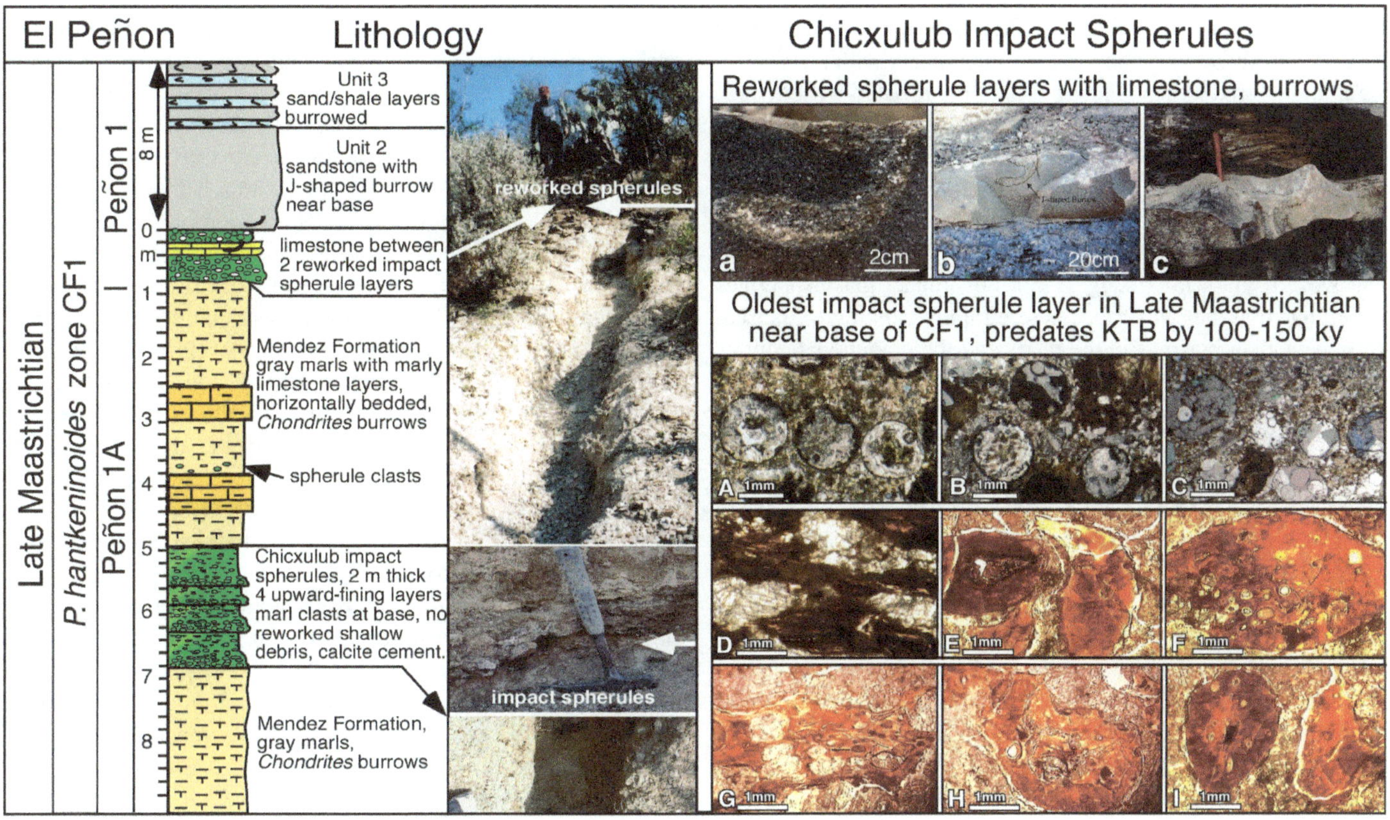

Figure 16. Composite section of the sandstone complex at El Peñon 1 and underlying trenched sequence at El Peñon 1A ~100 m southwest. At El Peñon 1A, sediments consist of horizontally bedded marls with two marly limestone layers and an impact spherule layer 4 m below the two reworked spherule layers at the base of the sandstone complex (dashed line). Peñon 1: (a) J-shaped burrow infilled with spherules and truncated from limestone (b) located between two reworked impact spherule layers, and (c) limestone layer with sculpted top between impact spherule layers at El Mimbral. Peñon 1A: A–B—Chicxulub impact spherules in marly matrix from the upper part of the 2-m-thick spherule unit; C—contact to overlying marl. D—melt-rock glass of welded spherules; E—spherules, some with concave-convex contact; F, I—compressed, vesicular spherules; G—vesicular glass shard; H—dumbbell. Deposition of this impact spherule layer occurred rapidly, possibly by raft-like accumulation of hot spherules at the sea surface and rapid sinking. KTB—K-T boundary.

the submarine channel at El Peñon. A trench was dug on the backside of the hill (Fig. 16; El Peñon 1A, location D). About 4–5 m below the base of the sandstone complex, a 2-m-thick impact spherule layer was discovered interbedded in undisturbed marls and marly limestones, thus confirming the hypothesis of an older pre–K-T boundary age for the Chicxulub impact. Subsequently, this older spherule layer was traced over ~100 m to El Peñon 1 (outcrop H) where strata dip below base level. Over the exposed distance, the spherule layer thins to 10–20 cm but remains parallel to the sandstone complex above (Fig. 15, dashed line). The 4–5 m of sediments consist of nearly horizontally layered marls and marly limestone with no signs of slumps, gravity flows, or tectonic disturbance (Keller et al., 2009d). This stratigraphically older spherule layer is near the base of zone CF1, similar to the oldest spherule layers at Loma Cerca and Mesa Juan Perez (Fig. 15; Keller et al., 2002b).

Nature of spherule deposition. Clues to the age and depositional nature of this spherule deposit can be gained from the spherules and matrix composition (Fig. 16; Keller et al., 2009d). A composite section based on El Peñon 1 and 1A shows the stratigraphy and deposition of the entire sequence in zone CF1, which spans the last 160 k.y. of the late Maastrichtian below the K-T boundary (Fig. 16). At the base, the spherule layer overlies an erosional surface with angular and rounded (2–5 cm, occasionally 10–20 cm) rip-up clasts from the underlying marls. Rip-up clasts decrease in abundance and size up section. The 2-m-thick spherule layer consists of four sublayers with upward-decreasing spherule abundance.

At the base, the more densely packed impact glass spherules form 5–10-cm-thick resistant layers of impact glass, and compressed and welded spherules, which are often amalgamated (Fig. 16D). Spherules are relatively large (2–5 mm), frequently compressed, oval (Figs. 16F and 16I), and dumbbell (Fig. 16H) vesicular, and some have concave/convex contacts (Fig. 16E). Vesicular glass shards are relatively common (Fig. 16G). The matrix is 80%–90% calcite with almost no clastic grains. Foraminifera are generally absent, except occasionally encased in glass (Keller et al., 2009d). In the upper parts of the unit, spherules are smaller and isolated in a marly matrix (Figs. 16A–16B). Above the spherule unit, the contact with the overlying marls is gradational (Fig. 16C) and bioturbated with diverse foraminiferal assemblages. Marls below the spherule layers are similar to those above (Keller et al., 2009d).

These spherule characteristics, including the distinct layers of welded glass, calcite cement, and near absence of detritus, indicate rapid settling of spherules through the water column. The welded glass rafts suggest that hot impact glass spherules accumulated at the sea surface before rapidly sinking to the seafloor. Abundant burrows in the marls and the gradational contact above the spherule unit reflect normal pelagic sedimentation after deposition of the spherule unit. Absence of exotic clasts and transported shallow-water debris indicates locally derived sediments. All these characteristics suggest that the late Maastrichtian spherule unit at El Peñon records the time of the Chicxulub impact.

Evidence from Brazos River, Texas. A pre–K-T boundary spherule age is also documented in the Brazos sections, where the K-T boundary mass extinction and evolution of Danian species occur up to 1 m above the sandstone complex of shallow submarine channels. Impact spherules, glauconite, and shell debris form an unconsolidated unit with rare lithified clasts at the base. Thin sections of the lithified clasts reveal the presence of impact spherules, and some clasts have burrows infilled with spherules (Keller et al., 2007, 2011b; Adatte et al., 2011). These spherule-rich clasts record a history of prior deposition, lithification, erosion, transport, and redeposition. Their age therefore predates the K-T boundary and also spherule deposition at the base of the channel. A yellow clay layer 45–60 cm below the channel unconformity was earlier thought to represent altered impact glass, but it has now been identified as a bentonite layer (Keller et al., 2011b; Adatte et al., 2011; Hart et al., 2012). Note that in the Brazos sections, Schulte et al. (2006) and Hart et al. (2011) place the K-T boundary at the spherule layer based on the redefinition of Chicxulub impact spherules as K-T boundary age, a practice that defies the primary K-T boundary identifying criteria—the mass extinction and evolution of Danian species up to 1 m above the top of the sandstone complex.

Evidence from the Chicxulub impact crater. A pre–K-T boundary age for the Chicxulub impact was also documented in the impact crater core Yaxcopoil-1 (Keller et al., 2004a, 2004b) and in two nearby Pemex cores (Ward et al., 1995). In the Yaxcopoil-1 core, the impact breccia (suevite) is separated from the K-T boundary by 50 cm of limestone with five glauconite clay layers, common burrows, a planktic foraminiferal assemblage of zone CF1, characteristic late Maastrichtian stable isotope values, and paleomagnetic chron 29r. Based on all these late Maastrichtian indicators, the age of the Chicxulub impact occurred sometime in zone CF1, preceding the K-T boundary mass extinction. Smit et al. (2004) argued that the 50 cm section of limestone represents backwash from the impact-generated tsunami and that the reported planktic foraminifera are just dolomite rhombs. Arz et al. (2004) and M. Caron (at the request of Smit) confirmed the presence of planktic foraminifera.

Probable age of Chicxulub impact. We can estimate the probable pre–Cretaceous boundary age of the Chicxulub impact based on Texas, NE Mexico, and the Chixculub crater core Yaxcopoil-1. In all localities, the spherule layers are in zone CF1, but only El Peñon and possibly Loma Cerca have the thick spherule unit suggesting primary deposition below the reworked spherules at the base of the sandstone complex. This impact spherule layer is near the base of zone CF1, although there is some erosion (undulating erosion surface, rip-up clasts). Zone CF1 spans the last 160 k.y. of the Maastrichtian below the K-T boundary (Fig. 3). A conservative estimate of spherule deposition is between 100 and 150 k.y. before the K-T boundary, which corresponds to the global warming (Li and Keller, 1998c; MacLeod et al., 2005; Abramovich et al., 2010; Keller et al., 2009d, 2011b; Punekar et al., this volume).

Probable age of reworked spherule layers. We can also estimate the depositional age of the reworked spherule layers at the base of the sandstone complex in the submarine channels, although this is more difficult because of downcutting and erosion. However, an age estimate can be made based on stable isotopes. Deposition occurred after the global warming, during a time of climate cooling and a sea-level fall in the upper half of zone CF1, as documented in the NE Mexico and Brazos River sections (Keller et al., 2007, 2009d, 2011b; Adatte et al., 1996, 2011; Punekar et al., this volume). This indicates that the sandstone complex (commonly called tsunami deposit) predates the K-T boundary by ~50–80 k.y. Near the end of the Maastrichtian, sea level rose again, and climate warmed across the K-T boundary, as observed in expanded sequences (e.g., El Kef and Elles, Tunisia, and Brazos River, Texas; Adatte et al., 2002, 2011; Abramovich et al., 2011; Keller et al., 2011b). No impact spherules are present at the K-T boundary in complete sequences (Keller et al., 2013).

Chicxulub and the K-T Boundary Mass Extinction

The stratigraphic position of the older impact spherule layers in undisturbed bedded marls and marly limestone at El Peñon and Loma Cerca permit evaluation of the environmental effects of the Chicxulub impact. Surprisingly, quantitative planktic foraminiferal analysis below and above the impact spherule layer at El Peñon reveals no significant change in species diversity or abundance, and all species survived with no long-term effects (Keller et al., 2009d). Similarly, no significant changes are observed in carbon isotopes or trace elements. The same results were earlier observed across the older impact spherule layers at Loma Cerca (Keller et al., 2002b). No mass extinction is associated with the spherules of the sandstone complex along the Brazos River, Texas, where the K-T boundary is up to 1.0 m up section (Keller et al., 2011b), or in the impact crater core Yaxcopoil-1, where the K-T boundary mass extinction is above a 0.5-m-thick limestone with five glauconitic clay horizons of zone CF1 age that overlies the impact breccia (Keller et al., 2004a, 2004b). These data suggest that the environmental effects of the Chicxulub impact were transient and not detectable within sample resolution of a few thousand years and that the kill effect of this impact has been overrated. Equally important is absence of impact spherules at the K-T boundary in complete sequences, which also indicates that this impact did not cause the mass extinction.

COINCIDENCE OF DECCAN VOLCANISM AND CHICXULUB IMPACT

Impact spherules are embedded in sediments at three different age intervals: (1) Most common are reworked spherule layers in the early Danian zone P1a, virtually always above a K-T boundary hiatus, as documented throughout the North Atlantic, Caribbean, and Central America (Fig. 14). (2) Impact spherules are restricted to the late Maastrichtian zone CF1 in over 80 sections of NE Mexico and Texas, where multiple spherule layers are reworked at the base of submarine channels (Figs. 14 and 15). (3) Also in NE Mexico (Loma Cerca and El Peñon), we find the stratigraphically oldest impact spherule layers in the late Maastrichtian near the base of zone CF1 (~100–150 k.y. prior to the K-T boundary), ~4–9 m below the submarine channels (Fig. 16). The late Maastrichtian has rarely been studied in NE Mexico, and a systematic investigation is likely to uncover more localities with the primary impact spherule deposits preserved similar to Loma Cerca and El Peñon.

We can now correlate the Chicxulub impact and Deccan volcanism, based on isotope stratigraphy and planktic foraminiferal biostratigraphy (Fig. 3). The onset of Deccan phase 2 volcanism began ~200 k.y. before the K-T boundary, coincident with rapid global warming that reached its maximum in the lower part of CF1. The stratigraphically oldest impact spherules near the base of zone CF1 occurred during this global warming. However, the 50 k.y. age uncertainty (between 100–150 k.y. prior to the K-T boundary) prevents evaluation of Chicxulub's contribution to the global warming. The impact may have exacerbated climate warming already under way due to Deccan volcanism, or it could have coincided with the onset of rapid global warming and resulted in intensified Deccan phase 2 volcanic eruptions.

The reworked spherules at the base of the sandstone complex with a depositional age of between 50 and 80 k.y. before the K-T boundary were deposited during global cooling and a sea-level fall. This appears to have been a time of decreased volcanic activity as suggested by geochemical data (Fig. 12; Keller et al., 2002b; Adatte et al., 1996). The single and multiple reworked impact spherule layers in early Danian sediments overlie a hiatus that juxtaposed Danian and late Maastrichtian zone CF3 (Keller et al., 2013). Erosion and redeposition occurred during a sea-level fall in the middle of zone P1a that has been documented in numerous deep-sea sections, including Haiti, Belize, Guatemala, and throughout the Caribbean (Keller et al., 1993, 2003a, 2003b, 2013). The fortuitous deposition of spherules overlying this hiatus (e.g., Bass River, New Jersey, ODP Sites 1049, 1259) is therefore not proof that the Chicxulub impact is precisely K-T boundary in age as frequently claimed.

REASSESSMENT OF THE MASS EXTINCTION SCENARIO

Over the past three decades, the hypothesis that a large extraterrestrial impact (Chicxulub) caused the mass extinction reached near-mythical status as the cause for the K-T boundary mass extinction and particularly the demise of the dinosaurs (Archibald, this volume). Evidence to the contrary was largely ignored or interpreted as the result of impact-induced giant earthquakes and tsunami waves creating massive gravity flows, slumps, margin collapse, faults, and folding that disturbed, mixed, or obliterated the sedimentary record. There is no doubt that a giant impact such as Chicxulub, creating a 175-km-diameter crater, would trigger tsunami and earthquakes, but are the resultant disturbed deposits preserved, and how are they identified?

Dypvik and Jansa (2003) observed that differentiating between gravity deposits triggered by rare impacts and those caused by common earthquakes is difficult and depends on the presence of melt-rock particles and iridium enrichments. For the Chicxulub impact, this is mainly the impact glass spherules surrounding the Chicxulub crater from the North Atlantic through the Caribbean and Central America (Fig. 14). However, Dypvik and Jansa (2003) observed a dearth of giant tsunami evidence associated with the Chicxulub impact outside the Gulf of Mexico, referring to the so-called tsunami deposits in the submarine channels of NE Mexico. As noted in this review, the sedimentary characteristics of these deposits demonstrate long-term deposition and repeated colonization of the seafloor that is incompatible with tsunami deposition.

In contrast to the popular Chicxulub impact hypothesis as cause for the K-T boundary mass extinction, Deccan volcanism—the other major catastrophe—has languished as an unlikely cause. However, over the past few years, researchers have made major advances in improved dating and correlation, identifying environmental changes based on basalt geochemistry and its weathering products, and based on micropaleontology documenting the direct link between the mass extinction and the massive flood basalt eruptions of phase 2 at the end of the Maastrichtian. Although much research remains to be done, the current data set for the first time permits evaluation of the age and faunal and environmental effects of Deccan volcanism in India and globally. Also for the first time, this database can be compared and correlated with the Chicxulub impact to evaluate the age, coincidence, and environmental effects of these two catastrophes. The following scenario integrates the major features of the Deccan and Chicxulub records in an effort to put the current data into context in order to better understand the events leading up to the K-T boundary mass extinction.

DECCAN VOLCANISM AND CHICXULUB IMPACT: AN INTEGRATED SCENARIO AND CONCLUSIONS

Deccan volcanism in palemagnetic chron C29r (planktic foraminiferal zones CF1–CF2) is linked to global climate and environmental changes ending with the K-T boundary mass extinction (Robinson et al., 2009; Keller et al., 2011a, 2012). Climate gradually cooled during the late Maastrichtian, reaching maximum cooling by the end of CF3, followed by rapid warming in the upper part of CF2 (Li and Keller, 1998c), beginning at ~200 k.y. before the K-T boundary (Fig. 17). Climate remained very warm but fluctuating until the middle of zone CF1, when it cooled and then warmed again near the end of the Maastrichtian (for high-resolution stable isotope records, see Punekar et al., this volume). Based on $^{187}Os/^{188}Os$ data, Robinson et al. (2009) speculated that a lower carbon burial rate during the last 200 k.y. of the Maastrichtian resulted from ocean acidification caused by Deccan volcanism. This age estimate is in agreement with the onset for Deccan phase 2 and warming in the upper part of zone CF2 but not with lower carbon burial, except near the end of the Maastrichtian (Font et al., 2011, this volume) (Fig. 3). Chenet et al. (2009) estimated that the entire phase 2 eruptions may have occurred within 100 k.y., with long periods of inactivity reducing active eruptions to possibly just 10 k.y. More precise age estimates must await better dating techniques, particularly the dating of single large eruptive events. Chenet et al. (2009) estimated the total gas output of Deccan phase 2 was up to 17,000 Gt of SO_2 and 35,000 Gt of CO_2, which is on the order of 20–200 times that of the Chicxulub impact (Courtillot and Fluteau, this volume).

The kill effect of volcanism depends not just on total gas output, but also on the rate of output and how fast eruptive events follow each other, preventing ecosystem recovery and causing runaway effects. The four largest lava flows of phase 2 near the end of the Maastrichtian followed each other in rapid succession and are separated by just 5 m, 10 m, and 15 m of sand and sandy shale, compared with marls in the infratrappean sediments of the Krishna-Godavari Basin wells (Fig. 10). This suggests a major change in sedimentation, with megaflow eruptions resulting in high clastic influx from the Krishna and Godavari Rivers over a short time span. If this was the case, then the rapid succession of massive volcanic eruptions at the end of phase 2 likely led to runaway effects, preventing ecosystem recovery and causing the mass extinction (Fig. 17).

Mass Extinction Crisis Onset: Stage 1

The crisis that led to the K-T boundary mass extinction began when Deccan volcanism reached critical levels during phase 2 in the latest Maastrichtian, depositing 80% of the entire

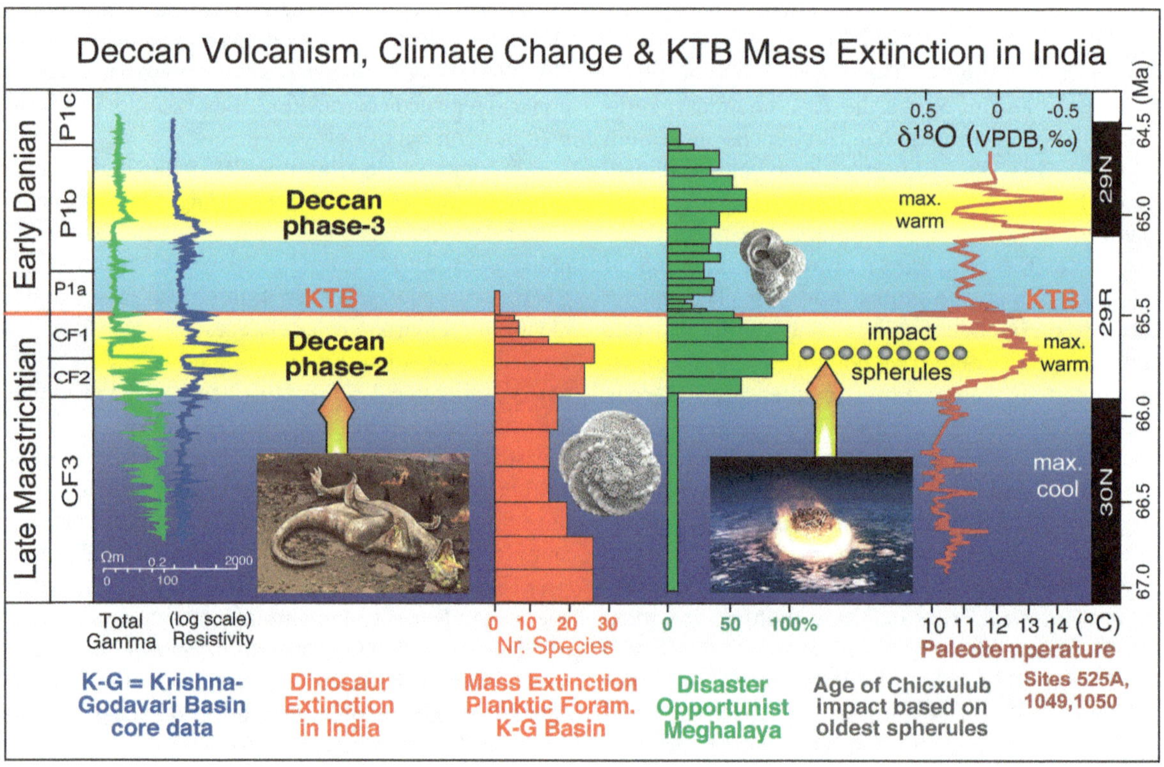

Figure 17. Summary of climatic, environmental, and biologic effects related to Deccan volcanism. All dinosaurs in India disappeared by the onset of the main eruptions in phase 2 (inset illustration at left: U.S. National Science Foundation, 2008); 50% of planktic foraminifera disappeared at this time, and nearly all went extinct by the fourth and last megaflow at the K-T boundary (KTB). Rapid climate changes and blooms of the disaster opportunist *Guembelitria cretacea* accompanied volcanic phase 2, in the early Danian after the mass extinction, and stress conditions intensified with eruptions of the last Deccan phase 3. The precise timing of the mass extinction is still uncertain. VPDB—Vienna Peedee belemnite.

3500-m-thick Deccan lava pile. With increasing volcanic intensity, SO_2 and CO_2 release led to acid rains on land and ocean acidification, which resulted in a carbonate crisis for marine organisms, as evident in dissolution effects and eventually extinctions (Robinson et al., 2009; Keller et al., 2011a, 2012; Font et al., 2011, this volume; Courtillot and Fluteau, this volume). The massive input of CO_2, beginning in the upper zone CF2 and continuing into the middle of CF1, led to the well-documented rapid global warming of 8 °C on land and 4 °C in the oceans worldwide (e.g., Li and Keller, 1998c; Wilf et al., 2003; Punekar et al., this volume).

During this global warming, high-stress marine environmental conditions resulted in the dominance (60%–80%) of the disaster opportunist *Guembelitria cretacea* worldwide, particularly in shallow-marine environments subjected to terrestrial, nutrient-rich, high runoff due to increased weathering (Fig. 13). On a global basis, the variable abundance of *Guembelitria* blooms and the decline in abundance of other species but no extinctions suggest that partial ecosystem recovery occurred between major volcanic eruptions, permitting species survival.

The Chicxulub impact on Yucatan occurred during this time, as indicated by the position of the stratigraphically oldest spherule layers in the lower part of CF1 in NE Mexico, Texas, and the pre–K-T boundary age of impact breccia in the crater. This impact likely contributed to and exacerbated the ongoing greenhouse warming and may have intensified volcanic eruptions. Environmental effects appear to have been short-term and transient, with no detectable changes in sample resolution of a few thousand years (Keller et al., 2004a, 2004b, 2009d, 2011b) (Fig. 17). No species extinctions are associated with this climate warming and impact. This indicates that the kill effect of the Chicxulub impact has been vastly overestimated.

Mass Extinction Crisis Middle: Stage 2

By the middle of zone CF1, greenhouse warming ended and climate cooled, accompanied by a falling sea level (Keller et al., 2007, 2011b). Intensified currents led to erosion, formation of submarine channels, and transport from nearshore areas into deeper waters, creating the sandstone complex known from submarine channels in NE Mexico and Texas and hiatuses elsewhere (Adatte et al., 1996). At this time, impact spherule deposits and shallow-water debris were repeatedly eroded from nearshore areas, transported seaward, and redeposited at the base of the sandstone complex. Subsequent sandstone deposition, with few spherule-filled burrows near the base, suggests gravity flows. The upper unit of fine sand, shale, and marl layers marks the rising sea level and return to normal sedimentation, with intensively burrowed horizons indicating colonization of the seafloor (Ekdale and Stinnesbeck, 1998). Popular scenarios interpret this sandstone complex as impact-generated tsunami deposits coeval with the K-T boundary and therefore place the mass extinction at the reworked impact spherule layers (e.g., Schulte et al., 2010).

Mass Extinction Crisis Ending: Stage 3

Near the end of the Maastrichtian, climate rapidly warmed, and sea level rose, eventually depositing the K-T boundary clay (Fig. 17). The largest four megaflow eruptions of the Krishna-Godavari Basin, spanning over 1500 km across India and out into the Bay of Bengal (Self et al., 2008), likely occurred during the last ~50 k.y. below the K-T boundary. This is suggested by (1) the maximum stress conditions below the K-T boundary in the Meghalaya section, where *Guembelitria* blooms reach 95%, ending with the mass extinction (Figs. 5 and 17; Gertsch et al., 2011a), (2) the Krishna-Godavari Basin wells where the onset of the mass extinction began just below the four megaflows, ending with the K-T boundary mass extinction (Figs. 10, 11, and 17), (3) the strong dissolution effects in intertrappean sediments (Figs. 5, 10, and 11; Keller et al., 2011a, 2012), and (4) strong dissolution effects in marine sections over an estimated ~30 k.y. below the K-T boundary mass extinction at Bidart, France (Font et al., 2011, this volume; Courtillot and Fluteau, this volume). Dissolution effects preceding the K-T boundary mass extinction have been observed by many paleontologists, but systematic documentation and evaluation on a global basis remain to be done.

Current data indicate that the mass extinction in planktic foraminifera occurred rapidly, beginning in infratrappean sediments just below the first of four megaflows, and was complete by the K-T boundary (Figs. 3, 10, 11, and 17). The last dinosaur remains are known from the infratrappean sediments below the megaflows, and their demise is interpreted as the direct result of Deccan volcanism (e.g., Samant and Mohabey, this volume; Prasad and Sahni, this volume). The immediate cause for the mass extinction was likely the rapid massive eruptions, increased volcanic intensity, and high SO_2 and CO_2 release, leading to runaway effects and preventing ecosystem recovery. The kill mechanism was likely arid climate and acid rains on land (Fig. 18), and ocean acidification in the marine realm, resulting in the observed carbonate crisis for marine organisms, as evident in dissolution effects and extinctions globally.

Environmental Effects in India

In the Deccan volcanic province, semiarid conditions known as "mock aridity" prevailed, whereas in the surrounding areas, abundant precipitation, acid rains, high chemical weathering, and continental runoff resulted in a major influx of silt and nutrients into the oceans (Fig. 12; Gertsch et al., 2011a). This led to increasingly toxic, turbid, mesotrophic to eutrophic waters, creating high-stress conditions that likely caused the mass extinction of nearly all planktic foraminifera. Enhanced weathering led to decreased CO_2 and pCO_2, reversing the global warming and leading to the rapid global cooling after the mass extinction. Parallel conditions have been observed associated with the Permian-Triassic volcanism and mass extinction. For example, Algeo and Twitchett (2010) reported high rates of physical and chemical erosion as a result of climate warming

Figure 18. Artist illustration of the end-Cretaceous mass extinction, including the dinosaur extinction, as a result of Deccan volcanism in India (U.S. National Science Foundation, 2008).

and acid rains at 16 Permian-Triassic mass extinction localities. They inferred that the resultant terrestrial ecosystem destruction, erosion, and runoff led to marine siltation and eutrophication, which contributed to the marine biotic crisis and delayed recovery after the mass extinction.

Mass Extinction After Effects: Stage 4

A few tiny, stress-tolerant planktic foraminiferal species evolved very quickly after the mass extinction. During the early Danian zone P1a (C29R), environmental stress conditions remained high, possibly because of continued high nutrient influx from weathering, as evident in the high total organic content of sediments (Gertsch et al., 2011a). As a result, marine productivity recovered only minimally, as evident by the very low species diversity, increasing from 5 to 12 species over ~300 k.y. (Figs. 5 and 10). All evolving species remained very small (<100 µm, frequently <63 µm), unornamented, and simple in morphology, reflecting the continued high-stress conditions. *Guembelitria* blooms prevailed, but generally of lower abundance compared with the latest Maastrichtian, and decreased shortly after the K-T boundary in India and globally (Figs. 13; Punekar et al., this volume). However, preservation of carbonate shells is very good, indicating no significant dissolution effects and absence of ocean acidification. This indicates that volcanic activity was largely absent in the early Danian (C29R, zone P1a) and that continued weathering of basalts resulted in the increased nutrient input and high-stress marine environment that delayed biotic recovery.

Early Danian Crisis—Deccan Volcanism Phase 3

The last Deccan phase 3 began at the base of C29N (zone P1b) accompanied by the extinction of *Parvularugoglobigerina eugubina* and *P. longiapertura* and decreased abundances of other earliest Danian species. No significant extinctions occurred during phase 3 volcanism, probably because eruptions were widely separated in time, permitting ecosystem recovery. Alternatively, the small early Danian species that evolved in the aftermath of the mass extinction were more stress tolerant than the specialized large Maastrichtian species and therefore able to survive the stress conditions associated with the Deccan phase 3. In contrast to the latest Maastrichtian, dissolution effects are minor, suggesting no significant change in marine pH conditions. However, species diversity remained low (<15 species), and simple, small morphotypes prevailed similar to zone P1a.

A gradual recovery to larger morphotypes and higher diversity began only after the Deccan phase 3 ended, and full recovery was restored with normal nutrient input and the return to oligotrophic conditions. This suggests that the long-delayed post–K-T boundary marine recovery that has remained an enigma for decades (e.g., Keller and Benjamini, 1991; Magaritz et al., 1992) was due to continued high-stress conditions induced by Deccan volcanism and its weathering products. Punekar et al. (this volume) detail the faunal event associated with Deccan phase 3 and its environmental and climatic consequences.

CONCLUSIONS

The proposed scenario for the K-T boundary mass extinction and delayed recovery is based on integration of large, multidisciplinary data sets on Deccan volcanism, the Chicxulub impact, and the mass extinction. It takes into consideration climate warming and cooling, sea-level changes, erosion, weathering, ocean acidification, high-stress environments with opportunistic species blooms, the mass extinction, and delayed postextinction recovery. The scenario based on all of these factors is necessarily complex, and no single event, such as the Chicxulub impact, can account for the observed long-term environmental and biotic changes. The Chicxulub impact coincided with the major late Maastrichtian rapid warming of 4 °C in the oceans and 8 °C on land that is commonly attributed to Deccan volcanism, and it may have significantly contributed to this warming and even intensified volcanic eruptions. This rapid warming triggered the observed cascade of climate changes and high environmental stress conditions that eventually led to the K-T boundary mass extinction and delayed recovery. Near the end of the Maastrichtian, another series of at least four massive volcanic eruptions in short order formed the longest lava flows on Earth and probably caused runaway effects, particularly ocean acidification, which resulted in the carbonate crisis commonly considered to be the prime cause for the mass extinction. The Chicxulub impact played no role in this mass extinction. The post–K-T boundary high-stress conditions and delayed recovery can be attributed to weathering of volcanic rocks followed by the last phase 3 of Deccan volcanism, delaying recovery.

ACKNOWLEDGMENTS

The review comments and suggestions of Richard Twitchett and Bill MacDonald are greatly appreciated and clarified various issues. The work presented in this review is the result of over two decades working on K-T boundary sequences with Chicxulub impact ejecta and 6 yr on Deccan volcanism, most of it in collaboration with Thierry Adatte and many other collaborators, as well as former and current students. Over the past 5 yr, many Indian colleagues have contributed to Deccan studies and field-based research, including Sunil Bajpai, Dhananjay Mohabey, Bandana Samant, Rahul Garg, and Nallamuthu Malarkodi. Research on deep wells in the Krishna-Godavari Basin would not have been possible without the permission and support of former directors D.K. Pande and S.V. Rao of the Oil and Natural Gas Corporation of India, the facilitation of Harinder Upadhyay, and the collaboration of its scientists, particularly B.C. Jaiprakash, A.N. Reddy, and D.S.N. Raju. This study is based upon work supported by the U.S. National Science Foundation (NSF) through the Continental Dynamics Program, Sedimentary Geology, and Paleobiology Program, and Office of International Science and Engineering's India Program under NSF grants EAR-0207407, EAR-0447171, and EAR-1026271.

REFERENCES CITED

Abramovich, S., and Keller, G., 2002, High stress late Maastrichtian paleoenvironment: Inference from planktonic foraminifera in Tunisia: Palaeogeography, Palaeoclimatology, Palaeoecology, v. 178, p. 145–164, doi:10.1016/S0031-0182(01)00394-7.

Abramovich, S., Almogi-Labin, A., and Benjamini, Ch., 1998, Decline of the Maastrichtian pelagic ecosystem based on planktic foraminifera assemblage changes: Implication for the terminal Cretaceous faunal crisis: Geology, v. 26, p. 63–66.

Abramovich, S., Yovel-Corem, S., Almogi-Labin, A., and Benjamini, C., 2010, Global climate change and planktic foraminiferal response in the Maastrichtian: Paleoceanography, v. 25, PA2201, doi:10.1029/2009PA001843.

Abramovich, S., Keller, G., Berner, Z., Cymbalista, M., and Rak, C., 2011, Maastrichtian planktic foraminiferal biostratigraphy and paleoenvironment of Brazos River, Falls County, Texas, *in* Keller, G., and Adatte, T., eds., The End-Cretaceous Mass Extinction and the Chicxulub Impact in Texas: Society for Sedimentary Geology (SEPM) Special Publication 100, p. 123–156.

Adatte, T., Stinnesbeck, W., and Keller, G., 1996, Lithostratigraphic and mineralogical correlations of near-K/T boundary clastic sediments in NE Mexico: Implications for origin and nature of deposition, *in* Ryder, G., Fastovsky, D., and Gartner, S., eds., The Cretaceous-Tertiary Event and Other Catastrophes in Earth History: Geological Society of America Special Paper 307, p. 211–226.

Adatte, T., Keller, G., Burns, S., Stoykova, K.H., Ivanov, M.I., Vangelov, D., Kramar, U., and Stüben, D., 2002, Paleoenvironment across the Cretaceous-Tertiary transition in eastern Bulgaria, *in* Koeberl, C., and MacLeod, K.G., eds., Catastrophic Events and Mass Extinctions: Impacts and Beyond: Geological Society of America Special Paper 356, p. 231–252.

Adatte, T., Keller, G., and Baum, G., 2011, Lithostratigraphy, sedimentology, sequence stratigraphy and the origin of the sandstone complex in the Brazos River KT sequences, *in* Keller, G., and Adatte, T., eds., The End-Cretaceous Mass Extinction and the Chicxulub Impact in Texas: Society for Sedimentary Geology (SEPM) Special Publication 100, p. 43–80.

Alegret, L., Molina, E., and Thomas, E., 2001, Benthic foraminifera at the Cretaceous-Tertiary boundary around the Gulf of Mexico: Geology, v. 29, no. 10, p. 891–894, doi:10.1130/0091-7613(2001)029<0891:BFATCT>2.0.CO;2.

Algeo, T.J., and Twitchett, R.J., 2010, Anomalous Early Triassic sediment fluxes due to elevated weathering rates and their biological consequences: Geology, v. 38, p. 1023–1026, doi:10.1130/G31203.1.

Archibald, D.J., 2014, this volume, What the dinosaur record says about extinction scenarios, *in* Keller, G., and Kerr, A.C., eds., Volcanism, Impacts, and Mass Extinctions: Causes and Effects: Geological Society of America Special Paper 505, doi:10.1130/2014.2505(10).

Archibald, D.J., Archibald, J.D., Clemens, W.A., Padian, K., Rowe, T., MacLeod, N., Barrett, P.M., Gale, A.S., Holroyd, P., Sues, H.-D., Arens, N.C., Horner, J.R., Wilson, G.P., Goodwin, M.B., Brochu, C.A., Lofgren, D.L., Hurlbert, S.H., Hartman, J.A., Eberth, D.A., Wignall, P.B., Currie, P.J., Weil, A., Prasad, G.V.R., Dingus, L., Courtillot, V., Milner, A., Milner, A., Bajpai, S., Ward, D.J., and Sahni, A., 2010, Cretaceous extinctions: Multiple causes: Science, v. 328, p. 973.

Arenillas, I., Arz, J.A., Grajales-Nishimura, J.M., Murillo-Muneton, G., Alvarez, W., Camargo-Zanguera, A., Molina, E., and Rosales-Dominguez, C., 2006, Chicxulub impact event is Cretaceous/Paleogene boundary in age: New micropaleontological evidence: Earth and Planetary Science Letters, v. 249, p. 241–257, doi:10.1016/j.epsl.2006.07.020.

Arz, J.A., Alegret, L., and Arenillas, I., 2004, Foraminiferal biostratigraphy and environmental reconstruction at the Yaxcopoil-1 drillcore, Chicxulub crater, Yucatan Peninsula: Meteoritics & Planetary Science, v. 39, no. 7, p. 1099–1111, doi:10.1111/j.1945-5100.2004.tb01131.x.

Badve, R.M., and Ghare, M.A., 1977, Palaeoecological aspects of the Bagh Beds, India: Recent Researches in Geology, v. 4, p. 388–402.

Bajpai, S., and Prasad, G.V.R., 2000, Cretaceous age for Ir-rich Deccan intertrappean deposits: Palaeontological evidence from Anjar, western India: Journal of the Geological Society of London, v. 157, p. 257–260, doi:10.1144/jgs.157.2.257.

Bajpai, S., Holmes, J., Bennett, C., Mandal, N., and Khosla, A., 2013, Paleoenvironment of northwestern India during the Late Cretaceous Deccan volcanic episode from trace element and stable isotope geochemistry of intertrappean ostracod shells: Global and Planetary Change, v. 107, p. 82–90, doi:10.1016/j.gloplacha.2013.04.011.

Baksi, A.K., 2005, Comment on "^{40}Ar/^{39}Ar dating of the Rajahmundry Traps, eastern India and their relations to the Deccan Traps" by Knight et al.: Earth and Planetary Science Letters, v. 239, p. 368–373, doi:10.1016/j.epsl.2005.03.029.

Baksi, A.K., Byerly, G.R., Chan, L.-H., and Farrar, E., 1994, Intracanyon flows in the Deccan province, India: Case history of the Rajahmundry Traps: Geology, v. 22, p. 605–608, doi:10.1130/0091-7613(1994)022<0605:IFITDP>2.3.CO;2.

Berggren, W.A., Kent, D.V., Swisher, C.C., III, and Aubry, M.-P., 1995, A revised Cenozoic geochronology and chronostratigraphy, *in* Berggren, W.A., Kent, D.V., Aubry, M.-P., and HardenboL, J., eds., Geochronology, Time Scales and Global Stratigraphic Correlation: Society for Sedimentary Geology (SEPM) Special Publication 54, p. 129–212.

Bhandari, A., 1995, Ostracodes from the inter-trappean beds near Duddukuru, Andhra Pradesh, and a note on their age and paleoecological significance: Indian Journal Petroleum Geology, v. 4, no. 1, p. 89–107.

Bhandari, N., Shukla, P.N., and Castognoli, C.G., 1993, Geochemistry of some K/T sections in India: Palaeogeography, Palaeoclimatology, Palaeoecology, v. 104, p. 199–211, doi:10.1016/0031-0182(93)90131-2.

Bhandari, N., Gupta, M., Panday, J., and Shukla, P.N., 1994, Chemical profiles in K/T boundary section of Meghalaya, India: Cometary, asteroidal or volcanic: Chemical Geology, v. 113, p. 45–60, doi:10.1016/0009-2541(94)90004-3.

Bhandari, N., Shukla, P.N., Ghevariya, Z.G., and Sundaram, S.M., 1996, K/T boundary layer in Deccan intertrappeans at Anjar, Kutch, *in* Ryder, G., Fastovsky, D., and Gartner, S., eds., The Cretaceous-Tertiary Event and Other Catastrophes in Earth History: Geological Society of America Special Paper 307, p. 417–424.

Blackburn, T.J., Olsen, P.E., Bowring, S.A., McLean, M.N., Kent, D.V., Puffer, J., McHone, G., Rasbury, T.E., and Et-Touhami, N.M., 2013, Zircon U-Pb geochronology links the end-Triassic extinction with the Central Atlantic magmatic province: Science, v. 340, p. 941–945, doi:10.1126/science.1234204.

Bodas, M.S., Khadri, S.F.R., and Subbarao, K.V., 1988, Stratigraphy of the Jawhar and Igatpuri Formations, Western Ghats Lava Pile, India, *in* Subbarao, K.B., ed., Deccan Flood Basalt: Geological Society of India Memoir 10, p. 235–252.

Bond, D.P.G., and Wignall, P.B., 2014, this volume, Large igneous provinces and mass extinctions: An update, *in* Keller, G., and Kerr, A.C., eds., Volcanism, Impacts, and Mass Extinctions: Causes and Effects: Geological Society of America Special Paper 505, doi:10.1130/2014.2505(02).

Bourgeois, J., Hansen, T.A., Wiberg, P.L., and Kauffman, E.G., 1988, A tsunami deposit at the Cretaceous-Tertiary boundary in Texas: Science, v. 241, p. 567–570, doi:10.1126/science.241.4865.567.

Chatterjee, S., Guven, N., Yoshenobu, A., and Donofrio, R., 2006, Shiva Structure: A Possible KT Boundary Impact Crater on the Western Shelf of India: Special Publications, Museum Texas Tech University, v. 50, 39 p.

Chenet, A.-L., Quidelleur, X., Fluteau, F., Courtillot, V., and Bajpai, S., 2007, ^{40}K/^{40}Ar dating of the main Deccan large igneous province: Further evidence of KTB age and short duration: Earth and Planetary Science Letters, v. 263, p. 1–15, doi:10.1016/j.epsl.2007.07.011.

Chenet, A.-L., Fluteau, F., Courtillot, V., Gerard, M., and Subbarao, K.V., 2008, Determination of rapid Deccan eruptions across the Cretaceous-Tertiary boundary using paleomagnetic secular variation: Results from a 1200-m-thick section in the Mahabaleshwar: Journal of Geophysical Research, v. 113, B04101, doi:10.1029/2006JB004635.

Chenet, A.-L., Courtillot, V., Fluteau, F., Gerard, M., Quidelleur, X., Khadri, S.F.R., Subbarao, K.V., and Thordarson, T., 2009, Determination of rapid Deccan eruptions across the Cretaceous-Tertiary boundary using paleomagnetic secular variation: 2. Constraints from analysis of eight new sections and synthesis for a 3500-m-thick composite section: Journal of Geophysical Research, v. 114, B06103, doi:10.1029/2008JB005644.

Coccioni, R., and Luciani, V., 2006, *Guembelitria irregularis* bloom at the K-T boundary: Morphological abnormalities induced by impact-related extreme environmental stress?, *in* Cockell, C., Koeberl, C., and Gilmour, I., eds., Biological Processes Associated with Impact Events: Impact Studies: Berlin, Springer, p. 179–196.

Colodner, D.C., Boyle, E.A., Edmond, J.M., and Thomson, J., 1992, Post-depositional mobility of platinum, iridium, and rhenium in marine sediments: Nature, v. 358, p. 402–404, doi:10.1038/358402a0.

Courtillot, V., and Fluteau, F., 2010, Cretaceous extinctions: The volcanic hypothesis: Science, v. 328, p. 973–974, doi:10.1126/science.328.5981.973-b.

Courtillot, V., and Fluteau, F., 2014, this volume, A review of the embedded time scales of flood basalt volcanism with special emphasis on dramatically short magmatic pulses, *in* Keller, G., and Kerr, A.C., eds., Volcanism, Impacts, and Mass Extinctions: Causes and Effects: Geological Society of America Special Paper 505, doi:10.1130/2014.2505(15).

Courtillot, V., and Renne, J., 2003, On the ages of flood basalt events: Comptes Rendus Geoscience, v. 335, p. 113–140, doi:10.1016/S1631-0713(03)00006-3.

Courtillot, V., Besse, J., Vandamme, D., Montigny, R., Jaeger, J.-J., and Cappetta, H., 1986, Deccan flood basalts at the Cretaceous/Tertiary boundary?: Earth and Planetary Science Letters, v. 80, p. 361–374, doi:10.1016/0012-821X(86)90118-4.

Courtillot, V., Feraud, G., Maluski, H., Vandamme, D., Moreau, M.G., and Besse, J., 1988, The Deccan flood basalts and the Cretaceous-Tertiary boundary: Nature, v. 333, p. 843–846, doi:10.1038/333843a0.

Courtillot, V., Gallet, Y., Rocchia, R., Feraud, G., Robin, E., Hofmann, C., Bhandari, N., and Ghevariya, Z.G., 2000, Cosmic markers, ^{40}Ar/^{39}Ar dating and paleomagnetism of the KT sections in the Anjar area of the Deccan large igneous province: Earth and Planetary Science Letters, v. 182, p. 137–156, doi:10.1016/S0012-821X(00)00238-7.

Cowie, J.W., Ziegler, W., and Remane, J., 1989, Stratigraphic Commission accelerates progress, 1984 to 1989: Episodes, v. 12, p. 79–83.

Crocket, J.H., and Paul, D.K., 2004, Platinum-group elements in Deccan mafic rocks: A comparison of suites differentiated by Ir contents: Chemical Geology, v. 208, p. 273–291, doi:10.1016/j.chemgeo.2004.04.017.

Dessert, C., Dupre, B., Gaillardet, J., Francois, L.M., and Allegre, C.J., 2003, Basalt weathering laws and the impact of basalt weathering on the global carbon cycle: Chemical Geology, v. 202, p. 257–273, doi:10.1016/j.chemgeo.2002.10.001.

Dypvik, H., and Jansa, L.F., 2003, Sedimentary signatures and processes during marine bolide impacts: A review: Sedimentary Geology, v. 161, p. 309–337, doi:10.1016/S0037-0738(03)00135-0.

Ekdale, A.A., and Stinnesbeck, W., 1998, Ichnology of Cretaceous-Tertiary (K/T) boundary beds in northeastern Mexico: Palaios, v. 13, p. 593–602, doi:10.2307/3515350.

Font, E., Nedelec, A., Ellwood, B.B., Mirao, J., and Silva, P.F., 2011, A new sedimentary benchmark for the Deccan Traps volcanism?: Geophysical Research Letters, v. 38, p. L24309, doi:10.1029/2011GL049824.

Font, E., Fabre, S., Nédélec, A., Adatte, T., Keller, G., Veiga-Pires, C., Ponte, J., Mirão, J., Khozyem, H., and Spangenberg, J., 2014, this volume, Atmospheric halogen and acid rains during the main phase of Deccan eruptions: Magnetic and mineral evidence, *in* Keller, G., and Kerr, A.C., eds., Volcanism, Impacts, and Mass Extinctions: Causes and Effects: Geological Society of America Special Paper 505, doi:10.1130/2014.2505(18).

Gale, A., 2006, The Cretaceous-Tertiary boundary on the Brazos River, Falls County, Texas; evidence for impact-induced tsunami sedimentation?: Proceedings of the Geologists' Association, v. 117, p. 1–13.

Garg, R., Khowaja-Ateequzzaman, and Prasad, V., 2006, Significant dinoflagellate cyst biohorizons in the Upper Cretaceous–Palaeocene succession of the Khasi Hills, Meghalaya: Journal Geological Society of India, v. 67, p. 737–747.

Gertsch, B., Keller, G., Adatte, T., Garg, R., Prasad, V., Fleitmann, D., and Berner, Z., 2011a, Environmental effects of Deccan volcanism across the Cretaceous-Tertiary transition in Meghalaya, India: Earth and Planetary Science Letters, v. 310, p. 272–285.

Gertsch, B., Keller, G., Adatte, T., and Berner, Z., 2011b, Platinum group element (PGE) geochemistry of Brazos sections, Texas, *in* Keller, G., and Adatte, T., eds., End-Cretaceous Mass Extinction and the Chicxulub Impact in Texas: Society for Sedimentary Geology (SEPM) Special Publication 100, p. 228–249.

Govindan, A., 1981, Foraminifera from the infra- and intertrappean subsurface sediments of Narsapur Well-1 and age of the Deccan Trap flows, *in* Khosla, S.C., and Kachara, R.P., eds, Proceedings of the 9th Indian Colloquium of Micropalaeontology and Stratigraphy, p. 81–93.

Gradstein, F., Ogg, J., and Smith, A., 2004, A Geologic Time Scale: Cambridge, UK, Cambridge University Press, 598 p.

Harris, J., and Van Couvering, J., 1995, Mock aridity and the paleoecology of volcanically influenced ecosystems: Geology, v. 23, p. 593–596, doi:10.1130/0091-7613(1995)023<0593:MAATPO>2.3.CO;2.

Hart, M.B., Searle, S., Feist, S.E., Leighton, A.D., Price, G.D., Smart, C.W., and Twitchett, R.J., 2011, The distribution of benthic foraminifera across the Cretaceous-Paleogene boundary in Texas (Brazos River) and Denmark (Stevns Klint), *in* Keller, G., and Adatte, T., eds., The End-Cretaceous Mass Extinction and the Chicxulub Impact in Texas: Society for Sedimentary Geology (SEPM) Special Publication 100, p. 179–196.

Hart, M.B., Yancey, T.E., Leighton, A.D., Miller, B., Liu, C., Smart, C.W., and Twitchett, R.J., 2012, The Cretaceous-Paleogene boundary on the Brazos River, Texas: New stratigraphic sections and revised interpretations: Gulf Coast Association of Geological Societies Journal, v. 1, p. 69–80.

Heymann, D., Yancey, T.E., Wolbach, W.S., Thiemens, M.H., Johnson, E.A., Roach, D., and Moecker, S., 1998, Geochemical markers of the Cretaceous-Tertiary boundary event at Brazos River, Texas, USA: Geochimica et Cosmochimica Acta, v. 62, no. 1, p. 173–181, doi:10.1016/S0016-7037(97)00330-X.

Hofmann, C., Feraud, G., and Courtillot, V., 2000, 40Ar/39Ar dating of mineral separates and whole rocks from the Western Ghats lava pile: Further constraints on duration and age of the Deccan Traps: Earth and Planetary Science Letters, v. 180, p. 13–27, doi:10.1016/S0012-821X(00)00159-X.

Jaiprakash, B.C., Singh, J., and Raju, D.S.N., 1993, Foraminiferal events across the K/T boundary and age of Deccan volcanism in Palakollu area, Krishna-Godavari Basin, India: Journal of the Geological Society of India, v. 41, p. 105–117.

Jay, A.E., and Widdowson, M., 2008, Stratigraphy, structure and volcanology of the south-east Deccan continental flood basalt province: Implications for eruptive extent and volumes: Journal of the Geological Society of London, v. 165, p. 177–188, doi:10.1144/0016-76492006-062.

Jay, A.E., MacNiocaill, C., Widdowson, M., Self, S., and Turner, W., 2009, New palaeomagnetic data from the Mahabaleshwar Plateau, Deccan flood basalt province, India: Implications for the volcanostratigraphic architecture of continental flood basalt provinces: Journal of the Geological Society of London, v. 166, p. 13–24, doi:10.1144/0016-76492007-150.

Keller, G., 2008, Impact Stratigraphy: Old principle, new reality, *in* Evans, K.R., Horton, J.W., and King, D.T., eds., The Sedimentary Record of Meteorite Impacts: Geological Society of America Special Paper 437, p. 147–178.

Keller, G., 2011, Defining the Cretaceous-Tertiary boundary: A practical guide and return to first principles, *in* Keller, G., and Adatte, T., eds., The End-Cretaceous Mass Extinction and the Chicxulub Impact in Texas: Society for Sedimentary Geology (SEPM) Special Publication 100, p. 23–42.

Keller, G., and Abramovich, S., 2009, Lilliput effect in late Maastrichtian planktic foraminifera: Response to environmental stress: Palaeogeography, Palaeoclimatology, Palaeoecology, v. 284, p. 47–62, doi:10.1016/j.palaeo.2009.08.029.

Keller, G., and Benjamini, C., 1991, Paleoenvironment of the eastern Tethys in the early Danian: Palaios, v. 6, p. 439–464, doi:10.2307/3514984.

Keller, G., and Pardo, A., 2004, Disaster opportunists Guembelitrinidae—Index for environmental catastrophes: Marine Micropaleontology, v. 53, p. 83–116, doi:10.1016/j.marmicro.2004.04.012.

Keller, G., Lyons, J.B., MacLeod, N., and Officer, C.B., 1993, Is there evidence for Cretaceous-Tertiary boundary impact deposits in the Caribbean and Gulf of Mexico?: Geology, v. 21, p. 776–780, doi:10.1130/0091-7613(1993)021<0776:ITEFCT>2.3.CO;2.

Keller, G., Li, L., and Macleod, N., 1995, The Cretaceous/Tertiary boundary stratotype section at El Kef, Tunisia: How catastrophic was the mass extinction?: Palaeogeography, Palaeoclimatology, Palaeoecology, v. 119, p. 221–254, doi:10.1016/0031-0182(95)00009-7.

Keller, G., Lopez-Oliva, J.G., Stinnesbeck, W., and Adatte, T., 1997, Age, stratigraphy and deposition of near-K/T siliciclastic deposits in Mexico: Relation to bolide impact?: Geological Society of America Bulletin, v. 109, p. 410–428, doi:10.1130/0016-7606(1997)109<0410:ASADON>2.3.CO;2.

Keller, G., Adatte, T., Stinnesbeck, W., Luciani, V., Karoui, N., and Zaghbib-Turki, D., 2002a, Paleoecology of the Cretaceous-Tertiary mass extinction in planktic foraminifera: Palaeogeography, Palaeoclimatology, Palaeoecology, v. 178, p. 257–297, doi:10.1016/S0031-0182(01)00399-6.

Keller, G., Adatte, T., Stinnesbeck, W., Affolter, M., Schilli, L., and Lopez-Oliva, J.G., 2002b, Multiple spherule layers in the late Maastrichtian of northeastern Mexico, *in* Koeberl, C., and MacLeod, K.G., eds., Catastrophic Events and Mass Extinctions: Geological Society of America Special Paper 356, p. 145–162.

Keller, G., Stinnesbeck, W., Adatte, T. and Stueben, D., 2003a, Multiple impacts across the Cretaceous-Tertiary boundary: Earth-Science Reviews, v. 62, p. 327–363.

Keller, G., Stinnesbeck, W., Adatte, T., Holland, B., Stueben, D., Harting, M., de Leon, C., and de la Cruz, J., 2003b, Spherule deposits in Cretaceous-Tertiary boundary sediments in Belize and Guatemala: Journal of the Geological Society of London, v. 160, p. 783–795, doi:10.1144/0016-764902-119.

Keller, G., Adatte, T., Stinnesbeck, W., Rebolledo-Vieyra, M., Urrutia Fuccugauchi, J., Kramar, G., and Stueben, D., 2004a, Chicxulub predates the K/T boundary mass extinction: Proceedings of the National Academy of Sciences of the United States of America, v. 101, p. 3753–3758, doi:10.1073/pnas.0400396101.

Keller, G., Adatte, T., and Stinnesbeck, W., 2004b, More evidence that Chicxulub predates KT boundary: Meteoritics & Planetary Science, v. 39, no. 6/7, p. 1127–1144, doi:10.1111/j.1945-5100.2004.tb01133.x.

Keller, G., Adatte, T., Berner, Z., Harting, M., Baum, G., Prauss, M., Tantawy, A.A., and Stüben, D., 2007, Chicxulub impact predates KT boundary: New evidence from Brazos, Texas: Earth and Planetary Science Letters, v. 255, no. 3–4, p. 339–356, doi:10.1016/j.epsl.2006.12.026.

Keller, G., Adatte, T., Gardin, S., Bartolini, A., and Bajpai, S., 2008, Main Deccan volcanism phase ends near the K-T boundary: Evidence from the Krishna-Godavari Basin, SE India: Earth and Planetary Science Letters, v. 268, p. 293–311, doi:10.1016/j.epsl.2008.01.015.

Keller, G., Adatte, T., Bajpai, S., Mohabey, D.M., Widdowson, M., Khosla, A., Sharma, R., Khosla, S.C., Gertsch, B., Fleitmann, D., and Sahni, A., 2009a, K-T transition in Deccan Traps and intertrappean beds in central India mark major marine seaway across India: Earth and Planetary Science Letters, v. 282, p. 10–23, doi:10.1016/j.epsl.2009.02.016.

Keller, G., Khosla, S.C., Sharma, R., Khosla, A., Bajpai, S., and Adatte, T., 2009b, Early Danian planktic foraminifera from Cretaceous-Tertiary intertrappean beds at Jhilmili, Chhindwara District, Madhya Pradesh, India: Journal of Foraminiferal Research, v. 39, no. 1, p. 40–55, doi:10.2113/gsjfr.39.1.40.

Keller, G., Sahni, A., and Bajpai, S., 2009c, Deccan volcanism, the KT mass extinction and dinosaurs: Journal of Biosciences, v. 34, p. 709–728, doi:10.1007/s12038-009-0059-6.

Keller, G., Adatte, T., Pardo Juez, A., and Lopez-Oliva, J., 2009d, New evidence concerning the age and biotic effects of the Chicxulub impact in NE Mexico: Journal of the Geological Society of London, v. 166, p. 393–411, doi:10.1144/0016-76492008-116.

Keller, G., Adatte, T., Pardo, A., Bajpai, S., Khosla, A., and Samant, B., 2010, Cretaceous extinctions: Evidence overlooked: Science, v. 328, p. 974–975, doi:10.1126/science.328.5981.974-a.

Keller, G., Bhowmick, P.K., Upadhyay, H., Dave, A., Reddy, A.N., Jaiprakash, B.C., and Adatte, T., 2011a, Deccan volcanism linked to the Cretaceous-Tertiary boundary (KTB) mass extinction: New evidence from ONGC wells in the Krishna-Godavari Basin, India: Journal of the Geological Society of India, v. 78, p. 399–428, doi:10.1007/s12594-011-0107-3.

Keller, G., Abramovich, S., Adatte, T., and Berner, Z., 2011b, Biostratigraphy, age of the Chicxulub impact, and depositional environment of the Brazos River KTB sequences, *in* Keller, G., and Adatte, T., eds., The End-Cretaceous Mass Extinction and the Chicxulub Impact in Texas: Society for Sedimentary Geology (SEPM) Special Publication 100, p. 81–122.

Keller, G., Adatte, T., Bhowmick, P.K., Upadhyay, H., Dave, A., Reddy, A.N., and Jaiprakash, B.C., 2012, Nature and timing of extinctions in Cretaceous-Tertiary planktic foraminifera preserved in Deccan intertrappean sediments of the Krishna-Godavari Basin, India: Earth and Planetary Science Letters, v. 341–344, p. 211–221, doi:10.1016/j.epsl.2012.06.021.

Keller, G., Malarkodi, N., Khozeym, H., Adatte, T., Spangenberg, J.E., and Stinnesbeck, W., 2013, Chicxulub impact spherules in the NW Atlantic and Caribbean: Age constraints and KTB hiatus: Geological Magazine, v. 150, no. 5, p. 885–907, doi:10.1017/S0016756812001069.

Khadkikar, A.S., Sant, D.A., Gogte, V., and Karanth, R.V., 1999, The influence of Deccan volcanism in climate: Insights from lacustrine intertrappean deposits, Anjar, western India: Palaeogeography, Palaeoclimatology, Palaeoecology, v. 147, p. 141–149, doi:10.1016/S0031-0182(98)00156-4.

Khadri, S.F.R., Subbarao, K.V., Hooper, P.R., and Walsh, J.N., 1988, Stratigraphy of Thakurvadi Formation, western Deccan basalt province, India, *in* Subbarao, K.B., ed., Deccan Flood Basalt: Geological Society of India Memoir 10, p. 281–304.

Khosla, S.C., and Nagori, M.L., 2002, Ostracodes from the intertrappean beds (early Paleocene) of the east coast of India: Palaeontological Research, v. 6, no. 2, p. 191–210.

Khosla, S.C., Nagori, M.L., Jakhar, S.R., and Rathore, A.S., 2009, Mixed marine brackish water and non-marine microfaunal association in the intertrappean beds (early Paleocene) of Jhilmili, Chhindwara District, Madhya Pradesh: Journal of the Geological Society of India, v. 73, p. 724–732, doi:10.1007/s12594-009-0056-2.

Knight, K.B., Renne, P.R., Halkett, A., and White, N., 2003, ^{40}Ar/^{39}Ar dating of the Rajahmundry Traps, eastern India, and their relationship to the Deccan Traps: Earth and Planetary Science Letters, v. 208, p. 85–99, doi:10.1016/S0012-821X(02)01154-8.

Knight, K.B., Renne, P.R., Baker, J., Waight, T., and White, N., 2005, Reply to "^{40}Ar/^{39}Ar dating of the Rajahmundry Traps, eastern India, and their relationship to the Deccan Traps: Discussion" by A.K. Baksi: Earth and Planetary Science Letters, v. 239, p. 374–382, doi:10.1016/j.epsl.2005.08.013.

Kuiper, K.F., Deino, A., Hilgen, F.J., Krijgsman, W., Renne, P.R., and Wijbrans, J.R., 2008, Synchronizing rock clocks of Earth history: Science, v. 320, p. 500–504, doi:10.1126/science.1154339.

Li, L., and Keller, G., 1998a, Maastrichtian climate, productivity and faunal turnovers in planktic foraminifera of South Atlantic DSDP Sites 525A and 21: Marine Micropaleontology, v. 33, no. 1–2, p. 55–86, doi:10.1016/S0377-8398(97)00027-3.

Li, L., and Keller, G., 1998b, Diversification and extinction in Campanian-Maastrichtian planktic foraminifera of northwestern Tunisia: Eclogae Geologica Helvetiae, v. 91, p. 75–102.

Li, L., and Keller, G., 1998c, Abrupt deep-sea warming at the end of the Cretaceous: Geology, v. 26, p. 995–998, doi:10.1130/0091-7613(1998)026<0995:ADSWAT>2.3.CO;2.

Luciani, V., 2002, High resolution planktonic foraminiferal analysis from the Cretaceous/Tertiary boundary at Ain Settara (Tunisia): Evidence of an extended mass extinction: Palaeogeography, Palaeoclimatology, Palaeoecology, v. 178, no. 3, p. 299–319, doi:10.1016/S0031-0182(01)00400-X.

MacLeod, K.G., Huber, B.T., and Isaza-Londono, C., 2005, North Atlantic warming during global cooling at the end of the Cretaceous: Geology, v. 33, p. 437–440, doi:10.1130/G21466.1.

MacLeod, K.G., Whitney, D.L., Huber, B.T., and Koeberl, C., 2007, Impact and extinction in remarkably complete Cretaceous-Tertiary boundary sections from Demerara Rise, tropical western North Atlantic: Geological Society of America Bulletin, v. 119, p. 101–115, doi:10.1130/B25955.1.

Magaritz, M., Benjamini, C., Keller, G., and Moshkovitz, S., 1992, Early diagenetic isotopic signal at the Cretaceous/Tertiary boundary, Israel: Palaeogeography, Palaeoclimatology, Palaeoecology, v. 91, no. 3, p. 291–304, doi:10.1016/0031-0182(92)90073-E.

Malarkodi, N., Keller, G., Fayazudeen, P.J., and Mallikarjuna, U.B., 2010, Foraminifera from the early Danian intertrappean beds in Rajahmundry Quarries, Andhra Pradesh, SE India: Journal of the Geological Society of India, v. 75, p. 851–863, doi:10.1007/s12594-010-0066-0.

McLean, D., 1978, A terminal Mesozoic "greenhouse": Lessons from the past: Science, v. 201, no. 4354, p. 401–406, doi:10.1126/science.201.4354.401.

McLean, D., 1985, Deccan Traps mantle degassing in the terminal Cretaceous marine extinctions: Cretaceous Research, v. 6, p. 235–259, doi:10.1016/0195-6671(85)90048-5.

Mehrotra, N.C., and Sargeant, W.A.S., 1987, Late Cretaceous to Early Tertiary dinoflagellate cysts from Narsapur Well-1, Godavari-Krishna Basin, south-India: Geobios, v. 20, p. 149–191, doi:10.1016/S0016-6995(87)80033-5.

Miller, K.G., Sherrell, R.M., Browning, J.V., Field, M.P., Gallagher, W., Olsson, R.K., Sugarman, P.J., Tuorto, S., and Wahyudi, H., 2010, Relationship between mass extinction and iridium across the Cretaceous-Paleogene boundary in New Jersey: Geology, v. 38, p. 867–870, doi:10.1130/G31135.1.

Misra, K.S., 2005, Distribution pattern, age and duration and mode of eruption of Deccan and associated volcanics: Gondwana Geological Magazine, v. 8, p. 53–60.

Molina, E., Alegret, L., Arenillas, I., Arz, J., Gallala, N., Hardenbol, J., von Salis, K., Steurbaut, E., Vanhenberghe, N., and Zaghbib-Turki, D., 2006, The global boundary stratotype section and point for the base of the Danian Stage (Paleocene, Paleogene, "Tertiary," Cenozoic) at El Kef, Tunisia—Original definition and revision: Episodes, v. 29, no. 4, p. 263–273.

Mussard, M., Le Hir, G., Fluteau, F., Lefebvre, B., and Godderis, Y., 2014, this volume, Modeling the carbon-sulfate interplays in climate changes related to the emplacement of continental flood basalts, *in* Keller, G., and Kerr, A.C., eds., Volcanism, Impacts, and Mass Extinctions: Causes and Effects: Geological Society of America Special Paper 505, doi:10.1130/2014.2505(17).

Nesbitt, H.W., and Young, G.M., 1982, Early Proterozoic climates and plate motions inferred from major element chemistry of lutites: Nature, v. 299, p. 715–717, doi:10.1038/299715a0.

Nesbitt, H.W., and Young, G.M., 1989, Formation and diagenesis of weathering profiles: The Journal of Geology, v. 97, p. 129–147, doi:10.1086/629290.

Norris, R.D., Huber, B.T., and Self-Trail, B.T., 1999, Synchroneity of the KT oceanic mass extinction and meteorite impact: Blake Nose, western North Atlantic: Geology, v. 27, p. 419–422, doi:10.1130/0091-7613(1999)027<0419:SOTKTO>2.3.CO;2.

Norris, R.D., Firth, J., Blusztajn, J., and Ravizza, G., 2000, Mass failure of the North Atlantic margin triggered by the Cretaceous-Paleogene bolide impact: Geology, v. 28, p. 1119–1122, doi:10.1130/0091-7613(2000)28<1119:MFOTNA>2.0.CO;2.

Olsson, R.K., Miller, K.G., Browning, J.V., Habib, D., and Sugarmann, P.J., 1997, Ejecta layer at the Cretaceous-Tertiary boundary, Bass River, New Jersey (Ocean Drilling Program Leg 174AX): Geology, v. 25, p. 759–762, doi:10.1130/0091-7613(1997)025<0759:ELATCT>2.3.CO;2.

Pande, K., Pattanayak, S.K., Subbarao, K.V., Navaneethakrishnan, P., and Venkatesan, T.R., 2004, 40Ar/39Ar age of a lava flow from the Bhimashankar Formation, Giravali Ghat, Deccan Traps: Proceedings of the Indian Academy of Sciences, v. 113, p. 755–758.

Pandey, J., 1990, Cretaceous/Tertiary boundary, iridium anomaly, and foraminifer breaks in the Um Sohryngkew River section, Meghalaya: Current Science, v. 59, p. 570–575.

Pardo, A., and Keller, G., 2008, Biotic effects of environmental catastrophes at the end of the Cretaceous: *Guembelitria* and *Heterohelix* blooms: Cretaceous Research, v. 29, no. 5/6, p. 1058–1073, doi:10.1016/j.cretres.2008.05.031.

Pardo, A., Ortiz, N., and Keller, G., 1996, Latest Maastrichtian and K/T boundary foraminiferal turnover and environmental changes at Agost, Spain, *in* MacLeod, N., and Keller, G., eds., The Cretaceous-Tertiary Mass Extinction: Biotic and Environmental Effects: New York, Norton Press, p. 157–191.

Prasad, B., and Pundeer, B.S., 2002, Palynological events and zones in Cretaceous-Tertiary boundary sediments of Krishna-Godavari and Cauvery Basins, India: Palaeontographica, abt. B, v. 262, p. 39–70.

Prasad, G.V.R., and Sahni, A., 2014, this volume, Vertebrate fauna from the Deccan volcanic province: Response to volcanic activity, *in* Keller, G., and Kerr, A.C., eds., Volcanism, Impacts, and Mass Extinctions: Causes and Effects: Geological Society of America Special Paper 505, doi:10.1130/2014.2505(09).

Premovic, P.I., Ilic, B.S., and Dordevic, M.G., 2012, Iridium anomaly in the Cretaceous-Paleogene boundary at Hojerup (Stevns Klint, Denmark) and Woodside Creek (New Zealand): The question of an enormous proportion of extraterrestrial component: Journal of the Serbian Chemical Society, v. 77, no. 2, p. 247–255, doi:10.2298/JSC110404178P.

Punekar, J., Mateo, P.M., and Keller, G., 2014, this volume, Effects of Deccan volcanism on paleoenvironment and planktic foraminifera: A global survey, *in* Keller, G., and Kerr, A.C., eds., Volcanism, Impacts, and Mass Extinctions: Causes and Effects: Geological Society of America Special Paper 505, doi:10.1130/2014.2505(04).

Quillévéré, F., Norris, R.D., Kroon, D., and Wilson, P.A., 2008, Transient ocean warming and shifts in carbon reservoirs during the early Danian: Earth and Planetary Science Letters, v. 265, p. 600–615, doi:10.1016/j.epsl.2007.10.040.

Raju, D.S.N., 2008, Stratigraphy of India: ONGC Bulletin Special Issue 43, no. 1, 44 p.

Raju, D.S.N., Jaiprakash, B.C., Kumar, A., Saxena, R.K., Dave, A., Chatterjee, T.K., and Mishra, C.M., 1994, The magnitude of hiatus and sea-level changes across K/T boundary in Cauvery and Krishna-Godavari Basins, India: Journal of the Geological Society of India, v. 44, p. 301–315.

Raju, D.S.N., Jaiprakash, B.C., Kumar, A., Saxena, R.K., Dave, A., Chatterjee, T.K., and Mishra, C.M., 1995, Age of Deccan volcanism across KTB in Krishna-Godavari Basin: New evidences: Journal of the Geological Society of India, v. 45, p. 229–233.

Raju, D.S.N., Jaiprakash, B.C., and Kumar, A., 1996, Paleoenvironmental setup and age of basin floor just prior to the spread of Deccan volcanism in the Krishna-Godavari Basin, India: Geological Society of India Memoir 37, p. 285–295.

Remane, J., Keller, G., Hardenbol, J., and Ben Haj Ali, M., 1999, Report on the International Workshop on Cretaceous-Paleogene Transitions: Episodes, v. 22, no. 1, p. 47–48.

Renne, P.R., Deino, A.L., Hilgen, F.J., Kuiper, F.G., Mark, D.F., Mitchell, W.S., III, Morgan, L.E., Mundil, R., and Smit, J., 2013, Time scales of critical events around the Cretaceous-Paleogene boundary: Science, v. 339, p. 684–687, doi:10.1126/science.1230492.

Rice, A., 1990, The role of volcanism in K/T extinctions, *in* Lockley, M.G., and Rice, A., eds., Volcanism and Fossil Biotas: Geological Society of America Special Paper 244, p. 39–56.

Robinson, N., Ravizza, G., Coccioni, R., Peucker-Ehrenbrink, B., and Norris, R., 2009, A high-resolution marine 187Os/188Os record for the late Maastrichtian: Distinguishing the chemical fingerprints of Deccan volcanism and the KP impact event: Earth and Planetary Science Letters, v. 281, p. 159–168, doi:10.1016/j.epsl.2009.02.019.

Sahni, A., 1983. Upper Cretaceous palaeobiogeography of Peninsular India and the Cretaceous-Paleocene transition: The vertebrate evidence, *in* Maheshwari, H.K., ed., Cretaceous of India: Lucknow, India, Indian Association of Palynostratigraphers, p. 128–140.

Samant, B., and Mohabey, D.M., 2014, this volume, Deccan volcanic eruptions and impact on flora: Palynological evidence, *in* Keller, G., and Kerr, A.C, eds., Volcanism, Impacts, and Mass Extinctions: Causes and Effects: Geological Society of America Special Paper 505, doi:10.1130/2014.2505(08).

Sant, D.A., Mathew, G., Khadkikar, A.S., Gogte, V., and Gundurao, T.K., 2003, Co-existent cristobalite and iridium at 65 Ma, Anjar intertrappeans, Kachchh, western India: Cretaceous Research, v. 24, p. 105–110, doi:10.1016/S0195-6671(03)00028-4.

Saxena, R.K., and Misra, C.M., 1994, Time and duration of Deccan volcanism in the Razole area, Krishna-Godavari Basin, India: Current Science, v. 66, no. 1, p. 73–76.

Saxena, R.K., and Misra, C.M., 1995, Campanian-Maastrichtian nannoplankton biostratigraphy of the Narsapur claystone Formation, Krishna-Godavari Basin, India: Journal of the Geological Society of India, v. 45, p. 323–329.

Schmitz, B., Keller, G., and Stenvall, O., 1992, Stable isotope changes across the Cretaceous-Tertiary boundary at Stevns Klint, Denmark: Arguments for long-term oceanic instability before and after bolide impact event: Palaeogeography, Palaeoclimatology, Palaeoecology, v. 96, p. 233–260, doi:10.1016/0031-0182(92)90104-D.

Schulte, P., Stinnesbeck, W., Stueben, D., Kramar, U., Berner, Z., Keller, G., and Adatte, T., 2003, Fe-rich and K-rich mafic spherules from slumped and channelized Chicxulub ejecta deposits in the northern La Sierrita area, NE Mexico: International Journal of Earth Sciences, v. 92, p. 114–142.

Schulte, P., Speijer, R., Mai, H., and Kontny, A., 2006, The Cretaceous-Paleogene (K-P) boundary at Brazos, Texas: Sequence stratigraphy, depositional events and the Chicxulub impact: Sedimentary Geology, v. 184, no. 1–2, p. 77–109, doi:10.1016/j.sedgeo.2005.09.021.

Schulte, P., Alegret, L., Arenillas, I., Arz, J.A., Barton, P.J., Bown, P.R., Bralower, T.J., Christeson, G.L., Claeys, P., Cockell, C.S., Collins, G.S., Deutsch, A., Goldin, T.J., Goto, K., Grajales-Nishimura, J.M., Grieve, R.A.F., Gulick, S.P.S., Johnson, K.R., Kiessling, W., Koeberl, C., Kring, D.A., MacLeod, K.G., Matsui, T., Melosh, J., Montanari, A., Morgan, J.V., Neal, C.R., Nichols, D.J., Norris, R.D., Pierazzo, E., Ravizza, G., Rebolledo-Vieyra, M., Reimold, W.U., Robin, E., Salge, T., Speijer, R.P., Sweet, A.R., Urrutia-Fucugauchi, J., Vajda, V., Whalen, M.T., and Willumsen, P.S., 2010, The Chicxulub asteroid impact and mass extinction at the Cretaceous-Paleogene boundary: Science, v. 327, p. 1214–1218, doi:10.1126/science.1177265.

Self, S., Widdowson, M., Thordarson, T., and Jay, A.E., 2006, Volatile fluxes during flood basalt eruptions and potential effects on the global environment: A Deccan perspective: Earth and Planetary Science Letters, v. 248, p. 518–532, doi:10.1016/j.epsl.2006.05.041.

Self, S., Jay, A.E., Widdowson, M., and Keszthelyi, L.P., 2008, Correlation of the Deccan and Rajahmundry Trap lavas: Are these the longest and largest lava flows on Earth?: Journal of Volcanology and Geothermal Research, v. 172, p. 3–19, doi:10.1016/j.jvolgeores.2006.11.012.

Sharma, R., and Khosla, A., 2009, Early Paleocene ostracoda from the Cretaceous-Tertiary Deccan intertrappean sequence at Jhilmili, Chhindwara District, Central India: Journal of the Palaeontological Society of India, v. 54, no. 2, p. 197–208.

Sheth, H.C., Pande, K., and Bhutani, R., 2001, $^{40}Ar/^{39}Ar$ age of a national geological monument: The Gilbert Hill basalt, Deccan Traps, Bombay: Current Science, v. 80, p. 1437–1440.

Smit, J., Montanari, A., Swinburne, N.H.M., Alvarez, W., Hildebrand, A.R., Margolis, S.V., Claeys, P., Lowrie, W., and Asaro, F., 1992, Tektite-bearing, deep-water clastic unit at the Cretaceous-Tertiary boundary in northeastern Mexico: Geology, v. 20, p. 99–103, doi:10.1130/0091-7613(1992)020<0099:TBDWCU>2.3.CO;2.

Smit, J., Roep, T.B., Alvarez, W., Montanari, A., Claeys, P., Grajales-Nishimura, J.M., and Bermudez, J., 1996, Coarse-grained clastic sandstone complex at the K-T boundary around the Gulf of Mexico: Deposition by tsunami waves induced by the Chicxulub impact, *in* Ryder, G., Fastovsky, D., and Gartner, S., eds., The Cretaceous-Tertiary Event and Other Catastrophes in Earth History: Geological Society of America Special Paper 307, p. 151–182.

Smit, J., van der Gaast, S., and Lustenhouwer, W., 2004, Is the transition to post-impact rock complete? Some remarks based on XRF scanning electron microprobe and thin section analyses of the Yaxcopoil-1 core: Meteoritics & Planetary Science, v. 39, p. 1113–1126, doi:10.1111/j.1945-5100.2004.tb01132.x.

Subbarao, K.V., and Pathak, S., 1993, Reversely magnetized flows, Rajahmundry, Andhra Pradesh: Journal of the Geological Society of India, v. 41, p. 71–72.

Tantawy, A.A., 2003, Calcareous nannofossil biostratigraphy and paleoecology of the Cretaceous-Tertiary transition in the western desert of Egypt: Marine Micropaleontology, v. 47, p. 323–356, doi:10.1016/S0377-8398(02)00135-4.

Vandamme, D., Courtillot, V., Besse, J., and Montigny, R., 1991, Paleomagnetism and age determinations of the Deccan Traps (India): Results of a Nagpur-Bombay traverse and review of earlier work: Reviews of Geophysics and Space Physics, v. 29, v. 159–190.

Venkatesan, T.R., Pande, K., and Gopalan, K., 1993, Did Deccan volcanism pre-date the Cretaceous-Tertiary transition?: Earth and Planetary Science Letters, v. 119, no. 1–2, p. 181–189, doi:10.1016/0012-821X(93)90015-2.

von Salis, K., and Saxena, R.K., 1998, Calcareous nannofossils across the K/T boundary and the age of the Deccan Trap volcanism in southern India: Journal of the Geological Society of India, v. 51, p. 183–192.

Ward, W., Keller, G., Stinnesbeck, W., and Adatte, T., 1995, Yucatan subsurface stratigraphy: Implications and constraints for the Chicxulub impact: Geology, v. 23, p. 873–876, doi:10.1130/0091-7613(1995)023<0873:YNSSIA>2.3.CO;2.

Watkins, D.K., and Self-Trail, J.M., 2005, Calcareous nannofossil evidence for the existence of the Gulf Stream during the late Maastrichtian: Paleoceanography, v. 20, PA3006, doi:10.1029/2004PA001121.

Wedepohl, K.H., 1971, Environmental influences on the chemical composition of shales and clays, *in* Ahrens, L.H., Press, F., Runcorn, S.K., and Urey, H.C., eds., Physics and Chemistry of the Earth: Oxford, UK, Pergamon, p. 305–333.

Wignall, P.B., 2001, Large igneous provinces and mass extinctions: Earth-Science Reviews, v. 53, p. 1–33, doi:10.1016/S0012-8252(00)00037-4.

Wilf, P., Johnson, K.R., and Huber, B.T., 2003, Correlated terrestrial and marine evidence for global climate changes before mass extinction at the Cretaceous-Paleogene boundary: Proceedings of the National Academy of Sciences of the United States of America, v. 100, no. 2, p. 599–604, doi:10.1073/pnas.0234701100.

Manuscript Accepted by the Society 31 January 2014

Printed in the USA

The Geological Society of America
Special Paper 505
2014

Effects of Deccan volcanism on paleoenvironment and planktic foraminifera: A global survey

Jahnavi Punekar*
Paula Mateo
Gerta Keller
Department of Geosciences, Princeton University, Princeton, New Jersey 08544, USA

ABSTRACT

Deccan volcanism, one of Earth's largest flood basalt provinces, erupted ~80% of its total volume (phase 2) during a relatively short time in the uppermost Maastrichtian paleomagnetic chron C29r and ended with the Cretaceous-Tertiary boundary mass extinction. Full biotic recovery in the marine realm was delayed at least 500 k.y. or until after the last Deccan eruptions in C29n (phase 3, 14% of the total Deccan volume). For over 30 yr, the mass extinction has been commonly attributed to the Chicxulub impact, and the delayed recovery remained an enigma. Here, we demonstrate that the two phases of Deccan volcanism can account for both the mass extinction and delayed marine recovery.

In India, a direct correlation between Deccan eruptions (phase 2) and the mass extinction reveals that ~50% of the planktic foraminifer species gradually disappeared during volcanic eruptions prior to the first of four lava megaflows, reaching ~1500 km across India, and out to the Bay of Bengal. Another 50% disappeared after the first megaflow, and the mass extinction was complete with the last megaflow. Throughout this interval, blooms of the disaster opportunist *Guembelitria cretacea* dominate shallow-marine assemblages in coeval intervals from India to the Tethys and the Atlantic Oceans to Texas. Similar high-stress environments dominated by blooms of *Guembelitria* and/or *Globoconusa* are observed correlative with Deccan volcanism phase 3 in the early Danian C29n, followed by full biotic recovery after volcanism ended. The mass extinction and high-stress conditions may be explained by the intense Deccan volcanism leading to rapid global warming and cooling in C29r and C29n, enhanced weathering, continental runoff, and ocean acidification, resulting in a carbonate crisis in the marine environment.

*jpunekar@princeton.edu

Punekar, J., Mateo, P., and Keller, G., 2014, Effects of Deccan volcanism on paleoenvironment and planktic foraminifera: A global survey, *in* Keller, G., and Kerr, A.C., eds., Volcanism, Impacts, and Mass Extinctions: Causes and Effects: Geological Society of America Special Paper 505, p. 91–116, doi:10.1130/2014.2505(04).

INTRODUCTION

For over three decades, a large extraterrestrial impact has been advocated as the sole cause for the Cretaceous-Tertiary (K-T) boundary mass extinction (also known as the Cretaceous-Paleogene [or K-Pg] mass extinction), and Deccan volcanism in India was dismissed as ineffectual (see Schulte et al., 2010). In this scenario, the late Maastrichtian planktic foraminiferal turnover and survivorship across the mass extinction, first noted by Keller (1988a; Keller et al., 1995) were attributed to artifacts of the sedimentary record. A blind test was devised to prove that the mass extinction was a sudden event and left no survivors but one species, consistent with the impact hypothesis (Smit, 1982). When the blind test did not confirm this assumption, this test was considered inconclusive (Ginsburg, 1997a, 1997b, 1997c; Lipps, 1997).

Over the past 15 yr, these contested extinction and survivorship patterns and associated environmental changes have been documented worldwide and increasingly linked to the late Maastrichtian global warming and Deccan volcanism (e.g., Li and Keller, 1998a, 1998b; Abramovich et al., 1998; Keller, 2001, 2008a, 2008b, 2011; Keller et al., 2009a; Abramovich and Keller, 2003; Tobin et al., 2012). However, the proposed link remained speculative and vague, referring to Deccan volcanism as a single undifferentiated event largely expressed by the latest Maastrichtian global warming. Only with the documentation of Deccan volcanism as three distinct phases (Chenet et al., 2007, 2008, 2009) and the discovery and documentation of the direct link between the mass extinction and Deccan volcanism (Keller et al., 2008, 2009, 2011a, 2012) has it become possible to directly link environmental and faunal changes with specific phases of volcanism (see Keller, this volume).

The objective of this paper is to test the twin hypotheses:

(1) The main phase 2 of Deccan volcanism in C29r near the end of the Maastrichtian caused global high-stress conditions for planktic foraminifera that resulted in the mass extinction.
(2) Deccan phase 3 in the early Danian C29n prolonged environmental stress, preventing full marine recovery for about a half million years.

We evaluate these hypotheses based on previously published work and global archival data on planktic foraminifera and carbon and oxygen stable isotopes that may reveal the devastating environmental and climatic effects of these massive continental flood basalt eruptions. It is well known that Deccan phase 2 corresponds to the rapid global warming and cooling near the end of the Maastrichtian (Li and Keller, 1998b; Ravizza and Puecker-Ehrenbrink, 2003; MacLeod et al., 2005; Robinson et al., 2009) and more specifically in planktic foraminiferal zones CF2 and CF1, which span the last 120 k.y. and 160 k.y. of the late Maastrichtian paleomagnetic chron C29r, respectively (Li and Keller, 1998a, 1998b; Abramovich et al., 2002, 2003, 2011; Keller et al., 2011a, 2011b, 2012). Deccan phase 3 corresponds to the rapid global warming in the early Danian known as the Dan-C2 event (Quillévéré et al., 2008; Coccioni et al., 2010) correlative with zone P1b of Keller et al. (1995, 2002a, 2011a, 2011b). Both climatic events are associated with major faunal changes, including blooms of the disaster opportunist *Guembelitria cretacea*, dwarfing of species, and decreased diversity in India and globally (Fig. 1). For this review, we choose localities that span from India through the Tethys and Atlantic Oceans to Texas, concentrating mainly on low to middle latitudes where the faunal effects were the most severe.

ENVIRONMENTAL EFFECTS OF VOLCANISM

Understanding the environmental and biological effects of large igneous province volcanism is nearly beyond comprehension. Recent studies are just beginning to provide a glimpse of the potentially devastating effects of sequential pulses of rapid, massive outpouring of lavas, and injection of ashes and gases (e.g., carbon dioxide, sulfur dioxide, chlorine, fluorine) during large igneous province emplacement (e.g., Columbia River, Deccan, Central Atlantic, and Siberian volcanic provinces; see Courtillot and Fluteau, this volume). For a basic understanding, we turn to a more recent example with fissure-type eruptions, such as Laki, Iceland, believed to be similar to Deccan volcanism.

Laki Eruptions of Iceland 1783–1784

From June 1783 to February 1784, Iceland was rocked by the largest volcanic eruption in human recorded history. About 130 craters opened along fissures in Earth's surface (Fig. 2A), and the rising magma interacted with groundwater, creating explosive lava fountains reaching heights of 900–1300 m. This volcanic eruption released an estimated 8 Megatons (Mt) of fluorine and ~122 Mt of sulfur dioxide (SO_2) into the atmosphere, ~0.91 km^3 of tephra (ash) blanketed Iceland and large parts of Europe, and 14.3 km^3 of basaltic lava spread over Iceland in 8 mo of volcanic activity (Thordarson and Self, 2003).

Convective currents of the eruption column carried ~20% of the sulfur mass released from the lava into the lower troposphere (~5 km), where it formed a local haze (Fig. 2B; Thordarson and Self, 2003). About 80% of the sulfur mass (~95 Mt of SO_2) released was carried to a height of 10–15 km into the upper troposphere–lower stratosphere, where sulfuric aerosols formed by reaction of SO_2 and atmospheric vapor produced ~200 Mt of H_2SO_4 aerosols that spread as distal haze across Europe (Thordarson and Self, 2003). About 175 Mt of the H_2SO_4 aerosols were removed as acid rain over Europe, causing extreme volcanic pollution and blanketing the region as dry fog (Stothers, 1996; Thordarson and Self, 2003), which resulted in severe climate extremes with dire consequences for humans and livestock. The excessive fluorine input resulted in fluorine poisoning, which along with the ensuing famine killed 20%–25% of the human population and 50%–80% of livestock on Iceland (Brayshay and Grattan, 1999; Grattan et al., 2003; Stone, 2004). Extreme summer heat in 1783 turned to extreme winter cold in 1784, and spring flooding

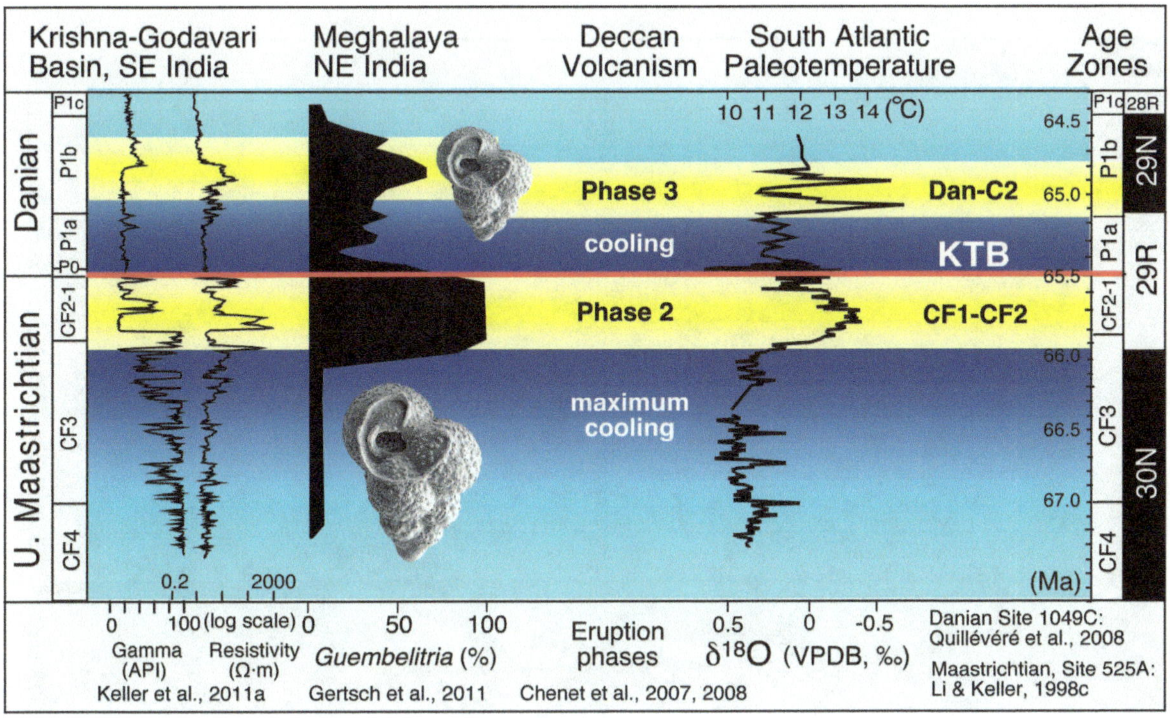

Figure 1. *Guembelitria* blooms in northeastern and southeastern India correlate with Deccan volcanic phase 2 and phase 3 lava megaflows from deep wells in the Krishna-Godavari Basin. Deccan phase 2 and phase 3 volcanism is correlative with global warming events in the late Maastrichtian zones CF1–CF2 and the early Danian Dan-C2 event in zone P1b. KTB—Cretaceous-Tertiary boundary; API—American Petroleum Institute units; VPDB—Vienna Peedee belemnite.

caused tens of thousands of additional deaths across Europe and North America. A decade of extreme climate disruption followed, destroying crops and leading to famines.

Deccan Eruptions of India

Compared with the Deccan flood basalt eruptions, the Laki eruptions were inconsequential. During Deccan eruptions, an estimated 1.5 million km^3 of lava (100,000 times the size of Laki) flooded the continent; these deposits currently cover an area the size of France and were probably three times larger prior to erosion (Fig. 3; Raja Rao et al., 1999). This lava outpouring occurred in three phases, with the initial and smallest phase 1 (~6% of the total volume) in the early late Maastrichtian chron C30r/C30n ca. 67.4 Ma (or ca. 68.33 Ma according to *The Geologic Time Scale* of Gradstein et al., 2012). The second and largest phase 2 erupted ~80% of the total lava pile (~80,000 times the Laki eruptions) near the end of the Maastrichtian in chron C29r (Subbarao et al., 2000; Jay and Widdowson, 2008; Chenet et al., 2007, 2008), ending with the K-T boundary mass extinction (Keller et al., 2011a, 2012). The last phase 3 (~14% of the total volume) erupted in the early Danian chron C29n.

However, even more important factors than the size of lava eruptions were the tempo of emplacement and the amounts of SO_2, CO_2, Cl, and other gases released into the atmosphere. For the Deccan Traps, Chenet et al. (2007, 2009) estimated a total of 15,000–35,000 Gigatons (Gt) of CO_2 and 6500–17,000 Gt of SO_2 (compared with 0.122 Gt of SO_2 at Laki), with 80% erupting during phase 2. These totals do not account for the gases released by massive magma intrusions, contact metamorphism with limestones, coal deposits, and organic carbon–rich sediments, which could exceed the amount of magmatic gases (Ganino and Arndt, 2009; Courtillot and Fluteau, this volume; Mussard et al., this volume). The tempo of eruptions for each of the three main phases is difficult to estimate because each consisted of several single eruptive events with lava volumes of 1000 km^3 to 20,000 km^3 and some megaflows reaching ~1500 km across India and out into the Bay of Bengal. Emplacement of each single eruptive event is estimated at less than a decade, with each of the three phases emplaced over ~100 k.y., including times of no volcanic activity between flows (Chenet et al., 2009; Courtillot and Fluteau, this volume). This age estimate is consistent with biostratigraphic data for phase 2, which ends with the rapid emplacement of the four largest lava megaflows across India coincident with the K-T boundary mass extinction (Keller et al., 2011a, 2012). A similar sequence of megaflows ends Deccan Traps phase 3 in the early Danian.

Deccan Volcanism and Chicxulub Impact

How does the gas emission of Deccan volcanism compare with the Chicxulub impact? Chenet et al. (2008, 2009) estimated that the SO_2 and CO_2 flux of the Chicxulub impact may be

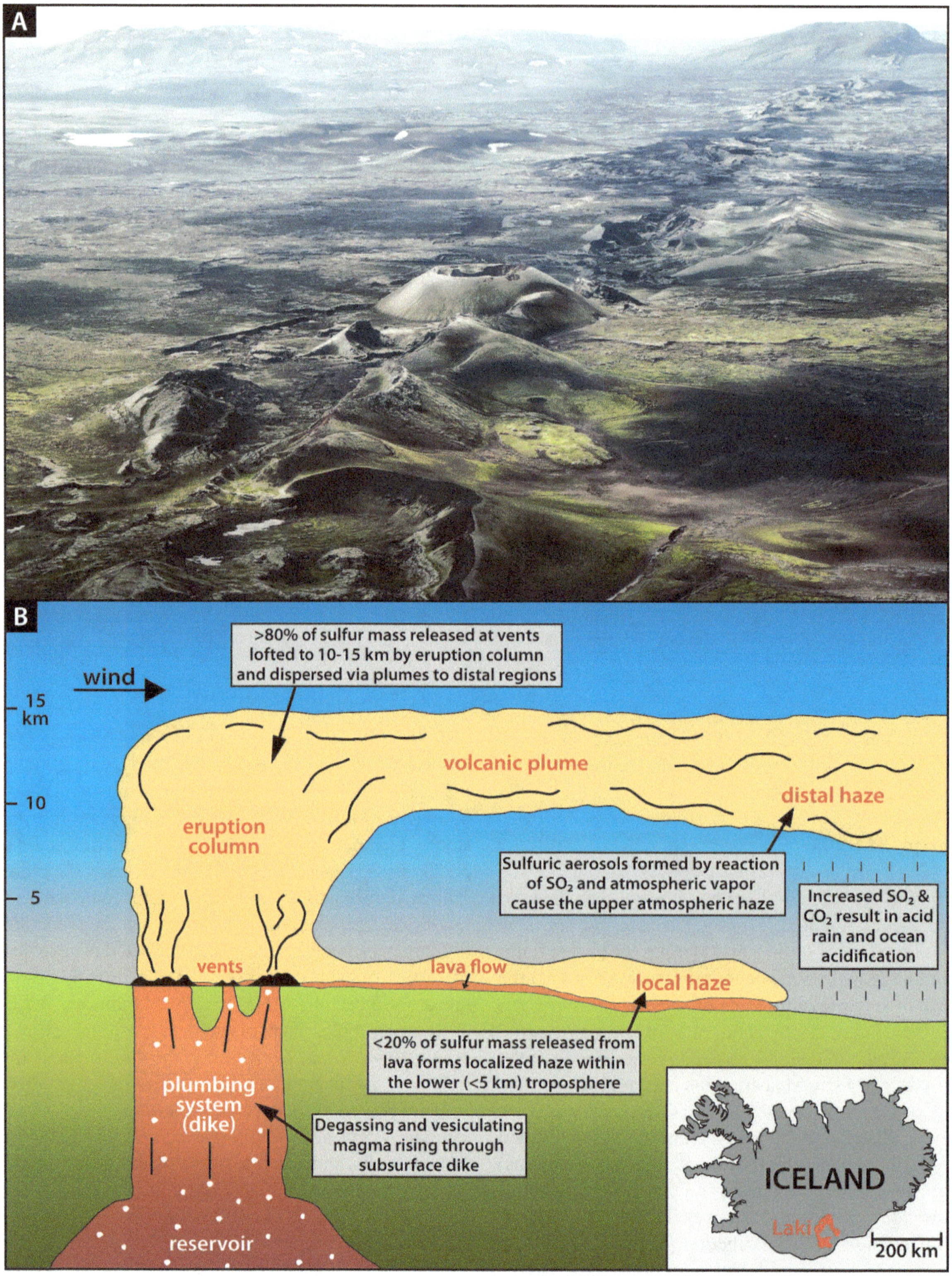

Figure 2. (A) Volcanoes dot the 1783–1784 Laki fissure eruptions of Iceland (photo by Klaus Ratzer, Germany, http://www.ratzer-art.de). (B) Schematic illustration of the Laki eruptions and distribution of volcanic gases (modified from Thordarson and Self, 2003). Similar fissure eruptions are believed to have characterized Deccan volcanism in India, although at a scale of ~100,000 times Laki today and ~300,000 times Laki prior to erosion.

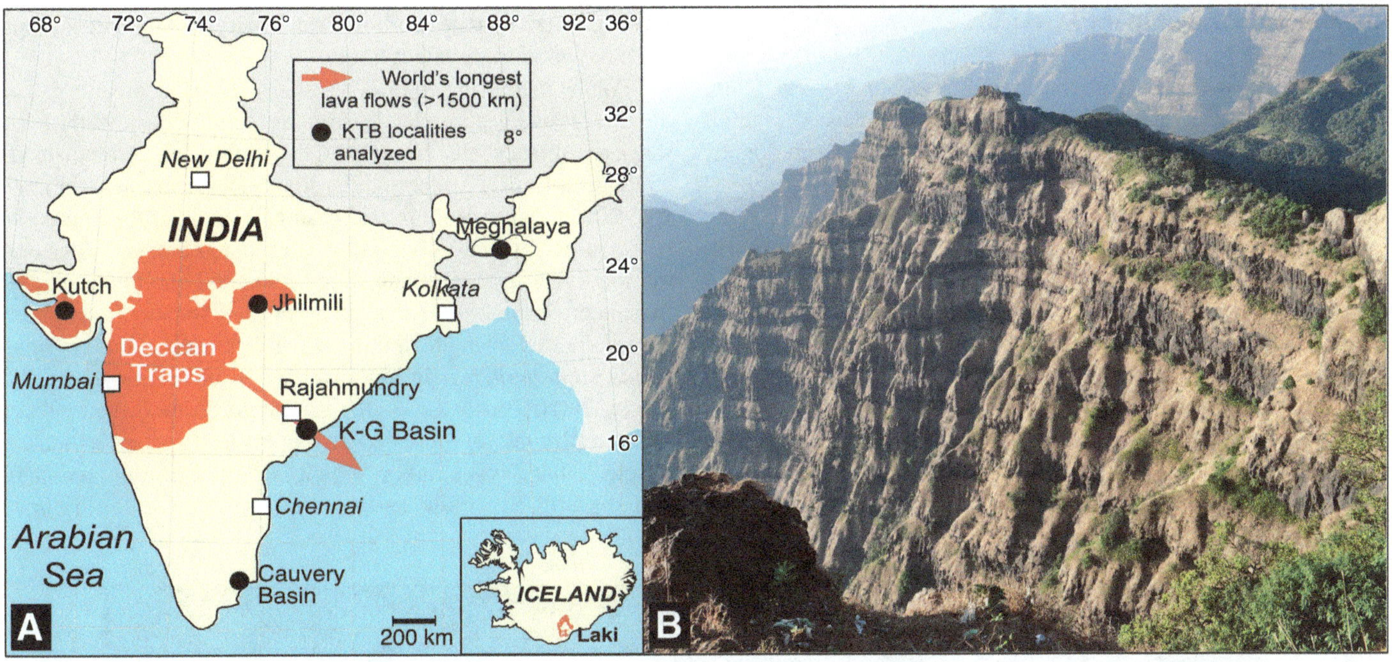

Figure 3. (A) Distribution of Deccan volcanic rocks in India today; prior to erosion, Deccan Trap basalts are believed to have covered about three times this area. Black circles mark sections studied. A map of Iceland with the area of the Laki eruptions is shown at the same relative size for comparison. (B) Photo shows the Deccan basalts forming 3000-m-high mountains near the Mahabaleshwar Hill Station in the Western Ghats. KTB—Cretaceous-Tertiary boundary; K-G—Krishna-Godavari Basin.

comparable to just one of the large single eruptive events of Deccan phase 2. However, the total Deccan gas output may have been on the order of 20–200 times that of the Chicxulub impact (Courtillot and Fluteau, this volume). Ultimately, the killing mechanism was not just the total gas output, but how rapidly eruptions followed each other. The environmental effects of a single event, such as the Chicxulub impact or one large Deccan eruption in phase 2, would have been relatively short-lived, and the environment would have recovered without necessarily causing species extinctions. Similarly, volcanic eruptions separated by long periods of quiet permit ecosystem recovery and the survival of most species. However, eruptions that follow each other in rapid succession prevent recovery and lead to runaway effects. This may have occurred at the end of the Maastrichtian, when lava eruptions followed in rapid succession with no evidence of long periods of inactivity, including the four lava megaflows that reached ~1500 km across India and into the Bay of Bengal, ending with the mass extinction (Keller et al., 2011a, 2012; Keller, this volume).

BIOTIC EFFECTS—PLANKTIC FORAMINIFERA

Planktic foraminifera occupy a wide range of habitats from the surface ocean to subthermocline depths, depending on the chlorophyll maximum zone, temperature, oxygen, and salinity constraints. Large-scale events, such as large igneous province volcanism and very large extraterrestrial impacts, can trigger global climate and environmental changes resulting in high-stress conditions for marine organisms derived primarily from increased precipitation and acid rain, enhanced weathering, and increased continental runoff (mesotrophy), particularly in nearshore environments, and changes in ocean water chemistry (ocean acidification; Keller, 2003, 2005; Gertsch et al., 2011; Font et al., 2011, this volume; Keller et al., 2011a, 2012). The response of planktic foraminifera depends on species tolerance to environmental changes, reproductive strategies to maximize survival, and the severity, duration, and frequency of perturbations.

Late Maastrichtian to early Paleocene (early Danian) climatic, environmental, and faunal changes have been documented, particularly in planktic foraminifera, for the past three decades and now form a considerable database revealing the biotic responses associated with the environmental stress that culminated in the K-T boundary mass extinction and the subsequent delayed recovery (e.g., Keller et al., 1993, 1995, 2002a, 2002b; Luciani, 1997, 2002; Li and Keller, 1998a, 1998b, 1998c; Kucera and Malmgren, 1998; Olsson et al., 2001; Abramovich and Keller, 2002, 2003; Nordt et al., 2003; Coccioni and Luciani, 2006; Pardo and Keller, 2008; Keller, 2002, 2004; Abramovich et al., 2010, 2011). This has led to the general acceptance that long-term environmental changes preceded and succeeded the K-T boundary mass extinction, despite decades of denial by impact proponents asserting that the Chicxulub impact was the sole cause with no significant ongoing environmental changes (e.g., Olsson et al., 1997; Norris et al., 1999; Lamolda et al., 2005; Arenillas et al., 2000, 2006; Molina et al., 2006; Schulte et al., 2010).

Late Maastrichtian Biotic Stress

Earlier studies suggest that during the late Maastrichtian, optimum planktic foraminiferal assemblages consisted of high species diversity, wide ranges of test sizes, diverse morphologies, and variable degrees of surface ornamentation, forming a continuum from small to large and simple to complex (e.g., Keller, 1988a, 2001; Pardo and Keller, 2008; Keller and Abramovich, 2009; Fig. 4). In these types of polytaxic communities, the diverse, large, complex specialized taxa (K-strategy species) utilize particular food sources, occupy particular ecological niches, have long life spans, and produce small numbers of offspring (Begon et al., 1996, 1998). In contrast, the low-diversity, smaller, less-ornamented species (R-strategists) tolerate a wide range of environments, utilize a variety of food sources, have shorter life spans, and produce larger numbers of offspring. R-strategists thus optimize chances for survival, whereas K-strategists optimize the good life while it lasts.

With increasing biotic stress, large specialized species are successively eliminated, survivors tend to be dwarfed, and the assemblages are dominated by ecological generalists (R-strategists; Fig. 5; Keller and Pardo, 2004; Keller and Abramovich, 2009). The result is decreasing diversity with increasing biotic stress, which may be due to climate change and/or open marine to successively shallower continental shelf and marginal marine environments. The most extreme biotic stress conditions are interpreted from the dominance of *G. cretacea* and/or *Globoconusa daubjergensis* in low-diversity assemblages (Keller and Pardo, 2004; Pardo and Keller, 2008; Keller and Abramovich, 2009). "Blooms" of these disaster opportunist species indicate an R-selected survival strategy, which is common in unpredictable and stressful environments where quick reproduction improves

Figure 4. Sizes and shapes of tests of Maastrichtian planktic foraminifera showing a continuum from small R-strategists to large complex K-strategists. R-strategy and opportunistic life of *Guembelitria* are inferred from minute test size, simple chamber arrangement, and isotopically light $\delta^{13}C$ values. K-strategy species are inferred from large and complex test morphology, small populations, and heavier $\delta^{13}C$ values (modified from Keller and Abramovich, 2009). 1, 2—*Guembelitria cretacea*, 3—*Heterohelix planata*, 4—*Heterohelix navarroensis*, 5—*Heterohelix globulosa*, 6—*Globigerinelloides aspera*, 7—*Pseudotextularia elegans*, 8—*Rugoglobigerina rugosa*, 9—*Globotruncana arca*, 10—*Pseudoguembelina costulata*, 11—*Pseudoguembelina kempensis*, 12—*Racemiguembelina fructicosa*, 13—*Contusotruncana walfischensis*, 14—*Globotruncanita stuarti*, 15—*Planoglobulina acervulinoides*, 16—*Abathomphalus mayaroensis*, 17—*Gublerina cuvillieri*, 18—*Planoglobulina multicamerata*.

the chances of survival. *Guembelitria* species are surface dwellers often associated with unstable environments, mesotrophic conditions of shallow continental shelves, upwelling areas, and regions under volcanic influence (Keller, 2003, 2005, 2011; Keller et al., 2007; reviews in Keller and Pardo, 2004; Pardo and Keller, 2008).

Species dwarfing, or the Lilliput effect, and other morphological deformities are the most easily identified signals of high-stress conditions. Dwarfing is likely a survival response to stress conditions by increasing reproduction rates through early sexual maturation (MacLeod et al., 2000; Keller and Abramovich, 2009). Such faunal responses have been linked to the global rapid climate warming in the latest Maastrichtian, decreased water mass stratification with resultant loss of habitat, high nutrient influx in shallow marine environments, leading to mesotrophic or eutrophic waters with reduced dissolved oxygen content, and reduced symbiotic activity (Abramovich and Keller, 2003; Keller and Abramovich, 2009).

Species richness reflects the actual number of species present in any given sample and is used as proxy for high variability and environmental stress as a function of climate, geography, and ecology (Abramovich and Keller, 2002). The lowest species richness has been consistently observed in areas of volcanic activity, such as in Deep Sea Drilling Project (DSDP) Site 216 on Ninetyeast Ridge, Indian Ocean, in the Neuquén Basin, Argentina, and in the Krishna-Godavari Basin and Meghalaya, India (Keller, 2003, 2005; Keller et al., 2007, 2011a, 2012; Gertsch et al., 2011a). In these regions, species richness varies between 6 and 12 species, but increases in intervals after volcanic activity ceased. We therefore identify extreme global environmental effects correlative with Deccan phase 2 by low species richness/diversity, dwarfing, dominance of disaster opportunists (*G. cretacea*), and variations among other environmentally tolerant small species.

Early Danian Biotic Stress

Significant numbers of small late Maastrichtian species appear to have survived the K-T boundary mass extinction into the early Danian zone P1a, as also testified by their Danian stable isotope signals (e.g., *Heterohelix globulosa*, *Heterohelix planata*, *Heterohelix navarroensis*, *Pseudoguembelina costulata*, *Hedbergella monmouthensis*, *Hedbergella holmdelensis*, *G. cretacea*; Keller, 1988a, 1989a, 1989b; Canudo et al., 1991; Keller et al., 1993, 1994; MacLeod and Keller, 1994; Barrera and Keller, 1990; Keller and Pardo, 2004; Pardo and Keller, 2008; Keller and Abramovich, 2009). However, only *G. cretacea* was a long-term survivor, and it may have been the progenitor of several Danian clades (Fig. 6; Olsson et al., 1999). This species thrived in high-stress environments of the late Maastrichtian as well as in the immediate aftermath of the mass extinction in zone P0 and

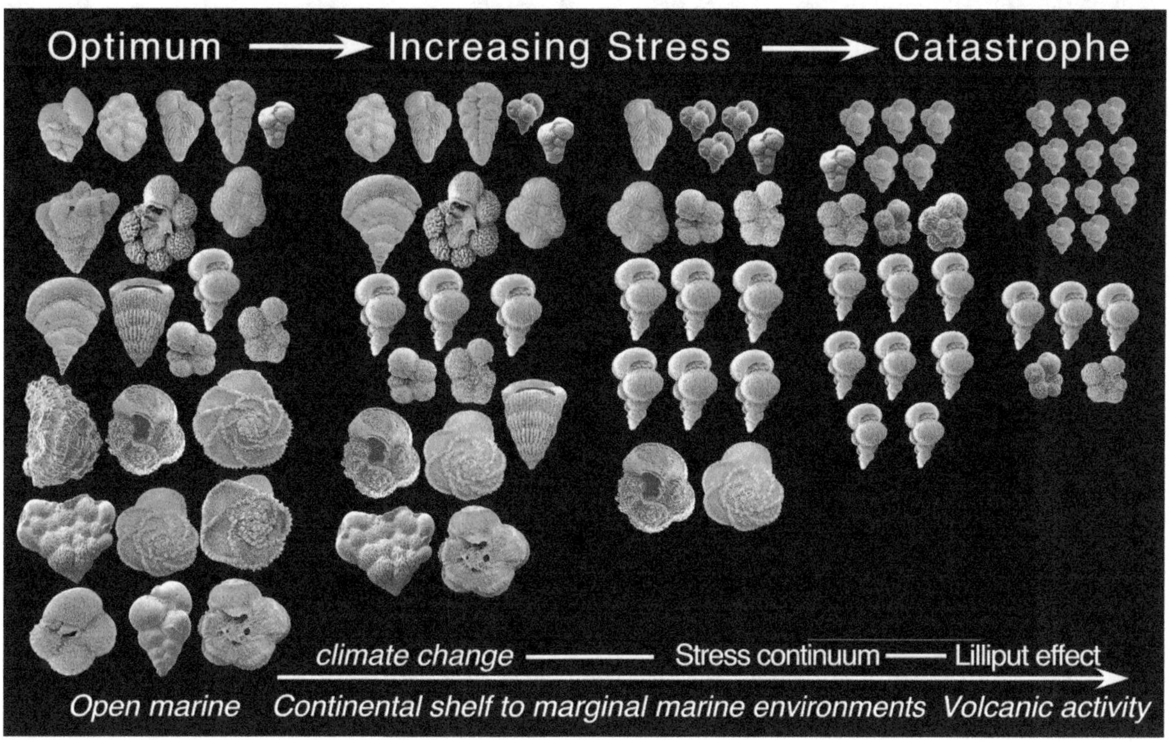

Figure 5. The effects of increasing environmental stress upon planktic foraminiferal assemblages from optimum to catastrophe conditions show the successive elimination of large, specialized K-strategy species, the survival of small R-strategy species, the overall dwarfing of these species, and their great abundance. Figure is modified from Keller and Abramovich (2009).

the lower part of zone P1a, but they decreased in the later part of P1a only to thrive again in P1b. Other survivor species were generally dwarfed in zones P0 and P1a, with *H. globulosa* abundant in shallow-water environments (e.g., Brazos River, Texas, and Nye Klov, Denmark). All survivors rapidly disappeared by the end of P1a (Fig. 6; MacLeod et al., 2000; Keller and Pardo, 2004; Keller and Abramovich, 2009).

All evolving new species found in the early Danian sections are small (generally <100 µm and frequently <63 µm) with thin test walls, small globular chambers, and simple biserial, triserial, or trochospiral morphologies (e.g., Keller, 1988a, 1989a; Keller et al., 1993, 1995, 2002a, 2002b; Luciani, 2002; Coccioni and Luciani, 2006; Pardo and Keller, 2008). The first new species evolved in zone P0 within a few thousand years of the mass extinction (e.g., *Woodringina hornerstownensis*, *Woodringina claytonensis*, *Parvularugoglobigerina extensa*, *Gl. daubjergensis*; Fig. 6). Shortly thereafter, in zone P1a, additional small species evolved with primarily trochospiral coiling (e.g., *Parvularugoglobigerina eugubina*, *Parvularugoglobigerina longiapertura*, *Eoglobigerina edita*, *Eoglobigerina eobulloides*, *Praemurica taurica*, *Globorotalia pentagona*, *Subbotina triloculionoides*, *Parasubbotina pseudobulloides*) and a few biserial morphotypes (*Chiloguembelina midwayensis*, *Chiloguembelina morsei*). Hence, early Danian species diversity remained low and only a few of these species dominated the assemblages in P1a (*P. eugubina*, *E. edita*, *W. hornerstownensis*, and other biserial species), suggesting continued high-stress conditions during the 100–200 k.y. after the mass extinction (Fig. 6).

The disappearance of *P. eugubina* and *P. longiapertura* marks the P1a/P1b zone boundary. Throughout zone P1b, species richness remains low with small simple test morphologies. *Guembelitria* species (*G. cretacea*, *G. irregularis*, *G. danica*, *G. dammula*) dominate yet again, along with *Gl. daubjergensis*, indicating a relapse to stressed environmental conditions. A notable recovery to higher diversity, larger test morphologies (>150 µm), and reduced abundance of stress-tolerant species does not take place until zone P1c—long after the K-T boundary mass extinction (Fig. 6).

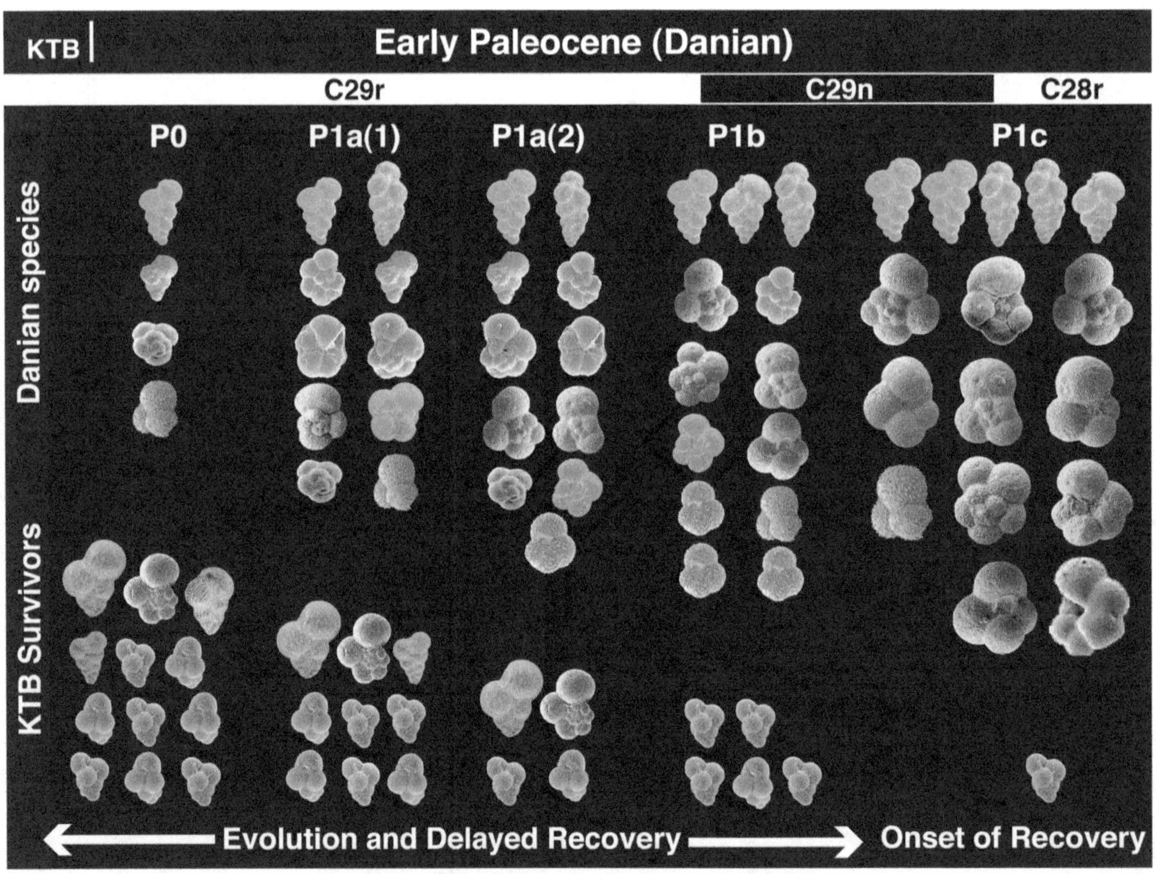

Figure 6. Early Danian evolution and Cretaceous survivor species illustrating the high-stress environment, small dwarfed species, and low diversity that mark the delayed marine recovery through zone P1b. The last phase 3 of Deccan volcanism occurred in zone P1b (C29n). The onset of marine recovery with higher diversity and larger species sizes begins only after Deccan volcanism ended. Note the decreasing abundance of survivor species with increasing evolution and diversity of early Danian assemblages. The number of same specimens qualitatively indicates relative abundance. KTB—Cretaceous-Tertiary boundary.

GLOBAL K-T BOUNDARY DATABASE

This study summarizes the faunal diversity and species abundance data that appear to link high-stress environments during the late Maastrichtian and early Danian to Deccan volcanism phase 2 and phase 3. Over 200 K-T boundary sequences from localities worldwide have been analyzed by one of the authors (Keller), students, and collaborators (Fig. 7). For many of these sections, oxygen and carbon stable isotopes and quantitative planktic foraminiferal analyses are available and make a formidable database with which to test various K-T boundary hypotheses regarding the nature, cause(s), and effects of the mass extinction. The analyzed sections encompass latest Maastrichtian zones CF3, CF2, and CF1 and early Danian zones P0, P1a (subzones P1a[1], P1a[2]), and P1b (correlative with P0, Pα, and P1a of Berggren et al., 1995) (Fig. 8). The database therefore permits evaluation of the Chicxulub and Deccan volcanism–related hypotheses on a global scale (Fig. 7).

We use the high-resolution biozonation that resulted from these studies and was first developed at the El Kef global stratotype section and point (GSSP), refined by additional studies at the nearby Elles section (Keller, 1988a; Li and Keller, 1998a, 1998b; Keller et al., 1995, 2002a; Fig. 8). This biozonation has been tested and successfully applied worldwide, including India (Keller et al., 2008, 2009, 2011a; Malarkodi et al., 2010; Gertsch et al., 2011a). Quantitative species analysis is an integral part of this biozonation and helps with recognition of zones, subzones, and intervals based on species acmes. The biozones and subzones were compared and integrated with lithological characteristics (black clay layer, red oxidized layer with iridium anomaly) and geochemical proxies ($CaCO_3$, total organic carbon [TOC], $\delta^{13}C$).

STABLE ISOTOPES

Late Maastrichtian to early Paleocene (early Danian) environmental changes can be inferred from oxygen and carbon isotopes of well-preserved monospecific planktic and benthic foraminifera and bulk rock sediments. For this review, we used the high-resolution records of the South Atlantic DSDP Site 525A (Li and Keller, 1998a), the Tethys record of Elles, Tunisia (Stüben

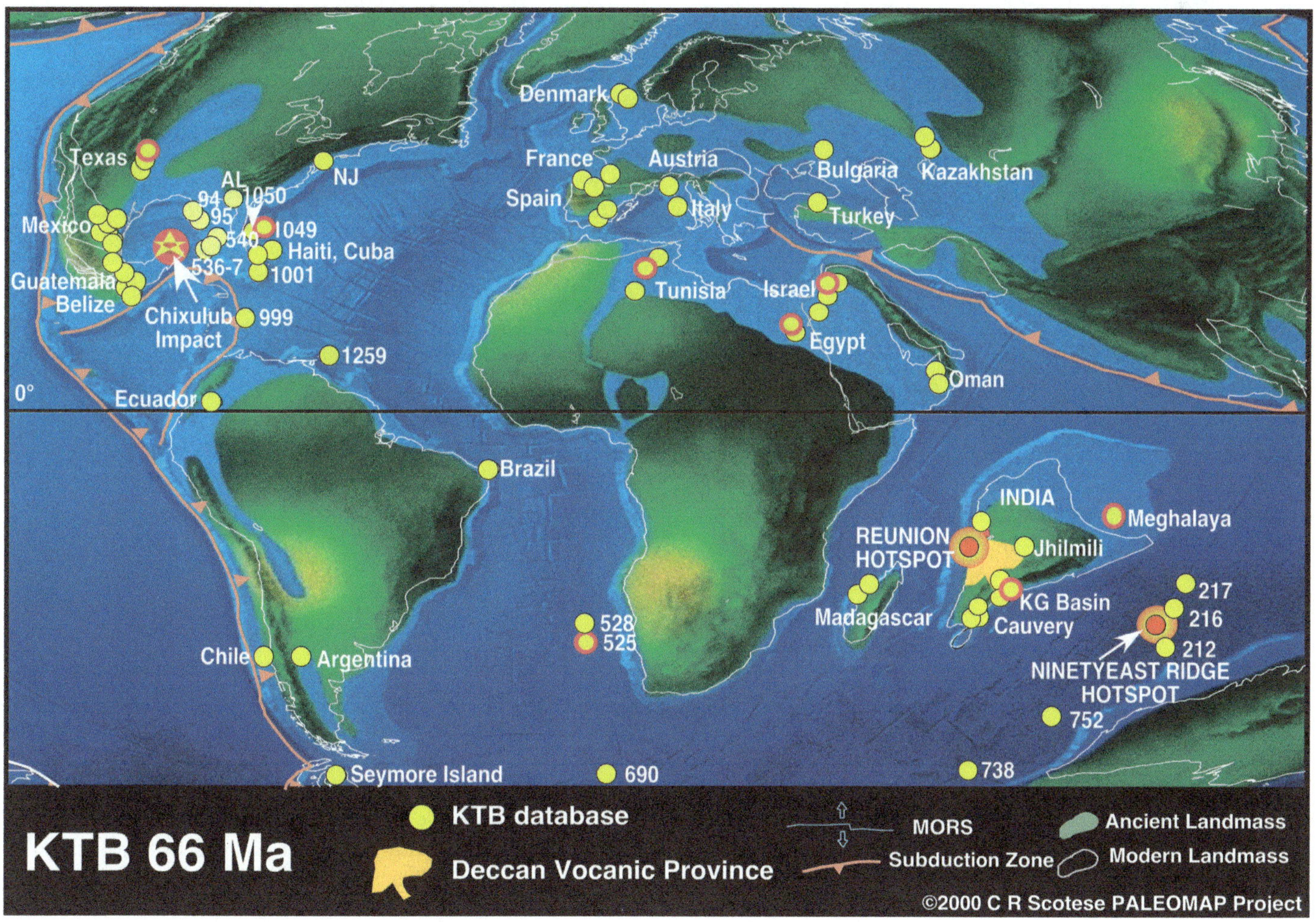

Figure 7. Global database of Cretaceous-Tertiary boundary (KTB) sections studied to date plotted on the paleomap of 66 Ma. Red-circled yellow dots mark sections discussed in this review. MORS—mid-ocean ridges.

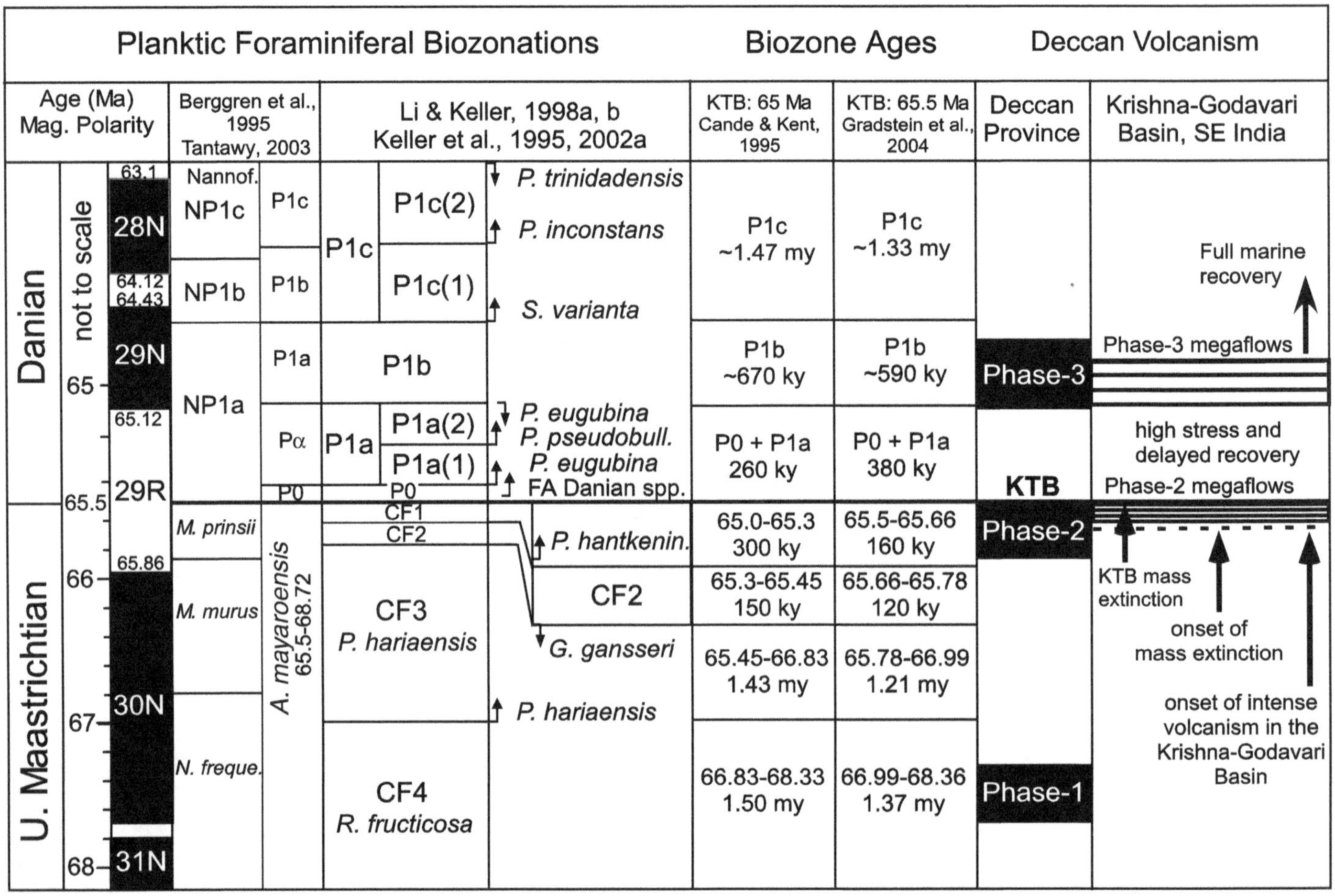

Figure 8. Late Maastrichtian and early Danian planktic foraminiferal zonal schemes and ages of biozones calculated based on two time scales with the Cretaceous-Tertiary boundary (KTB) at 65.0 and 65.5 Ma. The stratigraphic position of Deccan megaflows and the onset of the Cretaceous-Tertiary boundary mass extinction in the Krishna-Godavari Basin are based on Keller et al. (2011a). This review concentrates on the shaded interval. Index species: *Nephrolithus frequens, Micula murus, Micula prinsii* (nannofossils)*; Racemiguembelina fructicosa, Planoglobulina hariaensis, Gansserina gansseri, Plummerita hantkeninoides, Parvularugoglobigerina eugubina, Parasubbotina pseudobulloides, Subbotina varianta, Praemurica inconstans, Praemurica trinidadensis* (planktic foraminifera).

et al., 2003), and the shallow shelf record of Brazos, Texas (Mullinax-1 core; Abramovich et al., 2011). At all three localities, foraminiferal test preservation varies from excellent (Brazos, Texas) to well preserved (DSDP Site 525A, Elles, Tunisia).

Late Maastrichtian

South Atlantic DSDP Site 525A

Remarkable climate changes occurred during the last 280 k.y. (zones CF2 and CF1) of the Maastrichtian, as shown in the high-resolution records (samples analyzed at 2000–5000 yr intervals) of the middle-latitude South Atlantic (Walvis Ridge DSDP Site 525A; Fig. 9; Li and Keller, 1998a). At Site 525A, the long-term global cooling trend that began in the late Campanian ended with the maximum Cretaceous cooling near the base of chron C29r, correlative with the CF3-CF2 zones transition. In the following warm event, bottom waters warmed rapidly by 3–4 °C (0.8‰ decrease in benthic $\delta^{18}O$ values near the base of CF2). This deep-water warming trend was interrupted by a short cooling event (1 °C) and ended in the upper part of CF1, when climate cooled rapidly by 2–3 °C (0.5‰ increase in benthic $\delta^{18}O$ values; Fig. 9). In contrast to the more stable deep-water temperatures, planktic $\delta^{18}O$ values of surface waters (*Rugoglobigerina rugosa*) strongly fluctuated during the warm event (0.3‰–0.5‰), suggesting unstable climatic conditions, although some of this variation could also be attributed to diagenetic effects and/or variations in species habitats. Both surface- and deep-water $\delta^{13}C$ values remained relatively steady through the warm event, fluctuating ~0.3‰, but they significantly increased during the subsequent cooling in the upper zone CF1 (Fig. 9). The late Maastrichtian record at Site 525A ends at the K-T boundary hiatus with an indeterminate interval eroded (Li and Keller, 1998a, 1998b).

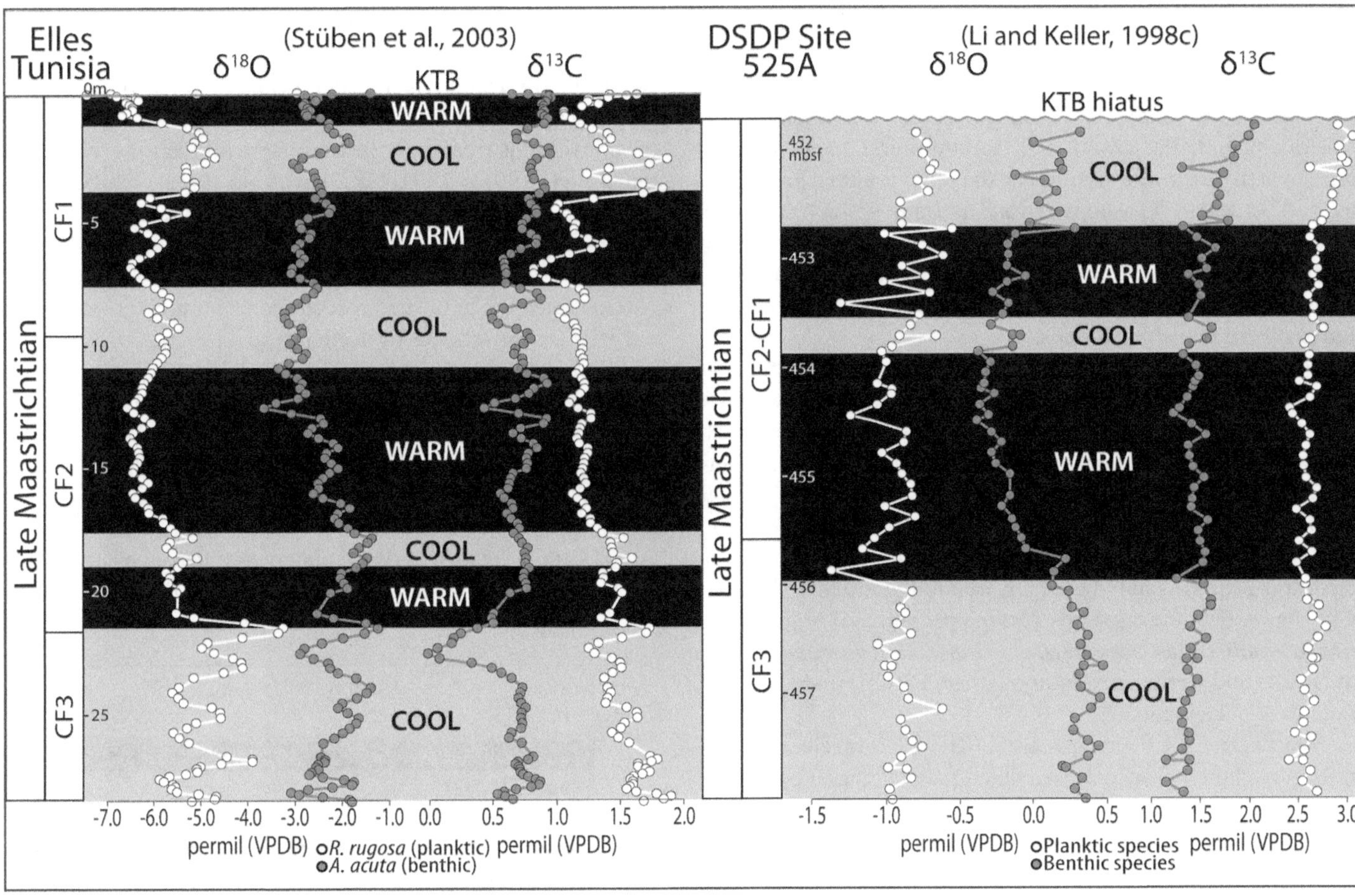

Figure 9. Carbon and oxygen isotopes of monospecific planktic and benthic foraminifera at Elles, Tunisia, and South Atlantic Deep Sea Drilling Project (DSDP) Site 525A. Gray marks cool intervals; black marks warm intervals. Note the broad climate warming of zones CF1–CF2 at Site 525A shows a more complex pattern at Elles, where two warm phases are separated by a short cool pulse. The late Maastrichtian warming at Elles is not present at Site 525A, probably because of the Cretaceous-Tertiary boundary (KTB) hiatus. Note that at Elles, oxygen isotope values are 3‰–4‰ lighter, suggesting diagenetic effects due to burial diagenesis and/or interaction with isotopically light meteoric water as a result of freshwater input. Data are from Li and Keller (1998a) and Stüben et al. (2003). VPDB—Vienna Peedee belemnite; planktic species—*Rugoglobigerina rugosa*; benthic species—*Anomalinoides acuta*, now known as *Cibicidoides pseudoacuta*.

Elles, Tunisia

In the middle-shelf section of Elles, Tunisia, samples were analyzed at 5–10 cm intervals, recording a sample resolution of ~2000 yr (Stüben et al., 2003). Oxygen isotope values are 3‰–4‰ lighter than at Site 525A (Fig. 9), suggesting freshwater influx and diagenetic effects due to burial diagenesis and/or interaction with isotopically light meteoric water (Stüben et al., 2003). This prevents the calculation of actual paleotemperatures, although climate trends are likely preserved, as indicted by their similarity to Site 525A. Significant differences are largely due to the expanded Elles record of 22 m, compared with just 4.2 m at Site 525A, and the lower-latitude location. As a result of this expanded record and detailed sampling, the climate changes are better resolved in both surface and bottom waters.

Three significant cool intervals can be recognized in planktic and benthic oxygen isotopes: lower part of CF2, base of CF1, and upper part of CF1 (Fig. 9). Only the upper two cool intervals are recognized in Site 525A. In addition, there is significant warming in the top meter immediately preceding the K-T boundary, which is not present at Site 525A due to a hiatus. Surface waters show less sample variability in $\delta^{18}O$ values and four distinct strong warm events (Fig. 9). The $\delta^{13}C$ values at Elles are significantly more variable than at Site 525A. In zone CF3, both surface and bottom waters show a gradual decrease, but the latter shows an abrupt 0.7‰ shift near the top of CF3. During the following warm event, the bottom-water $\delta^{13}C$ values gradually increase up to the K-T boundary. In contrast, surface $\delta^{13}C$ values decrease through the warm event, temporarily increase in the lower half of the subsequent cooling, and rapidly increase just below the K-T boundary (Fig. 9).

Brazos River, Texas

At the Brazos section (core Mullinax-1), samples were analyzed at 5–10 cm intervals (Abramovich et al., 2011). The sediment accumulation rate for the CF2–CF1 interval is relatively high (9 m) compared with Site 525A (4.2 m), though less than at Elles (22 m). However, this excludes erosion at the base of the sandstone complex, which is estimated to be several meters. Preservation of foraminiferal tests is excellent, which rules out significant diagenetic alteration. However, $\delta^{18}O$ and $\delta^{13}C$ values covary and are ~2‰–3‰ lighter than at Site 525A. This has been attributed to the shallow shelf location and influx of freshwater during warm intervals (Abramovich et al., 2011). Therefore, no temperatures calculations were made for Brazos.

Compared with deep-water environments, climate changes recorded in shallow nearshore environments can be expected to show more enhanced temperature changes modified by proximity to land. This is evident in the Mullinax-1 core, where $\delta^{18}O$ values vary by 2‰–3‰ (Fig. 10). The warm/cool climate spans the same interval from the CF3-CF2 transition to the K-T boundary, but there are significant differences in the timing relative to Elles. This may be largely due to the rarity of the zone CF1 and CF2 index species (first appearance of *Plummerita hantkeninoides*, last appearance of *Gansserina gansseri*; Fig. 8) at Brazos and hence uncertainty in the CF3-CF2 and CF2-CF1 boundary placements.

At Brazos, the first warm event (CF3-CF2 transition) is significantly shorter than at Elles, which may be due to erosion or nondeposition, which commonly accompanies the maximum cooling and sea-level regression near the end of CF3. The second warm event shows highly fluctuating values with freshwater influx (Fig. 10; Abramovich et al., 2011). The subsequent cool interval marks a sea-level fall, erosion, and deposition of the sandstone complex with reworked Chicxulub impact spherules (Adatte et al., 2011; Keller et al., 2011b). Claystone deposition resumes above the sandstone complex as sea level rises and temperatures gradually increase toward the K-T boundary (Fig. 10). Bulk $\delta^{13}C$ values decrease during the pre–K-T boundary warming (Keller et al., 2007), which is contrary to the increase observed at Elles.

Timing of Climate Change

We can derive best estimates for the age and duration of the major climatic changes based on sediment accumulation rates calculated from the expanded Elles section. We use the time scale of Gradstein et al. (2004), which places the K-T boundary at 65.5 Ma, and zones CF1 and CF2 span the last 160 k.y. and 120 k.y., respectively, corresponding to most of chron C29r at the top of the Maastrichtian. Rapid climate warming began ~280 k.y. before the K-T boundary and reached its maximum near the top of zone CF2, or within ~110 k.y. (including the short cooling in the lower part of CF2; Fig. 9). The subsequent cool and warm intervals span ~50 k.y. and 60 k.y., respectively, followed by another cool interval of ~40 k.y. and the latest Maastrichtian warming (~20 k.y.). Although these values are tentative, they demonstrate that the warm and cool intervals were of relatively short duration.

Early Danian Climate Change

The early Danian is a difficult time interval for evaluation of climatic and environmental changes because the K-T boundary transition is frequently incomplete with hiatuses (MacLeod and Keller, 1991a, 1991b; Keller et al., 2013), which makes comparison of isotope records difficult. In addition, few high-resolution Danian records exist, and most studies have concentrated only on the K-T boundary and its immediate aftermath. We examined the published records concentrating on zones P1a and P1b (zonal scheme by Keller et al., 1995, 2002a; Fig. 8). Because of frequent hiatuses at the K-T boundary, the earliest Danian stable isotope record is not well understood in zone P1a and at the P1a-P1b boundary, and the climatic and environmental history of this interval is still uncertain. However, for zone P1b, the sediment record is generally well represented, and we concentrate on this interval.

The most detailed early Danian record for zone P1b is from the North Atlantic Ocean Drilling Program (ODP) Site 1049C,

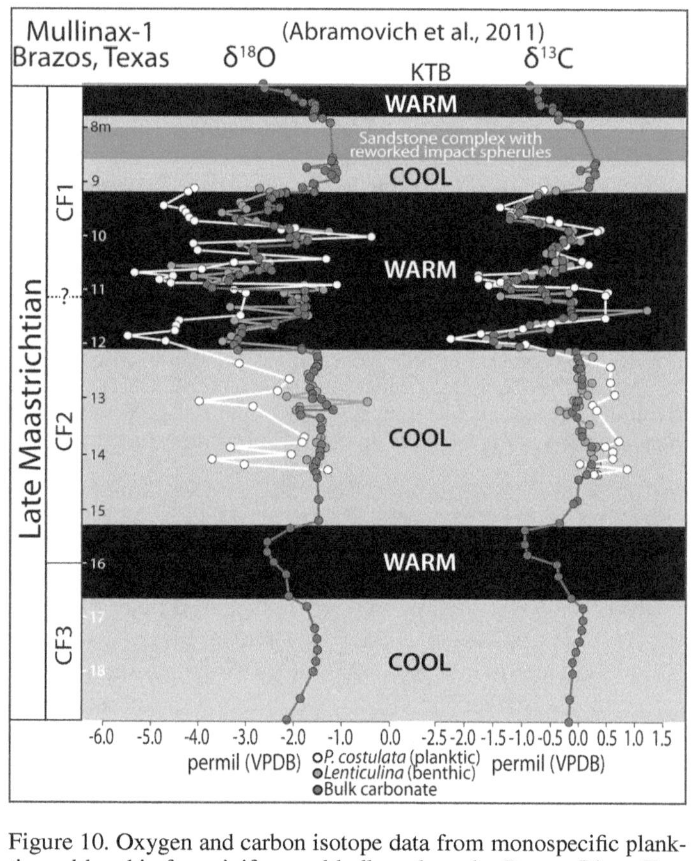

Figure 10. Oxygen and carbon isotope data from monospecific planktic and benthic foraminifera and bulk rock at the Brazos River, Texas, Mullinax-1 core. Note the late Maastrichtian CF2–CF1 warming phases are similar to Elles with two main warm intervals plus the end-Maastrichtian warming. Note: the very negative $\delta^{18}O$ values likely reflect freshwater influx during warm phases. Data are from Abramovich et al. (2011). KTB—Cretaceous-Tertiary boundary; VPDB—Vienna Peedee belemnite; planktic species—*Pseudoguembelina costulata;* benthic species—*Lenticulina.*

where Quillévéré et al. (2008) recorded two short-term negative $\delta^{18}O$ and $\delta^{13}C$ excursions that define the Dan-C2 event (~1.5‰; Fig. 11). Sea-surface temperature rose an estimated ~4 °C, as suggested by the ~1‰ shift in the planktic foraminifer *Praemurica taurica*, assuming that samples are not biased by postdepositional processes. Similar excursions were recorded at South Atlantic Sites 527, 528, Italy and Ukraine, suggesting a global event (Quillévéré et al., 2008; Coccioni et al., 2010; Gilmour et al., 2013).

However, available isotope records from the eastern Tethys (Israel, Egypt) do not mirror the structure or position of the Dan-C2 isotope event at Site 1049C, although distinct excursions are present (Fig. 11). In all eastern Tethys sections, the first $\delta^{18}O$ and $\delta^{13}C$ excursions are at or near the P1a-P1b boundary. The second excursion, which likely reflects the Dan-C2 event, occurs significantly above this interval, although the characteristic double-peak excursion of Site 1049C is not recognized. Several factors can explain these differences, including the presence of hiatuses and low sample resolution. All eastern Tethys sections, as well as North Atlantic Site 1049C, have multiple early Danian hiatuses, including at the K-T boundary, with most of zone P1a missing, and at the P1a-P1b boundary, with the lower part of zone P1b and upper part of zone P1a eroded (Keller and Benjamini, 1991; Keller et al., 2002a, 2013; Punekar et al., 2014). This can explain the variable stratigraphic position of the isotope excursion referred to as Dan-C2.

BIOTIC STRESS—LATE MAASTRICHTIAN

Biotic stress conditions associated with the late Maastrichtian climate warming and cooling have been documented in numerous localities, summarized in Pardo and Keller (2008) and Keller and Abramovich (2009). Here, we review the climatic and environmental effects on planktic foraminiferal species diversity and abundances based on high-resolution quantitative studies from the middle-latitude South Atlantic (Li and Keller, 1998a; Abramovich and Keller, 2003), the eastern Tethys (Abramovich and Keller, 2002; Stüben et al., 2003), and Western Interior Seaway (Texas; Abramovich et al., 2011; Keller, 2011).

DSDP Site 525A—South Atlantic

The biotic response to the late Maastrichtian climate warming in the South Atlantic Site 525A includes dwarfing, decreased diversity, temporary disappearance of species, decreased photosymbiotic activity, and depth migration (Fig. 12; Abramovich and Keller, 2003). The biotic stresses appear to be primarily associated with temperature-related changes in the water mass

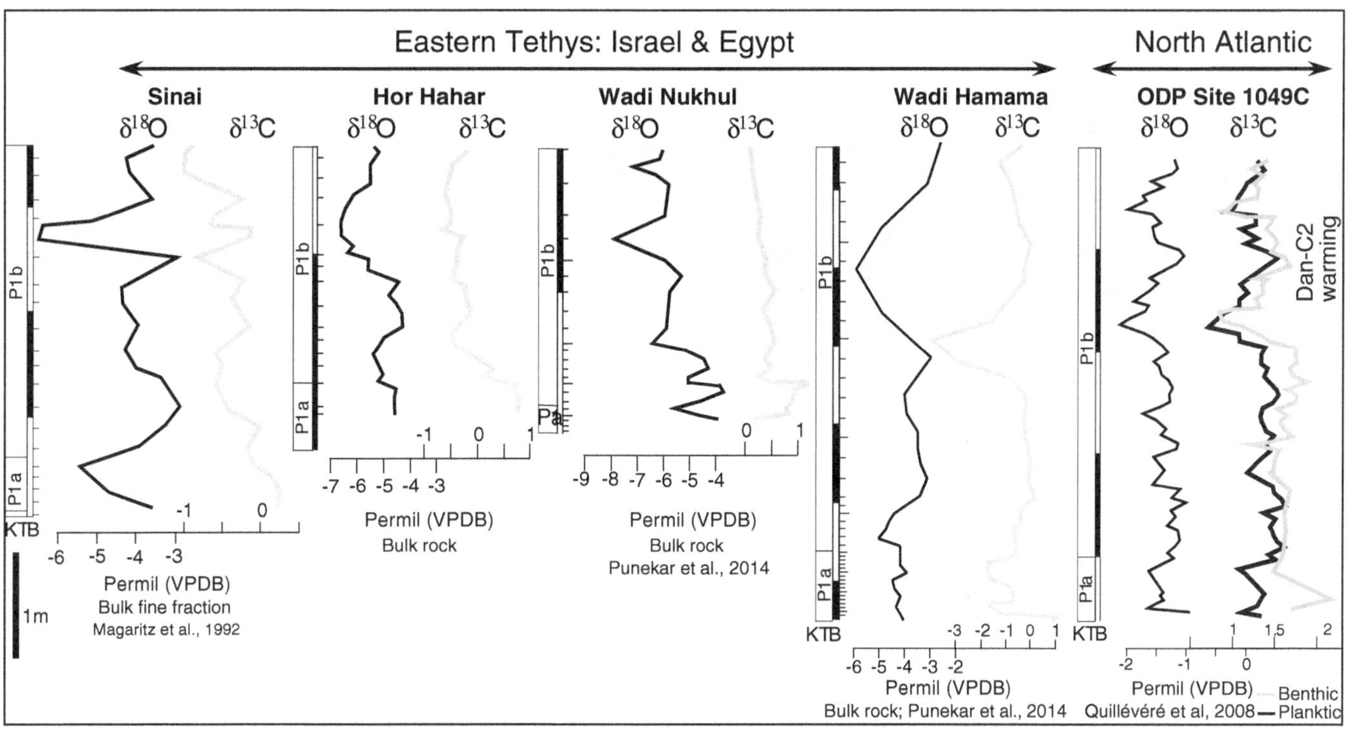

Figure 11. Oxygen and carbon isotope data (planktic, benthic, and bulk rock) in sections from the North Atlantic (Ocean Drilling Program [ODP] Site 1049C) and eastern Tethys (Israel and Egypt). Coeval zone P1b excursions (gray shaded area) originally identified at Site 1049C as Dan-C2 event (Quillévéré et al., 2008) are present but do not mirror the characteristic double-peak structure. The reason is likely low sample resolution. KTB—Cretaceous-Tertiary boundary; VPDB—Vienna Peedee belemnite.

stratification accompanied by decreased niche habitats for specialized large species.

The typical small (63–150 µm) middle-latitude species were least affected and retained their population sizes, with the exception of the dominant *Heterohelix planata* (formerly known as *Heterohelix dentata*), which decreased significantly during warm intervals and increased during cool intervals. *Heterohelix globulosa* is less abundant but also increased during the cooling near the end of the Maastrichtian. In contrast, the disaster opportunist *G. cretacea* is rare.

Dwarfing is the most striking response to the stress conditions associated with climate warming. All morphotypes are affected (Fig. 12), particularly the large (>150 µm), complex, specialized species living at intermediate depth (thermocline). Stressed assemblages show 30%–60% of the population dwarfed to one-third or half of their normal average size.

The Shannon-Weaver index (H′; Buzas and Gibson, 1969) for species diversity shows major changes at Site 525A. For large species (>150 µm), diversity drops dramatically at the onset of warming and recovers partly during cool intervals but never regains the prewarming high diversity (Fig. 12). The reason for this diversity drop is the high number and high abundance of dwarfed larger species to sizes below 150 µm.

Elles, Tunisia—Western Tethys

The faunal turnovers in the middle-latitude South Atlantic and Tethys Oceans are broadly similar, though *H. globulosa* dominates in the Tethys. *Heterohelix planata* mirrors the pattern at Site 525A, with greater abundance during cool intervals and decreased abundance (from ~30% to ~10%) during warm intervals (Fig. 13). *Heterohelix globulosa* shows a general decline during warm intervals, although this relationship is not very consistent. Increased abundances of *G. cretacea* are observed in zones CF3 and CF1, though abundances are too low to compare with shallower shelf sequences where this species dominates (Keller and Pardo, 2004; Pardo and Keller, 2008). Complex species, such as the intermediate-dwelling globotruncanids, decrease in their relative abundance during cool intervals with a slight recovery during warming.

Diversity trends at Elles are different from Site 525A. The values of the diversity index H′ for small (63–150 µm) and large (>150 µm) species show parallel trends. During cool intervals in zones CF3, CF2, and CF1, diversity of large species drops, and diversity of small species increases (Fig. 13). During the first two warm intervals, diversity of the large species is higher than that of small species. The main decrease begins in the

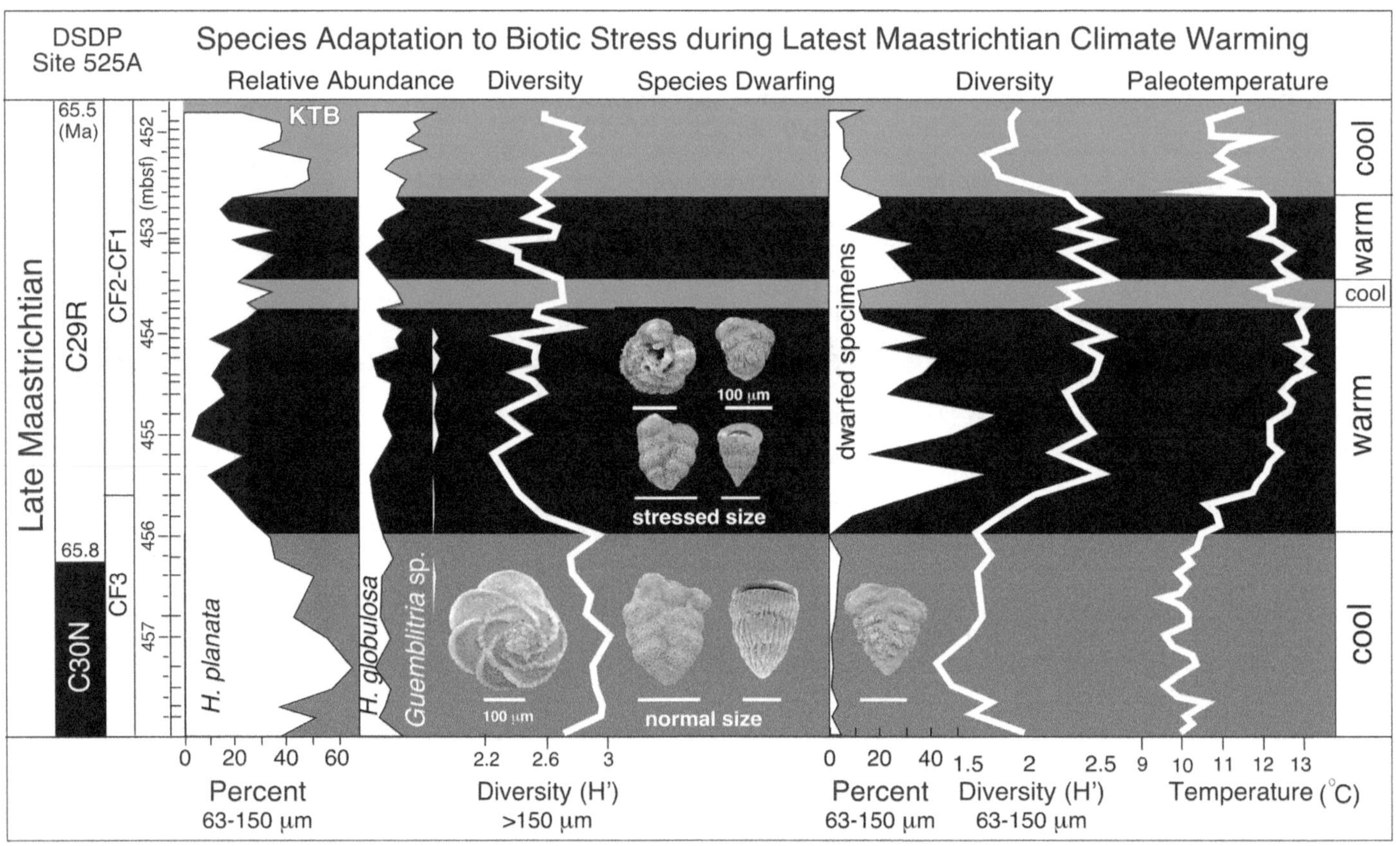

Figure 12. Species adaptation to biotic stress during the late Maastrichtian climate warming at Deep Sea Drilling Project (DSDP) Site 525A, modified from Li and Keller (1998a) and Abramovich and Keller (2003). KTB—Cretaceous-Tertiary boundary; *H. planata—Heterohelix planata, H. globulosa—Heterohelix globulosa.*

warm interval near the top of zone CF2 and continues into the first cooling of zone CF1, reaching minimum values along with small species. During the CF1 warm and cool intervals up to the K-T boundary, small species diversity is slightly higher than for large species. Diversity H′ for larger species thus indicates high-stress conditions beginning near the top of zone CF2, whereas for small species, diversity increases. In contrast, at Site 525A, dwarfing of larger species resulted in significant diversity reduction (Fig. 12). The diversity decrease in the CF1 cool intervals likely reflects the decreased abundance or absence of specialized large species living at thermocline depths, possibly due to upwelling effectively reducing habitat space. This can be evaluated based on species richness.

All variability in species richness is observed in intermediate-depth dwellers (the thermocline dwellers; Abramovich and Keller, 2003; Abramovich et al., 2003), which increase in diversity during warming and decrease during cooling, indicating major changes in the water mass stratification (Fig. 13). These changes may be related to upwelling and reduced thermocline depth during cooling and expansion of the oxygen minimum zone during warming. Both cooling and warming would have placed maximum stress on the intermediate-dwelling large specialized species, driving them to near extinction. Specialized species show normal combined species diversity (66%) but very low combined total species abundances (<10% overall, <5% globotruncanids). Surface- and deep-water dwellers experienced little variability.

Brazos River, Texas—Western Interior Seaway

For most of the late Maastrichtian, claystone deposition in the Brazos River area occurred in a middle shelf environment, which dropped to inner shelf depth in the upper part of zone CF1, marked by an unconformity and deposition of the sandstone complex with reworked Chicxulub impact spherules (Fig. 14; Keller et al., 2007, 2011b). Claystone deposition resumed with latest Maastrichtian zone CF1 assemblages and stable isotope signals marking deposition of the sandstone complex and the reworked Chicxulub impact spherules as latest Maastrichtian in age.

The same small heterohelicids (*H. planata*, *H. globulosa*) and *Guembelitria* that dominate at Elles and Site 525A dominate at Brazos. These three species account for 75%–95% of the

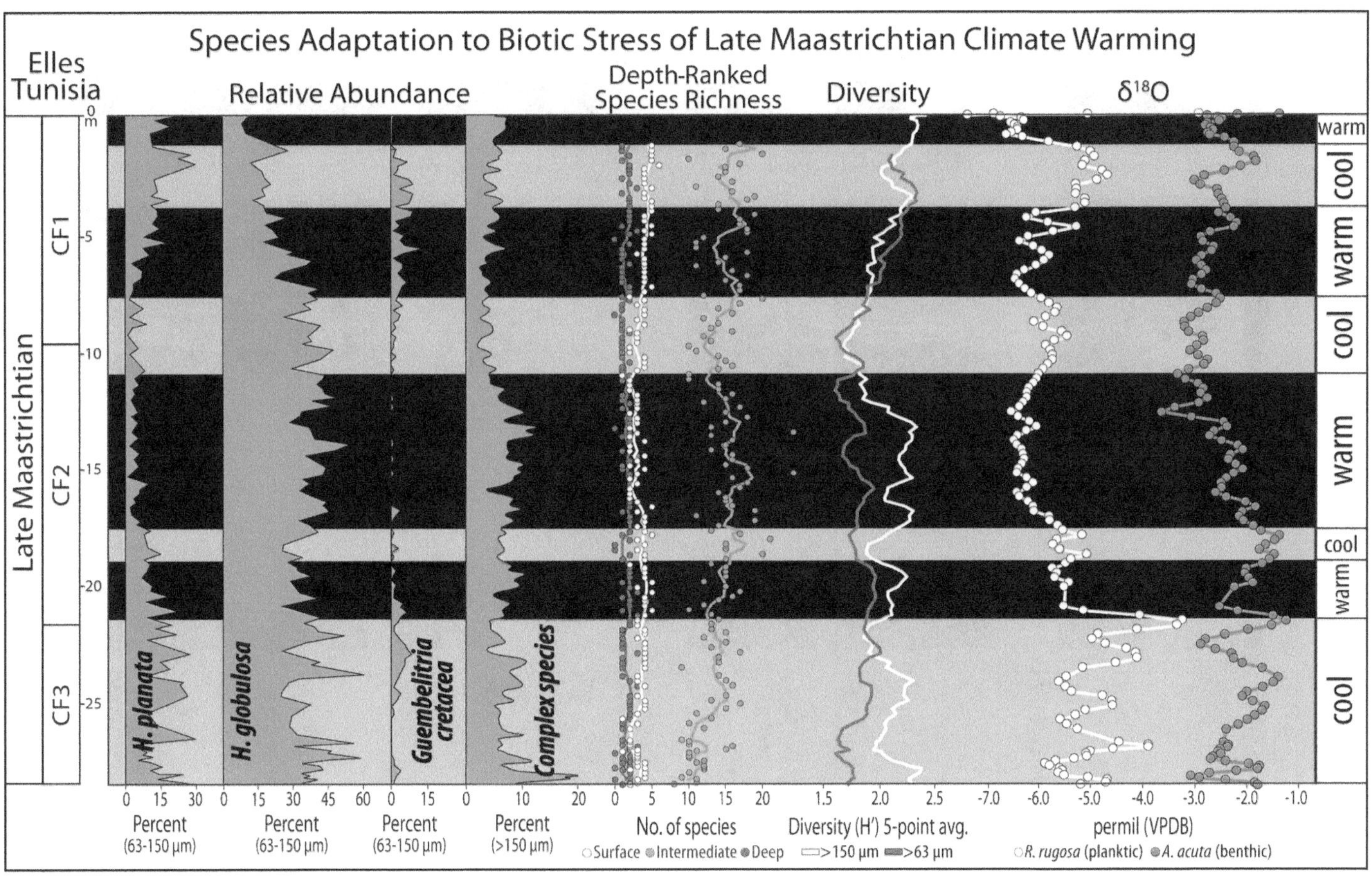

Figure 13. Species adaptation, species richness, and diversity compared with oxygen isotope trends at Elles, Tunisia. Data are from Abramovich and Keller (2002) and Stüben et al. (2003). VPDB—Vienna Peedee belemnite; *H. planata*—*Heterohelix planata*; *H. globulosa*—*Heterohelix globulosa*.

total planktic foraminiferal population (Fig. 14; Abramovich et al., 2011; Pardo and Keller, 2008). *Heterohelix planata* shows a distinct preference for cooler temperatures with major decreases in relative abundances in both warm phases, similar to Elles and Site 525A. In the first warm phase (CF3-CF2 transition), relative abundance drops from 50% to 25%, but it recovers during the subsequent cooling. Interestingly, the terminal population decrease begins in the middle warm interval (zone CF1), with no recovery in the subsequent cooling. *Heterohelix globulosa* shows the opposite trend, with higher abundances during the warm intervals and continuing high abundance across the K-T boundary into the early Danian. Species are dwarfed in zone CF1 as well as in the early Danian zones P0–P1a, where survivor populations of *H. globulosa* show Danian isotope signals (Barrera and Keller, 1990; MacLeod et al., 2000). The most abundant species during CF1 warming is *G. cretacea*, with pronounced peaks (>50%) at the warm-cool transitions.

The diversity index H′ shows little change except for a gradual decrease through the latest Maastrichtian. Species richness plotted as histograms for small and large species for each meter interval shows larger species disappearing in cool intervals and particularly during deposition of the sandstone complex (Fig. 14). This is due to the sea-level drop to inner neritic depths and even subaerial exposure within 2.5 km of Mullinax-1, which excluded all but the surface dwellers from the area (Keller et al., 2011b; Adatte et al., 2011). Species richness (5 point average) fluctuates with climate changes, showing maxima during warm phases and minima during cool phases, as also observed at Elles.

BIOTIC STRESS—EARLY DANIAN

The planktic foraminifera during the early Danian indicate prolonged environmental stress and delayed ecosystem recovery after the mass extinction. The most expanded early Danian sequences known to date are from Tunisia (Elles and El Kef sections) and Texas (Brazos River sections). However, high-resolution stable isotope records and faunal analyses for zones P1b and P1c are still lacking. In the condensed and less complete sequences from the North and South Atlantic (D'Hondt and Keller, 1991; Keller, 1993; Keller et al., 2013) and the eastern Tethys (Fig.

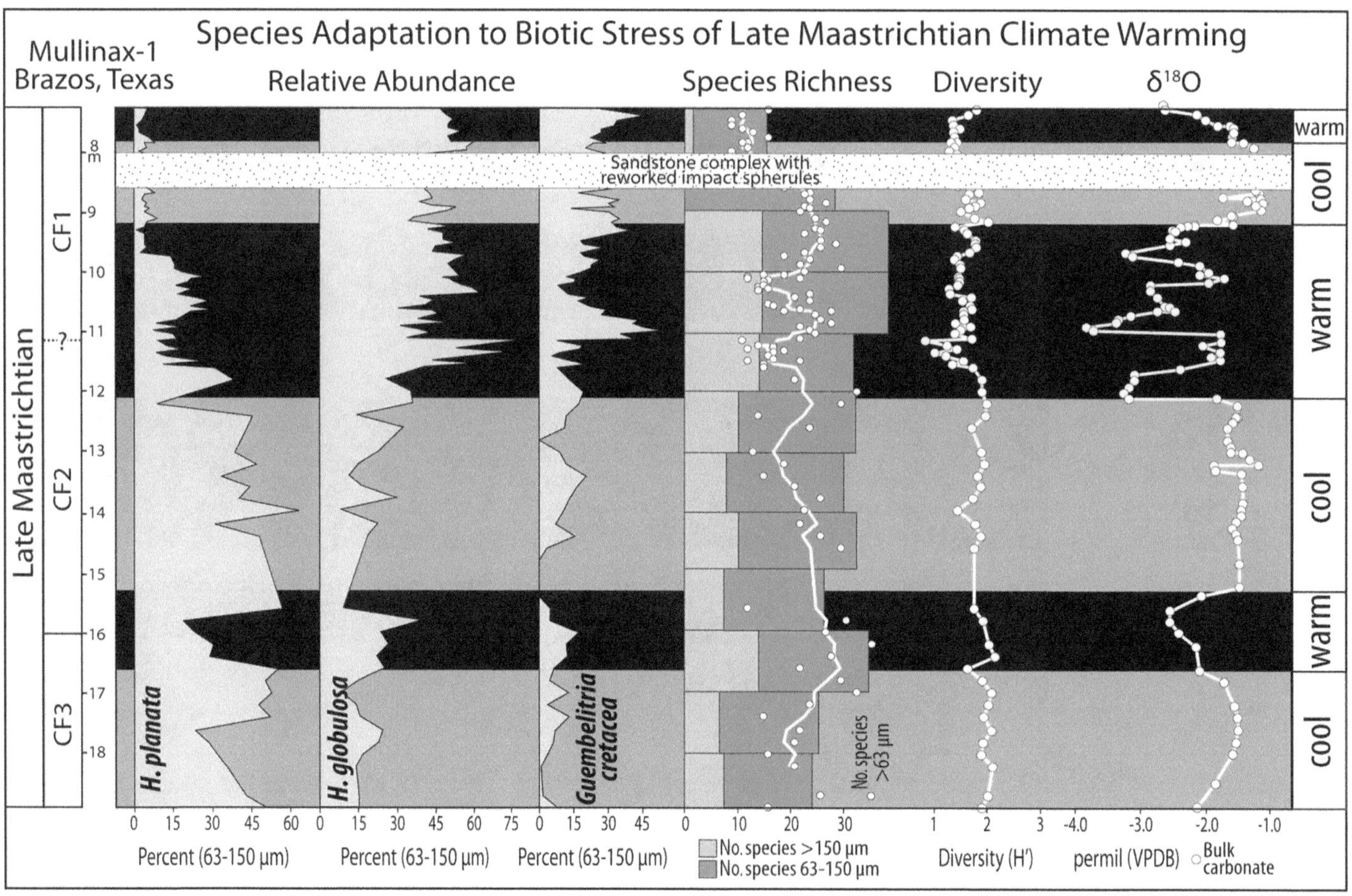

Figure 14. Species adaptation, species richness, and diversity compared with oxygen isotope trends at Brazos, Texas. Note the distinct changes with warm/cool events. Data are from Abramovich et al. (2011). VPDB—Vienna Peedee belemnite; *H. planata—Heterohelix planata; H. globulosa—Heterohelix globulosa.*

15), early Danian zones P0 and P1a contain multiple hiatuses (MacLeod and Keller, 1991a, 1991b; Keller and Benjamini, 1991; Keller et al., 2013; Punekar et al., 2014). For this reason, we concentrate on zone P1b, which is relatively complete.

Guembelitria blooms are well known from the aftermath of the K-T boundary mass extinction, but their reappearance in zone P1b (frequently with blooms of *Gl. daubjergensis*) is less well documented. In most sections in Israel and Egypt, the P1b *Guembelitria* blooms begin near the P1a-P1b hiatus, correlative with a major isotope excursion at Hor Hahar and Wadi Hamama (Figs. 11 and 15). Species richness remains low, and species are dwarfed through zone P1b (10–15 species; e.g., Keller and Benjamini, 1991; Keller, 1988a, 1988b, 2002, 2004; Punekar et al., 2014). Near the top of P1b, *Guembelitria* blooms decrease rapidly and are replaced by blooms of chiloguembelinids, which in turn give way to the first larger (>150 μm) trochospiral species morphologies and higher diversity in zone P1c. This marks the onset of full marine recovery after the mass extinction.

This pattern of *Guembelitria* blooms, frequently accompanied by blooms of *Gl. daubjergensis*, is observed from India to the Tethys and Atlantic to Texas, marking global high-stress environments in open marine and marginal marine settings (Figs. 16 and 17). These renewed high-stress conditions follow a period of recovery in the upper part of the *P. eugubina* zone (P1a[2]), as indicated by the low or nearly disappearing *Guembelitria* species (Fig. 17). Interestingly, the onset of the renewed high-stress environment is marked by the extinction of *P. eugubina* and its similar morphotype *P. longiapertura*, as well as changes in the relative abundances of other species (Fig. 15) documented in numerous publications (e.g., Keller, 1988a, 1989a, 1989b; Keller and Benjamini, 1991; Canudo et al., 1991; Keller et al., 2013; Punekar et al., 2014). The renewed high-stress environment and long-delayed marine recovery after the mass extinction have long been an enigma and largely ignored by paleontologists. Recent documentation of early Danian zone P1b assemblages in intertrappean sediments between Deccan lava megaflows in India provides clues to this mystery.

DECCAN VOLCANISM AND THE K-T BOUNDARY MASS EXTINCTION

Recent studies of deep wells in the Krishna-Godavari Basin of India have documented the K-T boundary mass extinction in marine sediments between Earth's largest megaflows, reaching >1000 km across India during the main Deccan eruption phase 2 (80% of total Deccan volume) in the latest Maastrichtian chron

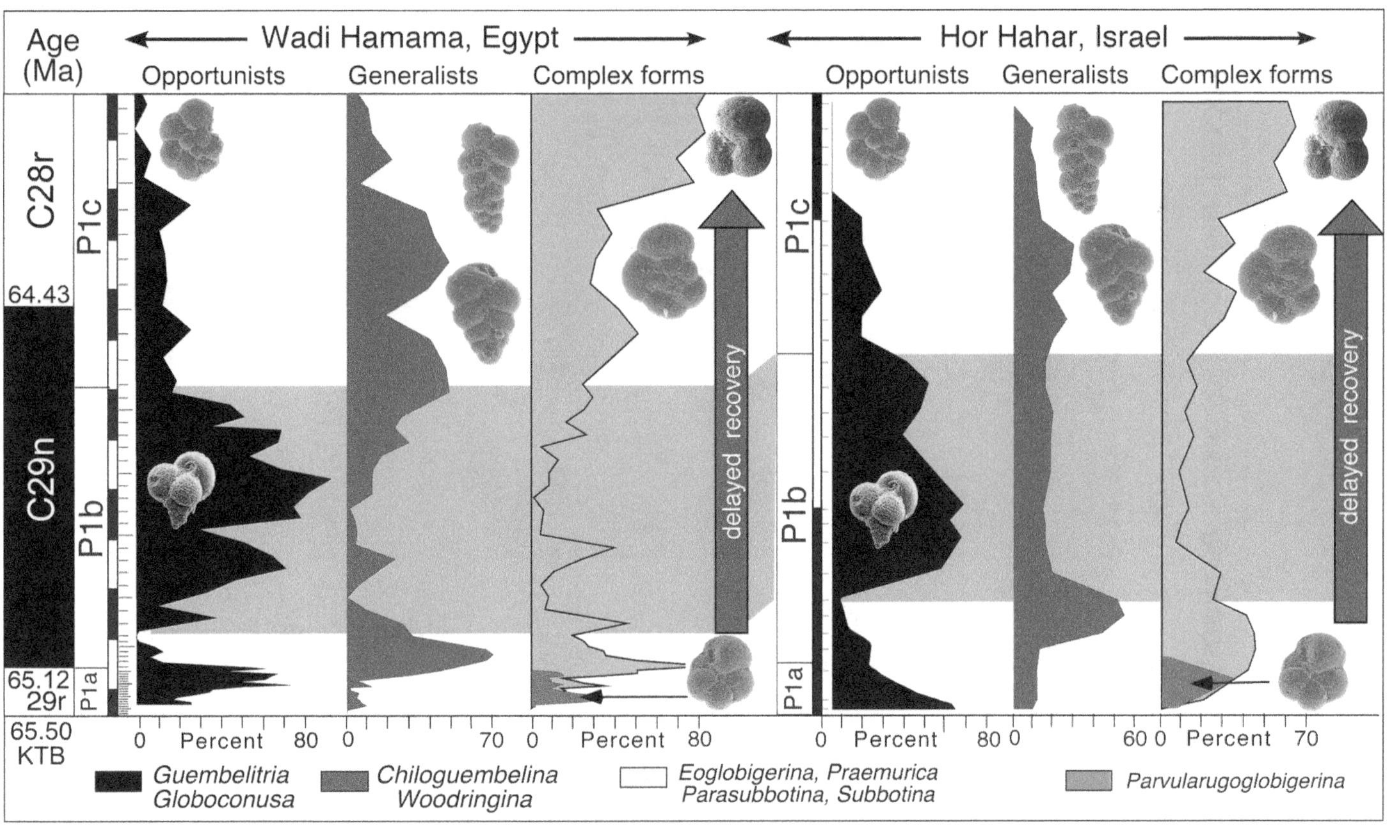

Figure 15. Dominant species populations in the early Danian zones P1b and P1c in the eastern Tethys. Note the blooms of ecological opportunists *Guembelitria* and *Globoconusa* during zone P1b, indicating maximum high-stress environments for planktic foraminifera correlative with Deccan volcanism phase 3. KTB—Cretaceous-Tertiary boundary.

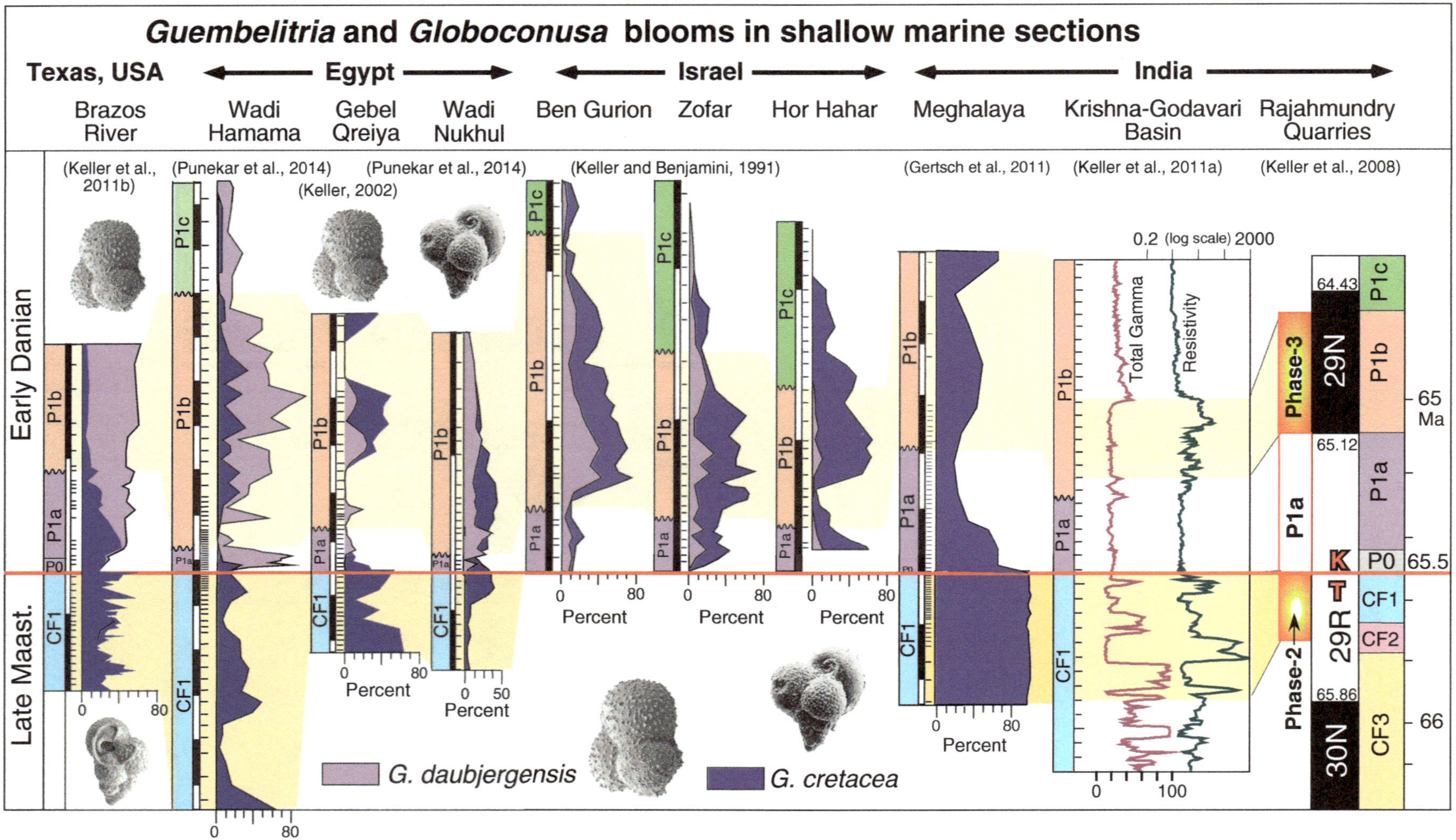

Figure 16. *Guembelitria* stress events in zone CF1 and P1b correlative with Deccan volcanic phase 2 and phase 3 of the late Maastrichtian and early Danian of continental shelf sequences from India through the Tethys, North Atlantic, and Texas. Maximum stress observed in the latest Maastrichtian of NE India (Meghalaya) decreases westward. In contrast, *Guembelitria* blooms in the early Danian zone P1b remain high across the oceans. K—Cretaceous; T—Tertiary; *G. daubjergensis—Globoconusa daubjergensis; G. cretacea—Guembelitria cretacea.*

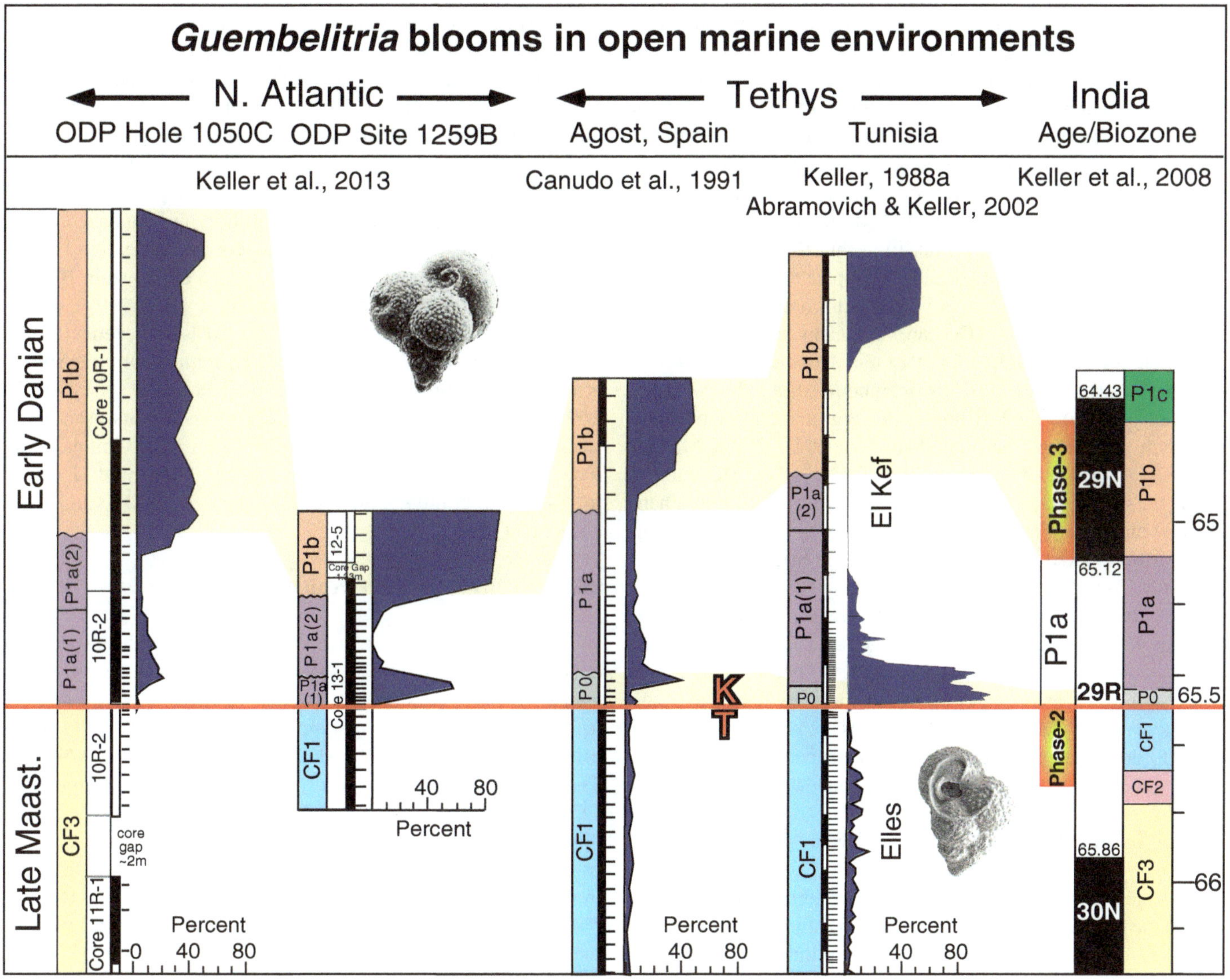

Figure 17. *Guembelitria* blooms in open marine environments show similar distributions and abundances in the early Danian zone P1b, but only minor *Guembelitria* blooms in the latest Maastrichtian zone CF1. K—Cretaceous; T—Tertiary.

C29r (zones CF1–CF2; Chenet et al., 2007, 2008; Self et al., 2008; Keller et al., 2011a, 2012). These data now permit re-evaluation of the climatic changes and global high-stress conditions in zones CF1–CF2 before the K-T boundary mass extinction that could never be explained by the Chicxulub impact hypothesis.

The expanded paleoclimatic records from Tunisia and Texas with distinct multiple rapid warm-cool events suggest that climatic changes in the late Maastrichtian zones CF1–CF2 were complex and episodic (Figs. 9 and 10), rather than a single global hyperthermal event in chron C29r, as originally identified at Site 525A and elsewhere (Fig. 9; Li and Keller, 1998a; Ravizza and Puecker-Ehrenbrink, 2003; Nordt et al., 2003; MacLeod et al., 2005; Robinson et al., 2009). The apparently single hyperthermal event can be explained by condensed sedimentation at Site 525A and elsewhere, as evident by comparison with Tunisia and Texas, where sediment accumulation is two to three times higher. The rapid paleoclimatic fluctuations evident in the carbon and oxygen isotope records point toward recurrent environmental perturbations, which likely resulted from the rapid succession of major Deccan eruptions, particularly the four lava megaflows that span planktic foraminiferal zones CF1–CF2 and most of the *Micula prinsii* nannofossil zone in the Krishna-Godavari Basin (Keller et al., 2011a, 2012). In the sediments below the first megaflow, planktic foraminifera gradually suffered a nearly 50% drop in species richness due to volcanic activity in the central magmatic province, as indicated by geochemical and mineralogical data (Keller et al., 2012). In the intertrappean sediments after the first megaflow eruption, survivors suffered another nearly 50% drop, leaving only 7–8 species. No recovery occurred between the next three megaflows, and the K-T boundary mass extinction was complete by the end of the last megaflow of phase 2. Whether this extinction pattern in India can be recognized globally remains to be examined in high-resolution expanded sediment records. Keys to this endeavor are complete K-T boundary sequences with high sediment accumulation rates, good carbonate preservation, and high-resolution stable isotope records, and ideally the recognition of unique chemical markers of Deccan volcanism.

Robinson et al. (2009) have linked a decline in $^{187}Os/^{188}Os$ beginning at the C30n-C29r transition and the late Maastrichtian global warming event with the main phase 2 of Deccan volcanism. Font et al. (2011, this volume) discovered a promising technique for identifying Deccan volcanism signals based on magnetic studies at Bidart (France) and Gubbio (Italy), where Cl-bearing iron oxides in low-magnetic-susceptibility (MS) intervals just below the K-T boundary suggest a reaction between HCl-rich volcanic gas and liquid-solid aerosols. Detailed faunal analysis of the Bidart section reveals that this low MS interval spans the last 30 cm of zone CF1. The planktic foraminifera here are few and poorly preserved due to dissolution, marking ocean acidification linked to Deccan volcanism (Font et al, 2011). Sial et al. (2013) used mercury concentrations as a proxy for volcanic activity and determined the atmospheric Hg and CO_2 buildup in three K-T boundary localities (Argentina, Brazil, Denmark); they discovered that the main species, Hg^{+2}, is of volcanic origin and may be used as indicator of volcanic activity during extreme climatic and biotic events. Similar studies are needed on a global basis to document the causes and consequences of extreme climatic and biotic events during the K-T boundary transition.

Ocean Acidification and Its Biotic Effects

Large-scale volcanism can severely perturb the carbon cycle due to the increased CO_2 emissions into the atmosphere. Loading of pCO_2 at a rate faster than the buffering time/capacity of the ocean (~1000 yr; Zeebe, 2012) can significantly affect seawater carbonate chemistry. Equilibration at the atmosphere-ocean interface would result in lowering of the CO_3^{2-} ion concentration and/or lowering of the surface ocean pH (Kump et al., 2009). Acidification would therefore lead to a calcification crisis for the carbonate-secreting planktons such as coccolithophores, foraminifera, and pteropods. The removal of huge volumes of volcanic SO_2 as acid rain would also lead to short-term lowering of surface ocean pH and cause severe biotic stress. Several faunal turnovers and mass extinction events that affected marine calcifiers in geological history may have involved ocean acidification events, including ocean anoxic events of the Paleozoic and the Paleocene-Eocene thermal maximum (see review in Hönisch et al., 2012). Dwarfism, malformed individuals, and R-strategist–dominated assemblages are recorded in calcareous nannoplankton during oceanic anoxic event 1a (OAE 1a), likely linked with stepwise accumulation of CO_2 in the ocean and progressive acidification (Erba et al., 2010).

In the past ~300 yr, the pH of the ocean has dropped by 0.1 units, and a further drop of 0.7 units is expected in the next ~300 yr due to anthropogenic CO_2 emissions. At this rate, models predict the southern oceans to become undersaturated with respect to aragonite in the next 50 yr and with respect to calcite in the next 100 yr (Orr et al., 2005; Zeebe et al., 2008). Experiments show that shell dissolution in live pteropods begins in less than 48 h on exposure to undersaturated waters (Fabry et al., 2008; Doney et al., 2009). Calcite-precipitating organisms are likely to suffer similar test dissolution on further acidification. Laboratory culture experiments confirm a decrease in the shell mass of planktic foraminifera with a decrease in $[CO_3^{2-}]$ of ambient water (Spero et al., 1997; Bijma et al., 1999). Acidification appears to have resulted in a 30%–35% reduction in calcification in modern *Globigerina bulloides* from the Southern Ocean as compared to Holocene specimens (Moy et al., 2009). Modern benthic foraminifera in sediments associated with shallow volcanic CO_2 seeps in Papua New Guinea show a decline in abundance and diversity correlative with decreasing pH (with complete absence in pH <7.9, >700 µatm pCO_2; Uthicke et al., 2013).

Deccan Volcanism and Ocean Acidification

Deccan volcanism phase 2 during the late Maastrichtian occurred as a series of at least 30 major eruptive events. The estimated CO_2 emission rate for each single eruptive event is up to 20 Gt/yr (assuming a maximum duration of ~10 yr for each event; Chenet et al., 2009; Courtillot and Fluteau, this volume). This rate is comparable with the magnitude of anthropogenic CO_2 loading (~30 Gt/yr) since the advent of the industrial revolution (Forster et al., 2007). A single Deccan eruptive event, or meteorite impact, of geologically short duration could result in ocean acidification that persisted for no more than 1000 yr (Gibbs et al., 2010), which may not be resolvable in the sedimentary record. However, the rapid succession (months, years, decades to 100,000 yr; Courtillot and Fluteau, this volume) of major Deccan eruptive events likely changed the seawater chemistry of the ocean, resulting in ocean acidification within a few hundred years of the onset of volcanism (analogous to the present-day anthropogenic situation of 300–600 yr) and may have lasted for 10,000–100,000 yr (the time scale of restoration of pre-perturbation marine $[CO_3^{2-}]$ by continental weathering processes; Archer, 2005; Hönisch et al., 2012). At this time scale, the effects of ocean acidification would have persisted long enough to be resolved in the sediment record. The episodic nature of Deccan eruptions leading to ocean acidification along with changes in water mass stratification caused by rapid global warming and cooling may explain the severe stress conditions for marine calcifiers near the end of the Maastrichtian.

Although this is a compelling hypothesis, acidification as a kill mechanism in the mass extinction of planktic foraminifera still requires rigorous evaluation. The intervals that record faunal stress must be tested for paleo-pH or $[CO_3^{2-}]$ changes using proxies such as $\delta^{11}B$, B/Ca, and Zn/Ca (Yu and Elderfield, 2007; Hönisch et al., 2012). Dissolution-based proxies such as $CaCO_3$ %

and foraminiferal shell weight may also be used as supporting evidence in sections that have not been significantly affected by postdepositional dissolution and recrystallization.

DECCAN VOLCANISM AND DELAYED RECOVERY AFTER THE MASS EXTINCTION

The delayed recovery in marine biota in the aftermath of the K-T boundary mass extinction has remained an enigma ever since the beginning of the K-T boundary debate (Hsü and McKenzie, 1985; Keller, 1988a, 1989a; Magaritz et al., 1992; review in Keller and Abramovich, 2009). Magaritz et al. (1992) first identified a major $\delta^{13}C$ anomaly associated with a black clay layer in the early Danian section in the Negev, Israel, and Keller and Benjamini (1991) documented very high abundances of *G. cretacea*, identifying this event as a major high-stress environment similar to the immediate aftermath of the K-T boundary mass extinction. More recently, Quillévéré et al. (2008) observed negative $\delta^{18}O$ and $\delta^{13}C$ anomalies in the North Atlantic ODP Site 1049C, and DSDP Sites 527 and 528, in zone P1a of Berggren et al. (1995), which corresponds to zone P1b of Keller et al. (1995; Fig. 8), and labeled this the hyperthermal Dan-C2 event. Coccioni et al. (2010) linked this warming to the last phase of Deccan volcanism, which began near the base of chron C29n coincident with the zone P1a-P1b boundary (Chenet et al., 2007, 2008, 2009; Keller et al., 2008, 2011a).

Keller et al. (2011a, 2012) documented zones P1a, P1b, and P1c assemblages in sediments below, between, and above the megaflows of Deccan phase 3 in nine deep wells of the Krishna-Godavari Basin. The early Danian planktic foraminiferal turnover associated with Deccan volcanism records the same low diversity, small species morphologies, and sequence of appearances/disappearances as observed in zones P1a and P1b worldwide. Biotic recovery to larger size and higher diversity followed in zone P1c after Deccan phase 3 ended. This study firmly links the early Danian high-stress environment and delayed recovery to the last phase 3 of Deccan volcanism. High-stress environments in zone P1b are evident by blooms of the opportunistic species *Guembelitria* and/or *Gl. daubjergensis* from India to the eastern Tethys (Israel and Egypt) and Texas (Figs. 16 and 17).

Interestingly, there are no extinctions associated with the Danian perturbations. Several factors can explain the differing responses between Deccan phase 2 and phase 3: (1) phase 3 volcanic eruptions were not as frequent or extreme as in phase 2 and therefore permitted ecosystem recovery between eruptions. (2) Climate changes associated with phase 3 were less extreme compared with phase 2 during the latest Maastrichtian, when temperatures on land rose by 5–8 °C and temperatures in the deep sea rose by 3–4 °C (Li and Keller, 1998b, 1998c; Nordt et al., 2003; Wilf et al., 2003). (3) Deccan phase 3 accounts for 14% of the total Deccan eruption volume as compared with 80% for phase 2. (4) The less catastrophic biotic response to phase 3 volcanism is at least partly related to faunal assemblages that evolved during post–K-T boundary high-stress environments and therefore were better adapted with a wider range of tolerance than the highly specialized Maastrichtian species.

CONCLUSIONS AND FUTURE WORK

Until a few years ago, the K-T boundary mass extinction was commonly attributed to the Chicxulub impact as the sole cause, and the long-term pre- and postextinction climatic, environmental, and faunal changes were either denied as nonexisting or ignored as unrelated to the mass extinction (see Schulte et al., 2010). Today, it is generally recognized that large igneous province volcanism played a major role in four out of five mass extinctions, including that at the K-T boundary (Bond and Wignall, this volume; Courtillot and Fluteau, this volume; Keller, this volume). Deccan volcanism is the only large igneous province event where a mass extinction has been documented in sediments between major basalt flows, providing an excellent case study for evaluation of the immediate effects of volcanic eruptions on marine life. This review summarizes the global climatic and faunal changes associated with the main Deccan phase 2 that led to the K-T boundary mass extinction and the last phase 3 that delayed recovery in the early Danian for at least 500 k.y. (Fig. 18). The initial Deccan phase 1, which accounts for ~6% of the total basalt eruptions, can also be recognized by *Guembelitria* blooms in India and Texas in zone CF4. However, there are little data available on this time interval to date.

There are still many challenges in working out the details of the faunal and climatic records globally and correlating these to Deccan volcanism. The major hurdle is the incomplete record in most localities, with hiatuses at the K-T boundary, within zone P1a, and at the P1a-P1b boundary. These short hiatuses are generally not recognized by routine biostratigraphic analysis based solely on the presence or absence of the index species, which frequently leads to erroneous conclusions that the record is complete when in fact just a small fraction of a biozone is preserved. High-resolution quantitative faunal analysis is necessary to determine the completeness of the sedimentation records, though such studies are often not done because they are very labor intensive.

Another challenge is the accurate evaluation of the age and tempo of the individual pulsed eruptions and megaflows, which requires chronologic dating using zircons for maximum age resolution. Equally important are accurate age estimates for the duration of quiescent intervals between eruptive events; it is possible that good age estimates can be obtained from red bole horizons between Deccan flows and from intertrappean marine sediments.

Multidisciplinary studies and use of a wide variety of geochemical proxies are required to understand the deleterious effects of Deccan volcanism on climate, the environment, and marine biota, with particular emphasis on potential stress and kill mechanisms, such as surface ocean acidification, marine calcification crisis, increased continental weathering, and nutrient and productivity changes.

Better atmospheric and oceanic global climate models for the late Maastrichtian–early Danian based on geochemical,

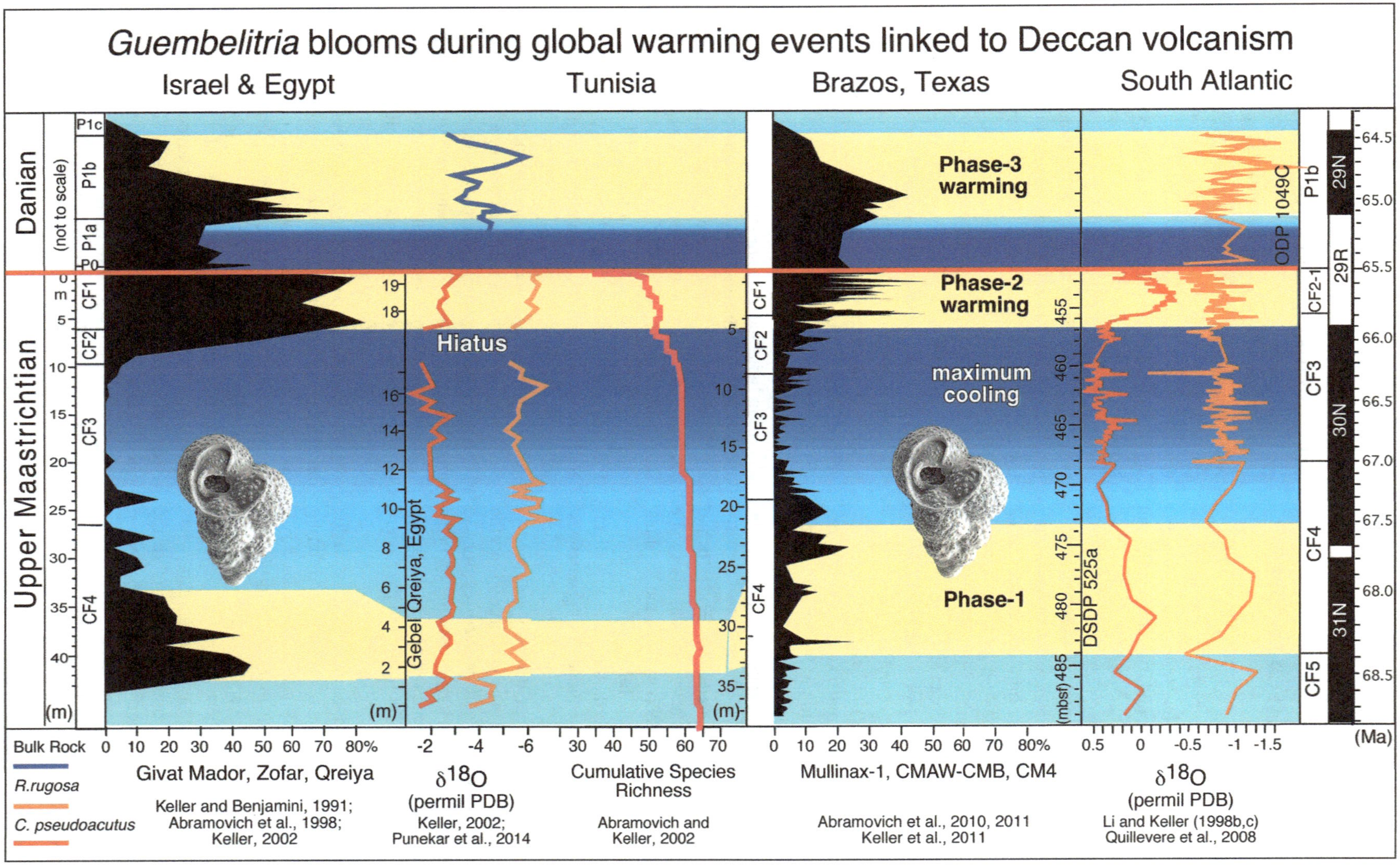

Figure 18. Global warming and faunal stress events during the late Maastrichtian and early Danian correlate with the three phases of Deccan volcanism. DSDP—Deep Sea Drilling Project; ODP—Ocean Drilling Program; PDB—Peedee belemnite; *R. rugosa*—*Rugoglobigerina rugosa*; *C. pseudoacuta*—*Cibicidoides pseudoacuta*.

paleoclimatic, and faunal records could improve our understanding of the runaway effects that likely caused the mass extinction.

ACKNOWLEDGMENTS

We are grateful for the reviews and constructive comments by Sigal Abramovich and Mark Leckie. This work is based upon work supported by the U.S. National Science Foundation (NSF) through the Continental Dynamics Program, Sedimentary Geology and Paleobiology Program, and Office of International Science & Engineering's India Program under NSF grants OISE-0912144, EAR-0207407, EAR-0447171, and EAR-1026271 and Princeton University's Scott and Tuttle Funds.

REFERENCES CITED

Abramovich, S., and Keller, G., 2002, High stress late Maastrichtian paleoenvironment: Inference from planktonic foraminifera in Tunisia: Palaeogeography, Palaeoclimatology, Palaeoecology, v. 178, p. 145–164, doi:10.1016/S0031-0182(01)00394-7.

Abramovich, S., and Keller, G., 2003, Planktonic foraminiferal response to the latest Maastrichtian abrupt warm event: A case study from South Atlantic DSDP Site 525A: Marine Micropaleontology, v. 48, no. 3–4, p. 225–249, doi:10.1016/S0377-8398(03)00021-5.

Abramovich, S., Almongi-Labin, A., and Benjamini, C., 1998, Decline of the Maastrichtian pelagic ecosystem based on planktic foraminifera assemblage changes: Implication for the terminal Cretaceous faunal crisis: Geology, v. 26, p. 63–66, doi:10.1130/0091-7613(1998)026<0063:DOTMPE>2.3.CO;2.

Abramovich, S., Keller, G., Adatte, T., Stinnesbeck, W., and Hottinger, L., 2002, Age and paleoenvironment of the Maastrichtian to Paleocene of the Mahajanga Basin, Madagascar : A multidisciplinary approach: Marine Micropaleontology, v. 47, p. 17–70, doi:10.1016/S0377-8398(02)00094-4.

Abramovich, S., Keller, G., Stüeben, D., and Berner, Z., 2003, Characterization of late Campanian and Maastrichtian planktonic foraminiferal depth habitats and vital activities based on stable isotopes: Palaeogeography, Palaeoclimatology, Palaeoecology, v. 202, no. 1–2, p. 1–29, doi:10.1016/S0031-0182(03)00572-8.

Abramovich, S., Yovel-Corem, S., Almogi-Labin, A., and Benjamini, C., 2010, Global climate change and planktic formaminiferal response in the Maastrichtian: Paleoceanography, v. 25, p. PA2201, doi:10.1029/2009PA001843.

Abramovich, S., Keller, G., Berner, Z., Cymbalista, M., and Rak, C., 2011, Maastrichtian planktic foraminiferal biostratigraphy and paleoenvironment of Brazos River, Falls County, Texas, *in* Keller, G., and Adatte, T., eds., The End-Cretaceous Mass Extinction and the Chicxulub Impact in Texas: Society for Sedimentary Geology (SEPM) Special Publication 100, p. 123–156.

Adatte, T., Keller, G., and Baum, G., 2011, Age and origin of the Chicxulub impact and sandstone complex, Brazos River, Texas: Evidence from lithostratigraphy and sedimentology, *in* Keller, G., and Adatte, T., eds., The End-Cretaceous Mass Extinction and the Chicxulub Impact in Texas: Society for Sedimentary Geology (SEPM) Special Publication 100, p. 43–80.

Archer, D., 2005, Fate of fossil fuel CO2 in geological time: Journal of Geophysical Research, v. 111, C09S05, doi:10.1029/2004JC002625.

Arenillas, I., Arz, J.A., Molina, E., and Dupuis, C., 2000, The Cretaceous/Tertiary boundary at Ain Settara, Tunisia: Sudden catastrophic mass extinction in planktic foraminifera: Journal of Foraminiferal Research, v. 30, p. 202–218, doi:10.2113/0300202.

Arenillas, I., Arz, J.A., Grajales-Nishimura, J.M., Murillo-Muneton, G., Alvarez, W., Camargo-Zanguera, A., Molina, E., and Rosales-Dominguez, C., 2006, Chicxulub impact event is Cretaceous/Paleogene boundary in age: New micropaleontological evidence: Earth and Planetary Science Letters, v. 249, p. 241–257, doi:10.1016/j.epsl.2006.07.020.

Barrera, E., and Keller, G., 1990, Foraminiferal stable isotope evidence for gradual decrease of marine productivity and Cretaceous species survivorship in the earliest Danian: Paleoceanography, v. 5, p. 867–890, doi:10.1029/PA005i006p00867.

Begon, M., Mortimer, M., and Thompson, D.J., 1996, Population Ecology: A Unified Study of Plants and Animals: Cambridge, UK, Blackwell, 247 p.

Begon, M., Harper, J.L., and Townsend, C.R., 1998, Ecology: Individuals, Populations and Communities: Boston, Blackwell Science, 1068 p.

Berggren, W.A., Kent, D.V., Swisher, C.C., III, and Aubry, M.-P., 1995, A revised Cenozoic geochronology and chronostratigraphy, *in* Berggren, W.A., Kent, D.V., Aubry, M.-P., and Hardenbol, J., eds., Geochronology, Time Scales and Global Stratigraphic Correlation: Society for Sedimentary Geology (SEPM) Special Publication 54, p. 129–212.

Bijma, J., Spero, H.J., and Lea, D.W., 1999, Reassessing foraminiferal stable isotope geochemistry: Impact of the oceanic carbonate system (experimental results), *in* Fischer, G., and Wefer, G., eds., Use of Proxies in Paleoceanography: Examples from the South Atlantic: Berlin, Springer, p. 489–512.

Bond, D.P.G., and Wignall, P.B., 2014, this volume, Large igneous provinces and mass extinctions: An update, *in* Keller, G., and Kerr, A.C., eds., Volcanism, Impacts, and Mass Extinctions: Causes and Effects: Geological Society of America Special Paper 505, doi:10.1130/2014.2505(02).

Brayshay, M., and Grattan, J., 1999, Environmental and social responses in Europe to the 1783 eruption of the Laki fissure volcano in Iceland: A consideration of contemporary documentary evidence, *in* Firth, C.R., and McGuire, W.J., eds., Volcanoes in the Quaternary: Geological Society of London Special Publication 161, p. 173–188.

Buzas, M.A., and Gibson, T.G., 1969, Species diversity: Benthonic foraminifera in the western North Atlantic: Science, v. 163, p. 72–75, doi:10.1126/science.163.3862.72.

Cande, S., and Kent, D.V., 1995, Revised calibration of the geomagnetic polarity timescale for the Late Cretaceous and Cenozoic: Journal of Geophysical Research, v. 100, p. 6093–6095, doi:10.1029/94JB03098.

Canudo, J.I., Keller, G., and Molina, E., 1991, Cretaceous/Tertiary boundary extinction pattern and faunal turnover at Agost and Caravaca, SE Spain: Marine Micropaleontology, v. 17, p. 319–341, doi:10.1016/0377-8398(91)90019-3.

Chenet, A.-L., Quidelleur, X., Fluteau, F., Courtillot, V., and Bajpai, S., 2007, $^{40}K/^{40}Ar$ dating of the main Deccan large igneous province: Further evidence of Cretaceous-Tertiary boundary age and short duration: Earth and Planetary Science Letters, v. 263, p. 1–15, doi:10.1016/j.epsl.2007.07.011.

Chenet, A.-L., Fluteau, F., Courtillot, V., Gerard, M., and Subbarao, K.V., 2008, Determination of rapid Deccan eruptions across the Cretaceous-Tertiary boundary using paleomagnetic secular variation: Results from a 1200-m-thick section in the Mahabaleshwar: Journal of Geophysical Research, v. 113, B04101, doi:10.1029/2006JB004635.

Chenet, A.-L., Courtillot, V., Fluteau, F., Gerard, M., Quidelleur, X., Khadri, S.F.R., Subbarao, K.V., and Thordarson, T., 2009, Determination of rapid Deccan eruptions across the Cretaceous-Tertiary boundary using paleomagnetic secular variation: 2. Constraints from analysis of eight new sections and synthesis for a 3500-m-thick composite section: Journal of Geophysical Research, v. 114, B06103, doi:10.1029/2008JB005644.

Coccioni, R., and Luciani, V., 2006, *Guembelitria irregularis* bloom at the K-T boundary: Morphological abnormalities induced by impact-related extreme environmental stress?, *in* Cockell, C., Koeberl, C., and Gilmour, I., eds., Biological Processes Associated with Impact Events: Impact Studies: Berlin, Springer, p. 179–196.

Coccioni, R., Frontalini, F., Bancalà, G., Fornaciari, E., Jovane, L., and Sprovieri, M., 2010, The Dan-C2 hyperthermal event at Gubbio (Italy): Global implications, environmental effects, and cause(s): Earth and Planetary Science Letters, v. 297, p. 298–305, doi:10.1016/j.epsl.2010.06.031.

Courtillot, V., and Fluteau, F., 2014, this volume, A review of the embedded time scales of flood basalt volcanism with special emphasis on dramatically short magmatic pulses, *in* Keller, G., and Kerr, A.C., eds., Volcanism, Impacts, and Mass Extinctions: Causes and Effects: Geological Society of America Special Paper 505, doi:10.1130/2014.2505(15).

D'Hondt, S., and Keller, G., 1991, Some patterns of planktic foraminiferal assemblage turnover at the Cretaceous-Tertiary boundary: Marine Micropaleontology, v. 17, p. 77–118, doi:10.1016/0377-8398(91)90024-Z.

Doney, S.C., Fabry, V.J., Feely, R.A., and Kleypas, J.A., 2009, Ocean acidification: The other CO2 problem: Annual Review of Marine Science, v. 1, p. 169–192, doi:10.1146/annurev.marine.010908.163834.

Erba, E., Bottini, C., Weissert, H.J., and Keller, C.E., 2010, Calcareous nannoplankton response to surface-water acidification around oceanic anoxic event 1a: Science, v. 329, no. 5990, p. 428–432, doi:10.1126/science.1188886.

Fabry, V.J., Seibel, B.A., Feely, R.A., and Orr, J.C., 2008, Impacts of ocean acidification on marine fauna and ecosystem processes: ICES Journal of Marine Science: Journal du Conseil, v. 65, no. 3, p. 414–432, doi:10.1093/icesjms/fsn048.

Font, E., Nedelec, A., Ellwood, B.B., Mirao, J., and Silva, P.F., 2011, A new sedimentary benchmark for the Deccan Traps volcanism?: Geophysical Research Letters, v. 38, L24309, doi:10.1029/2011GL049824.

Font, E., Fabre, S., Nédélec, A., Adatte, T., Keller, G., Veiga-Pires, C., Ponte, J., Mirão, J., Khozyem, H., and Spangenberg, J., 2014, this volume, Atmospheric halogen and acid rains during the main phase of Deccan eruptions: Magnetic and mineral evidence, *in* Keller, G., and Kerr, A.C., eds., Volcanism, Impacts and Mass Extinctions: Causes and Effects: Geological Society of America Special Paper 505, doi:10.1130/2014.2505(18).

Forster, P., Ramaswamy, V., Artaxo, P., Berntsen, T., Betts, R., Fahey, D.W., Haywood, J., Lean, J., Lowe, D.C., Myhre, G., Nganga, J., Prinn, R., Raga, G., Schultz, M., and Van Dorland, R., 2007, Changes in atmospheric constituents and in radiative forcing, *in* Solomon., S., et al., eds., Climate Change 2007: The Physical Science Basis. Contribution of Working Group I to the Fourth Assessment Report of the Intergovernmental Panel on Climate Change: Cambridge, UK, Cambridge University Press, p. 129–234.

Ganino, C., and Arndt, N.T., 2009, Climate changes caused by degassing of sediments during the emplacement of large igneous provinces: Geology, v. 37, p. 323–326, doi:10.1130/G25325A.1.

Gertsch, B., Keller, G., Adatte, T., Garg, R., Prasad, V., Fleitmann, D., and Berner, Z., 2011, Environmental effects of Deccan volcanism across the Cretaceous-Tertiary transition in Meghalaya, India: Earth and Planetary Science Letters, v. 310, p. 272–285, doi:10.1016/j.epsl.2011.08.015.

Gibbs, S.J., Stoll, H.M., Brown, P.R., and Bralower, T.J., 2010, Ocean acidification and surface water carbonate production across the Paleocene-Eocene thermal maximum: Earth and Planetary Science Letters, v. 295, p. 583–592, doi:10.1016/j.epsl.2010.04.044.

Gilmour, I., Gilmour, M., Jolley, D.W., Kelley, S.P., Kemp, D.B., Daly, R.J., and Watson, J.S., 2013, A high-resolution nonmarine record of an early Danian hyperthermal event, Boltysh Crater, Ukraine: Geology, v. 41, p. 783–786, doi:10.1130/G34292.1.

Ginsburg, R.N., 1997a, An attempt to resolve the controversy over the end-Cretaceous extinction of planktic foraminifera at El Kef, Tunisia, using a blind test; introduction; background and procedures: Marine Micropaleontology, v. 29, no. 2, p. 67–68, doi:10.1016/S0377-8398(96)00038-2.

Ginsburg, R.N., 1997b, Not so blind El Kef test: Marine Micropaleontology, v. 32, no. 3–4, p. 399, doi:10.1016/S0377-8398(97)89276-6.

Ginsburg, R.N., 1997c, Perspectives on the blind test: Marine Micropaleontology, v. 29, no. 2, p. 101–103, doi:10.1016/S0377-8398(96)00046-1.

Gradstein, F., Ogg, J., and Smith, A., 2004, A Geologic Time Scale: Cambridge, UK, Cambridge University Press, 598 p.

Gradstein, F.M., Ogg, G., and Schmitz, M., 2012, The Geologic Time Scale 2-Volume Set: Amsterdam, Elsevier, 92 p.

Grattan, J., Durand, M., and Taylor, S., 2003, Illness and elevated human mortality in Europe coincident with the Laki fissure eruption, *in* Oppenheimer C., Pyle, D.M., and Barclay J., eds., Volcanic Degassing: Geological Society of London Special Publication 213, p. 401–414.

Hönisch, B., Ridgwell, A., Schmidt, D.N., Thomas, E., Gibbs, S.J., Sluijs, A., Zeebe, R., Kump, L., Martindale, R.C., Greene, S.E., Kiessling, G., Ries, J., Zachos, J.C., Royer, D.L., Barker, S., Marchitto, T.M., Jr., Moyer, R., Pelejero, C., Ziveril, P., Foster, G.L., and Williams, B., 2012, The geological record of ocean acidification: Science, v. 335, no. 6072, p. 1058–1063, doi:10.1126/science.1208277.

Hsü, K.J., and McKenzie, J.A., 1985, A "strangelove" ocean in earliest Tertiary, *in* Sundquist, E.T., and Broecker, W., eds., The Carbon Cycle and Atmospheric CO_2: Natural Variations Archean to Present: American Geophysical Union Geophysical Monograph 32, p. 487–492.

Jay, A.E., and Widdowson, M., 2008, Stratigraphy, structure and volcanology of the south-east Deccan continental flood basalt province: Implications for eruptive extent and volumes: Journal of the Geological Society of London, v. 165, p. 177–188, doi:10.1144/0016-76492006-062.

Keller, G., 1988a, Extinction, survivorship and evolution of planktic foraminifers across the Cretaceous/Tertiary boundary at El Kef, Tunisia: Marine Micropaleontology, v. 13, p. 239–263, doi:10.1016/0377-8398(88)90005-9.

Keller, G., 1988b, Biotic turnover in benthic foraminifera across the Cretaceous-Tertiary boundary at El Kef, Tunisia: Palaeogeography, Palaeoclimatology, Palaeoecology, v. 66, p. 153–171, doi:10.1016/0031-0182(88)90198-8.

Keller, G., 1989a, Extended period of extinctions across the K/T boundary in planktic foraminifera of continental shelf sections: Implications for impact and volcanism theories: Geological Society of America Bulletin, v. 101, p. 1408–1419, doi:10.1130/0016-7606(1989)101<1408:EPOEAT>2.3.CO;2.

Keller, G., 1989b, Extended K/T boundary extinctions and delayed populational change in planktic foraminiferal faunas from Brazos River, Texas: Paleoceanography, v. 4, no. 3, p. 287–332, doi:10.1029/PA004i003p00287.

Keller, G., 1993, The Cretaceous-Tertiary boundary transition in the Antarctic Ocean and its global implications: Marine Micropaleontology, v. 21, p. 1–45, doi:10.1016/0377-8398(93)90010-U.

Keller, G., 2001, The end-Cretaceous mass extinction in the marine realm: Year 2000 assessment: Planetary and Space Science, v. 49, no. 8, p. 817–830, doi:10.1016/S0032-0633(01)00032-0.

Keller, G., 2002, *Guembelitria*-dominated late Maastrichtian planktic foraminiferal assemblages mimic early Danian in central Egypt: Marine Micropaleontology, v. 47, p. 71–99, doi:10.1016/S0377-8398(02)00116-0.

Keller, G., 2003, Biotic effects of impacts and volcanism: Earth and Planetary Science Letters, v. 215, no. 1–2, p. 249–264, doi:10.1016/S0012-821X(03)00390-X.

Keller, G., 2004, Low-diversity, late Maastrichtian and early Danian planktic foraminiferal assemblages of the eastern Tethys: Journal of Foraminiferal Research, v. 34, no. 1, p. 49–73, doi:10.2113/0340049.

Keller, G., 2005, Biotic effects of late Maastrichtian mantle plume volcanism: Implications for impacts and mass extinctions: Lithos, v. 79, no. 3–4, p. 317–341, doi:10.1016/j.lithos.2004.09.005.

Keller, G., 2008a, Cretaceous climate, volcanism, impacts and biotic effects: Cretaceous Research, v. 29, no. 5–6, p. 754–771, doi:10.1016/j.cretres.2008.05.030.

Keller, G., 2008b, Impact stratigraphy: Old principle, new reality, *in* Evans, K.R., Horton, J.W., and King, D.T., eds., The Sedimentary Record of Meteorite Impacts: Geological Society of America Special Paper 437, p. 147–178.

Keller, G., 2011, Cretaceous-Tertiary mass extinction in marginal and open marine environments: Texas, U.S.A., and Tunisia, *in* Keller, G., and Adatte, T., eds., The End-Cretaceous Mass Extinction and the Chicxulub Impact in Texas: Society for Sedimentary Geology (SEPM) Special Publication 100, p. 197–226.

Keller, G., 2014, this volume, Deccan volcanism, the Chicxulub impact, and the end-Cretaceous mass extinction: Coincidence? Cause and effect?, *in* Keller, G., and Kerr, A.C., eds., Volcanism, Impacts, and Mass Extinctions: Causes and Effects: Geological Society of America Special Paper 505, doi:10.1130/2014.2505(03).

Keller, G., and Abramovich, S., 2009, Lilliput effect in late Maastrichtian planktic foraminifera: Response to environmental stress: Palaeogeography, Palaeoclimatology, Palaeoecology, v. 284, p. 47–62, doi:10.1016/j.palaeo.2009.08.029.

Keller, G., and Benjamini, C., 1991, Paleoenvironment of the eastern Tethys in the early Danian: Palaios, v. 6, p. 439–464, doi:10.2307/3514984.

Keller, G., and Pardo, A., 2004, Disaster opportunists Guembelitrinidae—Index for environmental catastrophes: Marine Micropaleontology, v. 53, p. 83–116, doi:10.1016/j.marmicro.2004.04.012.

Keller, G., Barrera, E., Schmitz, B., and Matsson, E., 1993, Gradual mass extinction, species survivorship, and long term environmental changes across the Cretaceous-Tertiary boundary in high latitudes: Geological Society of America Bulletin, v. 105, no. 8, p. 979–997, doi:10.1130/0016-7606(1993)105<0979:GMESSA>2.3.CO;2.

Keller, G., Stinnesbeck, W., and Lopez-Oliva, J. G, 1994, Age, deposition and biotic effects of the Cretaceous/Tertiary boundary event at Mimbral, NE Mexico: Palaios, v. 9, no. 2, p. 144–157.

Keller, G., Li, L., and MacLeod, N., 1996, The Cretaceous/Tertiary boundary stratotype section at El Kef, Tunisia: How catastrophic was the mass extinction?: Palaeogeography, Palaeoclimatology, Palaeoecology, v. 119, p. 221–254, doi:10.1016/0031-0182(95)00009-7.

Keller, G., Adatte, T., Stinnesbeck, W., Luciani, V., Karoui, N., and Zaghbib-Turki, D., 2002a, Paleoecology of the Cretaceous-Tertiary mass extinction in planktic foraminifera: Palaeogeography, Palaeoclimatology, Palaeoecology, v. 178, p. 257–297, doi:10.1016/S0031-0182(01)00399-6.

Keller, G., Adatte, T., Burns, S.J., and Tantawy, A.A., 2002b, High-stress paleoenvironment during the late Maastrichtian to early Paleocene in central Egypt: Palaeogeography, Palaeoclimatology, Palaeoecology, v. 187, no. 1, p. 35–60, doi:10.1016/S0031-0182(02)00504-7.

Keller, G., Adatte, T., Tantawy, A.A., Berner, Z., Stinnesbeck, W., Stueben, D., and Leanza, H.A., 2007, High stress late Maastrichtian–early Danian palaeoenvironment in the Neuquén Basin, Argentina: Cretaceous Research, v. 28, no. 6, p. 939–960, doi:10.1016/j.cretres.2007.01.006.

Keller, G., Adatte, T., Gardin, S., Bartolini, A., and Bajpai, S., 2008, Main Deccan volcanism phase ends near the K-T boundary: Evidence from the Krishna-Godavari Basin, SE India: Earth and Planetary Science Letters, v. 268, p. 293–311, doi:10.1016/j.epsl.2008.01.015.

Keller, G., Khosla, S.C., Sharma, R., Khosla, A., Bajpai, S., and Adatte, T., 2009, Early Danian planktic foraminifera from Cretaceous-Tertiary intertrappean beds at Jhilmili, Chhindwara District, Madhya Pradesh, India: Journal of Foraminiferal Research, v. 39, no. 1, p. 40–55, doi:10.2113/gsjfr.39.1.40.

Keller, G., Bhowmick, P.K., Upadhyay, H., Dave, A., Reddy, A.N., Jaiprakash, B.C., and Adatte, T., 2011a, Deccan volcanism linked to the Cretaceous-Tertiary boundary (KTB) mass extinction: New evidence from ONGC wells in the Krishna-Godavari Basin, India: Journal of the Geological Society of India, v. 78, p. 399–428, doi:10.1007/s12594-011-0107-3.

Keller, G., Abramovich, S., Adatte, T., and Berner, Z., 2011b, Biostratigraphy, age of the Chicxulub impact, and depositional environment of the Brazos River KTB sequences, *in* Keller, G., and Adatte, T., eds., The End-Cretaceous Mass Extinction and the Chicxulub Impact in Texas: Society for Sedimentary Geology (SEPM) Special Publication 100, p. 81–122.

Keller, G., Adatte, T., Bhowmick, P.K., Upadhyay, H., Dave, A., Reddy, A.N., and Jaiprakash, B.C., 2012, Nature and timing of extinctions in Cretaceous-Tertiary planktic foraminifera preserved in Deccan intertrappean sediments of the Krishna-Godavari Basin, India: Earth and Planetary Science Letters, v. 341–344, p. 211–221, doi:10.1016/j.epsl.2012.06.021.

Keller, G., Khozyem, H.M., Adatte, T., Malarkodi, N., Spangenberg, J.E., and Stinnesbeck, W., 2013, Chicxulub impact spherules in the North Atlantic and Caribbean: Age constraints and Cretaceous-Tertiary boundary hiatus: Geological Magazine, v. 150, no. 5, p. 885–907, doi:10.1017/S0016756812001069.

Kucera, M., and Malmgren, B.A., 1998, Terminal Cretaceous warming event in the mid-latitude South Atlantic Ocean: Evidence from poleward migration of *Contusotruncana contusa* (planktonic foraminifera) morphotypes: Palaeogeography, Palaeoclimatology, Palaeoecology, v. 138, p. 1–15, doi:10.1016/S0031-0182(97)00124-7.

Kump, L., Bralower, T., and Ridgwell, A., 2009, Ocean acidification in deep time: Oceanography, v. 22, no. 4, p. 94–107.

Lamolda, M.A., Melinte, M.C., and Kaiho, K., 2005, Nannofloral extinction and survivors across the K/T boundary at Caravaca, southeastern Spain: Palaeogeography, Palaeoclimatology, Palaeoecology, v. 224, p. 27–52, doi:10.1016/j.palaeo.2005.03.030.

Li, L., and Keller, G., 1998a, Maastrichtian climate, productivity and faunal turnovers in planktic foraminifera of South Atlantic DSDP Sites 525A and 21: Marine Micropaleontology, v. 33, no. 1–2, p. 55–86, doi:10.1016/S0377-8398(97)00027-3.

Li, L., and Keller, G., 1998b, Abrupt deep-sea warming at the end of the Cretaceous: Geology, v. 26, p. 995–998, doi:10.1130/0091-7613(1998)026<0995:ADSWAT>2.3.CO;2.

Li, L., and Keller, G., 1998c, Diversification and extinction in Campanian-Maastrichtian planktic foraminifera of northwestern Tunisia: Eclogae Geologica Helvetiae, v. 91, p. 75–102.

Lipps, J.H., 1997, The Cretaceous-Tertiary boundary; the El Kef blind test: Marine Micropaleontology, v. 29, no. 2, p. 65–66, doi:10.1016/S0377-8398(96)00037-0.

Luciani, V., 1997, Planktonic foraminiferal turnover across the Cretaceous-Tertiary boundary in the Vajont valley (southern Alps, northern Italy): Cretaceous Research, v. 18, p. 799–821, doi:10.1006/cres.1997.0088.

Luciani, V., 2002, High resolution planktonic foraminiferal analysis from the Cretaceous/Tertiary boundary at Ain Settara (Tunisia): Evidence of an extended mass extinction: Palaeogeography, Palaeoclimatology, Palaeoecology, v. 178, no. 3, p. 299–319, doi:10.1016/S0031-0182(01)00400-X.

MacLeod, K.G., Huber, B.T., and Isaza-Londono, C., 2005, North Atlantic warming during global cooling at the end of the Cretaceous: Geology, v. 33, p. 437–440, doi:10.1130/G21466.1.

MacLeod, N., and Keller, G., 1991a, Hiatus distributions and mass extinctions at the Cretaceous/Tertiary boundary: Geology, v. 19, p. 497–501, doi:10.1130/0091-7613(1991)019<0497:HDAMEA>2.3.CO;2.

MacLeod, N., and Keller, G., 1991b, How complete are Cretaceous/Tertiary boundary sections? A chronostratigraphic estimate based on graphic correlation: Geological Society of America Bulletin, v. 103, p. 1439–1457, doi:10.1130/0016-7606(1991)103<1439:HCACTB>2.3.CO;2.

MacLeod, N., and Keller, G., 1994, Comparative biogeographic analysis of planktic foraminiferal survivorship across the Cretaceous/Tertiary (K/T) boundary: Paleobiology, v. 20, no. 2, p. 143–177.

MacLeod, N., Ortiz, N., Fefferman, N., Clyde, W., Schulter, C., and MacLean, J., 2000, Phenotypic response of foraminifera to episodes of global environmental change, *in* Culver, S.J., and Rawson, P., eds., Biotic Response to Global Environmental Change: The Last 145 Million Years: Cambridge, UK, Cambridge University Press, p. 51–78.

Magaritz, M., Benjamini, C., Keller, G., and Moshkovitz, S., 1992, Early diagenetic isotopic signal at the Cretaceous/Tertiary boundary, Israel: Palaeogeography, Palaeoclimatology, Palaeoecology, v. 91, no. 3, p. 291–304, doi:10.1016/0031-0182(92)90073-E.

Malarkodi, N., Keller, G., Fayazudeen, P.J., and Mallikarjuna, U.B., 2010, Foraminifera from the early Danian intertrappean beds in Rajahmundry Quarries, Andhra Pradesh, SE India: Journal of the Geological Society of India, v. 75, p. 851–863, doi:10.1007/s12594-010-0066-0.

Molina, E., Alegret, L., Arenillas, I., Arz, J., Gallala, N., Hardenbol, J., von Salis, K., Steurbaut, E., Vanhenberghe, N., and Zaghbib-Turki, D., 2006, The global boundary stratotype section and point for the base of the Danian Stage (Paleocene, Paleogene, "Tertiary," Cenozoic) at El Kef, Tunisia—Original definition and revision: Episodes, v. 29, no. 4, p. 263–273.

Moy, A.D., Howard, W.R., Bray, S.G., and Trull, T.W., 2009, Reduced calcification in modern Southern Ocean planktonic foraminifera: Nature Geoscience, v. 2, no. 4, p. 276–280, doi:10.1038/ngeo460.

Mussard, M., Le Hir, G., Fluteau, F., Lefebvre, V., and Goddéris, Y., 2014, this volume, Modeling the carbon-sulfate interplays in climate changes related to the emplacement of continental flood basalts, *in* Keller, G., and Kerr, A.C., eds., Volcanism, Impacts, and Mass Extinctions: Causes and Effects: Geological Society of America Special Paper 505, doi:10.1130/2014.2505(17).

Nordt, L., Atchley, S., and Dworkin, S., 2003, Terrestrial evidence for two greenhouse events in the latest Cretaceous: GSA Today, v. 13, no. 12, p. 4–9, doi:10.1130/1052-5173(2003)013<4:TEFTGE>2.0.CO;2.

Norris, R.D., Huber, B.T., and Self-Trail, B.T., 1999, Synchroneity of the KT oceanic mass extinction and meteorite impact: Blake Nose, western North Atlantic: Geology, v. 27, p. 419–422, doi:10.1130/0091-7613(1999)027<0419:SOTKTO>2.3.CO;2.

Olsson, R.K., Miller, K.G., Browning, J.V., Habib, D., and Sugarmann, P.J., 1997, Ejecta layer at the Cretaceous-Tertiary boundary, Bass River, New Jersey (Ocean Drilling Program Leg 174AX): Geology, v. 25, p. 759–762, doi:10.1130/0091-7613(1997)025<0759:ELATCT>2.3.CO;2.

Olsson, R.K., Hemleben, C., Berggren, W.A., and Huber, B.T., 1999, Atlas of Paleocene Planktonic Foraminifera: Smithsonian Contributions to Paleobiology 85: Washington, DC, Smithsonian Institution Press, 252 p.

Olsson, R.K., Wright, J.D., and Miller, K.G., 2001, Paleobiogeography of *Pseudotextularia elegans* during the latest Maastrichtian global warming event: Journal of Foraminiferal Research, v. 31, no. 3, p. 275–282, doi:10.2113/31.3.275.

Orr, J.C., Fabry, V.J., Aumont, O., Bopp, L., Doney, S.C., Feely, R.A., Gnanadesikan, A., Gruber, N., Ishida, A., Joos, F., Key, R.M., Lindsay, K., Maier-Reimer, E., Matear, R., Monfray, P., Mouchet, A., Najjar, R.G., Plattner, G-K., Rodgers, K.B., Sabine, C.L., Sarmiento, J.L., Schlitzer, R., Slater, R.D., Totterdell, I.J., Weirig, M-F., Yamanaka, Y., and Yool, A., 2005, Anthropogenic ocean acidification over the twenty-first century and its impact on calcifying organisms: Nature, v. 437, p. 681–686, doi:10.1038/nature04095.

Pardo, A., and Keller, G., 2008, Biotic effects of environmental catastrophes at the end of the Cretaceous: *Guembelitria* and *Heterohelix* blooms: Cretaceous Research, v. 29, no. 5–6, p. 1058–1073, doi:10.1016/j.cretres.2008.05.031.

Punekar, J., Keller, G., Khozyem, H.M., Hamming, C., Adatte, T., Tantawy, A.A., and Spangenberg, J., 2014, Late Maastrichtian–early Danian high-stress environments and delayed recovery linked to Deccan volcanism: Cretaceous Research, v. 49, p. 63–82.

Quillévéré, F., Norris, R.D., Kroon, D., and Wilson, P.A., 2008, Transient ocean warming and shifts in carbon reservoirs during the early Danian: Earth and Planetary Science Letters, v. 265, p. 600–615, doi:10.1016/j.epsl.2007.10.040.

Raja Rao, C.S., Sahasrabudhe, S.S., Deshmukh, S.S., and Raman, R., 1999, Distribution, structure and petrography of the Deccan Traps, India, *in* Subbarao, K.V., ed., Deccan Volcanic Province: Geological Society of India Memoir 43, p. 401–414.

Ravizza, G., and Peucker-Ehrenbrink, B., 2003, Chemostratigraphic evidence of Deccan volcanism from the marine osmium isotope record: Science, v. 302, p. 1392–1395, doi:10.1126/science.1089209.

Robinson, N., Ravizza, G., Coccioni, R., Peucker-Ehrenbrink, B., and Norris, R., 2009, A high-resolution marine 187Os/188Os record for the late Maastrichtian: Distinguishing the chemical fingerprints of Deccan volcanism and the KP impact event: Earth and Planetary Science Letters, v. 281, p. 159–168, doi:10.1016/j.epsl.2009.02.019.

Schulte, P., Alegret, L., Arenillas, I., Arz, J.A., Barton, P.J., Bown, P.R., Bralower, T.J., Christeson, G.L., Claeys, P., Cockell, C.S., Collins, G.S., Deutsch, A., Goldin, T.J., Goto, K., Grajales-Nishimura, J.M., Grieve, R.A.F., Gulick, S.P.S., Johnson, K.R., Kiessling, W., Koeberl, C., Kring, D.A., MacLeod, K.G., Matsui, T., Melosh, J., Montanari, A., Morgan, J.V., Neal, C.R., Nichols, D.J., Norris, R.D., Pierazzo, E., Ravizza, G., Rebolledo-Vieyra, M., Reimold, W.U., Robin, E., Salge, T., Speijer, R.P., Sweet, A.R., Urrutia-Fucugauchi, J., Vajda, V., Whalen, M.T., and Willumsen, P.S., 2010, The Chicxulub asteroid impact and mass extinction at the Cretaceous-Paleogene boundary: Science, v. 327, p. 1214–1218, doi:10.1126/science.1177265.

Self, S., Jay, A.E., Widdowson, M., and Keszthelyi, L.P., 2008, Correlation of the Deccan and Rajahmundry Trap lavas: Are these the longest and largest lava flows on Earth?: Journal of Volcanology and Geothermal Research, v. 172, p. 3–19, doi:10.1016/j.jvolgeores.2006.11.012.

Sial, A.N., Lacerda, L.D., Ferreira, V.P., Frei, R., Marquillas, R.A., Barbosa, J.A., Gaucher, C., Windmöller, C.C., and Pereira, N.S., 2013, Mercury as a proxy for volcanic activity during extreme environmental turnover: The Cretaceous–Paleogene transition: Palaeogeography, Palaeoclimatology, Palaeoecology, v. 387, p. 153–164.

Smit, J., 1982, Extinction and evolution of planktonic foraminifera after a major impact at the Cretaceous/Tertiary boundary, *in* Silver, L.T., and Schultz, P.H., eds., Geological Implications of Impacts of Large Asteroids and Comets on the Earth: Geological Society of America Special Paper 190, p. 329–352.

Spero, H.J., Bijma, J., Lea, D.W., and Bemis, B.E., 1997, Effect of seawater carbonate concentration on foraminiferal carbon and oxygen isotopes: Nature, v. 390, no. 6659, p. 497–500, doi:10.1038/37333.

Stone, R., 2004, Iceland's doomsday scenario: Science, v. 306, p. 1278–1281, doi:10.1126/science.306.5700.1278.

Stothers, R.B., 1996, Major optical depth perturbations to the stratosphere from volcanic eruptions: Pyrheliometric period, 1881–1960: Journal of Geophysical Research, v. 101, no. D2, p. 3901–3920, doi:10.1029/95JD03237.

Stüben, D., Kramar, U., Berner, Z., Leosson, M., Keller, G., Adatte, T., and Stinnesbeck, W., 2002, High resolution carbon and oxygen isotope profiles of foraminifera and the Ca-normalized Sr curve of the late Maastrichtian at Elles, Tunisia: Palaeogeography, Palaeoclimatology, Palaeoecology, v. 178, p. 321–345.

Stüben, D., Kramar, U., and Berner, Z.A., 2003, Late Maastrichtian paleoclimatic and paleoceanographic changes inferred from Sr/Ca ratio and stable isotopes: Palaeogeography, Palaeoclimatology, Palaeoecology, v. 199, p. 107–127, doi:10.1016/S0031-0182(03)00499-1.

Subbarao, K.V., Bodas, M.S., Khadri, S.F.R., and Beane, J.E., 2000, Penrose Deccan 2000 Field Excursion Guide to the Western Deccan Basalt Province, Penrose Field Guides, B: New Delhi, Geological Society of India, p. 241.

Tantawy, A.A.A., 2003, Calcareous nannofossil biostratigraphy and paleoecology of the Cretaceous–Tertiary transition in the central eastern desert of Egypt: Marine Micropaleontology, v. 47, no. 3, p. 323–356.

Thordarson, T., and Self, S., 2003, Atmospheric and environmental effects of the 1783–1784 Laki eruption: A review and reassessment: Journal of Geophysical Research, v. 108, no. D1, 4011, doi:10.1029/2001JD002042.

Tobin, T.S., Ward, P.D., Steig, E.J., Olivero, E.B., Hilburn, I.A., Mitchell, R.N., Diamond, M.R., Raub, T.D., and Kirschvink, J.L., 2012, Extinction patterns, $\delta^{18}O$ trends, and magnetostratigraphy from a southern high-latitude Cretaceous-Paleogene section: Links with Deccan volcanism: Palaeogeography, Palaeoclimatology, Palaeoecology, v. 350–352, p. 180–188, doi:10.1016/j.palaeo.2012.06.029.

Uthicke, S., Momigliano, P., and Fabricius, K.E., 2013, High risk of extinction of benthic foraminifera in this century due to ocean acidification: Scientific Reports, v. 3, doi:10.1038/srep01769.

Wilf, P., Johnson, K.R., and Huber, B.T., 2003, Correlated terrestrial and marine evidence for global climate changes before mass extinction at the Cretaceous-Paleogene boundary: Proceedings of the National Academy of Sciences of the United States of America, v. 100, no. 2, p. 599–604, doi:10.1073/pnas.0234701100.

Yu, J., and Elderfield, H., 2007, Benthic foraminiferal B/Ca ratios reflect deep water carbonate saturation state: Earth and Planetary Science Letters, v. 258, p. 73–86, doi:10.1016/j.epsl.2007.03.025.

Zeebe, R.E., 2012, History of seawater carbonate chemistry, atmospheric CO_2, and ocean acidification: Annual Review of Earth and Planetary Sciences, v. 40, p. 141–165, doi:10.1146/annurev-earth-042711-105521.

Zeebe, R.E., Zachos, J.C., Caldeira, K., and Tyrrell, T., 2008, Carbon emissions and acidification: Science, v. 321, p. 51–52, doi:10.1126/science.1159124.

Manuscript Accepted by the Society 31 January 2014

Printed in the USA

The Geological Society of America
Special Paper 505
2014

Plants and floral change at the Cretaceous-Paleogene boundary: Three decades on

Robert A. Spicer*
Centre for Earth, Planetary, Space and Astronomical Research, The Open University, Walton Hall, Milton Keynes MK7 6AA, UK

Margaret E. Collinson
Department of Earth Sciences, Royal Holloway University of London, Egham, Surrey TW20 0EX, UK

ABSTRACT

We review the extensive record of plant fossils before, at, and after the Cretaceous-Paleogene event horizons, recognizing that key differences between plants and other organisms have important implications for understanding the patterns of environmental change associated with the Cretaceous-Paleogene event. Examples are given of the breadth of prior environmental conditions and ecosystem states to place Cretaceous-Paleogene events in context. Floral change data across the Cretaceous-Paleogene are reviewed with new data from North America and New Zealand. Latest Cretaceous global terrestrial ecology was fire prone and likely to have been adapted to fire. Environmental stress was exacerbated by frequent climate variations, and near-polar vegetation tolerated cold dark winters. Numerous floristic studies across Cretaceous-Paleogene event horizons in North America attest to continent-wide ecological trauma, but elsewhere greater floral turnover is sometimes seen well before the Cretaceous-Paleogene boundary rather than at it. Data from the Teapot Dome site (Wyoming) indicate continued photosynthesis, but during or immediately after the Cretaceous-Paleogene event, growth was restricted sufficiently to curtail normal plant reproductive cycles. After the Cretaceous-Paleogene transition in New Zealand, leaf form appears to have been filtered for leaves adapted to extreme cold, but at other high-southern-latitude sites, as in the Arctic, little change in floral composition is observed. Although lacking high-resolution (millimeter level) stratigraphy and Cretaceous-Paleogene event horizons, gradual floral turnover in India, and survival there of normally environmentally sensitive taxa, suggests that Deccan volcanism was unlikely to have caused the short-term trauma so characteristic elsewhere but may have played a role in driving global environmental change and ecosystem sensitivity prior to and after the Cretaceous-Paleogene boundary.

*r.a.spicer@open.ac.uk

Spicer, R.A., and Collinson, M.E., 2014, Plants and floral change at the Cretaceous-Paleogene boundary: Three decades on, *in* Keller, G., and Kerr, A.C., eds., Volcanism, Impacts, and Mass Extinctions: Causes and Effects: Geological Society of America Special Paper 505, p. 117–132, doi:10.1130/2014.2505(05). For permission to copy, contact editing@geosociety.org.

INTRODUCTION

Understanding global events at the Cretaceous-Paleogene boundary must include terrestrial organisms—in particular those that underpin almost all nonmarine food chains, provide four-dimensional (structural changes through time) ecological space for the majority of other land-based organisms, and profoundly affect the chemical composition of the atmosphere, oceans, and geosphere. We refer, of course, to photosynthetic plants. In this paper, key differences between plants and other organisms are outlined that have important implications for understanding plant response to the Cretaceous-Paleogene event. Examples of the breadth of environmental conditions in which plants were living prior to the Cretaceous-Paleogene boundary are given to place Cretaceous-Paleogene boundary changes into the context of the prior state of ecosystems. Existing data on floral change across the Cretaceous-Paleogene boundary are reviewed, along with new data from North America and New Zealand.

Why Plants Are Different

Plants as Extinction Indicators

Using land plants to study extinction studies is not straightforward. The reasons for this are several. First, land plants are almost never preserved as whole entities. Leaves, pollen, spores, flowers, and seeds are all produced and shed in vast numbers throughout the lifetime of an individual (especially in deciduous trees), but each has different transport and preservation potential, and, in the case of leaves, often display high levels of phenotypic plasticity. These factors can differentially bias the apparent extinction record provided by each organ type (Mander et al., 2010). Second, different plant organs are exposed to different selection pressures and so evolve at different rates. For example, coevolution with pollinators may influence morphological change in floral morphology, while changes in leaf morphology are directly related to environmental factors that affect photosynthetic return in relation to structural costs, water management, and respiratory drain. Thus, the rates and directions in which floral and foliar organs evolve can be largely decoupled from one another. Biological drivers affecting the rate and direction of flower evolution cannot of course be divorced from changes in the physical environment, but the relationship is a loose one. Thus, within a genetic lineage, evolutionary rates in different plant organs are often independent of each other, a phenomenon known as mosaic evolution through the process of heterobathmy (Hill, 1986; Hill and Crane, 1982; Takhtajan, 1959). Combine mosaic evolution with differential shedding, transport, and preservation of organs, and the use of land plant fossils to assess extinction rates becomes complex to say the least. Pollen and spores in particular are comparable to marine planktic organisms in being abundant, having high preservation potential (in this case, sporopollenin walls), and having small size, enabling study of multiple assemblages at high temporal resolution. However, range truncations in individual organ morphologies (such as pollen, spore, or leaf morphotypes) cannot readily be equated to the termination of genetic lineages represented by whole organisms.

Despite these arguments, plant fossils do provide unique insights into biotic change that cannot be obtained from animal remains, even those with a strong fossil record. For example, the preservation of complex floral morphology can be directly linked to reproductive strategies, even if specific vectors are not preserved, while leaves provide an intimate record of atmospheric conditions, in particular climate.

An additional issue affects paleobotanical studies across the Cretaceous-Paleogene transition and that is the tendency for recent work on Late Cretaceous and earliest Paleocene floras to have focused on delimitation of morphotypes for use in biostratigraphy and paleoclimate analysis. Morphological intergradation is pronounced in many Cretaceous leaf forms, there has been little detailed morphometric work on large assemblages, and there are relatively few well-supported identifications of leaf morphotypes to named lineages or clades. The reasons for the morphotype approach are several and include the legacy of assigning names to leaves based on superficial resemblance (picture matching sensu Wolfe, 1973) without a well-developed evolutionary and developmental context, as well as high levels of hybridization and thus reticulate evolution early in angiosperm history (Spicer, 1986). Such reticulate evolution also appears in pollen (Hofmann et al., 2011). It is beyond the scope of this review to revise the existing systematics, so we have to proceed using the published nomenclature, taxonomic assignments, and opinions of the various authors cited.

Land Plant Reproduction and Survival in Adversity

Plants have a broad spectrum of reproductive strategies, including a diversity of asexual strategies not seen in animals. These range from simple sexual reproduction, involving both biotic and physical vectors and gamete fusion between separate individuals, through self-fertilization to vegetative growth from specialized propagules (e.g., adventitious buds) or fragmented parts of the plant body such as stolons, rhizomes, and roots. Self-fertilization and vegetative reproduction effectively render the plant independent of biotic vectors and thus decouple the lineage from the demise of coevolutionary partners, at least temporarily.

Recolonization of modern landscapes devastated by volcanic eruptions has often been used as a model for understanding past ecosystem traumas, particularly to explain the so-called "fern-spore spike" (= fern-spore abundance anomaly) associated with many Cretaceous-Paleogene event horizons (see reviews in Nichols and Johnson, 2008; Spicer, 1989). A good example is that of El Chichón, Chiapas, Mexico, (17°20′N, 93°12′W), which erupted on 28 March–4 April 1982. Here, collapse of a classic Plinian eruptive column produced base surges that locally obliterated the surrounding paratropical rain forest. Ash deposits covered the landscape, burying most of the soil and downed biomass to depths of several meters. Subsequent high rainfall led to the development of erosional gullies that, after several months, exposed buried, but still-living, fern (*Pityrogramma calomelanos* [L.] Link) rhizomes. These quickly produced fertile fronds that released vast numbers of spores, with the consequence that 2 yr after the eruption, even the most heavily impacted proximal landscapes on the slopes of the volcano were covered in large numbers of sporelings. Each spore has the ability to give rise to a sporeling by selfing at the gametophyte stage. In this tropical (17°N) high-rainfall zone, this resulted in typical sporeling densities of 40 per m^2 (Spicer et al., 1985). In this example, initial recolonization success was possible because the rhizomes could survive trauma and burial for several months. Rhizomes, along with underground stolons and roots, make a soil reservoir of plant genetic material, the longevity of which is hard to quantify because it is highly dependent on local conditions. What can be demonstrated is that in the case of *Pityrogramma*, rhizome viability persisted for several months, until they were exposed by erosion. Fern spores have limited viability in the soil spore bank, with proven longevities typically measured in months or at most years (Ballesteros et al., 2011; Dyer and Lindsay, 1992; Simabukuro et al., 1998). In the case of seeds, however, viabilities can be much longer, often exceeding a decade or more, more rarely centuries, and in extreme cases, millennia if conditions are appropriate (e.g., the Beal experiment reviewed *in* Bewley and Black, 1994; germination of *Nelumbo* dated at 1288 ± 271 yr reported by Shen-Miller et al., 1995, 2002; 2000-yr-old date palms *in* Sallon et al., 2008; for other examples, see Monaco et al., 2002). Soil seed banks, and if spores are included, "soil propagule banks," are well studied and provide an important mechanism by which plants can survive far longer periods (potentially centuries) of environmental adversity than animals.

Implications for Studies of the Cretaceous-Paleogene Boundary

It follows from the previous discussion that while the Cretaceous-Paleogene plant fossil record may be complex to interpret in terms of lineage (clade) extinction (e.g., a pollen morphotype range truncation that is useful for local to regional biostratigraphy may not equate to lineage extinction), it is a powerful tool for recognizing ecological disruption. Phenomena such as "fern-spore spikes" in the fossil record are indicative of short-term disturbance, while the scale and persistence of floral turnover (by implication reflecting recovery and recolonization) can indicate the severity and duration of an event. For these reasons, the plant fossil record across a boundary such as the Cretaceous-Paleogene provides information critical to understanding the nature of environmental change. However, knowledge of the state (e.g., stable or fluctuating or vulnerable) of the ecosystems immediately before the event is necessary to assess the effects of the event in the context of background conditions.

CRETACEOUS-PALEOGENE PREBOUNDARY ENVIRONMENTAL PERTURBATIONS AFFECTING PLANTS

Fire

Terrestrial ecosystems throughout the Cretaceous were heavily impacted by, and in some cases specifically adapted to, wildfire (Brown et al., 2012). This is evidenced by the globally widespread occurrence of charcoal, not only in sediments deposited under seasonally dry climates at low latitudes, but also in the humid polar regions. Charcoal occurs commonly even in the near-polar swamp forests represented today by northern Alaskan coals, but it shows an increase in frequency in the Campanian and Maastrichtian, interpreted as a consequence of regional drying (Spicer and Parrish, 1987). Most, if not all, terrestrial plant ecosystems were impacted by fire before the boundary events, and, using multiple lines of evidence, Belcher et al. (2009) demonstrated that wildfires across the Cretaceous-Paleogene event were not above background levels.

Temperature, Climate, and pCO_2 Shifts

As well as disturbance by fire, terrestrial ecosystems during the last 10 m.y. of the Cretaceous experienced considerable temperature variations. As ever, these variations operated over a range of time scales, with many short-term (<1 k.y.) fluctuations being lost in the often-limited temporal resolution offered by the geological record. Strong temperature fluctuations have been recorded in surface and intermediate-depth waters in the equatorial Pacific and middle and high latitudes in the South Atlantic, with a major cooling of sea-surface temperatures (SSTs) of 4–5 °C occurring between 73 and 70 Ma at middle and high latitudes. This was followed by a short period of warming before cooling resumed between 68.5 and 65.5 Ma (Li and Keller, 1999). Rapid warming (an SST rise of 3–4 °C) occurred ~450 k.y. before the Cretaceous-Paleogene boundary, although equatorial temperatures changed little, followed by another cooling episode (SSTs declined by 2–3 °C) in the last 200 k.y. of the Cretaceous (Li and Keller, 1999).

Similar fluctuations have been recorded from terrestrial deposits. Temperatures, as estimated from North American floras, warmed into the Santonian, cooled into the Campanian and Maastrichtian, and rose close to Santonian warm levels at the very end of the Maastrichtian (Wolfe and Upchurch, 1987a). This warming was examined in more detail by Wilf et al. (2003), who, using both foraminifera from a variety of Ocean Drilling Program (ODP) sites and leaf margin analysis of floras from the Williston Basin, North America, estimated that temperatures rose over an interval of some 200 k.y. before the Cretaceous-Paleogene, followed by cooling and a loss of plant diversity. A similar warming immediately prior to the Cretaceous-Paleogene event horizons was noted in North Dakota by Johnson et al. (1989) using megafloral remains (n = 11,503) collected from 87 localities. These temperature fluctuations are similar to those of Li and Keller (1999) and mirror changes in pCO_2 as determined from soil carbonates (Nordt et al., 2002). Specifically, there appears to have been a significant increase in, reportedly a doubling of, pCO_2 0.5 m.y. before the boundary, followed by a reduction to "normal" levels just before the boundary. Subsequently, a doubling of pCO_2 occurred, but not until 1.5 m.y. after the boundary (Nordt et al., 2002).

Spatial and temporal shifts among warm/wet, cool/dry, and cool/wet climates are documented in the sedimentological record through the late Maastrichtian of the Horseshoe Canyon Formation, Canada (Eberth and Braman, 2012). While it is not possible to correlate precisely the various temperature fluctuations, it is certain that the Maastrichtian was a time of dynamic climate change.

Winter Darkness and Cold

Despite the Maastrichtian cool interval between 73 and 70 Ma, seasonally deciduous forests extended to 82°N paleolatitude (Spicer and Herman, 2010). Exposed to continuous winter darkness of up to 3 months, the Arctic forests, present throughout the Late Cretaceous and into the Paleogene, survived extended periods of temperatures close to, and slightly below, freezing by entering dormancy. This is evidenced by the deciduous nature of the leaf litter as well as tree-ring characteristics (Parrish and Spicer, 1988; Spicer and Parrish, 1990a, 1990b). At ca. 70 Ma, the mean annual air temperature at 82°N is estimated to have been 6.3 ± 2.2 °C, using multivariate analysis of leaf form (Spicer and Herman, 2010). Trees were predominantly deciduous taxodioid conifers (Cupressaceae), with ferns and *Equisetites* making up most of the ground cover. Angiosperm pollen is present but shows high rates of species turnover, which may indicate they were understory disturbed-site herbaceous annuals and shrubs (Frederiksen et al., 1988). Leaves or wood of angiosperm trees are extremely rare at latitudes 80°N and above in the late Maastrichtian, and their absence persists well into the early Paleocene.

Seasonality

The predominantly deciduous nature of the Arctic vegetation (termed the polar broadleaved deciduous forest by Wolfe, 1985) was largely confined to latitudes above 70°N. Southward, there is an increase in the frequency of leaves and leafy shoots exhibiting evergreen characteristics (Godefroit et al., 2009; Golovneva, 1994, 2000; Tomsich et al., 2010; Wolfe and Upchurch, 1987a) associated with higher mean annual temperatures.

At paleolatitudes of 65°N in central Alberta, Maastrichtian conifers exhibit well-defined growth rings with a high ratio of early to late wood indicative of strong seasonality, but at 45°N, rings are often nonexistent, indicating weak, if any, seasonality. Between 35°N and 45°N, the vegetation was predominantly evergreen megathermal and subhumid, with angiosperms accounting for 79%–95% of the overall flora, conifers between 0% and 10%, and pteridophytes 0%–14% (Upchurch and Wolfe, 1993).

CRETACEOUS-PALEOGENE TRANSITION: EXAMPLES OF FLORAL CHANGE

Change in Floral Provinces

Herngreen et al. (1996) and Braman and Sweet (2012) mapped the Late Cretaceous *Aquilapollenites* and Normapolles palynofloral provinces. The *Aquilapollenites* province became established in the Campanian (Nichols and Jacobson, 1982), although the first occurrence of the triprojectate group of grains to which they belong is in the middle Coniacian (Braman and Sweet, 2012). The province is primarily located in western North America, Canada, Alaska, Greenland, the northern North Sea Basin, far eastern Russia, northern Japan, and northeast China, but *Aquilapollenites* grains and other triprojectate pollen are also found elsewhere, such as India (Cripps et al., 2005; Samant et al., 2013). The Normapolles province extended across southeastern North America, Europe and European Russia, and North Africa.

Across the Cretaceous-Paleogene boundary, the *Aquilapollenites* province all but disappeared in western North America, and overall 30%–45% of the Maastrichtian palynoflora ceased to exist in that area, while other taxa underwent a significant decline in abundance (Nichols, 1990; Nichols and Fleming, 1990; Nichols et al., 1990; Hotton, 2002; Nichols and Johnson, 2008). Most of the change was seen in the angiosperms. This loss of the *Aquilapollenites* province occurred suddenly, within a few centimeters of the event horizons. In Canada, Braman and Sweet (2012, p. 26) reported "a large number of extinctions" at the Maastrichtian-Danian boundary, including seven species of *Aquilapollenites*. A general characteristic of the pollen record is that large, thick-walled grains such as those of the triprojectates, typified by *Aquilapollenites*, are most affected across the Cretaceous-Paleogene interval (Sweet and Braman, 2001). These grains have all the characteristics of being zoophilous, especially entomophilous, and perhaps the demise of the pollinators could explain the disproportionate loss of these taxa (Bercovici et al., 2012). The Normapolles province appears to show a different pattern of change going into the Paleocene. The available records suggest no sudden disappearance but instead a gradual decline thought to be associated with evolutionary change into modern taxa belonging to the order Juglandales (Traverse, 2007), but without continuous successions across the Cretaceous-Paleogene interval in the Normapolles province, this pattern cannot be confirmed.

North America

Nichols and Johnson (2008) documented 105 terrestrial Cretaceous-Paleogene boundary sites worldwide, 85 of which are in North America. This is not surprising, because it reflects the occurrence of event horizons that provide a definitive stratigraphic marker. The widespread occurrence of impactite layers (event horizons) in North America has provided a common datum for correlation and has driven (1) a search for further sites and (2) a focus on the boundary itself, rather than a broader view of conditions leading up to the event horizons. We cannot do justice here to the large body of work that exists documenting plant fossil occurrences across the Cretaceous-Paleogene in North America; suffice it to say that the pattern of vegetation and environmental change across the Cretaceous-Paleogene transition is most marked in New Mexico, extends up into the Western Interior, and appears independent of differences in sedimentary facies (Nichols and Johnson, 2008). Section thicknesses and taxonomic details differ from site to site (Nichols and Johnson, 2008), and there is some suggestion of latitudinal control on variations in survival floras (upper event horizons) and overlying recovery floras (earliest Paleocene), shown by Sweet and Braman (1992, their fig. 12), and supported by subsequent high-resolution work by Sweet et al. (1999). Here, we refer to four examples that record important aspects of the changing floral composition across the Cretaceous-Paleogene event horizons in North America.

Raton Basin, New Mexico, USA

Despite being derived in part from the literature and lacking quantitative sampling on a regional scale (Nichols and Johnson, 2008), the pattern of floral changes across the Cretaceous-Paleogene interval in the Raton Basin, New Mexico, recorded by Wolfe and Upchurch (1987b) provides a comparator for terrestrial Cretaceous-Paleogene sections elsewhere.

Phase 1. In the late Maastrichtian, leaf characters, including cuticle morphology, suggest subhumid megathermal (mean annual temperature >20 °C) vegetation typified by palms, Laurales, Illicales, Euphorbiaceae, and Menispermaceae, together with some conifers.

Phase 2. Occurring in the Upper Cretaceous–Paleogene event horizon and immediately above, the "fern-spore spike" assemblage is dominated by fern spores (Tschudy et al., 1984) and a few sterile fern fronds similar to those of the extant *Stenocleana.* Dispersed cuticles are typically thin and indicative of herbaceous taxa, including fern rachises.

Phase 3. Overlying phase 2 for just a few meters, this low-diversity (8 megafossil and 10 cuticle taxa) assemblage contains early successional but large-leaved taxa with attenuate apices and is indicative of angiosperm recolonization in a wet climate. Palms are again present (although different taxa to phase 1), together with lobed leaves such as "*Cissites*" that may represent the Euphorbiaceae. Cuticles of herbaceous taxa decline, suggesting the re-establishment of woody vegetation.

Phase 4. More diverse than phase 3, phase 4 extends up to 200 m above the Cretaceous-Paleogene event horizon. Leaves here are larger, and other indicators suggest high precipitation. Palms and megathermal evergreen dicots predominate.

Phase 5. Although the taxa are mostly different to phase 1, the assemblage is indicative of a diverse (40–50 megafossil species) megathermal evergreen and deciduous rain forest. Palms and deciduous conifers are present.

Wolfe and Upchurch (1987b) reported that floral turnover across the Cretaceous-Paleogene transition is greater amongst leaf taxa than palynomorphs, especially affecting thermophilic taxa (a similar phenomenon to that seen elsewhere; e.g., Johnson et al., 1989; Wilf and Johnson, 2004), with only 12 of the 75 phase 1 evergreen dicots continuing into the Paleocene. This suggests that a selection for deciduousness (i.e., the capacity to enter dormancy for extended periods) was advantageous across the Cretaceous-Paleogene transition in this region. Using an early form of Climate Leaf Analysis Multivariate Program (CLAMP) and leaf fossil assemblages from the Mississippi Embayment across the Western Interior and into Alberta, Wolfe (1990) inferred paleolatitude-corrected temperature and precipitation changes across the Cretaceous-Paleogene interval. Phase 2 assemblages, devoid of angiosperms, cannot be used for CLAMP, and diversity in phase 3 is too low ($n < 10$) for reliable results; nevertheless, Wolfe (1990) argued for a sharp rise in temperature and precipitation. This hot and humid climate lasted for up to 1 m.y. after the event horizons, and although becoming cooler and drier in phases 4 and 5, temperatures and rainfall remained elevated with respect to levels of the late Maastrichtian. The reliability of Wolfes' quantitative estimates remains in question, however, because of low species diversity during phase 3 and the then-primitive CLAMP calibration and methodology.

Sugarite, New Mexico, USA

At Sugarite, the "fern-spore spike" is dominated (72% of the palynoflora) by *Cyathidites diaphana*, and in Maastrichtian coals, 20% of the palynomorphs are fern spores. Across the boundary, 20% of the 48 Maastrichtian palynomorph taxa disappear. File cells, regarded as indicative of herbaceous taxa, show a steady increase throughout the uppermost meter of the Maastrichtian section, followed by marked fluctuations immediately above the event horizons. The diversities of both monocots and dicots show large fluctuations, both before and after the boundary (Beerling et al., 2001).

Following on from pioneering work on carbon isotope fluctuations across the Cretaceous-Paleogene interval (Arens and Jahren, 2000; Schimmelmann and DeNiro, 1984), the carbon isotope signature in the carbonaceous and coal-rich Sugarite section indicates that "perturbations in the biogeochemical cycling of carbon by marine ecosystems continued well after the initial recovery of terrestrial ecosystem structure" (Beerling et al., 2001, p. 739). Thus, terrestrial recovery was rapid, even at this site proximal to the Chicxulub impact.

Teapot Dome, Wyoming, USA

At the Teapot Dome locality (U.S. Geological Survey [USGS] paleobotanical locality D 7402), Wolfe (1991) studied a succession of five beds across the Cretaceous-Paleogene interval underlain by a Cretaceous mud rock (Table 1), from what he interpreted to be a preserved "lily pond." The event horizons,

TABLE 1. SUMMARY AND INTERPRETATION OF THE TEAPOT DOME SECTION LITHOLOGIES AND ASSOCIATED FLORA

Bed no.[1]	Boundary lithology terminology[2]	Boundary event terminology[2–4]	Terminology in this paper	Duration	Age	Lithology	Thickness (mm)	Megafossils & mesofossils	Palynomorphs ≥ 3%	Dominant POM	Raw diversity palynomorph categories	POM/ palynomorph particle ratio
5			Tertiary mud rock		Tertiary	Gray claystone	30–50	*Kurtziflora* flowers, *Nelumbites* leaves, seed type 2, rare cuticle, rare charcoal	*Cyathidites* 75, *bisaccates* 3, *Schizophacus* 3, *Ulmipollenites* 3	Woody	16 (5 pollen)	20:1
4	Laminated shale		Laminated layer	Years[2]	Cretaceous-Paleogene boundary	Blackish-red claystone, clear bedding	4–5.4	Abundant *Azolla* soral clusters, rare *Costatheca*, cuticles, charcoal	*Cyathidites* 42, **tetrads** 22, ***Kurtzipites*** 13, ***Pandaniidites*** 9, *Azolla* 4, *Ghoshispora* 3, ***Ulmipollenites*** 3	Orange unstructured, pale cuticles	22 (15 pollen)	6:1
3b	Laminated shale		Laminated layer	Years[2]	Cretaceous-Paleogene boundary	Reddish-black claystone, clear bedding.	Few mm	*Azolla* at top, abundant cuticle, seed type 1, *Costatheca*, *Spermatites*, charcoal, fern sporangia				
3a	Satiny claystone	Fireball or impact or magic layer	Upper claystone	Hours to days[5] or days to months[2–4,6]		Reddish-black claystone. Contains clasts of bed below. "Coaly" in places.	6–10		*Cyathidites* 47, *Ghoshispora* 19, *Kurtzipites* 10, *Ulmipollenites* 7, *Pandaniidites* 3	Pale unstructured, pale cuticles	20 (12 pollen)	13:1
1, 2	Hackly claystone	Ejecta layer, boundary claystone	Lower claystone	Minutes to hours[2,7]	Cretaceous-Paleogene boundary	Grayish-pink claystone with spherules at top	7–10	Charcoal, cuticles, rare *Costatheca*	*Cyathidites* 65, *Ghoshispora* 8, *Kurtzipites* 4, *Ulmipollenites* 3, *Pandaniidites* 3	Woody, charcoal, dark unstructured	17 (7 pollen)	34:1
			Cretaceous mud rock		Cretaceous	Gray mudstone fining upward	100	Abundant charcoal	*Cyathidites* 85, punctate spore 6	Woody, charcoal, dark unstructured	13 (2 pollen)	34:1
			Cretaceous mud rock		Cretaceous		100	Charcoal	*Concentricites* 36, *Cyathidites* 24, algal cysts 8, monocolp. pollen 7, fungal spores 9	Woody, charcoal	13 (1 pollen)	34:1

Note: POM—palynological organic matter. References: 1—Wolfe (1991), 2—Sweet et al. (1999), 3—Izett (1990), 4—Hildebrand (1993), 5—Pope (2002), 6—Toon et al. (1997), 7—Sweet (2001). Palynomorph names in bold are those occurring in clumps as well as dispersed.

In the field, bed 3b could not be separated from the upper claystone layer (bed 3a). By contrast, a clear separation occurred within the basal few millimeters of the laminated layer (at junction of bed 3b with 4). The mesofossils, therefore, are from a bulk sample of bed 3a plus 3b. For palynology only the nonlaminated claystone (bed 3a) was used also avoiding clasts of underlying lower claystone and "coaly" areas.

with an iridium anomaly and shocked quartz grains, are present in the section, which locally has well-preserved mega- and mesofossils as well as a rich palynoflora. Wolfe's interpretation of not one but two impacts, one distant and one proximal, that occurred perhaps only weeks apart in June, accompanied by a rapid drop in the mean daily temperature of ~30 °C and freezing of the pond, proved controversial. Nichols (*in* Nichols et al., 1992) suggested that Wolfe had misidentified several of his palynomorphs to the extent that his interpretations were invalid. In our opinion, the pollen that Wolfe (1991) referred to as *Nuphar*-like is *Pandaniidites*, and the tetrad pollen that Wolfe referred to as *Nelumbo*-like is *Inaperturotetradites*. *Nelumbo*-like leaves are unlikely to represent *Nelumbo* (lotus) and may be better referred to the genus *Nelumbago*. Here, we present a re-evaluation of the site based on a detailed examination of an excavated section made in 1993.

Cretaceous mud rocks (base of the section; Table 1). The latest Cretaceous palynomorphs are dominated by spores with very few pollen types. There is a shift in palynomorph content from mixed algal, fungal, and fern spore to purely fern spore dominated just before the Cretaceous-Paleogene event horizons. This can be interpreted as background Cretaceous "pteridophyte-dominated" vegetation and varying within-pond ecology (e.g., algal growth reduced in the very latest Cretaceous), possibly associated with shallowing of the pond.

Cretaceous-Paleogene claystones (event horizons; Fig. 1A). In the field, the lowermost few millimeters of the overlying laminated layer could not be separated from the upper claystone layer. By contrast, a clear separation occurred within the basal few millimeters of the laminated layer, and this was used as the junction for the bulk samples (Table 1). For palynology, only nonlaminated claystone was used for the upper claystone sample, also avoiding clasts of underlying lower claystone and "coaly" areas.

In the lower claystone, there is no change in the dominant palynological organic matter (POM) or proportions of palynomorphs to POM compared with the underlying Cretaceous. However, the palynomorph composition changes drastically (Table 1), with an increase in diversity and abundance of angiosperm pollen. The lower Cretaceous-Paleogene claystone is judged to have been deposited in minutes to hours, resulting from ballistically distributed melted target rock.

The lower and upper claystones have almost identical palynomorph signatures, varying only in percentage values and overall pollen diversity. However, the upper claystone has very different POM (pale cuticles, nonwoody). The upper Cretaceous-Paleogene claystone is considered to have been deposited in hours to days or days to months, resulting from hydrodynamic collapse of an impact fireball where condensates and shocked minerals settled through the atmosphere.

The rapid depositional histories inferred for both Cretaceous-Paleogene claystones, especially the lower claystone, rule out a possible new pollen or spore source from newly dispersing plants that had colonized a postimpact disturbed landscape. Therefore, the change in palynomorph composition must reflect either (1) different depositional conditions (e.g., settling through atmosphere vs. runoff) or (2) changes in local vegetation or local ecology that had happened prior to impact. *Ghoshispora* is a large spore, with a spore body 150–200 µm in diameter (probably a small megaspore from a heterosporous plant; Batten et al., 2011), and it is most unlikely to have been dispersed in the atmosphere. The same applies to the seed-sized *Costatheca* and *Spermatites* mesofossils. Therefore, a change in local vegetation prior to the impact seems more likely.

Laminated layer. *Azolla* is not present in the Cretaceous-Paleogene claystones, but all of the laminated layer is characterized by the presence of abundant *Azolla* remains on lamina surfaces. The *Azolla* remains include megaspore apparatuses and soral clusters (multiple sporangia still grouped within sori). This combination of remains and their abundance indicate the growth of *Azolla* on the surface of the pond in which the remains were deposited. Normally in *Azolla*, microspore massulae, containing

Figure 1. (A) Polished block of the lower and upper claystone from the Teapot Dome locality, Wyoming. (B) Soral clusters of *Azolla* isolated from the laminated layers at Teapot Dome indicating a repeated failure to complete the sexual life cycle of this invasive water fern during the latest part of the Cretaceous-Paleogene event.

~8–20 microspores, are dispersed individually, and when surface hairs (glochida) are present, they radiate outward. Specialized glochidia tips become hooked and entwined in hairs on the megaspore apparatus (e.g., Collinson et al., 2013; their fig. 2), thus bringing sperm produced by microspores into close contact with the eggs inside the megaspore. For details of the numerous "normal" records of fossil *Azolla*, see Collinson et al. (2013) and van der Burgh et al. (2013, and references cited therein). For a normal example from another stratigraphic position in a Cretaceous-Paleogene section, see Bercovici et al. (2009, their fig. 14).

By striking contrast, the numerous large *Azolla* soral clusters in the laminated layer at Teapot Dome (Fig. 1B) indicate that microspore massulae were not dispersed, although the initial stages in separation of microsporangial contents had just begun, as evidenced by scanning electron microscope (SEM) imagery in van Bergen et al. (1993). Futhermore, even in the smaller groups (probably broken fragments) of microspore massulae (<125 μm) in palynological preparations, the glochida do not radiate from the surface. Therefore, the normal sexual life cycle of *Azolla* was not completed during deposition of this layer. *Azolla* is well known as an invasive species newly colonizing large areas within 1 yr (e.g., Aber et al., 2010). As spores survive at least 3 yr (Janes, 1998b) but were not present in the underlying horizons, new colonization at Teapot Dome could have been by biotically dispersed spores or small plant fragments. Plants can survive and increase biomass in air temperatures down to at least –5 °C in low light levels (Janes, 1998a); hence, they could persist, and potentially even spread through vegetative reproduction, in the absence of successful sexual reproduction.

The palynomorph signature of the laminated layers is similar to that of the Cretaceous-Paleogene claystones but with three striking differences. The first is the presence of *Azolla*, already discussed here. The second is the presence, in abundance, of a tetrad pollen type that was lacking in the Cretaceous-Paleogene claystones. SEM and light microscope (LM) observations show that this pollen can be placed in the species *Inaperturotetradites scabratus* Tschudy (as was suggested by Nichols and Johnson, 2008, p. 140, based on specimens from Wolfe, 1991). The unidentified tetrad in Bercovici et al. (2009) is very similar. The third is the occurrence of pollen clumps of four pollen types representing different flowering plants, namely, *Kurtzipites*, the tetrad pollen, *Pandaniidites*, and *Ulmipollenites*. Such pollen clumps were not recorded in underlying or overlying strata. The most likely explanation for these monotypic pollen clumps is failure of dispersal

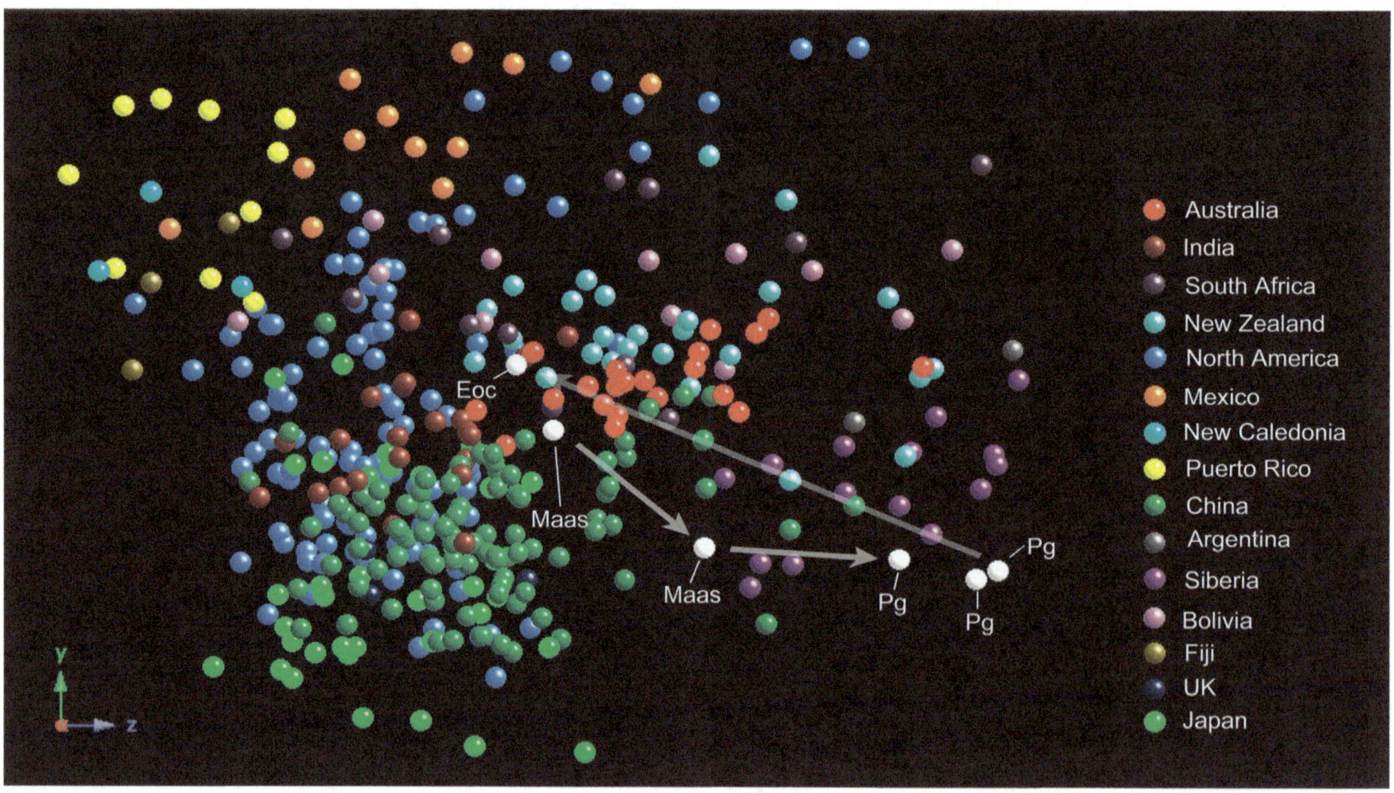

Figure 2. Axis 2/3 plot of Climate Leaf Analysis Multivariate Program (CLAMP) foliar physiognomic space for 378 modern vegetation sites from different geographic regions represented by colored spheres. Cooling during the Maastrichtian (Maas) drives the foliar physiognomic signature of New Zealand vegetation toward the lower right of the plot. The earliest Paleocene sites have similar leaf characteristics to those of modern extreme cold-adapted Siberian vegetation. For comparison, an Eocene (Eoc) New Zealand sample shows a return to warmer conditions in the center of the plot. For details of the CLAMP methodology, see http://clamp.ibcas.ac.cn. Pg—Paleogene.

from the anther in the flower such that anther fragments were shed from the plant or fell from the plant on death. Although other pollen of these plants, which were dispersed (and are documented as dispersed grains), would potentially have enabled completion of the sexual life cycles, the occurrences of these clumps indicate some disruption to the normal life cycle of these plants, perhaps at the end of the flowering season.

In summary, the laminated Cretaceous-Paleogene layer contains several indicators of ecological disturbance: (1) incoming of a new abundant flowering plant (tetrad pollen producer), which subsequently disappears, (2) incoming of abundant *Azolla*, an aquatic surface-floating heterosporous fern well known as one of the world's most invasive plants, (3) failure of *Azolla* to complete its sexual life cycle, and (4) disruption to the life cycles of four other types flowering plants.

Tertiary mud rocks. The palynomorph signature of the earliest Tertiary differs from that of the Cretaceous-Paleogene layers. *Kurtzipites* and *Pandaniidites* are no longer important, and bisaccates occur. Flowering plant abundance and diversity have fallen, perhaps suggesting a return to more stable fern-dominated vegetation. The POM assemblage is again dominated by woody components, not cuticular or pale components. A striking difference is the presence of numerous *Nelumbo*-like leaves and *Kurtziflora* flowers. *Schizophacus* resting cysts in a gray claystone indicate continuing pond deposition; however, *Azolla* is lacking. This can be interpreted as a subsequent successional stage in the pond where greater stability resulted in loss of the invasive *Azolla. Kurtziflora* is the staminate ("male") flowers that produce *Kurtzipites* pollen. The shedding of numerous intact staminate flowers but very little dispersed pollen (<3% palynoflora) of the same taxon is evidence of continuing disruption to the life cycles, at least of this one flowering plant. The fact that plants were able to produce reproductive structures, even if the life cycle was not completed, indicates continued photosynthesis but accompanied by a reduction in the length of the growing season (or perhaps another factor inhibiting growth). There is no evidence (e.g., mud cracks) for drying out of the pond, which could have affected the *Azolla*, and no indication of deformity such as might be caused by disease. The mutational effect of ultraviolet (UV) radiation on the gene initiating separation has been invoked as an explanation for presence of unseparated lycopsid tetrads at the Permian-Triassic boundary (Visscher et al., 2004). This factor cannot be ruled out, but it is not the most parsimonius explanation, especially when some dispersed pollen are also present in the same samples as the clumps. In the absence of a mechanism for seasonal reduction of light levels at midlatitudes, we consider that this curtailment in growth was most likely caused by low temperatures.

Canada

Sweet et al. (1999) undertook a high-resolution study and noted sequential palynofloral changes across the event horizons, which, like those further south, include an iridium anomaly and shocked quartz. Their summary (Sweet et al., 1999, their fig. 16) shows that some 5–10 cm below the event horizons, angiosperm pollen diversity is reduced and replaced by a gymnosperm-dominated assemblage. Within the event horizons, there are stepped reductions in diversity, and in the upper event horizons, there is codominance of *Cyathidites* fern spores and ulmaceous flowering plant pollen. A sharp change occurs ~3 mm above the event horizons, where the dominant spore type switches to *Laevigatosporites*, and *Kurtzipites* flowering plant pollen dominates over ulmaceous pollen. Thus, there are clear signs of floral reorganization prior to the boundary, followed by floral transitions through and after the event horizons.

Alaska

In Alaska, no impactite layer or geochemical anomalies have so far been encountered, but palynological evidence from shot holes across the projected subsurface location of the boundary suggests a change from predominantly conifer-dominated vegetation to one rich in wind-pollinated angiosperms (Frederiksen, 1989). The lack of event horizons inevitably confounds high-resolution stratigraphy, and only general patterns of vegetational change are discernible. To better understand the effects of Cretaceous-Paleogene events in the North Pacific region, it is useful to examine the Alaskan floral record alongside that from eastern Russia.

Russia and China

Russian accounts of changes across the Cretaceous-Paleogene interval were criticized by Nichols and Johnson (2008, p. 194) with the following words, "In general, it appears that the Russian paleobotanists have been too quick to assign global stage names and associated temporal significance to local and poorly dated terrestrial sections in the Russian Far East. These sections have clear potential, but to date it has not been realized, and it is premature to discuss the significance of these floras as if they were known to chron- or impactite-level resolution." While it is true that sections containing impactite horizons have so far not been discovered in northeastern Russia, this overlooks detailed stratigraphic work that demonstrates clearly a lack of vegetational change across the Cretaceous-Paleogene transition. Of particular relevance to terrestrial events at the Cretaceous-Paleogene transition is the Koryak phytostratigraphic interval (Herman, 2011; Moiseeva, 2012), which ranges from the Upper Maastrichtian through to the Danian. Conifers, mostly Cenozoic genera, are important components, often representing the dominant subgroup in addition to angiosperms. This floral type extends across NE Russia, and its age is constrained by marine megafossil biostratigraphy in northwestern Kamchatka, Ugol'naya Bay and Amaam Lagoon, and Khatyrka River (see details in Moiseeva, 2012). In the Pakul'nei region, dinosaur remains indicate a late Maastrichtian age. Numerous components of the Koryak phytostratigraphic unit in NE Russia also occur in the Sagwon flora of Selandian (Paleocene) age in northern Alaska (Moiseeva, 2012; Moiseeva et al., 2009). Clearly, these taxa, and the composition of the Koryak floral type overall, survived the Cretaceous-Paleogene transition intact, even if the impactite signature is missing.

The banks of Heilongjian River (known as the Amur River in Russia) have long been known to contain Maastrichtian dinosaurs (Godefroit et al., 2011, and references therein), and the Cretaceous-Paleogene transition is projected, on stratigraphic grounds, to occur in the river bed. No impactite signature is known to be exposed along the river banks. Joint Russian-Chinese drilling on the Chinese side has revealed only a gradual floral change across the Cretaceous-Paleogene transition, with almost half as many taxa disappearing before the inferred boundary (7 species) as at it (15 species), and 11 species continuing into the Paleocene (Sun et al., 2011). By contrast, the most profound floral change in the region took place between the middle and late Maastrichtian, roughly coincident with the start of the Koryak phytostratigraphic unit. The middle Maastrichtian palynoflora is dominated by gymnosperms (Pinaceae, taxodioid members of the Cupressaceae and *Ginkgocycadophytus*), while the angiosperms are represented by forms close to extant Betulaceae, Juglandaceae, Fagaceae, Ulmaceae, and Myricaceae. A cooler late Maastrichtian palynoflora is dominated by pollen close to the taxodioid members of the Cupressaceae and Ulmaceae, and among spores, *Laevigatosporites* (a fern) predominates (Markevich et al., 2012).

Japan

Saito et al. (1986) reported on a shallow-marine section off Hokkaido, northern Japan. The Cretaceous-Paleogene interval, recognized based on foraminifera, lacks an iridium anomaly and shocked quartz but does contain a "fern-spore spike" and occurs just below a detrital coal associated with a 2‰ change in $\delta^{13}C$. Within 30 cm below the Cretaceous-Paleogene interval, there occurs a progressive reduction in the relative proportions of both angiosperms and gymnosperms compared to fern spores. Thus, ferns, relative to seed plants, show an increase prior to the boundary, similar to that seen on the Russian-Chinese border. Fern abundance peaks at the boundary and then declines as gymnosperms, rather than angiosperms, increase. *Pinus* pollen shows a particular increase, extending at least 40 cm above the boundary. Angiosperms do not disappear altogether, and, as in North America, species of *Aquilapollenites* continue into the Paleogene.

New Zealand

There are several Cretaceous-Paleogene transition sections in New Zealand spanning terrestrial and shallow- and deep-marine sections. The nonmarine sections are typified by the section at Moody Creek Mine (Vajda et al., 2001; Vajda and McLoughlin, 2004). Here, there are no event horizons, but elevated iridium levels occur in a thin coal within a carbonaceous shale. Beneath the iridium-rich layer, gymnosperms increase to ~70% before declining dramatically (in relation to ferns) at the iridium layer itself. This increase spans 14 cm of section and is progressive across three successive samples. Unlike the gymnosperms, flowering plants show a steady decline in the same interval, with some variation in relation to fluctuations in tree fern abundance. One section revealed that the lowest part of the iridium-rich layer contained only fungal palynomorphs (Vajda and McLoughlin, 2004). Above this, still within the iridium-rich layer, there is an immediate increase in ground ferns, as represented initially by *Laevigatisporites*, followed in some sections by *Gleicheniidites*, and then by *Baculatisporites*. These taxa are interpreted to be tolerant of open and acidic soil conditions. Tree ferns are the last to show an increase (*Cyathidites/Cibotiidites*). Despite these increases in fern abundances, some gymnosperms represent a significant component of the palynoflora well into the early Paleocene, while the angiosperms remain at low abundances. From 40 cm above the iridium-rich layer, ferns decline in relation to gymnosperms, with an increase in *Phyllocladidites*. The presence of a "fern-spore spike" is clear and in some respects is similar to that seen in North America and Japan.

Also similar to Japan is the pattern seen at the shallow-marine section at mid-Waipara River (Vajda and Raine, 2003; Vajda et al., 2001), which bears a weak iridium anomaly (Brooks et al., 1986) marking the position of the Cretaceous-Paleogene boundary. Again, there is a marked decline in gymnosperm and angiosperm pollen abundance over a 5 m thickness of section prior to the anomaly. As in Japan, there is a significant continuance of both gymnosperm and angiosperm pollen across the boundary, and the gymnosperms rise more than other taxa in terms of relative abundance after the peak in the fern spores that follows immediately after the anomaly. Some of the continuance may be due to temporal smearing resulting from the inevitable short-term reworking within a shallow-marine environment. This phenomenon is unlikely to account for the decline in gymnosperm and angiosperm taxa prior to the iridium anomaly but may reduce the magnitude of that decline.

Material that is immune to reworking are the leaf fossil assemblages. These are known in Maastrichtian and Lower Paleogene sections, but because they are not in association with boundary markers such as an iridium anomaly, they are less precisely positioned with respect to event horizons. Nevertheless, some interesting environmental signals emerge from these assemblages. First described in this context by Kennedy et al. (2002), lowest in the succession is Pakawau Bush Road, an assemblage yielding 58 different dicot leaf forms. Based on palynology and regional stratigraphic relations, the next highest in the succession, but still in the latest Maastrichtian, is the Cameron's Pit assemblage (Kennedy, 2003). In the lowest Paleocene, close to but at an unknown stratigraphic level above the boundary, there are three assemblages: Ian's Tip, Pillar Point Track, and Spring Creek Mine (Kennedy et al., 2002; Kennedy, 2003), each with no more than approximately half the diversity of leaf forms seen at Pakawau Bush Road. Figure 2 shows the results of a multivariate analysis of leaf form (CLAMP; Kovach and Spicer, 1995; Spicer, 2008; Wolfe, 1993; Yang et al., 2011). Here, the New Zealand fossil sites are shown in physiognomic space defined by leaf form in 387 modern vegetation sites where climate is known. Pakawau Bush Road plots in an area occupied by humid temperate sites from modern New Zealand, Japan, and South Africa. Cameron's Pit plots with cooler sites from western Siberia, while all three Paleocene sites are positioned outside of modern physiognomic space but close to sites in central Siberia where the cold month mean temperatures are below –40 °C. This suggests that New Zealand leaf form in the early Paleocene was filtered in favor of features that today confer an adaptive advantage in extreme cold.

Table 2 shows the predicted climates for these samples, but values for the Paleocene should be regarded with caution. Testing using modern vegetation sites suggests that where the cold month mean temperatures are low, such as in central Siberia, CLAMP yields temperatures that are too warm (Spicer et al., 2004), because in this region physiognomic space is curved strongly, and the vectors used to derive the predictions do not reflect this well. So the earliest winter climates represented by the Ian's Tip, Pillar Point Track, and Spring Creek Mine assemblages may have been well below –15 °C.

Selection for foliar physiognomies characteristic of low temperatures does not imply wholesale changes in species composition or extinction of these already cold-adapted, high-latitude floras. Just as in the Arctic floral turnover as recorded by palynological studies on Seymour Island (Antarctic Peninsula), marine successions show no fern abundance anomaly and no significant change in pollen composition across the Cretaceous-Paleogene interval, as marked by a weak iridium anomaly and changes in dinoflagellate cysts (Askin, 1988, 1990; Elliot et al., 1994). Geochemical analyses indicate consistent cool marine temperatures (11.4 ± 5 °C) from the latest Cretaceous to the Paleogene (O'Brien et al., 2013). This implies that the duration and magnitude of any changes in air temperatures were insufficient to significantly overcome marine thermal inertia.

India

Understanding floral changes in the Late Cretaceous and into the Paleogene in India is critical to contextualizing the relative contribution of the Deccan Traps eruptions and the Chicxulub impact to worldwide biotic changes across the Cretaceous-Paleogene transition. Prior to the onset of Deccan volcanism, palynological evidence shows that vegetation in India was composed of gymnosperms (*Araucariacites*, *Classopollis*, *Cycadopites*), angiosperms (*Graminidites*, *Palmaepollenites*, *Retimonocolpites*, etc.), and pteridophytes (*Isoetes*, etc.). After the eruptions began, the fringes of newly created ponds and lakes in the lava fields were populated by angiosperms and pteridophytes: typical species of disturbed sites. Coincidental with the acme of eruptive activity, there is a decline in abundance of pollen grains and a rise in the abundance of *Glomus*-like mycorrhizal fungi. Centric- and pennate-type diatoms suggest eutrophic lakes. During the waning phase of volcanic activity in the early Paleocene, there is a return of a diverse forested landscape, but one distinctly different from that of the Maastrichtian in that it is closer in composition to that of the posteruption floras of the later Paleocene and Eocene (Samant and Mohabey, 2012), and there is a loss of diversity in the triprojectate pollen, including *Aquilapollenites* (Samant et al., 2013).

An early Paleocene assemblage at the heart of the Deccan eruption-influenced region shows that a diverse biota preserved in intertrappean lagoonal sediments near Mumbai characterizes the closing stages (ca. 64.5 Ma) of the Deccan flood basalt episode. Mumbai Island Formation intertrappean faunal and floral communities are distinct from those common to many late Maastrichtian intertrappean sediments across the Deccan province, but organisms usually sensitive to environmental perturbations, including frogs, turtles, and crocodiles, thrived, and while plant material deposition was responsive to environmental shifts, a vegetated landscape of palms, podocarp ferns, and bamboo occupied the lagoon margins (Cripps et al., 2005).

Thus, while the Deccan eruptions undoubtedly influenced the composition and productivity of terrestrial ecosystems within, and proximal to, the eruption centers, organisms normally regarded as environmentally sensitive (e.g., frogs) show no evidence of local, let alone global demise. It seems highly unlikely, therefore, that the Deccan continental flood basalt episode was the primary driver for extinctions across the Cretaceous-Paleogene boundary, but rather its influence was likely to have been felt as a driver for changes in atmospheric composition and longer-term perturbations to the climate system that sensitized ecosystems to trauma associated with the end-Cretaceous impact events.

TABLE 2. TEMPERATURES PREDICTED BY CLIMATE LEAF ANALYSIS MULTIVARIATE PROGRAM (CLAMP) FOR NEW ZEALAND LEAF FLORAS

Locality	New Zealand geological time scale	MAT (°C)	WMMT (°C)	CMMT (°C)
Pakawau Bush Road	Cretaceous; Mata Series; Haumurian	12.9	21.0	4.6
Cameron's Pit	Cretaceous; Mata Series; Haumurian	7.6	17.9	–1.4
Ian's Tip	Earliest Paleocene; Dannevirke Series; Teurian	5.1	15.7	–3.3
Pillar Point Track	Earliest Paleocene; Dannevirke Series; Teurian	3.9	15.6	–5.1
Spring Creek Mine	Earliest Paleocene; Dannevirke Series; Teurian	4.9	15.6	–3.5
Pikopiko	Late Eocene	21.4	24.9	16.9

Notes: MAT—mean annual temperature; WMMT—warm month mean temperature; CMMT—cold month mean temperature. The temperatures are only indicative. The earliest Paleocene sites are outside physiognomic calibration space, and thus uncertainties are not given.

Boltysh Impact Event

The Boltysh impact occurred just prior to that at Chicxulub at 65.59 ± 0.64 Ma (Kelley and Gurov, 2002), and in-filling of the 24-km-diameter crater records in detail vegetation changes across the Cretaceous-Paleogene interval in the Ukraine. Vegetation change across that interval appears almost identical to that seen in the Raton, Denver, and Powder River Basins, North America: a clear "fern-spore spike," followed by an influx of *Arecipites*, taken to indicate thermophilic palms, and an early post–Cretaceous-Paleogene event plant community recovery that exhibited a common posteruption seral successional pathway observed in igneous provinces, although enhanced by high moisture availability. Thereafter, palynological data from the more expanded and continuous lacustrine sedimentation of Boltysh suggest that wet-dry climate cycles, which were also a common feature of low to midlatitudes in the Late Cretaceous, extended the recovery period (Jolley et al., 2013) over a period of ~300 k.y. up to the Dan-C2 negative carbon isotope excursion (CIE), reminiscent of the extended recovery period reported by Wolfe and Upchurch (1987b). The frequent occurrences of dry episodes may have delayed the development of closed canopy forest until ~20 k.y. after the boundary event. The Dan-C2 CIE signature is greater than in the marine record, attests to a worldwide event, and is accompanied by a rise in thermophilic, predominantly Normapolles, taxa (Gilmour et al., 2013).

SYNTHESIS

Terrestrial Maastrichtian ecosystems worldwide were subject to considerable environmental instability (Eberth and Braman, 2012; Johnson et al., 1989; Nordt et al., 2002; Wilf et al., 2003) and fire disturbance (Brown et al., 2012). In some regions, floral compositional turnover was greater prior to the Cretaceous-Paleogene event than at it (Markevich et al., 2012; Spicer et al., 1994). The drivers of this change are at present unknown, but given the documented instability, climate may have played a role. The causes of climatic instability are poorly understood but could have included changes in sea level and seaway connections (Li and Keller, 1999), changes in atmospheric composition connected with large-scale and prolonged volcanism (Deccan Traps), and possibly a minor perturbation associated with the Boltysh impact (Gilmour et al., 2013; Jolley et al., 2013). It is not yet possible to identify ecological trauma attributable to the Boltysh impact against the background of ongoing latest Maastrichtian environmental change. Nevertheless, environmental changes undoubtedly filtered terrestrial ecosystem composition and structure prior to the Cretaceous-Paleogene boundary, and it is very likely that the nature of any Chicxulub impact–related ecological trauma was determined in part by the state of ecosystem resilience in a given area at the time of the event.

There can be little doubt that the Chicxulub impact induced profound terrestrial ecological trauma at low and middle latitudes in North America, but even there, proximal to the Chicxulub impact site, recovery was evidently rapid compared to that in marine ecosystems (Beerling et al., 2001). Although standing biomass was vulnerable to immediate impact-related trauma, considerable land plant resilience to all but the most extended period of adversity was conferred by soil propagule banks. Where there are marked changes in the abundance and diversity of particular pollen forms, such as the demise of the *Aquilapollenites* province, this may have been due to more vulnerable links in the life cycle being broken (e.g., the demise of pollinators; Bercovici et al., 2012).

Interpreting genuine extinctions (termination of genetic lineages) from plant fossil organs (leaves, pollen, etc.) is complicated by organ shedding, mosaic evolution, and varying transport and preservation potentials of different plant organs combined with a focus on morphotype recognition more suitable for biostratigraphy and paleoclimate studies rather than for phylogenetic analyses. A more tractable avenue is determining how extensive and prolonged that trauma was worldwide and how sensitive to trauma global terrestrial ecosystems were at the time of impact. Until we can better characterize ecosystem vulnerability before the Cretaceous-Paleogene event, we cannot extrapolate reliably the consequences for future possible perturbations, whether due to impacts or other causes.

Ecological trauma was most marked in North America, particularly at mid- and low latitudes (Nichols and Johnson, 2008; Spicer, 1986; Sweet et al., 1999). Typical Cretaceous-Paleogene event horizons are characterized by a fern-spore abundance anomaly, floral turnover, and a recovery pattern that mimics that seen in normal seral succession, albeit possibly on a longer time scale (Nichols and Johnson, 2008; Wolfe and Upchurch, 1987b). Based on modern fern-spore viability (even at low temperatures), the duration of any postimpact adversity where no growth was possible is unlikely to have been more than 1 or 2 yr. Detailed analysis of dispersed pollen and spores, as well as plant meso- and megafossil remains at the Teapot Dome site in Wyoming attests to vegetation changes immediately prior to the Cretaceous-Paleogene event horizons, followed by short-term (minutes to months) loss of diversity and seral recolonization during and after the deposition of those horizons. There is, however, no evidence of fire intensity different to background levels. In the earliest postevent interval, photosynthesis continued but growth was compromised, probably by a curtailed growing season. The survival of *Azolla* but failure to complete its sexual reproductive cycle, and the disruption to the sexual life cycles of several flowering plants suggest that the growth period was cool and/or short in relation to photosynthetic return.

Further afield, the effects of the impact were less pronounced. In New Zealand, an immediate pre–Cretaceous-Paleogene floral change was followed at the iridium-rich layer by fungal abundance and an obvious increase in fern abundance, accompanied by angiosperms and gymnosperms continuing up section (Vajda et al., 2001; Vajda and McLoughlin, 2004). The relative continuity of the vegetation records a low-temperature "filter" on leaf form, leaving a long-term foliar physiognomic spectrum similar to that seen in modern central Siberia. Immediate postimpact cooling explains the susceptibility of low- and midlatitude vegetation to the Cretaceous-Paleogene event, expressed strongly in the intensively studied North American record, where evergreen thermophilic taxa were particularly vulnerable (Wolfe and Upchurch, 1987b; Johnson et al., 1989; Wilf and Johnson, 2004), as well as the muted floral change observed at near-polar latitudes (Moiseeva, 2012; Askin, 1988; Elliot et al., 1994). At lower (warmer) latitudes in North America floral change was greatest, and cooler temperatures, or temperature fluctuations, must have persisted long enough to overcome the resilience conferred by the soil seed bank.

The Boltysh record (Gilmour et al., 2013; Jolley et al., 2013) provides further insights into the early Paleocene. Here, lacking studies of the leaf record, there is no evidence of low temperatures, and thermophilic taxa such as palms persist and even increase in relative abundance during the post–"fern-spore spike" recovery phase. Climatic fluctuations, seen previously in the Maastrichtian, continued, and repeated dry periods persisting thousands of years into the Paleocene extended the period during which an open rather than forested landscape prevailed. Such an extended period of vegetation and climate change cannot be directly ascribed to the Chicxulub impact alone but appears to be a continuation of environmental instability that began well before the arrival of the bolide.

Background environmental fluctuations close to the Cretaceous-Paleogene boundary, possibly driven by large-scale volcanism, determined ecosystem status at the time of the Cretaceous-Paleogene impacts. In the context of the Cretaceous-Paleogene boundary, it seems that asteroid impacts alone probably cause, at most, short-term ecological trauma limited in geographical extent, and result in minimal long-term global vegetation change. By itself, the ecological trauma across the Cretaceous-Paleogene boundary, being so short-lived and thus leaving the soil propagule reservoir more or less intact, could not have caused long-term land plant ecosystem changes. These, after all, were under way before the Chicxulub and Boltysh events. However, the effects that impacts have depend not only on the time, size, and place of the event, but on the ecosystem vulnerability at the time of the impact. In this case, the abundance and diversity of taxa capable of entering extended periods of dormancy and/or with propagules possessing long in-soil viabilities are critical factors in determining vulnerability. To properly evaluate susceptibility to impact events requires studies of ecosystem characteristics well before and after any event signature. The focus on high-resolution studies of event horizons at the millimeter scale, while crucial to understanding short-term responses, has to some extent distracted attention from the wider environmental context we consider critical to reliable generalizations on just how damaging to life on Earth impacts, or other major ecosystem traumas, are likely to be.

ACKNOWLEDGMENTS

We thank Liz Kennedy (Institute of Geological & Nuclear Sciences Ltd. [GNS] New Zealand) for making available the New Zealand leaf data, Ian Raine (GNS New Zealand) for providing information on the stratigraphic position of the New Zealand leaf assemblages, and Daphne Lee (University of Otago) and the William Evans Fellowship committee for supporting Spicer in New Zealand. For work at Teapot Dome, we thank National Geographic for funding helicopter transport; the late Jack A. Wolfe for his collaboration, enthusiasm, and friendship; and S. Davies, P. Lang, P. Vollum, and the late Jack Wolfe for help with field work. Katy Roucoux undertook work on Teapot Dome during an undergraduate project supervised by Collinson, and her contribution is gratefully acknowledged. Collinson would like to thank Art Sweet and the late Doug Nichols for helpful discussion on palynology and suggestions for identification of rarer palynomorphs at Teapot Dome.

REFERENCES CITED

Aber, J.S., Owens, L.C., Aber, S.W., Eddy, T., Schulenberg, J.H., Sundberg, M., and Penner, R.L., II, 2010, Recent *Azolla* bloom at Cheyenne Bottoms, Kansas: Transactions of the Kansas Academy of Science, v. 113, p. 56–63, doi:10.1660/062.113.0204.

Arens, N.C., and Jahren, A.H., 2000, Carbon isotope excursion in the atmospheric CO_2 at the Cretaceous-Tertiary boundary: Evidence from terrestrial sediments: Palaios, v. 15, p. 314–322, doi:10.1669/0883-1351(2000)015<0314:CIEIAC>2.0.CO;2.

Askin, R.A., 1988, The palynological record across the Cretaceous/Tertiary transition on Seymour Island, Antarctica, *in* Feldman, R.M., and Woodburne, M.O., eds., Geology and Paleontology of Seymour Island, Antarctic Peninsula: Geological Society of America Memoir 169, p. 155–62.

Askin, R.A., 1990, Campanian to Paleocene spore and pollen assemblages of Seymour Island, Antarctica, *in* Truswell, E.M., and Owen, J.A.K., eds., Proceedings of the 7th International Palynological Congress: Review of Palaeobotany and Palynology, v. 65, p. 105–113.

Ballesteros, D., Estrelles, E., Walters, C., and Ibars, A.M., 2011, Effect of storage temperature on green spore longevity for the ferns *Equisetum ramosissimum* and *Osmunda regalis*: Cryo Letters, v. 32, p. 89–98.

Batten, D.J., Zavattieri, A.M., and Collinson, M.E., 2011, Megaspores from the Upper Maastrichtian of the eastern Spanish Pyrenees and their biostratigraphic, palaeogeographic and palaeoenvironmental significance: Review of Palaeobotany and Palynology, v. 167, p. 156–172, doi:10.1016/j.revpalbo.2011.08.002.

Beerling, D.J., Lomax, B.H., Upchurch, G.R., Nichols, D.J., Pillmore, C.L., Handley, L.L., and Scrimgeour, C.M., 2001, Evidence for the recovery of terrestrial ecosystems ahead of marine primary production following a biotic crisis at the Cretaceous-Tertiary boundary: Journal of the Geological Society, v. 158, p. 737–740, doi:10.1144/jgs.158.5.737.

Belcher, C.M., Finch, P., Collinson, M.E., Scott, A.C., and Grassineau, N.V., 2009, Geochemical evidence for combustion of hydrocarbons during the K-T impact event: Proceedings of the National Academy of Sciences of the United States of America, v. 106, p. 4112–4117.

Bercovici, A., Pearson, D., Nichols, D.J., and Wood, C., 2009, Biostratigraphy of selected K/T boundary sections in southwestern North Dakota, USA: Towards refinement of palynological identification criteria: Cretaceous Research, v. 30, p. 632–658, doi:10.1016/j.cretres.2008.12.007.

Bercovici, A., Vajda, V., and Sweet, A., 2012, Pollen and spore stratigraphy of the Cretaceous-Paleogene mass extinction interval in the Northern Hemisphere: Journal of Stratigraphy, v. 36, p. 165–178.

Bewley, J.D., and Black, M., 1994, Seeds; Physiology of Dormancy and Germination (2nd ed.): London, Plenum Press, 445 p.

Braman, D.R., and Sweet, A.R., 2012, Biostratigraphically useful Late Cretaceous–Paleocene terrestrial palynomorphs from the Canadian

Western Interior Sedimentary Basin: Palynology, v. 36, p. 8–35, doi:10.1080/01916122.2011.642127.

Brooks, R.R., Strong, C.P., Lee, J., Orth, C.J., Gilmore, J.S., Ryan, D.E., and Holzbecher, J., 1986, Stratigraphic occurrences of iridium anomalies at four Cretaceous/Tertiary boundary sites in New Zealand: Geology, v. 14, p. 727–729, doi:10.1130/0091-7613(1986)14<727:SOOIAA>2.0.CO;2.

Brown, S.A.E., Scott, A.C., Glasspool, I.J., and Collinson, M.E., 2012, Cretaceous wildfires and their impact on the Earth system: Cretaceous Research, v. 36, p. 162–190, doi:10.1016/j.cretres.2012.02.008.

Collinson, M.E., Smith, S.Y., Van Konijnenburg-van Cittert, J.H.A., Batten, D.J., van der Burgh, J., Barke, J., and Marone, F., 2013, New observations and synthesis of Paleogene heterosporous water ferns: International Journal of Plant Sciences, v. 174, p. 350–363, doi:10.1086/668249.

Cripps, J.A., Widdowson, M., Spicer, R.A., and Jolley, D.W., 2005, Coastal ecosystem responses to late stage Deccan Trap volcanism: The post K-T boundary (Danian) palynofacies of Mumbai (Bombay), west India: Palaeogeography, Palaeoclimatology, Palaeoecology, v. 216, p. 303–332, doi:10.1016/j.palaeo.2004.11.007.

Dyer, A.F., and Lindsay, S., 1992, Spore banks of temperate ferns: American Fern Journal, v. 82, p. 89–122, doi:10.2307/1547792.

Eberth, D.A., and Braman, D.R., 2012, A revised stratigraphy and depositional history for the Horseshoe Canyon Formation (Upper Cretaceous), southern Alberta plains: Canadian Journal of Earth Sciences, v. 49, p. 1053–1086, doi:10.1139/e2012-035.

Elliot, D.H., Askin, R.A., Kyte, F.T., and Zinsmeister, W.J., 1994, Iridium and dinocysts at the Cretaceous-Tertiary boundary on Seymour Island, Antarctica: Implications for the K-T event: Geology, v. 22, p. 675–678, doi:10.1130/0091-7613(1994)022<0675:IADATC>2.3.CO;2.

Frederiksen, N.O., 1989, Changes in floral diversities, floral turnover rates, and climates in Campanian and Maastrichtian time: North Slope of Alaska: Cretaceous Research, v. 10, p. 249–266.

Frederiksen, N.O., Ager, T.A., and Edwards, L.E., 1988, Palynology of Maastrichtian and Paleocene rocks, lower Colville River region, North Slope of Alaska: Canadian Journal of Earth Sciences, v. 25, p. 512–527, doi:10.1139/e88-051.

Gilmour, I., Gilmour, M., Jolley, D.W., Kelley, S.P., Kemp, D., Daly, R., and Watson, J., 2013, A high-resolution nonmarine record of an early Danian hyperthermal event, Boltysh crater, Ukraine: Geology, v. 41, p. 783–786, doi:10.1130/G34292.1.

Godefroit, P., Golvneva, L.B., Shchepetov, S., Garcia, G., and Alekseev, P., 2009, The last polar dinosaurs: High diversity of latest Cretaceous arctic dinosaurs in Russia: Naturwissenschaften, v. 96, no. 4, p. 495–501, doi:10.1007/s00114-008-0499-0.

Godefroit, P., Lauters, P., Van Itterbeeck, J., Bolotsky, Y.L., Dong, Z., Jin, L., Wu, W., Bolotsky, I., Hai, S., and Yu, T., 2011, Recent advances on study of hardrosaurid dinosaurs in Helingjiang (Amur) River area between China and Russia: Global Geology, v. 14, p. 160–191.

Golovneva, L.B., 1994, The flora of the Maastrichtian–Danian deposits of the Koryak Uplands, northeast Russia: Cretaceous Research, v. 15, p. 89–100, doi:10.1006/cres.1994.1004.

Golovneva, L.B., 2000, The Maastrichtian (Late Cretaceous) climate in the Northern Hemisphere, *in* Hart, M.B., ed., Climates: Past and Present: Geological Society of London Special Publication 181, p. 43–54.

Herman, A.B., 2011, Al'bskaya–Paleotsenovaya Flora Severnoi Patsifiki (Albian–Paleocene Flora of the North Pacific Region): Moscow, GEOS, 230 p.

Herngreen, G.F.W., Kedves, M., Rovnina, L.V., and Smirnova, S.B., 1996, Cretaceous palynofloral provinces: A review, *in* Jansonius, J., and McGregor, D.C., eds., Palynology: Principles and Applications: Dallas, Texas, American Association of Stratigraphic Palynologists, p. 1157–1188.

Hildebrand, A.R., 1993, The Cretaceous/Tertiary boundary impact (or the dinosaurs didn't have a chance): The Journal of the Royal Astronomical Society of Canada, v. 87, p. 77–118.

Hill, C.R., 1986, The epidermis/cuticle and in situ spores and pollen in fossil plant taxonomy, *in* Spicer, R.A., and Thomas, B.A., eds., Systematic and Taxonomic Approaches in Palaeobotany: Oxford, UK, Oxford University Press, p. 123–136.

Hill, C.R., and Crane, P.R., 1982, Evolutionary cladistics and the origin of angiosperms, *in* Joysey, K.A., and Friday, A.E., eds., Problems in Phylogenetic Reconstruction: London, Academic Press, p. 269–361.

Hofmann, C.-C., Spicer, R.A., Ahlberg, A., and Herman, A.B., 2011, Scanning electron microscopy investigation of monads and tetrads of basal core eudicots from the Upper Cretaceous Vilui Basin, Siberia: Evidence for reticulate evolution: Review of Palaeobotany and Palynology, v. 167, p. 196–211, doi:10.1016/j.revpalbo.2011.08.007.

Hotton, C., 2002, Palynology of the Cretaceous-Tertiary boundary in central Montana: Evidence for extraterrestrial impact as a cause of the terminal Cretaceous extinctions, *in* Hartman, J.H., Johnson, K.R., and Nichols, D.J., eds., The Hell Creek Formation and the Cretaceous-Tertiary Boundary in the Northern Great Plains: An Integrated Continental Record of the End of the Cretaceous: Geological Society of America Special Paper 361, p. 473–501.

Izett, G.A., 1990, The Cretaceous/Tertiary Boundary Interval, Raton Basin, Colorado and New Mexico, and its Content of Shock-Metamorphosed Minerals: Evidence Relevant to the K/T Boundary Impact-Extinction Theory: Geological Society of America Special Paper 249, 100 p.

Janes, R.A., 1998a, Growth and survival of *Azolla filiculoides* in Britain: I. Vegetative reproduction: The New Phytologist, v. 138, p. 367–375, doi:10.1046/j.1469-8137.1998.00114.x.

Janes, R.A., 1998b, Growth and survival of *Azolla filiculoides* in Britain: II. Sexual reproduction: The New Phytologist, v. 138, p. 377–384, doi:10.1046/j.1469-8137.1998.00113.x.

Johnson, K.R., Nichols, D.J., Attrep, M., Jr., and Orth, C.J., 1989, High-resolution leaf fossil record spanning the Cretaceous-Tertiary boundary: Nature, v. 340, p. 708–711, doi:10.1038/340708a0.

Jolley, D.W., Daly, R., Gilmour, I., and Kelley, S.P., 2013, Climatic oscillations stall vegetation recovery from K/Pg event devastation: Journal of the Geological Society, v. 170, p. 477–482, doi:10.1144/jgs2012-088.

Kelley, S.P., and Gurov, E., 2002, Boltysh, another impact at the KT boundary: Meteoritics & Planetary Science, v. 37, p. 1031–1043, doi:10.1111/j.1945-5100.2002.tb00875.x.

Kennedy, E.M., 2003, Late Cretaceous and Paleocene terrestrial climates of New Zealand: Leaf fossil evidence from South Island assemblages: New Zealand Journal of Geology and Geophysics, v. 46, p. 295–306, doi:10.1080/00288306.2003.9515010.

Kennedy, E.M., Spicer, R.A., and Rees, P.M., 2002, Quantitative paleoclimate estimates from Late Cretaceous and Paleocene leaf floras in the northwest of the South Island, New Zealand: Palaeogeography, Palaeoclimatology, Palaeoecology, v. 184, p. 321–345, doi:10.1016/S0031-0182(02)00261-4.

Kovach, W.L., and Spicer, R.A., 1995, Canonical correspondence analysis of leaf physiognomy: A contribution to the development of a new palaeoclimatological tool: Palaeoclimates, v. 1, p. 125–173.

Li, L., and Keller, G., 1999, Variability in Late Cretaceous climate and deep waters: Evidence from stable isotopes: Marine Geology, v. 161, p. 171–190, doi:10.1016/S0025-3227(99)00078-X.

Mander, L., Kürschner, W.M., and McElwain, J.C., 2010, An explanation for conflicting records of Triassic-Jurassic plant diversity: Proceedings of the National Academy of Sciences of the United States of America, v. 107, p. 15,351–15,356, doi:10.1073/pnas.1004207107.

Markevich, V.S., Bugdaeva, E.V., Sun, G., and Ashraf, A.R., 2012, Paleofloristic and paleoenvironmental changes in Amur (Heilongjiang) River region throughout the Late Cretaceous and Paleocene: Palynological and paleobotanical evidence, *in* 13th International Palynological Congress and 9th International Organisation of Paleobotany Conference Abstracts: Tokyo, Japan, p. SS21–008(311).

Moiseeva, M.G., 2012, The Maastrichtian flora of the Amaam Lagoon area (northeastern Russia): Stratigraphy and Geological Correlation, v. 20, p. 579–679, doi:10.1134/S0869593812070027.

Moiseeva, M.G., Herman, A.B., and Spicer, R.A., 2009, Late Paleocene flora of the northern Alaska peninsula: The role of transberingian plant migrations and climate change: Paleontological Journal, v. 43, p. 1298–1308, doi:10.1134/S0031030109100116.

Monaco, T.J., Ashton, F.M., and Well, S.C., 2002, Weed Science: Practice and Principles: New York, Wiley Blackwell, 671 p.

Nichols, D.J., 1990, Geologic and biostratigraphic framework of the nonmarine Cretaceous-Tertiary boundary interval in western North America: Review of Palaeobotany and Palynology, v. 65, p. 75–84, doi:10.1016/0034-6667(90)90058-Q.

Nichols, D.J., and Fleming, R.F., 1990, Plant microfossil record of the terminal Cretaceous event in the western United States and Canada, *in* Sharpton, V.L., and Ward, P.D., eds., Global Catastrophes in Earth History: An Interdisciplinary Conference on Impacts, Volcanism, and Mass Mortality: Geological Society of America Special Paper 247, p. 95–143.

Nichols, D.J., and Jacobson, S.R., 1982, Palynostratigraphic framework for the Cretaceous (Albian-Maestrichtian) of the overthrust belt of Utah and

Wyoming: Palynology, v. 6, p. 119–147, doi:10.1080/01916122.1982.9989238.

Nichols, D.J., and Johnson, K.R., 2008, Plants and the K-T Boundary: Cambridge, UK, Cambridge University Press, 280 p.

Nichols, D.J., Fleming, R.F., and Frederiksen, N.O., 1990, Palynological evidence of effects of the terminal Cretaceous event in western North America, *in* Kauffman, E.G., and Walliser, O.H., eds., Extinction Events in Earth History: Lecture Notes in Earth Sciences: Berlin, Springer, p. 351–364.

Nichols, D.J., Hickey, L.J., McWeeney, L.J., and Wolfe, J.A., 1992, Plants at the K/T boundary; Discussion and reply: Nature, v. 356, p. 295–296, doi:10.1038/356295b0.

Nordt, L., Atchley, S., and Dworkin, S.I., 2002, Paleosol barometer indicates extreme fluctuations in atmospheric CO_2 across the Cretaceous-Tertiary boundary: Geology, v. 30, p. 703–706, doi:10.1130/0091-7613(2002)030<0703:PBIEFI>2.0.CO;2.

O'Brien, C.L., Robinson, S.A., Kemp, D.B., Crame, J.A., Francis, J.E., Lunt, D.L., Ineson, J.R., Whittle, R.J., and Bowman, V., 2013, Paleoclimatic changes across the Cretaceous-Paleogene boundary: Geochemical reconstructions from Seymour Island, Antarctica, *in* Goldschmidt 2013 Conference Abstracts, p. 1869, doi:10.1180/minmag.2013.077.5.15.

Parrish, J.T., and Spicer, R.A., 1988, Middle Cretaceous wood from the Nanushuk Group, central North Slope, Alaska: Palaeontology, v. 31, p. 19–34.

Pope, K.O., 2002, Impact dust not the cause of the Cretaceous-Tertiary mass extinction: Geology, v. 30, p. 99–102, doi:10.1130/0091-7613(2002)030<0099:IDNTCO>2.0.CO;2.

Saito, T., Tamanoi, T., and Kaiho, K., 1986, End-Cretaceous devastation of terrestrial flora in the boreal Far East: Nature, v. 323, p. 253–255, doi:10.1038/323253a0.

Sallon, S., Solowey, E., Cohen, Y., Korchinsky, R., Egli, M., Woodhatch, I., Simchoni, O., and Kislev, M., 2008, Germination, genetics, and growth of an ancient date seed: Science, v. 320, p. 1464, doi:10.1126/science.1153600.

Samant, B., and Mohabey, D.M., 2012, Deccan continental flood basalt induced floral changes in Late Cretaceous–early Paleocene: Evidences from palynomorph bearing interlava sediments of Indian subcontinent, *in* 13th International Palynological Congress and 9th International Organisation of Paleobotany Conference Abstracts: Tokyo, Japan, p. SS21–O10 (447).

Samant, B., Mohabey, D.M., and Paudayal, K.N., 2013, *Aquilapollenites* and other triprojectate pollen from the Late Cretaceous–early Paleocene Deccan intertrappean deposits of India: Palynology, v. 37, no. 2, p. 298–315, doi:10.1080/01916122.2013.787125.

Schimmelmann, A., and DeNiro, M.J., 1984, Elemental and stable isotope variations of organic matter from a terrestrial sequence containing the Cretaceous-Tertiary boundary at York Canyon: New Mexico: Earth and Planetary Science Letters, v. 68, p. 392–398, doi:10.1016/0012-821X(84)90124-9.

Shen-Miller, J., Mudgett, M.B., and Schopf, J.W., 1995, Exceptional seed longevity and robust growth: Ancient sacred lotus from China: American Journal of Botany, v. 82, p. 1367–1380, doi:10.2307/2445863.

Shen-Miller, J., Schopf, J.W., Harbottle, G., Cao, R.-J., Ouyang, S., Zhou, K.-S., Southon, J.R., and Liu, G.-H., 2002, Long-living lotus: Germination and soil gamma-irradiation of centuries-old fruits, and cultivation, growth, and phenotypic abnormalities of offspring: American Journal of Botany, v. 89, p. 236–247, doi:10.3732/ajb.89.2.236.

Simabukuro, E.A., Dyer, A.F., and Felippe, G.M., 1998, The effect of sterilization and storage conditions on the viability of the spores of *Cyathea delgadii*: American Fern Journal, v. 88, p. 72–80, doi:10.2307/1547226.

Spicer, R.A., 1986, Comparative leaf architectural analysis of Cretaceous radiating angiosperms, *in* Spicer, R.A., and Thomas, B.A., eds., Systematic and Taxonomic Approaches to Palaeobotany: Systematics Association Special Volume 31, p. 223–233.

Spicer, R.A., 1989, Plants at the Cretaceous-Tertiary boundary: Philosophical Transactions of the Royal Society of London, ser. B, Biological Sciences, v. 325, p. 291–305, doi:10.1098/rstb.1989.0090.

Spicer, R.A., 2008, CLAMP, *in* Gornitz, V., ed., Encycopedia of Paleoclimatology and Ancient Environments: Dordrecht, Netherlands, Springer, p. 156–158.

Spicer, R.A., and Herman, A.B., 2010, The Late Cretaceous environment of the Arctic: A quantitative reassessment using plant fossils: Palaeogeography, Palaeoclimatology, Palaeoecology, v. 295, p. 423–442, doi:10.1016/j.palaeo.2010.02.025.

Spicer, R.A., and Parrish, J.T., 1987, Plant megafossils, vertebrate remains, and paleoclimate of the Kogosukruk Tongue (Late Cretaceous), North Slope, Alaska, *in* Hamilton, T.D., and Galloway, J.P., eds., Geologic Studies in Alaska by the United States Geological Survey during 1986: U.S. Geological Survey Circular 998, p. 47–48.

Spicer, R.A., and Parrish, J.T., 1990a, Late Cretaceous–Early Tertiary palaeoclimates of northern high latitudes: A quantitative view: Journal of the Geological Society, v. 147, p. 329–341, doi:10.1144/gsjgs.147.2.0329.

Spicer, R.A., and Parrish, J.T., 1990b, Latest Cretaceous woods of the central North Slope Alaska: Palaeontology, v. 33, p. 225–242.

Spicer, R.A., Burnham, R.J., Grant, P., and Glicken, H., 1985, *Pityrogramma calomelanos*, the primary, post-eruption colonizer of volcan Chichonal, Chiapas, Mexico: American Fern Journal, v. 75, p. 1–5, doi:10.2307/1546571.

Spicer, R.A., Davies, K.S., and Herman, A.B., 1994, Circum-Arctic plant fossils and the Cretaceous-Tertiary transition, *in* Boulter, M.C., and Fisher, H.C., eds., Cenozoic Plants and Climates of the Arctic: Berlin, Springer, NATO ASI Series 1, v. 27, p. 161–174.

Spicer, R.A., Herman, A.B., and Kennedy, E.M., 2004, The foliar physiognomic record of climatic conditions during dormancy: CLAMP and the cold month mean temperature: The Journal of Geology, v. 112, p. 685–702, doi:10.1086/424579.

Sun, G., Akhmetiev, M., Markevich, V., Ashraf, A.R., Bugdaeva, E., Godefroit, P., Bolotsky, Y., Dong, Z.M., Golovneva, L., Yanf, H.X., Sun, C.L., Sun, Y.W., Quan, C., Kodrul, T., Kezina, T., Johnson, K., Nishida, H., Dilcher, D.L., Harding, I., and Chen, Y.J., 2011, Late Cretaceous biota and the Cretaceous-Paleogene boundary in Jiayin of Heilongjiang, China: Global Geology, v. 14, p. 115–143.

Sweet, A.R., 2001, Plants, a yardstick for measuring the environmental consequences of the Cretaceous-Tertiary boundary event: Geoscience Canada, v. 28, p. 127–138.

Sweet, A.R., and Braman, D.R., 1992, The K-T boundary and contiguous strata in western Canada: Interactions between paleoenvironments and palynological assemblages: Cretaceous Research, v. 13, p. 31–79, doi:10.1016/0195-6671(92)90027-N.

Sweet, A.R., and Braman, D.R., 2001, Cretaceous-Tertiary palynofloral perturbations and extinctions within the *Aquilapollenites* phytogeographic province: Canadian Journal of Earth Sciences, v. 38, p. 249–269.

Sweet, A.R., Braman, D.R., and Lerbekmo, J.F., 1999, Sequential palynological changes across the composite Cretaceous-Tertiary boundary claystone and contiguous strata, western Canada and Montana, U.S.A.: Canadian Journal of Earth Sciences, v. 36, p. 743–768, doi:10.1139/e98-061.

Takhtajan, A., 1959, Die Evolution der Angiospermen: Jena, Germany, VEB Gustav Fischer, 344 p.

Tomsich, C.S., McCarthy, P.J., Fowell, S.J., and Sunderlin, D., 2010, Paleofloristic and paleoenvironmental information from a Late Cretaceous (Maastrichtian) flora of the lower Cantwell Formation near Sable Mountain, Denali National Park, Alaska: Palaeogeography, Palaeoclimatology, Palaeoecology, v. 295, p. 389–408, doi:10.1016/j.palaeo.2010.02.023.

Toon, O.B., Zahnle, K., Morrison, D., Turco, R.P., and Covey, C., 1997, Environmental perturbations caused by the impacts of asteroids and comets: Reviews of Geophysics, v. 35, p. 41–78, doi:10.1029/96RG03038.

Tschudy, R.H., Pillmore, J.S., Orth, C.J., Gilmore, J.S., and Knight, J.D., 1984, Disruption of the terrestrial plant ecosystem at the Cretaceous-Tertiary boundary, Western Interior: Science, v. 225, p. 1030–1032, doi:10.1126/science.225.4666.1030.

Traverse, A., 2007, Paleopalynology: Boston, Springer, 772 p.

Upchurch, G.R., and Wolfe, J.A., 1993, Cretaceous vegetation of the Western Interior and adjacent regions of North America, *in* Caldwell, W.G.E., and Kauffman, E.G., eds., Cretaceous Evolution of the Western Interior Basin: Geological Society of Canada Special Paper 39, p. 243–281.

Vajda, V., and McLoughlin, S., 2004, Fungal proliferation at the Cretaceous-Tertiary boundary: Science, v. 303, p. 1489, doi:10.1126/science.1093807.

Vajda, V., and Raine, J.I., 2003, Pollen and spores in marine Cretaceous/Tertiary boundary sediments at mid-Waipara River, North Canterbury, New Zealand: New Zealand Journal of Geology and Geophysics, v. 46, p. 255–273, doi:10.1080/00288306.2003.9515008.

Vajda, V., Raine, J.I., and Hollis, C.J., 2001, Indication of global deforestation at the Cretaceous-Tertiary boundary by New Zealand fern spike: Science, v. 294, p. 1700–1702, doi:10.1126/science.1064706.

van Bergen, P.F., Collinson, M.E., and de Leeuw, J.W., 1993, Chemical composition and ultrastructure of fossil and extant salvinialean microspore massulae and megaspores: Grana, v. 32, S1, p. 18–30, doi:10.1080/00173139309427448.

van der Burgh, J., Collinson, M.E., van Konijnenburg-van Cittert, J.H.A., Barke, J., and Brinkhuis, H., 2013, The freshwater fern *Azolla* (Azollaceae) from the Eocene Arctic and Nordic Sea sediments: New species and their stratigraphic distribution: Review of Palaeobotany and Palynology, v. 194, p. 50–68, doi:10.1016/j.revpalbo.2013.02.002.

Visscher, H., Looy, C.V., Collinson, M.E., Brinkhuis, H., van Konijnenburg-van Cittert, J.H.A., Kurschner, W.M., and Sephton, M.A., 2004, Environmental mutagenesis during the end-Permian ecological crisis: Proceedings of the National Academy of Sciences of the United States of America, v. 101, p. 12,952–12,956, doi:10.1073/pnas.0404472101.

Wilf, P., and Johnson, K.R., 2004, Land plant extinction at the end of the Cretaceous: A quantitative analysis of the North Dakota megafloral record: Paleobiology, v. 30, p. 347–368, doi:10.1666/0094-8373(2004)030<0347:LPEATE>2.0.CO;2.

Wilf, P., Johnson, K.R., and Huber, B.T., 2003, Correlated terrestrial and marine evidence for global climate changes before mass extinction at the Cretaceous-Paleogene boundary: Proceedings of the National Academy of Sciences of the United States of America, v. 100, p. 599–604, doi:10.1073/pnas.0234701100.

Wolfe, J.A., 1973 (1974), Fossil forms of Amentiferae: Brittonia, v. 25, p. 334–355, doi:10.2307/2805639.

Wolfe, J.A., 1985, Distribution of major vegetational types during the Tertiary, *in* Sundquist, E.T., and Broecker, W.S., eds., The Carbon Cycle and Atmospheric CO_2: Natural Variations, Archean to Present: American Geophysical Union Geophysical Monograph 32, p. 357–375.

Wolfe, J.A., 1990, Palaeobotanical evidence for a marked temperature increase following the Cretaceous/Tertiary boundary: Nature, v. 343, p. 153–156, doi:10.1038/343153a0.

Wolfe, J.A., 1991, Palaeobotanical evidence for a June "impact winter" at the Cretaceous/Tertiary boundary: Nature, v. 352, p. 420–423, doi:10.1038/352420a0.

Wolfe, J.A., 1993, A Method of Obtaining Climatic Parameters from Leaf Assemblages: U.S. Geological Survey Bulletin 2040, p. 1–73.

Wolfe, J.A., and Upchurch, G.R., 1987a, North American nonmarine climates and vegetation during the Late Cretaceous: Palaeogeography, Palaeoclimatology, Palaeoecology, v. 61, p. 33–77, doi:10.1016/0031-0182(87)90040-X.

Wolfe, J.A., and Upchurch, G.R., 1987b, Leaf assemblages across the Cretaceous-Tertiary boundary in the Raton Basin, New Mexico and Colorado: Proceedings of the National Academy of Sciences of the United States of America, v. 84, p. 5096–5100, doi:10.1073/pnas.84.15.5096.

Yang, Jian, Spicer, R.A., Spicer, T.E.V., and Li, Cheng-Sen, 2011, "CLAMP Online": A new Web-based palaeoclimate tool and its application to the terrestrial Paleogene and Neogene of North America: Palaeobiodiversity and Palaeoenvironments, v. 91, p. 163–183, doi:10.1007/s12549-011-0056-2.

Manuscript Accepted by the Society 31 January 2014

Printed in the USA

The Geological Society of America
Special Paper 505
2014

The early Danian hyperthermal event at Boltysh (Ukraine): Relation to Cretaceous-Paleogene boundary events

Iain Gilmour*
Centre for Earth, Planetary, Space and Astronomical Research, The Open University, Milton Keynes MK7 6AA, UK

David Jolley
Department of Geology & Petroleum Geology, Kings College, Aberdeen AB9 2UE, UK

David Kemp
Simon Kelley
Mabs Gilmour
Centre for Earth, Planetary, Space and Astronomical Research, The Open University, Milton Keynes MK7 6AA, UK

Rob Daly
Department of Geology & Petroleum Geology, Kings College, Aberdeen AB9 2UE, UK

Mike Widdowson
Centre for Earth, Planetary, Space and Astronomical Research, The Open University, Milton Keynes MK7 6AA, UK

ABSTRACT

The Boltysh meteorite impact crater formed in the Ukrainian Shield on the margin of the Tethys Ocean a few thousand years before the Cretaceous-Paleogene boundary and was rapidly filled by a freshwater lake. Sediments filling the lake vary from early lacustrine turbidites and silts to ~300 m of fine silts, organic carbon–rich muds, oil shales, and lamenites that record early Danian terrestrial climate signals at high temporal resolution. Combined carbon isotope and palynological data show that the fine-grained organic carbon–rich lacustrine sediments preserve a uniquely complete and detailed negative carbon isotope excursion in an expanded section of several hundred meters.

The position of the carbon isotope excursion in the early Danian stage of the Paleogene period, around 200 k.y. above the Cretaceous-Paleogene boundary, leads us to correlate it to the Dan-C2 carbon isotope excursion recorded in marine sediments of the same age. The more complete Boltysh carbon isotope excursion record indicates a $\delta^{13}C$ shift of around −3‰, but also a more extended recovery period,

*iain.gilmour@open.ac.uk

Gilmour, I., Jolley, D., Kemp, D., Kelley, S., Gilmour, M., Daly, R., and Widdowson, M., 2014, The early Danian hyperthermal event at Boltysh (Ukraine): Relation to Cretaceous-Paleogene boundary events, *in* Keller, G., and Kerr, A.C., eds., Volcanism, Impacts, and Mass Extinctions: Causes and Effects: Geological Society of America Special Paper 505, p. 133–146, doi:10.1130/2014.2505(06). For permission to copy, contact editing@geosociety.org.

strikingly similar in pattern to the highest fidelity carbon isotope excursion records available for the Toarcian and Paleocene-Eocene hyperthermal events. Changes in floral communities through the carbon isotope excursion recorded at Boltysh reflect changing biomes caused by rapidly warming climate, followed by recovery, indicating that this early Danian hyperthermal event had a similar duration to the Toarcian and Paleocene-Eocene events.

INTRODUCTION

The Maastrichtian and Danian deposits that encompass the globally distributed Cretaceous-Paleogene boundary layer (66.04 ± 0.05 Ma; Renne et al., 2013) are some of the most intensively studied deposits in the geological record. Their association with physical and chemical markers of the Chicxulub impact and the end-Mesozoic mass extinction means that they have been subjected to extensive paleontologic, stratigraphic, geochemical, and petrographic investigation, and some 350 locations are now known worldwide that preserve this boundary to varying degrees of stratigraphic completeness (Smit, 1999). Thirty years after the original suggestion that an extraterrestrial impact was responsible for the end-Mesozoic mass extinctions (Alvarez et al., 1980; Smit and Hertogen, 1980), there is still debate as to whether a causal link between the Chicxulub impact and the Cretaceous-Paleogene mass extinction has been established and the role that large igneous provinces, such as the Deccan Traps, may have played in the late Maastrichtian and early Danian (Schulte et al., 2010; Keller et al., 2010; Archibald et al., 2010; Courtillot and Fluteau, 2010; Keller, 2012). Establishing the relationships between different events across the Cretaceous-Paleogene boundary is dependent on the completeness of the stratigraphic record and the time resolution that can be achieved within that record.

Following the Cretaceous-Paleogene boundary, the Paleocene geological record is characterized by a succession of hyperthermal events culminating in the prominent Paleocene-Eocene thermal maximum. Such hyperthermal events are interpreted as evidence for the addition of thousands of petagrams of ^{13}C-depleted carbon to the ocean-atmosphere system, recording major short-term perturbations of the global carbon cycle (McInerney and Wing, 2011). In the early Paleocene, several hyperthermal events have been identified, including the Dan-C2 event at ca. 65.2 Ma (Quillévéré et al., 2008), the latest Danian event at ca. 61.7 Ma (Westerhold et al., 2008; Bornemann et al., 2009), and the Danian-Selandian transition event at ca. 61 Ma (Speijer, 2003). The earliest of these, the Dan-C2 event, has been documented in marine sections near the top of magnetozone C29r, postdating the Cretaceous-Paleogene boundary by less than 300 k.y. To date, the Dan-C2 event has been found in marine carbonate δ^{13}C and δ^{18}O records from the Atlantic (Quillévéré et al., 2008; Westerhold et al., 2011) and western Tethys (Coccioni et al., 2010). However, a high-resolution benthic δ^{13}C record from the Pacific appeared to show no evidence for an early Danian event, leading Westerhold et al. (2011) to suggest that the event may have been restricted to the Atlantic. The Dan-C2 hyperthermal event recorded in marine sequences is coincident with a major phase of Deccan volcanism, which commenced in the upper part of magnetochron C29r and lasted until the lowermost part of magnetochron C29n, releasing an estimated 2×10^{14} kg of sulfur dioxide and 4×10^{14} kg of carbon dioxide (Self et al., 2006; Chenet et al., 2009; Punekar et al., this volume). Since Earth's oceans represent the major short-term carbon cycle reservoir, changes in the δ^{13}C of marine biogenic carbonate associated with a global Dan-C2 event should have affected atmospheric and continental carbon reservoirs, as is observed at the Paleocene-Eocene thermal maximum (Koch et al., 1992). Understanding the extent to which the Dan-C2 event is global and shares common driving mechanisms with other hyperthermal events, and its relation to late Maastrichtian and early Danian events, requires further high-resolution studies, particularly of terrestrial sequences.

A newly discovered Cretaceous-Paleogene record is preserved in the post-impact fill of the Boltysh meteorite crater, Ukraine (Fig. 1). As the only Eurasian terrestrial example of the Cretaceous-Paleogene section, Boltysh is a significant record of late Maastrichtian–early Danian events. This significance is enhanced by the high spatial resolution of post–Cretaceous-Paleogene crater fill, comprising 500 m of lacustrine sediments, which ensure a long-term continuous record. The Boltysh impact crater formed on the Proterozoic crystalline basement of the Ukrainian Shield and has yielded an Ar-Ar age from crystalline impact melt samples of 65.59 ± 0.64 Ma, an age within error of the Cretaceous-Paleogene boundary and the Chicxulub impact crater (Kelley and Gurov, 2002; age recalculated to the constants recommended by Renne et al., 2011). However, radiometric dating alone was not able to establish the precise age relationship between the Boltysh and Chicxulub impacts. The crater was drilled in the 1960s–1980s in a study of economic oil shale deposits. Unfortunately, the cores were not curated and have been lost; in consequence, we drilled a borehole in 2008 in the deepest part of the crater fill (Fig. 2). Jolley et al. (2010) established the close temporal relationship between the Boltysh and Chicxulub impacts, demonstrating that the Boltysh postimpact flora, formed on the ejecta layer, was in turn devastated by the Cretaceous-Paleogene event. The fern spike and negative carbon isotope shift that documents the Cretaceous-Paleogene boundary at Boltysh has strong similarities to the record in North America (e.g., Wolfe and Upchurch, 1987) and other global Cretaceous-Paleogene boundary locations (e.g., Japan—Saito et al., 1986; New Zealand—Vajda et al., 2001). This correlation is confirmed by paleontological constraints that record crater ejecta containing late Maastrichtian foraminifera overlain by calcareous nannoplankton

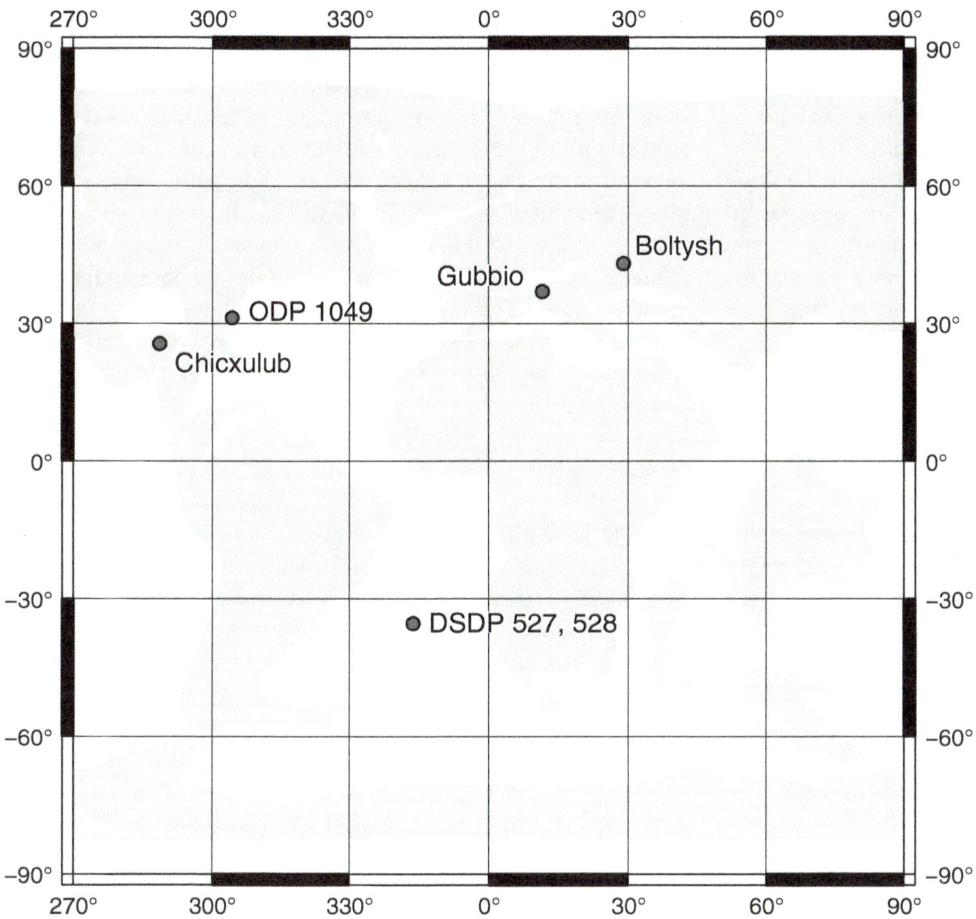

Figure 1. Cretaceous-Paleogene paleogeographic reconstruction of the Atlantic and Tethys Oceans showing the location of the Boltysh and Chicxulub impact craters and locations at which the Dan-C2 hyperthermal event has been reported. Paleoreconstruction of continental configuration was generated using Ocean Drilling Stratigraphic Network plate reconstruction (www.odsn.de/odsn/services/paleomap/paleomap.html). ODP—Ocean Drilling Program; DSDP—Deep-Sea Drilling Project.

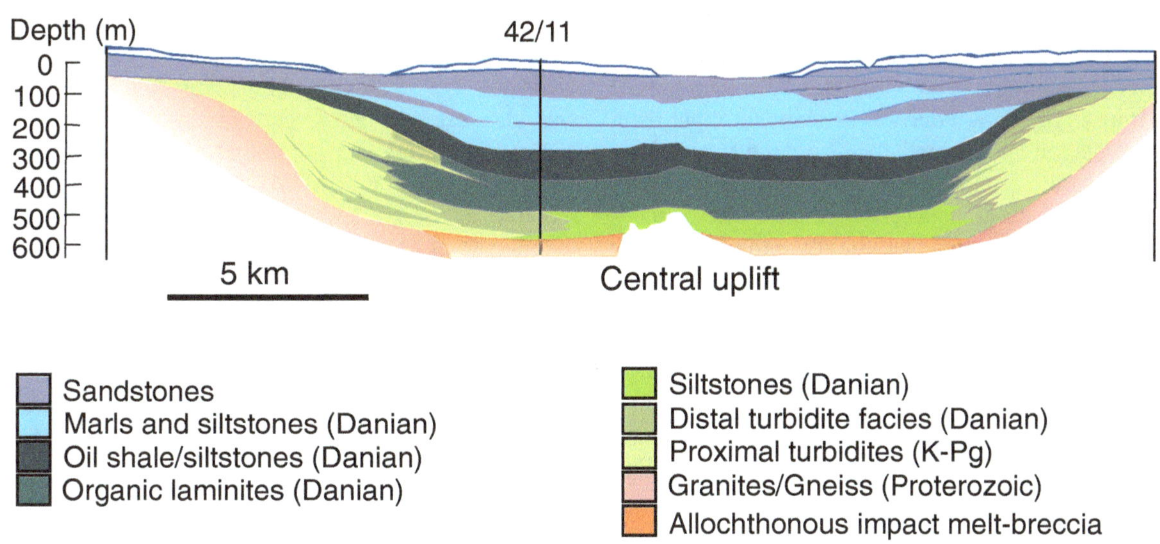

Figure 2. Schematic cross section through the Boltysh impact crater showing location of the 42/11 borehole and depths of sedimentary lithologies. K-Pg—Cretaceous-Paleogene.

zone NP1 of the early Danian (Valter and Plotnikova, 2003). A detailed study of the Cretaceous-Paleogene event recovery preserved in the Boltysh record is given in Jolley et al. (2013). Taken together, these lines of evidence indicate that the Boltysh impact occurred a few thousand years prior to the Cretaceous-Paleogene boundary event (Jolley et al., 2013; Gilmour et al., 2013).

Here, we examine in detail the geochemical and palynological evidence for an early Danian hyperthermal event recorded in the lacustrine sediments of the 2008 Boltysh drill core, and we place constraints on the timing and duration of the event and assess the role that the Deccan large igneous province may have played in its inception.

SAMPLES AND METHODS

Core Description

Previous descriptions of the Boltysh crater fill have mixed lithological and stratigraphic interpretations (Gurov et al., 2006). For the 596 m cored borehole (hole 42/11) drilled in 2008 to the west of the crater's central uplift at 48°54′N, 42°10′45″E, a lithological log was integrated with the gamma-ray log recorded in the open hole during the drilling of the borehole. The basal 20 m of the core comprise an allochthonous impact breccia containing glass, monomineralic (quartz, feldspar) fragments, and polymineralic fragments. These impact breccias are overlain by 395 m of Cenozoic crater-fill sediment. The oldest 90 m of these post-impact sedimentary rocks consist of fining-upward sequences of poorly sorted sands and sandy mudstones. These cyclic deposits are of varied thickness (<1–5 m) and are interpreted as having been deposited by turbidity currents. Within this turbidite interval, there are some units exhibiting contorted bedding, suggesting soft sediment deformation or slumping. Although the geometries of lake marginal deltas are not readily determined from the available 1970s Russian data (Fig. 2), turbidite fans are present around the crater margin. These turbidite fans back step toward the crater margin, implying a reduction in sediment input from ejecta blanket depletion and, potentially, decreased fluvial flow. Above 490 m, these turbidite cycles give way to increasingly fine-grained and finely laminated organic-rich shales. These shales are interpreted as having accumulated rapidly within the crater lake, preserving abundant plant macrofossil debris and, in the finest-grained intervals, submillimeter-scale lamination. Higher in the core (e.g., 315–300 m), intervals with abundant ostracods and gastropods in life assemblages are interbedded with gypsum laminations, which are interpreted as being deposited in a shallow, evaporative lake. The highest parts of the core are marked by the increased frequency of sandy strata and intervals of pale-colored marl deposition.

Palynology

To construct a high-resolution, long-term record of changes in both the crater and regional environment during the period of crater infill, samples were taken for palynology at 1 m intervals. These samples were crushed to small fragments, and a standard weight of sample was treated with 40% hydrofluoric acid to remove silicate minerals. The neutralized sample was then sieved using a 5 mm mesh nylon sieve, and the resultant residue was treated with dilute nitric acid for <5 min to remove any pyrite. Samples from intervals dominated by sapropellic kerogen were treated in Shultze's solution for up to 3 h. All neutralized residues were then pipetted onto glass coverslips using dilute polyvinyl acetate (PVA) and mounted on glass slides as permanent strew mounts using a two-part epoxy medium. Slides were examined under a transmitted light microscope, with counts of up to 250 specimens being undertaken where the abundance of the palynomorphs allowed.

Geochemistry

The lacustrine turbidites, sands, and organic-rich shales of the Boltysh core have variable carbonate contents with carbonate-rich intervals containing abundant ostracods and gastropods. This necessitated pre-analysis acid treatment of samples prior to determination of total organic carbon (TOC) contents and carbon stable isotope compositions for total organic matter. Following an initial rinsing of powdered sample with 0.1 *M* HCl, samples were treated with 1 *M* HCl for 24 h and then rinsed to neutrality with MilliQ water (18.2 MΩ). Each step involving a change of reagent or water was preceded by centrifugation (10 min at 1500*g*) to minimize the loss of fine material in suspension. TOC $\delta^{13}C$ values were determined using either a PDZ Europa elemental analyzer coupled to a PDZ Europa GEO 20-20 mass spectrometer or a Thermo Flash HT elemental analyzer coupled to a Thermo MAT 253 mass spectrometer (EA-irm-MS). The $\delta^{13}C$ values are reported relative to Vienna Peedee belemnite (VPDB), and analytical reproducibility for the within-run laboratory standard (sucrose/potassium nitrate), calibrated against IAEA CH-6 (sucrose), is –26.39‰ ± 0.09‰ (±1σ; $n = 43$); this standard was also used to check interinstrument consistency. In total, 453 samples were analyzed at ~1 m intervals from 596.4 m to 181.52 m. Organic carbon and nitrogen concentrations are reported as weight percent carbon and nitrogen, respectively (Table DR1[1]). The accuracy and precision of the TOC data obtained by elemental analysis were confirmed by comparison with Rock-Eval pyrolysis and TOC data obtained on 176 samples, $r^2(174) = 0.95$. The mean T_{max} from the Rock-Eval pyrolysis, 436 ± 12 °C (±3σ; $n = 176$), indicates that the organic matter in the Boltysh core samples is uniformly immature throughout the core section studied.

Statistical analyses of the isotopic data were performed using Aabel software (Gigawiz Ltd.), and correlations are reported as

[1]GSA Data Repository item 2014280, Table DR1, details, is available online at www.geosociety.org/pubs/ft2014.htm, or on request from editing@geosociety.org or Documents Secretary, GSA, P.O. Box 9140, Boulder, CO 80301, USA.

R^2 (degrees of freedom) and unpaired *t*-tests as *t* (degrees of freedom), *p* (two tailed).

RESULTS AND DISCUSSION

Carbon Isotope Stratigraphy

The $\delta^{13}C$ curve, TOC concentration, N concentration, and C/N ratio for the Boltysh 42/11 core are shown in Figure 3. Carbon contents are initially low (<0.5%) below 570 m core depth before becoming progressively more organic rich, with a mean carbon concentration of ~7% throughout the rest of the section. The $\delta^{13}C$ values are consistent with a predominantly freshwater algal and higher plant input to the lake; the maximum $\delta^{13}C$ is −19.4‰, with a minimum value of −37.1‰ and a mean of −29.0‰. C/N ratios vary significantly through the core, with both low C/N (~4–10) typical of phytoplankton inputs and C/N ratios of >20, typical of vascular plants (Meyers, 2003). The $\delta^{13}C$ curve records a significant carbon isotope excursion averaging −3‰ in magnitude between 537.4 and 271 m. The $\delta^{13}C$ records of TOC have proven to be reliable indicators of carbon cycle changes at the Paleocene-Eocene thermal maximum, for example, at Polecat Bench and Honeycombs, Wyoming (Magioncalda et al., 2004; Yans et al., 2006), and at sections in the south-central Pyrenees, Spain (Domingo et al., 2009), particularly when combined with other proxies. However, caution must be applied when interpreting bulk lacustrine $\delta^{13}C$ values, because they can be susceptible to bias due to selective preservation of different organic constituents, maturation-induced changes in $\delta^{13}C$, or the input of allochthonous material. It is not possible to completely exclude possible modification of the Boltysh $\delta^{13}C$ record due to these effects. However, organic matter in the 42/11 sediment samples is predominantly a uniformly dispersed and well-mixed amorphous sapropellic kerogen that is immature throughout the section studied, and thus the likelihood of selective preservation and maturation effects is minimal. No significant correlation exists between the organic carbon content of the samples and $\delta^{13}C$ (pre–carbon isotope excursion: $R^2_{[48]} = 0.36$, $p < 0.001$; carbon isotope

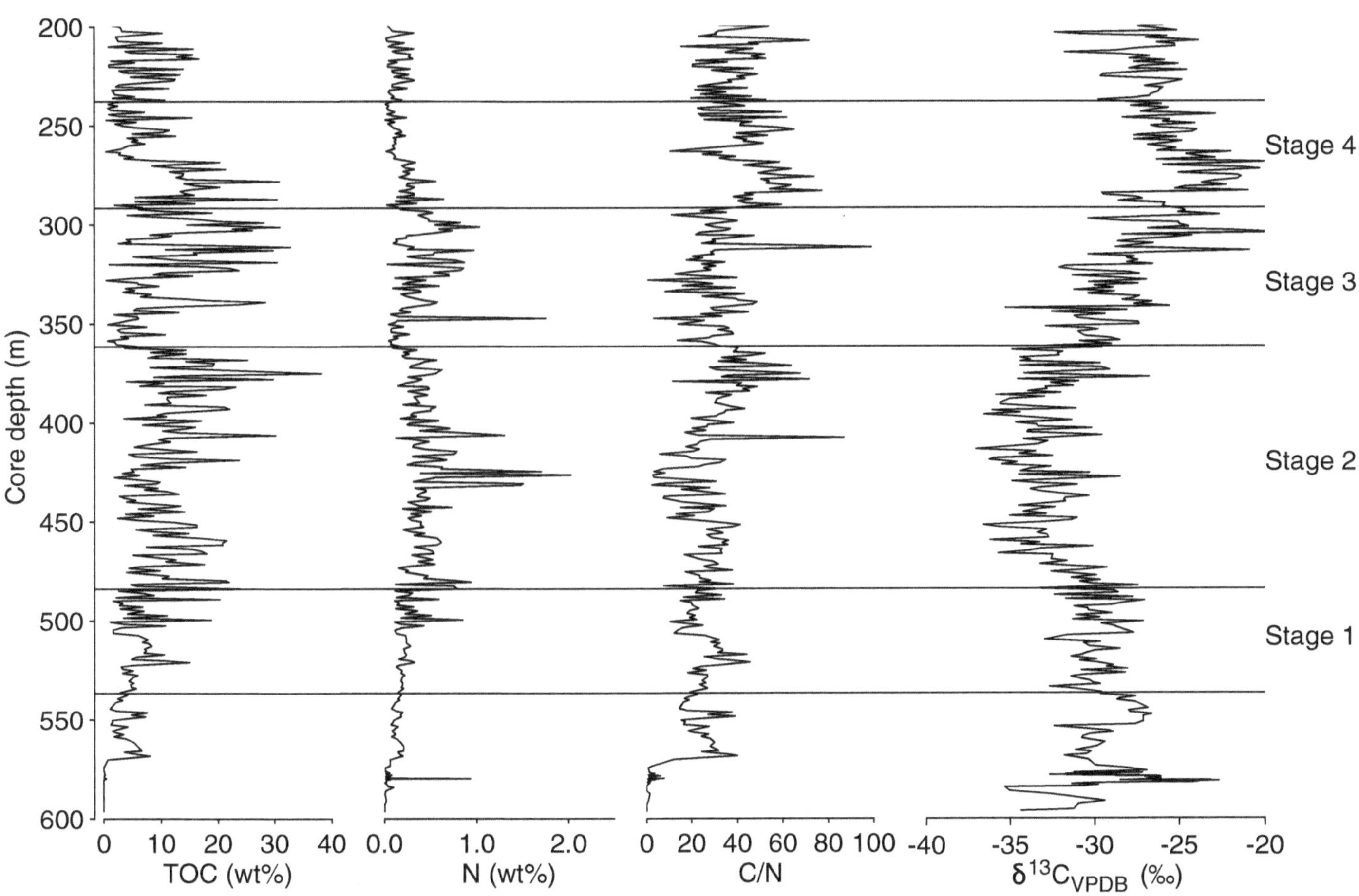

Figure 3. Total organic carbon (TOC, wt%), total nitrogen (wt%), C/N ratio, and $\delta^{13}C_{VPDB}$ (VPDB—Vienna Peedee belemnite) of organic matter through the Boltysh 42/11 core. By applying the same criteria as used by Cohen et al. (2007), we have identified a number of distinct stages in the Boltysh carbon isotope record, revealing broadly similar patterns of initiation and recovery to those for the Paleocene-Eocene thermal maximum and Toarcian hyperthermal.

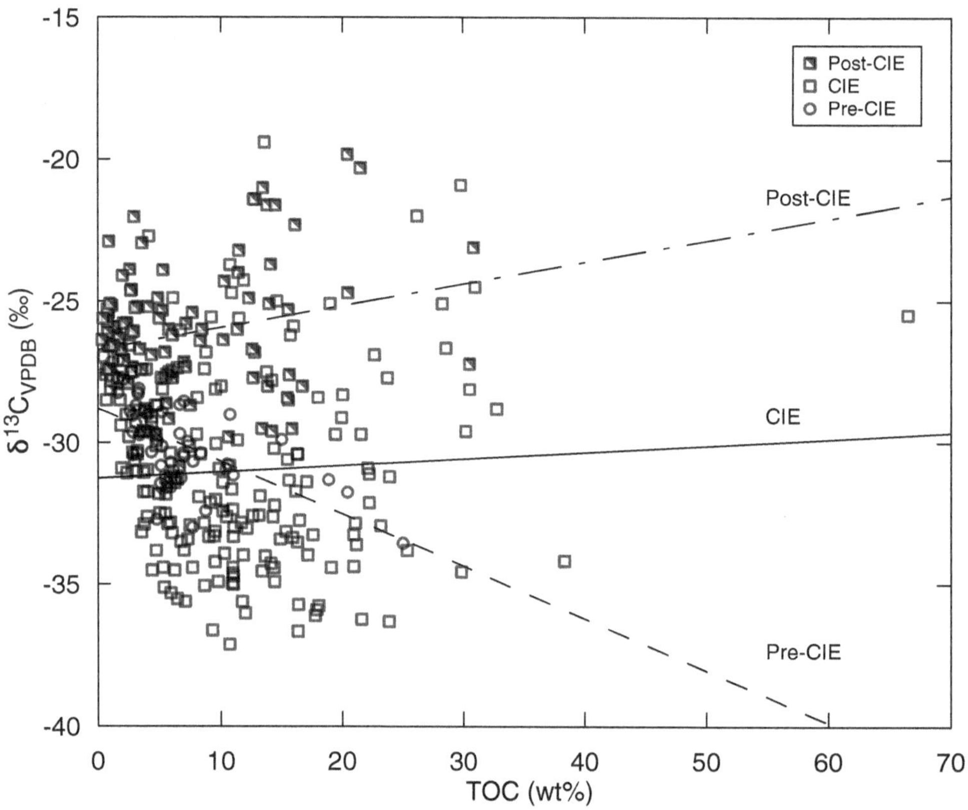

Figure 4. $\delta^{13}C_{VPDB}$ (VPDB—Vienna Peedee belemnite) vs. total organic carbon (TOC, wt%) for organic matter from the Boltysh core. Pre– and post–carbon isotope excursion (CIE) values have been separated from carbon isotope excursion values.

excursion: $R^2_{[109]} = 0.042$, $p = 0.031$; post–carbon isotope excursion: $R^2_{[120]} = 0.029$, $p = 0.06$; Fig. 4). This indicates that $\delta^{13}C$ values are not influenced by the amount of organic carbon preserved and that $\delta^{13}C$ values reflect the main vegetation inputs to the sediments, which in turn would reflect atmospheric CO_2 values.

The fine structure and variability observed in the Boltysh 42/11 $\delta^{13}C$ record are more analogous to the high-resolution $\delta^{13}C$ records obtained from Quaternary lake sediments (e.g., Talbot and Johannessen, 1992), where they reflect lacustrine cyclicity, than those typically obtained from the marine record of carbon isotope excursions, where lower sedimentation rates tend to smooth the isotopic record. Figure 3 shows that there is a clear carbon isotope excursion within the 42/11 core, and we have identified four distinct stages, which appear to mirror other carbon isotope excursion records. Mean $\delta^{13}C$ values for each of these stages are given in Table 1. The $\delta^{13}C$ record below 478.5 m shows some variability, with a slight decrease in mean $\delta^{13}C$ in the interval 537.4–479.6 m (stage 1; Fig. 3) compared with the underlying interval. Above 478.5 m, there is a clear carbon isotope excursion within the 42/11 core, and in the interval between 478.5 and 360.4 m, there are two abrupt shifts toward more negative $\delta^{13}C$ values (stage 2; Fig. 3). Subsequently, $\delta^{13}C$ values steadily become more positive, reaching stage 1 values at around 291 m (stage 3; Fig. 3). From 290 to 271 m, there is a distinct positive excursion in $\delta^{13}C$ values (stage 4; Fig. 3), above which values return to stage 1 levels in the overlying interval.

TABLE 1. DIFFERENCE BETWEEN MEAN $\delta^{13}C$ VALUES ACROSS THE CARBON ISOTOPE EXCURSION IN THE 42/11 CORE

Stage	Description*	Interval (m)	Cycles†	Duration§ (k.y.)	Mean % C	Mean $\delta^{13}C_{TOC}$ (‰)	Correlation# (R^2)	$\Delta\delta^{13}C_{TOC}$ to previous stage (‰)
Post CIE	Postexcursion	237.00–181.52		–	5.1 ± 4.7	–27.1 ± 1.7	0.089	–1.9 ($t_{104} = 5.2$, $p < 0.001$)
Stage 4	Positive excursion	290.00–238.00	~2	42/80	15.0 ± 7.1	–25.2 ± 2.5	0.069	+2.8 ($t_{118} = –5.9$, $p < 0.001$)
Stage 3	Recovery	360.40–289.00	~2.5	52.5/100	11.9 ± 10.9	–28.0 ± 2.9	0.073	+4.8 ($t_{177} = –13.1$, $p < 0.001$)
Stage 2	Negative excursion	482.45–361.32	~4.1	84/160	11.6 ± 6.7	–32.9 ± 2.1	0.042	–2.9 ($t_{157} = 8.9$, $p < 0.001$)
Stage 1	Cyclic variation	537.40–484.00		–	7.0 ± 5.5	–30.0 ± 1.5	0.360	–1.0 ($t_{71} = 2.5$, $p = 0.014$)
Pre-CIE	Pre-excursion	568.60–538.30		–	3.6 ± 2.2	–29.0 ± 1.7	0.134	

Note: TOC—total organic carbon.
*Stage descriptions follow those of Cohen et al. (2007).
†Estimated number of cycles from spectral analysis.
§Duration based on estimate of number of 21 k.y. precessional or 41 k.y. obliquity cycles.
#Correlation R^2 is linear regression coefficient for $\delta^{13}C$ on wt% C for each interval.

Palynofloras of the Boltysh Core

Palynofloras recovered from the borehole 42/11 samples are consistent with other terrestrial early Danian successions (see also Cavagnetto and Gaudant, 2000). These rich assemblages are dominated by pollen and spores, but they also include intervals of highly abundant algae. Monospecific, highly abundant occurrences of the chlorophycean algae *Botryococcus braunii* are recorded through the initial phase of turbiditic crater-fill sedimentation and continue to the base of the negative carbon isotope excursion. Following almost total absence of algae from the negative carbon isotope excursion interval, *B. braunii* is recorded in high frequencies from the uppermost 5 m of the negative carbon isotope excursion to the top of the cored section in the post–carbon isotope excursion Danian interval (Fig. 5).

Variations within the higher plant palynofloras appear to reflect changes in both crater marginal and regional vegetation, providing a record of early Danian environmental change. The cored interval from the post–Cretaceous-Paleogene boundary section to the inception of the carbon isotope excursion yielded higher plant palynofloras dominated by abundant Pteridophyta (ferns) and Cupressaceae (swamp cypresses). Angiosperm pollen is abundant, with equal frequencies of archaic *Normapolles* and post-*Normapolles* pollen (Fig. 5). *Normapolles* comprise an important part of many Late Cretaceous angiosperm assemblages from midlatitudes in the Northern Hemisphere within a region extending from West Siberia across Europe to eastern North America (Batten, 1981). The abundance of ferns and swamp cypresses, together with the abundant occurrences of platanaceous and fagaceous pollen, indicates that warm temperate regional vegetation dominated the interval from the post–Cretaceous-Paleogene boundary to the inception of the carbon isotope excursion (Fig. 5). However, this palynoflora shows a transition to archaic *Normapolles* angiosperm pollen dominance at the carbon isotope excursion inception, a transition that continues to the midpoint of the carbon isotope excursion (Fig. 5). The frequency of *Normapolles* pollen is of significance to climate change interpretation; primarily, it indicates the continued influence of the *Normapolles* phytogeographic province (Herngreen et al., 1996) from the Late Cretaceous into the Danian. Secondly, it signifies an increasingly dry and warming climate.

The virtual disappearance of crater lake algae (mainly *B. braunii*) at the inception of the carbon isotope excursion and the increasing dominance of the thermophylic *Normapolles* group indicate restriction of the lake within the crater in an increasingly warm, dry climate. Co-occurring with the dominant *Normapolles* in the carbon isotope excursion and early recovery phase are common occurrences of Juglandaceous pollen (*Platycaryapollenites*, particularly *P. anticyclus*). These pollen are derived from plants that colonized mires and abandoned watercourses (Wing and Hickey, 1984), supporting the interpretation that a wetland zone was established on areas of emergent crater floor as the crater lake reduced in size. Climatic drying is also reflected in evidence for the regional vegetation. For example, Pinaceae pollen (almost wholly *Pityosporites* haploxylonoid types) occurs commonly throughout the carbon isotope excursion and is interpreted here as being derived from windblown input from dry substrate regions distal to the crater.

The dominance of *Normapolles* pollen continues from the carbon isotope excursion into the recovery stage, before declining sharply in the upper part of the recovery (stage 3; Fig. 5). This simplification masks a major environmental change that immediately predates the end of the carbon isotope excursion. Alongside the first reappearance of *B. braunii*, there are occurrences of common aquatic marginal monocot pollen derived from *Sparganium*- and *Restio*-type plants. Taken together with the influx of chlorophycean algae, the monocot pollen can be interpreted as resulting from reflooding of the outer crater zone. The dominant Restionaceae species, *Milfordia hungarica*, has been associated with brackish marshes (Lenz and Riegel, 2001), which, together with the brackish water tolerance of *B. braunii*, indicates that the lake was evaporitic during the early part of the recovery phase. This interpretation is supported by the occurrences of the bivalve *Glycimeris* and gastropod life assemblages implying shallow freshwater to weakly brackish water, and the abundance of gypsum in submillimeter laminations in the sediments.

Post–carbon isotope excursion changes in the regional and crater vegetation are marked by increased abundances of Cupressaceae and Pinaceae pollen within the early recovery stage, marking a return to temperate climatic conditions. This cooling trend is further marked by influxes of Fagaceae and Myricaceae pollen, which increase to dominance in the late recovery stage. The increase to dominance of these temperate taxa coincides with a compositional change in the crater lake margin flora from brackish to freshwater. Freshwater algae continue to occur abundantly, including abundant *Pediastrum bifidites*.

Comparison with the Marine Dan-C2 Record

The ecological succession evident in the palynological record strongly suggests that the Boltysh 42/11 core represents an expanded and high-resolution record predating the Cretaceous-Paleogene boundary by a few thousand years and extending through the lowermost Danian with no significant breaks. This is supported by the sedimentological and paleontological evidence, which suggests that the lake did not dry out or shallow until above 300 m core depth. Precise temporal correlation with the Dan-C2 marine record is not straightforward, in part due to age uncertainties in the latter (e.g., Renne et al., 2013). The Dan-C2 event was first observed in the Atlantic (Quillévéré et al., 2008) and subsequently in western Tethys (Coccioni et al., 2010; Punekar et al., this volume) and is characterized by a pair of negative carbon isotope excursions of around −1.5‰ in marine carbonates, indicating that the event was at least supraregional with a carbon cycle perturbation. However, Westerhold et al. (2011) questioned whether this was a global event, since it was not observed in a high-resolution carbon isotope record from the Pacific, though at

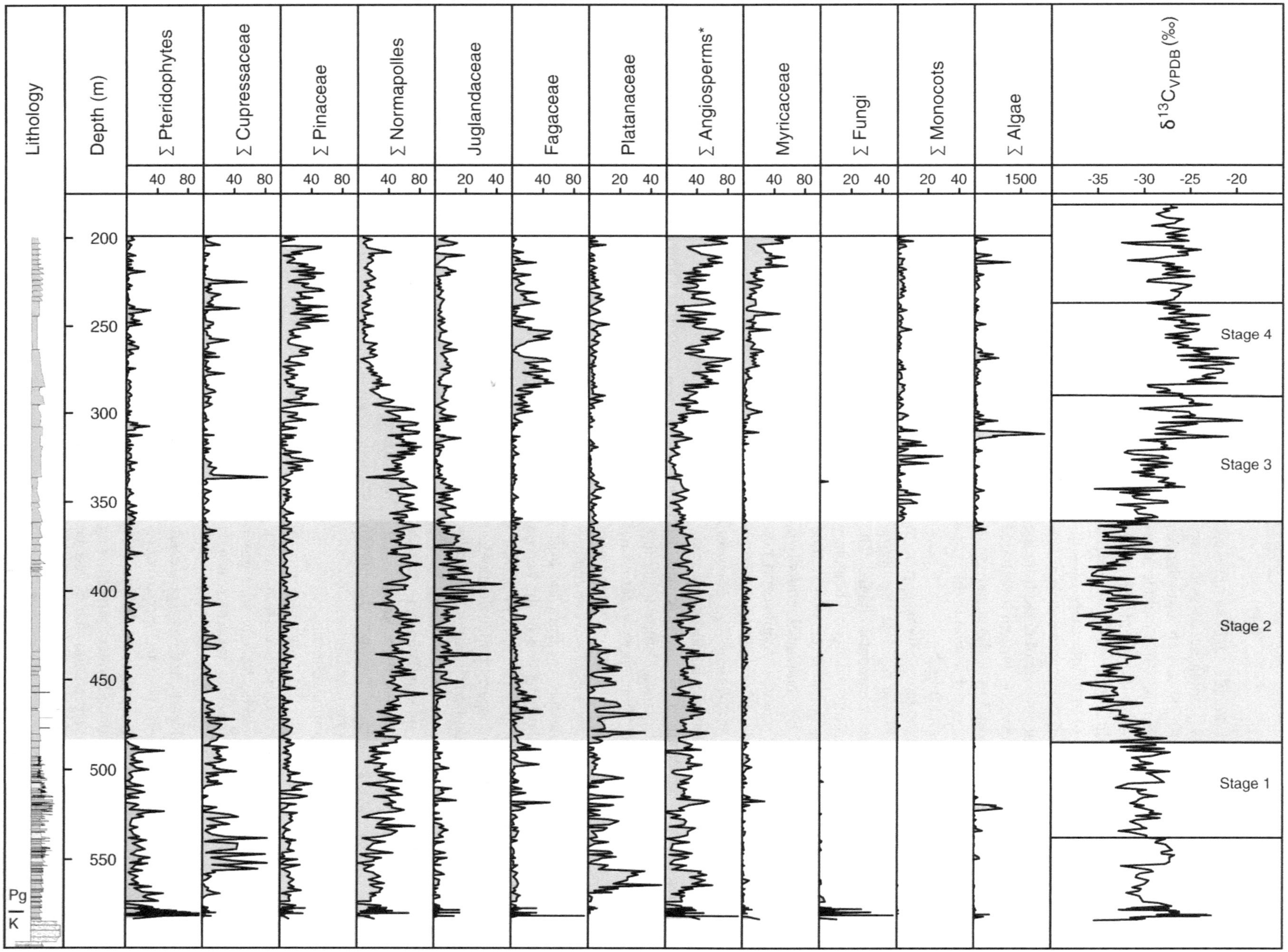

Figure 5. Palynofloral botanical affinity distribution data and $\delta^{13}C_{VPDB}$ (VPDB—Vienna Peedee belemnite) values for total organic carbon in the 42/11 core. The palynological data are summed by botanical affinity. *Selected post-*Normapolles* angiosperm families are shown separately. The shaded area indicates the interval of the Boltysh carbon isotope excursion that we interpret as corresponding to the Dan-C2 carbon isotope excursion recorded in marine carbonates. K—Cretaceous; Pg—Paleogene.

this site, the basal Danian is composed of a fining-upward sedimentary sequence, making a complete Cretaceous-Paleogene boundary record unlikely. The marine record places the Dan-C2 at the top of magnetozone C29r, within planktonic foraminiferal zone P1a, postdating the Cretaceous-Paleogene boundary by around 250 k.y. (Quillévéré et al., 2008). The onset of the carbon isotope excursion in the 42/11 drill core occurs at ~479.6 m, within the lowermost crater lake turbidite sequence and around 100 m above the fern spike that marks the Cretaceous-Paleogene boundary (Jolley et al., 2013).

The duration of the Dan-C2 carbon isotope excursion in the marine realm has been estimated as ~100 k.y. based on extrapolation of sedimentation rates between the independently dated Cretaceous-Paleogene boundary and overlying C29r/C29n magnetozone boundary (Quillévéré et al., 2008). The relatively short duration inferred for the event led Quillévéré et al. (2008) to suggest a possible astronomical control on the $\delta^{13}C$ changes. Indeed, a direct astronomical control on the pacing and structure of carbon isotope excursions is a previously documented feature of other hyperthermal events, notably the Toarcian and Paleocene-Eocene thermal maximum (e.g., Cohen et al., 2007). Spectral analysis of linearly interpolated raw $\delta^{13}C_{TOC}$ data over the entire excursion interval between 237 and 483 m (stages 2–4) reveals the presence of statistically significant cycles with a wavelength of ~30 m (Fig. 6A). The inferred ~30 m cyclicity is also apparent in the raw data and is highlighted using a 9.18 m moving average (thick black line, Fig. 6B). Cycle counting over the analyzed interval using both the raw and filtered

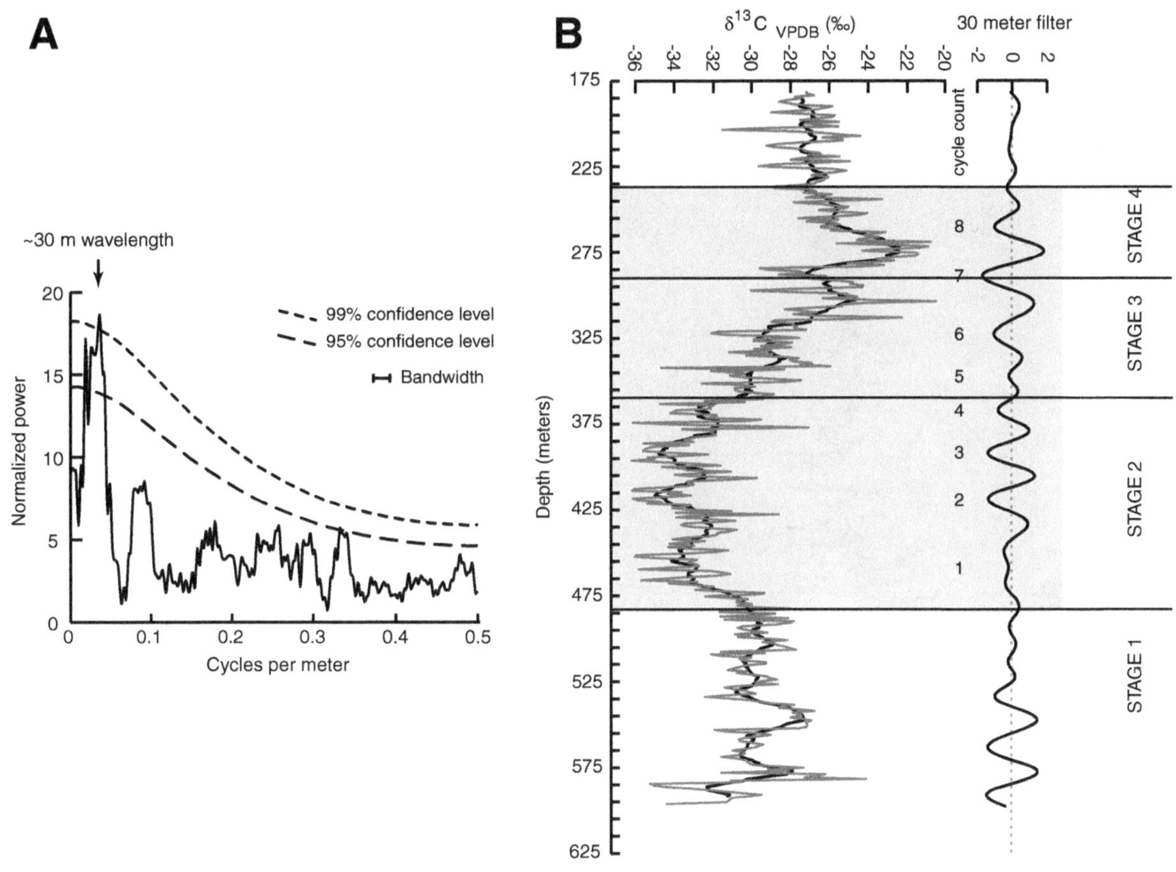

Figure 6. (A) Multitaper spectral analysis of linearly interpolated raw $\delta^{13}C$ data for total organic carbon over the entire excursion interval between 237 and 483 m (shaded region in B). This analysis highlights the presence of statistically significant cycles with a wavelength of ~30 m. Multitaper spectral analysis was conducted using five tapers. Data were detrended by removing the third-order polynomial fit prior to analysis. Significance levels were calculated based on a red noise background fitted to the spectrum (e.g., Weedon, 2003). (B) Smoothed and filtered data highlighting the ~30 m cyclicity over the analyzed interval. Filtering was conducted using a Gaussian filter centered at a frequency of 0.033 (0.016 bandwidth). The inferred ~30 m cyclicity is also apparent in the raw data and is highlighted using a 9.18 m moving average (thick black line). Cycle counting over the analyzed interval using both the raw and filtered data suggests the presence of around 8.5 ~30 m cycles. Note, however, the mismatch between cycles apparent in the raw data and the ~30 m filter output between 480 and 430 m. Assuming an astronomical precession (21 k.y. period) or obliquity (40 k.y. period) origin for the cycles, the duration of the entire excursion (i.e., stages 2–4) is putatively constrained to either ~178 k.y. or 340 k.y.

data suggests the presence of around 8.5 ~30 m cycles. Note, however, the mismatch between cycles apparent in the raw data and the ~30 m filter output between 480 and 430 m (Fig. 6B). Assuming an astronomical precession (21 k.y. period) or obliquity (40 k.y. period) origin for the cycles, the duration of the entire excursion (i.e., stages 2–4; Figs. 5 and 6B) is putatively constrained to either ~178 k.y. or 340 k.y., and the duration of the main phase of the negative excursion (i.e., stage 2; Figs. 5 and 6B) can be estimated as ~84 k.y. or 160 k.y., respectively. These estimates are comparable with estimates from the marine Dan-C2 records, which apparently lack the recovery and positive excursion stages observed at Boltysh and may also be subject to dissolution (Fig. 7). The marine-continental link provided by the Boltysh carbon isotope record together with a palynofloral record that reflects changing biomes caused by a rapidly warming climate are strongly indicative of a hyperthermal event of similar duration and contemporaneous with the Dan-C2 event recorded in marine sediments in Tethys and the Atlantic and suggests that that there were profound environmental changes occurring on a global scale less than 300 k.y. after the Cretaceous-Paleogene boundary.

Carbon Cycle Changes Associated with the Dan-C2 Hyptherthermal

Quillévéré et al. (2008) considered that the Dan-C2 event was of lesser magnitude, shorter duration, and lacked evidence of a recovery phase when compared with the Paleocene-Eocene thermal maximum. They therefore considered that the event was not associated with a large input of CO_2 from buried carbon reservoirs. However, this view is brought into question by the Boltysh isotope record, which shows similar patterns of initiation and recovery to those for the Paleocene-Eocene thermal maximum and Toarcian carbon isotope excursions (Table 1; Figs. 5 and 7). The magnitude of the Dan-C2 carbon isotope excursion in the Boltysh lacustrine record is around −3‰, about twice the size of that in the marine carbonate record. This parallels the relationship observed between terrestrial and deep-sea $\delta^{13}C$ records at the Paleocene-Eocene thermal maximum, where the carbon isotope excursion in terrestrial records is generally between −3‰ and −6‰ (e.g., Koch et al., 1992; Bowen et al., 2001; Magioncalda et al., 2004), compared to a marine carbonate carbon isotope excursion of ~−3‰ (McInerney and Wing, 2011).

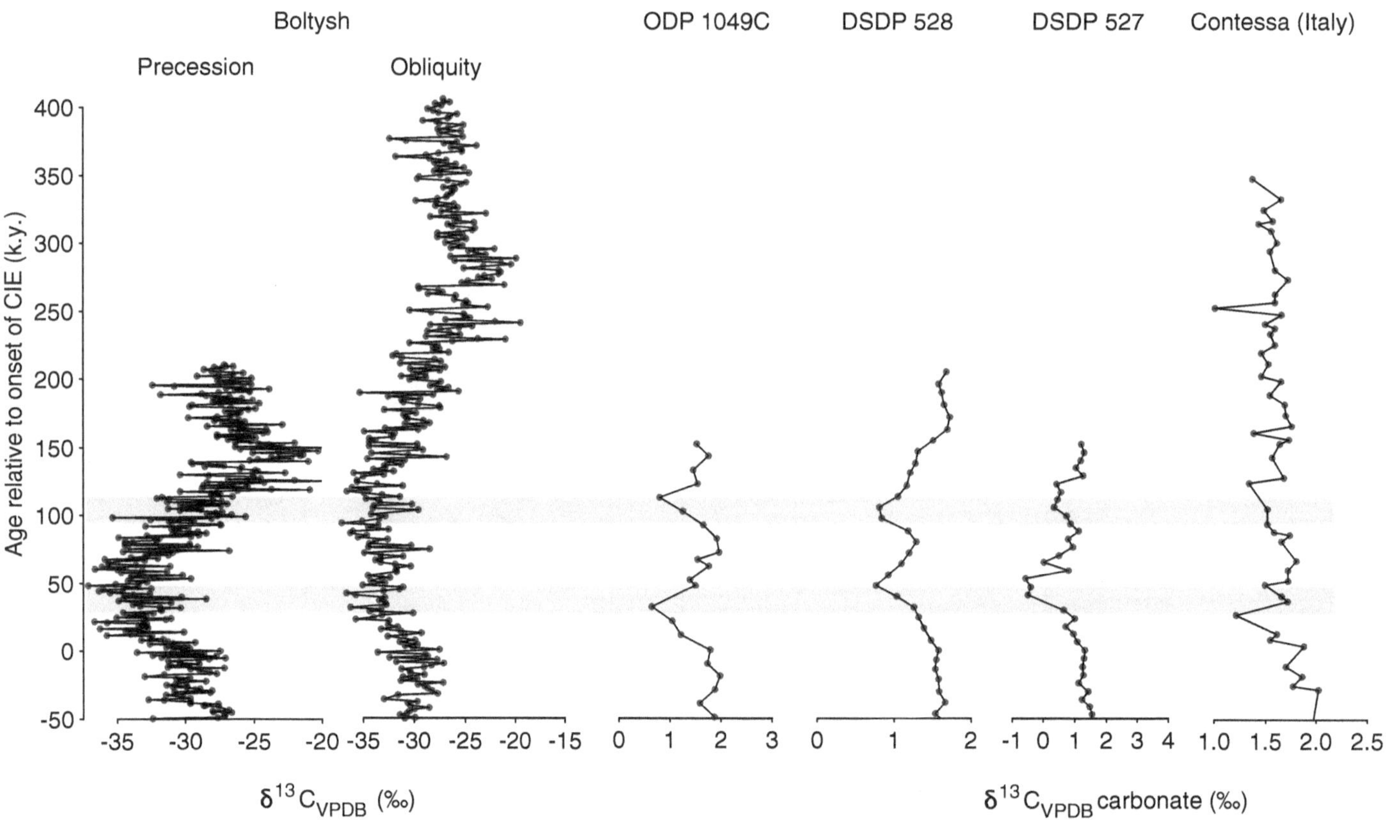

Figure 7. Comparison of the relative carbon isotope excursion (CIE) time scales of the Boltysh terrestrial record (based on spectral analysis of the $\delta^{13}C$ curve with precessional and obliquity drivers of cyclicity) and marine records of the Dan-C2 event (based on extrapolation of sedimentation rates between the independently dated Cretaceous-Paleogene boundary and overlying C29r-C29n magnetozone boundary). Shaded areas shows potential coincidence of marine record double carbon isotope excursion with cyclicity documented in the Boltysh record. ODP—Ocean Drilling Program; DSDP—Deep-Sea Drilling Project; VPDB—Vienna Peedee belemnite.

McInerney and Wing (2011) used mass balance estimates of the mass of carbon released from different sources over a range of carbon isotope excursion values to evaluate the effect of different sources of light carbon on the Paleocene-Eocene thermal maximum carbon isotope excursion. Using the same assumptions for the initial mass and carbon isotope composition of the early Paleocene surface reservoir of 5×10^{16} kg and −2.5‰, a mean terrestrial carbon isotope excursion for the Dan-C2 of −3‰ corresponds to masses of released ^{13}C-depleted carbon from 3×10^{15} kg of methane hydrate with a $\delta^{13}C$ of −60‰ source or 1×10^{16} kg from wildfires or epicontinental seas with $\delta^{13}C$ values of around −22‰. This suggests that the Dan-C2 carbon isotope excursion represents a massive input of ^{13}C-depleted carbon into the ocean and atmosphere on a scale comparable to the Paleocene-Eocene thermal maximum.

While the global nature and large carbon release associated with major hyperthermal events such as the Paleocene-Eocene thermal maximum and Toarcian hyperthermal are accepted, a central debate concerns the source and mass of ^{13}C-depleted carbon and whether there are common driving mechanisms. The nonmarine record at Boltysh provides strong evidence that the early Danian transient warming and carbon cycle perturbations were also global, requiring mechanisms that would operate on a global scale when assessing potential causes. Three potential mechanisms are commonly proposed: large-scale release of ^{13}C-depleted carbon from methane hydrate dissociation, the release of thermogenic methane from organic-rich sediments (e.g., Svensen et al., 2004), and biomass burning (Finkelstein et al., 2006). A combination of these mechanisms can also not be ruled out; however, given the temporal coincidence of the Dan-C2 with a major phase of Deccan volcanism, we next assess whether viable mechanisms exist for a direct link between the two.

Evaluating the Relationship between the Dan-C2 and Deccan Large Igneous Province

Temporal links between hyperthermal events and large igneous provinces exist in the coincidence of the Toarcian and Paleocene-Eocene thermal maximum events with the Karoo-Ferrar and North Atlantic large igneous provinces, respectively. Associations become more persuasive if we include other events with lower-fidelity carbon isotope excursions, such as the coincidence of the Triassic-Jurassic carbon isotope excursion and the Central Atlantic magmatic province around 200 Ma, and the Late Permian–Triassic carbon isotope excursion with the Siberian Traps large igneous province around 250 Ma (see review by Wignall, 2001). A potential link between the Dan-C2 hyperthermal and the Deccan large igneous province was suggested by Coccioni et al. (2010) based on magnetostratigraphic constraints that placed the Dan-C2 event at the top of chron 29r.

Recent work has begun to unravel the evolution of the Deccan large igneous province at a temporal resolution sufficient to separate the main Deccan eruptions into multiple eruptive phases (Chenet et al., 2008, 2009). Significant crustal extension preceded or was contemporaneous with the Deccan emplacement and began along the north Indian margin and propagated southward during chrons 29–27, culminating in the separation of the Seychelles microcontinent and onset of ocean lithosphere development between chrons 28 and 27 (Collier et al., 2008; Minshull et al., 2008). Plume-lithosphere interaction at the Cambay-Narmada rift zones of NW peninsular India resulted in the development of a series of offshore gulfs and basins where several kilometers of Maastrichtian–Danian sediments accumulated. Another major volcanic center, also associated with this early stage rifting, is located on the Saurashtra Platform (offshore northwest India), where it is intruded into Mesozoic rift and Late Cretaceous marine sediments (Carmichael et al., 2009; Corfield et al., 2010). In the Mumbai area, there is onshore evidence for the interaction between Deccan lavas and sediments in the Mumbai area of western India, where interbedded marly sandstones and organic-rich shales occur as intertrappean sequences (Cripps et al., 2005).

The marine records of the Dan-C2 event place the carbon isotope excursion entirely within the upper part of chron 29r (Quillévéré et al., 2008; Coccioni et al., 2012). However, in contrast to the marine records, the Boltysh carbon isotope excursion record shows both a clear recovery stage and positive excursion, reminiscent of the Toarcian and Paleocene-Eocene thermal maximum carbon isotope excursions, and it suggests a longer time scale for the event, crossing the chron 29r-29n boundary (Fig. 7). This would place the Dan-C2 coincident with an acceleration of volcanic output during the generation of the volumetrically important Mahabaleshwar Formation, which accounts for ~5% of the entire Deccan large igneous province output (Chenet et al., 2009; Mussard et al., this volume).

Deccan Volcanism as a Source of ^{13}C-Depleted Thermogenic Methane

Svensen et al. (2004) suggested that the climate impact of a large igneous province could be considerable if the melt is partly or completely emplaced into carbon-rich sedimentary successions. The production of thermogenic methane from sedimentary organic matter remains a poorly constrained process, although the injection of magma into organic-rich sediments resulting in the explosive release of methane is a proposed mechanism for both the Paleocene-Eocene thermal maximum and Toarcian carbon isotope excursions (McInerney and Wing, 2011; Cohen et al., 2007). Experimental evidence indicates that the production of thermogenic methane results in carbon isotopic fractionation, with the produced methane being more ^{13}C-depleted than the starting sedimentary organic matter. The fractionation is dependent on organic matter type and extent of gas generation, but it has been measured at between −1.4‰ and −17.5‰ in laboratory experiments (Clayton, 1991). Using mass balance calculations, we estimate that a mean carbon isotope excursion of −3‰ would correspond to a mass of released ^{13}C-depleted carbon at the Dan-C2 of between 5×10^{15} and 1×10^{16} kg C, for a thermogenic methane source with an isotopic composition of between −25‰ and −42‰.

Aarnes et al. (2010) attempted to quantify gas formation by the cracking of organic material and mineral devolatilization during contact metamorphism through numerical modeling and showed that in shales with a TOC content of >5 wt%, methane was the dominant volatile generated through organic cracking. Applying this model to the Dan-C2 carbon isotope excursion, it is possible to estimate the extent of contact metamorphism required to produce the calculated amount of carbon released. For example, a cumulative intrusion thickness of 100 m into a 5 wt% TOC shale would require an intruded area of 150,000 km^2 to produce 5.10^{15} kg C as CH_4. This equates to a single sill with a radius of ~200 km. This is larger than the extent of any likely interaction between Deccan lavas and onshore sedimentary basins. However, given that large rift-related, hydrocarbon-rich sedimentary basins do occur offshore of the Indian continent (Calvès et al., 2011), and that Deccan volcanism clearly interacted with these offshore deposits, it is entirely conceivable that the introduction of multiple large sills and feeder dike systems, together with a regionally elevated geothermal gradient, would affect these organic-rich successions and result in the release of substantial amounts of CH_4 into the atmosphere over a short time period.

The sudden injection of ^{13}C-depleted thermogenic methane is an attractive hypothesis for the initiation of hyperthermal events. However, much less is known about how hyperthermal events persist after their initial inception and about what drives their pacing and sustainability. This is largely due to the difficulties in quantifying the magnitude of the carbon isotope excursion and atmospheric pCO_2 levels. Further work on the Boltysh record will be required to evaluate the rates of environmental change and climate response signal to ascertain whether the inception of the Dan-C2 represents a sudden or gradual change.

CONCLUSIONS

Sediments filling the 24-km-diameter Boltysh impact crater in Ukraine record a negative carbon isotope excursion in the immediate post–Cretaceous-Paleogene period of Earth history. The −4‰ negative carbon isotope excursion appears around 100 m above the fern spike that marks the Cretaceous-Paleogene boundary. The carbon isotope excursion consists of a rapid decrease followed by slower recovery to pre-excursion values and a small positive excursion at the end, around 210 m above the inception. The form of the carbon isotope excursion maps very closely to the form of other high-resolution complete carbon isotope excursions in the Toarcian and at the Paleocene-Eocene thermal maximum, leading us to conclude that it reflects an early Danian hyperthermal event of similar proportion to the Paleocene-Eocene thermal maximum and Toarcian. The unusual occurrence of the Boltysh impact crater has created an extremely high-resolution record of the carbon isotope excursion, on a par with Quaternary lake deposits, and it thus represents a unique opportunity to study sudden climate change in an ice-free Earth system.

The Dan-C2 event occurred toward the top of chron 29r and probably outlasted the boundary into chron 29r. As such, it coincides with a major pulse of onshore Deccan eruption, and the development of large, offshore rift-related sedimentary basins. Evidence that the latter phase of Deccan eruptions was probably the only one to interact with carbon-rich sediments allowed us to estimate a potential outcome, and we conclude that it is possible that sufficient methane may have been thermogenically produced from sedimentary carbon to account for the Dan-C2 carbon isotope excursion.

ACKNOWLEDGMENTS

The drilling of Boltysh crater was funded by a grant from the Natural Environmental Research Council (NERC) (NE/D005043/1). Kemp thanks NERC for the award of an advanced fellowship. We would like to thank Eugene Gurov for provoking our interest in the Boltysh impact crater and Mr. Bratchuk and Mr. Panachenko of the Geologiska Drilling Company, Ukraine, for their help in recovering high-quality and almost complete core.

REFERENCES CITED

Aarnes, I., Svensen, H., Connolly, J.A.D., and Podladchikov, Y.Y., 2010, How contact metamorphism can trigger global climate changes: Modeling gas generation around igneous sills in sedimentary basins: Geochimica et Cosmochimica Acta, v. 74, no. 24, p. 7179–7195, doi:10.1016/j.gca.2010.09.011.

Alvarez, L.W., Alvarez, W., Asaro, F., and Michel, H.V., 1980, Extraterrestrial cause for the Cretaceous-Tertiary extinction: Science, v. 208, p. 1095–1108, doi:10.1126/science.208.4448.1095.

Archibald, J.D., Clemens, W.A., Padian, K., Rowe, T., Macleod, N., Barrett, P.M., Gale, A., Holroyd, P., Sues, H.D., and Arens, N.C., 2010, Cretaceous extinctions: Multiple causes: Science, v. 328, p. 973, doi:10.1126/science.328.5981.973-a.

Batten, D.J., 1981, Stratigraphic, palaeogeographic and evolutionary significance of late Cretaceous and Early Tertiary Normapolles pollen: Review of Palaeobotany and Palynology, v. 35, no. 2–4, p. 125–137, doi:10.1016/0034-6667(81)90104-4.

Bornemann, A., Schulte, P., Sprong, J., Steurbaut, E., Youssef, M., and Speijer, R.P., 2009, Latest Danian carbon isotope anomaly and associated environmental change in the southern Tethys (Nile Basin, Egypt): Journal of the Geological Society, v. 166, p. 1135–1142, doi:10.1144/0016-76492008-104.

Bowen, G.J., Koch, P., Gingerich, P.D., Norris, R.D., Bains, S., and Corfield, R.M., 2001, Refined isotope stratigraphy across the continental Paleocene-Eocene boundary on Polecat Bench in the northern Bighorn Basin, *in* Gingerich, P.D., ed., Paleocene-Eocene Stratigraphy and Biotic Change in the Bighorn and Clarks Fork Basins, Wyoming: University of Michigan Papers on Paleontology 33, p. 73–88.

Calvès, G., Schwab, A.M., Huuse, M., Clift, P.D., Gaina, C., Jolley, D., Tabrez, A.R., and Inam, A., 2011, Seismic volcanostratigraphy of the western Indian rifted margin: The pre-Deccan igneous province: Journal of Geophysical Research–Planets, v. 116, no. B1, B01101, doi:10.1029/2010JB000862.

Carmichael, S.M., Akhter, S., Bennet, J.K., Fatimi, M.A., Hosein, K., Jones, R.W., Longacre, M.B., Osborne, M.J., and Tozer, R.S.J., 2009, Geology and hydrocarbon potential of the offshore Indus Basin, Pakistan: Petroleum Geoscience, v. 15, p. 107–116, doi:10.1144/1354-079309-826.

Cavagnetto, C., and Gaudant, J., 2000, A palynoflora of Palaeocene age from the fossiliferous sapropels of the Boltyshka depression, central Ukraine: Newsletters on Stratigraphy, v. 38, no. 1, p. 39–56.

Chenet, A.-L., Fluteau, F., Courtillot, V., Gérard, M., and Subbarao, K.V., 2008, Determination of rapid Deccan eruptions across the

Cretaceous-Tertiary boundary using paleomagnetic secular variation: Results from a 1200-m-thick section in the Mahabaleshwar escarpment: Journal of Geophysical Research–Solid Earth, v. 113, B04101, doi:10.1029/2006JB004635.

Chenet, A.-L., Courtillot, V., Fluteau, F., Gérard, M., Quidelleur, X., Khadri, S.F.R., Subbarao, K.V., and Thordarson, T., 2009, Determination of rapid Deccan eruptions across the Cretaceous-Tertiary boundary using paleomagnetic secular variation: 2. Constraints from analysis of eight new sections and synthesis for a 3500-m-thick composite section: Journal of Geophysical Research–Planets, v. 114, no. B6, B06103, doi:10.1029/2008JB005644.

Clayton, C., 1991, Carbon isotope fractionation during natural gas generation from kerogen: Marine and Petroleum Geology, v. 8, no. 2, p. 232–240, doi:10.1016/0264-8172(91)90010-X.

Coccioni, R., Frontalini, F., Bancala, G., Fornaciari, E., Jovane, L., and Sprovieri, M., 2010, The Dan-C2 hyperthermal event at Gubbio (Italy): Global implications, environmental effects, and cause(s): Earth and Planetary Science Letters, v. 297, no. 1–2, p. 298–305, doi:10.1016/j.epsl.2010.06.031.

Coccioni, R., Bancala, G., Catanzarit, R., Fornaciari, E., Frontalini, F., Giusberti, L., Jovane, L., Luciani, V., Savian, J., and Sprovieri, M., 2012, An integrated stratigraphic record of the Palaeocene–lower Eocene at Gubbio (Italy): New insights into the early Palaeogene hyperthermals and carbon isotope excursions: Terra Nova, v. 24, no. 5, p. 380–386, doi:10.1111/j.1365-3121.2012.01076.x.

Cohen, A., Coe, A., and Kemp, D.B., 2007, The late Palaeocene–early Eocene and Toarcian (Early Jurassic) carbon isotope excursions: A comparison of their time scales, associated environmental changes, causes and consequences: Journal of the Geological Society, v. 164, no. 6, p. 1093, doi:10.1144/0016-76492006-123.

Collier, J.S., Sansom, V., Ishizuka, O., Taylor, R.N., Minshull, T.A., and Whitmarsh, R.B., 2008, Age of Seychelles–India break-up: Earth and Planetary Science Letters, v. 272, p. 264–277, doi:10.1016/j.epsl.2008.04.045.

Corfield, R.I., Scott, C., Bennett, J., Akhter, S, Fatimi, M., and Craig, T., 2010, Variability in the crustal structure of the west Indian continental margin in the northern Arabian Sea: Petroleum Geoscience, v. 16, p. 257–265, doi:10.1144/1354-079309-902.

Courtillot, V., and Fluteau, F., 2010, Cretaceous extinctions: The volcanic hypothesis: Science, v. 328, no. 5981, p. 973–974, doi:10.1126/science.328.5981.973-b.

Cripps, J., Widdowson, M., Spicer, R.A., and Jolley, D., 2005, Coastal ecosystem responses to late stage Deccan Trap volcanism: The post K-T boundary (Danian) palynofacies of Mumbai (Bombay), west India: Palaeogeography, Palaeoclimatology, Palaeoecology, v. 216, p. 303–332, doi:10.1016/j.palaeo.2004.11.007.

Domingo, L., López-Martínez, N., Leng, M.J., and Grimes, S.T., 2009, The Paleocene–Eocene thermal maximum record in the organic matter of the Claret and Tendruy continental sections (south-central Pyrenees, Lleida, Spain): Earth and Planetary Science Letters, v. 281, no. 3–4, p. 226–237, doi:10.1016/j.epsl.2009.02.025.

Finkelstein, D.B., Pratt, L.M., and Brassell, S.C., 2006, Can biomass burning produce a globally significant carbon isotope excursion in the sedimentary record?: Earth and Planetary Science Letters, v. 250, p. 501–510, doi:10.1016/j.epsl.2006.08.010.

Gilmour, I., Gilmour, M., Jolley, D.W., Kelley, S.P., Kemp, D.B., Daly, R.J., and Watson, J.S., 2013, A high-resolution nonmarine record of an early Danian hyperthermal event, Boltysh crater, Ukraine: Geology, v. 41, p. 783–786, doi:10.1130/G34292.1.

Gurov, E.P., Kelley, S.P., Koeberl, C., and Dykan, N.I., 2006, Sediments and impact rocks filling the Boltysh impact crater, *in* Cockell, C.S., Koeberl, C., and Gilmour, I., eds., Biological Processes Associated with Impact Events: Berlin, Springer, p. 335–358.

Herngreen, G., Kedves, M., Rovnina, I.V., and Smirnova, S.B., 1996, Cretaceous floral provinces: A review, *in* Jansonius, J., and McGregor, D.C., eds., Palynology, Principles, and Applications: Dallas, Texas, American Association of Stratigraphic Palynologists, Inc., p. 1157–1188.

Jolley, D.W., Gilmour, I., Gurov, E.P., Kelley, S.P., and Watson, J.S., 2010, Two large meteorite impacts at the Cretaceous-Paleogene boundary: Geology, v. 38, p. 835–838, doi:10.1130/G31034.1.

Jolley, D.W., Daly, R.J., Gilmour, I., and Kelley, S.P., 2013, Climatic oscillations stall vegetation recovery from K/Pg event devastation: Journal of the Geological Society, v. 170, p. 477–482, doi:10.1144/jgs2012-088.

Keller, G., 2012, The Cretaceous-Tertiary mass extinction, Chicxulub impact and Deccan volcanism, *in* Talent, J., ed., Earth and Life, Extinction Intervals and Biogeogrpahic Perturbations through Time: Dordrecht, Springer Science and Media, p. 759–793.

Keller, G., Adatte, T., Pardo, A., Bajpai, S., Khosla, A., and Samant, B., 2010, Cretaceous extinctions: Evidence overlooked: Science, v. 328, no. 5981, p. 974, doi:10.1126/science.328.5981.974-a.

Kelley, S.P., and Gurov, E.P., 2002, Boltysh, another end-Cretaceous impact: Meteoritics & Planetary Science, v. 37, no. 8, p. 1031–1043, doi:10.1111/j.1945-5100.2002.tb00875.x.

Koch, P., Zachos, J.C.C., and Gingerich, P.D., 1992, Correlation between isotope records in marine and continental carbon reservoirs near the Paleocene-Eocene boundary: Nature, v. 358, no. 6384, p. 319–322, doi:10.1038/358319a0.

Lenz, O.K., and Riegel, W., 2001, Isopollen maps as a tool for the reconstruction of a coastal swamp from the middle Eocene at Helmstedt (northern Germany): Facies, v. 45, no. 1, p. 177–194, doi:10.1007/BF02668111.

Magioncalda, R., Dupuis, C., Smith, T., Steurbaut, E., and Gingerich, P.D., 2004, Paleocene-Eocene carbon isotope excursion in organic carbon and pedogenic carbonate: Direct comparison in a continental stratigraphic section: Geology, v. 32, p. 553–556, doi:10.1130/G20476.1.

McInerney, F.A., and Wing, S.L., 2011, The Paleocene-Eocene thermal maximum: A perturbation of carbon cycle, climate, and biosphere with implications for the future: Annual Review of Earth and Planetary Sciences, v. 39, no. 1, p. 489–516, doi:10.1146/annurev-earth-040610-133431.

Meyers, P.A., 2003, Applications of organic geochemistry to paleolimnological reconstructions: A summary of examples from the Laurentian Great Lakes: Organic Geochemistry, v. 34, no. 2, p. 261–289, doi:10.1016/S0146-6380(02)00168-7.

Minshull, T.A., Lane, C.I., Collier, J.S., and Whitmarsh, R.B., 2008, The relationship between rifting and magmatism in the northeastern Arabian Sea: Nature Geoscience, v. 1, p. 463–467, doi:10.1038/ngeo228.

Mussard, M., Le Hir, G., Fluteau, F., Lefebvre, B., and Godderis, Y., 2014, this volume, Modeling the carbon-sulfate interplays in climate changes related to the emplacement of continental flood basalts, *in* Keller, G., and Kerr, A.C., eds., Volcanism, Impacts, and Mass Extinctions: Causes and Effects: Geological Society of America Special Paper 505, doi:10.1130/2014.2505(17).

Punekar, J., Mateo, P.M., and Keller, G., 2014, this volume, Effects of Deccan volcanism on paleoenvironment and planktic foraminifera: A global survey, *in* Keller, G., and Kerr, A.C., eds., Volcanism, Impacts, and Mass Extinctions: Causes and Effects: Geological Society of America Special Paper 505, doi:10.1130/2014.2505(04).

Quillévéré, F., Norris, R.D., Kroon, D., and Wilson, P.A., 2008, Transient ocean warming and shifts in carbon reservoirs during the early Danian: Earth and Planetary Science Letters, v. 265, no. 3–4, p. 600–615, doi:10.1016/j.epsl.2007.10.040.

Renne, P.R., Balco, G., Ludwig, K.R., Mundil, R., and Min, K., 2011, Response to the comment by W.H. Schwarz et al. on "Joint determination of ^{40}K decay constants and ^{40}Ar*/^{40}K for the Fish Canyon sanidine standard, and improved accuracy for ^{40}Ar/^{39}Ar geochronology" by P.R. Renne et al. (2010): Geochimica et Cosmochimica Acta, v. 75, p. 5097, doi:10.1016/j.gca.2011.06.021.

Renne, P.R., Deino, A.L., Hilgen, F.J., Kuiper, K.F., Mark, D.F., Mitchell, W.S., Morgan, L.E., Mundil, R., and Smit, J., 2013, Time scales of critical events around the Cretaceous-Paleogene boundary: Science, v. 339, no. 6120, p. 684–687, doi:10.1126/science.1230492.

Saito, T., Yamanoi, T., and Kaiho, K., 1986, End-Cretaceous devastation of terrestrial flora in the boreal Far East: Nature, v. 323, no. 6085, p. 253–255, doi:10.1038/323253a0.

Schulte, P., Alegret, L., Arenillas, I., Arz, J.A., Barton, P.J., Bown, P.R., Bralower, T.J., Christeson, G.L., Claeys, P., Cockell, C.S., Collins, G.S., Deutsch, A., Goldin, T.J., Goto, K., Grajales-Nishimura, J.M., Grieve, R.A.F., Gulick, S.P.S., Johnson, K.R., Kiessling, W., Koeberl, C., Kring, D.A., MacLeod, K.G., Matsui, T., Melosh, J., Montanari, A., Morgan, J.V., Neal, C.R., Nichols, D.J., Norris, R.D., Pierazzo, E., Ravizza, G., Rebolledo-Vieyra, M., Reimold, W.U., Robin, E., Salge, T., Speijer, R.P., Sweet, A.R., Urrutia-Fucugauchi, J., Vajda, V., Whalen, M.T., and Willumsen, P.S., 2010, The Chicxulub asteroid impact and mass extinction at the Cretaceous-Paleogene boundary: Science, v. 327, no. 5970, p. 1214–1218, doi:10.1126/science.1177265.

Self, S., Widdowson, M., Thordarson, T., and Jay, A.E., 2006, Volatile fluxes during flood basalt eruptions and potential effects on the global environment: A Deccan perspective: Earth and Planetary Science Letters, v. 248, no. 1–2, p. 518–532, doi:10.1016/j.epsl.2006.05.041.

Smit, J., 1999, The global stratigraphy of the Cretaceous-Tertiary boundary impact ejecta: Annual Review of Earth and Planetary Sciences, v. 27, p. 75–113, doi:10.1146/annurev.earth.27.1.75.

Smit, J., and Hertogen, J., 1980, An extraterrestrial event at the Cretaceous-Tertiary boundary: Nature, v. 285, p. 198–200, doi:10.1038/285198a0.

Speijer, R.P., 2003, Danian-Selandian sea-level change and biotic excursion on the southern Tethyan margin (Egypt), *in* Wing, S.L., Gingerich, P.D., Schmitz, B., and Thomas, E., eds., Causes and Consequences of Globally Warm Climates in the Early Paleogene: Geological Society of America Special Paper 369, p. 275–290.

Svensen, H., Planke, S., Malthe-Sorenssen, A., Jamtveit, B., Myklebust, R., Eidem, T., and Rey, S., 2004, Release of methane from a volcanic basin as a mechanism for initial Eocene global warming: Nature, v. 429, no. 6991, p. 542–545, doi:10.1038/nature02566.

Talbot, M.R., and Johannessen, T., 1992, A high resolution palaeoclimatic record for the last 27,500 years in tropical West Africa from the carbon and nitrogen isotopic composition of lacustrine organic matter: Earth and Planetary Science Letters, v. 110, no. 1–4, p. 23–37, doi:10.1016/0012-821X(92)90036-U.

Vajda, V., Raine, J.I., and Hollis, C.J., 2001, Indication of global deforestation at the Cretaceous-Tertiary boundary by New Zealand fern spike: Science, v. 294, no. 5547, p. 1700–1702, doi:10.1126/science.1064706.

Valter, A.A., and Plotnikova, L., 2003, Biostratigraphic indications of the age of the Boltysh impact crater in Ukraine, *in* Koeberl, C., and Martinez-Ruiz, F., eds., Impact Markers in the Stratigraphic Record: Impact Studies Volume 3: Heidelberg, Springer, p. 163–178, doi:10.1007/978-3-642-55463-6_7.

Weedon, G.P., 2003, Time-Series Analysis and Cyclostratigraphy: Examining Stratigraphic Records of Environmental Cycles: Cambridge, UK, Cambridge University Press, 259 p.

Westerhold, T., Röhl, U., Raffi, I., Fornaciari, E., Monechi, S., Reale, V., Bowles, J., and Evans, H.F., 2008, Astronomical calibration of the Paleocene time: Palaeogeography, Palaeoclimatology, Palaeoecology, v. 257, no. 4, p. 377–403, doi:10.1016/j.palaeo.2007.09.016.

Westerhold, T., Röhl, U., Donner, B., McCarren, H.K., and Zachos, J.C.C., 2011, A complete high-resolution Paleocene benthic stable isotope record for the central Pacific (ODP Site 1209): Paleoceanography, v. 26, no. 2, PA2216, doi:10.1029/2010PA002092.

Wignall, P.B., 2001, Large igneous provinces and mass extinctions: Earth-Science Reviews, v. 53, p. 1–33, doi:10.1016/S0012-8252(00)00037-4.

Wing, S.L., and Hickey, L., 1984, The Platycarya perplex and the evolution of the Juglandaceae: American Journal of Botany, v. 71, p. 388–411, doi:10.2307/2443497.

Wolfe, J.A., and Upchurch, G.R., 1987, Leaf assemblages across the Cretaceous-Tertiary boundary in the Raton Basin, New Mexico and Colorado: Proceedings of the National Academy of Sciences of the United States of America, v. 84, p. 5096–5100, doi:10.1073/pnas.84.15.5096.

Yans, J., Strait, S.G., Smith, T., Dupuis, C., Steurbaut, E., and Gingerich, P.D., 2006, High-resolution carbon isotope stratigraphy and mammalian faunal change at the Paleocene-Eocene boundary in the Honeycombs area of the southern Bighorn Basin, Wyoming: American Journal of Science, v. 306, no. 9, p. 712–735, doi:10.2475/09.2006.02.

Manuscript Accepted by the Society 31 January 2014

Printed in the USA

The Geological Society of America
Special Paper 505
2014

Plant macrofossils from Boltysh crater provide a window into early Cenozoic vegetation

Sergei V. Vikulin*
Laboratory of Paleobotany, V.L. Komarov Botanical Institute of the Russian Academy of Sciences (BIN RAS), 2 Prof. Popov ul., St. Petersburg 197376, Russia

Nadia Bystriakova
Harald Schneider
Department of Life Sciences, Natural History Museum, Cromwell Road, London SW7 5BD, UK

David Jolley
Department of Geology and Petroleum Geology, King's College, University of Aberdeen, Aberdeen AB24 3UE, Scotland, UK

ABSTRACT

We analyzed the plant macro- and mesofossil records deposited in the Paleocene oil shales of the Boltysh crater (Ukraine) in terms of leaf morphology and its implication for reconstruction of the vegetation and paleoecology of the region. During the early Cenozoic, the Boltysh astrobleme formed a geothermal crater lake that accumulated sediments, preserving a record from the Paleocene to the early middle Eocene. These sediments contain fossil leaf fragments of ferns and angiosperms that grew close to the lake. The occurrence of the Mesozoic fern *Weichselia reticulata* is of importance. This discovery suggests the survival of this Jurassic to Cretaceous fern into the early Paleogene in the refugial geothermal ecosystem of the Boltysh crater area. Our finding is the youngest record of this fern, although it was a widespread and common element of secondary vegetation during the Cretaceous. The local survival of this fern may have been fostered by the unique combination of edaphic environmental factors of the Boltysh hydrothermal area. Other plant fossils include fragments of leaves that represent ferns likely belonging to lineages that diversified in the shadow of angiosperms, as well as remains of the flowering plants *Pseudosalix*, *Sorbus*, *Comptonia*, and ?*Myrica* leaf morphotypes.

*vickulin@gmail.com

Vikulin, S.V., Bystriakova, N., Schneider, H., and Jolley, D., 2014, Plant macrofossils from Boltysh crater provide a window into early Cenozoic vegetation, *in* Keller, G., and Kerr, A.C., eds., Volcanism, Impacts, and Mass Extinctions: Causes and Effects: Geological Society of America Special Paper 505, p. 147–169, doi:10.1130/2014.2505(07). For permission to copy, contact editing@geosociety.org.

INTRODUCTION

Cretaceous terrestrial evolution continues to be a challenge to our understanding of the development of life on Earth (Song et al., 1986; Benton, 2009; Kerr, 2012). One of the most popular explanations for the mass extinction of life at the end of the Cretaceous period 65 m.y. ago has been a single asteroid impact at Chicxulub on the Yucatan Peninsula, Mexico, precisely at the Cretaceous-Paleogene boundary (Schulte et al., 2010), or predating the Cretaceous-Paleogene by 100–150 k.y. or 300 k.y., depending on the time scale used (Keller, 2012). A recent discovery of a second smaller crater at Boltysh in the Ukraine with a similar age raises the possibility that a shower of asteroids or comets impacted Earth close to the Cretaceous-Paleogene boundary (Kelley et al., 2009; Watson et al., 2010).

During the period from around 130 to 100 Ma, the taxonomic composition of the vast majority of terrestrial plant ecosystems underwent dramatic changes. The flowering plants evolved rapidly to become not only the most species-rich group of land plants but also to dominate in most habitats (e.g., Retallack and Dilcher, 1986; Crane and Lidgard, 1989; Fiz-Palacios et al., 2011). At the same time, other lineages declined in species numbers, distribution range, and ecological importance (Niklas et al., 1985; Crane and Lidgard, 1989; Lupia et al., 1999; Judd et al., 2002). However, recent studies have provided increasing evidence that the majority of terrestrial nonflowering vascular land plant lineages showed opportunistic diversification into habitats dominated by angiosperms (Schneider et al., 2004; Fiz-Palacios et al., 2011).

These observations support a hypothesis of intralineage replacement (Christenhusz et al., 2011; Schneider, 2007; Schneider et al., 2004, 2009) in most lineages by land plants and perhaps also other groups of organisms. This hypothesis is best illustrated by the fern lineage, because comprehensive evidence can be obtained by integrating the fossil record and data based on molecular diversification time estimates. Mesozoic fern families, such as Dipteridacae and Matoniaceae, are recognized as species-rich, widespread, and ecologically important families in terrestrial ecosystems of the Jurassic and Early Cretaceous, but they declined in the Late Cretaceous and/or early Paleogene. Today, Dipteridaceae and Matoniaceae are represented by a few extant species restricted to the botanical subkingdom that encompasses the islands of Southeast Asia and the Malay Peninsula. In contrast, the derived ferns rapidly diversified in the Late Cretaceous to early Eocene and contributed to more than 80% of the extant fern diversity. However, many of the details of this process are not fully understood due to the paucity of reports on the fern fossil record in the Late Cretaceous and early Paleogene (Schneider, 2007), and confirmation of some deoxyribonucleic acid (DNA)–based diversification time estimates is lacking (Schneider et al., 2004; Schuettpelz and Pryer, 2009; Fiz-Palacios et al., 2011).

Most importantly, the proposed decline of some lineages has only been sketched out, and it requires detailed testing. For example, Collinson (2001, 2002) argued that the Matoniaceae family has been restricted to its current distribution range in the Malesian biogeographic region since the earliest Paleogene, because only a few fossils are recorded in the Late Cretaceous and early Paleogene. In contrast, these ferns are commonly found in the fossil record of terrestrial ecosystems of the Jurassic and Early Cretaceous. During most of the Mesozoic, these ferns likely occurred on all continents and showed a notable range of ecological diversity (Collinson 2001, 2002; Brown et al., 2012).

The Paleogene was a time of profound reorganization of the biosphere, following the Cretaceous-Paleogene mass extinction (Vajda et al., 2001; Benton and Harper, 2009), which was followed by the loss of major Mesozoic lineages and low-diversity floras during the Paleocene (Wappler et al., 2009; Wing and DiMichele, 1995; Peppe, 2010). Here, we explore new evidence of ferns and flowering plants in the early Paleogene drill-core macrofossil record of the Boltysh meteorite impact crater (Kelley and Gurov, 2002; Gurov et al., 2006), which provides a unique window into Paleocene vegetation. The 24-km-diameter Boltysh crater in the Ukraine (Fig. 1) was formed ~65.17 m.y. ago by a meteorite impact similar to that of Chicxulub, but it is considered independent from the Chicxulub impact on Yucatan, Mexico (Jolley et al., 2010). Although today the Boltysh crater is located in the Ukraine, it was likely part of an island at the edge of a shallow sea at the southeastern range of an island chain that contained most of modern Europe (Jolley et al., 2010; Watson et al., 2009).

The importance of the Boltysh flora is enhanced because of the rarity of European Paleocene floras with finely preserved plant megafossils. A related middle Paleocene flora is known from the Menat Basin, Puy-de-Do me, France, dated at 60–61 Ma (Vincent et al., 1977), which is the Western European macroflora stratigraphically closest to the Cretaceous-Paleogene event. In addition to Menat, France, some leaf floras are known from the Paris Basin Vaugirard Formation (formerly Argiles Plastiques) at the sites of Bagneux, Cessoy, Meudon, and Vervins, and a similar

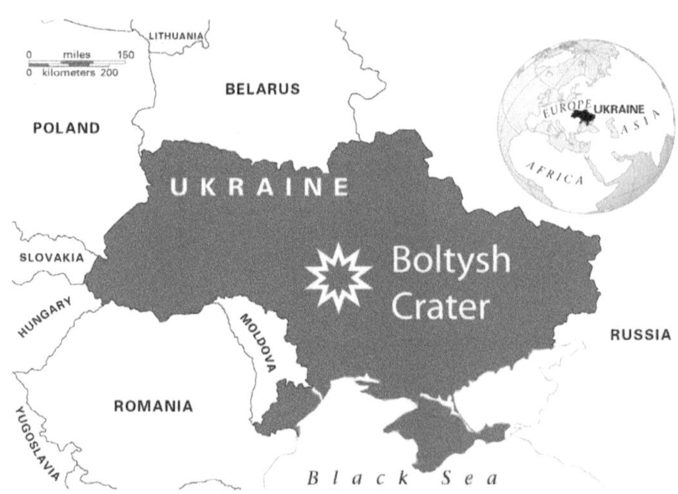

Figure 1. Map showing location of the Boltysh crater in central Ukraine, Eastern Europe and a globe showing where Ukraine is located in Europe. Coordinates of the Boltysh crater: 48.9°N, 32.25°E.

flora is known from the Travertin de Sézanne. These floras are difficult to interpret because they have not been recently revised and are frequently poorly preserved (Fritel, 1910; Feugueur, 1963). Mai (1995) has noted leaves in the Sézanne flora as being entire margined and megaphyllous. In the London Basin, age-equivalent floras come from the Upnor (late Thanetian), Woolwich, and Reading Formations (Paleocene-Eocene thermal maximum; Speijer et al., 2012). These are either fruit and seed (Upnor Formation) or mainly fruit, seed, and leaf floras (Hooker and Collinson, 2012). Until recently, the only Paleocene floras in Eastern Europe were derived from the Kanev Formation in the Ukraine (Kryshtofovich, 1952) and from sandstones from the Lower Volga River, European Russia (Krassilov et al., 1996).

Only the Paleocene fossil-lagerstätten flora of Menat, France (Laurent, 1912, 1919; Vincent et al., 1977; Kedves and Russell, 1982; Piton, 1940; Mai, 1995; Wappler et al., 2009), is comparable in age and taxonomic composition to the newly unearthed Boltysh leaf flora from Ukraine drill cores of oil shales. This Menat flora is found in fossil lake deposits with sedimentary infill interpreted as a maar lake created by volcanism (Vincent et al., 1977) or a meteorite impact. Plants from Menat and Boltysh are preserved as impressions and compressions in thinly laminated, bituminous shales. Most of the aforementioned Paleocene leaf floras are predominantly angiospermous, with rare fern records.

To date, paleobotanical research of the Boltysh crater has focused mainly on the palynomorph record, including pollen, spores, and remains of various groups of algae (Jolley et al., 2010), whereas the macrofossil leaf record from the Boltysh crater lake shale has been ignored. The existing studies have focused mainly on the vegetation change from the Cretaceous-Paleogene transition (Jolley et al., 2010, 2012). Here, we focus on the phylogenetic and taxonomic composition of vascular plants communities in the paratropical regions of Eastern Europe in the early Paleogene. On one hand, the report of macrofossils of the Boltysh crater will improve our understanding of the climatic and vegetation history of Eastern Europe (Kvaček, 2010). On the other hand, the records will contribute to the ongoing inference about the lineage replacement within the fern lineage in the Late Cretaceous and early Paleogene.

PREVIOUS STUDIES OF BOLTYSH FOSSILS

According to the latest findings, continental lake and lagoon sedimentary layers of organic mudrocks and oil shales correspond to the Pselian and Merlian regional stages of the Paleocene, and the Kanevian regional stage of the Lower Eocene of the Dnieper-Donets depression (Zosymovych and Ryabokon', 2010). The composition of palynomorphs (Jolley et al., 2010, 2012) and macro- and mesofossils provide evidence supporting an early Paleogene age. For example, drill cores in well 1715 (Stanislavsky, 1968) recovered ~300 m of shales, argillite, and oil-shale lake sapropelites. These sediments contain remnants of early Paleogene species, such as the conifer *Doliostrobus* sp., and angiosperms, similar to some convergent Proteaceae/Myricaceae morphotypes, dentate leaves designated as "*Hakea*" *exulata* Heer, "*Dryandroides*" *antique*, *Comptonia* sp., *Myrica* sp., and *Salix* sp., as well as two species of *Dryophyllum*, an archaic fagaceous representative, *Dryophyllum curticellense*, and *Dryophyllum furcinerve*, of the same morphotypes as in Thanetian sandstones of the Paris Basin (Fritel, 1910; Feugueur, 1963; Vincent et al., 1977). According to Stanislavsky (1968, p. 110) "...this [oil-shale, S.V.] thickness on all levels contains numerous vegetative remains. The latter are mainly dispersed, and unequivocally, were carried by streams to the places of deposition. In this sequence fern leaf fragments occur, among which three or four morphotypes may be recognized." The originally suggested Paleocene–Eocene origin of the sedimentary oil-shale fill of the Boltysh crater (Stanislavsky, 1968) was further confirmed by the results of subsequent drilling operations (Zosymovych and Ryabokon', 2010; Gurov, 1996).

GEOLOGICAL SETTING AND LOCATIONS

The crater lake formed by the Boltysh meteorite impact structure (48°9′N, 32°25′E) in central Ukraine (Fig. 1) represents an ideal depositional basin for the study of Paleocene and Eocene stratigraphy and paleoecology of the fossil plants preserved within. Up to the late early Eocene, the crater was filled by a geothermal lake (Gurov, 1996; Gurov et al., 2006). The structure is ~24 km in diameter and 570–600 m deep. According to ^{40}Ar-^{39}Ar dating estimates, the age is 65.17 ± 0.64 Ma, placing it within the error margin of the Cretaceous-Paleogene boundary (Kelley and Gurov, 2002; Gurov et al., 2006). However, the Cretaceous-Paleogene boundary is now placed at 66 Ma based on U-Pb dating, which places the Ar/Ar date of the crater sediments well into the Paleocene (Renne et al., 2013).

After the impact, a crater lake formed, and fine-grained lacustrine sediments accumulated through the Paleocene and early Eocene, in which a wealth of terrestrial pollen and spores and leaf compressions are preserved. An ~400 m section of shale, oil shale, marl, sand, and clay preserves the diversity of plant macrofossils, which are the focus of this study. Results so far (Daly and Jolley, 2007) show a dynamic, fluctuating assemblage at the bottom of the section, revealing the ecological effects in the aftermath of the meteorite impact. Higher up section, the environmental variability is probably a consequence of the Paleocene–Eocene thermal maximum, evident in a rapid drop in diversity and shift in the floral balance. The shales of Boltysh crater are overlain by strata of the middle Eocene Kiev Formation, dated on the basis of foraminifera (Ryabokon', 2002). The underlying sapropelite of the core has been dated as late Paleocene–early Eocene based on gastropods, ostracods, insects, fishes (Stanislavsky 1968; Sytchevskaya 1986), salamanders (Skutschas and Gubin, 2012), and flora (Stanislavsky, 1968). The section interval of the lithological column studied for this paper and the sample positions and names of plant fossils found are shown in Figure 2 (modified and updated from Gurov et al., 2006).

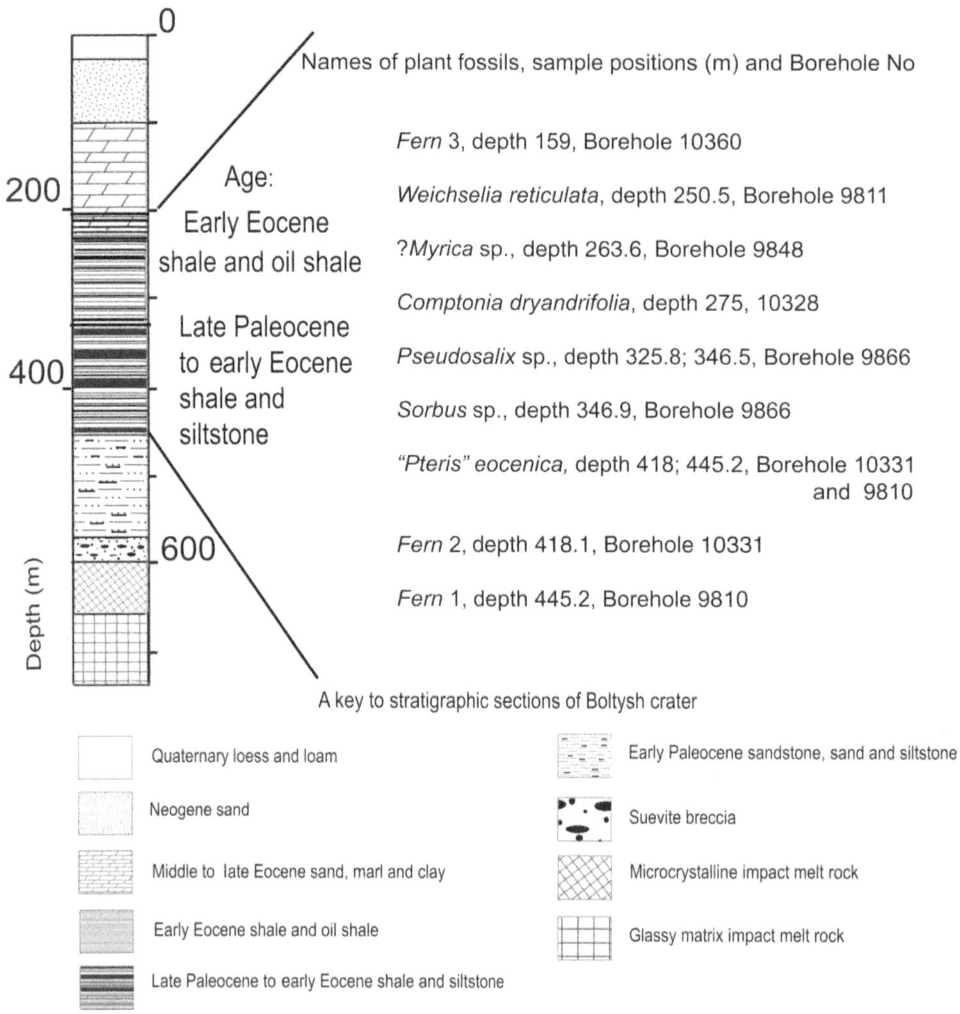

Figure 2. Lithological column of Boltysh crater sediments showing late Paleocene to early Eocene section interval with plant fossils studied for this paper (modified according to Gurov et al., 2006).

MATERIAL AND METHODS

The crater was drilled in the 1960s–1980s, but most of the cores were subsequently lost (Watson et al., 2010). A small, very interesting collection with Boltysh plant macrofossils was preserved in the Paleobotany Laboratory of the V.L. Komarov Botanical Institute of the Russian Academy of Sciences, BIN RAS. The material for this study consists of the Yu.I. Selin collection of drill-core samples, which have imprints of several ferns and angiosperms from the postimpact Boltysh crater sedimentary infill (shales; Kirovograd Province, Ukraine, Fig. 1). Samples were obtained from the wells drilled through the Paleocene–early Eocene section at depths of ~200–450 m (Fig. 2; Depository: collection 1128, Paleobotany Laboratory of BIN RAS). In 1967, geologist Y.I. Selin ("Kievgeologia") transferred plant remains from Boltysh to I.V. Vasiliev of the All-Union Geological Institute (VSEGEI) for identification, but nothing more than a brief summary species list across the whole Cenozoic borehole (not accompanied by illustrations) has been published (Vasiliev and Selin, 1970). Sedimentary rock samples collected by Selin ("Kievgeologia") in 1966–1967 from the Boltysh cores for further macrofossil plant analysis were sent to I.V. Vasiliev of VSEGEI (Leningrad). At the time, S.G. Zhilin was an assistant of Dr. Vasiliev and later on came to BIN RAS and brought a small part of the Selin-Vasiliev collection (the Paleocene-Eocene shales) to the Komarov Botanical Institute. Zhilin (1980) identified *Comptonia acutiloba* Brongn. from this material. In the mid-1980s, Zhilin provided several cores from the Boltysh (collection 1128, BIN) to the senior author (Vikulin) for review (Fig. 2).

Photography of fossil leaf impressions of ferns and angiosperms was conducted with a digital camera AxioCam ICc 3 stereomicroscope Zeiss Stemi 2000-CS at the Paleobotany Laboratory, BIN RAS (Figs. 3, 4, and 6–11). Modern plants for comparison with fossils were photographed in the Herbarium (LE) of the V.L. Komarov Botanical Institute with a digital camera Canon "PowerShot" SD 630 Digital Elph (SV); measurements (in millimeters) were taken either directly through a stereomicroscope using an optical reticle or from digital images using an AxioCam ICc 3 stereomicroscope Zeiss Stemi 2000-CS, viewed in Adobe Photoshop.

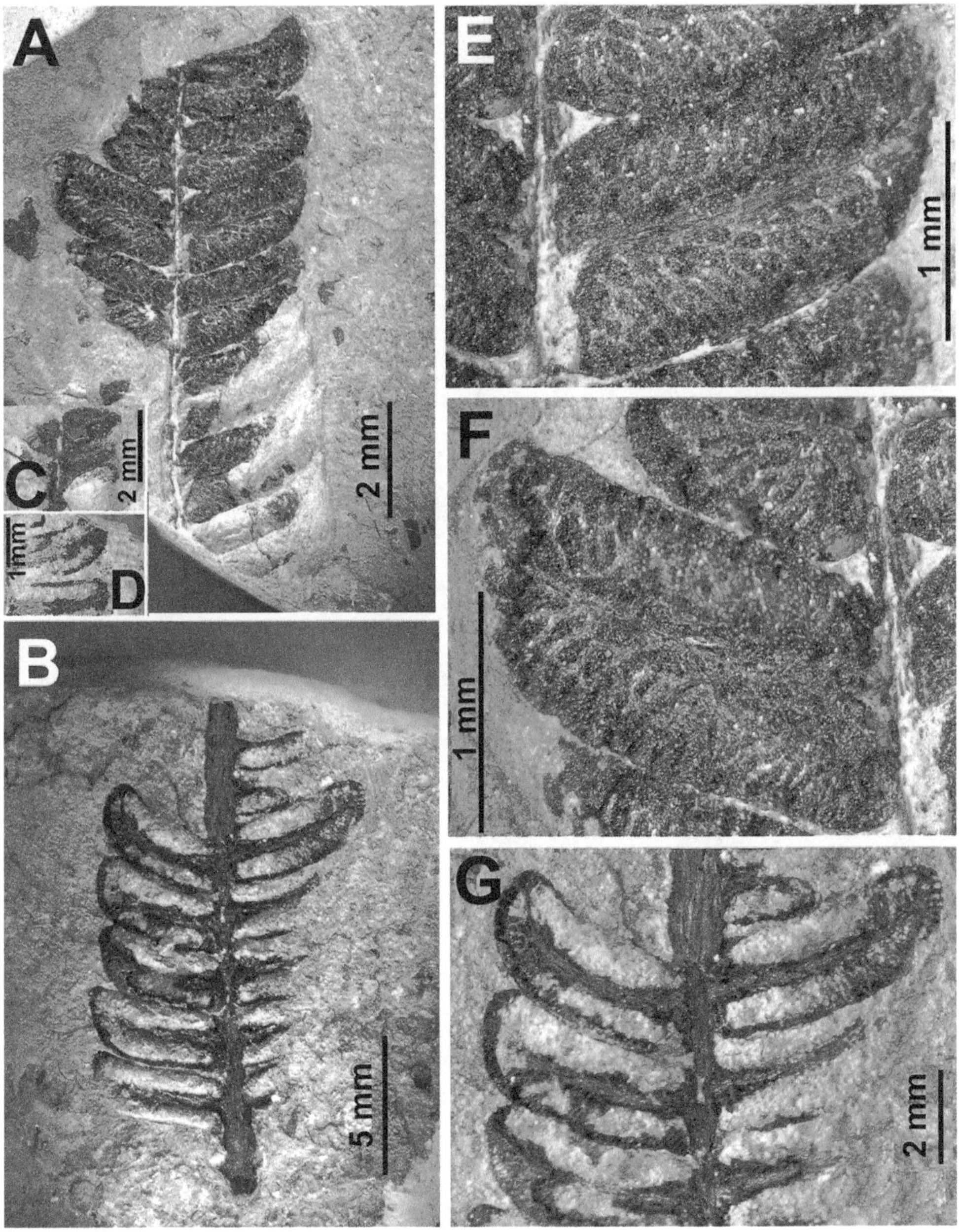

Figure 3. Photographs showing specimens of late Paleocene–early Eocene ferns from the Boltysh drill core; scales are in mm. (A) *Weichselia reticulata*, depth 250.5 m, borehole no. 9811/F 780 b, sample 12/1128 BIN RAS: an imprint of a pinna with a medium-prominent rachis. (B) *W. reticulata*, depth 250.5 m, borehole no. 9811/F 800, sample 11/1128 BIN RAS: imprint of a black pinna with a prominent rachis; the pinnules have a prominent venation; *Weichselia* remains preserved as fossil charcoal. (C) Small fragment of *W. reticulata*, depth 250.5 m, borehole no. 9811/F 800, sample 12a/1128 BIN RAS. (D) The same from sample 11a/1128 BIN RAS. (E–F) Close-up of a pinnula, the same as A. (G) The same as in B, close-up: detail showing charred pinnula of *Weichselia*; for sketches of venation, see Figure 4.

SYSTEMATIC PALEOBOTANY

Ferns

Weichselia reticulata **(Stokes et Webb 1824) Fontaine in Ward, 1899**

Class Polypodiopsida Cronquist, Takht. et Zimmerm. 1966

Order Gleicheniales Link 1825

Family Matoniaceae C. Presl 1847

Weichselia Stiehler 1857

Weichselia reticulata (Stokes et Webb 1824) Fontaine in Ward, 1899

For a detailed list of synonyms, see Seward (1894), Alvin (1971, 1974), El-Khayal (1986), and Diėguez and Melėndez (2000).

Holotype

An imprint of a fragment of a sterile frond, Sussex, Upper Weald Clay Formation, Barremian, Early Cretaceous (Stokes and Webb, 1824).

Material

Core samples from the Paleocene–early Eocene oil shale from the Boltysh crater: specimen 11/1128 BIN RAS: (1) imprint of a black pinnae with a prominent rachis, pinnules

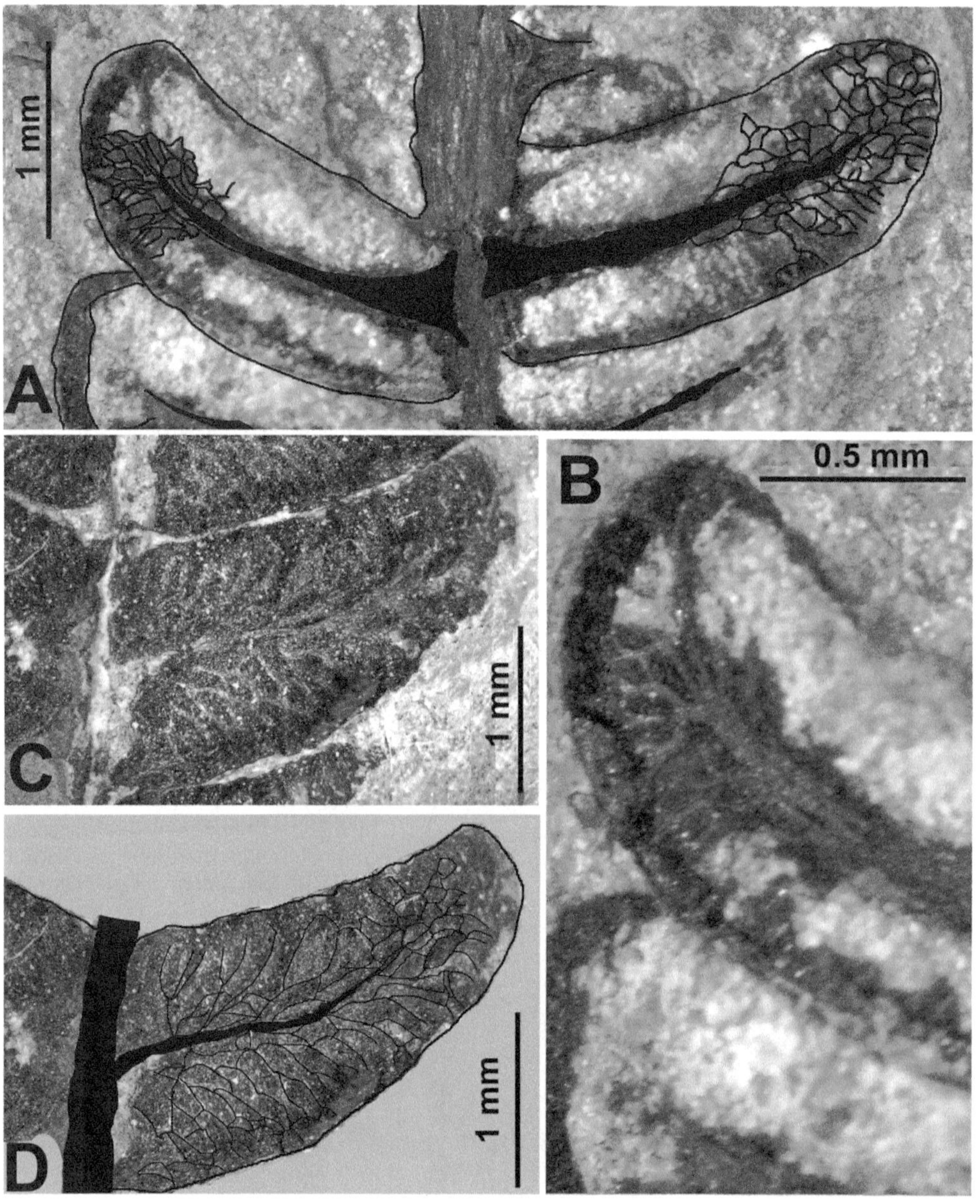

Figure 4. Photographs showing specimens of late Paleocene–early Eocene ferns from the Boltysh drill core; scales are in mm. (A–D) *Weichselia reticulata*: the reticulate venation is preserved in all the pinnules. As shown in sketches (A and D), it has 5–6 vein meshes between the midrib and the pinnulae margin, and the meshes tend to be elongated nearly parallel to the long axis near the midrib, but gradually more transverse near the margins. A and B: depth 250.5 m, borehole no. 9811/F 800, sample 11/1128 BIN RAS; C and D: depth 250.5 m, borehole no. 9811/F 780 b, sample 12/1128 BIN RAS.

with prominent venation; (2) specimen 12/1128 BIN RAS: imprint of a pinnae with a medium-prominent rachis; (3–4) small fragments of *W. reticulata*, samples 11a and 12a/1128 BIN (Figs. 2, 3A–3G, and 4A–4D).

Selected Fossil Occurrences

Geographic distribution of *Weichselia*: Berry (1928); Serlin (1982); El Khayal (1985); Vakhrameev (1964, 1988, 1991). For locality details, see Figure 5.

Description

Morphology of sterile fronds is presented in Figures 3A–3G and 4A–4D. Fragments of sterile fronds are clearly paripinnate. Convex, elliptical, slightly falcate pinnules, around 4 mm in length and 2.6–4.8 mm in width, are attached to a prominent, straight, expanding at the base rachis, ranging from 1 to 3 mm in width (2.3 mm on average). The angle between the midvein of a pinnula and the rachis ranges from 75° to 90°. From the point of attachment to the rachis, midveins bend and then branch out and stop short of the broadly rounded apex at three quarters of the length of a pinnula: The midvein fades into the net venation before reaching the apex. The upper surface of the pinnule is convex, with the midvein marked by a furrow (Figs. 3E–3G and 4A–4D). Prominent, thin veins of higher order, ranging from 0.2 to 0.3 mm in length, form a prominent reticulate structure consisting of elongated, pentagonal vein meshes, or anastomoses. The reticulate venation is preserved in all of the pinnules. It has 5–6 vein meshes between the midrib and the pinnule margin, and the meshes tend to be elongated nearly parallel to the long axis near the midrib, but gradually become more transverse near the

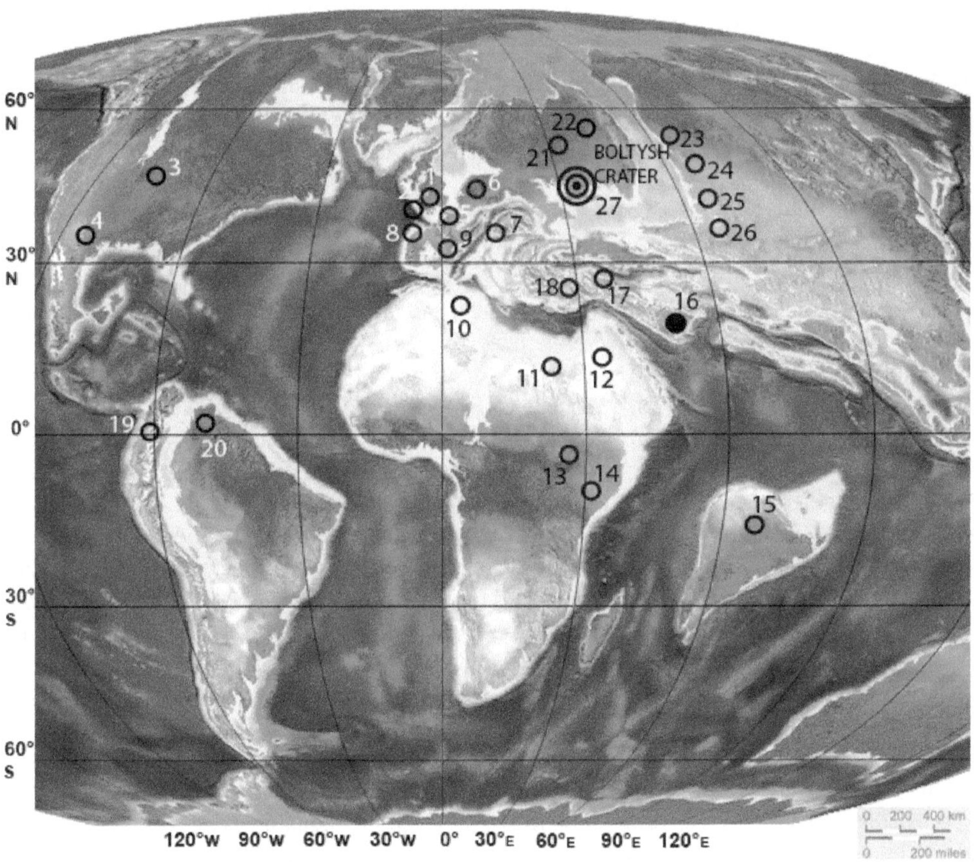

Figure 5. Paleogeographic map of the early Paleogene showing distribution of the fossil *Weichselia* localities. All localities are identified by numbers (1–27). Localities in Siberia and Russian Pacific are beyond range of the map (for details, see Vakhrameev, 1964, 1988, 1991). ○—Cretaceous and ●—Early Cretaceous localities; ◎—Boltysh crater locality, late Paleocene–early Eocene. Key to fossil *Weichselia* localities identified by numbers (1–27): 1—English Wealden, Sussex, UK, the Upper Weald Clay Formation, Barremian, Early Cretaceous (Stokes and Webb, 1824; Alvin, 1974). 2—Isle of Wight, UK, Shepherd's Chine ravine, 0.5 miles (0.8 km) from Atherfield Point, a lens of siltstones, the Upper Weald Clay Formation, Barremian, Early Cretaceous (Brown et al., 2012). 3—United States, Wyoming, Hay Creek coal field, Black Hills Formation, Cretaceous (Ward, 1899). 4—United States, northwest Texas (Nolan County), chert nodules associated with the Edwards Limestone (middle Albian) near Abilene, Early Cretaceous (Fontaine, 1893; Berry, 1928; Serlin, 1982). 5—Germany, Quedlinburg, Barremian-Albian, Early Cretaceous (Daber, 1968). 6—Wealden of Belgium, Hautrage, Mons Basin, Barremian-Aptian, Early Cretaceous (Alvin, 1971; Gomez et al., 2008). 7—Central Europe, Poland–Czech Republic border, Moravia-Silesia, Early Cretaceous (Reymanówna, 1965). 8—Spain, Cuenca, Las Hoyas, Ell Collado, and La Hue Arguina Formation, Barremian, Early Cretaceous (Barale, 1979). 9—NE Spain, Teruel province, Escucha Formation, Albian, Cretaceous (Diez, et al., 2005). 10—North Africa, southern Tunisia, Kimmeridgian and Portlandian, Late Jurassic–Early Cretaceous (Boureau and de Lapparent, 1951). 11—North Africa, southwestern desert of Egypt, on the border with Sudan, Gilf el Kebir plane, Abu Ballas Formation (= Nubian Series), late Tithonian–Berriasian, Early Cretaceous (Böttcher, 1982; Barthel and Böttcher, 1978). 12—Jordan, Nubian Sandstone: Wadi Jabbok, Wadi Kenetri, and Nahr es Zerka, uppermost part of the Early Cretaceous (Bender, 1974). 13—Western Sudan, Darfur, Nubian Sandstone in Jebel Dirra, Cretaceous (Edwards, 1932). 14—Northern Kenya, Kailta in the Wergudud area, "post-Jurassic" (Alvin, 1971). 15—Central India, Jabalpur series, Early Cretaceous (Sahni, 1936; Bose and Sukh Dev, 1959). 16—Saudi Arabia, Wasia Formation, Cenomanian-Turonian, Late Cretaceous (El-Khayal, 1985). 17—Israel, Makhtesh Ramon, Upper Hatira Formation, Early Cretaceous (Silantieva and Krasilov, 2006). 18—Lebanon, Madayrij, 24 km east of Beirut, Early Cretaceous (Edwards, 1929). 19—South America, Venezuela, Early Cretaceous (Schlagintweit, 1919). 20—South America, Colombia, Early Cretaceous (Berry, 1937). 21–26—Eastern Europe to Western Siberia, Kazakhstan, and Uzbekistan; for details, see Vakhrameev (1964, 1988, 1991). 27—Eastern Europe, Ukraine, Kirovograd Province. Oil-shale fill of the Boltysh crater, the Paleocene to the early Eocene, Thanetian-Ypresian, Pselian, and Merlian regional stages (Vikulin, 2013; Vikulin et al., 2013).

margins (Figs. 4A–4D). Total length of the leaf fragments varies from 17 to 18 mm, and their width is 8 mm.

Comparative Analysis

Although, there is some variation in the size and shape of the pinnules of *W. reticulata*, they are slightly sigmoid and parallel sided, except for the rounded end, and their venation is distinctly reticulate. Scleromorphic charred pinnules from the Boltysh crater are nearly identical to the fragments of *W. reticulata* from the Early Cretaceous fluvial sediments and seaside locations of North Africa, southwestern desert of Egypt (Böttcher, 1982; Barthel and Böttcher, 1978), as well as to the materials from the Late Cretaceous of the Wasia Formation, central Saudi Arabia (El-Khayal, 1985). All fragments are characterized by prominent reticulate venation preserved in pinnules and by a thick, well-developed rachis. These features enable a relatively straightforward determination. Reymanówna (1965) pointed out that identification of the genus *Weichselia* based on fragments presents no difficulties, as the members of the genus possess a very peculiar reticulate venation (Figs. 4A–4D). The midrib fades into the net venation before reaching the apex (Figs. 4C and 4D). There are 4–5–6 meshes between the midrib and the pinnule margin, and the meshes tend to be elongated nearly parallel to the long axis near the midrib, but gradually become more transverse near the margins (Figs. 4A and 4B). Therefore, the *Weichselia* fragments found in the Upper Cretaceous of Saudi Arabia (El-Khayal, 1985) as well as their supposed Paleocene descendants from Boltysh can be safely assigned to the genus *Weichselia*. Both occurrences of *Weichselia* belong to the coastal areas of the paleo-Tethys Ocean. In the protolog of *W. reticulata* (basionym *Pecopteris reticulata*: Stokes and Webb, 1824, p. 424), it was noted that "the ramifications and interlacing of the veins, at the tips of the pinnae, are peculiar, and serve to distinguish it from other species."

Discussion

The generic characters of *Weichselia* based on external morphology of the pinnae may be summarized briefly as follows: frond bipinnate, rachis broad and rigid, pinnae long and with prominent axes, pinnules entire with obtuse apex, attached by the whole of their bases, distinct, venation reticulate. It has 5–6 vein meshes between the midrib and the pinnule margin, and the meshes tend to be elongated nearly parallel to the long axis near the midrib, but gradually become more transverse near the margins (Figs. 4A–4D). Pinnulae lengths around 4 mm of *Weichselia* from Boltysh are comparable with the mean pinnulae lengths of the northern European *Weichselia*, indicated as 4.3 mm (Alvin, 1971) and 5 mm (Reymanówna, 1965). In southern European and African–Middle Eastern material, pinnulae length is ~7–9 mm (Edwards 1932; Dieguez and Melendez, 2000), except for the specimens from Saudi Arabia, for which El-Khayal (1985) gives the length range 4–13 mm, with a mean of 4.5 mm.

The type specimens from Tilgate forest, Sussex, pictured in Stokes and Webb (1824), are presumed lost (Alvin, 1971), and a neotype description has been proposed. Alvin (1971) offered an emended diagnosis of the *Weichselia* neotype, where attributive assignment of the vegetative pinnules to the extinct genus *Weichselia* is given. Alvin (1971) pointed out that they (i.e., vegetative pinnules) are typically 3–5 mm long, 1.3–2.3 mm wide, and oblong with rounded ends. They are attached along the whole breadth of the base, with adjacent pinnules becoming adnate and more pointed in extreme distal parts of secondary pinnae. The adaxial surface of the pinnule is convex with slightly depressed midrib and rolled-over margins. Midrib and veins are more prominent on the adaxial side. Typically 4–5–6 vein meshes are between the midrib and margin (see Figs. 4A–4D).

All these diagnostic features of *Weichselia* are quite straightforwardly observed in our materials of charred fern pinnules of *Weichselia* from Paleocene drill cores of the Boltysh crater (Figs. 3–4). Unequivocal attribution to the given *Weichselia* genus is evident according to the basic diagnostic principles of botanical systematics (see description of the *W. reticulata* type and neotype pinnae external morphology). *Weichselia reticulata* was widely distributed in the Cretaceous, and its presence in the Paleocene-Eocene sediments of the Boltysh crater is the first record in the early Paleogene. This species was common in both tropical and subtropical climatic zones in the Early Cretaceous, but it is also reported from the Late Cretaceous of Saudi Arabia (El-Khayal, 1985). A potential migration might have taken place in the Late Cretaceous–early Paleogene along the shores of the Tethys Ocean and via island land bridges (Fig. 5).

Weichselia reticulata was distributed along the coast of the Tethys Ocean and proto–Atlantic Ocean from North America to Europe, Siberia, and Japan, and from South America to North Africa and central India (see Fig. 5; see also Alvin, 1971, 1974; Barale, 1979; Vakhrameev, 1964, 1988, 1991). Since *Weichselia* is commonly found in monospecific assemblages in Cretaceous coastal sediments, in coastal swamps, matoniaceous ferns may have displaced angiosperms and conifers due to possibly greater competitiveness (Lacovara et al., 2003). Like the modern vegetation of estuarine banks and interdune swales, this genus probably preferred effectively xeric, coastal sedimentary environments.

The reported occurrence of *Weichselia* in the late Paleocene to early Eocene of Boltysh (Vikulin, 2013; Vikulin et al., 2013) is only the fourth report of the fern family in the Cenozoic. The fern family Matoniaceae was species rich, ecologically diverse, and widespread in the Mesozoic (Tidwell and Ash, 1994; Skog, 2001). Today, this family consists of four species in two genera with an occurrence restricted to a few places in the Malesian floristic region (Kato, 1998). The contrast of common fossils in the Mesozoic versus putative absence in the Cenozoic was emphasized in previous surveys of Matoniaceae diversity. Matoniaceae distribution through time is documented in reports from the Paleogene of Tasmania (Tidwell and Skog, 1992), the Paleocene to early Eocene of Eastern Europe (Vikulin, 2013), and the Eocene and Oligocene of Central Europe (Schmidt and Dörfelt, 2007). These fossils support the maintenance of a wider distribution despite a possible decline in abundance during the Paleogene. The reduced distribution range of these ferns may have occurred in the Neogene. However, these findings do not reject the hypothesis of a decline in matoniaceous fern species diversity and ecological disparity as a result of the rise of angiosperms.

***Pteris? eocenica* Ett. et Gardn. Gardner and Ettingshausen, 1879–1882**

Order Polypodiales Link 1833

Family Pteridaceae E.D.M. Kirchner 1831

Pteris L. 1753

Pteris? eocenica Ett. et Gardn. Gardner and Ettingshausen, 1879–1882 vs. '*Blechnum*' *dentatum* (Göppert 1836) A. Braun 1852

For a detailed list of synonyms, see Gardner and Ettingshausen (1879–1882), Barthel (1976), Frankenhauser and Wilde (1993), and Wilde (1989, 2001).

Material

Core sample locations and fossils identified from the Paleocene–early Eocene sediments of the Boltysh crater are shown in Figure 2 and Figures 6A–6C.

Selected Fossil Occurrences: Boltysh and Beyond

Boltysh, Ukraine, Paleocene–Eocene oil shales (Stanislavsky, 1968; Vikulin, 2013); Geiseltal, Messel, Kayna-Süd, Eckfeldt, Germany, Eocene, "im Farnflöz" fern layer (Barthel, 1976; Frankenhäuser and Wilde, 1993; Wilde, 1989, 2001); London clays, UK, late Eocene (Gardner and Ettingshausen, 1879–1882); Badhyz, Turkmenistan, Eocene (Vassilevskaya, 1957); Sarmatian, Miocene, Katar Mountain, interstream area of Kura and Iori, Transcaucasia, Azerbaijan (Fataliev, 1960).

Description

Morphology of vegetative fronds: The fern is represented by incomplete imprints of segments of sterile pinnae. The margins of the pinnae are parallel, or tend to be lobed or sinuous, and very finely toothed. The most complete segment has a length of 17 mm and is lancet-shaped, with a width of 9–10 mm (Fig. 6A). All segments have toothed margins with small and irregular teeth. The angle between teeth and margins ranges from 55° to 70°. The teeth are ~3 mm wide, and the base is ~0.33 mm long. The midrib is prominent and elongated, equal to 0.5 mm thick, and gets thinner toward the apex. The secondaries are numerous, closely spaced, and thin, ~2 mm apart at the point of attachment to the main vein, once or usually twice dichotomous. The first dichotomy occurs close to the main vein, and the second near the margin. Secondaries emerge from the main vein at acute angles, ~70° at the point of attachment, and ~23° within the remaining area of the segment. The secondary veins are parallel to each other, curving outward to the margin, terminating in teeth, most dichotomous once, some twice, and a very few remain undivided. At the margin of the segments, the distance between the neighbor teeth is ~1.8 mm; the whole 5-mm-long segment contains 10–11 veins.

Discussion

The two pinnae from the Komarov Boltysh collection at drill-core depths of 418 m and 445.2 m (Figs. 6A–6C) are comparable with various fern families, including the derived fern

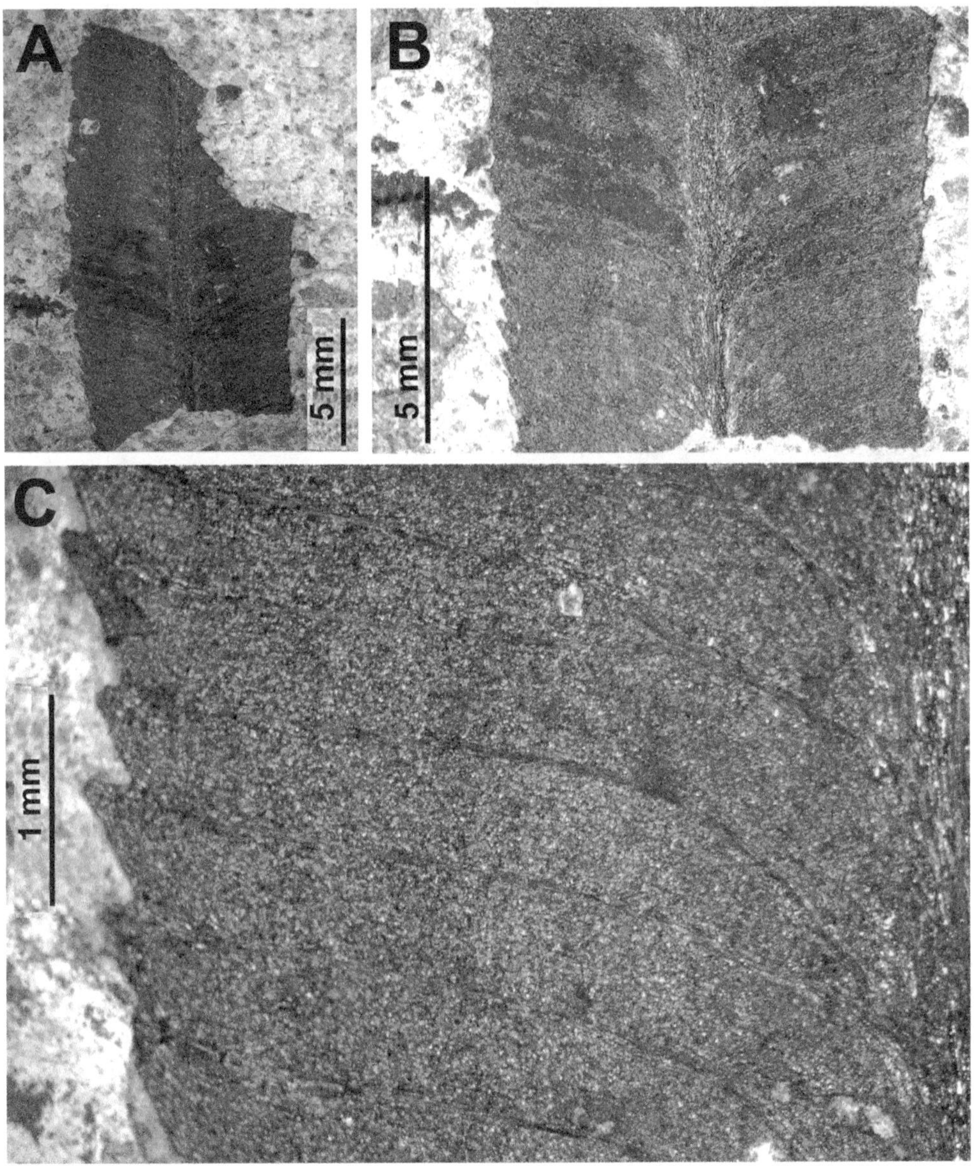

Figure 6. Photographs showing specimens of late Paleocene–early Eocene ferns from the Boltysh drill core; scales are in mm. (A–C) *Pteris eocenica* (syn. *"Blechnum" dentatum*), depth 445.2 m, borehole no. 9810/F1589a, specimen 21/1128 BIN RAS: a large fragment of a black pinnula. (C) Close-up image: note dentate margin and dichotomous venation.

families Blechnaceae and Pteridaceae (Copeland, 1947). Sterile pinnae of fern foliage with free secondary veins as in *Pteris eocenica* are indistinguishable from the *Blechnum dentatum* of Barthel (1976) from the Upper Eocene of Geiseltal. Both fossil taxa show the same type of venation and pinnae shape. Therefore, we argue that these two fossil taxa cannot be distinguished.

Based on the limited morphological characters preserved, these fossils cannot be unequivocally assigned to any extant lineages of ferns, nor can the assignment to *Blechnum* or *Pteris* be rejected. The assignment to *Pteris* (Tryon, and Lugardon, 1990; Walker, 1962) and relatives may (Tryon and Lugardon, 1990) be verified if the *Pteris*-like spores occurring in the palynomorph record of the same core layers (Jolley, 2013, personal observation) were formed by this fern. In the Paleocene deposits near the Boltysh (Central Ukraine) area of the Crimea (Rotman, 1983), the Kuban River near Cherkessk (southern Russia, North Caucasus) (Panova, 1968), and south Kazakhstan (Blyakhova, 1976), spores of *Pteris* sp. div. comprise up to 35% in a rich composition of fern palynospectra. This peculiar fern leaf morphotype with dentate/serrate margins of pinnae and bifurcating secondary veins is known from a number of European Eocene sites under the names of "*Blechnum*" *dentatum*, or "*Pteris*" *eocenica*. In the Boltysh macrofossil collection available to us, this fossil is known only from two sterile pinnae fragments. It is noteworthy that the primary vein of the pinnae is not curving. The presence of curving primary veins is sometimes considered of some diagnostic value.

Comparison of Fossil and Living Species

The type of venation described here is found in various ferns with rather different phylogenetic relationships (Schneider et al., 2009). These fossils resemble the extant ferns belonging to the *Pteris cretica* complex in aspects of venation and the dentate pinnae margin. However, other features, such as the presence of hydathodes and the epidermal morphology, need to be confirmed in future studies. Venation of this fossil is very similar to the widespread European Eocene fern assigned to "*Blechnum*" *dentatum*, also known under synonym name as "*Pteris*" *eocenica* (Gardner and Ettingshausen, 1879–1882; Barthel, 1976, p. 460). *Pteris eocenica* is abundant in the Eocene "fern-beds" of

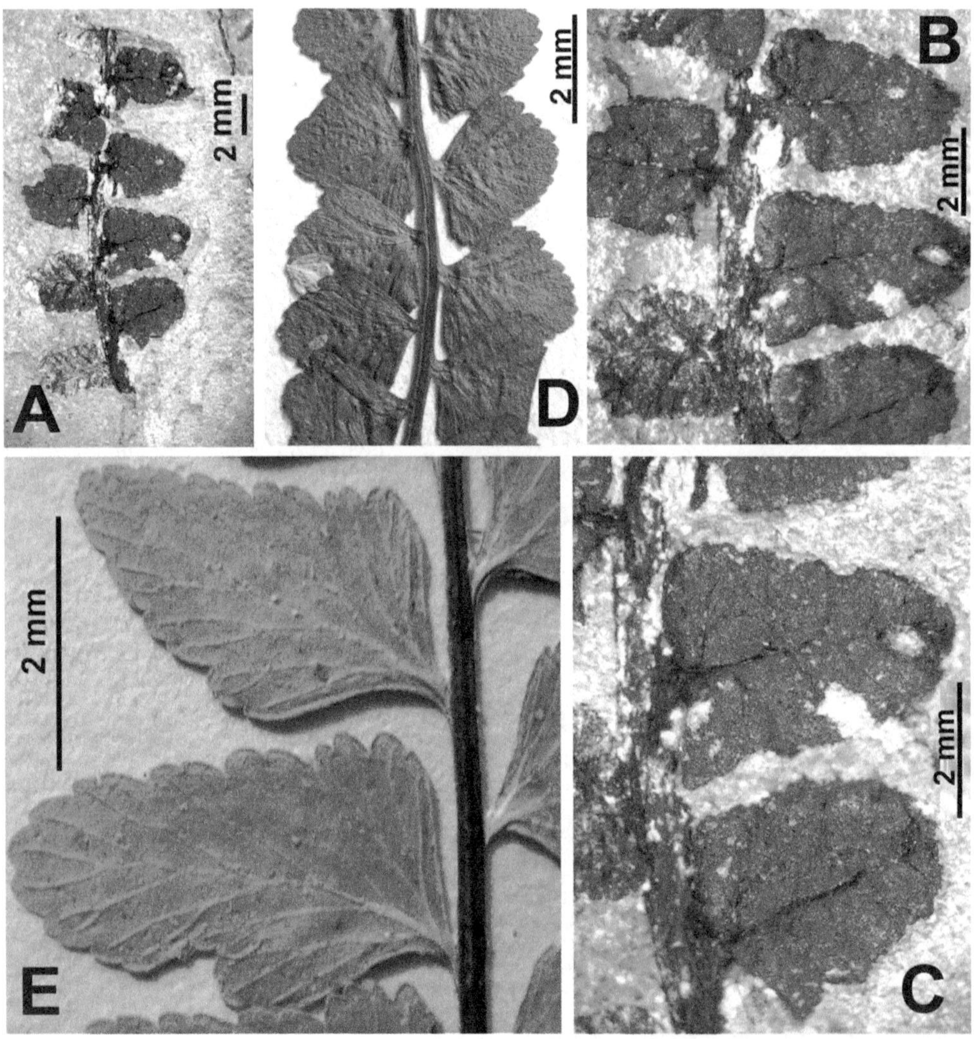

Figure 7. Photographs showing specimens of late Paleocene–early Eocene ferns from the Boltysh drill core and its nearest living relatives; scales are in mm. (A–C) Specimen from borehole no. 9810, depth 445.2 m, sample no. F1594, specimen 31/1128 BIN RAS, a sterile fossil frond fragment, fern 1, resembling a number of living ferns in the leaf dissections. These ferns belong mainly to derived fern families such as Aspleniaceae: e.g., *Asplenium adulterinum* (D), *Asplenium marinum* (E).

Bournemouth in association with swampy and marshy plants. The affinities of fossil samples with sterile pinnulae to *Blechnum dentatum* or to *Pteris eocenica* must remain open pending a detailed study of leaf cuticle and co-occurring spores and leaf cuticle, while leaf macro- and micromorphological features of the *Blechnum/Pteris* type are not distinctive enough to help in this respect.

Fern 1

Material

Borehole number 9810, depth 445.2 m, sample number F1594 (1967, Yu.I. Selin), specimen 31/1128 BIN RAS (Figs. 2 and 7A–7C).

Description

A sterile fossil frond fragment cf. *Asplenium* sp. is pinnate with a length of some 10 mm and width ~4.5 mm (Fig. 7A). The pinnae consist of five pairs of pinnules that are slightly lobed, rhombic or ovate in shape, and subopposite, growing in an alternating pattern down the black, distinct costae (Figs. 7B and 7C). Each pinnulae averages no more than 2 mm in length; the margins are crenate or entire, and the veins are free and forked.

Discussion

A sterile fossil frond fragment resembles a number of living ferns in the leaf dissections (Figs. 7D and 7E). These ferns belong mainly to derived fern families, such as Aspleniaceae: *Asplenium trichomanes* (Figs. 7D and 7E) and Pteridaceae, but similar leaf shapes are also known from ferns belonging to the schizaeoid fern family Anemiacae. Thus, the phylogenetic information provided by these fossils requires careful studies of epidermal features and the presence of hydathodes.

Fern 2

Material

Borehole number 10331, depth 418.1 m; sample number 1350a (1967, Yu.I. Selin); specimen 41/1128 BIN RAS (Figs. 2 and 8A).

Description

The Asplenioid? fern 2 is represented by one pinnate fragment of a dissected blade with linear filiform or lanceolate pinnules in opposite pairs (Fig. 8A). A dorsoventrally flattened rachis narrows to a point of verticil formed by opposite pairs of pinnules. The length of the whole fragment is 4 mm; the width of each pinnule is ~0.6–1.0 mm. Secondary veins in each pinnule emerge from the primary vein at an acute angle of ~64°.

Discussion

The fossil fern 2 is comparable in its morphology to derived ferns with highly dissected lamina such as *Asplenium fragrans* and *Asplenium viviparum* (Figs. 8B and 8C). It has a dorsoventrally flattened rachis narrowing to a point of verticil formed by opposite pairs of pinnules.

Fern 3

Material

Borehole number 10360; sample F747 (1967, Yu.I. Selin); specimen 51/1128 BIN RAS: imprint of a ruptured pinnae with a prominent rachis (Figs. 2 and 8D–8H).

Description

The sterile lamina is pinnate and ruptured, and the total length of the "pinnae segments" or "pinnules" equals 30 mm; their costa is strong and distinctly raised; the pinnules are variable in shape from triangular to suborbicular, usually obtuse, and adnate throughout. The average width and length of the pinnules are ~5–6 mm and 2.5 mm. Their texture is thickly sub coriaceous, succulent in appearance, and very dark; the pinnule veins are simple or once-furcated (i.e., bifurcated), the veinlets mesh is more clearly visible near the margins, and the basal pinnulae reduce gradually to small flanges at the base.

Discussion

Pinnae are sessile, variable in shape, and subopposite and vary from triangular to suborbicular. These fern fragments resemble the living fern *Blechnum blechnoides* (*Lomaria banksii* Hook. F.) (Figs. 8I and 8J). Sterile pinnae of the modern species *B. blechnoides* are less triangular, with a lower length/width ratio in comparison with the fossil species.

FLOWERING PLANTS

Representatives of flowering plants (Figs. 9–11) in the Boltysh sediments consist of a small assemblage of leaf fossils, related in composition to those obtained from the Paleocene to early Eocene flora of the south of European Russia (Krasnov, 1910; Baikovskaya and Koval', 1968; Zhilin and Vikulin, 1986), the Paris Basin (Saporta and Marion, 1873, 1878; Piton, 1940), and the Green River oil-shale formation flora from North America (Pigg and DeVore, 2010).

***Pseudosalix* sp.**

Family Salicaceae

Pseudosalix Boucher, Manchester et Judd, 2003

Material

Boltysh borehole number 9866, depth 325.8 m; sample 1546b (1967, Yu.I. Selin); specimen 61/1128 BIN RAS; borehole number 9866, depth 346.5 m, sample F1575 (1967, Yu.I. Selin), specimen 101/1128 BIN RAS (Figs. 2 and 9A–9E).

Description

The elongate slender leaf fragments are ~40 mm long and 10 mm wide, serrated in one specimen, cuneated in another;

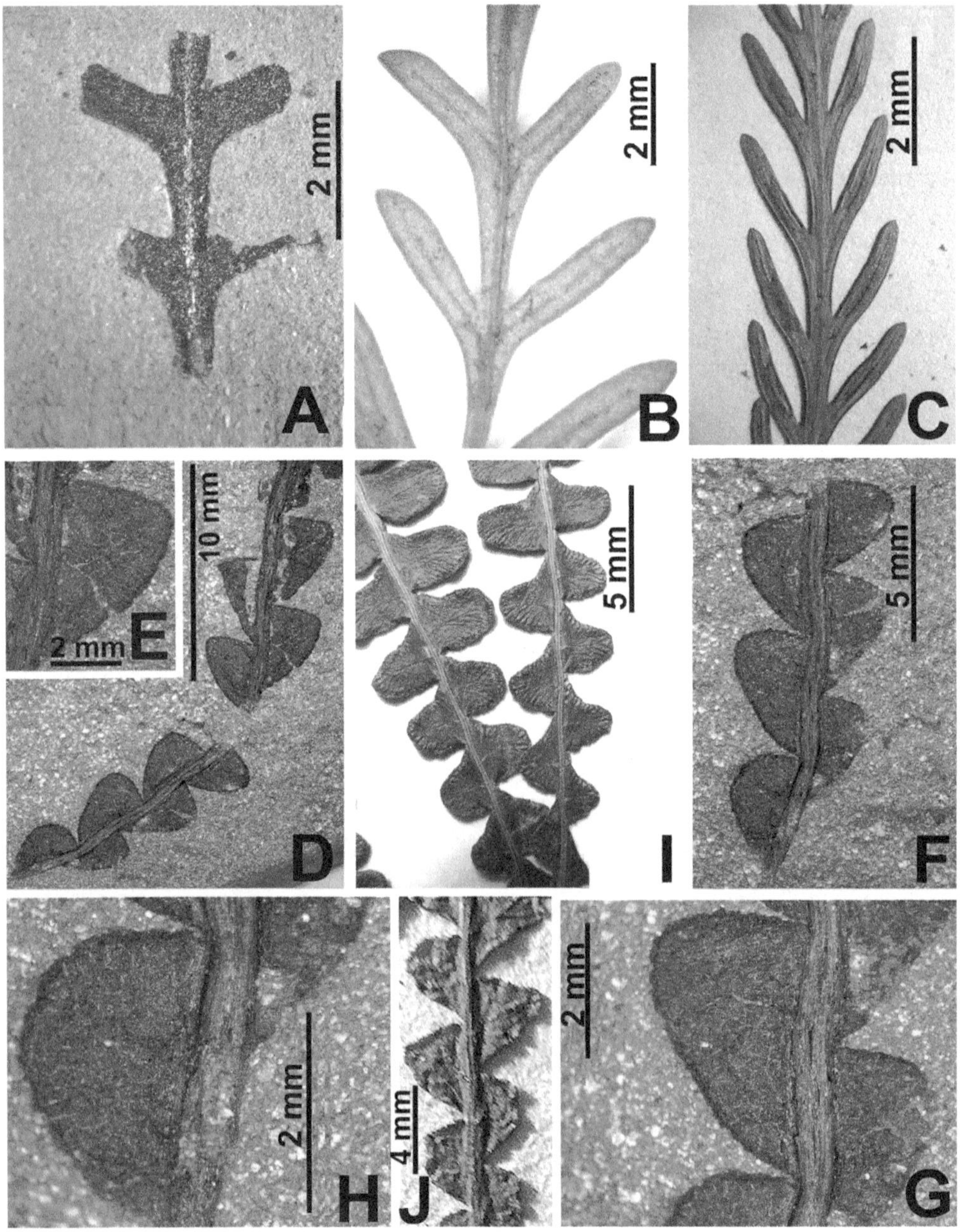

Figure 8. Photographs showing specimens of late Paleocene–early Eocene ferns from the Boltysh drill core and their nearest living relatives; scales are in mm. (A) Fern 2, borehole no. 10331, depth 418.1 m; sample no. 1350a; specimen 41/1128 BIN RAS. The fossil fern 2 is comparable in morphology to derived ferns with highly dissected lamina such as *Asplenium viviparum* (B), and *Asplenium flaccidum* (C). (D–H) Fern 3, borehole no. 10360, sample F747, specimen 51/1128 BIN RAS: imprint of a ruptured pinna with a prominent rachis; these fern fragments (D) resemble the living fern *Blechnum blechnoides* (I and J). Sterile pinnae of the modern species *B. blechnoides* are less triangular, with a lower length/width ratio in comparison with fossil species.

the base is not preserved in one leaf (Fig. 9A) and cuneated in another (Fig. 9E). The leaf fragment with a preserved cuneated base is 22 mm long and 10 mm wide (Fig. 9E). The margin is serrated with glandular teeth, and the tooth apex ends in a circular gland (Figs. 9C and 9D). Teeth are somewhat irregularly spaced, simple and ~1 mm from the base to the apex. The tooth shape is concave on the apical side and convex to straight on the basal side with a rounded sinus. Venation is pinnate with intersecondary veins and a moderately thick midvein. Secondary venation is semicraspedodromous with irregular spacing, the secondary veins are prominently looped closer to the leaf margin, and the angle between the primary and secondary veins is from 45° to 55° degrees, with weak intersecondary veins (Fig. 9C). The spacing and looping of the secondaries are well defined, and the tertiary venation is mixed with alternate-opposite percurrent. Secondary veins usually branch before reaching the margin, sending one branch to the tooth and the other to loop with the supra-adjacent secondary vein (Fig. 9D). Each tooth is generally enervated by the principal vein of the tooth, often extending directly from the secondary vein and entering the tooth medially, and a pair of adjacent, slightly thinner accessory veins on each side originate from veins above and below (Figs. 9C and 9D). Tertiary venation is random reticulate, with the veinlet angle variably obtuse to the primary vein.

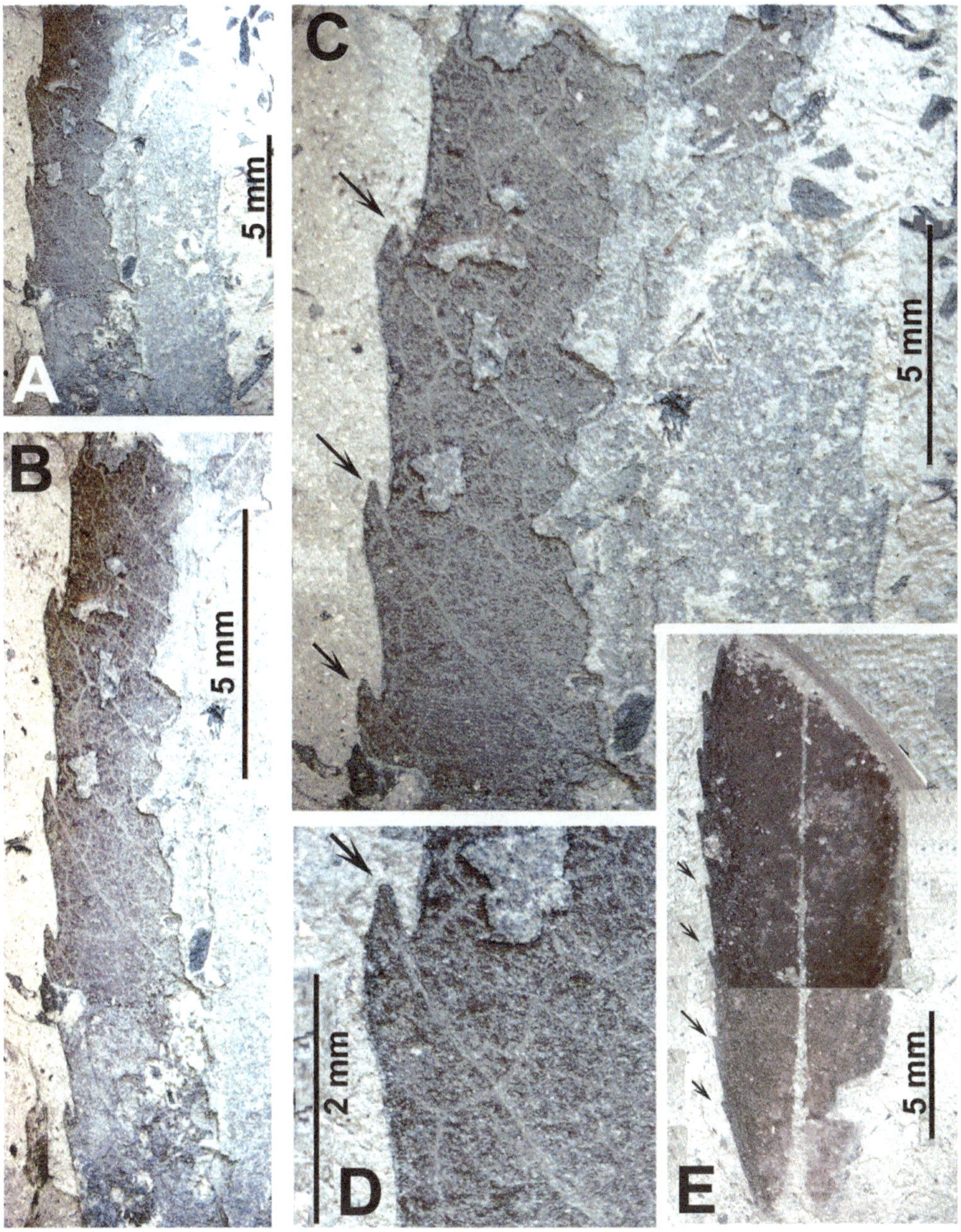

Figure 9. Photographs showing specimens of late Paleocene–early Eocene angiosperms from the Boltysh drill core; scales are in mm. (A–E) *Pseudosalix* sp., borehole no. 9866, depth 325.8 m, sample 1546b, specimen 61/1128 BIN RAS; borehole no. 9866, depth 346.5 m, sample F1575, specimen 101/1128 BIN RAS; leaves of *Pseudosalix* from Boltysh share characters: elongate lamina, semicraspedodromous venation, and salicoid teeth with middle Eocene *Pseudosalix handleyi* of Bonanza, Utah. The size of the marginal teeth both of Boltysh leaves and teeth of *Pseudosalix* from North America is equal to ~1 mm. The size of teeth in *Pseudosalix* and Boltysh *Salix*-like leaves is somewhat larger than in many of the modern willows. (C and D) The spherulate gland at the tooth apex (marked by arrows) is most similar to what has been termed a "salicoid tooth" (Hickey and Wolfe, 1975; Judd et al., 2002).

Discussion

Boltysh leaves with glandular teeth and pinnate venation resemble salicoid leaves of *Pseudosalix*, which would likely have been fused with those of *Salix* because of the shared characters of elongate lamina, semicraspedodromous venation, and salicoid teeth. The size of the marginal teeth of both Boltysh leaves and teeth of *Pseudosalix* from North America is ~1 mm. The size of teeth in *Pseudosalix* and Boltysh *Salix*-like leaves is somewhat larger than in many of the modern willows. The spherulate gland (Figs. 9C and 9D) at the tooth apex (Fig. 9D) is most similar to what has been termed a salicoid tooth (Hickey and Wolfe, 1975; Judd et al., 2002). The tooth apex with its gland is enervated by secondary or intersecondary accessory veins, both along the apical margin of the tooth and along the basal flank (Fig. 9D).

Leaves of *Pseudosalix* have somewhat larger teeth compared with those of modern species of *Salix*. Among fossil leaves similar to the Boltysh salicoid leaf are those previously described from the Paleogene of the Green River flora—*Pseudosalix handleyi* and *Salix cockerelli* Brown and *Salix longiacuminata* Brown (Brown, 1934; Boucher et al., 2003). The secondary veins and intersecondaries are both prominent and organized similarly to European fossil species *Salix varians* Goeppert, *Salix longinqua* Sap. et Mar, and *Salix schimperi* Lesquereux and North American fossil species *S. cockerelli* and *P. handleyi* leaves.

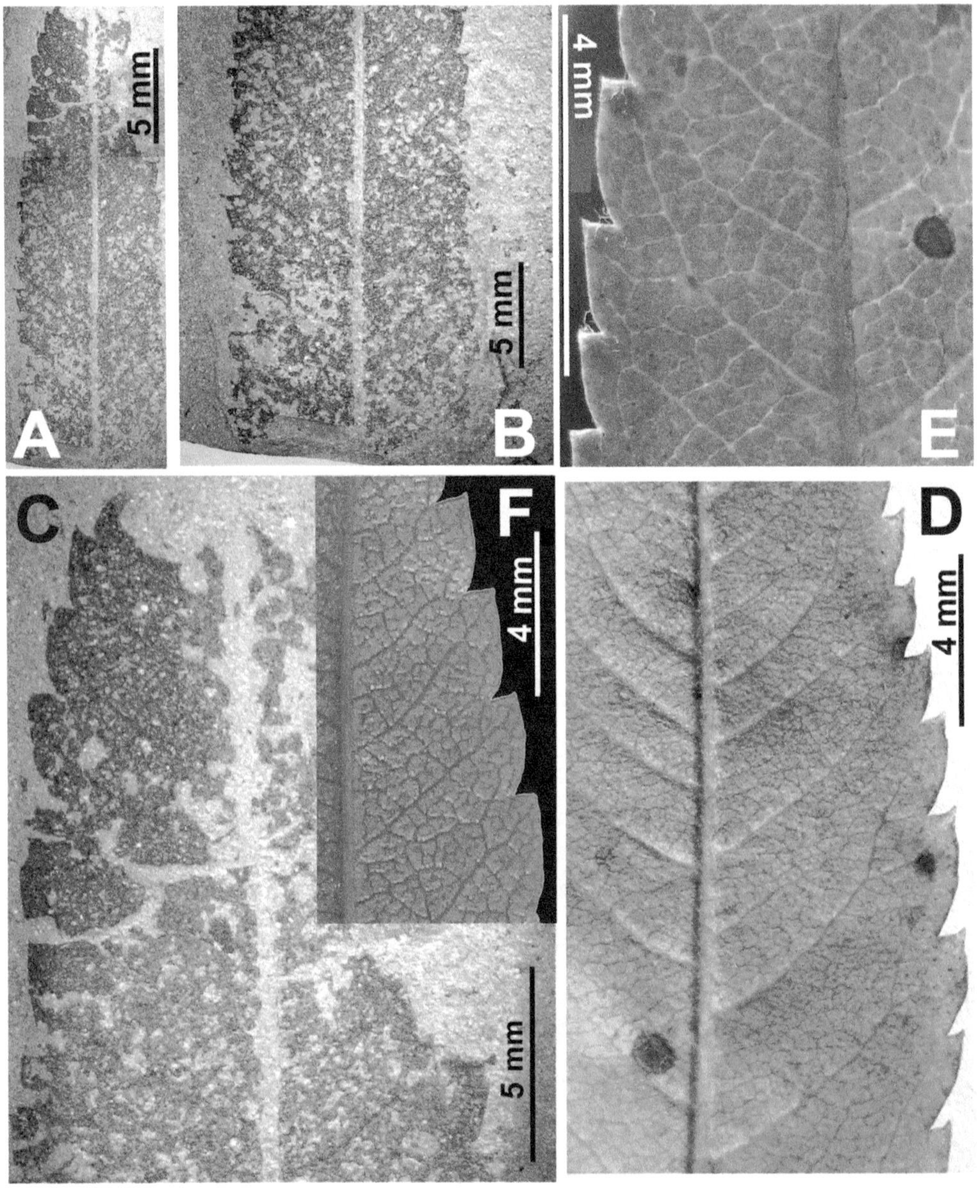

Figure 10. Photographs showing specimens of late Paleocene–early Eocene angiosperms from the Boltysh drill core and its nearest living relatives; scales are in mm. (A–C) *Sorbus* sp., borehole no. 9866, depth 346.9 m, sample F1577, specimen 71/1128 BIN RAS: the fossil leaves share combined characters of *Sorbus aucuparia* (D and E), a European species of mountain ash, and *Sorbus sambucifolia* (F), a Siberian mountain ash.

Among living species, *Salix* from Boltysh resemble *Salix exigua* and *Salix lasiandra* from North America. Boltysh salicaceous taxa based on fossil leaves are characterized by the presence of typical salicoid teeth, semicraspedodromous secondary venation, and lanceolate shape. The presence of circular glands on teeth of the leaf margin (i.e., salicoid teeth) distinguishes Salicaceae s.l. (Figs. 9A–9E) from the other candidate families. These distinctive teeth are likely a characteristic of all members of Salicaceae s.l. (Judd et al., 2002; see also Wilkinson, 2007). Their presence in the Boltysh Paleocene oil shale resembles Eocene salicaceous leaves from Green River Formation, described as *Pseudosalix*, providing strong support for placement of Boltysh fossil leaves with salicoid teeth in Salicaceae s.l. In view of the lack of *Salix* infructescences, an evaluation of whether Boltysh *Salix*-like leaves represent an extinct taxon, such as *Pseudosalix* (sharing features of *Salix*, *Populus*, and some Flacourtiaceae) or the extant genus *Salix*, requires more complete material.

Sorbus sp.

Family Rosaceae

Sorbus L., 1753

Material

Borehole number 9866, depth 346.9 m; sample F1577′ (1967, Yu.I. Selin); specimen 71/1128 BIN RAS (Figs. 2 and 10A–10C).

Figure 11. Photographs showing specimens of late Paleocene–early Eocene angiosperms from the Boltysh drill core and their nearest living relatives (NLR); scales are in mm. (A–E) *Comptonia dryandrifolia* of Myricaceae family (syn. *Comptonia schrankii*), borehole no. 10328, depth 275 m, sample F370a, specimen 81a/1128 BIN RAS; borehole no. 10328, depth 275 m, sample F370b, specimen 81b/1128 BIN RAS; diagnostic features of *C. dryandrifolia*: long, narrow leaves with deep pinnate lobes that might extend to the midvein, pinnate primary vein, secondary veins craspedodromous. Compound toothed margin occurs. (F–H) Compare with NLR of fossil species and modern North American *Comptonia peregrina.* (I) Phyllites cf. *Myrica longifolia*, borehole no. 9848, depth 263.6 m, sample F979, specimen 91/1128 BIN RAS; fossil impression resembles leaves that have been assigned to *M. longifolia*. Though modern affinities of Boltysh *Myrica*-like leaf are open, it resembles *Myrica lindeniana* (J) native to Mesoamerica.

Description

Leaflet is asymmetrical, 40 mm long and 13 mm wide (length/width ratio > 4:1). The shape of the leaflet is narrow, oblong, and lanceolate; the apex and the base are not preserved; the margin is coarsely serrate, the apical angle is acute, sinuses are angular, and spacing is irregular; and teeth are simple, regularly to irregularly spaced, forming segments with concave apical and basal sides. The venation is pinnate, semicraspedodromous. The midrib is strong, slightly curved. About 12 pairs of curved secondary veins arise at angles of 70°–50° to the midrib; secondary veins are straight, looping sharply near the margin to make an almost quadrangular intercostal area. The angle of the divergence may be more acute on one side of the midrib than on the other and initiate straight or curved, becoming more curved and running tangent to, and occurring quite close to, the margin. The external vein arises from the middle of the highly curved segment of the secondary loop and enters the tooth medially. External secondary branches enter teeth typically centrally but some apically; tertiary veins are orthogonal, forming irregularly polygonal reticulum; quaternary veins are orthogonal, forming a quadrangular to pentagonal reticulum.

Margin segments have concave apical and basal sides (cf. close-up Fig. 10C with nearest living relatives [NLR] Figs. 10D–10F); branch from secondary loop enters tooth (cf. close-up Fig. 10C with NLR Figs. 10D–10F); their secondary veins pronouncedly and sharply curve near the margin following a semicraspedodromous pattern; Tertiary venation forms a coarse, irregular mesh. Their sinuses are supplied with veins originating on the abmedial sides of the secondary vein loops and from craspedodromous external secondary veins.

Discussion

The single specimen observed from Boltysh has some asymmetry in the lamina, which is a strong indication of a leaflet from a pinnately compound leaf. Particularly peculiar is the consistently semicraspedodromous venation. The difficulty of separating isolated leaflets of extant species of *Sorbus* or *Rosa* makes determination of this one fossil specimen speculative. The asymmetrical leaflet from the Boltysh collection with pinnate venation and serrate margin also has a superficial resemblance to the leaflet called *Barghoornia oblongifolia* Wolfe and Wehr (Bursearceae) from the early-middle Eocene of northeastern Washington, United States (Wolfe and Werr, 1987). However, a more conservative interpretation places the Boltysh specimen in the Rosaceae, in particular *Sorbus*. Similarity of fossil leaf shape and venation architecture is observed with living species of European *Sorbus aucuparia* (Figs. 10D and 10E) and Siberian mountain ash *Sorbus sambucifolia* (Fig. 10F) of amphi-Pacific (both sides) distribution in temperate areas of the Russian Pacific, Kamchatka, Korea, Japan, and North American Pacific Alaska (Hulten, 1968; Krestov et al., 2008).

***Comptonia dryandrifolia* Brongn. (syn. *C. schrankii*)**

Family *Myricaceae*

Comptonia L'Heritier, 1789

Material

Borehole number 10328, depth 275 m; sample F370a (1967, Yu.I. Selin); specimen 81a/1128 BIN RAS (20 mm leaf fragment); borehole number 10328, depth 275 m; sample F370b (1967, Yu.I. Selin); specimen 81b/1128 BIN RAS (25 mm leaf fragment) (Figs. 2 and 11A–11E).

Selected Fossil Occurrences

East European Plain: Eocene, lava in vicinity of Putivl', Sumskaya Province, Ukraine (Krasnov, 1910); early mid-Eocene, Kamenka of Oboyanskiy district, Kursk Province, European Russia (Baikovskaya and Koval', 1968); early mid-Eocene, Visloe near Belgorod, Belgorod Province, European Russia (Zhilin and Vikulin, 1986), late Paleocene, Romankulsay, South Ural (Baikovskaya, 1984).

Description

Leaves are simple, narrow elliptical to mostly lanceolate (all fragmented) fragments (middle and basal parts of pinnate-laciniate leaves) of *C. dryandrifolia* observed in Boltysh cores. The basal pair of lobes is sometimes shortly decurrent on the petiole (Figs. 11A and 11C). The shape of lobes is mainly triangular, rarely obtuse-triangular, or falcated. Apical parts are not preserved. Laminae are deeply lobed to dissected, with alternate to opposite lobes. Sinuses between the lobes are regular and reach the midvein. Individual lobes are triangular, falcate, or rounded, and half-oval with rounded or acute tips (Figs. 11A–11E). Lobes are alternate or opposite, and sessile to midrib. Original length is more than 20–25 mm; width is 4–5 mm. The shape of lobes is regular, and apices are acute. The margin is entire, the midrib is stout, and secondary veins arise at angles of 30°–85°. Leaves are narrow, with an average width of 5 mm. Preserved length of the fragments is up to 25 mm. Most of the specimens of this species previously described from the areas of the former USSR, except for specimens 12 and 33 from the early Eocene of Visloe, Belgorod Province, Russia (coll. 1387, BIN RAS), lack venation of higher orders. These specimens, especially 12/1387 (Zhilin and Vikulin, 1986: Pl.1, fig. 1, 6), possess remained venation of the third of fourth orders, forming a reticulum of tiny areoles. Tertiary veins branch from secondary veins at an angle of 30°–48°.

Discussion

The only living representative of the genus *Comptonia* today inhabits eastern North America. Several species of *Comptonia* are known from the extratropical areas of the Northern Hemisphere during the Late Cretaceous, Paleogene, and Neogene, as indicated by the fossil records of leaves, carpoid remains, and pollen, which were especially numerous in Eurasia (Zhilin and Vikulin, 1986). An extensive updated list of geographical distributions along with paleobiogeographical implication of the genus *Comptonia* was recently considered (Liang et al., 2010). Most common late Paleogene and Neogene species (e.g., *Comptonia acutiloba* Brongn. and *Comptonia naumannii* [Nath.] Krysht) have similar leaf architecture to the modern *Comptonia peregrina* (L.) Coult. Its leaf blade is deeply pinnatifid into numerous oblique or sparingly dentate lobes, 3–15.5 × 0.3–2.9 cm, with the lobes alternate to nearly opposite, and the base truncate, cuneate to attenuate, or oblique and the apex acute. Characteristic of the older Paleogene, *Comptonia schrankii* (syn. *C. dryandrifolia* Brongn.) has smaller-size lobes, which distinguish these leaves from the later Oligocene and Neogene species *C. acutiloba.*

Comparison with Fossil and Living Species

Leaves of the fossil species *C. schrankii* were described and sketched by Sternberg as early as 1822. He thought that he was dealing with a fern and used the generic name *Aspleniopteris.* Meanwhile, the morphological resemblance of the leaf shape of the Angiospermous genus *Comptonia* to some ferns with a dissected lamina (e.g., some species of Aspleniaceae) is purely superficial. Contrary to ferns, living and fossil *Comptonia* species, also known as sweet-fern, or fern-gale, of the dicotyledonous angiosperm possess an arching micro-architecture of several secondary veins along the border of the leaf lobes. In all ferns, species with a dissected lamina, which superficially resemble *Comptonia*, have only one or no distinct vein per ultimate segment (i.e., a single vein per pinnule or segment often not reaching the margin). Their pinnae rarely show anastomosing, dichotomously netted venation.

Venation. Despite some similarities of leaf shapes of *Comptonia* and ferns, the latter have only one single vein per ultimate leaf segment, whereas venation of *Comptonia* is characterized by two or three secondaries arching in each lobe (Berry, 1906). The secondaries proceed from the midrib at a wide angle, often 90°, gradually curving upward, in greater extent in the lobes toward the tip of the leaf. The uppermost secondary in each lobe usually runs directly to the tip of the lobe when the latter is pointed, or directly to the margin in those leaves with rounded lobes; its position on the margin is indicated by a slight mucronate point (Figs. 11F–11H). The one or two secondaries below the upper one are inserted at equal distances apart and become more curved as they proceed outward, curving upward to join a short downwardly directed lateral veinlet branch from the secondary next above. The tertiary venation in the border region is looping along the margin of the leaf; the finer areolation shows three-, four-, or five-sided reticulations.

Fossil leaves of *C. dryandrifolia* possess the exact characters of the young leaves of existing *Comptonia*, which show near the growing tips thick midrib, acute lobes, and coriaceous texture. Not all leaves of living *Comptonia* are acutely lobed as in fossil *C. dryandrifolia*. However, this is seen in specimens that are not completely unfolded, or from herbarium specimens in which the lobes usually become somewhat involuted during drying (Berry, 1906). Many authors use the name of *Comptonia dryandrifolia* as a synonym for *C. schrankii* (Sternberg) Berry. However, the depository of the holotype *Aspleniopteris schrankii* Sternberg is unknown, and its image (Sternberg, 1825) is so ambiguous and obscure that the use of the name leads to errors. A specimen from the holotype location (Hoering, Austria) is stored in the Paleobotany Laboratory, BIN RAS, and labeled "*Dryandra schrankii* Sternb. spec." (apparently written by Ettingshausen) and has the same poor state of preservation (Zhilin, 1980) as the holotype. Venation on these samples is indistinguishable. Brongniart (1828) illustrated a whole leaf and a fragment of the leaf of *Comptonia dryandrifolia* with the details of venation of the third and fourth orders, allowing quite a clear idea of this species. The smaller size of the leaves and regular lobes distinguish these from *Comptonia difformis* (syn. of C. *acutiloba*).

However, some transitional leaf forms reveal the need for revision. Furthermore, leaves of *C. difformis* and *C. schrankii* fall in the morphological variability of leaves of modern *C. peregrina* (Berry, 1906). The smaller leaf size and regular lobes distinguish these from *C. difformis* (syn. *Comptonia acutiloba* Brongn.) and *C. naumannii.* The Boltysh material also represents a simple (single blade) leaf, and thus cannot represent the leaflet of the somewhat similar *Lyonothamnus* (Rosaceae). The foliage of Boltysh *Comptonia* with a double serration is rare in the genus (Figs. 11B and 11F); most of the described fossil species and the one extant species are mainly singly serrate. These small leaves have a distinctive narrow-lobed or deeply toothed outline with lateral veins, which tend to converge toward the apices (Figs. 11C and 11D). These leaves are common in some early Eocene outcrops near Belgorod, Visloe, Russia (Zhilin and Vikulin, 1986), ~300 km from the Boltysh sites. *Comptonia* is an extant North American genus of the Myricaceae (e.g., *C. peregrina* [L.] J.M. Coult.), but a similar leaf form is also found in *Dryandra*, an extant Australian genus of the Proteaceae, recently subsumed within *Banksia* (Mast and Thiele, 2007). Myricaceous pollen is present in the Boltysh Paleogene cores, although pollen comparable to *Dryandra/Banksia* of Proteaceae is not. Myricaceae pollen does occur elsewhere in the European Paleogene and Neogene. Pollen of the Proteaceae is not known unequivocally from the sediments of this age in the East European Plain.

Phyllites cf. *Myrica longifolia* Unger
Insertae Sedis
? *Myrica* L.

Material

Borehole number 9848, depth 263.6 m; sample F979 (1967, Yu.I. Selin); specimen 91/1128 BIN RAS (leaf fragment 8 mm width × 16 mm length) (Figs. 2 and 11I).

Description

The leaf shape is oblong-linear, the leaf fragments are 8–10 mm wide and 16–30 mm long, and the strong leaf lamina descends into the leaf base, which is symmetrical acute to cuneate. The blade tip is not preserved. The leaf margin is finely serrated with irregular teeth, partially pointed to slightly splay-spinose. In the basal region, teeth are less sharp, with almost smooth edges. Primary nerve is distinctly visible. Venation is semicraspedodromous. The secondary veins arise at acute angles up to 85°, forming decreasing loops near the leaf margin. Accessorial veinlets to the tooth tip are not visible.

Discussion

The fossil impression resembles leaves that have been assigned to *Myrica longifolia.* According to Ettingshausen (1853), these morphotypes of Paleogene leaves were identified earlier as belonging to the Proteaceae *Hakea*, *Banksia*, or *Dryandroides* but were later revised (Ettingshausen and Standfest, 1888) and identified as *Myrica lignitum* (Unger) Saporta. Based on leaf cuticular studies by Kvaček (2004), there are two different forms of the latter species, previously described as *Myrica lignitum*: (1) *M. longifolia* Unger, and (2) *M. lignitum* (Unger) Saporta sensu stricto. A detailed list of synonyms is given in Rüffle (1976, p. 343–347). Leaf impressions of *M. longifolia* are common in European Eocene and especially early Oligocene sediments, where this species dominated in wet coal floras of the Haselbach series and Haring (Mai and Walther, 1978).

In the tropics and subtropics, where *Myrica* spp. may occur in the mountains, they may also grow on dry slopes, ocean sand beaches, or heavy clay soils, and in wet environments. Although modern affinities of Boltysh *Myrica*-like leaves are open, they resemble *Myrica lindeniana* C. DC. (Fig. 11J), which are native to Mesoamerica.

GENERAL DISCUSSION

Mesozoic and Tertiary Fern Elements in Boltysh Paleogene Flora

Most of the ancient matoniaceous ferns, including the extinct genus *Weichselia*, were dominant elements of the vegetation in many parts of the world during the Early Cretaceous. They were thought to have disappeared at the end of the Cretaceous together with several other groups, such as seed-ferns (McLoughlin et al., 2008), that had occupied key positions in the terrestrial ecosystems (Collinson, 2001; Skog, 2001). Their demise was generally correlated with competition from diversifying flowering plants through the Cretaceous and the global environmental crisis attributed to the Chicxulub impact at the end of the Maastrichtian. Nevertheless, new finds of fossils of matoniaceous ferns[1] from Baltic/Central European amber (Schmidt and Dörfelt, 2007) and *Weichselia* from the Boltysh crater oil shale in the Ukraine, eastern Europe, along with similarly spectacular findings of the Tasmanian seed-fern *Komlopteris* (Corystospermaceae) (McLoughlin et al., 2008), show that these "Lazarus taxa" survived into the early Cenozoic.

Both microfossils and macrofossils suggest the occurrence of ferns belonging to Mesozoic lineages of ferns and derived ferns. Some are typical Mesozoic elements (*Weichselia*), whereas others (*Pteris eocenica*, "*Blechnum dentatum*," and *Asplenioid* and *Blechnoid* ferns) appear to represent groups that are frequently found in Tertiary fern flora and may belong to the derived ferns, a lineage rapidly diversifying in the Late Cretaceous. The recovery of *W. reticulata*–like fossils may be among the most interesting discoveries. These Mesozoic tree ferns were common and widespread during the Late Jurassic and Early Cretaceous (Diez et al., 2005; Brown et al., 2012; El-Khayal, 1985). *Weichselia reticulata* is generally assumed to be related to Matoniacae, a distinct family of Gleicheniales (Collinson, 2001; Diez et al., 2005; Silantieva and Krassilov, 2006). Similar to the sister family Dipteridacae, the Mationaceae were a diverse, globally distributed, and ecologically important fern family from the Late Triassic to the Middle Cretaceous. However, they likely decreased in species richness, distribution range, and ecological variation during the Late Cretaceous, as suggested by the few macrofossils recorded since the Cenomanian and complete absence in the earliest Tertiary (Collinson, 2001, 2002). Today, the family is only known from four species occurring in relict habitats in the Malay Archipelago (Kato, 1998). Collinson (2001, 2002) argued that the *Matoniaceae* were restricted to their modern range already in the Early Tertiary, but this view has been questioned by recent reports of Cenozoic amber fossils from the Baltic and Central Europe (Schmidt and Dörfelt, 2007). The recovery of *Weichselia* from the late Paleocene–early Eocene of Eastern Europe suggests that this fire-adapted fern (Brown et al., 2012) survived the rise of the angiosperms, at least in some areas.

Paleoecology of Ferns and Angiosperms Growing near Boltysh Crater Lake

Myricaceae and Salicaceae of the Boltysh ecosystem were presumably capable of nitrogen fixation similar to their living relatives, which enabled them to colonize old lava flows, regenerating freely under open canopy as pioneer colonizers. Salicaceae species—poplars (*Populus* spp.) and willows (*Salix* spp.)—are early succession trees, also nitrogen fixers, with rapid growth and deep roots adapted to nutrient-poor environments (i.e., riparian habitats with sandy soils; Wuehlisch, 2011). A large endophytic community resides in the stem tissue of Salicaceae species (Ulrich et al., 2008). Among these endophytes, several diazotrophic (nitrogen-fixing) bacteria[2] were recently identified (Taghavi et al., 2009; Kraiser et al., 2011).

The angiosperm vegetation cover of the Boltysh ecosystem was reinforced by a number of ferns, which occur any time that the environment gets destroyed, as they are the first colonizers. Representatives of both angiosperm families (Myricaceae and Salicaceae) are nitrogen-fixing pioneer trees in their native ranges and often become dominant in their ecosystems. Both spread rapidly through aggressive growth and fruit production. The myricaceous representative, the genus *Comptonia*, is considered a pioneer by Mai (1995). This genus was identified in Boltysh (Vikulin, 2013), Visloe (Zhilin and Vikulin, 1986), and Messel based on leaves (Wilde, 1989; Wilde and Frankenhäuser, 1998) and likely represents part of the pioneer succession. Leaf remains of myricaceous and salicaceous land plants in the Boltysh crater lake oil shales may be attributed to arborescent and shrubby pioneer vegetation invading the immediate vicinity and initial reforestation of the disturbed crater area. Within the initial herbaceous communities, ferns like *W. reticulata* (Matoniaceae) (Figs. 3–4), *Pteris eocenica* (Fig. 6), and other ferns resembling extant fern morphotypes (Figs. 7–8) played an important role in recolonizing the area. Low and rapidly growing trees and shrubs, including Rosaceae (Fig. 10), may have soon have followed and invaded the primary herbaceous vegetation in a subsequent succession.

Paleoecosystem of the Boltysh Geothermal Lake—An Early Paleogene Refuge

The Boltysh crater is compatible by size and ecosystem with a modern giant caldera of the ancient volcano Ngoro-Ngoro in eastern Africa. Nevertheless, geologists studying the Boltysh currently consider its depression to be an impact crater with

[1]The amber inclusions of *Matonia striata* show that the family was still present in Europe during the whole Paleogene and that the formation of the Southeast Asian relict area of the family happened much later (Schmidt and Dörfelt, 2007).

[2]Nitrogen-fixing bacteria occur in the living tissues of the stem and branches of Salicaceae and not in root nodules like in the legume family (Fabaceae) and a number of woody plant families (i.e., Betulaceae, Myricaceae, Rosaceae; Taghavi et al., 2009; Ulrich et al., 2008; Kraiser et al., 2011).

geothermal activity (e.g., Kelley and Gurov, 2002; Gurov et al., 2006). The crater formed on one of the islands near the northern shore of the tropical Tethys Ocean (Gavrilov et al., 2009; Jolley et al., 2010). After 15 m.y. of existence (Thanetian-Ypresian), the 24-km-diameter Boltysh crater lake disappeared as a result of a marine transgression at the end of the middle Eocene (Stanislavsky, 1968; Gurov et al., 2006; Zosymovych and Ryabokon', 2010). As the thermal phase of Boltysh lake decreased, a succession of pioneer vegetation took place, recently preliminarily confirmed by palynology (Jolley et al., 2010), which showed an increased content in fern spores in the Paleocene core drilling samples. Similar "ecological disruption and mass kill" (Wolfe and Upchurch, 1986, p. 148; Vajda et al., 2001) associated with the so-called "fern spikes" are confined not only to the Cretaceous-Paleogene boundary but also to other stratigraphic levels marked by volcanic activity (Krasilov, 2006). Such fern spikes are not unique to impacts; they occur any time that the environment gets destroyed, because they are the first colonizers. There are impressive examples of living tropical ferns (e.g., "*Pityrogramma austroamericana*" [sic! H.Sch.]) capable of survival in extreme environments, such as environments characterized by high acidity of fallen volcanic ashes and high concentrations of sulfur and carbonates (Kornas, 1978; Spicer et al., 1985; Riba and Reyes, 1990). Comparable local environments near Boltysh hydrothermal area might have created a unique combination of edaphic environmental factors, enabling survival of relict Cretaceous ferns (e.g., *W. reticulata*) during the Paleocene by easing competition with the evolutionarily younger species of flowering plants, for which adaptability to ecological extremes was still inadequate.

Further disappearance of the Boltysh refugial ecosystem inhabited by the Cretaceous relict xeric *Weichselia* fern, fossilized in sediments of the geothermal crater lake, was due to the marine transgression of the middle-late Eocene (Vikulin, 2013). As a result of this transgression, hydrothermal manifestations ceased and increased competition with rapidly evolving modern flowering plants (Wilf and Johnson, 2004), including representatives of a living prototype of coastal mangroves: Marine to brackish-water seashore, fluvial, and inlet habitats (Wilkinson, 1981; Collinson, 1993; Lacovara et al., 2003; Graham, 2011; Vikulin 2013) caused ancient *Weichselia* fern to disappear by the middle Eocene. Graham (2011) suggested that the mangrove habitats were "...present long before the Cretaceous but were occupied by plants other than modern *Rhizophora*, which did not appear until the mid-Eocene" (p. 337). *Nypa* palms were widespread in the Eocene coastal marshes of the southern Russian plain (Schmalhausen, 1883; Kryshtofovich, 1927, 1931; Gee, 2001) and could have been competitors of these ancient *Weichselia* ferns through the coastal Tethys areas. By the middle Eocene, the ancient Cretaceous xerophytic fern species *W. reticulata* disappeared (Vikulin, 2013), yielding the nearshore marine habitats to specialized modern angiosperms and younger species of ferns (e.g., *Acrostichum*), which evolved during the Eocene (Arnold and Daugherty, 1963). In the brackish sediments of the middle Eocene, the angiosperm seeds of the halophilous species *Potamogeton* and related extinct *Limnocarpus* are also observed (Stanislavsky, 1968; Collinson and Van Bergen, 2004).

CONCLUSIONS

1. The recovery of the postimpact plant associations in the Boltysh meteorite crater has a particularly refugial aspect, probably due to the influence of hydrothermal activity, which may have created a unique combination of edaphic environmental factors that contributed to survival of relict Cretaceous ferns (*W. reticulata*) into the Paleocene, competing with taxa of flowering plants with lower adaptability to extreme environments until the middle Eocene.

The discovery of *Weichselia* in the late Paleocene to early Eocene of Boltysh (Vikulin, 2013; Vikulin et al., 2013) is similar to the recent Baltic and Central European discoveries of Eocene and uppermost Oligocene Matoniaceae known from amber inclusions (Schmidt and Dörfelt, 2007), and it closes the gap in the fossil record of matoniaceous ferns, restricting Matoniaceae to the extant relict area in Southeast Asia and confining it to the Neogene. Thus, the Paleogene records of *W. reticulata* as well as *Matonia striata* document the latest matoniaceous ferns in Europe, which might have survived as relict species in Southeast Asia until the Neogene cooling.

2. The Boltysh angiosperms contain fossil species related to the North American Myricaceae (*Comptonia*) and Salicaceae (*Pseudosalix* sp.), which are extinct willow-like species that supported symbiotic nitrogen-fixing microorganisms, enabling them to thrive on riparian habitats with sandy soils, cooling lava, ashes, and other substrates inappropriate for the survival of non-symbiont-bearing flowering plants (Wuehlisch, 2011). Salicaceae—poplar (*Populus* spp.) and willow (*Salix* spp.) species—are early successional trees with rapid growth, deep roots, and nitrogen fixers adapted to nutrient-poor environments (i.e., riparian habitats with sandy soils; Wuehlisch, 2011). It is well known that a large endophytic community resides in the stem tissue of poplar and willow species (Ulrich et al., 2008; Taghavi et al., 2009). Among these endophytes of Salicaceae, several diazotrophic (nitrogen-fixing) bacteria were recently identified (Doty et al., 2009; Wuehlisch, 2011). The presence of these diazotrophic microorganisms may help to explain the ability of these Salicaceae pioneering tree species to grow under nitrogen limitation (Doty et al., 2009).

ACKNOWLEDGMENTS

We would like to thank Dr. Eugene Gurov for provoking our interest in the Boltysh impact crater. We thank Gary Upchurch for providing bibliographical references on occurrences of fossil *Weichselia* in North America, and we are very grateful to Peta Hayes (curator of paleobotany at the Natural History Museum, London) for acquaintance with topo-type material for *Weichselia*. Thanks are due to Bob Spicer, S.P. Kelley, and

M.A. Gilmour for helpful comments and suggestions. Two anonymous reviewers are thanked for their helpful reviews of the manuscript. This study was supported by V.L. Komarov Botanical Institute of the Russian Academy of Sciences, BIN RAS (Vikulin).

REFERENCES CITED

Alvin, K.L., 1971, *Weichselia reticulata* (Stokes et Webb) Fontaine from the Wealden of Belgium: Institut Royal Sciences Naturelles Belgique Mémoire 166, p. 3–31.

Alvin, K.L., 1974, Leaf anatomy of *Weichselia* based on fusainized material: Paleontology, v. 17, p. 587–598.

Arnold, C.A., and Daugherty, L.H., 1963, The fern genus *Acrostichum* in the Eocene Clarno Formation of Oregon: Contributions from the Museum of Paleontology, University of Michigan, v. 18, p. 205–227.

Baikovskaya, T.N., 1984, The Paleocene Flora of the Romankulsay (South Ural) (A.L. Takhtajan ed.): Leningrad, Nauka, 79 p. (In Russian.)

Baikovskaya, T.N., and Koval' Ya.M., 1968, On the Eocene flora of Kursk Province: Botanicheskii Zhurnal, v. 53, p. 1251–1262.

Barale, G., 1979, Découverte de *Weichselia reticulata* (Stokes & Webb) Fontaine emend. Alvin, filicinée leptosporangiée, dans le Crétacé Inférieur de la Province de Lérida (Espagne): Implications stratigraphiques et paléoécologiques: Geobios, v. 12, p. 313–319, doi:10.1016/S0016-6995(79)80087-X.

Barthel, M., 1976, Farne und Cycadeen: Eozäne Floren des Geiseltales: Abhandlungen des Zentralen Geologischen Instituts, v. 26, p. 439–498.

Barthel, W., and Böttcher R., 1978, Abu Ballas formation (Tithonian/Berriasian; southwestern desert, Egypt), a significant lithostratigraphic unit of the former "Nubian series": Bayerische Staatssammlung für Paläontologie und Historische Geologie München, v. 18, p. 153–166.

Bender, F., 1974, Geology of Jordan. Contribution of the Regional Geology of the Earth: Berlin, Borntraeger, 196 p.

Benton, M.J., 2009, The red queen and the court jester: Species diversity and the role of biotic and abiotic factors through time: Science, v. 323, p. 728–732, doi:10.1126/science.1157719.

Benton, M.J., and Harper, D.A.T., 2009, Introduction to Paleobiology and the Fossil Record: Chichester, UK, John Wiley and Sons, 592 p.

Berry, E.W., 1906, Living and fossil species of *Comptonia*: American Naturalist, v. 40, p. 485–520, doi:10.1086/278644.

Berry, E.W., 1928, *Weichselia* from the Lower Cretaceous of Texas: Journal of the Washington Academy of Sciences, v. 18, p. 1–5.

Berry, E.W., 1937, On the presence of the fern *Weichselia* in Colombia, South America: Journal of the Washington Academy of Sciences, v. 37, p. 458–461.

Blyakhova, S.M., 1976, Palynology and stratigraphy of the Cretaceous-Paleogene border deposits of southern Kazakhstan, *in* Palynology of Kazakhstan: Alma-Ata, Nauka Publishing House, p. 69–78. (In Russian.)

Bose, M., and Sukh Dev, S., 1959, Occurrence of two characteristic Wealden ferns in the Jabalpur Series: Nature, v. 183, p. 130–131, doi:10.1038/183130a0.

Böttcher, R., 1982, Die Abu Ballas Formation (Lingula Shale) (Apt?) der Nubischen Gruppe Südwest-Ägyptens. Eine Beschreibung der formation unter besonderer Berücksichtigung der paläontologie: Berlin: Geowissenschaftliche Abhandlungen Reihe A, v. 39, p. 145s.

Boucher, L.D., Manchester, S.R., and Tudd, W.S., 2003, An extinct genus of Salicaceae based on twigs with attached flowers, fruits, and foliage from the Eocene Green River Formation of Utah and Colorado, USA: American Journal of Botany, v. 90, p. 1389–1399, doi:10.3732/ajb.90.9.1389.

Boureau, E., and de Lapparent, A.F., 1951, Découverte de structures de *Weichselia reticulata* Stokes et Webb = *Paradoxopteris Stromeri* Hirmer, dans le Jurassique du Sud de la Tunisie: Compte Rendu Sommaire des Séances de la Société Géologique de France, v. 7, p. 107–109.

Brongniart, Ad., 1828, Notice sur les plantes d'Armissan, près Narbonne: Annales des Sciences Naturelles, ser. 1, v. 15, p. 43–51.

Brown, R.W., 1934, The recognizable species of the Green River flora: U.S. Geological Survey Professional Paper 185-C, p. 45–77.

Brown, S.A.E, Scott, A.C., Glasspool, I.J., and Collinson, M.E., 2012, Cretaceous wildfires and their impact on the Earth system: Cretaceous Research, v. 36, p. 162–190.

Christenhusz, M.J.M., Zhang, X.C., and Schneider, H., 2011, A linear sequence of extant families and genera of lycophytes and ferns: Phytotaxa, v. 19, p. 7–54.

Collinson, M.E., 1993,Taphonomy and fruiting biology of recent and fossil Nypa: Special Papers in Paleontology, v. 49, p. 165–180.

Collinson, M.E., 2001, Cainozoic ferns and their distribution: Brittonia, v. 53, p. 173–235, doi:10.1007/BF02812700.

Collinson, M.E., 2002, The ecology of Cainozoic ferns: Review of Palaeobotany and Palynology, v. 119, p. 51–68, doi:10.1016/S0034-6667(01)00129-4.

Collinson, M.E., and Van Bergen, P.F., 2004, Evolution of angiosperm fruit and seed dispersal biology and ecophysiology: Morphological, anatomical and chemical evidence from fossils, *in* Hemsley, A.R., and Poole, I., eds., The Evolution of Plant Physiology: London, The Linnean Society of London, p. 344–377.

Copeland, E.B., 1947, Genera Filicum: Waltham, Massachusetts, Chronica Botanica, 247 p.

Crane, P.R., and Lidgard, S., 1989, Angiosperm diversification and paleolatitudinal gradients in Cretaceous floristic diversity: Science, v. 246, p. 675–678, doi:10.1126/science.246.4930.675.

Daber, R., 1968, A *Weichselia-Stiehleria*-Matoniaceae community within the Quedlinburg Estuary of Lower Cretaceous age: Journal of the Linnean Society of London: Botany, v. 61, p. 75–85.

Daly, R., and Jolley, D., 2007, The palaeoecology of the eastern European Palaeogene from deposits of the Boltysh post-impact crater, Ukraine, *in* Paleontological Association of Sweden, 51st Annual Meeting Programme with Abstracts: Uppsala, Sweden, Uppsala University, p. 64.

Dieguez, C., and Melendez, N., 2000, Early Cretaceous ferns from lacustrine limestones at Las Hoyas, Cuenca Province, Spain: Palaeontology, v. 43, p. 1113–1141, doi:10.1111/1475-4983.00163.

Diez, J.B., Sender, L.M., Villanueva-Amadoz, U., Ferrer, J., and Rubio, C., 2005, New data regarding *Weichselia reticulata*: Soral clusters and the spore development process: Review of Palaeobotany and Palynology, v. 135, p. 99–107, doi:10.1016/j.revpalbo.2005.02.007.

Doty, L.D., Oakley, B., Xin, G., Kang, J.W., Singleton, G., Khan, Z., Vajzovic, A., and Staley, J.T., 2009, Diazotrophic endophytes of native black cottonwood and willow: Symbiosis, v. 47, p. 23–33, doi:10.1007/BF03179967.

Edwards, W.N., 1929, Lower Cretaceous plants from Syria and Transjordania: Annals & Magazine of Natural History, v. 4, p. 394–405, doi:10.1080/00222932908673072.

Edwards, W.N., 1932, Some Mesozoic plants from Africa: Annals & Magazine of Natural History, v. 10, p. 406–411, doi:10.1080/00222933208673589.

El-Khayal, A.A., 1985, Occurrence of a characteristic Wealden fern (*Weichselia reticulata*) in the Wasia Formation, central Saudi Arabia: Scripta Geologia, v. 79, p. 75–87.

Ettingshausen, C., 1853, Die Tertiarfloren der Oesterreichischen Monarchie. Tertiäre Flora von Häring in Tirol: Abhandlungen der Kaiserlichen Königlichen Geologischen Reichsanstalt, v. 2, p. 1–118.

Ettingshausen, C., and Standfest, F., 1888, Über *Myrica lignitum* Unger und ihre Beziehung zu den lebenden Myrica Arten: Denkschriften der kaiserlichen Akademie der Wissenschaften, Mathematisch Naturwissenschaftliche Klasse, v. 54, p. 255–260.

Fataliev, R.A., 1960, Fossil ferns from Sarmatian deposits of the Rivers Kura and Iori interfluve: Botanicheskii Zhurnal, v. 45, p. 1213–1220.

Feugueur, L., 1963, L'Yprésien du Bassin de Paris. Essai de Monographie Stratigraphique. Mémoire pour Servir à l'Explication de la Carte Géologique de France: Paris, Imprimerie nationale, 568 p.

Fiz-Palacios, O., Schneider, H., Heinrichs, J., and Savolainen, V., 2011, Diversification of land plants: Insights from a family-level phylogenetic analysis: BioMed Central Evolutionary Biology, v. 11, p. 341.

Fontaine, W.M., 1893, Notes on some fossil plants from the Trinity Division of Comanche Series of Texas: Proceedings of the United States National Museum, Washington, v. 16, p. 261–292.

Frankenhäuser, H., and Wilde, V., 1993, Farne aus der Mitteleozanen Maarfulling von Eckfeld bei Manderscheid in der Eifel: Mainzer Naturwissenschaftliches Archiv, v. 31, p. 149–167.

Fritel, P.-H., 1910, Etude sur les végétaux fossiles de l'étage Sparnacien du Bassin de Paris: Mémoires de la Société Géologique de France, v. 40, p. 1–37.

Gardner, J.S., and Ettingshausen, C., 1879–1882, The British Eocene Flora I–III: London, Palaeontographical Society, 86 p.

Gavrilov, Y., Shcherbinina, E., Golovanova, O., and Pokrovsky, B., 2009, A variety of PETM records in different settings, northeastern peri-Tethys, *in* Crouch, E.M., Strong, C.P., and Hollis, C.J., eds., Climatic and Biotic

Events of the Early Paleogene (CBEP 2009), Extended abstracts from an international conference in Wellington, New Zealand, 12–15 January 2009: GNS Science Miscellaneous Series 18, p. 67–70.

Gee, C.T., 2001, The mangrove palm *Nypa* in the geologic past of the New World: Wetlands Ecology and Management, v. 9, p. 181–203, doi:10.1023/A:1011148522181.

Gomez, B., Gillot, T., Daviero-Gomez, V., Spagna, P. and Yans, J., 2008, Palaeoflora from the Wealden facies strata of Belgium—Mega- and mesofossils of Hautrage: Bulletin of the Geological Survey of Canada, v. 55, p. 53–59.

Göppert, H.R., 1836, Die fossilen Farnkräuter [Systema filicum fossilium]. Verhandlungen der kaiserlichen leopoldisch-carolinischen Akademie der Naturforscher [Novorum actorum Academiae Caesareae Leopoldino - Carolinae Naturae Curiosorum, voluminis septimi decimi supplementum], vol. 17 (suppl.), 486 p.

Graham, A., 2011, The age and diversification of terrestrial New World ecosystems through Cretaceous and Cenozoic times: American Journal of Botany, v. 98, p. 336–351, doi:10.3732/ajb.1000353.

Gurov, E.P., 1996, The Boltysh impact crater: Lake basin with heated bottom, *in* International Workshop Abstracts: Planet Earth: Role of the Impact Processes in Geology and Biological Evolution, Postojna, September–2 October 1996, Abstr. Geol. West Sloven.: Field Guide 27: Ljubljana, Slovenia, p. 30–31.

Gurov, E.P., Kelley, S.P., Koeberl, C., and Dykan, N.I., 2006, Sediments and impact rocks filling the Boltysh impact crater, *in* Cockell, C., Koeberl, C., and Gilmour, I., eds., Biological Processes Associated with Impact Events: Berlin, Springer, p. 335–354.

Hickey, L.J., and Wolfe, J.A., 1975, The bases of angiosperm phylogeny: Vegetative morphology: Annals of the Missouri Botanical Garden, v. 62, p. 538–589, doi:10.2307/2395267.

Hooker, J.J., and Collinson, M.E., 2012, Mammalian faunal turnover across the Paleocene-Eocene boundary in NW Europe: The roles of displacement, community evolution and environment: Austrian Journal of Earth Sciences, Vienna, v. 105, no. 1, p. 17–18.

Hulten, E., 1968, Flora of Alaska and Neighboring Territories: A Manual of the Vascular Plants: Stanford, California, Stanford University Press, 1008 p.

Jolley, D.W., Gilmour, I., Gurov, E., Kelley, S.P., and Watson, J.S., 2010, Two large meteorite impacts at the Cretaceous-Paleogene boundary: Geology, v. 38, p. 835–838, doi:10.1130/G31034.1.

Jolley, D.W., Daly, R., Gilmour, I., Gilmour, M., and Kelley, S.P., 2012, The Boltysh crater record of rapid vegetation change during the Dan-C2 hypothermal event: European Geosciences Union, EGU, General Assembly, Vienna: Geophysical Research Abstracts, v. 14, p. 4635.

Judd, W.S., Campbell, C.S., Kellogg, E.A., Stevens, P.F., and Donoghue M.J., 2002, Plant Systematics: A Phylogenetic Approach (2nd ed.): Sunderland, Massachusetts, Sinauer Associates Inc., 576 p.

Kato, M., 1998, Matoniaceae: Flora Melesiana, ser. 2, v. 3, p. 289–294.

Kedves, M., and Russell, D.E., 1982, Palynology of the Thanetian layers of Menat. The geology of the Menat Basin, France: Palaeontographica B, v. 182, p. 87–150.

Keller, G., 2012, The Cretaceous-Tertiary mass extinction, Chicxulub impact, and Deccan volcanism, *in* Talent, J.A., ed., Earth and Life, International Year of Planet Earth: Berlin, Springer Science, p. 759–793.

Kelley, S.P., and Gurov, E., 2002, Boltysh, another end-Cretaceous impact: Meteoritics & Planetary Science, v. 37, p. 1031–1043, doi:10.1111/j.1945-5100.2002.tb00875.x.

Kelley, S.P., Jolley, D., Gurov, E., Gilmour, I., and Watson, J., 2009, Relationship between the Boltysh and Chicxulub asteroid impacts: Implications for celestial mechanics at the K/Pg boundary: Washington, D.C., American Geophysical Union, Fall Meeting 2009, abstract no. NH31D–06.

Kerr, R.A., 2012, Before the dinosaurs' demise, a clambake extinction?: Science, v. 337, p. 1280, doi:10.1126/science.337.6100.1280.

Kornas, J., 1978, Fire resistance in the Pteridophytes of Zambia: The Fern Gazette, v. 11, p. 373–384.

Kraiser, T., Gras, D.E., Gutierrez, A.G., Gonzalez, B., and Gutierrez, R.A., 2011, A holistic view of nitrogen acquisition in plants: Journal of Experimental Botany, v. 62, p. 1455–1466, doi:10.1093/jxb/erq425.

Krasilov, V.A., 2006, Paleontology and the Paradigm of Modern Science: Evolution of the Biosphere and Biodiversity: On the 70th Anniversary of A.Yu. Rozanov: Moscow, KMK Scientific Press Ltd., p. 7–19.

Krasnov, A.N., 1910, Fundamentals of the Tertiary Flora of Southern Russia: Proceedings of the Society of Naturalists at the Imperial Khar'kov University: Khar'kov, Governorate of the Russian Empire, Pechatnik Publishing, v. 44, p. 148–253. (Also published separately: Krasnov, A.N., 1911, Fundamentals of the Tertiary Flora of Southern Russia. Primitiae Florae Tertiariae Rossiae Meridionalis. Kharkov, Khar'kov Governorate of the Russian Empire: Pechatnik Publishing, 108 p.)

Krassilov, V.A., Makulbekov, N.M., and Maslova, N.P., 1996, *Ushia*, a Palaeocene angiosperm of *Nothofagus* affinities from the Lower Volga and western Kazakhstan: Palaeontographica B, v. 239, p. 137–145.

Krestov, P.V., Omelko A.M., and Nakamura, Y., 2008, Vegetation and natural habitats of Kamchatka: Hannover, Germany, Reinhold Tüxen Gesellschaft, v. 20, p. 195–218.

Kryshtofovich, A.N., 1927, Remains of the palm (*Nipadites burtinii* Brongn.) from the Eocene near the city Voznesensk of Odessa Province: Leningrad: Proceedings of the Geological Committee, v. 45, p. 639–642.

Kryshtofovich, A.N., 1931, New data on vegetation of the mangrove bogs of Ukraine during Tertiary: Proceedings of the Scientific Conference of the Academy of Sciences of the Ukrainian SSR in memory of academician P.A. Tutkovskiy, v. 2, p. 273–276.

Kryshtofovich, A.N., 1952, Tertiary Plants from Kanev Formation: Proceedings of VSEGEI (All Union Geological Institute, Leningrad): Paleontology and Stratigraphy Series, p. 198–203 (In Russian).

Kvaček, Z., 2010, Forest flora and vegetation of the European early Palaeogene—A review: Bulletin of Geosciences, v. 85, p. 63–76, doi:10.3140/bull.geosci.1146.

Lacovara, K.J., Smith, J.R., Smith, J.B., and Lamanna, M.C., 2003, The Ten Thousand Islands coast of Florida: A modern analog to low-energy mangrove coasts of Cretaceous epeiric seas, *in* Davis, R.A., Jr., ed., Proceedings of the 5th International Conference on Coastal Sediments, Clearwater Beach, Florida: Singapore, World Scientific Publishing Company, p. 1773–1784.

Laurent, L., 1912, Flore fossile des schistes de Menat (Puy-de-Do^me): Annales du Museé d'Histoire Naturelle de Marseille, v. 14, p. 1–246.

Laurent, L., 1919, Addition à la flore des schistes de Menat: Annales du Musée d'Histoire Naturelle de Marseille, v. 17, p. 1–8.

Liang, X.-Q., Wilde, V., Ferguson, D.K., Kvaček, Z., Ablaev, A.G., Wang, Y.-F., and Li, Ch.-S., 2010, *Comptonia naumannii* (Myricaceae) from the early Miocene of Weichang, China, and the palaeobiogeographical implication of the genus: Review of Palaeobotany and Palynology, v. 163, p. 52–63, doi:10.1016/j.revpalbo.2010.09.004.

Lupia, R., Lidgard, S., and Crane, P.R., 1999, Comparing palynologia abundance and diversity: Implications for biotic replacement during the Cretaceous angiosperm radiation: Paleobiology, v. 25, p. 305–340.

Mai, D.H., 1995, Tertiäre Vegetationsgeschichte Europas: Jena, Germany, Gustav Fischer, 691 p.

Mai, D.H., and Walther, H., 1978, Die Floren der Haselbacher Serie im Weisselster-Becken (Bezirk, Leipzig, DDR): Abhandlungen des Staatlichen Museums für Mineralogie und Geologie zu Dresden, v. 28, p. 1–200.

Mast, A.R., and Thiele, K., 2007, The transfer of *Dryandra* R.Br. to *Banksia* L.f. (Proteaceae): Australian Systematic Botany, v. 20, p. 63–71, doi:10.1071/SB06016.

McLoughlin, S., Carpenter, R.J., Jordan, G.J., and Hill, R.S., 2008, Seed ferns survived the end-Cretaceous mass extinction in Tasmania: American Journal of Botany, v. 95, p. 465–471, doi:10.3732/ajb.95.4.465.

Niklas, K.J., Tiffany, B.H., and Knoll, A.H., 1985, Patterns in vascular plant diversification: An analysis at the species level, *in* Valentin, J.W., ed., Phanerozoic Diversity Patterns: Profiles in Macroevolution: Princeton, New Jersey, Princeton University Press, p. 97–128.

Panova, L.A., 1968, Paleocene and Eocene spore-pollen assemblages of the southern and central West-Siberian lowland, *in* Paleo-Palynological Methods in Stratigraphy: 2nd International Palynological Conference: Leningrad, VSEGEI, p. 206–218 (translation in Russian).

Peppe, D.J., 2010, Megafloral change in the early and middle Paleocene in the Williston Basin, North Dakota, USA: Palaeogeography, Palaeoclimatology, Palaeoecology, v. 298, p. 224–234, doi:10.1016/j.palaeo.2010.09.027.

Pigg, K.B., and DeVore, M.L., 2010, Floristic composition and variation in late Paleocene to early Eocene floras in North America: Bulletin of Geosciences, v. 85, p. 135–154, doi:10.3140/bull.geosci.1136.

Piton, L.E., 1940, Paléontologie du gisement éocène de Menat (Puy-de-Dôme) (flore et faune): Mémoires de la Société d'Histoire Naturelle de Auvergne, v. 1, p. 1–303.

Renne, P.R., Deino, A.L., Hilgen, F.J., Kuiper, K.F., Mark, D.F., Mitchell III, W.S., Morgan, L.E., Mundil, R., Smit, J., 2013, Time scales of critical events around the Cretaceous-Paleogene boundary: Science, v. 339, no. 6120, p. 684–687, doi:10.1126/science.1230492.

Retallack, G.J., and Dilcher, D.L., 1986, Cretaceous angiosperm invasion of North America: Cretaceous Research, v. 7, p. 227–252, doi:10.1016/0195-6671(86)90027-3.

Reymanówna, M., 1965, On *Weichselia reticulata* and *Frenelopsis hoheneggeri* from the western Carpathians: Acta Palaeobotanica, v. 6, p. 15–26.

Riba, R., and Reyes, J.I., 1990, *Pityrogramma calomelanos* (L.) Link (Adiantaceae) on layers of volcanic ash in Los Tuxtlas, State of Veracruz, Mexico: Annals of the Missouri Botanical Garden, v. 77, p. 287–289, doi:10.2307/2399542.

Rotman, R.N., 1983, Palynological characteristics of the Upper Maastrichtian deposits of Crimea and the Cretaceous-Paleogene boundary: Proceedings of the Western Siberian Geology and Oil Research Institute (NIGNI), v. 179, p. 6–10.

Rüffle, L., 1976, Myricaceae, Leguminosae, Icacinaceae, Sterculiaceae, Nymphaeaceae, Monocotyledones, Coniferae: Abhandlungen des Zentralen Geologischen Instituts, Paläontologische Abhandlungen, v. 26, p. 337–438. (In German.)

Ryabokon', T.S., 2002, Biostratigraphy of Kiev Formation type section (mid-Eocene) of Dnieper–Donets depression by the data of studying foraminiferes: Geologo–Mineralogičnyj Visnik, v. 2, p. 39–50. (In Ukrainian.)

Sahni, B., 1936, The occurrence of *Matonidium* and *Weichselia* in India: Records of the Geological Survey of India, v. 71, p. 152–165.

Saporta, G., and Marion, A.F., 1873, Essai sur l'état de la végétation à l'époque des marnes heersiennes de Gelinden: Mémoires Couronnés et Mémoires des Savants Étrangers Publiés par l'Académie Royale des Sciences, des Lettres et des Beaux-Arts de Belgique, v. 37, p. 3–94.

Saporta, G., and Marion, A.F., 1878, Révision de la flore Heersiennes de Gelinden: Mémoires Couronnés et Mémoires des Savants Étrangers Publiés par l'Académie Royale des Sciences, des Lettres et des Beaux-Arts de Belgique, v. 41, p. 1–112.

Schlagintweit, O., 1919, *Weichselia mantelli* im Nordostlichen Venezuela: Zentralblatt für Mineralogie: Geologie und Paläontologie, v. 17–20, p. 315–319.

Schmalhausen, J., 1883, Tertiarflora Sud-West-Russlands: Palæontologische Abhandlungen, v.1, p. 1–53.

Schmidt, A.R., and Dörfelt, H., 2007, Evidence of Cenozoic Matoniaceae from Baltic and Bitterfeld amber: Review of Palaeobotany and Palynology, v. 144, p. 145–156, doi:10.1016/j.revpalbo.2006.07.009.

Schneider, H., 2007, Plant morphology as the cornerstone to the integration of fossils and extant taxa in phylogenetic systematics: Species, Phylogeny and Evolution, v. 1, p. 65–74.

Schneider, H., Schuettpelz, E., Pryer, K.M., Cranfill, R., Magallon, S., and Lupia, R., 2004, Ferns diversified in the shadow of angiosperms: Nature, v. 428, p. 553–557, doi:10.1038/nature02361.

Schneider, H., Smith, A.R., and Pryer, K.M., 2009, Is morphology really at odds with molecules in estimating fern phylogeny?: Systematic Botany, v. 34, p. 455–475, doi:10.1600/036364409789271209.

Schuettpelz, E., and Pryer, K.M., 2009, Evidence for a Cenozoic radiation of ferns in an angiosperm-dominated canopy: Proceedings of the National Academy of Sciences of the United States of America, v. 106, p. 11,200–11,205, doi:10.1073/pnas.0811136106.

Schulte, P., Alegret, L., Arenillas, I., Arz, J.A., Barton, P.J., Bown, P.R., and Willumsen, P.S., 2010, The Chicxulub asteroid impact and mass extinction at the Cretaceous-Paleogene boundary: Science, v. 327, no. 5970, p. 1214–1218, doi:10.1126/science.1177265.

Serlin, B.S., 1982, An Early Cretaceous fossil flora from northwest Texas: Its composition and implications: Palaeontographica B, v. 182, p. 52–86.

Silantieva, N., and Krassilov, V., 2006, *Weichselia* Stiehler from Lower Cretaceous of Makhtesh Ramon, Israel: New morphological interpretation and taxonomical affinities: Acta Palaeobotanica, v. 46, p. 119–136.

Skog, J.E., 2001, Biogeography of Mesozoic leptosporangiate ferns related to extant ferns: Brittonia, v. 53, p. 236–269, doi:10.1007/BF02812701.

Skutschas, P.P., and Gubin, Y.M., 2012, A new salamander from the late Paleocene–early Eocene of Ukraine: Acta Palaeontologica Polonica, v. 57, p. 135–148, doi:10.4202/app.2010.0101.

Song, Z.-C., Li, M.-Y., and Zhong, L., 1986, Cretaceous and Early Tertiary spore-pollen assemblages from the Sanshui basin, Guandong Province: Paleontologia Sinica, no. 171: New Series A, v. 10, p. 1–169.

Speijer, R.P., Scheibner, Ch., Stassen, P., and Mors, M., 2012, Response of marine ecosystems to deep-time global warming: A synthesis of biotic patterns across the Paleocene-Eocene thermal maximum (PETM): Austrian Journal of Earth Sciences, Vienna, v. 105, no. 1, p. 6–16.

Spicer, R.A., Burnham, R.J., Grant, P., and Glicken, H., 1985, *Pityrogramma calomelanos*, the primary, post-eruption colonizer of volcán Chichonal, Chiapas, Mexico: American Fern Journal, v. 75, p. 1–5, doi:10.2307/1546571.

Stanislavsky, F.A., 1968, The age and stratigraphy of the sapropelites of the Boltysh depression: Geological Zhurnal of Ukraine, Kiev, v. 28, p. 109–115.

Sternberg, K., von, 1820–1838, Versuch einer Geognostisch-Botanischen Darstellung der Flora der Vorwelt: Kommission im Deutschen Museum, Fr. Fleischer in Leipzig, Leipzig und Prag, 220 s.

Stokes, C., and Webb, P.B., 1824, Description of some fossil vegetables of the Tilgate forest in Sussex: Transactions of the Geological Society of London, ser. 2, v. 1, p. 421–424.

Sytchevskaya, E.K., 1986, Paleogene freshwater fish fauna of the USSR and Mongolia: Trudy Sovmestnoj Sovetsko–Mongol'skoj Paleontologičeskoj Ekspedicii, v. 29, p. 1–157. (In Russian.)

Taghavi, S., Garafola, C., Monchy, S., Newman, L., Hoffman, A., Weyens, N., Barac, T., Vangronsveld, J., and van der Lelie, D., 2009, Genome survey and characterization of endophytic bacteria exhibiting a beneficial effect on growth and development of poplar trees: Applied and Environmental Microbiology, v. 75, p. 748–757, doi:10.1128/AEM.02239-08.

Tidwell, W.D., and Ash, S.R., 1994, A review of selected Triassic to Early Cretaceous ferns: Journal of Plant Research, v. 107, p. 417–442, doi:10.1007/BF02344066.

Tidwell, W.D., and Skog, J.E., 1992, Two new fossil matoniaceous stem genera from Tasmania, Australia. Review of Palaeobotany and Palynology, v. 70, p. 263–277.

Tryon, A.F., and Lugardon, B., 1990, Spores of the Pteridophyta: Surface, Wall Structure, and Diversity Based on Electron Microscope Studies: New-York, Springer-Verlag, 670 p.

Ulrich, K., Ulrich, A., and Ewald, D., 2008, Diversity of endophytic bacterial communities in poplar grown under field conditions: Federation of European Microbiological Societies: Microbiology Ecology, v. 63, p. 169–180.

Vajda, V., Raine, J.I., and Hollis, C.J., 2001, Indication of global deforestation at the Cretaceous-Tertiary boundary by New Zealand fern spike: Science, v. 294, p. 1700–1702, doi:10.1126/science.1064706.

Vakhrameev, V.A., 1964, Yurskie i rannemelovye Flory Evrazii i paleofloristicheskie provintsii etogo vremeni (Jurassic and early Cretaceous floras of Eurasia and the palaeofloristic provinces of this period): Trudy Geologicheskogo Instituta Akademii Nauk S.S.S.R., v. 102, p. 1–262. (In Russian.)

Vakhrameev, V.A., 1988, Jurassic and Cretaceous floras and climates of Earth, Proceedings of the Geological institute of the Academy of Sciences of the USSSR, Moscow, Nauka (publisher), v. 430, 214 p.

Vakhrameev, V.A., 1991, Jurassic and Cretaceous Floras and Climates of the Earth: Cambridge, UK, Cambridge University Press, 338 p. (Originally published in Russian by Nauka, Moscow, 1988.)

Vasiliev, I.V., and Selin, Yu.I., 1970, New data on paleontological characteristics of the productive thickness of the Boltysh oil-shale deposit: Proceedings of the Ukraine Academy of Sciences, ser. B, v. 12, p. 1059–1061.

Vassilevskaya, N.D., 1957, The Eocene Flora of Badhyz in Turkmenia: Collection of Scientific Works in Memory of A.N. Kryshtofovich: Moscow-Leningrad, USSR Academy of Sciences Publisher, p. 103–176. (In Russian.)

Vikulin, S.V., 2013, Ferns *Weichselia* (Matoniaceae) and *Blechnum* (Blechnaceae) in Paleocene–early Eocene of Boltysh Depression (Ukraine, Kirovograd Province): Botanicheskii Zhurnal, v. 98, no. 4, p. 453–467.

Vikulin, S.V., Bystriakova, N., and Schneider, H., 2013, Post impact plant macrofossils from the Boltysh Crater illustrate survival of a Cretaceous fern relict Weichselia reticulata into the Paleogene within a refugial geothermal ecosystem, *in* Keller, G., and Kerr, A., eds., Abstracts: International Conference on Volcanism, Impacts and Mass Extinctions: Causes and Effects, Flett Lecture Theatre, The Natural History Museum (NHM), London, 27–29 March 2013, p. 80.

Vincent, P.M., Aubert, M., Boivin, P., Cantagrel, J.M., and Lenat, J.-F., 1977, Decouverte d'un volcanisme Paleocène en Auvergneles maars de Menat et leurs annexes: Etude geologique et géophysique: Bulletin de la Société Géologique de France, v. 19, p. 1067–1070.

Walker, T.G., 1962, Cytology and evolution in the fern genus *Pteris* L: Evolution; International Journal of Organic Evolution, v. 16, p. 27–43, doi:10.2307/2406264.

Wappler, T., Currano, E.D., Wilf, P., Rust, J., and Labandeira, C.C., 2009, No post-Cretaceous ecosystem depression in European forests? Rich insect-feeding damage on diverse middle Palaeocene plants, Menat, France: Royal Society of London Proceedings, ser. B, v. 276, p. 4271–4277, doi:10.1098/rspb.2009.1255.

Ward, L.F., 1899, The Cretaceous Formation of the Black Hills as Indicated by the Fossil Plants: Annual Report of the U.S. Geological Survey 1897–1898, 651 p.

Watson, J.S., Gilmour, I., Kelley, S.P., and Jolley, D.W., 2009, Scientific drilling of the Boltysh impact crater, Ukraine, *in* Proceedings of the 40th Lunar and Planetary Science Conference: Houston, Texas, Lunar and Planetary Institute.

Watson, J.S., Gilmour, I., Jolley, D.W., Kelley, S.P., Gilmour, M.A., and Gurov, E.P., 2010, Molecular parameters of post impact cooling in the Boltysh impact structure, *in* 41st Lunar and Planetary Science Conference, 1–5 March 2010, Houston: Lunar and Planetary Institute Contribution 1533, p. 2296.

Wilde, V., 1989, Untersuchungen zur Systematik der Blattreste aus dem Mitteleozän der Grube Messel bei Darmstadt (Hessen, Bundesrepublik Deutschland): Courier Forschungsinstitut Senckenberg, v. 15, 213 p.

Wilde, V., 2001, Ein Farnhorizont aus dem Mitteleozän des Geiseltales, Sachsen-Anhalt, Deutschland: Hallesches Jahrbuch für Geowissenschaften in Halle (Saale), ser. B, v. 13, p. 69–75.

Wing, S.L., and DiMichele, W. A., 1995, Conflict between local and global changes in plant diversity through geological time: Palaios, v. 10, p. 551–564.

Wilde, V., and Frankenhäuser, H., 1998, The middle Eocene plant taphocoenosis from Eckfeld (Eifel, Germany): Review of Palaeobotany and Palynology, v. 101, p. 7–28, doi:10.1016/S0034-6667(97)00067-5.

Wilf, P., and Johnson, K.R., 2004, Land plant extinction at the end of the Cretaceous: A quantitative analysis of the North Dakota megafloral record: Paleobiology, v. 30, p. 347–368, doi:10.1666/0094-8373(2004)030<0347:LPEATE>2.0.CO;2.

Wilkinson, H.P., 1981, The anatomy of the hypocotyls of *Ceriops* Arnott (Rhizophoraceae), recent and fossil: Botanical Journal of the Linnean Society, London, v. 82, p. 139–164.

Wilkinson, H.P., 2007, Leaf teeth in certain Salicaceae and 'Flacourtiaceae': Botanical Journal of the Linnean Society, v. 155, p. 241–256, doi:10.1111/j.1095-8339.2007.00695.x.

Wolfe, J.A., and Upchurch, G.R., Jr., 1986, Vegetation, climatic and floral changes at the Cretaceous-Tertiary boundary: Nature, v. 324, p. 148–152, doi:10.1038/324148a0.

Wolfe, J.A., and Wehr, W.C., 1987, Mid-Eocene dicotyledonous plants from Republic, northeastern Washington: U.S. Geological Survey Bulletin 1597, p. 1–25.

Wuehlisch, G., 2011, Evidence for nitrogen-fixation in the Salicaceae Family: Tree Planters' Notes, v. 54, p. 38–41.

Zhilin, S.G., 1980, Notes on the systematics of fossil plants: Myricaceae, *in* Zhilin, S.G., ed., Systematics and Evolution of Higher Plants: Leningrad, Nauka Publishing House, p. 9–20.

Zhilin, S.G., and Vikulin, S.V., 1986, *Comptonia dryandrifolia* (Myricaceae) from the Eocene of the south of Middle-Russian upland: Botanicheskii Zhurnal, v. 71, p. 148–153.

Zosymovych, V.Yu, and Ryabokon', T.S., 2010, Preliminary investigations of the bore hole 42/11 Paleogene section (Boltyshka depression, Ukrainian shield): Regional problems of lithology, *in* Gozhyk, P.F., ed., Collection of Scientific Works of Institute of Geological Sciences of National Academy of Sciences: Kiev, Ukraine, Institute of Geological Sciences of National Academy of Sciences, v. 3, p. 38–42. (In Ukrainian.)

Manuscript Accepted by the Society 31 January 2014

Printed in the USA

The Geological Society of America
Special Paper 505
2014

Deccan volcanic eruptions and their impact on flora: Palynological evidence

Bandana Samant*
Postgraduate Department of Geology, Rashtrasant Tukadoji Maharaj Nagpur University, Nagpur-440001, India

Dhananjay M. Mohabey
Ex. Geological Survey of India, C-3, HIG, Giripeth, Nagpur-440010, India

ABSTRACT

Palynological studies of the late Maastrichtian infratrappean and intertrappean sedimentary beds from various stratigraphic levels in the Deccan continental flood basalt of the Nand-Dongargaon Basin in central India show that infratrappean sediments (Maastrichtian C30n-C29r) are characterized by gymnosperm (*Araucariacites, Classopollis, Cycadopites, Podocarpidites*) and angiosperm (*Cretacaeiporites, Compositoipollenites, Graminidites, Longapertites, Palmaepollenites*) palynomorphs. A distinct floral turnover is observed in intertrappean sediments with the initiation of volcanic activity in the basin. At the lowest stratigraphic level, the earliest floral change is recorded by the appearance of angiosperm-pteridophyte–dominated association (*Aquilapollenites, Azolla, Gabonisporis, Tricolpites, Triporopollenites,* and *Normapolles* group pollen) and the appearance of peridinoid dinoflagellates. At higher stratigraphic levels in C29r, this palynofloral association continues with the appearance of new forms, such as *Scabrastephanocolpites* spp., *Scollardia conferta,* and *Triporopollenites cracentis.* A sharp decline in diversity of titanosauriform-abelisaurid–dominated dinosaurian fauna is also observed with the onset of Deccan volcanic activity. At this time, out of at least eight dinosaur species, only a single taxon of titanosauriform dinosaurs survived the onset of volcanism. This suggests that the floral change and decreased dinosaur diversity are strongly linked with the initiation of Deccan volcanism in C29r in India. The peak eruption of this volcanic phase resulted in the Deccan's largest volume of lava flows, which played a significant role in the global mass extinction at the Cretaceous-Paleogene boundary.

*bandanabhu@gmail.com

Samant, B., and Mohabey, D.M., 2014, Deccan volcanic eruptions and their impact on flora: Palynological evidence, *in* Keller, G., and Kerr, A.C., eds., Volcanism, Impacts, and Mass Extinctions: Causes and Effects: Geological Society of America Special Paper 505, p. 171–191, doi:10.1130/2014.2505(08). For permission to copy, contact editing@geosociety.org.

INTRODUCTION

The Cretaceous-Paleogene boundary (also known as KTB for Cretaceous-Tertiary) marks one of the five major mass extinction events in Earth's history. The relative importance of the most often cited major causes—the Chicxulub impact (Alvarez et al., 1980) and the massive eruption of Deccan continental flood basalts (Courtillot et al., 1986, 2000)—has been debated for over 30 yr (Schulte et al., 2010; Archibald et al., 2010; Keller, 2012). To date, in the context of chronostratigraphic framework, there has been little assessment of the biotic changes recorded in sediments proximal to the Deccan continental flood basalts. However, where Deccan lava flows interfinger with shallow marine sediments, such as in Jhilmili (Chhindwara) central India, in Rajahmundry quarries, and deep wells of the Krishna-Godavari Basin in SE India, the Cretaceous-Paleogene boundary has been identified in the intertrappean sediments above the main phase of Deccan volcanism (Keller et al., 2008, 2009a, 2009b). Moreover, the rapid nature of the mass extinction in planktic foraminifera in C29r has been documented between the lava flows in the deep wells drilled by the Oil and Natural Gas Corporation of India (ONGC; Keller et al., 2011, 2012).

The Deccan continental flood basalt province is one of the largest and best-studied continental flood basalt provinces in the world. It presently covers an area over of 500,000 km^2 in peninsular and western India, and it produced a large eruptive volume, estimated at ~1.3×10^6 km^3 (Jay and Widdowson, 2008), between 69 and 62 Ma (Sheth et al., 2001; Pande, 2002; Sheth and Pande 2014). Deccan volcanism was associated with rifting and rapid migration of the Indian plate over the Reunion Island hotspot, which may have caused changes at the surface that affected both topography and drainage (Cox, 1989) and consequently affected the composition of the sediments as well as biota (Tandon et al., 1995; Tandon, 2002; Tandon and Andrews, 2001). The established lava stratigraphy of the Deccan continental flood basalt sequence in central and southern peninsular India has been based on the systematic mapping of lava flows, and analysis of their magnetostratigraphy, petrography, and chemistry by the Geological Survey of India and others (Deshmukh, 1984; Deshmukh et al., 1996; Cox and Hawkesworth, 1984; Beane et al., 1986; Courtillot et al., 1986; Nair et al., 1996; Jay and Widdowson, 2008). These sequences are classified as Sahyadri, Amarkantak, Malwa, and Satpura groups (Geological Survey of India, 2001a, 2001b, 2001c, 2001d; Nair and Bhusari, 2001). The lava flows of the Deccan continental flood basalt sequences of western India, including Kutch and Saurashtra, are still not formally classified (Fig. 1).

Sediments associated with Deccan continental flood basalts occur as "infratrappean" and "intertrappean" sediments, which are positioned below and between the flows, respectively. Although it is typically inferred that deposition of infratrappean sediments took place before initiation of the Deccan continental flood basalt, whereas intertrappeans were deposited during pauses in volcanic activity, Sahni (1984) suggested that in some cases they may actually be syndepositional. Based on magnetostratigraphy, Milankovitch stratigraphy, and organic carbon isotope study, Hansen et al. (2005) suggested that the infratrappean (Lameta Formation) and intertrappean sediments are time-transgressive.

Infratrappean sediments, collectively referred to as the Lameta Formation, are considered to have been deposits in five inland basins (Fig. 1): (1) Balasinor-Jhabua, (2) Sagar, (3) Jabalpur, (4) Amarkantak-Ambikapur, and (5) Nand-Dongargaon (Mohabey, 1996a). Lameta sediments in the Kheda area of the Balasinor-Jhabua Basin, which contain abundant remains of dinosaurs and their nesting sites, were deposited during C30n, as were the overlying intertrappean beds (Hansen et al., 2001, 2005). In contrast, in the Jabalpur Basin, the dinosaur-bearing sediments in the lower parts of the Lameta Formation were deposited during Maastrichtian chron C29r (Vandamme and Courtillot, 1992). Infratrappean and intertrappean sediments at different stratigraphic levels within the Deccan volcanic sequence in geographically separated areas are rich in fauna and flora. The fauna represents various groups of vertebrates (e.g., anurans, turtles, titanosaurid and abelisaurid dinosaurs, crocodylomorphs, lizards, snakes, fishes, and mammals) and invertebrates (mostly molluscs and ostracods). However, except for in situ dinosaur remains, the fauna does not provide a precise stratigraphic age owing to their extended stratigraphic range (Chatterjee, 1978; Prasad et al., 1994; Prasad and Khajuria, 1995; Bajpai and Prasad, 2000; Rana and Wilson, 2008; Samant and Mohabey, 2009; Mohabey et al., 2011; D'Emic et al., 2009).

Previous paleontological investigations of the Lameta sediments of the Nand-Dongargaon Basin reported a diverse terrestrial to semi-aquatic and aquatic fauna including dinosaurs, turtles, crocodylomorphs, pelobatids, frogs, fishes, and eutharian mammals (Mohabey and Udhoji, 1996; Mohabey et al., 1993; Mohabey 1996a, 1996b; Mohabey and Udhoji, 1993). A palynological study by Samant and Mohabey (2005) of the sediments associated with the lower part of the Sahyadri Group of the Deccan volcanic sequence in the Nand-Dongargaon Basin revealed the presence of a gymnosperm-angiosperm–rich palynoflora in the Lameta Formation and a characteristic Maastrichtian palynoflora in the overlying intertrappean sediments. It was suggested that the onset of volcanic activity in the basin changed the existing gymnosperm-angiosperm flora to a pteridophyte-angiosperm flora, which continued with some diversification in higher stratigraphic levels (Samant and Mohabey, 2009).

The present paper documents changes in floral diversity and turnover based on quantitative and qualitative assessments of the palynomorphs in sediments at different stratigraphic levels within Deccan continental flood basalt sequences of the Nand-Dongargaon Basin (Fig. 1). In order to investigate a possible link between the observed palynofloral change and Deccan volcanism, we extended our studies to intertrappean sediments higher up in the sequence within the Ajanta Formation of the Sahyadri Group in the Nand-Dongargaon Basin and also included sediments associated with the Amarkantak Group of the Deccan volcanic sequence in the Chhindwara-Mandla-Jabalpur sector in Madhya Pradesh and Saurashtra (Fig. 1).

MATERIAL AND METHODS

Systematic samples for palynological analysis were collected from the Lameta Formation in the Nand-Dongargaon Basin. After preliminary analysis of all the samples from different lithofacies of the Lameta Formation, it was found that only paludal and lacustrine facies exhibit palynomorphs. About 50 samples were taken from the palustrine deposits of Polgaon, Pisdura, and Piraya and the lacustrine facies at Dongargaon (Fig. 2). The intertrappean sections were sampled from different stratigraphic levels, lithotypes, and their facies variants. Samples were also taken from surface and available subsurface intertrappean sediments in adjoining areas (Table 1). Thirty samples of type A plant–bearing coprolites, 20 of type B coprolites, and 20 of type Ba and type C coprolites (Matley, 1939) were also macerated for palynological study. Other intertrappean samples collected include 10 samples from the Mohgaon Kalan and Jhilmili section of the Amarkantak Group of the Chhindwara area, and 50 samples from intertrappean deposits from the Saurashtra area, which include 15 samples from the Ninama Hill and Ninama River sections.

The samples were macerated using standard techniques (HCl, HF, HNO_3, KOH). The concentration and duration of the chemical treatment were dependent on the sample type. For the extraction of phytoliths from the sediments, the samples were first treated with H_2O_2, followed by HNO_3 and KOH treatments. Later on, the residues were sieved using a 15 μm sieve. Slides were prepared using polyvinyl alcohol and mounted with distyrene plasticize xylene (DPX).

The preservation of organic matter in the intertrappean sediments proved to be better than that of the Lameta sediments. The slides are housed in the museum of the Postgraduate Department of Geology, Rashtrasant Tukadoji Maharaj Nagpur University, Nagpur, India.

STRATIGRAPHY OF THE NAND-DONGARGAON BASIN

In the Nand-Dongargaon Basin, Lameta sediments crop out across a >700 km^2 area, attaining a total thickness of up to 20 m (Fig. 2). Sediment deposition occurred during the late Maastrichtian paleomagnetic chrons C30n to C29r (Hansen et al., 2005). The Lameta sediments are unconformably underlain by

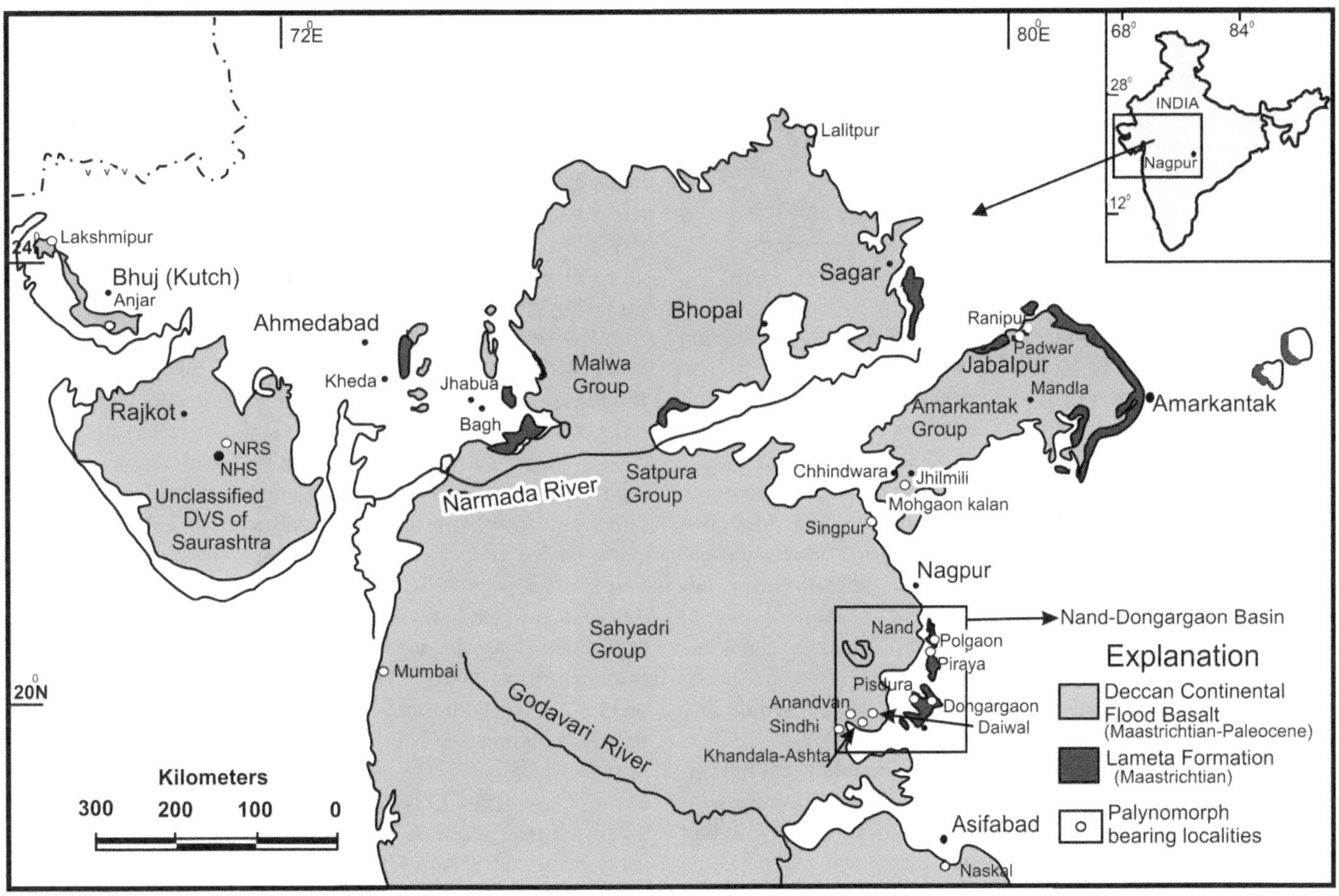

Figure 1. Map of Deccan continental flood basalt province of central and western India showing Nand-Dongargaon Basin and major palynomorph-bearing localities. DVS—Deccan volcanic sequence; NRS—Ninama River section; NHS—Ninama Hill section.

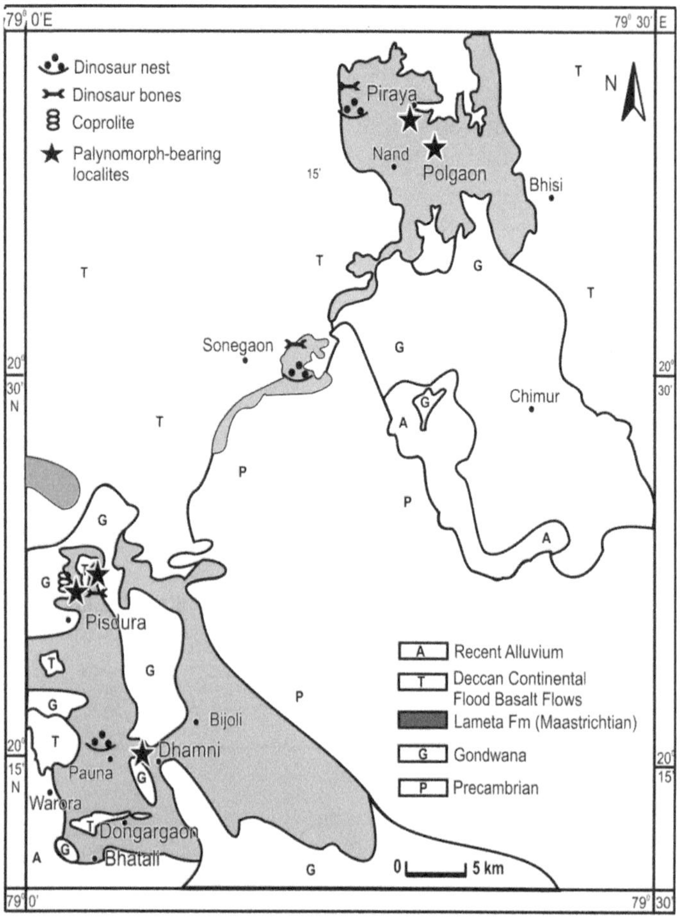

Figure 2. Geological map of the Lameta Formation of Nand-Dongargaon Basin showing palynomorph-bearing localities.

Precambrian and Gondwana sediments and overlain by Deccan continental flood basalt lava fields of the Sahyadri Group, having a composite thickness of ~500 m with 29 basaltic flows and associated intertrappean horizons (Nair and Bhusari, 2001; Geological Survey of India, 2001a, 2001b, 2001c). The Lameta sediments were deposited in overbank, channel, paludal/palustrine, and lacustrine settings (Mohabey and Udhoji, 1993; Mohabey et al., 1993). They are associated with fossils of titanosaur and abelisaurid dinosaurs, as well as their eggs and coprolites (Matley, 1939; Mohabey, 1998, 2005; Ghosh et al., 2003; Wilson et al., 2003, 2009; Wilson and Mohabey, 2006). Biofacies and lithofacies studies indicate that the sediments were deposited under semiarid to arid climatic conditions with strong seasonality (Mohabey et al., 1993; Mohabey and Samant, 2005).

Palynological samples were collected from different lithofacies of the Lameta Formation in the Nand-Dongargaon Basin at Polgaon, Piraya, Pisdura, and Dongargaon localities (Fig. 2) and from intertrappean horizons at different stratigraphic levels in the Sahyadri Group, the adjoining area to the north in the Amarkantak Group, and unclassified Deccan volcanic sequences in Saurashtra (Fig. 3). The Lameta deposits at Polgaon (Fig. 3A) comprise a succession of 5–6-m-thick red and green clays with thin bands of fossiliferous palustrine marls at two levels. The marls and clays contain a rich, diverse assemblage of gastropods, ostracods, and plant fossils (angiosperm seeds and fruits, monocot and dicot leaf impressions, conifers [*Araucarites*], *Isoetes*, and charophytes) associated with freshwater fishes and turtles (Mohabey and Udhoji, 1996). Algae, fungal remains, phytoliths, spores, and pollen are recorded mostly from the marls.

In the section at Piraya (Fig. 3B), a thin palustrine carbonate band occurs in silty overbank clays with occasional calcareous concretion nodules. Gastropods, ostracods, and charophytes are present in the clays. The concretionary nodules have yielded pollen, algae, fungal remains, and phytoliths.

At Pisdura (Fig. 3C), the palynomorph-yielding palustrine marls occur at two levels in a sequence of overbank red and green clays and sandstone nearly 5 m thick. The sediments contain a variety of fossils: vertebrates (including dinosaurs, turtles, snakes, mammals, and fishes), invertebrates (molluscs and ostracods), charophytes, plant megafossils, and plant-bearing coprolites (Von Huene and Matley, 1933; Matley, 1939; Jain and Sahni, 1986; Jain, 1989; Mohabey and Udhoji, 1996; Mohabey, 2001a; Ghosh et al., 2003; Mohabey et al., 2011).

At Dongargaon (Fig. 3D), a nearly 11-m-thick lake sequence in the lower part of the Lameta Formation contains freshwater diatoms *(Aulacoseira)* and sponge spicules in varved clays. Septaria concretions in the upper part of the succession yielded algae, fungal spores, phytoliths, and pollen. The fine-grained siliciclastic facies are rich in freshwater ostracods, molluscs, and fishes (Mohabey et al., 1993). A partially articulated skeleton of *Isisaurus colberti* (= *Titanosaurus colberti*; Jain and Bandyopadhyay, 1997; Wilson and Upchurch, 2003) has been described from the green clays in the lower part of the section. The sediments are capped by the basal volcanic flow of the Sahyadri Group.

The overlying Sahyadri Group of the Deccan continental flood basalt sequence is subdivided into the Ajanta, Chikhli, Buldhana, and Karanja formations in ascending stratigraphic order (Nair and Bhusari, 2001; Geological Survey of India 2001a, 2001b, 2001c). Intertrappean beds with thicknesses ranging from 1 m to more than 5 m occur in these formations at different stratigraphic levels. The sediments include cherts, tuffs, clays, shales, porcellanitic clays, shales, marls, and limestones. The intertrappean beds of the Ajanta Formation are exposed at different levels at Daiwal, Anandvan, Kharwad, Chikni, Sindhi, Mahalgaon, Narsala, and Gargoti. The sediments at Daiwal indicate deposition during C29r (Hansen et al., 2005). Lithological sections of palynomorph-bearing intertrappean sediments are shown in Figures 3E–3H.

The unclassified Deccan volcanic sequence of the Saurashtra area of Gujarat has 13 lava flows varying in thickness from 10 m to 80 m; intertrappean beds occur at five stratigraphic levels (Shekhavat and Sharma, 1996). Although we collected a series of samples from nine intertrappean localities, only the section at Ninama Hill yielded palynomorphs (Fig. 4). This intertrappean deposit is ~4 m thick and composed of marl and calcareous

TABLE 1. PALYNOLOGICAL AND FAUNAL ANALYSES OF INTERTRAPPEAN SECTIONS IN THE NAND-DONGARGAON (N-D) BASIN, CHHINDWARA, AND SAURASHTRA AREAS

Name of intertrappean section	Lithology	Fauna (invertebrate)	Palynomorphs/megaflora
A) Surface intertrappean exposures in the N-D basin			
1. Ramnagar	Green clay	Molluscs	Absent
2. Pander Puri	Green clay	Molluscs	Absent
3. Telankhedi	Green clay	–	Absent
4. Mahurzari	Chert	Molluscs	Absent (fossil wood, humic matter)
5. Nawargaon	Chert	–	Absent (fossil wood, humic matter)
6. Kalmeshwar	Chert, p. clay	Molluscs	Absent (humic matter)
7. Anandvan	Chert, p. clay	Molluscs	**Present**
8. Daiwal	Chert, shale	Ostracods	**Present**
9. Kharwad	Chert, lst	Molluscs	Absent (humic matter)
10. Sindhi	Chert, lst	Ostracod	**Present**
11. Narsala	Chert, lst	–	Absent
12. Gargoti	Chert, lst	–	Absent
13. Buranda	Chert	Sponges	Absent
14. Kolampad	Chert	–	Absent
15. Kutki	Chert	–	Absent
B) Subsurface intertrappean beds in the N-D basin			
16. Regional borehole KA-1 Khandala-Ashta	Black shale	–	**Present**
17. Regional borehole WSR-1 Selu 5 sections,	Green-gray clay	–	Absent
18. Khadkura borehole N-192, 194, 2 sections,	Green clay	–	Absent
19. Makardhokla MR-152-156, 5 sections,	Green clay	–	Absent
20. Wadki WAK M-1, 2; 4 sections	Black-gray shale	–	Absent
C) Intertrappeans from Chhindwara			
21. Mohgaon Kalan Well	Shale	–	**Present**
22. Mohgaon Kalan Fossil Forest	Chert	Molluscs	**Present** (fossil wood)
23. Jhilmili	Shales/clay	Ostracods, foraminifera	Absent
D) Intertrappeans from Saurashtra			
24. Ninama Hill section	Marl/chert	Absent	**Present**
25. Ninama River section	Marl/shale	Molluscs Ostracods	Absent

Note: lst—limestone: p. clay—porcellanitic clay.

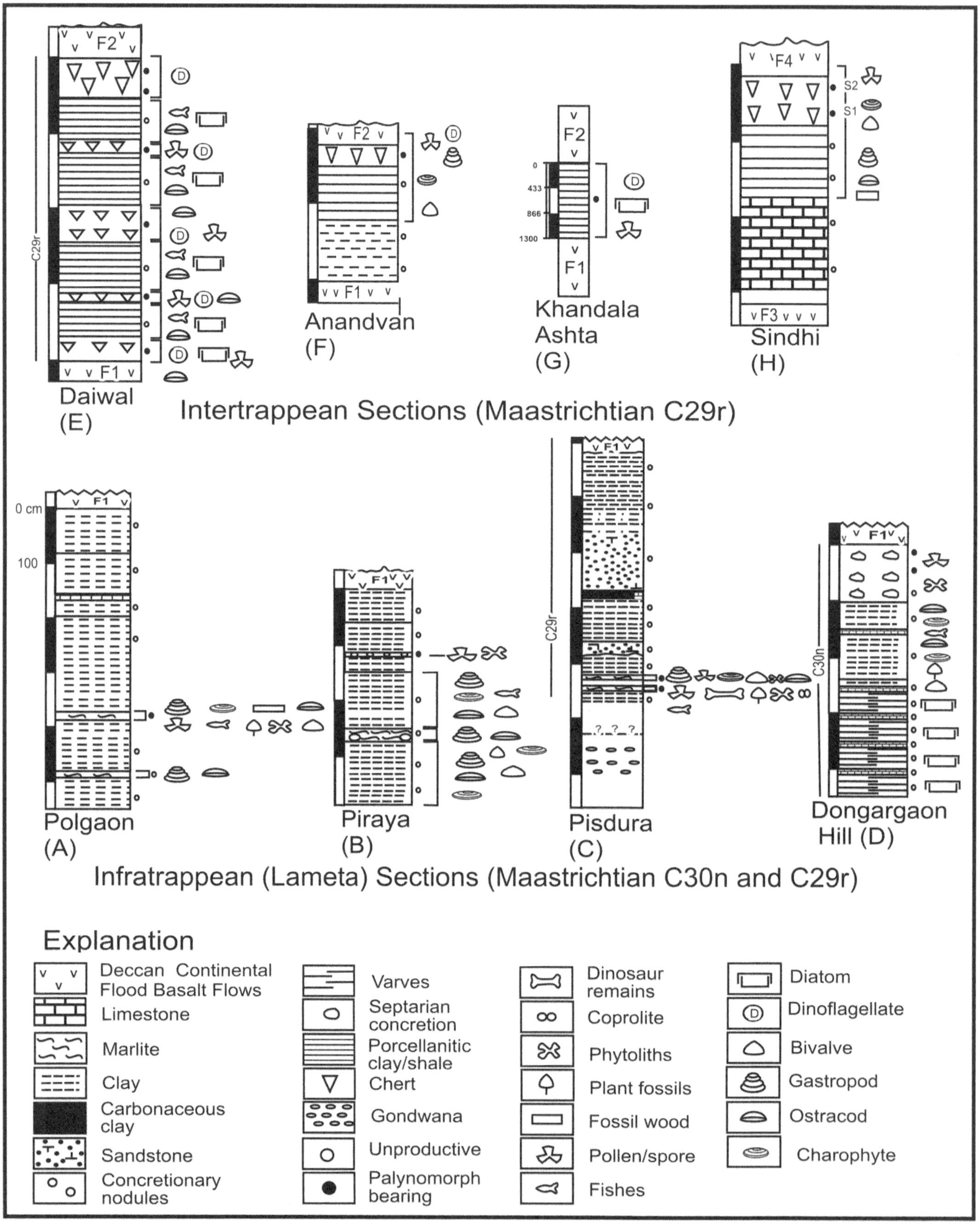

Figure 3. Lithologs of palynomorph-bearing sections of (A–D) Lameta and (E–H) intertrappean deposits in Nand-Dongargaon Basin.

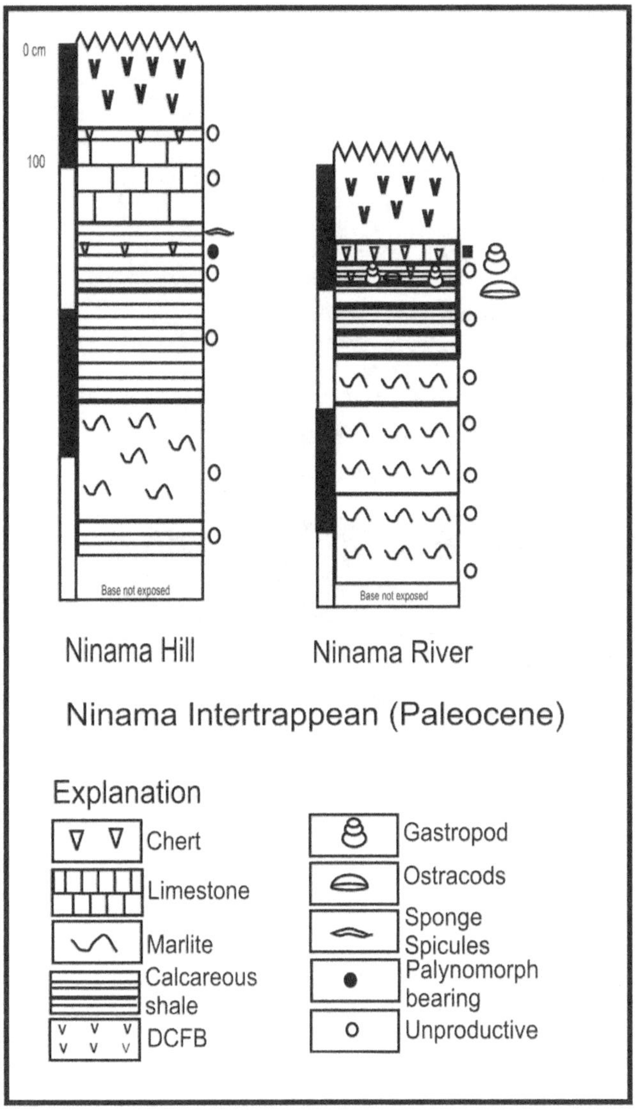

Figure 4. Lithologs of intertrappean deposits of Ninama Hill and Ninama River sections. DCFB—Deccan continental flood basalt.

shale in the lower part that proved to be either unfossiliferous or yielded only biodegraded organic matter or mycorrhizal fungi. However, a thin chert band in the section overlying the marl yielded a good palynomorph assemblage (Fig. 4). A few poorly preserved sponge spicules were recovered from the shale band overlying the chert. The uppermost limestone is unfossiliferous.

MEGAFLORA AND PALYNOFLORA: LAMETA FORMATION OF NAND-DONGARGAON BASIN

Megaflora

The megaflora record from the Lameta Formation of the Nand-Dongargaon Basin is meager and reported only from the localities at Polgaon and Dongargaon. The reported megaflora include charophyte plants (leaf whorls, nodes, and gyrogonites), pteridophytes (*Isoetes*), seeds (*Araucarites*), leaf impressions of gymnosperms and angiosperms (e.g., palms and dicots; Mohabey and Udhoji, 1993, 1996), monocotyledonous and dicotyledonous fruits (*Viracarpon*—Biradar and Bonde, 1979; Capparidaceae—Dutta and Ambwani, 2007; *Phoenix* seed—Ambwani and Dutta, 2005), small pieces of wood attributable to palm (Bonde and Birader, 1981; Dutta et al., 2011), Lecythidaceae (*Barringtonioxylon*), and Sapindaceae (*Euphorioxylon*; Kar et al., 2004).

Palynoflora

The palynological assemblage from the Lameta sediments consists of pollen and spores, including *Araucariacites*, *Callialasporites*, and *Trilobosporites* (Mohabey and Udhoji, 1996), *Classopollis*, *Cycadopites*, *Compositoipollenites*, *Cretacaeiporites*, *Graminidites* spp., *Longapertites*, *Palmaepollenites*, *Periporopollenites*, *Retimonosulcites*, and a few tri- and tetracolporate grains. Apart from these palynomorphs, grass phytoliths (bilobate, cross-shaped), silicified leaf cuticles, and trichomes of dicots are also recorded from the sediments (Fig. 5). Grass phytoliths (bilobate, cross-shaped, polylobate saddle, trapezoidal bilobate) were reported earlier from plant-bearing coprolites collected from Lameta sediments at Pisdura (Prasad et al., 2005, 2011). The pollen record of *Compositoipollenites* (Compositae) from the Late Cretaceous of Nand-Dongargaon Basin suggests that this family may have originated in India. To date, the oldest record of the Compositae family is from Eocene sediments of South America (Barreda et al., 2010). Based on phylogenetic analysis, Barres et al. (2013) suggested that this family originated in west Asia. Freshwater sponge spicules (Megascleres) associated with diatoms (*Aulacoseira*) and fungal spores (mostly *Inapertisporites* and *Dicellaesporites*) have been observed in sediments. *Lecaniella* and *Botryococcus* algae are commonly present in the lake sediments.

The macerated concentrate of sediments yielded large quantities of biodegraded organic matter, but the preservation of the palynomorphs is generally poor, possibly due to their degradation in semiarid to arid climatic conditions under oxidizing environments. Compared with palynomorphs, the preservation and concentration of the chlamydospores of *Glomus*, like mycorrhizal fungi, are better in many samples. This could be due to their association with plant roots, which might have prevented their aerial exposure, consequently arresting biodegradation. Maceration of various coprolite ichnomorphotypes (types A, B, Ba, and C of Matley, 1939) also yielded palynomorphs. Of the coprolites sampled, types A and Ba were the most productive for palynomorphs. Palynomorphs recovered represent pteridophytes (*Azolla*, *?Gabonisporis*; Ghosh et al., 2003), gymnosperms (*Araucariacites*, *Cycadopites*, *Podocarpidites*), and angiosperms (*Graminidites*, *Multiareolites*, *Palmaepollenites*). Additionally, diatom frustules (*Aulacoseira*), alga (*Oedogonium*-like alga), tracheal and woody material of gymnosperms and angiosperms, silicified plant remains, bacterial remains, mycorrhizal fungi, and

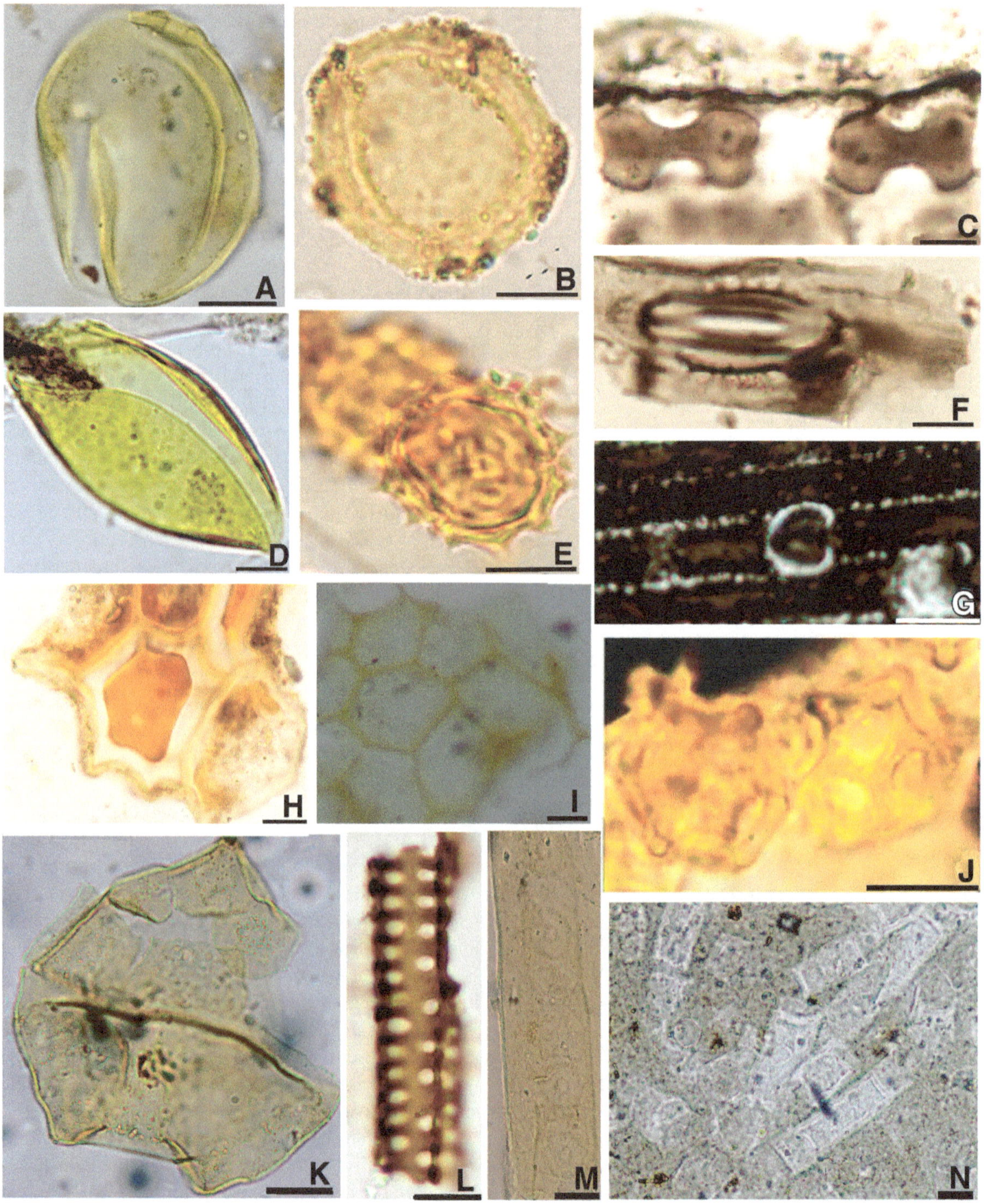

Figure 5. Microphotographs of selected palynomorphs and other organic matter; scale bars represent 10 µm; EF—England finder reading of palynomorphs. (A) *Graminidites annulatus* Potonié, 1960; slide no. Pol. 3(8); EF F27. (B) *Palmaepollenites* sp. A; slide no. Pol. 15(4); EF Q49. (C) Bilobate phytolith of Poaceae from Lameta sediments; slide no. Pis. (02); EF L36. (D) *Cycadopites* sp. from Lameta sediments; slide no. Pis. 6; EF W42. (E) *Compositoipollenites* sp.; slide no. Pol. 3(12); EF G48. (F) Grass cuticle with stomate phytolith from the Lameta sediments; slide no. Piraya–II; EF H41. (G) Cuticle with cuneiform bulliform phytolith with occluded carbon from type A coprolite; slide no. a 22(6); EF M54. (H) Phytolith from Lameta sediments; slide no. Piraya 10-II; EF H28. (I) Cuticular material from type A coprolite; slide no. a 25 (6); EF L38. (J) *Cretacaeiporites* sp.; slide no. Pol. 18(3); EF H53. (K) Dinoflagellate from intertrappean sediments of Surla; slide no. Surla 1; EF H47. (L) Plant remain from type A coprolite; slide no. a 2(6); EF J56. (M) Tracheary material from type A coprolite; slide no. Pis A (1); EF T37. (N) *Aulacoseira* in varved clays in Lameta sediments; slide no. V2; M47.

fungal remains, such as *Dicellaesporites*, *Pluricellaesporites*, *Multicellaesporites*, *Meliolinites*, *Quilonia*, and *Frasnacritetrus*, are also recorded from the coprolites. Some of these fungal spores show similarity with extant fungi (*Alternaria*, *Aspergillus*, *Helminthosporum*, and *Meliola*). Type Ba coprolites contain a good concentration of plant-derived organic matter, sponge spicules (Megascleres), and diatoms *(Aulacoseira)*. The overall quantitative analysis of palynological data from the Lameta sediments and associated coprolites indicates predominance of angiosperm pollen over gymnosperm pollen. In total, 12 genera and 15 species of pollen and spores were recovered (Table 2). The overall preservation and concentration of palynomorphs and other plant-derived material are better in coprolites than in sediment, possibly due to their concentration in guts of plant-eating sauropods. The fecal mass possibly prevented the direct aerial exposure of the palynomorphs and other plant material.

Some plant families are represented both as mega and microfossils. Araucareaceae is represented by pollen *Araucariacites* and seeds *Araucarites* (Mohabey and Udhoji, 1996), Arecaceae by pollen *Palmaepollenites*, *Longapertites*, fossil wood *Palmoxylon* (Bonde and Birader, 1981; Dutta et al., 2011), leaves (Mohabey and Udhoji, 1996), and *Phoenix* seed (Ambwani and Dutta, 2005), and the Poaceae family by pollen *Graminidites*, phytoliths, and permineralized leaf tissues.

PALYNOFLORA FROM INTERTRAPPEAN SEDIMENTS: NAND-DONGARGAON BASIN

In the Nand-Dongargaon Basin, detailed quantitative assessment of the palynological data of the successive intertrappean sediments indicates that the lowest stratigraphic level (i.e., Daiwal, Anandvan, and Khandala-Ashta; Figs. 3E–3G) has good concentrations of well-preserved palynomorphs.

The Daiwal intertrappean sediments (Fig. 3E) are mostly porcellanitic clays and black chert, dominated by diatoms in porcellanitic clays and dinoflagellates in the interlayered black cherts. The palynomorphs are mostly present in the marginally deposited black chert. The black chert of Anandvan (Fig. 3F) contains pollen, spores, and dinoflagellates. At Surla, which is an extension of Anandvan, high numbers of algae *Pediastrum* along with dinoflagellates are present. Khandala-Ashta (Fig. 3G) also contains a good concentration of spores and pollen assemblages, along with the dinoflagellates. These intertrappean beds are dominated by *Azolla cretacea*, *Aquilapollenites bengalensis*, and *Triporopollenites* sp. (Table 2; Fig. 6).

The overall data indicate that after the onset of volcanic activity in the basin, the characteristic palynomorphs of the infratrappean sediments (e.g., *Araucariacites*, *Classopollis*, *Cycadopites*, *Podocarpidites*, *Compositoipollenites*, *Graminidites*, and *Palmaepollenites*) either disappeared or were drastically reduced in number. A new palynoassemblage evolved, dominated by *Aquilapollenites bengalensis*, *Azolla cretacea*, *Gabonisporis vigourouxii*, *Tricolpites reticulatus*, *Triporoletes reticulatus*, *Triporopollenites* sp., and *Normapolles* group pollen. This assemblage is observed in varying abundances in all lowermost intertrappean beds (C29r) of the Nand-Dongargaon Basin. The palynomorphs of the lower intertrappean sediments (Daiwal, Khandala-Ashta, and Anandvan) are also associated with blooms of monogeneric peridinoid freshwater dinoflagellates. In total, 27 species belonging to 24 genera were identified in the assemblage (quantitative data in Table 2; Fig. 7). Quantitative analysis of freshwater dinocysts was not carried out in these sediments because their blooms dominated the studied samples.

The overlying intertrappean sediments at Sindhi between lava flow 3 and flow 4 also have a good concentration of pollen and spores, with a predominance of *Gabonisporis vigourouxii* (Fig. 3H). In total, 16 genera and 20 species of pollen and spores are present at this level (Table 2), which is marked by the characteristic appearance of the polyaperturate pollen *Scabrastephanocolpites* spp., *Scollardia conferta*, and *Triporopollenites cracentis*. Dinocysts and diatoms are not recorded.

Overall, all intertrappean beds at different stratigraphic levels from the late Maastrichtian C29r in the Nand-Dongargaon Basin are dominated by angiosperm pollen and pteridophytic spores. Gymnosperm pollen grains commonly present in Lameta sediments are rare in the intertrappean assemblage, despite good preservation (Fig. 7). Some taxa, such as *Gabonisporis vigourouxii*, *Azolla cretacea*, and *Aquilapollenites bengalensis*, are present in varying abundances in all intertrappean beds. Distinct floral assemblages present in different stratigraphic levels in the Lameta and intertrappean beds have been quantified based on relative abundances of major groups (Fig. 7).

Diatoms and Dinoflagellate Cysts

A monogeneric assemblage of *Aulacoseira* and a few sponge spicules are observed in four levels of the varved clays in the Lameta lake sequence at Dongargaon, which was deposited during C30n (Fig. 3D). In the immediately overlying intertrappean sediments of Daiwal, Anandvan, and Khandala-Ashta, diatoms are reduced in abundance, with the first appearance of monogeneric peridinoid dinoflagellates. Both diatoms and dinoflagellates disappear higher up in the sequence and are not recorded in the overlying intertrappean sediments.

The C29r intertrappean sediments of the Nand-Dongargaon Basin record the oldest monogeneric freshwater dinoflagellate cyst assemblage, which so far is known only from the Miocene sediments of Idaho, United States (Batten et al., 1999), and Lake Enspel, Germany (Kohler and Clausing, 2000). The increased concentration of diatoms in Columbia River Basalt sediments (Jolley et al., 2008) has been attributed to volcanogenic input. Dinoflagellate blooms in Lake Enspel and maar lake of Messel, Germany, have been linked to increased nutrient availability (Clausing et al., 2000; Lenz et al., 2006). It is possible that the appearance and blooms of a new peridinoid monogeneric freshwater dinoflagellate genus in the basal intertrappean lake sediments, which were deposited on a fresh Deccan lava surface, were also linked to the volcanically produced nutrient availability.

TABLE 2. PALYNOMORPHS RECOVERED FROM THE STUDIED LAMETA FORMATION AND INTERTRAPPEAN (IT) SEDIMENTS OF NAND-DONGARGAON (N-D) BASIN

Lameta sediments and coprolites		Intertrappean beds →	Daiwal IT	Anandvan IT	Anandvan IT (Surla)	Khandala-Ashta IT	Sindhi IT S1	Sindhi IT S2
A) Megaflora		Palynomorphs/taxa						
1) Fossil woods		*Pediastrum diffuses* Singh and Khanna, 1978	0	0	**46**	0	0	0
a) Palms		*Ariadnaesporites* sp.	0	0	0	1	0	0
b) Sapindaceae		*Azolla cretacea* Stanley, 1965	1	0	0	**94**	4	3
c) Euphorbiaceae		*Cyathidites australis* Couper ,1953	0	0	0	5	8	9
2) Leaves		*Gabonisporis vigourouxii* Boltenhagen, 1967	2	3	1	2	**140**	**176**
a) Palms		*Lygodiumsporites* spp.		0	0	6	9	9
b) Dicots		*Marsilea*-like pollen	0	0	0	**12**	7	**10**
c) *Isoetes*		*Todisporites* spp.		0	0	6	2	2
3) Fruits		*Triporoletes reticulatus* Playford, 1971	0	0	0	2	0	0
a) Capparidaceae		*Triplanosporites* sp.		0	0	0	2	1
b) *Viracarpon*		*Cycadopites* sp.		0	0	0	0	0
4) Seeds a) *Phoenix*		*Equisetosporites* spp.		0	0	0	3	5
B) Palynoflora		*Podocarpidites* sp.	0	0	1	0	0	0
Taxa		*Aquilapollenites bengalensis* Baksi & Deb, 1976	**16**	4	1	**76**	4	5
		Graminidites sp.	0	1	0	0	0	0
Lecaniella sp.	2	*Echitricolpites* sp. A		0	0	2	0	0
Oedogonium sp.	**10**	*Farabeipollis* spp.		0	0	2	1	1
Azolla sp.	1	*Jiangsupollis* sp. A		1	0	0	0	2
Araucariacites australis Cookson, 1947	8	*Neocouperipollis* sp.	1	2	0	0	0	0
Cycadopites sp.	3	*Palmaepollenites* sp.	0	0	0	1	0	0
Classopollis sp.	3	*Proxapertites operculatus* Van der Hammen,1956	1	3	0	0	0	0
Podocarpidites sp.	1	*Psilodiporites* sp.	0	0	0	1	0	1
Cretacaeiporites sp.	2	*Recemonocolpites maximus* Prasad et al., 1995	1	0	0	0	0	0
Compositoipollenites sp. A	6	*Retitricolpites crassireticulatus* Samant & Phadtare, 1997	**12**	3	2	1	0	0
Graminidites annulatus Potonie ,1960	5	*Tricolpites reticulatus* Couper, 1953	2	**11**	1	2	0	0
Graminidites assamicus Sah & Dutta, 1974	2	Tetracolporate striate pollen	4	0	0	0	0	0
Longapertites sp.	2	*Triporopollenites* sp. A	**24**	**60**	0	0	0	0
Multiareolites sp. A	5	*Turonipollis* sp. A		0	0	9	0	1
Palmaepollenites sp. A	**11**	*Normapolles* group pollen	0	0	0	2	2	2
Palmaepollenites sp.	1	Other small pollen	0	**10**	0	0	0	0
Periporopollenites sp.	1	*Scabrastephanocolpites singpurensis* Samant et al., 2008	0	0	0	0	8	8
Retimonosulcites sp.	2	*S. intertrappea* Samant et al., 2008	0	0	0	0	3	2
Tetracolporate pollen	1	*Scollardia conferta* Venkatachala & Sharma, 1974	0	0	0	0	1	0
Tricolporate pollen	1	*Triporopollenites cracentis* Frederiksen, 1994	0	0	0	0	7	**15**
Fungal spores	**25**	Fungal spores	**25**	3	0	**76**	**99**	**50**
		Phragmothyrites	1	0	0	0	0	0
Total	92	Total	100	100	52	300	300	300
		Numbers represent specimens recovered; 0 = none present	(excluding dinocysts)	(excluding dinocysts)	(excluding dinocysts)	(excluding dinocysts)		

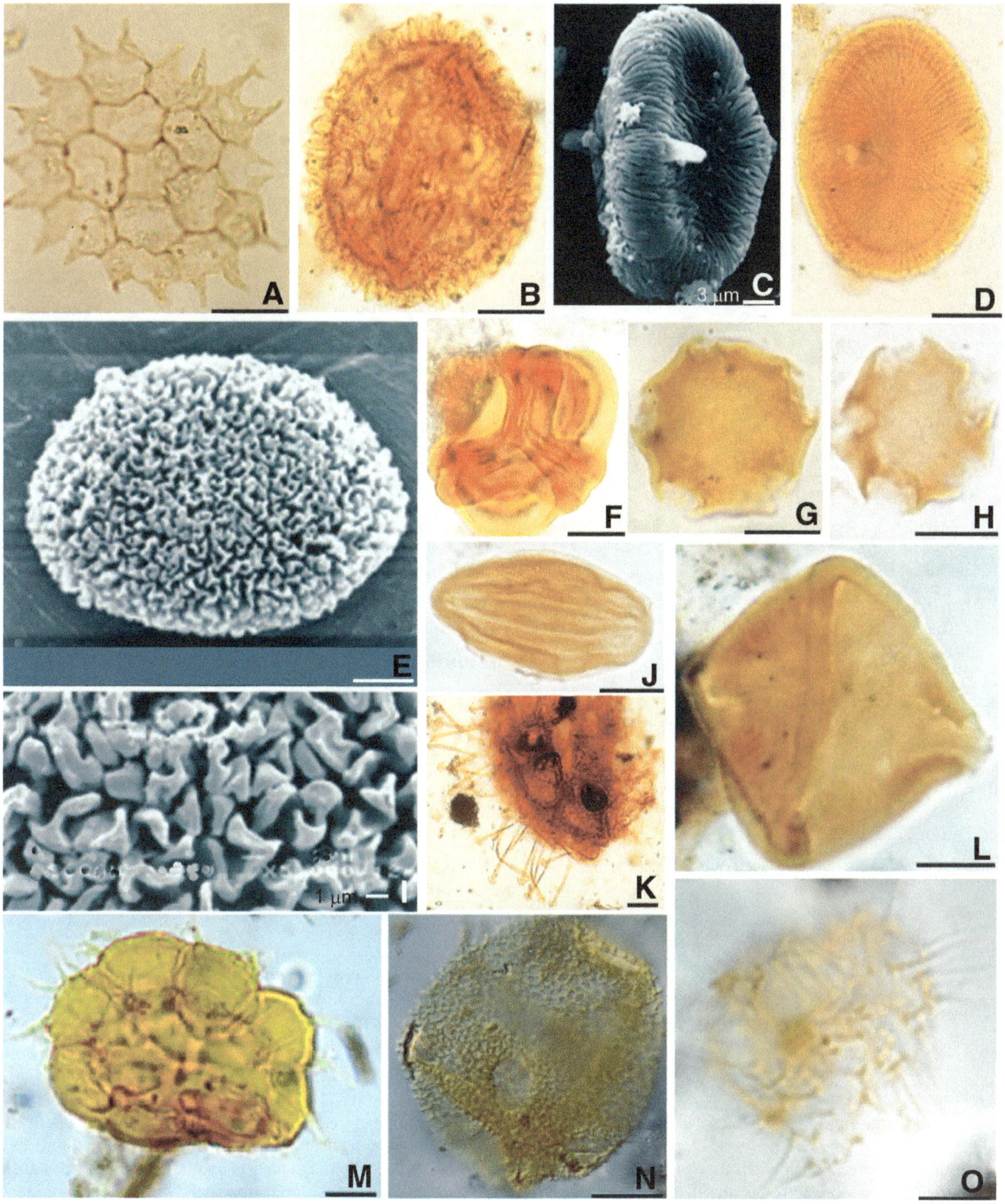

Figure 6. Microphotographs of selected microfossils; scale bars represent 10 µm unless otherwise noted; EF—England finder reading of palynomorphs. (A) *Pediastrum diffuses* Singh and Khanna, 1978; slide no. SCh-2; EF R25. (B) *Gabonisporis vigourouxii* Boltenhagen, 1967; slide no. IJ(10); EF N18. (C) Scanning electron microscope (SEM) photographs of *Aquilapollenites bengalensis* Baksi and Deb, 1976. (D) *Aquilapollenites bengalensis* Baksi and Deb, 1976; slide no. PCH 1; EF W25. (E and I) SEM photograph of *Gabonisporis* spore. (F) *Normapolles* group pollen; slide no. P-15(11); EF L19. (G) *Scabrastephanocolpites singpurensis* Samant et al., 2008; slide no. Si 17(4); EF P58. (H) *S. singpurensis* Samant et al., 2008; slide no. Si. 17(4); EF Q31. (J) *Equisetosporites ovatus* Singh (Pierce) Singh, 1964; slide no. Si(7); EF J20. (K) *Azolla cretacea* Stanley, 1965; slide no. J 10; EF T49. (L) *Jiangsupollis* sp. A; slide no. Si14(1); EF N41. (M) Foraminiferal lining from the Mohgaon Kalan Well section (MKWS); slide no. MKWS 6; EF O23. (N) *Typha*-like pollen from intertrappean section; slide no. Sau. 5; EF Y34. (O) Dinoflagellate from Ninama intertrappean section; slide no. NM1 A; EF D39.

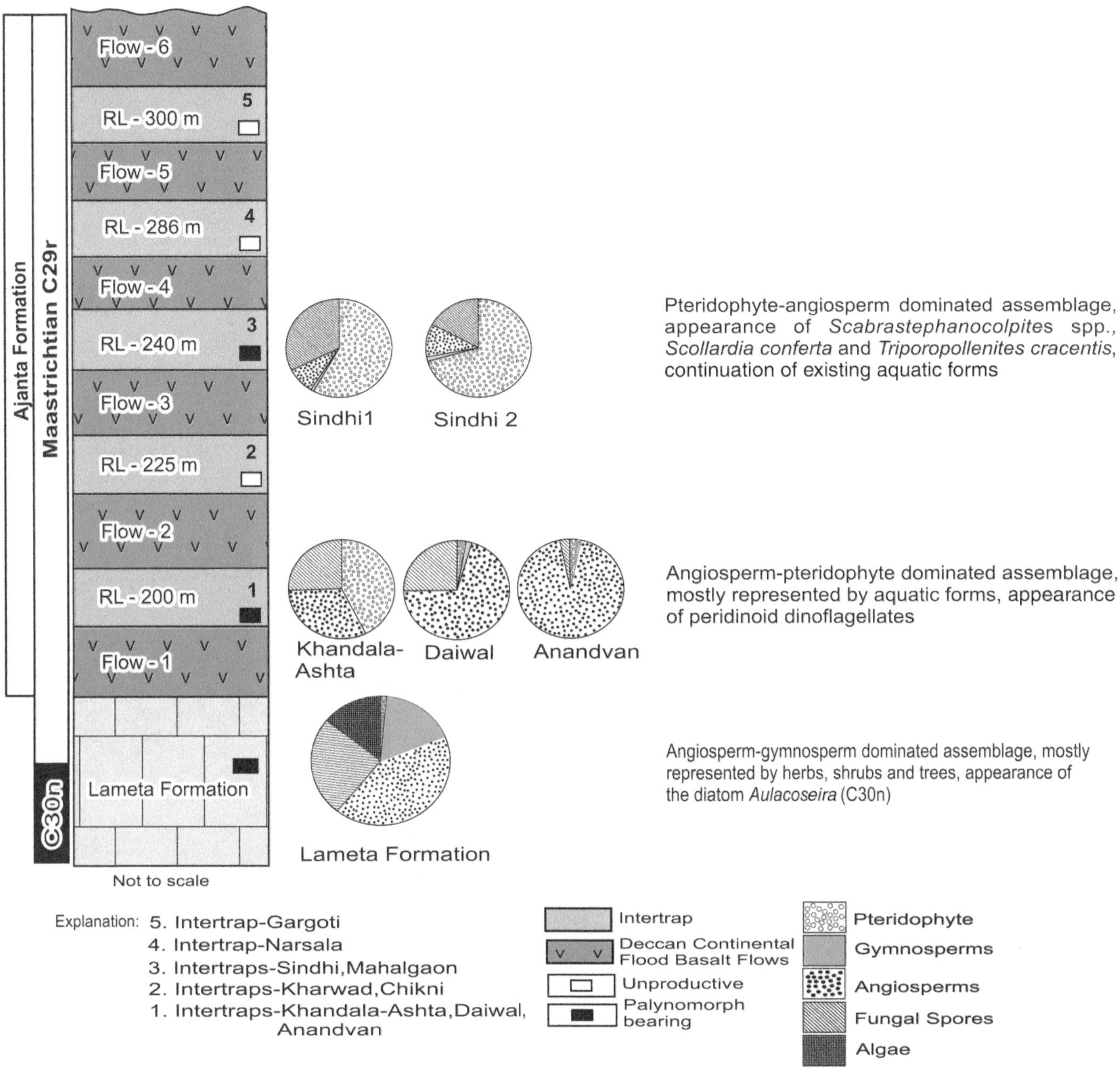

Figure 7. Schematic stratigraphic sequence of the Lameta Formation and intertrappean beds at different stratigraphic levels in the Deccan continental flood basalt province of the Nand-Dongargaon Basin showing palynological events associated with Lameta and intertrappean beds. Percentages of the major floral groups represented by palynofloral assemblages of the Lameta Formation and intertrappean beds are shown for two stratigraphic levels (lower intertrappean—Khandala-Ashta, Daiwal, Anandvan, and higher intertrappean—Sindhi). RL—reduced level.

PALYNOSTRATIGRAPHIC CORRELATION: DECCAN CONTINENTAL FLOOD BASALT PROVINCES

Late Maastrichtian

Deccan continental flood basalt–associated intertrappean sediments from the central and southern peninsula of India have yielded rich faunal remains from a number of localities (summary in Khosla and Sahni, 2003; Prasad and Sahni, this volume). The megafloral record from the intertrappean sediments is limited to a few localities from the Sahyadri and Amarkantak Groups of central India (Kapgate, 2005; Bonde, 2008). Palynoflora from the intertrappeans are limited mostly to the Amarkantak Group of central India and an unclassified Deccan sequence in Kutch, western India. In the course of this study, many palynomorph-bearing intertrappean localities from different stratigraphic levels in the Sahyadri Group were analyzed. These data were compared and correlated with those from the Amarkantak Group to the north of the Nand-Dongargaon Basin, Kutch, and Saurashtra in order to evaluate the qualitative variations in palynofloral composition in these geographically separated areas.

The Deccan volcanic sequence of the Amarkantak Group in Chhindwara, Mandla, and Jabalpur areas is classified into eight formations, namely, Mandla, Dhuma, Piparadhi, Linga, Multai, Amarwara, Khampa, and Kuleru Formations, in ascending stratigraphic order (Geological Survey of India, 2001d). Many intertrappean deposits are present in these formations. However, good palynomorph-bearing intertrappean sediments are present only in the Mandla and Dhuma Formations. The lowermost part of these formations at Mohgaon Kalan, Chhindwara area, has two palynomorph-bearing intertrappean deposits: the Mohgaon Kalan Well section and the Mohgaon Kalan Fossil Forest (Samant et al., 2008). The lower Mohgaon Kalan Well section is associated with dinosaur eggshells and Maastrichtian palynomorphs (Srinivasan, 1996; Kar and Srinivasan, 1997; Kumaran et al., 1997), whereas the upper Mohgaon Kalan Fossil Forest is rich in megafloral remains, including angiosperms, gymnosperms, pteridophytes, algae, and fungi. The megaflora is dominated by angiosperms (Kapgate, 2005; Bonde, 2008).

In this study, a palynomorph-rich layer in the Mohgaon Kalan Fossil Forest chert contains the Maastrichtian marker palynotaxa *Aquilapollenites bengalensis*, *Azolla cretacea*, and *Gabonisporis vigourouxii*. This palynoassemblage is significant because it resolves the controversy over Maastrichtian or Paleocene age for these deposits and firmly places the Mohgaon Kalan Fossil Forest in the Maastrichtian. The Mohgaon Kalan Well section contains palynomorphs of *Spinizonocolpites* (*Nypa*), *Proxapertites* (Arecaceae), and foraminiferal linings, whereas Mohgaon Kalan Fossil Forest contains *Spinizonocolpites* (*Nypa*) and *Nypa* megaflora (Kapgate, 2005; Bonde, 2008), which suggest marine influence during Maastrichtian chron C30n and C29r. In the Chhindwara area, a Paleocene marine influence was documented by Keller et al. (2009a) based on early Danian zone P1a planktic foraminifera in the Jhilmili intertrappean sediments that stratigraphically overlie the Mohgaon Kalan Fossil Forest. A palynological study of the Jhilmili intertrappean sediments revealed only biodegraded organic matter and fungal spores. In the Jabalpur area to the north of Chhindwara, a Maastrichtian palynoassemblage was reported from Ranipur and Padwar intertrappean sediments associated with the Dhuma Formation (Mathur and Sharma, 1990; Prakash et al., 1990). The Maastrichtian palynoassemblages from the Mohgaon Kalan Well section and Mohgaon Kalan Fossil Forest, as well as from the Dhuma Formation at Ranipur and Padwar, are similar to those recorded from the intertrappean sediments of the Nand-Dongargaon Basin.

In the Deccan volcanic province in Kutch, Maastrichtian palynomorph assemblages are reported from the Lakshmipur and Anjar localities (Samant and Bajpai, 2005; Dogra et al., 2004). The Lakshmipur intertrappean assemblage is dominated by gymnosperm and angiosperm pollen, whereas the Anjar assemblage is dominated by pteridophytes and angiosperms. Overall, the palynoassemblage is similar to the Nand-Dongargaon Basin. A late Maastrichtian palynoassemblage is also reported from the Naskal intertrappean locality in the southern Deccan volcanic province (Singh et al., 2006).

Palynomorphs are recorded from the Lameta sediments of the Nand-Dongargaon Basin (Samant and Mohabey, 2005, 2009) and the Jabalpur Basin (Dogra et al., 1994), whereas the megaflora is exclusively known from the Lameta Formation of the Nand-Dongargaon Basin (Mohabey and Udhoji, 1996). Paleontological investigations in the last decade have not produced additional megafloral fossil localities in the Lameta sediments. On the contrary, there is an increasing record of megaflora and palynoflora from the intertrappean sediments. The correlation of palynological data from intertrappean sediments in different provinces shows that Maastrichtian sediments are characterized by dominance of *Azolla cretacea*, *Aquilapollenites bengalensis*, *Gabonisporis vigourouxii*, *Triporoletes reticulatus*, and *Normapolles* group pollen. This assemblage is remarkably different from the assemblage recorded from the underlying Lameta sediments.

Paleocene

We report palynomorphs from the intertrappean bed at the Ninama locality in Saurashtra (Fig. 1). This intertrappean bed contains palynomorphs *Intrareticulites brevis*, *Striacolporites striatus*, *Retitricolpites crassimarginatus*, and *Rhombipollis* sp. (Table 3), which have so far been recorded only from the Paleocene and early Eocene of the Indian subcontinent (Saxena, 1990; Frederiksen, 1994). Except for a single grain of *Aquilapollenites bengalensis*, no other Maastrichtian marker palynomorphs of the Deccan volcanic sequence are reported from the Ninama intertrappean bed. Dominance of *Neocouperipollis* in the Ninama palynoassemblage, along with non-peridinoid dinoflagellates, suggests marine influence (Fig. 8). To date, the only Paleocene palynomorph-bearing intertrappean assemblage from the Deccan continental flood basalt province is known from Lalitpur in Uttar Pradesh (Singh and Kar, 2002).

Intertrappean beds of Paleocene age are also known from Mumbai (Sukheshwala, 1956; Bande et al., 1988; Sethna, 2003). However, their Paleocene age is based on the absolute age dating of associated Deccan volcanic flows. The palynomorph recovery from these intertrappean beds is poor, and except for *Aquilapollenites* sp., other forms are identifiable only at the family level. Cripps et al. (2005) suggested that the Mumbai intertrappean sediments associated with pyroclastic discharges were deposited in a post–Cretaceous-Paleogene boundary lagoon system.

INFLUENCE OF DECCAN VOLCANISM ON FLORA AND CLIMATE

During the long Deccan flood basalts eruptive history (ca. 69–62 Ma; Sheth et al., 2001; Pande, 2002; Sheth and Pande, 2014), volcanism produced a large volume of magma and

TABLE 3. PALYNOMORPHS RECOVERED FROM NINAMA HILL INTERTRAPPEAN SEDIMENTS

Palynomorph/taxa	Number of specimens recovered
Pediastrum boryanum Maeghini	4
Frasnacritetrus sp. A	1
Dinoflagellate	1
Fungal spore	**17**
Pteridacidites sp.	1
Araucariacites sp.	2
Aquilapollenites bengalensis Baksi & Deb, 1976	1
A. ovatus Hofmann & Zetter, 2007	4
Arecipites communis Mathur & Chopra, 1987	1
Crototricolpites densus Salard-Cheboldaeff, 1978	5
Graminidites sp.	1
Liliacidites sp.	1
Intrareticulites brevis (Sah & Kar) Kar, 1985	1
Neocouperipollis sp. A	**132**
Retitricolpites crassimarginatus Samant & Phadtare, 1997	1
Retitricolpites sp.	1
Rhombipollis sp. A	7
Striacolporites striatus Sah & Kar, 1970	4
Proteacidites sp.	2
Triporopollenites sp. A	8
Typha-like pollen	1
Tricolpate reticulate pollen	3
Total	200

associated gases (~1.3×10^6 km^3; Cox, 1989; Self et al., 2005; Jay and Widdowson, 2008). Deposition of lava flows now classified into various formations in different Deccan volcanic subprovinces is in conformity with the northward migration of the Indian plate over a hotspot. Most of the lava piles in widely separated areas are considered to have different sites and sources of eruptions (Sen and Cohen, 1994; Nair et al., 1996). The contemporary local fauna and flora in the vicinity of active eruptive sites of different provinces would be affected by the direct and indirect activity of the Deccan volcanism (early Maastrichtian–Paleocene). The history of plants existing just before the arrival of the first Deccan flows (Sahyadri Group) in the Nand-Dongargaon Basin is well preserved in the Lameta sediments. The overlying intertrappean beds higher up in the sequence record the history of palynofloral changes that may have been induced by the active volcanism. The present study suggests a possible link between palynofloral changes in the Deccan volcanic sequence–associated sediments and the Deccan volcanism phase 2 eruptive event (Fig. 9; Chenet et al., 2007, 2008, 2009).

In the Nand-Dongargaon Basin, the Lameta sediments are characterized by well-developed calcrete profiles and gypsum-rich clays. The flora is composed mainly of angiosperm-gymnosperm plants. The gymnosperm plant association consists of evergreen flora (*Araucaria*, *Cycas*, *Podocarpus*, and *Classopollis*) with an understory of angiosperms, including Arecaceae (*Palmaepollenites*, *Longapertites*), Poaceae (*Graminidites*), Caryophyllaceae (*Cretacaeiporites*, *Periporopollenites*), Asteraceae (*Compositoipollenites*), Acanthaceae (*Multiareolites*), and other dicots. Autochthonous *Botryococcus*, *Oedogonium*, and *Lecaniella* algae thrived in local water bodies.

The lithofacies, biofacies, and oxygen isotope studies of the Lameta sediments (Sarkar et al., 1991; Tandon and Andrews, 2001; Mohabey and Samant, 2005; Ghosh et al., 2003) suggest semiarid to arid climatic conditions. The chlamydospores of *Glomus*-like mycorrhizal fungus and a large amount of biodegraded organic matter in the lacustrine and paludal sediments also suggest intermittent desiccation, evaporation, and fluctuating lake levels (Turner and Friese, 1998; Lodge, 1989). The presence of gypsum and thin limestone-carbonate mud facies is indicative of alkaline conditions and excessive evaporation (Mohabey and Samant, 2005). Alkaline lake conditions are also indicated by the presence of algal gyrogonites of Chara (*Platycharaperlata*, *Microchara*) and other algae (*Oedogonium*, *Lecaniella*). The record of the pollen of Cheirolepidiaceae (*Classopollis*), and to some extent Asteraceae (*Compositoipollenites*) and Poaceae (phytoliths of mostly cross and dumb-bell type; Vartika et al., 2007), suggests dry climatic conditions at the time of deposition (Srivastava, 1976).

With the onset of volcanism in the Nand-Dongargaon Basin, the approaching lava fronts not only came in direct contact with the existing vegetation, but the outpouring of voluminous lava flows also brought physiographic changes. The volcanic effusions and associated ash falls influenced the existing flora and sediments. Lithologically, the intertrappean sediments are dominated by cherts and porcellanitic clays. The large amount of silica mobilization in the intertrappean lakes could be mostly from acid rains caused by volcanism. Outpouring of the lava flows destroyed the existing vegetation, and new vegetation appeared in the lakes and ponds that developed over the fresh lava fields and on the volcanic substrate. Such observations are also known from the middle Eocene maar lake sediments of Germany (Lenz et al., 2006) and recent volcanic sites at Mount St. Helens (USA) and Usu (Japan) (Tsuyuzaki, 1987; Tsuyuzaki and del Moral, 1995; del Moral and Grishin, 1999).

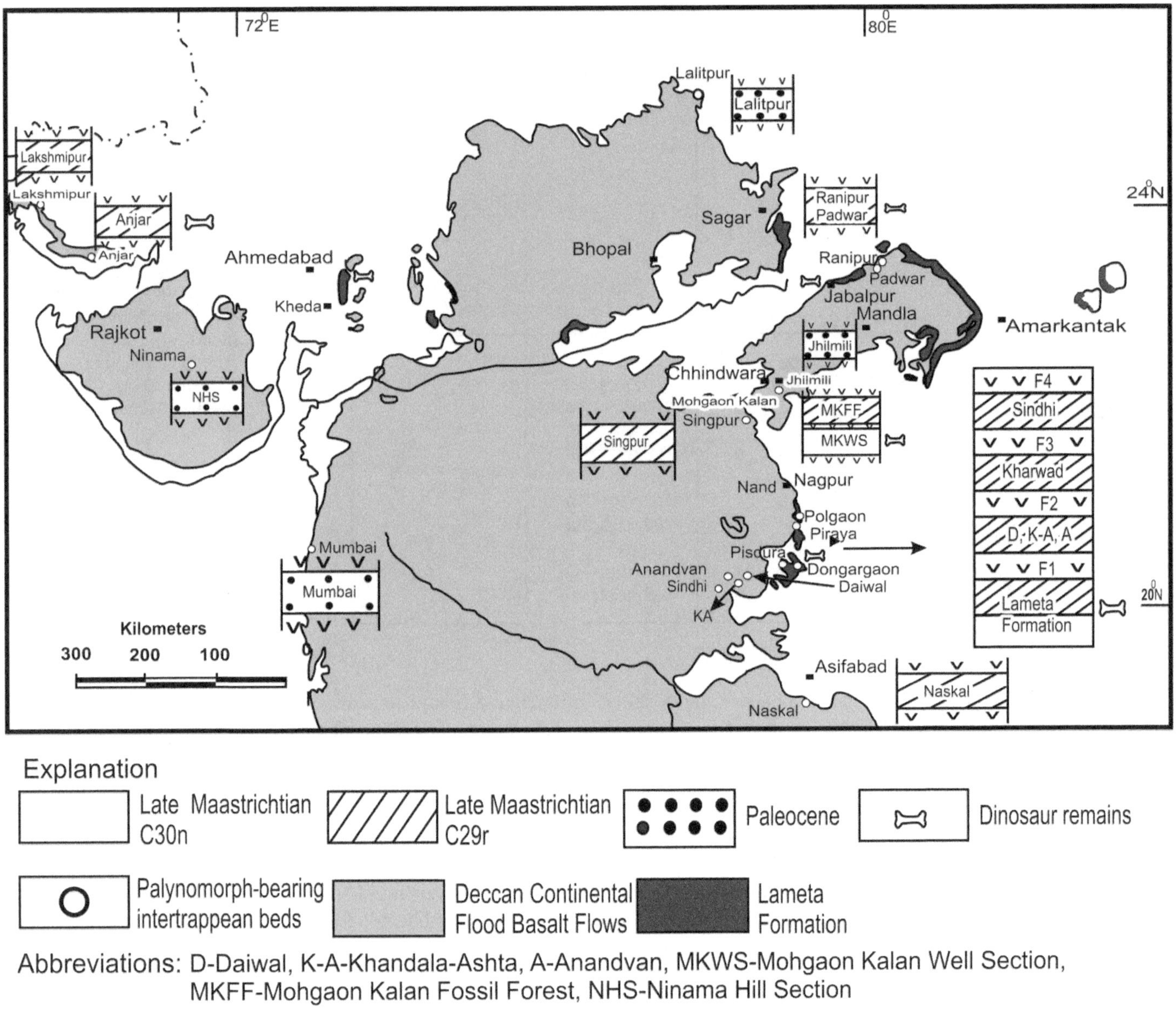

Figure 8. Stratigraphic correlation of geographically separated intertrappean beds based on palynomorphs and dinosaur fossils.

In the lower intertrappean sediments (Daiwal, Anandvan, Khandala-Ashta; Fig. 9), the overall palynofloral composition of the lakes is distinctly different from the underlying Lameta sediments. These intertrappean also show the appearance and domination of peridinoid dinoflagellates (Fig. 7). Importantly, Araucareaceae, Podocarpaceae, Cycadeoidae, Cheirolepidiaceae, Asteraceae, Caryophyllaceae, and Poaceae pollen are either absent or poorly represented, despite good preservation conditions in the intertrappean sediments (Table 2). The abundance of dinocysts, as well as the presence of a large concentration of *Pediastrum* algae in the Surla lower intertrappean sediments, could be due to eutrophication of the lakes by volcanic effusions (Tappan, 1980; Yi, 1997). The palynoflora of the earliest intertrappean sediments is mostly represented by aquatic floral elements (*Pediastrum*, dinocysts, *Azolla*, *Gabonisporis*) and taxa like *Aquilapollenites bengalensis*, *Tricolpites reticulatus*, *Triporopollenites* sp., and *Normapolles* group pollen. Fossil wood is rare in these sediments (Table 2). These evidences suggest that the vegetation community consisted exclusively of aquatic plants, small herbs, and shrubs. A similar vegetation succession has been observed in Lake Messel, Germany (Lenz et al., 2006). At the higher stratigraphic level (Sindhi) in the Deccan volcanic sequence, the flora diversified, and some new forms appeared, including *Scabrastephanocolpites* spp., *Scollardia conferta*, and *Triporopollenites cracentis*. Dinoflagellates and diatoms are absent, and fossil wood fragments are present.

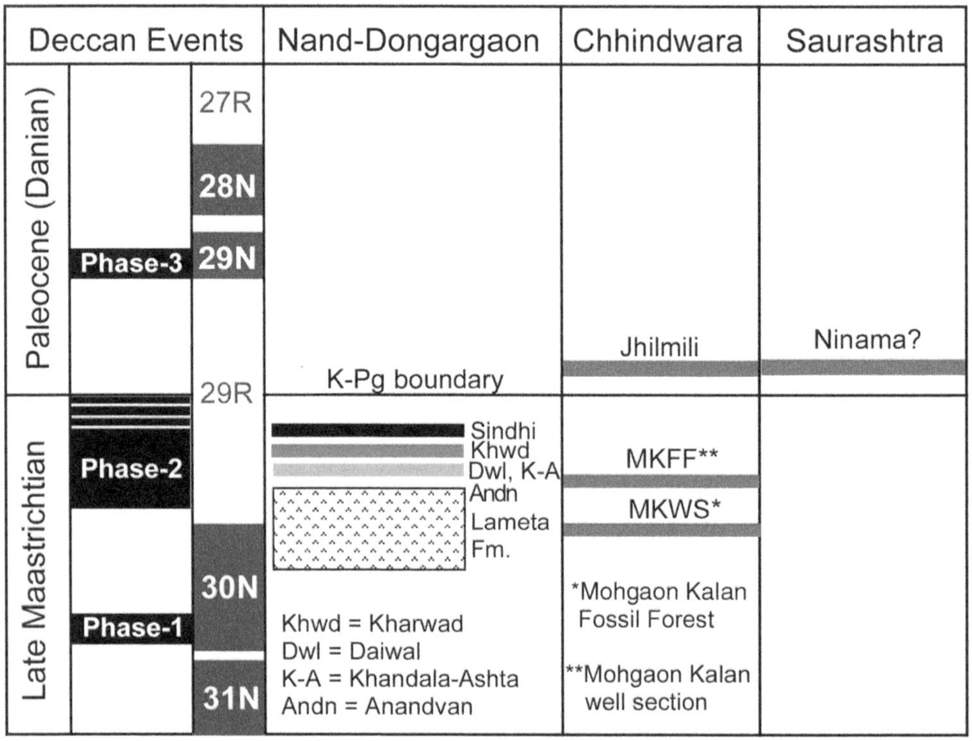

Figure 9. Summary of the age and stratigraphic position of the Lameta and intertrappean sections analyzed and their relative positions in the Deccan volcanic sequence of the Nand-Dongargaon Basin, Chhindwara and Saurashtra (Ninama). The three eruptive phases are defined by Chenet et al. (2007, 2008, 2009). K-Pg—Cretaceous-Paleogene.

In the Amarkantak Group, in the Mohgaon Kalan locality, Chhindwara area (Figs. 1 and 9), the lowermost intertrappean bed of the Mohgaon Kalan Well section is associated with black carbonaceous, lignitic to sapropelic shales. These shales contain a diverse and well-preserved palynoassemblage and high concentrations of fungal spores and fruiting bodies, suggesting a humid environment. The immediately overlying Mohgaon Kalan Fossil Forest intertrappean section contains a diverse megaflora of *Elaeocarpoxylon anticum* (Elaeocarpaceae), *Alianthoxylon ghiarense* (Simaroubaceae), *Barringtonioxylon deccanensis*, *Barringtonioxylon eopterocarpon* (Lecythidaceae), *Tetrameleoxylon prenudiflora* (Datiscaceae), *Grewioxylon mahurzariensis* (Tiliaceae), *Palmoxylon* of cf. *Phoenix* (Arecaceae), *Rhamnoxylon intertrappea* (Rhamnaceae), *Musa cardiosperma* (Musaceae), and *Ebenoxylon mohgaonense* (Ebenaceae). This flora indicates a tropical evergreen to semi-evergreen forest with some moist deciduous plants (Kapgate, 2005). The fungal fruiting bodies of Microthyriaceae (*Phragmothyrites*) also indicate prevailing warm humid conditions and heavy rainfall (Cookson, 1947; Dilcher, 1965; Selkirk, 1975). The flora as well as sediments of both the Mohgaon Kalan Well section and Mohgaon Kalan Fossil Forest intertrappean sediments suggest the prevalence of humid climatic conditions during their deposition.

The subsequent climatic shift from humid to semiarid and arid conditions is indicated in the immediately overlying Paleocene sediments at Jhilmili. Lithology, clay mineralogy, and oxygen isotopes of this 14-m-thick intertrappean section suggest deposition in mostly semiarid to arid climatic condition (Keller et al., 2009a). The overall data of the Amarkantak Group indicate fluctuating humid to semiarid to arid climates from the Maastrichtian into the Paleocene.

DINOSAURS: NAND-DONGARGAON BASIN

Historic Record and Diversity

The dinosaur fossil record in sediments of the Deccan volcanic province has not been assessed from a taphonomic perspective. This record may be incomplete as the last two decades of exploration in Late Cretaceous continental sediments and museum research have failed to measurably increase the nonavian dinosaurs biodiversity. In fact, restudy of described and unidentified specimens in museum collections has reduced the known biodiversity of Indian Late Cretaceous dinosaurs. In 1828, W.H. Sleeman (Sleeman, 1844) discovered the first dinosaur in India from the Lameta Formation of Jabalpur. Based on this material, Lydekker (1877) recognized a new genus and species of the sauropod *Titanosaurus indicus*. He named a second species *Titanosaurus blandfordi* (Lydekker, 1879) based on the material from the Lameta of Pisdura in the Nand-Dongargaon Basin. Von Huene and Charles Matley provided an exhaustive description of the dinosaur skeletal remains excavated by Mately in 1917–1919 and 1932–1933 from the Lameta Formation at Bara Simla and Chhota Simla, Jabalpur (Carrano et al., 2010; Wilson et al., 2011; Mohabey, 2011; Mohabey et al., 2013). Von Huene and Matley (1933) described at least three genera and four species of titanosaur sauropods, nine new theropod species, including three small-sized theropods, and one species of Thyreophora ornithichia, *Lametasaurus indicus*. The latter was redescribed as a large-sized theropod by Chakravarthi (1935).

Currently, many of these Indian dinosaur taxa are considered *nomen dubia* or invalid species. Recent analyses have suggested that the Indian Late Cretaceous dinosaur fauna consists of at least two sauropod dinosaurs, *Titanosaurus colberti* and *Jainosaurus septentrionalis*, and several theropod dinosaurs, including *Laevisuchus indicus*, *Rajasaurus narmadensis*, and possibly *Indosuchus raptorius*, *Indosaurus matleyi*, *Lametasaurus indicus*, and *Rahiolisaurus gujaratensis* (Wilson et al., 2003, 2011; Novas et al., 2010). However, evaluating the taphonomical biases (Lyman, 2001) in the currently known biodiversity of Indian dinosaurs requires much information from the undescribed collections in the repositories of the Geological Survey of India and the Indian Museum, Kolkata. Discovery of additional dinosaur fossil localities in Kheda area in Gujarat in 1981 (Dwivedi et al., 1982; Mohabey, 1983) produced two new abelisaurid taxa *Rajasaurus narmadensis* (Wilson et al., 2003) and *Rahiolisaurus gujaratensis* (Novas et al., 2010), which expanded the database of Indian Late Cretaceous dinosaurs.

Dinosaurs in Lameta Formation (C30n-C29r)

The Nand-Dongargaon Basin is one of the three main inland Lameta basins that have yielded most of the Late Cretaceous (Maastrichtian) dinosaur fossils in India (Lydekker, 1877; Von Huene and Matley, 1933; Matley, 1939; Berman and Jain, 1982; Jain and Sahni, 1986; Jain and Bandyopadhyay, 1997; Mohabey, 1998; Wilson and Mohabey, 2006; D'Emic et al., 2009). The other two basins are Jabalpur and Balasinor-Jhabua (Dwivedi et al., 1982; Mohabey, 1983). The infratrappean sediments in the Nand-Dongargaon Basin have dinosaur fossils mainly from the Pisdura and Dongargaon Hill. The skeleton of *T. colberti* (Jain and Bandyopadhyay, 1997; renamed *Isisaurus colberti* by Wilson and Upchurch, 2003) was discovered at Dongargaon Hill. The green silty clays of this locality with *T. colberti* were deposited during C30n. The Pisdura sediments yielded remains of *J. septentrionalis* (Wilson et al., 2009). Based on the study of plant-bearing coprolites from Pisdura, Mohabey (2005) suggested that the plant-eating sauropods were feeding upon soft tissues of angiosperm and gymnosperm plants. Prolific dinosaur egg sites in the infratrappean sediments of central and western India (Mohabey, 1996b, 1998, 2001a, 2001b, 2005) indicate that the Lameta environments offered ideal habitats for dinosaur nesting.

Dinosaurs in Intertrappeans (C29r)

The magnetic polarity of the flow overlying the Lameta sediments at Dongargaon Hill was interpreted as latest Maastrichtian C29r by Vandamme et al. (1991). The sauropod eggs and nests found at Kholdoda and Pavna were also deposited in C29r (Hansen et al., 2005). The diversity of the dinosaurs abruptly diminished with the arrival of the initial lava flow. Only a single species of titanosauriform dinosaur has been discovered to date from the intertrappean sediments associated with the basalmost part of the Deccan continental flood basalt sequences. The intertrappean localities that have yielded scarce remains of dinosaurs are Anjar in Kutch (Ghevariya, 1988), associated with unclassified basalts; Ranipur and Mohgaon Kalan (Mohgaon Kalan Well section) in the Chhindwara-Mandla-Jabalpur sector in Madhya Pradesh, associated with the lowermost Mandla and Dhuma Formations of Amarkantak Group; and a recently documented locality at Ukala in the Dhar area of Madhya Pradesh, associated with the basal Mandleshwar Formation of the Malwa Group (DMM personal observation 2010). In all these localities, the dinosaur fossils represent only one or two individuals of titanosauriforms. The last level of appearance of dinosaurs is recorded in the intertrappean sediments deposited during Maastrichtian C29r (Hansen et al., 2005).

Of these localities, the dinosaur-bearing sediments of Ranipur are associated with Maastrichtian marker palynomorphs *Gabonisporis vigourouxii*, *Aquilapollenites bengalensis*, *Azolla cretacea*, and *Triporoletes reticulatus* (Mathur and Sharma, 1990), whereas at Anjar (Kutch), the palynomorphs *Gabonisporis bacricumulus* and *Aquilapollenites bengalensis* are present (Dogra et al., 2004). The sharp decline in the diversity and abundance of dinosaur fauna in intertrappean sediments suggests that the initial volcanic activity adversely affected the dinosaur fauna, leading to their extinction. In contrast, other associated vertebrates, including turtles and snakes, successfully survived the environmental changes caused by volcanism close to the Cretaceous-Paleogene boundary.

CONCLUSIONS

Prior to the initiation of Deccan volcanism in the Nand-Dongargaon Basin, Lameta sedimentation in Maastrichtian chron C30n occurred in terrestrial environments with lakes populated by diverse flourishing vegetation including gymnosperms, angiosperms, pteridophytes, and algae. With the onset of phase 2 Deccan volcanic activity, the existing ecosystems changed dramatically in physiographic and atmospheric aspects. Over the newly developed lava fields, small lakes and ponds developed, providing new ecological niches. A new angiosperm-pteridophyte–dominated palynoassemblage developed preferentially in the fringe areas of water bodies and in volcanic substrate. Megaflora is absent, and blooms of the monogeneric peridinoid dinoflagellate first appeared in C29r. The same palynofloral associations continue into the overlying intertrappean sections, with the addition of some new Maastrichtian forms (*Scabrastephanocolpites* spp., *Scollardia conferta*, and *Triporopollenites cracentis*) and a few fossil dicot wood species, although diatoms and dinoflagellates are absent. Diversity thus dramatically decreased in intertrappean sections, and floral assemblages are dominated by just one or two taxa of mostly aquatic forms, with terrestrial vegetation impoverished.

The prolific dinosaur fauna of the Lameta deposits (skeletal remains, eggs, and nest sites) sharply decreased in diversity and abundance with the onset of Deccan volcanic activity, and almost all species disappeared (a single titanosauriform dinosaur is known from intertrappeans). In contrast, snakes and turtles, mollusks, ostracods, and charophytes survived the initial phase 2 volcanic onslaught and continued to the Cretaceous-Paleogene boundary in the late phase of volcanic activity. A distinct lithologic change from calcretes and gypsum-rich clays of the Lameta beds to cherts and porcellanitic clays in the lower intertrappean beds marks the observed changing floral and faunal environmental conditions as a result of Deccan volcanism. Detrimental effects of volcanism may have included direct contact with approaching lava fronts, pyroclastic input, ash fall, acid rains, toxic gases, and changing surface topography and drainage in and around the active eruptive volcanic sites. The changes in the flora and the dinosaurs predate the Cretaceous-Paleogene boundary in India. The indirect effect of the main Deccan volcanic eruptions away from the Indian subcontinent could be a relatively late event that contributed to the global mass extinction at the Cretaceous-Paleogene boundary.

ACKNOWLEDGMENTS

The present work was carried out as part of the Joint Collaborative project under Memorandum of understanding involving Rashtrasant Tukadoji Maharaj (RTM) Nagpur University and the Geological Survey of India (GSI). Samant is thankful to the Department of Science and Technology, New Delhi (No. SR/SY/ES-32/2003 and SR/S4/ES-619/2012) and University Grant Commission-Special Assistance Programme-I grant for financial assistance. Samant is also thankful to the head of the PG Department of Geology, RTM Nagpur University, Nagpur, for providing working facilities. Mohabey is thankful to the Director General, GSI, Kolkata, and Deputy Director General, Central Region, for excellent cooperation for pursuing the work under the collaborative program, and to Jeffrey Wilson for his contributions to the study of Indian dinosaurs and discussion therein. Both authors are thankful to Gerta Keller for fruitful discussions during her visit to India for field work in some intertrappean localities, which helped in building up some ideas presented in the paper, and to three reviewers and the editor for comments, corrections, and suggestions, which immensely helped in improving the manuscript.

REFERENCES CITED

Alvarez, L.W., Alvarez, W., Asaro, F., and Michel, H.V., 1980, Extraterrestrial cause for the Cretaceous-Tertiary extinction: Science, v. 208, p. 1095–1108, doi:10.1126/science.208.4448.1095.

Ambwani, K., and Dutta, D., 2005, Seed-like structure in dinosaurian coprolite of Lameta Formation (Upper Cretaceous) at Pisdura, Maharashtra, India: Current Science, v. 88, p. 352–354.

Archibald, J.D., Clemens, W.A., Padian, K., Rowe, T., MacLeod, N,. Barret, P.M., Gale, A., Holroyd, P., Sues, H.-D., Arens, N.C., Horner, J.R., Wilson, G.P., Goodwin, M.K., Brochu, C.A., Lofgren, D.L., Hurlbert, S.H., Hartman, J.H., Eberth, D.A., Wignall, P.B., Currie, P.J., Weil, A., Prasad, G.V.R., Dingus, L., Courtillot, V., Milner, A., Milner, A., Bajpai, S., Ward, D.J., and Sahni, A., 2010, Cretaceous extinction: Multiple causes: Science, v. 21, p. 973.

Bajpai, S., and Prasad, G.V.R., 2000, Cretaceous age for Ir-rich Deccan intertrappean deposits: Paleontological evidence from Anjar, western India: Journal of the Geological Society, v. 157, p. 257–260, doi:10.1144/jgs.157.2.257.

Baksi, S.K., and Deb, U., 1976, On new occurrence of *Aquilapollenites* from eastern India: Pollen et Spores, v. 18, no. 3, p. 399–406.

Bande, M.B., Chandra, A., Venkatachala, B.S., and Mehrotra, R.C., 1988, Deccan intertrappean floristics and its stratigraphic implications, *in* Maheshwari, H.K., ed., Paleocene of India: Lucknow, India, Indian Association of Palynostratigraphers, p. 83–123.

Barreda, V.D., Palazzesi, L., Tellería, M.C., Katinas, L., Crisci, J.V., Bremer, K., Passalia, M.G., Corosolini, R., Rodriquez Brizuela, R., and Bechis, F., 2010, Eocene Patagonia fossils of the daisy family: Science, v. 329, p. 1621, doi:10.1126/science.1193108.

Barres, L., Sanmartín, I., Anderson, C.L., Susanna, A., Buerki, S., Casals, M.G., and Vilatersana, R., 2013, Reconstructing the evolution and biogeographic history of tribe Cardueae (Compositae): American Journal of Botany, v. 100, no. 5, p. 867–882, doi:10.3732/ajb.1200058.

Batten, D.J., Gray, J., and Harland, R., 1999, Paleoenvironmental significance of monospecific assemblage of dinoflagellate cysts from the Miocene Clarkia beds, Idaho, USA: Palaeogeography, Palaeoclimatology, Palaeoecology, v. 153, p. 161–177, doi:10.1016/S0031-0182(99)00103-0.

Beane, J.E., Turner, C.A., Hooper, P.R., Subbarao, K.V., and Walsh, J.N., 1986, Stratigraphy, composition and form of the Deccan basalt, Western Ghats, India: Bulletin of Volcanology, v. 48, no. 1, p. 61–83, doi:10.1007/BF01073513.

Berman, D.S., and Jain, S.L., 1982, Braincase of a small sauropod dinosaur (Reptilia: Saurischia) from the Upper Cretaceous Lameta Group, central India, with review of Lameta Group localities: Annals of the Carnegie Museum, v. 51, p. 405–422.

Birader, N.V., and Bonde, S.D., 1979, On a fossil palm peduncle from Dongargaon District, Chandrapur Maharashtra, India: Geophytology, v. 9, p. 132–138.

Boltenhagen, E., 1967, Pollen et spores du Crétacé Supérieur du Gabon: Pollen et Spores, v. 9, p. 335–355.

Bonde, S.D., 2008, Indian fossil monocotyledons: Current status, recent developments and future direction: Palaeobotanist, v. 57, p. 141–164.

Bonde, S.D., and Birader, N.V., 1981, On two palm woods from the Deccan intertrappean beds of Dongargaon District, Chandrapur, Maharashtra (India): Journal of the University of Poona, v. 54, p. 247–257.

Carrano, M.T., Wilson, J.A., and Barret, P.M., 2010, The history of dinosaur collecting in central India, 1828–1947, *in* Moody, R.T.J., Buffetaut, E., Naish, D., and Martill, D.M., eds., Dinosaurs and Other Extinct Saurians: A Historical Perspective: Geological Society of London Special Publication 343, p. 161–173, doi:10.1144/SP343.9.

Chakravarthi, D.K., 1935, Is *Lametasaurus indicus* an armoured dinosaur?: American Journal of Science, v. 30, p. 138–141.

Chatterjee, S., 1978, *Indosuchus* and *Indosaurus*, Cretaceous carnosaurs from India: Journal of Paleontology, v. 52, p. 570–580.

Chenet, A.L., Quidelleu, X., Fluteau, F., Courtillot, V., and Bajpai, S., 2007, ^{40}K-^{40}Ar dating of the main Deccan large igneous province: Further evidence of KTB age and short duration: Earth and Planetary Science Letters, v. 263, p. 1–15, doi:10.1016/j.epsl.2007.07.011.

Chenet, A.-L., Fluteau, F., Courtillot, V., Gerard, M., and Subbarao, K.V., 2008, Determination of rapid Deccan eruptions across the KTB using paleomagnetic secular variation: 1. Results from 1200 m thick section in the Mahabaleshwar escarpment: Journal of Geophysical Research, v. 113, B04101, doi:10.1029/2006JB004635.

Chenet, A.-L., Courtillot, V., Fluteau, F., Gerard, M., Quidelleur, X., Khadri, S.F.R., Subbarao, K.V., and Thordarson, T., 2009, Determination of rapid Deccan eruptions across the Cretaceous-Tertiary boundary using paleomagnetic secular variation: 2. Constraints from analysis of eight new sections and synthesis for a 3500 m thick composite section: Journal of Geophysical Research, v. 114, B06103, doi:10.1029/2008JB005644.

Clausing, A., Kohler, J., and Schudack, U., 2000, Dinoflagellatenzysten aus oberoligozan en Susswassersedimenten des Enspel See, Westerwald, Deutschland: International Maar Conference: Terra Nostra, v. 6, p. 93–98.

Cookson, I.C., 1947, Plant microfossils from the lignite of Kerulen Archipelago: Banzare, ser. A, v. 2, p. 127–148.

Couper, R.A., 1953, Upper Mesozoic and Cainozoic Spores and Pollen from New Zealand: New Zealand Geological Survey Paleontological Bulletin 22, p. 1–77.

Courtillot, V., Besse, J., Vandamme, D., Montigny, R., Jaeger, J.J., and Cappetta, H., 1986, Deccan flood basalt at the Cretaceous/Tertiary boundary:

Earth and Planetary Science Letters, v. 80, p. 361–374, doi:10.1016/0012-821X(86)90118-4.

Courtillot, V., Gallet, Y., Rocchia, R., Feraud, G., Robin, E., Hoffman, C., Bhanadari, Z., and Ghevariya, Z., 2000, Cosmic markers, $^{40}Ar/^{39}Ar$ dating and palaeomagnetism of the K/T sections in the Anjar area of the Deccan large igneous province: Earth and Planetary Science Letters, v. 182, p. 137–156, doi:10.1016/S0012-821X(00)00238-7.

Cox, K.G., 1989, The role of mantle plumes in the development of continental drainage patterns: Nature, v. 342, p. 873–877, doi:10.1038/342873a0.

Cox, K.G., and Hawkesworth, C.J., 1984, Relative contribution of crest and mantle to flood basalt magmatism, Mahabaleshwar area, Deccan Traps: Philosophical Transactions of the Royal Society of London, v. 310, p. 627–641, doi:10.1098/rsta.1984.0011.

Cripps, J.A., Widdowson, M., Spicer, R.A., and Jolley, D.W., 2005, Coastal ecosystem response to late stage Deccan Trap volcanism: The post K-T boundary (Danian) palynofacies of Mumbai (Bombay), west India: Palaeogeography, Palaeoclimatology, Palaeoecology, v. 216, p. 303–332, doi:10.1016/j.palaeo.2004.11.007.

del Moral, R., and Grishin, S.Y., 1999, Consequences of volcanic eruptions, *in* Walker, R.L., ed., Ecosystem of Disturbed Ground: Amsterdam, Netherlands, Elsevier Science, p. 137–160.

D'Emic, M.D., Wilson, J.A., and Chatterjee, S., 2009, The titanosaur (Dinosauria: Sauropoda) osteoderm record: Review and first definitive specimen from the Upper Cretaceous of India: Journal of Vertebrate Paleontology, v. 29, p. 165–177, doi:10.1671/039.029.0131.

Deshmukh, S.S., 1984, Petrography and petrochemistry of the Deccan basalts in Maharashtra, *in* Proceedings of Symposium on Deccan Traps and bauxite, Pune, Maharashtra: Geological Survey of India Special Publication 12, p. 56–76.

Deshmukh, S.S., Sano, T., and Nair, K.K., 1996, Geology and chemical stratigraphy of the Deccan basalts of Chikaldara and Behramghat sections from the eastern part of the Deccan Trap province, India: Gondwana Geological Magazine, v. 2, Special Issue, p. 1–22.

Dilcher, D.L., 1965, Epiphyllous fungi from Eocene deposits in western Tennessee, USA: Palaeontographica B, v. 116, p. 1–156.

Dogra, N.N., Singh, R.Y., and Kulshreshta, S.K., 1994, Palynostratigraphy of infratrappean Jabalpur and Lameta Formations (Lower and Upper Cretaceous) in Madhya Pradesh, India: Cretaceous Research, v. 15, p. 205–215, doi:10.1006/cres.1994.1011.

Dogra, N.N., Raghumani, Y., and Singh, R.Y., 2004, Palynological assemblage from the Anjar intertrappean, Kachchh district, Gujarat: Age implication: Current Science, v. 86, p. 1596–1597.

Dutta, D., and Ambwani, K., 2007, Capers: A food for Upper Cretaceous dinosaurs of Pisdura, India: Current Science, v. 92, p. 897–899.

Dutta, D., Ambwani, K., and Ruiz, E.E., 2011, Late Cretaceous palm stem *Palmoxylon lametaei* sp. nov. from Bhisi Village, Maharashtra, India: Revista Mexicana de Ciencias Geológicas, v. 28, p. 1–9.

Dwivedi, G.N., Mohabey, D.M., and Bandyopadhyay, S., 1982, On the discovery of vertebrate fossils from infratrappean Lameta beds, Kheda District, Gujarat: Current Trends in Geology, v. 7, p. 79–87.

Frederiksen, N.O., 1994, Middle and late Paleocene angiosperm pollen from Pakistan: Palynology, v. 18, p. 91–137, doi:10.1080/01916122.1994.9989442.

Geological Survey of India, 2001a, District Resource Map of Yeotmal District, Maharashtra: Kolkata, Geological Survey of India, scale 1:125,000.

Geological Survey of India, 2001b, District Resource Map of Nanded District, Maharashtra: Kolkata, Geological Survey of India, scale 1:125,000.

Geological Survey of India, 2001c, District Resource Map of Chandrapur District, Maharashtra: Kolkata, Geological Survey of India, scale 1:125,000.

Geological Survey of India, 2001d, District Resource Map of Chhindwara, District, Madhya Pradesh: Kolkata, Geological Survey of India, scale 1:125,000.

Ghevariya, Z.G., 1988, Intertrappean dinosaur fossils from Anjar area, Kachcha District, Gujarat: Current Science, v. 57, p. 248–251.

Ghosh, P., Bhattacharya, S.K., Sahni, A., Kar, R.K., Mohabey, D.M., and Ambwani, K., 2003, Dinosaur coprolites from the Late Cretaceous (Maastrichtian) Lameta Formation of India: Isotopic and other markers suggesting a C3 plant diet: Cretaceous Research, v. 24, p. 743–750, doi:10.1016/j.cretres.2003.08.002.

Hansen, H.J., Mohabey, D.M., and Toft, P., 2001, No K-T boundary at Anjar, Gujarat: Evidence from magnetic susceptibility and carbon isotopes: Proceedings of the Indian Academy of Sciences, v. 11, no. 92, p. 1–10.

Hansen, H.J., Mohabey, D.M., Lojen, S., Toft, P., and Sarkar, A., 2005, Orbital cycles and stable carbon isotopes of sediments associated with Deccan volcanic suite, India: Implications for the stratigraphic correlation and Cretaceous/Tertiary boundary: Gondwana Geological Magazine, v. 8, Special Issue, p. 5–28.

Hofmann, C., and Zetter, R., 2007, Upper Cretaceous pollen flora from the Vilui Basin, Siberia: Circumpolar and endemic *Aquilapollenites*, *Manicorpus*, and *Azonia* species: Grana, v. 46, p. 227–249, doi:10.1080/00173130701763142.

Jain, S.L., 1989, Recent dinosaur discoveries in India, including eggshells, nests and coprolites, *in* Gillet, D.D., and Lockley, M.G., eds., Dinosaur Tracks and Traces: Cambridge, UK, Cambridge University Press, p. 99–108.

Jain, S.L., and Bandyopadhyay, S., 1997, New titanosurid (Dinosauria: Sauropoda) from the Late Cretaceous of central India: Journal of Vertebrate Paleontology, v. 17, p. 114–136, doi:10.1080/02724634.1997.10010958.

Jain, S.L., and Sahni, A., 1986, Some Cretaceous vertebrates from central India: Their palaeogeographical implications: Lucknow, Indian Association of Palynostratigraphers, Symposium, Birbal Sahni Institute of Palaeobotany, p. 66–83.

Jay, A.E., and Widdowson, M., 2008, Stratigraphy, structure and volcanology of SE Deccan continental flood basalt province: Implications for eruptive extent and volumes: Journal of the Geological Society, v. 165, p. 177–188, doi:10.1144/0016-76492006-062.

Jolley, D.W., Widdowson, M., and Shelf, S., 2008, Volcanogenic nutrient fluxes and plant ecosystem in a large igneous province: An example from the Columbia River basalt group: Journal of the Geological Society, v. 165, p. 955–966, doi:10.1144/0016-76492006-199.

Kapgate, D.K., 2005, Megafloral analysis of intertrappean sediments with focus on diversity and abundance of flora of Mohgaon Kalan, Mandla, and adjoining areas of Madhya Pradesh: Gondwana Geological Magazine, v. 20, no. 1, p. 31–45.

Kar, R.K., and Srinivasan, S., 1997, Late Cretaceous palynofossils from the Deccan intertrappean beds of Mohgaon Kalan, Chhindwara District, Madhya Pradesh: Geophytology, v. 27, p. 17–22.

Kar, R.K., Mohabey, D.M., and Shrivastava, R., 2004, First occurrence of angiospermous (dicot) fossil woods from the Lameta Formation (Maastrichtian), Maharashtra, India: Geophytology, v. 33, p. 21–27.

Keller, G., 2012, The Cretaceous-Tertiary mass extinction, Chicxulub impact and Deccan volcanism, *in* Talent, J.A., ed., Earth and Life, Extinction Intervals and Biogeographic Perturbations through Time: International Year of Planet Earth: Dordrecht, Netherlands, Springer Science and Media, p. 759–793.

Keller, G., Adatte, T., Gardin, S., Bartolini, A., and Bajpai, S., 2008, Main Deccan volcanism phase ends near the K-T boundary: Evidence from the Krishna-Godavari Basin, SE India: Earth and Planetary Science Letters, v. 268, p. 293–311, doi:10.1016/j.epsl.2008.01.015.

Keller, G., Adatte, T., Bajpai, S., Mohabey, D.M., Widdowson, M., Khosla, A., Sharma, R., Khosla, S.C., Gertsch, B., Fleitman, D., and Sahni, A., 2009a, K-T transition in Deccan Traps of central India marks major marine seaway across India: Earth and Planetary Science Letters, v. 282, p. 10–23, doi:10.1016/j.epsl.2009.02.016.

Keller, G., Sahni, A., and Bajpai, S., 2009b, Deccan volcanism, the K-T mass extinction and dinosaurs: Journal of Biosciences, v. 34, p. 709–728, doi:10.1007/s12038-009-0059-6.

Keller, G., Bhowmick, P.K., Upadhyay, H., Dave, A., Reddy, A.N., Jaiprakash, B.C., and Adatte, T., 2011, Deccan volcanism linked to the Cretaceous-Tertiary boundary (KTB) mass extinction: New evidence from ONGC wells in the Krishna-Godavari Basin, India: Journal of the Geological Society of India, v. 78, p. 399–428, doi:10.1007/s12594-011-0107-3.

Keller, G., Adatte, T., Bhowmick, P.K., Upadhyay, H., Dave, A., Reddy, A.N., and Jaiprakash, B.C., 2012, Nature and timing of extinctions in Cretaceous-Tertiary planktic foraminifera preserved in Deccan intertrappean sediments of the Krishna-Godavari Basin, India: Earth and Planetary Science Letters, v. 341–344, p. 211–221, doi:10.1016/j.epsl.2012.06.021.

Khosla, A., and Sahni, A., 2003, Biodiversity during Deccan volcanic eruptive episode: Journal of Asian Earth Sciences, v. 21, p. 895-908.

Kohler, J., and Clausing, A., 2000, Taxonomy and paleoecology of dinoflagellate cysts from Upper Oligocene freshwater sediments of Lake Enspel, Westerwald area, Germany: Review of Palaeobotany and Palynology, v. 112, p. 39–49, doi:10.1016/S0034-6667(00)00034-8.

Kumaran, K.P.N., Bonde, S.D., and, Kanitkar, M.D., 1997, An *Aquilapollenites* associated palynoflora from Mohgaon Kalan and its stratigraphic implications for age and stratigraphic correlation of Deccan intertrappean beds: Current Science, v. 72, p. 590–592.

Lenz, O.K., Wilde, V., and Riegel, W., 2006, Recolonization of a middle Eocene volcanic site: Quantitative palynology of the initial phase of the maar lake of Messel (Germany): Review of Palaeobotany and Palynology, v. 145, p. 217–242, doi:10.1016/j.revpalbo.2006.11.001.

Lodge, D.J., 1989, The influence of soil moisture and flooding on formation of VA-endo- and ectomycorrhizae in *Populus* and *Salix*: Plant and Soil, v. 117, p. 243–253, doi:10.1007/BF02220718.

Lydekker, R., 1877, Notices of new and other vertebrates from Indian Tertiary and Secondary rocks: Records of the Geological Survey of India, v. 10, p. 30–43.

Lydekker, R., 1879, Indian pre-Tertiary vertebrata. 3. Fossil Reptilia and Batrachia: Memoirs of the Geological Survey of India, Palaeontologia Indica, ser. 4, v. 1, p. 1–35.

Lyman, R.L., 2001, Vertebrate Taphonomy: Cambridge Manuals in Archeology: Cambridge, UK, Cambridge University Press, 557 p.

Mathur, N.S., and Sharma, S.D., 1990, Palynofossils and age of the Ranipur intertrappean beds, Gaur River, Jabalpur M.P., *in* Sahni, A., and Jolley, A., eds., Cretaceous Event Stratigraphy and Correlation of Indian Non Marine Strata: Chandigarh, International Geological Correlation Project 216–245, p. 58–59.

Mathur, Y.K., and Chopra, A.S., 1987, Palynofossils from the Cenozoic subsurface sediments of the Bengal Basin, India: Geoscience Journal, v. 18, p. 109–152.

Matley, C.A., 1939, The coprolites of Pijdura, Central Province: Records of the Geological Survey of India, v. 74, p. 535–547.

Mohabey, D.M., 1983, On the occurrence of dinosaurian fossil eggs from infratrappean limestone, Kheda District, Gujarat: Current Science, v. 52, p. 1194–1195.

Mohabey, D.M., 1996a, Depositional environment of Lameta Formation (Late Cretaceous) of Nand-Dongargaon Basin, Maharashtra: The fossil and lithofacies evidence: Memoirs of the Geological Society of India, v. 37, p. 363–386.

Mohabey, D.M., 1996b, A new oospecies *Megaloolithus matleyi* from the Lameta Formation (Upper Cretaceous) of Chandrapur District, Maharashtra, and general remarks on paleoenvironment and nesting behavior of dinosaurs: Cretaceous Research, v. 17, p. 183–196, doi:10.1006/cres.1996.0014.

Mohabey, D.M., 1998, Systematics of Indian Upper Cretaceous dinosaur and Chelonian eggshells: Journal of Vertebrate Paleontology, v. 18, p. 348–362, doi:10.1080/02724634.1998.10011063.

Mohabey, D.M., 2001a, Dinosaurs eggs and dung (fecal mass) from Late Cretaceous of central India: Dietary implications, *in* Recent Advances in the Field of Earth Sciences and their Implications in National Development: Geological Survey of India Special Publication 64, p. 605–615.

Mohabey, D.M., 2001b, Indian dinosaur eggs: A review: Journal of the Geological Society of India, v. 58, p. 479–508.

Mohabey, D.M., 2005, Late Cretaceous (Maastrichtian) nests, eggs, and dung mass (coprolites) of sauropods (titanosaurs) from India, *in* Tidwell, V., and Carpenter, K., eds., Thunder Lizards: The Sauropodomorph Dinosaurs: Bloomington, Indiana, Indiana University Press, p. 466–489.

Mohabey, D.M. 2011, History of Late Cretaceous dinosaur finds in India and current status of their study: Journal of Palaeontological Society of India, v. 56, no. 2, p. 127–135.

Mohabey, D.M., and Samant, B., 2005, Lacustrine facies association of a Maastrichtian lake (Lameta Formation) from Deccan volcanic terrain, central India: Implications to depositional history, sediment cyclicity and climates: Gondwana Geological Magazine, v. 8, Special Issue, p. 37–52.

Mohabey, D.M., and Udhoji, S.G., 1993, Palaeoenvironment interpretation of Lameta Formation (Late Cretaceous) of Nand area, Nagpur District Maharashtra: Gondwana Geological Magazine, v. 4, p. 13–22.

Mohabey, D.M., and Udhoji, S.G., 1996, Fauna and flora from Late Cretaceous (Maastrichtian) nonmarine Lameta sediments associated with Deccan volcanic episodes, Maharashtra: Its relevance to the K/T boundary problem, palaeoenvironment and palaeoclimate: Gondwana Geological Magazine, v. 2, Special Issue, p. 349–364.

Mohabey, D.M., Udhoji, S.G., and Verma, K.K., 1993, Palaeontological and sedimentological observations on non-marine Lameta Formation (Upper Cretaceous) of Maharashtra, India: Their palaeoecological and palaeoenvironmental significance: Palaeogeography, Palaeoclimatology, Palaeoecology, v. 105, p. 83–94, doi:10.1016/0031-0182(93)90108-U.

Mohabey, D.M., Head, J.J., and Wilson, J.A., 2011, A new species of the snake *Madtsoia* from the Upper Cretaceous of India and its palaeobiogeographic implications: Journal of Vertebrate Paleontology, v. 31, p. 588–595, doi:10.1080/02724634.2011.560220.

Mohabey, D.M., Sen, S., and Wilson, J.A., 2013, India's first dinosaur, rediscovered: Current Science, v. 104, no. 1, p. 34–37.

Nair, K.K.K., and Bhusari, B., 2001, Stratigraphy of Deccan Traps: A review, *in* Recent Advances in the Field of Earth Sciences and their Implications in National Development: Geological Survey of India Special Publication 64, p. 477–491.

Nair, K.K.K., Chatterjee, A.K., and Sano, T., 1996, Stratigraphy and geochemistry of the Deccan basalt along Toranmal section, western Satpura region: Gondwana Geological Magazine, v. 2, Special Issue, p. 23–48.

Novas, F.E., Chhatterjee, S., Rudra, D.K., and Dutta, P.M., 2010, *Rahiolisaurus gujaratensis*, n. gen. sp., A new abelisaurid theropod from the Late Cretaceous of India, *in* Novas, F.E., and Bandyopadyay, S., eds., New Aspect of Mesozoic Biodiversity: Berlin, Springer Verlag, Lecture Notes in Earth Sciences 132, p. 45–61, doi:10.1007/978-3-642-10311-7_3.

Pande, K., 2002, Age and duration of the Deccan Trap, India: Constraints from radiometric and paleomagnetic data: Proceedings of the Indian Academy of Sciences, v. 111, p. 115–123.

Playford, R.L., 1971, Lower Upper Cretaceous plant microfossils from Minnesota: Minnesota Geological Survey Bulletin, v. 42, p. 1–86.

Potonié, R., 1960, Sporologie der ecozanen khole von Kalewa in Burma: Senckenbergiana Lethaea, v. 41, p. 451–481.

Prakash, T., Singh, R.Y., and Sahni, A., 1990, Palynofloral assemblage from the Padwar Deccan intertrappean (Jabalpur) M.P., *in* Sahni, A., and Jolly, A., eds., Cretaceous Event Stratigraphy and Correlation of Indian Non Marine Strata: Chandigarh, International Geological Correlation Project 216–245, p. 68–69.

Prasad, B., Jain, A.K., and Mathur, Y.K., 1995, A standard palynozonation scheme for the Cretaceous and pre-Cretaceous subsurface sediments of Krishna-Godavari Basin, India: Geoscience Journal, v. 16, p. 155–232.

Prasad, G.V.R., and Khajuria, C.K., 1995, Implications of the infra- and inter-trappean biota from the Deccan, India, for the role of volcanism in Cretaceous-Tertiary boundary extinctions: Journal of the Geological Society, v. 152, p. 289–296, doi:10.1144/gsjgs.152.2.0289.

Prasad, G.V.R., and Sahni, A., 2014, this volume, Vertebrate fauna from the Deccan volcanic province: Response to volcanic activity, *in* Keller, G., and Kerr, A.C., eds., Volcanism, Impacts, and Mass Extinctions: Causes and Effects: Geological Society of America Special Paper 505, doi:10.1130/2014.2505(09).

Prasad, G.V.R., Jaeger, J.J., Sahni, A., Gheerbrandt, E., and Khajuria, C.K., 1994, Eutherian mammals from the Upper Cretaceous (Maastrichtian) intertrappean beds of Naskal, Andhra Pradesh, India: Journal of Vertebrate Paleontology, v. 14, p. 260–277, doi:10.1080/02724634.1994.10011556.

Prasad, V., Stromberg, C.A.E., Alimonammadian, H., and Sahni, A., 2005, Dinosaurian coprolites and the early evolution of grasses and grazers: Science, v. 310, p. 1177–1180, doi:10.1126/science.1118806.

Prasad, V., Stromberg, C. A. E., Leach, A.D., Samant, B., Patnaik, R., Tang, L., Mohabey, D. M., and Sahni, A., 2011, Late Cretaceous origin of the rice tribe provides evidence for early diversification in Poaceae: Nature Communication, v. 2, p. 480, doi:10.1038/ncomms1482.

Rana, R.S., and Wilson, G.P., 2008, New Late Cretaceous mammals from the intertrappean beds of Rangapur, India, and paleobiogeographic framework: Acta Palaeontologica Polonica, v. 53, p. 1–14, doi:10.4202/app.2008.0101.

Sah, S.C.D., and Dutta, S.K., 1974, Palynostratigraphy of the sedimentary formation of Assam: 3. Biostratigraphic zonation of south Shillong Plateau: Palaeobotanist, v. 21, p. 42–46.

Sah, S.C.D., and Kar, R.K., 1970, Palynology of the Laki sediments in Kutch: 3. Pollen from the bore-holes around Jhulrai, Baranda and Panandhro: Palaeobotanist, v. 18, p. 127–142.

Sahni, A., 1984, Cretaceous-Paleocene terrestrial fauna in India: Lack of endemism during drifting of India: Science, v. 226, p. 441–443, doi:10.1126/science.226.4673.441.

Salard-Cheboldaeff, S., 1978, Sur la palynoflore Maestrichtienne et Tertiaire du bassin sedimentaire littoral du Cameroun. Pollen et spores, v. 20, p. 215–260.

Samant, B., and Bajpai, S., 2005, Palynoflora from Lakshmipur intertrappean deposits of Kutch, Gujarat: Age implications: Journal Paleontological Society of India, v. 50, no. 2, p. 177–182.

Samant, B., and Mohabey, D.M., 2005, Response of flora to Deccan volcanism: A case study from Nand-Dongargaon Basin of Maharashtra: Implications to environment and climate: Gondwana Geological Magazine, v. 8, Special Issue, p. 151–164.

Samant, B., and Mohabey, D.M., 2009, Palynoflora from Deccan volcano-sedimentary sequence (Cretaceous-Paleogene transition) of central India: Implication for spatio-temporal correlation: Journal of Biosciences, v. 34, no. 5, p. 811–823, doi:10.1007/s12038-009-0064-9.

Samant, B., and Phadtare, N.R., 1997, Stratigraphic palynoflora of the early Eocene Rajpardi lignite, Gujarat, and the lower age limit of the Tarkeswar Formation of South Cambay Basin, India: Palaeontographica, v. B245, p. 1–108.

Samant, B., Mohabey, D.M., and Kapgate, D.K., 2008, Palynoassemblage from Singpur intertrappean beds, Chhindwara District, Madhya Pradesh: Implications for stratigraphic resolution within Late Cretaceous Deccan intertrappean beds: Journal of the Geological Society of India, v. 71, p. 851–858.

Sarkar, A., Bhattacharya, S.K., and Mohabey, D.M., 1991, Stable isotope analysis of dinosaur eggshells: Paleoenvironmental implications: Geology, v. 19, p. 1068–1071, doi:10.1130/0091-7613(1991)019<1068:SIAODE>2.3.CO;2.

Saxena, R.K., 1990, A Catalogue of Fossil Plants from India. Part 5A. Cenozoic (Tertiary) A. Spores and Pollen: Lucknow, India, Birbal Sahni Institute of Palaeobotany, 147 p.

Schulte, P., Alegret, L., Arenillas, I., Arz, J.A., Barton, P.J., Bown, P.R., Bralower, T.J., Christeson, G.L., Clays, P., Cockett, C.S., Collins, G.S., Deutsch, A., Goldin, T.J., Goto, K., Grajales-Nishimura, J.m., Grieve, R.A.F., Gulick, S.P.S., Johnson, K.R., Kiessling, W., Koebert, C., Kring, D.A., MacLeod, K.G., Matsui, T., Melosh, J., Montanari, A., Morgan, J.V., Neal, C.R., Nichols, D.J., Norris, R.D., Pierazzo, E., Ravizza, G., Rebolledo-Vieyra, M., Reimold, W U., Robin, E., Salge, T., Speijer, R.P., Sweet, A.R., Urrutia-Fucugauchi, J., Vajda, V., Whalen, M.T., and Willumsen, P.S., 2010, The Chicxulub asteroid impact and mass extinction at the Cretaceous-Paleogene boundary: Science, v. 327, no. 5970, p. 1214–1218, doi:10.1126/science.1177265.

Self, S., Thordarson, T., and Widdowson, M., 2005, Gas fluxes from flood basalt eruptions: Elements, v. 1, p. 283–287, doi:10.2113/gselements.1.5.283.

Selkirk, D.R., 1975, Tertiary fossil fungi from Kiandra, New South Wales: Proceedings of the Linnean Society of New South Wales, v. 100, no. 1, p. 70–94.

Sen, G., and Cohen, T.H., 1994, Deccan intrusion, crustal extension, doming and the size of Deccan Reunion plume head, *in* Subbarao, K.V., ed., Volcanism: Bombay, Wiley Eastern Limited, p. 201–216.

Sethna, S.F., 2003, The occurrence of acid and intermediate rocks in the Deccan volcanic province with associated high positive gravity anomalies and their probable significance: Journal of the Geological Society of India, v. 51, p. 813–815.

Shekhavat, L.S., and Sharma, V.P., 1996, Deccan basalts of Wakaner-Rajkot area, Gujarat: Gondwana Geological Magazine, v. 2, Special Issue, p. 89–100.

Sheth, H.C., and Pande, K., 2014, Geological and ^{40}Ar/^{39}Ar age constraints on late-stage Deccan rhyolitic volcanism, inter-volcanic sedimentation, and the Panvel flexure from the Dongri area, Mumbai: Journal of Asian Earth Sciences, v. 84, p. 167–175.

Sheth, H.C., Pande, K., and Bhutani, R., 2001, ^{40}Ar/^{39}Ar ages of Bombay trachytes: Evidence for a Paleocene phase of Deccan volcanism: Geophysical Research Letters, v. 28, no. 18, p. 3513–3516, doi:10.1029/2001GL012921.

Singh, C., 1964, Microflora of the Lower Cretaceous Manville Group, east central Alberta: Research Council of Alberta Bulletin, v. 166, p. 1–233.

Singh, H.P., and Khanna, A.K., 1978, Some fossil species of *Pediastrum* and their palaeoecological significance in the Subathu Formation of Himachal Pradesh: Palaeobotanist, v. 25, p. 466–473.

Singh, R.S., and Kar, R., 2002, Paleocene palynofossils from the Lalitpur intertrappean beds, Uttar Pradesh: Journal of the Geological Society of India, v. 60, no. 2, p. 213–216.

Singh, R.S., Kar, R., and Prasad, G.V.R., 2006, Deccan intertrappean beds of Naskal, Rangareddi district, Andhra Pradesh: Current Science, v. 90, p. 1281–1285.

Sleeman, W.H., 1844, Rambles and Recollection of an Indian Official: London, I.J. Hatchard and Son, 478 p.

Srinivasan, S., 1996, Late Cretaceous egg shells from the Deccan volcano-sedimentary sequences of central India, *in* Sahni, A., ed., Cretaceous Stratigraphy and Environment. Rama Rao Volume: Memoir of the Geological Society of India, v. 37, p. 321–336.

Srivastava, S.K., 1976, The fossil pollen grains *Classopollis*: Lethaia, v. 9, p. 437–457, doi:10.1111/j.1502-3931.1976.tb00985.x.

Stanley, E.A., 1965, Upper Cretaceous and Paleocene plant microfossils and Paleocene dinoflagellates and hystrichosphaerids from northwestern South Dakota: Bulletins of American Paleontology, v. 49, p. 179–384.

Sukheswala, R.N., 1956, Notes on the field occurrences and petrography of the rocks of the Bombay Island, Bombay: Transactions of the Mining, Geological, and Metallurgical Institute of India, v. 50, p. 101–126.

Tandon, S.K., 2002, Record of the influence of Deccan volcanism on contemporary sedimentary environment in central India: Sedimentary Geology, v. 147, p. 177–192, doi:10.1016/S0037-0738(01)00196-8.

Tandon, S.K., and Andrews, J.E., 2001, Lithofacies association and stable isotopes of palustrine and calcrete carbonates, examples from an Indian Maastrichtian regolith: Sedimentology, v. 48, p. 339–355, doi:10.1046/j.1365-3091.2001.00367.x.

Tandon, S.K., Sood, A., Andrews, J.E., and Davis, P.I., 1995, Paleoenvironments of dinosaur bearing Lameta beds (Maastrichtian) Narmada Valley, central India: Palaeogeography, Palaeoclimatology, Palaeoecology, v. 117, p. 153–184, doi:10.1016/0031-0182(94)00128-U.

Tappan, H., 1980, The Paleobiology of Plant Protists: San Francisco, California, W.H. Freeman, 1028 p.

Tsuyuzaki, S., 1987, Origin of plant recovery on the volcano Usu northern Japan, since the eruption of 1997 and 1998: Vegetato, v. 73, p. 53–58, doi:10.1007/BF00031851.

Tsuyuzaki, S., and del Moral, R., 1995, Species attributes in early primary succession on volcanoes: Journal of Vegetation Science, v. 6, p. 517–522, doi:10.2307/3236350.

Turner, S.D., and Friese, C.F., 1998, Plant mycorrhizal community dynamics associated with a moisture gradient within a rehabilitated prairie fern: Restoration Ecology, v. 6, p. 44–51, doi:10.1046/j.1526-100x.1998.00616.x.

Vandamme, D., and Courtillot, V., 1992, Paleomagnetic constraints on the structure of the Deccan Taps: Physics of the Earth and Planetary Interiors, v. 74, p. 241–261, doi:10.1016/0031-9201(92)90013-L.

Vandamme, D., Courtillot, V., Besse, J., and Montigny, R., 1991, Paleomagnetism and age determinations of the Deccan Traps (India): Results of a Nagpur-Bombay traverse and review of earlier work: Reviews of Geophysics and Space Physics, v. 29, p. 159–190, doi:10.1029/91RG00218.

Van Der Hammen, T., 1956, A palynological systematic nomenclature: Boletín Geológico Bogotá, v. 4, no. 2–3, p. 63–101.

Vartika, V., Prasad, V., and Chakroborty, S., 2007, Phytoliths as indicators of monsoonal variability during mid–late Holocene in mainland Gujarat, western India: Current Science, v. 92, no. 12, p. 1754–1759.

Venkatachala, B.S., and Sharma, K.D., 1974, Palynology of the Cretaceous sediments from the subsurface of Vridhachalam area: Cauvery Basin: Geophytology, v. 4, no. 2, p. 153–183.

Von Huene, F.B., and Matley, C.A., 1933, Cretaceous Saurischia and Ornithischia of the Central Province of India: Memoirs of the Geological Survey of India, Paleontologica Indica, v. 21, no. 1, p. 1–74.

Wilson, J.A., and Mohabey, D.M., 2006, A titanosauriform (Dinosauria: Sauropoda) axis from the Lameta Formation (Upper Cretaceous: Maastrichtian) of Nand, central India: Journal of Vertebrate Paleontology, v. 26, no. 2, p. 471–479, doi:10.1671/0272-4634(2006)26[471:ATDSAF]2.0.CO;2.

Wilson, J.A., and Upchurch, P., 2003, A revision of the genus *Titanosaurus* Lydekker (Dinosauria: Sauropoda), the first dinosaur genus with a Gondwanan distribution: Journal of Systematic Palaeontology, v. 1, p. 125–160, doi:10.1017/S1477201903001044.

Wilson, J.A., Serano, P.C., Srivastava, S., Bhatte, D.K., Khosla, A., and Sahni, A., 2003, A new abelisaurid (Dinosauria, Theropoda) from the Lameta Formation (Cretaceous, Maastrichtian) of India: Contributions of the Museum of Paleontology, University of Michigan, v. 31, p. 1–32.

Wilson, J.A., D'Emic, M., Curry, C.A., Mohabey, D.M., and Sen, S., 2009, Reassessment of sauropod dinosaur *Jainosaurus* (=*Antarctosaurus*) *septentrionalis* from Upper Cretaceous of India: Contributions of the Museum of Paleontology, University of Michigan, v. 32, no. 2, p. 17–40.

Wilson, J.A., Barret, P.M., and Carrano, M.T., 2011, An associated partial skeleton of *Jainosaurus* cf. *septentrionalis* (Dinosauria: Sauropoda) from the Late Cretaceous of Chhota Simla, central India: Palaeontology, v. 54, no. 5, p. 981–998, doi:10.1111/j.1475-4983.2011.01087.x.

Yi, S., 1997, Zygnematacean zygospores and other freshwater algae from the Upper Cretaceous of the Yellow Sea Basin, southwest coast of Korea: Cretaceous Research, v. 18, p. 515–544, doi:10.1006/cres.1996.0072.

Manuscript Accepted by the Society 31 January 2014

Printed in the USA

The Geological Society of America
Special Paper 505
2014

Vertebrate fauna from the Deccan volcanic province: Response to volcanic activity

Guntupalli V.R. Prasad*
Department of Geology, Centre for Advanced Studies, University of Delhi, Delhi 110007, India

Ashok Sahni
98, Mahatma Gandhi Marg, Lucknow 226001, India

ABSTRACT

During the last two decades, extensive paleontological research in the main Deccan volcanic province has led to a better understanding of biodiversity close to the Cretaceous-Paleogene boundary. Several infratrappean localities exposed in Jabalpur, Kheda, Balasinor, Rahioli, Dohad, and Bagh in the Narmada Valley (India) preserve one of the most geographically widespread dinosaur nesting sites known in the world. The well-studied intertrappean beds, such as those of Naskal on the southern margin, Asifabad and Nagpur on the eastern margin, Kisalpuri and Mohgaon Kalan on the northeastern margin, and Anjar on the northwestern margin of the main Deccan volcanic province, have yielded Maastrichtian fish (*Igdabatis*) and dinosaur remains and palynofossils (*Aquilapollenites-Gabonisporites-Ariadnaesporites*), either separately or in association, that suggest a Maastrichtian age for these beds. Only two intertrappean sections, Papro on the northern margin and Jhilmili on the northeastern margin of the main Deccan volcanic province, have produced Paleocene fossils. The fossil record from the infratrappean and intertrappean beds demonstrates that the dinosaurs survived the early phase of volcanism, though there was an apparent decline in their diversity, and that freshwater vertebrate fauna was least affected by the initial volcanic activity. The episodic nature of Deccan volcanism may possibly explain the survival of many freshwater and terrestrial communities during the periods of quiescence. In addition, as in the case of the late Maastrichtian sections in eastern Montana, North America, detritus-feeding freshwater vertebrate communities possibly had greater potential for survival than the terrestrial communities dependent on primary productivity. A close examination of the vertebrate faunal distribution across the two stratigraphic intervals (infratrappean and intertrappean) suggests that sampling bias in the infratrappean beds may have also masked the actual diversity of these beds.

*guntupalli.vrprasad@gmail.com

Prasad, G.V.R., and Sahni, A., 2014, Vertebrate fauna from the Deccan volcanic province: Response to volcanic activity, *in* Keller, G., and Kerr, A.C., eds., Volcanism, Impacts, and Mass Extinctions: Causes and Effects: Geological Society of America Special Paper 505, p. 193–211, doi:10.1130/2014.2505(09). For permission to copy, contact editing@geosociety.org.

INTRODUCTION

The Deccan Traps, representing one of the largest continental flood basalts on Earth's surface, cover about two thirds of peninsular India. At many places on the eastern, southern, northern, and southwestern margins of the main Deccan volcanic province, the Deccan volcanic flows are found in association with sedimentary beds that have been designated as infratrappean beds (or Lameta Formation), intertrappean, and supratrappean beds, depending on their physical position with respect to the volcanic flows (Fig. 1). The infratrappean beds (= Lameta Formation) occur below the local basal flows, whereas the intertrappean beds, deposited during the dormant stages of the volcanic activity, occur intercalated within the lava flows. The sediments immediately overlying the youngest volcanic flow are referred as supratrappean beds and are generally not encountered in outcrops.

Recent radiometric and paleomagnetic studies of basalt flows in the Western Ghats indicate that Deccan volcanism took place in three short phases interspersed with quiescent periods of considerable time (Chenet et al., 2007, 2008, 2009). Phase I, with an estimated volume of ~6%, occurred close to base of C30n at ca. 67.4 Ma in the late Maastrichtian (Chenet et al., 2007, 2009). Phase II, consisting of several eruptive events of considerable volume (~80% volume of total Deccan Traps), took place within C29r below the Cretaceous-Paleogene transition over a short period of time ranging from thousands to tens of thousands of years (Chenet et al., 2008; Jay and Widdowson, 2008; Jay et al., 2009). Phase III eruptions, representing ~14% of Deccan Traps volume, occurred at or near the base of C29n within the early Danian (Jay and Widdowson, 2008; Jay et al., 2009; Keller et al., 2012).

Currently, there are two competing and widely debated hypotheses to explain the cause of mass extinction at the Cretaceous-Paleogene (also known as KTB after Cretaceous-Tertiary) boundary: (1) the asteroid impact hypothesis of Alvarez et al. (1980) and (2) the volcanic hypothesis of McLean (1985), Courtillot et al. (1986), and Officer et al. (1987). Ever since the Deccan volcanic activity of India was suggested as a causal link for this extinction event (McLean, 1985; Officer et al., 1987; Courtillot et al., 1986, 1988), the sedimentary beds occurring in association with the Deccan Traps (infratrappean, intertrappean, and supratrappean beds) and yielding plant and animal fossils have received wide attention from the geoscientific community worldwide. In the last three decades, application of bulk screen-washing techniques on the infratrappean and intertrappean beds to recover fossils from different stratigraphic levels of the main Deccan volcanic province has yielded very promising results. A variety of vertebrate microfossils, ostracods, molluscs, foraminifers, and charophytes previously unknown from the main Deccan volcanic province have been documented. In the following sections, we present a brief history of paleontological research in the main Deccan volcanic province, the distribution, depositional environment, and age of infratrappean and intertrappean beds, and finally an analysis of the response of vertebrate fauna to volcanic activity.

HISTORY OF PALEONTOLOGICAL RESEARCH IN THE MAIN DECCAN VOLCANIC PROVINCE

There are three phases during which paleontological research was carried out on the Deccan infratrappean and intertrappean beds. During the first phase, which spans India's pre-independence era, most of the early studies were carried out by British geologists, army men, and medical officers. W.H. Sleeman made the first discovery of fossils from the main Deccan volcanic province in 1828 (Sleeman, 1844). These fossils, represented by caudal vertebrae, were later referred to a sauropod dinosaur *Titanosaurus indicus* by Lydekker (1877), which is now regarded as an invalid species (Wilson and Upchurch, 2003). Following this, Hislop (1860) published a detailed account on the molluscan fauna of the Deccan infratrappean and intertrappean beds. Later on, Woodward (1908) documented the presence of fishes (*Lepisosteus indicus*, *Eoserranus hislopi*, *Pycnodus lametae*) from the infratrappean beds of Dongargaon in Chandrapur District, Maharashtra State (Fig. 2). Additional fish remains represented mainly by scales were described by Hora (1938) from the intertrappean beds of Deothan and Kheri in Madhya Pradesh (Fig. 3). Many skull bones, dentitions, and postcranial bones of dinosaurs were also reported from the infratrappean beds of

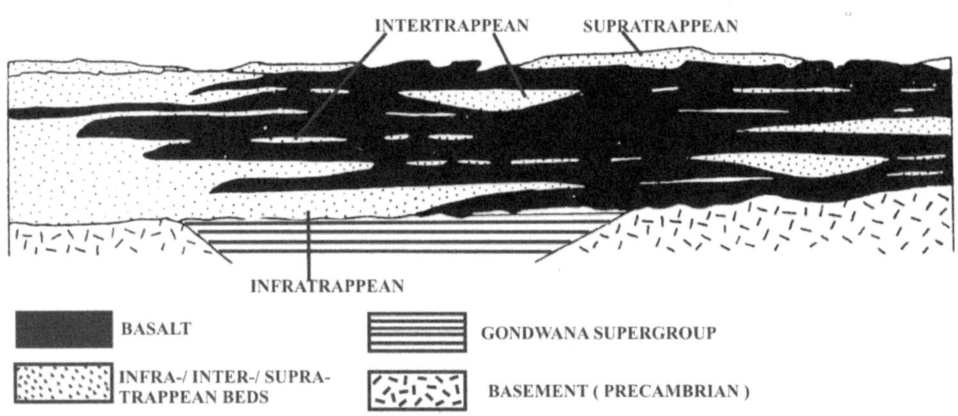

Figure 1. Schematic diagram showing the physical position of infratrappean, intertrappean, and supratrappean beds with respect to Deccan volcanic flows (modified after Sahni et al., 1994).

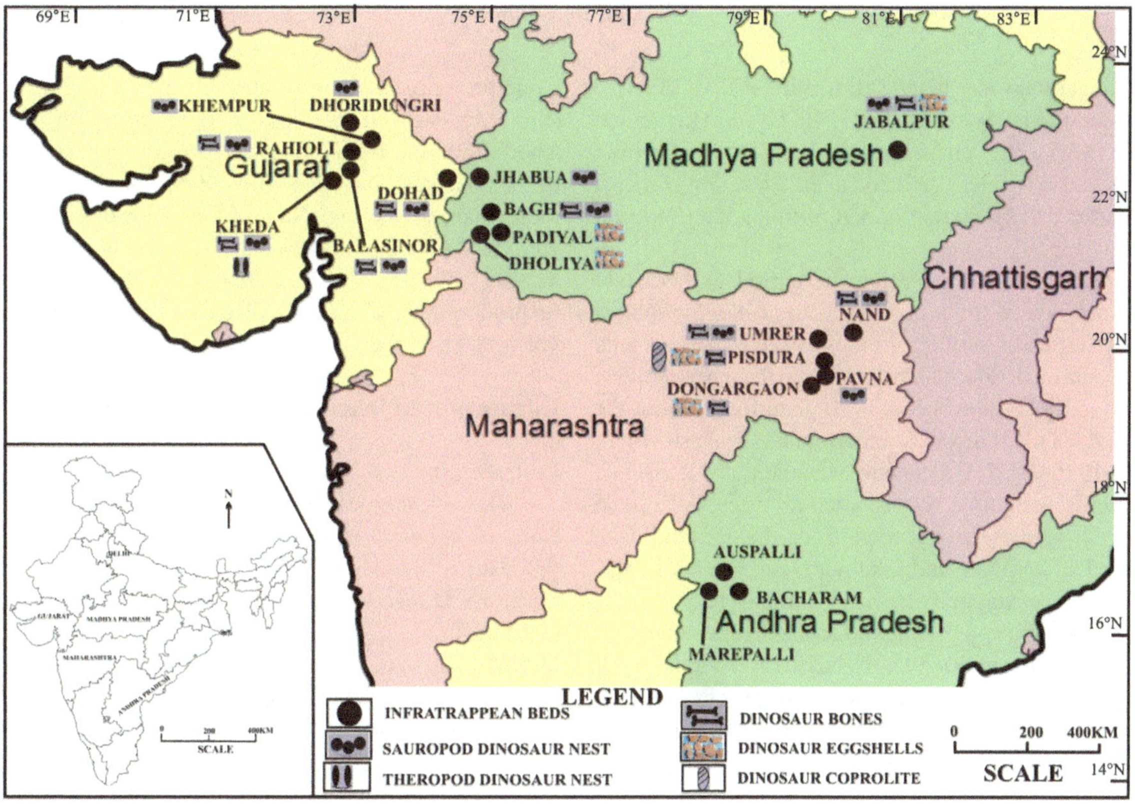

Figure 2. Map showing the infratrappean vertebrate fossil sites along the Narmada Valley and in Central and South India. Inset shows outline map of India highlighting the states in which the fossil sites are located.

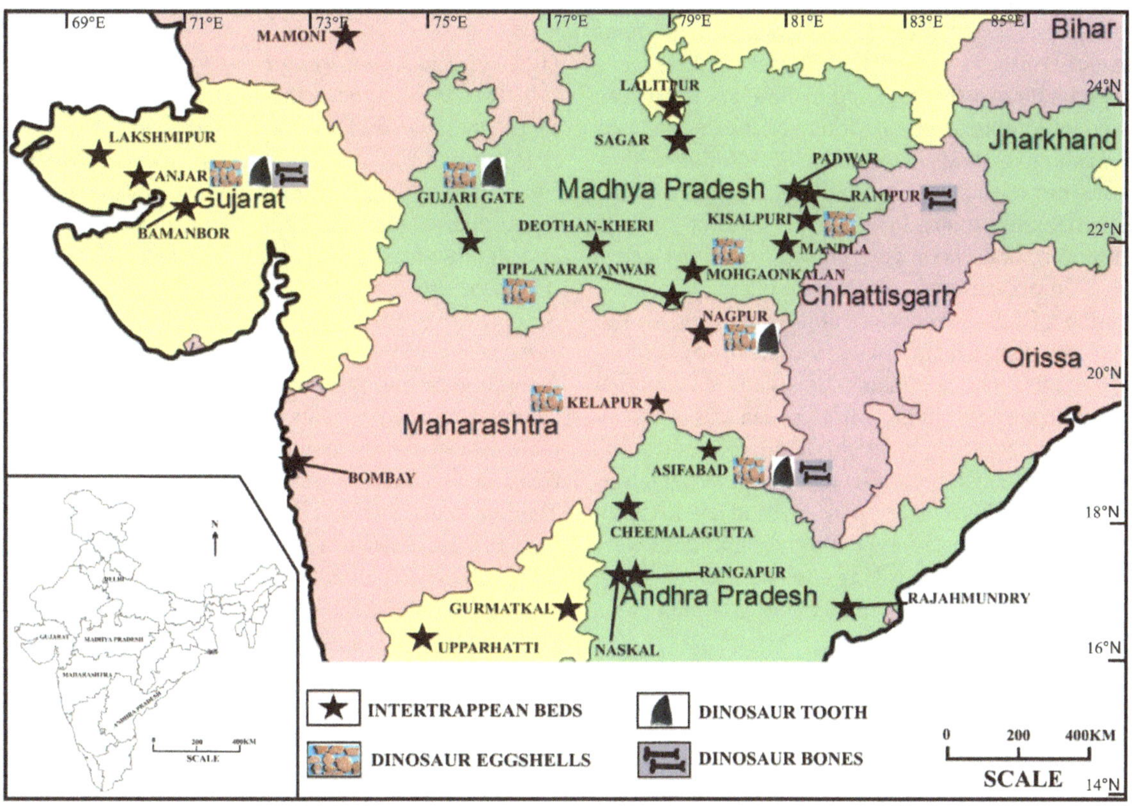

Figure 3. Map showing vertebrate fossil–yielding intertrappean sites of the Deccan volcanic province. The inset map of India highlights the states in which these fossil sites are located.

Pisdura and Dongargaon in Maharashtra, and Jabalpur (Fig. 2) in Madhya Pradesh (Matley, 1921, 1923, 1929; von Huene and Matley, 1933). During this phase, frog remains (*Indobatrachus pusillus* [Owen, 1847] Noble, 1930) and a turtle plastron (*Carteremys leithii* [Carter, 1852]) were reported from the intertrappean beds of Bombay (Fig. 3).

In the second phase, representing the period immediately after the independence of India in 1947 to ca. 1980, the majority of the studies on fossil biota from the main Deccan volcanic province were confined to foraminifers from the intertrappean beds of Rajahmundry (Bhalla, 1967) or to plant fossils from the intertrappean beds of Central and Southern India (Prakash, 1960; Shivarudrappa, 1976–1977; Bhatia and Mannikeri, 1976; Bande et al., 1986). Some vertebrate fossils, such as frogs, turtle, and dinosaur remains, were also recorded from the infratrappean beds of Pisdura, Dongargaon, and Jabalpur (Jain, 1977; Chatterjee, 1978) and the intertrappean beds of Bombay (Chiplonkar, 1940; Verma, 1965). Prior to 1980, the Deccan Traps were considered to span a duration of 30–50 m.y., based on K-Ar dating of the basaltic flows (Alexander, 1981) and molluscs, charophytes, and other plant fossils from the intertrappean beds (Hislop, 1860; Sahni, 1934; Hora, 1938; Prakash, 1960; Bhatia and Mannikeri, 1976; Shivarudrappa, 1976–1977).

The third phase of research on the Deccan infratrappean and intertrappean biota started in the beginning of 1980. Following the publications by McLean (1985) linking Deccan Traps to the Cretaceous-Paleogene boundary mass extinction and Courtillot et al.'s (1986) interpretation of Deccan volcanism as a short-duration event (<1.0 m.y.) close to the Cretaceous-Paleogene boundary, renewed interest was generated in the study of Deccan infratrappean and intertrappean biota. Moreover, the introduction of bulk screen-washing methods in the early 1980s for the collection of vertebrate microfossils substantially increased the diversity of various vertebrate groups. Prior to this phase of research activity, very few vertebrate taxa were known from the Lameta Formation and intertrappean beds of eruptive phase II. Following the application of bulk screen-washing techniques, almost all vertebrate groups (fishes, amphibians, turtle, snakes, lizards, turtles, crocodiles, dinosaurs, and mammals), with the exception of birds, have been reported from the infratrappean and intertrappean beds (Table 1). In addition to vertebrate microfossils, ~100 species of ostracods have also been documented from the infratrappean and intertrappean beds during phases II and III (see Khosla et al. [2011] for references), and planktic foraminifera were discovered in the intertrappean beds between phase II and phase III basalt flows at Jhilmili and Rajahmundry (Fig. 3; Keller et al., 2008, 2009a, 2009b).

DISTRIBUTION, DEPOSITIONAL ENVIRONMENT, AND AGE OF THE INFRATRAPPEAN AND INTERTRAPPEAN BEDS

The infratrappean beds and phase II and phase III intertrappean beds are widely known throughout the main Deccan volcanic province in outcrops and subsurface sections, whereas the supratrappean beds are known only from subsurface sections. Though marine infratrappean and intertrappean strata spanning the Cretaceous-Paleogene boundary have been delineated in more than 50 Oil and Natural Gas Corporation (ONGC) deep wells in Krishna-Godavari Basin (Raju et al., 1991; Jaiprakash et al., 1993; Keller et al., 2011, 2012), the present paper is restricted to predominantly continental outcrops from the main Deccan volcanic province.

Infratrappean Beds or Lameta Formation

Distribution

The infratrappean beds are exposed in discontinuous patches from Balasinor (Gujarat) in the west to Ambikapur (Madhya Pradesh) in the east, and from Sagar (Madhya Pradesh) in the north to Dongargaon, Pisdura (Maharashtra), and Marepalli (Andhra Pradesh) in the south (Fig. 2), and they reach a maximum thickness of 75 m, with a thick section known from the type area at Jhiraghat near Jabalpur in Central India. In the type section near Jabalpur, this formation consists of a Green Sandstone, Lower Limestone, Mottled Nodular Bed, Upper Limestone, and Upper Sandstone (Matley, 1921). The latter two units were redesignated as Upper (calcified) Sandstone by Tandon et al. (1995). Not all these units are preserved in the western part and in the Nand-Dongargaon Basin, where the Upper (calcified) Sandstone occurs as the most common lithologic unit.

Depositional Environment

The infratrappean beds (Lameta Formation) consist of sandstones, clays, limestones, and marls that are underlain by Precambrian basement or rocks of the Gondwana Supergroup. From time to time, the infratrappean beds have been interpreted as continental (Matley, 1921; Sahni and Mehrotra, 1974), fluvial and shallow-marine (Kumar and Tandon, 1979), shallow-marine (Chanda, 1967), tidal and estuarine (Singh and Srivastava, 1981; Singh, 1981; Saha et al., 2010), and pedogenically modified palustrine mudflat deposits (Brookfield and Sahni, 1987). Sedimentological and isotopic analyses of pedogenic carbonates and calcretes favored an alluvial-plain setting, with braided streams, palustrine flats, and sheetfloods under semiarid climatic conditions for the Lameta outcrops of Jabalpur (Ghosh et al., 1995; Tandon et al., 1995). The Lameta Formation of the Kheda and Dongargaon areas was interpreted as pedogenically altered overbank, channel, paludal, and lacustrine deposits of an alluvial-plain setting (Mohabey, 1996). Oxygen and carbon isotope studies of sauropod dinosaur eggshells from the infratrappean beds of Balasinor led Sarkar et al. (1991) to conclude that the Lameta dinosaurs consumed water from freshwater pools in a semiarid climate and ate C3 plants. On the whole, the burden of evidence, both paleontological and sedimentological, favors a pedogenically altered, paludal/lacustrine depositional environment in an alluvial-plain setting under semiarid climate conditions during deposition of the Lameta Formation.

TABLE 1. LIST OF VERTEBRATE FOSSILS DESCRIBED FROM THE DECCAN INFRA- AND INTERTRAPPEAN BEDS

Stratigraphic position	Taxon	Locality	State
Intertrappean beds phase II volcanism	***Fishes:***		
	Igdabatis indicus Prasad and Cappetta, 1993	Asifabad, Naskal	Andhra Pradesh
		Gurmatkal	Karnataka
		Nagpur	Maharashtra
		Kisalpuri	Madhya Pradesh
		Piplanarayanwar	Madhya Pradesh
	Rhombodus sp.	Asifabad	Andhra Pradesh
		Piplanarayanwar	Madhya Pradesh
	Raja sudhakari Prasad and Cappetta, 1993	Asifabad	Andhra Pradesh
	Rajiforme indet. (Prasad and Cappetta, 1993)	Asifabad	Andhra Pradesh
	Siluriformes indet. (Cione and Prasad, 2002)	Naskal	Andhra Pradesh
	Clupeidae (Hora, 1938; Prasad and Srinivasan, 1990)	Gurmatkal	Karnataka
		Deothan-Kheri	Madhya Pradesh
	Cyprinidae (Hora, 1938)	Deothan-Kheri	Madhya Pradesh
	Polycanthidae (Hora, 1938)	Deothan-Kheri	Madhya Pradesh
	Serranidae (Hora, 1938	Deothan-Kheri	Madhya Pradesh
	Nandidae (Hora, 1938)	Deothan-Kheri	Madhya Pradesh
		Gurmatkal	Karnataka
	Pristolepidae (Hora, 1938)	Deothan-Kheri	Madhya Pradesh
		Gurmatkal	Karnataka
	Horaclupea intertrappea Borkar, 1973	Bamanbor	Gujarat
	Palaeopristolepis feddeni Borkar, 1973	Bamanbor	Gujarat
	Palaeopristolepis chiplonkari Borkar, 1984	Bamanbor	Gujarat
	Indiaichthys bamanborensis Arratia et al., 2004	Bamanbor	Gujarat
	Percomorpha indet. (Arratia et al., 2004)	Bamanbor	Gujarat
	Pycnodontidae indet.	Naskal, Asifabad	Andhra Pradesh
		Rajahmundry	Andhra Pradesh
		Nagpur	Maharashtra
		Piplanarayanwar	Madhya Pradesh
		Gurmatkal	Karnataka
	Lepidotes sp. (Mohabey, 1996)	Nagpur	Maharashtra
	Lepisosteus indicus Woodward, 1908	Nagpur	Maharashtra
		Piplanarayanwar	Madhya Pradesh
		Kisalpuri	Madhya Pradesh
		Gujri Gate	Madhya Pradesh
		Naskal, Asifabad	Andhra Pradesh
	?*Belonostomus* sp. (Gayet et al., 1984)	Nagpur	Maharashtra
	Phareodus sp., Osteoglossidae indet.	Naskal, Asifabad	Andhra Pradesh
		Nagpur, Kelapur	Maharashtra
		Kisalpuri	Madhya Pradesh
		Piplanarayanwar	Madhya Pradesh
		Deothan-Kheri	Madhya Pradesh
	Enchodus ferox Leidy, 1855	Nagpur	Maharashtra
	Enchodus sp. (Gayet et al., 1984)	Nagpur	Maharashtra
	Palaeolabrus cf. *dormaalensis* (Prasad and Sahni, 1987)	Asifabad	Andhra Pradesh
	Palaeolabrus sp. (Gayet et al., 1984)	Nagpur	Maharashtra
	?*Sphyraena* (Gayet et al., 1984)	Nagpur	Maharashtra
	Percoidei indet. (Gayet et al., 1984)	Nagpur	Maharashtra
	Ostracion sp. (Gayet et al., 1984)	Nagpur	Maharashtra
	Otoliths:		
	Heterotidinarum heterotoides Nolf et al., 2008	Naskal	Andhra Pradesh
	Osteoglossidarum deccanensis Rana, 1988	Naskal, Rangapur	Andhra Pradesh
	Osteoglossidarum intertrappus Rana, 1988	Naskal, Rangapur	Andhra Pradesh
	Notopteridarum nolfi Rana, 1988	Naskal, Rangapur	Andhra Pradesh
		Cheemalagutta	Andhra Pradesh
	Clupeidarum valdiyai Rana and Sahni, 1989	Nagpur	Maharashtra
	Clupeidarum sahnii (Rana, 1996)	Rangapur	Andhra Pradesh
	Gonorynchidarum rectangulus (Rana, 1988)	Naskal, Rangapur	Andhra Pradesh
		Nagpur	Maharashtra
	Gonorynchidarum sp. (Nolf et al., 2008)	Rangapur	Andhra Pradesh
		Nagpur	Maharashtra
	?Ariidae (Nolf et al., 2008)	Naskal, Rangapur	Andhra Pradesh
	Ariidae indet. (Bajpai and Srinivasan, 1996)	Anjar	Gujarat
	Anthracoperca bhatiai (Rana, 1996)	Naskal, Rangapur	Andhra Pradesh
	Percoideorum citreum Nolf et al., 2008	Naskal	Andhra Pradesh
	Percoideorum nagpurensis (Rana and Sahni, 1989)	Nagpur	Maharashtra
	Percoideorum rangapurensis Rana, 1988	Naskal, Rangapur	Andhra Pradesh

(*Continued*)

TABLE 1. LIST OF VERTEBRATE FOSSILS DESCRIBED FROM THE DECCAN INFRA- AND INTERTRAPPEAN BEDS (*Continued*)

Stratigraphic position	Taxon	Locality	State
Intertrappean beds phase II volcanism	Percoidei sp. 1 (Nolf et al., 2008)	Naskal, Rangapur	Andhra Pradesh
	Percoidei sp. 2 (Nolf et al., 2008)	Rangapur	Andhra Pradesh
	Percoidei indet. (Bajpai and Srinivasan, 1996)	Anjar	Gujarat
	?*Centropomidarum takliensis* (Rana and Sahni, 1989)	Nagpur	Maharashtra
	Dapalis erici Nolf et al., 2008	Rangapur	Andhra Pradesh
	Ambassidarum cappettai (Rana and Sahni, 1989)	Nagpur	Maharashtra
	Apogonidarum curvatus (Rana, 1996)	Naskal, Rangapur	Andhra Pradesh
	?*Pristolepidinarum jaegeri* (Rana and Sahni, 1989)	Nagpur	Maharashtra
	Blenniidarum sp. (Nolf et al., 2008)	Rangapur	Andhra Pradesh
	Serranidarum sp. 1 (Bajpai and Srinivasan, 1996)	Anjar	Gujarat
	Serranidarum sp. 2 (Bajpai and Srinivasan, 1996)	Anjar	Gujarat
	Amphibians:		
	Indobatracus pusillus (Owen, 1847) Noble, 1930	Bombay	Maharashtra
	Gobiatinae (Discoglossidae, ?Hylidae)	Naskal	Andhra Pradesh
	Ranoidea (Ranidae or Rhacophoridae) (Prasad and Rage, 2004).	Naskal	Andhra Pradesh
	Leptodactylidae or Hemisotidae (Prasad and Rage, 2004)	Naskal	Andhra Pradesh
		Kisalpuri	Madhya Pradesh
	Pelobatidae indet. (Sahni et al., 1982)	Nagpur	Maharashtra
	Lizards:		
	Anguidae indet. (Prasad and Rage, 1995)	Naskal	Andhra Pradesh
	Litakis sp. (Rana, 2005)	Nagpur	Maharashtra
	Pristiguana sp. (Rana, 2005)	Nagpur	Maharashtra
	Iguanidae indet. (Rana, 2005)	Nagpur	Maharashtra
	Agama sp. (Rana, 2005)	Rangapur	Andhra Pradesh
	?*Contogenys* sp. (Rana, 2005)	Nagpur	Maharashtra
	Eumeces sp. (Rana, 2005)	Rangapur	Andhra Pradesh
	Exostinus estesai Rana, 2005	Nagpur	Maharashtra
	Gekkonid eggshells (Sahni et al., 1984)	Nagpur	Maharashtra
		Anjar	Gujarat
	Turtles:		
	Carteremys leithii (Carter, 1852) Williams, 1953	Bombay	Maharashtra
	cf. *Carteremys leithii* (de Lapperent de Broin et al., 2009)	Lakshmipur	Gujarat
	Carteremys sp.(de Lapperent de Broin et al., 2009)	Kisalpuri	Madhya Pradesh
	Sankuchemys sethnai Gaffney et al., 2003	Bombay	Maharashtra
	Cf. Bothremydidae (de Lapperent de Broin et al., 2009)	Upparhatti	Karnataka
	Crocodiles:		
	Crocodylia indet. (Rana, 1990; Prasad and de Lapparent de Broin, 2002)	Naskal, Asifabad	Andhra Pradesh
		Nagpur, Kelapur	Maharashtra
		Kisalpuri,	Madhya Pradesh
		Piplanarayanwar	Madhya Pradesh
		Gujri Gate	Madhya Pradesh
		Anjar	Gujarat
		Bombay	Maharashtra
	Crocodylidae indet. (Singh et al., 1998)	Bombay	Maharashtra
	Dyrosauridae indet. (Khosla et al., 2009a)	Kisalpuri	Madhya Pradesh
	Eggshells (Singh et al., 1998).	Bombay	Maharashtra
	Snakes:		
	Indophis sahnii Rage and Prasad, 1992	Naskal	Andhra Pradesh
		Kelapur	Maharashtra
		Anjar	Gujarat
		Gujri Gate	Madhya Pradesh
		Piplanarayanwar	Madhya Pradesh
	Madtsoiidae indet. (Rage et al., 2004)	Kelapur	Maharashtra
	Serpentes *incertae sedis* (Rage and Prasad, 1992)	Naskal	Andhra Pradesh
	Coniophis sp. (Rage et al., 2004)	Naskal	Andhra Pradesh
	Dinosaurs:		
	Indeterminate sauropod bones (Rao and Yadagiri, 1981; Ghevariya, 1988; Mathur and Sharma, 1990)	Asifabad	Andhra Pradesh
		Ranipur	Madhya Pradesh
		Anjar	Gujarat
	Theropod teeth (*Massospondylus rawesi*, Lydekker, 1890)	Nagpur	Maharashtra
		Anjar	Gujarat

(*Continued*)

TABLE 1. LIST OF VERTEBRATE FOSSILS DESCRIBED FROM THE DECCAN INFRA- AND INTERTRAPPEAN BEDS (*Continued*)

Stratigraphic position	Taxon	Locality	State
Intertrappean beds phase II volcanism		Gujri Gate	Madhya Pradesh
		Asifabad	Andhra Pradesh
	Sauropod eggshells (Sahni et al., 1984)	Asifabad	Andhra Pradesh
		Nagpur, Kelapur	Maharashtra
		Piplanarayanwar	Madhya Pradesh
	?*Hypselosaurus* sp. (Srinivasan, 1996)	Mohgaon Kalan	Madhya Pradesh
	Subtiliolithus kachchhensis (Khosla and Sahni, 1995)	Anjar	Gujarat
	Ornithoid eggshells (Bajpai et al., 1993)	Anjar	Gujarat
		Gujri Gate	Madhya Pradesh
		Kisalpuri	Madhya Pradesh
	Mammals:		
	Deccanolestes hislopi Prasad and Sahni, 1988	Naskal, Rangapur	Andhra Pradesh
	Deccanolestes robustus Prasad et al., 1994	Naskal, Rangapur	Andhra Pradesh
	Deccanolestes narmadensis Prasad et al., 2010	Kisalpuri	Madhya Pradesh
	Sahnitherium rangapurensis Rana and Wilson, 2003	Rangapur	Andhra Pradesh
	Kharmerungulatum vanvaleni Prasad et al., 2007a	Kisalpuri	Madhya Pradesh
	Bharattherium bonapartei Prasad et al., 2007b (=*Dakshina jederi* Wilson et al., 2007)	Kisalpuri	Madhya Pradesh
		Naskal	Andhra Pradesh
Infratrappean beds below phase I and/or phase II volcanism	***Fishes:***		
	Igdabatis indicus Prasad and Cappetta, 1993	Pisdura	Maharashtra
		Jabalpur	Madhya Pradesh
		Marepalli	Andhra Pradesh
		Nand-Dongargaon	Maharashtra
	Rhombodus sp.	Marepalli	Andhra Pradesh
	Lepisosteus indicus Woodward, 1908	Dongargaon	Maharashtra
		Marepalli	Andhra Pradesh
		Jabalpur	Madhya Pradesh
	Pycnodus lametae Woodward, 1908	Dongargaon	Maharashtra
	Pycnodontidae indet.	Jabalpur	Madhya Pradesh
	Eoserranus hislopi Woodward, 1908	Dongargaon	Maharashtra
	Lepidotes deccanensis Sykes, 1851	Nand-Dongargaon	Maharashtra
	Clupea sp. (Mohabey, 1996)	Nand-Dongargaon	Maharashtra
	Nandidae indet. (Mohabey, 1996)	Marepalli	Andhra Pradesh
	Osteoglossidae indet. (Mohabey, 1996)	Marepalli	Andhra Pradesh
		Dongargaon	Maharashtra
		Jabalpur	Madhya Pradesh
	Enchodus sp. (Jain and Sahni, 1983; Mohabey, 1996)	Nand-Dongargaon	Maharashtra
		Pisdura	Maharashtra
		Marepalli	Andhra Pradesh
	Arius sp. (Jain and Sahni, 1983)	Pisdura	Maharashtra
	Crocodiles:		
	Crocodylia indet.	Marepalli	Andhra Pradesh
	Dyrosauridae indet (Rana, 1987)	Auspalli,	Andhra Pradesh
		Marepalli	Andhra Pradesh
	Turtles:		
	Shweboemys pisdurensis (Jain, 1986)	Pisdura	Maharashtra
	Eggshells (Bajpai et al., 1997)	Rajahmundry	Andhra Pradesh
	Snakes:		
	Sanajeh indicus (Madtsoiidae) Wilson et al., 2010	Kheda	Gujarat
	Madtsoia pisdurensis (Madtsoiidae) Mohabey et al., 2011	Pisdura	Maharashtra
	Dinosaurs:		
	Jainosaurus (=*Antarctosaurus*) *septentrionalis* (von Huene and Matley, 1933) Hunt et al., 1994	Pisdura	Maharashtra
		Jabalpur	Madhya Pradesh
	Jainosaurus cf. *septentrionalis* (Wilson et al., 2011)	Jabalpur	Madhya Pradesh
	Isisaurus colberti (Jain and Bandyopadhyay, 1997) Wilson and Upchurch, 2003	Dongargaon	Maharashtra
	Titanosauriformes indet. (Wilson and Mohabey, 2006)	Nand	Maharashtra
	Indosuchus raptorius von Huene and Matley, 1933	Jabalpur	Madhya Pradesh
	Indosaurus matleyi von Huene and Matley, 1933	Jabalpur	Madhya Pradesh
	Laevisuchus indicus von Huene and Matley, 1933	Jabalpur	Madhya Pradesh
	Lametasaurus indicus von Huene and Matley, 1933	Jabalpur	Madhya Pradesh

(*Continued*)

TABLE 1. LIST OF VERTEBRATE FOSSILS DESCRIBED FROM THE DECCAN INFRA- AND INTERTRAPPEAN BEDS (*Continued*)

Stratigraphic position	Taxon	Locality	State
Infratrappean beds below phase I and/or phase II volcanism	*Rajasaurus narmadensis* Wilson et al., 2003	Rahioli	Gujarat
	Rahiolisaurus gujaratensis Novas et al., 2010	Rahioli	Gujarat
	Oospecies:		
	Megaloolithus cylindricus Khosla and Sahni, 1995	Jabalpur	Madhya Pradesh
	M. mohabeyi Khosla and Sahni, 1995	Dholiya	Madhya Pradesh
	M. padiyalensis Khosla and Sahni, 1995	Padiyal	Madhya Pradesh
	M. jabalpurensis Khosla and Sahni, 1995	Jabalpur	Madhya Pradesh
	M. dholiyaensis Khosla and Sahni, 1995	Dholiya	Madhya Pradesh
	M. dhoridungriensis Mohabey, 1998	Dhoridungri	Gujarat
	M. khempurensis Mohabey, 1998	Khempur	Gujarat
	M. megadermus Mohabey, 1998	Dohad	Gujarat
	M. baghensis Khosla and Sahni, 1995	Bagh	Madhya Pradesh
	Sauropod eggshells (Jain and Sahni, 1985)	Pisdura	Maharashtra
	Ellipsoolithus khedaensis (Family Elongatoolithidae) (Mohabey, 1998; Loyal et al., 1998).	Kheda	Gujarat
	Mammals:		
	Avashishta bacharamensis Anantharaman et al., 2006	Bacharam	Andhra Pradesh

Age

For a long time, the Lameta Formation was regarded as Turonian in age, based on the similarity of dinosaur fauna between the Lameta Formation and the Maevarano Formation of Madagascar, which was initially assigned a Turonian age (von Huene and Matley, 1933; Chiplonkar, 1986). While reviewing the Lameta dinosaur fauna, Buffetaut (1987) rejected the Turonian age for the Lameta Formation and favored a Maastrichtian age. The latter age was confirmed by the subsequent discoveries of a Maastrichtian fish (*Igdabatis*, *Rhombodus*; Jain and Sahni, 1983; Courtillot et al., 1986) and a palynological assemblage consisting of *Ariadnaesporites* and *Aquilapollenites* (Dogra et al., 1988). The Maastrichtian age for the Lameta Formation was further supported by a $^{40}Ar/^{39}Ar$ age of 66.4 ± 1.9 Ma (Courtillot et al., 1988) for the basal flow overlying the Maastrichtian sediments at Dongargaon (Fig. 2). However, paleomagnetic studies of the volcanic flows overlying the Lameta Formation at Jabalpur had placed these flows in magnetochron C30n (Courtillot et al., 1986) and thus in phase I of Deccan volcanism. Hence, the fossils and radiometric data from the Lameta Formation underlying the basal volcanic flows indicate a late Maastrichtian lower age limit for the Deccan volcanic eruptions.

Intertrappean Beds

Distribution

The intertrappean sedimentary beds are found mainly along the southern, eastern, northeastern, and northwestern margins of the main Deccan volcanic province. The well-known intertrappean sections are from Gurmatkal and Upparhatti (Karnataka State), Naskal, Rangapur, and Asifabad (Andhra Pradesh State), Nagpur and Bombay (Maharashtra State), Mohgaon Kalan, Jhilmili, Kisalpuri, Mandla, Padwar, Ranipur, Piplanarayanwar, and Gujri Gate (Madhya Pradesh State), Anjar and Bamanbor (Gujarat State), Mamoni (Rajasthan State), and Papro (Uttar Pradesh state) (Fig. 3). At present, there are no paleomagnetic or geochronological controls on the position of the intertrappean beds between the lava flows. However, based on geochemical characteristics, Jay and Widdowson (2008) correlated the basaltic flows on the southeastern, eastern, northeastern, and northern margins of the main Deccan volcanic province with the Poladpur and Ambenali Formations of the Wai Group in the Western Ghats. In view of this and the presence of Maastrichtian fossils and dinosaur remains in the intertrappean beds of Gurmatkal, Naskal, Rangapur, Asifabad, Nagpur, Mohgaon Kalan, Kisalpuri, Mandla, Padwar, Ranipur, Piplanarayanwar, Gujri Gate, Anjar, and Mamoni (Fig. 3), we prefer to place these beds in phase II volcanism.

The intertrappean beds of Jhilmili (Fig. 3), with early Paleocene fossils and lower and upper traps placed within chrons C29r and C29r/C29n, respectively, were considered to lie between phases II and III (Keller et al., 2009b). The intertrappean beds of Papro (Fig. 3), with Paleocene palynofossils, may also be placed between volcanic phases II and III. The intertrappean beds of Bombay were placed within phase III (Cripps et al., 2005). Currently, there are limited fossil data to pinpoint the position of Upparhatti and Bamanbor (Fig. 3) intertrappean beds within the lava flow stratigraphy of the main Deccan volcanic province. Compared with the Lameta Formation, the intertrappean beds of phases II and III are relatively thin, generally reaching a few centimeters to 30 cm in thickness, and rarely measuring up to 5 m, except at Jhilmili, where these intertrappean beds measure 13.5 m (Keller et al., 2009a, 2009b).

Depositional Environment

Since the pioneering work of Hislop (1860), many intertrappean outcrops have been investigated from the paleontological point of view. Of these, the intertrappean beds of phase II, such as Rangapur, Naskal, Asifabad, Nagpur, Kisalpuri, Piplanarayanwar, and Anjar (Fig. 3), have been intensively studied for vertebrate microfossils. Generally, they are composed of siltstones, mudstones, sandstones, clays, marls, and shales, frequently associated with a prominent fossiliferous chert and occasionally with a limestone bed. The Deccan intertrappean beds are traditionally considered freshwater lacustrine deposits that filled low-lying areas on the surface of the lava flows during the dormant stages of volcanism.

The phase II intertrappean beds of Naskal and Anjar (Fig. 3) were studied in some detail in the context of their depositional environment. Based on sedimentological, paleontological, and taphonomic studies, the Naskal intertrappean beds were interpreted as alkaline lake deposits in a floodplain setting intermittently subjected to subaerial exposure, leading to the development of two phases of paleosols (Khajuria and Prasad, 1998). Sedimentological studies of the iridium-bearing Anjar intertrappean section indicated a freshwater lacustrine environment in a semiarid climate setting (Khadkikar et al., 1999). However, the mixed presence of marine and nonmarine biota in the intertrappean beds of phase II at Asifabad and Nagpur (Fig. 3) indicates some marine influence in these areas (Sahni, 1983; Prasad and Sahni, 1987). This has been proven by the recent discovery of planktonic foraminifers from the intertrappean beds occurring between phase II and phase III basalt flows at Jhilmili in Central India (Keller et al., 2009b).

Age

Early paleontological studies overestimated the age of Deccan intertrappean beds as Early Tertiary (with the majority favoring an Eocene age) based on fish, molluscs, and plant fossils, including charophytes and plant megafossils (Hislop, 1860; Sahni, 1934; Hora, 1938; Bhatia and Mannikeri, 1976; Shivarudrappa, 1976–1977; Bande et al., 1986), and the absence of dinosaurs as compared to their frequent occurrence in the Lameta Formation (Matley, 1921, 1929; von Huene and Matley, 1933, Chatterjee, 1978). Khajuria et al. (1994) discussed in detail the inadequacy of these fossils for assigning an Eocene age for the intertrappean beds. Further, dinosaur remains (a few bones and teeth, and several eggshell fragments) have been found in many intertrappean beds (Rao and Yadagiri, 1981; Vianey-Liaud et al., 1987; Prasad, 1989; Sahni and Bajpai, 1988; Ghevariya, 1988; Mathur and Sharma, 1990).

Subsequent findings of Maastrichtian fishes (Prasad and Cappetta, 1993) and the *Ariadnaesporites-Aquilapollenites-Gabonisporites* palynological assemblage (Mathur and Sharma, 1990; Sahni et al., 1996; Kumaran et al., 1997; Dogra et al., 2004; Singh et al., 2006) from the majority of the intertrappean beds favor a Maastrichtian age. A couple of intertrappean outcrops from the main Deccan volcanic province have been assigned a younger early Paleocene age. The northernmost intertrappean section near Papro in Lalitpur District (Uttar Pradesh State) has been assigned an early Paleocene age based on a palynological assemblage consisting of *Dandotiaspora dilata*, *Dandotiaspora pseudoauriculata*, *Dandotiaspora plicata*, *Spinizonocolpites echinatus*, *Matanomadhiasulcites* sp., and *Lakiapollis ovatus* (Singh and Kar, 2002).

The Anjar intertrappean section in Kachchh, western India, was initially designated as Cretaceous-Paleogene age based on anomalous concentrations of iridium in the sediments sandwiched between volcanic flows III and IV (Bhandari et al., 1996). However, later paleontological studies demonstrated that the intertrappean beds overlying the iridium-bearing levels yield Maastrichtian fossils, including dinosaur remains (Bajpai and Prasad, 2000), and the high iridium concentration was attributed to leaching from the Deccan volcanic flows (Khadkikar et al., 1999). As a result, the iridium-enriched levels of the Anjar intertrappean beds are no longer regarded as representing the Cretaceous-Paleogene boundary. Therefore, no fossiliferous intertrappean outcrop spanning the Cretaceous-Paleogene transition has been recorded from the main Deccan volcanic province to date.

More recently, Keller et al. (2009a, 2009b) reported the presence of early Danian planktic foraminifers in the intertrappean beds between volcanic phases II and III at Jhilmili in Chhindwara District (Madhya Pradesh State). However, the Maastrichtian to Danian transition has yet to be discovered in this area.

On the whole, the fossil record from the infratrappean and intertrappean outcrops is in agreement with the fossil record from the subsurface intertrappean sections of ONGC wells in the Krishna-Godavari Basin, where the shallow-marine foraminiferal assemblages from the infratrappean beds and phase II intertrappean beds indicate a late Maastrichtian age, and phase III intertrappean and supratrappean beds indicate an early Paleocene (Danian) age for the Deccan volcanism (Raju et al., 1991; Jaiprakash et al., 1993; Keller et al., 2011, 2012). The paleontological data from infratrappean and intertrappean beds from the main Deccan volcanic province outcrops and from shallow-marine subsurface sections are in agreement with radiometric dates from a thick Deccan volcanic pile from the Western Ghats, which indicate that Deccan volcanism began ca. 67.4 Ma and terminated around 62 Ma (Venkatesan et al., 1993; Chenet et al., 2007, 2008).

RESPONSE OF VERTEBRATE FAUNA TO DECCAN VOLCANISM

One frequently asked question is whether Deccan volcanism was capable of affecting the ecosystems at the global level, and, if so, what were its effects at the site of eruption? To seek answers to these questions, one should examine the fossil evidence from the sediments deposited prior to the initiation of volcanism (infratrappean) and those deposited during (intertrappean) or immediately after Deccan volcanic activity.

Fishes, Crocodiles, Turtles, and Snakes

A comparison of the infratrappean and intertrappean vertebrate fauna demonstrates that at least 10 infratrappean fish taxa (five each at species and family level) survived into the intertrappean beds of phase II (Table 1; Fig. 4). Although some indeterminate anurans and lizards were reported by Jain and Sahni (1983) and Mohabey (1996), no recognizable amphibian and lizard remains have been recorded to date from the infratrappean beds. On the other hand, the intertrappean amphibians and lizards are represented by at least four and six families, respectively. Two crocodilian groups (Crocodylia indet. and Dyrosauridae) of the infratrappean beds survived into the intertrappean beds (Fig. 4). The lone turtle species (*Shweboemys pisdurensis* Jain, 1986) known from the infratrappean beds belongs to the family Podocnemididae. In comparison, the intertrappean turtles are identified as *Carteremys leithii* (Carter, 1852), *Sankuchemys sethnai* Gaffney et al., 2003 of the family Bothremydidae, and an indeterminate form distinct from all these taxa (de Lapparent de Broin et al., 2009). The infratrappean snake fauna is represented

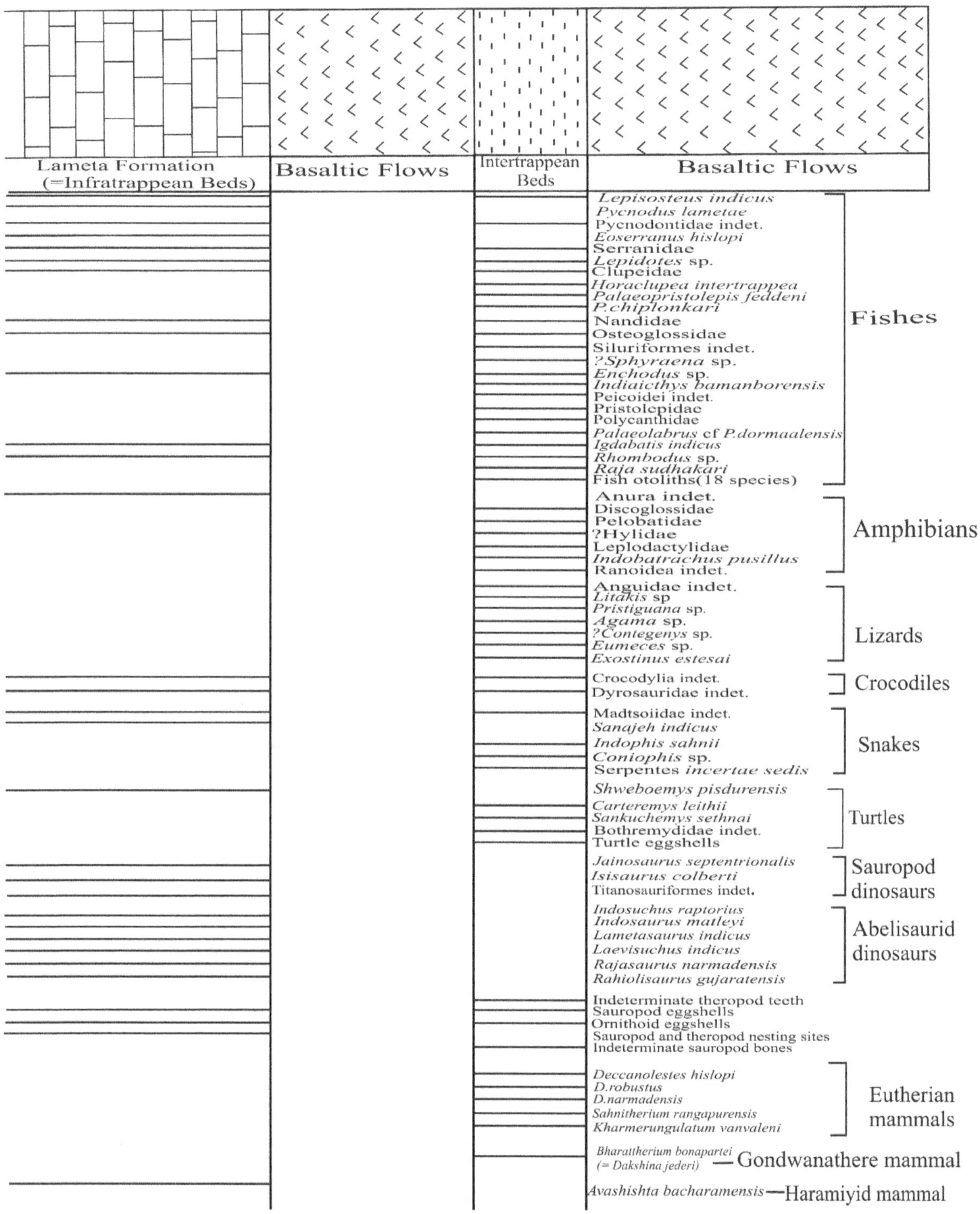

Figure 4. Range chart of various vertebrate groups across the Deccan volcano-sedimentary sequences.

by two taxa: *Sanajeh indicus* Wilson et al., 2010 and *Madtsoia pisdurensis* Mohabey et al., 2011 of the family Madtsoiidae. Not only did the madtsoiid snakes survive into the intertrappean beds at the family level, but so did three new taxa: *Indophis sahnii* (?Nigerophiidae) Rage and Prasad, 1992, Serpentes *incertae sedis* (Rage and Prasad, 1992), and *Coniophis* (Aniliidae) (Rage et al., 2004) appear at this stratigraphic level.

Dinosaur Bones, Nesting Sites, Eggshells, and Teeth

Lameta Formation

One of the most important fossil findings from the main Deccan volcanic province in the last three decades is the discovery of geographically extensive nesting sites and eggshells in the Lameta Formation (below phase II volcanism) over a stretch of 1000 km from Jabalpur in the east to Balasinor in the west and in the Nand-Dongargaon Basin in Central India (Fig. 2). These nesting sites are located near Jabalpur in the upper Narmada Valley, Bagh, Jobat, Dohad, Kheda, Balasinor, and Rahioli in the lower Narmada Valley, and Pavna and Dongargaon in the Nand-Dongargaon Basin (Fig. 2). In all these sites, nests and eggshells belonging to nine oospecies of *Megaoolithus* (Table 1) were documented from a fine- to medium-grained calcretized sandstone (Srivastava et al., 1986; Mohabey, 1998; Vianey-Liaud et al., 2003). Only one oospecies of *Ellipsoolithus* (Table 1) has so far been described from the Lameta Formation near Kheda (Mohabey, 1998). The occurrence of nests in a similar lithofacies has been attributed to strong nesting site selectivity, and extensiveness of nesting sites with morphologically similar eggs was shown as evidence for colonial nesting (Sahni et al., 1994; Mohabey, 2001).

Besides nesting sites and eggshells, the Lameta Formation at Jabalpur, Balasinor, Pisdura, Dongargaon, and Nand has been the source for all known Cretaceous dinosaur taxa from India named on the basis of skull and postcranial bones (Sleeman, 1844; Lydekker, 1877; Matley, 1921, 1923, 1929; von Huene and Matley, 1933; Chatterjee, 1978; Berman and Jain, 1982; Srivastava et al., 1986; Mathur and Srivastava, 1987; Jain and Bandyopadhyay, 1997; Wilson et al., 2003; Novas et al., 2010; Wilson and Mohabey, 2006; Wilson et al., 2011). To date, three titanosaurid sauropods and six abelisaurid theropods have been described from the Lameta Formation.[1]

Intertrappean Beds

Another significant finding from the main Deccan volcanic province is the occurrence of dinosaur remains in the intertrappean beds. All the early reports of skeletal remains of dinosaurs, coprolites, and dinosaur nesting sites were from the Lameta Formation (Matley, 1921; Matley, 1929; von Huene and Matley, 1933; Chatterjee, 1978; Berman and Jain, 1982; Srivastava et al., 1986; Mathur and Srivastava, 1987). Prior to 1980s, dinosaurs were considered to have been restricted to the Lameta Formation, and the intertrappean beds were generally regarded as lacking dinosaur fossils. This was one of the reasons for assigning an Early Tertiary (Eocene) age for the intertrappean beds. However, since 1980, several dinosaur fossils, including a few bones and teeth, and many eggshell fragments, have been documented from phase II intertrappean beds. Moreover, they have been reported from widely separated intertrappean sections such as those of Asifabad (Rao and Yadagiri, 1981; Sahni et al., 1984; Prasad, 1989), Nagpur (Vianey-Liaud et al., 1987), Anjar (Ghevariya, 1988; Sahni and Bajpai, 1988; Bajpai et al., 1993; Bajpai and Prasad, 2000), Ranipur (Mathur and Sharma, 1990), Mohgaon Kalan (Srinivasan, 1996), and Kisalpuri (Khosla et al., 2004) (Fig. 3). More recently, dinosaur eggshell fragments have also been found in the intertrappean beds of Kelapur (Maharashtra) and Piplanarayanwar and Gujri Gate (Madhya Pradesh) (Fig. 3) (work in progress at Delhi University).

These intertrappean eggshells are generally thin as compared to those of the Lameta Formation. Although they have not been studied as extensively and in as much detail as those of the Lameta Formation, limited ultrastructural studies indicate that eggshells from Asifabad, Nagpur (Sahni et al., 1984; Vianey-Liaud et al., 1987), Mohgaon Kalan (Srinivasan, 1996), Kelapur, and Piplanarayanwar belong to sauropod dinosaurs, and those from Anjar (Bajpai et al., 1993), Kisalpuri, and Gujri Gate are assignable to the ornithoid type. Until now, no parataxonomic classification has been attempted for the intertrappean dinosaur eggshells.

Depositional Environment of Dinosaur Fossils: Lameta versus Intertrappean

The Lameta Formation is generally thick, reaching a maximum thickness of 75 m, and the dinosaur nests and eggs are preferentially preserved in a sandy calcretized unit of the carbonate facies at Jabalpur, Rahioli, and Dongargaon (Mohabey, 2001). At Jabalpur, the dinosaur skeletal remains were recovered from the base and within the lower limestone. At Rahioli, dinosaur bones occur within a conglomerate horizon underlying the dinosaur egg-bearing sandy calcrete (Mohabey, 2001). In the Nand-Dongargaon Basin, dinosaur bones are reworked and transported elements within the overbank red and green clays, whereas the eggs are exclusively found in sandy calcrete and channel sandstone (Mohabey, 2001). In the area near the Bagh Caves in the lower Narmada Valley, dinosaur-egg yielding levels of the Lameta Formation are confined to the sandy and nodular calcretes of massive cherty limestone (Mohabey, 2001).

In contrast, the intertrappean beds are relatively thin (generally a few centimeters to 30 cm thickness, rarely reaching 5 m) with widely varying lithologies consisting of soft siltstone,

[1]Titanosaurid sauropods: *Jainosaurus* (= *Antarctosaurus*) *septentrionalis* (von Huene and Matley, 1933) Hunt et al., 1994; *Jainosaurus* cf. *septentrionalis* (Wilson et al., 2011); *Isisaurus colberti* (Jain and Bandyopadhyay, 1997) Wilson and Upchurch, 2003; Titanosauriformes indet. (Wilson and Mohabey, 2006). Abelisaurid theropods: *Indosuchus raptorius* von Huene and Matley, 1933; *Indosaurus matleyi* von Huene and Matley, 1933; *Laevisuchus indicus* von Huene and Matley, 1933; *Lametasaurus indicus* von Huene and Matley, 1933; *Rajasaurus narmadensis* Wilson et al., 2003; and *Rahiolisaurus gujaratensis* (Novas et al., 2010) (see Table 1).

shale, mudstone, marl, and clay. Sporadic dinosaur bones have been reported from the phase II intertrappean beds at Asifabad, near Hyderabad (Rao and Yadagiri, 1981), and Ranipur (Mathur and Sharma, 1990). However, the comparatively thick (~3.5 m) intertrappean beds of Anjar have yielded several large, yet to be described dinosaur bones (Ghevariya, 1988).

A major difference between the dinosaur fauna of the Lameta Formation and phase II intertrappean beds is the abundance of bones and nesting sites and fully preserved eggs in the former as compared to relatively limited presence of bones and no complete eggs or nesting sites in the latter. In view of the fact that the Lameta and phase II intertrappean dinosaur fossils were preserved in two distinctly different lithofacies, it is difficult to say whether there was a true decline of dinosaurs from the Lameta Formation to phase II intertrappean beds or whether this is an artifact of preservation. Depositional environments seem to have played an important role in the preservation of dinosaur fossils. The hard, sandy carbonate facies of the Lameta Formation deposited in a coastal-plain environment and proximal to the dinosaur habitat might have ensured better preservation of bones and complete eggs. In comparison, the intertrappean depositional environments represented by thin lacustrine facies distal to the dinosaur habitat may have received transported eggshell fragments and occasional large bones.

Mammals

At least two major groups of mammals have been documented from three widely separated phase II intertrappean sites. The intertrappean beds of Naskal near Hyderabad (Fig. 3) yielded the most diversified microvertebrate assemblage, including the first Cretaceous mammals of India. *Deccanolestes hislopi* was the first eutherian mammal recorded from the intertrappean beds of Naskal (Prasad and Sahni, 1988). Following this report, a few more taxa, such as *Deccanolestes robustus* (Prasad et al., 1994) and *Sahnitherium rangapurensis* (Rana and Wilson, 2003), were also documented from this area. Later on, one more species of *Deccanolestes* (*Deccanolestes narmadensis* Prasad et al., 2010) and a possible ungulate (*Kharmerungulatum vanvaleni* Prasad et al., 2007a) were described from the intertrappean beds of Kisalpuri in Central India (Fig. 3). Phylogenetic analysis of *Deccanolestes* has shown that it belongs to an insectivorous group of mammals (Adapisoriculidae), and this group possibly originated in India and dispersed to Africa and Europe some time close to the Cretaceous-Paleogene boundary (Goswami et al., 2011; de Bast et al., 2012).

These discoveries of eutherian mammals from the main Deccan volcanic province assume great significance in light of recent molecular phylogenies suggesting the former Gondwanaland as a center of origin for certain placental mammal orders (Springer, 1997; Waddell et al., 1999). As compared to a moderately diverse mammalian assemblage known from the intertrappean beds, only one possible haramiyidan mammal tooth (*Avashishta bacharamensis* Anantharaman et al., 2006) has been described from the infratrappean beds. The latter was recovered from the infratrappean beds of Bacharam, Rangareddi District, Andhra Pradesh (Anantharaman et al., 2006). The poor mammalian record from the infratrappean beds can be attributed to sampling bias. So far, samples from 10 phase II intertrappean sections have been bulk screen-washed, out of which only three turned out to be mammal-bearing. Out of four infratrappean sections (i.e., Pisdura, Jabalpur, Bacharam, and Marepalli; Fig. 2) subjected to bulk screen-washing, only one has produced a mammalian fossil. Thus, future works focused on washing of large numbers of samples may improve the diversity of infratrappean mammalian fauna.

SURVIVAL PATTERNS: LAMETA VERSUS INTERTRAPPEAN

The Lameta (infratrappean) fauna is represented by 12 fish, two crocodile, one turtle, two snake, and nine dinosaur taxa (three sauropod and six theropod) and 10 dinosaur oospecies (Table 1). In comparison, the phase II intertrappean vertebrate fauna is represented by 47 fish species (including 21 otolith species), five amphibian, eight lizard, four snake, three crocodile, four turtle, and six mammalian taxa. Besides these, sauropod and ornithoid eggshells, theropod teeth, and a few large sauropod bones are also known from the intertrappean beds. The diversity of intertrappean vertebrate fauna more than doubled, at least in the case of fish, snakes, and turtles at species and higher taxonomic levels.

This diversity increase may be due to sampling bias or the availability of more nutrients in the ecosystem following the initial phase of volcanism. Although the infratrappean beds have been extensively prospected for dinosaurs, very few sections have been explored for vertebrate microfossils by means of bulk screen-washing methods. In contrast, large numbers of intertrappean sections have been investigated based on this method. In addition, following erosion and transportation of materials from the newly deposited mafic volcanic rocks, many new nutrients might have been added to the lacustrine bodies, which may have played an important role in increasing their biodiversity. To discriminate between these two plausible explanations for the increased biodiversity in the intertrappean beds, more infratrappean sections need to be sampled on a larger scale.

Overall, the infratrappean vertebrate fauna survived Deccan volcanism phase I and, with the exception of dinosaurs, proliferated in the intertrappean ecosystems. This pattern of vertebrate survival from the infratrappean to intertrappean beds has a close parallel with the continental late Maastrichtian record of the western interior of North America (eastern Montana). There, the uppermost Cretaceous (late Maastrichtian) Hell Creek Formation yielded 107 vertebrates species, of which 52 species (49%) survived into the lower part of the overlying Lower Paleocene Tullock Formation (Archibald, 2011). Compared to 28% survival in land-dwelling animals, freshwater animals such as bony fishes, amphibians, turtles, crocodiles, and champsosaurs had a high survival rate of 76%. Sheehan and Fastovsky (1992) inferred a higher rate of survival for freshwater faunas (90%) as compared to land-dwellers (12%). This differential pattern of extinction of freshwater and terrestrial vertebrates has been explained in terms of dependence of land-based communities on primary productivity in contrast to buffering of freshwater communities from extinction by detritus feeding when there was a temporary drop in primary productivity (Sheehan and Fastovsky, 1992).

A similar explanation can be offered for the preferential survival of the predominantly freshwater aquatic community compared with the terrestrial community from the time of deposition of infratrappeans to intertrappean beds. There is a notable increase in the diversity of the intertrappean freshwater vertebrates such as fishes, turtles, and snakes as compared to that of the infratrappean vertebrates (Fig. 5). The freshwater vertebrates constitute 52%, while the terrestrial vertebrates form 49% of the infratrappean vertebrate fauna (Fig. 5). In contrast, the intertrappean vertebrate fauna is dominated by freshwater vertebrates (70%), compared to terrestrial vertebrates (30%). Additionally, it is quite possible that depositional environments also played a significant role in the preservation of fossils. The humid, tropical, lacustrine environments of the intertrappean beds had possibly greater potential for the preservation of vertebrate fossils as compared to overbank, paludal environments of the semiarid climate during deposition of the infratrappean beds. Prasad and Khajuria (1995) suggested that the survival of Deccan phase I volcanism by freshwater communities might be related to the episodic nature of the volcanic eruptions, and the fauna could have recovered from the environmental stress during the repose periods.

Invertebrates also seem to have responded to the initial Deccan volcanic activity in this manner. Khosla et al. (2011) described 41 species of freshwater ostracods from the Lameta Formation of Jabalpur, Pisdura, and Dongargaon. According to these authors, of the 41 species, five are indeterminate, six are new, and 30 are common to the intertrappean beds. Besides their common occurrence in Maastrichtian infratrappean and intertrappean beds, many of these Maastrichtian species are found in association with Danian planktic foraminifers and brackish-water ostracod species in Jhilmili intertrappean beds (Keller et al., 2009b). This demonstrates that the ostracods of the Maastrichtian intertrappean beds survived into the early Paleocene, apparently without major extinctions due to volcanism (Sharma and Khosla, 2009; Khosla et al., 2011).

According to Cripps et al. (2005), the third and last phase of Deccan volcanism did not affect life significantly. Cripps et al. (2005) considered the intertrappean beds of Bombay as part of the Mumbai Island Formation (Salsette Subgroup) overlying the Wai Subgroup of main Deccan volcanic province and hence representative of the final phase III of Deccan volcanism that followed the Cretaceous-Paleogene boundary mass extinction. Singh and Sahni (1996) assigned a Maastrichtian age for the intertrappean beds of Bombay on the basis of an ostracod assemblage similar to that of other intertrappean outcrops of the main Deccan volcanic province. However, the recent report of similar ostracod assemblage in association with early Paleocene (Danian) planktic foraminifers from the intertrappean beds of Jhilmili argues against the use of ostracods as age markers for the intertrappean beds. Palynofacies analysis of Bombay intertrappean beds led Cripps et al. (2005) to conclude that these beds were deposited in a lagoonal environment proximal to land, sheltered from a strong marine influence and more organic rich in comparison to those of the main Deccan volcanic province. Because of the absence of evidence for wildfires that would cause mass mortality and increase the availability of plant material to the lagoonal ecosystem throughout the numerous pyroclastic eruptions in this region, Cripps et al. (2005) argued that the late-stage explosive Deccan volcanism might not have caused a major ecosystem collapse.

However, based on palynofloral analysis of the Lameta Formation and succeeding intertrappean beds in Nand-Dongargaon Basin in Maharashtra, Samant and Mohabey (2005) concluded that the initial Deccan volcanic activity affected the plant community. The floral transition from the Lameta Formation to the overlying intertrappean stratigraphic levels is marked by a decline in diatom diversity and abundance, an increase in angiosperm and pteridophyte diversity at the expense of gymnosperms, and the first appearance of dinoflagellates (Samant and Mohabey, this volume). They attributed these floral changes to initial physiographic and chemical changes caused by Deccan volcanism. Though Samant and Mohabey (2005) observed total absence of diatoms and dinoflagellates in spore-pollen–rich intertrappean beds at higher stratigraphic levels, the spore-pollen assemblages from the intertrappean beds of different stratigraphic levels are generally more or less similar. Even the pollen-spore assemblage represented by *Ariadnaesporites*, *Aquilapollenites*, *Gabonisporites*, *Azolla*, etc. is broadly similar to Lameta and intertrappean beds. A marked difference in the palynofloral assemblage has only been noticed between the Maastrichtian outcrops and the Paleocene intertrappean beds of Papro, Lalitpur (Singh and Kar, 2002).

SUMMARY

Intensified vertebrate faunal studies in recent years have revealed the presence of widespread nesting sites over an area of 1000 km in an east-west direction along the Narmada Valley and in Central India. Following the application of bulk screen-washing methods, there has been a remarkable improvement in the record of diversity of vertebrate fauna of the Deccan infratrappean and intertrappean beds. As a consequence, all vertebrate groups, with the exception of birds, have been recovered from the intertrappean beds. In comparison to the common occurrence of dinosaur skeletal remains and nesting sites and eggs in the Lameta Formation, the phase II intertrappean beds yielded only isolated teeth and bones in a few sites and eggshell fragments at many localities. This apparent decline of the dinosaur fauna subsequent to the eruption of the first phase of Deccan volcanism might be attributed to taphonomic bias, as the infratrappean and intertrappean dinosaur-bearing sedimentary facies are distinct from each other. A comparison of vertebrate fauna from the

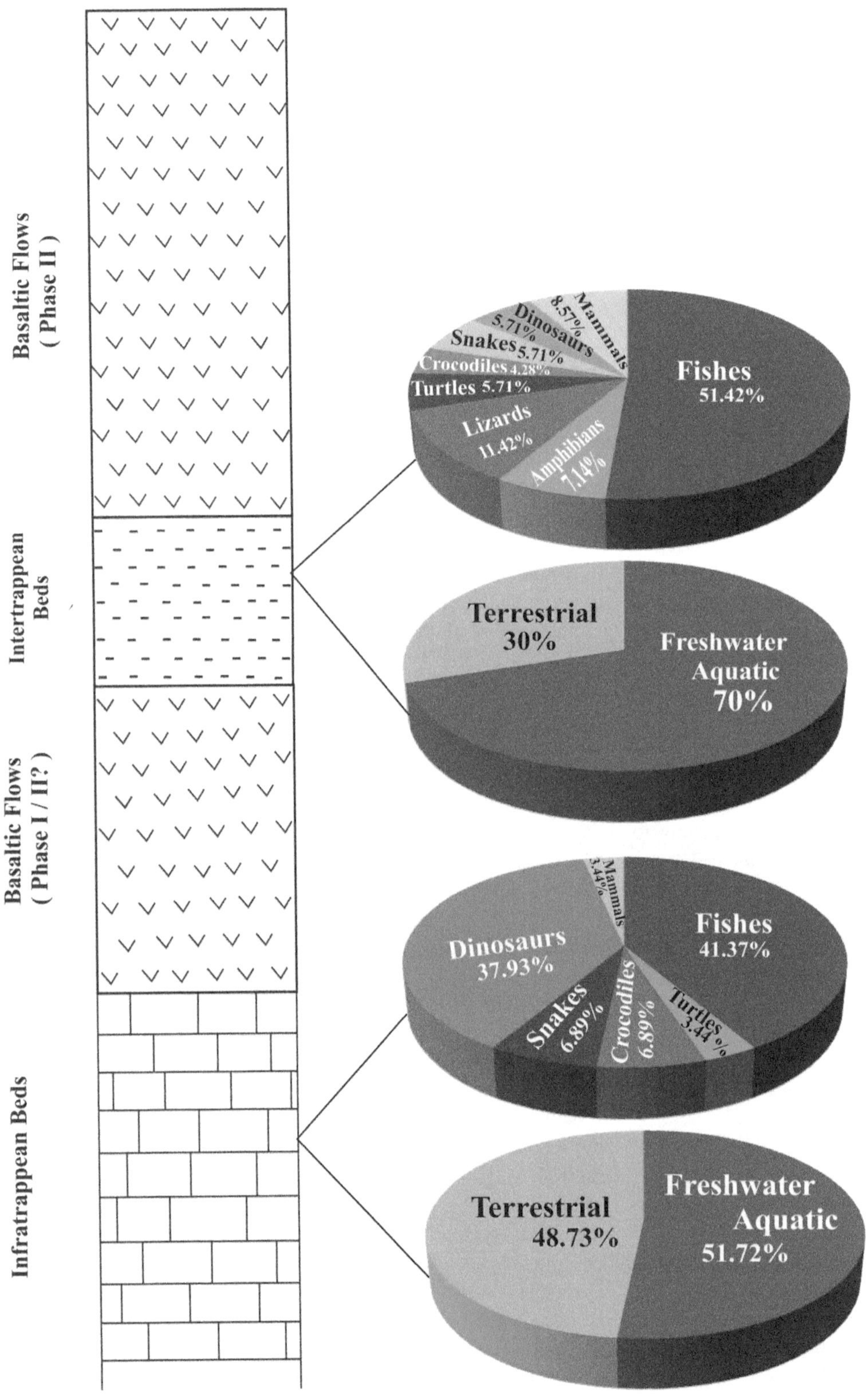

Figure 5. Relative composition of freshwater aquatic and terrestrial vertebrates of the infratrappean and intertrappean beds.

infratrappean and intertrappean beds revealed that many freshwater taxa survived the first phase of Deccan volcanism. The reasons one could offer for their survival are: (1) ability of the fauna to bounce back during the quiescent periods of volcanism; (2) the different feeding strategies adapted by freshwater and terrestrial communities; or (3) sampling bias—the observed increase in the vertebrate diversity from the infratrappean to intertrappean beds could be a consequence of limited sampling from the infratrappean beds in comparison to large-scale screen-washing of intertrappean sediments. If Deccan volcanism had any impact on life, it was possibly during the second phase, when 80% of the total volume erupted. Although a linkage has been suggested between Deccan volcanism and foraminiferal extinction in Cretaceous-Paleogene boundary subsurface sections of ONGC wells in the Krishna-Godavari Basin (Keller et al., 2011, 2012) and in the Um Sohryngkew section of Meghalaya, ~800 km northeast of the main Deccan volcanic province (Gertsch et al., 2011), so far no well-defined Cretaceous-Paleogene boundary section has been discovered in continental outcrops of the main Deccan volcanic province. Therefore, future studies should focus on identifying clearly marked Cretaceous-Paleogene boundary sections in the main Deccan volcanic province outcrops in order to understand the effects of volcanism on physical environments and biota in a terrestrial environment.

ACKNOWLEDGMENTS

Prasad acknowledges the research grant from J.C. Bose National Fellowship of Department of Science and Technology, New Delhi. We are thankful to Gerta Keller for inviting us to contribute to this volume. This manuscript has benefited from critical reviews by Gerta Keller, W.A. Clemens, and Jeffrey A. Wilson. Help from L. Ranjit Singh, Department of Geology, University of Delhi, in making the figures is duly acknowledged.

REFERENCES CITED

Alexander, P.O., 1981, Age and duration of Deccan volcanism: K-Ar evidence, *in* Subba Rao, K.V., and Sukheshwala, R.N., eds., Deccan Volcanism and Related Basalt Provinces in Other Parts of the World: Geological Society of India Memoir 3, p. 244–258.

Alvarez, L.W., Alvarez, W., Asaro, F., and Michel, H.V., 1980, Extraterrestrial cause for the Cretaceous-Tertiary extinction: Science, v. 208, p. 1095–1108, doi:10.1126/science.208.4448.1095.

Anantharaman, S., Wilson, G.P., Das Sarma, D.C., and Clemens, W.A., 2006, A possible Late Cretaceous "haramiyidan" from India: Journal of Vertebrate Paleontology, v. 26, p. 488–490, doi:10.1671/0272-4634(2006)26[488:APLCHF]2.0.CO;2.

Archibald, J.D., 2011, Extinction and Radiation: How the Fall of Dinosaurs Led to the Rise of Mammals: Baltimore, Maryland, John Hopkins University Press, 108 p.

Arratia, G., López-Arbarello, A., Prasad, G.V.R., Parmar, V., and Kriwet, J., 2004, Late Cretaceous-Paleocene percomorphs (Teleostei) from India—Early radiation of Perciformes, *in* Arratia, G., Wilson, M.V.H., and Cloutier, R., eds., Recent Advances in the Origin and Early Radiation of Vertebrates: Münich, Germany, Verlag Dr. Friedrich Pfeil, p. 635–663.

Bajpai, S., and Prasad, G.V.R., 2000, Cretaceous age for Ir-rich Deccan intertrappean deposits: Paleontological evidence from Anjar, western India: Journal of the Geological Society of London, v. 157, p. 257–260.

Bajpai, S., and Srinivasan, S., 1996, Fish otoliths from the Deccan intertrappean beds of Kachchh, *in* Pandey, J., Azmi, R.J., Bhandari, A., and Dave, A., eds., Contributions to XV Indian Colloquium on Micropalaeontology and Stratigraphy: Dehradun, India, Wadia Institute of Himalayan Geology, p. 471–475.

Bajpai, S., Sahni, A., and Srinivasan, S., 1993, Ornithoid eggshells from Deccan intertrappean beds near Anjar (Kachchh), western India: Current Science, v. 64, p. 42–45.

Bajpai, S., Srinivasan, S., and Sahni, A., 1997, Fossil turtle eggshells from infratrappean beds of Duddukuru, Andhra Pradesh: Journal of the Geological Society of India, v. 49, p. 209–213.

Bande, M.B., Chandra, A., Venkatachala, B.S., and Mehrotra, R.C., 1986, Deccan intertrappean floristics and its stratigraphic implications, *in* Venkatachala, B.S., ed., Proceedings of the Symposium on Paleocene of India: Limits and Subdivisions: Lucknow, India, Birbal Sahni Institute of Palaeobotany, p. 83–123.

Berman, D.S., and Jain, S.L., 1982, The brain-case of a small sauropod dinosaur (Reptilia: Saurischia) from the Upper Cretaceous Lameta Group, Central India, with review of Lameta Group localities: Annals of the Carnegie Museum, v. 51, p. 405–422.

Bhalla, S.N., 1967, Foraminifera from the intertrappean beds (Lower Eocene) of the Pangadi area, India: Micropalaeontology, v. 13, p. 351–368, doi:10.2307/1484837.

Bhandari, N., Shukla, P.N., Ghevariya, Z.G., and Sundaram, S.M., 1996, K/T boundary layer in Deccan intertrappeans at Anjar, Kutch, *in* Ryder, G., Fastovsky, D.E., and Gartner, S., eds., The Cretaceous-Tertiary Event and Other Catastrophes in Earth History: Geological Society of America Special Paper 307, p. 417–424.

Bhatia, S.B., and Mannikeri, M.S., 1976, Some charophytes from the Deccan inter-trappean beds near Nagpur, Central India: Geophytology, v. 6, p. 75–81.

Borkar, V.D., 1973, Fossil fishes from the intertrappean beds of Surendranagar District, Gujarat State: Proceedings of the Indian Academy of Sciences, v. 78, p. 181–193.

Borkar, V.D., 1984, *Palaeopristolepis chiplonkari*, a new fossil fish from the inter-trappean bed of Bamanbor, Surendranagar District, Gujarat: Biovigyanam, v. 10, p. 65–67.

Brookfield, M.E., and Sahni, A., 1987, Paleoenvironments of the Lameta Beds (Late Cretaceous) at Jabalpur, Madhya Pradesh, India: Soils and biotas of a semi-arid alluvial plain: Cretaceous Research, v. 8, p. 1–14, doi:10.1016/0195-6671(87)90008-5.

Buffetaut, E., 1987, On the age of the dinosaur fauna from the Lameta Formation (Upper Cretaceous of Central India): Newsletters on Stratigraphy, v. 18, p. 1–6.

Carter, H.J., 1852, Geology of the island of Bombay: Journal of the Bombay Branch of Royal Asiatic Society, v. 21, p. 161–215.

Chanda, S.K., 1967, Petrogenesis of the calcareous constituents of the Lameta Group around Jabalpur, M.P., India: Journal of Sedimentary Petrology, v. 37, p. 425–437, doi:10.1306/74D716E5-2B21-11D7-8648000102C1865D.

Chatterjee, S., 1978, *Indosuchus* and *Indosaurus*, Cretaceous carnosaurs from India: Journal of Paleontology, v. 52, p. 570–580.

Chenet, A.-L., Quidelleur, X., Fluteau, F., Courtillot, V., and Bajpai, S., 2007, 40K-40Ar dating of the main Deccan large igneous province; further evidence of KTB age and short duration: Earth and Planetary Science Letters, v. 263, p. 1–15, doi:10.1016/j.epsl.2007.07.011.

Chenet, A.-L., Fluteau, F., Courtillot, V., Gérard, M., and Subbarao, K.V., 2008, Determination of rapid Deccan eruptions across the Cretaceous-Tertiary boundary using paleomagnetic secular variation: Results from a 1200 m thick section in the Mahabaleshwar escarpment: Journal of Geophysical Research, v. 113, B04101, doi:10.1029/2006JB004635.

Chenet, A.-L., Courtillot, V., Fluteau, F., Gérard, M., Quidelleur, X., Khadri, S.F.R., Subbarao, K.V., and Thordarson, T., 2009, Determination of rapid Deccan eruptions across the Cretaceous-Tertiary boundary using paleomagnetic secular variation: 2. Constraints from analysis of eight new sections and synthesis for a 3500 m thick composite section: Journal of Geophysical Research, v. 114, B06103, doi:10.1029/2008JB005644.

Chiplonkar, G.W., 1940, A new species of fossil frog from the intertrappean beds of Worli Hill, Bombay: Journal of the Bombay Branch of Royal Asiatic Society, v. 40, p. 799–804.

Chiplonkar, G.W., 1986, Age of the Deccan Traps: Publication of the Centre of Advanced Study in Geology, Panjab University: New Series, v. 2, p. 1–22.

Cione, A.L., and Prasad, G.V.R., 2002, The oldest known catfish (Teleostei: Siluriformes) from Asia (India, Late Cretaceous): Journal of Paleontology, v. 76, p. 190–193, doi:10.1666/0022-3360(2002)076<0190:TOKCTS>2.0.CO;2.

Courtillot, V., Besse, J., Vandamme, D., Montigny, R.J., Jaeger, J.-J., and Cappetta, H., 1986, Deccan flood basalts at the Cretaceous-Tertiary boundary?: Earth and Planetary Science Letters, v. 80, p. 361–374, doi:10.1016/0012-821X(86)90118-4.

Courtillot, V., Feraud, G., Maluski, H., Vandamme, D., Moreau, M.G., and Besse, J., 1988, The Deccan flood basalts and the Cretaceous-Tertiary boundary: Nature, v. 333, p. 843–846, doi:10.1038/333843a0.

Cripps, J.A., Widdowson, M., Spicer, R.A., and Jolley, D.W., 2005, Coastal ecosystem responses to late stage Deccan Trap volcanism: The post K-T boundary (Danian) palynofacies of Mumbai (Bombay), west India: Palaeogeography, Palaeoclimatology, Palaeoecology, v. 216, p. 303–332, doi:10.1016/j.palaeo.2004.11.007.

de Bast, E., Sigé, B., and Smith, T., 2012, Diversity of the adapisoriculid mammals from the early Paleocene of Hainin, Belgium: Acta Palaeontologica Polonica, v. 57, p. 35–52, doi:10.4202/app.2010.0115.

de Lapparent de Broin, F., Prasad, G.V.R., Bajpai, S., and Verma, O., 2009, Chelonian remains from the Upper Cretaceous Deccan intertrappean beds of peninsular India: Review, new material and comparisons, *in* Abstract Volume, Gaffney Turtle Symposium: Alberta, Canada, Royal Tyrrell Museum of Paleontology, p. 95–97.

Dogra, N.N., Singh, R.Y., and Kulshrestha, S.K., 1988, Palynological evidence on the age of Jabalpur and Lameta Formations in the type area: Current Science, v. 57, p. 954–956.

Dogra, N.N., Singh, Y.R., and Singh, R.Y., 2004, Palynological assemblage from the Anjar intertrappean beds, Kutch, District Gujarat: Current Science, v. 86, p. 1596–1597.

Gaffney, E.S., Sahni, A., Schleich, H., Singh, S.D., and Srivastava, R., 2003, *Sankuchemys*, a new side-necked turtle (Pelomedusoides: Bothremydidae) from the Late Cretaceous of India: American Museum Novitates, v. 3405, p. 1–10, doi:10.1206/0003-0082(2003)405<0001:SANSTP>2.0.CO;2.

Gayet, M., Rage, J.-C., and Rana, R.S., 1984, Nouvelles ichthyofaune et herpéto faune de Gitti Khdan, le plus ancien gisement connu du Deccan (Crétacé/Paléocène) à microvertébrés: Implications paléogéographiques: Memoirs Société Géologique de France, v. 147, p. 55–65.

Gertsch, B., Keller, G., Adatte, T., Garg, R., Prasad, V., Fleitmann, D., and Berner, Z., 2011, Environmental effects of Deccan volcanism across the Cretaceous-Tertiary transition in Meghalaya, India: Earth and Planetary Science Letters, v. 310, p. 272–285, doi:10.1016/j.epsl.2011.08.015.

Ghevariya, Z.G., 1988, Intertrappean dinosaurian fossils from Anjar area, Kachchh District, Gujarat: Current Science, v. 57, p. 248–251.

Ghosh, P., Bhattacharya, S.K., and Jani, R.A., 1995, Paleoclimate and paleovegetation in Central India during the Upper Cretaceous based on the stable isotope composition of the paleosol carbonates: Palaeogeography, Palaeoclimatology, Palaeoecology, v. 114, p. 285–296, doi:10.1016/0031-0182(94)00082-J.

Goswami, A., Prasad, G.V.R., Upchurch, P., Boyer, D.M., Seiffert, E.R., Verma, O., Gheerbrant, E., and Flynn, J.J., 2011, A radiation of basal arboreal mammals beginning in the Late Cretaceous of India: Proceedings of the National Academy of Sciences of the United States of America, v. 108, p. 16,333–16,338, doi:10.1073/pnas.1108723108.

Hislop, S., 1860, On the Tertiary deposits associated with trap rock in the East Indies with description of fossil shells: Quarterly Journal of the Geological Society of London, v. 16, p. 154–166, doi:10.1144/GSL.JGS.1860.016.01-02.22.

Hora, S.L., 1938, On some fossil fish scales from the intertrappean beds at Deothan and Kheri, Central Provinces: Records of the Geological Survey of India, v. 73, p. 267–294.

Hunt, A.G., Lockley, M.G., Lucas, S.G., and Meyer, C.A., 1994, The global sauropod fossil record: Gaia, v. 10, p. 261–279.

Jain, S.L., 1977, A new fossil pelomedusid turtle from the Upper Cretaceous Pisdura sediments, Central India: Journal of the Palaeontological Society of India, v. 20, p. 360–365.

Jain, S.L., 1986, New pelomedusid turtle (Pleurodira: Chelonia) remains from Lameta Formation (Maastrichtian) at Dongargaon, Central India, and a review of pelomedusids from India: Journal of the Palaeontological Society of India, v. 31, p. 63–75.

Jain, S.L., and Bandyopadhyay, S., 1997, New titanosaurid (Dinosauria: Sauropoda) from the Late Cretaceous of Central India: Journal of Vertebrate Paleontology, v. 17, p. 114–136, doi:10.1080/02724634.1997.10010958.

Jain, S.L., and Sahni, A., 1983, Some Upper Cretaceous vertebrates from Central India and their paleogeographical implications, *in* Maheshwari, H.K., ed., Cretaceous of India: Indian Association of Palynostratigraphers Symposium: Lucknow, India, Birbal Sahni Institute of Palaeobotany, p. 66–83.

Jain, S.L., and Sahni, A., 1985, Dinosaurian eggshell fragments from the Lameta Formation at Pisdura, Chandrapur District, Maharashtra: Geoscience Journal, v. 6, p. 211–220.

Jaiprakash, B.C., Singh, J., and Raju, D.S.N., 1993, Foraminiferal events across K/T boundary and age of Deccan volcanism in Palakollu area, Krishna-Godavari Basin, India: Journal of the Geological Society of India, v. 41, p. 105–117.

Jay, A.E., and Widdowson, M., 2008, Stratigraphy, structure and volcanology of the south-east Deccan continental flood basalt province: Implications for eruptive extent and volumes: Journal of the Geological Society of London, v. 166, p. 1–12.

Jay, A.E., Mac Niocaill, C., Widdowson, M., Self, S., and Tuner, W., 2009, New palaeomagnetic data from the Mahabaleshwar Plateau, Deccan flood basalt province, India: Implications from the volcanostratigraphic architecture of continental flood basalt provinces: Journal of the Geological Society of London, v. 166, p. 13–24, doi:10.1144/0016-76492007-150.

Keller, G., Adatte, T., Gardin, S., Bartolini, A., and Bajpai, S., 2008, Main Deccan volcanism phase ends near the K-T boundary: Evidence from the Krishna-Godavari Basin, SE India: Earth and Planetary Science Letters, v. 268, p. 293–311, doi:10.1016/j.epsl.2008.01.015.

Keller, G., Khosla, S.C., Sharma, R., Khosla, A., Bajpai, S., and Adatte, T., 2009a, Early Danian planktic foraminifera from Cretaceous-Tertiary intertrappean beds at Jhilmili, Chhindwara District, Madhya Pradesh, India: Journal of Foraminiferal Research, v. 39, p. 40–55, doi:10.2113/gsjfr.39.1.40.

Keller, G., Adatte, T., Bajpai, S., Mohabey, D.M., Widdowson, M., Khosla, A., Sharma, R., Khosla, S.C., Gertsch, B., Fleitmann, D., and Sahni, A., 2009b, K-T transition in Deccan Traps of Central India marks major marine seaway across India: Earth and Planetary Science Letters, v. 282, p. 10–23, doi:10.1016/j.epsl.2009.02.016.

Keller, G., Bhowmick, P.K., Upadhyay, H., Dave, A., Reddy, A.N., Jaiprakash, B.C., and Adatte, T., 2011, Deccan volcanism linked to the Cretaceous-Tertiary boundary mass extinction: New evidence from ONGC wells in the Krishna-Godavari Basin: Journal of the Geological Society of India, v. 78, p. 399–428, doi:10.1007/s12594-011-0107-3.

Keller, G., Adatte, T., Bhowmick, P.K., Upadhyay, H., Dave, A., Reddy, A.N., and Jaiprakash, B.C., 2012, Nature and timing of extinctions in Cretaceous-Tertiary planktic foraminifera preserved in Deccan intertrappean sediments of the Krishna-Godavari Basin, India: Earth and Planetary Science Letters, v. 341–344, p. 211–221, doi:10.1016/j.epsl.2012.06.021.

Khadkikar, A.S., Sant, D.A., Gogte, V., and Karanth, R.V., 1999, The influence of Deccan volcanism on climate: Insights from lacustrine intertrappean deposits, Anjar, western India: Palaeogeography, Palaeoclimatology, Palaeoecology, v. 147, p. 141–149, doi:10.1016/S0031-0182(98)00156-4.

Khajuria, C.K., and Prasad, G.V.R., 1998, Taphonomy of a Late Cretaceous mammal-bearing microvertebrate assemblage from the Deccan intertrappean beds of Naskal, Peninsular India: Palaeogeography, Palaeoclimatology, Palaeoecology, v. 137, p. 153–172, doi:10.1016/S0031-0182(97)00098-9.

Khajuria, C.K., Prasad, G.V.R., and Manhas, B.K., 1994, Paleontological constraints on the age of Deccan Traps, peninsular India: Newsletters on Stratigraphy, v. 31, p. 21–32.

Khosla, A., and Sahni, A., 1995, Parataxonomic classification of Late Cretaceous dinosaur eggshells from India: Journal of the Palaeontological Society of India, v. 40, p. 87–102.

Khosla, A., Prasad, G.V.R., Verma, O., Jain, A.K., and Sahni, A., 2004, Discovery of a micromammal-yielding Deccan intertrappean site near Kisalpuri, Dindori District, Madhya Pradesh: Current Science, v. 87, p. 380–383.

Khosla, A., Sertich, J.J.W., Prasad, G.V.R., and Verma, O., 2009a, Dyrosaurid remains from the intertrappean beds of India and the Late Cretaceous distribution of Dyrosauridae: Journal of Vertebrate Paleontology, v. 29, p. 1321–1326, doi:10.1671/039.029.0416.

Khosla, S.C., Nagori, M.L., Jakhar, S.R., and Rathore, A.S., 2009b, Mixed marine, brackish water and non-marine microfaunal association in the intertrappean beds (early Paleocene) of Jhilmili, Chhindwara District, Madhya Pradesh: Journal of the Geological Society of India, v. 73, p. 724–732, doi:10.1007/s12594-009-0056-2.

Khosla, S.C., Rathore, A.S., Nagori, M.L., and Jakhar, S.R., 2011, Non-marine Ostracoda from the Lameta Formation (Maastrichtian) of Jabalpur (Madhya Pradesh) and Nand-Dongargaon Basin (Maharashtra), India: Their correlation, age and taxonomy: Revista Española de Micropaleontologia, v. 43, p. 209–260.

Kumar, S., and Tandon, K.K., 1979, Trace fossils and environment of deposition of the sedimentary succession of Jabalpur, M.P.: Journal of the Geological Society of India, v. 20, p. 103–106.

Kumaran, K.P.N., Bonde, S.D., and Kanitkar, M.D., 1997, An *Aquilapollenites* associated palynoflora from Mohgaonkalan and its stratigraphic implications for age and stratigraphic correlation of Deccan intertrappean beds: Current Science, v. 72, p. 590–592.

Leidy, J., 1855, Indications of twelve species of fossil fishes: Proceedings of the Academy of Natural Sciences of Philadelphia, v. 7, p. 395–397.

Loyal, R.S., Mohabey, D.M., Khosla, A., and Sahni, A., 1998, Status and paleobiology of the Late Cretaceous Indian theropods with description of a new theropod eggshell oogenus and oospecies, *Ellipsoolithus khedaensis*, from the Lameta Formation, district Kheda, Gujarat, western India: Gaia, v. 15, p. 379–387.

Lydekker, R., 1877, Notices of new and other Vertebrata from Indian Tertiary and Secondary rocks: Records of the Geological Survey of India, v. 10, p. 30–43.

Lydekker, R., 1890, Note on certain vertebrate remains from the Nagpur District: Records of the Geological Survey of India, v. 23, p. 21–24.

Mathur, U.B., and Srivastava, S., 1987, Dinosaur teeth from Lameta Group (Upper Cretaceous) of Kheda District, Gujarat: Journal of the Geological Society of India, v. 29, p. 554–566.

Mathur, Y.K., and Sharma, K.D., 1990, Palynofossils and age of Ranipur intertrappean beds, Gaur River, Jabalpur, M.P., *in* Sahni, A., ed., Cretaceous Event Stratigraphy and Correlation of the Indian Non-Marine Strata: Contributions to Seminar cum Workshop (IGCP 216 & 245): Chandigarh, India, CAS in Geology, Panjab University, p. 58–59.

Matley, C.A., 1921, On the stratigraphy, fossils and geological relationships of the Lameta beds of Jubbulpore: Records of the Geological Survey of India, v. 53, p. 142–169.

Matley, C.A., 1923, Note on an armored dinosaur from the Lameta beds of Jubbulpore: Records of the Geological Survey of India, v. 55, p. 105–109.

Matley, C.A., 1929, Stratigraphy, fossils and geological relationship of the Lameta Beds of Jabalpur: Records of the Geological Survey of India, v. 53, p. 142–164.

McLean, D.M., 1985, Deccan Traps mantle degassing in the terminal Cretaceous marine extinctions: Cretaceous Research, v. 6, p. 235–259, doi:10.1016/0195-6671(85)90048-5.

Mohabey, D.M., 1996, *Pycnodus lametae* (Pycnodontidae), a holostean fish from freshwater Upper Cretaceous Lameta Formation of Maharashtra: Journal of the Geological Society of India, v. 47, p. 593–598.

Mohabey, D.M., 1998, Systematics of Indian Upper Cretaceous dinosaur and chelonian eggshells: Journal of Vertebrate Paleontology, v. 18, p. 348–362, doi:10.1080/02724634.1998.10011063.

Mohabey, D.M., 2001, Indian dinosaur eggs: A review: Journal of the Geological Society of India, v. 58, p. 479–508.

Mohabey, D.M., Head, J.J., and Wilson, J.A., 2011, A new species of the snake *Madtsoia* from the Upper Cretaceous of India and its paleobiogeographic implications: Journal of Vertebrate Paleontology, v. 31, p. 588–595, doi:10.1080/02724634.2011.560220.

Noble, G.K., 1930, The fossil frogs of the intertrappean beds of Bombay, India: American Museum Novitates, v. 401, p. 1–13.

Nolf, D., Rana, R.S., and Prasad, G.V.R., 2008, Late Cretaceous (Maastrichtian) fish otoliths from the Deccan intertrappean beds, India: A revision: Bulletin de l'Institut Royal des Sciences Naturelles de Belgique, v. 78, p. 239–259.

Novas, F.E., Chatterjee, S., Rudra, D.K., and Datta, P.M., 2010, *Rahiolisaurus gujaratensis* n. gen. n. sp., a new abelisaurid theropod from the Late Cretaceous of India, *in* Bandhyopadhyay, S., ed., New Aspects of Mesozoic Biodiversity: Berlin, Springer-Verlag, Lecture Notes in Earth Sciences, v. 132, p. 45–62.

Officer, C.B., Hallam, A., Drake, C.L., and Devine, J.D., 1987, Late Cretaceous and paroxysmal Cretaceous/Tertiary extinctions: Nature, v. 326, p. 143–149, doi:10.1038/326143a0.

Owen, F.G.S., 1847, On the batracholites, indicative of a small species of frogs (*Rana pusilla* Owen): Quarterly Journal of the Geological Society of London, v. 3, p. 224–225, doi:10.1144/GSL.JGS.1847.003.01-02.24.

Prakash, U., 1960, A survey of the Deccan intertrappean flora of India: Journal of Paleontology, v. 34, p. 1027–1040.

Prasad, G.V.R., 1989, Vertebrate fauna from the infratrappean and intertrappean beds of Andhra Pradesh: Age implications: Journal of the Geological Society of India, v. 34, p. 161–173.

Prasad, G.V.R., and Cappetta, H., 1993, Late Cretaceous selachians from India and the age of Deccan Traps: Palaeontology, v. 36, p. 231–248.

Prasad, G.V.R., and de Lapparent de Broin, F., 2002, Late Cretaceous crocodile remains from Naskal (India): Comparisons and bio-geographic affinities: Annales de Paléontologie, v. 88, p. 19–71, doi:10.1016/S0753-3969(02)01036-4.

Prasad, G.V.R., and Khajuria, C.K., 1995, Implications of the infratrappean and intertrappean biota from the Deccan, India, for the role of volcanism in Cretaceous-Tertiary boundary extinctions: Journal of the Geological Society of London, v. 152, p. 289–296, doi:10.1144/gsjgs.152.2.0289.

Prasad, G.V.R., and Rage, J.-C., 1995, Amphibians and squamates from the Maastrichtian of Naskal, India: Cretaceous Research, v. 16, p. 95–107, doi:10.1006/cres.1995.1006.

Prasad, G.V.R., and Rage, J.-C., 2004, Fossil frogs (Amphibia: Anura) from the Upper Cretaceous intertrappean beds of Naskal, Andhra Pradesh: Revue de Paléobiologie, v. 23, p. 99–116.

Prasad, G.V.R., and Sahni, A., 1987, Coastal-plain microvertebrate assemblage from the terminal Cretaceous of Asifabad, Andhra Pradesh: Journal of the Palaeontological Society of India, v. 32, p. 5–19.

Prasad, G.V.R., and Sahni, A., 1988, First Cretaceous mammal from India: Nature, v. 332, p. 638–640, doi:10.1038/332638a0.

Prasad, G.V.R., and Srinivasan, S., 1990, Gurmatkal intertrappean beds—Age reassessment, *in* Sahni, A., ed., Cretaceous Event Stratigraphy and Correlation of the Indian Non-Marine Strata: Contributions to Seminar cum Workshop (IGCP 216 & 245): Chandigarh, India, CAS in Geology, Panjab University, p. 48–49.

Prasad, G.V.R., Jaeger, J.-J., Sahni, A., Gheerbrant, E., and Khajuria, C.K., 1994, Eutherian mammals from the Upper Cretaceous (Maastrichtian) intertrappean beds of Naskal, Andhra Pradesh, India: Journal of Vertebrate Paleontology, v. 14, p. 260–277, doi:10.1080/02724634.1994.10011556.

Prasad, G.V.R., Verma, O., Sahni, A., Parmar, V., and Khosla, A., 2007a, A Cretaceous hoofed mammal from India: Science, v. 318, p. 937, doi:10.1126/science.1149267.

Prasad, G.V.R., Verma, O., Sahni, A., Krause, D.W., Khosla, A., and Parmar, V., 2007b, A new Late Cretaceous Gondwanatherian mammal from Central India: Proceedings of the Indian National Science Academy, v. 73, p. 17–24.

Prasad, G.V.R., Verma, O., Gheerbrant, E., Goswami, A., Khosla, A., Parmar, V., and Sahni, A., 2010, First mammal evidence from the Late Cretaceous of India for biotic dispersal between India and Africa at the K/T transition: Comptes Rendus Palévol, v. 9, p. 63–71, doi:10.1016/j.crpv.2009.12.003.

Rage, J.-C., and Prasad, G.V.R., 1992, New snakes from the Late Cretaceous (Maastrichtian) of Naskal, India: Neues Jahrbuch für Geologie und Palaontologie, Abhandlungen, v. 187, p. 83–97.

Rage, J.-C., Prasad, G.V.R., and Bajpai, S., 2004, Additional snakes from the uppermost Cretaceous (Maastrichtian) of India: Cretaceous Research, v. 25, p. 425–434, doi:10.1016/j.cretres.2004.02.003.

Raju, D.S.N., Ravichandran, C.N., Dave, A., Jaiprakash, B.C., and Singh, J., 1991, K/T boundary events in the Cauvery and Krishna-Godavari Basins and the age of Deccan volcanics: Geoscience Journal, v. 12, p. 177–190.

Rana, R.S., 1987, Dyrosaurid crocodile (Mesosuchia) from the infratrappean beds of Vikarabad, Hyderabad District, Andhra Pradesh: Current Science, v. 56, p. 532–534.

Rana, R.S., 1988, Freshwater fish otoliths from the Deccan Trap associated sedimentary (Cretaceous-Tertiary transition) beds of Rangapur, Hyderabad District, Andhra Pradesh: Geobios, v. 21, p. 465–493, doi:10.1016/S0016-6995(88)80045-7.

Rana, R.S., 1990, Alligatorine teeth from the Deccan intertrappean beds near Rangapur, Andhra Pradesh, India: Further evidence of Laurasiatic elements: Current Science, v. 59, p. 49–51.

Rana, R.S., 1996, Additional fish otoliths from the Deccan Trap associated sedimentary beds exposed near Rangapur, Rangareddi District, Andhra Pradesh, *in* Pandey, J., Azmi, R.J., Bhandari, A., and Dave, A., eds., Contributions of XV Indian Colloquium on Micropalaeontology and Stratigraphy: Dehradun, India, Wadia Institute of Himalayan Geology, p. 477–492.

Rana, R.S., 2005, Lizard fauna from intertrappean (Late Cretaceous–early Paleocene) beds of peninsular India, *in* Mohabey, D.M., ed., Indian Non-Marine Late Cretaceous: Advances and Challenges: Gondwana Geological Magazine, v. 8, supplement, p. 123–132.

Rana, R.S., and Sahni, A., 1989, Fish otoliths from the Takli Formation (inter-trappean beds) of Nagpur, India: Geoscience Journal, v. 10, p. 145–164.

Rana, R.S., and Wilson, G.P., 2003, New Late Cretaceous mammals from the intertrappean beds of Rangapur, India, and paleobiogeographic framework: Acta Palaeontologica Polonica, v. 48, p. 331–348.

Rao, B.R.J., and Yadagiri, P., 1981, Cretaceous intertrappean beds from Andhra Pradesh and their stratigraphic significance, *in* Subbarao, K.V., and Sukheshwala, R.N., eds., Deccan Volcanism and Related Basalt Provinces in Other Parts of the World: Geological Society of India Memoir 3, p. 287–291.

Saha, O., Shukla, U.K., and Rani, R., 2010, Paleoenvironmental implication of trace fossils in the Upper Cretaceous Lameta Beds of Jabalpur area, M.P., India: Journal of the Geological Society of India, v. 76, p. 607–620, doi:10.1007/s12594-010-0121-x.

Sahni, A., 1983, Upper Cretaceous paleobiogeography of peninsular India and the Cretaceous-Paleocene transition: The vertebrate evidence, *in* Maheshwari, H.K., ed., Cretaceous of India: Lucknow, India, Indian Association of Palynostratigraphers, p. 128–140.

Sahni, A., and Bajpai, S., 1988, Cretaceous-Tertiary boundary events: The fossil vertebrate, paleomagnetic and radiometric evidence from peninsular India: Journal of the Geological Society of India, v. 32, p. 382–397.

Sahni, A., and Mehrotra, D.K., 1974, Turonian terrestrial communities of India: Geophytology, v. 4, p. 102–105.

Sahni, A., Kumar, K., Hartenberger, J.-L., Jaeger, J.-J., Rage, J.-C., Sudre, J., and Vianey-Liaud, M., 1982, Microvertébrés nouveau des Traps du Deccan (Inde): Mise en evidence d'une voie de communication terrestre probable entre la Laurasie et l'Inde à la limite Crétacé-Tertiaire: Bulletin de la Société Géologique de France, v. 24, p. 1093–1099.

Sahni, A., Rana, R.S., and Prasad, G.V.R., 1984, SEM studies of thin eggshell fragments from the intertrappeans (Cretaceous-Tertiary transition) of Nagpur and Asifabad, peninsular India: Journal of the Palaeontological Society of India, v. 29, p. 26–33.

Sahni, A., Tandon, S.K., Jolly, A., Bajpai, S., Sood, A., and Srinivasan, S., 1994, Upper Cretaceous dinosaur eggs and nesting sites from the Deccan volcano-sedimentary province of peninsular India, *in* Carpenter, K., Hirsch, K.F., and Horner, J.R., eds., Dinosaur Eggs and Babies: Cambridge, UK, Cambridge University Press, p. 204–226.

Sahni, A., Venkatachala, B.S., Kar, R.K., Rajnikanth, A., Prakash, T., Prasad, G.V.R., and Singh, R.Y., 1996, New paleontological data from the intertrappean beds: Implications for the latest record of dinosaurs and synchronous initiation of volcanic activity in India, *in* Sahni, A., ed., Cretaceous Stratigraphy and Paleoenvironments: Geological Society of India Memoir 37, p. 267–203.

Sahni, B., 1934, The Deccan Traps: Are they Cretaceous or Tertiary?: Current Science, v. 3, p. 134–136.

Samant, B., and Mohabey, D.M., 2005, Response of flora to Deccan volcanism: A case study from Nand-Dongargaon Basin of Maharashtra, implications to environment and climate: Gondwana Geological Magazine, v. 8, p. 151–164.

Samant, B., and Mohabey, D.M., 2014, this volume, Deccan volcanic eruptions and their impact on flora: Palynological evidence, *in* Keller, G., and Kerr, A.C., eds., Volcanism, Impacts, and Mass Extinctions: Causes and Effects: Geological Society of America Special Paper 505, doi:10.1130/2014.2505(08).

Sarkar, A., Bhattacharya, S.K., and Mohabey, D.M., 1991, Stable isotope analyses of dinosaur eggshells: Paleoenvironmental implications: Geology, v. 19, p. 1068–1107, doi:10.1130/0091-7613(1991)019<1068:SIAODE>2.3.CO;2.

Sharma, R., and Khosla, A., 2009, Early Paleocene Ostracoda from the Cretaceous-Tertiary (K-T) Deccan intertrappean sequence at Jhilmili, District, Chhindwara, Central India: Journal of the Palaeontological Society of India, v. 54, p. 197–208.

Sheehan, P.M., and Fastovsky, D.E., 1992, Major extinctions of land dwelling vertebrates at the Cretaceous-Tertiary boundary, eastern Montana: Geology, v. 20, p. 556–560, doi:10.1130/0091-7613(1992)020<0556:MEOLDV>2.3.CO;2.

Shivarudrappa, T.V., 1976–1977, Discovery of fossil reproductive organs of *Chara* from intertrappeans of Gurmatkal, Gulbarga District, Karnataka: Journal of Mysore University, sec. B–Science, v. 27, p. 104–110.

Singh, I.B., 1981, Paleoenvironment and paleogeography of Lameta Group sediments (Late Cretaceous) in Jabalpur area, India: Journal of the Palaeontological Society of India, v. 26, p. 38–53.

Singh, R.S., and Kar, R.K., 2002, Paleocene palynofossils from the Lalitpur inter-trappean beds, Uttar Pradesh, India: Journal of the Geological Society of India, v. 60, p. 213–216.

Singh, R.S., Kar, R., and Prasad, G.V.R., 2006, Palynological constraints on the age of mammal yielding Deccan intertrappean beds of Naskal, Rangareddi District, Andhra Pradesh: Current Science, v. 90, p. 1281–1285.

Singh, S.D., and Sahni, A., 1996, Bombay intertrappeans: New data on age and faunal affinities, *in* Pandey, J., Azmi, R.J., Bhandari, A., and Dave, A., eds., Contributions of XV Indian Colloquium on Micropaleontology and Stratigraphy: Dehradun, India, Wadia Institute of Himalayan Geology, p. 465–469.

Singh, S.D., Sahni, A., Gaffney, E., and Schleich, H.H., 1998, Reptilia from the inter-trappean beds of Bombay (India), *in* Schleich, H.H., and Kästle, W., eds., Contributions to the Herpetology of South Asia (Nepal, India): Veröffentlichungen aus dem Fuhlrott-Museum, v. 4, p. 307–321.

Singh, S.K., and Srivastava, H.K., 1981, Lithostratigraphy of Bagh Beds and its correlation with the Lameta Beds: Journal of the Palaeontological Society of India, v. 26, p. 77–85.

Sleeman, W.H., 1844, Rambles and Recollections of an Indian Official, Volume 1: London, J. Hatchard and Son, 478 p.

Springer, M.S., 1997, Molecular clocks and the timing of the placental and marsupial radiations in relation to the Cretaceous-Tertiary boundary: Journal of Mammalian Evolution, v. 4, p. 285–302, doi:10.1023/A:1027378615412.

Srinivasan, S., 1996, Late Cretaceous eggshells from the Deccan volcano-sedimentary sequences of Central India, *in* Sahni, A., ed., Cretaceous Stratigraphy and Palaeoenvironments: Geological Society of India Memoir 37, p. 321–336.

Srivastava, S., Mohabey, D.M., Sahni, A., and Pant, S.C., 1986, Upper Cretaceous dinosaur egg clutches from Kheda District, Gujarat, India: Their distribution, shell structure and paleoecology: Palaeontographica Abteilung A, v. 193, p. 219–233.

Sykes, C., 1851, On a fossil fish from the table land of the Deccan, in the peninsula of India, with a description of specimens by P.M.G. Egerton: Quarterly Journal of the Geological Society of London, v. 7, p. 272–273, doi:10.1144/GSL.JGS.1851.007.01-02.49.

Tandon, S.K., Sood, A., Andrews, J.E., and Dennis, P.F., 1995, Paleoenvironments of dinosaur-bearing Lameta Beds (Maastrichtian), Narmada Valley, Central India: Palaeogeography, Palaeoclimatology, Palaeoecology, v. 117, p. 153–184, doi:10.1016/0031-0182(94)00128-U.

Venkatesan, T.R., Pande, K., and Gopalan, K., 1993, Did Deccan volcanism pre-date the Cretaceous/Tertiary transition?: Earth and Planetary Science Letters, v. 119, p. 181–189, doi:10.1016/0012-821X(93)90015-2.

Verma, K.K., 1965, Note on a new species of fossil frog from the intertrappean beds of Malabar Hill, Bombay: Current Science, v. 34, p. 182–183.

Vianey-Liaud, M., Jain, S.L., and Sahni, A., 1987, Dinosaur eggshells (Saurischia) from the Late Cretaceous intertrappean and Lameta Formations (Deccan India): Journal of Vertebrate Paleontology, v. 7, p. 408–424, doi:10.1080/02724634.1988.10011673.

Vianey-Liaud, M., Khosla, A., and Garcia, G., 2003, Relationships between European and Indian dinosaur eggs and eggshells of the oofamily Megaloolithidae: Journal of Vertebrate Paleontology, v. 23, p. 575–585, doi:10.1671/0272-4634(2003)023[0575:RBEAID]2.0.CO;2.

von Huene, F., and Matley, C.A., 1933, The Cretaceous Saurischia and Ornithischia of the Central Provinces of India: Memoirs Geological Survey of India, Palaeontologica Indica: New Series, v. 21, p. 1–74.

Waddell, P.J., Okada, N., and Hasegawa, M., 1999, Progress in resolving the interordinal relationships of placental mammals: Systematic Biology, v. 48, p. 1–5, doi:10.1093/sysbio/48.1.1.

Williams, E., 1953, Fossils and distribution of cheylid turtles. 1. "*Hydraspis*" *leithii* (Carter) in the Eocene of India is a pelomedusid: Breviora, v. 1, p. 1–8.

Wilson, G.P., Das Sarma, D.C., and Anantharaman, S., 2007, Late Cretaceous sudamericid gondwanatherian from India with paleobiogeographic considerations of Gondwanan mammals: Journal of Vertebrate Paleontology, v. 27, p. 521–531, doi:10.1671/0272-4634(2007)27[521:LCSGFI]2.0.CO;2.

Wilson, J.A., and Mohabey, D.M., 2006, A titanosauriform (Dinosauria: Sauropoda) axis from the Lameta Formation (Upper Cretaceous, Maastrichtian) of Nand, Central India: Journal of Vertebrate Paleontology, v. 26, p. 471–479, doi:10.1671/0272-4634(2006)26[471:ATDSAF]2.0.CO;2.

Wilson, J.A., and Upchurch, P., 2003, A revision of Titanosaurus Lydekker (Dinosauria: Sauropoda), the first dinosaur genus with a 'Gondwanan' distribution: Journal of Systematic Palaeontology, v. 1, p. 125–160, doi:10.1017/S1477201903001044.

Wilson, J.A., Sereno, P.C., Srivastava, S., Bhatt, D.K., Khosla, A., and Sahni, A., 2003, A new abelisaurid (Dinosauria, Theropoda) from the Lameta Formation (Cretaceous, Maastrichtian) of India: Contributions from the Museum of Paleontology: University of Michigan, v. 31, p. 1–42.

Wilson, J.A., Mohabey, D.M., Peters, S.E., and Head, J.J., 2010, Predation upon hatchling dinosaurs by a new snake from the Late Cretaceous of India: PLoS Biology, v. 8, p. e1000322, doi:10.1371/journal.pbio.1000322.

Wilson, J.A., Barrett, P.M., and Carrano, M.T., 2011, An associated partial skeleton of *Jainosaurus* cf. *septentrionalis* (Dinosauria: Sauropoda) from the Late Cretaceous of Chhota Simla, Central India: Palaeontology, v. 54, p. 981–998, doi:10.1111/j.1475-4983.2011.01087.x.

Woodward, A.S., 1908, On some fish-remains from the Lameta Beds at Dongargaon, Central Provinces: Palaeontologica Indica, v. 3, p. 1–6.

Manuscript Accepted by the Society 31 January 2014

Printed in the USA

The Geological Society of America
Special Paper 505
2014

What the dinosaur record says about extinction scenarios

J. David Archibald*
Department of Biology, San Diego State University, San Diego, California 92182-4614, USA

ABSTRACT

The record of dinosaurs over the last 10 m.y. of the Cretaceous, as well as surrounding the Cretaceous-Paleogene boundary, helps to define extinction scenarios. Although Late Cretaceous dinosaur fossils occur on all present-day continents, only in North America do we find a terrestrial vertebrate fossil record spanning the Cretaceous-Paleogene boundary, although promising work may yield comparable records in South America, India, China, and Europe. For the present then, the North American record represents the proxy for our knowledge of dinosaur extinction. Over the last 10 m.y. of the Cretaceous (late Campanian to late Maastrichtian) in the northern part of the western interior of North America, the number of nonavian dinosaur species dropped from 49 to 25, almost a 50% reduction, even though a 16% greater extent of fossil-bearing exposures record the last dinosaurs in the latest Cretaceous in the western interior. Important, but less-well-exposed, nonavian-dinosaur–bearing units suggest this drop occurred around, or at least commenced by, the Campanian-Maastrichtian boundary. These losses began during climatic fluctuations, occurring during and possibly in part caused by the last major regressive cycle of the Cretaceous, which also reduced the expanse of the low coastal plains inhabited by nonavian dinosaurs. The pulse of Deccan Trap emplacement that began some time later in the latest Cretaceous was also likely a major driver of climatic change. As for the dinosaur record near the Cretaceous-Paleogene boundary, even the best-known records from North America remain enigmatic and open to interpretation. Newer studies suggest some decline in at least relative abundance approaching the Cretaceous-Paleogene boundary, but the cause (or causes) for the final extinction (if it was the case) of nonavian dinosaurs remains unresolved, although the Chicxulub impact undoubtedly played a major role.

INTRODUCTION

The best-preserved and most-studied sequences of Late Cretaceous and Paleocene fossil vertebrates occur in western North America. From the latest Cretaceous of western North America alone, over 100 vertebrate species are known, representing all major vertebrate clades. Although nonavian dinosaurs represent at most only 20% of this taxonomic diversity, interest in the possible causes and patterns of their extinction far outweighs the interest for the extinction of other species, vertebrate or not (Archibald, 1996). The study of dinosaur extinction is certainly not the most conducive for general understanding of the processes

*darchibald@mail.sdsu.edu

Archibald, J.D., 2014, What the dinosaur record says about extinction scenarios, *in* Keller, G., and Kerr, A.C., eds., Volcanism, Impacts, and Mass Extinctions: Causes and Effects: Geological Society of America Special Paper 505, p. 213–224, doi:10.1130/2014.2505(10). For permission to copy, contact editing@geosociety.org.

of extinction, quite simply because dinosaur population sizes and the environments in which they lived did not greatly favor preservation in the fossil record, with some very notable exceptions. Unfortunately, there are no such notable exceptions for any sites near the end of the Cretaceous.

Can any evidence be offered relating to the question of what the dinosaur record says about extinction scenarios, which the organizers of this volume posed as a topic to be addressed. The answer is a qualified "yes." Specifically, the two major issues to be explored are: (1) What is the nonavian dinosaur record throughout the later Cretaceous, approximately the last 10 m.y.? (2) What is the nonavian dinosaur record like near the end of the Cretaceous, say the last half-million years leading up to the Cretaceous-Paleogene boundary? The emphasis here is by necessity on these records and less on the possible extinction scenarios, quite simply because it is easier to examine what the nonavian dinosaur record is rather than why it is this way. The questions of why are examined in the instances in which the patterns of change in nonavian dinosaur taxonomic diversity seem to address extinction scenarios.

THE LAST TEN MILLION YEARS

Most exposures in Dinosaur Provincial Park, Alberta, Canada, range from 76 to 74 m.y. old, representing the later part of the late Campanian portion of the Late Cretaceous (Currie and Koppelhus, 2005). In western North America, terrestrial, geological sections based on vertebrate fossils of this age also are known as the Judithian after the Judith River Formation, first recognized in Montana, a partial equivalent of the Dinosaur Park Formation in Canada.

Ornithischian and saurischian dinosaurs both occur commonly in the late Campanian (Judithian)–aged sediments of Dinosaur Park. Saurischians are limited to theropods, the group including both birds and the mostly large to extremely large ground-dwelling, bipedal, mostly carnivorous dinosaurs. Sauropod saurischians do not make it this far north in the Late Cretaceous, being known only as far north as southern Wyoming. Thus, the Dinosaur Park nonavian dinosaur fauna can best be grouped as ornithischians and theropods, in which ornithischians outnumber theropods by four to one. Among ornithischians, ~50% are duckbilled hadrosaurs, 25% are horned ceratopsians, and the remaining 25% mostly represent armored ankylosaurs and bone-headed pachycephalosaurs. In the park, the ornithischians comprise three genera of ankylosaurs, each with one species, three species of the pachycephalosaur genus *Stegoceras*, six species of ceratopsians belonging to four different genera, and eight hadrosaurs arrayed in six different genera. Hypsilophodontidae (a paraphyletic group including basal neornithischians and basal ornithopods) is known only from isolated teeth in the park and thus cannot be clearly assigned to a genus or species (Table 1; references in Currie and Koppelhus, 2005; Archibald, 2011).

Theropods make up less than 10% of recovered fossils but are taxonomically quite diverse. Of the 14 families (39 genera) of dinosaurs recognized in the park, 42% of these families (40% of the genera) are theropods (excluding birds). Thus, what theropods lack in numbers of specimens some 75 m.y. ago (Ma), they make up in diversity (Currie, 2005). This pattern of low specimen abundance but higher taxonomic diversity is echoed some 10 m.y. later, just before nonavian dinosaur extinction.

It would be best to possess good samples of nonavian dinosaur taxa at half-million-year intervals, but this is unfortunately not the case in North America. To have comparably well-sampled dinosaur faunas, we must jump to the latest Cretaceous, or the late Maastrichtian, which in North America is also called the Lancian, the terrestrial vertebrate equivalent of this age, named after the type Lance Formation of eastern Wyoming. The taxonomically most diverse Lancian-aged nonavian dinosaur fauna, with 18 species, is perhaps known from the type Lance Formation. Its upper limit is essentially the Cretaceous-Paleogene boundary at roughly 66 Ma. The base of the Lancian at 69 Ma is less well constrained (see Cifelli et al., 2004; Wilson, 2005).

Comparing the Dinosaur Park and Lance Formations, the former represents about 2 m.y. and the latter between 2 and 3 m.y., respectively. Being terrestrial, largely fluvially derived units, we must keep in mind that although both formations span the times indicated, they do not preserve events through the entirety of these interval, but rather represent time-averaged packets. Most terrestrially derived fossils are from such time-averaged occurrences. Both formations represent similar depositional histories, preserving floodplain and coastal fluvial deposits draining easterly into the Pierre or Western Interior Seaway. The western continent of Laramidia lay to the west, and the eastern continent of Appalachia lay to the east of this seaway (Archibald, 1996), transgressing and regressing some four times during the last 35 m.y. of the Late Cretaceous (Fig. 1). The last incarnation of this epicontinental sea at its greatest width might have reached 1000 km wide and up to five times as long. Such seaways inundated almost all continents during the Late Cretaceous, equivalent to an area probably approximately the size of modern Africa. Nothing comparable in size exists today, although Hudson Bay, the North Sea, the Baltic Sea, and the Red Sea all have been proposed as smaller examples.

In a comparison of the nonavian dinosaur faunas, the Lance Formation, although quite rich, does not match the older Dinosaur Park Formation nonavian dinosaur fauna: Dinosaur Park has 39 species, whereas the Lance Formation has 18 species. This means a drop of over 50% species abundance during the last 10 m.y. of the Cretaceous in North America (Table 1). Additional Judithian- and Lancian-aged nonavian dinosaur faunas in western North America add further support in showing this decline is real. Other major Judithian-aged formations in the northern western interior are the Judith River and Two Medicine Formations in central Montana; the other major formation of Lancian age is the Hell Creek Formation of eastern Montana and western North and South Dakota. Adding these additional records to augment the Dinosaur Park and Lance Formations, we find a loss of dinosaur species of almost 50%, still a considerable decline (Table 1).

TABLE 1. GENERIC AND SPECIES COUNTS OF LATE CAMPANIAN THROUGH LATE MAASTRICHTIAN NONAVIAN DINOSAURS FROM NORTHERN LARAMIDIA (CURRIE, 2005; EBERTH ET AL., 2005; RYAN AND EVANS, 2005; LONGRICH AND CURRIE, 2009; WEISHAMPEL ET AL., 2004)

Dinosaur Park Formation (Judithian, late Campanian)*		Horseshoe Canyon Formation (Edmontonian, Campanian–Maastrichtian)		Lance Formation (Lancian, late Maastrichtian)[#]		Change in no. of species (with J, T, M, and D) between late Camp. and late Maast.
Genus	No. species (J and T)	Genus	No. species	Genus	No. species (M and D)	
Theropoda		**Theropoda**		**Theropoda**		**–9 (–4)**
Albertosaurus (J)	(1)	Albertonykus	1	Albertosaurus	1	
Bambiraptor (T)	(1)	Albertosaurus	1	*Caenagnathus* (D)	(1)	
Caenagnathus	2	Atrociraptor	1	*Chirostenotes* (M, D)[§]	(1)	
Chirostenotes	1	cf. *Dromaeosaurus*	1	Dromaeosaurus	1	
Daspletosaurus	2	*Epichirostenotes*[§]	1	*Elmisaurus* (M, D)[§]	(1)	
Dromaeosaurus	1	Ornithomimus	1	*Nanotyrannus* (M)	(1)	
Dromiceiomimus	1	cf. *Paronychodon*[†]	1	Ornithomimus	1	
Elmisaurus[§]	1	cf. *Richardoestesia*[†]	1	*Paronychodon*[†]	1	
Gorgosaurus	1	Struthiomimus	1	*Richardoestesia*[†]	1	
Hesperonychus	1	Troodon	1	Saurornitholestes	1	
Ornithomimus	1			*Struthiomimus* (D)	(1)	
Paronychodon[†]	1			Troodon	1	
Richardoestesia[†]	1			Tyrannosaurus	1	
Saurornitholestes	1					
Struthiomimus	1					
Therizinosaurid	1					
Troodon	1					
Ankylosauria		**Ankylosauria**		**Ankylosauria**		**–1(–2)**
Edmontonia (1 is from J)	1(+1)	Edmontonia	1	Ankylosaurus	1	
Euoplocephalus	1	Euoplocephalus	1	Edmontonia	1	
Panoplosaurus	1					
Euornithopoda		**Euornithopoda**		**Euornithopoda**		**+1(0)**
Hypsilophodontid	1	Parksosaurus	1	Bugenasaura	1	
Orodromeus (T)	(1)			Thescelosaurus	1	
Hadrosauridae		**Hadrosauridae**		**Hadrosauridae**		**–6 (–9)**
Brachylophosaurus	1	Edmontosaurus	1	Edmontosaurus	2	
Corythosaurus	1	Hypacrosaurus	1			
Gryposaurus	2	Saurolophus	1			
Hypacrosaurus (T)	(1)					
Lambeosaurus	2					
Maiasaura (T)	(1)					
Parasaurolophus (1 is from T)	1(+1)					
Prosaurolophus	1					

(*Continued*)

TABLE 1. GENERIC AND SPECIES COUNTS OF LATE CAMPANIAN THROUGH LATE MAASTRICHTIAN NONAVIAN DINOSAURS FROM NORTHERN LARAMIDIA (CURRIE, 2005; EBERTH ET AL., 2005; RYAN AND EVANS, 2005; LONGRICH AND CURRIE, 2009; WEISHAMPEL ET AL., 2004) (*Continued*)

Dinosaur Park Formation (Judithian, late Campanian)*		Horseshoe Canyon Formation (Edmontonian, Campanian–Maastrichtian)		Lance Formation (Lancian, late Maastrichtian)[#]		Change in no. of species (with J, T, M, and D) between late Camp. and late Maast.
Genus	No. species (J and T)	Genus	No. species	Genus	No. species (M and D)	
Pachycephalosauria		**Pachycephalosauria**		**Pachycephalosauria**		**−1 (+1)**
Stegoceras	3	cf. *Colepiocephale*	1	Pachycephalosaurus	1	
		Sphaerotholus	1	Stegoceras (M, D)	(1)	
				Sphaerotholus (M)	(1)	
				Stygimoloch	1	
Ceratopsia		**Ceratopsia**		**Ceratopsia**		**−3 (−6)**
Achelosaurus (T)	(1)	Anchiceratops	1	Diceratops (or Diceratus)	1	
Avaceratops (J)	(1)	Arrhinoceratops	1	Leptoceratops	1	
*Anchiceratops***	1	Eotriceratops	1	Triceratops	2[††]	
Centrosaurus	1	Montanoceratops	1			
Chasmosaurus	3	Pachyrhinosaurus	1			
Einiosaurus (T)	(1)					
Leptoceratops	1					
Styracosaurus	1					
Total numbers of species	**39 (49)**		**23**		**18 (25)**	**−19 (−24)**

*Additional genera from the Judith River (J) and Two Medicine (T) Formations of central Montana.
[†]*Paronychodon* and *Richardoestesia* are tooth taxa and accordingly not recognized by some workers.
[§]There is no consensus as to how many genera are represented by *Chirostenotes*, *Elmisaurus*, and *Epichirostenotes*. All are recognized here.
[#]Additional genera from the Hell Creek Formation of eastern Montana (M) and western part of the Dakotas (D).
**Found south, outside of Dinosaur Park.
[††]The number of species of *Triceratops* is uncertain. I use the higher number of two. Further, according to Scannella (2009), *Torosaurus latus* from the Lance Formation represents older individuals of *Triceratops*.

Some have argued that the Dinosaur Park Formation and Lance Formation are not ecologically equivalent but have offered no data to support this claim, going so far as to call such an assessment absurd (Fastovsky et al., 2005). Others have even attempted without success to statistically discount this very real decline (Fastovsky et al., 2005; Wang and Dodson, 2006; Lloyd et al., 2008). In agreement with the results presented here, a study by Barrett et al. (2009) compiled genus-richness estimates for three Mesozoic dinosaur clades, noting a marked decline in dinosaur generic richness in the final stages of the Cretaceous, which indicates that the clade decreased in diversity for several million years before its final extinction at the Cretaceous-Paleogene boundary. The authors of this study show that many fluctuations in dinosaur diversity through the Mesozoic found in previous studies are likely not real but are artifacts of fossil preservation. These authors further found that when biases in fossil preservation are taken into account, several genuine diversity signals are present; the most intriguing of these, according to these authors, shows that both ornithischian and theropod dinosaur lineages began to decline several million years prior to the Cretaceous-Paleogene extinction event, again in accord with the results discussed here.

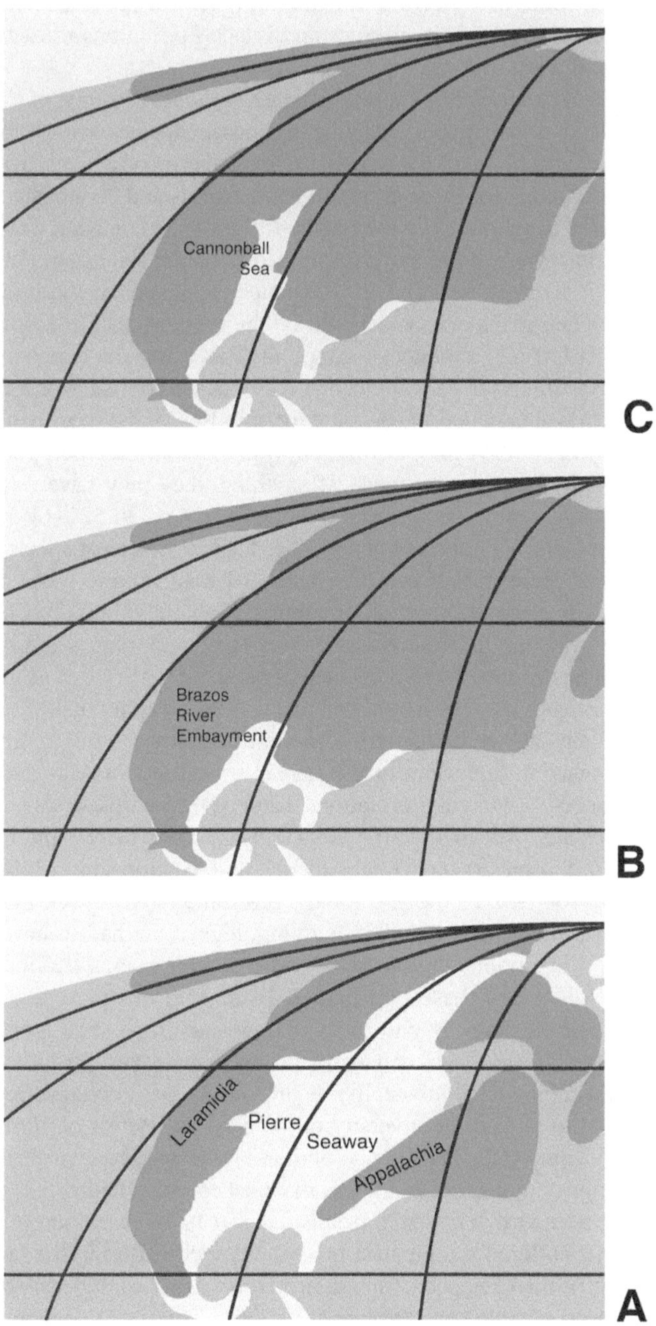

In 2012, Brusatte et al. (2012) calculated what they termed the morphological disparity or anatomical variability found in seven major dinosaur groups during the latest Cretaceous. They found both geographic and clade-specific heterogeneity. The morphological disparity declined for large-bodied herbivores (ceratopsids and hadrosauroids) and some North American taxa during the final two stages of the Cretaceous. At the same time, carnivorous dinosaurs, mid-sized herbivores, and some Asian taxa did not decrease in morphological disparity. These authors argued that the pattern of Late Cretaceous dinosaur evolution was complex and that there was no universal biodiversity trend. As with earlier studies, Brusatte et al. (2012) found that some dinosaur groups endured long-term declines in morphological variability before their final extinction at the end of the Cretaceous.

The data presented here as well as by Barrett et al. (2009) and Brusatte et al. (2012) indicate that the Late Cretaceous vertebrate fossil record in western North America is quite good and by all measures is reliable in providing proportional changes, if not absolute changes, between Judithian and Lancian, and Lancian and Cretaceous-Paleogene boundary sites. Nevertheless, Fastovsky et al. (2005, e75) repeated one of the common canards against comparison of the Dinosaur Park and Lance Formation nonavian dinosaur faunas: "There is no reason to suppose that the richest formations of each time interval are spatially equivalent or preserve faunas with equal fidelity." If by a possible lack of "spatial equivalence" of the rock units yielding these dinosaurs bones, they mean the two are not equivalent in areal exposure, they are correct, but certainly not in the way they implied. The argument here basically is that if you have more exposure, everything else being equal, you should also have more species of dinosaurs preserved. Following their reasoning, if the much greater taxonomic richness of the Dinosaur Park Formation could be attributed to a greater geologic exposure, these authors might have a case. Such is in fact, not the case.

Eberth (2005, p. 58) noted that "the bedrock exposures in the [Dinosaur Provincial] Park comprise Canada's largest area of badlands, covering approximately 75 km^2." The Dinosaur Park

Figure 1. (A) North American biogeography during the latter part of the Cretaceous showing the major western continent of Laramidia and the eastern continent of Appalachia separated by the epicontinental Pierre Seaway. (B) The same region near the Cretaceous-Paleogene boundary showing the Brazos River Embayment. (C) The same region near the mid-Paleocene showing the Cannonball Sea (after Archibald, 2011).

Formation comprises a large portion of this area, and as noted above, has produced 39 species of nonavian dinosaur. More broadly, the Dinosaur Park Formation in Alberta and Saskatchewan plus the equivalently aged Judith River and Two Medicine Formations to the south in Montana are mapped as covering something like 48,000 km^2 (Archibald, 2011). Doing this, an additional 10 nonavian dinosaur taxa can be added the 39 already known in and near Dinosaur Provincial Park, giving a total of 49 late Campanian (Judithian) nonavian dinosaurs from these three formations. By comparison, the late Maastrichtian (Lancian)-aged Lance Formation of Wyoming and the Hell Creek Formation of eastern Montana and western North and South Dakota are mapped as covering at least 57,000 km^2; yet at most five additional dinosaur species can be added, yielding perhaps 25 species of nonavian dinosaurs for this interval (Archibald, 2011). Clearly, the late Campanian (Judithian) nonavian-dinosaur–bearing formations in western North America have less exposure than late Maastrichtian (Lancian) nonavian dinosaur-bearing formations in western North America (Fig. 2). Although the late Campanian faunas are known from 16% fewer exposures than those faunas from the late Maastrichtian faunas, the former have produced almost double the number of species. Barrett et al. (2009) echoed such a percentage drop in dinosaur diversity.

The second claim by Fastovsky et al. (2005), that the dinosaur faunas of the Judithian and Lancian are not preserved with equal fidelity, cannot be justified other than possibly by statistical

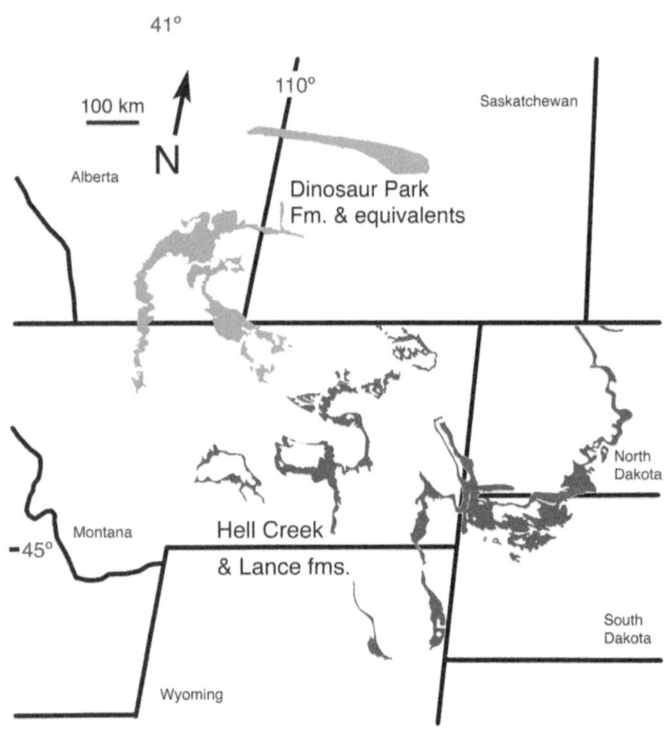

Figure 2. The mapped area of the late Campanian (ca. 75 Ma) Dinosaur Park Formation and its equivalents (48,000 km^2) compared to the mapped areas of late Maastrichtian (ca. 66 Ma) Lance and Hell Creek Formations (with 57,000 km^2; after Archibald, 2011).

machinations. The issue of whether the Dinosaur Park and age-equivalent formations and the Lance and age-equivalent formations are equally rich in the number of identifiable specimens that each has yielded cannot be addressed as definitively as the question concerning "spatial equivalence." There are no complete lists of the identifiable specimens that have been removed over the past 100 yr and more from these units. Many museums around the world have nonavian dinosaurs from especially the Lance and Hell Creek Formations. It is fair to say that both sets of formations have yielded countless nonavian dinosaur specimens, thus it is up to naysayers such as Fastovsky and his colleagues to demonstrate that there is a significant preservational bias between these two sets of formations. Barring this, the argument of preservational bias between the Dinosaur Park and Lance Formations must be treated as a red herring.

Yet another clear pattern of change in the last 10 m.y. of the Cretaceous shows that the decline in dinosaur species abundance is most certainly real. This concerns the patterns of change over the last 10 m.y. for more common hadrosaurids and ceratopsians compared to the far rarer theropods. If the Lance nonavian dinosaur fauna were more poorly sampled than the Dinosaur Park fauna, it would show first in a depauperate nonavian theropod fauna. The rarer theropods should be the first to disappear from the record, if it is a matter of sampling. This is decidedly not the case. For Dinosaur Park, nonavian theropods comprise ~44% of the species abundance and account for only 10% of the specimens. By the time of the Lance Formation, whereas nonavian theropods are equally rare as specimens, if anything, they may have relatively increased taxonomically, ranging from 44% to 52% of the nonavian dinosaur species abundance. Their numbers of species, however, declined between 32% and 53% compared to those of theropods from the Dinosaur Park faunas (Table 1). This is further strong evidence indicating that comparative sampling is not an issue. The decline of nonavian theropods in the last 10 m.y. of the Cretaceous in western North America is most certainly real.

The most-striking pattern of decline concerns the most common nonavian dinosaurs in the faunas from the Dinosaur Park and Lance–Hell Creek Formations, namely, the hadrosaurids and ceratopsians. Recall that for the Dinosaur Park fauna, among the ornithischians, hadrosaurs (50%) and ceratopsians (25%) account for 75% of the specimens. For Lance–Hell Creek Formations, this percentage is, if anything, higher, yet hadrosaurids declined taxonomically by 80% and the ceratopsians declined by 40% between these two faunas. Even with this precipitous taxonomic decline, by end of the Cretaceous in western North America, hadrosaurids and ceratopsians remain very common as specimens. Unequivocally, hadrosaurids and ceratopsians declined in taxonomic diversity in the western interior of North America during the last 10 m.y. of the Cretaceous. These patterns of change demonstrate that it was the most common hadrosaurids and, to a lesser extent, ceratopsians rather than the rarer theropod taxa that suffered the greatest taxonomic decline during the last 10 m.y. of the Cretaceous in western North America. No statistical special pleading can explain away these patterns of decline.

Determining the causes of these taxonomic declines is harder to do. A possibility noted earlier concerns the fact that the deposition of the Lance and Hell Creek Formations occurred as part of the last and most massive regression of the Pierre Seaway leading to near the Cretaceous-Paleogene boundary. The fossil vertebrate evidence available for the four transgressions and regressions in North America during the 35 m.y. of the Late Cretaceous suggests that during times of transgressions, freshwater vertebrates suffered but land-dwelling species, including nonavian dinosaurs, fared better, even showing a taxonomic increase. The opposite occurred during regressions, and, at best, land-dwelling species show taxonomic stasis, such as for the Dinosaur Park Formation, or decline during the much greater regression that occurred when the Lance–Hell Creek vertebrate faunas lived (Archibald, 1996).

Some parallels can be drawn between the changes that occurred between the Dinosaur Park and Lance–Hell Creek nonavian dinosaur faunas and differences we see in extant mammalian biotas. The single Lance–Hell Creek genus of hadrosaurid, *Edmontosaurus*, is arguably represented by more identifiable remains than any of the seven hadrosaurids from Dinosaur Park. Although the number of hadrosaurid species dramatically declined in the last 10 m.y. of the Cretaceous, the one or two species that replaced them were quite common. This is similar to the pattern of African savannah ungulates compared to ungulates of pre-European human invasion of North America. In Africa, there are as many as 60 medium (above 25 kg) to large grazers and browsers (elephants, rhinoceroses, zebra, pigs, hippopotamuses, deer, giraffes, antelope, sheep, goats, cattle-like bovids; Haltenorth and Diller, 1992), whereas in North America, there are only 12 such mammals (peccaries, deer, elk, moose, caribou, sheep, goats, bison, pronghorns; Whitaker, 1980). In Africa, some large migrating mammals such as zebras and wildebeest have prodigious numbers, but other herbivorous species are not uncommon. The migratory populations of wildebeest in the Serengeti-Mara have been measured as high as 1.3 million individuals, and single concentrations can range from 10,000 to 20,000 animals but are usually smaller (Kingdon, 1982). Whereas in North America, the bison was far and away the most numerous large grazing mammal when Europeans arrived some 500 yr ago. In the fifteenth century, 60 million bison roamed North America in massive migrations. Hornaday (1886) estimated in one area that a migrating herd reached over four million animals! By 1900, only 1000 animals remained, but this has now grown to over 30,000 individuals (Whitaker, 1980). North America had a large biomass of large ungulates, but it was dominated by a single species. It is possible that a trend toward lower diversity, but still with a large biomass, transpired for large herbivorous nonavian dinosaurs in the waning 10 m.y. of the Cretaceous in North America.

This likely scenario does not explain why the diversity of the carnivorous and scavenging theropods did not decline as dramatically over the same 10 m.y. Although the numbers of potential prey species clearly declined, as suggested in the comparison of Lancian hadrosaurs and North American bison, likely the total numbers of individuals remained similar over the same 10 m.y. Another possibility is that because the vast majority of these theropods are ostrich size or smaller, with a diversity of diets, a decline in the larger hadrosaurid and ceratopsian diversity would not have affected these smaller theropods as dramatically.

Given that the decline in dinosaur taxonomic diversity between the late Campanian (Judithian) and the late Maastrichtian (Lancian) is real, when did these declines occur and why? These questions may be addressed as more intermediate-aged nonavian-dinosaur–bearing sites are studied. One such study by Eberth et al. (2013) examined the dinosaur biostratigraphy of the Edmonton Group in Alberta, Canada, in considerable detail, encompassing beds ranging in age from ca. 72.5 to 65.5 Ma, or the latest Campanian into the early Paleogene. The upper portion, the Scollard Formation, includes the Cretaceous-Paleogene boundary, with the late Maastrichtian (Lancian) portion including the same dinosaur species known further to the south in the Lance and Hell Creek Formations. Below this, there is the Horseshoe Canyon Formation, which spans the Campanian-Maastrichtian (Edmontonian) boundary, including the best-known nonavian dinosaur fauna of this age in the northern part of present-day North America. The Eberth et al. (2013) study included occurrences and biostratigraphic distributions of the nonavian dinosaurs from the Horseshoe Canyon and Scollard Formations, which record the regressive phase of the Bearpaw transgressive-regressive (T-R) cycle and the Fox Hills T-R cycle. Comparing the areas of the Horseshoe Canyon Formation outcrops shown in their figure 1 to those of the older and younger beds shown here in Figure 1, it is clear the Horseshoe Canyon Formation represents considerably less exposure yet has produced 23 genera of nonavian dinosaurs. The middle two columns in Table 1 list these genera. As seen in this table, the taxonomic diversity of nonavian dinosaurs from the Horseshoe Canyon Formation rivals that of the younger Lance–Hell Creek Formations.

Two factors, however, suggest that nonavian dinosaurs are underrepresented from this intermediate interval. First, as noted, the Horseshoe Canyon Formation exposures, although extensive, are far less so than the earlier and later nonavian-dinosaur–producing exposures in northern Laramidia, likely in part yielding this lower taxonomic diversity. Second, when taxa are compared across the three time intervals shown in Table 1, one finds that there may be as many as six genera found both in the older Dinosaur Park and the younger Lance–Hell Creek Formations not known from the intermediate Horseshoe Canyon Formation. This follows the same pattern found for all of North America by Archibald and MacLeod (2007). If these Lazarus taxa are added to those known from the Horseshoe Canyon Formation, this interval would have upwards of 29 nonavian dinosaur genera, surpassing that of the much more greatly exposed late Maastrichtian Lance–Hell Creek Formations, but still considerably less than the late Campanian Dinosaur Park Formation.

Although the dip in dinosaurian taxonomic diversity in the latest Campanian and early Maastrichtian is almost certainly in part artifactual, caused by the nature of the fossil record, some of it is certainly real, suggesting that whatever the cause(s) of the decline between the late Campanian and late Maastrichtian, much of the decline may have begun or occurred by the Campanian-Maastrichtian transition. Eberth et al. (2013) noted that there is no evidence of significant climatic changes in the older, but relatively shorter expanse of the Dinosaur Park Formation, whereas each of the members of the Horseshoe Canyon Formation, being as long as the entire duration of the Dinosaur Park Formation, shows considerable climatic change. Compared to the Dinosaur Park Formation ornithischians, those in the Horseshoe Canyon Formation show longer-term morphological stasis and major climate-change–driven nonavian dinosaur migration into and out of the area. In the Horseshoe Canyon Formation interval, the authors detail climatic shifts involving a shift from warm and wet to cool and seasonally dry and back again, and finally to a warm and seasonally dry climate in the overlying latest Cretaceous Scollard Formation.

Some of these regional climatic changes appear to have been tectonically driven, but some changes may also have been more global in scale. What the results of these global changes might have been is not yet clear, but the longer-term pattern of regression commencing in earnest ca. 73 Ma, with a much shorter interval of transgression during the latest Campanian into the Maastrichtian (Eberth et al., 2013), might have been in part responsible for these climatic fluctuations. This regression also diminished the expanse of low coastal plains that the nonavian dinosaurs inhabited, as well as provided new avenues of intercontinental migrations (Archibald, 1996, 2011). Still later, ca. 67.4 Ma, the first but smallest of the three pulses of Deccan volcanism may have also contributed to climatic fluctuations, with the second, much larger pulse near the Cretaceous-Paleogene boundary contributing even further to climatic change. A smaller third pulse occurred in early Paleogene (Keller et al., 2012).

APPROACHING THE CRETACEOUS-PALEOGENE BOUNDARY

Although there is little question that at least in North America, nonavian dinosaur species abundance declined by at least 50% over the last 10 m.y. of the Cretaceous, the record becomes more problematic as one approaches the Cretaceous-Paleogene boundary. In order to examine this boundary interval, we must utilize the latest Cretaceous Hell Creek Formation and the overlying early Paleocene Tullock Formation of eastern Montana as well as parts of western North and South Dakota. This is because the records of all vertebrates, not just nonavian dinosaurs, have been more thoroughly studied in these areas, the Cretaceous-Paleogene boundary is better known, and at least in eastern Montana, there is a reasonably good vertebrate record in the overlying Paleocene beds.

Several studies have attempted to determine the diversity of nonavian dinosaurs in the Hell Creek Formation. One study (Sheehan et al., 1991) examined what happened to familial-level diversity of nonavian dinosaurs in a three-part partitioning of the Hell Creek Formation, arguing that the pattern was commensurate with sudden extinction. Unfortunately, one could not detect changes below the family level, which was a level too coarse to detect if there had been any decline within the families (Hurlbert and Archibald, 1995).

In some recent studies, Pearson et al. (2001; see also Pearson et al., 2002) identified over 10,000 specimens of very small to very large vertebrates of 61 vertebrate taxa from a 100 m section of Hell Creek Formation exposed mostly in southwestern North Dakota. One of their major conclusions seems well justified, namely, that there is no evidence for a decline in overall vertebrate diversity through the vertical extent of the formation. Problems arise with their second conclusion; there is no evidence for a decline of nonavian dinosaur diversity in the 3 m below the Cretaceous-Paleogene boundary. There is scarce evidence for any vertebrate fossils in at least the last 1.77 m of their section (Fig. 3). The single specimen they identified that constitutes this record is identified as an indeterminate ceratopsid. Of the over 10,000 specimens they collected, there are only three nonavian dinosaurs identified between 2 and 3 m below the Cretaceous-Paleogene boundary. These consist of an indeterminate ceratopsid, an indeterminate hadrosaurine, and the theropod tooth taxon *Richardoestesia.* Moving down section, according to their figures, one encounters five nonavian dinosaur taxa just below 3 m and eight taxa just below 8 m. Does this mean that, counter to their arguments, in fact we see a decline from eight to five to three to one and finally to zero by 1.77 m below the Cretaceous-Paleogene boundary. I think not.

The left side of Figure 3 is after their figure 2, whereas the right side shows their record for nonavian dinosaurs recast as a spindle or minaret-shaped diagram. In the spindle diagram, numbers show the nonavian dinosaur taxa either collected at that level or implied by their lines connecting occurrences below and above. If this record is to be believed, there is relative taxonomic stability until about the 4 m mark, when numbers start to drop, reaching zero before the Cretaceous-Paleogene boundary. This record would seem to indicate a gradual disappearance of dinosaurs in these sections. The authors argue that this is artifactual and that in reality this dinosaur record is commensurate with a catastrophic extinction. This argument mostly derives from the so-called Signor-Lipps effect, which argues that because the fossil record is never complete, we should not expect to have found the first or the last fossil records of any particular species or higher group of organisms. Invoking the Signor-Lipps effect means, however, that one can claim that catastrophic, gradual, or some other pattern of extinction is commensurate with the fossil record. The authors note that possibly the gap near the top of the Hell Creek Formation is a function of depositional conditions associated with the transitions between formations rather than reflecting true terminal Cretaceous faunal diversity, an assessment with which I agree.

If the single indeterminate ceratopsid noted above is discounted, the 1.77 m becomes a 2.37 m fossil gap in this data set. This gap has also sometimes been referred to as the 3 m or 10 ft gap. Some 30 yr ago, this notoriously unfossiliferous gap

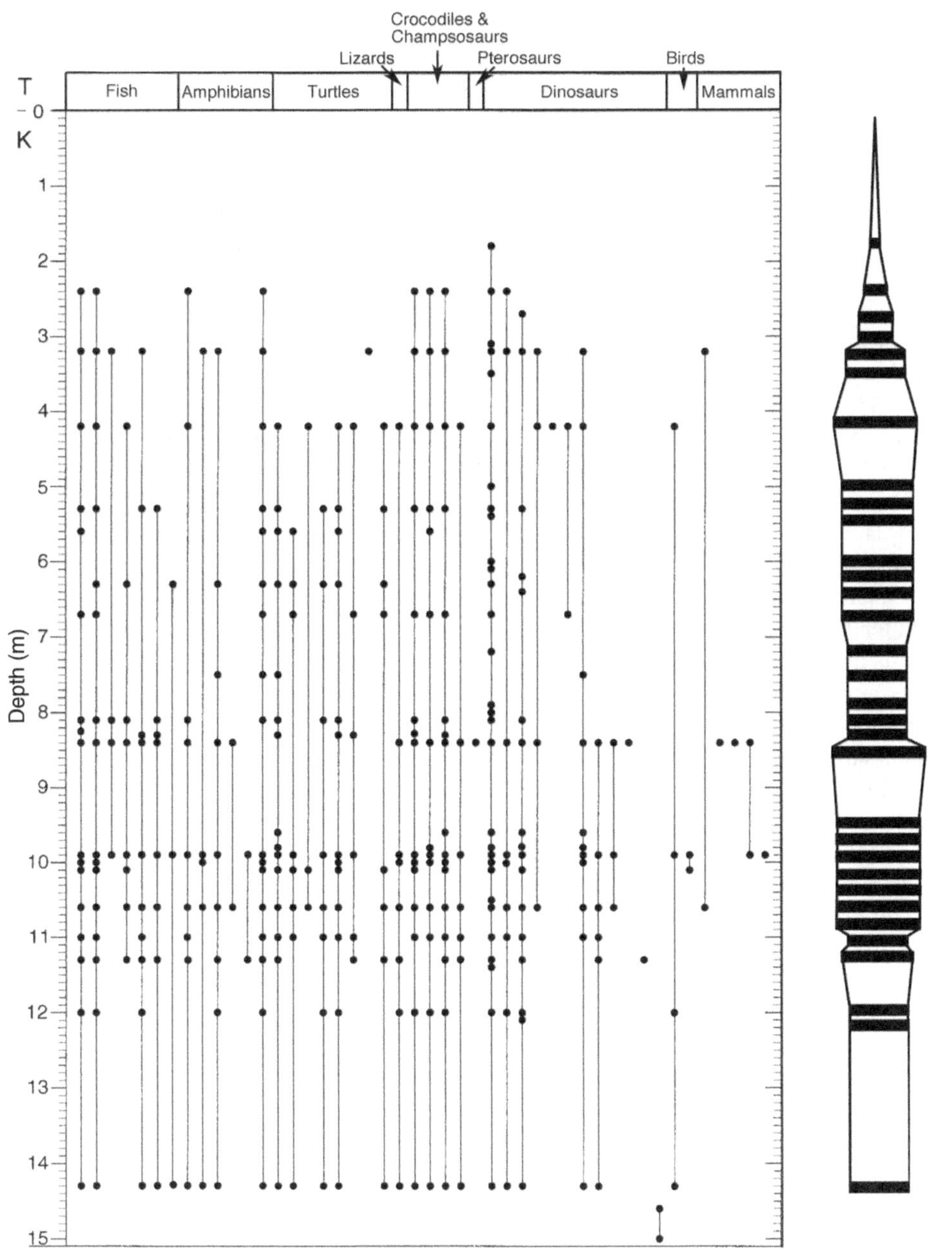

Figure 3. Isolated occurrences of taxa (dots) are identified within clades connected by vertical lines within the upper 15 m of the Hell Creek Formation, exposed mostly in southwestern North Dakota (after Pearson et al., 2001). Spindle diagram to the right shows repeating occurrences of dinosaurs at the various levels and interpolates the presence of a taxon if it was recovered above and below that level (after Archibald, 2011). Note the decline from seven to zero taxa in the last 4 m and the total lack of any vertebrates in the uppermost 1.77 m. K—Cretaceous; T—Tertiary.

was first thought possibly to represent the level at which dinosaur extinction occurred. This gap occurs below the Cretaceous-Paleogene boundary, based on the much more easily and better-sampled plant pollen record and the iridium-enriched level thought to mark the Chicxulub impact. Even before the advent of the impact theory of extinction, it was realized that although not completely barren, it is a generally unfossiliferous zone for vertebrate fossils. Thus, for over 30 yr, no one seriously argued that the base of this gap represented the level of final extinction for nonavian dinosaurs.

Unfortunately, this fact was lost on the authors of a paper in 2011 (Lyson et al., 2011) claiming to show that a chunk of a ceratopsian brow horn found 13 cm below what they identified as the Cretaceous-Paleogene boundary proved (p. 14) "that a gap devoid of non-avian dinosaur fossils does not exist and is inconsistent with the hypothesis that non-avian dinosaurs were extinct prior to the K-T boundary impact event." Even taking the conclusions of their field evidence for this occurrence at face value, it simply represents one more rare occurrence of a bone, dinosaur or otherwise, in this quite unfossiliferous gap in fossil representation near the top of the Hell Creek Formation.

Not surprisingly, this very real gap creates considerable problems in gauging what was happening not just to dinosaurs, but all vertebrates during this nearly barren interval. Some, such as Pearson et al. (2001) and Lyson et al. (2011), have tried to argue this gap away, basically saying that because no decline in vertebrate diversity can be detected, no such decline, especially for dinosaurs, occurred in the 3 m below the Cretaceous-Paleogene boundary. Given that this 3 m (or 2.37 m) gap in the records compiled by Pearson et al. (2001) is by far the largest gap in the upper 12 m of the Hell Creek Formation, such a claim is not warranted. There is no evidence supporting a catastrophic (bang) or gradual (whimper) extinction in these last 2 to 3 m of the Hell Creek Formation.

This unfossiliferous gap is vexing. Why is there such a gap at all? The views vary, and frankly all explanations seem to have some merit, but choosing one over another is problematic. One suggestion is that a lack of channel deposits within this 3 m interval may explain the absence of fossil localities. The vast majority of both small and large fossil vertebrates in the Hell Creek Formation are associated with fluvial or riverlain deposits. These can be the channels themselves, sand bars that breached the channels during flood events, or more rarely floodplain deposits on the margins of the streams. Especially compared to the overlying Paleocene Tullock (or Fort Union as it is referred to in the Dakotas) Formation, the Hell Creek Formation tends to preserve smaller streams or at least muddier streams in its upper reaches, although very large channels certainly are not present. The upper reaches of the Hell Creek Formation for the most part lack larger channel deposits, although most of the sediment still appears to be fluvially related, and thus the rarity of channel deposits to explain the absence of fossil localities in the last 3 m seems to have merit.

Another suggestion is that as a result of the Chicxulub impact at the Cretaceous-Paleogene boundary, the resulting global acid rain fallout likely caused the destruction of microfossils immediately underlying the impact layer (Bailey et al., 2005). At first an appealing idea, the survival of most aquatic vertebrate species belies the idea of a very low pH acid rain. The possibility of the dissolution of most teeth and bones in the upper 3 m of the Hell Creek Formation may yet prove to be true, but not from acid rain. Rather, the carbonaceous and lignitic coals that most often overlie the top of the Hell Creek Formation could have caused the leaching near the top of the Hell Creek Formation. This admittedly more prosaic explanation seems more likely to have been the cause.

Finding and thoroughly exploring other terrestrial Cretaceous-Paleogene sections will help to determine if this gap is a regionally driven phenomenon or is more global in extent. Finding a single bone in a single section, such as done by Lyson et al. (2011), does not begin to address the issue or answer any question about the rapidity of dinosaur extinction. Similarly, although not concerned with the subject of this paper (i.e., nonavian dinosaur extinction), two recent papers dealing with extinctions of other vertebrate groups misrepresent the nature of the vertebrate fossil record near and at the Cretaceous-Paleogene boundary. One study shows the fossil record of lizards and snakes extending exactly to the Cretaceous-Paleogene boundary (Longrich et al., 2012). The second, and more egregious in its misrepresentation of the fossil record, shows the record for birds as single dots immediately below the Cretaceous-Paleogene boundary (Longrich et al., 2011). Although some species of snakes, lizards, and birds may have become extinct at the Cretaceous-Paleogene boundary, to my knowledge, there is no record of any of these taxa within 2 to 3 m of the Cretaceous-Paleogene boundary. The problem with all of these studies is the impression of a sudden extinction, when in fact such evidence is completely lacking.

Another approach to examining what was occurring leading up to and across the Cretaceous-Paleogene boundary has used large samples of vertebrate microfossils amassed over almost 40 yr, mostly by crews often under the direction of W.A. Clemens from the Museum of Paleontology at the University of California, Berkeley, working in the late Maastrichtian (Lancian) Hell Creek Formation and early Paleocene Tullock Member of northeastern Montana. One more recent approach by Wilson and colleagues (Hutchison et al., 2004; Wilson, 2005) placed vertebrate localities into a temporal framework utilizing stratigraphic position. Although their work did not address dinosaur records, it is relevant to understanding possible broader climatic changes that certainly would have affected dinosaurs. These authors noted changes in the better-sampled amphibian, turtle, and mammalian paleocommunities approaching the Cretaceous-Paleogene boundary. They compared these data with those for published paleofloral and paleoclimatic data. Echoing the findings of Pearson and colleagues, Wilson and colleagues did not find any evidence of long-term changes within paleocommunities leading up to the Cretaceous-Paleogene boundary, although they did find some changes in relative abundances, taxonomic composition, and body sizes that likely reflect normal responses to background levels of climate change. However, unlike Pearson and colleagues, Wilson and colleagues did find dramatic changes in turtle and mammalian paleocommunities within the last 100 k.y. of the Cretaceous. These authors went on to write that while their results are consistent with a sudden extinction, such as from an extraterrestrial impact, the response was nonlinear, so that long-term causes or multiple short-term causes cannot be rejected. They further wrote that their assessments will improve with increased temporal resolution of the chronostratigraphic framework, improvement of the density of fossil sampling, the incorporation of other fossil taxa such as plants and mollusks, and comparisons with other study areas.

This work shows that during the last 2 m.y. of the Cretaceous, maximum diversities of turtles, mammals, and plants correlate with the maximum latest Cretaceous warming trend, while the drop in taxonomic richness of these three groups correlates to a rapid drop in paleotemperature in the last 100,000 yr of the Cretaceous of the same region (Hutchison et al., 2004; Wilson, 2005). If verified by further studies, these results clearly point to factors driving climatic change well before an extraterrestrial impact that would be expected to have affected dinosaur relative abundances and/or taxonomic diversity in some fashion.

Finally, a more recent approach by Horner et al. (2011) examined in considerable detail the stratigraphic distribution of nonavian dinosaur skeletons and individual bones in the Hell

Creek Formation of northeastern Montana. Whereas up to six genera are known in the lower two thirds based on skeletons and individual bones (*Triceratops*, *Tyrannosaurus*, *Edmontosaurus*, *Thescelosaurus*, *Ornithomimus*, *Pachycephalosaurus*, and *Ankylosaurus*), only three genera (*Triceratops*, *Tyrannosaurus*, *Edmontosaurus*) are known in the upper one third, and this portion is dominated by a single taxon—*Triceratops*. The evidence then, if it shows anything, is that there was a decline in relative abundance, if not dinosaur diversity, in the upper Hell Creek Formation.

CONCLUSIONS

The lack of well-studied, fossiliferous Cretaceous-Paleogene terrestrial sections on all continents except western North American means that the records from these North American locations must for now act as proxy for our knowledge of dinosaur extinction. During the last 10 m.y. of the Cretaceous, in at least the northern part of the western interior of North America, nonavian dinosaur species dropped from 49 to 25, almost a 50% decline. This is the case even though the latest Cretaceous vertebrate-bearing beds are represented by a 16% greater areal extent than beds some 10 m.y. earlier in the same region. This decrease in nonavian dinosaur taxonomic diversity commenced near the Campanian-Maastrichtian boundary and likely continued during the greatest marine regression to occur in the past 250 m.y. and later during phases of Deccan Trap emplacement. Resulting climatic changes and loss of habitat likely contributed to this decline. The nonavian dinosaur record near the Cretaceous-Paleogene boundary, even for the best-known records from North America, remains enigmatic and open to interpretation. Newer studies suggest some decline in at least relative abundance of nonavian dinosaurs as one approaches the Cretaceous-Paleogene boundary, but the cause (or causes) for their final extinction (if it was the case) remains unresolved, although the Chicxulub impact undoubtedly played a major role following upon the effects of longer-term regression and volcanism.

ACKNOWLEDGMENTS

The organizers of the International Conference on Volcanism, Impacts and Mass Extinctions: Causes and Effects—Norman MacLeod, Gerta Keller, and Andrew Kerr—are thanked for their kind invitation to participate and for arranging partial financial support. Alexander Averianov, Hans Dieter-Sues, and Gerta Keller are thanked for reading and providing helpful comments on the manuscript.

REFERENCES CITED

Archibald, J.D., 1996, Dinosaur Extinction and the End of an Era: What the Fossils Say: New York, Columbia University Press, 237 p.

Archibald, J.D., 2011, Extinction and Radiation: How the Fall of the Dinosaurs Led to the Rise of the Mammals: Baltimore, Maryland, Johns Hopkins University Press, 108 p.

Archibald, J.D., and MacLeod, N., 2007, Dinosaurs, extinction theories for, *in* Levin, S.A., ed., Encyclopedia of Biodiversity: Amsterdam, Netherlands, Elsevier, p. 1–9.

Bailey, J.V., Cohen, A.S., and Kring, D.A., 2005, Lacustrine fossil preservation in acidic environments: Implications of experimental and field studies for the Cretaceous-Paleogene boundary acid rain trauma: Palaios, v. 20, p. 376–389, doi:10.2110/palo.2003.p03-88.

Barrett, P.M., McGowan, A.J., and Page, V., 2009, Dinosaur diversity and the rock record: Proceedings of the Royal Society, ser. B, Biological Sciences, v. 276, p. 2667–2674, doi:10.1098/rspb.2009.0352.

Brusatte, S.L., Butler, R.J., Prieto-Márquez, A., and Norell, M.A., 2012, Dinosaur morphological diversity and the end-Cretaceous extinction: Nature Communications, v. 3, p. 804, doi:10.1038/ncomms1815.

Cifelli, R.L., Eberle, J.J., Lofgren, D.L., Lillegraven, J.A., and Clemens, W.A., 2004, Mammalian biochronology of the latest Cretaceous, *in* Woodburne, M.O., ed., Late Cretaceous and Cenozoic Mammals of North America: New York, Columbia University Press, p. 21–42.

Currie, P.J., 2005, Theropods including birds, *in* Currie, P., and Koppelhus, E.B., eds., Dinosaur Provincial Park: A Spectacular Ancient Ecosystem Revealed: Bloomington, Indiana, Indiana University Press, p. 367–397.

Currie, P.J., and Koppelhus, E.B., eds., 2005, Dinosaur Provincial Park: A Spectacular Ancient Ecosystem Revealed: Bloomington, Indiana, Indiana University Press, 647 p.

Eberth, D., 2005, The geology, *in* Currie, P., and Koppelhus, E.B., eds., Dinosaur Provincial Park: A Spectacular Ancient Ecosystem Revealed: Bloomington, Indiana, Indiana University Press, p. 54–82.

Eberth, D., Evans, D.C., Brinkman, D.B., Therrien, F., Tanke, D.H., and Russell, L.S., 2013, Dinosaur biostratigraphy of the Edmonton Group (Upper Cretaceous), Alberta, Canada: Evidence for climate influence: Canadian Journal of Earth Sciences, v. 50, p. 701–726, doi:10.1139/cjes-2012-0185.

Fastovsky, D.E., Huang, Y., Hsu, J., Martin-McNaughton, J., Sheehan, P.M., and Weishampel, D.B., 2005, The shape of Mesozoic dinosaur richness: Reply: Geology, p. e75, doi:10.1130/0091-7613-33.1.e75.

Haltennorth, T., and Diller, H., 1992, Mammals of Africa including Madagascar: London, Collins, 400 p.

Hornaday, W.T., 1886, The extermination of the American bison: Report of the National Museum, 1886–1887, p. 369–548.

Horner, J.R., Goodwin, M.B., and Myhrvold, N., 2011, Dinosaur census reveals abundant *Tyrannosaurus* and rare ontogenetic stages in the Upper Cretaceous Hell Creek Formation (Maastrichtian), Montana, USA: PLoS ONE, v. 6, p. 1–9, doi:10.1371/journal.pone.0016574.

Hurlbert, S.H., and Archibald, J.D., 1995, No evidence of sudden (or gradual) dinosaur extinction at the K/T boundary: Geology, v. 23, p. 881–884, doi:10.1130/0091-7613(1995)023<0881:NSSFSO>2.3.CO;2.

Hutchison, J.H., Holroyd, P.A., Gregory, P., and Wilson, G.P., 2004, The possible role of climate on successive turtle assemblages from the Upper Cretaceous Hell Creek Formation: Journal of Vertebrate Paleontology, v. 24, supplement to 3, p. 73A.

Keller, G., Adatte, T., Bhowmick, P.K., Upadhyay, H., Dave, A., Reddy, A.N., and Jaiprakash, B.C., 2012, Nature and timing of extinctions in Cretaceous-Tertiary planktic foraminifera preserved in Deccan intertrappean sediments of the Krishna-Godavari Basin, India: Earth and Planetary Science Letters, v. 341–344, p. 211–221, doi:10.1016/j.epsl.2012.06.021.

Kingdon, J., 1982, East African Mammals: Volume IIID. Bovids: Chicago, Illinois, University of Chicago Press, 746 p.

Lloyd, G.T., Davis, K.E., Pisani, D., Tarver, J.E., Ruta, M., Sakamoto, M., Hone, D.W.E., Jennings, R., and Benton, M.J., 2008, Dinosaurs and the Cretaceous terrestrial revolution: Proceedings of the Royal Society, ser. B, Biological Sciences, v. 275, p. 2483–2490, doi:10.1098/rspb.2008.0715.

Longrich, N.R., and Currie, P.J., 2009, A microraptorine (Dinosauria-Dromaeosauridae) from the Late Cretaceous of North America: Proceedings of the National Academy of Sciences of the United States of America, v. 106, p. 5002–5007, doi:10.1073/pnas.0811664106.

Longrich, N.R., Tokaryk, T.T., and Field, D.J., 2011, Mass extinction of birds at the Cretaceous-Paleogene (K-Pg) boundary: Proceedings of the National Academy of Sciences of the United States of America, v. 108, p. 15,253–15,257, doi:10.1073/pnas.1110395108.

Longrich, N.R., Bhullar, B.-A.S., and Gauthier, J.A., 2012, Mass extinction of lizards and snakes at the Cretaceous-Paleogene boundary: Proceedings of the National Academy of Sciences of the United States of America, v. 109, p. 21,396–21,401, doi:10.1073/pnas.1211526110.

Lyson, T.R., Bercovici, A., Chester, S.G.B., Sargis, E.J., Pearson, D., and Joyce, W.G., 2011, Dinosaur extinction: Closing the '3 m gap': Biology Letters, v. 7, p. 925–928, doi:10.1098/rsbl.2011.0470.

Pearson, D.A., Schaefer, T., Johnson, K.R., and Nichols, D.J., 2001, Palynologically calibrated vertebrate record from North Dakota consistent with abrupt dinosaur extinction at the Cretaceous-Tertiary boundary: Geology, v. 29, p. 39–42, doi:10.1130/0091-7613(2001)029<0039:PCVRFN>2.0.CO;2.

Pearson, D.A., Schaefer, T., Johnson, K.R., Nichols, D.J., and Hunter, J.P., 2002, Vertebrate biostratigraphy of the Hell Creek Formation in southwestern North Dakota and northwestern South Dakota, *in* Hartman, J.H., Johnson, K.R., and Nichols, D.J., eds., The Hell Creek Formation and the Cretaceous Tertiary Boundary in the Northern Great Plains: An Integrated Continental Record at the End of the Cretaceous: Geological Society of America Special Paper 361, p. 145–168.

Ryan, M.J., and Evans, D.C., 2005, Ornithischian dinosaurs, *in* Currie, P., and Koppelhus, E.B., eds., Dinosaur Provincial Park: A Spectacular Ancient Ecosystem Revealed: Bloomington, Indiana, Indiana University Press, p. 312–348.

Scannella, J., 2009, And then there was one: Synonymy consequences of *Triceratops* cranial ontogeny: Journal of Vertebrate Paleontology, v. 29, supplement to 3, p. 177A.

Sheehan, P.M., Fastovsky, D.E., Hoffman, R.G., Berghaus, C.B., and Gabriel, D.L., 1991, Sudden extinction of the dinosaurs: Latest Cretaceous, upper Great Plains, USA: Science, v. 254, p. 835–839, doi:10.1126/science.11536489.

Wang, S.C., and Dodson, P., 2006, Estimating the diversity of dinosaurs: Proceedings of the National Academy of Sciences of the United States of America, v. 103, p. 13,601–13,605, doi:10.1073/pnas.0606028103.

Weishampel, D.B., Dodson, P., and Osmólska, H., eds., 2004, The Dinosauria (2nd ed.): Berkeley, California, University of California Press, 861 p.

Whitaker, J.O., Jr., 1980, The Audubon Society Field Guide to North American Mammals: New York, Alfred A. Knopf, 937 p.

Wilson, G.P., 2005, Mammalian faunal dynamics during the last 1.8 million years of the Cretaceous in Garfield County, Montana: Journal of Mammalian Evolution, v. 12, p. 53–76, doi:10.1007/s10914-005-6943-4.

Manuscript Accepted by the Society 31 January 2014

Printed in the USA

The Geological Society of America
Special Paper 505
2014

The Pliensbachian–Toarcian (Early Jurassic) extinction: A North American perspective

Andrew H. Caruthers*
Paul L. Smith*
Department of Earth, Ocean and Atmospheric Sciences, University of British Columbia, Room 2020 Earth Sciences Building, 2207 Main Mall, Vancouver, British Columbia V6T 1Z4, Canada

Darren R. Gröcke*
Department of Earth Sciences, Durham University, South Road, Durham DH1 3LE, UK

ABSTRACT

Currently, it is believed that volcanogenic outgassing of CO_2 during Karoo-Ferrar igneous province eruptions caused prolonged global warming and a multiphased extinction event during the Pliensbachian–Toarcian interval of the Early Jurassic. Warmer water temperatures are thought to have caused a release from the methane hydrate reservoir and global marine anoxia in the Early Toarcian (dubbed the Toarcian oceanic anoxic event). Recently, the timing and geographic extent of these events have been questioned, with emphasis placed on regional conditions in the Tethys Ocean area rather than global controls.

Our study compares paleontological and geochemical data from western North America with previously established correlative data in Europe to provide a perspective on the duration, extent, and controlling mechanisms of this Early Jurassic extinction event. Our data indicate that during Pliensbachian–Toarcian time, concurrent with Karoo-Ferrar magmatism: (1) there were six globally correlative intervals of taxonomic diversity decline that constitute evidence of a multiphased extinction event; (2) there was a major disruption in the long-term $\delta^{13}C$ profile (~3‰–7‰), suggesting volcanogenic CO_2 outgassing as a preeminent factor driving global warming and mass extinction; (3) there was a large negative excursion in the Early Toarcian $\delta^{13}C$ profile in two successions on Haida Gwaii (formerly known as Queen Charlotte Islands), Canada, suggesting global methane release; and (4) the northeast Panthalassa Ocean did not contain anoxic water during the Early Toarcian. Although the Toarcian oceanic anoxic event does not appear to have globally affected every marine environment in the same manner, it is possible that local anoxic water masses occurred in restricted basins outside the Tethys Ocean area.

*caruthers.andrew@gmail.com; psmith@eos.ubc.ca; d.r.grocke@durham.ac.uk.

Caruthers, A.H., Smith, P.L., and Gröcke, D.R., 2014, The Pliensbachian–Toarcian (Early Jurassic) extinction: A North American perspective, *in* Keller, G., and Kerr, A.C., eds., Volcanism, Impacts, and Mass Extinctions: Causes and Effects: Geological Society of America Special Paper 505, p. 225–243, doi:10.1130/2014.2505(11).

INTRODUCTION

Our understanding of the Pliensbachian–Toarcian second-order mass extinction is heavily biased toward Europe, but the resolution of remaining questions requires a global perspective. In Europe, the seemingly instantaneous acme of the extinction occurred during the Early Toarcian, at a level correlative with the Tenuicostatum/Serpentinum (Falciferum) Zone boundary (Hallam, 1961, 1987; Benton, 1993; Little and Benton, 1995; Harries and Little, 1999; Aberhan and Baumiller, 2003; Cecca and Macchioni, 2004; Zakharov et al., 2006; Caswell et al., 2009; Dera et al., 2010; Bilotta et al., 2010). One calculation suggests a species extinction intensity that exceeds 90% among ammonites, belemnites, bivalves, and gastropods (Harries and Little, 1999). Numerous benthic and pelagic groups in addition to the molluscs were affected, including radiolarians, foraminifera, calcareous nannoplankton, brachiopods, ostracods, dinocysts, palynomorphs, crinoids, crustaceans, marine reptiles, and fish (Hallam, 1986, 1987; Little and Benton, 1995; Hallam and Wignall, 1997; Harries and Little, 1999; Bucefalo Palliani et al., 2002; Vörös, 2002; Cecca and Macchioni, 2004; Zakharov et al., 2006; Mattioli et al., 2008; Caswell et al., 2009; García Joral et al., 2011; Fraguas et al., 2012; Barrón et al., 2013; Arias, 2013; Goričan et al., 2013). Dera et al. (2010) recently calculated that ammonite species in the northwest Tethyan and Arctic domains suffered losses of 40%–65% at the subzone level and 70%–90% at the zone level. Dera et al. (2010) concluded that the interval from the Pliensbachian–Toarcian boundary to the Tenuicostatum/Serpentinum (Falciferum) Zone was, in fact, the main phase of a multiphased extinction in a sequence of five distinct intervals of biodiversity decline that occurred during the Late Pliensbachian to Late Toarcian (Fig. 1). They also recognized an additional diversity drop in the Early Pliensbachian. Because the main phase of the extinction has been observed across the Tethyan and northwest European Boreal regions, many have postulated that it was global in extent (Zakharov et al., 2006; fig. 4 in Caswell et al., 2009; Dera et al., 2010). However, Dera et al. (2010) suggested that the other, less-intense extinction phases were restricted to the Tethys Ocean area.

It is becoming increasingly accepted that large igneous provinces, and the profound environmental changes induced by their eruptions, have played a central role in causing extinctions throughout the Phanerozoic (Courtillot, 1999; Pálfy and Smith, 2000; Wignall, 2001, 2005; Courtillot and Renne, 2003; Suan

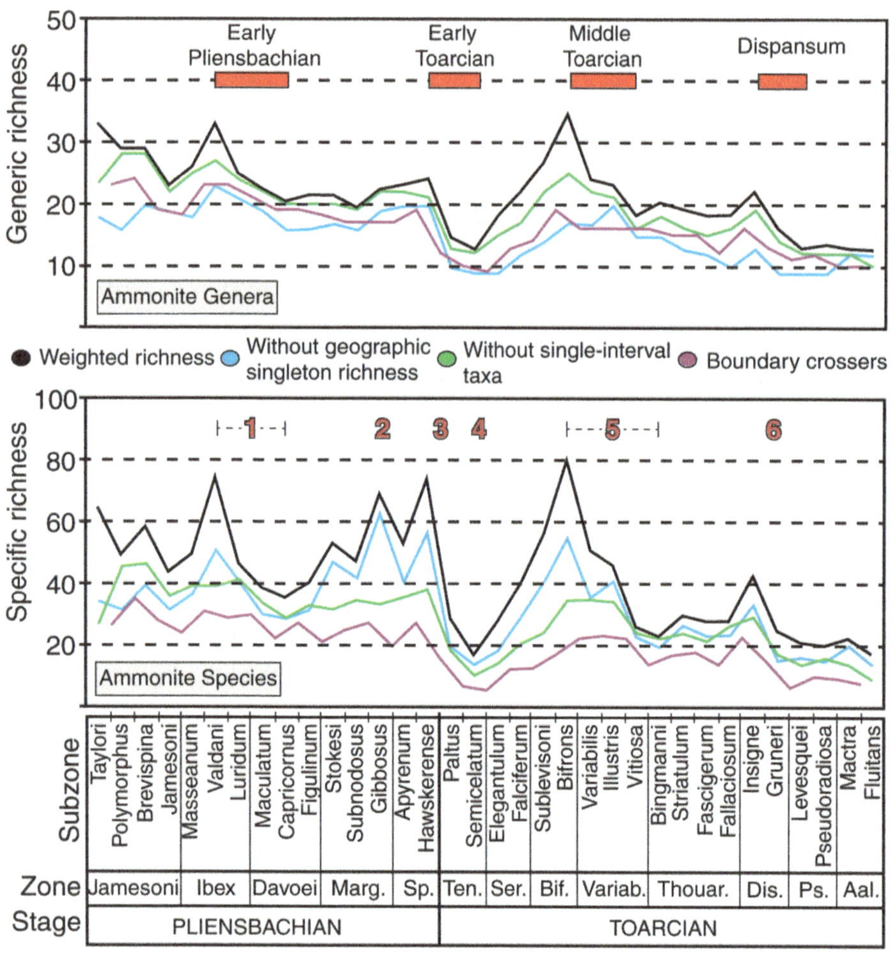

Figure 1. Species and generic ammonite diversity in the Pliensbachian and Toarcian of Europe (modified from Dera et al., 2010; Caruthers et al., 2013). Six major declines in biodiversity are recognized, with the largest being in the Lower Toarcian at the Tenuicostatum/Serpentinum Zone boundary. Major declines in biodiversity are numbered and in red. "Weighted richness" are the total number of taxa measured in that time interval; geographic singletons are taxa that are confined to a specific geographic area; "single-interval taxa" are confined to a specific subzone; and "boundary crossers" are taxa that occur in two separate time intervals. Marg.—Margaritatus, Sp.—Spinatum, Ten.—Tenuicostatum, Ser.—Serpentinum, Bif.—Bifrons, Variab.—Variabilis, Thouar.—Thouarsense, Dis.—Dispansum, Ps.—Pseudoradiosa, Aal.—Aalensis.

et al., 2008; Caswell et al., 2009). Volcanogenic greenhouse gas emissions, particularly CO_2, can cause prolonged global warming that results in warmer water temperatures which in turn are thought to initiate a series of events, as summarized in Figure 2. Such events include: (1) the release of methane from continental shelf hydrates or possibly from thermogenic sources; (2) a disruption in ocean water dynamics involving the thermohaline circulation pattern; (3) increased weathering and erosion on the continents; and (4) ocean stagnation and marine anoxia. This model, dubbed the "Early Jurassic volcanic greenhouse scenario" by Wignall (2005), is partly evidenced in the sedimentary record through elevated concentrations of organic carbon and major perturbations in many isotopic and geochemical systems.

Rift-related tectonic activity during the breakup of Gondwana initiated the eruption of the Karoo-Ferrar large igneous province, which today covers parts of South Africa, Antarctica, South America, Australia, and New Zealand (Riley and Knight, 2001; Fig. 3D). It is this igneous province that has been causally linked to the Pliensbachian–Toarcian extinction (Pálfy and Smith, 2000). Outgassing of volcanogenic CO_2 subsequently destabilized ~5000 Gt of methane hydrate from continental shelf sediments in the Early Toarcian, causing a large ~5‰–7‰ negative carbon-isotope excursion, or CIE (Hesselbo et al., 2000; Beerling et al., 2002). This CIE is recognizable in many successions representing the epicontinental seaway of northwest Europe and the western Tethys Ocean (Fig. 3C). The disruption in seawater geochemistry is recorded in many materials, including bulk organic carbon, carbonate, organic biomarkers, benthic shells, fossil wood, and phytoclasts, which suggests a direct interplay between the oceanic and atmospheric carbon reservoirs (Hesselbo et al., 2000, 2007; Schouten et al., 2000; McElwain et al., 2005; van Breugel et al., 2006; Suan et al., 2008, 2010; Hesselbo and Pienkowski, 2011).

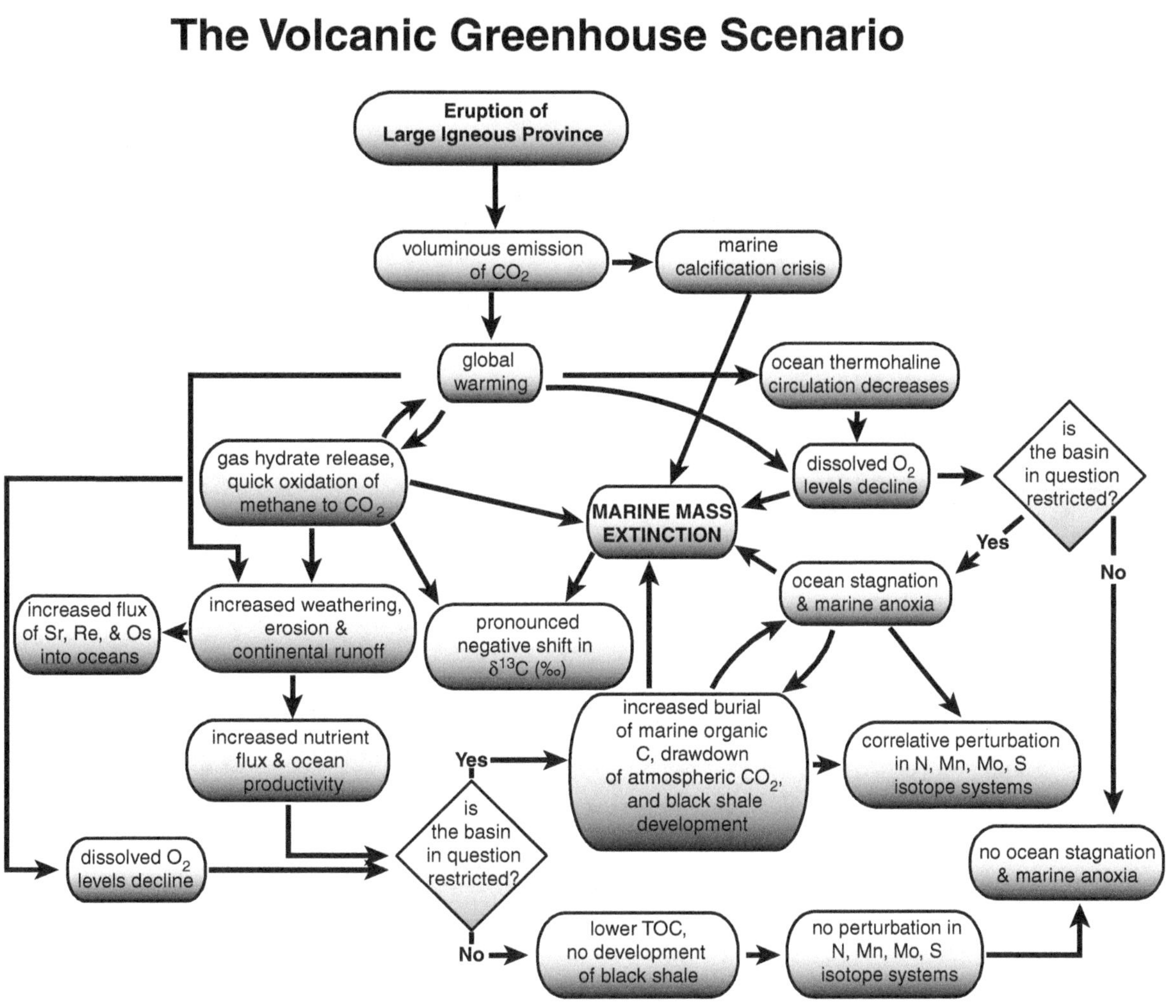

Figure 2. The volcanic greenhouse scenario: a model showing the potential chain of events leading to environmental change and mass extinction (modified from Wignall, 2001, 2005). Note: diamonds are the potential effects of global warming in restricted versus unrestricted basins (as proposed herein). Figure is adapted from Wignall (2005). TOC—total organic carbon.

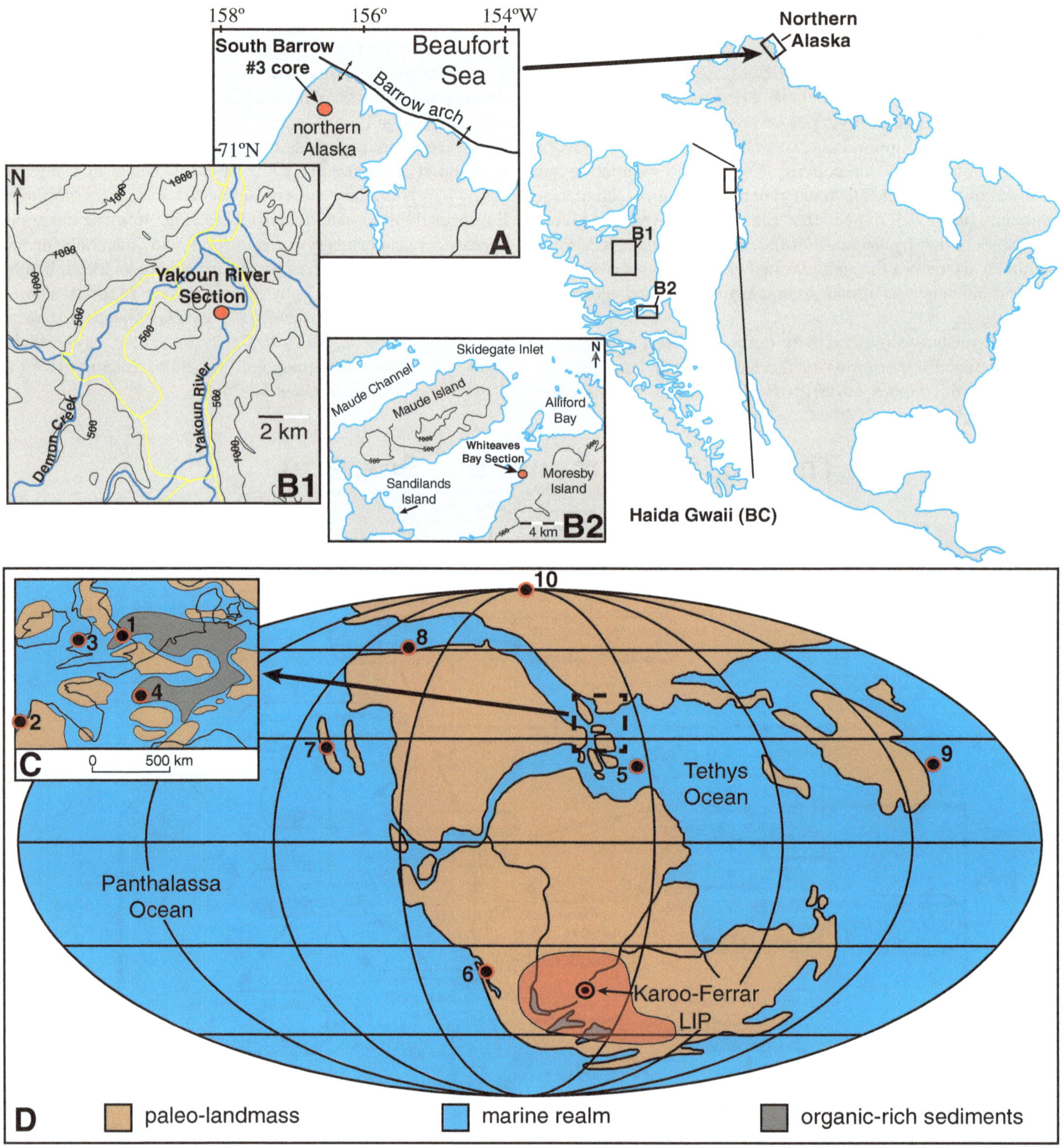

Figure 3. (A–B) Map of present-day western North America showing the locations of stratigraphic sections sampled in this study. (C–D) Early Toarcian paleogeography showing the location of the Tethys and Panthalassa Oceans, the Karoo-Ferrar large igneous province (LIP), and various localities in northwest Europe (after Hesselbo et al., 2007; Hermoso et al., 2009; Caswell et al., 2009), the Mediterranean (after Sabatino et al., 2009), South America (after Al-Suwaidi et al., 2010), and this study. 1—Yorkshire (UK), 2—Peniche (Portugal), 3—Mochras (Wales), 4—Sancerre (France), 5—Monte Mangart (Italian-Slovenian border), 6—Neuquén Basin (Argentina), 7—Wrangellia (Smith, 2006), 8—northern Alaska (Smith, 2006; Imlay, 1981), 9—Japan (Gröcke et al., 2011; Izumi et al., 2012), 10—northern Siberia (Suan et al., 2011).

McElwain et al. (2005) suggested that the large negative CIE was caused by the intrusion of Karoo-Ferrar magmas into coalfields rather than from the destabilization of methane hydrates. Svensen et al. (2007) built on this interpretation, inferring that contact metamorphism by dikes and sills caused a transformation of organic-rich sediments into isotopically light CO_2 and CH_4. However, Gröcke et al. (2009) challenged the thermogenic methane release hypothesis by showing no significant carbon-isotope changes in Karoo Basin coals that have been intruded by dikes. Gröcke et al. (2009) postulated that the transformation of organic-rich sediments and coal into isotopically lighter CO_2 and CH_4 should be reflected in the carbon-isotope ratio of the metamorphosed coals. They also disputed the thermogenic methane hypothesis by suggesting that, because of moisture content and the depth of the coal at the time of intrusion, the Karoo dikes would not have had sufficient energy to generate thermogenic methane.

Although there is geographically extensive carbon-isotope data from the Lower Toarcian, a record for the entire Pliensbachian–Toarcian duration of Karoo-Ferrar magmatism is currently not available. A geochemical record of this scope should shed light on the debate between a continental margin versus thermogenic methane release hypothesis. In addition, a long-term record of the carbon cycle will reveal the impact of volcanogenic outgassing of CO_2 and other greenhouse gases on the climatic, environmental, and oceanic systems. Global warming during the Early Jurassic is thought to have caused: (1) seawater temperatures to increase by ~3–5 °C in the subtropical seas of the Tethys and 8–10 °C in the polar domains (Dera et al., 2011; Dera and Donnadieu, 2012); (2) significant increases in continental weathering rates, as interpreted using oceanic strontium-isotope (Fig. 4; Jones and Jenkyns, 2001; McArthur et al., 2000) and osmium-isotope values (Fig. 4; Cohen et al., 2004); and (3) a major reduction in biocalcification (Mattioli et al., 2004, 2009; Tremolada et al., 2005; Suan et al., 2010) that further reinforced the main phase of marine extinction (Caswell et al., 2009; Bilotta et al., 2010; Dera et al., 2010).

The Early Toarcian interval was also a time of black shale deposition and perturbations in a variety of other geochemical systems in the Tethys Ocean area (Jenkyns, 2010). These are summarized in Figure 4 and include: nitrogen (Jenkyns et al., 2001), manganese (Sabatino et al., 2011), sulfur (McArthur et al., 2008; Gill et al., 2011), and molybdenum (Pearce et al., 2008; McArthur et al., 2008). Jenkyns (2010, and references therein) suggested that the disruption in these geochemical systems is evidence for rapid injection and oxidation of methane, which reduced oceanic O_2 levels and increased organic carbon burial (Fig. 2), referred to as the Toarcian ocean anoxic event. In modern marine environments, boundaries of oxygenation regimes are identified by changes to microbially driven oxidation reactions at various water depths, specific to individual basins (Breck, 1974; Piper and Calvert, 2009). During the Toarcian oceanic anoxic event, the isotopic perturbations in the epicontinental seaway of northwest Europe are thought to have resulted from bacterial degradation of organic matter by nitrates and sulfates in a suboxic-anoxic water column (Jenkyns et al., 2001; McArthur et al., 2008; Jenkyns, 2010).

Recent studies, though, have questioned the timing, evidence for, and geographic extent of black shale deposition and isotopic perturbations associated with the Toarcian oceanic anoxic event (Wignall et al., 2005; van de Schootbrugge et al., 2005; McArthur et al., 2008; Gómez et al., 2008; Bilotta et al., 2010; Gómez and Goy, 2011; García Joral et al., 2011). Specifically, McArthur et al. (2008) dispute the geographic extent and concentration (wt%) of organic carbon in "globally distributed black shale units" based on an interpretation of molybdenum concentration (Mo, ppm) and molybdenum isotope ratios presented in Pearce et al. (2008). These concerns have prompted an alternative hypothesis, referred to here as the "restricted basin scenario," which suggests that the Toarcian oceanic anoxic event was caused by local controls operating in the epicontinental seaway of northwest Europe, where several silled basins were located (McArthur et al., 2008). It is proposed that the deposition of black shales, the presence of ^{12}C-enriched bottom waters, and the generation of geochemical perturbations were in response to euxinic conditions caused by a salinity-driven pycnocline resulting from increased freshwater input (Küspert, 1982; Jenkyns, 1988; Sælen et al., 1996; Schouten et al., 2000; McArthur et al., 2008). In the restricted basin scenario, the negative CIE occurred by diachronous upwelling of ^{12}C-rich bottom water and was therefore not a global event. In relation to the Toarcian oceanic anoxic event, it is possible that both hypotheses are applicable and that simultaneous regional and global controls could have reinforced each other. The formation of Early Toarcian anoxic marine water might not have occurred in the entire global marine environment but may have been limited to restricted basins (diamonds in Fig. 2), which could, of course, have been distributed globally.

This paper summarizes recent research on the Pliensbachian–Toarcian interval of western North America and covers four main objectives:

(1) confirm that the main extinction phase was potentially global and associated with a negative carbon-isotope anomaly in western North America;
(2) determine whether the multiphased extinction evident in Europe is recognizable in North America;
(3) understand the role of Karoo-Ferrar magmatism in shaping the dynamics of the Pliensbachian–Toarcian extinction, specifically with respect to: (a) what the North American record says about the timing of extinction relative to Karoo-Ferrar igneous activity; (b) what the long-term influence of CO_2 outgassing is on the carbon-isotope cycle; and (c) whether the North American record can shed light on the discussion regarding methane hydrate and/or thermogenic methane causes with respect to the negative CIE in the Early Toarcian; and
(4) test the global geographic extent of marine anoxia (of the Toarcian oceanic anoxic event) by determining whether anoxia occurred in the northeastern part of the Panthalassa Ocean.

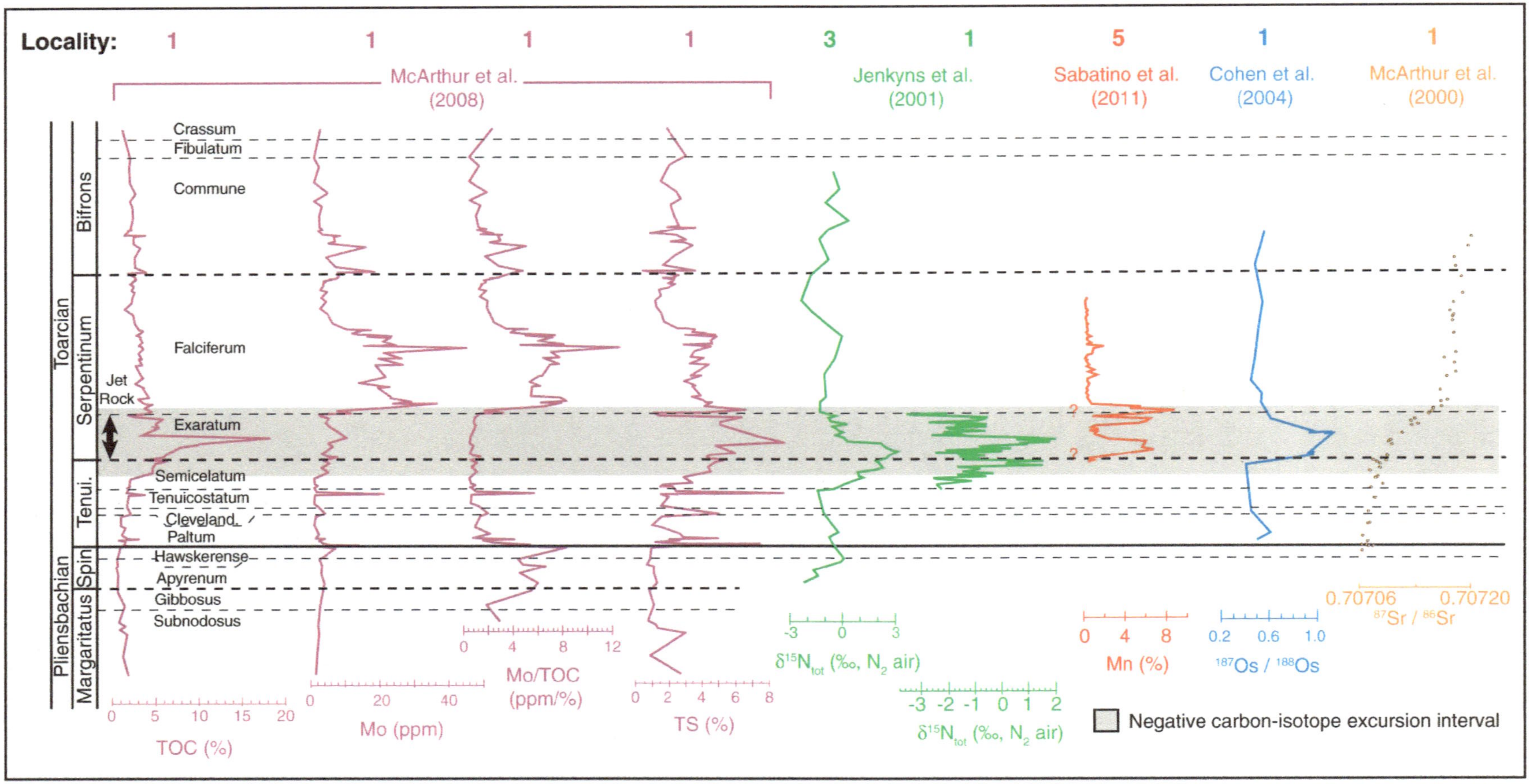

Figure 4. Correlative chemostratigraphy throughout the Late Pliensbachian–Middle Toarcian successions of the Cleveland (locality 1) and Cardigan Bay (locality 3) Basins (northwest Europe) showing geochemical profiles and concentrations of (from left to right): organic carbon, molybdenum, sulfur, nitrogen isotopes, manganese, osmium isotopes, and strontium-isotope values (modified with supplementary data from McArthur et al., 2008). TOC—Total Organic Carbon, TS—Total Sulphur, Spin.—Spinatum, Tenui.—Tenuicostatum, Cleveland.—Clevelandicum. Locality numbers at top refer to paleogeographic locations in Figure 3C and 3D. Note that ages of manganese concentrations from Monte Mangart (locality 5 Italy-Slovenia border) were not given at the zone level by Sabatino et al. (2011).

METHODS

This is a review of our paleontological and geochemical results (Caruthers et al., 2011, 2013; Caruthers, 2013), which were calibrated using the geologic time scale for the Pliensbachian and Toarcian presented by Pálfy et al. (2000b) and subsequently in Gradstein et al. (2012). In western North America, previous studies have established an ammonite-based zonal scheme for Pliensbachian and Toarcian successions (Smith et al., 1988; Jakobs et al., 1994; Smith and Tipper, 1996; Jakobs, 1997). This is a secondary standard zonal scheme (in the sense of Callomon, 1985) that has been calibrated with the primary standard of northwest Europe. In addition, the North American scheme has been calibrated with U-Pb and Ar-Ar dates from interbedded zircon-bearing ash beds and lava flows (Pálfy et al., 1997, 2000a, 2000b), as summarized in Figure 5. However, recent age determinations have indicated that the base of the Toarcian is at 182.7 ± 0.7 Ma (Gradstein et al., 2012) rather than 183.6 +1.7/–1.1 Ma (Pálfy et al., 2000b).

Ammonite and foraminiferal species-level diversities for western North America are described using the metrics established by Foote (2000) and Hammer and Harper (2006) based on data compiled from numerous stratigraphic range charts (Imlay, 1955, 1981; Tappan, 1955; Smith et al., 1988; Jakobs, 1992, 1995, 1997; Thomson and Smith, 1992; Jakobs et al., 1994; Smith and Tipper, 1996; Jakobs and Smith, 1996; Smith et al., 2001; Kottachchi et al., 2002; 2003; Caruthers and Smith, 2012). Stratigraphic range charts are presented as supplemental data in Caruthers et al. (2013).

Geochemical samples presented in Caruthers et al. (2011) and Caruthers (2013) were collected from three successions in western North America. Two successions are located in the Haida Gwaii (formerly the Queen Charlotte Islands, Canada), namely, Whiteaves Bay and Yakoun River (Figs. 3B1 and B2). The third is located in northern Alaska, namely, the South Barrow #3 well core (SB #3 core herein; Fig. 3A). The Jurassic stratigraphy of these areas has been described in a number of publications (Imlay, 1981; Cameron and Tipper, 1985; Tipper et al., 1991; Smith and Tipper, 1996; Jakobs, 1997; Sherwood et al., 2002; Houseknecht and Bird, 2004; Mickey et al., 2006). In this study, samples were collected from the Ghost Creek, Fannin, and Whiteaves Formations in Haida Gwaii and from the Lower Kingak Formation in northern Alaska.

EXTINCTION DYNAMICS

Analysis of 206 ammonite and 242 foraminiferal species ranges was presented in Caruthers et al. (2013) based on numerous Pliensbachian–Toarcian successions in western North America. Our analysis reveals a multiphased extinction that is similar to that observed in the northwest European and Arctic domains (Dera et al., 2010). In western North America, species diversity in both the ammonites and foraminifera declined and reached low points in six separate intervals (Fig. 6A): (1) middle Whiteavesi–middle Freboldi Zones; (2) upper Kunae–lower Carlottense Zones; (3) upper Carlottense–middle Kanense Zones; (4) upper Planulata–lower Crassicosta Zones; (5) middle Crassicosta–Hillebrandti Zones; and (6) lower–middle Yakounensis Zone (Caruthers et al., 2013). These intervals of species decline also seem to correlate approximately with genera-level declines seen in a lower-resolution data set from the Neuquén Basin of South America (Fig. 6C, using data in Riccardi, 2008).

Dera et al. (2010) suggested that the decline in the Early Pliensbachian of Europe was most likely a regional phenomenon, linked to the disappearance of endemic taxa as a result of sporadic northward incursions by Mediterranean taxa (Dommergues and Meister, 2009). They postulated that it was potentially connected to changes in water circulation in the Tethys Ocean (Dera et al., 2009, 2010). However, in both western North America (Caruthers et al., 2013) and in the generic data from South America (Riccardi, 2008), there are similar modest declines in diversity that are coeval with the Early Pliensbachian decline in the Tethys Ocean area (Figs. 6A–6C). This suggests a global controlling mechanism.

Of the five episodes of diversity decline that constitute the multiphased Late Pliensbachian–Toarcian extinction described by Dera et al. (2010), all are distinguishable in western North America (numbers 2–6 in Fig. 6A), and they are also potentially recorded in the low-resolution data set from South America (Riccardi, 2008; Fig. 6C). However, a recent study concerning radiolarian diversity from the Tethys and northeastern Panthalassa Oceans by Goričan et al. (2013) showed a somewhat different pattern. Goričan et al. (2013) showed radiolarian diversity reaching maximum levels in the middle part of the Early Pliensbachian and then declining gradually throughout Late Pliensbachian–Middle Toarcian time (Fig. 6D). This pattern of radiolarian diversity correlates broadly with phases 2–4 of the multiphased extinction scenario and supports the hypothesis of long-term environmental change as a potential controlling mechanism.

Within the Tethys Ocean area and western North America, the main phase of extinction is described as a progressive decline in species diversity that begins just before the Pliensbachian–Toarcian boundary and extends into the Lower Toarcian (at an interval correlative with the Tenuicostatum/Serpentinum Zone boundary of northwest Europe), where diversity reaches minimum values (number 3 in Figs. 6A and 6B). The main phase has therefore now been recorded in three ocean basins, namely, the Panthalassa, Arctic, and Tethys Oceans. During the main phase of extinction in western North America, ammonite and foraminiferal diversity reached minimum values of 5 and 25 species, respectively, in the middle part of the Kanense Zone at an interval that is correlative with the negative CIE recognized by Caruthers et al. (2011; Fig. 6A). The apparent synchronicity between the negative CIE and the most profound diversity minima of the main phase of the extinction suggests that methane release played an important role in escalating the extinction. In addition, recognition of four other correlative intervals of taxonomic decline in three ocean basins occurring throughout the Late Pliensbachian–Toarcian interval also suggests global controls related to the

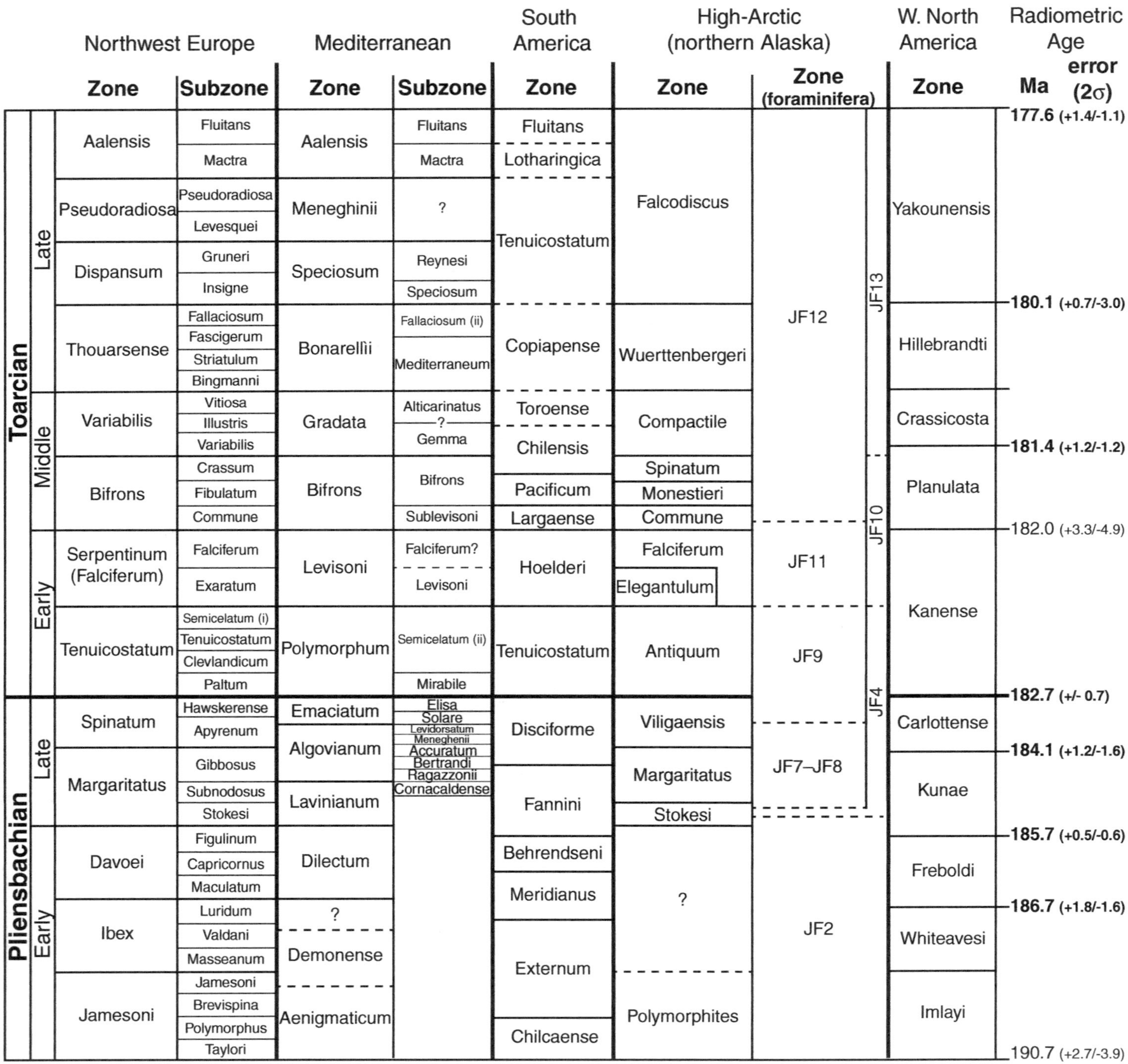

Figure 5. Correlative Pliensbachian–Toarcian time scales for northwest Europe and the Mediterranean (Dean et al., 1961; Schlatter, 1980; Braga et al., 1982; Howarth, 1992; Cariou and Hantzpergue, 1997; Page, 2003); High-Arctic (Zakharov et al., 1997; Nikitenko et al., 2008); western North America (Smith et al., 1988; Jakobs et al., 1994); and South America (von Hillebrandt, 2006). Zones are based on ammonites unless indicated otherwise. Absolute U-Pb and Ar-Ar age determinations are from Pálfy et al. (1997, 2000b) and Gradstein et al. (2012). Absolute age determinations in boldface font have an error range that is less than 5 m.y. and are interpreted as good quality. The Pliensbachian zonal biostratigraphy of the Mediterranean Province refers to areas of Spain (Page, 2003).

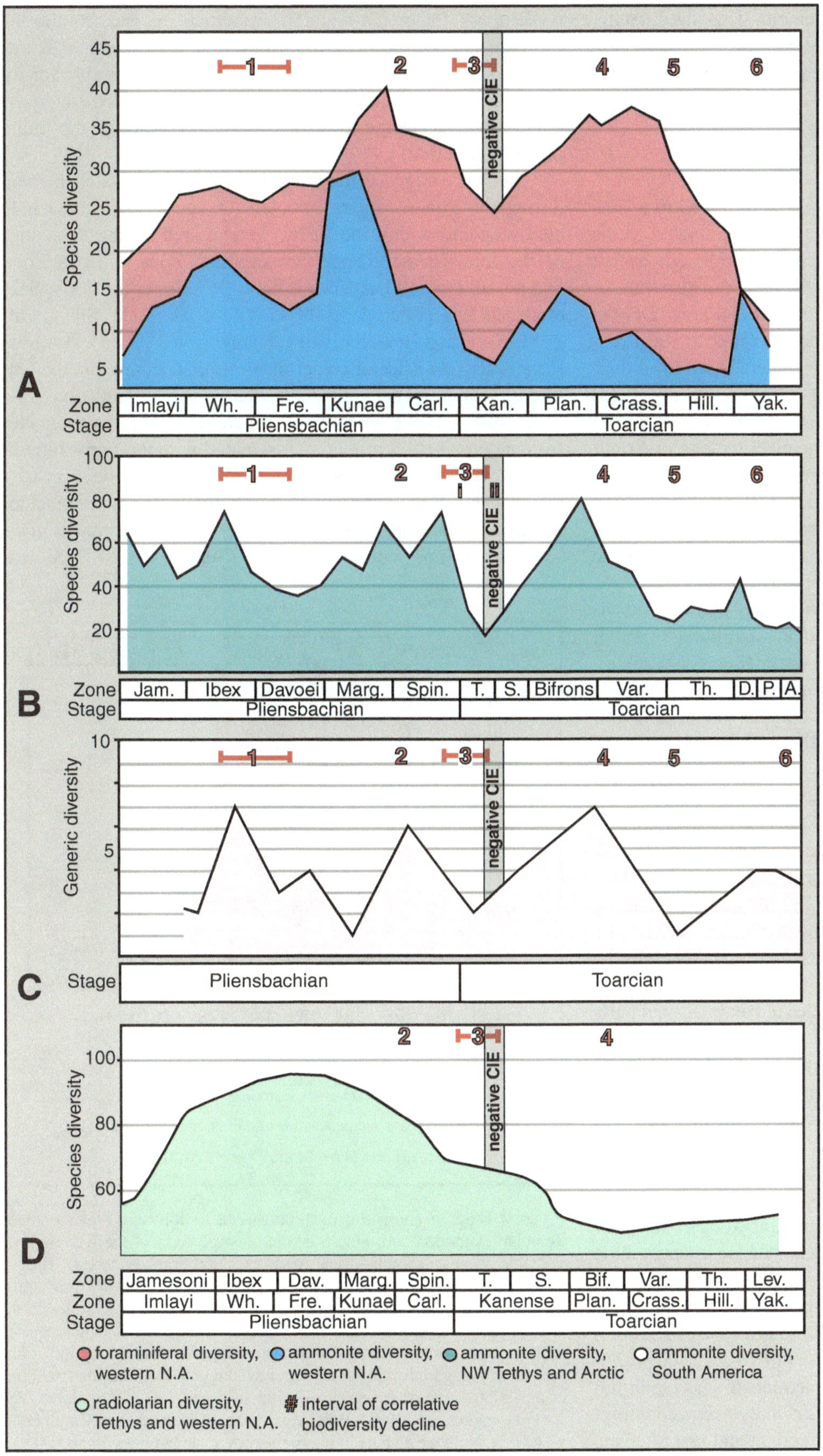

Figure 6. Pliensbachian–Toarcian ammonite, foraminiferal, and radiolarian biodiversity in (A) western North America (Caruthers et al., 2013), (B) northwest Tethys and Arctic domains (Dera et al., 2010), (C) South America (Riccardi, 2008), and (D) Tethys and western North America (Goričan et al., 2013). Figure shows a multiphased event with major declines occurring over six correlative intervals. The main phase of extinction is a large progressive decline that begins just before the Pliensbachian–Toarcian boundary and extends into the Lower Toarcian at an interval coeval with the negative carbon-isotope excursion (CIE). Note: the Middle Toarcian event in Dera et al. (2010) is illustrated here as two separate events (4, 5). Intervals I and II in (B) show approximately the extinction intervals identified by Harries and Little (1999) and Caswell et al. (2009). N.A.—North America, Wh.—Whiteavesi, Fre.—Freboldi, Carl.—Carlottense, Kan.—Kanense, Plan.—Planulata, Crass.—Crassicosta, Hill.—Hillebrandti, Yak.—Yakounensis, Jam.—Jamesoni, Marg.—Margaritatus, Spin.—Spinatum, T.—Tenuicostatum, S.—Serpentinum, Bif.—Bifrons, Var.—Variabilis, Th.—Thouarsense, D.—Dispansum, P.—Pseudoradiosa, A.—Aalensis, Lev.—Levisoni. Negative CIE—negative carbon-isotope excursion interval in Caruthers et al. (2011, and references therein). Figure is adapted from Caruthers et al. (2013).

Early Jurassic volcanic greenhouse scenario. This significantly expands the suspected duration of the influence of the Early Jurassic volcanic greenhouse scenario and underlines the complexity of the associated environmental changes (Fig. 2).

KAROO-FERRAR MAGMATISM

Currently, the multiphased extinction is attributed to a variety of paleoenvironmental changes that could be related to the Early Jurassic volcanic greenhouse scenario, although except for the main phase (number 3 in Fig. 6), the extinction phases were thought to be restricted to the Tethys Ocean area (Dera et al., 2010). Paleoenvironmental influences related to the Early Jurassic volcanic greenhouse scenario include: (1) sea-level fluctuation affecting restricted basins; (2) saline stratification; (3) warming (or cooling) of seawater; (4) changing water current dynamics; and (5) perturbation in geochemical cycles (McArthur et al., 2000, 2008; Bailey et al., 2003; Suan et al., 2008, 2010; Dera et al., 2010; Dera and Donnadieu, 2012). However, if the six extinction phases were global in extent, it is plausible that the eruption of the Karoo-Ferrar large igneous province was the underlying trigger and controlling factor.

The approximate timing of the six phases of extinction is shown in Figure 6. The middle Whiteavesi–middle Freboldi Zone phase occurred at ca. 186 Ma, the upper Kunae–lower Carlottense Zone phase at ca. 184 Ma, the upper Carlottense–middle Kanense Zone phase at ca. 183 Ma, the upper Planulata–lower Crassicosta Zone phase between ca. 182 and ca. 181 Ma, the middle Crassicosta–Hillebrandti Zone phase between ca. 181 and ca. 180 Ma, and the Yakounensis Zone phase at ca. 179–178 Ma. A recent study by Jourdan et al. (2008) showed eruption ages for magmatism in the Karoo Basin that occurred between 186 and 176 Ma, with the main pulses of magmatism taking place over an interval of ~3 m.y. from 184 to 181 Ma (green histogram in Fig. 7). Riley and Knight (2001) showed eruption ages for the Ferrar province that occurred between 182 and 172 Ma, with the majority of magmatism occurring at ca. 180 Ma (dark-gray histogram in Fig. 7). However, based on studies in Patagonia and parts of the Antarctic Peninsula, Pankhurst et al. (2000) suggested that magmatism occurred as three distinct individual episodes (V1–V3) that altogether spanned more than 30 m.y., from 188 to 153 Ma. Their initial V1 episode occurred between 188 and 178 Ma, and it is this episode that is correlative with the Pliensbachian–Toarcian extinction (orange outlined histogram in Fig. 7). This combined data set suggests that much of the Karoo-Ferrar volcanic activity occurred over an interval of ~10 m.y. of the Early Jurassic and would have had a protracted impact on the global environment.

A comparison of these eruption ages for the Karoo-Ferrar igneous province and the timing of the multiphased extinction reveals that: (1) four of the six extinction pulses occurred within the main phase of Karoo and Ferrar magmatism; (2) the initial Early Pliensbachian decline in species diversity occurred within error range of the onset of Karoo magmatism and just after the peak in V1 Patagonia-Antarctic magmatism; and (3) the latest Toarcian decline occurred within the later stages of Karoo and Ferrar magmatism (Fig. 7). This makes it possible for the Karoo-Ferrar large igneous province and associated long-term environmental change to have been responsible for the multiphased extinction.

Evidence suggests that there were numerous warm intervals during the Pliensbachian–Toarcian. Oxygen-isotope data indicate warming within the Tethys Ocean area during deposition of the Ibex–Davoei Zones, Serpentinum Zone, Bifrons Zone, and possibly in the Dispansum Zone (Rosales et al., 2004; Suan et al., 2008; Dera et al., 2009; Gómez et al., 2008; Silva et al., 2011; Korte and Hesselbo, 2011; Harazim et al., 2013). However, there is also evidence of colder intervals that occurred in the Late Pliensbachian and at the Pliensbachian–Toarcian transition (Suan et al., 2008). These colder intervals are thought to have resulted from a combination of factors associated with the Early Jurassic volcanic greenhouse scenario that worked to lower the short-term atmospheric pCO_2 during greenhouse conditions. These include increases in primary productivity and forest development under warmer and more humid conditions (Suan et al., 2010; Silva et

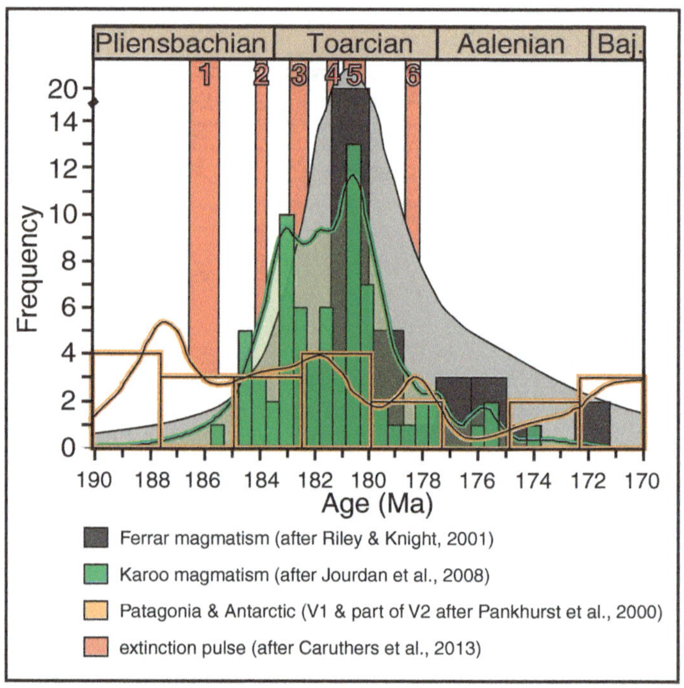

Figure 7. Diagram comparing the timing of the multiphased Pliensbachian–Toarcian extinction with emplacement of three parts of the Karoo-Ferrar volcanic province (South Africa, Antarctica, and South America). The Ar-Ar, U-Pb zircon, and Rb-Sr whole-rock age probability density distribution diagrams and frequency histograms show a correlation in timing with the known phases of extinction. All six phases of diversity decline occurred within the duration of Karoo-Ferrar magmatism. V1 and V2 are individual episodes of magmatism in Antarctica and Patagonia identified by Pankhurst et al. (2000), occurring on the diagram from ca. 190–177 ma and 175–170 ma respectively. Baj.—Bajocian. Figure is adapted from Caruthers et al. (2013). Note the change in the Frequency scale between 14 and 20.

al., 2011; Korte and Hesselbo, 2011), as well as declining carbonate production along platforms and platform margins (Donnadieu et al., 2011).

NORTH AMERICAN GEOCHEMICAL DATA

A Pliensbachian–Toarcian geochemical record from North America could contribute to our understanding of: (1) the potential effect of CO_2 outgassing on the long-term carbon-isotope record; (2) the timing and source of methane release; and (3) the geographic extent of the Toarcian oceanic anoxic event.

In the case of the Permian–Triassic Siberian Traps, coeval carbon profiles show an instability in the carbon reservoir seen as numerous smaller negative and larger positive excursions with magnitudes of up to ~8‰ (fig. 1 in Payne and Kump, 2007). Payne and Kump (2007) argued that multiple episodes of CO_2 outgassing during large igneous province formation could have produced small-magnitude negative CIEs and that increased primary productivity (related to elevated CO_2 input during several pulses of volcanism) produced the large positive excursions. Payne and Kump (2007) discounted the possibility of multiple releases of methane from hydrate as a potential cause of the negative CIEs because the time scale for methane clathrate regeneration is too long to produce multiple events consistent with the carbon-isotope record. In addition, release from a methane hydrate reservoir should produce a much larger-magnitude negative excursion, although this is dependent on many factors, including where the methane is oxidized and the turnover time (i.e., productivity) of the oceans. This is primarily based on the fact that $\delta^{13}C_{methane}$ is ≈ –60‰ (Hesselbo et al., 2000), whereas $\delta^{13}C$ of mantle-derived CO_2 is bimodal at –5‰ and –25‰ (Deines, 2002).

The Early Jurassic carbon-isotope curves for western North America show that from the upper part of the Hettangian into the Sinemurian, $\delta^{13}C_{org}$ values are relatively consistent at ~–31‰ (fig. 1 in Williford et al., 2007). In the Lower Pliensbachian, there is a steady positive trend, where $\delta^{13}C_{org}$ values reach ~–28‰ in the lower part of the Whiteavesi Zone (Fig. 8A). From the middle Whiteavesi Zone to the Middle Toarcian, a unique pattern develops, in that $\delta^{13}C_{org}$ values fluctuate frequently, and small negative CIEs are followed by larger positive excursions that have a magnitude ranging ~3‰ to 7‰. The onset of the Early Pliensbachian perturbation, near the basal Whiteavesi Zone, occurs between 190.7 (+2.7/–3.9) and 186.7 (+1.8/–1.6) Ma. The onset of Karoo magmatism occurred at 186 ± 3.6 Ma (Jourdan et al., 2008), and the onset of V1 magmatism in Patagonia-Antarctica occurred at ca. 188 Ma (Pankhurst et al., 2000); therefore, it is possible that this perturbation in the carbon cycle resulted from the onset of Karoo-Ferrar magmatism. In the Middle Toarcian Planulata Zone, values begin a gradual return to more negative values of –29‰. The top of the Planulata Zone is calibrated at ca. 181.4 Ma, and the top of the Hillebrandti Zone at ca. 180.1 Ma (Fig. 5), which is near the termination of main-phase Karoo and Ferrar magmatism.

Interestingly, a study by Silva et al. (2011) generated a $\delta^{13}C$ profile from the Lusitanian Basin of Portugal that shows a similar pattern of perturbation within the Lower to Upper Pliensbachian (Fig. 9). A comparison of this lower-resolution data set from Portugal with our data from western North America demonstrates a series of three small positive and five small negative shifts in the $\delta^{13}C$ profiles. Positive CIEs occur within the Whiteavesi (Ibex) Zone; lower part of the Freboldi (Davoei) Zone; and upper part of the Freboldi Zone. Negative CIEs occur at the Whiteavesi/Freboldi Zone boundary, within the Freboldi Zone, and within the Kunae Zone. During the Pliensbachian, Portugal was located in the western reaches of the Tethys Ocean, Haida Gwaii was located in the eastern part of the Panthalassa Ocean, and northern Alaska was located in the paleo–Arctic Ocean (Figs. 3C and 3D). Therefore, similar perturbation in the Pliensbachian carbon reservoir is evident in coeval strata from three geographically far-removed ocean basins and should be considered global in extent.

In addition to the Lower Toarcian negative CIE in the Yakoun River section that is described by Caruthers et al. (2011), there is also another coeval occurrence in a correlative succession at Whiteaves Bay (Figs. 8 and 10; Caruthers, 2013). In the Toarcian succession along the Yakoun River, the negative CIE averages –31‰ for 11 m, whereas at Whiteaves Bay, the negative CIE averages –28‰ for 6 m in the section. The North American negative CIEs (Fig. 10) occur over a similar time-stratigraphic interval as the large negative CIE in Europe, supporting the idea of methane release producing a widespread global signal. Furthermore, the isotope data from Haida Gwaii help to support the broad correlation of the North American Kanense Zone with the Tenuicostatum and Serpentinum ammonite zones of the primary standard northwest European scheme (Fig. 5). If the CIE is isochronous, it suggests that the Tenuicostatum/Serpentinum Zone boundary occurs approximately within the middle part of the Kanense Zone. The large negative CIE occurs well within the middle of the main phase of Karoo-Ferrar magmatism (using age determinations in Riley and Knight, 2001; Jourdan et al., 2008), which supports the idea of destabilization of clathrates along continental margins rather than contact metamorphism of coal and organic-rich sediments by intrusive dikes and sills in the Karoo Basin. A thermogenic source would have produced a negative CIE at the onset of Karoo magmatism (near the Whiteavesi/Freboldi Zone zonal boundary in Fig. 8A), which is not evident in our data.

A more recent study by Svensen et al. (2012) provides an updated data set of 14 U-Pb age dates from Karoo Basin sills. Using these new data, they suggested that total large igneous province emplacement took place over a much shorter interval of ~0.5 m.y., beginning at ca. 183.0 to ca. 182.3 Ma. If correct, then only the main phase of extinction would be correlative with this large igneous province eruption. This would mean that the other globally correlative events within the multiphased extinction scenario and the previously mentioned correlative geochemical perturbations could not be attributed to the Karoo-Ferrar Early Jurassic volcanic greenhouse scenario. However, in the study by

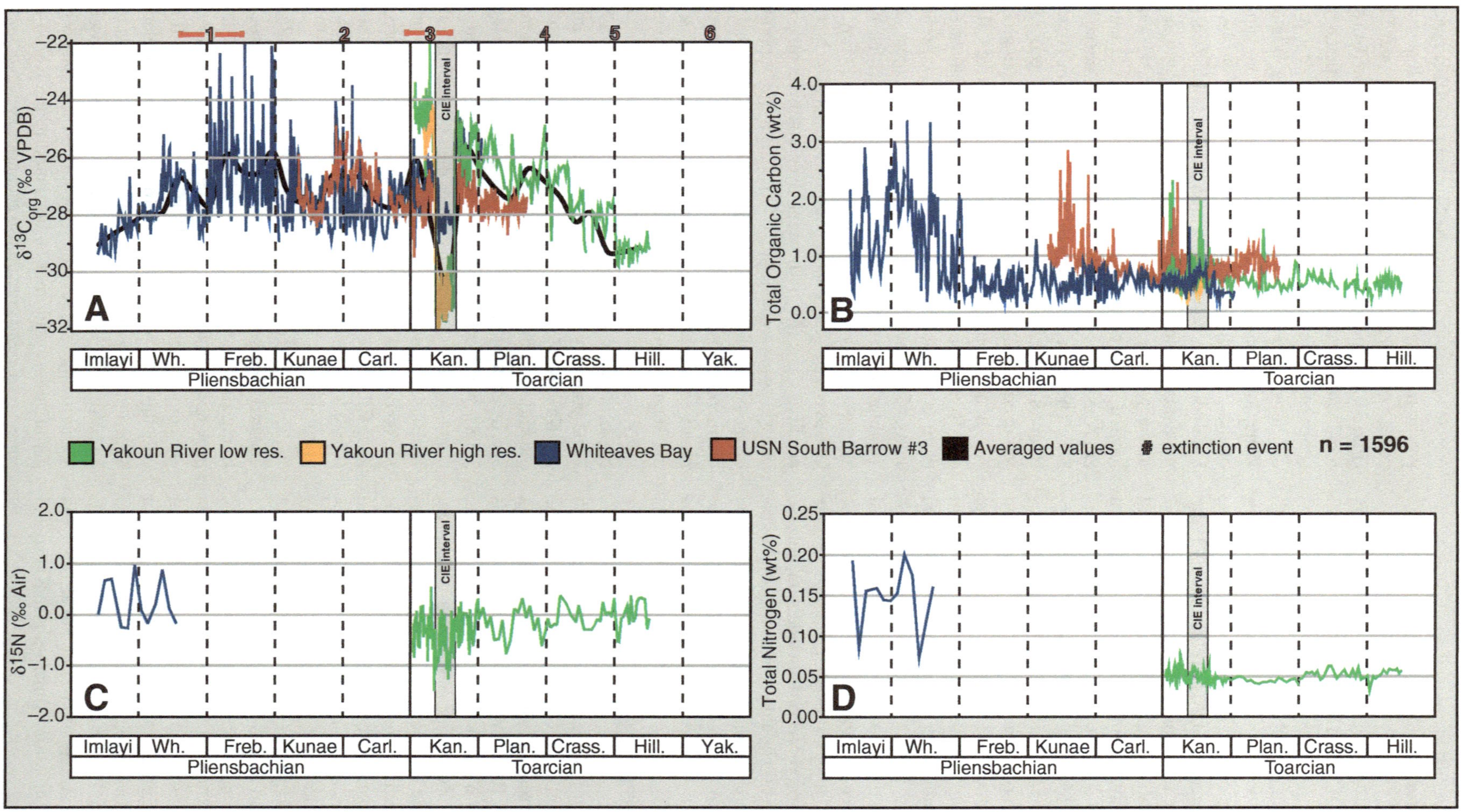

Figure 8. Combined Pliensbachian–Toarcian geochemical data from Haida Gwaii and Arctic Alaska. (A) $\delta^{13}C_{org}$ values; data from the Yakoun River section (green line) are from Caruthers et al. (2011), and extinction events 1–6 refer to correlative intervals of Dera et al. (2010) and Caruthers et al. (2013). (B) Total organic carbon (TOC) data; Yakoun River section data (green line) are from Caruthers et al. (2011). (C) $\delta^{15}N$ values. (D) Total organic nitrogen (TN) values. VPDB—Vienna Peedee belemnite; see caption for Figure 6 for other abbreviations.

Svensen et al. (2012), there is no discussion of the timing of the onset of Karoo magmatism at ca. 186 Ma that is recorded by Jourdan et al. (2008). Therefore, it remains unclear why the duration of Karoo magmatism was drastically shortened. It seems that these new sill age determinations could very well fall within the main pulses of Karoo magmatism as described by Jourdan et al. (2008).

OCEAN ANOXIA

The global distribution of organic-rich black shales co-occurring with a positive carbon-isotope excursion in the Early Toarcian is considered to be one of the primary indicators of anoxic marine waters (Jenkyns, 1988, 2010, and references therein). However, several studies have questioned the timing, evidence for, and geographic extent of black shale deposition and isotopic perturbation associated with the Toarcian oceanic anoxic event (Bucefalo Palliani et al., 2002; Wignall et al., 2005; van de Schootbrugge et al., 2005; McArthur et al., 2008; Gómez et al., 2008; Bilotta et al., 2010; Gómez and Goy, 2011; García Joral et al., 2011). In addition to perturbations in the carbon reservoir, correlative Early Toarcian perturbations in other isotopic systems include (Fig. 4) nitrogen, manganese, sulfur, and molybdenum, which are thought to have resulted from bacterial degradation of organic matter by nitrates and sulfates in a suboxic-anoxic water column (Jenkyns et al., 2001; McArthur et al., 2008; Jenkyns, 2010).

In our study, total organic carbon (TOC) values are generally low and mostly range from 0.5 to 1.5 wt% (Fig. 8B) throughout much of the Pliensbachian and Toarcian (excluding the Early Pliensbachian). These values are more typical of modern hemipelagic shelf sediments (McIver, 1975; table 1 in Tyson and Pearson, 1991). However, over the following intervals, TOC appears to be slightly elevated, reaching ~1.5%–3.5%: (1) Early Pliensbachian (Imlayi and Whiteavesi Zones), (2) Late Pliensbachian (Kunae Zone), (3) the earliest Toarcian (below the negative CIE interval), and (4) within the Early Toarcian negative CIE interval (Fig. 8B). Within the negative CIE interval, concentrations of TOC are much lower than those seen in northwest and central Europe (e.g., ~15–20 wt% in Jenkyns, 1985, 1988, 2010; van Breugel et al., 2006), and the organic carbon maximum (2 wt%) consists of only one data point and is not a consistent trend throughout the interval.

Sabatino et al. (2009), Mattioli et al. (2004), and Kafousia et al. (2011) reported TOC values from within the negative CIE interval in pelagic shales of the western Tethys that are locally elevated (e.g., up to 2–3 wt%) but still much lower than those in northern Europe, where values reach 15–20 wt%. Sabatino et al. (2009) and Kafousia et al. (2011) suggested that in deeper-water environments, a variety of factors could have led to a degradation of organic matter in the water column prior to deposition, resulting in lower organic carbon values. In addition, Bucefalo Palliani et al. (2002), Mattioli et al. (2008), and Kafousia et al. (2011) suggested that environmental and topographical differences between the northern and southern margins of the Tethys Ocean could have contributed toward the observed lower concentrations of organic carbon. Furthermore, a recent comparison of carbon-isotope data from the Paris (France) and Lusitanian (Portugal) basins demonstrates a negative CIE that is decoupled from black shale deposition (Hermoso et al., 2009). These studies, together with TOC data from the Panthalassa Ocean, therefore suggest that organic carbon concentration seems to be

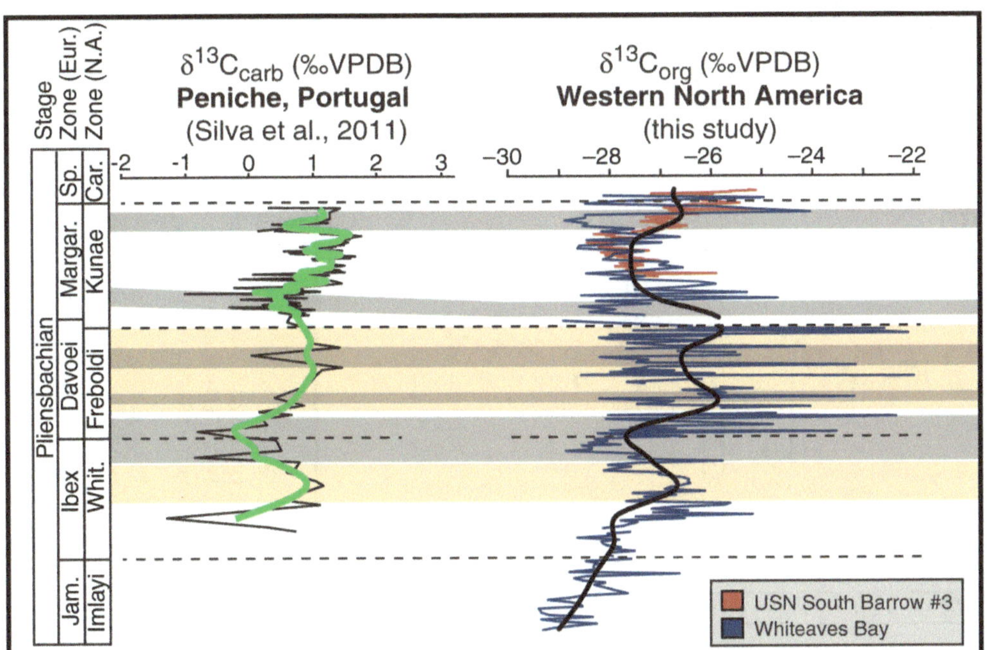

Figure 9. Comparative carbon-isotope profiles for the Pliensbachian of Peniche, Portugal (from Silva et al., 2011), and western North America (Fig. 8A). Data show similar trends in $\delta^{13}C$ values among the western Tethys, Panthalassa Ocean, and paleo–Arctic Ocean carbon reservoirs. Jam.—Jamesoni, Whit.—Whiteavesi, Car.—Carlottense, Sp.— Spinatum, VPDB—Vienna Peedee belemnite, N.A.—North America, Eur.—Europe.

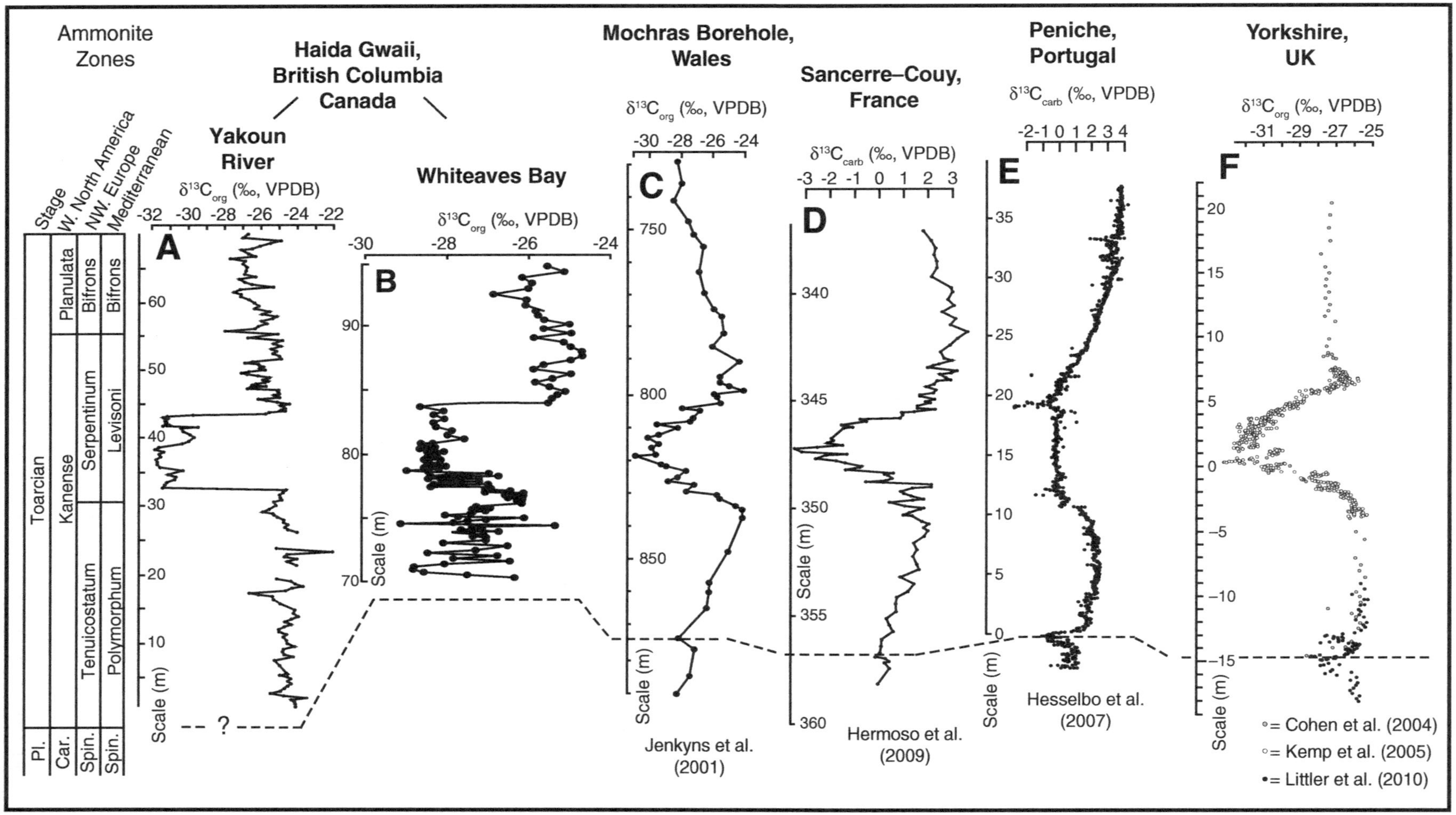

Figure 10. Correlative Lower to Middle Toarcian carbon-isotope data from western North America, northwest Europe, and the Mediterranean (modified from Caruthers et al., 2011). (A–B) Kanense–Planulata Zone data from Whiteaves Bay and the Yakoun River sections on Haida Gwaii, British Columbia, Canada; (C) Spinatum–Bifrons Zone data from the Mochras borehole, Wales; (D) Spinatum–Falciferum Zone data from Sancerre-Couy borehole, France; (E) Spinatum–Levisoni Zone data from Peniche, Portugal; (F) Spinatum–Falciferum Zone data from Yorkshire, UK. Pl.—Pliensbachian, Car.—Carlottense Zone, Spin.—Spinatum Zone.

controlled locally and should not be used to characterize a global Toarcian oceanic anoxic event. While North American TOC values do not rule out marine anoxia in the Panthalassa Ocean, they also do not support it.

Aside from organic carbon concentrations, excursions in nitrogen (and other) isotopes have been used as a more definitive indicator of changing redox conditions in the Tethys Ocean area (Jenkyns et al., 2001; McArthur et al., 2008; Jenkyns, 2010). Early Toarcian $\delta^{15}N$ data from northern Europe (British Isles) and southern Europe (Italy) show a pronounced positive excursion in $\delta^{15}N$ that generally ranges in magnitude from ~3‰ to 6‰ (Fig. 4) and also correlates with elevated TOC (~15 wt%) in the negative CIE interval (Jenkyns et al., 2001). This enrichment in $\delta^{15}N$ was interpreted by Jenkyns et al. (2001) to represent upwelling of a deoxygenated water mass that had undergone partial denitrification and is therefore evidence of the Toarcian oceanic anoxic event.

Our data from western North America do not show a positive excursion in $\delta^{15}N$ within the negative CIE interval (Fig. 8C). In fact, $\delta^{15}N$ values show a general fluctuation between approximately −1.0‰ and +0.25‰ throughout much of the Toarcian, with occasional (rare) shifts of ±1‰ magnitude and total nitrogen (TN) concentrations that are very low, generally ranging between 0.05% and 0.10% (Fig. 8D). Throughout the negative CIE interval, $\delta^{15}N$ values show two successive (albeit of small magnitude) negative excursions that reach minimum values of −1.50‰ and −1.27‰, respectively. As discussed by Jenkyns et al. (2001), typical Jurassic and Cretaceous shales deposited in normal marine conditions have $\delta^{15}N$ values that range between −2.5‰ and +4‰, which encompasses the observed Toarcian values from western North America. In modern marine sediments, $\delta^{15}N$ ranges approximately from +4‰ to +10‰ (Altabet and Francois, 1994; fig. 2 in Robinson et al., 2012), with denitrification ($\delta^{15}N$ of NO_3^-) within anoxic water masses producing mean values that reach +19‰ in the eastern equatorial Pacific Ocean (Tesdal et al., 2013).

Therefore, it seems that this part of the northeast Panthalassa Ocean was not experiencing denitrification during the Toarcian. If denitrification was not occurring, then it is possible that the Toarcian sediments were deposited under normal redox conditions, and there was no global marine anoxia during the Early Toarcian. If correct, then it can be surmised that the release of methane hydrate and the consequent acceleration in greenhouse conditions at this time created anoxic water only in certain marine environments (Fig. 2) through mechanisms related to the "restricted basin scenario" described by McArthur et al. (2008). There is no reason to think that such restricted basins were limited to the Tethys Ocean area. They could have been distributed globally, wherever appropriate basin architecture occurred.

In certain circumstances, the right environmental parameters and conditions could have been present allowing for the development of black shale deposition (from water-column anoxia) during the negative CIE interval (i.e., in the deep-water basins of the Panthalassa Ocean; see Hori, 1997; Wignall et al., 2010; Gröcke et al., 2011; Dera and Donnadieu, 2012), whereas other environments not in the right window for water-column anoxia and black shale deposition may have led to normal sedimentary deposition during the CIE interval.

CONCLUSIONS

This study reviews paleontological and geochemical data from western North America to provide a perspective on the duration, extent, and controlling mechanisms of the Pliensbachian–Toarcian extinction. Analysis of the stratigraphic ranges of 206 ammonite and 242 foraminiferal species (presented in Caruthers et al., 2013) in western North America indicates declines in diversity during six separate intervals throughout Pliensbachian–Toarcian time. These intervals of extinction can be correlated with ammonite diversity declines in northwest European and Arctic domains (Dera et al., 2010), South America (Riccardi, 2008), and declines 2–4 with radiolarian data from Europe and western North America (Goričan et al., 2013). This suggests that the Pliensbachian–Toarcian multiphased extinction occurred across the Panthalassa, Arctic, and Tethys Ocean basins and, consequently, was almost certainly global. Phases of extinction are evident in the: (1) middle of the Lower Pliensbachian (middle Whiteavesi–middle Freboldi Zones); (2) middle of the Upper Pliensbachian (upper Kunae–lower Carlottense Zones); (3) interval from the Pliensbachian–Toarcian boundary into the Lower Toarcian (upper Carlottense–middle Kanense Zones); (4) Middle Toarcian (upper Planulata–lower Crassicosta Zones); (5) upper Middle–lower Upper Toarcian (middle Crassicosta–Hillebrandti Zones); and (6) Upper Toarcian (lower–middle Yakounensis Zone).

Data from western North America also support previous conclusions suggesting that the main phase of extinction was a global event during which diversity declined in a gradual fashion beginning at the Pliensbachian–Toarcian boundary and extending into the Lower Toarcian (Harries and Little, 1999; Caswell et al., 2009). In western North America, species diversity in both ammonite and foraminifera reaches minimum values in the middle part of the Lower Toarcian Kanense Zone.

Previously, it was argued that volcanogenic outgassing of CO_2 during the eruption of the Karoo-Ferrar large igneous province initiated greenhouse conditions and caused the extinction of marine organisms (Pálfy and Smith, 2000). Our data show a correlation between the duration of Karoo-Ferrar eruptions (Pankhurst et al., 2000; Riley and Knight, 2001; Jourdan et al., 2008) and the duration of the multiphased, Pliensbachian–Toarcian extinction. Four of the six extinction pulses occur within the main phase of Karoo and Ferrar magmatism, the Early Pliensbachian diversity decline occurred within error range of the onset of Karoo and Patagonia/Antarctica magmatism, and the Late Toarcian decline coincided with the later stages of Karoo and Ferrar magmatism.

Pliensbachian–Toarcian geochemical data from western North America (presented in Caruthers, 2013) show a distinctive pattern in the Lower Pliensbachian–Middle Toarcian interval, with $\delta^{13}C$ values becoming more positive and fluctuating abruptly and frequently, with a magnitude of ~3‰–7‰. This pattern is consistent with the model proposed by Payne and Kump (2007) to explain the Late Permian–Middle Triassic influence of Siberian Trap magmatism. It suggests that volcanogenic outgassing of CO_2 during large igneous province eruption produces small negative CIEs that are followed by larger (higher magnitude) positive excursions brought on by increasing primary productivity. In western North America, the initial part of the $\delta^{13}C$ perturbation in the Early Pliensbachian is correlative with the onset of Karoo-Ferrar magmatism (using age determinations of Pankhurst et al., 2000; Riley and Knight, 2001; Jourdan et al., 2008), and values show a return to more negative levels during the waning of magmatism in the Middle Toarcian. In both the Siberian and Karoo-Ferrar large igneous province eruptions, repeated magmatism could have produced the observed pattern in $\delta^{13}C$ fluctuation (Payne and Kump, 2007; this study). Our new long-term $\delta^{13}C$ profile suggests that outgassing of CO_2 during Karoo-Ferrar magmatism could have had a significant influence on the carbon-isotope profile.

Our $\delta^{13}C$ data presented in Caruthers et al. (2011) and Caruthers (2013) also show good evidence of the Early Toarcian negative CIE interval, which we interpret as being produced by release of methane from the hydrate reservoir along continental shelf settings (Hesselbo et al., 2000) rather than from thermogenic sources (McElwain et al., 2005; Svensen et al., 2007). This is because a large negative CIE does not occur at the onset of Karoo-Ferrar magmatism, as one would expect if contact metamorphism were creating a significant amount of thermogenic methane.

Lastly, we used organic carbon, $\delta^{15}N$, and total nitrogen data to test the debated global extent of the Toarcian oceanic anoxic event (Jenkyns, 1988, 2010). Our North American data (Caruthers, 2013) show concentrations of TOC that do not exceed 2% within the negative CIE interval, a value that is more typical of modern hemipelagic shelf sediments (McIver, 1975; Tyson and Pearson, 1991). Within the time interval in which the Toarcian oceanic anoxic event occurred in Europe, and throughout the majority of the Toarcian, North American $\delta^{15}N$ values are consistently low and do not show any positive excursion. Therefore, the TOC and $\delta^{15}N$ data do not suggest denitrification within the water column for this part of the northeast Panthalassa Ocean during the Early Toarcian. Unlike the significantly negative $\delta^{13}C$ excursion, which seems to have been globally isochronous, TOC concentration is dependent on local factors and therefore should not be used to characterize global marine anoxia.

ACKNOWLEDGMENTS

Funding for this work was provided by Natural Sciences and Engineering Research Council research grants to Paul Smith (8493) and Darren Gröcke (288321), a Natural Environment Research Council grant to Darren Gröcke (NE/H021868/1), and a grant from the Alaska Geological Society (2008) to Andrew Caruthers. We express our sincerest gratitude to Ken Papp and the Geologic Materials Center in Eagle River, Alaska, for their help with the SB #3 core samples, Luke Beranek for assistance in the field, Robert B. Blodgett for logistical support, Boris Nikitenko, Ken Bird, and John Reeder for correspondence, and Joanne Peterkin at the Stable-Isotope Biogeochemistry Laboratory (SIBL) at Durham University for analytical assistance. Lastly, we thank Guillaume Dera and Emanuela Mattioli for their insightful comments and helpful suggestions for improving this manuscript.

REFERENCES CITED

Aberhan, M., and Baumiller, T.K., 2003, Selective extinction among Early Jurassic bivalves: A consequence of anoxia: Geology, v. 31, p. 1077–1080, doi:10.1130/G19938.1.

Al-Suwaidi, A.H., Angelozzi, G.N., Baudin, F., Damborenea, S.E., Hesselbo, S.P., Jenkyns, H.C., Manceñido, M.O., and Riccardi, A.C., 2010, First record of the early Toarcian oceanic anoxic event from the Southern Hemisphere, Neuquén Basin, Argentina: Journal of the Geological Society, v. 167, p. 633–636, doi:10.1144/0016-76492010-025.

Altabet, M.A., and Francois, R., 1994, Sedimentary nitrogen isotopic ratio as a recorder for surface nitrate utilization: Global Biogeochemical Cycles, v. 8, p. 103–116, doi:10.1029/93GB03396.

Arias, C., 2013, The early Toarcian (Early Jurassic) ostracod extinction events in the Iberian Range: The effect of temperature changes and prolonged exposure to low dissolved oxygen concentrations: Palaeogeography, Palaeoclimatology, Palaeoecology, v. 387, p. 40–55, doi:10.1016/j.palaeo.2013.07.004.

Bailey, T.R., Rosenthal, Y., McArthur, J.M., van de Schootbrugge, B., and Thirlwall, M.F., 2003, Paleoceanographic changes of the late Pliensbachian–early Toarcian interval: A possible link to the genesis of an oceanic anoxic event: Earth and Planetary Science Letters, v. 212, p. 307–320, doi:10.1016/S0012-821X(03)00278-4.

Barrón, E., Comas-Rengifo, M.J., and Duarte, L.V., 2013, Palynomorph succession of the Upper Pliensbachian–Lower Toarcian of the Peniche section (Portugal): Comunicações Geológicas, v. 100, Edição Especial I, p. 55–61.

Beerling, D.J., Lomas, M.R., and Gröcke, D.R., 2002, On the nature of methane gas-hydrate dissociation during the Toarcian and Aptian oceanic anoxic events: American Journal of Science, v. 302, p. 28–49, doi:10.2475/ajs.302.1.28.

Benton, M.J., ed., 1993, The Fossil Record 2: London, UK, Chapman and Hall, 846 p.

Bilotta, M., Venturi, F., and Sassaroli, S., 2010, Ammonite faunas, OAE and the Pliensbachian-Toarcian boundary (Early Jurassic) in the Apennines: Lethaia, v. 43, p. 357–380.

Braga, J.C., Comas-Rengifo, M.J., Goy, A., and Rivas, P., 1982, Comparaciones faunisticas y correlaciones en el Pliensbachiense de la Zona Subbética y Cordillera Ibérica: Bolletin de la Reale Sociedad Española de Historia Naturale: Seccion Geologica, v. 80, p. 133–152.

Breck, W.G., 1974, Redox levels in the sea, *in* Goldberg, E.D., ed., The Sea: Volume 5: New York, Wiley, p. 153–179.

Bucefalo Palliani, R., Mattioli, E., and Riding, J.B., 2002, The response of marine phytoplankton and sedimentary organic matter to the early Toarcian (Lower Jurassic) oceanic anoxic event in northern England: Marine Micropaleontology, v. 46, p. 223–245, doi:10.1016/S0377-8398(02)00064-6.

Callomon, J.H., 1985, Biostratigraphy, chronostratigraphy and all that—Again!, *in* Michelsen, O., and Zeiss, A., eds., International Symposium on Jurassic Stratigraphy, Erlangen 1984: Copenhagen, Geological Survey of Denmark, v. 3, p. 611–624.

Cameron, B.E.B., and Tipper, H.W., 1985, Jurassic Stratigraphy of the Queen Charlotte Islands, British Columbia: Geological Survey of Canada Bulletin 365, p. 1–49.

Cariou, E., and Hantzpergue, P., 1997, Groupe Français d'études du Jurassique, biostratigraphie du Jurassique ouest-européen et méditerranen. Zonations parallèles et distribution des invertébrés et microfossiles: Bulletin des Centres de Recherches Exploration-Production Elf-Aquitaine, v. 17, p. 1–400.

Caruthers, A.H., 2013, Pliensbachian–Toarcian (Early Jurassic) Extinction in Western North America [Ph.D. thesis]: Vancouver, Canada, University of British Columbia, 305 p.

Caruthers, A.H., and Smith, P.L., 2012, Pliensbachian ammonoids from the Talkeetna Mountains (Peninsular terrane) of southern Alaska: Review de Paléobiologie, Volume Spécial 11, p. 365–378.

Caruthers, A.H., Gröcke, D.R., and Smith, P.L., 2011, The significance of an Early Jurassic (Toarcian) carbon-isotope excursion in Haida Gwaii (Queen Charlotte Islands), British Columbia, Canada: Earth and Planetary Science Letters, v. 307, p. 19–26.

Caruthers, A.H., Smith, P.L., and Gröcke, D.R., 2013, The Pliensbachian-Toarcian (Early Jurassic) extinction, a global multi-phased event: Palaeogeography, Palaeoclimatology, Palaeoecology, v. 386, p. 104–118, doi:10.1016/j.palaeo.2013.05.010.

Caswell, B.A., Coe, A.L., and Cohen, A.S., 2009, New range data for marine invertebrate species across the early Toarcian (Early Jurassic) mass extinction: Journal of the Geological Society, v. 166, p. 859–872, doi:10.1144/0016-76492008-0831.

Cecca, F., and Macchioni, F., 2004, The two early Toarcian (Early Jurassic) extinction events in ammonoids: Lethaia, v. 37, p. 35–56, doi:10.1080/00241160310008257.

Cohen, A.S., Coe, A.L., Harding, S.M., and Schwark, L., 2004, Osmium isotope evidence for the regulation of atmospheric CO_2 by continental weathering: Geology, v. 32, p. 157–160, doi:10.1130/G20158.1.

Courtillot, V., 1999, Evolutionary Catastrophes: The Science of Mass Extinction: Cambridge, UK, Cambridge University Press, 237 p.

Courtillot, V., and Renne, P.R., 2003, On the ages of flood basalt events: Compte Rendus– Académie des Sciences, v. 335, p. 113–140.

Dean, W.T., Donovan, D.T., and Howarth, M.K., 1961, The Liassic ammonite zones and subzones of northwest European province: Bulletin of the British Museum (Natural History): Geology, v. 4, p. 435–505.

Deines, P., 2002, The carbon isotope geochemistry of mantle xenoliths: Earth-Science Reviews, v. 58, p. 247–278, doi:10.1016/S0012-8252(02)00064-8.

Dera, G., and Donnadieu, Y., 2012, Modeling evidences for global warming, Arctic seawater freshening, and sluggish oceanic circulation during the early Toarcian anoxic event: Paleoceanography, v. 27, PA2211, doi:10.1029/2012PA002283.

Dera, G., Pucéat, E., Pellenard, P., Neige, P., Delsate, D., Joachimski, M., Reisberg, L., and Martinez, M., 2009, Water mass exchange and variations in seawater temperature in the NW Tethys during the Early Jurassic: Evidence from neodymium and oxygen isotopes of fish teeth and belemnites: Earth and Planetary Science Letters, v. 286, p. 198–207, doi:10.1016/j.epsl.2009.06.027.

Dera, G., Neige, P., Dommergues, J.-L., Fara, E., Laffont, R., and Pellenard, P., 2010, High-resolution dynamics of Early Jurassic marine extinctions: The case of Pliensbachian–Toarcian ammonites (Cephalopoda): Journal of the Geological Society, v. 167, p. 21–33, doi:10.1144/0016-76492009-068.

Dera, G., Brigaud, B., Monna, F., Laffont, R., Pucéat, E., Deconinck, J.-F., Pellenard, P., Joachimski, M., and Durlet, C., 2011, Climatic ups and downs in a disturbed Jurassic world: Geology, v. 39, p. 215–218, doi:10.1130/G31579.1.

Dommergues, J.-L., and Meister, C., 2009, Area of mixed marine faunas between two major paleogeographical realms, exemplified by the Early Jurassic (Late Sinemurian and Pliensbachian) ammonites in the Alps: Palaeogeography, Palaeoclimatology, Palaeoecology, v. 86, p. 265–282.

Donnadieu, Y., Dromart, G., Goddéris, Y., Pucéat, E., Brigaud, B., Dera, G., Dumas, C., and Oliver, N., 2011, A mechanism for brief glacial episodes in the Mesozoic greenhouse: Paleoceanography, v. 26, PA3212, doi:10:1029/2010PA002100.

Foote, M., 2000, Origination and extinction components of taxonomic diversity: General problems: Paleobiology, v. 26, supplement, p. 74–102, doi:10.1666/0094-8373(2000)26[74:OAECOT]2.0.CO;2.

Fraguas, Á., Comas-Rengifo, M.J., Gómez, J.J., and Goy, A., 2012, The calcareous nannofossil crisis in northern Spain (Asturias province) linked to the early Toarcian warming-driven mass extinction: Marine Micropaleontology, v. 94–95, p. 58–71, doi:10.1016/j.marmicro.2012.06.004.

García Joral, F., Gómez, J.J., and Goy, A., 2011, Mass extinction and recovery of the early Toarcian (Early Jurassic) brachiopods linked to climate change in northern and central Spain: Palaeogeography, Palaeoclimatology, Palaeoecology, v. 302, p. 367–380, doi:10.1016/j.palaeo.2011.01.023.

Gill, B.C., Lyons, T.W., and Jenkyns, H.C., 2011, A global perturbation to the sulfur cycle during the Toarcian oceanic anoxic event: Earth and Planetary Science Letters, v. 312, p. 484–496, doi:10.1016/j.epsl.2011.10.030.

Gómez, J.J., and Goy, A., 2011, Warming-driven mass extinction in the early Toarcian (Early Jurassic) of northern and central Spain: Correlation with other time-equivalent European sections: Palaeogeography, Palaeoclimatology, Palaeoecology, v. 306, p. 176–195, doi:10.1016/j.palaeo.2011.04.018.

Gómez, J.J., Goy, A., and Canales, M.L., 2008, Seawater temperature and carbon isotope variations in belemnites linked to mass extinction during the Toarcian (Early Jurassic) in central and northern Spain: Comparison with other European sections: Palaeogeography, Palaeoclimatology, Palaeoecology, v. 258, p. 28–58, doi:10.1016/j.palaeo.2007.11.005.

Goričan, Š., Carter, E.S., Guex, J., O'Dogherty, L., De Wever, P., Dumitrica, P., Hori, R.S., Matsuoka, A., and Whalen, P.A., 2013, Evolutionary patterns and palaeobiogeography of Pliensbachian and Toarcian (Early Jurassic) radiolaria: Palaeogeography, Palaeoclimatology, Palaeoecology, v. 386, p. 620–636, doi:10.1016/j.palaeo.2013.06.028.

Gradstein, F.M., Ogg, J.G., Schmitz, M.D., and Ogg, G.M., 2012, The Geologic Time Scale 2012: Oxford, UK, Elsevier, doi:10.1016/B978-0-444-59425-9.00004-4.

Gröcke, D.R., Rimmer, S.M., Yoksoulian, L.E., Cairncross, B., Tsikos, H., and van Hunen, J., 2009, No evidence for thermogenic methane release in coal from the Karoo-Ferrar large igneous province: Earth and Planetary Science Letters, v. 277, p. 204–212.

Gröcke, D.R., Hori, R.S., Trabucho-Alexandre, J., Kemp, D.B., and Schwark, L., 2011, An open ocean record of the Toarcian oceanic anoxic event: Solid Earth, v. 2, p. 245–257, doi:10.5194/se-2-245-2011.

Hallam, A., 1961, Cyclothems, transgressions and faunal change in the Lias of north west Europe: Transactions of the Edinburgh Geological Society, v. 18, p. 124–174, doi:10.1144/transed.18.2.124.

Hallam, A., 1986, The Pliensbachian and Tithonian extinction events: Nature, v. 319, p. 765–768, doi:10.1038/319765a0.

Hallam, A., 1987, Radiations and extinctions in relation to environmental change in the marine Jurassic of north west Europe: Paleobiology, v. 13, p. 152–168.

Hallam, A., and Wignall, P.B., 1997, Mass Extinctions and Their Aftermath: Oxford, UK, Oxford University Press, 320 p.

Hammer, Ø., and Harper, D., 2006, Paleontological Data Analysis: Malden, Massachusetts, Blackwell Publishing, 351 p.

Harazim, D., van de Schootbrugge, B., Sorichter, K., Fiebig, J., Weug, A., Suan, G., and Oschmann, W., 2013, Spatial variability of watermass conditions within the European epicontinental seaway during the Early Jurassic (Pliensbachian–Toarcian): Sedimentology, v. 60, p. 359–390, doi:10.1111/j.1365-3091.2012.01344.x.

Harries, P.J., and Little, C.T.S., 1999, The early Toarcian (Early Jurassic) and the Cenomanian–Turonian (Late Cretaceous) mass extinctions: Similarities and contrasts: Palaeogeography, Palaeoclimatology, Palaeoecology, v. 154, p. 39–66, doi:10.1016/S0031-0182(99)00086-3.

Hermoso, M., Minoletti, F., Le Callonnec, L., Jenkyns, H.C., Hesselbo, S.P., Rickaby, R.E.M., Renard, M., de Rafélis, M., and Emmanuel, L., 2009, Global and local forcing of early Toarcian seawater chemistry: A comparative study of different paleoceanographic settings (Paris and Lusitanian Basins): Paleoceanography, v. 24, PA4208, doi:10.1029/2009PA001764.

Hesselbo, S.P., and Pienkowski, G., 2011, Stepwise atmospheric carbon-isotope excursion during the Toarcian oceanic anoxic event (Early Jurassic, Polish Basin): Earth and Planetary Science Letters, v. 301, p. 365–372, doi:10.1016/j.epsl.2010.11.021.

Hesselbo, S.P., Gröcke, D.R., Jenkyns, H.C., Bjerrum, C.J., Farrimond, P., Morgans Bell, H.S., and Green, O.R., 2000, Massive dissociation of gas hydrate during a Jurassic oceanic anoxic event: Nature, v. 406, p. 392–395, doi:10.1038/35019044.

Hesselbo, S.P., Jenkyns, H.C., Duarte, L.V., and Oliveira, L.C.V., 2007, Carbon-isotope record of the Early Jurassic (Toarcian) oceanic anoxic event from fossil wood and marine carbonate (Lusitanian Basin, Portugal): Earth and Planetary Science Letters, v. 253, p. 455–470, doi:10.1016/j.epsl.2006.11.009.

Hori, R.S., 1997, The Toarcian radiolarian event in bedded cherts from southwestern Japan: Marine Micropaleontology, v. 30, p. 159–169, doi:10.1016/S0377-8398(96)00024-2.

Houseknecht, D.W., and Bird, K.J., 2004, Sequence stratigraphy of the Kingak Shale (Jurassic–Lower Cretaceous): National Petroleum Reserve in Alaska: American Association of Petroleum Geologists Bulletin, v. 88, p. 279–302.

Howarth, M.K., 1992, The Ammonite Family Hildoceratidae in the Lower Jurassic of Britain: London, Palaeontographical Society Monograph 145, 200 p.

Imlay, R.W., 1955, Characteristic Jurassic mollusks from northern Alaska: U.S. Geological Survey Professional Paper 274-D, p. 69–96.

Imlay, R.W., 1981, Early Jurassic Ammonites from Alaska: U.S. Geological Survey Professional Paper 1190, 40 p.

Izumi, K., Miyaji, T., and Tanabe, K., 2012, Early Toarcian (Early Jurassic) oceanic anoxic event recorded in the shelf deposits in the northwestern Panthalassa: Evidence from the Nishinakayama Formation in the Toyora area, west Japan: Palaeogeography, Palaeoclimatology, Palaeoecology, v. 315–316, p. 100–108, doi:10.1016/j.palaeo.2011.11.016.

Jakobs, G.K., 1992, Toarcian (Lower Jurassic) Ammonite Biostratigraphy and Ammonite Fauna of North America [Ph.D. thesis]: Vancouver, Canada, University of British Columbia, 682 p.

Jakobs, G.K., 1995, New occurrences of *Leukadiella* and *Paroniceras* (Ammonoidea) from the Toarcian (Lower Jurassic) of the Canadian Cordillera: Journal of Paleontology, v. 69, p. 89–98.

Jakobs, G.K., 1997, Toarcian (Early Jurassic) Ammonoids from Western North America: Geological Survey of Canada Bulletin 428, p. 1–137.

Jakobs, G.K., and Smith, P.L., 1996, Latest Toarcian ammonoids from the North American Cordillera: Palaeontology, v. 39, p. 97–147.

Jakobs, G.K., Smith, P.L., and Tipper, H.W., 1994, An ammonite zonation for the Toarcian (Lower Jurassic) of the North American Cordillera: Canadian Journal of Earth Sciences, v. 31, p. 919–942, doi:10.1139/e94-083.

Jenkyns, H.C., 1985, The early Toarcian and Cenomanian-Turonian anoxic events in Europe: Comparisons and contrasts: International Journal of Earth Sciences, v. 74, p. 505–518.

Jenkyns, H.C., 1988, The early Toarcian (Jurassic) event: Stratigraphy, sedimentary and geochemical evidence: American Journal of Science, v. 288, p. 101–151, doi:10.2475/ajs.288.2.101.

Jenkyns, H.C., 2010, Geochemistry of oceanic anoxic events: Geochemistry Geophysics Geosystems, v. 11, Q03004, doi:10.1029/2009GC002788.

Jenkyns, H.C., Gröcke, D.R., and Hesselbo, S.P., 2001, Nitrogen isotope evidence for water mass denitrification during the early Toarcian (Jurassic) ocean anoxic event: Paleoceanography, v. 16, p. 593–603, doi:10.1029/2000PA000558.

Jones, C.E., and Jenkyns, H.C., 2001, Seawater strontium isotopes, oceanic anoxic events, and seafloor hydrothermal activity in the Jurassic and Cretaceous: American Journal of Science, v. 301, p. 112–149, doi:10.2475/ajs.301.2.112.

Jourdan, F., Féraud, G., Bertrand, H., Watkeys, M.K., and Renne, P.R., 2008, The $^{40}Ar/^{39}Ar$ ages of the sill complex of the Karoo large igneous province: Implications for the Pliensbachian–Toarcian climate change. Geochemistry Geophysics Geosystems, v. 9, p. 1–20, Q06009, doi:10.1029/2008GC001994.

Kafousia, N., Karakitsios, V., Jenkyns, H.C., and Mattioli, E., 2011, A global event with a regional character: The early Toarcian oceanic anoxic event in the Pindos Ocean (northern Peloponnese, Greece): Geological Magazine, v. 148, p. 619–631, doi:10.1017/S0016756811000082.

Kemp, D.B., Coe, A.L., Cohen, A.S., and Schwark, L., 2005, Astronomical pacing of methane release in the Early Jurassic period: Nature, v. 437, p. 396–399.

Korte, C., and Hesselbo, S.P., 2011, Shallow-marine carbon- and oxygen-isotope and elemental records indicate icehouse-greenhouse cycles during the Early Jurassic: Paleoceanography, v. 26, PA4219, doi:10.1029/2011PA002160.

Kottachchi, N., Schröder-Adams, C.J., Haggart, J.W., and Tipper, H.W., 2002, Jurassic foraminifera from the Queen Charlotte Islands, British Columbia, Canada: Biostratigraphy, paleoenvironments and paleogeographic implications: Palaeogeography, Palaeoclimatology, Palaeoecology, v. 180, p. 93–127, doi:10.1016/S0031-0182(01)00424-2.

Kottachchi, N., Schröder-Adams, C.J., Haggart, J.W., and Page, J.E., 2003, Lower and Middle Jurassic Foraminifera of Queen Charlotte Islands, British Columbia: Raw Data and Preliminary Results: Geological Survey of Canada Open-File 1739, 47 p.

Küspert, W., 1982, Environmental changes during oil shale deposition as deduced from stable isotope ratios, *in* Einsele, G., and Seilacher, A., eds., Cyclic and Event Stratification: Berlin, Springer, p. 482–501.

Little, C.T.S., and Benton, M.J., 1995, Early Jurassic mass extinction: A global long-term event: Geology, v. 23, p. 495–498, doi:10.1130/0091-7613(1995)023<0495:EJMEAG>2.3.CO;2.

Littler, K., Hesselbo, S.P., and Jenkyns, H.C., 2010, A carbon-isotope perturbation at the Pliensbachian–Toarcian boundary: Evidence from the Lias Group, NE England: Geological Magazine, v. 147, p. 181–192.

Mattioli, E., Pittet, B., Bucefalo Palliani, R., Röhl, H.J., Schmid-Röhl, A., and Morettini, E., 2004, Phytoplankton evidence for timing and correlation of palaeoceanographical changes during the early Toarcian oceanic anoxic event (Early Jurassic): Journal of the Geological Society, v. 161, p. 685–693, doi:10.1144/0016-764903-074.

Mattioli, E., Pittet, B., Suan, G., and Mailliot, S., 2008, Calcareous nannoplankton changes across the early Toarcian oceanic anoxic event in the western Tethys: Paleoceanography, v. 23, PA3208, doi:10.1029/2007PA001435.

Mattioli, E., Pittet, B., Petitpierre, L., and Mailliot, S., 2009, Dramatic decrease of pelagic carbonate production by nannoplankton across the early Toarcian anoxic event (T-OAE): Global and Planetary Change, v. 65, p. 134–145, doi:10.1016/j.gloplacha.2008.10.018.

McArthur, J.M., Donovan, D.T., Thirlwall, M.F., Fouke, B.W., and Mattey, D., 2000, Strontium isotope profile of the early Toarcian (Jurassic) oceanic anoxic event, the duration of ammonite biozones, and belemnite palaeotemperatures: Earth and Planetary Science Letters, v. 179, p. 269–285, doi:10.1016/S0012-821X(00)00111-4.

McArthur, J.M., Algeo, T.J., van de Schootbrugge, B., Li, Q., and Howarth, R.J., 2008, Basinal restriction, black shales, Re-Os dating, and the early Toarcian (Jurassic) oceanic anoxic event: Paleoceanography, v. 23, p. 1–22, doi:10.1029/2008PA001607.

McElwain, J.C., Wade-Murphy, J., and Hesselbo, S.P., 2005, Changes in carbon dioxide during an oceanic anoxic event linked to intrusion into Gondwana coals: Nature, v. 435, p. 479–482, doi:10.1038/nature03618.

McIver, R., 1975, Hydrocarbon occurrences from JOIDES Deep Sea Drilling Project, *in* Proceedings of the 9th World Petroleum Congress (Tokyo), 2: Barking, UK, Applied Science Publishers, p. 269–280.

Mickey, M.B., Haga, H., and Bird, K.J., 2006. Micropaleontology of Selected Wells and Seismic Shot Holes, Northern Alaska: U.S. Geological Survey Open-File Report 2006–1055, 11 p.

Nikitenko, B., Shurygin, B., and Mickey, M., 2008, High resolution stratigraphy of the Lower Jurassic and Aalenian of Arctic regions as the basis for detailed palaeobiogeographic reconstructions: Norwegian Journal of Geology, v. 88, p. 267–277.

Page, K.N., 2003, The Lower Jurassic of Europe: Its subdivision and correlation: Geological Survey of Denmark and Greenland Bulletin, v. 1, p. 23–59.

Pálfy, J., and Smith, P.L., 2000, Synchrony between Early Jurassic extinction, oceanic anoxic event, and the Karoo-Ferrar flood basalt volcanism: Geology, v. 28, p. 747–750, doi:10.1130/0091-7613(2000)28<747:SBEJEO>2.0.CO;2.

Pálfy, J., Parrish, R.R., and Smith, P.L., 1997, A U-Pb age from the Toarcian (Lower Jurassic) and its use for time scale calibration through error analysis of biochronologic dating: Earth and Planetary Science Letters, v. 146, p. 659–675, doi:10.1016/S0012-821X(96)00241-5.

Pálfy, J., Mortensen, J.K., Smith, P.L., Friedman, R.M., McNicoll, V., and Villeneuve, M., 2000a, New U-Pb zircon ages integrated with ammonite biochronology from the Jurassic of the Canadian Cordillera: Canadian Journal of Earth Sciences, v. 37, p. 549–567, doi:10.1139/e99-115.

Pálfy, J., Smith, P.L., and Mortensen, J.K., 2000b, A U-Pb and ^{40}Ar-^{39}Ar time scale for the Jurassic: Canadian Journal of Earth Sciences, v. 37, p. 923–944, doi:10.1139/e00-002.

Pankhurst, R.J., Riley, T.R., Fanning, C.M., and Kelley, S.P., 2000, Episodic silicic volcanism in Patagonia and the Antarctic Peninsula: Chronology of magmatism associated with the break-up of Gondwana: Journal of Petrology, v. 41, p. 605–625, doi:10.1093/petrology/41.5.605.

Payne, J.L., and Kump, L.R., 2007, Evidence for recurrent Early Triassic massive volcanism from quantitative interpretation of carbon isotope fluctuations: Earth and Planetary Science Letters, v. 256, p. 264–277, doi:10.1016/j.epsl.2007.01.034.

Pearce, C.R., Cohen, A.S., Coe, A.L., and Burton, K.W., 2008, Molybdenum isotope evidence for global ocean anoxia coupled with perturbations to the carbon cycle during the Early Jurassic: Geology, v. 36, p. 231–234, doi:10.1130/G24446A.1.

Piper, D.Z., and Calvert, S.E., 2009, A marine biogeochemical perspective on black shale deposition: Earth-Science Reviews, v. 95, p. 63–96, doi:10.1016/j.earscirev.2009.03.001.

Riccardi, A., 2008, The marine Jurassic of Argentina: A biostratigraphic framework: Episodes, v. 31, no. 3, p. 326–335.

Riley, T.R., and Knight, K.B., 2001, Age of pre-break-up Gondwana magmatism: Antarctic Science, v. 13, p. 99–110, doi:10.1017/S0954102001000177.

Robinson, R.S., Kienast, M., Albuquerque, A.L., Altabet, M., Contreras, S., Holz, R.D.P., Dubois, N., Francois, R., Galbraith, E., Hsu, T.-C., Ivanochko, T., Jaccard, S., Kao, S.-J., Kiefer, T., Kienast, S., Lehmann, M.F., Philippe, M., McCarthy, M., Möbius, J., Pedersen, T., Quan, T.M., Ryabenko, E., Schmittner, A., Schneider, R., Schneider-Mor, A.,

Shigemitsu, M., Sinclair, D., Somes, C., Studer, A., Thunell, R., and Yang, J.-Y., 2012, A review of nitrogen isotopic alteration in marine sediments: Paleoceanography, v. 27, PA4203, doi:10.1029/2012PA002321.

Rosales, I., Quesada, S., and Robles, S., 2004, Paleotemperature variations of Early Jurassic seawater recorded in geochemical trends of belemnites from the Basque-Cantabrian Basin, northern Spain: Palaeogeography, Palaeoclimatology, Palaeoecology, v. 203, p. 253–275, doi:10.1016/S0031-0182(03)00686-2.

Sabatino, N., Neri, R., Bellanca, A., Jenkyns, H.C., Baudin, F., Parisi, G., and Masetti, D., 2009, Carbon-isotope records of the Early Jurassic (Toarcian) oceanic anoxic event from the Valdorbia (Umbria–Marche Apennines) and Monte Mangart (Julian Alps) sections: Palaeoceanographic and stratigraphic implications: Sedimentology, v. 56, p. 1307–1328, doi:10.1111/j.1365-3091.2008.01035.x.

Sabatino, N., Neri, R., Bellanca, A., Jenkyns, H.C., Masetti, D., and Scopelliti, G., 2011, Petrography and high-resolution geochemical records of Lower Jurassic manganese-rich deposits from Monte Mangart, Julian Alps: Palaeogeography, Palaeoclimatology, Palaeoecology, v. 299, p. 97–109, doi:10.1016/j.palaeo.2010.10.039.

Sælen, G., Doyle, P., and Talbot, M.R., 1996, Stable isotope analyses of belemnite rostra from the Whitby Mudstone Fm., England: Surface water conditions during deposition of a marine black shale: Palaios, v. 11, p. 97–117, doi:10.2307/3515065.

Schlatter, R.V., 1980, Biostratigraphie und Ammonitenfauna des Unter-Pliensbachium im Typusgebiet (Pliensbach, Holzmaden und Nürtingen; Württemberg, SW.-Deutschland): Stuttgarter Beitrage zur Naturkunde, ser. B, Geologie und Palaontologie, v. 65, 251 p.

Schouten, S., Van Kaam-Peters, H.M.E., Rijpstra, W.I.C., Schoell, M., and Sinninghe Damste, J.S., 2000, Effects of an oceanic event on the stable carbon isotopic composition of early Toarcian carbon: American Journal of Science, v. 300, p. 1–22, doi:10.2475/ajs.300.1.1.

Sherwood, K.W., Johnson, P.P., Craig, J.D., Zerwick, S.A., Lothamer, R.T., Thurston, D.K., and Hurlbert, S.B., 2002, Structure and stratigraphy of the Hanna Trough, U.S., Chukchi Shelf, Alaska, *in* Miller, E.L., Grantz, A., and Klemperer, S.L., eds., Tectonic Evolution of the Bering Shelf–Chukchi Sea–Arctic Margin and Adjacent Landmasses: Geological Society of America Special Paper 360, p. 39–66.

Silva, R.L., Duarte, L.V., Comas-Rengifo, M.J., Mendonça Filho, J.G., and Azerêdo, A.C., 2011, Update of the carbon and oxygen isotopic records of the early–late Pliensbachian (Early Jurassic, ~187 Ma): Insights from the organic-rich hemipelagic series of the Lusitanian Basin (Portugal): Chemical Geology, v. 283, p. 177–184, doi:10.1016/j.chemgeo.2011.01.010.

Smith, P.L., 2006, Paleobiogeography and Early Jurassic molluscs in the context of terrane displacement in western Canada, *in* Haggart, J.W., Enkin, R.J., and Monger, J.W.H., eds., Geological Association of Canada Special Paper 46, p. 81–94.

Smith, P.L., and Tipper, H.W., 1996, Pliensbachian (Lower Jurassic) ammonites of the Queen Charlotte Islands, British Columbia: Bulletins of American Paleontology, v. 108, p. 1–122.

Smith, P.L., Tipper, H.W., Taylor, D.G., and Guex, J., 1988, An ammonite zonation for the Lower Jurassic of Canada and the United States: The Pliensbachian: Canadian Journal of Earth Sciences, v. 25, p. 1503–1523, doi:10.1139/e88-142.

Smith, P.L., Tipper, H.W., and Ham, D.M., 2001, Lower Jurassic Amaltheidae (Ammonitina) in North America: Paleobiogeography and tectonic implications: Canadian Journal of Earth Sciences, v. 38, p. 1439–1449, doi:10.1139/e01-034.

Suan, G., Mattioli, E., Pittet, B., Mailliot, S., and Lécuyer, C., 2008, Evidence for major environmental perturbation prior to and during the Toarcian (Early Jurassic) oceanic anoxic event from the Lusitanian Basin: Paleoceanography, v. 23, p. 1–14, doi:10.1029/2007PA001459.

Suan, G., Mattioli, E., Pittet, B., Lécuyer, C., Suchéras-Marx, B., Duarte, L.V., Philippe, M., Reggiani, L., and Martineau, F., 2010, Secular environmental precursors to early Toarcian (Jurassic) extreme climate changes: Earth and Planetary Science Letters, v. 290, p. 448–458, doi:10.1016/j.epsl.2009.12.047.

Suan, G., Nikitenko, B.L., Robov, M.A., Baudin, F., Spangenberg, J.E., Knyazev, V.G., Glinskikh, L.A., Goryacheva, A.A., Adatte, T., Riding, J.B., Föllmi, K.B., Pittet, B., Mattioli, E., and Lécuyer, C., 2011, Polar record of Early Jurassic massive carbon injection: Earth and Planetary Science Letters, v. 312, p. 102–113, doi:10.1016/j.epsl.2011.09.050.

Svensen, H., Planke, S., Chevallier, L., Malthe-Sørenssen, A., Corfu, F., and Jamtveit, B., 2007, Hydrothermal venting of greenhouse gases triggering Early Jurassic global warming: Earth and Planetary Science Letters, v. 256, p. 554–566, doi:10.1016/j.epsl.2007.02.013.

Svensen, H., Corfu, F., Polteau, S., Hammer, Ø., and Planke, S., 2012, Rapid magma emplacement in the Karoo large igneous province: Earth and Planetary Science Letters, v. 325–326, p. 1–9, doi:10.1016/j.epsl.2012.01.015.

Tappan, H., 1955, Jurassic Foraminifera: Part 2. Foraminifera from the Arctic Slope of Alaska: U.S. Geological Survey Professional Paper 236–B, p. 21–90.

Tesdal, J.E., Galbraith, E.D., and Kienast, M., 2013, Nitrogen isotopes in bulk marine sediment: Linking seafloor observations with subseafloor records: Biogeosciences, v. 10, p. 101–118, doi:10.5194/bg-10-101-2013.

Thomson, R.C., and Smith, P.L., 1992, Pliensbachian (Lower Jurassic) Biostratigraphy and Ammonite Fauna of the Spatsizi Area, North-Central British Columbia: Geological Survey of Canada Bulletin 437, 87 p.

Tipper, H.W., Smith, P.L., Cameron, B.E.B., Carter, E.S., Jakobs, G.K., and Johns, M.J., 1991, Biostratigraphy of the Lower Jurassic formations of the Queen Charlotte Islands, British Columbia, *in* Woodsworth, G.J., ed., Evolution and Hydrocarbon Potential of the Queen Charlotte Basin, British Columbia: Geological Survey of Canada Paper 90–10, p. 203–235.

Tremolada, F., van de Schootbrugge, B., and Erba, E., 2005, Early Jurassic schizosphaerellid crisis in Cantabria, Spain: Implications for calcification rates and phytoplankton evolution across the Toarcian oceanic anoxic event: Paleoceanography, v. 20, PA2011, doi:10.1029/2004PA001120.

Tyson, R.V., and Pearson, T.H., 1991, Modern and ancient continental shelf anoxia: An overview, *in* Tyson, R.V., and Pearson, T.H., eds., Modern and Ancient Continental Shelf Anoxia: Geological Society Special Publication 58, p. 1–24.

van Breugel, Y., Bass, M., Schouten, S., Mattioli, E., and Sinninghe Damsté, J.S., 2006, Isorenieratane record in black shales from the Paris Basin, France: Constraints on recycling of respired CO_2 as a mechanism for negative carbon isotope shifts during the Toarcian oceanic anoxic event: Paleoceanography, v. 21, PA4220, 8 p., doi:10.1029/2006PA001305.

van de Schootbrugge, B., Bailey, T.R., Rosenthal, Y., Katz, M.E., Wright, J.D., Miller, K.G., Feist-Burkhardt, S., and Falowski, P., 2005, Early Jurassic climate change and the radiation of organic-walled phytoplankton in the Tethys Ocean: Paleobiology, v. 31, p. 73–97, doi:10.1666/0094-8373(2005)031<0073:EJCCAT>2.0.CO;2.

von Hillebrandt, A., 2006, Ammoniten aus dem Pliensbachium (Carixium und Domerium) von Südamerika: Revue de Paléobiologie, v. 25, p. 1–403.

Vörös, A., 2002, Victims of the early Toarcian anoxic event: The radiation and extinction of Jurassic Koninckinidae (Brachiopoda): Lethaia, v. 35, p. 345–357, doi:10.1111/j.1502-3931.2002.tb00093.x.

Wignall, P.B., 2001, Large igneous provinces and mass extinctions: Earth-Science Reviews, v. 53, p. 1–33, doi:10.1016/S0012-8252(00)00037-4.

Wignall, P.B., 2005, The link between large igneous province eruptions and mass extinctions: Elements, v. 1, p. 293–297, doi:10.2113/gselements.1.5.293.

Wignall, P.B., Newton, R.J., and Little, C.T.S., 2005, The timing of paleoenvironmental change and cause-and-effect relationships during the Early Jurassic mass extinction Europe: American Journal of Science, v. 305, p. 1014–1032, doi:10.2475/ajs.305.10.1014.

Wignall, P.B., Bond, D.P.G., Kuwahara, K., Kakuwa, Y., Newton, R.J., and Poulton, S.W., 2010, An 80 million year oceanic redox history from Permian to Jurassic pelagic sediments of the Mino-Tamba terrane, SW Japan, and the origin of four mass extinctions: Global and Planetary Change, v. 71, p. 109–123, doi:10.1016/j.gloplacha.2010.01.022.

Williford, K.H., Ward, P.D., Garrison, G.H., and Buick, R., 2007, An extended organic carbon-isotope record across the Triassic-Jurassic boundary in the Queen Charlotte Islands, British Columbia, Canada: Palaeogeography, Palaeoclimatology, Palaeoecology, v. 244, p. 290–296, doi:10.1016/j.palaeo.2006.06.032.

Zakharov, V.A., Bogomolov, Y.I., Il'ina, V.I., Konstantinov, A.G., Kurushin, N.I., Lebedeva, N.K., Meledina, S.V., Nikitenko, B.L., Sobolev, E.S., and Shurygin, B.N., 1997, Boreal zonal standard and biostratigraphy of the Siberian Mesozoic: Russian Geology and Geophysics, v. 38, p. 965–993.

Zakharov, V.A., Shurygin, B.N., Il'ina, V.I., and Nikitenko, B.L., 2006, Pliensbachian–Toarcian biotic turnover in north Siberia and the Arctic region: Stratigraphy and Geological Correlation, v. 14, p. 399–417, doi:10.1134/S0869593806040046.

Manuscript Accepted by the Society 31 January 2014

Printed in the USA

The Geological Society of America
Special Paper 505
2014

Volcanism of the Central Atlantic magmatic province as the trigger of environmental and biotic changes around the Triassic-Jurassic boundary

József Pálfy*
Department of Physical and Applied Geology, Eötvös University, Budapest, H-1117, Hungary and
MTA–MTM–ELTE Research Group for Paleontology, P.O. Box 137, Budapest, H-1431, Hungary

Ádám T. Kocsis
MTA–MTM–ELTE Research Group for Paleontology, P.O. Box 137, Budapest, H-1431, Hungary and
Department of Physical and Applied Geology, Eötvös University, Budapest, H-1117, Hungary

ABSTRACT

In the last decade, major advances have been made in our understanding of the end-Triassic mass extinction, related environmental changes, and volcanism of the Central Atlantic magmatic province. Studies of various fossil groups and synoptic analyses of global diversity document the extinction and subsequent recovery. The concomitant environmental changes are manifested in a series of carbon isotope excursions (CIE), suggesting perturbations in the global carbon cycle. Besides the earlier-recognized initial and main negative anomalies, a more complex picture is emerging with other CIEs, both negative and positive, prior to and following the Triassic-Jurassic boundary. The source of isotopically light carbon remains debated (methane from hydrate dissociation vs. thermogenic methane), but either process is capable of amplifying an initial warming, resulting in runaway greenhouse conditions. Excess CO_2 entering the ocean causes acidification, an effective killing mechanism for heavily calcified marine biota that appears implicated in the reef crisis. The spatial and temporal extent of Central Atlantic magmatic province volcanism is established through a growing data set of radiometric ages. Since the Central Atlantic magmatic province is one of the largest Phanerozoic large igneous provinces, volcanic CO_2-driven warming is plausible as a key factor in the chain of Triassic-Jurassic boundary events. Greenhouse warming may have been punctuated by short-term cooling episodes due to H_2S emission and production of sulfate aerosols, a process more difficult to trace in the stratigraphic record. Taken together, recently generated data significantly increase the support for Central Atlantic magmatic province volcanism as a viable trigger for the environmental and biotic changes around the Triassic-Jurassic boundary.

*palfy@nhmus.hu

Pálfy, J., and Kocsis, Á.T., 2014, Volcanism of the Central Atlantic magmatic province as the trigger of environmental and biotic changes around the Triassic-Jurassic boundary, *in* Keller, G., and Kerr, A.C., eds., Volcanism, Impacts, and Mass Extinctions: Causes and Effects: Geological Society of America Special Paper 505, p. 245–261, doi:10.1130/2014.2505(12). For permission to copy, contact editing@geosociety.org.

INTRODUCTION

Since mass extinctions in the Phanerozoic history of life were recognized through modern synoptic analyses of the fossil record (Raup and Sepkoski, 1982), understanding these events and searching for their causes have been major targets of multidisciplinary research in earth sciences. The proposal of an extraterrestrial impact as the cause for the end-Cretaceous extinction (Alvarez et al., 1980) triggered an unprecedented amount of research into this event. However, because the end-Permian extinction stands out as the largest crisis (Fig. 1; Alroy, 2008), some of the attention has shifted toward this event (Benton and Twitchett, 2003; Erwin, 1994). The end-Triassic extinction, long recognized as another one of the "Big Five" mass extinctions, had been relatively neglected in early research (Hallam, 1990). Ranked third by measures of extinction rate and loss of taxonomic diversity (Alroy, 2008) and also by ecological severity (McGhee et al., 2004), the end-Triassic event has increasingly come into focus, recently helped by international collaborative research efforts such as an International Geoscience Programme (IGCP) project in the early 2000s (Hesselbo et al., 2007).

A common conclusion of studies of extinction events is that biotic turnover was concomitant with severe environmental change detrimental to life. Growing evidence supporting the Alvarez hypothesis (Schulte et al., 2010) led to proposals that extraterrestrial impacts are uniformly responsible for extinction crises (Rampino et al., 1997; Raup, 1992), through a cascade of environmental perturbations. The impact that generated the large Manicouagan crater was suggested to have caused the end-Triassic extinction (Olsen et al., 1987), and an iridium anomaly was also reported to support this case (Olsen et al., 2002). However, U-Pb dating of Manicouagan melt rock yielded an age that predates the Triassic-Jurassic boundary by more than 10 m.y. (Hodych and Dunning, 1992), and the Ir anomaly is more than an order of magnitude less than that measured at the Cretaceous-Paleogene boundary; thus, an impact scenario for the end-Triassic events remains unsubstantiated. Alternatively, volcanism of large igneous provinces has also been proposed as the recurring ultimate cause of extinctions, well supported by temporal coincidence (Courtillot and Renne, 2003; Wignall, 2001). Similar to the Deccan Traps, which are implicated in the end-Cretaceous event (Courtillot et al., 1988; Courtillot and Fluteau, 2010; Keller et al., 2008, 2009, 2012), and the Siberian Traps, which are now widely held to trigger the end-Permian extinction (Renne and Basu, 1991; Sobolev et al., 2011), the Central Atlantic magmatic province is linked to the end-Triassic event. Marzoli et al. (1999) demonstrated the large areal extent and determined the age of Central Atlantic magmatic province, which was shown to broadly coincide with the Triassic-Jurassic boundary (Pálfy et al., 2000a), although details of the correlation remain debated. Synchronous environmental change was exemplified by perturbation of the carbon cycle, which was also linked to Central Atlantic magmatic province volcanism (Hesselbo et al., 2002).

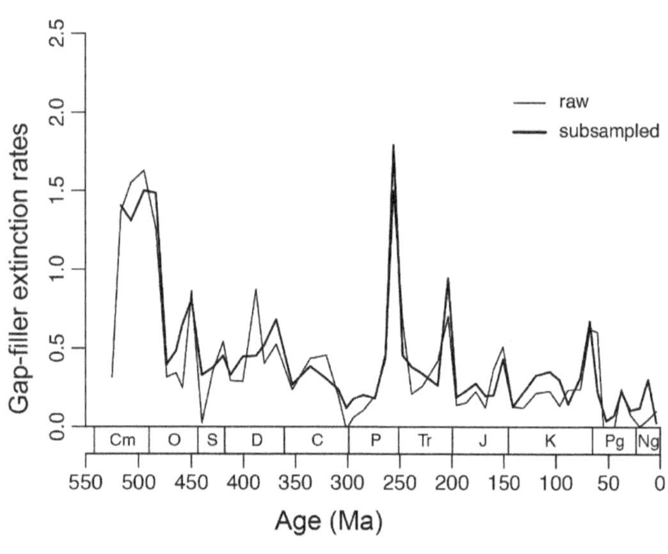

Figure 1. Corrected gap-filler raw and sampling standardized extinction rates of marine invertebrate genera calculated using the gap-filler metrics from Paleobiology Database (PBDB) occurrence data (from Alroy, 2014). Note that the end-Triassic extinction is among the most severe mass extinctions of the Phanerozoic. Cm—Cambrian, O—Ordovician, S—Silurian, D—Devonian, C—Carboniferous, P—Permian, Tr—Triassic, J—Jurassic, K—Cretaceous, Pg—Paleogene, NG—Neogene.

The role of the Central Atlantic magmatic province as a driving force for the end-Triassic extinction was reviewed earlier by Pálfy (2003), within a monograph devoted to multifaceted research on the Central Atlantic magmatic province (Hames et al., 2003). Another in-depth treatment of the topic is found in Tanner et al. (2004). Much new data generated and major advances made in the past decade warrant a new summary. Thus, the main goal of this paper is to provide an updated review of the role of Central Atlantic magmatic province in triggering environmental and biotic change at the end of the Triassic. Building on the synthesis of Hesselbo et al. (2007), particular emphasis is given to the most recent results. The "Volcanism, Impacts and Mass Extinctions: Causes and Effects" conference, held in London in 2013, showcased an emerging view of strong and well-supported temporal and causal linkages among large igneous provinces, episodes of global environmental change, and major biotic crises. An overview of the Central Atlantic magmatic province and end-Triassic events may present a key case study in how the Earth system reacts to episodic perturbations created by magmatic events manifest in the geological record as large igneous provinces. Our aim is to test the agreement of various types of data sets and predictions of the Central Atlantic magmatic province–induced environmental change and extinction scenario. Such critical assessment will also point out those critical pieces of evidence that are still lacking and thus may help steer future research.

KEY FEATURES OF THE END-TRIASSIC BIOTIC AND ENVIRONMENTAL CHANGE

The end of the Triassic stands out in both the fossil and the stratigraphic records as a brief interval of significant change. Its two most salient aspects are the biotic extinction, ranking among the largest Phanerozoic mass extinctions, and a major perturbation of the global carbon cycle, preserved as a series of anomalies in the carbon isotope ratios. Current scenarios for these profound and correlative changes in the Earth system consider the role of Central Atlantic magmatic province volcanism as the common forcing factor for both phenomena.

Global Diversity Patterns and Rate Changes

Mass extinctions are global events of profound changes in the global ecosystem, recorded by substantial temporary increase in extinction rates and commonly by decreased global biodiversity as shown by the fossil record. The global nature of such events requires synoptic analyses of large databases, which have become an integral part of modern paleobiological studies. The pioneering, single-handed effort of database compilation (Sepkoski et al., 2002) has been replaced by the community-driven Paleobiology Database (PBDB), which has grown enormously over the past decade. At the time of writing this paper, PBDB contains over 1.1 million occurrences form nearly 150,000 fossil collections. The growth of the analytical toolkit is also impressive: novel standardization methods (Alroy, 2010; Alroy et al., 2008) and a number of robust metrics have been developed (Alroy, 2008; Foote, 2000) to assess the dynamical aspects of taxonomic diversity. Contrary to earlier, range-based compilations, the PBDB allows the application of sampling-standardization methods, which correct for some significant biases in the known fossil record. Although earlier analyses lumped the Rhaetian into the Norian stage as the terminal Triassic unit, as long advocated by Tozer (1967), the Rhaetian is now universally accepted, and its global boundary stratotype section and point (GSSP) designation is under way (Krystyn et al., 2007). Unfortunately, several earlier studies failed to distinguish the Rhaetian, a problem exacerbated by the recently recognized long duration of the Norian stage (Mundil et al., 2010). Although most recent works (Alroy, 2014) treat the Rhaetian separately, there is still a need for higher stratigraphic resolution in global analyses. Even if PBDB-based studies employ stage-level time slices, they provide the only means by which to compensate for biases in the fossil record introduced by the poor stratigraphic record and sampling around the Triassic-Jurassic boundary. Recent comprehensive analyses of the PBDB confirm the severity of the end-Triassic extinction (Fig. 1; Alroy et al., 2008, Alroy, 2014).

Carbon Cycle Perturbation from the $\delta^{13}C$ Record

Recurrent coincidence of mass extinction events and carbon isotope excursions (CIE) in the geological record suggests that both could be driven by environmental change. CIEs were already well established for the end-Cretaceous and the end-Permian events (Holser et al., 1996) when the first robust data sets were published by the early 2000s to demonstrate a CIE for the end-Triassic event (Hesselbo et al., 2002; Pálfy et al., 2001; Ward et al., 2001). The pattern of a sharp and prominent negative anomaly (termed initial CIE) that was followed by another, more protracted one (the main CIE) was subsequently reproduced from numerous localities worldwide. Local differences, deviation from the most common observations, and additional complexities were also discovered. Prominent issues in the focus of current research pertain to those isotope shifts that have not been unambiguously recognized in all or most studied sections (1) the negative precursor excursion in the Rhaetian (Ruhl and Kurschner, 2011), (2) the validity of a positive excursion in the Hettangian (Williford et al., 2007), and (3) the reproducibility of some smaller variations in between the larger anomalies (Bartolini et al., 2012). Details of the carbon isotope events will be discussed later, in the context of the way in which the emerging picture of carbon cycle history across the Triassic-Jurassic transition can be explained by perturbations triggered by Central Atlantic magmatic province magmatism.

BRIEF OUTLINE OF CENTRAL ATLANTIC MAGMATIC PROVINCE MAGMATISM

The extensive occurrence of continental flood basalts that formed close to the Triassic-Jurassic boundary was coined the Central Atlantic magmatic province by Marzoli et al. (1999). The name has gained widespread acceptance as it captures the fact that lava flows, dikes, and sills in North and South America, northeast Africa, and southwest Europe are cogenetic, and they are related to the opening of the Central Atlantic Ocean basin (Fig. 2), which represents the first episode of the breakup of Pangea (Schlische et al., 2003). Large igneous provinces are now recognized as records of brief and extremely voluminous and spatially extensive episodes of mostly extrusive magmatism, including continental flood basalts (Saunders, 2005). Although early compilations (Coffin and Eldholm, 1994) did not include volcanic provinces around the Central Atlantic, $^{40}Ar/^{39}Ar$ ages obtained by Marzoli et al. (1999) from Brazil led to the recognition of the true extent of ca. 200 Ma magmatism in South America and allowed definition of the Central Atlantic magmatic province. Although it is now split among four continents, and its deep erosion hinders estimates of its original size, the original area of Central Atlantic magmatic province may have exceeded 1.1×10^7 km^2, whereas the total volume of erupted lavas approached 2.4×10^6 km^3 (McHone, 2003). These estimates are now in general use by authors of reviews and new studies on the Central Atlantic magmatic province (e.g., Tanner et al., 2004; Blackburn et al., 2013).

Lithologically, most Central Atlantic magmatic province rocks are tholeiitic basalts, but magmatic evolution resulted in

Figure 2. Global paleogeographic map at the Triassic-Jurassic transition (ca. 200 Ma). Surface distribution of Central Atlantic magmatic province rocks is in dark red; reconstructed total area of Central Atlantic magmatic province is shown in light and dark red (after McHone, 2003). Numbered areas and localities include the important rift basins on the eastern North American and the northwest African side of the future Central Atlantic, and key sections with marine biostratigraphic data, $\delta^{13}C$ data, or radiometric ages mentioned in the text. 1—Fundy Basin (Blackburn et al., 2013; Schoene et al., 2006); 2—Hartford Basin (Whiteside et al., 2010); 3—Newark Basin (Blackburn et al., 2013; Marzoli et al., 2011; Schaller et al., 2011; Whiteside et al., 2010); 4—Culpeper Basin (Marzoli et al., 2011); 5—Argana Basin (Blackburn et al., 2013; Deenen et al., 2010); 6—High Atlas Basin (Marzoli et al., 2004); 7—Northern Calcareous Alps (Kürschner et al., 2007; Ruhl and Kürschner, 2011; Ruhl et al., 2009, 2011); 8—Pelso Unit, Hungary (Pálfy et al., 2001, 2007); 9—Western Carpathians (Michalík et al., 2007, 2010); 10—southern Alps (Bachan et al., 2012; Galli et al., 2007; van de Schootbrugge et al., 2008); 11—Apennines (Bachan et al., 2012; van de Schootbrugge et al., 2008); 12—southern Germany (van de Schootbrugge et al., 2008; Ruhl and Kürschner, 2011); 13—Polish Trough (Pienkowski et al., 2012); 14—northern Germany (van de Schootbrugge et al., 2013); 15—Danish Basin (Lindström et al., 2012); 16—southwest Britain (Clémence et al., 2010; Hesselbo et al., 2002; Korte et al., 2009; Ruhl et al., 2010); 17—East Greenland (McElwain et al., 1999; Hesselbo et al., 2002); 18—Queen Charlotte Islands (Friedman et al., 2008; Pálfy et al., 2000a; Ward et al., 2001; Williford et al., 2007); 19—New York Canyon, Nevada (Bartolini et al., 2012; Guex et al., 2004, 2012; Schoene et al., 2010; Ward et al., 2004); 20—Utcubamba Valley, Peru (Schaltegger et al., 2008; Schoene et al., 2010). Base map is from Ruiz-Martínez et al. (2012).

temporally significant geochemical differences that permit correlation between now widely separated parts of the province, as well as matching of cogenetic shallow intrusive units (dikes and sills) and extrusive units. The eruptive sequence of the Moroccan section of the Central Atlantic magmatic province consists of lower, intermediate, upper, and recurrent basalts, where the SO_2- and TiO_2-rich lower unit evolves into an SO_2- and TiO_2-poor upper unit, and the younger recurrent unit is characterized by the highest TiO_2 and low SO_2 content (Marzoli et al., 2004). The lower basalt in the Argana Basin may be older than the lowest flows in eastern North America, where the major- and trace-element signature of the North Mountain Basalt is transitional between the Moroccan lower and intermediate basalts, and the Orange Mountain Basalt, the lowest eruptive unit in the Newark Basin, is geochemically similar to the intermediate unit (Marzoli et al., 2004, 2011; Blackburn, 2013). Several basins in eastern North America have a tripartite Central Atlantic magmatic province eruptive complex, e.g., in the Newark Basin succession, the Orange Mountain Basalt is overlain, with intercalated sedimentary formations, by flows of the Preakness and Hook Mountain Basalts. The typical thickness of individual lava units there ranges from 50 to 200 m each, but the North Mountain Basalt in the center of Fundy Basin may reach up to 1000 m (McHone, 2003). The Palisades sill, which has a complex history and appears to be geochemically related to both the Orange Mountain and Preakness basalts, reaches 350 m in thickness along the Hudson Valley in New Jersey (Marzoli et al., 2011).

The origin of Central Atlantic magmatic province magmatism is subject to ongoing debate, with conflicting views favoring either a mantle plume origin (Ruiz-Martínez et al., 2012) or partial melting of the lithospheric mantle under the central part of the Pangea supercontinent (Callegaro et al., 2013). Discussing the processes involved in generating the Central Atlantic magmatic province is beyond the scope of this review; instead, our emphasis is on the volume and fast production rate of its volcanic products and released gases that are relevant to assess its possible impact on the environment and indirectly on biota.

ESTABLISHING SYNCHRONY: DATING AND DURATION OF END-TRIASSIC EVENTS AND THE CENTRAL ATLANTIC MAGMATIC PROVINCE

To establish a cause-and-effect relationship between the Central Atlantic magmatic province and environmental and biotic crises, first and foremost their synchrony needs to be demonstrated. Neither radiometric dating of primarily basaltic volcanics, nor numeric calibration of the chronostratigraphic Triassic-Jurassic boundary are straightforward, and both indeed have been fraught with problems. However, recent progress has led to convincing proofs for temporal coincidence, a prerequisite of causality. In addition, dating efforts have also yielded constraints on the duration of events, where abruptness and brevity are important factors in assessing catastrophic scenarios.

U-Pb Dating of Marine Boundary Sections

Before the millennium, the age of the Triassic-Jurassic boundary was estimated through interpolation using a sparse radiometric database that contained ages of dubious quality, obtained using now-outdated methods (K-Ar, Rb-Sr) on low-temperature minerals. Successive geological time scales prior to 2000 quoted Triassic-Jurassic boundary ages that differed greatly and carried large uncertainties (Pálfy, 1995). Major improvements were achieved through direct radiometric dating of volcanogenic layers within marine Triassic-Jurassic boundary sections (Pálfy et al., 2000a) and the selective use of U-Pb and $^{40}Ar/^{39}Ar$ dates of superior accuracy and precision (Pálfy et al., 2000b). Thus, a zircon U-Pb age of 199.6 ± 0.3 Ma (Queen Charlotte Islands or Haida Gwaii, Canada) was obtained from a volcanic ash layer from Kunga Island near the Triassic-Jurassic boundary, defined by radiolarian biostratigraphy, which later anchored the boundary age in the 2004 *Geological Time Scale* (GTS) (Ogg, 2004). The first decade of the twenty-first century saw major methodological advances in U-Pb geochronology, as dating chemically abraded single grains emerged as the method of choice for time-scale calibration due to its superior performance in achieving concordance. Schaltegger et al. (2008) obtained a single-crystal zircon U-Pb age of 201.58 ± 0.28 Ma for a volcanic ash layer in the Utcubamba Valley (Peru), from within the ammonite-constrained Triassic-Jurassic boundary interval. The redating of the Kunga Island Triassic-Jurassic boundary ash sample using the new methods yielded a comparable age of 201.7 ± 0.6 Ma (Friedman et al., 2008). Further analyses of the ash sample dated by Schaltegger et al. (2008) revised its age to 201.40 ± 0.18 Ma, whereas two stratigraphically somewhat younger ashes from the Peruvian section yielded dates of 201.36 ± 0.13 and 201.29 ± 0.16 Ma (Schoene et al., 2010). Notably, the latter dates bracket the first occurrence of the ammonoid *Psiloceras spelae*, the biostratigraphic marker event of the recently defined GSSP for the Triassic-Jurassic boundary (Hillebrandt et al., 2013). The official system boundary thus corresponds to a point in the early biotic recovery phase, rather than the earlier mass extinction event itself. Another ash bed, stratigraphically immediately above the first occurrence of *Psiloceras spelae*, was dated from the New York Canyon section in Nevada as 201.33 ± 0.13 Ma (Schoene et al., 2010). Seven additional ash beds from the uppermost Triassic Rhaetian and lowermost Jurassic Hettangian Stages in the Utcubamba Valley in Peru were reported by Guex et al. (2012). Internally consistent with their stratigraphic order and in agreement with the other ages, these allow the duration of the Hettangian Stage to be constrained, and hence the initial recovery phase, to ~2 m.y. by a 199.43 ± 0.10 Ma date close to the Hettangian-Sinemurian Stage boundary.

This set of new U-Pb ages from chemically abraded single zircons allowed a significant improvement in the 2012 GTS for the Triassic-Jurassic boundary age: Ogg and Hinnov (2012) picked 201.3 ± 0.2 Ma for the boundary age, which is thus 1.7 m.y. older than that in the 2004 GTS. Critically, although the end-Triassic extinction preceded the GSSP-defined system boundary, this tight set of biostratigraphically constrained U-Pb dates suggests that the time difference may be close to 100 k.y.

Age and Duration of Central Atlantic Magmatic Province from $^{40}Ar/^{39}Ar$ and U-Pb Dating

Attempts to firmly link the Central Atlantic magmatic province to the Triassic-Jurassic boundary events also stimulated much geochronological research on the flood basalts. In a Central Atlantic magmatic province monograph, Baksi (2003) presented a critical compilation of 21 $^{40}Ar/^{39}Ar$ ages that were deemed accurate and showed a significant concentration around 200 Ma. Subsequent work provided a large number of new $^{40}Ar/^{39}Ar$ ages from Morocco (Knight et al., 2004; Marzoli et al., 2004; Verati et al., 2007) and the eastern United States (Jourdan et al., 2009; Marzoli et al., 2011). A critical summary based on selected $^{40}Ar/^{39}Ar$ ages available up to 2006 that met stringent quality criteria concluded that intrusive phases of Central Atlantic magmatic province magmatism commenced at 202 Ma, onset of extrusive activity was abrupt at 200 Ma and peaked at 199 Ma, and minor later phases centered at 195 and 192 Ma were also recognized, although those did not significantly add to the area and volume of the province (Nomade et al., 2007). However, comparison of $^{40}Ar/^{39}Ar$ and U-Pb dates proved difficult with increasing realization of the inaccurately known decay constant of ^{40}K (Begemann et al., 2001), until a newly determined constant became available (Renne et al., 2010). Significantly, the newest sets of $^{40}Ar/^{39}Ar$ ages from eastern North America together with previously published and recalculated ages identify peak volcanic activity at 201.4 ± 0.9 Ma both in eastern North American basins (Newark, Culpeper, Fundy) and the Moroccan High Atlas and Argana Basin, indistinguishable in age from the Triassic-Jurassic boundary (Marzoli et al., 2011). Intrusive activity was more extended, as exemplified by the complex Palisades sill, for which ages range from 195.1 ± 2.1 Ma to 202.8 ± 1.8 Ma (Marzoli et al., 2011).

Although flood basalts generally lack or only rarely contain zircon as an accessory mineral, exceptions are known, and the applicability of U-Pb dating to Central Atlantic magmatic province rocks has been demonstrated (Dunning and Hodych, 1990; Hodych and Dunning, 1992). The first chemically abraded single-crystal ages were reported from the North Mountain Basalt by Schoene et al. (2006; 201.27 ± 0.06 Ma), followed by another suite of analyses from this flow unit yielding an age of 201.38 ± 0.02 Ma (Schoene et al., 2010). Recently, Blackburn et al. (2013) dated the North Mountain Basalt as 201.566 ± 0.031 and the Preakness Basalt (the second extrusive unit in the Newark Basin) as 201.274 ± 0.032 Ma. These authors also obtained an age of 201.520 ± 0.034 Ma for part of the Palisades sill, which is thought to have fed the Orange Mountain Basalt (i.e., the lowermost extrusive unit in the Newark Basin; Puffer et al., 2009), in addition to a coherent set of five other high-precision U-Pb ages from Central Atlantic magmatic province intrusive units.

In summary, recent advances in both $^{40}Ar/^{39}Ar$ and U-Pb geochronology have resulted in a remarkable consistency of increasingly precise radiometric ages on Central Atlantic magmatic province products. The latest results confirm an abrupt onset of extrusive volcanism and rapid succession of major lava flows, the oldest of which appears coeval with the end-Triassic extinction level immediately preceding the system boundary in marine sections.

Cyclostratigraphy and Astrochronology of Triassic-Jurassic Boundary Sections

Independent of radiometric ages, important constraints on the duration of environmental and biotic events have been obtained from astrochronology of cyclic sedimentary successions. Mostly lacustrine Upper Triassic to lowermost Jurassic strata in the Newark Basin display prominent cyclicity, interpreted as precession and eccentricity Milankovitch cycles (Olsen and Kent, 1996). The long, continuous, and highly resolved record from a series of stratigraphically overlapping cores serves as a basis of a Late Triassic astrochronologic scale (Olsen and Kent, 1999). Studies of complementary outcrop sections have been partly aimed at constraining the duration of Central Atlantic magmatic province eruptive units, and the time gap between them, exploiting the 20 k.y. resolution afforded by precession cycles. In the Newark Basin, duration of Central Atlantic magmatic province volcanism is established as 610 k.y. (Whiteside et al., 2010), assuming completeness of the sedimentary successions between the flows. The astrochronologically estimated time separating eruptions of the Orange Mountain and Preakness basalts is 250 ± 20 k.y. (Whiteside et al., 2010), which is in agreement with U-Pb geochronological results of Blackburn et al. (2013). Astrochronologic constraints on the Preakness Basalt suggest an eruption interval of ~80 k.y. (Whiteside et al., 2007).

Continental deposits below and above Central Atlantic magmatic province basalts in the Argana Basin in Morocco also display cyclicity. Deenen et al. (2010) combined cyclo- and magnetostratigraphic correlation to suggest that the oldest basalt there predates the Orange Mountain Basalt by ~20 k.y., thus representing the oldest known eruptive product of the Central Atlantic magmatic province.

Turning to marine strata, cyclicity of lowermost Jurassic formations on the Somerset coast of SW England was used earlier to provide a minimum estimate of 1.29 m.y. for the duration of the Hettangian Stage (Weedon et al., 1999). Analyses of the more complete composite St. Audrie's Bay and East Quantoxhead sections allowed a more robust estimate of 1.8 m.y. (Ruhl et al., 2010), which is in agreement with U-Pb age constraints on the Hettangian Stage boundaries from Peru (Guex et al., 2012; Schaltegger et al., 2008). Although not directly relevant to the Central Atlantic magmatic province, such constraints are useful for assessing the temporal framework of carbon isotope anomalies and the biotic extinction and recovery.

Additional Data for Correlation

Central Atlantic magmatic province volcanics are embedded in continental sedimentary strata; thus, palynostratigraphy is the primary means of biostratigraphic correlation and establishing a direct link to mass extinction. A long-held view that floral turnover preceded the emplacement of the oldest extrusives was based on palynological studies in the Newark Basin (Fowell and Olsen, 1993; Olsen et al., 2002; Whiteside et al., 2007). This view was challenged by Cirilli et al. (2009), who reported a Late Triassic pollen assemblage from above a flow unit of North Mountain Basalt and claimed that the last occurrence of the palynomorph index species *Patinasporites densus*, used to infer the position of the Triassic-Jurassic boundary, was actually much older in the Rhaetian.

Significantly, the conchostracan biostratigraphic data of Kozur and Weems (2007, 2010) suggest that sedimentary strata overlying the basal Central Atlantic magmatic province basalt units in different basins of the Newark Supergoup contain a latest Rhaetian fauna; hence, these authors imply a Rhaetian emplacement for the first Central Atlantic magmatic province flows.

A different line of evidence to link the Central Atlantic magmatic province with the end-Triassic events is based on direct observation of distal Central Atlantic magmatic province ejecta in marine sections. At Kendlbachgraben, in a classical Triassic-Jurassic boundary section that is well correlated with the GSSP also in the Northern Calcareous Alps, occurrence of mafic minerals, their characteristic clay alteration products (vermiculite and low-charged smectite), anomalous trace-element patterns, and clay-altered volcanic glass spherules were interpreted to record distal fallout of Central Atlantic magmatic province ejecta (Pálfy and Zajzon, 2012). This horizon occurs immediately below the extinction level (McRoberts et al., 2012) and the initial negative CIE in a marine section.

CAUSATION AND POSSIBLE SCENARIOS

The Central Atlantic magmatic province volcanism is not unique as a large igneous province implicated in causing a major extinction and environmental perturbation. In fact, repeated temporal coincidences of such events have been noted by several workers (Courtillot and Renne, 2003; Rampino and Stothers, 1988), leading to the proposal of a general model for the environmental and biotic consequences of large igneous province formation (Wignall, 2005, 2001). Wignall's model is readily adaptable to the Central Atlantic magmatic province–end-Triassic extinction linkage (Pálfy, 2003). It builds a systemic network of several testable cause-and-effect relationships for which much observational evidence has been collected recently.

The key component of the model is the atmospheric loading from volcanic emission. CO_2 and SO_2 are major climate-forcing gases, although to opposing effects and at different time scales, and halogens are also of environmental significance. McHone (2003) estimated that the total volatile emission from Central Atlantic magmatic province volcanism introduced 5.19×10^{18} g CO_2, 2.31×10^{18} g SO_2, 1.58×10^{18} g Cl, and 1.11×10^{18} g F into the atmosphere. Volume, duration, and frequency of individual eruptions are key factors that control the actual volatile fluxes, but their determination from the geological record carries significant uncertainty. On the basis of studies on Deccan flood basalts, Self et al. (2006) emphasized the effect of SO_2 flux, formation of sulfate aerosols, and transient cooling. On the other hand, these authors suggested that large igneous province–induced CO_2 loading is inadequate to force major climate change, because it represents a relatively small contribution to the total atmospheric CO_2 budget, especially at times when pCO_2 is already relatively high. Turning to the better-studied Deccan Traps for analogy, it is instructive to note that its eruption history was strongly pulsed, with the main phase that produced the bulk of the lava volume consisting of short mega-eruptions (lasting only for decades or a few hundred years), constrained by paleomagnetic secular variation (Chenet et al., 2009). The deeply eroded Central Atlantic magmatic province is not amenable to studies of such detail, but similarly pulsed eruptive activity can be reasonably assumed.

Apart from direct volcanogenic CO_2 emission, oxidation of methane, either from gas hydrate dissociation or thermogenically generated, may provide an alternative source of extra CO_2. Central in Wignall's model is the CO_2-induced global warming, amplified by the positive feedback of hydrate dissociation, which is expected to be reflected in the carbon isotope record. Increased weathering is a predicted consequence of hot and humid climate; its geochemical tests include Sr and Os isotopic ratios (Cohen, 2004). Fresh surfaces of lava flows exposed to intense weathering enhance nutrient delivery to the ocean, stimulating eutrophication of surface waters. Warming also promotes water-mass stratification and stagnation in the ocean, which favor development of anoxia. The model also posits that SO_2 and halogen emissions result in brief episodes of cooling and acid rain, although these are more difficult to trace in the stratigraphic record.

Ocean acidification is an important additional environmental consequence (Greene et al., 2012; Hönisch et al., 2012) of volcanogenic CO_2 emission, with deleterious effects for the marine biota, especially calcifiers. Several other novel proxy records have been recognized in recent studies. Next, we will review how compatible the observed patterns of environmental and biotic change are with Central Atlantic magmatic province forcing.

ENVIRONMENTAL CHANGE AROUND THE TRIASSIC-JURASSIC BOUNDARY

Elevated Atmospheric CO_2 and Greenhouse Warming

Searches for proxy records for both atmospheric CO_2 and paleoclimate were stimulated by an original study of macroplant fossils from the Astartekløft section in East Greenland, which demonstrated changes in leaf stomata and inferred changes in pCO_2 and super-greenhouse warming across the Triassic-Jurassic boundary (McElwain et al., 1999). Subsequent stomatal analyses of separate, taxonomically and physiologically distinct leaves from taphonomically different and geographically distant localities confirmed a pattern of steep [CO_2] rise and doubling of values from 1000 ppm in the Rhaetian to 2000–2500 ppm at the Triassic-Jurassic boundary, remaining elevated during the Hettangian (Steinthorsdottir et al., 2011).

Atmospheric [CO_2] can also be inferred from $\delta^{13}C$ of pedogenic carbonates. Although an earlier study based on low-resolution sampling found no significant [CO_2] increase across the Triassic-Jurassic boundary (Tanner et al., 2001), recent work from the Newark Basin with much increased sampling resolution revealed large-amplitude, pulsed rises after each main eruptive unit in Central Atlantic magmatic province (Schaller et al., 2011). Calculated absolute values are about twice as high as those inferred from stomatal density (McElwain et al., 1999; Steinthorsdottir et al., 2011), but the doubling from pre-eruptive values (2000 ppm) to posteruptive, perturbed values (up to 4400 ppm) is similar (Schaller et al., 2011). However, modeling suggests that the [CO_2] increase was heavily dependent on the eruptive time scale, and the estimates of Schaller et al. (2011) may be too high (Rampino and Caldeira, 2011). Return to background level appears to have taken place on a time scale of a few hundred thousand years, suggestive of drawdown of CO_2 through increased continental silicate weathering.

Greenhouse warming forced by rising CO_2 is expected to drive diverse changes in climate; thus, in addition to temperature rise, humidity and storminess are also expected to increase in some regions. These changes could lead to increased frequency of wildfires. Indeed, fire indicators have been repeatedly reported from several continental Triassic-Jurassic boundary sections. A peak in the abundance of charcoal was found in East Greenland (Belcher et al., 2010) and the Danish Basin (Petersen and Lindström, 2012), in both cases accompanying changes in floral composition. While the increases in fire intensity and frequency point to climate warming, it is not necessarily related to the biotic turnover, as modern studies, at least at regional scale, have found that wildfires promote rather than suppress plant diversity (e.g., Keeley et al., 2005). Polycyclic aromatic hydrocarbons (PAHs) are organic geochemical tracers of combustion, and these have been found to occur at the Triassic-Jurassic boundary in the Danish Basin (Petersen and Lindström, 2012) and the Polish Trough (Marynowski and Simoneit, 2009; Pienkowski et al., 2012). However, PAHs may not only derive from wildfires, as van de Schootbrugge et al. (2009) reported their occurrence without a coeval charcoal peak and interpreted their origin from combustion of sedimentary organic matter by ascending Central Atlantic magmatic province lavas.

Clay minerals are also useful indicators of paleoclimate. There are numerous, geographically widespread localities where sudden dominance of kaolinite, a weathering product under hot and humid climate, occurs consistently in the uppermost Rhaetian, at or immediately after the Triassic-Jurassic boundary: in the Northern Calcareous Alps (Zajzon et al., 2012), Western Carpathians (Michalík et al., 2010), Germanic Basin (van de Schootbrugge et al., 2009), Polish Trough (Pienkowski et al., 2012), and Danish Basin (Ahlberg et al., 2003). Soil acidification, induced by acid rains from volcanogenic SO_2, may have also contributed to accelerated weathering.

Climatically enhanced continental weathering leaves a geochemical signature in marine sediments, as the ratio of Sr and Os isotopes changes toward more radiogenic values when weathering of old, crustally derived rocks increases. On the contrary, the isotopic balance could shift toward nonradiogenic values from the contribution of freshly exposed tholeiitic flood basalt provinces. First applied to investigate the link between the Central Atlantic magmatic province and the Triassic-Jurassic boundary by Cohen and Coe (2002), their later, more detailed study (Cohen and Coe, 2007) confirmed that a shift to unusually nonradiogenic Os isotope ratios is best interpreted to reflect rapid weathering of Central Atlantic magmatic province lavas at equatorial latitudes.

More controversial are the findings of Kuroda et al. (2010) from a pelagic section in Japan, where the shift toward nonradiogenic Os ratios starts relatively early in the Rhaetian, suggesting for these authors an early onset of Central Atlantic magmatic province volcanism. Analyzing the Sr isotopic composition of conodont apatite, similar conclusions were reached by Callegaro et al. (2012). These tantalizing hints for possible early onset of Central Atlantic magmatic province volcanism are at odds with known age distributions of Central Atlantic magmatic province extrusive units and need validation from further studies.

Details of the $\delta^{13}C$ Record

Major CIEs near the Triassic-Jurassic boundary have received much attention since their initial discoveries (Hesselbo et al., 2002; Pálfy et al., 2001; Ward et al., 2001) for two principal reasons: They reflect carbon cycle perturbations that are key to understanding the environmental history, and the $\delta^{13}C$ curve is increasingly used for stratigraphic correlation, among and between both marine and terrestrial sections. Currently, the carbon isotope stratigraphy for over 30 Triassic-Jurassic boundary sections results in global coverage (Fig. 2). The most important marine reference sections are concentrated in the western Tethys (Northern Calcareous Alps: Kuhjoch, Tiefengraben, Kendlbachgraben and others [Kuerschner et al., 2007; Ruhl et al., 2009; Ruhl and Kurschner, 2011], southern Alps: Val Adrara [van de Schootbrugge et al., 2008], Csővár [Pálfy et al., 2007], parts of the European epicontinental seaway, SW Britain: St. Audrie's Bay, Doniford, Lavernock [Clémence et al., 2010; Hesselbo et al., 2002; Korte et al., 2009; Ruhl et al., 2010], and eastern paleo-Pacific in western North America such as Queen Charlotte Islands [Ward et al., 2004; Williford et al., 2007] and New York Canyon, Nevada [Guex et al., 2004; Ward et al., 2007]). However, only a few section have been studied so far from large swaths of Panthalassa (New Zealand—Akikuni et al., 2010; Japan—Kuroda et al., 2010). Important terrestrial records of $\delta^{13}C$ evolution from Pangea are available from East Greenland (Hesselbo et al., 2002) and rift basins at the subsequent opening of Central Atlantic (Deenen et al., 2010; Whiteside et al., 2010).

The majority of carbon isotope ratio data have been measured on either bulk carbonate or bulk organic matter. Interpretation of the primary signal (i.e., inferred changes in the exchangeable atmospheric and oceanic carbon reservoirs) is not always straightforward, because carbonate might be prone to diagenetic alteration, whereas isotopic composition of organic matter may be affected by changing proportions of the source material. To circumvent the problems for carbonates, oyster shell material of low-Mg calcite was selectively analyzed by Korte et al. (2009). The reliability of organic carbon analyses was enhanced by selective analysis of wood particles or compound-specific measurement of alkanes (Ruhl et al., 2011; Whiteside et al., 2010). Results from the use of these advanced methods are in general agreement with established CIE patterns.

The carbon isotopic evolution through the Triassic was recently reviewed by Tanner (2010). Focusing here to the Rhaetian to Hettangian interval, the first CIE to note is the recently recognized late Rhaetian precursor excursion, a 2‰–3‰ negative shift observed in $\delta^{13}C_{org}$ in multiple sections (Ruhl and Kürschner, 2011). The most pronounced feature is a CIE termed the initial isotope excursion (Hesselbo et al., 2002). Now recognized in nearly 20 sections worldwide, it registers as a brief, large-amplitude negative shift of ~5‰–6‰ in organic carbon and also preserved in carbonate carbon (Clémence et al., 2010; Pálfy et al., 2007), coincident with the most prominent pulse of marine extinction (Ward et al., 2004). After a transient return to previous background values, many (but not all) sections exhibit a prolonged negative shift, termed the main isotope excursion (Hesselbo et al., 2002), which spans the GSSP-defined Triassic-Jurassic boundary as demonstrated in the stratotype section of Kuhjoch (Ruhl et al., 2009).

The most prominent feature of the Hettangian $\delta^{13}C$ curve is a large positive CIE, first reported from the Queen Charlotte Islands, as a shift of up to 5‰ in organic carbon (Williford et al., 2007). Although not universally present, it has been verified from $\delta^{13}C_{carb}$ data from the Southern Alps (van de Schootbrugge et al., 2008). Bartolini et al. (2012) recorded smaller and greater positive anomalies in the well-dated New York Canyon section in Nevada and argued that the larger, younger anomaly in the late Hettangian is correlated with the Queen Charlotte Islands. The pattern of two separate Hettangian positive CIEs agrees with the findings of Bachan et al. (2012) from Italian sections in the southern Alps and the Apennines.

Besides the problem of global reproducibility of locally or regionally observed $\delta^{13}C$ curves, causes for this series of carbon cycle perturbations also remain controversial. Central Atlantic magmatic province degassing has been invoked to directly account for the negative CIEs (Hesselbo et al., 2002), but modeling results and mass balance calculations require involvement of an additional source of light carbon (Beerling and Berner, 2002). Similar to other prominent negative CIEs at major events (e.g., Early Jurassic oceanic anoxic event, Paleocene-Eocene thermal maximum), methane-derived ^{12}C-enriched carbon may originate from either dissociation of gas hydrates triggered by global warming (Pálfy et al., 2001) or from thermal metamorphism of organic-rich sediments intruded by Central Atlantic magmatic province magmas (Ruhl et al., 2011; van de Schootbrugge et al., 2009). Changes in the source of organic matter and a greater proportion of terrestrially derived organic matter may also be reflected in the negative excursion of $\delta^{13}C_{org}$ (van de Schootbrugge et al., 2008). Positive CIEs are attributed to enhanced organic carbon burial (Bachan et al., 2012; Bartolini et al., 2012).

Sulfur Dioxide Emission and Cooling

Contrary to volcanogenic CO_2, the other major volatile component, SO_2, is expected to force short-term climate cooling due to formation of sulfate aerosols (Tanner et al., 2004). Proxies of atmospheric SO_2 are difficult to obtain. Floral change observed across the Triassic-Jurassic boundary has been hypothesized to result partly from soil acidification caused by sulfuric acid washed out from the atmosphere (van de Schootbrugge et al., 2009). Experimentally backed results from analyses of gymnosperm macroplant remains from East Greenland demonstrated leaf physiognomic changes across the Triassic-Jurassic boundary, where increased roundness of fossil leaves is compatible with those of nearest living relatives grown under experimental atmospheric conditions of increased SO_2 (Bacon et al., 2013). Stomatal density and the derived measure of stomatal index of plant leaves, a key indicator for [CO_2], was found to be also affected by [SO_2]. Experimental studies on modern plants indicate different responses by different plant species, where *Gingko biloba*, the modern analogue of fossil Ginkgoales, widely used in Triassic-Jurassic boundary studies, was found to show a parallel reaction of decrease of stomatal density to both [SO_2] and [CO_2] increase, rendering deconvolution of the forcing of these two atmospheric gases problematic (Haworth et al., 2012). Compared to paleobotanical evidence, a more hypothetical inference is made from the sea-level drop interpreted in the New York Canyon section in Nevada as a result of glaciation caused by Central Atlantic magmatic province–derived SO_2-forced cooling (Schoene et al., 2010).

Changes in Ocean Chemistry: Anoxia and Acidification

Marine anoxia has been frequently cited as a causal agent of extinctions in the marine realm. Although there are sections (e.g., in the British Isles) where organic-rich sedimentary rocks are intimately associated with the Triassic-Jurassic boundary (Wignall and Bond, 2008), anoxia no longer appears to have played a major direct role in the end-Triassic biotic crisis. However, biomarker evidence from lowermost Hettangian strata in the northwest European epicontinental seaway suggests abundant green sulfur bacteria in the upper water column, indicating euxinic conditions (absence of O_2 and presence of free H_2S) in the photic zone (Richoz et al., 2012). Organic-rich dark mudrocks are widespread in the Hettangian deposits in the epicontinental seaway bordering the western Tethys, suggesting that hothouse climate and nutrient-rich continental runoff favored the development of stratified water masses and persistent anoxic conditions for some time after the main perturbation immediately before the Triassic-Jurassic boundary (van de Schootbrugge et al., 2013).

Because the ocean acts as a sink for excess atmospheric CO_2, large increases of pCO_2 have led to ocean acidification at discrete intervals in Earth history (Hönisch et al., 2012). Volcanogenic outgassing of SO_2 also contributes to a change in ocean chemistry (Greene et al., 2012). Ocean acidification induced by Central Atlantic magmatic province degassing was first proposed by Hautmann (2004; Hautmann et al., 2008) as the dominant factor in the end-Triassic extinctions, supported by the relatively high ratio of extinct taxa having aragonite or high-Mg calcite skeletal mineralogy in the compendium of Sepkoski et al. (2002). Additional evidence is the presence of a gap in carbonate sedimentation in numerous stratigraphic sections worldwide. A comprehensive review by Greene et al. (2012) found the sedimentary evidence for Triassic-Jurassic boundary ocean acidification compelling, the Central Atlantic magmatic province to be a viable trigger, and the biotic extinction record consistent with the ocean acidification model.

However, in a sampling standardized analysis of the PBDB, Kiessling et al. (2007) found no significant relationship between the magnitude of extinctions and skeletal mineralogy of calcifying taxa. Bivalves demonstrated greater chance of survival, which may have been related to higher calcite content in their shells and their epifaunal habitat preference; this was used to rule out ocean acidification as the primary cause of the extinctions. On the other hand, Hautmann et al. (2008) argued that brachiopods were generally more affected than bivalves by the end-Triassic event, and the high amount of extinct calcitic brachiopods might bias the result of aragonite/calcite selectivity in the entire data set. This is further supported by the significantly high ratio of surviving brachiopods with phosphatic shells.

Unbuffered organisms such as corals and calcareous sponges were among the primary victims of the end-Triassic crisis, which was one of the most severe reef crises in the Phanerozoic, and which manifested in a dramatic loss of reef volume (Kiessling and Simpson, 2011), reduction in number and size of reef sites (Fig. 3), and reduction of within-reef diversities (Kiessling et al., 2007). The reef crisis is regarded as manifestation of ocean acidification at the Triassic-Jurassic boundary crisis (Kiessling and Simpson, 2011), but latitudinal patterns of diversity dynamics were different in the Rhaetian from any other time during the Triassic and Jurassic, and extinctions were significantly higher in the tropics than in nontropical areas (Fig. 4; Kiessling and Aberhan, 2007). Because ocean acidification preferentially affects colder waters that are less saturated with respect to CO_3^{2-}, the extinctions should have hit groups living in deeper and higher-latitude environments harder, if ocean acidification was the dominant cause of the extinction.

These results suggest that both ocean acidification and global warming affected the biota, but the greater severity of tropical extinctions suggests that elevated seawater temperature

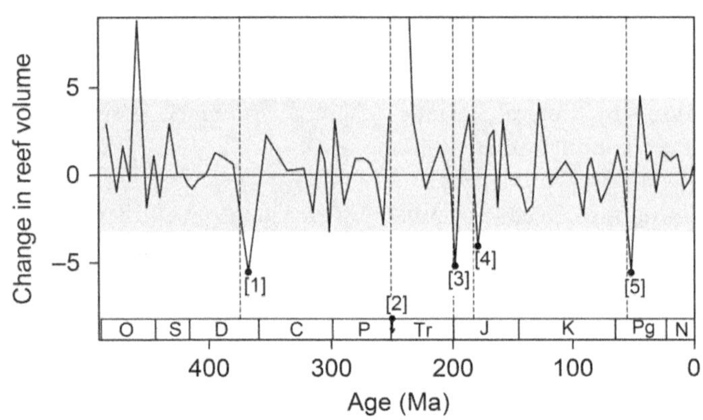

Figure 3. Changes (log-return) of global metazoan reef volume as recorded in The PaleoReefs Database (after Kiessling and Simpson, 2011). Background fluctuations are delineated by gray box; significant crises are marked by black dots; and the pre-crisis intervals are marked by dashed lines. The end-Triassic [3] mass extinction was associated with significant reef volume depletion. See caption for Figure 1 for abbreviations.

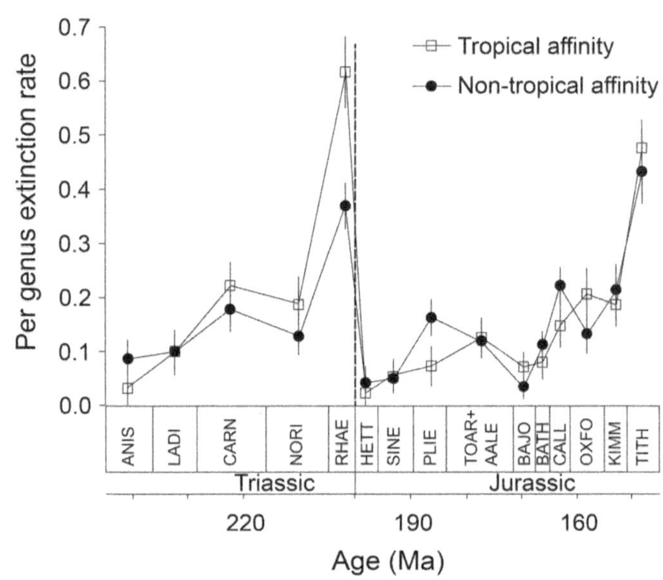

Figure 4. Comparison between the extinction rates of marine invertebrate genera with significant affinities for tropical (latitudes lower than 30°) and nontropical environments in the Triassic-Jurassic interval (redrawn after Kiessling and Aberhan, 2007). Sampling standardized to 130 Paleobiology Database (PBDB) occurrences. The high Rhaetian extinction spike of tropical genera implies a strong climatic component in the end-Triassic mass extinction. ANIS—Anisian, LADI—Ladinian, CARN—Carnian, NORI—Norian, RHAE—Rhaetian, HETT—Hettangian, SINE—Sinemurian, PLIE—Pliensbachian, TOAR—Toarcian, AALE—Aalenian, BAJO—Bajocian, BATH—Bathonian, CALL—Callovian, OXFO—Oxfordian, KIMM—Kimmeridgian, TITH—Tithonian.

due to global warming played a crucial role in the extinction. This is further supported by the pattern of reef reoccurrence in the Early Jurassic as well, which implies escape from heat rather than acidity (Kiessling and Simpson, 2011) and points to long-term hot climate conditions after an initial, transient extreme greenhouse event.

BIOTIC CHANGE AROUND THE TRIASSIC-JURASSIC BOUNDARY

Changes in the Marine Realm

Research of global diversity in the past decade has confirmed that a profound mass extinction took place at the end of the Triassic period (Alroy, 2008; Kiessling et al., 2007), coupled with an abrupt drop of global diversity (Alroy, 2010). Use of the PBDB and a modern analytical toolkit confirmed Raup and Sepkoski's (1982) original view that the end-Triassic event is one of the "Big Five" marine mass extinctions of the Phanerozoic. The drop in diversity is partially a result of lower originations in the latest Triassic, as first noted by Bambach et al. (2004) using Sepkoski's compendium and later confirmed by Alroy et al. (2008) from the PBDB. These analyses were performed at the stage level of stratigraphic resolution, but the issue of finer temporal resolution is increasingly pressing. Research priorities must also shift from purely taxon-based to process-based ecological and biological assessments of the mass extinction, to better understand the causality and consequences of the environmental perturbations.

Plankton Groups

Marine plankton are at the base of the marine food web; therefore, their fate during environmental and biotic crises is of critical importance. However, relatively little is known about the planktonic biota of the early Mesozoic, because the major groups that dominate this realm of the modern oceans did not rise to dominance until the later part of the Mesozoic (Falkowski et al., 2004).

The history of nannoplakton and organic-walled phytoplankton across the Triassic-Jurassic boundary is known from a few boundary sections. Van de Schootbrugge et al. (2007) observed severe extinctions among dinoflagellates and coccolithophorids in the St. Audrie's Bay section. These groups are absent in the crisis interval and gave way to blooms of chlorophytes such as acritarchs and prasinophytes. At Doniford, a correlative section in SW England, Clémence et al. (2010) also found that calcareous nannoplankton productivity was drastically reduced at the Triassic-Jurassic boundary. They suggested that the decreased carbonate production contributed to the early Hettangian warmth as oceanic uptake of volcanogenic CO_2 was lessened by a less efficient biological carbonate pump. The inferences of these studies are somewhat compromised by facies changes in the boundary interval recorded in the St. Audrie's Bay and Doniford sections.

Radiolaria are the only extant planktonic group with an excellent fossil record from the Upper Triassic. Although analyses of individual stratigraphic sections in Canada and Japan suggest a major species turnover at the Triassic-Jurassic boundary (Carter and Hori, 2005), using Sepkoski's compendium (2002), Hautmann et al. (2008) found that the group was relatively unaffected by the end-Triassic extinction, which they explained by the group's lesser vulnerability to ocean acidification. A PBDB-based, sampling standardized analysis of radiolarian occurrence data indeed failed to show significant extinctions at the stage-level resolution (Kiessling and Danelian, 2011). However, our analyses using a refined stratigraphic resolution of the time series yielded quantitative evidence for significantly higher radiolarian extinctions in the late Rhaetian (Kocsis et al., in press).

Nekton Groups

Although conodonts are justifiably regarded as victims of the end-Triassic mass extinction (Orchard, 2010), the effect of the end-Permian extinctions on the waning of the group has not been properly assessed. Studies about the diversity changes of conodonts in the Triassic are hindered by the taxonomic challenges posed by the multi-element conodont apparatus; hence, no up-to-date study is available on Late Triassic conodont origination and extinction rate changes and diversity trajectory. Reviews of the older literature (e.g., Tanner et al., 2004) conclude that conodont standing diversity was low throughout the Rhaetian, and the group was already depauperate when hit by its final extinction.

It is widely held that ammonoids were heavily affected by the end-Triassic extinction and underwent major taxonomic restructuring at the Triassic-Jurassic boundary and during subsequent recovery (Guex et al., 2012). However, quantitative studies of diversity changes of this group have so far been restricted to analyses of nonpublic databases (Dera et al., 2010). Although a recent range-based analysis (Brayard et al., 2009) from the Ammon and Goniat databases suggests an incremental diversity drop at the end of the Triassic period, this decline is possibly biased by the Signor-Lipps effect. Although their broad sampling would make ammonoids an ideal target group for spatial analysis, so far the neglect of PBDB by ammonite taxonomers prevents a sampling standardized analysis.

Benthic Groups

Because higher-rank taxonomic groups of the Mesozoic benthic communities are well represented in modern ecosystems, their physiologies, habitat preferences, and ecological requirements are much better known than those of the mostly extinct major nektonic groups, and these ecological attributes are important for unraveling extinction processes.

Bivalves are a group with an excellent fossil record, and their end-Triassic extinction (McRoberts and Newton, 1995) and ecological selectivity (McRoberts and Newton, 1995) are well studied. Detailed investigation of a classical section in the Northern Calcareous Alps revealed local changes of assemblages compatible with acidification at the Triassic-Jurassic boundary (McRoberts et al., 2012). However, a synoptic analysis of PBDB found no clear support for selectivity by shell mineralogy and ecological guild structure of bivalve assemblages across the Triassic-Jurassic boundary (Kiessling et al., 2007). This study also pointed out that inshore faunas and reefs were more seriously affected. The end-Triassic reef crisis has already been discussed as a test case for the ocean acidification scenario (Kiessling and Simpson, 2011). An independent analysis by Lathuilière and Marchal (2009) confirms the severity of extinction among corals and argues for the role of hothouse climate.

Terrestrial Floral and Faunal Change

At the end of the Triassic, the terrestrial and marine ecosystems were affected equally. Floral turnover and vegetation change are evident from both palynological and paleobotanical data. As discussed already, a sharp palynological turnover is closely (albeit controversially) associated with the onset of Central Atlantic magmatic province volcanism in rift basins of eastern North America and Morocco (Cirilli, 2010). An intriguing feature is the Triassic-Jurassic boundary spore spike, first emphasized in the Newark Basin record (Fowell et al., 1994) and later recognized at multiple locations. Van de Schootbrugge et al. (2009) attached great significance to the fern spike as a record of replacement of gymnosperm forests by pioneer fern successions in disturbed ecosystems. Co-occurrence of a spore spike with peak abundance of prasynophytes, regarded as disaster taxa in marine phytoplankton, was used to argue for synchronous end-Triassic crises in the terrestrial and marine ecosystems (Götz et al., 2009). However, Bonis et al. (2010) found multiple, cyclical increases of spore abundance in the St. Audrie's Bay section and called for caution in regarding it as a single, global event, but maintained that climate change was the likely trigger of this vegetation response. This view of the Triassic-Jurassic boundary palynological event is not universally shared. A review by Cirilli (2010) found mostly gradual change in the Late Triassic, whereas some detailed local studies suggest that climatically controlled cyclical vegetation changes characterize the Triassic-Jurassic boundary interval (Ruckwied, 2009).

Paleobotanical data do not show significant extinctions at the higher taxonomic levels, but turnover at the Triassic-Jurassic boundary is evident from local sections, most notably in East Greenland, as first discussed by Harris (1937). Modern studies of the rich flora, started by McElwain et al. (1999), have yielded a wealth of new data. Thousands of leaf remains from a series of plant beds across the Triassic-Jurassic boundary record ecosystem collapse. The recovery is seen by macroecological changes of the terrestrial vegetation in response to climate change (McElwain et al., 2007). Moreover, the observed patterns of relative abundance distribution point to sudden loss of biodiversity (McElwain et al., 2009) exactly where stomatal data suggest massive CO_2 increase, confirming the key role of global warming.

The terrestrial vertebrate record is stratigraphically well resolved in North America and also records a major turnover across the Triassic-Jurassic boundary. This faunal change appears so sudden that it has been suggested to be compatible with an extraterrestrial impact-related extinction scenario (Olsen et al., 2002), yet it can be similarly explained by Central Atlantic magmatic province–induced catastrophic environmental change. Global analysis supports that the end-Triassic extinction was a turning point in archosaur evolutionary history, where previously dominant crurotarsans went extinct, and dinosaurs exploited the ecological opportunity and rose to dominance (Brusatte et al., 2008). The global terrestrial vertebrate record is temporally rather poorly resolved, as even the Rhaetian stage is often left unrecognized (Brusatte et al., 2010), preventing detailed analyses of diversity trajectories. A bias caused by the uneven continental stratigraphic record introduces another problem that hinders diversity studies (Barrett et al., 2009).

CONCLUSIONS

A survey of recent literature reveals that forcing of end-Triassic environmental and biotic change by Central Atlantic magmatic province volcanism is currently the leading hypothesis. Previous doubts about temporal coincidence are removed by increasingly precise radiometric dates and new biostratigraphic data. Cyclostratigraphy, together with high-resolution geochronology, confirms the brevity of peak changes, confined within the 20 k.y. range of a precession cycle. Carbon isotope stratigraphy and the carbon cycle history of the Triassic-Jurassic transition are not yet fully understood, but the presence of major perturbations and complexity of changes are evident. Strong evidence exists for global warming and ocean acidification at the Triassic-Jurassic boundary. The biotic response to the Central Atlantic magmatic province–induced environmental forcing was a first-order extinction event globally, but one that selectively affected both marine and terrestrial groups.

A review published 10 yr ago (Pálfy, 2003) implicated Central Atlantic magmatic province volcanism as the potential driving force in end-Triassic events. Research in the last decade has provided much new evidence to corroborate and refine the overarching scenario; thus, Central Atlantic magmatic province volcanism is now regarded as the most likely driver of Earth system changes across the Triassic-Jurassic transition.

However, there remain several outstanding issues for further research.

1. The early eruptive history of the Central Atlantic magmatic province and its consequences remain inadequately known. The timing and environmental effects of Rhaetian onset of magmatism, as hinted by precursor CIEs and some Sr and Os geochemical data, need more evidence.
2. Lava volumes and durations of major individual Central Atlantic magmatic province eruptions and their volatile fluxes are still poorly constrained, leaving room for speculation about their environmental consequences.
3. The geological record of brief SO_2-driven cooling episodes is patchy, and the degree and duration of cooling are controversial.

4. Details of the carbon isotope record often disagree among sections, and forcing mechanisms of the individual negative and positive anomalies are still debated.
5. Ocean acidification is increasingly recognized, but disentangling its effects from that of extreme greenhouse warming is not straightforward.
6. High-resolution sampling standardized analyses of the global fossil record are needed to better understand extinction rate changes and selectivity among major fossil groups.

Further study of the end-Triassic events is stimulated by the recognition of important similarities of this environmental and biotic crisis in deep time with that humanity will face in the coming centuries (Kiessling et al., 2008).

ACKNOWLEDGMENTS

The organizers and sponsors of the "Volcanism, Impacts, and Mass Extinctions: Causes and Effects" conference (London, 2013) are thanked for the opportunity to present there a contribution that formed the basis of this paper. Constructive reviews by Paul Wignall and an anonymous reviewer helped improve the manuscript significantly. Research was financially supported by the Hungarian Scientific Research Fund (grant K72633). This is Hungarian Academy of Sciences–Hungarian Natural History Museum–Eötvös University Paleo contribution 188.

REFERENCES CITED

Ahlberg, A., Olsson, I., and Simkevicius, P., 2003, Triassic-Jurassic weathering and clay mineral dispersal in basement areas and sedimentary basins of southern Sweden: Sedimentary Geology, v. 161, no. 1–2, p. 15–29, doi:10.1016/S0037-0738(02)00381-0.

Akikuni, K., Hori, R.S., Vajda, V., Grant-Mackie, J.A., and Ikehara, M., 2010, Stratigraphy of Triassic-Jurassic boundary sequences from the Kawhia coast and Awakino gorge, Murihiku terrane, New Zealand: Stratigraphy, v. 7, no. 1, p. 7–24.

Alroy, J., 2008, Dynamics of origination and extinction in the marine fossil record: Proceedings of the National Academy of Sciences of the United States of America, v. 105, supplement 1, p. 11,536–11,542.

Alroy, J., 2010, The shifting balance of diversity among major marine animal groups: Science, v. 329, no. 5996, p. 1191–1194, doi:10.1126/science.1189910.

Alroy, J., 2014, Accurate and precise estimates of origination and extinction rates: Paleobiology, v. 40, no. 3 (in press).

Alroy, J., Aberhan, M., Bottjer, D.J., Foote, M., Fürsich, F.T., Harries, P.J., Hendy, A.J.W., Holland, S.M., Ivany, L.C., Kiessling, W., Kosnik, M.A., Marshall, C.R., McGowan, A.J., Miller, A.I., Olszewski, T.D., Patzkowsky, M.E., Peters, S.E., Villier, L., Wagner, P.J., Bonuso, N., Borkow, P.S., Brenneis, B., Clapham, M.E., Fall, L.M., Ferguson, C.A., Hanson, V.L., Krug, A.Z., Layou, K.M., Leckey, E.H., Nürnberg, S., Powers, C.M., Sessa, J.A., Simpson, C., Tomašových, A., and Visaggi, C.C., 2008, Phanerozoic trends in the global diversity of marine invertebrates: Science, v. 321, no. 5885, p. 97–100, doi:10.1126/science.1156963.

Alvarez, L., Alvarez, W., Asaro, F., and Michel, H., 1980, Extraterrestrial cause for the Cretaceous-Tertiary extinction: Science, v. 208, p. 1095–1108, doi:10.1126/science.208.4448.1095.

Bachan, A., van de Schootbrugge, B., Fiebig, J., McRoberts, C.A., Ciarapica, G., and Payne, J.L., 2012, Carbon cycle dynamics following the end-Triassic mass extinction: Constraints from paired delta C-13(carb) and delta C-13(org) records: Geochemistry Geophysics Geosystems, v. 13, p. Q09008, doi:10.1029/2012GC004150.

Bacon, K.L., Belcher, C.M., Haworth, M., and McElwain, J.C., 2013, Increased atmospheric SO_2 detected from changes in leaf physiognomy across the Triassic-Jurassic boundary interval of East Greenland: PLoS ONE, v. 8, no. 4, p. e60614, doi:10.1371/journal.pone.0060614.

Baksi, A.K., 2003, Critical evaluation of $^{40}Ar/^{39}Ar$ ages for the Central Atlantic magmatic province: Timing, duration and possible migration of magmatic centers, *in* Hames, W.E., McHone, J.G., Renne, P.R., and Ruppel, C., eds., The Central Atlantic Magmatic Province: Insights from Fragments of Pangea: American Geophysical Union Geophysical Monograph 136, p. 77–90.

Bambach, R.K., Knoll, A.H., and Wang, S.C., 2004, Origination, extinction, and mass depletions of marine diversity: Paleobiology, v. 30, no. 4, p. 522–542, doi:10.1666/0094-8373(2004)030<0522:OEAMDO>2.0.CO;2.

Barrett, P.M., McGowan, A.J., and Page, V., 2009, Dinosaur diversity and the rock record: Proceedings of the Royal Society, ser. B, Biological Sciences, v. 276, no. 1667, p. 2667–2674.

Bartolini, A., Guex, J., Spangenberg, J., Schoene, B., Taylor, D., Schaltegger, U., and Atudorei, V., 2012, Disentangling the Hettangian carbon isotope record: Implications for the aftermath of the end-Triassic mass extinction: Geochemistry Geophysics Geosystems, v. 13, p. Q01007, doi:10.1029/2011GC003807.

Beerling, D.J., and Berner, R.A., 2002, Biogeochemical constraints on the Triassic-Jurassic boundary carbon cycle event: Global Biogeochemical Cycles, v. 16, no. 3, 1036, doi:10.1029/2001GB001637.

Begemann, F., Ludwig, K.R., Lugmair, G.W., Min, K., Nyquist, L.E., Patchett, P.J., Renne, P.R., Shih, C.-Y., Villa, I.M., and Walker, R.J., 2001, Call for an improved set of decay constants for geochronological use: Geochimica et Cosmochimica Acta, v. 65, no. 1, p. 111–121, doi:10.1016/S0016-7037(00)00512-3.

Belcher, C.M., Mander, L., Rein, G., Jervis, F.X., Haworth, M., Hesselbo, S.P., Glasspool, I.J., and McElwain, J.C., 2010, Increased fire activity at the Triassic/Jurassic boundary in Greenland due to climate-driven floral change: Nature Geoscience, v. 3, no. 6, p. 426–429, doi:10.1038/ngeo871.

Benton, M.J., and Twitchett, R.J., 2003, How to kill (almost) all life: The end-Permian extinction event: Trends in Ecology & Evolution, v. 18, no. 7, p. 358–365, doi:10.1016/S0169-5347(03)00093-4.

Blackburn, T.J., Olsen, P.E., Bowring, S.A., McLean, N.M., Kent, D.V., Puffer, J., McHone, G., Rasbury, E.T., and Et-Touhami, M., 2013, Zircon U-Pb geochronology links the end-Triassic extinction with the Central Atlantic magmatic province: Science, v. 340, no. 6135, p. 941–945, doi:10.1126/science.1234204.

Bonis, N.R., Ruhl, M., and Kurschner, W.M., 2010, Milankovitch-scale palynological turnover across the Triassic-Jurassic transition at St. Audrie's Bay, SW UK: Journal of the Geological Society, v. 167, no. 5, p. 877–888, doi:10.1144/0016-76492009-141.

Brayard, A., Escarguel, G., Bucher, H., Monnet, C., Bruhwiler, T., Goudemand, N., Galfetti, T., and Guex, J., 2009, Good genes and good luck: Ammonoid diversity and the end-Permian mass extinction: Science, v. 325, no. 5944, p. 1118–1121, doi:10.1126/science.1174638.

Brusatte, S.L., Benton, M.J., Ruta, M., and Lloyd, G.T., 2008, Superiority, competition, and opportunism in the evolutionary radiation of dinosaurs: Science, v. 321, no. 5895, p. 1485–1488, doi:10.1126/science.1161833.

Brusatte, S.L., Nesbitt, S.J., Irmis, R.B., Butler, R.J., Benton, M.J., and Norell, M.A., 2010, The origin and early radiation of dinosaurs: Earth-Science Reviews, v. 101, no. 1–2, p. 68–100, doi:10.1016/j.earscirev.2010.04.001.

Callegaro, S., Rigo, M., Chiaradia, M., and Marzoli, A., 2012, Latest Triassic marine Sr isotopic variations, possible causes and implications: Terra Nova, v. 24, no. 2, p. 130–135, doi:10.1111/j.1365-3121.2011.01046.x.

Callegaro, S., Marzoli, A., Bertrand, H., Chiaradia, M., Reisberg, L., Meyzen, C., Bellieni, G., Weems, R.E., and Merle, R., 2013, Upper and lower crust recycling in the source of CAMP basaltic dykes from southeastern North America: Earth and Planetary Science Letters, v. 376, p. 186–199, doi:10.1016/j.epsl.2013.06.023.

Carter, E.S., and Hori, R.S., 2005, Global correlation of the radiolarian faunal change across the Triassic-Jurassic boundary: Canadian Journal of Earth Sciences, v. 42, no. 5, p. 777–790, doi:10.1139/e05-020.

Chenet, A.-L., Courtillot, V., Fluteau, F., Gérard, M., Quidelleur, X., Khadri, S.F.R., Subbarao, K.V., and Thordarson, T., 2009, Determination of rapid Deccan eruptions across the Cretaceous-Tertiary boundary using paleomagnetic secular variation: 2. Constraints from analysis of eight new sections and synthesis for a 3500-m-thick composite section: Journal of Geophysical Research–Solid Earth, v. 114, no. B6, B06103, doi:10.1029/2008JB005644.

Cirilli, S., 2010, Upper Triassic–lowermost Jurassic palynology and palynostratigraphy: A review, *in* Lucas, S.G., ed., The Triassic Timescale: Geological Society of London Special Publication 334, p. 285–314.

Cirilli, S., Marzoli, A., Tanner, L., Bertrand, H., Buratti, N., Jourdan, F., Bellieni, G., Kontak, D., and Renne, P.R., 2009, Latest Triassic onset of the Central Atlantic magmatic province (CAMP) volcanism in the Fundy Basin (Nova Scotia): New stratigraphic constraints: Earth and Planetary Science Letters, v. 286, no. 3–4, p. 514–525, doi:10.1016/j.epsl.2009.07.021.

Clémence, M.-E., Bartolini, A., Gardin, S., Paris, G., Beaumont, V., and Page, K.N., 2010, Early Hettangian benthic-planktonic coupling at Doniford (SW England): Palaeoenvironmental implications for the aftermath of the end-Triassic crisis: Palaeogeography, Palaeoclimatology, Palaeoecology, v. 295, no. 1–2, p. 102–115, doi:10.1016/j.palaeo.2010.05.021.

Coffin, M.F., and Eldholm, O., 1994, Large igneous provinces: Crustal structure, dimensions and external consequences: Reviews of Geophysics, v. 32, p. 1–36, doi:10.1029/93RG02508.

Cohen, A.S., 2004, The rhenium-osmium isotope system: Applications to geochronological and palaeoenvironmental problems: Journal of the Geological Society, v. 161, p. 729–734, doi:10.1144/0016-764903-084.

Cohen, A.S., and Coe, A.L., 2002, New geochemical evidence for the onset of volcanism in the Central Atlantic magmatic province and environmental change at the Triassic-Jurassic boundary: Geology, v. 30, no. 3, p. 267–270, doi:10.1130/0091-7613(2002)030<0267:NGEFTO>2.0.CO;2.

Cohen, A.S., and Coe, A.L., 2007, The impact of the Central Atlantic magmatic province on climate and on the Sr- and Os-isotope evolution of seawater: Palaeogeography, Palaeoclimatology, Palaeoecology, v. 244, no. 1–4, p. 374–390, doi:10.1016/j.palaeo.2006.06.036.

Courtillot, V., and Fluteau, F., 2010, Cretaceous extinctions: The volcanic hypothesis: Science, v. 328, no. 5981, p. 973–974 (author reply p. 975–976), doi:10.1126/science.328.5981.973-b.

Courtillot, V., and Renne, P.R., 2003, On the ages of flood basalt events: Comptes Rendus Geoscience, v. 335, no. 1, p. 113–140, doi:10.1016/S1631-0713(03)00006-3.

Courtillot, V., Feraud, G., Maluski, H., Vandamme, D., Moreau, M.G., and Besse, J., 1988, Deccan flood basalts and the Cretaceous/Tertiary boundary: Nature, v. 333, p. 843–846, doi:10.1038/333843a0.

Deenen, M.H.L., Ruhl, M., Bonis, N.R., Krijgsman, W., Kuerschner, W.M., Reitsma, M., and van Bergen, M.J., 2010, A new chronology for the end-Triassic mass extinction: Earth and Planetary Science Letters, v. 291, no. 1–4, p. 113–125, doi:10.1016/j.epsl.2010.01.003.

Dera, G., Neige, P., Dommergues, J.L., Fara, E., Laffont, R., and Pellenard, P., 2010, High-resolution dynamics of Early Jurassic marine extinctions: The case of Pliensbachian-Toarcian ammonites (Cephalopoda): Journal of the Geological Society, v. 167, p. 21–33, doi:10.1144/0016-76492009-068.

Dunning, G.R., and Hodych, J.P., 1990, U/Pb zircon and baddeleyite ages for the Palisades and Gettysburg sills of the northeastern United States: Implications for the age of the Triassic/Jurassic boundary: Geology, v. 18, p. 795–798, doi:10.1130/0091-7613(1990)018<0795:UPZABA>2.3.CO;2.

Erwin, D.H., 1994, The Permo-Triassic extinction: Nature, v. 367, p. 231–236, doi:10.1038/367231a0.

Falkowski, P.G., Katz, M.E., Knoll, A.H., Quigg, A., Raven, J.A., Schofield, O., and Taylor, F.J.R., 2004, The evolution of modern eukaryotic phytoplankton: Science, v. 305, no. 5682, p. 354–360, doi:10.1126/science.1095964.

Foote, M., 2000, Origination and extinction components of taxonomic diversity: General problems: Paleobiology, v. 26, no. 4, p. 74–102, doi:10.1666/0094-8373(2000)26[74:OAECOT]2.0.CO;2.

Fowell, S.J., and Olsen, P.E., 1993, Time calibration of Triassic/Jurassic microfloral turnover, eastern North America: Tectonophysics, v. 222, p. 361–369, doi:10.1016/0040-1951(93)90359-R.

Fowell, S.J., Cornet, B., and Olsen, P.E., 1994, Geologically rapid Late Triassic extinctions: Palynological evidence from the Newark Supergroup, *in* Klein, G.D., ed., Pangea: Paleoclimate, Tectonics, and Sedimentation During Accretion, Zenith, and Breakup of a Supercontinent: Geological Society of America Special Paper 288, p. 197–206.

Friedman, R., Mundil, R., and Palfy, J., 2008, Revised zircon U-Pb ages for the Triassic-Jurassic boundary and the earliest Jurassic employing the chemical abrasion pretreatment (CA-TIMS) technique: Geochimica et Cosmochimica Acta, v. 72, no. 12, p. A284.

Galli, M.T., Jadoul, F., Bernasconi, S.M., Cirilli, S., and Weissert, H., 2007, Stratigraphy and palaeoenvironmental analysis of the Triassic-Jurassic transition in the western Southern Alps (Northern Italy): Palaeogeography, Palaeoclimatology, Palaeoecology, v. 244, no. 1-4, p. 52–70.

Götz, A.E., Ruckwied, K., Pálfy, J., and Haas, J., 2009, Palynological evidence of synchronous changes within the terrestrial and marine realm at the Triassic/Jurassic boundary (Csővar section, Hungary): Review of Palaeobotany and Palynology, v. 156, no. 3–4, p. 401–409, doi:10.1016/j.revpalbo.2009.04.002.

Greene, S.E., Martindale, R.C., Ritterbush, K.A., Bottjer, D.J., Corsetti, F.A., and Berelson, W.M., 2012, Recognising ocean acidification in deep time: An evaluation of the evidence for acidification across the Triassic-Jurassic boundary: Earth-Science Reviews, v. 113, no. 1–2, p. 72–93, doi:10.1016/j.earscirev.2012.03.009.

Guex, J., Bartolini, A., Atudorei, V., and Taylor, D., 2004, High-resolution ammonite and carbon isotope stratigraphy across the Triassic-Jurassic boundary at New York Canyon (Nevada): Earth and Planetary Science Letters, v. 225, no. 1–2, p. 29–41, doi:10.1016/j.epsl.2004.06.006.

Guex, J., Schoene, B., Bartolini, A., Spangenberg, J., Schaltegger, U., O'Dogherty, L., Taylor, D., Bucher, H., and Atudorei, V., 2012, Geochronological constraints on post-extinction recovery of the ammonoids and carbon cycle perturbations during the Early Jurassic: Palaeogeography, Palaeoclimatology, Palaeoecology, v. 346, p. 1–11, doi:10.1016/j.palaeo.2012.04.030.

Hallam, A., 1990, The end-Triassic mass extinction event, *in* Sharpton, V.L., and Ward, P.D., eds., Global Catastrophes in Earth History; An Interdisciplinary Conference on Impacts, Volcanism, and Mass Mortality: Geological Society of America Special Paper 247, p. 577–583.

Hames, W.E., McHone, J.G., Renne, P., and Ruppel, C., 2003, The Central Atlantic Magmatic Province: Insights from Fragments of Pangea: American Geophysical Union Geophysical Monograph 136, 267 p.

Harris, T.M., 1937, The fossil flora of Scoresby Sound East Greenland. Part 5: Stratigraphic relations of the plant beds: Meddelelser om Groenland, v. 112, no. 2, p. 1–112.

Hautmann, M., 2004, Effect of end-Triassic CO_2 maximum on carbonate sedimentation and marine mass extinction: Facies, v. 50, p. 257–261, doi:10.1007/s10347-004-0020-y.

Hautmann, M., Benton, M.J., and Tomasovych, A., 2008, Catastrophic ocean acidification at the Triassic-Jurassic boundary: Neues Jahrbuch für Geologie und Palaontologie, Abhandlungen, v. 249, no. 1, p. 119–127, doi:10.1127/0077-7749/2008/0249-0119.

Haworth, M., Elliott-Kingston, C., Gallagher, A., Fitzgerald, A., and McElwain, J.C., 2012, Sulphur dioxide fumigation effects on stomatal density and index of non-resistant plants: Implications for the stomatal palaeo-[CO_2] proxy method: Review of Palaeobotany and Palynology, v. 182, p. 44–54, doi:10.1016/j.revpalbo.2012.06.006.

Hesselbo, S.P., Robinson, S.A., Surlyk, F., and Piasecki, S., 2002, Terrestrial and marine mass extinction at the Triassic-Jurassic boundary synchronized with major carbon-cycle perturbation: A link to initiation of massive volcanism?: Geology, v. 30, no. 3, p. 251–254, doi:10.1130/0091-7613(2002)030<0251:TAMEAT>2.0.CO;2.

Hesselbo, S.P., McRoberts, C.A., and Pálfy, J., 2007, Triassic-Jurassic boundary events: Problems, progress, possibilities: Palaeogeography, Palaeoclimatology, Palaeoecology, v. 244, no. 1–4, p. 1–10, doi:10.1016/j.palaeo.2006.06.020.

Hillebrandt, A.v., Krystyn, L., Kürschner, W.M., Bonis, N.R., Ruhl, M., Richoz, S., Schobben, M.A.N., Urlichs, M., Bown, P.R., Kment, K., McRoberts, C.A., Simms, M., and Tomãsových, A., 2013, The global stratotype sections and point (GSSP) for the base of the Jurassic System at Kuhjoch (Karwendel Mountains, Northern Calcareous Alps, Tyrol, Austria): Episodes, v. 36, no. 3, p. 162–198.

Hodych, J.P., and Dunning, G.R., 1992, Did the Manicouagan impact trigger end-of-Triassic mass extinction?: Geology, v. 20, p. 51–54, doi:10.1130/0091-7613(1992)020<0051:DTMITE>2.3.CO;2.

Holser, W.T., Magaritz, M., and Ripperdan, R.L., 1996, Global isotopic events, *in* Walliser, O.H., ed., Global Events and Event Stratigraphy in the Phanerozoic: Berlin, Springer, p. 63–88.

Hönisch, B., Ridgwell, A., Schmidt, D.N., Thomas, E., Gibbs, S.J., Sluijs, A., Zeebe, R., Kump, L., Martindale, R.C., Greene, S.E., Kiessling, W., Ries, J., Zachos, J.C., Royer, D.L., Barker, S., Marchitto, T.M., Moyer, R., Pelejero, C., Ziveri, P., Foster, G.L., and Williams, B., 2012, The geological record of ocean acidification: Science, v. 335, no. 6072, p. 1058–1063, doi:10.1126/science.1208277.

Jourdan, F., Marzoli, A., Bertrand, H., Cirilli, S., Tanner, L.H., Kontak, D.J., McHone, G., Renne, P.R., and Bellieni, G., 2009, Ar-40/Ar-39 ages of

CAMP in North America: Implications for the Triassic-Jurassic boundary and the K-40 decay constant bias: Lithos, v. 110, no. 1–4, p. 167–180, doi:10.1016/j.lithos.2008.12.011.

Keeley, J.E., Fotheringham, C., and Baer-Keeley, M., 2005, Factors affecting plant diversity during post-fire recovery and succession of Mediterranean-climate shrublands in California, USA: Diversity & Distributions, v. 11, no. 6, p. 525–537, doi:10.1111/j.1366-9516.2005.00200.x.

Keller, G., Adatte, T., Gardin, S., Bartolini, A., and Bajpai, S., 2008, Main Deccan volcanism phase ends near the K–T boundary: Evidence from the Krishna–Godavari Basin, SE India: Earth and Planetary Science Letters, v. 268, no. 3, p. 293–311.

Keller, G., Sahni, A., and Bajpai, S., 2009, Deccan volcanism, the KT mass extinction and dinosaurs: Journal of Biosciences, v. 34, no. 5, p. 709–728.

Keller, G., Adatte, T., Bhowmick, P., Upadhyay, H., Dave, A., Reddy, A., and Jaiprakash, B., 2012, Nature and timing of extinctions in Cretaceous-Tertiary planktic foraminifera preserved in Deccan intertrappean sediments of the Krishna–Godavari Basin, India: Earth and Planetary Science Letters, v. 341, p. 211–221.

Kiessling, W., and Aberhan, M., 2007, Geographical distribution and extinction risk: Lessons from Triassic-Jurassic marine benthic organisms: Journal of Biogeography, v. 34, p. 1473–1489, doi:10.1111/j.1365-2699.2007.01709.x.

Kiessling, W., and Danelian, T., 2011, Trajectories of Late Permian–Jurassic radiolarian extinction rates: No evidence for an end-Triassic mass extinction: Fossil Record, v. 14, no. 1, p. 95–101, doi:10.1002/mmng.201000017.

Kiessling, W., and Simpson, C., 2011, On the potential for ocean acidification to be a general cause of ancient reef crises: Global Change Biology, v. 17, no. 1, p. 56–67, doi:10.1111/j.1365-2486.2010.02204.x.

Kiessling, W., Aberhan, M., Brenneis, B., and Wagner, P.J., 2007, Extinction trajectories of benthic organisms across the Triassic-Jurassic boundary: Palaeogeography, Palaeoclimatology, Palaeoecology, v. 244, p. 201–222, doi:10.1016/j.palaeo.2006.06.029.

Kiessling, W., Aberhan, M., and Villier, L., 2008, Phanerozoic trends in skeletal mineralogy driven by mass extinctions: Nature Geoscience, v. 1, no. 8, p. 527–530, doi:10.1038/ngeo251.

Knight, K.B., Nomade, S., Renne, P.R., Marzoli, A., Bertrand, H., and Youbi, N., 2004, The Central Atlantic magmatic province at the Triassic-Jurassic boundary: Paleomagnetic and $^{40}Ar/^{39}Ar$ evidence from Morocco for brief, episodic volcanism: Earth and Planetary Science Letters, v. 228, no. 1–2, p. 143–160, doi:10.1016/j.epsl.2004.09.022.

Kocsis, T.Á., Kiessling, W., and Pálfy, J., 2014, Radiolarian biodiversity dynamics through the Triassic and Jurassic: Implications for proximate causes of the end-Triassic mass extinction: Paleobiology (in press).

Korte, C., Hesselbo, S.P., Jenkyns, H.C., Rickaby, R.E.M., and Spotl, C., 2009, Palaeoenvironmental significance of carbon- and oxygen-isotope stratigraphy of marine Triassic-Jurassic boundary sections in SW Britain: Journal of the Geological Society, v. 166, p. 431–445, doi:10.1144/0016-76492007-177.

Kozur, H., and Weems, R.E., 2007, Upper Triassic conchostracan biostratigraphy of the continental rift basins of eastern North America: Its importance for correlating Newark Supergroup events with the Germanic Basin and the international geologic time scale, *in* Lucas, S.G., and Spielmann, J.A., eds., The Global Triassic: New Mexico Museum of Natural History and Science Bulletin, v. 41, p. 137–188.

Kozur, H., and Weems, R.E., 2010, The biostratigraphic importance of conchostracans in the continental Triassic of the Northern Hemisphere, *in* Lucas, S.G., ed., The Triassic Timescale: Geological Society of London Special Publication 334, p. 315–417.

Krystyn, L., Bouquerel, H., Kuerschner, W., Richoz, S., and Gallet, Y., 2007, Proposal for a candidate GSSP for the base of the Rhaetian Stage, *in* Lucas, S.G., and Spielman, J.A., eds., The Global Triassic: New Mexico Museum of Natural History and Science Bulletin, v. 41, p. 189–199.

Kürschner, W.M., Bonis, N.R., and Krystyn, L., 2007, Carbon-isotope stratigraphy and palynostratigraphy of the Triassic-Jurassic transition in the Tiefengraben section—Northern Calcareous Alps (Austria): Palaeogeography, Palaeoclimatology, Palaeoecology, v. 244, no. 1–4, p. 257–280, doi:10.1016/j.palaeo.2006.06.031.

Kuroda, J., Hori, R.S., Suzuki, K., Gröcke, D.R., and Ohkouchi, N., 2010, Marine osmium isotope record across the Triassic-Jurassic boundary from a Pacific pelagic site: Geology, v. 38, no. 12, p. 1095–1098, doi:10.1130/G31223.1.

Lathuilière, B., and Marchal, D., 2009, Extinction, survival and recovery of corals from the Triassic to Middle Jurassic time: Terra Nova, v. 21, no. 1, p. 57–66, doi:10.1111/j.1365-3121.2008.00856.x.

Lindstrom, S., van de Schootbrugge, B., Dybkjaer, K., Pedersen, G.K., Fiebig, J., Nielsen, L.H., and Richoz, S., 2012, No causal link between terrestrial ecosystem change and methane release during the end-Triassic mass extinction: Geology, v. 40, no. 6, p. 531–534.

Marynowski, L., and Simoneit, B.R.T., 2009, Widespread Upper Triassic to Lower Jurassic wildfire records from Poland: Evidence from charcoal and pyrolytic polycyclic aromatic hydrocarbons: Palaios, v. 24, no. 12, p. 785–798, doi:10.2110/palo.2009.p09-044r.

Marzoli, A., Renne, P.R., Piccirillo, E.M., Ernesto, M., Bellieni, G., and De Min, A., 1999, Extensive 200-million-year-old continental flood basalts of the Central Atlantic magmatic province: Science, v. 284, p. 616–618, doi:10.1126/science.284.5414.616.

Marzoli, A., Bertrand, H., Knight, K.B., Cirilli, S., Buratti, N., Verati, C., Nomade, S., Renne, P.R., Youbi, N., Martini, R., Allenbach, K., Neuwerth, R., Rapaille, C., Zaninetti, L., and Bellieni, G., 2004, Synchrony of the Central Atlantic magmatic province and the Triassic-Jurassic boundary climatic and biotic crisis: Geology, v. 32, no. 11, p. 973–976, doi:10.1130/G20652.1.

Marzoli, A., Jourdan, F., Puffer, J.H., Cuppone, T., Tanner, L.H., Weems, R.E., Bertrand, H., Cirilli, S., Bellieni, G., and De Min, A., 2011, Timing and duration of the Central Atlantic magmatic province in the Newark and Culpeper Basins, eastern U.S.: Lithos, v. 122, no. 3–4, p. 175–188, doi:10.1016/j.lithos.2010.12.013.

McElwain, J.C., Beerling, D.J., and Woodward, F.I., 1999, Fossil plants and global warming at the Triassic-Jurassic boundary: Science, v. 285, p. 1386–1390, doi:10.1126/science.285.5432.1386.

McElwain, J.C., Popa, M.E., Hesselbo, S.P., Haworth, M., and Surlyk, F., 2007, Macroecological responses of terrestrial vegetation to climatic and atmospheric change across the Triassic/Jurassic boundary in East Greenland: Paleobiology, v. 33, no. 4, p. 547–573, doi:10.1666/06026.1.

McElwain, J.C., Wagner, P.J., and Hesselbo, S.P., 2009, Fossil plant relative abundances indicate sudden loss of Late Triassic biodiversity in East Greenland: Science, v. 324, no. 5934, p. 1554–1556, doi:10.1126/science.1171706.

McGhee, G.R., Sheehan, P.M., Bottjer, D.J., and Droser, M.L., 2004, Ecological ranking of Phanerozoic biodiversity crises: Ecological and taxonomic severities are decoupled: Palaeogeography, Palaeoclimatology, Palaeoecology, v. 211, no. 3–4, p. 289–297, doi:10.1016/j.palaeo.2004.05.010.

McHone, J.G., 2003, Volatile emissions from Central Atlantic magmatic province basalts: Mass assumptions and environmental consequences, *in* Hames, W.E., McHone, J.G., Renne, P.R., and Ruppel, C., eds., The Central Atlantic Magmatic Province: Insights from Fragments of Pangea: American Geophysical Union Geophysical Monograph 136, p. 241–254.

McRoberts, C.A., and Newton, C.R., 1995, Selective extinction among end-Triassic European bivalves: Geology, v. 23, no. 2, p. 102–104, doi:10.1130/0091-7613(1995)023<0102:SEAETE>2.3.CO;2.

McRoberts, C.A., Krystyn, L., and Hautmann, M., 2012, Macrofaunal response to the end-Triassic mass extinction in the west-Tethyan Kössen Basin, Austria: Palaios, v. 27, no. 9, p. 607–616, doi:10.2110/palo.2012.p12-043r.

Michalík, J., Lintnerová, O., Gazdzicki, A., and Soták, J., 2007, Record of environmental changes in the Triassic-Jurassic boundary interval in the Zliechov Basin, Western Carpathians: Palaeogeography, Palaeoclimatology, Palaeoecology, v. 244, no. 1-4, p. 71–88.

Michalík, J., Biroň, A., Lintnerová, O., Götz, A.E., and Ruckwied, K., 2010, Climate change at the Triassic/Jurassic boundary in the northwestern Tethyan realm, inferred from sections in the Tatra Mountains (Slovakia): Acta Geologica Polonica, v. 60, no. 4, p. 535–548.

Mundil, R., Pálfy, J., Renne, P.R., and Brack, P., 2010, The Triassic timescale: New constraints and a review of geochronological data, *in* Lucas, S.G., ed., The Triassic Timescale: Geological Society of London Special Publication 334, p. 41–60, doi:10.1144/SP334.3.

Nomade, S., Knight, K.B., Beutel, E., Renne, P.R., Verati, C., Feraud, G., Marzoli, A., Youbi, N., and Bertrand, H., 2007, Chronology of the Central Atlantic magmatic province: Implications for the Central Atlantic rifting processes and the Triassic-Jurassic biotic crisis: Palaeogeography, Palaeoclimatology, Palaeoecology, v. 244, no. 1–4, p. 326–344, doi:10.1016/j.palaeo.2006.06.034.

Ogg, J.G., 2004, The Jurassic Period, *in* Gradstein, F.M., Ogg, J.G., and Smith, A.B., eds., A Geologic Time Scale 2004: Cambridge, UK, Cambridge University Press, p. 307–343.

Ogg, J.G., and Hinnov, L.A., 2012, Jurassic, *in* Gradstein, F.M., Ogg, J.G., Schmitz, M.D., and Ogg, G., eds., The Geologic Time Scale 2012: Amsterdam, Elsevier, p. 731–791.

Olsen, P.E., and Kent, D.V., 1996, Milankovitch climate forcing in the tropics of Pangaea during the Late Triassic: Palaeogeography, Palaeoclimatology, Palaeoecology, v. 122, p. 1–26, doi:10.1016/0031-0182(95)00171-9.

Olsen, P.E., and Kent, D.V., 1999, Long-period Milankovitch cycles from the Late Triassic and Early Jurassic of eastern North America and their implications for the calibration of the early Mesozoic time scale and the long-term behavior of the planets: Philosophical Transactions of the Royal Society, ser. A, Mathematical and Physical Sciences, v. 357, p. 1761–1786, doi:10.1098/rsta.1999.0400.

Olsen, P.E., Shubin, N.H., and Anders, M.H., 1987, New Early Jurassic tetrapod assemblages constrain Triassic-Jurassic tetrapod extinction event: Science, v. 237, p. 1025–1029, doi:10.1126/science.3616622.

Olsen, P.E., Kent, D.V., Sues, H.-D., Koeberl, C., Huber, H., Montanari, A., Rainforth, E.C., Fowell, S.J., Szajna, M.J., and Hartline, B.W., 2002, Ascent of dinosaurs linked to an iridium anomaly at the Triassic-Jurassic boundary: Science, v. 296, p. 1305–1307, doi:10.1126/science.1065522.

Orchard, M.J., 2010, Triassic conodonts and their role in stage boundary definition, *in* Lucas, S.G., ed., The Triassic Timescale: Geological Society of London Special Publication 334, p. 139–161, doi:10.1144/SP334.7.

Pálfy, J., 1995, Development of the Jurassic geochronologic scale: Hantkeniana, v. 1, p. 13–25.

Pálfy, J., 2003, Volcanism of the Central Atlantic magmatic province as a potential driving force in the end-Triassic mass extinction, *in* Hames, W.E., McHone, J.G., Renne, P.R., and Ruppel, C., eds., The Central Atlantic Magmatic Province: Insights from Fragments of Pangea: American Geophysical Union Geophysical Monograph 136, p. 255–267.

Pálfy, J., and Zajzon, N., 2012, Environmental changes across the Triassic-Jurassic boundary and coeval volcanism inferred from elemental geochemistry and mineralogy in the Kendlbachgraben section (Northern Calcareous Alps, Austria): Earth and Planetary Science Letters, v. 335, p. 121–134, doi:10.1016/j.epsl.2012.01.039.

Pálfy, J., Mortensen, J.K., Carter, E.S., Smith, P.L., Friedman, R.M., and Tipper, H.W., 2000a, Timing the end-Triassic mass extinction: First on land, then in the sea?: Geology, v. 28, no. 1, p. 39–42, doi:10.1130/0091-7613(2000)28<39:TTEMEF>2.0.CO;2.

Pálfy, J., Smith, P.L., and Mortensen, J.K., 2000b, A U-Pb and $^{40}Ar/^{39}Ar$ time scale for the Jurassic: Canadian Journal of Earth Sciences, v. 37, no. 6, p. 923–944, doi:10.1139/e00-002.

Pálfy, J., Demény, A., Haas, J., Hetényi, M., Orchard, M., and Vetö, I., 2001, Carbon isotope anomaly and other geochemical changes at the Triassic-Jurassic boundary from a marine section in Hungary: Geology, v. 29, no. 11, p. 1047–1050, doi:10.1130/0091-7613(2001)029<1047:CIAAOG>2.0.CO;2.

Pálfy, J., Demény, A., Haas, J., Carter, E.S., Görög, Á., Halász, D., Oravecz-Scheffer, A., Hetényi, M., Márton, E., Orchard, M.J., Ozsvárt, P., Vető, I., and Zajzon, N., 2007, Triassic-Jurassic boundary events inferred from integrated stratigraphy of the Csovar section, Hungary: Palaeogeography, Palaeoclimatology, Palaeoecology, v. 244, no. 1–4, p. 11–33, doi:10.1016/j.palaeo.2006.06.021.

Petersen, H.I., and Lindström, S., 2012, Synchronous wildfire activity rise and mire deforestation at the Triassic-Jurassic boundary: PLoS ONE, v. 7, no. 10, p. e47236, doi:10.1371/journal.pone.0047236.

Pienkowski, G., Niedzwiedzki, G., and Waksmundzka, M., 2012, Sedimentological, palynological and geochemical studies of the terrestrial Triassic-Jurassic boundary in northwestern Poland: Geological Magazine, v. 149, no. 2, p. 308–332, doi:10.1017/S0016756811000914.

Puffer, J.H., Block, K.A., and Steiner, J.C., 2009, Transmission of flood basalts through a shallow crustal sill and the correlation of sill layers with extrusive flows: The Palisades intrusive system and the basalts of the Newark Basin, New Jersey, U.S.A.: The Journal of Geology, v. 117, no. 2, p. 139–155, doi:10.1086/595663.

Rampino, M.R., and Caldeira, K., 2011, Comment on "Atmospheric pCO_2 Perturbations associated with the Central Atlantic magmatic province": Science, v. 334, no. 6056, p. 594, doi:10.1126/science.1208653.

Rampino, M.R., and Stothers, R.B., 1988, Flood basalt volcanism during the last 250 million years: Science, v. 241, p. 663–668, doi:10.1126/science.241.4866.663.

Rampino, M.R., Haggerty, B.M., and Pagano, T.C., 1997, A unified theory of impact crises and mass extinctions: Quantitative tests: Annals of the New York Academy of Sciences, v. 822, no. 1, p. 403–431, doi:10.1111/j.1749-6632.1997.tb48358.x.

Raup, D.M., 1992, Large-body impact and extinction in the Phanerozoic: Paleobiology, v. 18, no. 1, p. 80–88.

Raup, D.M., and Sepkoski, J.J., Jr., 1982, Mass extinctions in the marine fossil record: Science, v. 215, p. 1501–1503, doi:10.1126/science.215.4539.1501.

Renne, P.R., and Basu, A.R., 1991, Rapid eruption of the Siberian Traps flood basalts at the Permo-Triassic boundary: Science, v. 253, p. 176–179, doi:10.1126/science.253.5016.176.

Renne, P.R., Mundil, R., Balco, G., Min, K.W., and Ludwig, K.R., 2010, Joint determination of K-40 decay constants and Ar-40*/K-40 for the Fish Canyon sanidine standard, and improved accuracy for Ar-40/Ar-39 geochronology: Geochimica et Cosmochimica Acta, v. 74, no. 18, p. 5349–5367, doi:10.1016/j.gca.2010.06.017.

Richoz, S., van de Schootbrugge, B., Pross, J., Puttmann, W., Quan, T.M., Lindstrom, S., Heunisch, C., Fiebig, J., Maquil, R., Schouten, S., Hauzenberger, C.A., and Wignall, P.B., 2012, Hydrogen sulphide poisoning of shallow seas following the end-Triassic extinction: Nature Geoscience, v. 5, p. 662–667, doi:10.1038/ngeo1539.

Ruckwied, K., 2009, Palynology of Triassic/Jurassic Boundary Key Sections of the NW Tethyan Realm (Hungary and Slovakia) [Ph.D. thesis]: Darmstadt, University of Technology, 95 p.

Ruhl, M., and Kurschner, W.M., 2011, Multiple phases of carbon cycle disturbance from large igneous province formation at the Triassic-Jurassic transition: Geology, v. 39, no. 5, p. 431–434, doi:10.1130/G31680.1.

Ruhl, M., Kuerschner, W.M., and Krystyn, L., 2009, Triassic-Jurassic organic carbon isotope stratigraphy of key sections in the western Tethys realm (Austria): Earth and Planetary Science Letters, v. 281, no. 3–4, p. 169–187, doi:10.1016/j.epsl.2009.02.020.

Ruhl, M., Deenen, M.H.L., Abels, H.A., Bonis, N.R., Krijgsman, W., and Kurschner, W.M., 2010, Astronomical constraints on the duration of the Early Jurassic Hettangian Stage and recovery rates following the end-Triassic mass extinction (St. Audrie's Bay/East Quantoxhead, UK): Earth and Planetary Science Letters, v. 295, no. 1–2, p. 262–276, doi:10.1016/j.epsl.2010.04.008.

Ruhl, M., Bonis, N.R., Reichart, G.J., Damste, J.S.S., and Kurschner, W.M., 2011, Atmospheric carbon injection linked to end-Triassic mass extinction: Science, v. 333, no. 6041, p. 430–434, doi:10.1126/science.1204255.

Ruiz-Martínez, V.C., Torsvik, T.H., van Hinsbergen, D.J., and Gaina, C., 2012, Earth at 200 Ma: Global palaeogeography refined from CAMP palaeomagnetic data: Earth and Planetary Science Letters, v. 331, p. 67–79, doi:10.1016/j.epsl.2012.03.008.

Saunders, A.D., 2005, Large igneous provinces: Origin and environmental consequences: Elements, v. 1, no. 5, p. 259–263, doi:10.2113/gselements.1.5.259.

Schaller, M.F., Wright, J.D., and Kent, D.V., 2011, Atmospheric pCO_2 perturbations associated with the Central Atlantic magmatic province: Science, v. 331, no. 6023, p. 1404–1409, doi:10.1126/science.1199011.

Schaltegger, U., Guex, J., Bartolini, A., Schoene, B., and Ovtcharova, M., 2008, Precise U-Pb age constraints for end-Triassic mass extinction, its correlation to volcanism and Hettangian post-extinction recovery: Earth and Planetary Science Letters, v. 267, no. 1–2, p. 266–275, doi:10.1016/j.epsl.2007.11.031.

Schlische, R.W., Withjack, M.O., and Olsen, P.E., 2003, Relative timing of CAMP, rifting, continental breakup, and basin inversion: tectonic significance, *in* Hames, W.E., McHone, J.G., Renne, P.R., and Ruppel, C., eds., The Central Atlantic Magmatic Province: Insights from Fragments of Pangea: American Geophysical Union Geophysical Monograph 136, p. 33–59.

Schoene, B., Crowley, J.L., Condon, D.J., Schmitz, M.D., and Bowring, S.A., 2006, Reassessing the uranium decay constants for geochronology using ID-TIMS U-Pb data: Geochimica et Cosmochimica Acta, v. 70, p. 426–445, doi:10.1016/j.gca.2005.09.007.

Schoene, B., Guex, J., Bartolini, A., Schaltegger, U., and Blackburn, T.J., 2010, Correlating the end-Triassic mass extinction and flood basalt volcanism at the 100 ka level: Geology, v. 38, no. 5, p. 387–390, doi:10.1130/G30683.1.

Schulte, P., Alegret, L., Arenillas, I., Arz, J.A., Barton, P.J., Bown, P.R., Bralower, T.J., Christeson, G.L., Claeys, P., and Cockell, C.S., 2010, The Chicxulub asteroid impact and mass extinction at the Cretaceous-Paleogene boundary: Science, v. 327, no. 5970, p. 1214–1218, doi:10.1126/science.1177265.

Self, S., Widdowson, M., Thordarson, T., and Jay, A.E., 2006, Volatile fluxes during flood basalt eruptions and potential effects on the global environment: A Deccan perspective: Earth and Planetary Science Letters, v. 248, no. 1–2, p. 518–532, doi:10.1016/j.epsl.2006.05.041.

Sepkoski, J.J., Jablonski, D., and Foote, M., 2002, A Compendium of Fossil Marine Animal Genera: Bulletins of American Paleontology 363, 560 p.

Sobolev, S.V., Sobolev, A.V., Kuzmin, D.V., Krivolutskaya, N.A., Petrunin, A.G., Arndt, N.T., Radko, V.A., and Vasiliev, Y.R., 2011, Linking mantle plumes, large igneous provinces and environmental catastrophes: Nature, v. 477, no. 7364, p. 312–316, doi:10.1038/nature10385.

Steinthorsdottir, M., Jeram, A.J., and McElwain, J.C., 2011, Extremely elevated CO_2 concentrations at the Triassic/Jurassic boundary: Palaeogeography, Palaeoclimatology, Palaeoecology, v. 308, no. 3–4, p. 418–432, doi:10.1016/j.palaeo.2011.05.050.

Tanner, L.H., 2010, The Triassic isotope record, *in* Lucas, S.G., ed., The Triassic Timescale: Geological Society of London Special Publication 334, p. 103–118.

Tanner, L.H., Hubert, J.F., Coffey, B.P., and McInerney, D.P., 2001, Stability of atmospheric CO_2 levels across the Triassic/Jurassic boundary: Nature, v. 411, p. 675–677, doi:10.1038/35079548.

Tanner, L.H., Lucas, S.G., and Chapman, M.G., 2004, Assessing the record and causes of Late Triassic extinctions: Earth-Science Reviews, v. 65, no. 1–2, p. 103–139, doi:10.1016/S0012-8252(03)00082-5.

Tozer, E.T., 1967, A Standard for Triassic Time: Geological Survey of Canada Bulletin 156, 103 p.

van de Schootbrugge, B., Tremolada, F., Rosenthal, Y., Bailey, T.R., Feist-Burkhardt, S., Brinkhuis, H., Pross, J., Kent, D.V., and Falkowski, P.G., 2007, End-Triassic calcification crisis and blooms of organic-walled 'disaster species': Palaeogeography, Palaeoclimatology, Palaeoecology, v. 244, no. 1–4, p. 126–141, doi:10.1016/j.palaeo.2006.06.026.

van de Schootbrugge, B., Payne, J.L., Tomasovych, A., Pross, J., Fiebig, J., Benbrahim, M., Follmi, K.B., and Quan, T.M., 2008, Carbon cycle perturbation and stabilization in the wake of the Triassic-Jurassic boundary mass-extinction event: Geochemistry Geophysics Geosystems, v. 9, no. 4, p. 1–16, doi:10.1029/2007GC001914.

van de Schootbrugge, B., Quan, T.M., Lindstrom, S., Puttmann, W., Heunisch, C., Pross, J., Fiebig, J., Petschick, R., Rohling, H.G., Richoz, S., Rosenthal, Y., and Falkowski, P.G., 2009, Floral changes across the Triassic/Jurassic boundary linked to flood basalt volcanism: Nature Geoscience, v. 2, p. 589–594, doi:10.1038/ngeo577.

van de Schootbrugge, B., Bachan, A., Suan, G., Richoz, S., and Payne, J.L., 2013, Microbes, mud and methane: Cause and consequence of recurrent Early Jurassic anoxia following the end-Triassic mass extinction: Palaeontology, v. 56, no. 4, p. 685–709.

Verati, C., Rapaille, C., Feraud, G., Marzoli, A., Bertrand, H., and Youbi, N., 2007, $^{40}Ar/^{39}Ar$ ages and duration of the Central Atlantic magmatic province volcanism in Morocco and Portugal and its relation to the Triassic-Jurassic boundary: Palaeogeography, Palaeoclimatology, Palaeoecology, v. 244, no. 1–4, p. 308–325, doi:10.1016/j.palaeo.2006.06.033.

Ward, P.D., Haggart, J.W., Carter, E.S., Wilbur, D., Tipper, H.W., and Evans, T., 2001, Sudden productivity collapse associated with the Triassic-Jurassic boundary mass extinction: Science, v. 292, p. 1148–1151, doi:10.1126/science.1058574.

Ward, P.D., Garrison, G.H., Haggart, J.W., Kring, D.A., and Beattie, M.J., 2004, Isotopic evidence bearing on Late Triassic extinction events, Queen Charlotte Islands, British Columbia, and implications for the duration and cause of the Triassic/Jurassic mass extinction: Earth and Planetary Science Letters, v. 224, no. 3–4, p. 589–600, doi:10.1016/j.epsl.2004.04.034.

Ward, P.D., Garrison, G.H., Williford, K.H., Kring, D.A., Goodwin, D., Beattie, M.J., and McRoberts, C.A., 2007, The organic carbon isotopic and paleontological record across the Triassic-Jurassic boundary at the candidate GSSP section at Ferguson Hill, Muller Canyon, Nevada, USA: Palaeogeography, Palaeoclimatology, Palaeoecology, v. 244, no. 1–4, p. 281–289, doi:10.1016/j.palaeo.2006.06.042.

Weedon, G.P., Jenkyns, H.C., Coe, A.L., and Hesselbo, S.P., 1999, Astronomical calibration of the Jurassic time-scale from cyclostratigraphy in British mudrock formations: Philosophical Transaction of the Royal Society, ser. A, v. 357, no. 1757, p. 1787–1813.

Whiteside, J.H., Olsen, P.E., Kent, D.V., Fowell, S.J., and Et-Touhami, M., 2007, Synchrony between the Central Atlantic magmatic province and the Triassic-Jurassic mass-extinction event?: Palaeogeography, Palaeoclimatology, Palaeoecology, v. 244, no. 1–4, p. 345–367, doi:10.1016/j.palaeo.2006.06.035.

Whiteside, J.H., Olsen, P.E., Eglinton, T., Brookfield, M.E., and Sambrotto, R.N., 2010, Compound-specific carbon isotopes from Earth's largest flood basalt eruptions directly linked to the end-Triassic mass extinction: Proceedings of the National Academy of Sciences of the United States of America, v. 107, no. 15, p. 6721–6725, doi:10.1073/pnas.1001706107.

Wignall, P.B., 2001, Large igneous provinces and mass extinctions: Earth-Science Reviews, v. 53, no. 1–2, p. 1–33, doi:10.1016/S0012-8252(00)00037-4.

Wignall, P.B., 2005, The link between large igneous province eruptions and mass extinctions: Elements, v. 1, no. 5, p. 293–297, doi:10.2113/gselements.1.5.293.

Wignall, P.B., and Bond, D.P.G., 2008, The end-Triassic and Early Jurassic mass extinction records in the British Isles: Proceedings of the Geologists' Association, v. 119, p. 73–84, doi:10.1016/S0016-7878(08)80259-3.

Williford, K.H., Ward, P.D., Garrison, G.H., and Buick, R., 2007, An extended organic carbon-isotope record across the Triassic-Jurassic boundary in the Queen Charlotte Islands, British Columbia, Canada: Palaeogeography, Palaeoclimatology, Palaeoecology, v. 244, no. 1–4, p. 290–296, doi:10.1016/j.palaeo.2006.06.032.

Zajzon, N., Kristály, F., Pálfy, J., and Németh, T., 2012, Detailed clay mineralogy of the Triassic-Jurassic boundary section at Kendlbachgraben (Northern Calcareous Alps, Austria): Clay Minerals, v. 47, p. 177–189, doi:10.1180/claymin.2012.047.2.03.

Manuscript Accepted by the Society 31 January 2014

Printed in the USA

The Geological Society of America
Special Paper 505
2014

Climatic reversals related to the Central Atlantic magmatic province caused the end-Triassic biotic crisis—Evidence from continental strata in Poland

Grzegorz Pieńkowski*
Polish Geological Institute–National Research Institute, Rakowiecka 4, PL-00-975 Warszawa, Poland

Grzegorz Niedźwiedzki*
Subdepartment of Evolution and Development, Department of Organismal Biology, Evolutionary Biology Centre, Uppsala University, Norbyvägen 18A, 752 36 Uppsala, Sweden

Paweł Brański
Polish Geological Institute–National Research Institute, Rakowiecka 4, PL-00-975 Warszawa, Poland

ABSTRACT

Eight climatic events can be distinguished in the Triassic–Jurassic (ca. 201 Ma) continental strata of Poland. These events are distinguished by kaolinite/illite ratio, chemical index of alteration (CIA), color of sediments, and palynomorphs. The first transition to wetter climate, evidenced by a shift from smectite- to kaolinite-dominated mudrocks, coincides with the earlier ("precursor") Rhaetian negative $\delta^{13}C_{org}$ excursion, which means that the beginning of climate perturbations predates the oldest known Central Atlantic magmatic province flood basalts by some 100–200 k.y. The later global, late Rhaetian "initial" negative $\delta^{13}C_{org}$ excursion is divided into two subpeaks, each corresponding to hot and humid events, separated by a cooler and drier event. The upper subpeak is also associated with perturbation of the osmium isotope system (attributed to volcanic fallout), and darkened miospores, pointing to acid rains. Between the "initial" excursion and the Triassic-Jurassic boundary interval, five climatic fluctuations are inferred from the changing kaolinite/illite ratio, the last two of which are also associated with an Os isotope perturbation, polycyclic aromatic hydrocarbon (PAH) occurrences, a "spore peak," and darkened miospores. A series of periodic atmospheric loading events by CO_2, CH_4, or alternatively by SO_2, sulfate aerosols, and toxic compounds, is inferred to have caused this series of rapid climatic reversals and resulting extinction of many less-adapted forms. Just above the

*grzegorz.pienkowski@pgi.gov.pl; grzegorz.niedzwiedzki@ebc.uu.se.

Pieńkowski, G., Niedźwiedzki, G., and Brański, P., 2014, Climatic reversals related to the Central Atlantic magmatic province caused the end-Triassic biotic crisis—Evidence from continental strata in Poland, *in* Keller, G., and Kerr, A., eds., Volcanism, Impacts, and Mass Extinctions: Causes and Effects: Geological Society of America Special Paper 505, p. 263–286, doi:10.1130/2014.2505(13). For permission to copy, contact editing@geosociety.org.

palynofloral extinction level, appearance of new forms commenced Jurassic palynofloral recovery. Tetrapod evolution events in the end-Triassic–earliest Jurassic were related to the extinction of the Pseudosuchia, Dicynodontia, Capitosauroidea, Plagiosaroidea, and Rhynchosauria, while appearance of highly diversified tetrapod ichnofauna in the earliest Jurassic strata indicates a rapid recovery and refill of ecological niches by dinosaurs.

INTRODUCTION

At least four of the five Phanerozoic mass extinction events can be associated with large igneous province formation. The eruption of continental flood basalts of the Central Atlantic magmatic province at the end of the Triassic (ca. 201 Ma) was approximately coeval with the breakup of Pangea, sea-level changes, abrupt negative carbon isotope excursions (CIEs) triggered by volcanism and/or methane hydrate dissociation, climatic and environmental changes, and mass extinction (e.g., McHone, 1996, 2003; Hesselbo and Jenkyns, 1998; Marzoli et al., 2004, 2011; McElwain et al., 1999, 2007; Olsen et al., 2002; Hesselbo et al., 2002, 2004, 2007; Hounslow et al., 2004; Guex et al., 2004; Pieńkowski, 2004; Tanner et al., 2004; Galli et al., 2005; Jourdan et al., 2009; Ruhl et al., 2009, 2010; Korte and Kozur, 2011; Pieńkowski et al., 2012). Central Atlantic magmatic province ages gathered so far from the North American basins and Morocco (Jourdan et al., 2009; Cirilli et al., 2009; Deenen et al., 2010; Marzoli et al., 2011; Blackburn et al., 2013) suggest a short (~1.5 m.y.) duration for the main magmatic activity, extending from the latest Triassic to the earliest Jurassic.

Mass extinctions are thought to reflect significant interactions among volcanic/geochemical cycles, climate, and biota. Recently, Central Atlantic magmatic province volcanism has been shown to coincide with the end-Triassic extinction (Cirilli et al., 2009; Deenen et al., 2010; Ruhl et al., 2010; Blackburn et al., 2013), although some inconsistency of chronology occurs between the studies of Deenen et al. (2010) and Blackburn et al. (2013) from one side and Marzoli et al. (2011) on the other side. Nevertheless, there remain uncertainties concerning cause and effect, particularly in the terrestrial realm. Global anoxic events, regression, primary productivity crash, methane hydrate dissociation, and rapid global warming have all figured in recent proposed kill mechanisms (Hallam and Wignall, 1997, 1999; Pálfy et al., 2000, 2001; Hesselbo et al., 2000, 2002, 2007; Ward et al., 2001, 2004; Galli et al., 2005; McElwain et al., 1999, 2007).

Series of perturbations in the global carbon cycle in particular have proved useful in evaluating the paleoenvironmental changes during the biotic crisis, and also in discussion of the separate issue of the placement of the Triassic-Jurassic boundary and its global correlation. After recognition of an "initial" and a "main" negative carbon isotope excursion (CIE), a more complex picture is emerging with other excursions, both negative and positive, prior to and following the Triassic-Jurassic boundary (e.g., Pálfy et al., 2001; Galli et al., 2005; Hesselbo et al., 2007; Hillebrandt et al., 2007; Korte and Kozur, 2011; Pieńkowski et al., 2012).

The origin of the huge amount of CO_2 or methane enriched in ^{12}C (^{13}C depleted) is still a matter of debate; it could have been derived directly from the Central Atlantic magmatic province basalts, subvolcanic thermal metamorphism of subsurface organic-rich strata, or from methane hydrates. None of these mechanisms is mutually exclusive, and all three may have contributed to the release of ^{12}C-enriched carbon at the end of the Triassic (Ruhl et al., 2011). Relative contributions of these end members remain unknown, but given the very short duration and magnitude of the observed end-Triassic ^{12}C release, a significant contribution from methane hydrates is likely. Nevertheless, cumulative carbon release might have been capable of amplifying an initial warming, resulting in runaway greenhouse conditions. Support for this scenario comes from fossil plants through stomatal density studies and other paleobotanical work (McElwain et al., 1999, 2007; Bacon et al., 2013).

Resolution of the extinction timing and the recognition of an ecological selectivity for extinctions of different groups of plants and animals have also helped to clarify the nature of the crisis (McRoberts and Newton, 1995; Ward et al., 2001, 2004; McElwain et al., 1999, 2007; Pieńkowski et al., 2012). The direct causal link between carbon-cycle disturbance (methane release) and onset of terrestrial biotic disturbances has been recently challenged, on the basis that the onset of the biotic crisis predates the initial negative $\delta^{13}C$ excursion (Lindström et al., 2012). Furthermore, it is unclear what caused the quite conspicuous (Hesselbo et al., 2002; Ruhl et al., 2009; Lindström et al., 2012) earlier Rhaetian $\delta^{13}C$ excursions, which predate the onset of Central Atlantic magmatic province volcanism. There are also reports pointing out possible cooling and/or release of toxic compounds as a cause of extinction (Hubbard and Boutler, 2000; Schoene et al., 2010; van de Schootbrugge et al., 2009). Furthermore, Korte et al. (2009) provided $\delta^{18}O$ data from fossil oysters that argue for cool ocean temperatures immediately after the initial $\delta^{13}C$ excursion, followed by ~8 °C early Hettangian warming.

The present study is based on data sets obtained from two boreholes (Kamień Pomorski IG-1 and Niekłań PIG-1, referred to herein as Kamień Pomorski and Niekłań) and two outcrops, Lisowice and Sołtyków (Fig. 1), which yielded records across the Triassic-Jurassic transition in continental deposits (lacustrine–alluvial plain). Many of the sedimentological, geochemical, and palynological results from the Kamień Pomorski borehole have been published (Pieńkowski et al., 2012). In this paper, we complement these results with clay mineral studies and discuss the results in the context of currently expanded knowledge of the Triassic-Jurassic boundary events, including mass extinction. A temporal sequence of events at the end of the Triassic and beginning of Jurassic is presented, and we also attempt to link environmental/climate changes with tetrapod biota evolution, based on data from Lisowice and Sołtyków, in order to arrive at a more comprehensive picture of the continental end-Triassic–earliest Jurassic climate, mass extinction of plants and tetrapods, and their subsequent recovery.

GEOLOGICAL SETTING

Tectonics

In the late Rhaetian and early Hettangian, the Polish Basin was located at about 40 degrees of the northern latitude, some 1200–1500 km from the nearest Central Atlantic magmatic province volcanic extrusions in Europe (Fig. 1). The Kamień Pomorski and Niekłań boreholes, ~500 km away of each other, are both located in the Mid-Polish Trough, as is the Sołtyków outcrop, situated close to the Niekłań borehole, while the Lisowice outcrop is located further to the south-west, outside the Mid-Polish Trough (Fig. 1). The Mid-Polish Trough is one of many inverted basins in Western and Central Europe (Ziegler, 1990), running generally along the Teisseyre-Tornquist zone and the Trans-European suture zone. The Mid-Polish Trough is the largest of these basins, with a length of more than 700 km, and it contains an accumulated sedimentary thickness of the order of 10 km. Rapid subsidence in the trough commenced in the early Hettangian (Pieńkowski, 2004), and in the Rhaetian subsidence was less marked (Fig. 1). Additionally, in Rhaetian-Hettangian times, the area around Kamień Pomorski was affected by a number of N-S–trending faults, grabens, and semigrabens (Dadlez, 1969; Dadlez et al., 1995). These synsedimentary faults and grabens acted as "catchment traps" for sandy sediments delivered from the Kaszuby Land, located eastward from the Koszalin-Chojnice fault zone (Fig. 1).

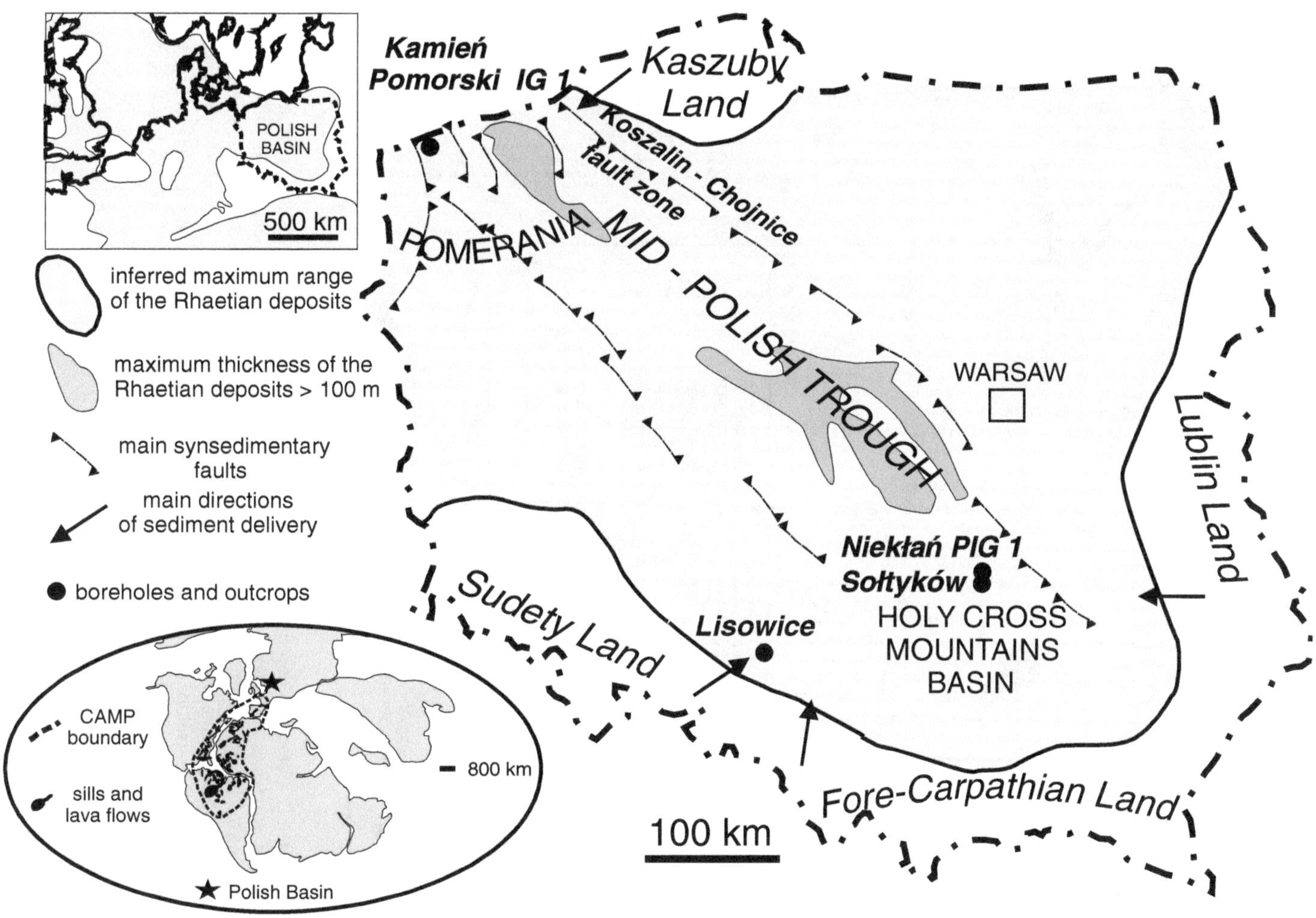

Figure 1. Late Rhaetian Polish Basin, its location in the European Basin, and its position on Earth ca. 201 Ma, in relation to the Central Atlantic magmatic province (CAMP).

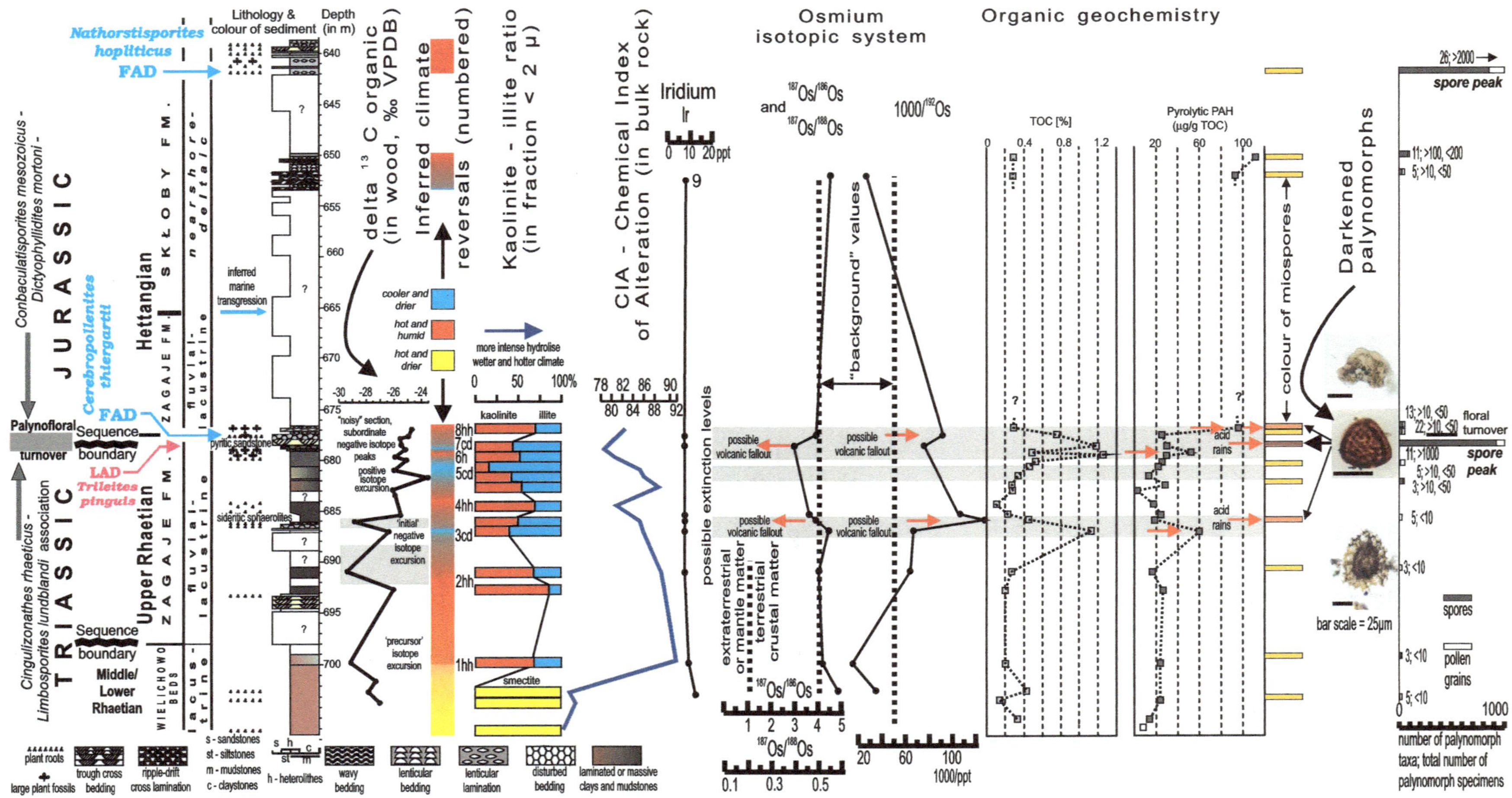

Figure 2. Kamień Pomorski IG-1 profile showing integrated stratigraphy, lithology, kaolinite-illite ratio in the <2 μm fraction, chemical index of alteration (CIA) in bulk rock, iridium content, carbon and osmium isotopes, total organic carbon (TOC), organic geochemistry (pyrolytic polycyclic aromatic hydrocarbons [PAH] and total organic carbon [TOC]), and miospore characteristics (color, diversity, and abundance). Inferred climate changes are based on clay mineralogy, and to lesser extent on lithology (sediment color), palynology, and other indices. Note increasing frequency of these changes above the "initial" carbon isotope excursion (CIE) toward the Triassic-Jurassic boundary, established on the basis of biostratigraphy. Osmium isotope ratios $^{187}Os/^{186}Os$ and $^{187}Os/^{188}Os$ show almost identical changes; therefore, they are presented as one plot with two scales, respectively. Stratigraphy, lithology, C and Os isotopes, and miospore characteristics are after Pieńkowski et al. (2012); pyrolytic PAH is after Marynowski and Simoneit (2009). FAD—first appearance datum; LAD—last appearance datum; VPDB—Vienna Pee Dee Belemnite standard.

Stratigraphy and Sedimentology

The Triassic-Jurassic boundary is located between the last appearance datum (LAD) of megaspore *Trileites pinguis* (Marcinkiewicz, 1962, 1971) and first appearance datum (FAD) of *Cerebropollenites thiergartii* in the Kamień Pomorski profile (Fig. 2; Pieńkowski et al., 2012). This placement of the Triassic-Jurassic boundary in Kamień Pomorski is further supported by major palynofloral turnover from the *Cingulizonathes rhaeticus–Limbosporites lundblandii* association, which corresponds to the *Rhaetipollis-Ricciisporites* (= *Rhaetipollis-Limbosporites*) zone of Rhaetian age, and the typically Hettangian *Conbaculatisporites mesozoicus–Dictyophyllidites mortoni–Cerebropollenites thiergartii* association (with the age-diagnostic pollen *C. thiergartii*), which corresponds to the *Pinuspollenites-Trachysporites* (= *Trachysporites-Heliosporites*) zone (Pieńkowski et al., 2012). This boundary is correlated with the base-Jurassic global stratotype section and point (GSSP) profile in Kuhjoch (Kürschner et al., 2007; Hillebrandt et al., 2007). A factor of crucial significance for stratigraphy is chemostratigraphic correlation between Kamień Pomorski and other profiles, based on carbon isotopes (Pieńkowski et al., 2012, their fig. 7).

Due to local tectonic conditions, the majority of coarser sediments were deposited east of Kamień Pomorski, and further to the west, the channel energy is inferred to have decreased on a "sand-starved" alluvial plain (Pieńkowski et al., 2012). Thus, muddy overbank deposits dominate in the Kamień Pomorski profile (Fig. 2). Fault activity in Rhaetian times periodically created additional accommodation space, allowing temporarily enhanced deposition of thicker packages of sediments.

Subsequent Early Jurassic sedimentation in Poland, including Pomerania and the Holy Cross Mountains, was characterized in detail by Pieńkowski (2004). Earliest Hettangian alluvial and lacustrine deposition (although much steadier and faster) was a continuation of Rhaetian deposition (Figs. 2 and 3). Therefore, both upper Rhaetian and lowermost Hettangian alluvial and lacustrine deposits are assigned to the same lithoformation—Zagaje Formation (Pieńkowski, 2004).

Preliminary geological and sedimentological observations from Lisowice outcrop (spanning the Triassic-Jurassic boundary) were presented by Szulc et al. (2006), Dzik et al. (2008a, 2008b), Niedźwiedzki and Sulej (2008, 2010), and Niedźwiedzki et al. (2012). We interpret these sediments as having been deposited on an alluvial plain (see Fig. 4 and the "Results" section). The whole 12-m-thick section at Lisowice is assigned herein to the informal lithostratigraphic units of the Zbąszynek Beds (lower red beds) and Wielichowo Beds (gray sandstone-mudstone bone-bearing interval plus heterolith unit). In previous work, the section in Lisowice was included in the other informal lithological unit, Woźniki Limestones or Woźniki Formation (Szulc et al., 2006). However, this diachronous red bed–carbonate unit does not fit into the general lithological character of the beds exposed in Lisowice, particularly in regard to the middle, gray-colored part of the outcrop (Fig. 4). Presumably, the Woźniki Limestones can be correlated laterally only with palustrine carbonate deposits (crenogenic in origin; see Szulc et al., 2006). These carbonates are also exposed in a small, abandoned quarry located ~600 m to the north of the clay pit.

Based on the palynomorph assemblages (Staneczko, 2007; Dzik et al., 2008a, 2008b; Świło et al., 2013), these strata would represent the subzone IVc (defined by the appearance of *Rhaetipollis germanicus* in Orłowska-Zwolińska 1983, 1985) of the *Corollina meyeriana* zone, and the *Ricciisporites tuberculatus* zone, which was later incorporated into the Rhaetian *Rhaetipollis germanicus* zone (Lund, 1977; Herngreen, 2005; Kürschner and Herngreen, 2010). According to the existing stratigraphical interpretation (Fig. 4), the Lisowice strata correlate with latest Norian Zbąszynek Beds and Rhaetian Wielichowo Beds (Deczkowski, 1997; Franz et al., 2007a, 2007b; Franz, 2008). The presence of Rhaetian strata is further supported by cuticle fragments of the typical Rhaetian seed-fern *Lepidopteris ottonis* (Goeppert, 1836) (Staneczko, 2007; Ociepa et al., 2008; Dzik et al., 2008a, 2008b; Wawrzyniak and Ziaja, 2009; Wawrzyniak, 2010a, 2010b, 2010c, 2011), the conifer *Stachyotaxus septentrionalis* (this genus has a distribution strongly restricted to the Rhaetian deposits of Scania and Greenland; see Taylor et al., 2009), isoëtalean macrospores *Trileites* cf. *pinguis* (Harris, 1935), and *Horstisporites bertelseni* Fuglewicz, 1977 (Fuglewicz and Śnieżek, 1980). At Lisowice, two horizons with conchostracans were identified, the lower with numerous *Shipingia* sp. and rare small *Euestheria* sp., and the upper with *Gregoriusella polonica* Kozur et Weems, 2010. These two conchostracan zones identify the presence of the latest Norian and early (possible also younger) Rhaetian deposits (Kozur and Weems, 2005, 2007, 2010), consistent with palynological data.

An outcrop of the Hettangian alluvial-plain deposits in Sołtyków, located close to the Niekłań borehole (Fig. 1), was described by Pieńkowski (2004). The strata there represent a siliciclastic coal-bearing lithofacies association of an alluvial wetland, with deposition controlled by high-sinuosity, anastomosing streams. The climate was warm to temperate, with wetter and drier seasons. An earliest Hettangian age is indicated by sequence stratigraphic correlation (Pieńkowski, 2004), macroflora dominated by the conifer *Hirmeriella muensteri* (Schenk) Jung and the benettite *Pterophyllum alinae* Barbacka, 2010 (Reymanówna, 1992; Wcisło-Luraniec, 1991; Barbacka et al., 2010), palynomorphs (Ziaja, 2006), and conchostracans *Bulbilimnadia kilianorum* Kozur, Weems, et Lucas, 2010).

MATERIAL AND METHODS

Recovery of core in the Kamień Pomorski borehole was ~42% in the Rhaetian and Hettangian, but, fortunately, a more complete core was recovered from the most important Triassic-Jurassic boundary interval. Moreover, the Kamień Pomorski profile represents a fairly continuous succession through the Triassic-Jurassic transition, which is unique in the Polish Basin. The new Niekłań core is 100% complete, but the Rhaetian profile is much more fragmentary due to hiatuses. However, despite

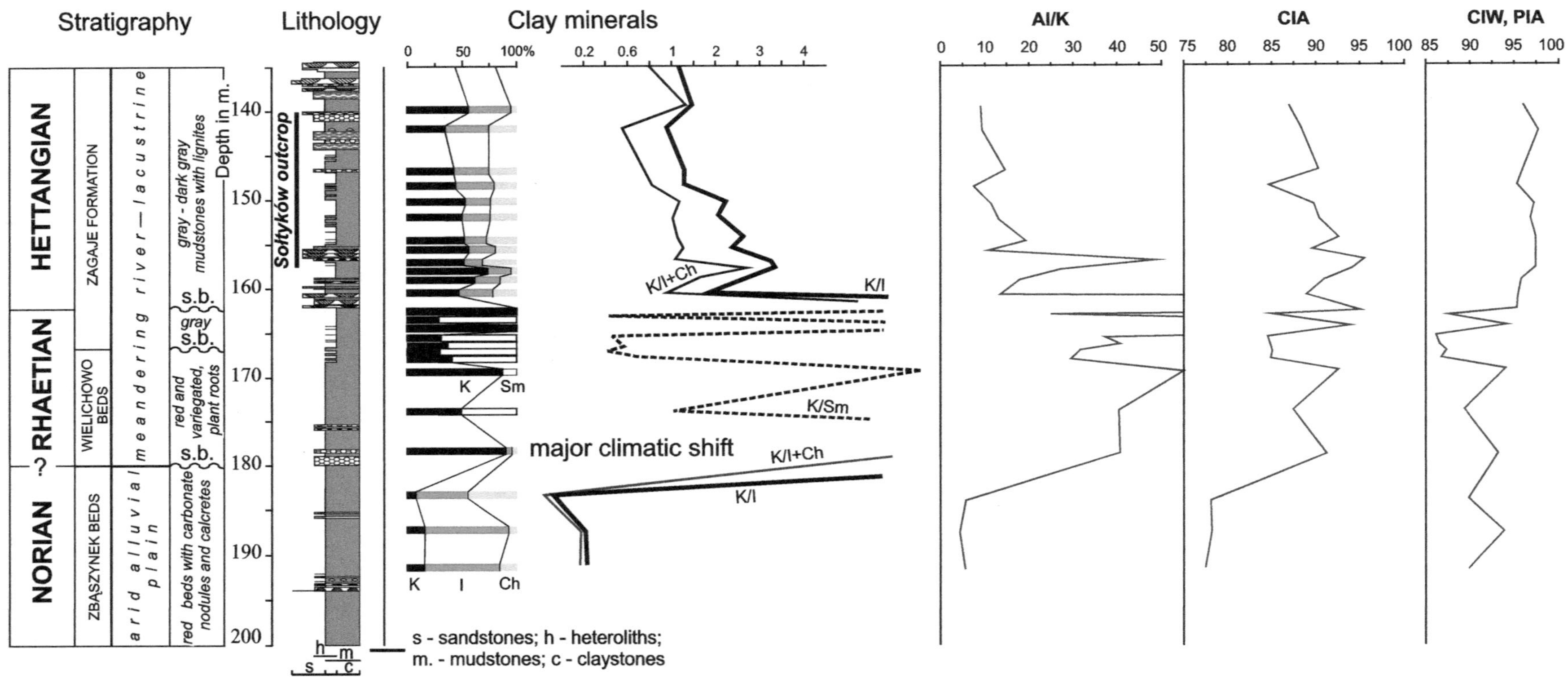

Figure 3. Niekłań PIG 1 borehole Triassic-Jurassic continental transition section showing stratigraphy, lithology, clay mineralogy, and weathering indices established in continental mudrocks (s.b.—sequence boundary). Note the mineralogical and geochemical record of a major climatic shift in the Rhaetian and series of frequent and abrupt changes in the clay mineral composition and weathering indices values below and at the Rhaetian-Hettangian (= Triassic-Jurassic) boundary, which correspond to inferred climate reversals. Explanations: Clay minerals: K—kaolinite, I—illite, Ch—chlorite, Sm—smectite; weathering indices: Al/K—aluminum/potash ratio; chemical index of alteration (CIA) = $(Al_2O_3/[Al_2O_3 + CaO^* + Na_2O + K_2O]) \times 100$ (Nesbitt and Young, 1982); chemical index of weathering (CIW) = $(Al_2O_3/[Al_2O_3 + CaO^* + Na_2O]) \times 100$ (Harnois, 1988); plagioclase index of weathering (PIA) = $\{(Al_2O_3 - K_2O)/([Al_2O_3 - K_2O] + CaO^* + Na_2O)\} \times 100$ (Fedo et al., 1995; for lithology column legend see Fig. 2). Stratigraphical position of the nearby (~4 km) Sołtyków outcrop is shown.

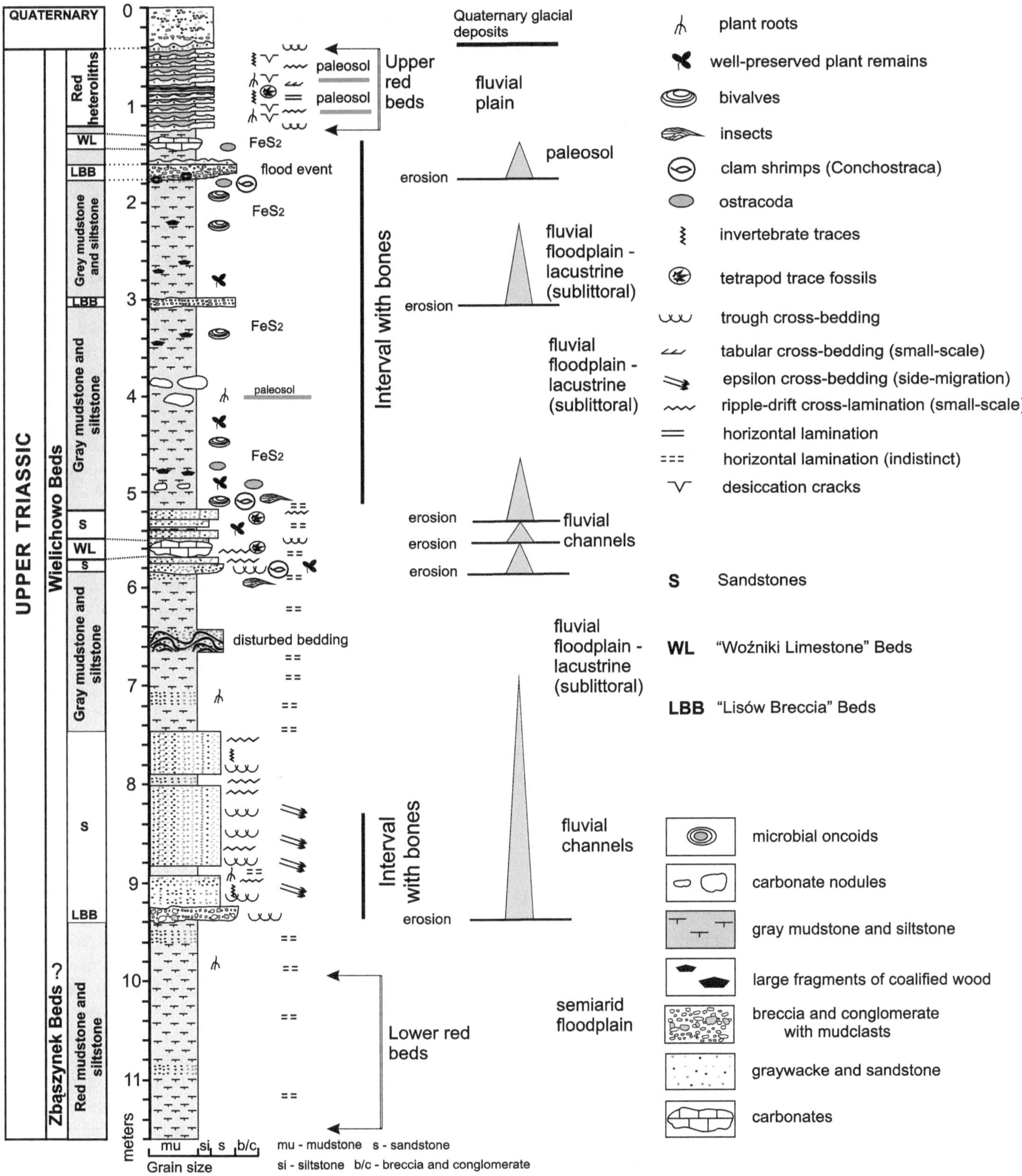

Figure 4. Lithologic, stratigraphic, and sedimentologic profile of the main exposure in the Lisowice (= "Lipie Śląskie" clay pit at Lisowice, Silesia) with position of the bone-bearing interval, tetrapod trace fossils, other faunal and floral remains, and intervals with red-bed deposits.

fragmentary preservation of the Rhaetian section, this core yielded a valuable, continuous profile through the Hettangian, together with clay mineral data reflecting important Rhaetian climatic changes, and giving insight into the mineralogy of underlying upper Norian red beds, deposited under dry climate conditions. Moreover, the Niekłań borehole is located very close to the Sołtyków outcrop, allowing integrated interpretation with very rich tetrapod ichnofossils (Gierliński et al., 2004; Niedźwiedzki, 2006, 2011b).

We acknowledge that the analysis of 24 samples (Kamień Pomorski) for $\delta^{13}C_{org}$, 15 samples for palynology, and even fewer for Os isotopes, spanning >60 m of section, is a relatively small number of data on which to report. However, 35 m of those 60 m section were not cored at all. We sampled the remaining 25 m for $\delta^{13}C_{org}$ as densely as possible (depending on the core preservation), and the average density of sampling in cored profile is close to 1 sample per meter, although sampling density is irregular, again due to core preservation and lithology (only mudstones yielded sufficient palynomaceral material).

Palynology, Carbon and Osmium Isotopes, Iridium Content, and Organic Geochemistry in the Kamień Pomorski Profile

Palynological results, carbon isotopes, osmium isotopes, and organic geochemistry in the Kamień Pomorski profile are taken from the previous paper by Pieńkowski et al. (2012), and a detailed description of material and methods is included therein. Fifteen rock samples from the Kamień Pomorski core were selected for quantitative palynological analysis, 12 of which yielded palynomorphs (Fig. 2).

Samples for $\delta^{13}C_{org}$ from the Kamień Pomorski borehole (Fig. 2) were separated manually under a binocular microscope from palynomorphs, and only woody phytoclast separates were taken for the carbon isotope analysis (homogenized samples). The carbon isotope ratios were measured using an elemental analyzer Carlo-Erba 1110 connected online to a Thermo Finnigan Delta Plus mass spectrometer (see Pieńkowski et al., 2012).

Analyses of osmium, iridium, and rhenium content, as well as analyses of the following isotopes, ^{186}Os, ^{187}Os, ^{188}Os, ^{192}Os, and ^{187}Re, were carried out from nine samples (Fig. 2). The analytical procedure used here followed that described in detail by Brauns (2001). The isotopic composition of Os in the blank is very close to the natural composition and has a ratio of 0.112 for $^{187}Os/^{188}Os$. All data were blank corrected on the basis of these measurements in combination with a yield of 90%, and an Os blank of 0.10 pg. For more information, see Pieńkowski et al. (2012).

The total organic carbon (TOC) and polycyclic aromatic hydrocarbons (PAH) were determined in 24 mudstone samples using a chromatographic, coulometric method (procedure PB-23) in an automated LECO analyzer. The gas chromatography–mass spectrometry (GC-MS) analyses of the aliphatic and aromatic fractions were performed with an Agilent 6890 Series Gas Chromatograph interfaced to an Agilent 5973 Network Mass Selective Detector and Agilent 7683 Series Injector (Agilent Technologies, Palo Alto, California) by Marynowski and Simoneit (2009).

Clay Mineralogy

The bulk-rock mineralogy, clay mineralogy, and major-element geochemistry were determined from 40 claystone and mudstone samples from Kamień Pomorski and the lower part of the Niekłań borehole. Bulk-rock mineralogy and clay minerals (in the <2 µm fraction) were identified by X-ray diffraction (XRD) using a Phillips PW 3020 X'Pert diffractometer with CuKα radiation. Major-element amounts were measured by X-ray fluorescence using a Phillips PW 2400 spectrometer. All samples were analyzed at the Polish Geological Institute–National Research Institute laboratories. For more details, see Brański (2014).

Tetrapods

The vertebrate remains and trace fossils were described from the uppermost Norian–Rhaetian strata of Lisowice, while a rich assemblage of the earliest Hettangian tetrapod trace fossils comes from the Sołtyków outcrop. In order to obtain the sedimentological and stratigraphical framework (Fig. 4) for the most important part of the profile in Lisowice with spectacular bone-bearing deposits, a few short trenches were excavated through the site (2007–2013). A detailed description of the taphonomic processes recognized for all bone-bearing horizons at this site is presented in Niedźwiedzki (2014).

RESULTS

Lithology

The Upper Norian deposits in the Niekłań borehole (Fig. 3) and uppermost Norian Lisowice outcrop (Fig. 4, the lower red bed) represent a red-bed association (mostly mudstones and siltstones) with rare calcrete, other carbonate nodules, and oxidized plant roots, and are assigned to the Zbąszynek Beds. The superposed Rhaetian sediments generally show bipartite development (Figs. 2 and 3). The lower part (belonging to the Wielichowo Beds) is represented by red-brownish, yellow-greenish, or variegated mudstone, with poorly preserved (oxidized) rootlets and scattered calcium carbonate concretions or carbonate calcretes in its lower part. In the Kamień Pomorski and Niekłań boreholes, the Upper Rhaetian (belonging together with the overlying Lower Hettangian strata to the Zagaje Formation) commences with sandstone resting on an erosional boundary, identified with the sequence boundary (depth of 698 m in Kamień Pomorski [Fig. 2]; 180 m in Niekłań [Fig. 3]). Upper Rhaetian mudrock of the Zagaje Formation (678.4–693 m [Fig. 2]; 162–180 m [Fig. 3]) shows a conspicuous change in color into gray/dark gray in Kamień Pomorski. However, there are thin intervals showing an intermittent return to a variegated or reddish gray color (Fig. 2).

In Niekłań, the color change from red to variegated colors occurs in the bottom of the 162–180 m interval, and gray sediments appear in the top of this interval (Fig. 3).

This change of color into gray/dark gray is associated with the appearance of frequent paleosols with coalified rootlets, occasionally topped with thin lignite beds, which further point to a general humidification of climate. At the top of the Rhaetian deposits, the second sequence boundary occurs (678.4 m in Kamień Pomorski; 161.9 m in Niekłań), slightly below or at the Triassic-Jurassic boundary. This boundary, linked to another sea-level lowstand at the Triassic-Jurassic boundary (Pieńkowski, 2004), is usually marked by regional erosion in the whole sedimentary basin of Poland (quite often, the Rhaetian deposits are missing in the whole basin or for a significant part). Only in Kamień Pomorski, due to a local tectonic regime, is this erosion less significant (fine-grained sandstone above the erosional boundary does not contain coarser grains or mud clasts, and bedding at the bottom of pyritic sandstone is characteristic of a rhythmic, lower-flow regime; see Pieńkowski et al., 2012). On the other hand, the Niekłań profile (Fig. 3) shows significant erosion and hiatuses at the inter-Rhaetian and Triassic-Jurassic sequence boundaries. Judging from clay mineral composition and its comparison to the more complete Kamień Pomorski profile, the whole uppermost Rhaetian section is missing in Niekłań. The Sołtyków outcrop exposes only the lowermost Hettangian strata (Fig. 3; for detailed profile of the outcrop, see Pieńkowski, 2004).

Above the Norian red beds of the clay pit at Lisowice (Fig. 4), the major part of the section is represented by gray mudstones and cross-bedded sandstones (graywackes), mainly with trough cross-bedding (general transport of material was to NE; Fig. 1) and ripple-drift cross-lamination, including climbing ripples. The epsilon cross-bedding produced by lateral migration of channels is common. Both sandstone and mudstone lithofacies contain numerous drifted plant fossils and pedogenic horizons. The mudstones and sandstones are slightly calcareous, likely because of redeposition and destruction of older carbonate rocks. Layers of palustrine carbonates or dolomitic mudstones also occur (called regionally the "Woźniki Limestone," a diachronous and genetically inhomogeneous lithological unit). Conglomerates and breccia-like coarse sediments appear mostly at the bottoms of sandstone layers (= bottoms of fluvial cycles) and are composed of reworked carbonate nodules from older Upper Triassic sediments and rock fragments (quartzites of shales, probably of Paleozoic age). These conglomerates are traditionally known as the "Lisów breccia" bed (diachronous lithological unit). A factor of note is a layer of disturbed bedding in the middle part of the profile (Fig. 4), possibly representing a tectonic shake effect. A red-greenish heterolith unit (interbedded mudstone, siltstone, and sandstone beds) occurs in the top of the outcrop, and it contains carbonate nodules, often infilling plant roots (Fig. 5).

There are two major intervals with bones in Lisowice (Fig. 4), and the upper (main) one has four discrete bone-bearing beds, separated by bone-barren mudstone-siltstone levels with numerous plant remains and casts of bivalve shells. Lithological characteristics, geometry of the lithic units, and sedimentary structures (Fig. 4) suggest that the gray (middle) strata in Lisowice were formed in a mosaic of alluvial-plain environments—swamps with ephemeral ponds and laterally shifting river channels. In particular, texturally immature sediments (graywackes), lack of stable channels, and presence of climbing ripples point to an increased rainfall and runoff, combined with fast deposition. This interpretation is supported by macrofloral and microfloral remains (Fuglewicz and Śnieżek, 1980; Wawrzyniak, 2010a, 2010b, 2010c, 2011). Both red bed units at the bottom and at the top of the outcrop were also deposited on an alluvial plain, but in semiarid climate conditions.

Clay Mineralogy

The examined bulk-rock samples of claystones and mudstones from the Zbąszynek Beds (Norian), Wielichowo Beds, and Zagaje Formation (late Rhaetian–early Hettangian) of Kamień Pomorski (Fig. 2) and Niekłań borehole (Fig. 3) are mainly composed of phyllosilicates (50%–90%). Quartz is less abundant, but occurs in variable amounts (7%–43%). In the Wielichowo Beds (early-middle Rhaetian), hematite, goethite, feldspars, calcite, and locally dolomite were observed with proportions of 0%–8%. In contrast, in the Zagaje Formation, calcite and dolomite are absent, but instead siderite is present in some samples with proportions of 0%–7%. Feldspar, hematite, and goethite appear sporadically, mostly in accessory amounts.

Figure 5. Top 3 m of the Lisowice outcrop (see Fig. 4), showing the gray mudstones (the upper part of the interval with bones), capped by the red-greenish heterolithic unit with pedogenic horizons. Note numerous plant roots associated with carbonate nodules (inset). The heterolithic unit probably reflects climatic fluctuation toward aridity. Such fluctuations might have caused biotic crises, including extinction of vertebrates. Scale bar is 1 m.

The clay mineral composition is very diverse in the <2 μm fraction, which is reflected in kaolinite/illite, and kaolinite/illite + chlorite, and in places kaolinite/smectite ratios (Figs. 2 and 3). In three samples taken from the Upper Norian strata (Zbąszynek Beds; Fig. 3), illite (with subsidiary chlorite) totally predominates, and kaolinite content is insignificant. In the overlying Wielichowo Beds (Rhaetian), the composition of clay mineral assemblage is different (Figs. 2 and 3). With exception of a few samples (see following), an abundance of smectite is observed; kaolinite appears mostly in lesser amounts (Fig. 3) or is absent (Fig. 2). In Kamień Pomorski, smectite makes up 100% of clay minerals in claystones of the Wielichowo Beds (Fig. 2). Small amounts of illite and chlorite were detected only in a single sample in Niekłań. In the superposed Zagaje Formation, kaolinite generally predominates (with an average content of 54% in Kamień and 49% in Niekłań) over illite (average content of 46% and 33%, respectively). Chlorite is still absent in the late Rhaetian part of the Zagaje Formation. It should be noted that in the lowest part of this formation in Niekłań section, a significant content of smectite (up to 70%) was still observed (Fig. 3). However, smectite is absent in the Zagaje Formation in Kamień Pomorski, while the kaolinite phase is accompanied by significant amounts of berthierine or serpentine. In the Hettangian part of Zagaje Formation (Fig. 3), the clay mineral assemblage consists of kaolinite and illite, with significant subsidiary chlorite, and smectite is almost absent.

Clay mineral data are reflected in the major-element geochemistry and in the values of alteration indices, which are high in Upper Rhaetian and Lower Hettangian strata. The chemical index of alteration (CIA) ranges from 79 to 96, but in the vast majority of cases, it exceeds 85, and these are due to residual clays (Nesbitt and Young, 1982). CIA shows the loss of Ca^{2+}, K^{+}, and Na^{+} in relation to Al^{2+} (Nesbitt and Young, 1982; see caption of Fig. 2). Therefore, CIA data are very useful in recognizing climate fluctuations.

Clay minerals in the studied sections are largely detrital and show generally insignificant diagenetic overprint (Brański, 2009), due to moderate burial and to the fairly closed hydrologic system. There is no distinct and systematic evolution of clay-mineral assemblages from top to bottom of the studied boreholes that would be indicative of burial diagenesis. Importantly, smectite is abundant in Lower-Middle Rhaetian Wielichowo Beds despite its sensitivity to diagenetic alteration. Low thermal alteration of the Rhaetian-Hettangian strata in the Polish Basin was confirmed by palynomorph color (Pieńkowski and Waksmundzka, 2009) and biomarkers (Marynowski and Simoneit, 2009). Moreover, a focus of this study is kaolinite content, which is resistant under moderate diagenetic conditions. Moreover, the scanning electron microscope (SEM) observations show that clay minerals are mostly detrital, and authigenic kaolinite appears rarely. In the studied <2 μm fraction, most diagenetic kaolinite should be excluded, because this has a typical particle size of 5–10 μm (when it has replaced feldspar) and an order of magnitude greater when it pseudomorphs mica (e.g., Hesselbo et al., 2009). On the other hand, breaking of kaolinite crystallites may occur during sample preparation (Hesselbo et al., 2009).

The kaolinite peaks in some intervals in the Rhaetian and at the beginning of the Hettangian on clay mineral ratio curves are of particular interest (Figs. 2 and 3). It should be noted that the first distinct rise of kaolinite content already appeared earlier in the Rhaetian Wielichowo Beds (Figs. 2 and 3; see the 1 hh climatic step at 700 m in Fig. 2, where the sequences of frequent and marked climatic reversals are named as 1hh, 2hh, 3 cd, 4 hh, 5 cd, 6 h, 7 cd, and 8 hh events [hh—hot and humid, cd—cold and dry, h—humid]; Fig. 2; Table 1). Similar peaks are observed in the Zagaje Formation, and they correspond in the Kamień Pomorski profile to the three rapid increases of kaolinite through the late Rhaetian section of the Zagaje Formation, and the last one at the beginning of Hettangian section of this formation. These peaks are described (Fig. 2; Table 1) as 2 hh, 4 hh, 6 hh, and 8 hh climatic steps.

Higher in the section, the kaolinite content significantly decreases and stabilizes (Fig. 3). The kaolinite peaks usually correlate with the highest values of weathering indices (in the most cases, CIA values are greater than 90). A few discrepancies may relate to the fact that clay minerals were analyzed in the <2 μm fraction, but the major-element content was measured on bulk-rock samples, and most probably some of them (in Kamień Pomorski borehole; Fig. 2) may have been somewhat influenced by Na metasomatism (Brański, 2014).

Interestingly, recent studies focused on the weathering proxies (CIA, chemical index of weathering [CIW] = $[Al_2O_3/\{Al_2O_3 + CaO^* + Na_2O\} \times 100$ [Harnois, 1988]; plagioclase index of weathering [PIA] = $[(Al_2O_3 - K_2O)/\{(Al_2O_3 - K_2O) + CaO^* + Na_2O\}] \times 100$ [Fedo et al., 1995]) are indicative of intense continental weathering associated with pulsed end-Cretaceous Deccan eruptions and abundant rainfall (acid rains), which resulted in mesotrophic waters (Gertsch et al., 2013).

Carbon Isotopes, Osmium Isotopes, Organic Geochemistry, and Miospores

The values of $\delta^{13}C_{org}$ (see Pieńkowski et al., 2012; Fig. 2; Table 1) show significant fluctuations through the Rhaetian. The most conspicuous negative $\delta^{13}C_{org}$ excursion is identified with the Rhaetian "precursor" and double-peaked "initial" CIE, followed by a positive excursion and again by slightly more negative values, probably representing subordinate fluctuations within a longer positive excursion, where the Triassic-Jurassic boundary is now inferred (Korte et al., 2009; Korte and Kozur, 2011). The double-peaked initial CIE shows negative values of -29.38‰ $\delta^{13}C_{org}$ at 691 m and -28.85‰ $\delta^{13}C_{org}$ at 686 m, separated by a more positive value (~-26‰ at 687 m). However, the interpretation of "subpeaks" should be treated with some caution, because there are only a few carbon-isotope data, and each feature is represented only by a single data point, with significant data gaps separating each inferred excursion. The amplitude of negative values indicates that these disturbances can be compared only to

TABLE 1. SEQUENCE OF $\delta^{13}C$ EXCURSIONS, LITHOLOGICAL, CLAY MINERAL, AND MINERALOGICAL CHANGES, OSMIUM ISOTOPE AND POLYCYCLIC AROMATIC HYDROCARBON (PAH) EVENTS, CHARACTERISTICS OF MIOSPORE APPEARANCE, AND INFERRED SUCCESSION OF MAIN CLIMATE AND EVENTS IN THE KAMIEŃ POMORSKI BOREHOLE SECTION ACROSS THE TRIASSIC-JURASSIC (T-J) BOUNDARY

Depth (m), event no. (Fig. 2)	$\delta^{13}C$ excursions	Sediment color, kaolinite/illite/smectite, CIA	Os isotopes	PAH and TOC	Color of miospores	Environment, climate
676.9 8hh	Slightly more positive	Gray, kaolinite >> illite		PAH increase	Darkened miospores	Intermittent drop in palynofloral diversity, final disappearance of Triassic miospores
677.5 7 cd–8hh	Slightly more negative	Gray	Increase of ^{192}Os			Appearance of new miospores, high diversity, onset of Jurassic palynofloral recovery
678.6 (T-J boundary) 7cd	Slightly more positive	Sequence boundary, dark-gray, illite > kaolinite, drop in CIA, pyrite	Drop in $^{187}Os/^{186}Os$	PAH and TOC increase	Strongly darkened miospores, fern peak	CAMP volcanic fallout, acid rains, cool episode, extirpation of Triassic palynoflora
679.5 6 h	Slightly more negative	Dark gray-olive gray, illite > kaolinite, kaolinite rise				Intermittent warming
680–682.5 5 cd	Positive excursion	Dark gray–**reddish**, illite > kaolinite (rapid kaolinite drop)			Pollen grain>spores	Drier and cooler
684.5, 4hh	Return to background values	Dark gray, rapid rise of kaolinite, kaolinite >> illite				Hot and humid
686.0 3 cd–4 hh	"Initial" negative excursion (upper subpeak of the INE)	Dark gray, kaolinite = illite	Os isotope disturbance, decrease in $^{187}Os/^{186}Os$, $^{187}Os/^{188}Os$ and ^{192}Os rise		Darkened miospores	CAMP volcanic fallout, acid rains, beginning of floral turnover
687.0 3 cd	Positive subpeak within the INE	Gray, illite > kaolinite	Stabilization	High PAH, high TOC		Cooler and drier, forest fires and/or volcanogenic PAH release
689–692 2 hh	"Initial" negative excursion (lower subpeak of the INE)	Dark gray, kaolinite >> illite	Slight disturbance in $^{187}Os/^{186}Os$, $^{187}Os/^{188}Os$, and ^{192}Os			Hot and humid, ? oldest known CAMP volcanism
699.3–700 1 hh	"Precursor" negative excursion	Variegated, kaolinite >> illite; 698 m - sequence boundary	Slight decrease in $^{187}Os/^{186}Os$ and $^{187}Os/^{188}Os$		Spores>pollen grains	Major climate turnover into more humid, intensified hydrolysis
>700		Reddish, 100% smectite, carbonate nodules				Semidry, seasonal climate

Note: INE—"initial" negative excursion; CIA—chemical index of alteration; TOC—total organic carbon; CAMP—Central Atlantic magmatic province.

the "initial" CIE, because there are not such light values known in the Rhaetian terrestrial organic matter (Hesselbo et al., 2002; Whiteside et al., 2010; Ruhl et al., 2010, 2011).

A subsequent positive excursion follows (-23.67‰ $\delta^{13}C_{org}$ at 681.5 m, although this is again based on a single data point), which is in turn followed by more negative values with subordinate fluctuations. At the Triassic-Jurassic boundary (677.5 m), there is a slight trend toward more positive values at the interval 676.9–677.5 m, followed by an uncored section between 653.4 and 676.5 m for which there are no data. Noticeably, the sediments spanning the "initial" double-peaked $\delta^{13}C_{org}$ excursion are markedly expanded in comparison to other known profiles (~6 m thick; Fig. 2), and assuming that the "initial" CIE lasted for only some 10,000–20,000 yr (Ruhl et al., 2011) and the precompaction thickness of this mud-sandy interval could be some 8–9 m (assuming compaction factor of 1.4), we obtain an average sedimentation rate of some 80–90 cm/1000 yr (taking 10,000 yr duration, or half of this figure in case of 20,000 yr). Such a sedimentation rate is reasonable during an intensified hydrologic cycle for an alluvial-plain environment, under steady subsidence rate.

The first suites of osmium/rhenium isotope data from the continental deposits across the Triassic-Jurassic boundary were obtained by Pieńkowski et al. (2012); here, we use these results to further investigate geochemical events occurring at the Triassic-Jurassic boundary. The data set, obtained from nine samples (Fig. 2), represents corrected, initial Re/Os isotopic ratios of $^{187}Os/^{186}Os$ and $^{187}Os/^{188}Os$, as well as content of unradiogenic osmium isotope ^{192}Os (Fig. 2; Table 1).

In the lowermost part of the section, $^{187}Os/^{188}Os$ and $^{187}Os/^{186}Os$ ratios slightly decrease upward, along with the stable ^{192}Os, up to the sequence boundary at 678.4 m. Conspicuous decreases of $^{187}Os/^{188}Os$ and $^{187}Os/^{186}Os$, linked with a sharp increase in ^{192}Os, are associated with the CIE (higher "subpeak" of the "initial" CIE at 686 m) and are interpreted as the result of a volcanic fallout event (Pieńkowski et al., 2012). Another decrease of $^{187}Os/^{188}Os$ and $^{187}Os/^{186}Os$ at 678.6 m, just below the Triassic-Jurassic boundary (though linked with a decrease in ^{192}Os), could also reflect volcanic fallout (association with darkened miospores). Higher in the profile, the $^{187}Os/^{188}Os$ and $^{187}Os/^{186}Os$ ratios in the Hettangian section increase. The content of ^{192}Os drops back to Lower/Middle Rhaetian values (one sample at depth 652.1 m). All measured values of the initial ratio $^{187}Os/^{186}Os$ are around 3 or higher (Fig. 2), which is indicative of a crustal origin (Koeberl and Shirey, 1997; Koeberl, 1998). Additionally, iridium content was measured, and all the values were very low, below 10 ppt (Fig. 2). The disturbances in the Os isotopic system are coeval or almost coeval with two levels showing elevated polycyclic aromatic hydrocarbon (PAH) contents (Marynowski and Simoneit, 2009). Their first increase of PAH content slightly predates (is 1 m below) the major Os excursions at the depth of 686 m, the second PAH abundance peak coincides with the less-marked excursions at the Triassic-Jurassic boundary (677.5–678.6 m), and the third at 652 m (Hettangian) is not related to any Os system disturbance (Fig. 2).

Systematic description of miospores was given by Pieńkowski et al. (2012). There are three levels of darkened miospores (Fig. 2; Table 1; 686 m, 678.6 m [Rhaetian], and 676.6 m [Hettangian] section), indicating acid rains (van Schootbrugge et al., 2009; Pieńkowski et al., 2012). Darkened miospores at 686 m, within the "initial" negative isotope excursion, coincide with carbon and osmium isotope as well as with PAH disturbances. Between 682.5 and 700 m in Kamień Pomorski (Fig. 2), the miospores are very scarce, 3–5 specimens per sample. Above 682.5 m, there is an increase in abundance and diversity of spores, and the changes of spore/pollen grain proportions generally follow the changes in kaolinite/illite ratio (Fig. 2). Just below the Triassic-Jurassic boundary, the abundance of miospores is very high (1355 specimens), linked with their marked darkening, $^{187}Os/^{186}Os$ and $^{187}Os/^{188}Os$ drop, as well as a significant drop in kaolinite and CIA (Fig. 2; Table 1). Both diversity and abundance of miospores are generally much higher in the Lower Hettangian strata, and spores dominate over pollen grains (Pieńkowski et al., 2012; Fig. 2). Another spore peak occurs higher up, within the Hettangian section (Fig. 2, depth 641.8 m), but within this peak, we observe much more diversified palynomorphs (i.e., number of palynofloral taxa), and also many more pollen grains produced by gymnosperm plants.

Tetrapod and Other Vertebrate Fossils

The recovered accumulation of bones in Lisowice includes over 400 large tetrapod bones and over 500 small vertebrate remains (fish teeth and scales, small tetrapod bones and teeth), representing up to 15 species (Niedźwiedzki, 2014). The vertebrate assemblage contains typical Late Triassic components and is dominated by bones of a giant dicynodont (Fig. 6D) and remains of actinopterygian fishes. Other vertebrate skeletal remains are rare and include small to large archosauromorphs, a large capitosaur, and a small plagiosaur. Well-preserved bones occur only in the lenticular body of gray mudstone and claystone deposits (Figs. 4 and 5). The Lisowice clay pit is dominated by terrestrial rather than amphibious or aquatic tetrapods. In two layers with large wood fragments, remains of huge dicynodonts are associated with remains of *Smok wawelski* (a basal predatory dinosaur), small- to medium-size archosaurs (pterosaur, dinosauromorph, dinosaur, and poposaurid), and other more non-archosaur archosauromorphs (a choristodere-like animal and rhynchosaur). Identification of many of the disarticulated, often fragmentary bones of other tetrapods is difficult and usually problematic (e.g., Dzik et al., 2008a, 2008b; Niedźwiedzki et al., 2011, 2012).

The temnospondyl amphibians (*Cyclotosaurus* sp. and *Gerrothorax* sp.) are known from an isolated, partially preserved skull (Fig. 6B), skull bones, jaw bones, and numerous long bones collected in a layer less than a meter above the principal bone-bearing bed of the clay-pit exposure. A few isolated long bones of capitosaur and plagiosaur were also found in the main bone-bearing horizon. Such "amphibian assemblages" are typical of a frequently flooded alluvial floodplain. Numerous macroremains of coelacanth and dipnoan fishes, hybodont sharks (Świło, 2010a, 2010b, 2010c; Świło and Kowalski, 2011), and palaeonisciform fishes were also found. A very rare mammaliaform or mammal tooth (Morganucodontidae indet.) was also recorded (Świło et al., 2013).

The tetrapod ichnofauna of the Zagaje Formation (lower Hettangian) exposed at the Sołtyków clay pit and also other sites (Figs. 7A–7F) shows a high ichnotaxonomic diversity (Gierliński and Pieńkowski, 1999; Gierliński et al., 2004; Niedźwiedzki, 2011b). Rare tetrapod bones have also been found (Fig. 7G). After re-examination of the collected material, specimens observed in the field, and the study of extended, as-yet-undescribed material from more recent collecting activities (Niedźwiedzki, 2011b), the ichnotaxa assemblage includes 12 ichnotaxa of dinosaurs representing: predatory dinosaur footprints; trackways and tracks of herbivorous sauropodomorphs; and isolated prints of early ornithischia (*Stenonyx* isp., *Grallator* isp., *Anchisauripus* isp., *Kayentapus soltykovensis*, *Eubrontes* isp., cf. *Megalosauripus* isp., *Anomoepus* isp., *Delatorrichnus* isp., *Parabrontopodus* isp., cf. *Otozoum* isp., cf. *Tetrasauropus* isp.). Dinosaur swimming traces are represented by *Characichnos* isp. The assemblage reflects notable dinosaur diversity, thus far not known from other Hettangian or pre–Early Jurassic localities. By the morphology of imprints, the most abundant large, tridactyl theropod dinosaur ichnotaxa (*Kayentapus soltykovensis*, *Eubrontes* isp., cf. *Megalosauripus* isp.) can be attributed to three species of early predatory dinosaurs. *Parabrontopodus* isp., cf. *Otozoum* isp., and cf. *Tetrasauropus* isp. might possibly represent footprints of early sauropodomorphs ("prosauropods" and early sauropods). The numerous gigantic and large theropod dinosaur footprints were discovered in the Sołtyków track site. These finds provide evidence for the occurrence of gigantic and large predatory dinosaurs in the earliest Jurassic (early Hettangian) times (Gierliński et al., 2001,

Figure 6. Characteristic tetrapod fossils from the Late Triassic of Poland: (A) *Metoposaurus diagnosticus krasiejovensis* Sulej, 2002, skull, Krasiejów; (B) part of mandible of *Cyclotosaurus* sp., Lisowice; (C) maxilla of *Polonosuchus silesiacus* (Sulej, 2005), Krasiejów (Silesia); (D) femur, tibia, and fibula of the new species of large dicynodont, Lisowice (Silesia); (E) cf. *Grallator* isp., tridactyl dinosaur footprint, Lisowice (Silesia); (F) Chirotheriidae indet., pentadactyl pseudosuchian footprint, Woźniki (Silesia); (G) cf. *Pseudotetrasauropus* isp., tetradactyl pseudosuchian footprint, Skarszyny (Holy Cross Mountains). Scale bars: A and B = 10 cm; C, E–G = 5 cm; D = geological hammer 32 cm long.

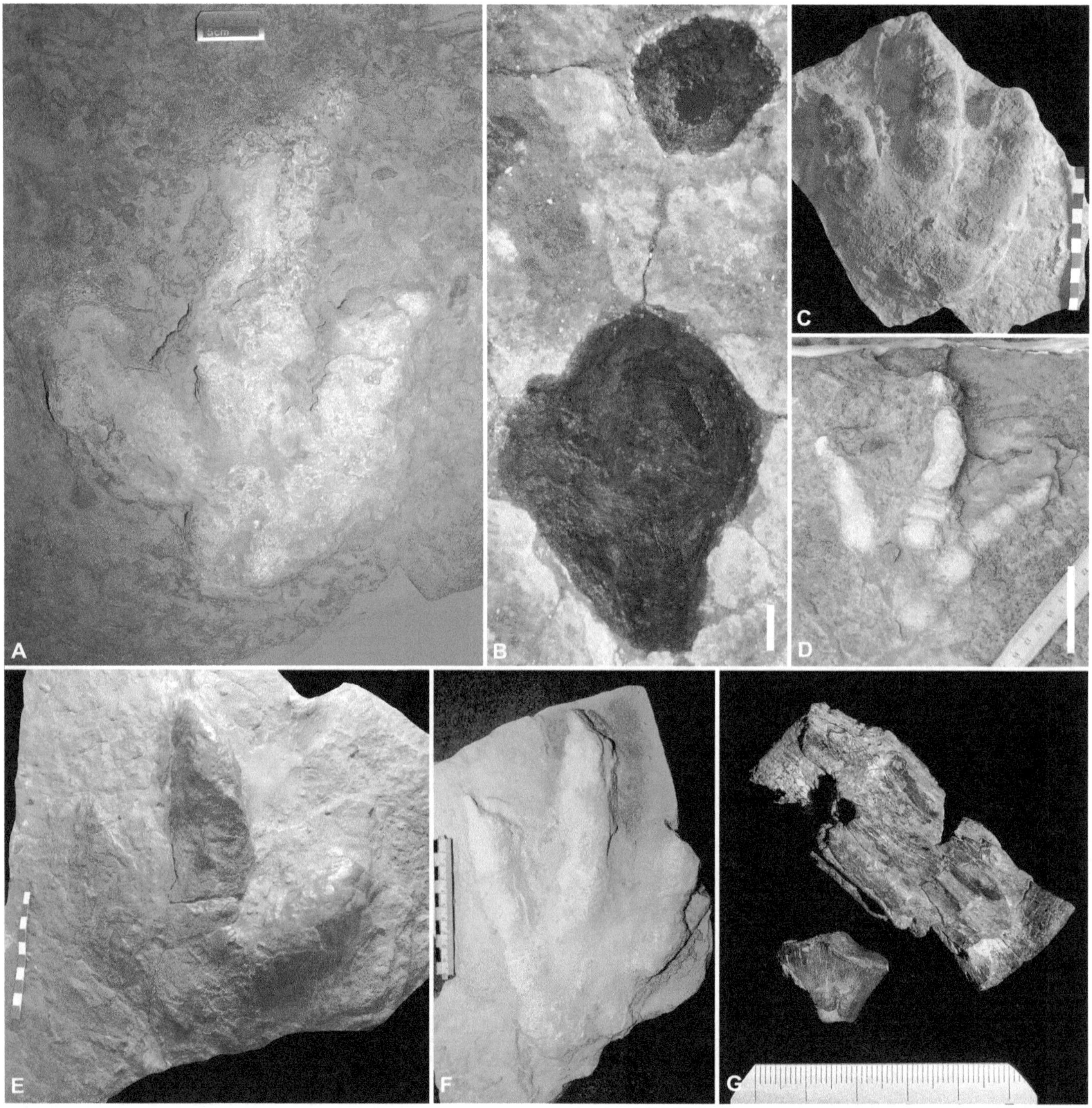

Figure 7. Characteristic tetrapod fossils from the Early Jurassic of Poland: (A) cf. *Megalosauripus* isp., footprint of large theropod dinosaur, Sołtyków; (B) *Parabrontopodus* isp., manus and pes imprints of the early sauropod, Sołtyków; (C) *Moyenisauropus* cf. *natator* Ellenberger, 1974, footprint of the early ornithischian dinosaur, Gromadzice; (D) *Anomoepus* isp., footprint of the early ornithischian dinosaur, Sołtyków; (E) *Kayentapus* isp., footprint of the large theropod dinosaur, Sołtyków; (F) *Eubrontes* isp., footprint of the large theropod dinosaur, Gromadzice; (G) dinosaur bones, Sołtyków. Scale bars: A, B, D, and G = 5 cm; C, E, and F = 10 cm.

2004). The largest footprints were described as cf. *Megalosauripus* isp. (50–65 cm long). Another large theropod footprints (30–45 cm) identified at Sołtyków resemble *Kayentapus* -like ichnites (*Kayentapus soltykovensis*) and classic ichnotaxa of the Newark Supergroup, i.e., *Eubrontes giganteus* (Olsen et al., 2002). Of note are also the finds of nondinosaur footprints: *Ameghinichnus* isp., left by a mammal-like animal, and small tetra- or pentadactyl footprints left by small reptiles (pterosaurs and small diapsids) and basal crocodylomorphs.

CLIMATIC AND ENVIRONMENTAL CHANGES VERSUS TERRESTRIAL EXTINCTION AND RECOVERY—DISCUSSION

Climatic Changes and Palynofloral Crisis

Results presented herein allow characterization of climatic and other environmental changes at the end of the Triassic and beginning of the Jurassic and link them with the end-Triassic terrestrial extinction and ensuing Early Jurassic recovery. The most continuous (though not completely recovered) profile from Kamień Pomorski shows conspicuous fluctuations in clay mineral composition, carbon and osmium isotopes, PAHs, and palynomorph frequency and preservation (Fig. 2; Table 1).

Our climatic interpretation is mostly based on clay mineralogy and to a lesser extent on color of sediments and floral remains (Figs. 2, 3, 4, and 5; Table 1). Much of clay deposits in sedimentary basins represent a final product of continental weathering processes, and they may reveal climatic fluctuations on continents, if diagenetic transformations were not very significant (e.g., Singer, 1984; Chamley, 1989; Ruffell et al., 2002; Ahlberg et al., 2003; Deconinck et al., 2003; Raucsik and Varga, 2008; Hesselbo et al., 2009; Dera et al., 2009; Brański, 2009, 2010, 2012). In particular, variations in the detrital kaolinite content of the clay fraction are considered as reliable proxy for humidity.

Abundance of major elements is intimately related to mineralogy of the mudrocks. A distinct domination of illite (with chlorite) in the Zbąszynek Beds and comparatively low values of chemical indices (Fig. 3) are clearly related to the dominance of physical weathering due to the (semi-) arid climate in Norian time. The abundance of smectite in part of the Wielichowo Beds (Figs. 2 and 3) indicates some increase in precipitation and a distinct seasonality in the early-middle Rhaetian compared with the Norian. The crucial change in the clay mineral assemblage is observed between the Wielichowo Beds and Zagaje Formation (Fig. 2, event 1 hh; Fig. 3), which reflects the climate humidification during Rhaetian time, with a shift from smectite- to kaolinite-dominated mudrocks. Moreover, the clay mineral change was associated with the earlier Rhaetian "precursor" $\delta^{13}C$ excursion of Ruhl and Kürschner (2011), which, according to Lindström et al. (2012), commenced the series of environmental perturbations.

In the upper part of the Wielichowo Beds and in the Zagaje Formation, the kaolinite-illite association prevails and signifies the predominance of warm climate with high year-round rainfall, even though initially some seasonality was still possible, as indicated by the appearance of smectite in the Niekłań section (Fig. 3) but not in Kamień Pomorski (Fig. 2). The values of weathering indices are similar to modern residual clays or muds from rivers draining strongly weathered tropical areas like central Uganda, where the CIA values from 87 to 96 have been recorded (Nyakairu and Koeberl, 2001). In general, progressive weathering caused the loss of Ca, Na, and Mg and finally, K, which reflect the decrease in both nonclay silicate minerals (plagioclases, K-feldspars) and compositionally immature clay minerals such as smectite (cf. Weaver, 1989).

A significant admixture of berthierine and serpentine in the Kamień section is probably linked to the co-occurrence of weathered mafic rocks in the source area (Kaszuby Land of the Baltic Shield; Fig. 1). Importantly, some beds with very high kaolinite/illite ratio and highly elevated values of chemical indices were observed in the sections studied (Figs. 2 and 3). The sharp increase in kaolinite content records the abrupt change in weathering regime. These layers, particularly rich in kaolinite, with usually very high CIA value, were developed as a result of extreme chemical weathering in the aftermath of rapid warming and abundant rainfall. Strong leaching has almost removed the mobile alkali and alkaline earth elements while immobile elements such as Al and Ti are enriched. As noted already, the first such greenhouse event was recorded in the earlier Rhaetian Wielichowo Beds. The last distinct kaolinite episode was detected in the lowermost Hettangian. High amounts of kaolinite are also reported from the Rhaetian-Hettangian sections in southern Sweden (Ahlberg et al., 2002, 2003), from the Tatra Mountains in Slovakia (Michalik et al., 2010), and from topmost Triassic *Triletes* beds in the German Basin (van de Schootbrugge et al., 2009). Recently, distinct kaolinite enrichment was also documented from the topmost Triassic Kossen Formation (Eiberg Basin) in Austria (Pálfy and Zajzon, 2012; Zajzon et al., 2012). Regardless of this, the periods of less intense chemical weathering during cooler and less humid conditions were recorded in some layers characterized by lesser kaolinite content and lower CIA values.

Rapid and episodic fluctuations in the composition of clay minerals and in the values of weathering indices point to the sequence of frequent and marked climatic reversals, named as 1 hh, 2 hh, 3 cd, 4 hh, 5 cd, 6 h, 7 cd, and 8 hh events (Fig. 2; Table 1). These changes above the "initial" CIE become conspicuously more frequent as the Triassic-Jurassic boundary is approached, assuming a relatively constant average sedimentation rate (Fig. 2).

The data provided by the palynomorphs from Triassic-Jurassic transitional beds in Poland show conspicuous changes in composition and character of plant biodiversity (see Pieńkowski et al., 2012). The extinction processes and replacement in the latest Triassic floras in Poland probably began along with an increase in humidity in the middle to late Rhaetian time, as it is associated with a marked clay mineral and CIA shift (Fig. 2; Table 1; depth ~700 m, the 1 hh climate event). This is the shift that likely correlates (at least partly) with a series of negative

$\delta^{13}C_{org}$ excursions (so-called "precursor" negative $\delta^{13}C_{org}$ excursion [Ruhl and Kürschner, 2011], or Neg-1 excursion [Lindström et al., 2012]), also observed in Kamień Pomorski $\delta^{13}C_{org}$ excursions. This somewhat neglected CIE event is also associated with spore peak and bivalve extinction, as well as foraminiferal and marine plankton change (Deenen et al., 2010; Ruhl et al., 2010; Lindström et al., 2012). Interestingly, this CIE seems to be unrelated to any known magmatic/volcanic event, as it predates the oldest known volcanic events of the Central Atlantic magmatic province (Deenen et al., 2010; Lindström et al., 2012; Blackburn et al., 2013).

Increased rainfall did not result in an increase of miospore diversity and abundance, although dominance of spores at 700 m contrasts with the lower, pollen grain–dominated sample (703 m), which seems to support humidification. The scarcity of miospores between 682.5 and 700 m in Kamień Pomorski (Fig. 2) is most probably caused by a low preservational potential, given the lithology, which is unfavorable to organic matter preservation (although a floral crisis might have also been involved). Darkened miospores appear at 686 m, coincident within the "initial" CIE and osmium isotope as well as with PAH disturbances. This coincidence of indices points to marked environmental disturbances, which could be a reflection of the disturbances that are generally regarded to mark the onset of the end-Triassic extinction (Cirilli et al., 2009; Deenen et al., 2010; Ruhl et al., 2010; Blackburn et al., 2013). Above 682.5 m, there is an increase in abundance and diversity of miospores, and the changes of spore/pollen grain proportions follow the changes in kaolinite/illite ratio (Fig. 2). The super-abundances of spores just below the Triassic-Jurassic boundary, linked with their marked darkening, decreases in $^{187}Os/^{186}Os$ and $^{187}Os/^{188}Os$, as well as significant drops in kaolinite and CIA (Fig. 2; Table 1), point to a marked cooling, acid rains, and domination of "plant disaster taxa" associated with this "spore peak" (Fig. 2; event 7 cd at 678.4 m). The spore spike could be explained by a short but severe crisis for land plants, generated by an eruption (or an impact), in which all adult photosynthesis organs died off for lack of light, or in a prolonged frost, or in acid rain, or other toxic compounds, or all of them. Abundance of spores might also be partly related to better preservational conditions (i.e., diminishing rate of biogenic decay of miospores, associated with acidification of soil). Perhaps ferns were the first plants to recolonize the debris, and higher plants returned later.

This event has been noticed in many parts of the world at the Triassic-Jurassic boundary (Deenen et al., 2010; Whiteside et al., 2010; Blackburn et al., 2013). In North America, high-diversity pollen assemblages composed of monosulcates and monosaccates give way to lower-diversity assemblages dominated by *Classopollis*, a pollen type normally associated with hot and/or arid climate conditions, and palynofloral diversity loss is estimated at ~60% (Fowell and Olsen, 1993). It should be noted, that Cirilli et al. (2009) cast some doubts on the existence of both widespread palynological turnover and a "fern peak" in North America (particularly in the Newark Basin), based on palynological results from the Fundy Basin (Nova Scotia, Canada); they attributed abundance of spores to rather diachronous and short-term ecological perturbations that were regional in scope (see also Lucas and Tanner, 2007b). However, it should be noted that Cirilli et al. (2009) were dealing with entirely Triassic material. Some current studies from Europe (Kürschner and Herngreen, 2010; Cirilli, 2010) show that palynofloral composition between the Late Triassic and Hettangian was relatively steady and without abrupt changes and consequently claim that the end-Triassic biotic crisis appears to have little affected palynofloral species diversity, at least in Europe. Also estimates of diversity loss based on macrofossils are typically much higher than estimates of diversity loss based on miospores (Mander et al., 2010).

As the Triassic-Jurassic palynofloral turnover in Poland is more conspicuous than changes observed in other regions in Europe, it may also be interpreted that the extinction rate was related to the original paleolatitudinal position, implying that the associated plant migration could also have been involved in the Triassic-Jurassic palynofloral turnover (Pieńkowski et al., 2012). In that respect, is also noticeable that that the Rhaetian palynofloral assemblage from Pomerania (Pieńkowski et al., 2012) shows differences compared to Lisowice (Staneczko, 2007; Dzik et al., 2008a, 2008b; Świło et al., 2013). The floral crisis observed in Poland was relatively short, and the next humid event sparked the re-composition of the flora, as indicated by the increased proportion of hygrophilous plants (and spores) in the earliest Hettangian ecosystems. Another "spore peak" (Fig. 2, depth 641.8 m), much more diversified in terms of miospore taxa and containing many gymnosperm pollen grains, could be related to a similar event (i.e., volcanic eruption) or to localized change of environment associated with very favorable hydrologic and preservational conditions.

Osmium Isotopic System, Iridium Content, and Organic Geochemistry—Impact versus Volcanic Scenario

The $^{187}Os/^{186}Os$ ratio and very low iridium content lend no support for a role for asteroid impact at the Triassic-Jurassic boundary (Pieńkowski et al., 2012). Furthermore, the recently redated 201 ± 2 Ma impact structure in Rochechouart in France (Schmieder et al., 2010) seems to be too small (some tens of kilometers) to have caused global extinction, despite its original size being still a matter of debate (Smith, 2011). A hypothetical comet impact or "comet shower," leaving behind no geochemical traces, is still a theoretically possible explanation, but it seems that volcanic volatiles from the Central Atlantic magmatic province continental flood basalts remain the most plausible explanation for end-Triassic continental mass extinction. Noteworthy factors are increased abundances of ^{192}Os, observed twice in the Kamień Pomorski profile (Fig. 2). Content of ^{192}Os, the most common unradiogenic osmium isotope, is thought to be mainly derived from igneous activity (Cohen and Coe, 2002, 2007; Kuroda et al., 2010). Also, the negative shift in $^{187}Os/^{188}Os$ values suggests input of unradiogenic Os of mantle (or extraterrestrial) origin or a reduction of continentally derived Os or both (Cohen and Coe, 2002;

Kuroda et al., 2010). Observed changes in osmium isotopes in the Kamień Pomorski profile (the higher value of $^{187}Os/^{188}Os$ at the bottom) resemble, to some extent, trends noted by Cohen and Coe (2002). Higher in the profile, values of this ratio are stable, followed by the marked drop concomitant with the upper "subpeak" of the "initial" CIE, and lower values continue until the top of Triassic strata (Fig. 2). Because these are terrestrial deposits, the higher $^{187}Os/^{188}Os$ values can be tentatively linked to volcanic fallout coming from the Central Atlantic magmatic province (Fig. 1). Such fallout could be associated with other volcanic impacts, as acid rain and subvolcanic PAH releases (triggered when hot magma interacts with organic-rich rocks), which could lead both to observed darkening of spores in several horizons (Fig. 2) and defoliation, which in turn might have led to intensification of forest fires.

The PAHs were produced also by local wildfires (Marynowski and Simoneit, 2009; Pieńkowski and Waksmundzka, 2009), which should also be associated with the presence of charcoal in palynomacerals. In some samples, this is the case (i.e., Fig. 2, 687 m, 3 cd event; many Hettangian samples in Kamień Pomorski and Sołtyków), but elsewhere it is not. Lack of charcoal linked with elevated PAHs would rather support a subvolcanic origin of these compounds (Pieńkowski et al., 2012; see also van Schootbrugge et al., 2009). Moreover, charcoal, even if present, must be taken with some caution as an "in situ" indicator of wildfire frequency because it can be widely redeposited due to its resistance to biogenic degradation and buoyancy (Pieńkowski and Waksmundzka, 2009).

Anyhow, release of toxic pollutants such as SO_2, sulfate aerosols, and PAHs certainly led to defoliation, which increased forest flammability and resulting fire activity (see also Pieńkowski et al., 2012), similar to the climate-driven shift from broad-leaved to narrow-leaved taxa at the Triassic-Jurassic boundary (McElwain et al., 2009; Belcher et al., 2010).

Tetrapods across the Triassic-Jurassic Boundary in Poland

The Lisowice and Sołtyków localities, linked with other data from older or younger sites (Figs. 6, 7, 8, and 9), provide paleontological data valuable for determining biodiversity and evolutionary changes of the terrestrial tetrapod fauna across the

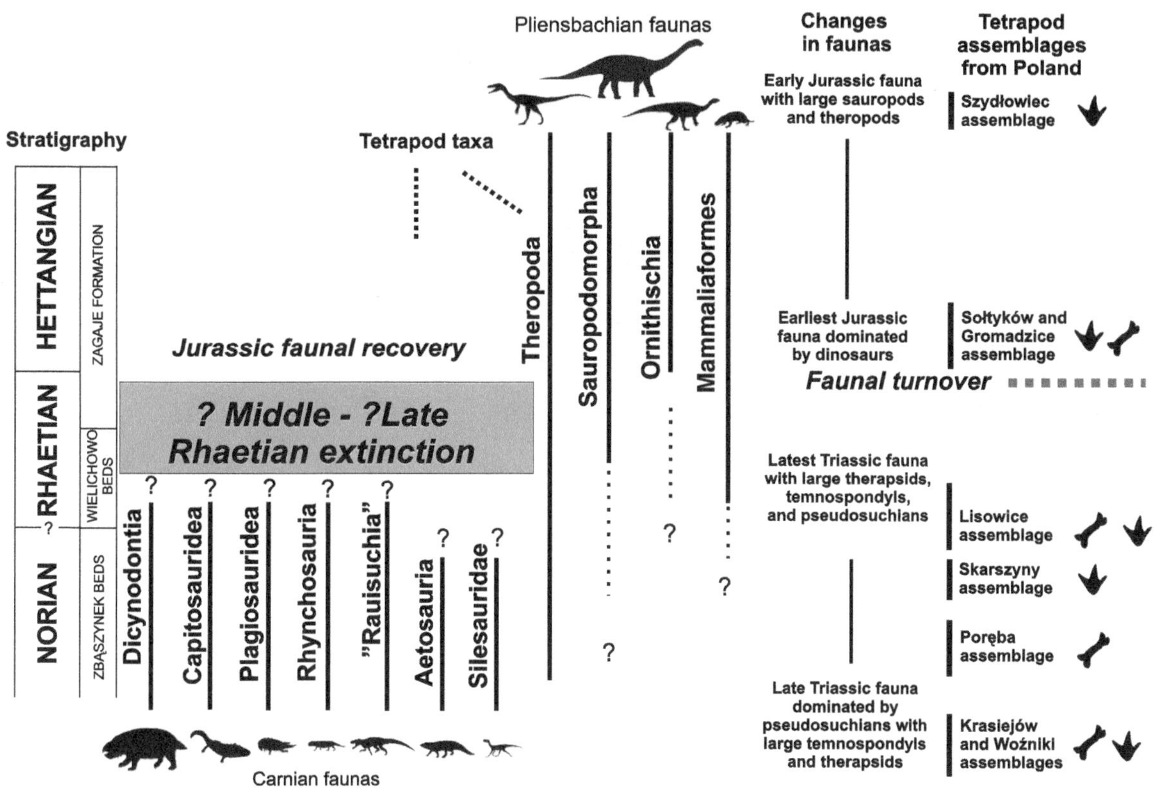

Figure 8. Distribution of terrestrial tetrapods and tetrapod assemblages (based on both body and trace fossils record) across the Late Triassic–Early Jurassic time interval in Poland. Although extinction period cannot be precisely indicated due to the lack of continuous outcrops spanning the Triassic-Jurassic boundary, the dramatic faunal turnover following the extinction period between the middle and end Rhaetian is visible, as is rapid tetrapod recovery in earliest Jurassic time (Sołtyków-Gromadzice assemblage). Figure is based on data from Gierliński et al. (2004), Dzik and Sulej (2007), Sulej (2009), Sulej et al. (2011, 2012), Dzik et al. (2008a, 2008b), Niedźwiedzki (2006, 2011a, 2011b, 2014), Niedźwiedzki et al. (2011, 2012, 2014).

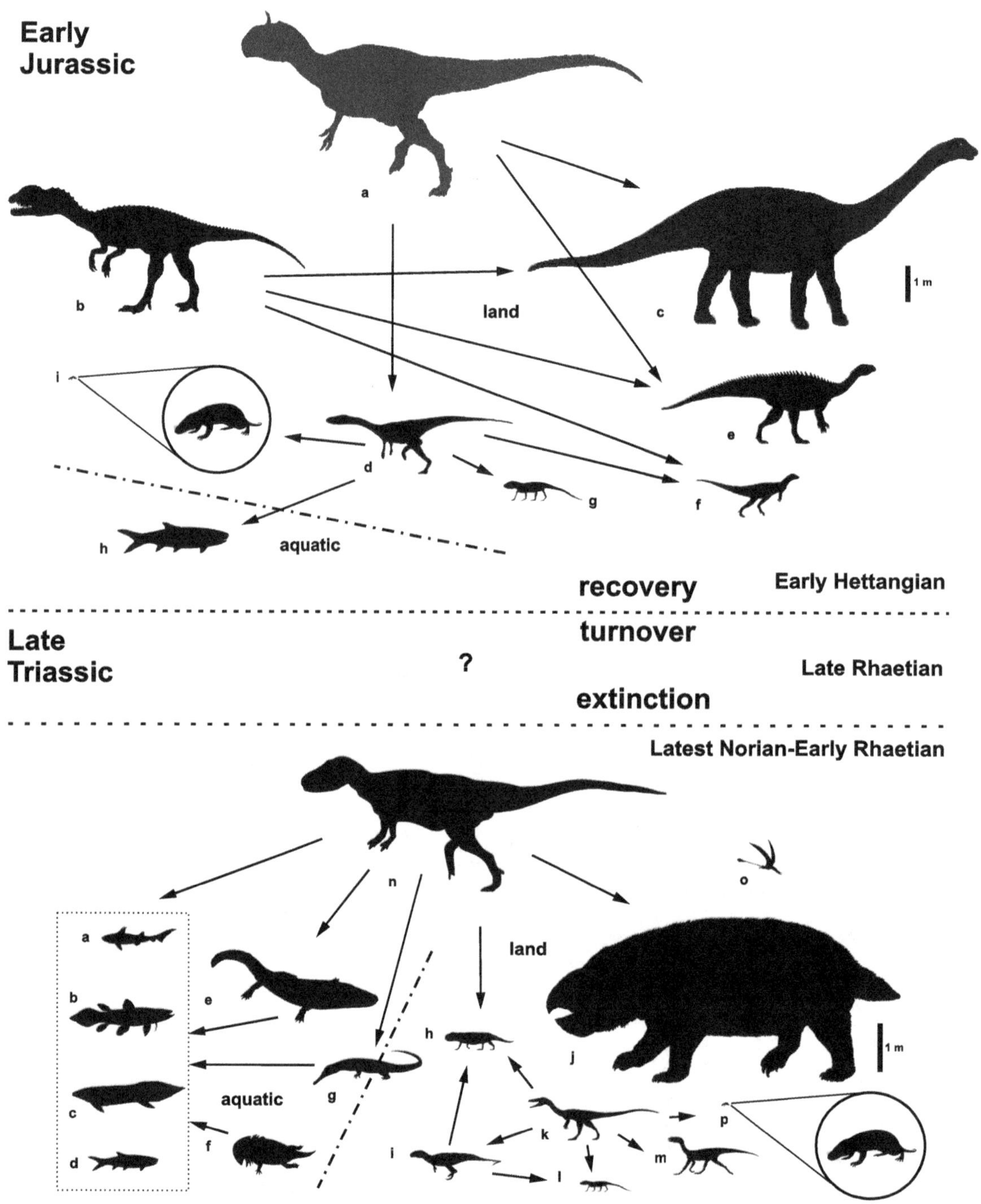

Figure 9. Reconstructed vertebrate fauna and simplified food web for the Lisowice community (Late Triassic: a—hybodont sharks; b—coelacanth fish; c—dipnoan fish; d—palaeonisciform fishes; e—capitosaur; f—plagiosaur; g—choristodere-like archosauromorph; h—rhynchosaur; i—poposaur; j—dicynodont; k—small predatory dinosaur; l—sphenodontid; m—small dinosauromorph; n—large predatory dinosaur; o—pterosaur; p—mammal or mammaliaform) and Sołtyków community (Early Jurassic: a—large theropod dinosaur; b—medium theropod dinosaur; c—sauropod dinosaur; d—small theropod dinosaur; e—medium ornithischian dinosaur; f—small ornithischian dinosaur; g—small crocodylomorph; h—palaeonisciform fishes; i—mammal or mammaliaform). Solid lines with arrows show feeding pathways.

Triassic-Jurassic boundary in Poland (Fig. 8). Reconstructions of assemblages from both localities reveal a rather complex trophic array, which today is characterized as stable and not disturbed ecosystems (Fig. 9).

The two well-characterized assemblages of tetrapod fossils (Lisowice and Sołtyków assemblages) from the southern Poland bracket the Triassic-Jurassic boundary. Those two assemblages encompass body fossils and footprints from a variety of lithofacies that represent similar depositional systems. This makes it easiest to simply compare each assemblage to the other because the differences between the assemblages in large part arose from taphonomic and not paleoenvironmental factors. In addition, the similar ages and paleogeographical positions of both sites are a good indicator of temporal succession and evolution of tetrapods across Triassic-Jurassic boundary in Poland.

In the Lisowice site, both body and trace fossil records of tetrapods are rich, and both are comparable from a taxonomic point of view. The Sołtyków site shows only an ichnological record (bones are extremely rare), but it is very diverse. Despite these differences, clear events in tetrapod evolution in the end of Triassic (or across Triassic-Jurassic boundary) in Poland are visible. The first event is the extinction of the pseudosuchians in the end of Triassic. This extinction, usually referred to as the extinction of "thecodonts" (or pseudosuchians), was identified as the principal tetrapod extinction at the end of the Triassic (see Colbert, 1958; Olsen et al., 2002; Lucas and Tanner, 2007a; Brusatte et al., 2008; Langer et al., 2010; Sues and Fraser, 2010). Pseudosuchian body fossils and footprints ("Rauisuchia") are present in the Lisowice sites, and also older Triassic sites in Poland (Fig. 6C), but are absent in Sołtyków. We take this to indicate pseudosuchian extinction somewhere between the Lisowice and Sołtyków assemblages, thus very close to the Triassic-Jurassic boundary. The second event is the extinction of typical Triassic fauna elements such as Dicynodontia, Capitosauroidea, Plagiosauroidea, and Rhynchosauria (Fig. 8). All these tetrapods disappear from the fossil record between the middle Rhaetian and earliest Hettangian. The third trend in tetrapod evolution across the Triassic-Jurassic boundary in Poland worth additional comment is the dramatic latest Triassic change in dinosaur diversity. The tetrapod assemblage from the earliest Hettangian of Poland shows that a sudden increase in numbers, diversity, and body sizes of dinosaurs took place during the latest Triassic (probably late Rhaetian) or just after the Triassic-Jurassic boundary. Thus, earliest Hettangian tetrapod assemblage of Poland is dominated by dinosaur tracks, with relatively rare tracks of other small tetrapods (basal crocodylomorphs, early mammals, small reptiles, and pterosaurs). The large-sized theropod dinosaurs (~7–8 m in length) might have been the top predators hunting the sauropodomorphs (probably early sauropods or last "prosauropods"; Fig. 9). The tetrapod ichnofauna from the Zagaje Formation supports the presence of a locally distributed theropod-sauropodomorph assemblage in the earliest Jurassic, preceding characteristic associations with numerous tracks of early ornithischia occurring in the late Hettangian and Sinemurian.

Numerous workers (e.g., Benton, 1986; Hunt, 1991; Olsen et al., 2002; Brusatte et al., 2008, 2010) have drawn attention to a relatively sudden increase in dinosaur abundance, diversity, and body size during the latest Triassic–earliest Jurassic, and this event is geographically widespread and not lithofacies correlated, so probably it is a record of evolutionary events connected with terminal Triassic extinction of therapsid-pseudosuchian faunas on the land. Additionally, according to paleontological records from Poland (and also from Italy, United States, Argentina, South Africa), a few evolutionary lines of dinosaurs were already widespread in the latest Triassic and earliest Jurassic. The relatively high diversified dinosaur ichnofauna from Sołtyków indicates to us a rapid recovery (or radiation) and refill of ecological niches after the end-Triassic extinction and within ~0.5–1.5 m.y. during the latest Rhaetian and earliest Hettangian.

The first Central Atlantic magmatic province pulse (latest Rhaetian in age; see Blackburn et al., 2013) probably coincided with the major extinction event of plants and other organisms in the continental realm. As recognized in the Newark and Fundy Basins (North America), the extinction of pseudosuchians and the rise of dinosaur-dominated fauna (Olsen et al., 2002) were also linked to Central Atlantic magmatic province volcanism (Blackburn et al., 2013). The record of extinction of the Triassic elements in the tetrapod fauna across Triassic-Jurassic boundary in Poland is probably connected with initial Central Atlantic magmatic province volcanism (i.e., according to the existing knowledge, also with the "initial" CIE). Some of the later (post–"initial" CIE, pre–Triassic-Jurassic boundary; 5 cd and 7 cd events; Fig. 2) releases of toxic pollutants (SO_2, sulfate aerosols, and other gases) were most likely the final blow to the end-Triassic pseudosuchian-therapsid–dominated ecosystems. We agree with the conclusion of Langer et al. (2010) that the Rhaetian stage was probably associated also with a crisis in dinosaurs.

The end-Triassic extinction event also affected many groups of large aquatic tetrapods and was perhaps an ecological event linked with changes in water chemistry of lakes and ponds (poisoning of surface water). Successive Central Atlantic magmatic province eruptions contributed both CO_2 (and CH_4 and PAHs) and sulfur injections (of note is the presence of highly pyritic sandstone just below the Triassic-Jurassic boundary in the Kamień Pomorski profile, different in its high intensity from other forms of pyritic mineralization in the profiles studied; see Fig. 2) with cooling lasting years or decades. Relatively longer periods of greenhouse events (2 hh, 4 hh, 8 hh; Fig. 2) should have allowed some migration of "vulnerable" taxa (temnospondyls, pseudosuchians, therapsids) to the lower latitudes, but abrupt cooling led to less time for those animals to survive (either by migration or accommodation). In contrast, dinosaurs, early mammals, small crocodylomorphs, pterosaurs, and also their insulated relatives could better withstand the cold challenge. Although still speculative (because of insufficient data), this scenario is consistent with the post-Triassic recovery in which dinosaurs took over the world (Figs. 8 and 9).

INTEGRATED STRATIGRAPHIC FRAMEWORK

The existing chronostratigraphic framework based on zircon U-Pb geochronologic and astronomical constraints from marine and continental sections (Deenen et al., 2010; Ruhl et al., 2010; Blackburn et al., 2013) can be matched to our profiles based on carbon isotope stratigraphy and palynoflora. All the aforementioned authors correlate the end-Triassic extinction with the onset of Central Atlantic magmatic province volcanism and the marked "initial" CIE (see also Ruhl et al., 2011). The following recovery at the Triassic-Jurassic boundary occurred just ~120 k.y. afterwards, spanning six precession cycles (Ruhl et al., 2010). We inferred four climatic events (reversals) in this time interval, but some part of sedimentary record may be missing at the sequence boundary, close to the Triassic-Jurassic boundary (Fig. 2).

Interestingly, Bonis et al. (2010) recognized four pronounced spore peaks in the St. Audrie's Bay section of the UK, in the end-Rhaetian Lilstock Formation. They attributed these peaks to precession-induced increased runoff, which would require the duration of the initial CIE to have been at least 20,000 yr, which is just on the maximum limit indicated by Ruhl et al. (2011). Bonis et al. (2010) also linked spore peaks to climate change rather than to any catastrophic events. Linking our climatic steps inferred from clay mineralogy with astronomical cycles (precession cycles) is not impossible. However, our study and other reports (i.e., Korte and Kozur, 2011; Ruhl et al., 2011) show that the initial CIE is bipartite, with two negative subpeaks. If the duration of the whole initial CIE was between 10,000 and 20,000 yr, these changes (reflected also in climate; Fig. 2; Table 1) are of too high frequency to be linked with astronomical forcing. Instead, they could be associated with episodes of Central Atlantic magmatic province eruption, which is further supported by differences of certain geochemical and palynological properties (Os isotope system disturbances as well as darkening of miospores and, circumstantially, PAH content) of these two peaks (Fig. 2; Table 1). It is possible that those independent mechanisms (astronomical forcing and volcanic activity) could amplify or oppositely, alleviate adverse effects, but Central Atlantic magmatic province–related processes seem to be the most probable explanation for the biotic crisis at the end of the Triassic. A series of periodic atmospheric loadings by CO_2, CH_4, or alternatively by SO_2, sulfate aerosols, and toxic compounds, is inferred to have caused this series of rapid climatic reversals and resulting biota crisis.

Our study confirms the results of Lindström et al. (2012), which indicated that the climatic and biotic changes commenced earlier than the initial CIE, somewhere at the level of the "precursor" CIE (Ruhl and Kürschner, 2011), judging from the astrochronological scale of Ruhl et al. (2010), some 100–200 k.y. earlier than the initial CIE and hitherto known onset of Central Atlantic magmatic province volcanism (Blackburn et al., 2013). The direct cause of this CIE is still unclear; perhaps an earlier phase of Central Atlantic magmatic province volcanism is still to be discovered.

The only other event that seems to be approximately correlatable with this shift is a marked erosional surface (698 m in Kamień Pomorski; Fig. 2), identified with the sequence boundary, which is probably concomitant with the Rhaetian lowstand, inferred to be one of the lowest in Phanerozoic (Hallam, 2001). This sequence boundary can be identified with emergence surfaces within the Lilstock Formation, occurring both in the St. Audrie's (Hesselbo et al., 2002) and Larne (Simms and Jeram, 2006) sections and a lowstand (correlative with a sequence boundary) at Kuhjoch, located at the top of the Koessen Formation (Hillebrandt et al., 2007) and in Csövár (Pálfy et al., 2001, 2007). However, the lowstand and exposure of vast areas would rather lead to steepened latitudinal temperature gradients and increased environmental extremes due to continentality, rather than to marked humidification.

Our conclusion is that the extinction period was more prolonged than the 20,000 yr duration of the initial CIE. Slight palynofloral change commenced at the initial $\delta^{13}C$ excursion, correlated with the main onset of Central Atlantic magmatic province volcanism, but the following positive carbon isotope excursion, accompanied by a rapid drop in kaolinite content and intermittent reappearance of a thin layer of red beds (5 cd event; Fig. 2; Table 1; possibly also red-greenish strata in Lisowice [Fig. 5], although existing stratigraphical resolution in Lisowice does not allow precise correlation with the Kamień Pomorski profile), was of equal, if not greater significance, causing the deepening crisis for Triassic flora and tetrapod fauna. Increased aridity might have been unfavorable, particularly for amphibians (i.e., Plagiosauridae). The next three steps, particularly 7 cd (Fig. 2; Table 1) at the Triassic-Jurassic boundary, associated with acid rains, "spore peaks," and darkened miospores, probably dealt the "final blow" to the otherwise weakened Triassic ecosystem, which is reflected by the disappearance of most of the Triassic palynomorphs and a number of characteristic tetrapod taxa. Obtained values of initial $^{187}Os/^{186}Os$ between 2.905 and 4.873 and very low iridium content (Fig. 2) lend no support for a role of an asteroid impact at the Triassic-Jurassic boundary event. However, Central Atlantic magmatic province–related volcanic volatiles causing climatic disturbances and influencing the ecosystem in many ways (Self et al., 2006) seem to be a much better substantiated cause for the end-Triassic extinction.

ACKNOWLEDGMENTS

We are grateful to Stephen Hesselbo and Simonetta Cirilli for their constructive reviews. Gerta Keller, Andrew Kerr, and Gina Harlow are thanked for their editorial assistance. This paper is a part of a project financed by the Polish National Science Centre, granted on the basis of decision no. DEC-2012/06/M/ST10/00478. Niedźwiedzki is currently funded by a Wallenberg scholarship grant awarded to P.E. Ahlberg (Uppsala University).

REFERENCES CITED

Ahlberg, A., Arndorff, L., and Guy-Ohlsson, D., 2002, Onshore climate change during the Late Triassic marine inundation of the Central European Basin: Terra Nova, v. 14, p. 241–248, doi:10.1046/j.1365-3121.2002.00416.x.

Ahlberg, A., Olsson, I., and Simkevicius, P., 2003, Triassic-Jurassic weathering and clay mineral dispersal in basement areas and sedimentary basins of southern Sweden: Sedimentary Geology, v. 161, p. 15–29, doi:10.1016/S0037-0738(02)00381-0.

Bacon, K.L, Belcher, C.M., Haworth, M., and McElwain, J.C., 2013, Increased atmospheric SO_2 detected from changes in leaf physiognomy across the Triassic-Jurassic boundary interval of East Greenland: PLoS ONE, v. 8, no. 4, e60614, doi:10.1371/journal.pone.0060614.

Barbacka, M., Ziaja, J., and Wcisło-Luraniec, E., 2010, Taxonomy and palaeoecology of the Early Jurassic macroflora from Odrowąż, central Poland: Acta Geologica Polonica, v. 60, p. 373–392.

Belcher, C.M., Mander, L., Rein, G., Jervis, F.X., Haworth, M., Hesselbo, S.P., Glasspool, I.J., and McElwain, J.C., 2010, Increased fire activity at the Triassic/Jurassic boundary in Greenland due to climate-driven floral change: Nature Geoscience, v. 3, p. 426–429, doi:10.1038/ngeo871.

Benton, M.J., 1986, The Late Triassic tetrapod extinction events, *in* Padian, K., ed., The Beginning of the Age of Dinosaurs: Faunal Change Across the Triassic-Jurassic Boundary: Cambridge, UK, Cambridge University Press, p. 303–320.

Blackburn, T.J., Olsen, P.E., Bowring, S.A., McLean, N.M., Kent, D.V., Puffer, J., McHone, G., and Rasbury, T., 2013, Zircon U-Pb geochronology links the end-Triassic extinction with the Central Atlantic magmatic province: Science, v. 340, p. 941–945, doi:10.1126/science.1234204.

Bonis, N.R., Ruhl, M., and Kürschner, W., 2010, Milankovitch-scale palynological turnover across the Triassic-Jurassic transition at St. Audrie's Bay, SW UK: Journal of the Geological Society, v. 167, p. 877–888, doi:10.1144/0016-76492009-141.

Brański, P., 2009, Influence of palaeoclimate conditions and greenhouse effect on the Hettangian clay mineral assemblages (Holy Cross Mts. area, Polish Basin): Geological Quarterly, v. 53, p. 363–368.

Brański, P., 2010, Kaolinite peaks in early Toarcian profiles from the Polish Basin—An inferred record of global warming: Geological Quarterly, v. 54, p. 15–24.

Brański, P., 2012, The mineralogical record of the early Toarcian stepwise climate changes and other environmental variation (Ciechocinek Formation, Polish Basin): Volumina Jurassica, v. 10, p. 1–24.

Brański, P., 2014, Climatic disaster at the Triassic-Jurassic boundary—A clay minerals and major elements record from the Polish Basin: Geological Quarterly, v. 58 (in press).

Brauns, C.M., 2001, A rapid, low blank technique for the extraction of osmium from geological samples: Chemical Geology, v. 176, p. 379–384.

Brusatte, S.L., Benton, M., Ruta, M., and Lloyd, G., 2008, Superiority, competition, and opportunism in the evolutionary radiation of dinosaurs: Science, v. 321, p. 1485–1488, doi:10.1126/science.1161833.

Brusatte, S.L., Nesbitt, S.J., Irmis, R.B., Butler, R.J., Benton, M.J., and Norell, M.A., 2010, The origin and early radiation of dinosaurs: Earth-Science Reviews, v. 101, p. 68–100, doi:10.1016/j.earscirev.2010.04.001.

Chamley, H., 1989, Clay Sedimentology: Berlin, Springer-Verlag, 623 p.

Cirilli, S., 2010, Upper Triassic–lowermost Jurassic palynology and palynostratigraphy: A review, *in* Lucas, S.G., ed., The Triassic Timescale: Geological Society of London Special Publication 334, p. 285–314, doi:10.1144/SP334.12.

Cirilli, S., Marzoli, A., Tanner, L., Bertrand, H., Buratti, N., Jourdan, F., Bellieni, G., Kontak, D., and Renne, P.R., 2009, Latest Triassic onset of the Central Atlantic magmatic province (CAMP) volcanism in the Fundy Basin (Nova Scotia): New stratigraphic constraints: Earth and Planetary Science Letters, v. 286, p. 514–525, doi:10.1016/j.epsl.2009.07.021.

Cohen, A.S., and Coe, A.L., 2002, New geochemical evidence for the onset of volcanism in the Central Atlantic magmatic province and environmental change at the Triassic-Jurassic boundary: Geology, v. 30, p. 267–270, doi:10.1130/0091-7613(2002)030<0267:NGEFTO>2.0.CO;2.

Cohen, A.S., and Coe, A.L., 2007, The impact of the Central Atlantic Magmatic province on climate and on the Sr- and Os-isotope evolution of sea water: Palaeogeography, Palaeoclimatology, Palaeoecology, v. 244, p. 374–390, doi:10.1016/j.palaeo.2006.06.036.

Colbert, E.H., 1958, Tetrapod extinction at the end of the Triassic Period: Proceedings of the National Academy of Sciences of the United States of America, v. 44, p. 973–977, doi:10.1073/pnas.44.9.973.

Dadlez, R., 1969, Stratigraphy of the Lias in western Poland: Proceedings of the Institute of Geology, Poland, v. 57, p. 1–92.

Dadlez, R., Narkiewicz, M., Stephenson, R.A., Visser, M.T.M., and Ven Wees, J.-D., 1995, Tectonic evolution of the Mid-Polish Trough: Modelling implications and significance for central European geology: Tectonophysics, v. 252, p. 179–195, doi:10.1016/0040-1951(95)00104-2.

Deconinck, J.-F., Hesselbo, S.P., Debuisser, N., Averbuch, O., Baudin, F., and Bessa, J., 2003, Environmental controls on clay mineralogy of an Early Jurassic mudrock (Blue Lias Formation, southern England): International Journal of Earth Sciences, v. 92, p. 255–266.

Deczkowski, Z., 1997, Noryk–Retyk, Jura dolna, *in* Marek, S., and Pajchlowa, M., eds., Epikontynentalny perm i mezozoik w Polsce: Prace Państwowego Instytutu Geologicznego, v. 153, p. 174–235.

Deenen, M.H.L., Ruhl, M., Bonis, N.R., Krijgsman, W., Kürschner, W.M., Reitsma, M., and Van Bergen, M.J., 2010, A new chronology for the end-Triassic mass extinction: Earth and Planetary Science Letters, v. 291, p. 113–125, doi:10.1016/j.epsl.2010.01.003.

Dera, G., Pellenard, P., Neige, P., Deconinck, J.-F., Puceat, E., and Dommergues, J.-L., 2009, Distribution of clay minerals in Early Jurassic Peritethyan seas: Palaeoclimatic significance inferred from multiproxy comparisons: Palaeogeography, Palaeoclimatology, Palaeoecology, v. 271, p. 39–51, doi:10.1016/j.palaeo.2008.09.010.

Dzik, J., and Sulej, T., 2007, A review of the early Late Triassic Krasiejów biota from Silesia, Poland: Palaeontologia Polonica, v. 64, p. 3–27.

Dzik, J., Sulej, T., and Niedźwiedzki, G., 2008a, A dicynodont-theropod association in the latest Triassic of Poland: Acta Palaeontologica Polonica, v. 53, p. 733–738, doi:10.4202/app.2008.0415.

Dzik, J., Niedźwiedzki, G., and Sulej, T., 2008b, Zaskakujące uwieńczenie ery gadów ssakokształtnych: Ewolucja, v. 3, p. 2–21.

Fedo, C.M., Nesbitt, H.W., and Young, G.M., 1995, Unraveling the effects of potassium metasomatism in sedimentary rocks and paleosols, with implications for paleoweathering conditions and provenance: Geology, v. 23, p. 921–924, doi:10.1130/0091-7613(1995)023<0921:UTEOPM>2.3.CO;2.

Fowell, S.J., and Olsen, P.E., 1993, Time calibration of Triassic–Jurassic microfloral turnover, eastern North America: Tectonophysics, v. 222, p. 361–369, doi:10.1016/0040-1951(93)90359-R.

Franz, M., 2008, Litho- und Leitflächenstratigraphie, Chronostratigraphie, Zyklo- und Sequenzstratigraphie des Keupers im östlichen Zentraleuropäischen Becken (Deutschland, Polen) und Dänischen Becken (Dänemark, Schweden) [Dissertation zur Erlangung des Akademischen Grades doctor rerum naturalium (Dr. rer. nat.)]: Wittenberg, Germany, Naturwissenschaftlichen Fakultät III der Martin-Luther-Universität Halle-Wittenberg, 198 p.

Franz, M., Bachmann, G.H., and Beutler, G., 2007a, Retyk sensu polonico versus Rhaet sensu germanico—New results: Schriftenreihe der Deutschen Gesellschaft für Geowissenschaften, v. 53, p. 99–100.

Franz, M., Bachmann, G.H., and Beutler, G., 2007b, Sedimentology and facies of the Polish Retyk and the German Arnstadt and Exter Formations (Norian, Rhaetian) in the eastern Central European Basin (CEB): Schriftenreihe der Deutschen Gesellschaft für Geowissenschaften, v. 53, p. 101.

Fuglewicz, R., 1977, New species of megaspores from the Trias of Poland: Acta Palaeontologica Polonica, v. 22, p. 405–431.

Fuglewicz, R., and Śnieżek, P., 1980, Upper Triassic megaspores from Lipie Śląskie near Lubliniec: Przegląd Geologiczny, v. 28, p. 459–461.

Galli, M.T., Jadoul, F., Bernasconi, S.M., and Weissert, H., 2005, Anomalies in global carbon cycling and extinction at the Triassic/Jurassic boundary: Evidence from a marine C-isotope record: Palaeogeography, Palaeoclimatology, Palaeoecology, v. 216, p. 203–214, doi:10.1016/j.palaeo.2004.11.009.

Gertsch, B., Keller, G., and Adatte, T., 2013, Geochemical proxies (major, trace and platinum group elements) across the KT boundary: What can they tell about the role of impacts and volcanism?, *in* Keller, G., Kerr, A., and McLeod, N., eds., Volcanism, Impacts and Mass Extinctions: Causes and Effects, International Conference Abstract Volume: London, The Natural History Museum, p. 38.

Gierliński, G., and Pieńkowski, G., 1999, Dinosaur track assemblages from Hettangian of Poland: Geological Quarterly, v. 43, p. 329–346.

Gierliński, G., Niedźwiedzki, G., and Pieńkowski, G., 2001, Gigantic footprint of a theropod dinosaur in the Early Jurassic of Poland: Acta Palaeontologica Polonica, v. 46, p. 441–446.

Gierliński, G., Pieńkowski, G., and Niedźwiedzki, G., 2004, Tetrapod track assemblage in the Hettangian of Sołtyków, Poland, and its paleoenvironmental background: Ichnos, v. 11, p. 195–213, doi:10.1080/10420940490444861.

Guex, J., Bartolini, A., Atudorei, V., and Taylor, D., 2004, High-resolution ammonite and carbon isotope stratigraphy across the Triassic-Jurassic boundary at New York Canyon (Nevada): Earth and Planetary Science Letters, v. 225, p. 29–41, doi:10.1016/j.epsl.2004.06.006.

Hallam, A., 2001, A review of the broad pattern of Jurassic sea-level changes and their possible causes in the light of current knowledge: Palaeogeography, Palaeoclimatology, Palaeoecology, v. 167, p. 23–37, doi:10.1016/S0031-0182(00)00229-7.

Hallam, A., and Wignall, P.B., 1997, Mass Extinctions and their Aftermath: Oxford, UK, Oxford University Press, 320 p.

Hallam, A. and Wignall, P.B., 1999, Mass extinctions and sea-level changes: Earth-Science Reviews, v. 48, p. 217–250.

Harnois, L., 1988, CIW Index: A new chemical index of weathering: Sedimentary Geology, v. 55, p. 319–322, doi:10.1016/0037-0738(88)90137-6.

Harris, T.M., 1935, The fossil flora of Scoresby Sound, East Greenland, Part 4. Ginkgoales, Lycopodiales and isolated fructifications: Meddelelser om Grønland, v. 112, p. 1–121.

Herngreen, G.F.W., 2005, Triassic sporomorphs of NW Europe: Taxonomy, morphology and ranges of marker species with remarks on botanical relationship and ecology and comparison with ranges in the Alpine Triassic: Kenniscentrum Biogeology (UU/TNO)-TNO report, NITG 04-176-C, Utrecht, 83 p.

Hesselbo, S.P., and Jenkyns, H.C., 1998, British Lower Jurassic sequence stratigraphy, *in* de Graciansky, P.C., Hardenbol, J., Jacquin, T., Farley, M., and Vail, P.R., eds., Mesozoic–Cenozoic Sequence Stratigraphy of European Basins: Special Publication of the Society for Sedimentary Geology, v. 60, p. 561–581.

Hesselbo, S.P., Gröcke, D.R., Jenkyns, H.C., Bjerrum, C.J., Farrimond, P., Bell, H.S.M., and Green, O.R., 2000, Massive dissociation of gas hydrate during a Jurassic oceanic anoxic event: Nature, v. 406, p. 392–395, doi:10.1038/35019044.

Hesselbo, S.P., Robinson, S.A., Surlyk, F., and Piasecki, S., 2002, Terrestrial and marine extinction at the Triassic-Jurassic boundary synchronized with major carbon-cycle perturbation: A link to initiation of massive volcanism?: Geology, v. 30, p. 251–254, doi:10.1130/0091-7613(2002)030<0251:TAMEAT>2.0.CO;2.

Hesselbo, S.P., Robinson, S.A., and Surlyk, F., 2004, Sea-level change and facies development across potential Triassic-Jurassic boundary horizons, SW Britain: Journal of the Geological Society, v. 161, p. 365–379, doi:10.1144/0016-764903-033.

Hesselbo, S.P., McRoberts, C.A., and Pálfy, J., 2007, Triassic/Jurassic boundary events: Problems, progress, possibilities: Palaeogeography, Palaeoclimatology, Palaeoecology, v. 244, p. 1–10, doi:10.1016/j.palaeo.2006.06.020.

Hesselbo, S.P., Deconinck, J.-F., Huggett, J.M., and Morgans-Bell, H.S., 2009, Late Jurassic palaeoclimatic change from clay mineralogy and gamma-ray spectrometry of the Kimmeridge Clay, Dorset, UK: Journal of the Geological Society, v. 166, p. 1123–1133, doi:10.1144/0016-76492009-070.

Hillebrandt, A.V., Krystyn, I. and Kürschner, W.M., with contribution by Bown, P., McRoberts, M., Uhl, M., Simms, M., Tomasovych, A., and Ulrichs M., 2007, A candidate GSSP for the base of the Jurassic in the Northern Calcareous Alps (Kuhjoch section; Karwendel Mountains, Tyrol, Austria): ISJS Newsletter, v. 34, p. 1–38.

Hounslow, M.W., Posen, P.E., and Warrington, G., 2004, Magnetostratigraphy and biostratigraphy of the Upper Triassic and lowermost Jurassic succession, St. Audrie's Bay, UK: Palaeogeography, Palaeoclimatology, Palaeoecology, v. 213, p. 331–358, doi:10.1016/j.palaeo.2004.07.018.

Hubbard, R.N.B.L., and Boutler, M.C., 2000, Phytogeography and palaeoecology in western Europe and eastern Greenland near the Triassic-Jurassic boundary: Palaios, v. 15, p. 120–131, doi:10.1669/0883-1351(2000)015<0120:PAPIWE>2.0.CO;2.

Hunt, A.P., 1991, The early diversification pattern of dinosaurs in the Late Triassic: Modern Geology, v. 16, p. 43–60.

Jourdan, F., Marzoli, A., Bertrand, H., Cirilli, S., Tanner, L., Kontak, D.J., McHone, G., Renne, P.R., and Bellieni, G., 2009, 40Ar/39Ar ages of CAMP in North America: Implications for the Triassic-Jurassic boundary and the 40K decay constant bias: Lithos, v. 110, p. 167–180, doi:10.1016/j.lithos.2008.12.011.

Koeberl, C., 1998, Identification of meteoritical components in impactites, *in* Grady, M.M., Hutchison, R., McCall, G.J.H., and Rothery, D.A., eds., Meteorites: Flux with Time and Impact Effects: Geological Society of London Special Publication 140, p. 133–153.

Koeberl, C., and Shirey, S.B., 1997, Re–Os isotope systematics as a diagnostic tool for the study of impact craters and ejecta: Palaeogeography, Palaeoclimatology, Palaeoecology, v. 132, p. 25–46.

Korte, C., and Kozur, W., 2011, Bio- and chemostratigraphic assessment of carbon isotope records across the Triassic-Jurassic boundary at Csővár quarry (Hungary) and Kendlbachgraben (Austria) and implications for global correlations: Bulletin of the Geological Society of Denmark, v. 59, p. 101–115.

Korte, C., Hesselbo, S.P., Jenkyns, H.C., Rickaby, R.E.M., and Spötl, C., 2009, Palaeoenvironmental significance of carbon- and oxygen-isotope stratigraphy of marine Triassic-Jurassic boundary sections in SW Britain: Journal of the Geological Society, v. 166, p. 431–445, doi:10.1144/0016-76492007-177.

Kozur, H.W., and Weems, R.E., 2005, Conchostracan evidence for a late Rhaetian to early Hettangian age for the CAMP volcanic event in the Newark Supergroup, and a Sevatian (late Norian) age for the immediately underlying beds: Hallesches Jahrbuch für Geowissenschaften, v. B27, p. 21–51.

Kozur, H.W., and Weems, R.E., 2007, Upper Triassic conchostracan biostratigraphy of the continental basins of eastern North America: Its importance for correlating Newark Supergroup events with the Germanic Basin and the International Geologic Time Scale, *in* Lucas, S.G., and Spielmann, J.A., eds., The Global Triassic: New Mexico Museum of Natural History and Science Bulletin 41, p. 137–188.

Kozur, H.W., and Weems, R.E., 2010, The biostratigraphic importance of conchostracans in the continental Triassic of the northern hemisphere, *in* Lucas, S.G., ed., The Triassic Timescale: Geological Society of London Special Publication 334, p. 315–417.

Kuroda, J., Hori, R.S., Suzuki, K., Gröcke, D.R., and Ohkouchi, N., 2010, Marine osmium isotope record across the Triassic-Jurassic boundary from a Pacific pelagic site: Geology, v. 38, p. 1095–1098, doi:10.1130/G31223.1.

Kürschner, W.M., and Herngreen, G.F.W., 2010, Triassic palynology of central and northwestern Europe: A review of palynofloral diversity patterns and biostratigraphic subdivisions, *in* Lucas, S.G., ed., The Triassic Timescale: Geological Society of London Special Publication 334, p. 263–283.

Kürschner, W.M., Bonis, N.R., and Krystyn, L., 2007, Carbon-isotope stratigraphy of the Triassic-Jurassic transition in the Tiefengraben section, Northern Calcareous Alps: Palaeogeography, Palaeoclimatology, Palaeoecology, v. 244, p. 257–280, doi:10.1016/j.palaeo.2006.06.031.

Langer, M.C., Ezcurra, M.D., Bittencourt, J.S., and Novas, F.E., 2010, The origin and early evolution of dinosaurs: Biological Reviews of the Cambridge Philosophical Society, v. 84, p. 1–56.

Lindström, S., van de Schootbrugge, B., Dybkjær, K., Pedersen, G.K., Fiebig, J., Nielsen, L.H., and Richoz, S., 2012, No causal link between terrestrial ecosystem change and methane release during the end-Triassic mass extinction: Geology, v. 40, p. 531–534, doi:10.1130/G32928.1.

Lucas, S.G., and Tanner, L.H., 2007a, Tetrapod biostratigraphy and biochronology of the Triassic-Jurassic transition on the southern Colorado Plateau, USA: Palaeogeography, Palaeoclimatology, Palaeoecology, v. 244, p. 242–256, doi:10.1016/j.palaeo.2006.06.030.

Lucas, S.G., and Tanner, L.H., 2007b, The nonmarine Triassic-Jurassic boundary in the Newark Supergroup of eastern North America: Earth-Science Reviews, v. 84, p. 1–20, doi:10.1016/j.earscirev.2007.05.002.

Lund, J., 1977, Rhaetic to Lower Liassic palynology of the onshore south-eastern North Sea Basin: Danmarks Geologiske Undersøgelse II, v. 109, p. 1–81.

Marcinkiewicz, T., 1962, Rhaetian and Lias megaspores from borehole Mechowo near Kamień Pomorski and their stratigraphical value: Instytut Geologiczny Prace, v. 30, p. 469–493.

Marcinkiewicz, T., 1971, The stratigraphy of the Rhaetian and Lias in Poland based on megaspore investigations: Instytut Geologiczny Prace, v. 65, p. 1–58.

Mander, L., Kürschner, W.M., and McElwain, J.C., 2010, An explanation for conflicting records of Triassic–Jurassic plant diversity: Proceedings of the National Academy of Sciences of the United States of America, v. 107, p. 15351–15356.

Marynowski, L., and Simoneit, B.R.T., 2009, Widespread Late Triassic to Early Jurassic wildfire records from Poland: Evidence from charcoal and pyrolytic polycyclic aromatic hydrocarbons: Palaios, v. 24, p. 785–798, doi:10.2110/palo.2009.p09-044r.

Marzoli, A., Bertrand, H., Knight, K.B., Cirilli, S., Buratti, N., Verati, C., Nomade, S., Renne, P.R., Youbi, N., Martini, R., Allenbach, K., Neuwerth, R., Rapaille, C., Zaninetti, L., and Bellieni, G., 2004, Synchrony of the Central Atlantic magmatic province and the Triassic-Jurassic boundary climatic and biotic crisis: Geology, v. 32, p. 973–976, doi:10.1130/G20652.1.

Marzoli, A., Jourdan, F., Puffer, J.H., Cuppone, T., Tanner, L.H., Weems, R.E., Bertrand, H., Cirilli, S., Bellieni, G., and De Min, A., 2011, Timing and duration of the Central Atlantic magmatic province in the Newark and Culpeper Basins, eastern U.S.A.: Lithos, v. 122, p. 175–188, doi:10.1016/j.lithos.2010.12.013.

McElwain, J.C., Beerling, D.J., and Woodward, F.I., 1999, Fossil plants and global warming at the Triassic-Jurassic boundary: Science, v. 285, p. 1386–1390, doi:10.1126/science.285.5432.1386.

McElwain, J.C., Popa, M.E., Hesselbo, S.P., Haworth, M., and Surlyk, F., 2007, Macroecological responses of terrestrial vegetation to climatic and atmospheric change across the Triassic/Jurassic boundary in East Greenland: Paleobiology, v. 33, p. 547–573, doi:10.1666/06026.1.

McElwain, J.C., Wagner, P., and Hesselbo, S.P., 2009, Fossil plant relative abundances indicate sudden loss of Late Triassic biodiversity in East Greenland: Science, v. 324, p. 1554–1556, doi:10.1126/science.1171706.

McHone, J.G., 1996, Broad-terrane Jurassic flood basalts across northeastern North America: Geology, v. 24, p. 319–322.

McHone, J.G., 2003, Volatile emissions from Central Atlantic magmatic province basalts; mass assumptions and environmental consequences, *in* Hames, W.E., et al., eds., The Central Atlantic Magmatic Province; Insights from Fragments of Pangea: American Geophysical Union Geophysical Monograph 136, p. 241–254.

McRoberts, C.A., and Newton, C.R., 1995, Selective extinction among end-Triassic European bivalves: Geology, v. 23, p. 102–104, doi:10.1130/0091-7613(1995)023<0102:SEAETE>2.3.CO;2.

Michalik, J., Biron, A., Lintnerova, O., Götz, A., and Ruckwied, K., 2010, Climate change at the Triassic-Jurassic boundary in the northwestern Tethyan realm, inferred from sections in the Tatra Mountains (Slovakia): Acta Geologica Polonica, v. 60, p. 535–548.

Nesbitt, H.W., and Young, G.M., 1982, Early Proterozoic climates and plate motions inferred from major element chemistry of lutites: Nature, v. 299, p. 715–717, doi:10.1038/299715a0.

Niedźwiedzki, G., 2006, Ślady wielkich teropodów z wczesnojurajskich osadów Gór Świętokrzyskich: Przegląd Geologiczny, v. 54, p. 615–621.

Niedźwiedzki, G., 2011a, Footprints of large theropod dinosaurs from the Upper Triassic of Poland, *in* Kear, B.P., and Streng, M., eds., 2nd Wiman Meeting on Scandinavian-Baltic Palaeontology, 17–18 November 2011: Uppsala, Sweden, Uppsala University, p. 19.

Niedźwiedzki, G., 2011b, Tropy dinozaurów z wczesnojurajskiego ekosystemu z Sołtykowa w Górach Świętokrzyskich: Biuletyn Państwowego Instytutu Geologicznego, v. 447, p. 49–98.

Niedźwiedzki, G., 2014, The large predatory archosaur *Smok wawelski* from the latest Triassic of Poland: Palaeontologia Polonica, v. 67 (in press).

Niedźwiedzki, G., and Sulej, T., 2008, Lipie Śląskie – okno na późnotriasowy ekosystem lądowy (artykuł informacyjny): Przegląd Geologiczny, v. 56, p. 821–822.

Niedźwiedzki, G., and Sulej, T., 2010, Wycieczki terenowe. Lisowice–cegielnia Lipie Śląskie (punkt 2). Kopalne Biocenozy w czasie i przestrzeni XXI Konferencja Naukowa Sekcji Paleontologicznej PTG, 13-16 września 2010, Żarki-Letnisko, p. 102–107.

Niedźwiedzki, G., Gorzelak, P., and Sulej, T., 2011, Bite traces on dicynodont bones and the early evolution of large terrestrial predators: Lethaia, v. 44, p. 87–92, doi:10.1111/j.1502-3931.2010.00227.x.

Niedźwiedzki, G., Sulej, T., and Dzik, J., 2012, A large predatory archosaur from the Late Triassic of Poland: Acta Palaeontologica Polonica, v. 57, p. 267–276, doi:10.4202/app.2010.0045.

Niedźwiedzki, G., Brusatte, S.L., Sulej, T., and Butler, R.J., 2014, Basal dinosauriform and theropod dinosaurs from the mid–late Norian (Late Triassic) of Poland: Implications for Triassic dinosaur evolution and distribution: Palaeontology, doi:10.1111/pala.12107.

Nyakairu, G.W.A., and Koeberl, C., 2001, Mineralogical and chemical composition and distribution of rare earth elements in clay-rich sediments from central Uganda: Geochemical Journal, v. 35, p. 13–28, doi:10.2343/geochemj.35.13.

Ociepa, A.-M., Staneczko, K., Feldman-Olszewska, A., and Barbacka, M., 2008, Nowe stanowiska *Lepidopteris ottonis* (Goeppert) Schimper w Polsce, *in* Utwory przełomu jury i kredy w zachodnich Karpatach fliszowych polsko-czeskiego pogranicza, Jurassica VII, Abstrakty: Geologia, v. 34, p. 199–200.

Olsen, P.E., Kent, D.V., Sues, H.D., Koeberl, C., Huber, H., Montanari, A., Rainforth, E.C., Fowell, S.J., Szajna, M.J., and Hartline, B.W., 2002, Ascent of dinosaurs linked to an iridium anomaly at the Triassic-Jurassic boundary: Science, v. 296, p. 1305–1307, doi:10.1126/science.1065522.

Orłowska-Zwolińska, T., 1983, Palynostratigraphy of the upper part of Triassic epicontinental sediments in Poland: Prace Instytutu Geologicznego, v. 104, p. 1–88.

Orłowska-Zwolińska, T., 1985, Palynological zones of the Polish epicontinental Triassic: Bulletin of the Polish Academy of Sciences–Earth Sciences, v. 33, p. 107–117.

Pálfy, J., and Zajzon, N., 2012, Environmental changes across the Triassic-Jurassic boundary and coeval volcanism inferred from elemental geochemistry and mineralogy in the Kendlbachgraben section (Northern Calcareous Alps, Austria): Earth and Planetary Science Letters, v. 335–336, p. 121–134, doi:10.1016/j.epsl.2012.01.039.

Pálfy, J., Smith, P.L., and Mortensen, J.K., 2000, A U-Pb and 40Ar/39Ar time scale for the Jurassic: Canadian Journal of Earth Sciences, v. 37, p. 923–944, doi:10.1139/e00-002.

Pálfy, J., Demeny, A., Haas, J., Hetenyi, M., Orchard, M.J., and Veto, I., 2001, Carbon isotope anomaly and other geochemical changes at the Triassic-Jurassic boundary from a marine section in Hungary: Geology, v. 29, p. 1047–1050, doi:10.1130/0091-7613(2001)029<1047:CIAAOG>2.0.CO;2.

Pálfy, J., Demeny, A., Haas, J., Carter, E.S., Gorog, A., Halasz, D., Oravecz-Scheffer, A., Hetenyi, M., Marton, E., Orchard, M.J., Ozsvart, P., Veti, I., and Zajzon, N., 2007, Triassic-Jurassic boundary events inferred from integrated stratigraphy of the Csővár section, Hungary: Palaeogeography, Palaeoclimatology, Palaeoecology, v. 244, p. 11–33, doi:10.1016/j.palaeo.2006.06.021.

Pieńkowski, G., 2004, The Epicontinental Lower Jurassic of Poland: Polish Geological Institute Special Paper 12, 154 p.

Pieńkowski, G., and Waksmundzka, M., 2009, Palynofacies in Lower Jurassic epicontinental deposits of Poland: Tool to interpret sedimentary environments: Episodes, v. 32, p. 21–32.

Pieńkowski, G., Niedźwiedzki, G., and Waksmundzka, M., 2012, Sedimentological, palynological, and geochemical studies of the terrestrial Triassic-Jurassic boundary in north-western Poland: Geological Magazine, v. 149, p. 308–332, doi:10.1017/S0016756811000914.

Raucsik, B., and Varga, A., 2008, Climato-environmental controls on clay mineralogy of the Hettangian-Bajocian successions of the Mecsek Mountains, Hungary: An evidence for extreme continental weathering during the early Toarcian oceanic anoxic event: Palaeogeography, Palaeoclimatology, Palaeoecology, v. 265, p. 1–13, doi:10.1016/j.palaeo.2008.02.004.

Reymanówna, M., 1992, Two conifers from the Liassic flora of Odrowaz in Poland, *in* Kovar-Eder, J., ed., Palaeovegetational Development in Europe and Regions Relevant to its Palaeofloristic Evolution, Proceedings of the Pan-European Palaeobotanical Conference, Vienna, 19–23 September 1991: Vienna, Austria, Museum of Natural History, p. 307–311.

Ruffell, A., McKinley, J.M., and Worden, R.H., 2002, Comparison of clay mineral stratigraphy to other proxy palaeoclimate indicators in the Mesozoic of NW Europe: Philosophical Transactions of the Royal Society of London, ser. A, v. 360, p. 675–693, doi:10.1098/rsta.2001.0961.

Ruhl, M., and Kürschner, W.M., 2011, Multiple phases of carbon cycle disturbance from large igneous province formation at the Triassic-Jurassic transition: Geology, v. 39, p. 431–434, doi:10.1130/G31680.1.

Ruhl, M., Kürschner, W.M., and Krystyn, L., 2009, Triassic-Jurassic organic carbon isotope stratigraphy of key sections in the western Tethys realm (Austria): Earth and Planetary Science Letters, v. 281, p. 169–187, doi:10.1016/j.epsl.2009.02.020.

Ruhl, M., Deenen, M.H.L., Abels, H.A., Bonis, R.N., Krijgsman, W., and Kürschner, W.M., 2010, Astronomical constraints on the duration of the Early Jurassic Hettangian Stage and recovery rates following the end-Triassic mass extinction (St. Audrie's Bay/East Quantoxhead, UK): Earth and Planetary Science Letters, v. 295, p. 262–276, doi:10.1016/j.epsl.2010.04.008.

Ruhl, M., Bonis, R.N., Reichart, G.J., Sinninghe Damsté, J.S., and Kürschner, W.M., 2011, Atmospheric carbon injection linked to end-Triassic mass extinction: Science, v. 333, p. 430–434, doi:10.1126/science.1204255.

Schmieder, M., Buchner, E., Schwarz, W.H., Trieloff, M., and Lambert, P., 2010, A Rhaetian ^{40}Ar/^{39}Ar age for the Rochechouart impact structure (France) and implications for the latest Triassic sedimentary record: Meteoritics & Planetary Science, v. 45, no. 8, p. 1225–1242.

Schoene, B., Guex, J., Bartolini, A., Schaltegger, U., and Blackburn, T., 2010, Correlating the end-Triassic mass extinction and flood basalt volcanism at the 100 ka level: Geology, v. 38, p. 387–390, doi:10.1130/G30683.1.

Self, S., Widdowson, M., Thordarson, T., and Jay, A.E., 2006, Volatile fluxes during flood basalt eruptions and potential effects on the global environment: A Deccan perspective: Earth and Planetary Science Letters, v. 248, p. 518–532, doi:10.1016/j.epsl.2006.05.041.

Simms, M.J., and Jeram, A.J., 2006, Waterloo Bay, Larne, Northern Ireland: A potential GSSP for the base of the Jurassic System: Volumina Jurassica, v. 4, p. 297–298.

Singer, A., 1984, The paleoclimatic interpretation of clay minerals in sediment. A review: Earth-Science Reviews, v. 21, p. 251–293.

Smith, R., 2011, Dark days of the Triassic: Lost world: Nature, v. 479, p. 287–289, doi:10.1038/479287a.

Staneczko, K., 2007, Nowe dane paleobotaniczne na temat górnego triasu z Lipia Śląskiego koło Lublińca (południowa Polska): Geo-Sympozjum Młodych Badaczy Silesia, v. 2007, p. 155–168.

Sues, H.D., and Fraser, N.C., 2010, Triassic Life on Land: The Great Transition: New York, Columbia University Press, 224 p.

Sulej, T., 2009, Early dinosaurs within the vertebrate assemblages of the Germanic Basin, and the paleoecology of the Late Triassic ecosystems of Poland: Journal of Vertebrate Paleontology, v. 329, no. 3, supplement, p. 187A.

Sulej, T., Bronowicz, R., Tałanda, M., and Niedźwiedzki, G., 2011, A new dicynodont-archosaur assemblage from the Late Triassic (Carnian) of Poland: Earth & Environmental Science: Transactions of the Royal Society of Edinburgh, v. 101, p. 261–269, doi:10.1017/S1755691011020123.

Sulej, T., Niedźwiedzki, G., and Bronowicz, R., 2012, A new Late Triassic vertebrate fauna with turtles, aetosaurs and coelophysoid dinosaurs from Poland: Journal of Vertebrate Paleontology, v. 32, p. 1033–1041, doi:10.1080/02724634.2012.694384.

Świło, M., 2010a, Znaleziska szczątków rekinów ze stanowiska Lipie Śląskie, *in* Materiały do II Konferencji "Młodzi w Paleontologii": Warzawa, Instytut Paleobiologii PAN, p. 14.

Świło, M., 2010b, Szczątki rekinów triasowych na Śląsku, ich uzytecznosc w stratygrafii i paleoekologii [M.Sc. thesis]: Wrocław, Poland, Wydział Nauk o Ziemi i Kształtowania Srodowiska, Uniwersytet Wrocławski, 59 p.

Świło, M., 2010c, Znaleziska szczątków rekinów z Lipia Śląskiego, *in* Materiały z XXI Konferencji Naukowej Sekcji Paleontologicznej PTG "Kopalne biocenozy w czasie i przestrzeni": Żarki–Letnisko, Wydział Nauk o Ziemi UŚ, abstract book, p. 84.

Świło, M., and Kowalski, J., 2011, Mikroskamieniałości kręgowców z Lipia Śląskiego, *in* Materiały do III Konferencji "Młodzi w Paleontologii": Kraków, Instytut Botaniki Polskiej Akademii Nauk, p. 11.

Świło, M., Niedźwiedzki, G., and Sulej, T., 2013, Mammal-like tooth from the Late Triassic of Poland: Acta Palaeontologica Polonica, doi:10.4202/app.00016.2013.

Szulc, J., Gradziński, M., Lewandowska, A., and Heunisch, C., 2006, The Upper Triassic crenogenic limestones in Upper Silesia (southern Poland) and their paleoenvironmental context, *in* Alonso-Zarza, A.M., and Tanner, L.H., eds., Paleoenvironmental Record and Applications of Calcretes and Palustrine Carbonates: Geological Society of America Special Paper 416, p. 133–151.

Tanner, L.H., Lucas, S.G., and Chapman, M.G., 2004, Assessing the timing and causes of Late Triassic extinction: Earth-Science Reviews, v. 65, p. 103–139.

Taylor, T.N., Taylor, E.L., and Krings, M., 2009, Paleobotany—The biology and evolution of fossil plants: Burlington/Amsterdam, Academic Press/Elsevier, p. 1230.

van de Schootbrugge, B., Quan, T.M., Lindström, S., Püttmann, W., Heunisch, C., Pross, J., Fiebig, J., Petschick, R., Röhling, H.-G., Richoz, S., Rosenthal, Y., and Falkowski, P.G., 2009, Floral changes across the Triassic/Jurassic boundary linked to flood basalt volcanism: Nature Geoscience, v. 2, p. 589–594, doi:10.1038/ngeo577.

Ward, P.D., Haggart, J.W., Carter, E.S., Wilbur, D., Tipper, H.W., and Evans, T., 2001, Sudden productivity collapse associated with the Triassic-Jurassic boundary mass extinction: Science, v. 292, p. 1148–1151, doi:10.1126/science.1058574.

Ward, P.D., Garrison, G.H., Haggart, J.W., Kring, D.A., and Beattie, M.J., 2004, Isotopic evidence bearing on Late Triassic extinction events, Queen Charlotte Islands, British Columbia, and implications for the duration and the cause of the Triassic/Jurassic mass extinction: Earth and Planetary Science Letters, v. 224, p. 589–600, doi:10.1016/j.epsl.2004.04.034.

Wawrzyniak, Z., 2010a, Tafonomia późnotriasowej flory z Lipia Śląskiego, *in* Materiały do II Konferencji "Młodzi w Paleontologii": Warszawa, Instytut Paleobiologii PAN, p. 16.

Wawrzyniak, Z., 2010b, Późnotriasowa flora z Lipia Śląskiego [M.Sc. thesis]: Sosnowiec, Poland, Wydział Nauk o Ziemi, Uniwersytet Śląski, 76 p.

Wawrzyniak, Z., 2010c, Późnotriasowe kutykule z Lipia Śląskiego k. Lublińca, *in* Materiały z XI Konferencji Naukowej Sekcji Paleontologicznej PTG "Kopalne biocenozy w czasie i przestrzeni": Żarki–Letnisko, Wydział Nauk o Ziemi UŚ, abstract book, p. 85.

Wawrzyniak, Z., 2011, The Upper Triassic cuticles from Lipie Śląskie (South Poland), *in* Bąk, M., Kaminski, M.A., and Waśkowska, A., eds., Integrating Microfossil Records from the Oceans and Epicontinental Seas: Grzybowski Foundation Special Publication 17, p. 140–141.

Wawrzyniak, Z., and Ziaja, J., 2009, Wstepne wyniki badan gornotriasowej makroflory Lipia Śląskiego, Polska: Geologia, v. 35, p. 105–106.

Wcisło-Luraniec, E., 1991, The lower liassic flora from Odrowąż in Poland and its ecosystem. Fifth symposium on Mesozoic Terrestrial Ecosystems and Biota. Extended abstracts: Contributions from the Paleontological Museum University of Oslo, v. 364, p. 69–70.

Weaver, C.E., 1989, Clays, Muds and Shales: Amsterdam, Elsevier, 820 p.

Whiteside, J., Olsen, P.E., Eglinton, T., Brookfield, M.E., and Sambrotto, R.N., 2010, Compound-specific carbon isotopes from Earth's largest flood basalt eruptions directly linked to the end-Triassic mass extinction: Proceedings of the National Academy of Sciences of the United States of America, v. 107, p. 6721–6725, doi:10.1073/pnas.1001706107.

Zajzon, N., Kristaly, F., Palfy, J., and Nemeth, T., 2012, Detailed clay mineralogy of the Triassic-Jurassic boundary section at Kendlbachgraben (Northern Calcareous Alps, Austria): Clay Minerals, v. 47, p. 177–189, doi:10.1180/claymin.2012.047.2.03.

Ziaja, J., 2006, Lower Jurassic spores and pollen grains from Odrowąż, Mesozoic margin of the Holy cross Mountains, Poland: Acta Palaeobotanica, v. 46, p. 3–83.

Ziegler, P., 1990, Geological Atlas of Western and Central Europe (2nd ed.): Shell International Petroleum Maatschappij B.V.: Geological Society of London, 256 p.

Manuscript Accepted by the Society 31 January 2014

Printed in the USA

The Geological Society of America
Special Paper 505
2014

An earth system approach to understanding the end-Ordovician (Hirnantian) mass extinction

Howard A. Armstrong
David A.T. Harper
Department of Earth Sciences, Palaeoecosystems Group, Durham University, Lower Mountjoy, Durham DH1 3LE, UK

ABSTRACT

The Hirnantian mass extinction is recognized as the first of the "big three" extinctions and, along with the end-Permian and end-Cretaceous events, is the result of an acceleration in biotic extinctions concomitant with a rise in originations. The Hirnantian mass extinction is characterized by high taxonomic impact and within-community extinctions. The Hirnantian mass extinction is also unusual in that (1) it is associated with glaciation, but there is little evidence elsewhere in the younger Phanerozoic that glaciations have been a cause of mass extinction, and (2) there is limited understanding of how glaciation could directly cause mass extinction, particularly in the marine realm. In this review, we argue that coordinated extinctions occurred at the onset and termination of glaciation and were due to climatically induced changes in relative sea level, ocean redox stratification, and sea-surface temperature gradients. These earth system changes resulted in a reduction in prospective niche space, both in the water column and on the seafloor, which in turn led to increased competition and selection pressures, resulting in extinctions where the carrying capacities of particular ecological niches were exceeded. The long-term ventilation of the oceans broke the link between glaciation and mass extinction.

INTRODUCTION

The Hirnantian mass extinction is recognized as the first of the "big three" extinctions and, along with the end-Permian and end-Cretaceous events, is the result of an acceleration in biotic extinctions concomitant with a rise in originations (Bambach et al., 2004; Fig. 1). Estimates indicate that in the marine realm, 20% of families, 40% of genera (Sepkoski, 1996), and by extrapolation ~85% of marine species went extinct at this time (Jablonski, 1991), making the Hirnantian mass extinction second only in scale to the end-Permian extinction in taxonomic impact. This extinction terminated the Great Ordovician Biodiversification Event and was followed by the Paleozoic plateau in biodiversity, which continued until the end of the Paleozoic Era (Harper et al., 2014).

The Hirnantian mass extinction is unusual in that (1) it is associated with glaciation, but there is little evidence elsewhere in the Phanerozoic that glaciations have been a cause of mass extinction, and (2) there is limited understanding of how glaciation could directly cause mass extinction, particularly in the

Armstrong, H.A., and Harper, D.A.T., 2014, An earth system approach to understanding the end-Ordovician (Hirnantian) mass extinction, *in* Keller, G., and Kerr, A., eds., Volcanism, Impacts, and Mass Extinctions: Causes and Effects: Geological Society of America Special Paper 505, p. 287–300, doi:10.1130/2014.2505(14).

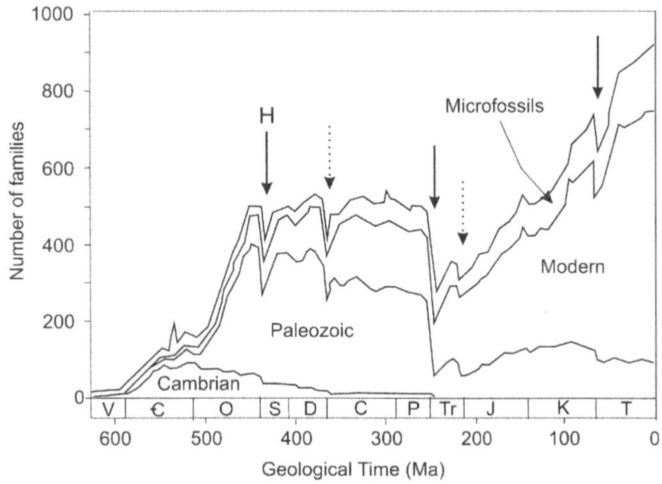

Figure 1. Global biodiversity changes through the Phanerozoic. The plot shows the family diversity of marine animals through the Phanerozoic indicating the three evolutionary faunas and microfossil record. The major extinction events (end-Ordovician, late Devonian, end-Permian, end-Triassic, and end-Cretaceous) are shown (after Sepkoski, 1981). Solid arrows indicate mass extinctions where extinction rates exceeded small increases in originations; dashed arrows indicate mass extinctions where reductions in origination exceeded extinction. H—Hirnantian; V—Vendian; Cm—Cambrian; O—Ordovician; S—Silurian; D—Devonian; C—Carboniferous; P—Permian; Tr—Triassic; J—Jurassic; K—Cretaceous; T—Tertiary.

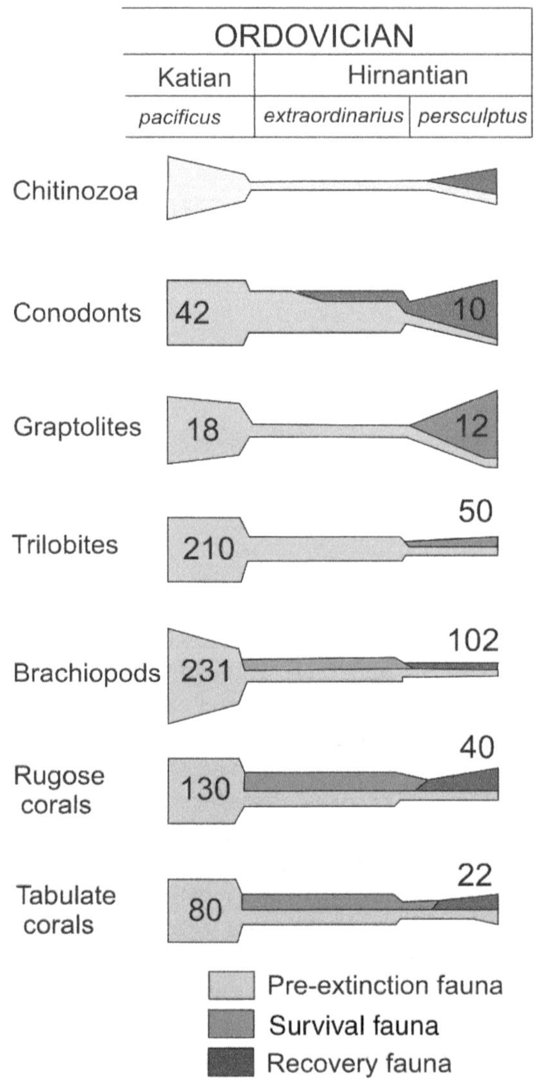

Figure 2. Biotic change through the Hirnantian mass extinction. The brachiopod data indicate the generic loss that appears to have been initiated already within the *Parakidograptus pacificus* graptolite biozone (after Brenchley et al., 2001; with modified brachiopod data from Rasmussen and Harper, 2011b).

marine realm. Pleistocene sea-level changes did not much affect the marine biosphere at any ecological level (Valentine and Jablonski, 1991), and habitat tracking appears to have been an important process in insulating the biosphere from mass extinction at this time (Brett et al., 2007). Glacially induced cooling, falling sea level, and chemical recycling in the oceans are three of the many suggested kill mechanisms for end-Ordovician extinction, but a general consensus is lacking (Finnegan et al., 2012; Harper et al., 2014).

The Hirnantian mass extinction consisted of two extinction phases, which are known in some detail (e.g., Brenchley and Cullen, 1984; Brenchley et al., 2006; Fig. 2). The "first strike" occurred at or just below the *Normalograptus extraordinarius* graptolite biozone and coincided with the onset of glaciation and a fall in global sea level (Fig. 3). Continental ice was centered on the supercontinent of Gondwana, located over the South Pole during the Late Ordovician (Brenchley and Cullen, 1984; Brenchley et al., 2001). During this phase, benthic organisms in deep- and shallow-water environments were more affected than organisms occupying midshelfal depths. Planktonic/nektonic organisms, particularly graptolites and nektonic groups, were differentially more greatly affected (Rasmussen and Harper, 2011a, 2011b, and references herein). The second strike started at the base of the *Normalograptus persculptus* graptolite biozone and coincided with a rise in sea level and the widespread deposition of black shale in continental shelf settings (Fortey, 1989). During this phase, coordinated extinctions occurred across the mid- to outer shelf, and particularly in midshelf settings.

Global Climate Context

Estimates of the duration of the Hirnantian Glacial Maximum vary, but it may have been less than 1 m.y. (e.g., Armstrong, 2007; Holmden et al., 2012). A "short, sharp glaciation" cause for mass extinction can now be re-evaluated against a better understanding of longer-term Ordovician climate change and a large, relatively new database of environmental proxies.

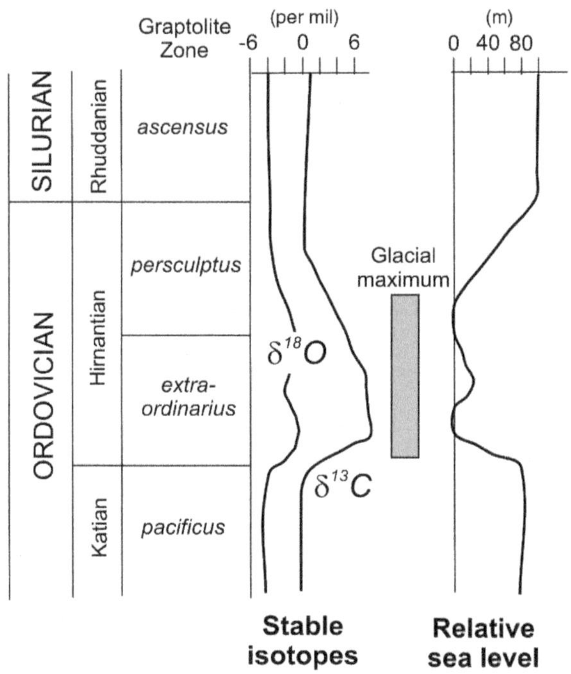

Figure 3. Late Ordovician–Early Silurian stratigraphy, stable isotope stratigraphy, and relative sea-level change. The base of the Hirnantian is placed at the base of the *Normalograptus extraordinarius* graptolite biozone (after Underwood et al., 1997). The carbon isotope profile and the sea level curve are modified from Brenchley et al. (1994). Figure is modified from Brenchley et al. (2001).

Emerging new climate scenarios for the Ordovician indicate that a modern-style "cool world" existed in the Early Ordovician (ca. 472 Ma; Trotter et al., 2008; Vandenbroucke et al., 2009), with successive glaciations during the Floian, Darriwilian, Katian (Guttenburg), and Hirnantian (Turner et al., 2011, 2012). These occurred against a backdrop of long-term declining pCO_2 (Godderis et al., 2001; Figs. 4 and 5), probably initiated by changes in plate configuration that resulted in increased weathering and nutrient cycling into the oceans, changes in volcanic outgassing of greenhouse gases, and the redirection of ocean currents. The rapid expansion of ice volume during successive intervals of Ordovician glaciation was triggered by orbitally induced cooling (Turner et al., 2011, 2012). During each glaciation, the Gondwana ice sheet grew large enough to be affected by eccentricity pacing of ice-margin processes embedded in obliquity, which largely controlled their size (Armstrong, 2007). Composite $\delta^{13}C_{carb}$ records for the entire Ordovician show a long-term broadly positive trend with increasingly large positive excursions through successive Ordovician glaciations (Bergström et al., 2009; Fig. 5).

Positive isotope excursions (base Floian, Darriwilian, Guttenburg, and Hirnantian carbon isotope excursions) coincide with intervals characterized by ~1.2-m.y.-long obliquity cycles, interpreted to indicate icehouse conditions (Boulila et al., 2011; Turner et al., 2011, 2012; Fig. 5). The $\delta^{13}C$ data suggest an ever-present Gondwana ice sheet that grew in a stepwise fashion to reach maximum size in the Hirnantian. The Hirnantian glaciation and associated mass extinction may thus be viewed as a "tipping point" in the Earth surface system, when the polar ice sheet was

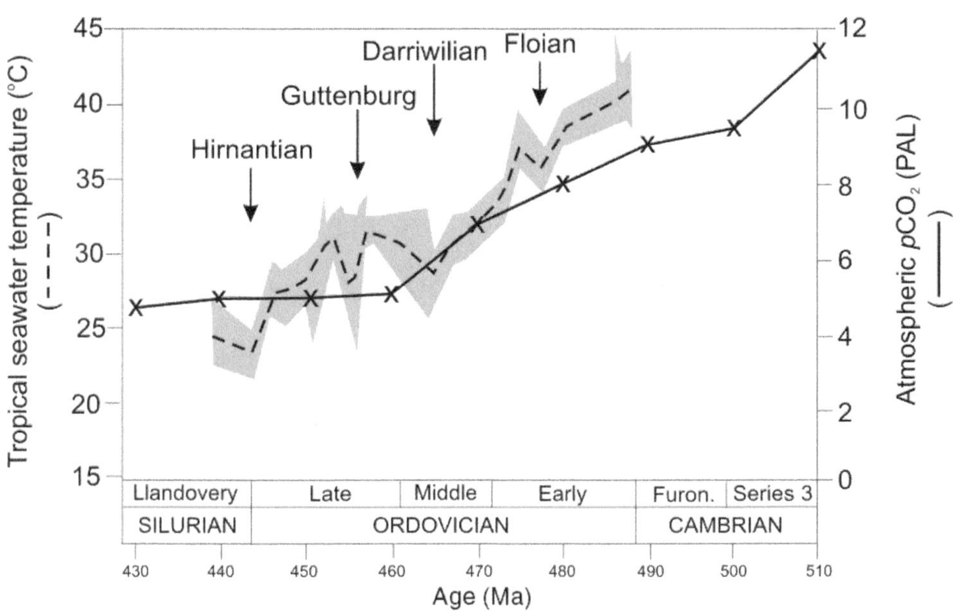

Figure 4. Modeled atmospheric carbon dioxide and tropical sea-surface temperatures through the Cambrian to Silurian. Tropical sea-surface temperatures are from Nardin et al. (2011), with the gray area indicating the error envelope. Atmospheric carbon dioxide is after Godderis et al. (2001).

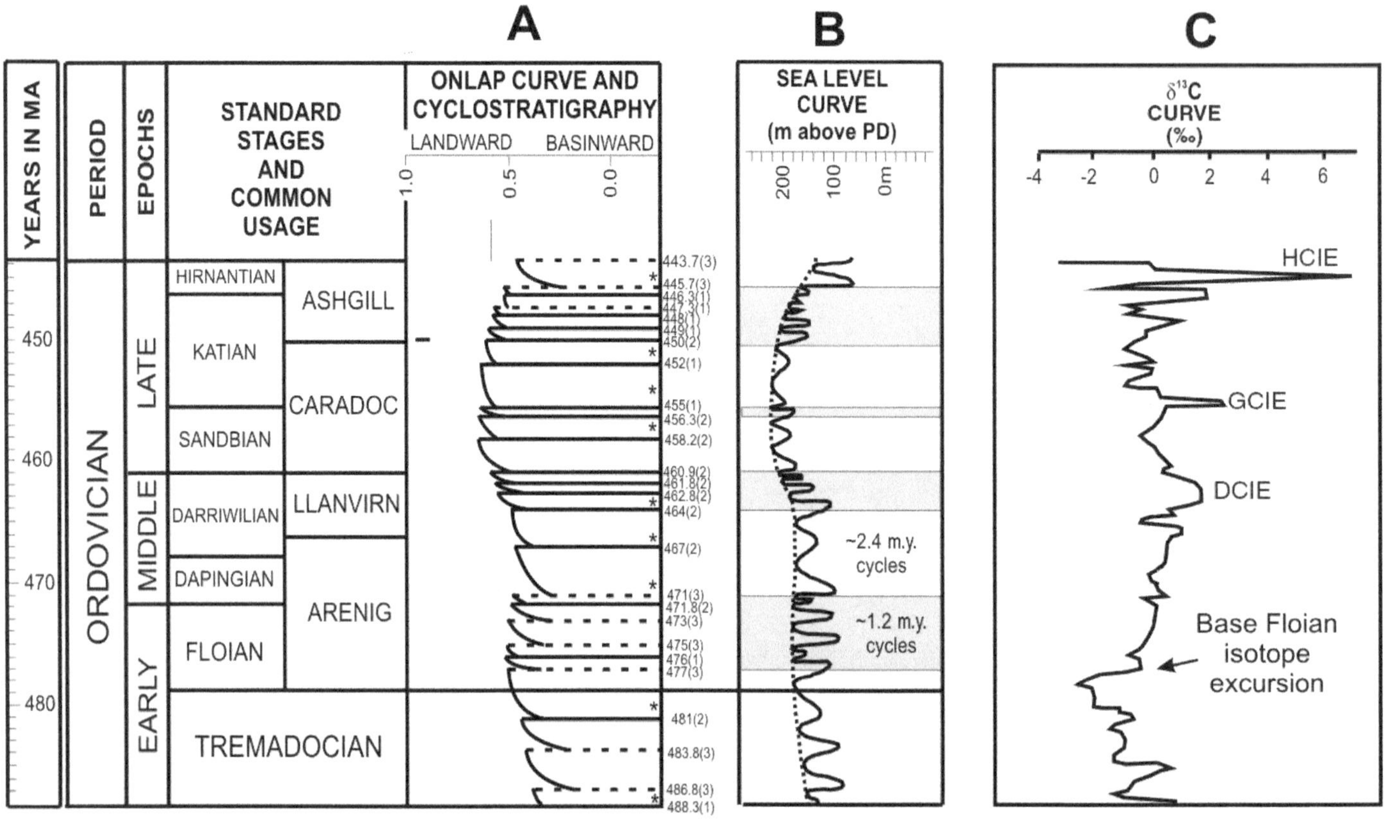

Figure 5. Time stratigraphic diagram showing chronostratigraphy and global composite proxy data. (A–B) Data for relative sea level (after Haq and Schulter, 2008). (C) Stable carbon (after Bergström et al., 2009). Gray horizontal boxes in B highlight the time periods of ~1.2 m.y. cyclic changes in global sea level, interpreted as indicative of icehouse periods (after Turner et al., 2011). DCIE—Darriwilian carbon isotope excursion; GCIE—Guttenburg carbon isotope excursion; HCIE—Hirnantian carbon isotope excursion; PD—present day.

large enough to have a maximal effect on the ocean-atmosphere system (Armstrong, 2007).

Biotic Patterns during Mass Extinction

Pattern 1. High Taxonomic Impact but Low Ecological Impact

No major taxonomic groups or trophic groups disappeared during the Hirnantian mass extinction, and the event has been classified as being of high taxonomic impact but low ecological impact (Droser et al., 2000; McGhee et al., 2013). It is acknowledged that the Hirnantian biotic record is far from complete, largely documented at the level of geological stage and biased toward continental shelf settings in low paleolatitudes. Despite this, the patterns and trends of extinction are uneven across continents, but these are as-yet poorly documented and understood (Harper and Rong, 2008). The uneven pattern of extinctions likely provided refugia during peak glaciation, as evidenced by a high proportion of Lazarus taxa (Brett et al., 2007; Rong et al., 2006). The relatively rapid recovery of the major clades and their benthic communities indicates minimal ecological disruption and largely within-community species-level extinction (Droser et al., 2000); this is particularly well demonstrated by tropical communities. Despite significant global cooling, tropical bioherm communities remained largely unaffected by the Hirnantian mass extinction, and rugose and tabulate corals continued to radiate into the Silurian (Fig. 6A). Bioherms expanded rapidly during the Llandovery and went on to dominate many Silurian carbonate environments, but these communities had similar taxonomic compositions, species richness, and trophic structure to those of the Katian (Copper and Jin, 2012).

Pattern 2. Variance in Patterns of Extinction between the Benthic and Pelagic Realms

Phytoplankton, including the acritarchs (chromophyte algae) and, in deeper environments, cyanobacteria, formed the base of the food chain in Late Ordovician ecosystems (Servais et al., 2008). Patterns of diversity decline are similar in phytoplankton, zooplankton, and nektonic groups (Fig. 6B). All groups reach peak diversity in the late Darriwilian (equivalent to the Llandeilo in the old British chronostratigraphy; see also Fig. 6 caption) and show declining diversity through the Late Ordovician, recovering at a variety of rates until the mid-Silurian. The onset of diversity decline is significantly earlier than in benthic groups. Acritarch diversity declined dramatically from the late Darriwilian (Llandeilo), with a similarly spectacular recovery in the Llandovery (Figs. 2 and 6B). Chitinozoa and graptolites formed the preserved zooplankton, and both groups declined markedly in diversity during the first strike of extinction. Graptolites were already in a state of decline from the Darriwilian (late Middle Ordovician); the group was reduced to only a few genera and less than 20 species by the first strike of the Hirnantian mass extinction (Bapst et al., 2012). Graptolite disparity, as evidenced by a marked reduction in the range of graptolite rhabdosome and theca morphologies, also decreased during the mass extinction interval. However, many of the thecal morphotypes present in the Ordovician are found in the Silurian monograptids, suggesting these were hydrodynamic adaptations as the graptolites reoccupied specific niches (M. Williams, 2013, personal commun.). Over 200 species have been reported from the Upper Llandovery (Lower Silurian; Zalasiewicz et al., 2009).

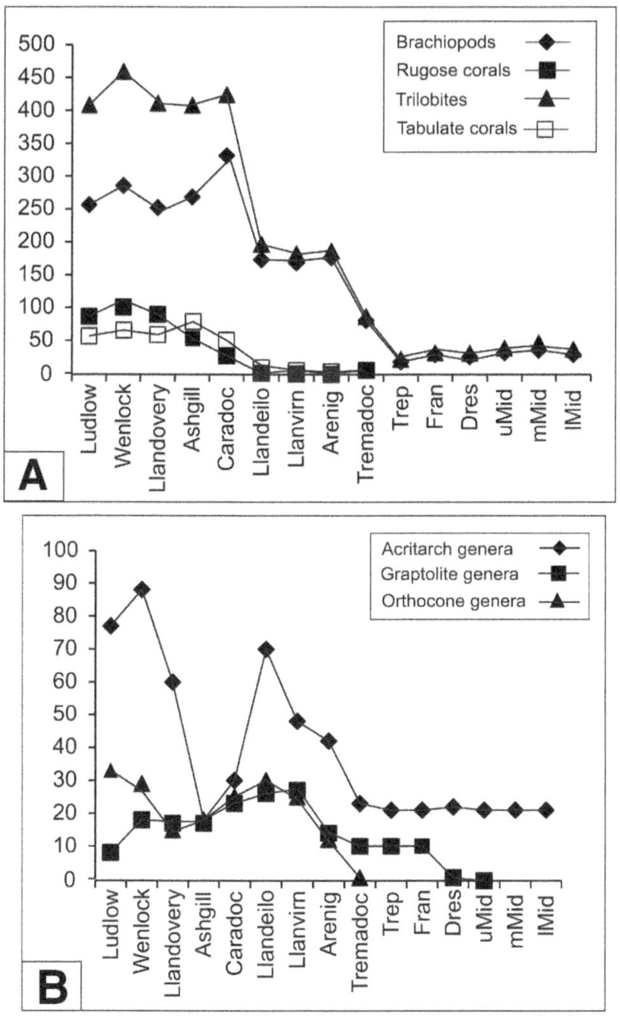

Figure 6. Biotic diversity during the Cambrian to Silurian. (A) Benthic generic diversity. (B) Pelagic and nektonic taxonomic diversity. (Data from Sepkoski Online, downloaded March 2013.) Detailed Ordovician curves can also be found in Servais et al. (2010). Note the data are assembled in the database using the British series names. The Llanvirn was extended by Fortey et al. (1995) to include part of the classical Llandeilo Series as a stage. For convenience, these are retained where appropriate in the text. British and globally recognized divisions are compared in Figure 5. Trep—Trempealeanan; Fran—Franconian; Dres—Dresbachian; uMid—upper Middle Cambrian; mMid—middle Middle Cambrian; lMid—lower Middle Cambrian.

Species distribution studies indicate graptolites lived in two primary depth zones of the ocean waters. A shallow, epipelagic zone biotope contained species found preserved in all depth facies, whereas a deep, mesopelagic zone biotope contained species found only in deep-water facies (isograptid biofacies; Cooper et al., 1991). Mesopelagic taxa thrived in continental-margin upwelling zones and within the oxygen minimum zone, where their rhabdosomes were readily preserved (Finney and Berry, 1997). Epipelagic biotope species were affected by sea-surface temperature (see following). Extinctions in the mesopelagic biotope are poorly constrained but would have reflected oceanographic changes and feedbacks (especially upwelling and redox conditions) that were driven by climatic and tectonic events (Finney and Berry, 1997).

In the nekton, trilobites, including the cyclopygids, disappeared entirely. The top predators of the Ordovician, the nautiloid cephalopods, were significantly reduced in diversity during the extinction interval, from nearly 300 species at the height of the great Ordovician biodiversification event to some 50 during the Hirnantian. This decline in diversity has been attributed to early Hirnantian regression, draining many of the epicontinental seas, and destroying habitats for potential prey (Harper et al., 2014).

In contrast, benthic groups, both sessile and motile, show diversity peaks at different times (e.g., see papers cited in Webby et al., 2004; Fig. 6A). Global analyses of Upper Ordovician and Lower Silurian brachiopods indicate that 18.6% and 12.5% of families and 51.0% and 41.3% of genera were eliminated in the first and second phases of the mass extinction, respectively, with the total loss of 28.4% of families and 69.0% of genera in the crisis (Rong et al., 2006). Although various major groups of brachiopods suffered substantially during the extinction, there was clearly phylogenetic and ecological continuity between the Late Ordovician and Early Silurian shelf faunas (Droser et al., 1997). The extinction of brachiopods did not cease at the Hirnantian mass extinction—there is evidence of a relict Ordovician association. Evidence from the Oslo region and South China suggests that part of the initial Silurian fauna consisted of the more eurytopic taxa of the regressive succession that survived the extinction in the deeper parts of the intracratonic basin (e.g., Baarli and Harper, 1986; Rong and Zhan, 2006; Owen et al., 2008). These species were subsequently able to create and participate within new community structures during the Early Silurian transgression (Rong and Harper, 1999).

Through the Hirnantian mass extinction, trilobites suffered losses in the region of 70% at the generic level, and all unequivocally pelagic taxa became extinct by the end of the Hirnantian mass extinction. Of the main trophic groups, the filter feeders were differentially affected but did not disappear, and these changes had little effect on the overall community structure (Adrain et al., 2004). In detail, there was a marked, two-stage reduction in diversity at the base of the *N. extraordinarius* and *N. persculptus* graptolite biozones (Brenchley et al., 2001; see following). Benthic taxa that survived the first phase of extinction commonly succumbed to the second.

Pattern 3. Extinctions in the Epipelagic Zone Were Associated with a Narrowing of Temperature-Defined Provinces

Biogeographical provinces are aggregations of communities, and their number and individual compositions affect global or gamma biodiversity levels. Today, temperature is the most important factor in defining the boundaries of biogeographical provinces, which broadly follow the climate belts in the marine realm (Kucera, 2007). Paleobiogeographical studies of epipelagic organisms through the Late Ordovician indicate that during the Hirnantian glaciation, the latitudinal temperature gradient steepened, and provinces shifted toward the equator (Vandenbroucke et al., 2010a, 2010b; Fig. 7). These data suggest that the geographical extent of the tropical province remained largely unaffected by the change from interglaciation (Sandbian distribution) to Hirnantian glaciation. The low-diversity polar province expanded in geographical extent. Extinction was highest in the more diverse subpolar province, associated with the narrowing of the zone (Vandenbroucke et al., 2010a).

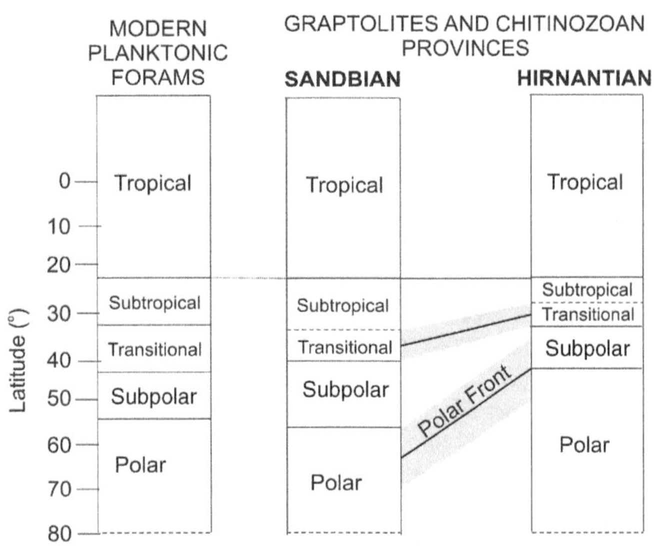

Figure 7. Late Ordovician polar front migration (after Vandenbroucke et al., 2010b). The figure compares the spatial distribution of Sandbian and Hirnantian chitinozoan and graptolite provinces and modern planktonic foraminiferan provinces. The changing positions of graptolite and chitinozoan provinces show an equatorward shift in the position of the polar front from 55°S–70°S to likely 40°S. This involves an equator-ward incursion of polar water and a narrowing of the subpolar province. The subtropical province moves slightly northward. The migration of the Hirnantian polar front compares well with known patterns from late Cenozoic interglacial to glacial transitions.

A link between species richness and provinciality is also recognized in brachiopods. Sheehan and Coorough (1990) recognized 10 Katian brachiopod provinces, which decreased to nine in the Hirnantian and only five in the Early to Middle Silurian. In contrast, Rong and Harper (1988) argued that only three provinces could be recognized during the Hirnantian, the Edgewood, Kosov (broadly equivalent to the distribution of the typical *Hirnantia* Fauna), and Bani provinces, suggesting that this difference in interpretation reflects the relative grouping of increasingly endemic faunas during the glaciation. Brachiopod diversity declined from the tropics into high latitudes (Harper and Mac Niocaill, 2002). In the early part of the Hirnantian, the relatively low-diversity *Hirnantia* brachiopod fauna expanded toward the equator (Temple, 1965), replacing the generally more diverse Edgewood Province, which dominated the tropics (Harper, 1981; Rong, 1979). A similar pattern is reported with the widely distributed *Mucronapsis* trilobite fauna (Owen, 1986).

Pattern 4. Benthic Extinction Was Associated with Depth

Detailed analyses of brachiopod data indicate the loss in alpha (within-community) and beta (between-community) biodiversity through the extinction events (Brenchley et al., 2001; Fig. 8). During the first strike, percentage extinctions were high across the shelf but were higher in the outer shelf; benthic assemblage 6 (BA 6), characteristic of the outer shelf and upper slope fauna (including the *Foliomena* brachiopod fauna), was completely eradicated. The second strike further reduced diversity in the mid- and outer-shelf biotas (BA 3 and BA 4/5), including the widespread cool-water *Hirnantia* brachiopod fauna (Rong and Harper, 1988).

Geochemical Evidence for Changing Ocean Conditions during Mass Extinction

The Ordovician oceans were very different from those of today (e.g., Munnecke et al., 2010). There is no direct evidence for the ventilation of the deep oceans. During the period, anoxic ocean bottom waters were widespread, and dissolved O_2 concentrations in the mixed layer are hypothesized to have been much reduced compared to the present day (Dahl et al., 2010); however, this is difficult to reconcile with the presence of large, complex marine organisms and diverse multitiered communities. The fact that a large reservoir of anoxic deep water persisted below the storm- and wave-influenced mixed layer is corroborated by a number of geochemical and sedimentary proxies. Brenchley and Marshall (1999) argued that the positive $\delta^{13}C$ isotopic excursion during the Hirnantian, and now reported for earlier events in the Ordovician, could only be realistically sustained with the disposal of ^{12}C into anoxic deep oceans. However, part of the excursion may have been in part due to increased carbonate precipitation rates in epeiric seas, where high photosynthesis rates in calcifying bacteria, maintained by high nutrient fluxes, locally increased the local carbonate-saturation state of the waters (LaPorte et al., 2009). Geochemical proxies such as iron speciation, molybdenum concentrations, and sulfur isotopes reveal complex patterns of changing regional and global redox conditions throughout the mass extinction (Figs. 9 and 10).

Sedimentary evidence indicates oxygenated deposition at least at shelfal depths through the peak glaciation (*N. extraordinarius* graptolite biozone); for example, there is a widespread shift from deposition of black shale to gray, occasionally bioturbated, shale at specific localities (e.g., Dob's Linn; Armstrong and Coe, 1997). This is coincident with pyrite that is significantly enriched in ^{34}S in a number of sections (e.g., Goodfellow and Jonasson, 1984; Zhang et al., 2009; Hammarlund et al., 2012; Fig. 9). The last suggests a widespread reduction in marine sulfate concentrations that can be attributed to an increase in pyrite burial during the early Hirnantian. The ^{34}S-isotope excursion coincides with a major positive carbon isotope excursion recorded globally in marine carbonates (Fig. 9) and indicates increased photic zone photosynthetic productivity and organic carbon flux to the seafloor (Armstrong and Coe, 1997; Bergström et al., 2006; Brenchley et al., 1994; Finney et al., 1999; see also

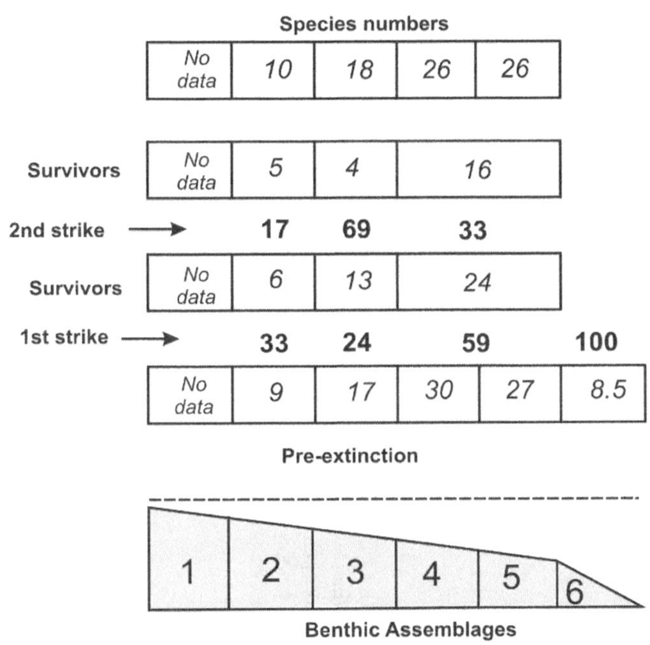

Figure 8. Changes within brachiopod community (alpha) diversity across the continental shelf and upper slope as calculated by Brenchley et al. (2001). Numbers are an average taken from communities within the benthic assemblage zones. There are no data for benthic assemblage 1. Numbers in bold indicate the percentage decrease in mean alpha diversity at the first and second strikes. The fall in alpha diversity after the first phase of extinction also includes the addition of new taxa belonging to the Hirnantian recovery fauna; no similar recovery fauna buffers the effect of the second phase of extinction. (Figure is modified from Brenchley et al., 2001.)

Kump et al., 1999). Together, these proxies indicate that as global sea level fell, increased nutrient flux and photic zone bioproductivity intensified and widespread anoxic, and in places euxinic, conditions prevailed or expanded as the chemocline rose through the water column (Briggs et al., 1988; Fortey, 1989; Goodfellow and Jonasson, 1984; Zhang et al., 2009; though for an alternative view, see Hammarlund et al., 2012).

Rapid sea-level fall can profoundly affect ocean chemistry, consistent with a pattern of increasing pyrite and organic carbon burial and a rising chemocline. As Hammarlund et al. (2012) observed, as sea level falls, organic carbon produced by primary producers moves farther offshore and sinks through a deeper water column (Wallmann, 2003). The longer settling time of organic carbon leads to a more complete organic carbon decomposition and release and recycling of dissolved inorganic phosphate (DIP). Models show that a sea-level drop of 100 m can result in a more than 50% increase in marine DIP concentration, at steady state (Bjerrum et al., 2006; Wallmann, 2003). Consequently, more phosphorus becomes available for primary production, which in turn feeds back into increased organic matter production, oxygen consumption at depth, and carbon burial. During a period of anoxic conditions, enhanced phosphorus regeneration from the sediment furthermore creates a positive feedback of phosphorus availability (Colman et al., 1997). The last Pleistocene glaciation provides an analogue for conditions when deep ocean waters contained less oxygen than during the Holocene (de Boer et al., 2007; Sigman et al., 2004; Toggweiler and Russell, 2008).

Associated with these changes at depth, an intensification of water-column stratification would also steepen the oxygen gradient, reducing the habitable space for the plankton and nekton occupying the dysoxic zone (Berry et al., 1990; Fig. 11).

The *N. persculptus* graptolite biozone is characterized by a major sea-level rise. Redox proxies from a limited number of sections suggest that anoxic and euxinic conditions, and even ferruginous waters, encroached onto the continental shelves (Hammarlund et al., 2012; Fig. 10). This change coincided with declining productivity and carbon and pyrite burial, all complexly linked to declining ocean nutrient cycling during the postglaciation period.

TOWARD A UNIFIED EARTH SYSTEM MECHANISM FOR GLACIALLY INDUCED MASS EXTINCTION

Figure 11 shows the results of a thought experiment, designed to explain the complex pattern of extinction within the benthos and motivated by notions of habitat loss as a driver of mass extinction (e.g., Sheehan, 2001). The model shows the distribution of benthic assemblages from nearshore to offshore. Ecological boundaries within the water column, e.g., fair weather wave base, storm wave base, and the chemocline, are pinned at the boundaries of the benthic assemblages and maintain their depths during successive phases of the glaciation. This underlying aspect of the model can be tested using paleoecological and sedimentological evidence.

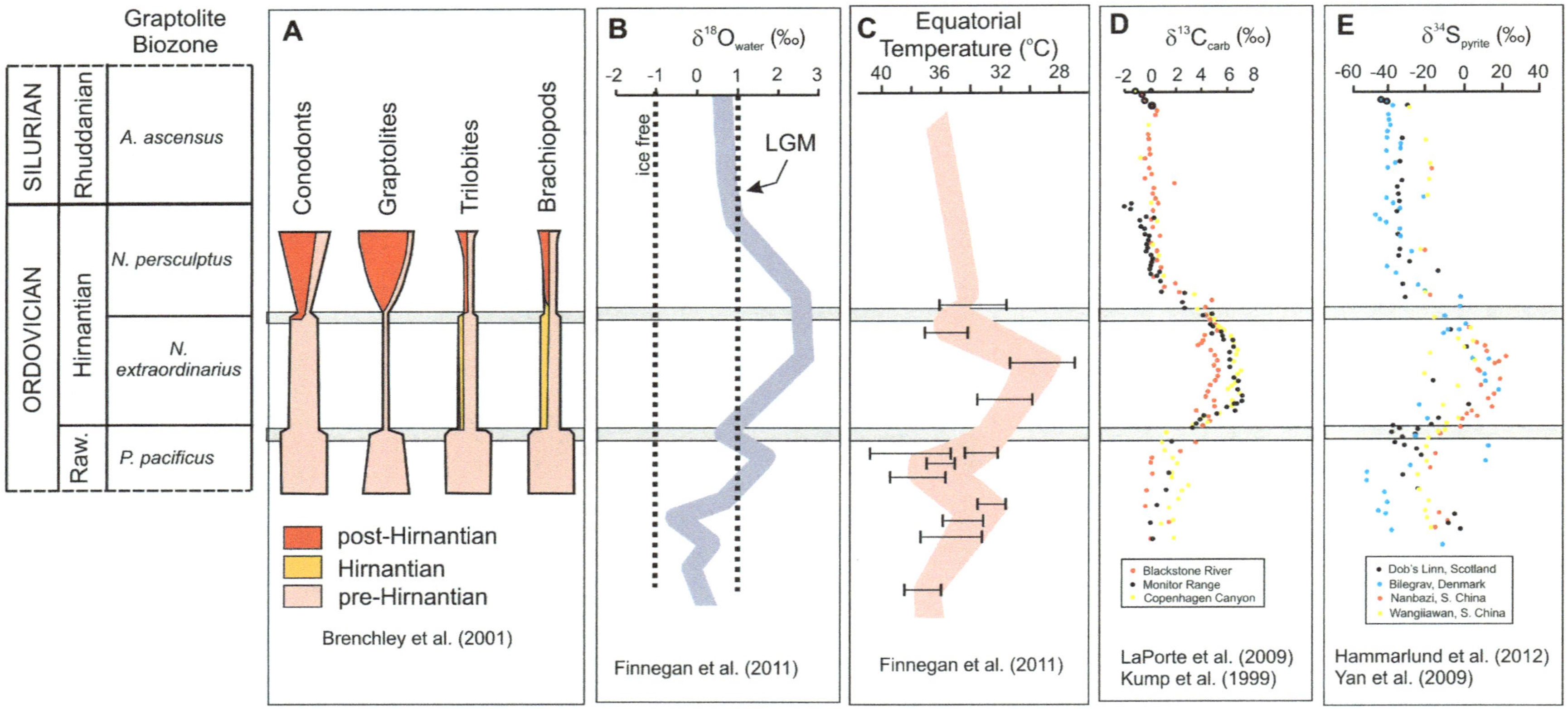

Figure 9. The Hirnantian mass extinction, glaciation, and isotope excursions (from Hammarlund et al., 2012). (A) Four major marine groups affected by the two-phased end-Ordovician extinction; pre-Hirnantian, Hirnantian, and post-Hirnantian refer to faunas (Brenchley et al., 1994, 2001). (B) Interpolated $\delta^{18}O$ values reflect fluctuations of Rawtheyan (Raw.) and Hirnantian sea level (Finnegan et al., 2011); LGM—Last Glacial Maximum. (C) Equatorial temperature fluctuations broadly parallel the $\delta^{18}O$ curve (Finnegan et al., 2011). (D) A compilation of three profiles of inorganic $\delta^{13}C$ shows a significant perturbation of oceanic carbon cycle dynamics during the Hirnantian (Kump et al., 1999; LaPorte et al., 2009). (E) A compilation of sulfur isotope data shows a major perturbation during the Hirnantian that parallels the $\delta^{13}C$ curve (Yan et al., 2009; Hammarlund et al., 2012).

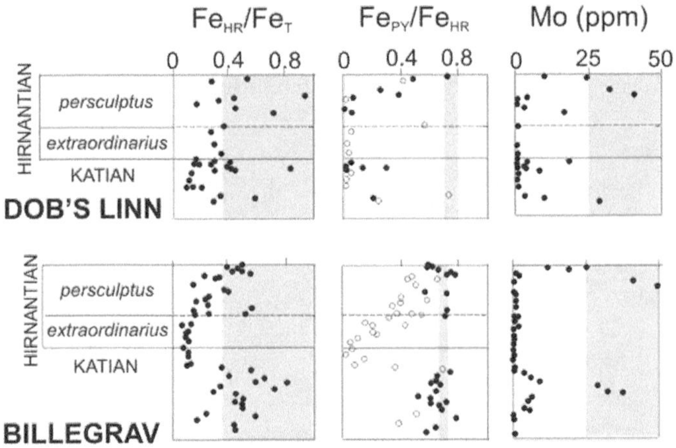

Figure 10. Iron and molybdenum data for the sections at Dob's Linn, Scotland (global stratotype section and point [GSSP] for the base of the Silurian), and Billegrav, Denmark, reveal extensive euxinic, and occasional ferruginous, conditions (after Hammarlund et al., 2012). Fe_{HR}/Fe_T (highly reactive iron over total iron) has a threshold at 0.38. For Fe_{PY}/Fe_{HR}, the ratio of pyrite over highly reactive iron, filled circles indicate values above 0.7, and open circles indicate values above below 0.7. The gray zone indicates the interval that is discussed as a lowermost threshold for euxinic conditions.

Brett et al. (1993) reviewed a variety of sedimentary and fossil evidence that could be used to estimate the absolute depth of the well-established depth gradient of Silurian onshore-to-offshore benthic assemblages (BA 1–5). This constrains much of the spectrum of benthic fossil communities to a narrow depth range, within the photic zone. They placed the depth of Silurian BA 1 through 4 at between 0 and 60 m. The absence of storm-disturbed beds over large parts of several major platforms below about the BA 3–4 boundary constrains the position of storm wave base. This is consistent with reports of Ordovician BA 4 through 6 being found in dysoxic waters below the storm wave base (Potter and Boucot, 1992). A more detailed study of the paleoecological controls of Ordovician benthic assemblages would provide an elegant test of the hypothesis.

Figure 11A shows the preglaciation template. During the first strike of extinction, the expansion of the ice sheet resulted in an ~80 m fall in global sea level (Loi et al., 2010), which was coincident with a rise in photosynthetic productivity and shallowing of the chemocline through the water column. The complete extinction of taxa within BA 6 indicates that the chemocline rose to shallow continental slope depths. The widths of the benthic assemblage have been adjusted to maintain the depths of the ecological boundaries, as described already. The result is to significantly reduce the widths of the benthic assemblages as compared to the preglaciation mode, particularly in BA 4–5. As shown in Figure 8, the extinctions during the first strike are found in all benthic assemblages but are preferentially higher in BA4–6.

During the second strike, rapid sea-level rise to +100 m (Brenchley et al., 1994; see also Fig. 3) shifts the remaining benthic assemblages onshore and coincides with the rise of the chemocline onto the shelf and spread of anoxic/euxinic waters, supported by the redox proxy data. Again, the widths of the benthic assemblages have been adjusted to maintain the constant depth of the ecological boundaries. The consequence is a further reduc-

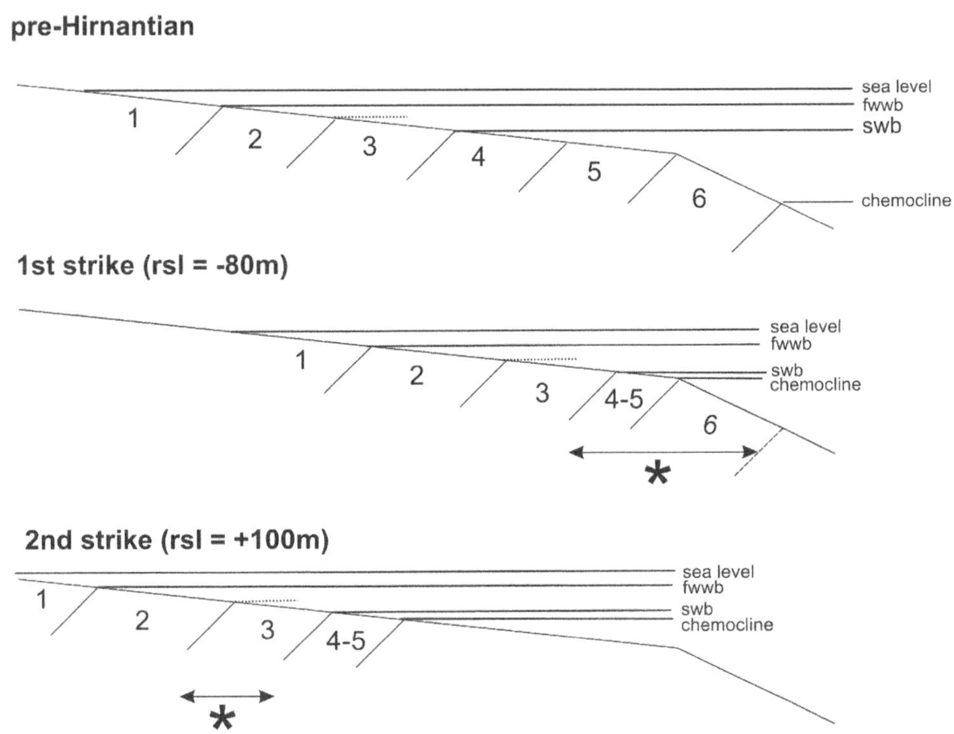

Figure 11. Conceptual model illustrating the paleoenvironmental changes associated with evolving sea-level changes (rsl—relative sea level). Stars indicate the areas with greatest biodiversity decline. Fair weather wave base (fwwb), storm wave base (swb), and pycnocline are fixed at the benthic assemblage boundaries and maintain their depths during sea-level change. The consequence is to change the habitable area occupied by the benthic assemblages. Changes in habitable area are consistent with areas of maximum species loss as shown in Figure 8.

tion in the widths of BA3 and BA 4–5, but significantly in BA3, a feature associated with higher percentage extinction in this benthic assemblage. The model predicts that reduction in prospective niche space at the seafloor is the primary cause of diversity fall. Only BA6 is removed completely, and community structures are maintained, though with reduced species richness.

DISCUSSION

Species go extinct for two main reasons: (1) population sizes are reduced to a point where chance events result in extinction, or (2) the niches species occupy disappear. Background extinction occurs all the time, but can these mechanisms be extrapolated to explain mass extinctions? At the current resolution of the available data sets, it is difficult to test for reduced population sizes. The complex patterns of extinctions during the Hirnantian mass extinction are set against a long-term decline in global diversity, from at least the mid-Ordovician. The underlying cause of this decline is currently not understood and may reflect a reduction of ecospace as a response to the gradual cooling prior to the glacial maximum. Alternately, reduction in ecospace may have been the result of the destruction of terranes and microcontinents themselves. Detailed analyses of locality-based databases suggest that the diverse habitats of the island arcs of the Iapetus Ocean were not sustained. This is particularly true for the peri-Laurentian terranes, which were accreted to the Laurentian craton during the Appalachian–Caledonian orogeny. Terrane accretion was already under way prior to and during the Hirnantian (Rasmussen and Harper, 2011a).

The first strike of the Hirnantian mass extinction has been attributed to: (1) cooling, reduced shelf areas as a result of sea-level drop, and increased ventilation of the oceans (Berry and Wilde, 1978; Brenchley et al., 1995), (2) expanding anoxia (Briggs et al., 1988; Fortey, 1989; Goodfellow and Jonasson, 1984; Hammarlund et al., 2012; Zhang et al., 2009), and (3) the amalgamation of microcontinents (Rasmussen and Harper, 2011a). For the second strike, anoxia is widely considered as the cause, as indicated by the widespread deposition of black shale following a rise in sea level (Brenchley et al., 2001; Rasmussen and Harper, 2011a; Rong and Harper, 1988). These hypotheses largely fail to explicitly link the inferred mechanism to the recognized biological explanations for extinction.

We hypothesize that coordinated extinctions occurred as a consequence of climatically induced changes in relative sea level, sea-surface temperatures, and ocean redox, which resulted in the declining availability of prospective niche space for both pelagic/nektonic and benthic species (Fig. 12).

This is a modification of the "shelf area hypothesis," which posits a relationship between species richness and habitable area; Phanerozoic species richness appears to have been in equilibrium with habitable area (Sepkoski, 1976). In the latest manifestation of this hypothesis, Finnegan et al. (2012) reported, based on sections from Laurentia, a link between tropical cooling and habitat loss. Sclafani and Holland (2013) also showed, using census

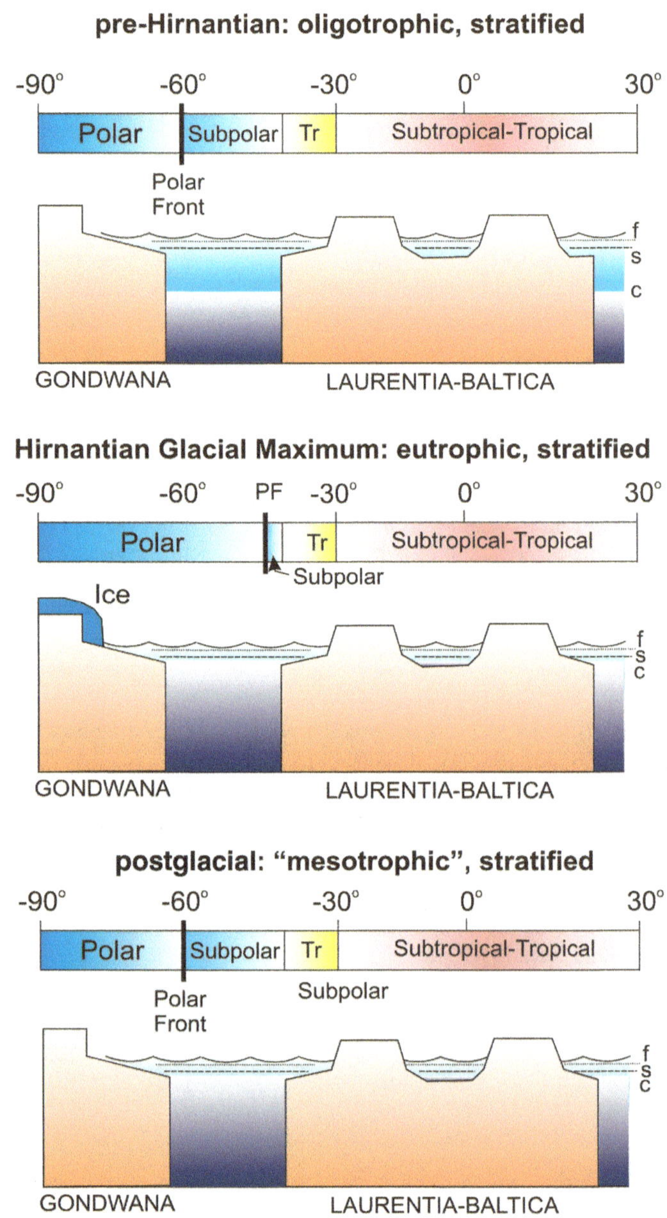

Figure 12. Model showing the combined effects of oceanographic and climatic changes during the mass extinction. Changing latitudinal temperature gradients affect the width and distribution of the planktonic provinces (polar, subpolar, transitional [Tr], subtropical-tropical). Simultaneously, oceanographic changes associated with changing sea level and ocean redox gradients affect benthic communities (f—fair weather wave base; s—storm wave base; c—chemocline). The chemocline marks the top of anoxic deep-ocean water. The combined effects of both these processes was to reduce the habitable areas/prospective niche space and hence increase competition between species, leading to extinctions where the carrying capacities of particular ecological niches were exceeded. Note: there are no continents on the North Pole to support a major ice sheet. PF—polar front.

data from Upper Ordovician strata in Laurentia, a weak positive relationship between province area, as defined by geochemical parameters, and biodiversity, though this was based on an untested assumption that early Paleozoic communities were species saturated.

During the first phase of the Hirnantian mass extinction, major eustatic sea-level fall removed significant habitable area on the continental shelves, which led to the down-shelf displacement of benthic assemblages, which in turn coincided with a rise in the chemocline and an increase in oxygen gradient, reducing the habitable space for the plankton and nekton occupying the dysoxic zone in the oceans. Within the mixed layer, declining diversity in planktonic provinces during the glacial maximum resulted from the steepening of latitudinal temperature gradients, which in turn resulted in the equator-ward spread of the low-diversity polar province and the narrowing of high-diversity provinces in the midlatitudes.

During the eustatic sea-level rise in the *N. persculptus* graptolite biozone, the potential increase of habitable shelf area was balanced by a narrowing of the shelf area occupied by BA 3–4/5 as ferruginous/euxinic water masses encroached onto the outer shelves, to close to storm wave base. During this interval, the rise of the chemocline onto the shelves reduced habitable space, coincident with a decline in diversity (Brenchley et al., 2001). The long-term ventilation of the oceans (Saltzman, 2005) and the deepening of the chemocline broke the link between glaciation and mass extinction.

A reduction in the area/volume occupied by a community, at the seafloor, within the water column, or at the sea surface, would increase competition and selection pressures, leading to extinctions where the carrying capacities of particular ecological niches were exceeded (Harper et al., 2014; Saltzman, 2005; Sheehan, 1975, 2008; Valentine, 1969). In this way, within-community extinctions would be the predicted pattern of diversity change during the Hirnantian mass extinction.

The availability of prospective niche space has implications not only for elevated extinction but postextinction recovery. In the allopatric model of speciation, incipient new species arise continually as a consequence of gene mutation and allopatry. The constraint on the fixing of new species is the availability of expanding ecospace (Erwin, 2001). In this model, recovery from the Hirnantian mass extinction could not be initiated until ocean stratification and chemistry had returned to the preglacial equilibrium state.

CONCLUSIONS

1. The Hirnantian mass extinction coincided with the glacial maximum of the early Paleozoic icehouse.
2. Existing causative hypotheses fail to provide a biological context for the Hirnantian mass extinction.
3. We hypothesize that coordinated extinctions occurred as a consequence of glacially induced changes in sea-surface temperatures and ocean oxygen stratification, which resulted in the declining availability of prospective niche space in both the pelagic and benthic realms.
4. During the first phase of the Hirnantian mass extinction, major eustatic sea-level fall coincident with a rise in the chemocline and a steepening of the water-column oxygen gradient displaced benthic communities offshore into reduced habitable areas. Declining diversity in the plankton resulted from the steepening of latitudinal temperature gradients, resulting in the equator-ward spread of the low-diversity polar province and the narrowing of high-diversity provinces in the midlatitudes.
5. Following glacial termination and during eustatic sea-level rise in the *N. persculptus* graptolite biozone, a potential increase of habitable shelf area was balanced by a narrowing of the shelf area occupied by BA 3 through 4/5 as ferruginous/euxinic water masses encroached onto the outer shelves, to close to storm wave base.
6. Together, each phase of mass extinction can be related to a loss of habitable area and a reduction in prospective niche space. Elevated density-dependent competition resulted in high levels of within-community species extinction.
7. Hirnantian mass extinction was contingent on the unique nature of the early Paleozoic oceans, with deep ocean anoxia. The effects of glaciations during the earlier Ordovician have yet to be studied in detail.
8. The progressive ventilation of the oceans during the later Paleozoic and Mesozoic meant that the biosphere would never again be subject to glacially induced mass extinction.

ACKNOWLEDGMENTS

We thank the conference conveners, who provided an opportunity to present and discuss our ideas. The Mineralogical Society provided financial support. Harper is funded by the Danish Council for Independent Research. We acknowledge the helpful comments of our referees, M. Williams and A.W. Owen, and the editor, whose work greatly improved the manuscript.

REFERENCES CITED

Adrain, J., Edgecombe, G.D., Fortey, R.A., Hammer, Ø., Laurie, J.R., McCormick, T., Owen, A.W., Waisfeld, B.G., Webby, B.D., Westrop, S.R., and Zhou, Z.-y., 2004, Trilobites, *in* Webby, B.D., Paris, F., Droser, M.L., and Percival, I.G., eds., The Great Ordovician Biodiversification Event: New York, Columbia University Press, p. 231–254.

Armstrong, H.A., 2007, On the cause of the Ordovician glaciation, *in* Williams, M., Haywood, A., and Gregory, J., eds., Deep Time Perspectives on Climate Change: Marrying the Signal from Computer Models and Biological Proxies: Micropalaeontological Society and Special Publication TMS002, p. 101–121.

Armstrong, H.A., and Coe, A.L., 1997, Deep sea sediments record the geophysiology of the end Ordovician glaciation: Journal of the Geological Society, v. 154, p. 929–934, doi:10.1144/gsjgs.154.6.0929.

Baarli, B.G., and Harper, D.A.T., 1986, Relict Ordovician brachiopod faunas in the Lower Silurian of Asker, Oslo region, Norway: Norsk Geologisk Tidsskrift, v. 66, p. 87–91.

Bambach, R.K., Knoll, A.H., and Wang, S., 2004, Origination, extinction, and mass depletions of marine diversity: Paleobiology, v. 30, no. 4, p. 522–542, doi:10.1666/0094-8373(2004)030<0522:OEAMDO>2.0.CO;2.

Bapst, D.W., Bullock, P.C., Melchin, M.J., Sheets, H.D., and Mitchell, C.E., 2012, Graptoloid diversity and disparity became decoupled during the Ordovician mass extinction: Proceedings of the National Academy of Sciences of the United States of America, v. 109, p. 3428–3433, doi:10.1073/pnas.1113870109.

Bergström, S.M., Saltzman, M.R., and Schmitz, B., 2006, First record of the Hirnantian (Upper Ordovician) delta C-13 excursion in the North American Midcontinent and its regional implications: Geological Magazine, v. 143, p. 657–678.

Bergström, S.M., Chen, X.U., Gutiérez-Marco, J.C., and Dronov, A., 2009, The new chronostratigraphic classification of the Ordovician System and its relations to major regional series and stages and to δ^{13}C chemostratigraphy: Lethaia, v. 42, no. 1, p. 97–107, doi:10.1111/j.1502-3931.2008.00136.x.

Berry, W.B.N., and Wilde, P., 1978, Progressive ventilation of the oceans—An explanation for the distribution of the Lower Paleozoic black shales: American Journal of Science, v. 278, p. 257–275, doi:10.2475/ajs.278.3.257.

Berry, W.B.N., Wilde, P., and Quinby-Hunt, M., 1990, Late Ordovician graptolite mass mortality and subsequent Early Silurian re-radiation, *in* Kauffman, E.G., and Walliser, O.H., eds., Extinction Events in Earth History: Berlin, Springer Verlag, Lecture Notes in Earth Sciences, v. 30, p. 115–123.

Bjerrum, C.J., Bendtsen, J., and Legarth, J.J.F., 2006, Modeling organic carbon burial during sea level rise with reference to the Cretaceous 1–24: Geochemistry Geophysics Geosystems, v. 7, p. 1–24, doi:10.1029/2005GC001032.

Boulila, S., Galbrun, B., Miller, K.G., Pekar, S.F., Browning, J.V., Laskar, J., and Wright, J.D., 2011, On the origin of Cenozoic and Mesozoic "third-order" eustatic sequences: Earth-Science Reviews, v. 109, p. 94–112, doi:10.1016/j.earscirev.2011.09.003.

Brenchley, P.J., and Cullen, B., 1984, The environmental distribution of associations belonging to the *Hirnantia* fauna—Evidence from North Wales and Norway, *in* Bruton, D.L., ed., Aspects of the Ordovician System: Paleontological Contributions from the University of Oslo 295, p. 113–126.

Brenchley, P.J., and Marshall, J.D., 1999, Relative timing of critical events during the Late Ordovician mass extinction—New data from Oslo: Acta Universitatis Carolinae: Geologica, v. 43, p. 187–191.

Brenchley, P.J., Marshall, J.D., Carden, G.A.F., Robertson, D.B.R., Long, D.G., Meidla, T., Hints, L., and Anderson, T.F., 1994, Bathymetric and isotopic evidence for a short-lived Late Ordovician glaciation in a greenhouse period: Geology, v. 22, p. 295–298, doi:10.1130/0091-7613(1994)022<0295:BAIEFA>2.3.CO;2.

Brenchley, P.J., Carden, G.A.F., and Marshall, J.D., 1995, Environmental changes associated with the 'first strike' of the Late Ordovician mass extinction: Modern Geology, v. 20, p. 69–82.

Brenchley, P.J., Marshall, J.D., and Underwood, C.J., 2001, Do all mass extinctions represent an ecological crisis? Evidence from the Late Ordovician: Geological Journal, v. 36, no. 329–340.

Brenchley, P.J., Marshall, J.D., Harper, D.A.T., Buttler, C.J., and Underwood, C.J., 2006, A Late Ordovician (Hirnantian) karstic surface in a submarine channel, recording glacioeustatic sea-level changes: Meifod, central Wales: Geological Journal, v. 41, p. 1–22, doi:10.1002/gj.1029.

Brett, C.E., Boucot, A.J., and Jones, B., 1993, Absolute depths of Silurian benthic assemblages: Lethaia, v. 26, no. 1, p. 25–40, doi:10.1111/j.1502-3931.1993.tb01507.x.

Brett, C.E., Hendy, A.J.W., Bartholomew, A.J., Bonelli, J.R., and McLaughlin, P.I., 2007, Response of shallow marine biotas to sea-level fluctuations: A review of faunal replacement and the process of habitat tracking: Palaios, v. 22, no. 3, p. 228–244, doi:10.2110/palo.2005.p05-028r.

Briggs, D.E.G., Fortey, R.A., and Clarkson, E.N.K., 1988, Extinction and the fossil record of the arthropods, *in* Larwood, G., ed., Systematics Association, Special Volume 34: Oxford, Clarendon Press, p. 171–209.

Colman, A., Mackenzie, F., Holland, H., Van Cappellen, P., and Ingall, E., 1997, Redox stabilization of the atmosphere and oceans and marine productivity: Science, v. 275, no. 5298, p. 406–408.

Copper, P., and Jin, J., 2012, Early Silurian (Aeronian) East Point coral patch reefs of Anticosti Island, eastern Canada: First reef recovery from the Ordovician/Silurian Mass extinction in eastern Laurentia: Geosciences, v. 2, p. 64–89, doi:10.3390/geosciences2020064.

Cooper, R.A., Fortey, R.A., and Lindholm, K., 1991, Latitudinal and depth zonation of early Ordovician graptolites: Lethaia, v. 24, no. 2, p. 199–218, doi:10.1111/j.1502-3931.1991.tb01468.x.

Dahl, T.W., Hammarlund, E.U., Anbar, A.D., Bond, D.P.G., Gill, B.C., Gordon, G.W., Knoll, A.H., Nielsen, A.T., Schovsbo, N.H., and Canfield, D.E., 2010, Devonian rise in atmospheric oxygen correlated to the radiations of terrestrial plants and large predatory fish: Proceedings of the National Academy of Sciences of the United States of America, v. 107, p. 17,911–17,915, doi:10.1073/pnas.1011287107.

de Boer, A.M., Sigman, D.M., Toggweiler, J.R., and Russell, J.L., 2007, Effect of global ocean temperature change on deep ocean ventilation: Paleoceanography, v. 22, p. PA2210, doi:10.1029/2005PA001242.

Droser, M.L., Bottjer, D.J., and Sheehan, P.M., 1997, Evaluating the ecological architecture of major events in the Phanerozoic history of marine invertebrate life: Geology, v. 25, p. 167–170, doi:10.1130/0091-7613(1997)025<0167:ETEAOM>2.3.CO;2.

Droser, M.L., Bottjer, D.J., Sheehan, P.M., and McGhee, G.R., 2000, Decoupling of taxonomic and ecologic severity of Phanerozoic marine mass extinctions: Geology, v. 28, no. 8, p. 675–678, doi:10.1130/0091-7613(2000)28<675:DOTAES>2.0.CO;2.

Erwin, D.H., 2001, Lessons from the past: Evolutionary impacts of mass extinctions: Proceedings of the National Academy of Sciences of the United States of America, v. 98, p. 5399–5403, doi:10.1073/pnas.091092698.

Finnegan, S., Bergmann, K., Eiler, J.M., Jones, D.S., Fike, D.A., Eisenman, L., Hughes, N.C., Tipati, A.K., and Fischer, W.W., 2011, The magnitude and duration of the Late Ordovician–Early Silurian glaciation: Science, v. 331, p. 903–906, doi:10.1126/science.1200803.

Finnegan, S., Heim, N.A., Peters, S.E., and Fischer, W.W., 2012, Climate change and the selective signature of the Late Ordovician mass extinction: Proceedings of the National Academy of Sciences of the United States of America, v. 109, no. 18, p. 6829–6834, doi:10.1073/pnas.1117039109.

Finney, S.C., and Berry, W.B.N., 1997, New perspectives on graptolite distributions and their use as indicators of platform margin dynamics: Geology, v. 25, no. 10, p. 919–922, doi:10.1130/0091-7613(1997)025<0919:NPOGDA>2.3.CO;2.

Finney, S.C., Berry, W.B.N., Cooper, J.D., Ripperdan, R.L., Sweet, W.C., Jacobson, S.R., Soufiane, A., Achab, A., and Noble, P.J., 1999, Late Ordovician mass extinction: A new perspective from stratigraphic sections in central Nevada: Geology, v. 27, p. 215–218, doi:10.1130/0091-7613(1999)027<0215:LOMEAN>2.3.CO;2.

Fortey, R.A., 1989, There are extinctions and extinctions—Examples from the Lower Palaeozoic: Philosophical Transactions of the Royal Society of London, ser. B, Biological Sciences, v. 325, p. 327–355, doi:10.1098/rstb.1989.0092.

Fortey, R.A., Harper, D.A.T., Ingham, J.K., Owen, A.W., and Rushton, A.W.A., 1995, A revision of Ordovician series and stages from the historical type area: Geological Magazine, v. 132, p. 15–30, doi:10.1017/S0016756800011390.

Godderis, Y., Francois, L.M., and Veizer, J., 2001, The early Paleozoic carbon cycle: Earth and Planetary Science Letters, v. 190, p. 181–196, doi:10.1016/S0012-821X(01)00377-6.

Goodfellow, W.D., and Jonasson, I.R., 1984, Ocean stagnation and ventilation defined by δ^{34}S secular trends in pyrite and barite, Selwyn Basin, Yukon: Geology, v. 12, p. 583–586, doi:10.1130/0091-7613(1984)12<583:OSAVDB>2.0.CO;2.

Hammarlund, E.U., Dahl, T.W., Harper, D.A.T., Bond, D.P.G., Nielsen, A.T., Bjerrum, C.J., Schovsbo, N.H., Schonlaub, H.P., Zalasiewicz, J.A., and Canfield, D.E., 2012, A sulfidic driver for the end-Ordovician mass extinction: Earth and Planetary Science Letters, v. 331–332, p. 128–139, doi:10.1016/j.epsl.2012.02.024.

Haq, B.U., and Schulter, S.R., 2008, A chronology of Paleozoic sea-level changes: Science, v. 322, p. 64–68, doi:10.1126/science.1161648.

Harper, D.A.T., 1981, The stratigraphy and faunas of the Upper Ordovician High Mains Formation of the Girvan district: Scottish Journal of Geology, v. 17, p. 247–255, doi:10.1144/sjg17040247.

Harper, D.A.T., and MacNiocaill, C., 2002, Early Ordovician rhynchonelliformean brachiopod biodiversity: Comparing some platforms, margins and intra-oceanic sites around the Iapetus Ocean, *in* Crame, J.A., and Owen, A.W., eds., Palaeobiogeography and Biodiversity Change: The Ordovician and Mesozoic-Cenozoic Radiations: Geological Society of London Special Publication 194, p. 25–34, doi:10.1144/GSL.SP.2002.194.01.03.

Harper, D.A.T., and Rong, J., 2008, Completeness of the Hirnantian brachiopod record: Spatial heterogeneity through the end Ordovician extinction event: Lethaia, v. 41, p. 195–197, doi:10.1111/j.1502-3931.2008.00098.x.

Harper, D.A.T., Hammarlund, E.U., and Rasmussen, C.M.Ø., 2014, End Ordovician extinctions: A coincidence of causes: Gondwana Research, v. 25, p. 1294–1307.

Holmden, C., Panchuk, K., and Finney, S.C., 2012, Tightly coupled records of Ca and C isotope changes during the Hirnantian glaciation event in an epeiric sea setting: Geochimica et Cosmochimica Acta, v. 98, p. 94–106, doi:10.1016/j.gca.2012.09.017.

Jablonski, D., 1991, Extinctions: A paleontological perspective: Science, v. 253, p. 754–757, doi:10.1126/science.253.5021.754.

Kucera, M., 2007, Planktonic foraminifera as traces of past oceanic environments in proxies in late Cenozoic palaeoceanography, *in* Hillaire-Marce, C., and De Vernal, A., eds., Developments in Marine Geology, Volume 1: Amsterdam, Elsevier, p. 213–262.

Kump, L.R., Arthur, M.A., Patzkowsky, M.E., Gibbs, M.T., Pinkus, D.S., and Sheehan, P.M., 1999, A weathering hypothesis for glaciation at high atmospheric *p*CO2 during the Late Ordovician: Palaeogeography, Palaeoclimatology, Palaeoecology, v. 152, p. 173–187, doi:10.1016/S0031-0182(99)00046-2.

LaPorte, D.F., Holmden, C., Patterson, W.P., Loxton, J.D., Melchin, M.J., Mitchell, C.E., Finney, S.C., and Sheets, H.D., 2009, Local and global perspectives on carbon and nitrogen cycling during the Hirnantian glaciation: Palaeogeography, Palaeoclimatology, Palaeoecology, v. 276, p. 182–195, doi:10.1016/j.palaeo.2009.03.009.

Loi, A., Ghienne, J.F., Dabard, M.P., Paris, F., Botquelen, A., Christ, N., Elaouad-Debbaj, Z., Gorini, A., Vidal, M., Videt, B., and Destombes, J., 2010, The Late Ordovician glacio-eustatic record from a high-latitude storm-dominated shelf succession: The Bou Ingarf section (Anti-Atlas, southern Morocco): Palaeogeography, Palaeoclimatology, Palaeoecology, v. 296, no. 3–4, p. 332–358, doi:10.1016/j.palaeo.2010.01.018.

McGhee, G.R., Jr., Clapham, M.E., Sheehan, P.M., Bottjer, D.J., and Droser, M.L., 2013, A new ecological-severity ranking of major Phanerozoic biodiversity crises: Palaeogeography, Palaeoclimatology, Palaeoecology, v. 370, p. 260–270, doi:10.1016/j.palaeo.2012.12.019.

Munnecke, A., Calner, M., Harper, D.A.T., and Servais, T., 2010, Ordovician and Silurian seawater chemistry and climate: A synopsis: Palaeogeography, Palaeoclimatology, Palaeoecology, v. 296, p. 389–413, doi:10.1016/j.palaeo.2010.08.001.

Nardin, E., Goddéris, Y., Donnadieu, Y., Hir, G.L., Blakey, R.C., Pucéat, E., and Aretz, M., 2011, Modeling the early Paleozoic long-term climatic trend: Geological Society of America Bulletin, v. 123, no. 5–6, p. 1181–1192, doi:10.1130/B30364.1.

Owen, A.W., 1986, The uppermost Ordovician (Hirnantian) trilobites of Girvan, SW Scotland, with a review of coeval trilobite faunas: Transactions of the Royal Society of Edinburgh–Earth Sciences, v. 77, p. 231–239, doi:10.1017/S0263593300010865.

Owen, A.W., Harper, D.A.T., and Heath, R.A., 2008, A route to recovery: The Early Silurian shallow-water shelly fauna in the northern Oslo Basin: Lethaia, v. 41, no. 2, p. 173–184, doi:10.1111/j.1502-3931.2008.00109.x.

Potter, A.W., and Boucot, A.J., 1992, Middle and Late Ordovician brachiopod benthic assemblages of North America, *in* Webby, B.D., and Laurie, J.R., eds., Global Perspectives on Ordovician Geology: Rotterdam, Netherlands, Balkema, p. 307–323.

Rasmussen, C.M.Ø., and Harper, D.A.T., 2011a, Did the amalgamation of continents drive the end Ordovician mass extinctions?: Palaeogeography, Palaeoclimatology, Palaeoecology, v. 311, p. 48–62.

Rasmussen, C.M.Ø., and Harper, D.A.T., 2011b, Interrogation of distributional data for the end-Ordovician crisis interval: Where did the disaster strike?: Geological Journal, v. 46, p. 478–500.

Rong, J., 1979, The Hirnantia fauna of China with comments on the Ordovician-Silurian boundary: Journal of Stratigraphy, v. 3, p. 1–29.

Rong, J., and Harper, D.A.T., 1988, A global synthesis of the latest Ordovician Hirnantian brachiopod faunas: Transactions of the Royal Society of Edinburgh–Earth Sciences, v. 79, p. 383–402, doi:10.1017/S026359330001436X.

Rong, J., and Harper, D.A.T., 1999, Brachiopod survival and recovery from the latest Ordovician mass extinctions in South China: Geological Journal, v. 34, no. 4, p. 321–348, doi:10.1002/(SICI)1099-1034(199911/12)34:4<321::AID-GJ809>3.0.CO;2-I.

Rong, J.-Y., and Zhan, R.-B., 2006, Surviving the end-Ordovician extinctions: Evidence from the earliest Silurian brachiopods of northeastern Jiangxi and western Zhejiang Provinces, East China: Lethaia, v. 39, no. 1, p. 39–48, doi:10.1080/00241160500443164.

Rong, J.-Y., Boucot, A.J., Harper, D.A.T., Zhan, R., and Neuman, R.B., 2006, Global analyses of brachiopods faunas through the Ordovician and Silurian transition: Reducing the role of the Lazarus effect: Canadian Journal of Earth Sciences, v. 43, p. 23–39, doi:10.1139/e05-089.

Saltzman, M.R., 2005, Phosphorus, nitrogen and the redox evolution of the Paleozoic oceans: Geology, v. 33, p. 573–576, doi:10.1130/G21535.1.

Sclafani, J.A., and Holland, S.M., 2013, The species-area relationship in the Late Ordovician: A test using neutral theory: Diversity, v. 5, p. 240–262, doi:10.3390/d5020240.

Sepkoski, J.J., Jr., 1976, Species diversity in the Phanerozoic: Species-area effects: Paleobiology, v. 2, no. 4, p. 298–303.

Sepkoski, J.J., Jr., 1981, A factor analysis description of the marine fossil record: Paleobiology, v. 7, no. 36–53.

Sepkoski, J.J., Jr., 1996, Patterns of Phanerozoic extinction: A perspective from global databases, *in* Walliser, O.H., ed., Global Events and Event Stratigraphy in the Phanerozoic: Berlin, Springer, p. 35–51.

Servais, T., and Owen, A.W., 2010, Early Palaeozoic palaeoenvironments and the 'explosion' of diversity of marine species, genera and families: Palaeogeography, Palaeoclimatology, Palaeoecology, v. 294, no. 3–4, p. 95–98, doi:10.1016/j.palaeo.2010.05.044.

Servais, T., Lehnert, O., Li, J., Miullins, G.L., Munnecke, A., Nutzel, A., and Vecoli, M., 2008, The Ordovician biodiversification: Revolution in the oceanic trophic chain: Lethaia, v. 41, p. 99–109, doi:10.1111/j.1502-3931.2008.00115.x.

Servais, T., Owen, A.W., Harper, D.A.T., Kröger, B., and Munnecke, A., 2010, The Great Ordovician Biodiversification Event (GOBE): The palaeoecological dimension: Palaeogeography, Palaeoclimatology, Palaeoecology, v. 294, no. 3–4, p. 99–119.

Sheehan, P.M., 1975, Brachiopod synecology in a time of crisis (Late Ordovician–Early Silurian): Paleobiology, v. 1, p. 205–212.

Sheehan, P.M., 2001, History of marine biodiversity: Geological Journal, v. 36, p. 231–249.

Sheehan, P.M., 2008, Did incumbency play a role in maintaining boundaries between Late Ordovician brachiopod realms?: Lethaia, v. 41, p. 147–153, doi:10.1111/j.1502-3931.2008.00116.x.

Sheehan, P.M., and Coorough, P.J., 1990, Brachiopod zoogeography across the Ordovician-Silurian extinction event, *in* McKerrow, W.S., and Scotese, C.R., eds., Palaeozoic Palaeogeography and Biogeography: Geological Society of London Memoir 12, p. 181–187.

Sigman, D.M., Jaccard, S.L., and Haug, G.H., 2004, Polar ocean stratification in a cold climate: Nature, v. 428, p. 59–63, doi:10.1038/nature02357.

Temple, J.T., 1965, Upper Ordovician brachiopods from Poland and Britain: Acta Palaeontologica Polonica, v. 10, p. 379–427.

Toggweiler, J.R., and Russell, J., 2008, Ocean circulation in a warming climate: Nature, v. 451, p. 286–288, doi:10.1038/nature06590.

Trotter, J.A., Williams, I.S., Barnes, C.R., Lecuyer, C., and Nicoll, R.S., 2008, Did cooling oceans trigger Ordovician biodiversification? Evidence from conodont thermometry: Science, v. 321, p. 550–554, doi:10.1126/science.1155814.

Turner, B.R., Armstrong, H.A., and Holt, P., 2011, Visions of ice sheets in the Early Ordovician greenhouse world: Evidence from the Peninsula Formation, Cape Peninsula, South Africa: Sedimentary Geology, v. 236, p. 226–238, doi:10.1016/j.sedgeo.2011.01.009.

Turner, B.R., Armstrong, H.A., Wilson, C.R., and Makhlouf, I.M., 2012, High frequency eustatic sea-level changes during the Middle to early Late Ordovician of southern Jordan: Indirect evidence for a Darriwilian ice age in Gondwana: Sedimentary Geology, v. 251–252, p. 34–48, doi:10.1016/j.sedgeo.2012.01.002.

Underwood, C.J., Crowley, S.F., Marshall, J.D., and Brenchley, P.J., 1997, High resolution carbon isotope stratigraphy of the basal Silurian stratotype (Dob's Linn, Scotland) and its global correlation: Journal of the Geological Society, v. 154, p. 709–718, doi:10.1144/gsjgs.154.4.0709.

Valentine, J.W., 1969, Patterns of taxonomic and ecological structure of the shelf benthos during Phanerozoic time: Palaeontology, v. 12, p. 684–709.

Valentine, J.W., and Jablonski, D., 1991, Biotic effects of sea level change: The Pleistocene test: Journal of Geophysical Research–Solid Earth, v. 96, no. B4, p. 6873–6878, doi:10.1029/90JB00602.

Vandenbroucke, T.R.A., Armstrong, H.A., Williams, M., Zalasiewicz, J.A., and Sabbe, K., 2009, Ground-truthing Late Ordovician climate models using the palaeobiology of graptolites: Palaeoceanography, v. 24, p. PA4202, doi:10.1029/2008PA001720.

Vandenbroucke, T.R.A., Armstrong, H.A., Williams, M., Paris, F., Sabbe, K., Zalasiewicz, J.A., Nõlvak, J., and Verniers, J., 2010a, Epipelagic chitinozoan biotopes map a steep latitudinal temperature gradient for earliest Late Ordovician seas: Implications for a cooling Late Ordovician climate: Palaeogeography, Palaeoclimatology, Palaeoecology, v. 294, no. 3–4, p. 202–219, doi:10.1016/j.palaeo.2009.11.026.

Vandenbroucke, T.R.A., Armstrong, H.A., Williams, M., Paris, F., Zalasiewicz, J.A., Sabbe, K., Nõlvak, J., Challands, T.J., Verniers, J., and Servais, T., 2010b, Polar front shift and atmospheric CO_2 during the glacial maximum of the early Paleozoic icehouse: Proceedings of the National Academy of Sciences of the United States of America, v. 107, no. 34, p. 14,983–14,986, doi:10.1073/pnas.1003220107.

Wallmann, K., 2003, Feedbacks between oceanic redox states and marine productivity: A model perspective focused on benthic phosphorus cycling: Global Biogeochemical Cycles, v. 17, p. 1084, doi:10.1029/2002GB001968.

Webby, B.D., Paris, F., Droser, M.L., and Percival, I.G., 2004, The Great Ordovician Biodiversification Event: New York, Columbia University Press, 496 p.

Yan, D., Chen, D., Wang, Q., and Wang, J., 2009, Geochemical changes across the Ordovician-Silurian transition on the Yangtze Platform, South China: Science in China, ser. D: Earth Science, v. 52, no. 1, p. 38–54.

Zalasiewicz, J.A., Taylor, L., Rushton, A.W.A., Loydell, D.K., Rickards, R.B., and Williams, M., 2009, Graptolites in British stratigraphy: Geological Magazine, v. 146, p. 785–850, doi:10.1017/S0016756809990434.

Zhang, T.G., Shen, Y.N., Zhan, R.B., Shen, S.Z., and Chen, X., 2009, Large perturbations of the carbon and sulfur cycle associated with the Late Ordovician mass extinction in South China: Geology, v. 37, p. 299–302, doi:10.1130/G25477A.1.

Manuscript Accepted by the Society 31 January 2014

Printed in the USA

The Geological Society of America
Special Paper 505
2014

A review of the embedded time scales of flood basalt volcanism with special emphasis on dramatically short magmatic pulses

Vincent Courtillot
Frédéric Fluteau
Equipe de Paléomagnétisme, Institut de Physique du Globe, UMR 7154, Sorbonne Paris Cité, F-75005, Paris, France, and
Sciences de la Terre, de l'Environnement et des Planètes, Université Paris Diderot, Sorbonne Paris Cité, F-75013, Paris, France

ABSTRACT

There are at least a dozen Phanerozoic continental flood basalts and oceanic plateaus (large igneous provinces) that roughly obey a "rule of two times one million" (volume of extruded lava of one million cubic kilometers and duration of volcanic activity ~1 m.y.). The correlation between large igneous province ages and mass extinctions (and oceanic anoxia events) is excellent, but quantitative scenarios are still wanting. We hypothesize that the temporal sequences of extrusions determine the severity of extinction: Volcanic pulses separated by thousands of years allow the ocean-atmosphere system time to recover, whereas large volcanic pulses occurring in a shorter sequence may result in a runaway effect and cause a mass extinction. Detailed flow-by-flow magnetic stratigraphies of thick sections have identified directional groups (sequences of superimposed lava flows with the same paleomagnetic direction that cooled in a time too short to record secular variation). With help of this simple tool, many single eruptive events with a volume larger than 1000 km^3, some in excess of 10,000 km^3, emplaced in possibly less than a decade, have been identified. We review this evidence, mainly for the following flood basalt provinces: Columbia, Brito-Arctic, Deccan, Karoo-Ferrar, Central Atlantic magmatic province, and Siberian Traps. Large igneous province volcanism occurs in a highly discontinuous way, on embedded time scales, on the order of 10 m.y., 1 m.y., 100 k.y., and 10 yr. This provides constraints for models of plume-lithosphere interaction and magma production. A next step is to model the consequences of massive injection of gases that can be derived from these time and duration estimates. Early attempts are reviewed in a companion paper in this volume.

Courtillot, V., and Fluteau, F., 2014, A review of the embedded time scales of flood basalt volcanism with special emphasis on dramatically short magmatic pulses, *in* Keller, G., and Kerr, A., eds., Volcanism, Impacts, and Mass Extinctions: Causes and Effects: Geological Society of America Special Paper 505, p. 301–317, doi:10.1130/2014.2505(15). For permission to copy, contact editing@geosociety.org.

INTRODUCTION

Twenty-five years ago, Rampino and Stothers (1988) showed that initiation of volcanism in major continental flood basalt provinces correlated well with the estimated dates of mass extinctions of marine organisms over the past 250 m.y. Subsequent analyses by various groups have since then increasingly confirmed this correlation and its significance (i.e., Courtillot and Renne, 2003; White and Saunders, 2005). Based on joint consideration of paleomagnetic, paleontological, and geochronological evidence, Courtillot et al. (1986) suggested that the 1-m.y.-long volcanic activity of the Deccan flood basalts straddled the Cretaceous-Tertiary boundary crisis. Since then, temporal coincidences between large igneous provinces and mass extinctions have been extended back to 360 Ma (Courtillot and Renne, 2003; see also Courtillot, 1994, 1999), confirming the quality and high degree of significance of this correlation. More recently, Kravchinsky (2012) extended the linkage possibly back to 550 Ma, though with large remaining uncertainties in exact chronology.

A significant amount of new chronological data has appeared in the literature in the past decade, and a new compilation and review would be useful. We do not attempt this in the present paper, which has another main goal as explained below. However, we can confidently summarize the following conclusions that are now generally accepted: Most of the Phanerozoic, major biotic depletions that can be labeled as "mass extinctions" (e.g., Sepkoski, 1990; Hallam and Wignall, 1999; Bambach et al., 2004) have now been connected to a continental flood basalt emplacement event or to an oceanic equivalent, all jointly designated as large igneous provinces. In some cases, these provinces were not even known or seriously considered as recently as the 1980s, e.g., the Central Atlantic magmatic province or the Emeishan Traps.

In the case of the Central Atlantic magmatic province, early studies (Olsen et al., 1996, 2002; Olsen, 1997) rejected the idea of a connection with the end-Triassic mass extinction, though Palfy (2003) and Courtillot and Renne (2003) argued otherwise. The latter view has now been largely vindicated (e.g., Blackburn et al., 2013). The case of the Emeishan Traps is interesting in another respect: It demonstrates the predictive power of the "correlated trap-extinction" hypothesis. Stanley and Yang (1994) pointed out that the Permian-Triassic extinction actually consisted of two separate events, one at the Permian-Triassic boundary itself (clearly and early associated with the Siberian Traps) and an earlier event at the Guadalupian-Tatarian boundary. Holser and Magaritz (1987) had emphasized a double anomaly in carbon isotopes and in sea level at precisely these two horizons. This led Courtillot et al. (1999) and Hallam and Wignall (1999) to propose that the recently rediscovered Emeishan Traps should be associated with the Guadalupian-Tatarian boundary and therefore have the same age. This has been supported by U-Pb ages on zircons from Zhou et al. (2002), and many other papers since (e.g., Wignall et al., 2009). Ali et al. (2002) made the connection but suggested that the eruptions were actually slightly older than the end-Guadalupian. The timing and duration of the felsic magmatism in the Emeishan large igneous province, however, are still the subject of debate (see Emeishan Flood Basalt Province section). A more recent candidate as a trap/extinction pair is the Viluy volcanics and Frasnian-Famennian and end-Devonian extinctions (Kravchinsky et al., 2002; Courtillot et al., 2010; Ricci et al., 2013). Absolute dating (K-Ar and $^{40}Ar/^{39}Ar$) is in progress and supportive of the hypothesis, though still with significant uncertainties. The correlation is now such that we can reasonably confidently, following Occam's razor, suggest that an as-yet-undiscovered flood basalt could be associated with the end-Ordovician glaciation and extinction (see also Courtillot and Olson [2007], who link magnetic superchrons with killer plumes, the end-Ordovician being one example). This could be the topic of a joint international effort: Are remnants of a 440 Ma large igneous province to be found in an orogen or allochthonous terranes, or has it been subducted and lost?

Altogether, observational and laboratory evidence is now overwhelming that all known Phanerozoic large igneous provinces were huge basaltic edifices with a typical original size of extruded material on the order of one million cubic kilometers (within a factor 10 at most) and a duration of emplacement that could have spanned several million years but in which most of the activity was confined within one or two "megapulses" or "phases" of 1 m.y. duration or less. This can for instance be accounted for with models of thermochemical mantle plumes (e.g., Lin and van Keken, 2005; see also Sobolev et al., 2012).

Despite this evidence, the possible effects of continental flood basalts on the environment and biota have been repeatedly challenged over the past decades (e.g., Schulte et al., 2010), largely because a duration of emplacement on the order of 1 m.y. or longer implied instantaneous volcanic fluxes that were too small, on the order of 1 km^3/yr. In reality, the total duration of emplacement represents the interval of time between the ages of the first and last lava flow and does not reflect the actual tempo of emplacement (Courtillot and Fluteau, 2010). Thordarson and Self (1998) estimated that a single eruptive event, such as the Roza flow in the Columbia River flood basalt province (>1000 km^3), could have been emplaced in a few years or even months. This implies effusive rates in excess of 100 km^3/yr, which is at least one order of magnitude larger than in the 1783 Laki eruption, possibly the largest balsaltic eruption in the past millennium (e.g., Thordarson and Self, 2003; Chenet et al., 2005).

Stothers et al. (1986) calculated the height of eruptive columns as a function of eruption flux rate; in the case of the Roza flow, they found that the fire fountain would have exceeded 1.5 km in height and would have allowed convective transport of volatiles up to the stratosphere. Even in the absence of explosive volcanism or of fire fountains, the heat released over the huge surface tops of trap lava flows would have induced penetrative convection, a powerful mechanism that will inject volcanic gases into the stratosphere (Kaminski et al., 2011). The study of recent volcanic eruptions has shown that massive injection of volcanic sulfur into the stratosphere can lead to regional, hemispheric, and probably even global surface cooling (Robock, 2000).

"Normal" (e.g., historical) volcanic eruptions also release carbon dioxide, but in amounts too small to have a major impact on climate. In the case of trap volcanism (i.e., on volume and time scales that have not occurred since 15 Ma), the total amounts of sulfur, carbon dioxide, and chlorine released to the atmosphere could significantly affect the climate system (see f.i. Dessert et al., 2001, for previously disregarded effects on the longer term, 1 m.y., time scale).

All this is not sufficient to understand why different large igneous provinces have had such different environmental and biological consequences. Differences might arise from height, size, and duration of the eruptive column, composition of erupted gases (e.g., carbon dioxide, sulfur, chlorine, etc.), nature of erupted ashes, geochemical composition of intruded crust, latitude of eruption, geography of continents and oceans, and pre-eruption composition and state of the atmosphere. None of these allows us to differentiate unambiguously the consequences of one trap from another.

Our working hypothesis is that the tempo of outpouring of huge amounts of magma over short durations is the key feature of trap volcanism. The study of huge past eruptions, such as Toba (ca. 74 ka) and Yellowstone (2.1 Ma, 1.3 Ma, and 0.64 Ma), teaches us that a single eruptive event can release a few hundred megatons to a few gigatons of SO_2 into the atmosphere and indeed cause a global surface cooling of 3–4 °C, which can disrupt the climate system for a few years (Timmreck, 2012). The amplitude of cooling does not vary linearly with the amount of sulfur injected, because of limiting effects such as the increase in aerosol size and radiation transmission (Timmreck, 2012). Using a three-dimensional (3-D) atmosphere–ocean–land-surface–marine biogeochemical coupled model, Segschneider et al. (2012) showed that supereruptions, such as those in Yellowstone 1–2 m.y. ago, can destabilize the carbon cycle. Because Yellowstone is located at midlatitudes, a globally asymmetric climate change with severe cooling in the Northern Hemisphere but only a minor response in the Southern Hemisphere is favored. The climate response leads to a destabilization of the carbon cycle (mainly the terrestrial cycle), leading to a decrease of atmospheric pCO_2 by ~3–7 ppm; a single eruptive event of "normal" (historical) size would be insufficient to produce severe climate and environmental perturbations.

However, in large igneous provinces, where huge, numerous lava flows can be successively emplaced in only a decade or so, we hypothesized (e.g., Chenet et al., 2008, and papers referenced therein) that the details of the time sequence of trap flows may be the key factor that determines the severity of environmental changes and subsequent extinctions. Indeed, the amounts of sulfur and carbon dioxide released by trap volcanism are many orders of magnitude larger than the supereruptions at Toba and Yellowstone. In the case of the Deccan Traps, Chenet et al. (2009) estimated that the total amount of SO_2 would have been between 6500 and 17,000 Gt, and for CO_2, 15,000–35,000 Gt.

In addition, the release of gases due to massive intrusions of magma through sediments could significantly increase the burden of pollution. The gases released by such contact metamorphism depend on the nature of the sedimentary rocks: gypsum and anhydrite in evaporites and sulfide-rich sediments release SO_2, limestone and coal release CO_2, and organic carbon–rich shales release CH_4 (Ganino and Arndt, 2009). The amounts of such sediment-derived gases could be even larger than the amounts of magmatic gases. However, it is difficult to estimate the exact amounts of these gases because of the generally rather limited knowledge of the volcanic intrusive systems in large igneous provinces.

A central idea of this paper is that for the same numbers, sizes, and total volumes of flows, the response of the coupled ocean-atmosphere system would be very different if flows erupted at intervals longer than the time needed for the ocean to re-equilibrate or in a faster sequence, leading to runaway effects. This requires a temporal resolution that is unattainable with current isotopic/geochronologic methods (despite major improvements in the past decade).

We have attempted to attain a resolution on the order of a decade to a century, using geomagnetic paleosecular variation recorded by flow magnetization as a tool. Indeed, the magnetic field at Earth's surface varies considerably, both in direction and intensity, over time scales ranging from years to millennia and beyond. These variations are irregular in time and space, due to the evolution of the dipole and nondipole components of the geomagnetic field. Different methods have been developed to reconstruct paleosecular variation.

Historical measurements from land-based stations and aboard ships yield a reasonable knowledge of paleosecular variation over the past four centuries. Before the seventeenth century, archeomagnetism provides a record of paleosecular variation: It is based on analyses of thermoremanent magnetization fossilized in situ by materials dated by archaeologists (e.g., kilns and potteries). Over the past three millennia, magnetic directions have varied by as much as 50° in declination and 20° in inclination (Fig. 1; Gallet et al., 2002). The distribution of secular variation over this period is log-normal, with a strong mode at 2°/century. Paleosecular variation can also be reconstructed 10,000 yr back and more, based on paleomagnetic results from volcanics (Hagstrum and Champion, 2002) and lacustrine sediments (Lund, 1996). These two methods yield similar results: While ages and magnetic directions are more reliable in volcanic rocks than in lacustrine sediments, details of paleosecular variation are not recorded because of the age distribution of sampled flows. Using the smoothed Holocene virtual geomagnetic poles (VGPs) from volcanic rocks from Hagstrum and Champion (2002), the distribution of secular variation over the past 14,000 yr has a Gaussian shape with mean at 7°/century.

As first noted by Mankinen et al. (1985), successive flows in some cases yield statistically identical magnetic remanence directions, i.e., they occur in groups that have recorded the same field direction. Samples from such flows should be averaged together, with the resulting directions being assigned to a "directional group" (see sample illustration in Fig. 2). Strict statistical

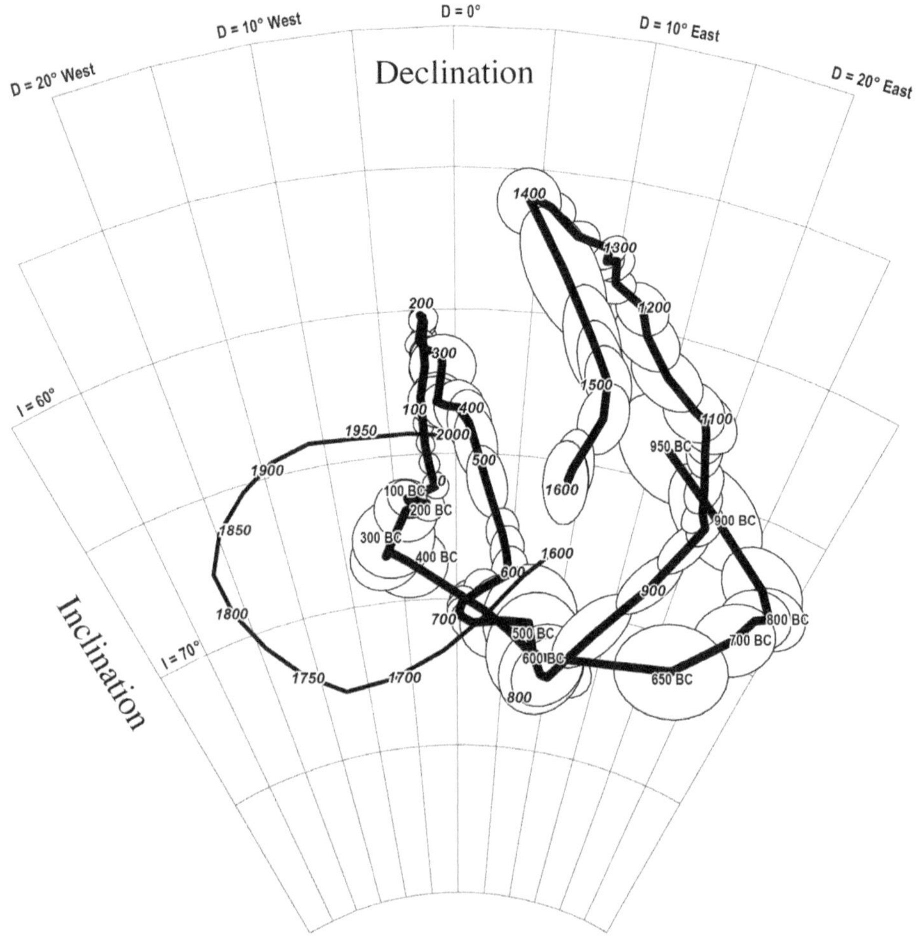

Figure 1. Directional evolution of Earth's magnetic field in France during the past three millennia (redrawn after Gallet et al., 2002). The thin black line represents the directional geomagnetic variations over the past four centuries measured from land-based stations. The thick black line is deduced from archeomagnetic data from France. A 160-yr-long sliding window is applied between 1000 B.C. and 50 B.C., and an 80-yr-long sliding window is applied between A.D. 0 and A.D. 1600.

criteria can be used to determine whether successive lava flows have indeed recorded the same magnetization direction. As noted by Mankinen et al. (1985, p. 10400), "if the mean magnetization directions of two or more flows showed no systematic trends and their ovals of 95% confidence overlapped, they were considered to record the same field direction. If there was an overlap of confidence ovals but the mean direction of one flow seemed to be progressing toward the next distinct directional group, it was considered to be different and was accorded a directional group number of its own. In most cases, the choices to be made are obvious, and few differences would occur if someone else were to make the decisions." Mankinen et al. (1985) applied this method to lavas from the Columbia flood basalt province, the smallest and youngest of all continental flood basalts, though they were not primarily interested in the dynamic aspects of volcanism. Riisager et al. (2003) were next to use the directional group method: They established a magnetic stratigraphy of the West Greenland continental flood basalt that included directional groups.

Following this work, Knight et al. (2004) and Chenet et al. (2008) applied the same method for the Central Atlantic magmatic province, and the Deccan Traps, respectively. Chenet et al. (2008, 2009) not only used the concept of directional groups to evaluate the cooling time (on the order of a decade to a century) of what were found to be huge flow fields, but also the time scales of magnetic reversals and secular variation, as documented by paleomagnetism and archeomagnetism (on the order of centuries to a few millennia; e.g., Valet et al., 2012; Gallet et al., 2002) and so defined a relative chronometer of the volcanic activity for large igneous provinces. Since then, this approach has been used on other traps (Karoo-Ferrar Traps—Moulin et al., 2011; Siberian Traps—Pavlov et al., 2011) and should be applied to all continental flood basalts (if possible) to more accurately assess their eruptive sequences. The present paper attempts to summarize the current main findings of the "directional group and paleosecular variation method" from the youngest to the older large igneous provinces.

COLUMBIA RIVER FLOOD BASALT PROVINCE

The remarkably thick (up to 1000 m) and numerous successions of lava flows (over 200) from the Steens Mountain sections of the Columbia River continental flood basalt have been the focus of a large number of breakthrough papers, mainly in relation to detailed recordings of the single reverse (R) to normal (N) magnetic field reversal found in the province (Mankinen et al., 1987). More recently, Jarboe et al. (2008) demonstrated

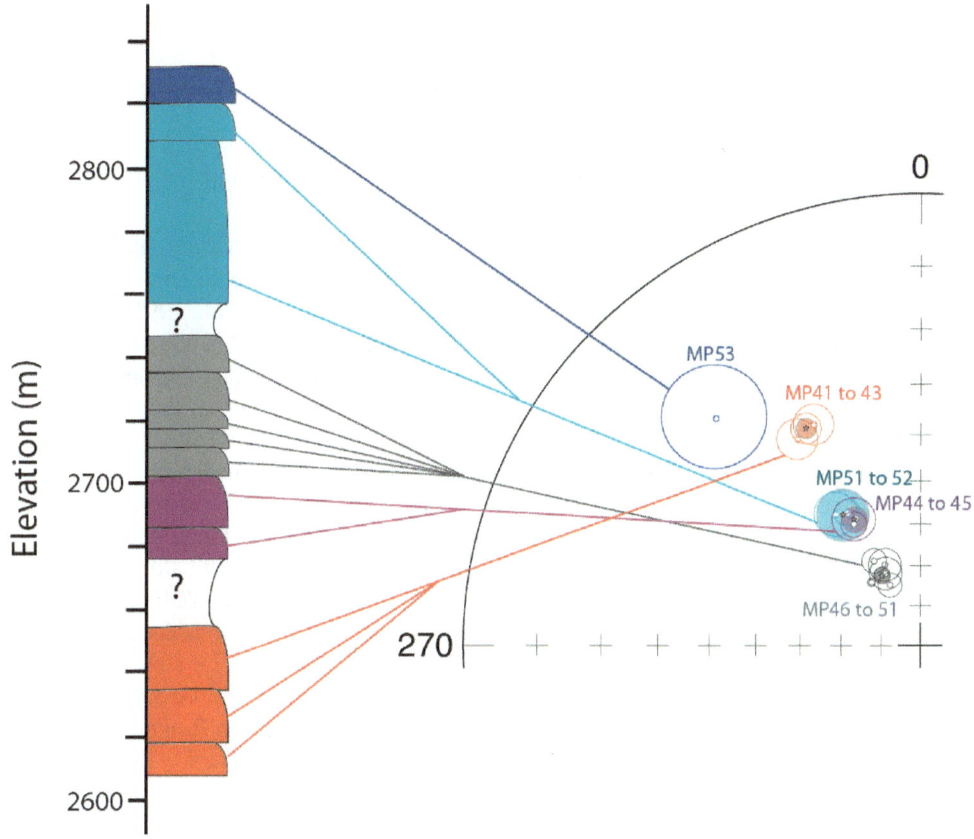

Figure 2. An example of paleomagnetic directional groups (DG) in a Karoo, Lesotho, section (redrawn after Moulin, 2011).

that the Steens basalts were all emplaced within a few hundred thousand years at ca. 16.7 Ma. The four sections they studied were erupted in "short bursts" of 1–3 k.y. within ~300 k.y. of the Steens reversal, as revealed by the lack of averaging and small loops of secular variation (13 directional groups and 37 individual flow directions). More isolated reverse polarity lavas were erupted within 1 m.y. of the reversal. Jarboe et al. (2008) apparently could not correlate directions from one section to another and did not use the paleomagnetic results to analyze volumes and dynamics of eruptions.

Barry et al. (2010) obtained new $^{40}Ar/^{39}Ar$ dates on the Grande Ronde Basalt, the most voluminous pulse of the Columbia River volcanism (15,000 km^3): They found that a series of ~110 flow fields had erupted in less than 420 ± 180 k.y. Many flows were greater than 1000 km^3, with some reaching ~10,000 km^3 (Reidel et al., 1989). Barry et al. (2010) observed that eruptive events were likely to be unevenly spaced, but the lack of significant sediments between lava flows precluded more detailed conclusions. However, it was clear to these authors that the amount of time taken for the emplacement of the Grande Ronde Basalt had been very short, on the order of 10,000 years, i.e., less than 3% ("and quite probably much less than 1%," Barry et al., 2010, p. 221) of the total duration of the main magmatic pulse. Barry et al. (2010, p. 213) concluded that the volumes and emplacement durations of Columbia River basalts might have had "obvious implications for climatic effects."

ETHIOPIAN AND YEMEN FLOOD BASALT PROVINCE

There is general agreement that the Ethiopian-Yemeni continental flood basalt erupted in a relatively short period (~1 m.y.) around 30 Ma (e.g., Baker et al., 1996; Hofmann et al., 1997; Arndt and Menzies, 2005). Rochette et al. (1998) suggested correlations of magnetic directions from 65 sites with the reversal time scale, with the bulk of magmatism occurring in chron C11r and therefore being linked with the Oi2 global cooling event. Coulié et al. (2003) showed that the K/Ar and $^{40}Ar/^{39}Ar$ ages from the Ethiopian and Yemeni parts of the province were indeed indistinguishable, 30.6 ± 0.4 and 30.2 ± 0.4 Ma, respectively. This continental flood basalt is different from several other major continental flood basalt provinces in that, rather than being entirely composed of basalts, it is bimodal, with a lower basaltic half and an upper silicic half (volcanism may have continued until 26 Ma [Baker et al., 1996; Coulié et al., 2003], possibly in relation to the continental flood basalt–rift–drift transition). A more recent study (Riisager et al., 2005) of 73 sites in Yemen has confirmed earlier conclusions, and shown that the volcanic stratigraphy extends from C11r to C9n, with five polarity zones recognized. The basal basaltic part erupted in less than 1 m.y. and is likely to have preceded the Oi2 cooling event, whereas the upper silicic part may have continued for some 3 m.y. as punctuated events after Oi2. To our knowledge, there are no high-resolution paleomagnetic studies of the type required for the analysis of directional groups on thick, continuous, and multiple stratigraphic lava successions/sections for the Oligocene Ethiopian-Yemeni continental flood basalt.

BRITO-ARCTIC FLOOD BASALT PROVINCE

Early detailed magnetostratigraphy of the Faroe Islands flood volcanics, part of the Brito-Arctic flood basalt province (together with the East Greenland traps), also known as the North Atlantic igneous province, was provided by Riisager et al. (2002). These authors found polarity chrons C26n to C24r (59–56 Ma) in the lower 700 m of the composite section, whereas the 2300 m of the middle and upper formations were all emplaced within C24r (56–55.5 Ma). Riisager et al. (2002) recognized the presence of 21 directional groups (that they interpreted as having cooled over less than a few hundred years), some 50 m thick. Combining directional groups with field observations and geochemical compositions, they were able to correlate cooling units between sections and to estimate eruption rate as being at least one flow per 70 k.y. Previous authors (e.g., Storey et al., 2007) distinguished two phases in the North Atlantic igneous province: phase 1 in the range ca. 62–58 Ma and phase 2 ca. 56–54 Ma, the latter coinciding with North Atlantic breakup and the Paleocene-Eocene thermal maximum (formerly known as the late Paleocene thermal maximum). Larsen and Tegner (2006) reconstructed the cooling history and pressure conditions under which the Skaergaard intrusion was formed and concluded that 5.3–6.3 ± 2.7 km of the East Greenland flood basalts were emplaced above the intrusion in less than 300 k.y.

In West Greenland, Riisager and Abrahamsen (1999) sampled a composite 1600-m-thick section of the lower part of the flood basalt, in which they found two polarity zones that they correlated with C27n-C26r. Riisager et al. (2003) combined paleomagnetic directional group analysis with stereo-photogrammetry and noted that (particularly in the ~300-m-thick Vaigat formation) consecutive flows often had similar mean remanence directions, a prominent example being a sequence of 31 lavas totaling 83 m in thickness. These authors concluded that directional groups implied flow fields having accumulated in less than a century. Finally, Tegner et al. (2012) suggested that the whole volume of the southern part of the North Atlantic igneous province might also have erupted in less than 300 k.y. (later revised to 40 k.y.; C. Tegner, 2013, oral communication at London meeting) and that the yearly fluxes of SO_2 and HCl released to the atmosphere might have reached those of the 1783 Laki eruption in Iceland. Svensen et al. (2007) suggested that intrusive volcanic systems related to the North Atlantic igneous province phase 2 in carbon-rich sedimentary strata may have caused a massive release of methane that initiated the Paleocene-Eocene thermal maximum (Svensen et al., 2007). However, more recently, Tegner et al. (2012) showed that the East Greenland flood basalt volcanism postdated the Paleocene-Eocene thermal maximum and may have initiated global cooling that would have ended the Paleocene-Eocene thermal maximum. The relationship between the second phase of North Atlantic igneous province emplacement and the Paleocene-Eocene thermal maximum needs more work to be fully understood.

DECCAN FLOOD BASALT PROVINCE

The Deccan has been one of the most extensively and intensively studied continental flood basalts, including by our group (Courtillot et al., 1986, 2000; Vandamme et al., 1991; Vandamme and Courtillot, 1992; Chenet et al., 2007, 2008, 2009). Courtillot et al. (1986) combined paleomagnetic, paleontological, and geochronological data and suggested that the Deccan large igneous province had erupted in less than 1 m.y., spanning the Cretaceous-Tertiary boundary. Despite much debate and thanks to numerous supplementary studies, these conclusions are now essentially vindicated (see the recent exchange among the following authors: Schulte et al., 2010; Archibald et al., 2010; Courtillot and Fluteau, 2010; Keller et al., 2010). In order to achieve some of the goals defined in the introduction, namely, attain much better time resolution, at least for some flow fields, Chenet et al. (2008, 2009) resampled several Deccan sections of the Main Province in great detail (the Main Province represents ~80% of the total volume of the Deccan Traps). This was made possible thanks to the stratigraphic logging and structural and volcanological analysis of Subbarao et al. (2000), Jay (2005), and Jay and Widdowson (2008). Chenet et al. (2008) studied the upper third, and Chenet et al. (2009) studied the lower two thirds of a 3500-m-thick composite section of the Main Province, combining petrology, volcanology, chemostratigraphy, morphology, geochronology, and analysis of sedimentary alteration horizons (boles). However, their main tool was the one central to the present paper, i.e., the identification of directional groups associated with cooling units (also called single eruptive events or SEE by Chenet et al., 2009).

Some thirty directional groups, representing major eruptive events, up to 180 m in thickness, were identified, and correlations of directions between sections allowed identification of flow fields more than 1000 km long (Self et al., 2008) and up to 10,000 km^3 in volume (e.g., Chenet et al., 2009; Bryan et al., 2010). The section consisted of some 40 individual lava units with volumes typically >1000 km^3. The ^{40}K-^{40}Ar dating carried out by Chenet et al. (2007) yielded indistinguishable ages of 64.5 ± 0.5 and 64.8 ± 0.6 Ma for the uppermost and some of the lower formations, respectively. A lowermost, volumetrically minor series of flows was dated at ca. 67.5 ± 1 Ma. Keller's group (e.g., Keller et al., 2010, 2011, 2012) has been instrumental in correlating flow sequences with microfossils found in remote marine sediment sections in which the flows were interbedded.

A synthesis of all this information led Chenet et al. (2009) to propose that Deccan activity had taken place over some 3 m.y. in three successive, separate, and much shorter phases (Fig. 3; after fig. 12b *in* Chenet et al., 2009), each having likely lasted only ~100 k.y. or less. Each phase itself consisted of a series of pulses; some of these individual pulses would have been truly gigantic and erupted remarkably fast. The total time of volcanic activity (sum of durations when flows were actually being emplaced) could have been (much) less than 10 k.y., i.e., less than a fraction of a percent of the total emplacement time of the province. Well above 99% of the time must have simply gone unrecorded. Special attention was given to the sedimentary or alteration levels (boles) between flows. Nine out of 23 thin boles were shown (using geomagnetic paleosecular variation) to have formed in no more than a few decades. Only two thicker boles were estimated to have formed in up to 50 k.y.

In a study of samples from wells in the Krishna-Godavari Basin, Keller et al. (2011a) were able to document the biological effects of Deccan volcanism. Phase 2 of Chenet et al. (2009), the main volcanic event by far, was entirely within C29r and ended just before the paleontologically defined Cretaceous-Tertiary boundary. It coincided with rapid global warming and subsequent cooling. Phase 3 began in the Danian and straddled the C29r-C29n reversal (Chenet et al., 2008). Full marine biotic recovery had to wait until after that last phase. Keller et al. (2011a, p. 399) concluded that "Deccan phase-2 alone could have caused the KTB mass extinction and (...) impacts may have had secondary effects." We believe that indeed the Deccan volcanism was the starting agent of the Cretaceous-Tertiary crisis (and the agent of most, if not all, other mass extinctions), but we acknowledge that an impact event may have added a significant blow to an already weakened and stressed environment (Courtillot, 2009; Chenet et al., 2009; Courtillot and Fluteau, 2010).

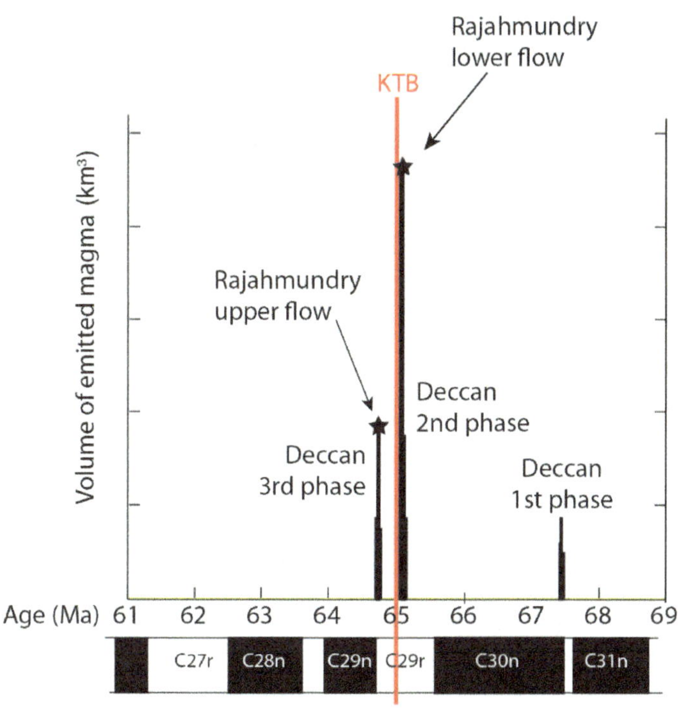

Figure 3. The scenario of embedded time scales of eruption in the Deccan Traps (redrawn after Chenet et al., 2009). Solid black columns represent the three phases of Deccan volcanism. Black stars indicate the time of emplacement of Rajahmundry lava flows (Keller et al., 2008). Geomagnetic time scale, showing normal and reverse chrons, is given along the bottom. KTB—Cretaceous-Tertiary boundary.

Chenet et al. (2008, 2009) evaluated the amounts and fluxes of SO_2 and CO_2 implied by these results. For SO_2, fluxes of up to 1 Gt/yr over durations of a century for each single eruptive event were obtained. The flux ranges estimated for the largest single eruptive events and for the Chicxulub impact overlap; the impact would have added stress to the biota, but not out of range compared to the Deccan volcanism, the total gas output of which (up to 17,000 Gt of SO_2 and 35,000 Gt of CO_2) could have been on the order of 20–200 Chicxulub impactors. Chenet et al. (2008) hypothesized that the detailed time sequence of single eruptive events might be the crucial parameter in determining the severity of a mass extinction in relation to trap magmatism. Jay et al. (2009) provided additional paleomagnetic and stratigraphic control and, following early work by Vandamme and Courtillot (1992), pointed out that the single magnetic reversal and the geochemical Ambenali-Mahabaleshwar Formation boundary provided two isochronous surfaces. They also found that the geochemical stratigraphy of the Deccan "could mask local and subregional detail in lava stacking patterns" (Jay et al., 2009, p. 13). Chenet et al. (2009) noted that the geochemical differences used to define formations were so subtle that sometimes stratigraphically implausible successions of formations occurred, or that two different formation names were assigned to flows that were in the same directional group.

A high-resolution marine $^{187}Os/^{188}Os$ record for the late Maastrichtian (C29r) led Robinson et al. (2009) to speculate an ~200-k.y.-long diminished carbon burial rate, having resulted from ocean acidification caused by Deccan volcanism, which would be our phase 2. In a study of extinction patterns, $\delta^{18}O$ trends, and magnetostratigraphy from an extended Cretaceous-Tertiary boundary section in the Antarctic peninsula, Tobin et al. (2012, p. 180) "find two prominent warming events synchronous with the three main phases of Deccan Traps flood volcanism, and the onset of the second is contemporaneous with a local extinction that pre-dates the bolide impact," therefore providing additional support to the findings of Chenet et al. (2008, 2009). Font et al. (2011) identified an interval of low magnetic susceptibility carried by a Cl-bearing iron oxide below the Cretaceous-Tertiary boundary that they interpreted as being due to heterogeneous reaction between HCl-rich volcanic gas and liquid-solid aerosols within plumes formed above the Deccan flows; this would also correspond to our phase 2.

Finally, Renne et al. (2013) produced an update and confirmation of earlier work by Swisher et al. (1992) establishing synchronicity of the Cretaceous-Tertiary boundary and impact to within 32 k.y. This is in no way in contradiction with the data reviewed in this section. As concluded by Renne et al. (2013, p. 687; see also Paelike, 2013; KPB—Cretaceous-Paleogene boundary),

"The cause of the precursory climate perturbations that pushed some ecosystems to the tipping point is unclear, but a leading candidate is volcanogenic volatile emissions from early pulses of the episodically erupted Deccan Traps. The magmatic event producing the Deccan Traps was clearly initiated prior to the KPB, and the most voluminous middle pulse of volcanism may be linked to either (i) the inception of a two-staged decline in marine $^{187}Os/^{188}Os$ beginning about 300 kyr before the KPB or (ii) the KPB itself. Existing geochronological data are insufficiently precise to constrain these relationships with age resolution comparable to that presented here for the KPB and the Chicxulub impact. Refining the timing and tempo of Deccan volcanism remains a considerable challenge whose resolution is key to evaluating the role of this event in the causes of biotic and environmental change at the KPB."

Unfortunately, the K content of Deccan lavas is very low, and there are not that many occurrences of minerals that would allow one to use the U/Pb method with its improved precision.

KAROO-FERRAR FLOOD BASALT PROVINCE

Jourdan et al. (2008) summarized a considerable amount of (mainly) $^{40}Ar/^{39}Ar$ dating of the main Karoo province and sill complex of the Karoo large igneous province. They concluded that (1) the main sill events lasted ~3 m.y. (181–184 Ma), (2) the main pulses of large igneous province magmatism lasted 4 ± 1 m.y., and (3) the entire province was emplaced in 8–10 m.y. At the same time, extinction of 5% of marine families occurred during the Pliensbachian and Toarcian, and extinction of 2.4% to 12.8% of continental organisms occurred during the Toarcian (Benton, 1995). The carbon cycle was strongly affected—a crisis of carbonate production (Dromart et al., 1996), global marine anoxia, and a sharp negative $\delta^{13}C$ that affected both carbon reservoirs (Hesselbo et al., 2000; Hermoso et al., 2009; Caruthers et al., 2011). In addition, rapid cooling and warming events have also been recorded (Dera et al., 2011). Close temporal correlation between the late Pliensbachian–early Toarcian environmental crisis (extinctions, oceanic anoxia, and climate change) and the emplacement of the Karoo traps suggests a causal relationship between the two (Palfy and Smith, 2000; Guex et al., 2001, 2012; Courtillot and Renne, 2003).

The probability density distribution (PDD) of age determinations of Karoo trap emplacement (see fig. 7c *in* Jourdan et al., 2008) is reminiscent of similar figures for other large igneous provinces published in recent decades. In many cases, exemplified by the Deccan continental flood basalt, it has been found that these early PDDs did not reflect the actual duration or phases of volcanism, and that they tended to shrink with improved data. We therefore believe that this PDD may not (yet?) reflect the actual time history of volcanism. The main peak of the PDD and emplacement of sills do, within uncertainty, match the mass extinction and isotopic excursions recorded at the base of the Toarcian. Jourdan et al. (2008) proposed that the low eruption rate of the Karoo large igneous province is one of the main reasons explaining why its biotic impact has remained comparatively low: The Pliensbachian–Toarcian period is marked by environmental perturbations, mostly during the early Toarcian. Caruthers et al. (2013) have recently shown that six episodes of decline of species diversity were coeval in Europe, western America, and the Arctic Sea. The synchronicity of extinctions suggests that the controlling mechanism acted on a global scale. Caruthers et al. (2013) argued that this multiphased extinction could be linked with peaks in Karoo magmatism.

We have undertaken a recent field and laboratory program and applied the methods, so successfully implemented in the Deccan region, to the Karoo case. We first analyzed an ~800-m-thick section at Naude's Nek in South Africa (Moulin et al., 2011). Dating using K/Ar suggests a succession of two separate volcanic phases, an older volumetrically minor one at ca. 185 Ma, and the main one at ca. 180 Ma (uncertainties on these absolute ages are ~2 m.y.). The lower 25 flows (115 m thick) have recorded a reversed polarity, the next 135 m are transitional, and the section ends with 500 m in normal polarity. Nineteen directional groups and 10 individual lava flows have been identified, implying a total duration of activity possibly less than 3 k.y., i.e., less than one thousandth of the total duration over which the section may have been emplaced. We also sampled a composite section (exposures at Oxbow and Moteng Pass; combined thickness 1500 m) some 200 km to the north in Lesotho (Moulin, 2011, and our as-yet-unpublished work).

New age determinations (both ^{40}K-^{40}Ar Cassignol-Gillot technique and $^{40}Ar/^{39}Ar$) show that the whole lava pile was emplaced in a relatively short time around the Pliensbachian-Toarcian boundary. Detailed flow-by-flow magnetostratigraphy shows that the eruptive sequences can be divided into 21 pulses (directional groups) and 16 interbedded individual lava flows. The total duration of activity was less than 4 k.y. Additional information comes from the remarkable high-resolution recording of the so-called "van Zijl" reversal (Moulin et al., 2012). Our two records can be combined with an earlier one from Prévot et al. (2003) into a single consistent reversal path. As suggested by Valet et al. (2012), reversals that occurred over the past 180 m.y. may have dynamical similarities, consisting of an ~2500-yr-long precursor, an ~1000-yr-long polarity switch, and an ~2500-yr-long rebound, with a total duration in the order of ~6 k.y. This provides yet another strong constraint on the duration of the central part of the sections in the 100–1000 yr time ranges.

The sections we studied (Oxbow-Moteng Pass and Naude's Nek) were emplaced in two volcanic phases: a minor one possibly leading to a first extinction during the Pliensbachian, and a second that could have led to a major climatic transition with marine anoxia and the main marine extinction near the Pliensbachian-Toarcian boundary (Dera et al., 2011; Moulin, 2011, and our as-yet-unpublished work). Assuming concentrations of sulfur and carbon dioxide similar to those in the Deccan Traps (Self et al., 2006), one can estimate that the whole Karoo traps emitted 15,000 to 20,000 Gt of SO_2 and 30,000 to 40,000 Gt of CO_2. The intrusion of magma into organic-rich sedimentary layers and ensuing metamorphism could have released another 27,000 Gt of CO_2 (Svensen et al., 2007), doubling the initial amount of volcanic CO_2. One still has to determine the amounts of SO_2 and CO_2 released during any given phase: The major volcanic phase is associated with the emplacement of the Lesotho Formation and would have led to the emission of large amounts of gases over a time scale at least as short as 500 k.y. Younger magmatic events of minor importance are known within the Karoo traps and would have been the possible cause of the last extinctions observed in marine species during the Toarcian (Caruthers et al., 2013; Moulin, 2011, and our as-yet-unpublished work). Thus, the emplacement of the Karoo was apparently the result of multiple (2–6) events with varying intensities, spread over a 10-m.y.-long time interval. Due to uncertainties, isotopic dating is unable to isolate the successive volcanic phases. The major pulse of volcanic activity likely emplaced the thickest section in Lesotho and can be associated with the main extinction event at the Pliensbachian-Toarcian boundary.

CENTRAL ATLANTIC MAGMATIC (FLOOD BASALT) PROVINCE

The Central Atlantic magmatic province was not recognized on most global maps of continental flood basalts until the 1990s. Olsen et al. (1996) and Olsen (1997) emphasized the extent, age, and short duration of basalts trapped in rifted basins on the Atlantic coast of the United States. Based on magneto- and cyclostratigraphy, these authors estimated total duration of emplacement in the Newark and nearby basins at 580 ± 100 k.y. Because the onset of volcanism appears some 30 m above the palynologically defined Triassic-Jurassic boundary, they rejected the possibility of a cause-effect linkage. In their review of this question, Courtillot and Renne (2003, p. 125) pointed out that "the fact that the first flows occur just slightly above the palynological T/J boundary (...) in the Newark basin are [is] no proof that volcanism had not started earlier somewhere outside of the small graben within the huge CAMP."

This has been vindicated by several studies that have appeared since. Knight et al. (2004) compiled all $^{40}Ar/^{39}Ar$ plateau ages that were then available. After filtering, they concluded that there was a major magmatic event at ca. 200 Ma that lasted less than 2 m.y. (at least in Morocco); they identified five volcanic pulses corresponding to paleomagnetic directional groups (which, based on Holocene archeomagnetism, they supposed to be less, or even much less, than 450 yr) and one reversal (estimated duration between 1000 and 8000 yr), providing an upper estimate of <20 k.y. for the duration of volcanic activity (note that Font et al. [2011] questioned some of the magnetic components found by Knight et al. [2004], but without contradicting the conclusions listed here). Marzoli and 14 coauthors (2004) combined data from stratigraphy, palynology, geochronology, paleomagnetism, and geochemistry to demonstrate that development of the Central Atlantic magmatic province straddled the Triassic-Jurassic boundary. Jourdan et al. (2009) measured 13 $^{40}Ar/^{39}Ar$ plateau ages from the American northeast and found that the duration of magmatic activity there was less than 2 or 3 m.y., and that the onset of magmatism preceded the Triassic-Jurassic boundary by a few hundred thousand years.

More recently, Schoene et al. (2010) obtained new high-precision U/Pb ages from volcanic ashes in two marine sections and showed that the end-Triassic biological crisis correlated with the onset of terrestrial flood volcanism in the Central Atlantic magmatic province to better than 150 k.y. They showed that a concomitant drop and rise in sea level and a negative $\delta^{13}C$ spike in the latest Triassic occurred in less than 290 k.y., associated with global cooling and glaciation (potentially driven by Central Atlantic magmatic province volcanism). Ruhl et al. (2011) further produced carbon isotope data of long-chain *n*-alkanes from waxes of land plants and found marked ^{13}C depletion of the end-Triassic atmosphere within 10–20 k.y. These authors linked the end-Triassic events to methane-derived massive carbon release and associated climate change (see also Deenen et al., 2010). The controversy between various groups has therefore diminished and seems basically resolved.

Blackburn et al. (2013[1], p. 942) presented zircon U-Pb data from seven sites in North America and one in Morocco. Using evidence of shared geochemical evolution of the lava units on both sides of the Atlantic, in addition to the geochronologic data, they concluded that the end-Triassic extinction began at the earliest phase of trap emplacement and that the recovery was under way before the end of the main volcanic phase. Their conclusion was that a causal relationship could be inferred "at a high level of precision."[1] Blackburn et al. (2013) further confirmed the existence of four discrete Central Atlantic magmatic province pulses based on the geochemical and geochronological data. This is in good agreement with the five paleomagnetic directional groups and pulses found by Knight et al. (2004), although more detailed work remains to be done. Furthermore, climate modeling by Paris et al. (2012) agrees with observed decreases in carbonate production only when a succession of intense degassing peaks is included in the model, as opposed to quasi-continuous eruption scenarios.

SIBERIAN FLOOD BASALT PROVINCE

As early as 1992, Wignall and Hallam (1992) pointed out anoxia as a cause of the (most intense) Permian-Triassic extinction. The Courtillot and Renne (2003) review noted the close

[1]Several coauthors of the Blackburn et al. (2013) study were also the main authors of the original Olsen et al. (1996) and Olsen (1997) papers that concluded there had been no such causal correlation, a good reason to believe the controversy is over. Olsen et al. (2002) still believed that a bolide impact was responsible for the Triassic-Jurassic mass extinction and rise of the dinosaurs, a now discounted hypothesis (see, e.g., Palfy, 2013).

temporal link of the (largest?) Siberian traps and Permian-Triassic extinction, that coincided to better than 400 k.y. (Renne et al., 1995) at ca. 250 Ma. The magnetic stratigraphy then available was limited to the work of Lind et al. (1994) and Gurevitch et al. (1995). The former found a single reversal within a composite 3500-m-thick succession of lavas in the Noril'sk area, the lower 8% of the section (Ivakinskaya Formation) being reversed and the rest normal. The thickness of the transitional part of the section is on the order of 1200 m, which likely erupted in less than 5–10 k.y. Gurevitch et al. (1995) found a reverse to normal reversal in a 1560-m-thick section in western Taymir, and then a more complex reverse-normal-reverse succession in the top 150 m of the section. Both groups concluded that most of this huge continental flood basalt had been emplaced in less than 1 m.y. at the Permian-Triassic boundary. Bowring et al. (1998, p. 1039) used U/Pb zircon ages to deduce that the late pulse of the end-Permian extinction lasted less than 1 m.y., and a negative excursion in $\delta^{13}C$ lasted on the order of 165 k.y., "suggesting a catastrophic addition of light carbon."

Gurevitch et al. (2004) have since confirmed that a single reversal is recorded in the traps of northwest central Siberia (directional groups can be suspected in their figs. 6 and 8 but were not pointed out then), and Heunemann et al. (2004) noted directional clustering in a series of 86 lava flows. This has been exploited and complemented recently by Pavlov et al. (2011), who identified 23 directional groups (and 12 individual directions) in the ~3000-m-thick section of Heunemann et al. at Noril'sk, and 17 directional groups (and 13 individual directions) in the new ~2000-m-thick section at Kotuy (updated values from V. Pavlov, 2013, personal commun.). The largest pulses reach more than 100 m (up to 12 flows) in Noril'sk and 57 m (up to 10 flows) in Kotuy. Pavlov et al. (2011) concluded that the total time of volcanic activity in the two sections may have been <8 k.y., whereas the total duration of emplacement (including periods of quiescence between flows and directional groups) may have been 10–100 k.y. Note that, as found in several other continental flood basalts, there are few or even no sedimentary layers or weathered crusts between flows described in the literature, further supporting the notion that the quiescent periods were brief. This work is ongoing and V. Pavlov (2013, personal commun.) further confirms the short durations of volcanism based on the "non-random-ordering factor" method of Biggin et al. (2008), where high correlation between directions of consecutive flows indicates brief time gaps. They have also been able to correlate a characteristic transitional direction in three areas separated by up to 100 km and to estimate the volume of a single eruptive event at more than 1000 km^3 (though not such an impressive amount for a continental flood basalt flow). Estimates of paleosecular variation, which are on the low side, led V. Pavlov (2013, personal commun.) to suggest an upper limit of 10 k.y. for the duration of the 1000-m-thick section at Kotuy.

Using paleontological identification of 537 species, Song et al. (2013) have recently shown that the marine extinction pattern consisted of two pulses separated by 180 k.y. The first phase occurred in the latest Permian and the second in the earliest Triassic, and the two phases show different extinction selectivity, possibly implying different environmental causes. Black et al. (2012) measured S, Cl, and F in melt inclusions from the Siberian Traps and found anomalously high concentrations with respect to other continental flood basalts. They estimated the amounts of S, Cl, and F released (in the range of 3000–13,000 Gt) and believe injection of these huge volumes into the stratosphere may have contributed to a drastic deterioration of the environment. These volatiles were "in part sweated out of the crustal rocks by the chambered magmas, and in part assimilated from the crust by the magmas and released upon eruption" (Elkins-Tanton et al., 2013, p. 35; see also Ganino and Arndt, 2009).

EMEISHAN FLOOD BASALT PROVINCE

The Emeishan basalt has long remained relatively un-noticed outside of the Chinese geological literature. It was mentioned in the large igneous province review of Coffin and Eldholm (1994) and recognized as a mantle plume–related continental flood basalt notably by Chung et al. (1998). We recalled in the introduction how the paleontological and geochemical findings of Stanley and Yang (1994) and Holser and Magaritz (1987), respectively, led Courtillot et al. (1999) and Hallam and Wignall (1999) to propose that the Emeishan continental flood basalt should be associated with the end-Guadalupian extinction event. This was soon supported by U-Pb ages on zircons from Zhou et al. (2002). Numerous studies of the Emeishan basalt have been published since. For instance, Wignall et al. (2009) found field evidence that the onset of volcanism coincided with large phreatomagmatic eruptions and extinctions among fusulinacean foraminifers and calcareous algae, supporting a cause and effect link.

Bond et al. (2010) recently reviewed the available evidence (notably collapse of species origination, a major regression [though somewhat postdating the marine extinction], evidence of a link between marine and terrestrial extinctions, coincidence with the onset of a major negative carbon excursion in South China and a cooling event). Critically, though, they redated the extinction/volcanism event as being not end-Guadalupian but middle Capitanian, i.e., occurring a couple of million years and two conodont zones earlier, as noted by Ali et al. (2002). However, Zhong et al. (2011) summarized the available geochronology and concluded that the mafic magmatism in the intrusive complexes of Emeishan occurred between 255.4 ± 3.1 and 259.5 ± 2.7 Ma. This would be consistent with the main stage of Emeishan magmatic activity at 259–262 ± 3 Ma, inferred from a previous study of silicic ignimbrites, the postvolcanic Xuanwei Formation and a clay tuff (He et al., 2007).

We focus here on the paleomagnetic and magnetostratigraphic evidence. This was reviewed by Ali et al. (2005). In most (rather thin; two on the order of 250 m and one up to 550 m) sections, most of the volcanism occurred in a normal interval and sometimes ended in a reversed one, with only one reversal (at most) indicated, implying a short duration for the volcanism, on the order of 1 m.y. Thompson et al. (2001, p. 597) also noted the absence of weathered flow tops and interflow paleosols. Bond et al. (2010) further pointed out that the Emeishan continental flood basalt was emplaced in two phases spanning part of two or three conodont zones, the first showing normal polarity and the second showing reversed polarity (this is what allows calibration of the marine extinction with the base and top of the Emeishan basalts). So the middle Capitanian–Emeishan continental flood basalt connection is firmly established, but, as concluded by Bond et al. (2010), improved magnetostratigraphy is still required to allow higher-resolution correlations.

The recent publication of Liu et al. (2012) agrees with previous work. A lower section (300 m thick with 30 flows) yields only 12 independent normal magnetic directions (a hint of directional grouping), whereas another section, 25 km away and stratigraphically above the former, is 50 m thick with five flows of reversed polarity. Liu et al. (2012) concluded that the whole continental flood basalt likely encompassed no more than two successive normal and reverse chrons (and therefore a single reversal). Although the end-Permian magnetostratigraphic time scale remains uncertain, if one follows the 2008 geologic time scale (Gradstein et al., 2008), the Emeishan magnetic stratigraphy could be correlated with the first of a pair of normal chrons occurring in the late Capitanian; the maximum duration of volcanism then would be ~1 m.y.

VILUY FLOOD BASALT PROVINCE

The major extinction events preceding the middle and end-Permian events are the end-Frasnian and end-Famennian extinctions in the Devonian. These were not associated with any particular continental flood basalt until as recently as a decade ago. Wilson and Lyashkevitch (1996) noted the Late Devonian magmatic activity in the Prypiat-Dnieper-Donets rift system, one dated near the end-Frasnian and the other near the end-Famennian (i.e., end-Devonian). The total volume of the lavas, less than 10,000 km^3, seems too small to associate them with the extinctions. In a paleomagnetic study of basaltic sills and dikes and kimberlitic pipes from the eastern Siberian platform (Viluy continental flood basalt, also known as Yakutsk large igneous province), Kravchinsky et al. (2002) found an unusual pole position associated with lavas that could not be dated better than the 377–350 Ma range, encompassing the ages of the Devonian events. Courtillot et al. (2010) made a first attempt at refining these age estimates. Using the K/Ar and ^{40}Ar/^{39}Ar techniques in the Orsay and Berkeley geochronology laboratories, they found a range from 344 to 367 Ma, and the best determined ages were from two plateaus at 360.3 ± 0.9 and 370.0 ± 0.7 Ma.

Redating yielded for the latter an age of 373.4 ± 0.7 Ma, compatible with then-available determinations of the end-Frasnian at 376 ± 3 Ma. In a continuation of the previous preliminary analysis, Ricci et al. (2013) analyzed seven samples from the Viluy rift. The ^{40}Ar/^{39}Ar ages cluster in two groups with means at 364.4 ± 1.7 and 376.7 ± 1.7 Ma. This confirms that the end-Devonian magmatism occurred in two major phases separated by some 7 m.y. Kaufmann (2006) placed the end-Frasnian boundary at 376.1 ± 3.6 Ma and the end-Famennian at 360.7 ± 2.7 Ma. Gradstein et al. (2012) used 358.9 ± 0.4 Ma for the latter. We note (1) that absolute ages for geological stage boundaries are unlikely to be defined to better than 0.5%–1% (i.e., 2–3 m.y.) and (2) that the age estimates of several of these boundaries have experienced worryingly large changes in the past decades (the end-Frasnian has fluctuated between 360 and 376 Ma).

We therefore conclude that much work remains to be done before we can confidently link the Viluy continental flood basalt with the Devonian extinctions. However, the few data that have accumulated since the suggestion of Kravchinsky et al. (2002) support the idea of two large separate phases of volcanism with ages compatible with the end-Frasnian and end-Famennian extinctions, a pair that could be compared with the Middle and Late Permian extinctions (though in that case, there were two separate large igneous provinces). Anoxic to euxinic conditions led to widespread black shale facies in the latest Frasnian (lower Kellwasser event) and end-Frasnian (upper Kellwasser event), both associated with positive carbon isotope excursions and significant climate cooling, implying that the latest Frasnian events were themselves a sequence of two pulses (e.g., Joachimski, 2013). The end-Devonian Hangenberg mass extinction may also have been a double event (Sallan, 2013). Once more, we encounter the possibility of embedded time scales of volcanism, a "pulse-in-pulse" structure, but for the Viluy Traps, resolution is still far from what one would like.

DISCUSSION AND CONCLUSION

We are not aware of similarly advanced studies for the remaining known Phanerozoic continental flood basalts, notably the Caribbean, Ontong-Java, Kerguelen, and Parana-Etendeka large igneous provinces. In the case of Ontong-Java, we note a recent controversy with Yan et al. (2008) arguing for a succession of up to 12 polarity intervals in a 600-m-thick section and a separate origin for the Ontong-Java Plateau and Nauru Basin, whereas Doubrovine et al. (2009) argued that there were numerous unidentified overprints and proposed a single normal polarity and a coeval formation for basalts from the two locations. In the case of the Parana-Etendeka continental flood basalt, recent work yields improved estimates for the onset of volcanism at ca. 134–135 Ma and a duration estimated at 3 m.y. by Janasi et al. (2011), but only 1 m.y. by Pinto Magalhães et al. (2011). Peate (2009; see also Chavagnac et al., 2008) argued from a study of Pb in sediments from Ocean Drilling Program (ODP) Site 1149B that Parana-Etendeka volcanism was indeed associated with the Valanginian δ^{13}C excursion but that only plumes linked to silicic, not basaltic, volcanism could reach the stratosphere and ensure global dispersal of Pb. However, we note the work of Thordarson and Self (2003) and Kaminski et al. (2011), who show, based both on observation and modeling, that plumes above continental flood basalts are able to reach the stratosphere.

Kravchinsky (2012) has reviewed the case of Paleozoic large igneous provinces and their possible correlation with mass extinctions. We have described some of them herein; some others do not have sufficient information for the type of analysis attempted in this paper. Such is the case for (1) the Skagerrak-centered volcanism in northern Europe and the Barguzin-Vitim magmatism near Baykal and Mongolia, possibly linked to Carboniferous events, (2) the Altay-Sayan event south of the Siberian platform and end-Silurian, (3) the Central Asian intraplate (large igneous province?) and the end–Late Cambrian, (4) the Kalkarindji/Antrim magmatism in Australia and the end–Early Cambrian, and (5) the Volyn large igneous province in Ukraine and the end-Ediacaran. To our knowledge, only the Kalkarindji is a serious contender (Glass and Phillips, 2006): This is an extensive basaltic province in northern and western-central Australia, cropping out over more than a million square kilometers, that is geochemically distinctive with a rather well-constrained age that places it near the end–Early Cambrian (early Toyonian) faunal mass extinction.

We are left with only one major mass extinction event (the first one of the "Big Five") in the Phanerozoic that to date has no clearly associated magmatic counterpart, namely, the end-Ordovician. Buggisch et al. (2010) have suggested that intense volcanism could indeed have triggered the first Late Ordovician icehouse, prior to the Hirnantian. They reconstructed the Late Ordovician paleotemperature history using oxygen isotopes from conodont apatite and found evidence of an abrupt and short-lived cooling that could have initiated a first short-term glacial episode, associated stratigraphically with the remains of huge volcanic ash eruptions (K-bentonites) that would have erupted large amounts of SO_2. Furthermore, they hypothesize that this may have been the first glacial interval of a longer-term Late Ordovician icehouse. This work has been criticized by Herrmann et al. (2011), who found that cooling had started before the K-bentonite layer. However, Young et al. (2009) have linked the large drop in seawater $^{87}Sr/^{86}Sr$ spanning the Middle-Late Ordovician boundary to a major volcanic episode.

Finnegan et al. (2012) have reviewed the selective nature and signature of the end-Ordovician mass extinction in connection with the climate changes that are recorded at this time. The extinction consisted of two phases/pulses at the start and in the middle of the Hirnantian Stage, i.e., separated by ~1 m.y. at ca. 446 Ma; the two pulses have been linked to intense glaciation (e.g., Armstrong and Harper, 2013). Finnegan et al. (2012, p. 6829) argued that "glacioeustatic sea-level fall and tropical ocean cooling played important roles in the first pulse of the Late Ordovician mass extinction in Laurentia." The ultimate cause of the end-Ordovician events remains to be found. Courtillot and Renne (2003) suggested that a trap might be related to these events. Application of Occam's razor and consideration of the fact that almost all major mass extinctions, and a number of "smaller" ones, and global oceanic anoxia events have been associated with a continental flood basalt (large igneous province) means that the "null hypothesis" to be tested is that a large continental flood basalt did exist and that this led to the end-Ordovician events some 450 m.y. ago. Finding the culprit could be the focus of a concerted international effort. Clearly, there is a possibility that if it ever existed, such a continental flood basalt could have been eroded away (but then feeder dike systems should have survived), or it could have been subducted or largely destroyed in a continental collision zone or within allochthonous terranes.

There are now six continental flood basalt provinces for which sufficient data are available to propose that a simple model of embedded time scales of volcanism could apply to all, including other provinces for which such data are not yet available (but available data are compatible). The model draws in particular from the work of Chenet et al. (2008, 2009) on the Deccan Traps, supported by similar observations from the Columbia flood basalt province, North Atlantic, Karoo-Ferrar, Central Atlantic magmatic province, and Siberian Traps. Continental flood basalt volcanism appears to occur over up to a few million years. It is not always easy to distinguish the time corresponding to material related to impingement of the plume head from early rifting to drifting stages in the many cases when opening of an ocean basin occurs (Courtillot et al., 1999), followed by seamount chains linked to the plume tail (e.g., Richards et al., 1989). However, in all cases, when appropriate quality geochronological data are available, volcanism seems to be concentrated in one or two (sometimes three) major phases, each on the order of a million years in duration or less. Finer analysis (based, for instance, on identification of $\delta^{18}O$ or $\delta^{13}C$ isotopic anomalies in correlative marine sections) shows that the main phases are themselves composed of a number of ~100 k.y. pulses.

Finally, these pulses contain a number of single eruptive events, many with volumes on the order of 1000 km^3, some reaching probably up to 10,000 km^3 (with flows at least in one instance in the Deccan Traps reaching 1000 km in length). Failure to record significant secular variation implies cooling times of flow fields on the order of a decade to at most a century, whereas recognition of small secular variation loops or structure in the path of a reversing field implies time constants on the order of centuries to a few millennia at most. The concept and recognition of directional groups allow us to propose that the total time during which much of a whole continental flood basalt was erupted (i.e., the total time of actual volcanic activity without intervening quiescence periods) was on the order of 10 k.y. or less. This is ~1% of the total duration of the main volcanic phase. We have noted that, in most cases, authors point out that there is a lack of sediment layers, weathered crusts, or paleosols between flows, possibly indicating that the total time of activity might have been on the same order as the total duration of a series of pulses or even a whole phase. There may, however, be climatic reasons for this lack of altered levels, and this certainly warrants further field studies. For instance, Chenet et al. (2008, 2009) pointed out the potential importance of an ~2-m-thick red bole in the Mahabaleshwar section that would warrant further field and laboratory work: Could it contain traces of the Chixculub impactor as found further north in Kutch Province (Bhandari et al., 1995; Courtillot et al., 2000)?

The volcanic time sequence with embedded time scales described herein is somewhat reminiscent of the fractal nature of "Cantor dust" (e.g., Helmberg, 2007). We have as many as four embedded scales with characteristic durations of ~10 m.y., 1 m.y., 100 k.y., and 10 yr, the latter three being the most significant. Interpretation of these patterns is important for plume dynamics, melt generation and eruption, and eruptive tectonics (faults and fissures), and it could also be a key element in evaluating the "killing potential" of a continental flood basalt eruption in relation to the characteristics of the associated mass extinction. The working hypothesis is that the same numbers *N*, volumes *V*, and CO_2 and SO_2 contents of two continental flood basalts could lead to vastly different biotic and environmental consequences, depending on whether the *N* flows or flow fields (single eruptive events) were erupted thousands of years apart, allowing the ocean to re-equilibrate the ocean-atmosphere system, or only decades or centuries apart, possibly leading to a runaway effect and a crisis. It is difficult to see how this hypothesis can be tested. Most absolute geochronology methods or isotope geochemistry profiles in sedimentary sections do not have enough resolution yet.

The injection of sulfur dioxide into the atmosphere during trap emplacement and its subsequent conversion to sulfate aerosols lead to climate changes on short time scales, shorter than the temporal resolution of most records. Interestingly, Bacon et al. (2013) have recently observed that SO_2 fumigation of leaves leads to increased leaf roundness in modern taxa. They observed a similar change in Triassic-Jurassic fossil leaves, providing direct evidence for the first time that SO_2 played a key role at the time of the mass extinction. The directional group method is unfortunately highly irregular in that it allows directional groups, and therefore single eruptive events, to be identified at the decadal time scale, but not the time between two successive flows with distinctly different directions (having recorded snapshots of secular variation separated by an unknown duration).

The emplacement of traps is also marked by the release of large amounts of carbon dioxide to the atmosphere. In the case of the Deccan Traps, we have estimated the flux of CO_2 released by magmas at 2 Gt per year (Chenet et al., 2009), i.e., 15 times less that the present-day anthropogenic flux (including biomass fire and fossil fuel rejections). However, the release may have lasted tens of years, and the elapsed time between two successive volcanic pulses would likely have been shorter than the time needed for the carbon cycle to reach equilibrium. During trap emplacement, atmospheric pCO_2 may have increased by a few tens of parts per million (Mussard et al., this volume). The rise of pCO_2 would have been enhanced by a feedback mechanism linked to the presence of sulfate aerosols in the atmosphere: During cooling events driven by these sulfate aerosols, the efficiency of silicate weathering would have been reduced, and the rise of pCO_2 would have been amplified by 25% (Mussard et al., this volume). Finally, the study of intertrappean levels represents a very tantalizing goal, though so far success has been limited. Also, a powerful tool could be climate modeling, which is the subject of the companion paper by Mussard et al. (this volume).

ACKNOWLEDGMENTS

We thank Gerta Keller and Andrew Kerr for permitting us to contribute to the Geological Society of America Special Paper on Volcanism, Impacts, and Mass Extinctions. We thank J. Ali and an anonymous reviewer for helpful comments on the original manuscript. Institut de Physique du Globe Contribution number 3216.

REFERENCES CITED

Ali, J.R., Thompson, G.M., Song, X.Y., and Wang, Y.L., 2002, Emeishan basalts (SW China) and the "end-Guadalupian" crisis: Magnetobiostratigraphic constraints: Journal of the Geological Society, v. 159, p. 21–29, doi:10.1144/0016-764901086.

Ali, J.R., Thompson, G.R., Zhou, M.F., and Song, X.Y., 2005, Emeishan large igneous province, SW China: Lithos, v. 79, p. 475–489, doi:10.1016/j.lithos.2004.09.013.

Archibald, J.D., et al., 2010, Cretaceous extinctions: Multiple causes: Science, v. 328, p. 973–974, doi:10.1126/science.328.5981.973-a.

Armstrong, H.A., and Harper, D.A., 2013, End-Ordovician mass extinction, *in* Abstracts, Conference on Volcanism, Impacts and Mass Extinctions: Causes and Effects: London, The Natural History Museum, p. 23.

Arndt, N., and Menzies, M.A., 2005, The Ethiopian large igneous province: http://www.largeigneousprovinces.org/05jan (accessed June 2013).

Bacon, K.L., Belcher, C.M., Haworth, M., and McElwain, J.C., 2013, Increased atmospheric SO_2 detected from changes in leaf physiognomy across the Triassic-Jurassic boundary interval of East Greenland: PLoS ONE, v. 8, p. e60614, doi:10.1371/journal.pone.0060614.

Baker, J., Snee, L., and Menzies, M., 1996, A brief Oligocene period of flood volcanism in Yemen: Implications for the duration and rate of continental flood volcanism at the Afro-Arabian triple junction: Earth and Planetary Science Letters, v. 138, p. 39–55, doi:10.1016/0012-821X(95)00229-6.

Bambach, R.K., Knoll, A.H., Steve, C., and Wang, S.C., 2004, Origination, extinction, and mass depletions of marine diversity: Paleobiology, v. 30, p. 522–542, doi:10.1666/0094-8373(2004)030<0522:OEAMDO>2.0.CO;2.

Barry, T.L., Self, S., Kelley, S.P., Reidel, S., Hooper, P., and Widdowson, M., 2010, New $^{40}Ar/^{39}Ar$ dating of the Grande Ronde lavas, Columbia River Basalts, USA: Implications for duration of flood basalt eruption episodes: Lithos, v. 118, p. 213–222, doi:10.1016/j.lithos.2010.03.014.

Benton, M.J., 1995, Diversification and extinction in the history of life: Science, v. 268, p. 52–58, doi:10.1126/science.7701342.

Bhandari, N., Shukla, P.N., Ghevariya, Z.G., and Sundaram, S., 1995, Impact did not trigger Deccan volcanism: Evidence from Anjar K/T boundary intertrappean sediments: Geophysical Research Letters, v. 22, p. 433–436, doi:10.1029/94GL03271.

Biggin, A.J., van Hinsbergen, D.J.J., Langereis, C.G., Straathof, G.B., and Deenen, M.H.L., 2008, Geomagnetic secular variation in the Cretaceous normal superchron and in the Jurassic: Physics of the Earth and Planetary Interiors, v. 169, p. 3–19, doi:10.1016/j.pepi.2008.07.004.

Black, B.A., Elkins-Tanton, L.T., Rowe, M.C., and Ukstins Peate, I., 2012, Magnitude and consequences of volatile release from the Siberian Traps: Earth and Planetary Science Letters, v. 317–318, p. 363–373, doi:10.1016/j.epsl.2011.12.001.

Blackburn, T.J., Olsen, P.E., Bowring, S.A., McLean, N.M., Kent, D.V., Puffer, J., McHone, G., Rasbury, E.T., and Et-Touhami, M., 2013, Zircon U-Pb geochronology links the end-Triassic extinction with the Central Atlantic magmatic province: Science, v. 340, no. 6135, p. 941–945, doi:10.1126/science.1234204.

Bond, D.P.G., Hilton, J., Wignall, P.B., Ali, J.R., Stevens, L.G., Sun, Y.D., and Lai, X.L., 2010, The Middle Permian (Capitanian) mass extinction on land and in the oceans: Earth-Science Reviews, v. 102, p. 100–116, doi:10.1016/j.earscirev.2010.07.004.

Bowring, S.A., Erwin, D.H., Jin, Y., Martin, M.W., Davidek, K., and Wang, W., 1998, U/Pb zircon geochronology and tempo of the end-Permian mass extinction: Science, v. 280, p. 1039–1045, doi:10.1126/science.280.5366.1039.

Bryan, S.E., Peate, I.U., Peate, D.W., Self, S., Jerram, D.A., Mawby, M.R., Marsh, J.S., and Miller, J.A., 2010, The largest volcanic eruptions on Earth: Earth-Science Reviews, v. 102, p. 207–229, doi:10.1016/j.earscirev.2010.07.001.

Buggisch, W., Joachimski, M.M., Lehnert, O., Bergström, S.M., Repetski, J.E., and Webers, G.F., 2010, Did intense volcanism trigger the first Late Ordovician icehouse?: Geology, v. 38, p. 327–330, doi:10.1130/G30577.1.

Caruthers, A.H., Gröcke, D.R., and Smith, P.L., 2011, The significance of an Early Jurassic (Toarcian) carbon-isotope excursion in Haida Gwaii (Queen Charlotte Islands), British Columbia, Canada: Earth and Planetary Science Letters, v. 307, p. 19–26, doi:10.1016/j.epsl.2011.04.013.

Caruthers, A.H., Smith, P.L., and Gröcke, D.R., 2013, The Pliensbachian–Toarcian (Early Jurassic) extinction, a global multi-phased event: Palaeogeography, Palaeoclimatology, Palaeoecology, v. 386, p. 104–118.

Chavagnac, V., German, C.R., and Taylor, R.N., 2008, Global environmental effects of large volcanic eruptions on ocean chemistry: Evidence from "hydrothermal" sediments (ODP Leg 185, Site 1149B): Journal of Geophysical Research, v. 113, no. B6, p. B06201, doi:10.1029/2007JB005333.

Chenet, A.-L., Fluteau, F., and Courtillot, V., 2005, Modelling massive sulphate aerosol pollution, following the large 1783 Laki basaltic eruption: Earth and Planetary Science Letters, v. 236, p. 721–731, doi:10.1016/j.epsl.2005.04.046.

Chenet, A.-L., Quidelleur, X., Fluteau, F., and Courtillot, V., 2007, ^{40}K-^{40}Ar dating of the main Deccan large igneous province: Further evidence of KTB age and short duration: Earth and Planetary Science Letters, v. 263, p. 1–15, doi:10.1016/j.epsl.2007.07.011.

Chenet, A.-L., Fluteau, F., Courtillot, V., Gérard, M., and Subbarao, K.V., 2008, Determination of rapid Deccan eruptions across the Cretaceous-Tertiary boundary using paleomagnetic secular variation: Results from a 1200-m-thick section in the Mahabaleshwar escarpment: Journal of Geophysical Research, v. 113, p. B04101, doi:10.1029/2006JB004635.

Chenet, A.-L., Courtillot, V., Fluteau, F., Gérard, M., Quidelleur, X., Khadri, S.F.R., Subbarao, K.V., and Thodarson, T., 2009, Determination of rapid Deccan eruptions across the Cretaceous-Tertiary boundary using paleomagnetic secular variation: 2. Constraints from analysis of eight new sections and synthesis for a 3500-m-thick composite section: Journal of Geophysical Research, v. 114, p. B06103, doi:10.1029/2008JB005644.

Chung, S.L., Jahn, B.M., Wu, G., Lo, C.H., and Cong, B., 1998, The Emeishan flood basalt in SW China: A mantle plume initiation model and its connection with continental breakup and mass extinction at the Permian-Triassic boundary, *in* Flower, M.F.J., Chung, S.L., Lo, C.H., and Lee, T.Y., eds., Mantle Dynamics and Plate Interactions in East Asia: American Geophysical Union Geodynamics Monograph 27, p. 47–58.

Coffin, M.F., and Eldholm, O., 1994, Large igneous provinces—Crustal structure, dimensions, and external consequences: Reviews of Geophysics, v. 32, p. 1–36, doi:10.1029/93RG02508.

Coulié, E., Quidelleur, X., Gillot, P.Y., Courtillot, V., Lefèvre, J.C., and Chiesa, S., 2003, Comparative K-Ar and Ar/Ar dating of Ethiopian and Yemenite Oligocene volcanism: Implications for timing and duration of the Ethiopian Traps: Earth and Planetary Science Letters, v. 206, p. 477–492, doi:10.1016/S0012-821X(02)01089-0.

Courtillot, V., 1994, Mass extinctions in the last 300 million years: One impact and seven flood basalts: Israel Journal of Earth Sciences, v. 43, p. 255–266.

Courtillot, V., 1999, Evolutionary Catastrophes: The Science of Mass Extinction: Cambridge, UK, Cambridge University Press, 171 p.

Courtillot, V., 2009, Nouveau Voyage au Centre de la Terre: Paris, Editions Odile Jacob, 348 p.

Courtillot, V., and Fluteau, F., 2010, Cretaceous extinctions: The volcanic hypothesis: Science, v. 328, no. 5981, p. 973–974, doi:10.1126/science.328.5981.973-b.

Courtillot, V., and Olson, P., 2007, Mantle plumes link magnetic superchrons to Phanerozoic mass depletion events: Earth and Planetary Science Letters, v. 260, p. 495–504, doi:10.1016/j.epsl.2007.06.003.

Courtillot, V., and Renne, P.R., 2003, On the ages of flood basalt events: Comptes Rendus Geoscience, v. 335, no. 1, p. 113–140, doi:10.1016/S1631-0713(03)00006-3.

Courtillot, V., Besse, J., Vandamme, D., Montigny, R., Jaeger, J.-J., and Cappetta, H., 1986, Deccan flood basalts at the Cretaceous/Tertiary boundary?: Earth and Planetary Science Letters, v. 80, p. 361–374, doi:10.1016/0012-821X(86)90118-4.

Courtillot, V., Jaupart, C., Manighetti, I., Tapponnier, P., and Besse, J., 1999, On causal links between flood basalts and continental breakup: Earth and Planetary Science Letters, v. 166, p. 177–195, doi:10.1016/S0012-821X(98)00282-9.

Courtillot, V., Gallet, Y., Rocchia, R., Féraud, G., Robin, E., Hofmann, C., Bhandari, N., and Ghevariya, Z.G., 2000, Cosmic markers, ^{40}Ar/^{39}Ar dating and paleomagnetism of the KT sections in the Anjar area of the Deccan large igneous province: Earth and Planetary Science Letters, v. 182, p. 137–156, doi:10.1016/S0012-821X(00)00238-7.

Courtillot, V., Kravchinsky, V.A., Quidelleur, X., Renne, P.R., and Gladkochub, D.P., 2010, Preliminary dating of the Viluy Traps (eastern Siberia): Eruption at the time of Late Devonian extinction events?: Earth and Planetary Science Letters, v. 300, p. 239–245, doi:10.1016/j.epsl.2010.09.045.

Deenen, M., Ruhl, H.L., Bonis, M., Krijgsman, N.R., Reitsma, W.M., and van Bergen, M., 2010, A new chronology for the end-Triassic mass extinction: Earth and Planetary Science Letters, v. 291, p. 113–125, doi:10.1016/j.epsl.2010.01.003.

Dera, G., Brigaud, B., Monna, F., Laffont, R., Pucéat, E., Deconinck, J.-F., Pellenard, P., Joachimski, M.M., and Durlet, C., 2011, Climatic ups and downs in a disturbed Jurassic world: Geology, v. 39, p. 215–218, doi:10.1130/G31579.1.

Dessert, C., Dupré, B., François, L., Schott, J., Gaillardet, G., Chakrapani, J., and Badjope, S., 2001, Erosion of Deccan Traps determined by river geochemistry: Impact on the global climate and oceanic ^{87}Sr/^{86}Sr: Earth and Planetary Science Letters, v. 188, p. 459–474, doi:10.1016/S0012-821X(01)00317-X.

Doubrovine, P.V., Tarduno, J.A., and Zhao, X.X., 2009, On the magnetostratigraphic age of Nauru Basin basalts of the western Pacific Ocean and timing of Ontong Java volcanism: Earth and Planetary Science Letters, v. 287, no. 1–2, p. 175–184, doi:10.1016/j.epsl.2009.07.040.

Dromart, G., Allemand, P., Garcia, J.P., and Robin, C., 1996, Variation cyclique de la production carbonatée au Jurassique le long d'un transect Bourgogne-Ardèche, E-France: Bulletin de la Société Géologique de France, v. 167, p. 423–433.

Elkins-Tanton, L.T., Black, B., Lamarque, J.F., Shields, C., and Kiehl, J., 2013, The Siberian flood basalts: Connecting the mantle, the continental crust, and the atmosphere, *in* Abstracts, Conference on Volcanism, Impacts and Mass Extinctions: Causes and Effects: London, The Natural History Museum, p. 35.

Finnegan, S., Heim., N.A., Peters, S.A., and Fisher, W.W., 2012, Climate change and the selective signature of the Late Ordovician mass extinction: Proceedings of the National Academy of Sciences of the United States of America, v. 109, no. 18, p. 6829–6834, doi:10.1073/pnas.1117039109.

Font, E., Nedelec, A., Ellwood, B.B., Mirao, J., and Silva, P.F., 2011, A new sedimentary benchmark for the Deccan Traps volcanism?: Geophysical Research Letters, v. 38, p. L24309, doi:10.1029/2011GL049824.

Gallet, Y., Genevey, A., and Goff, M.L., 2002, Three millennia of directional variation of the Earth's magnetic field in western Europe as revealed by archeological artifacts: Physics of the Earth and Planetary Interiors, v. 131, p. 81–89, doi:10.1016/S0031-9201(02)00030-4.

Ganino, C., and Arndt, N.T., 2009, Climate changes caused by degassing of sediments during the emplacement of large igneous provinces: Geology, v. 37, p. 323–326, doi:10.1130/G25325A.1.

Glass, L.M., and Phillips, D., 2006, The Kalkarindji continental flood basalt province: A new Cambrian large igneous province in Australia with possible links to faunal extinctions: Geology, v. 34, p. 461–464, doi:10.1130/G22122.1.

Gradstein, F.M., Ogg, J.G., and van Kranendonk, M., 2008, On the Geologic Time Scale 2008: www.stratigraphy.org/bak/gts2008.pdf (accessed June 2013).

Gradstein, F.M., Ogg, J.G., Schmitz, M., and Ogg, G., 2012, The Geologic Time Scale 2012: Boston, Massachusetts, Elsevier, 2 volumes plus chart, 1176 p.

Guex, J., Morard, A., Bartolini, A., and Morettini, E., 2001, Découverte d'une importante lacune stratigraphique à la limite Domérien-Toarcien: Implications paléo-océanographiques: Bulletin de la Société Vaudoise des Sciences Naturelles, v. 87, p. 277–284.

Guex, J., Bartolini, A., Spangenberg, J., Vicente, J.-C., and Schaltegger, U., 2012, Ammonoid multi-extinction crises during the late Pliensbachian–Toarcian and carbon cycle instabilities: Solid Earth, Discuss, v. 4, p. 1205–1228, doi:10.5194/sed-4-1205-2012.

Gurevitch, E., Westphal, M., Daragan-Suchov, J., Feinberg, H., Pozzi, J.-P., and Khramov, A.N., 1995, Paleomagnetism and magnetostratigraphy of the traps from western Taymir (northern Siberia) and the Permo-Triassic crisis: Earth and Planetary Science Letters, v. 136, p. 461–473, doi:10.1016/0012-821X(95)00183-D.

Gurevitch, E., Heunemann, C., Rad'ko, V., Westphal, M., Bachtadse, V., Pozzi, J.P., and Feinberg, H., 2004, Palaeomagnetism and magnetostratigraphy of the Permian-Triassic Siberian Trap basalts: Tectonophysics, v. 379, p. 211–226, doi:10.1016/j.tecto.2003.11.005.

Hagstrum, J.T., and Champion, D.E., 2002, A Holocene paleosecular variation record from ^{12}C-dated volcanic rocks in western North America: Journal of Geophysical Research, v. 107, p. 1–14.

Hallam, A., and Wignall, P.B., 1999, Mass extinctions and sea-level changes: Earth-Science Reviews, v. 48, p. 217–250, doi:10.1016/S0012-8252(99)00055-0.

He, B., Xu, Y.G., Huang, X.L., Luo, Z.Y., Shi, Y.R., Yang, Q.J., and Yu, S.Y., 2007, Age and duration of the Emeishan flood volcanism, SW China: Geochemistry and SHRIMP zircon U–Pb dating of silicic ignimbrites, post-volcanic Xuanwei Formation and clay tuff at the Chaotian section: Earth and Planetary Science Letters, v. 255, p. 306–323.

Helmberg, G., 2007, Getting Acquainted With Fractals: Berlin, Walter de Gruyter, 177 p.

Hermoso, M., Minoletti, F., Le Callonnec, L., Jenkyns, H.C., Hesselbo, S.P., Rickaby, R.E., Renard, M., de Rafélis, M., and Emmanuel, L., 2009, Global and local forcing of early Toarcian seawater chemistry: A comparative study of different paleoceanographic settings (Paris and Lusitanian Basins): Paleoceanography, v. 24, p. PA4208, doi:10.1029/2009PA001764.

Herrmann, A.D., Leslie, S.A., and MacLeod, K.G., 2011, Did intense volcanism trigger the first Late Ordovician icehouse?: Comment: Geology, v. 39, p. e237, doi:10.1130/G31758C.1.

Hesselbo, S.P., Gröcke, D.R., Jenkyns, H.C., Bjerrum, C.J., Farrimond, P., Morgans Bell, H.S., and Green, O.R., 2000, Massive dissociation of gas hydrate during a Jurassic oceanic event: Nature, v. 406, p. 392–395, doi:10.1038/35019044.

Heunemann, C., Krasa, D., Soffel, H.C., Gurevitch, E., and Bachtad, V., 2004, Directions and intensities of the Earth's magnetic field during a reversal: Results from the Permo-Triassic Siberian Trap basalts, Russia: Earth and Planetary Science Letters, v. 218, p. 197–213, doi:10.1016/S0012-821X(03)00642-3.

Hofmann, C., Courtillot, V., Feraud, G., Rochette, P., Yirgu, G., Ketefo, E., and Pik, R., 1997, Timing of the Ethiopian flood basalt event and implication for plume birth and global change: Nature, v. 369, p. 838–841.

Holser, W.T., and Magaritz, M., 1987, Events near the Permian-Triassic boundary: Modern Geology, v. 11, p. 155–180.

Hong, Zhong, Campbell, I.H., Wei-Guang, Zhue, Allen, C.M., Rui-Zhong, Hu, Lie-Wen, Xie, and De-Feng, He, 2011, Timing and source constraints on the relationship between mafic and felsic intrusions in the Emeishan large igneous province: Geochimica et Cosmochimica Acta, v. 75, p. 1374–1395.

Janasi, V. de A., de Freitas, V.A., and Heaman, L.H., 2011, The onset of flood basalt volcanism, northern Parana Basin, Brazil: A precise U-Pb baddeleyite/zircon age for a Chapeco-type dacite: Earth and Planetary Science Letters, v. 302, p. 147–153, doi:10.1016/j.epsl.2010.12.005.

Jarboe, N.A., Coe, R.S., Renne, P.R., Glen, J.M.G., and Mankinen, E.A., 2008, Quickly erupted volcanic sections of the Steens Basalt, Columbia River Basalt Group: Secular variation, tectonic rotation, and the Steens Mountain reversal: Geochemistry Geophysics Geosystems, v. 9, p. Q11010, doi:10.1029/2008GC002067.

Jay, A.E., 2005, Volcanic Architecture of the Deccan Traps, Western Maharashtra, India: An Integrated Chemostratigraphic and Palaeomagnetic Study [Ph.D. thesis]: Milton Keynes, UK, Open University, 346 p.

Jay, A.E., and Widdowson, M., 2008, Stratigraphy, structure, and volcanology of the south-east Deccan continental flood basalt province: Implications for eruptive extent and volumes: Journal of the Geological Society, v. 165, p. 177–188, doi:10.1144/0016-76492006-062.

Jay, A.E., Mac Niocaill, C., Widdowson, M., Self, S., and Turner, W., 2009, New palaeomagnetic data from the Mahabaleshwar Plateau, Deccan flood basalt province, India: Implications for the volcano-stratigraphic architecture of continental flood basalt provinces: Journal of the Geological Society, v. 166, p. 13–24, doi:10.1144/0016-76492007-150.

Joachimski, M., 2013, Causes of the Late Devonian mass extinction: Extraterrestrial or Earth-bound?, *in* Abstracts, Conference on Volcanism, Impacts and Mass Extinctions: Causes and Effects: London, The Natural History Museum, p. 47.

Jourdan, F., Féraud, G., Bertrand, H., Watkeys, M.K., and Renne, P.R., 2008, The ^{40}Ar/^{39}Ar ages of the sill complex of the Karoo large igneous province: Implications for the Pliensbachian-Toarcian climate change: Geochemistry Geophysics Geosystems, v. 9, p. Q06009, doi:10.1029/2008GC001994.

Jourdan, F., Marzoli, A., Bertrand, H., Cirilli, S., Tanner, L.H., Kontak, D.J., McHone, G., Renne, P.R., and Bellieni, G., 2009, ^{40}Ar/^{39}Ar ages of CAMP in North America: Implications for the Triassic-Jurassic boundary and the ^{40}K decay constant bias: Lithos, v. 110, p. 167–180, doi:10.1016/j.lithos.2008.12.011.

Kaminski, E., Chenet, A.-L., Jaupart, C., and Courtillot, V., 2011, Rise of volcanic plumes to the stratosphere aided by penetrative convection above large lava flows: Earth and Planetary Science Letters, v. 301, p. 171–178, doi:10.1016/j.epsl.2010.10.037.

Kaufmann, B., 2006, Calibrating the Devonian time scale: A synthesis of U-Pb ID-TIMS ages and conodont stratigraphy: Earth-Science Reviews, v. 76, p. 175–190, doi:10.1016/j.earscirev.2006.01.001.

Keller, G., Adatte, T., Gardin, S., Bartolini, A., and Bajpai, S., 2008, Main Deccan volcanism phase ends near the K-T boundary: Evidence from the Krishna Godavari Basin, SE India: Earth and Planetary Science Letters, v. 268, p. 293–311, doi:10.1016/j.epsl.2008.01.015.

Keller, G., Adatte, T., Pardo, Bajpai, S., Khosla, A., and Samant, B., 2010, Cretaceous extinctions: Evidence overlooked: Science, v. 328, p. 974–975, doi:10.1126/science.328.5981.974-a.

Keller, G., Bhowmick, P.K., Upadhyay, H., Dave, A., Reddy, A.N., Jaiprakash, B.C., and Adatte, T., 2011, Deccan volcanism linked to the Cretaceous-Tertiary boundary (KTB) mass extinction: New evidence from ONGC wells in the Krishna-Godavari Basin: Journal of the Geological Society of India, v. 78, p. 399–428, doi:10.1007/s12594-011-0107-3.

Keller, G., Adatte, T., Bhowmick, P.K., Upadhyay, H., Dave, A., Reddy, A.N., and Jaiprakash, B.C., 2012, Nature and timing of extinctions in Cretaceous-Tertiary planktic foraminifera preserved in Deccan intertrappean sediments of the Krishna-Godavari Basin, India: Earth and Planetary Science Letters, v. 341–344, p. 211–221, doi:10.1016/j.epsl.2012.06.021.

Knight, K.B., Nomade, S., Renne, P.R., Marzoli, A., Bertrand, H., and Youbi, N., 2004, The Central Atlantic magmatic province at the Triassic-Jurassic boundary: Paleomagnetic and ^{40}Ar/^{39}Ar evidence from Morocco for brief, episodic volcanism: Earth and Planetary Science Letters, v. 228, p. 143–160, doi:10.1016/j.epsl.2004.09.022.

Kravchinsky, V.A., 2012, Paleozoic large igneous provinces of northern Eurasia: Correlation with mass extinction events: Global and Planetary Change, v. 86–87, p. 31–36, doi:10.1016/j.gloplacha.2012.01.007.

Kravchinsky, V.A., Konstantinov, K.M., Courtillot, V., Savrasov, J.I., Valet, J.-P., Chernyi, S., Mishenin, S., and Parasotka, B., 2002, Paleomagnetism of East Siberian Traps and kimberlites: Two new poles and paleogeographic reconstructions at about 360 and 250 Ma: Geophysical Journal International, v. 148, p. 1–33, doi:10.1046/j.0956-540x.2001.01548.x.

Larsen, R.B., and Tegner, C., 2006, Pressure conditions for the solidification of the Skaergaard intrusion: Eruption of East Greenland flood basalts in less than 300,000 years: Lithos, v. 92, p. 181–197, doi:10.1016/j.lithos.2006.03.032.

Lin, S.-C., and van Keken, P.E., 2005, Multiple volcanic episodes of flood basalts caused by thermochemical mantle plumes: Nature, v. 436, p. 250–252, doi:10.1038/nature03697.

Lind, E.N., Kropotov, S.V., Czamanske, G.K., Grommé, S.C., and Fedorenko, V.A., 1994, Paleomagnetism of the Siberian flood basalts of the Noril'sk area: A constraint on eruption duration: International Geology Review, v. 36, p. 1139–1150, doi:10.1080/00206819409465508.

Liu Chengying, Pan Yongxin, and Zhu Rixiang, 2012, New paleomagnetic investigations of the Emeishan basalts in NE Yunnan, southwestern China: Constraints on eruption history: Journal of Asian Earth Sciences, v. 52, p. 88–97, doi:10.1016/j.jseaes.2012.02.014.

Lund, S.P., 1996, A comparison of Holocene paleomagnetic secular variation records from North America: Journal of Geophysical Research, v. 101, p. 8007–8024, doi:10.1029/95JB00039.

Mankinen, E.A., Prévot, M., Grommé, C.S., and Coe, R.S., 1985, The Steen's Mountain (Oregon) geomagnetic polarity transition: 1. Directional history, duration of episodes, and rock magnetism: Journal of Geophysical Research, v. 90B, p. 10,393–10,416, doi:10.1029/JB090iB12p10393.

Mankinen, E.A., Larson, E.L., Grommé, C.S., Prévot, M., and Coe, R.S., 1987, The Steens Mountain (Oregon) geomagnetic polarity transition: 3. Its regional significance: Journal of Geophysical Research, v. 92B, p. 8057–8076, doi:10.1029/JB092iB08p08057.

Marzoli, A., Bertrand, H., Knight, K.B., Cirilli, S., Buratti, N., Verati, C., Nomade, S., Renne, P.R., Youbi, N., Martini, R., Allenbach, K., Neuwerth, R., Rapaille, C., Zaninetti, L., and Bellieni, G., 2004, Synchrony of the Central Atlantic magmatic province and the Triassic-Jurassic boundary climatic and biotic crisis: Geology, v. 32, no. 11, p. 973–976, doi:10.1130/G20652.1.

Moulin, M., 2011, Les Traps du Karoo et les Extinctions du Jurassique Inférieur: Dynamique Éruptive et Perturbations de l'Environnement [Ph.D. thesis]: Paris, Institut de Physique du Globe de Paris, 390 p.

Moulin, M., Fluteau, F., Courtillot, V., Marsh, J., Delpech, G., Quidelleur, X., Gérard, M., and Jay, A.E., 2011, An attempt to constrain the age, duration, and eruptive history of the Karoo flood basalt: Naude's Nek section (South Africa): Journal of Geophysical Research, v. 116, p. B07403, doi:10.1029/2011JB008210.

Moulin, M., Courtillot, V., Fluteau, F., and Valet, J.-P., 2012, The "van Zijl" Jurassic geomagnetic reversal revisited: Geochemistry Geophysics Geosystems, v. 13, p. Q03010, doi:10.1029/2011GC003910.

Mussard, M., Le Hir, G., Fluteau, F., Lefebvre, V., and Godderis, Y., 2014, this volume, Chapter 17, Modeling the carbon-sulfate interplays in climate changes related to the emplacement of continental flood basalts, *in* Keller, G., and Kerr, A., eds., Volcanism, Impacts, and Mass Extinctions: Causes and Effects: Geological Society of America Special Paper 505, doi:10.1130/2014.2505(17).

Olsen, P.E., 1997, Stratigraphic record of the early Mesozoic breakup of Pangea in the Laurasia–Gondwana rift system: Annual Review of Earth and Planetary Sciences, v. 25, p. 337–401, doi:10.1146/annurev.earth.25.1.337.

Olsen, P.E., Schlische, R.W., and Fedosh, M.S., 1996, 580 ky duration of the Early Jurassic flood basalt event in eastern North America estimated using Milankovitch cyclostratigraphy: Bulletin of Museum of Northern Arizona, v. 60, p. 11–22.

Olsen, P.E., Kent, D.V., Sues, H.-D., Koeberl, C., Huber, H., Montanari, A., Rainforth, E.C., Fowell, S.J., Szajana, M.J., and Hartline, B.W., 2002, Ascent of dinosaurs linked to iridium anomaly at the Triassic-Jurassic boundary: Science, v. 296, p. 1305–1307, doi:10.1126/science.1065522.

Paelike, H., 2013, Impact and extinction: Science, v. 339, no. 6120, p. 655–656, doi:10.1126/science.1233948.

Palfy, J., 2003, Volcanism of the Central Atlantic magmatic province as a potential driving force in the end-Triassic mass extinction, *in* Hames, W., Mchone, J.G., Renne, P., and Ruppel, C., eds., The Central Atlantic Magmatic Province: Insights from Fragments of Pangea: American Geophysical Union Geophysical Monograph 136, p. 255–267.

Palfy, J., 2013, Volcanism of the Central Atlantic magmatic province as the trigger of environmental and biotic changes around the Triassic-Jurassic boundary, *in* Abstracts, Conference on Volcanism, Impacts and Mass Extinctions: Causes and Effects: London, The Natural History Museum, p. 64.

Palfy, J., and Smith, P.L., 2000, Synchrony between Early Jurassic extinction, oceanic anoxic event, and the Karoo-Ferrar flood basalt volcanism: Geology, v. 28, p. 747–750, doi:10.1130/0091-7613(2000)28<747:SBEJEO>2.0.CO;2.

Paris, G., Donnadieu, Y., Beaumont, V., Fluteau, F., and Godderis, Y., 2012, Modeling the consequences on Late Triassic environment of intense pulse-like degassing during the Central Atlantic magmatic province using the GEOCLIM model: Climate of the Past, Discussions, v. 8, p. 2075–2110, doi:10.5194/cpd-8-2075-2012.

Pavlov, V.E., Fluteau, F., Veselovskiy, R.V., Fetisova, A.M., and Latyshev, A.V., 2011, Secular geomagnetic variations and volcanic pulses in the Permian-Triassic traps of the Norilsk and Maimecha-Kotui provinces: Izvestiya, Physics of the Solid Earth, v. 47, p. 402–417, doi:10.1134/S1069351311040070.

Peate, D.W., 2009, Global dispersal of Pb by large-volume silicic eruptions in the Paraná–Etendeka large igneous province: Geology, v. 37, p. 1071–1074, doi:10.1130/G30338A.1.

Pinto Magalhães, V., Afraneo Hartmann, L., Orestes, S., Santos, J., McNaughton, N.J., and Wildner, W., 2011, Zircon U-Pb geochronology from the Paraná bimodal volcanic province supports a brief eruptive cycle at ~135 Ma: Chemical Geology, v. 281, p. 93–102, doi:10.1016/j.chemgeo.2010.11.031.

Prévot, M., Roberts, N., Thompson, J., Faynot, L., Perrin, M., and Camps, P., 2003, Revisiting the Jurassic geomagnetic reversal recorded in the Lesotho Basalt (southern Africa): Geophysical Journal International, v. 155, p. 367–378, doi:10.1046/j.1365-246X.2003.02029.x.

Rampino, M.R., and Stothers, R.B., 1988, Flood basalt volcanism during the past 250 million years: Science, v. 241, p. 663–668, doi:10.1126/science.241.4866.663.

Reidel, S.P., Tolan, T.L., Hooper, P.R., Beeson, M.H., Fecht, K.R., Bentley, R.D., and Anderson, J.L., 1989, The Grande Ronde Basalt, Columbia River Basalt Group; stratigraphic descriptions and correlations in Washington, Oregon, and Idaho, *in* Reidel, S.P., and Hooper, P.R., eds., Volcanism and Tectonism in the Columbia River Flood-Basalt Province: Geological Society of America Special Paper 239, p. 21–54, doi:10.1130/SPE239-p21.

Renne, P.R., Zichao, Z., Richards, M.A., Black, M.T., and Basu, A., 1995, Synchrony and causal relations between Permian-Triassic boundary crises and Siberian flood volcanism: Science, v. 269, p. 1413–1416, doi:10.1126/science.269.5229.1413.

Renne, P.R., Deino, A.L., Hilgen, F.J., Kuiper, K.F., Mark, D.F., Mitchell, W.S., III, Morgan, L.E., Mundil, R., and Smit, J., 2013, Time scales of critical events around the Cretaceous-Paleogene boundary: Science, v. 339, no. 6120, p. 684–687, doi:10.1126/science.1230492.

Ricci, J., Quidelleur, X., Pavlov, V., Orlov, S., and Courtillot, V., 2013, New ^{40}Ar/^{39}Ar and K/Ar ages of the Viluy Traps (eastern Siberia): Further evidence for a relationship with the Frasnian-Famennian mass extinction: Palaeogeography, Palaeoclimatology, Palaeoecology, v. 386, p. 531–540, doi:10.1016/j.palaeo.2013.06.020.

Richards, M.A., Duncan, R.A., and Courtillot, V., 1989, Flood basalts and hot spot tracks: Plume heads and tails: Science, v. 246, p. 103–107.

Riisager, J., Riisager, P., and Pedersen, A.K., 2003, Paleomagnetism of large igneous provinces: Case study from west Greenland, North Atlantic igneous province: Earth and Planetary Science Letters, v. 214, p. 409–425, doi:10.1016/S0012-821X(03)00367-4.

Riisager, P., and Abrahamsen, N., 1999, Magnetostratigraphy of Paleocene basalts from the Vaigat Formation of West Greenland: Geophysical Journal International, v. 137, p. 774–782, doi:10.1046/j.1365-246x.1999.00830.x.

Riisager, P., Riisager, J., Abrahamsen, N., and Waagstein, R., 2002, New paleomagnetic pole and magnetostratigraphy of Faroe Islands flood volcanics, North Atlantic igneous province: Earth and Planetary Science Letters, v. 201, p. 261–276, doi:10.1016/S0012-821X(02)00720-3.

Riisager, P., Knight, K.B., Baker, J.A., Peate, I.U., Al-Kadasi, M., Al-Subbary, A., and Renne, P.R., 2005, Paleomagnetism and ^{40}Ar/^{39}Ar geochronology of Yemeni Oligocene volcanics: Implications for timing and duration of Afro-Arabian traps and geometry of the Oligocene paleomagnetic field: Earth and Planetary Science Letters, v. 237, p. 647–672, doi:10.1016/j.epsl.2005.06.016.

Robinson, N., Ravizza, G., Coccioni, R., Peucker-Ehrenbrink, B., and Norris, R., 2009, A high-resolution marine ^{187}Os/^{188}Os record for the late Maastrichtian: Distinguishing the chemical fingerprints of Deccan volcanism and the KP impact event: Earth and Planetary Science Letters, v. 281, p. 159–168, doi:10.1016/j.epsl.2009.02.019.

Robock, A., 2000, Volcanic eruptions and climate: Reviews of Geophysics, v. 38, p. 191–219, doi:10.1029/1998RG000054.

Rochette, P., Tamrat, E., Feraud, G., Pik, R., Courtillot, V., Ketofo, E., Coulon, C., Hofmann, C., Vandamme, D., and Yirgu, G., 1998, Magnetostratigraphy and timing of the Oligocene Ethiopian traps: Earth and Planetary Science Letters, v. 164, p. 497–510, doi:10.1016/S0012-821X(98)00241-6.

Ruhl, M., Bonis, N.R., Reichart, G.-J., Sinninghe Damsté, J.S., and Kürschner, W.M., 2011, Atmospheric carbon injection linked to end-Triassic mass extinction: Science, v. 333, p. 430–434, doi:10.1126/science.1204255.

Sallan, L.C., 2013, The end-Devonian Hangenberg event: Causes and consequences of a major bottleneck in vertebrate evolution, *in* Abstracts, Conference on Volcanism, Impacts and Mass Extinctions: Causes and Effects: London, The Natural History Museum, p. 71.

Schoene, B., Guex, J., Bartolini, A., Schaltegger, U., and Blackburn, T.J., 2010, Correlating the end-Triassic mass extinction and flood basalt volcanism at the 100 ka level: Geology, v. 38, p. 387–390, doi:10.1130/G30683.1.

Schulte, P., and 40 others, 2010, The Chicxulub asteroid impact and mass extinction at the Cretaceous-Paleogene boundary: Science, v. 327, p. 1214–1218, doi:10.1126/science.1177265.

Segschneider, J., Beitsch, A., Timmreck, C., Brovkin, V., Ilyina, T., Jungclaus, J., Lorenz, S.J., Six, K.D., and Zanchettin, D., 2012, Impact of an extremely large magnitude volcanic eruption on the global climate and carbon cycle estimated from ensemble Earth System Model simulations:

Biogeosciences Discussions, v. 9, p. 8693–8732, doi:10.5194/bgd-9-8693-2012.

Self, S., Widdowson, M., Thordarson, T., and Jay, A.E., 2006, Volatile fluxes during flood basalt eruptions and potential effects on the global environment: A Deccan perspective: Earth and Planetary Science Letters, v. 248, p. 518–532, doi:10.1016/j.epsl.2006.05.041.

Self, S., Jay, A.E., Widdowson, M., and Keszthelyi, L.P., 2008, Correlation of the Deccan and Rajahmundry Trap lavas: Are these the longest and largest lava flows on Earth?: Journal of Volcanology and Geothermal Research, v. 172, p. 3–19, doi:10.1016/j.jvolgeores.2006.11.012.

Sepkoski, J.J., 1990, The taxonomic structure of periodic extinction, *in* Sharpton, V.L., and Ward, P.D., eds., Global Catastrophes in Earth History: Geological Society of America Special Paper 247, p. 33–44.

Sobolev, S., Sobolev, A., Kuzmin, D., Krivolutskaya, N., Petrunin, A., Arndt, N., Radko, V., and Vasiliev, Y., 2012, Linking mantle plumes, large igneous provinces and environmental catastrophes: Nature, v. 477, no. 7364, p. 312–316, doi:10.1038/nature10385.

Song, H., Wignall, P.B., Tong, J., and Yin, H., 2013, Two pulses of extinction during the Permian-Triassic crisis: Nature Geoscience, v. 6, p. 52–56, doi:10.1038/ngeo1649.

Stanley, S.M., and Yang, X.A., 1994, Double mass extinction at the end of the Paleozoic era: Science, v. 266, p. 1340–1344, doi:10.1126/science.266.5189.1340.

Storey, M., Duncan, R.A., and Tegner, C., 2007, Timing and duration of volcanism in the North Atlantic igneous province: Implications for geodynamics and links to the Iceland hotspot: Chemical Geology, v. 241, p. 264–281, doi:10.1016/j.chemgeo.2007.01.016.

Stothers, R.B., Wolff, J.A., Self, S., and Rampino, M.R., 1986, Basaltic fissure eruptions, plume heights, and atmospheric aerosols: Geophysical Research Letters, v. 13, p. 725–728, doi:10.1029/GL013i008p00725.

Subbarao, K.V., Bodas, M.S., Khadri, S.F.R., and Beane, J. E., 2000, Penrose Deccan 2000, Field Excursion Guide to the Western Deccan Basalt Province: Penrose Field Guides: Bangalore, India, Geological Society of India, 243 p.

Svensen, H., Planke, S., Chevallier, L., Malthe-Sörenssen, A., Corfu, F., and Jamtveit, B.R., 2007, Hydrothermal venting of greenhouse gases triggering Early Jurassic global warming: Earth and Planetary Science Letters, v. 256, p. 554–566, doi:10.1016/j.epsl.2007.02.013.

Swisher, C.C., III, Grajales-Nishimura, J.M., Montanari, A., Margolis, S.V., Claeys, P., Alvarez, W., Renne, P., Cedillo-Pardo, E., Maurrasse, M.R., Curtis, G.H., Smit, J., and McWilliams, M.O., 1992, Coeval $^{40}Ar/^{39}Ar$ ages of 65.0 million years ago from Chicxulub crater melt rock and Cretaceous-Tertiary boundary tektites: Science, v. 257, p. 954–958, doi:10.1126/science.257.5072.954.

Tegner, C., Heilmann-Clausen, C., Larsen, R.B., and Kent, A.J.R., 2012, East Greenland flood basalt volcanism: Duration, volatile flux and correlation to the Paleocene-Eocene thermal maximum, *in* European Geosciences Union General Assembly 2012: Vienna, Austria, European Geosciences Union, p. 11296.

Thompson, G.M., Ali, J.R., Song, X., and Jolley, D.W., 2001, Emeishan Basalts, SW China: Reappraisal of the formation's type area stratigraphy and a discussion of its significance as a large igneous province: Journal of the Geological Society, v. 158, p. 593–599, doi:10.1144/jgs.158.4.593.

Thordarson, T., and Self, S., 1998, The Roza Member, Columbia River basalt group: A gigantic pahoehoe lava flow field formed by endogenous processes?: Journal of Geophysical Research, v. 103, p. 27,411–27,445, doi:10.1029/98JB01355.

Thordarson, T., and Self, S., 2003, Atmospheric and environmental effects of the 1783–1784 Laki eruption: A review and reassessment: Journal of Geophysical Research, v. 108, no. D1, p. 4011, doi:10.1029/2001JD002042.

Timmreck, C., 2012, Modeling the climatic effects of large explosive volcanic eruptions: Wiley Interdisciplinary Reviews: Climate Change, v. 3, no. 6, p. 545–564, doi:10.1002/wcc.192.

Tobin, T.S., Ward, P.D., Steig, E.J., Olivero, E.B., Hilburn, I.A., Mitchell, R.N., Diamond, M.R., Raub, T.D., and Kirschvink, J.L., 2012, Extinction patterns, $\delta^{18}O$ trends, and magnetostratigraphy from a southern high-latitude Cretaceous–Paleogene section: Links with Deccan volcanism: Palaeogeography, Palaeoclimatology, Palaeoecology, v. 350–352, p. 180–188, doi:10.1016/j.palaeo.2012.06.029.

Valet, J.-P., Fournier, A., Courtillot, V., and Herrero-Bervera, E., 2012, Dynamical similarity of geomagnetic field reversals: Nature, v. 490, p. 89–93, doi:10.1038/nature11491.

Vandamme, D., and Courtillot, V., 1992, Paleomagnetic constraints on the structure of the Deccan Traps: Physics of the Earth and Planetary Interiors, v. 74, p. 241–261, doi:10.1016/0031-9201(92)90013-L.

Vandamme, D., Courtillot, V., Besse, J., and Montigny, R., 1991, Paleomagnetism and age determinations of the Deccan traps (India): Results of a Nagpur-Bombay traverse and review of earlier work: Reviews of Geophysics, v. 29, p. 159–190.

White, R.V., and Saunders, A.D., 2005, Volcanism, impact and mass extinctions: Incredible or credible coincidences?: Lithos, v. 79, p. 299–316, doi:10.1016/j.lithos.2004.09.016.

Wignall, P.B., and Hallam, A., 1992, Anoxia as a cause of the Permian/Triassic extinction: Facies evidence from northern Italy and the western United States: Palaeogeography, Palaeoclimatology, Palaeoecology, v. 93, p. 21–46, doi:10.1016/0031-0182(92)90182-5.

Wignall, P.B., Sun, Y., Bond, D.P.G., Izon, G., Newton, R.J., Védrine, S., Widdowson, M., Ali, J.R., Lai, X., Jiang, H., Cope, H., and Bottrell, S.H., 2009, Fluctuations in the Middle Permian of China: Science, v. 324, p. 1179, doi:10.1126/science.1171956.

Wilson, M., and Lyashkevitch, Z.M., Magmatism and the geodynamics of rifting of the Prypiat–Dnieper–Donets rift, 1996, East European Platform: Tectonophysics, v. 268, p. 65–81.

Yan, M.D., Zhao, X.X., and Riisager, P., 2008, New paleomagnetic results from basaltic drill cores of the Nauru Basin, western Pacific: Tectonic and magnetostratigraphic implications: Earth and Planetary Science Letters, v. 267, p. 175–187, doi:10.1016/j.epsl.2007.11.047.

Young, S.A., Saltzman, M.R., Foland, K.A., and Kump, L.R., 2009, A major drop in seawater $^{87}Sr/^{86}Sr$ during the Middle Ordovician (Darriwilian): Links to volcanism and climate?: Geology, v. 37, p. 951–954, doi:10.1130/G30152A.1.

Zhou, M.F., Malpas, J., Song, X.Y., Robinson, P.T., Sun, M., Kennedy, A.K., Lesher, C.M., and Keays, R.R., 2002, A temporal link between the Emeishan large igneous province (SW China) and the end-Guadalupian mass extinction: Earth and Planetary Science Letters, v. 196, p. 113–122, doi:10.1016/S0012-821X(01)00608-2.

Manuscript Accepted by the Society 31 January 2014

Printed in the USA

The Geological Society of America
Special Paper 505
2014

Emplacement characteristics, time scales, and volcanic gas release rates of continental flood basalt eruptions on Earth

S. Self*
Department of Environment, Earth & Ecosystems, The Open University, Walton Hall, Milton Keynes MK7 6AA, UK, and
Department of Earth and Planetary Science, University of California, Berkeley, California 94720-9980, USA

A. Schmidt
School of Earth and Environment, University of Leeds, Leeds LS2 9JT, UK

T.A. Mather
Department of Earth Sciences, University of Oxford, South Parks Road, Oxford OX1 3AN, UK

ABSTRACT

Continental flood basalt provinces are the subaerial expression of large igneous province volcanism. The emplacement of a continental flood basalt is an exceptional volcanic event in the geological history of our planet with the potential to directly impact Earth's atmosphere and environment. Large igneous province volcanism appears to have occurred episodically every 10–30 m.y. through most of Earth history. Most continental flood basalt provinces appear to have formed within 1–3 m.y., and within this period, one or more pulses of great magma production and lava eruption took place. These pulses may have lasted from 1 m.y. to as little as a few hundred thousand years. Within these pulses, tens to hundreds of volumetrically large eruptions took place, each producing 10^3–10^4 km^3 of predominantly pāhoehoe lava and releasing unprecedented amounts of volcanic gases and ash into the atmosphere. The majority of magmatic gas species released had the potential to alter climate and/or atmospheric composition, in particular during violent explosive phases at the eruptive vents when volcanic gases were lofted into the stratosphere. Aside from the direct release of magmatic gases, magma-sediment interactions featured in some continental flood basalt provinces could have released additional carbon, sulfur, and halogen-bearing species into the atmosphere. Despite their potential importance, given the different nature of the country rock associated with each continental flood basalt province, it is difficult to make generalizations about these emissions from one province to another. The coincidence of continental flood basalt volcanism with periods of major biotic change is well substantiated, but the actual mechanisms by which

*steve.self1815@gmail.com

Self, S., Schmidt, A., and Mather, T.A., 2014, Emplacement characteristics, time scales, and volcanic gas release rates of continental flood basalt eruptions on Earth, *in* Keller, G., and Kerr, A.C., eds., Volcanism, Impacts, and Mass Extinctions: Causes and Effects: Geological Society of America Special Paper 505, p. 319–337, doi:10.1130/2014.2505(16). For permission to copy, contact editing@geosociety.org.

the volcanic gases might have perturbed the environment to this extent are currently not well understood, and have been little studied by means of atmospheric modeling. We summarize current, albeit rudimentary, knowledge of continental flood basalt eruption source and emplacement characteristics to define a set of eruption source parameters in terms of magmatic gases that could be used as inputs for Earth system modeling studies. We identify our limited knowledge of the number and length of non-eruptive phases (hiatuses) during continental flood basalt volcanism as a key unknown parameter critical for better constraining the severity and duration of any potential environmental effects caused by continental flood basalt eruptions.

INTRODUCTION

Continental flood basalt provinces are a subset of the broader group of large igneous provinces. The generation of a large igneous province is an exceptional volcanic event in Earth history because of the large total volume of predominantly basaltic magma generated, and lava erupted, over brief periods of geological time (usually less than 1–2 m.y.). Individual continental flood basalt eruptions, of which hundreds to thousands constitute a province, have produced lava volumes on the order of 10^3–10^4 km^3, likely at mean eruption rates of tens up to 10^2 km^3 per year (Self et al., 2006, 2008b; Chenet et al., 2008). By contrast, 30 yr of almost constant volcanic activity at Kīlauea volcano (Hawai`i) from 1983 to 2013 has produced ~4 km^3 of basaltic lava (USGS, 2013), and the largest basaltic eruption of the past ~1000 yr (Eldgjá, Iceland, 934 CE, probably lasting up to 6 yr) produced an eruptive volume of ~20 km^3 (Thordarson et al., 2001). Thus, volumetrically, continental flood basalt eruptions were huge compared to basaltic volcanism on Earth today.

Through time, large igneous province volcanism has led to the construction of extensive lava plateaus, from 1 to 3 km in thickness at any one location on land, and considerably thicker on the seafloor (Fig. 1). The extrusive and intrusive products of large igneous provinces collectively cover areas in excess of hundreds of thousands of square kilometers (sometimes more than 1 million km^2 [1×10^6 km^2]), and, typically, the aggregate extruded volcanic deposit and lava volumes are greater than 1×10^6 km^3. Research between the late 1980s to the turn of the millennium (summarized by Courtillot and Renne, 2003) showed that during large igneous province volcanism, the climax or peak in terms of erupted lava volume was often much less than 1 m.y. in duration (Rampino and Stothers, 1988; Courtillot, 1999; Chenet et al., 2009).

Large igneous provinces have formed throughout much of Earth's history and are a manifestation of Earth's normal, but in this case episodic, internal processes. The processes of generation and emplacement can be thought of as unusual transient events in Earth's volcanic history when compared with magma production at mid-ocean ridges and subduction zones (for a monograph on large igneous province characteristics, see Mahoney and Coffin, 1997). The most common hypothesis to explain the formation of large igneous provinces invokes decompressional melting of mantle plume heads (Campbell et al., 1990; Ernst and Buchan, 2001; Campbell, 2005). The frequency of large igneous province events since the Archean eon (ca. 2.5 Ga) has been estimated at one every 20 m.y. (Ernst and Buchan, 2001), but when just considering the last 270 m.y. (i.e., including the oceanic large igneous province record, some of which was destroyed by subduction earlier than 270 Ma), the frequency may have been up to one every 10 m.y. (Coffin and Eldholm, 2001). However, Rampino and Prokoph (2013) recently proposed an average frequency of 30–35 m.y. for the past 500 Ma.

Large igneous province volcanism has occurred in two broad compositional varieties: basaltic, which are the most numerous, and silicic (Bryan, 2007), which are few in number and will not be further discussed here. In this paper, we focus on continental flood basalt provinces, which are the subaerial (i.e., on-land) expression of large igneous provinces. Continental flood basalt volcanism is characterized by the combination of large erupted volumes and relatively frequent, huge basaltic eruptions (Coffin and Eldholm, 1994; Bryan and Ernst, 2008). In detail, some continental flood basalts, which are dominated by vast piles of basalt lava flows (as shown on Fig. 2), have silicic igneous rocks amongst their sequences, and vice versa for silicic large igneous provinces.

There is a striking age correlation with mass extinction events for at least four continental flood basalt provinces emplaced in the past 300 m.y. (for reviews, see Officer et al., 1987; Wignall, 2001; Courtillot and Renne, 2003; Kelley, 2007). One of the most famous examples is the mass extinction event at the end-Cretaceous (ca. 66 Ma) and its coincidence with the emplacement of the Deccan Traps continental flood basalt province in India (e.g., Vogt, 1972; Courtillot, 1999; Keller et al., 2008, 2012). Moreover, this is the only well-substantiated case of an asteroid impact coinciding with a continental flood basalt event and a mass extinction event (e.g., Keller, 2014, this volume). Perhaps more strikingly, when the eruptions of the Siberian Traps were first proposed as a trigger for the end-Permian mass extinction (252 Ma), this presented a possible correlation of the emplacement of the largest continental flood basalt province with the greatest loss of floral and faunal diversity in Earth's history (Erwin, 1994; for a review, see Wignall, 2001; Reichow et al., 2009). There has also been important recent work on the possible association between the Central Atlantic magmatic

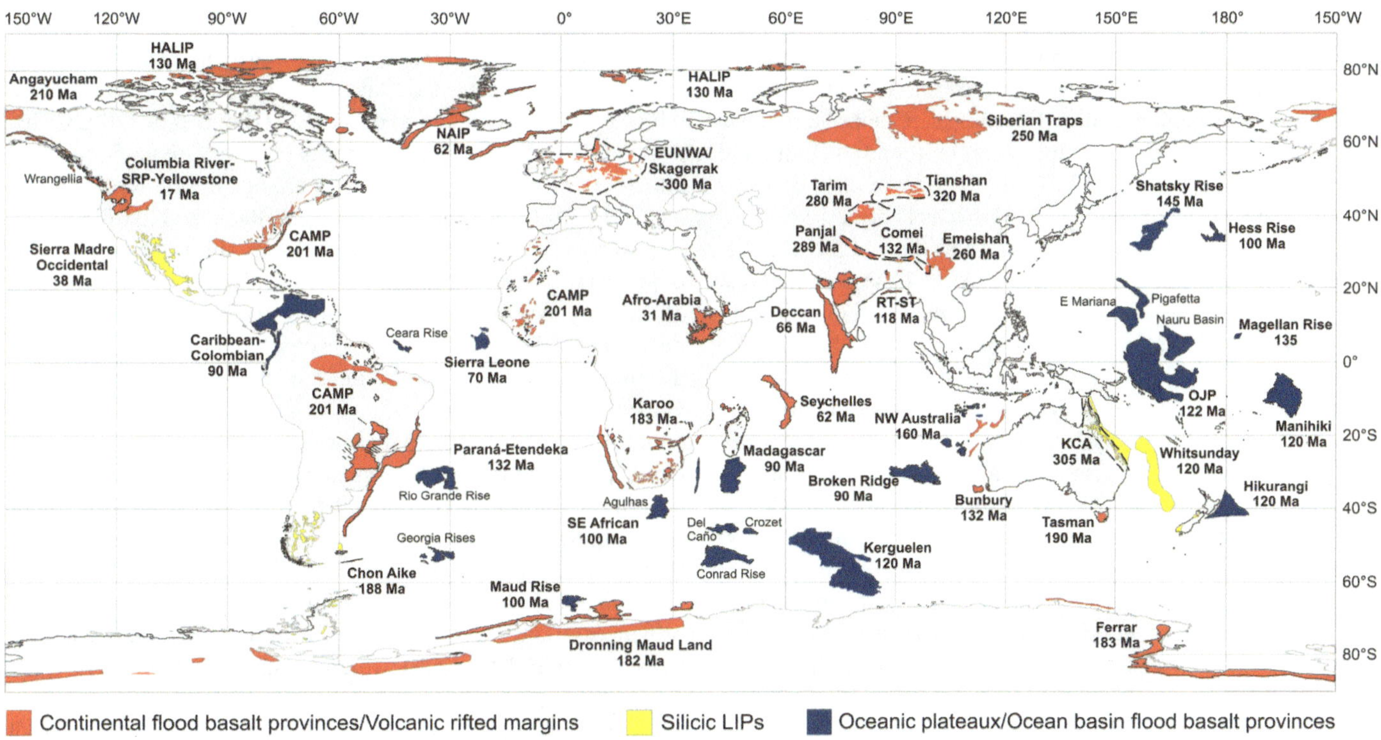

Figure 1. Large igneous provinces of the world, including continental flood basalt provinces, oceanic plateaus, and silicic large igneous provinces (LIPs; after Bryan and Ferrari, 2013). Ages are those of the main phase or pulse of volcanism, according to Bryan and Ferrari (2013). Inferred extent of some provinces shown by dashed lines. Discussion in this paper mainly focuses on the Columbia River Basalt Group and the Deccan Traps, as well as the Siberian Traps and the Emeishan provinces. Other LIPs not mentioned in this chapter; see Bryan and Ferrari, 2013.

Figure 2. A stack of lava flows, typical of the "trap" or stepped topography characteristic of flood basalt lavas, exposed along the Grande Ronde River in Washington State (USA). The >600-m-thick pile of lavas seen here is part of the Columbia River Basalt Group.

province and the extinction events at the Triassic-Jurassic boundary (e.g., Whiteside et al., 2007; Blackburn et al., 2013). Overall, however, there is still debate about whether and how continental flood basalt volcanism is causally linked to periods of biotic crisis throughout geologic time. Recent comprehensive reviews of large igneous province volcanism can be found in Bryan and Ferrari (2013) and Bryan et al. (2010). The topic of large igneous province volcanism and mass extinctions is discussed extensively in Courtillot (1999) and Wignall (2001), and the age dating of continental flood basalt provinces and mass extinctions events has been discussed in detail elsewhere (e.g., Courtillot and Renne, 2003), including meteorite impact events (Kelley, 2007).

In this paper, we review emplacement characteristics, time scales, and gas release rates of continental flood basalt eruptions. One of our aims is to compile a set of eruption source parameters for continental flood basalt eruptions. These eruption source parameters could be used as inputs for atmospheric modeling efforts to test whether, and to what degree, continental flood basalt eruptions could have caused or triggered environmental changes in the geological past. By necessity, we use information from those continental flood basalt provinces for which some geological evidence exists with regard to the size, duration, and style of the eruptions—that is the Columbia River Basalt Group and the Deccan Traps in particular. We focus on the gases released directly from erupted lavas (for which more information is available in the literature and where conclusions from one continental flood basalt province are more likely to be applicable to others), although we also briefly review other emissions sources associated with magma–country-rock (sediment) interactions. We aim to highlight areas of uncertainty that currently limit our abilities to gain a better understanding of the Earth system impacts of continental flood basalt volcanism. Because continental flood basalt provinces consist of distinct but volumetrically significant lava flow units, it is important to consider both the emplacement characteristics of the province as a whole, but also of the individual lava flow units, as discussed in the following two sections of this paper. In the final three sections of the paper, we discuss what we know about volcanic gas emissions and their potential environmental impacts, and we compile a set of eruption source parameters for typical magmatic sulfur releases from such events.

EMPLACEMENT CHARACTERISTICS OF CONTINENTAL FLOOD BASALT PROVINCES

Emplacement characteristics are key factors for understanding gas release from continental flood basalt provinces, both in terms of magma emplacement rates (and hence gas release rates) and also in terms of the duration of pulses of activity and the intervening hiatuses. A feature of all large igneous provinces is the high magma emplacement rates (e.g., Storey et al., 2007; Svensen et al., 2012), with aggregate extruded lava volumes of 1×10^6 km^3 or more erupted over periods of one to a few million years, corresponding to a time-averaged lava emplacement volume of at least 1 km^3/yr. The Siberian Traps formed ca. 250 Ma (e.g., Reichow et al., 2005, 2009) and they represent volumetrically the largest continental flood basalt province. The Siberian Trap province covers about 5×10^6 km^2 and totals about 4×10^6 km^3 of mainly basaltic lava and associated volcanic rocks, which erupted over a period of ~1 m.y. (Kamo et al., 2003). This province also has voluminous intrusive rocks (e.g., Svensen et al., 2009), as well as lava types that are not strictly basalt (Sobolev et al., 2011; for a recent review see Black et al., 2012).

Recent studies have shown that the bulk erupted lava volumes from continental flood basalt provinces only give us part of the story, because continental flood basalt volcanism is highly pulsed (i.e., alternating phases of activity and non-activity) over the time scale of the emplacement of the province as a whole. Understanding the timing, duration, and intensity as well as the length of episodes of activity and episodes of non-activity is a key requirement for the assessment of any potential environmental effects caused by continental flood basalt volcanism. Studies using the ^{40}Ar/^{39}Ar geochronological method (Courtillot and Renne, 2003), and, to a lesser extent, uranium-lead (U-Pb) age determinations, have shown that many continental flood basalt provinces erupted over quite brief periods of geologic time (for a review, see Kelley, 2007), which is especially true for the main pulse of lava production.

The best-understood and age-dated continental flood basalt province, the Columbia River Basalt Group (Fig. 3), consists of a total extrusive volume of 250,000 km^3, erupted between 17 and 10 Ma (Miocene), with minor volumes of lava erupted to as young as 6 Ma. However, as shown in Figure 3, a volume of ~150,000 km^3 (or 72% of the total volume) was emplaced over ~400,000 yr (Barry et al., 2010, 2013) at ca. 16 Ma in the Grande Ronde Basalt Formation (it should be remembered that this time interval is predicated upon the errors of the ^{40}Ar/^{39}Ar ages). For another example, the Deccan Traps flood basalts were previously believed to have erupted over a period of ~5 m.y., based on the large range of early age determinations (see Vandamme et al., 1991). Redating of the basalts using refined Ar/Ar techniques, combined with paleomagnetic studies of the lavas, however, indicates that much of the Deccan lavas were erupted over a period of only ~500,000–800,000 yr (Chenet et al., 2009). The entire duration of Deccan volcanism may have been ~3 m.y., although this still remains to be thoroughly documented. Some continental flood basalt provinces may show evidence of two or more peaks in magma output; it has been suggested that the Deccan had three pulses, with the main pulse emplacing a considerable portion of the total volume in as little as 500,000 yr or less (Chenet et al., 2009; Keller et al., 2008, 2012).

Constraining the temporal occurrence and exact duration of the main activity phases for all continental flood basalt provinces is difficult because of the precision of geochronological determinations. Volumetrically larger continental flood basalt provinces, such as the Siberian Traps, originally contained in total twenty times as much lava as the Columbia River Basalt Group, but little robustly supported detail is known about when the major pulses of volcanism occurred in the Siberian Traps. The volcanism seems to have occurred over a 2-m.y.-long period from 252 to 250 Ma. Most of the lava ages seem to fall within a 1-m.y.-long period (Kamo et al., 2003; Reichow et al., 2009), but large parts of the province remain relatively unstudied.

In summary, we stress again that the acme or main pulse(s) of volcanism in a continental flood basalt province is a very important parameter when considering environmental effects of flood basalt eruptions. Only two continental flood basalt provinces are presently well enough studied to say where in the lava pile, and when geochronologically, these main volumetric pulses of lava production might have occurred: the Columbia River Basalt Group province and the Deccan Traps province. It is not even certain yet that the Deccan's main volumetric pulse came late (i.e., at the time of the extinction peak) in the eruptive history of the province (Self et al., 2006; Chenet et al., 2007, 2009). Generalizing from these two relatively well-constrained examples, continental flood basalt volcanism appears to last for 1–3 m.y., with the pulse or pulses of main lava-volume emplacement being as brief as a few hundred thousand years. Not every continental flood basalt province may have followed this model of formation; therefore, caution is needed where part of a continental flood basalt province is poorly known and/or spatially very widespread and buried (e.g., the Siberian Traps; Reichow et al., 2005, 2009).

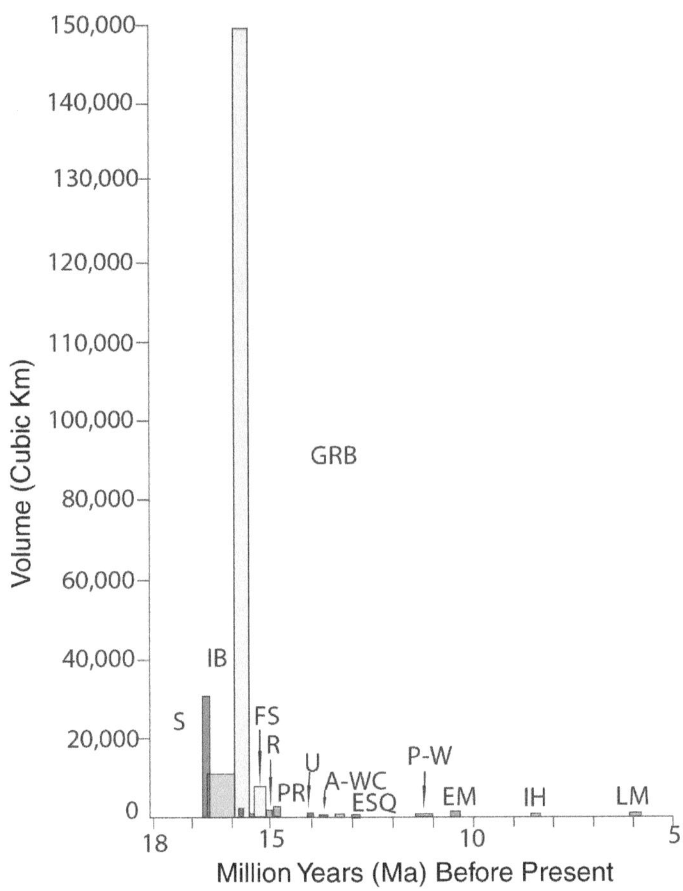

Figure 3. Volume of lava plotted against time (based on latest $^{40}Ar/^{39}Ar$ age dates; after Barry et al., 2013) for the Columbia River Basalt Group, showing the large "spike" of the main pulse of volcanism, the Grande Ronde Basalt Formation (GRB). Other formations of Columbia River lavas are identified by the initials in Barry et al. (2013) and are not mentioned in this chapter.

EMPLACEMENT CHARACTERISTICS OF INDIVIDUAL CONTINENTAL FLOOD BASALT ERUPTIONS

Eruption characteristics and style of individual eruptions are further important considerations when attempting to constrain the flux of volcanic gases from continental flood basalt provinces. By building up databases concerning these individual eruptive units, we will understand more in terms of the degree to which we can generalize from one eruption to another (both within the same provinces and between different provinces). This is important regarding generalizing their typical volumes, durations, and intensities, but also in order to allow us to make the most appropriate decisions regarding the heights of injection into the atmosphere to choose as model inputs.

During the emplacement of continental flood basalt provinces, hundreds to possibly thousands of eruptions produce immense lava flow fields, each the product of a single or cluster of vents, or a group of vents along a fissure (i.e., a line of volcanic vents; Self et al., 1997, 1998). Consequently, all flood basalt lava fields are dominated by basaltic lava flows (see Fig. 2), and, perhaps rather surprisingly, all continental flood basalt provinces examined in detail are dominated by compound pāhoehoe lava flow fields (e.g., White et al., 2009; Bryan et al., 2010). The term lava flow field designates the entire lava products of one prolonged eruption. Overall, this is a reflection of similar eruption conditions and magma types in most provinces. A certain amount of volcanological knowledge exists about the subaerial volcanism that occurred in continental flood basalt provinces (e.g., the Deccan province; Bondre et al., 2004), but this does not extend to the intrusive components, except in a couple of cases, i.e., the Karoo-Ferrar province (Svensen et al., 2012; Moulin et al., 2011; Elliot et al., 1999) and the Siberian province (e.g., Svensen et al., 2009; Sobolev et al., 2011).

For the purposes of estimating the gas output of a flood basalt eruption, it is desirable to break a province down to its individual eruptive units. Therefore, in the next two subsections, we discuss the general characteristics of individual flood basalt lava flow fields (see also Bryan et al., 2010), and the challenges of determining the timing and volumes produced by a main pulse of magma effusion during emplacement of a province. Resolving the duration of individual eruptions within continental flood basalt provinces is an even greater challenge. Only for the Roza flow field (1300 km^3 of basalt lava) of the Columbia River Basalt Group has an independent, first-order estimate been made based on the characteristics of the fissure-fed pāhoehoe lava flows that inflate during emplacement, which constrains the eruptive phase to have lasted a minimum of ~15 yr (Thordarson and Self, 1998).

In terms of understanding the impacts of continental flood basalt province emplacement on our planet, constraining the timing and length of hiatus periods within an eruption sequence is as important as knowing the durations of eruptions. For the Grande Ronde Formation of the Columbia River Basalt Group, there were ~100 separate eruptions within ~400,000 yr or less (Reidel and Tolan, 2013), but both of these values still have quite a degree of uncertainty. In the Columbia River Basalt Group case, the average hiatus was 3900 yr if we assume that each eruption lasted 100 yr, but in reality, this average hiatus may not be typical of the durations between eruptions.

Unfortunately, our understanding of most other continental flood basalt provinces is poorer by far than our knowledge about the Columbia River Basalt Group (Reidel et al., 2013). We have little to guide us when interpreting any eruption scenarios based on eruption frequency and hiatus length between eruptions, due the present inability of radiometric age-dating methods applied to basalts to be more accurate than about ±0.1 Ma at tens of millions of years, and about ±0.5–1.0 Ma at fifty to hundreds of millions of years. Some studies have attempted to define groups of lavas erupted over shorter periods of time; these are discussed later.

Eruption Products and Eruption Style

Flood basalt eruptions simply make pāhoehoe lavas, which are the smooth, ropey-surfaced flows that are common on all Hawaiian volcanoes, and are Earth's most common eruptive product (Self et al., 1998). All continental flood basalt provinces, from the very ancient geologic eras to the youngest on Earth (i.e., the Columbia River Basalt Group), consist almost entirely of pāhoehoe lavas. Vents for the lavas are rarely preserved, but the vent system for the Roza flow field of the Columbia River Basalt Group was discovered in 1975 and has recently been restudied (Swanson et al., 1975; Brown et al., 2014). The proximal vent facies show evidence of periodic explosive (fire-fountain) activity (Self et al., 2005) recurring many times during the eruption; thus, the style of activity was somewhat like Hawaiian effusions with accompanying fire-fountain activity, but the volumes of lava produced were magnified many times in scale, and the resulting flow fields are extensive and voluminous.

Flow fields are composed of thousands of lava sheet lobes (Fig. 4). Sheet lobes, and innumerable smaller lobes, stacked and superposed in some places and laterally arranged in others, make up whole flood basalt provinces. Evidence of hiatuses between the eruptions is presented by weathered surfaces on tops of lava flow fields, such as the red boles of the Deccan Traps (Widdowson et al., 1997), but we must remember that flow fields, per se, cannot be recognized in many continental flood basalt provinces, where only subvertical sections through the lavas are exposed. Difficulties in correlating individual lava flows between outcrops (e.g., Jay et al., 2009) presently preclude the recognition of the geographic extent of single flow fields in provinces such as the Deccan Traps.

Emplacement Rates and Lava Volumes

Insights on the lava volume produced and duration of individual eruptions can be gained by studying the physical characteristics of their deposits. Flow fields are composed of many lava flows, which never form a single stretch of lava but consist of a succession of lobes that represent the sequential emplacement of the lava flow (Fig. 4; see Hon et al., 1994; Self et al., 1998). These lava lobes can occur on several scales. For instance, kilometer-scale lobes can be found to include centimeter- to meter-scale lava lobes in the field. One common type in flood basalt provinces is sheet lobes, which are typically several kilometers in lateral dimensions and tens of meters thick. The sheet lobes inflate to a considerable thicknesses during the time that they reach their great extent (Self et al., 1997; Vye-Brown et al., 2013). This suggests a relatively gradual emplacement of the flows over a period of time, most likely years to decades (Thordarson and Self, 1998; Jerram and Widdowson, 2005). Yet, the average lava output rates during an individual continental flood basalt eruption must still have been higher than in any historic eruption except perhaps the extremes of some Icelandic lava outpourings. For example, the time required to emplace a 1000–2000 km^3 lava flow field at the range of peak output rates of the Laki eruption (2000–4000 m^3/s; Thordarson and Self, 2003) would be ~10–20 yr. If the eruptions were much faster (higher magma output rates), then there would be different types of lava in the flow fields, such as ʻaʻā (lava characterized by a rough or rubbly surface), and channeled flows with evidence of thermal erosion at the base of flows (Keszthelyi and Self, 1998; Greeley et al., 1998); if eruptions were much slower, then there would be insufficient time to accommodate the number of lava-producing eruptions, and hiatuses between eruptions, during the geochronologically constrained duration of a main continental flood basalt pulse.

Despite the minimum volumes of individual eruptive units being known for parts of the Columbia River Basalt Group, it is only recently that some understanding has been gained on the possible eruptive volume or mass of flood basalt eruptions from other continental flood basalt provinces, mainly the Deccan Traps of India (e.g., Chenet et al., 2008; Self et al., 2008a). Assumptions about typical flow thicknesses can also give us insights into potential flow volumes. For the Deccan example, if areas from 70,000 to 175,000 km^2 were covered by pāhoehoe lava flow fields during emplacement of some formations in the Wai Subgroup (Self et al., 2006, 2008a), then adopting a range of flow-field thicknesses from 30 to 50 m (Jay et al., 2009) indicates eruptive volumes in the range 2100–8750 km^3. Further, the Mesozoic to Cenozoic continental flood basalt provinces are the best studied and preserved, and the information is biased toward the younger examples. Constraints on the dimensions of almost all eruptive lava flow field units in continental flood basalt provinces are hampered by the effects of burial, tectonism, and erosion since emplacement.

One thing that seems to be exceptional about basaltic large igneous provinces is that most individual eruptions in continental

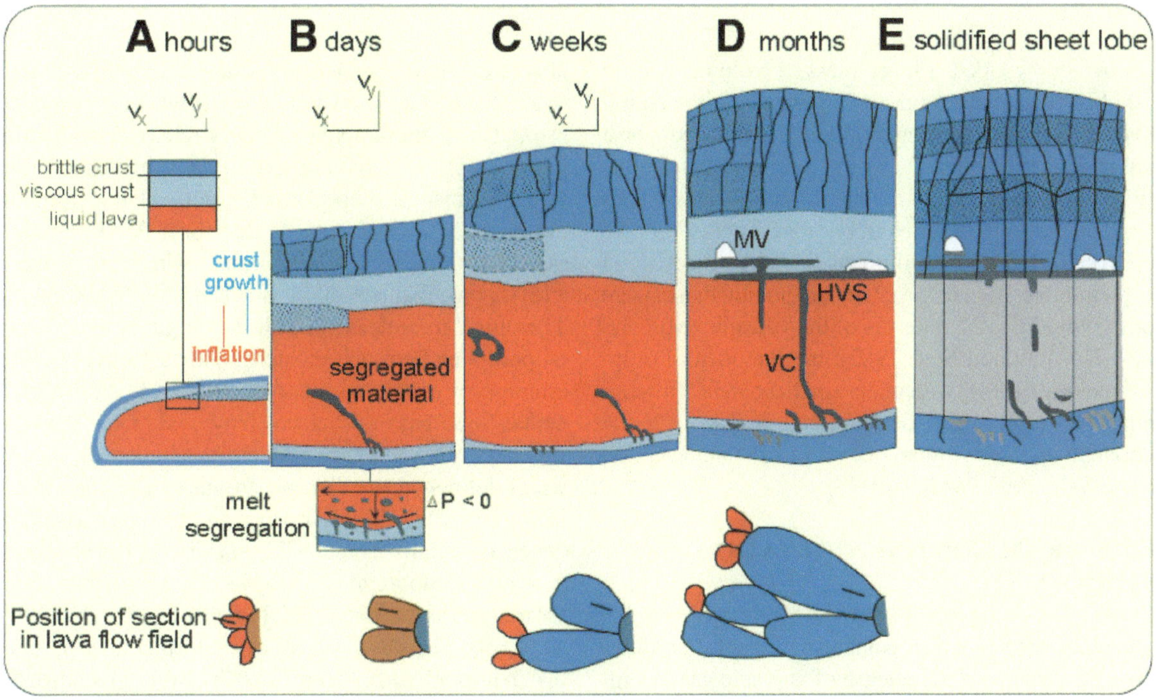

Figure 4. Cartoon showing the development of a pāhoehoe basalt lava flow field over time as the lava sheet lobes grow outward and downslope by advance. Lobes also coalesce and thicken by inflation. Vx and Vy show proportions of lateral and vertical expansion of flow lobe. Inset shows localized separation of flow lines leading to drops in internal pressure (P) and vesiculation at the basal crust-core boundary. Schematic plan view of growing lava flow field is given beneath each stage. MV—megavesicles; HVS—horizontal vesicular sheets; VC—vesicle cylinders. This scheme applies to flood basalt lava flow fields (after Thordarson and Self, 1998).

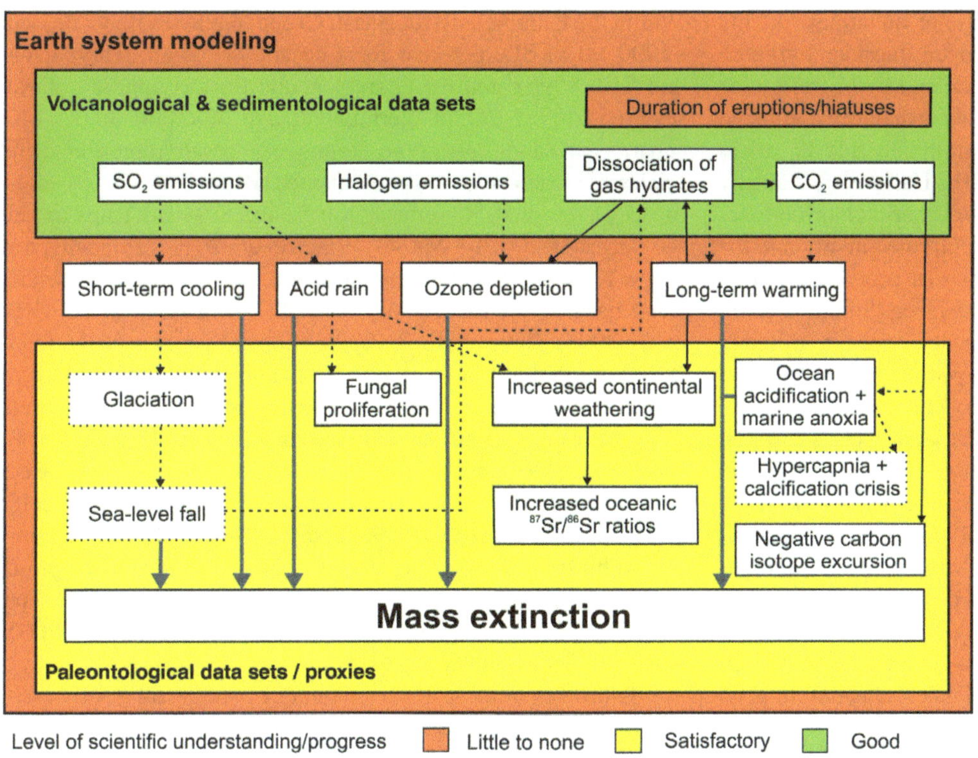

Figure 5. Schematic illustrating some of the proposed environmental effects and Earth system feedbacks of continental flood basalt (CFB) volcanism (adapted from Wignall [2001, 2007], who summarized the chain of feedbacks for the Siberian Traps). Not all volcanic gas emissions and therefore feedbacks shown here are applicable to all continental flood basalt provinces. Dashed lines and box outlines indicate effects and feedbacks that are possible in theory but have not yet been fully quantified or are still debated in the authors' opinion. Light gray arrows indicate main mechanisms thought to cause a mass extinction (either terrestrial or marine or both). The colors indicate the level of scientific understanding and progress of each discipline, illustrating that while the quantification of gas fluxes from continental flood basalt volcanism and paleontological proxies are available (green and yellow), some key mechanistic understanding that links the two is missing (orange). For example, lack of knowledge of the magnitude of short-term cooling from volcanic sulfur, or the amount of acid deposition of volcanic sulfur species, means that some of the proposed cause-and-effect relationships are weak and speculative. Earth system modeling is one of the means by which to quantify these linkages, but there is little scientific progress to date.

flood basalt provinces yielded a huge mass or volume; with a limit of 1×10^{15} kg for a super-eruption (Self, 2006), this is equivalent to ~360 km^3 of basaltic lava, and only a few of the smallest of the Columbia River Basalt Group units may fall below this class. Thus, it seems that when continental flood basalt provinces were forming, they did so by a series of basaltic super-eruptions, each one far exceeding the lower limit of this class (see Bryan et al., 2010, their table 2). However, even though these were huge eruptions volumetrically, in terms of the volcanic explosivity index (VEI; Newhall and Self, 1982), the majority of eruptions may have been VEI 3–4, with occasional explosive (intense fire-fountaining) phases of VEI 5–6 magnitude, analogous to the violent Strombolian and sub-Plinian phases of the Laki eruption (Thordarson and Self, 1993; Woods, 1993).

GAS RELEASE FROM CONTINENTAL FLOOD BASALT ERUPTIONS

One of the main ways in which volcanism impacts our atmosphere and environment is via the release of volcanic gases. Figure 5 summarizes some of the ways in which these gases may have affected the atmosphere and the environment in the geological past, including feedbacks that have been proposed to lead to sufficient ecosystem stress to cause a mass extinction.

In general, water vapor (H_2O) is the most abundant volatile species released during an eruption (contributing between 50% and 90% by volume of the gas phase) followed by carbon dioxide (CO_2), which contributes between 1% and 40% by volume (Gerlach, 2004). Halogens are released in minor quantities, mainly in the form of hydrogen chloride, hydrogen fluoride, and, with even lower abundance, hydrogen bromide and hydrogen iodide (Pyle and Mather, 2009). Despite their low abundance compared with other species, halogens play an important role in volcanic plume chemistry (von Glasow, 2010). Halogens efficiently destroy ozone; however, under low stratospheric chlorine loadings, an eruption must deliver sufficient chlorine directly into the stratosphere for heterogeneous reactions that promote chlorine activation on liquid sulfate aerosol to take place (Solomon et al., 1998). Whether or not volcanic eruptions can deliver sufficient halogens to stratospheric altitudes is, however, still a matter of debate (e.g., Tabazadeh and Turco, 1993; Kutterolf et al., 2013). To date, sulfur dioxide (SO_2) is the sole volatile species released during volcanic eruptions that, following conversion to sulfate aerosol, has been shown to alter the radiative balance of the atmosphere to a measurable extent (e.g., Robock, 2000). Sulfur (S) species contribute between 2% and 35% by volume of the gas phase, and SO_2 and hydrogen sulfide (H_2S) are most abundant (Gerlach, 2004; Textor et al., 2003, 2004).

Estimates of the volatile content of magmas from past eruptions usually come from melt (now glass) inclusions trapped within crystals, which is feasible but challenging for continental flood basalt volcanism, because both crystals and preserved degassed matrix glass are rare in continental flood basalt lavas, and both crystals and glassy lava are often (but not always) altered due to the great age of many continental flood basalt lavas (e.g., Blake et al., 2010). In Table 1, we have compiled published estimates of the amount of gases released by continental flood basalt eruptions and compared these estimates to those for explosive volcanism and flood basalt eruptions in historic times. Basaltic magmas, including those that form flood basalts, are usually rich in dissolved sulfur (commonly with sulfur concentrations of ≥1500 ppm; Wallace, 2005; Black et al., 2012; Zhang et al., 2013). Therefore, the release of sulfur-rich gases from a large basaltic eruption can be much greater than that from an explosive silicic eruption of equal volume. Not all volatile emissions from continental flood basalt provinces may come from the basalts. Minor volatile emissions may arise from silicic volcanism associated with continental flood basalt provinces (Scaillet and Macdonald, 2006; Bryan et al., 2010; Bryan and Ferrari, 2013). Recent evidence, especially from the Siberian Traps, has also highlighted the potentially important role of gas released via interactions between the magma and country rock (Ganino and Arndt, 2009; Svensen et al., 2009). These will be discussed further later herein.

Although data are still sparse (Table 1), comparisons can be made in terms of the gas release budgets from the magmas associated with different continental flood basalt provinces. Self et al. (2008b) measured major-element and volatile composition of melt inclusions and groundmass glass in the quenched products of the Deccan Trap lavas and estimated that individual decade-long continental flood basalt eruptions could have released up to 5000 Mt of SO_2, which equates to a total SO_2 release of up to 6.5×10^6 Mt estimated for the Deccan Trap province as a whole. Within the Columbia River Basalt Group, the Roza flow released ~1200 Mt of SO_2 per year for a decade or longer (Thordarson and Self, 1996), and the Grande Ronde Basalts appear to have released ~10^6 Mt SO_2 in intermittent bursts of <1000–30,000 Mt, separated by long-lasting non-eruptive intervals represented by thick soil horizons in the lava sequence (Blake et al., 2010). Estimates of the total SO_2 emission from the Siberian Traps are in the range $10–20 \times 10^6$ Mt SO_2 (Black et al., 2012). Typically, for every 1 km^3 of basaltic magma emplaced during a continental flood basalt eruption, 3.5–6.5 Mt of SO_2 will be released (Self et al., 2005; Blake et al., 2010). For context, studies of the Laki lava flows and ash indicate that the magma originally contained ~1700 ppm of sulfur, and that the eruption could have released ~120 Mt of SO_2 from the vent and lava flows (~15 km^3 in volume) over the course of 8 mo (two-thirds of which were released during the first 2 mo of intense activity; Thordarson and Self, 1993; Thordarson et al., 1996). The 1991 Mount Pinatubo eruption (Philippines) was the largest explosive volcanic eruption since 1912 and released ~20 Mt of SO_2 into the stratosphere. For context, the anthropogenic SO_2 flux to the atmosphere in 2005 was ~118 Mt (Smith et al., 2011).

Estimates of the halogen content of continental flood basalt magmas are more sparse than for sulfur. In some cases, the halogens appear to stay in the melt phase of the magma, suggesting limited degassing during eruption. This is probably controlled by details of the oxidation and degassing history as the magmas rise.

TABLE 1. SUMMARY OF PUBLISHED ERUPTION SOURCE PARAMETERS FOR DIFFERENT ERUPTION TYPES AND ERUPTION STYLES

Mode	Age of activity	Duration total	Estimate of total subaerial gases released (Mt)				Height of injection	References
			SO_2	CO_2	HCl	HF		
Present-day persistent emissions (Mt/yr)	Present day	Continuous	13 ± 5	65–540	4.3 ± 1	0.5 ± 0.2	Passive emissions vary with volcano summit height (most >1 km). In terms of SO_2: ~10 Mt/yr passive to troposphere, ~1–4 Mt/yr on average to stratosphere from "large" eruptions and rest from sporadic "small" eruptions to troposphere.	Pyle and Mather (2009); Stoiber et al. (1987); Andres and Kasgnoc (1998); Halmer et al. (2002); Pyle and Mather (2003); Mather et al. (2003); Burton et al. (2013); Graf et al. (1997).
Pinatubo (largest eruption of the satellite era)	1991	5 h (climactic phase)	12–26[*]	42	3–4.5		>30	Krueger et al. (1995); McCormick et al. (1995); Westrich and Gerlach (1992); Gerlach et al. (1996)
Tambora (largest historic eruption)	1815	5–6 d (climax ~24 h)	20–200	Not estimated	100		43	Oppenheimer (2003)
Youngest Toba Tuff (largest Quaternary eruption)	ca. 74 ka	9–14 d	25–3300	Not estimated	1000	1000	32–40	Oppenheimer (2002); Chesner and Luhr (2010)
Continental flood basalt volcanism (limited to cases where magmatic gas fluxes are published)								
Columbia River Basalt Group (CRBG, ca. 17–6 Ma)[†]								
Grande Ronde Basalts (CRBG)	ca. 16–15.6 Ma	0.4 m.y.	1×10^{6}[§]	Not estimated	Not estimated	Not estimated		Blake et al. (2010)
Roza Member (CRBG)	ca. 14.7 Ma	~10 yr	~12,420	Not estimated	~710	~1780	~9620 Mt SO_2, ~400 Mt HCl, and ~1450 Mt HF released at vents into 7–13 km altitude.	Thordarson and Self (1996)
Deccan Traps[#]	ca. 67–63 Ma	4 m.y. period, with most activity within a 1 m.y. period. Pulses of tens–thousands years with hiatuses of 10^3–10^4 yr.	3.5–6.5 × 10^6[**]	1.4 × 10^7[††]	1 × 10^6	Not estimated	Volatiles released from vents to altitudes between 7–15 km.	Self et al. (2006, 2008b)
Paraná-Etendeka Province[§§]	ca. 132 Ma	~1 m.y.	6.2–10.8 × 10^6[##]	Not estimated	7.2–48 × 10^4	6.3–12.6 × 10^5		Marks et al. (2014)
Siberian Traps[***]	ca. 252–248 Ma	Bulk of the eruption occurred during <1 m.y. period.	~12.6–15.6 × 10^6[†††]	Not estimated[§§§]	~3.4–8.7 × 10^6	~7.1–13.6 × 10^6	Uncertain but injection into the stratosphere is likely during more violent explosive phases. Early fragmental deposits of uncertain eruptive style.	Black et al. (2012, 2013)
Emeishan province[###]	ca. 262–256 Ma		1.5 × 10^6[****]	16.8 × 10^6[††††]	Not estimated	Not estimated	Volatiles released from vents into altitudes between 7–15 km. Violent phreatomagmatic phases at the onset of activity might have injected volatiles to altitudes >15 km.	Zhang et al. (2013); Wignall et al. (2009); Ganino and Arndt (2009)
Laki	June 1783 to February 1784 CE	8 mo	122[§§§§]	304	~7	~15	6–13 km emitting ~95 Tg Mt SO_2, ~3.5 Mt HCl, and 8.3 Tg Mt HF at the vents to the upper troposphere/lower stratosphere.	Thordarson and Self (2003); Thordarson et al. (1996); Hartley et al. (2014)

Note: Estimates of continental flood basalt ages, durations, and paleolatitude are from Kelley (2007), Rampino and Self (2000), and sources therein. Estimates of the nature of associated country rock are from Ganino and Arndt (2009).

[*]20 usually accepted value.

[†]Total volume ~0.25×10^6 km³ with a duration of the peak pulse ≤0.5 m.y. (for 72%). Its paleolatitude was 45°N. Associated country rock thought to be granitoid/gneiss, sandstone/conglomerate, and shale/silt.

[§]Delivered in intermittent bursts of <1 to 30 Gt separated by long noneruptive intervals.

[#]Total volume >1.3×10^6 km³ with a duration of the peak pulse ≤1 m.y. Its paleolatitude was ~20°S. Associated country rock is granitoid/gneiss, sandstone/conglomerate, and shale/silt.

[**]This is based on an estimate of 10^6 km³ of lava for the Deccan Trap province. Evidence suggests that fluxes of 10^2–10^3 Mt of SO_2 per year might have been sustained during eruptive phases for a decade.

[††]Based on Self et al. (2006), who estimated for 10^3 km³ and scaled up for 10^6 km³. They estimated 220–1100 Mt/yr for 10–50 yr pulses of 10^3 km³ of lava.

[§§]Its paleolatitude was 40°S. Associated country rock thought to be granitoid/gneiss, sandstone/conglomerate and shale/silt.

[##]Based on a volume of extruded magma in the province of 2.2 to 2.35×10^6 km³ and previously published degassing efficiencies. Scaillet and Macdonald (2006) suggest a further 9.2×10^3 Mt SO_2 for the metaluminous silicic magmas associated with the flood basalts.

[***]Total volume >2×10^6 km³ with a duration of the peak pulse ≤1 m.y. Its paleolatitude was 60–65°N. Associated country rock thought to be limestone, coal, shale/silt, and evaporite.

[†††]Based on an estimate of 4×10^6 km³ of lava for the Siberian Trap province.

[§§§]Estimates of up to >100,000 Gt CO_2 from metamorphism of organic matter and petroleum (Svensen et al., 2009) as well as large quantities of CH_4 (from organic-rich shale and coal) and CH_3Cl (from limestones and rock salt) (Aarnes et al., 2011).

[###]Total volume ~0.4×10^6 km³ with a duration of the peak pulse 1–2 m.y. Its paleolatitude was 25°N. Associated country rock thought to be limestone, dolomite, sandstone/conglomerate, and shale/silt.

[****]Based on a total lava volume of ~0.3×10^6 km³ (Zhang et al., 2013, and references therein).

[††††]Estimates of additional CO_2 emissions of 61,600–145,600 Gt from contact aureoles with country rock sediments.

[§§§§]Based on an estimate of 14.7 ± 1.0 km³ of lava (Thordarson and Self, 1993).

However, estimates for the Deccan Traps and the Siberian Traps suggest that emissions of chlorine and fluorine might be of the same order of magnitude as those of sulfur (Table 1; Self et al., 2008b; Black et al., 2012), or possibly higher (Sobolev et al., 2011). Further, there has been speculation that chlorine emissions from the Deccan Traps might account for a Cl-rich layer of sediments identified in France and Italy (Font et al., 2011). Volcanic bromine has greater potential ozone depletion efficiency than other volcanic halogen species; however, due to its low concentrations in the melt and analytical challenges, our understanding of volcanic bromine release is in its relative infancy (Pyle and Mather, 2009; Mather et al., 2012). A recent study by Kutterolf et al. (2013) suggested that new techniques such as synchrotron radiation might one day hold promise for studying bromine emissions from continental flood basalt volcanism.

Direct determinations of the amounts of CO_2 that might be degassed from flood basalt magmas upon eruption are not available. However, it is known that CO_2 is relatively insoluble in basaltic melts and that the mantle is undersaturated with respect to CO_2. Self et al. (2006) adopted 0.5 wt% as a high, but reasonable, value for pre-eruptive CO_2 concentration in flood basalt magmas and considered that magma degassing is a highly efficient process, with 70%–80% of the gas released at the eruptive vents (Thordarson et al., 2003). Self et al. (2006) estimated that ~1.4×10^{10} kg, or 14 Mt of CO_2, could be released for every 1 km^3 of basaltic lava erupted (assuming a density of 2750 kg m^{-3}). Therefore, the total release from an erupted lava volume of 1000 km^3 (the approximate volume for one eruption in the Deccan Traps) might be ~14×10^3 Mt of CO_2 (or 1.4×10^7 Mt CO_2 from the emplacement of the entire Deccan Traps). Ganino and Arndt (2009) used similar reasoning to estimate an emission of 16.8×10^6 Mt magmatic CO_2 for the Emeishan Province. These are the only estimates of CO_2 released from a continental flood basalt province that we are aware of to date (Table 1). While the emission from a single Deccan eruption is a very large mass, it should be noted that it represents less than 1/200th of the amount of CO_2 present in the modern atmosphere (~3×10^6 Mt, or 3×10^{15} kg). For comparison, the current anthropogenic emissions of CO_2 to the atmosphere are ~35 Gt/yr (Le Quéré et al., 2013). Thus, the annual CO_2 flux from a single Deccan eruption equates to only 40% of the current anthropogenic CO_2 flux under the assumption that all 14 Gt of magmatic CO_2 is emitted in an unrealistically short period of a year. If released over a more realistic decade time scale, these volcanic emissions represent ~4% of the current anthropogenic flux only. However, CO_2 has long atmospheric residence times, and the weathering of huge areas of basaltic rock could also have affected atmospheric CO_2 concentrations (Dessert et al., 2001) and/or ocean acidity, which could be tested in the future using state-of-the-art carbon cycle models that account for the "long tail" of CO_2 emissions.

As mentioned already, the volatile load carried by the basalts might only account for part of the release of gases to the atmosphere associated with continental flood basalt volcanism. Recent studies have suggested that gases released from nonmagmatic sources may make an important additional contribution to the gas release budget of continental flood basalt volcanism or indeed might dominate. For example, Ganino and Arndt (2009) have correlated the degree of environmental stress caused by continental flood basalt volcanism with the types of sediments beneath the lavas and estimated that emission of CO_2 from contact aureoles with the sedimentary country rock could have been 4–9 times larger than that released from the magma for the Emeishan Traps. In the case of the Siberian Traps, it has been suggested that magma-sediment interactions could dissociate vast quantities of hydrocarbons and halocarbons such as methane from heating of organic-rich shale and petroleum-bearing evaporates (basin-scale gas production potential estimates suggest that metamorphism of organic matter and petroleum could have generated >100,000 Gt CO_2), resulting in global warming and ozone depletion (Aarnes et al., 2011; Svensen et al., 2009). The suggested mechanism of gas release is through hydrothermal vent complexes, which are now being recognized as a characteristic component of large igneous provinces (for a review, see Svensen and Jamtveit, 2010). Iacono-Marziano et al. (2012a, 2012b) suggested that the release of vast amounts of reduced gases due to magma intrusion in the coaliferous sediments beneath the Siberian province contributed to global warming and to the negative carbon isotopic shift observed at the end-Permian. Grasby et al. (2011) suggested that deposition of coal fly-ash generated by magma-coal pyrometamorphism in the Siberian province resulted in toxic marine conditions. Black et al. (2013) presented modeling to suggest that the halocarbon release from crustal reservoirs associated with Siberian Traps volcanism (Aarnes et al., 2011) could have had significant effects on stratospheric ozone levels. It should be noted that the tropospheric lifetime of halocarbons such as CH_3Cl is sufficiently long (>1 yr) that even those released in the lower troposphere reach the stratosphere, reducing the sensitivity of their environmental effects to atmospheric injection.

GAS INJECTION HEIGHTS OF FLOOD BASALT ERUPTIONS

Basaltic eruptions are commonly more effusive (i.e., lava producing) in character, last much longer, and are much less explosive than silicic eruptions of comparable volume. Basaltic eruptions are also commonly considered to be much less effective in lofting gases into the stratosphere, where gas and aerosol particle lifetimes are longer than in the troposphere. Observations and plume model simulations (Woods, 1993; see Self et al., 2005) suggest, however, that the high eruption rates during the CE 1783–1784 Laki eruption in Iceland produced tall fire-fountains (600–1450 m), so that the convective plume rising above the fountains could have attained altitudes of up to 15 km above sea level (Thordarson et al., 1996; see Fig. 6), which is above the Icelandic tropopause during the summer months. By analogy, at the maximum eruption rates for Laki, the Roza flow eruption of the Columbia River Basalt Group would have produced lava fountains more than 1.5 km in height and may have created a convective column rising ~13–15 km above the volcanic vents (Table 1; Self et al., 2005; Brown et al., 2014).

Extensive ash deposits are associated with Icelandic basaltic fissure eruptions, such as Eldgjá, and the occurrences of deposits of spatter, spatter-fed lava, and scoria mounds along Columbia River Basalt Group eruptive fissures suggest violent to sometimes mild fire-fountaining during flood basalt eruptions. In support of this, observations of a widespread sulfurous haze over Europe in 1783 (Thordarson and Self, 2003; Witham and Oppenheimer, 2005; Schmidt et al., 2011), as well as a sulfuric acidity peak in Greenland ice in 1783 suggest that the Laki eruption column reached the lower stratosphere. For the much larger-volume lava flows typical of continental flood basalt eruptions, the models of convective plume height also indicate that sulfur-rich eruption plumes could reach the lower stratosphere by direct injection in an eruption column (Stothers et al., 1986; Thordarson and Self, 1996), or by injection within a volcanic plume helped by convection above a large, active lava field (Kaminski et al., 2011), or perhaps by a lofting of gases into the lower stratosphere in deep convective systems (e.g., Bourassa et al., 2012). Future efforts in this area should also account for paleo-atmospheric temperatures and composition when using plume rise models because the tropopause height is likely to differ significantly compared to the present day in, for example, a much warmer Late Cretaceous atmosphere.

POTENTIAL ENVIRONMENTAL EFFECTS OF CONTINENTAL FLOOD BASALT VOLCANISM

Evidence of the environmental impacts of continental flood basalt volcanism and the processes leading to these effects comes mainly from three sources: (1) direct evidence from the proxy record, (2) historical records of the effects of large-scale volcanic activity such as the Laki flood basalt eruption in 1783 CE, and (3) by scaling up from observations and measurements of the environmental impacts of present-day volcanism, which not only is on a different scale in terms of the mass eruption rate compared with continental flood basalt volcanism but also often differs fundamentally in terms of eruption style. Figure 5 summarizes the processes and feedbacks that have been proposed to have led to severe environmental changes at the time of continental flood basalt volcanism. The details of these processes are reviewed elsewhere (e.g., Officer et al., 1987; Wignall, 2001, 2007). Proposed cause-and-effect mechanisms include, among others, a short-term cooling effect from sulfate aerosol lasting years to decades and the deposition of sulfur species leading to acidification of

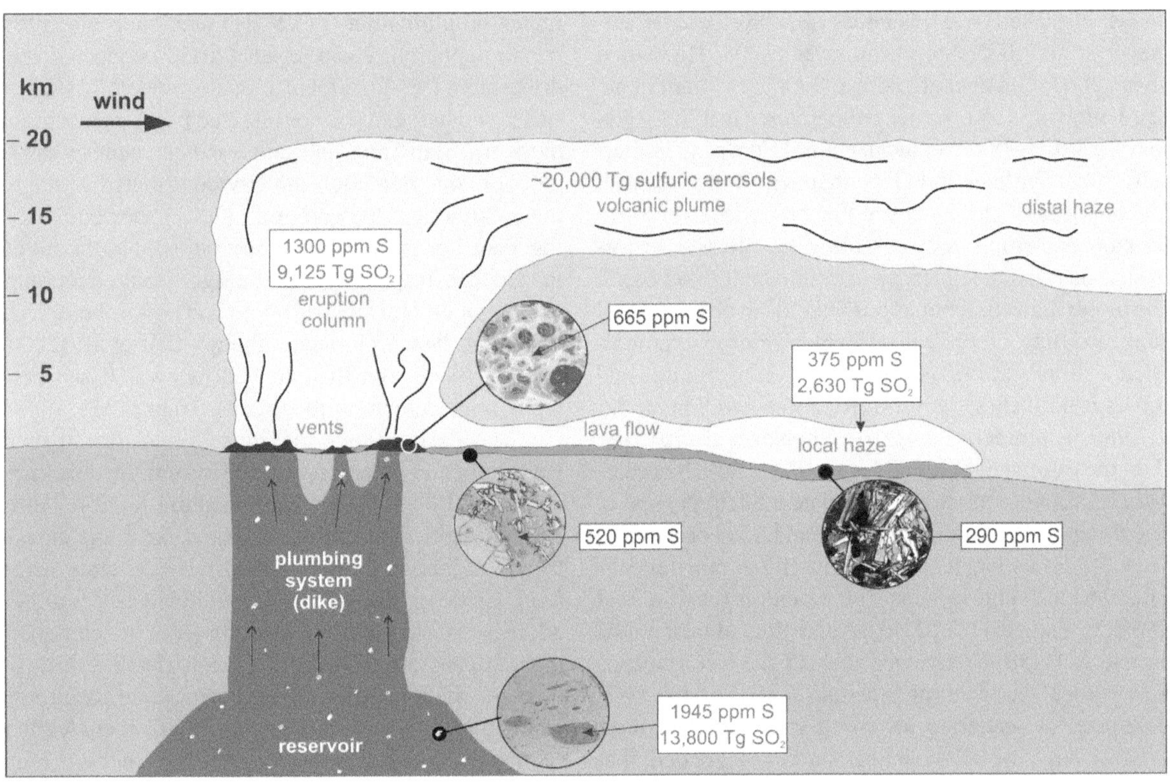

Figure 6. Cartoon showing generalized eruptive style and atmospheric dispersal of gas and aerosol particles (with minor ash) associated with basaltic fissure eruptive activity such as a flood basalt eruption, based on studies of the Roza eruption of the Columbia River Basalt province, Washington, USA (after Thordarson et al., 2009; for a more detailed explanation, see also Self et al., 2005). Tg is terragrams (10^{12} g). Total amount of SO_2 released into the atmosphere is estimated to have been almost 12,000 Tg, ~12 Gt (gigatons), with ~9000 Tg (9 Gt) coming from the vents with the eruption columns.

soils (Black et al., 2013; see the recent review by Bryan and Ferrari, 2013). Proposed long-term effects include global warming from the greenhouse gas forcing of volcanic CO_2 (released by the eruptions) lasting tens to thousands of years (Chenet et al., 2008; Sobolev et al., 2011; Joachimski et al., 2012), which was originally disputed by Caldeira and Rampino (1990) and is still a matter of debate (Self et al., 2006).

The proxy record, including the mass extinction events themselves, clearly suggests significant environmental perturbation around the time of continental flood basalt volcanism (e.g., Li and Keller, 1998; Courtillot and Renne, 2003; Wilf et al., 2003; Visscher et al., 2004; Wignall, 2005; Thibault and Gardin, 2006; Wignall et al., 2009; Keller et al., 2012; Sun et al., 2012; Joachimski et al., 2012; Bryan and Ferrari, 2013). However, one of the challenges that remains is to find a primary marker for the volcanism within these same archives, allowing a clear causal relationship to the environmental perturbation to be developed, rather than purely markers indicating environmental change. Recent efforts in this area have included the use of osmium isotopes (Ravizza and Peucker-Ehrenbrink, 2003; Robinson et al., 2009) and mercury (Sanei et al., 2012; Silva et al., 2014).

The Laki eruption might serve as a key historical analogue in terms of eruption style, but it was much shorter-lived (8 mo) compared with the average length of a continental flood basalt eruption. Modeling studies have demonstrated that its gas and aerosol clouds dispersed widely across the Northern Hemisphere (Chenet et al., 2005; Stevenson et al., 2003; Highwood and Stevenson, 2003; Oman et al., 2006a, 2006b; Schmidt et al., 2010, 2012). Following Laki, winter temperatures in 1783–1784 dropped up to -0.5 °C below average in central Europe (Thordarson and Self, 2003; Witham and Oppenheimer, 2005; Oman et al., 2006a; Schmidt et al., 2012). Given the difference in eruptive volume, duration, and eruption style, scaling the effects of historical and present-day volcanic activity to continental flood basalt scale may be highly flawed, if not impossible. Both observational evidence and modeling of short-lived explosive eruptions have suggested that a simple scaling between eruption magnitude and climatic impact is unlikely to be valid (Rampino and Self, 1982; Pinto et al., 1989; Timmreck et al., 2010; English et al., 2013).

Based on the estimated SO_2 flux for the Roza flow, Thordarson and Self (1996) were the first to estimate a likely change in aerosol optical depth (AOD) of between 7 and 13. For context, this is at least 2.5 times larger than current climate model estimates for the 74 ka Youngest Toba Tuff eruption (peak global mean AOD of 2.6; English et al., 2013). At the time of their study, Thordarson and Self (1996) assumed that all the SO_2 released forms sulfate aerosol particles of a certain size (i.e., the optimum particle size for scattering incoming solar radiation back to space). However, several modeling studies suggested that when considering aerosol microphysical processes, the impact of explosive volcanism on climate becomes self-limited with increasing SO_2 release, because the coagulation of particles causes particles to grow to large sizes, which have a lower optical depth per unit mass and fall out of the atmosphere faster than smaller particles (Pinto et al., 1989; Timmreck et al., 2010; English et al., 2013). Both Stevenson et al. (2003) and Schmidt et al. (2010) have shown that for a Laki-scale eruption, oxidants are heavily depleted, and so sulfate aerosol yields are lower (range 71–173 Mt of sulfate aerosol) than most previous estimates that assumed all SO_2 released was converted to sulfate aerosol (see tables 1 *in* Oman et al., 2006a and Schmidt et al., 2010).

Given the likely nonlinearities in the Earth system response to a continental flood basalt eruption, modeling studies are one of the key tools to allow us to interrogate the impact of volcanism on the environment. There have only been a handful of modeling studies directly tailored to continental flood basalt volcanism to date. This includes a carbon cycle modeling study, which found a negligible effect of magmatic CO_2 emissions from the Deccan Traps on Late Cretaceous temperatures (Caldeira and Rampino, 1990). Iacono-Marziano et al. (2012b) suggested that magma-sediment interactions during the emplacement of the Siberian Traps released vast amounts of reduced carbon monoxide and methane (CH_4), which resulted in enhanced atmospheric lifetimes of these species and a greenhouse gas forcing of climate on the order of 1.9 W/m^2 to 4.2 W/m^2 for a decade-long eruption. Using a two-dimensional (2-D) chemical transport model, Beerling et al. (2007) suggested that the release of hydrogen chloride and methyl chloride during the Siberian Traps eruptions could have resulted in mutagenic effects on plants through increased ultraviolet B (UV-B) radiation as a result of enhanced ozone depletion. Black et al. (2013) conducted the first three-dimensional (3-D) global climate modeling study of volcanic sulfur and halogen species released by a single magmatic episode of the Siberian Traps and suggested that widespread acid rain and global ozone depletion directly contributed to the end-Permian mass extinction. Dessert et al. (2001) showed, based on modeling seawater strontium isotope ratios, that the emplacement of the Deccan Traps basalts had a significant effect on geochemical cycling of carbon and strontium and atmospheric CO_2 concentrations, and hence climate, although Self et al. (2006) suggested that the model run that showed this was based on implausibly rapid emplacement of the Deccan Traps.

The dashed lines in Figure 5 illustrate that many of the environmental effects and Earth system feedbacks proposed for continental flood basalt volcanism are feasible but have not yet been quantified or are still debated, as is the case for volcanic CO_2 emissions and their impact on Late Cretaceous temperatures (e.g., Caldeira and Rampino, 1990; Chenet et al., 2008). Overall, there is a good to satisfactory level of scientific understanding and progress made by volcanologists estimating the amount of gases released during continental flood basalt volcanism and paleontologists gathering proxy records that clearly imply environmental changes. However, some key mechanistic understanding that links volcanic gas release and environmental effects is still missing. Earth system modeling may be one of the means by which to quantify these linkages, but there is little scientific progress to date. Eruption source parameter scenarios are key to meaningful future modeling efforts. In the next section, we therefore attempt to summarize eruption source parameters based on the currently available data discussed herein, which could be used as inputs for Earth system modeling studies.

Work described here has also highlighted the potential importance of magma–country-rock interactions. While Ganino and Arndt (2009) summarized the broad characteristics of the country rock into which the major known continental flood basalt provinces intruded, we are not yet in a position to generalize from better-studied continental flood basalt provinces to others, and so we do not include these emissions in the source parameters suggested for Earth system modeling detailed in the following section. It is clear, however, from the results of more detailed studies of the Emeishan province (Ganino and Arndt, 2009) and Siberian Traps (e.g., Svensen et al., 2009), that additional detailed work focused on each individual continental flood basalt province will further illuminate these processes and their potential impacts in each case. For example, shale metamorphism in the Karoo has been estimated to have emitted ~27,400 Gt CO_2 through breccia pipes (Svensen et al., 2007). The presence of coal deposits in the Karoo Basin (Cadle et al., 1993) also offers the intriguing possibility of interactions with the intrusive activity in this setting, as well as the Siberian Traps, although further investigation is needed to determine the potential size of this effect (Gröcke et al., 2009).

ERUPTION SOURCE PARAMETERS FOR EARTH SYSTEM MODELING

The Roza lava flow field in the Columbia River Basalt Group is the only individual continental flood basalt eruption for which information about the eruption source parameters is reasonably well constrained based on the geological record (see earlier). At the minimum, information on the volcanic gas flux and the eruption duration is needed as inputs for atmospheric modeling studies. For the majority of continental flood basalt provinces, these eruption source parameters are highly uncertain; yet they are crucial for the assessment of the length and severity of any potential environmental effects and Earth system feedbacks. For instance, if the non-eruptive phases (hiatuses) outlasted the duration of the volcanic sulfate aerosol forcing, then the Earth system would have had ample time to recover after an eruptive phase, which clearly weakens the feedback chain through to mass extinctions in Figure 5. Even hiatuses as short as 50 yr may leave ample time for the Earth system to recover from the cooling of climate caused by volcanic sulfur released during a decade-long continental flood basalt eruption.

Based on published data (Table 1) of the magmatic volatile release per cubic kilometer of lava emplaced (Fig. 7A) and the volume of lava emplaced per year (Fig. 7B), we calculate a mean SO_2 release of ~670 Mt per year for flood basalt eruptions (Fig. 7C). Individual continental flood basalt eruptions could have lasted between 10 and 50 yr (longer eruptions might be possible, but there is no recent analogue or any evidence to make informed estimates). Non-eruptive phases could have lasted many thousands of years, but overall we have little knowledge about their length and frequency. Red boles (soils) that are found sandwiched between the lava flow fields of the Deccan Traps (Widdowson et al., 1997) are thought to form on time scales of several hundreds to thousands of years (Fig. 7D), although in parts of the Deccan lavas, Chenet et al. (2008) made a case for soils forming in much less than 100 yr.

Other parameters needed for a model simulation are the injection height of the gases and the eruption style. The injection height in particular is difficult to constrain directly from the volcanic deposits. However, based on the similarity of the volcanic deposits and the chemical composition of the volcanic products, we can infer that large-scale flood basalt eruptions in historic times such as Laki provide a feasible and well-constrained analogue of the eruption style in the Columbia River Basalt Group and the Deccan Traps (Thordarson et al., 1996); that is a quasi-continuous injection of volcanic gases into the upper troposphere–lower stratosphere.

In summary, based on currently available data, eruption scenarios can be developed with some confidence for the SO_2 released from the Columbia River Basalt Group and the Deccan Traps lavas. By analogy, scenarios for the CO_2 release of the Deccan Traps or the release of halogen species by the Siberian Traps could follow this "template." However, for the Siberian Traps, detailed information on individual eruption volumes is lacking. Based on the volcanic deposits, the eruption type and style of the Columbia River Basalt Group and the Deccan Traps appear to be most commonly associated with emplacement of continental flood basalt provinces. Therefore, the suggested volume of lava emplaced per year and the gas flux per year for the Columbia River Basalt Group and Deccan Traps (Figs. 7B and 7C) may be a reasonable starting point for atmospheric modeling. Other continental flood basalt provinces such as the Siberian Traps may have exhibited large-magnitude explosive phases and also liberated substantial amounts of halogens, both directly via magmatic gases (Sobolev et al., 2011; Black et al., 2012, 2013) and via magma–country-rock interactions (Svensen et al., 2009; Iacono-Marziano et al., 2012a, 2012b; Ganino and Arndt, 2009). Black et al. (2013) addressed these uncertainties by modeling 27 perturbation and recovery scenarios to approximate gas release from plausible magmatic episodes, including a large pyroclastic eruption, metamorphism of hydrocarbon-bearing evaporite salts, and one or more explosive pipes, as well as stratospheric versus tropospheric injection. Overall, there is no simple way to generalize these scenarios for continental flood basalt volcanism as a whole; rather, when it comes to interactions with sedimentary rocks in the crust, we very much need to assess continental flood basalt provinces on a case-by-case basis.

SUMMARY

Continental flood basalt provinces are the subaerial expression of large igneous province volcanism, and their emplacements are exceptional volcanic events in the geological history of

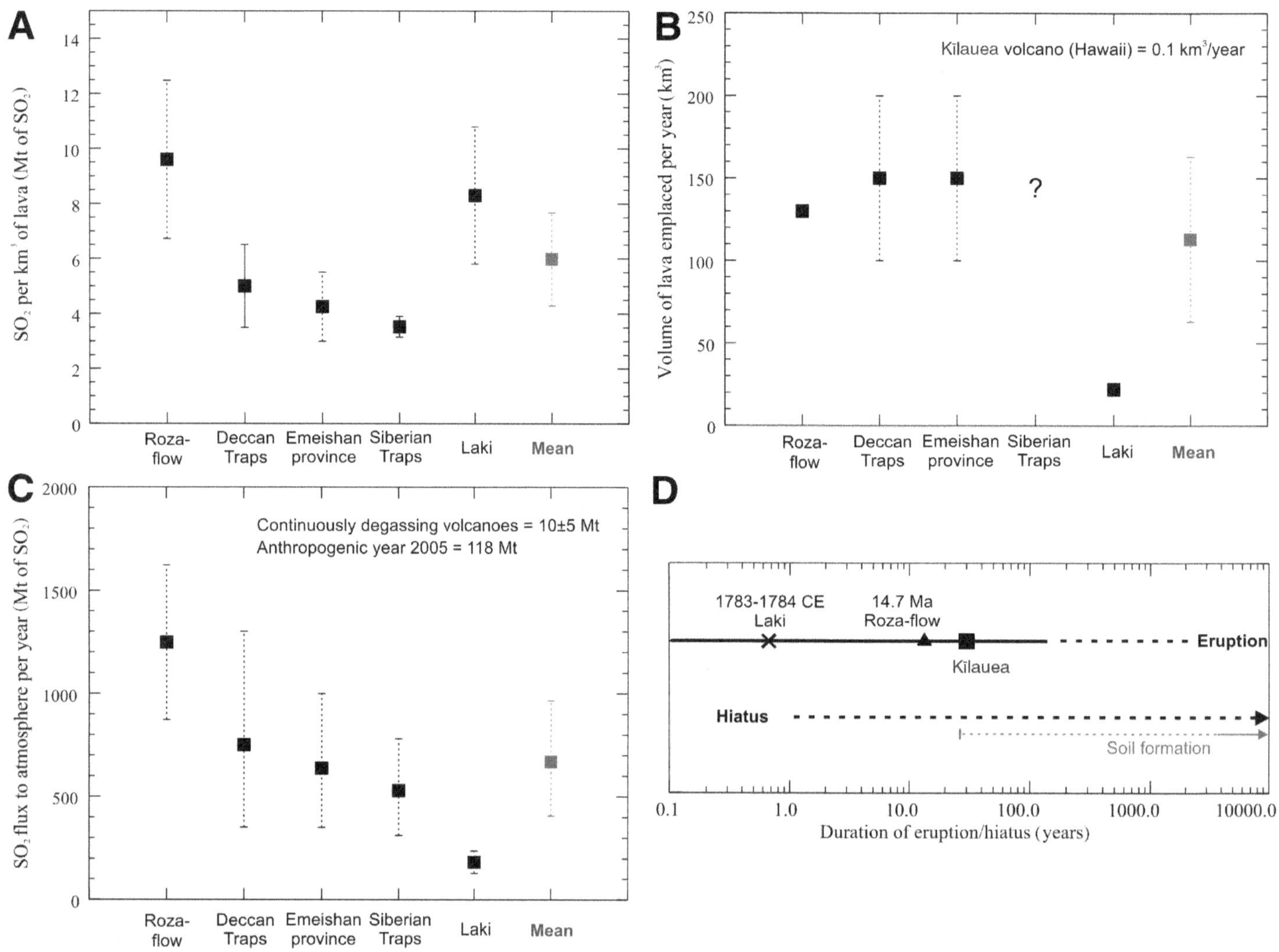

Figure 7. Compilation of eruption source parameters for flood basalt eruptions (see also Table 1 and references listed there), which could be used as inputs for Earth system modeling studies. (A) The amount of sulfur dioxide (SO_2) released per cubic kilometer (km^3) of lava erupted based on measurements of glass inclusions and groundmass glass in the quenched lava products. Dashed error bars indicate that a ±30% uncertainty is assumed, whereas solid error bars represent a reported analytical error. (B) The volume of lava erupted per year during an eruption based on estimates of the volume and duration of an average flow field in each province or, in the case of 1783–1784 CE Laki, a flood basalt eruption in historic times. Dashed error bars indicate that these estimates are highly uncertain. For the Siberian Traps, there is no estimate of the annual lava emplacement rate available as indicated by the question mark. For context, Kīlauea volcano on Hawai`i emplaced ~0.1 km^3 of lava per year averaged over the years 1983–2012 (USGS, 2013). (C) Combining data from (A) and (B), we derived the annual flux of SO_2 to the atmosphere. For context, the annual global mean flux of SO_2 from anthropogenic sources was ~118 Mt of SO_2 in the year 2005 (Smith et al., 2011), and continuously degassing volcanoes release ~10 ± 5 Mt of SO_2 per year (Graf et al., 1997, and references therein). Upper and lower limits shown in (C) are derived by accounting for both the error on the SO_2 measurements as shown in (A) and the error on the emplacement volume of a flow field as shown in (B). The mean is derived by averaging all flood basalt eruptions for which we have data. (D) Estimates of the duration of eruptions and non-eruptive phases for flood basalt volcanism. The latter are highly uncertain as indicated by the dashed lines. Time scales for formation of soils (and/or red boles) on lava flow tops, as observed in the Deccan Traps, have been estimated to be between tens and thousands of years (Chenet et al., 2008; Widdowson et al., 1997).

our planet, forming up to 4 × 10^6 km^3 of lava flows in as little as 2 m.y. For context, this volume would cover an area the size of the United States of America in 400 m of lava.

Continental flood basalt provinces appear to have formed within 1–3 m.y., and within this period, one or more pulses of great magma production and lava eruption took place, lasting from 1 m.y. to as little as a few hundred thousand years. The frequency and length of these voluminous lava-producing pulses are ill-defined. Each of these pulses featured tens to hundreds of individual eruptions, each producing up to possibly 10,000 km^3, or more usually 1000–5000 km^3, of predominantly pāhoehoe lava. The eruption of the 14.7 Ma Roza flow field of the Columbia

River Basalt Group (the youngest continental flood basalt province on Earth) produced a 1300 km^3 lava flow field and is the only eruptive phase for which a duration of between 10–15 yr has been estimated (Thordarson and Self, 1998). For other continental flood basalt provinces, we can only infer that eruptions lasted decades to probably centuries, based on analogy to the Roza flow and the dominance of pāhoehoe lava flow fields and the similarity of the chemical composition (i.e., tholeiitic melts) of the volcanic products between continental flood basalt provinces.

By measuring major-element and volatile composition of melt inclusions and groundmass glass in the quenched products of the Deccan Trap lavas, Self et al. (2008b) estimated that a decade-long eruption producing a 1000 km^3 lava flow field could have released 5000 Mt of SO_2 into the atmosphere. The Roza flow released ~12,000 Mt of SO_2 over the course of 10–15 yr (Thordarson and Self, 1996). Based on published estimates of the volcanic gas flux, it appears that for every 1 km^3 of basaltic magma emplaced during a continental flood basalt eruption, 6 ± 1.7 Mt of SO_2 could be released (Fig. 7C). Compared to an explosive silicic eruption of equal volume, the typical mass of sulfur-rich gases released from a large basaltic eruption is therefore much greater.

The majority of magmatic gas species released during continental flood basalt volcanism had the potential to alter climate and/or atmospheric composition, in particular during violent explosive phases at the eruptive vents when the gases were lofted into the stratosphere. However, the fate of these gases is more difficult to understand at present. For example, we have little knowledge about the atmospheric burdens, the atmospheric lifetimes, and the climatic impact of the magmatic sulfur- and halogen-bearing species released during a typical continental flood basalt eruption, but recent efforts in this direction include the use of global climate models (Black et al., 2013). Aside from the direct release of magmatic gases, the interaction of magma with the sedimentary bedrock during the emplacement of some continental flood basalt provinces could have released additional carbon and possibly sulfur and halogen-bearing species into the atmosphere (e.g., Svensen et al., 2009; Iacono-Marziano et al., 2012a, 2012b; Ganino and Arndt, 2009). Given the particular nature of these interactions in different provinces, it is difficult to generalize from one province to another, and further work is needed to constrain the magnitude of these emissions on a case-by-case basis.

The general coincidence of continental flood basalt volcanism with times of major biotic change is well substantiated to within the limits of radiometric age determinations; however, the mechanisms by which continental flood basalt volcanism might have perturbed the environment to such a great extent are not well understood. Figure 5 not only summarizes proposed mechanisms by which continental flood basalt volcanism could have triggered biotic crises, but it also gives our opinion concerning the level of current scientific understanding and progress made by specific scientific disciplines. Greater research efforts are to be encouraged in terms of the processes represented throughout Figure 5. Very little scientific progress has been made by means of atmospheric and (paleo-)Earth system modeling, which may help to, for instance, quantify the magnitude of the short-term reduction of temperatures or the effects of acid rain from volcanic sulfur deposition (Black et al., 2013). Given the likely differences in the Earth system response during different periods of Earth history, this is best done for each continental flood basalt province on a case-by-case basis. Notwithstanding the above, our knowledge of the number and length of non-eruptive phases (hiatuses) during continental flood basalt volcanism is far from complete—yet it is critical for quantifying the severity and duration of any environmental effects potentially caused by continental flood basalt volcanism.

ACKNOWLEDGMENTS

We thank Scott Bryan and Thor Thordarson for providing figures. Self was funded by UK Natural Environment Research Council (NERC) grants NER/B/S/2003/00246 and GR3/11474 for part of the work described herein, and also acknowledges support from The Open University Research Development Fund. Schmidt is funded by an Academic Research Fellowship from the School of Earth and Environment, University of Leeds. Mather acknowledges the Leverhulme Trust and NERC (NE/G01700X/1) for financial support and Lawrence Percival for very helpful discussions. We thank Gerta Keller for encouragement to submit this paper, and Nicholas Arndt and an anonymous reviewer for their timely and helpful reviews, which improved the manuscript.

REFERENCES CITED

Aarnes, I., Fristad, K., Planke, S., and Svensen, H., 2011, The impact of host-rock composition on devolatilization of sedimentary rocks during contact metamorphism around mafic sheet intrusions: Geochemistry Geophysics Geosystems, v. 12, p. Q10019, doi:10.1029/2011GC003636.

Andres, R.J., and Kasgnoc, A.D., 1998, A time-averaged inventory of subaerial volcanic sulfur emissions: Journal of Geophysical Research–Atmospheres, v. 103, p. 25,251–25,261, doi:10.1029/98JD02091.

Barry, T.L., Self, S., Kelley, S.P., Reidel, S., Hooper, P., and Widdowson, M., 2010, New $^{40}Ar/^{39}Ar$ dating of the Grande Ronde lavas, Columbia River Basalts, USA: Implications for duration of flood basalt eruption episodes: Lithos, v. 118, p. 213–222, doi:10.1016/j.lithos.2010.03.014.

Barry, T.L., Kelley, S.P., Reidel, S.P., Camp, V.E., Self, S., Jarboe, N.A., Duncan, R.A., and Renne, P.R., 2013, Eruption chronology of the Columbia River Basalt Group, *in* Reidel, S.P., Camp, V.C., Ross, M.E., Wolff, J.A., Martin, B.S., Tolan, T.L., and Wells, R.E., eds., The Columbia River Basalt Province: Geological Society of America Special Paper 497, p. 117–154.

Beerling, D.J., Harfoot, M., Lomax, B., and Pyle, J.A., 2007, The stability of the stratospheric ozone layer during the end-Permian eruption of the Siberian Traps: Philosophical Transactions of the Royal Society, ser. A, v. 365, p. 1843–1866, doi:10.1098/rsta.2007.2046.

Black, B.A., Elkins-Tanton, L.T., Rowe, M.C., and Peate, I.U., 2012, Magnitude and consequences of volatile release from the Siberian Traps: Earth and Planetary Science Letters, v. 317–318, p. 363–373, doi:10.1016/j.epsl.2011.12.001.

Black, B.A., Lamarque, J.-F., Shields, C.A., Elkins-Tanton, L., and Kiehl, J.T., 2013, Acid rain and ozone depletion from pulsed Siberian Traps magmatism: Geology, v. 42, p. 67–70, doi:10.1130/G34875.1.

Blackburn, T.J., Olsen, P.E., Bowring, S.A., McLean, N.M., Kent, D.V., Puffer, J., McHone, G., Rasbury, E.T., and Et-Touhami, M., 2013, Zircon U-Pb geochronology links the end-Triassic extinction with the Central Atlantic magmatic province: Science, v. 340, no. 6135, p. 941–945, doi:10.1126/science.1234204.

Blake, S., Self, S., Sharma, K., and Sephton, S., 2010, Sulfur release from the Columbia River Basalts and other flood lava eruptions constrained by a model of sulfide saturation: Earth and Planetary Science Letters, v. 299, p. 328–338, doi:10.1016/j.epsl.2010.09.013.

Bondre, N.R., Duraiswami, R.A., and Dole, G., 2004, Morphology and emplacement of flows from the Deccan volcanic province: Bulletin of Volcanology, v. 66, p. 29–45, doi:10.1007/s00445-003-0294-x.

Bourassa, A.E., Robock, A., Randel, W.J., Deshler, T., Rieger, L.A., Lloyd, N.D., Llewellyn, E.J., and Degenstein, D.A., 2012, Large volcanic aerosol load in the stratosphere linked to Asian monsoon transport: Science, v. 337, p. 78–81, doi:10.1126/science.1219371.

Brown, R.J., Blake, S., Thordarson, T., and Self, S., 2014, Pyroclastic edifices record vigorous lava fountains during the emplacement of a flood basalt flow field, Roza Member, Columbia River Basalt Province, USA: Geological Society of America Bulletin, v. 126, p. 875-891, doi:10.1130/B30857.1.

Bryan, S.E., 2007, Silicic large igneous provinces: Episodes, v. 30, p. 20–31.

Bryan, S.E., and Ernst, R.E., 2008, Revised definition of large igneous provinces (LIPs): Earth-Science Reviews, v. 86, p. 175–202, doi:10.1016/j.earscirev.2007.08.008.

Bryan, S.E., and Ferrari, L., 2013, Large igneous provinces and silicic large igneous provinces: Progress in our understanding over the past 25 years: Geological Society of America Bulletin, v. 125, p. 1053–1078, doi:10.1130/B30820.1.

Bryan, S.E., Ukstins Peate, I.A., Self, S., Peate, D., Jerram, D.A., Mawby, M.R., Miller, J., and Marsh, J.S., 2010, The largest volcanic eruptions on Earth: Earth-Science Reviews, v. 102, p. 207–229, doi:10.1016/j.earscirev.2010.07.001.

Burton, M.R., Sawyer, G.M., and Granieri, D., 2013, Deep carbon emissions from volcanoes: Reviews in Mineralogy and Geochemistry, v. 75, p. 323–354, doi:10.2138/rmg.2013.75.11.

Cadle, A.B., Cairncross, B., Christie, A.D.M., and Roberts, D.L., 1993, The Karoo Basin of South Africa: Type basin for the coal-bearing deposits of southern Africa: International Journal of Coal Geology, v. 23, p. 117–157, doi:10.1016/0166-5162(93)90046-D.

Caldeira, K., and Rampino, M.R., 1990, Carbon dioxide emissions from Deccan volcanism and a K/T boundary greenhouse effect: Geophysical Research Letters, v. 17, p. 1299–1302, doi:10.1029/GL017i009p01299.

Campbell, I.H., 2005, Large igneous provinces and the mantle plume hypothesis: Elements, v. 1, p. 265–269, doi:10.2113/gselements.1.5.265.

Campbell, I.H., Ross, W., and Griffiths, R.W., 1990, Implications of mantle plume structure for the evolution of flood basalts: Earth and Planetary Science Letters, v. 99, p. 79–93, doi:10.1016/0012-821X(90)90072-6.

Chenet, A.L., Fluteau, F., and Courtillot, V., 2005, Modelling massive sulphate aerosol pollution, following the large 1783 Laki basaltic eruption: Earth and Planetary Science Letters, v. 236, no. 3–4, p. 721–731, doi:10.1016/j.epsl.2005.04.046.

Chenet, A.L., Quidelleur, X., Fluteau, F., Courtillot, V., and Bajpai, S., 2007, K-40-Ar-40 dating of the Main Deccan large igneous province: Further evidence of KTB age and short duration: Earth and Planetary Science Letters, v. 263, p. 1–15, doi:10.1016/j.epsl.2007.07.011.

Chenet, A.L., Fluteau, F., Courtillot, V., Gerard, M., and Subbarao, K.V., 2008, Determination of rapid Deccan eruptions across the Cretaceous-Tertiary boundary using paleomagnetic secular variation: Results from a 1200-m-thick section in the Mahabaleshwar escarpment: Journal of Geophysical Research–Solid Earth, v. 113, p. B04101, doi:10.1029/2006JB004635.

Chenet, A.L., Fluteau, F., Courtillot, V.E., Gerard, M., Quidelleur, X., Khadri, S.F.R., Subbarao, K.V., and Thordarson, T., 2009, Determination of rapid Deccan eruptions across the Cretaceous–Tertiary boundary using paleomagnetic secular variation: 2. Constraints from analysis of eight new sections and synthesis for a 3500-m-thick composite section: Journal of Geophysical Research–Solid Earth, v. 114, p. B06103, doi:10.1029/2008JB005644.

Chesner, C.A., and Luhr, J.F., 2010, A melt inclusion study of the Toba Tuffs, Sumatra, Indonesia: Journal of Volcanology and Geothermal Research, v. 197, p. 259–278, doi:10.1016/j.jvolgeores.2010.06.001.

Coffin, M.F., and Eldholm, O., 1994, Large igneous provinces: Crustal structure, dimensions, and external consequences: Reviews of Geophysics, v. 32, no. 1, p. 1–36, doi:10.1029/93RG02508.

Coffin, M.F., and Eldholm, O., 2001, Large igneous provinces: Progenitors of some ophiolites?, *in* Ernst, R.E., and Buchan, K.L., eds., Mantle Plumes: Their Identification Through Time: Geological Society of America Special Paper 352, p. 59–70.

Courtillot, V.E., 1999, Evolutionary Catastrophes: The Science of Mass Extinctions (Translated by Joe McClinton): New York, Cambridge University Press, 188 p.

Courtillot, V.E., and Renne, P.R., 2003, On the ages of flood basalt events: Comptes Rendus Geoscience, v. 335, no. 1, p. 113–140, doi:10.1016/S1631-0713(03)00006-3.

Dessert, C., Dupre, B., Francois, L.M., Schott, J., Gaillardet, J., Chakrapani, G., and Bajpai, S., 2001, Erosion of Deccan Traps determined by river geochemistry: Impact on the global climate and the Sr-87/Sr-86 ratio of seawater: Earth and Planetary Science Letters, v. 188, p. 459–474, doi:10.1016/S0012-821X(01)00317-X.

Elliot, D.H., Fleming, T.H., Kyle, P.R., and Foland, K.A., 1999, Long-distance transport of magmas in the Jurassic Ferrar large igneous province, Antarctica: Earth and Planetary Science Letters, v. 167, p. 89–104, doi:10.1016/S0012-821X(99)00023-0.

English, J.M., Toon, O.B., and Mills, M.J., 2013, Microphysical simulations of large volcanic eruptions: Pinatubo and Toba: Journal of Geophysical Research–Atmospheres, v. 118, p. 1880–1895.

Ernst, R.E., and Buchan, K.L., 2001, Large mafic magmatic events through time and links to mantle-plume heads, *in* Ernst, R.E., and Buchan, K.L., eds., Mantle Plumes: Their Identification Through Time: Geological Society of America Special Paper 352, p. 483–575.

Erwin, D.H., 1994, The Permo-Triassic extinction: Nature, v. 367, p. 231–236, doi:10.1038/367231a0.

Font, E., Nedelec, A., Ellwood, B.B., Mirao, J., and Silva, P.F., 2011, A new sedimentary benchmark for the Deccan Traps volcanism?: Geophysical Research Letters, v. 38, p. L24309, doi:10.1029/2011GL049824.

Ganino, C., and Arndt, N.T., 2009, Climate changes caused by degassing of sediments during the emplacement of large igneous provinces: Geology, v. 37, no. 4, p. 323–326, doi:10.1130/G25325A.1.

Gerlach, T.M., 2004, Volcanic sources of tropospheric ozone-depleting trace gases: Geochemistry Geophysics Geosystems, v. 5, p. Q09007, doi:10.1029/2004GC000747.

Gerlach, T.M., Westrich, H.R., and Symonds, R.B., 1996, Preeruption vapor in magma of the climactic Mount Pinatubo eruption: Source of the giant stratospheric sulfur dioxide cloud, *in* Newhall, C.G., and Punongbayan, R.S., eds., Fire and Mud: Eruptions and Lahars of Mount Pinatubo, Philippines: Seattle, Washington, University of Washington Press, p. 415–433.

Graf, H.-F., Feichter, J., and Langmann, B.R., 1997, Volcanic sulfur emissions: Estimates of source strength and its contribution to the global sulfate distribution: Journal of Geophysical Research, v. 102, p. 10,727–10,738, doi:10.1029/96JD03265.

Grasby, S.E., Sanei, H., and Beauchamp, B., 2011, Catastrophic dispersion of coal fly ash into oceans during the latest Permian extinction: Nature Geoscience, v. 4, no. 2, p. 104–107, doi:10.1038/ngeo1069.

Greeley, R., Fagents, S.A., Harris, R.S., Kadel, S.D., Williams, D.A., and Guest, J.E., 1998, Erosion by flowing lava: Field evidence: Journal of Geophysical Research–Solid Earth, v. 103, no. B11, p. 27,325–27,345, doi:10.1029/97JB03543.

Gröcke, D.R., Rimmer, S.M., Yoksoulian, L.E., Cairncross, B., Tsikos, H., and van Hunen, J., 2009, No evidence for thermogenic methane release in coal from the Karoo-Ferrar large igneous province: Earth and Planetary Science Letters, v. 277, p. 204–212, doi:10.1016/j.epsl.2008.10.022.

Halmer, M.M., Schmincke, H.U., and Graf, H.F., 2002, The annual volcanic gas input into the atmosphere, in particular into the stratosphere: A global data set for the past 100 years: Journal of Volcanology and Geothermal Research, v. 115, p. 511–528, doi:10.1016/S0377-0273(01)00318-3.

Hartley, M.E., Maclennan, J., Edmonds, M., and Thordarson, T., 2014, Reconstructing the deep CO_2 degassing behaviour of large basaltic fissure eruptions: Earth and Planetary Science Letters, v. 393, p. 120–131.

Highwood, E.J., and Stevenson, D.S., 2003, Atmospheric impact of the 1783–1784 Laki Eruption: Part II. Climatic effect of sulphate aerosol: Atmospheric Chemistry and Physics, v. 3, p. 1177–1189, doi:10.5194/acp-3-1177-2003.

Hon, K., Kauahikaua, J., Denlinger, R., and Mackay, K., 1994, Emplacement and inflation of pahoehoe sheet flows: Observations and measurements of active lava flows on Kilauea volcano, Hawaii: Geological Society of America Bulletin, v. 106, p. 351–370, doi:10.1130/0016-7606(1994)106<0351:EAIOPS>2.3.CO;2.

Iacono-Marziano, G., Gaillard, F., Scaillet, B., Polozov, A.G., Marecal, V., Pirre, M., and Arndt, N.T., 2012a, Extremely reducing conditions reached during basaltic intrusion in organic matter–bearing sediments: Earth and Planetary Science Letters, v. 357–358, p. 319–326, doi:10.1016/j.epsl.2012.09.052.

Iacono-Marziano, G., Marecal, V., Pirre, M., Gaillard, F., Arteta, J., Scaillet, B., and Arndt, N.T., 2012b, Gas emissions due to magma–sediment interactions during flood magmatism at the Siberian Traps: Gas dispersion and environmental consequences: Earth and Planetary Science Letters, v. 357–358, p. 308–318, doi:10.1016/j.epsl.2012.09.051.

Jay, A.E., Mac Niocaill, C., Widdowson, M., Self, S., and Turner, W., 2009, New palaeomagnetic data from the Mahabaleshwar Plateau, Deccan Flood Basalt Province, India: Implications for the volcanostratigraphic architecture of continental flood basalt provinces: Journal of the Geological Society, v. 166, p. 13–24, doi:10.1144/0016-76492007-150.

Jerram, D.A., and Widdowson, M., 2005, The anatomy of continental flood basalt provinces: Geological constraints on the processes and products of flood volcanism: Lithos, v. 79, p. 385–405, doi:10.1016/j.lithos.2004.09.009.

Joachimski, M.M., Lai, X., Shen, S., Jiang, H., Luo, G., Chen, B., Chen, J., and Sun, Y., 2012, Climate warming in the latest Permian and the Permian-Triassic mass extinction: Geology, v. 40, no. 3, p. 195–198, doi:10.1130/G32707.1.

Kaminski, E., Chenet, A.-L., Jaupart, C., and Courtillot, V., 2011, Rise of volcanic plumes to the stratosphere aided by penetrative convection above large lava flows: Earth and Planetary Science Letters, v. 301, p. 171–178, doi:10.1016/j.epsl.2010.10.037.

Kamo, S.L., Czamanske, G.K., Amelin, Y., Fedorenko, V.A., Davis, D.W., and Trofimov, V.R., 2003, Rapid eruption of Siberian flood-volcanic rocks and evidence for coincidence with the Permian-Triassic boundary and mass extinction at 251 Ma: Earth and Planetary Science Letters, v. 214, p. 75–91, doi:10.1016/S0012-260 821X(03)00347-9.

Keller, G., 2014, this volume, Deccan volcanism, the Chicxulub impact, and the end-Cretaceous mass extinction: Coincidence? Cause and effect?, *in* Keller, G., and Kerr, A.C., eds., Volcanism, Impacts, and Mass Extinctions: Causes and Effects: Geological Society of America Special Paper 505, doi:10.1130/2014.2505(03).

Keller, G., Adatte, T., Gardin, S., Bartolini, A., and Bajpai, S., 2008, Main Deccan volcanism phase ends near the K-T boundary: Evidence from the Krishna-Godavari Basin, SE India: Earth and Planetary Science Letters, v. 268, p. 293–311, doi:10.1016/j.epsl.2008.01.015.

Keller, G., Adatte, T., Bhowmick, P.K., Upadhyay, H., Dave, A., Reddy, A.N., and Jaiprakash, B.C., 2012, Nature and timing of extinctions in Cretaceous-Tertiary planktic foraminifera preserved in Deccan intertrappean sediments of the Krishna-Godavari Basin, India: Earth and Planetary Science Letters, v. 341–344, p. 211–221, doi:10.1016/j.epsl.2012.06.021.

Kelley, S.P., 2007, The geochronology of large igneous provinces, terrestrial impact craters, and their relationship to mass extinctions on Earth: Journal of the Geological Society, v. 164, p. 923–936, doi:10.1144/0016-76492007-026.

Keszthelyi, L., and Self, S., 1998, Some physical requirements for the emplacement of long lava flows: Journal of Geophysical Research–Solid Earth, v. 103, p. 27,447–27,464, doi:10.1029/98JB00606.

Krueger, A.J., Walter, L.S., Bhartia, P.K., Schnetzler, C.C., Krotkov, N.A., Sprod, I., and Bluth, G.J.S., 1995, Volcanic sulfur dioxide measurements from the total ozone mapping spectrometer instruments: Journal of Geophysical Research, v. 100, p. 14,057–14,076, doi:10.1029/95JD01222.

Kutterolf, S., Hansteen, T., Appel, K., Freundt, A., Krüger, K., Pérez, W., and Wehrmann, H., 2013, Combined bromine and chlorine release from large explosive volcanic eruptions: A threat to stratospheric ozone?: Geology, v. 41, p. 707–710, doi:10.1130/G34044.1.

Le Quéré, C., Andres, R.J., Boden, T., Conway, T., Houghton, R.A., House, J.I., Marland, G., Peters, G.P., van der Werf, G.R., Ahlström, A., Andrew, R.M., Bopp, L., Canadell, J.G., Ciais, P., Doney, S.C., Enright, C., Friedlingstein, P., Huntingford, C., Jain, A.K., Jourdain, C., Kato, E., Keeling, R.F., Klein Goldewijk, K., Levis, S., Levy, P., Lomas, M., Poulter, B., Raupach, M.R., Schwinger, J., Sitch, S., Stocker, B.D., Viovy, N., Zaehle, S., and Zeng, N., 2013, The global carbon budget 1959–2011: Earth System Science Data, v. 5, no. 1, p. 165–1853.

Li, L., and Keller, G., 1998, Maastrichtian climate, productivity and faunal turnovers in planktic foraminifera in South Atlantic DSDP Sites 525A and 21: Marine Micropaleontology, v. 33, no. 1–2, p. 55–86, doi:10.1016/S0377-8398(97)00027-3.

Mahoney, J.J., and Coffin, M.F., eds., 1997, Large Igneous Provinces: Continental, Oceanic, and Planetary Flood Volcanism: American Geophysical Union Geophysical Monograph Series 100, 438 p.

Marks, L., Keiding, J., Wenzel, T., Trumbull, R.B., Veksler, I., Wiedenbeck, M., and Markl G., 2014, F, Cl, and S concentrations in olivine-hosted melt inclusions from mafic dikes in NW Namibia and implications for the environmental impact of the Paraná–Etendeka Large Igneous Province: Earth and Planetary Science Letters, v. 392, p. 39–49.

Mather, T.A., Pyle, D.M., and Oppenheimer, C., 2003, Tropospheric volcanic aerosol, *in* Robock, A., and Oppenheimer, C., eds., Volcanism and the Earth's Atmosphere: American Geophysical Union Geophysical Monograph 139, p. 189–212.

Mather, T.A., Witt, M.L.I., Pyle, D.M., Quayle, B.M., Aiuppa, A., Bagnato, E., Martin, R.S., Sims, K.W.W., Edmonds, M., Sutton, A.J., and Ilyinskaya, E., 2012, Halogens and trace metal emissions from the ongoing 2008 summit eruption of Kīlauea volcano: Hawai'i: Geochimica et Cosmochimica Acta, v. 83, p. 292–323, doi:10.1016/j.gca.2011.11.029.

McCormick, M.P., Thomason, L.W., and Trepte, C.R., 1995, Atmospheric effects of the Mt. Pinatubo eruption: Nature, v. 373, p. 399–404, doi:10.1038/373399a0.

Moulin, M., Fluteau, F., Courtillot, V., Marsh, J., Delpech, G., Quidelleur, X., Gérard, M., and Jay, A.E., 2011, An attempt to constrain the age, duration, and eruptive history of the Karoo flood basalt: Naude's Nek section (South Africa): Journal of Geophysical Research, v. 116, p. B07403, doi:10.1029/2011JB008210.

Newhall, C.G., and Self, S., 1982, The volcanic explosivity index (VEI): An estimate of explosive magnitude for historical volcanism: Journal of Geophysical Research, v. 87, p. 1231–1238, doi:10.1029/JC087iC02p01231.

Officer, C.B., Hallam, A., Drake, C.L., and Devine, J.D., 1987, Late Cretaceous and paroxysmal Cretaceous/Tertiary extinctions: Nature, v. 326, p. 143–149, doi:10.1038/326143a0.

Oman, L., Robock, A., Stenchikov, G.L., Thordarson, T., Koch, D., Shindell, D.T., and Gao, C., 2006a, Modeling the distribution of the volcanic aerosol cloud from the 1783–1784 Laki eruption: Journal of Geophysical Research, v. 111, p. D12209, doi:10.1029/2005JD006899.

Oman, L., Robock, A., Stenchikov, G.L., and Thordarson, T., 2006b, High-latitude eruptions cast shadow over the African monsoon and the flow of the Nile: Geophysical Research Letters, v. 33, p. L18711, doi:10.1029/2006GL027665.

Oppenheimer, C., 2002, Limited global change due to the largest known Quaternary eruption, Toba ~74 kyr BP?: Quaternary Science Reviews, v. 21, p. 1593–1609.

Oppenheimer, C., 2003, Climatic, environmental and human consequences of the largest known historic eruption: Tambora volcano (Indonesia) 1815: Progress in Physical Geography, v. 27, p. 230–259, doi:10.1191/0309133303pp379ra.

Pinto, J.P., Turco, R.P., and Toon, O.B., 1989, Self-limiting physical and chemical effects in volcanic eruption clouds: Journal of Geophysical Research, v. 94, no. D8, p. 11,165–11,174.

Pyle, D.M., and Mather, T.A., 2003, The importance of volcanic emissions in the global atmospheric mercury cycle: Atmospheric Environment, v. 37, p. 5115–5124, doi:10.1016/j.atmosenv.2003.07.011.

Pyle, D.M., and Mather, T.A., 2009, Halogens in igneous processes and their fluxes to the atmosphere and oceans from volcanic activity: A review: Chemical Geology, v. 263, p. 110–121, doi:10.1016/j.chemgeo.2008.11.013.

Rampino, M.R., and Prokoph, A., 2013, Are mantle plumes periodic?: Eos [Transactions, American Geophysical Union], v. 94, p. 113–114, doi:10.1002/2013EO120001.

Rampino, M.R., and Self, S., 1982, Historic eruptions of Tambora (1815), Krakatau (1883), and Agung (1963), their stratospheric aerosols, and climatic impact: Quaternary Research, v. 18, no. 2, p. 127–143, doi:10.1016/0033-5894(82)90065-5.

Rampino, M.R., and Stothers, R.B., 1988, Flood basalt volcanism during the past 250 million years: Science, v. 241, p. 663–668.

Rampino, M.R., Self, S., and Stothers, R.B., 1988, Volcanic winters: Annual Review of Earth and Planetary Sciences, v. 16, p. 73–99, doi:10.1146/annurev.ea.16.050188.000445.

Ravizza, G., and Peucker-Ehrenbrink, B., 2003, Chemostratigraphic evidence of Deccan volcanism from the marine osmium isotope record: Science, v. 302, p. 1392–1395, doi:10.1126/science.1089209.

Reichow, M.K., Saunders, A.D., White, R.V., Al'Mukhamedov, A.I., and Medvedev, A.Y., 2005, Geochemistry and petrogenesis of basalts from the

West Siberian Basin: An extension of the Permo-Triassic Siberian Traps, Russia: Lithos, v. 79, p. 425–452, doi:10.1016/j.lithos.2004.09.011.

Reichow, M.K., Pringle, M.S., Al'Mukhamedov, A.I., Allen, M.B., Andreichev, V.L., Buslov, M.M., Davies, C.E., Fedoseev, G.S., Fitton, J.G., Inger, S., Medvedev, A.Ya., Mitchell, C., Puchkov, V.N., Safonova, I.Yu., Scott, R.A., and Saunders, A.D., 2009, The timing and extent of the eruption of the Siberian Traps large igneous province: Implications for the end-Permian environmental crisis: Earth and Planetary Science Letters, v. 277, p. 9–20, doi:10.1016/j.epsl.2008.09.030.

Reidel, S.P., and Tolan, T.L., 2013, The Grande Ronde Basalt, Columbia River Basalt Group, *in* Reidel, S.P., Camp, V.C., Ross, M.E., Wolff, J.A., Martin, B.S., Tolan, T.L., and Wells, R.E., eds., The Columbia River Basalt Province: Geological Society of America Special Paper 497, p. 117–154.

Reidel, S.P., Camp, V.C., Ross, M.E., Wolff, J.A., Martin, B.S., Tolan, T.L., and Wells, R.E., eds., 2013, The Columbia River Basalt Province: Geological Society of America Special Paper 497, 440 p.

Robinson, N., Ravizza, G., Coccioni, R., Peucker-Ehrenbrink, B., and Norris, R., 2009, A high-resolution marine $^{187}Os/^{188}Os$ record for the late Maastrichtian: Distinguishing the chemical fingerprints of Deccan volcanism and the KP impact event: Earth and Planetary Science Letters, v. 281, p. 159–168, doi:10.1016/j.epsl.2009.02.019.

Robock, A., 2000, Volcanic eruptions and climate: Reviews of Geophysics, v. 38, no. 2, p. 191–219, doi:10.1029/1998RG000054.

Sanei, H., Grasby, S.E., and Beauchamp, B., 2012, Latest Permian mercury anomalies: Geology, v. 40, no. 1, p. 63–66, doi:10.1130/G32596.1.

Scaillet, B., and Macdonald, R., 2006, Experimental and thermodynamic constraints on the sulphur yield of peralkaline and metaluminous silicic flood eruptions: Journal of Petrology, v. 47, p. 1413–1437, doi:10.1093/petrology/egl016.

Schmidt, A., Carslaw, K.S., Mann, G.W., Wilson, M., Breider, T.J., Pickering, S.J., and Thordarson, T., 2010, The impact of the 1783–1784 AD Laki eruption on global aerosol formation processes and cloud condensation nuclei: Atmospheric Chemistry and Physics, v. 10, p. 6025–6041, doi:10.5194/acp-10-6025-2010.

Schmidt, A., Ostro, B., Carslaw, K.S., Wilson, M., Thordarson, T., Mann, G.W., and Simmons, A.J., 2011, Excess mortality in Europe following a future Laki-style Icelandic eruption: Proceedings of the National Academy of Sciences of the United States of America, v. 108, no. 38, p. 15,710–15,715, doi:10.1073/pnas.1108569108.

Schmidt, A., Thordarson, T., Oman, L.D., Robock, A., and Self, S., 2012, Climatic impact of the long-lasting 1783 Laki eruption: Inapplicability of mass-independent sulfur isotopic composition measurements: Journal of Geophysical Research–Atmospheres, v. 117, no. D23, p. D23116, doi:10.1029/2012JD018414.

Self, S., 2006, The effects and consequences of very large explosive volcanic eruptions: Philosophical Transactions of the Royal Society, Series A, v. 364, p. 2073–2097.

Self, S., Thordarson, T., and Keszthelyi, L., 1997, Emplacement of continental flood basalt lava flows, *in* Mahoney, J.J., and Coffin, M.F., eds., Large Igneous Provinces: Continental, Oceanic, and Planetary Flood Volcanism: American Geophysical Union Geophysical Monograph 100, p. 381–410.

Self, S., Keszthelyi, L., and Thordarson, T., 1998, The importance of pāhoehoe: Annual Review of Earth and Planetary Sciences, v. 26, p. 81–110, doi:10.1146/annurev.earth.26.1.81.

Self, S., Thordarson, T., and Widdowson, M., 2005, Gas fluxes from flood basalt eruptions: Elements, v. 1, p. 283–287, doi:10.2113/gselements.1.5.283.

Self, S., Widdowson, M., Thordarson, T., and Jay, A.E., 2006, Volatile fluxes during flood basalt eruptions and potential effects on the global environment: A Deccan perspective: Earth and Planetary Science Letters, v. 248, p. 518–532, doi:10.1016/j.epsl.2006.05.041.

Self, S., Jay, A.E., Widdowson, M., and Keszthelyi, L.P., 2008a, Correlation of the Deccan and Rajahmundry Trap lavas: Are these the longest and largest lava flows on Earth?: Journal of Volcanology and Geothermal Research, v. 172, p. 3–19, doi:10.1016/j.jvolgeores.2006.11.012.

Self, S., Blake, S., Sharma, K., Widdowson, M., and Sephton, S., 2008b, Sulfur and chlorine in Late Cretaceous Deccan magmas and eruptive gas release: Science, v. 319, p. 1654–1657, doi:10.1126/science.1152830.

Silva, M.V.N., Sial, A.N., Barbosa, J.A., Ferreira, V.P., Neumann, V.H., and De Lacerda, L.D., 2014, Carbon isotopes, rare-earth elements and mercury geochemistry across the K-T transition of the Paraíba Basin, northeastern Brazil, *in* Bojar, A.-V., Melinte-Dobrinescu, M.C., and Smit, J., eds., Isotopic Studies in Cretaceous Research: Geological Society of London Special Publication 382, doi:10.1144/SP382.2.

Smith, S.J., van Aardenne, J., Klimont, Z., Andres, R.J., Volke, A., and Delgado Arias, S., 2011, Anthropogenic sulfur dioxide emissions: 1850–2005: Atmospheric Chemistry and Physics, v. 11, p. 1101–1116, doi:10.5194/acp-11-1101-2011.

Sobolev, S.V., Sobolev, A.V., Kuzmin, D.V., Krivolutskaya, N.A., Petrunin, A.G., Arndt, N.T., Radko, V.A., and Vasiliev, Y.R., 2011, Linking mantle plumes, large igneous provinces and environmental catastrophes: Nature, v. 477, no. 7364, p. 312–316, doi:10.1038/nature10385.

Solomon, S., Portmann, R.W., Garcia, R.R., Randel, W., Wu, F., Nagatani, R., Gleason, J., Thomason, L., Poole, L.R., and McCormick, M.P., 1998, Ozone depletion at mid-latitudes: Coupling of volcanic aerosols and temperature variability to anthropogenic chlorine: Geophysical Research Letters, v. 25, no. 11, p. 1871–1874, doi:10.1029/98GL01293.

Stevenson, D.S., Johnson, C.E., Highwood, E.J., Gauci, V., Collins, W.J., and Derwent, R.G., 2003, Atmospheric impact of the 1783–1784 Laki eruption: Part I. Chemistry modelling: Atmospheric Chemistry and Physics, v. 3, p. 487–507, doi:10.5194/acp-3-487-2003.

Stoiber, R.E., Williams, S.N., and Huebert, B., 1987, Annual contribution of sulfur dioxide to the atmosphere by volcanoes: Journal of Volcanology and Geothermal Research, v. 33, p. 1–8, doi:10.1016/0377-0273(87)90051-5.

Storey, M., Duncan, R.A., and Tegner, C., 2007, Timing and duration of volcanism in the North Atlantic igneous province: Implications for geodynamics and links to the Iceland hotspot: Chemical Geology, v. 241, no. 3–4, p. 264–281, doi:10.1016/j.chemgeo.2007.01.016.

Stothers, R.B., Wolff, J.A., Self, S., and Rampino, M.R., 1986, Basaltic fissure eruptions, plume heights, and atmospheric aerosols: Geophysical Research Letters, v. 13, no. 8, p. 725–728, doi:10.1029/GL013i008p00725.

Sun, Y., Joachimski, M.M., Wignall, P.B., Yan, C., Chen, Y., Jiang, H., Wang, L., and Lai, X., 2012, Lethally hot temperatures during the Early Triassic greenhouse: Science, v. 338, no. 6105, p. 366–370, doi:10.1126/science.1224126.

Svensen, H., and Jamtveit, B., 2010, Metamorphic fluids and global environmental changes: Elements, v. 6, no. 3, p. 179–182, doi:10.2113/gselements.6.3.179.

Svensen, H., Planke, S., Chevallier, L., Malthe-Sørenssen, A., Corfu, B., and Jamtveit, B., 2007, Hydrothermal venting of greenhouse gases triggering Early Jurassic global warming: Earth and Planetary Science Letters, v. 256, p. 554–566, doi:10.1016/j.epsl.2007.02.013.

Svensen, H., Planke, S., Polozov, A.G., Schmidbauer, N., Corfu, F., Podladchikov, Y.Y., and Jamtveit, B., 2009, Siberian gas venting and the end-Permian environmental crisis: Earth and Planetary Science Letters, v. 277, no. 3–4, p. 490–500, doi:10.1016/j.epsl.2008.11.015.

Svensen, H., Corfu, F., Polteau, S., Hammer, Ø., and Planke, S., 2012, Rapid magma emplacement in the Karoo large igneous province: Earth and Planetary Science Letters, v. 325–326, p. 1–9, doi:10.1016/j.epsl.2012.01.015.

Swanson, D.A., Wright, T.L., and Helz, R.T., 1975, Linear vent systems and estimated rates of magma production and eruption for the Yakima Basalt on the Columbia Plateau: American Journal of Science, v. 275, no. 8, p. 877–905, doi:10.2475/ajs.275.8.877.

Tabazadeh, A., and Turco, R.P., 1993, Stratospheric chlorine injection by volcanic eruptions: HCl scavenging and implications for ozone: Science, v. 260, no. 5111, p. 1082–1086, doi:10.1126/science.260.5111.1082.

Textor, C., Graf, H.-F., Herzog, M., and Oberhuber, J.M., 2003, Injection of gases into the stratosphere by explosive volcanic eruptions: Journal of Geophysical Research–Atmospheres, v. 108, no. D19, p. 4606, doi:10.1029/2002JD002987.

Textor, C., Graf, H.F.T.C., and Robock, A., 2004, Emissions from volcanoes, *in* Granier, C., Artaxo, P., and Reeves, C., eds., Emissions of Chemical Compounds and Aerosols in the Atmosphere, Volume 6: Dordrecht, Netherlands, Kluwer, p. 269–303.

Thibault, N., and Gardin, S., 2006, Maastrichtian calcareous nannofossil biostratigraphy and paleoecology in the equatorial Atlantic (Demerara Rise, ODP Leg 207 Hole 1258A): Revue de Micropaleontologie, v. 49, no. 4, p. 199–214, doi:10.1016/j.revmic.2006.08.002.

Thordarson, T., and Self, S., 1993, The Laki (Skaftar fires) and Grimsvötn eruptions in 1783–1785: Bulletin of Volcanology, v. 55, no. 4, p. 233–263, doi:10.1007/BF00624353.

Thordarson, T., and Self, S., 1996, Sulfur, chlorine and fluorine degassing and atmospheric loading by the Roza eruption, Columbia River Basalt Group, Washington, USA: Journal of Volcanology and Geothermal Research, v. 74, no. 1–2, p. 49–73, doi:10.1016/S0377-0273(96)00054-6.

Thordarson, T., and Self, S., 1998, The Roza Member, Columbia River Basalt Group: A gigantic pāhoehoe lava flow field formed by endogenous processes?: Journal of Geophysical Research–Solid Earth, v. 103, no. B11, p. 27,411–27,445, doi:10.1029/98JB01355.

Thordarson, T., and Self, S., 2003, Atmospheric and environmental effects of the 1783–1784 Laki eruption: A review and reassessment: Journal of Geophysical Research–Atmospheres, v. 108, no. D1, p. 4011, doi:10.1029/2001JD002042.

Thordarson, T., Self, S., Oskarsson, N., and Hulsebosch, T., 1996, Sulfur, chlorine, and fluorine degassing and atmospheric loading by the 1783–1784 AD Laki (Skaftar fires) eruption in Iceland: Bulletin of Volcanology, v. 58, no. 2–3, p. 205–225, doi:10.1007/s004450050136.

Thordarson, T., Miller, D.J., Larsen, G., Self, S., and Sigurdsson, H., 2001, New estimates of sulfur degassing and atmospheric mass-loading by the 934 AD Eldgja eruption, Iceland: Journal of Volcanology and Geothermal Research, v. 108, no. 1–4, p. 33–54, doi:10.1016/S0377-0273(00)00277-8.

Thordarson, T., Self, S., Miller, D.J., Larsen, G., and Vilmundardottir, E.G., 2003, Sulphur release from flood lava eruptions in the Veidivotn, Grimsvotn and Katla volcanic systems, Iceland: Volcanic Degassing, v. 6, no. 213, p. 103–121.

Thordarson, T., Rampino, M., Keszthelyi, L.P., and Self, S., 2009, Effects of megascale eruptions on Earth and Mars, *in* Chapman, M.G., and Keszyhelyi, L.P., eds., Preservation of Random Megascale Events on Mars and Earth: Geological Society of America Special Paper 453, p. 37–53.

Timmreck, C., Graf, H.-F., Lorenz, S.J., Niemeier, U., Zanchettin, D., Matei, D., Jungclaus, J.H., and Crowley, T.J., 2010, Aerosol size confines climate response to volcanic super-eruptions: Geophysical Research Letters, v. 37, no. 24, p. L24705, doi:10.1029/2010GL045464.

USGS, 2013, Kīlauea's East Rift Zone (Pu'u'Ō'ō) Eruption: 1983 to Present: http://hvo.wr.usgs.gov/kilauea/summary/ (accessed 30 September 2013).

Vandamme, D., Courtillot, V., Besse, J., and Montigny, R., 1991, Paleomagnetism and age determinations of the Deccan Traps (India): Results of a Nagpur-Bombay traverse and review of earlier work: Reviews of Geophysics, v. 29, no. 2, p. 159–190, doi:10.1029/91RG00218.

Visscher, H., Looy, C.V., Collinson, M.E., Brinkhuis, H., van Konijnenburg-van Cittert, J.H.A., Kurschner, W.M., and Sephton, M.A., 2004, Environmental mutagenesis during the end-Permian ecological crisis: Proceedings of the National Academy of Sciences of the United States of America, v. 101, no. 35, p. 12,952–12,956, doi:10.1073/pnas.0404472101.

Vogt, P.R., 1972, Evidence for global synchronism in mantle plume convection, and possible significance for geology: Nature, v. 240, no. 5380, p. 338–342, doi:10.1038/240338a0.

von Glasow, R., 2010, Atmospheric chemistry in volcanic plumes: Proceedings of the National Academy of Sciences of the United States of America, v. 107, no. 15, p. 6594–6599, doi:10.1073/pnas.0913164107.

Vye-Brown, C.L., Self, S., and Barry, T.L., 2013, Architecture and emplacement of flood basalt flow fields: Case studies from the Columbia River flood basalts, USA: Bulletin of Volcanology, v. 75, p. 697, doi:10.1007/s00445-013-0697-2.

Wallace, P.J., 2005, Volatiles in subduction zone magmas: Concentrations and fluxes based on melt inclusion and volcanic gas data: Journal of Volcanology and Geothermal Research, v. 140, no. 1–3, p. 217–240, doi:10.1016/j.jvolgeores.2004.07.023.

Westrich, H.R., and Gerlach, T.M., 1992, Magmatic gas source for the stratospheric SO_2 cloud from the June 15, 1991, eruption of Mount Pinatubo: Geology, v. 20, no. 10, p. 867–870, doi:10.1130/0091-7613(1992)020<0867:MGSFTS>2.3.CO;2.

White, J.D.L., Bryan, S.B., Ross, P.-S., Self, S., and Thordarson, T., 2009, Physical volcanology of large igneous provinces: Update and review, *in* Thordarson, T., Self, S., Larsen, G., Rowland, S.K., and Hoskuldsson, A., eds., Advances in Volcanology (The Legacy of George Walker): Special Publications of IAVCEI Volume 2: London, Geological Society of London, p. 291–321.

Whiteside, J.H., Olsen, P.E., Kent, D.V., Fowell, S.J., and Et-Touhami, M., 2007, Synchrony between the Central Atlantic magmatic province and the Triassic-Jurassic mass-extinction event?: Palaeogeography, Palaeoclimatology, Palaeoecology, v. 244, no. 1–4, p. 345–367, doi:10.1016/j.palaeo.2006.06.035.

Widdowson, M., Walsh, J.N., and Subbarao, K.V., 1997, The geochemistry of Indian bole horizons: Palaeoenvironmental implications of Deccan intravolcanic palaeosurfaces, *in* Widdowson, M., ed., Palaeosurfaces: Recognition, Reconstruction and Palaeoenvironmental Interpretation: Geological Society of London Special Publication 120, p. 269–281.

Wignall, P.B., 2001, Large igneous provinces and mass extinctions: Earth-Science Reviews, v. 53, p. 1–3, doi:10.1016/S0012-8252(00)00037-4.

Wignall, P.B., 2005, The link between large igneous province eruptions and mass extinctions: Elements, v. 1, no. 5, p. 293–297, doi:10.2113/gselements.1.5.293.

Wignall, P.B., 2007, The end-Permian mass extinction—How bad did it get?: Geobiology, v. 5, no. 4, p. 303–309, doi:10.1111/j.1472-4669.2007.00130.x.

Wignall, P.B., Sun, Y., Bond, D.P.G., Izon, G., Newton, R.J., Védrine, S., Widdowson, M., Ali, J.R., Lai, X., Jiang, H., Cope, H., and Bottrell, S.H., 2009, Volcanism, mass extinction, and carbon isotope fluctuations in the Middle Permian of China: Science, v. 324, no. 5931, p. 1179–1182, doi:10.1126/science.1171956.

Wilf, P., Johnson, K.R., and Huber, B.T., 2003, Correlated terrestrial and marine evidence for global climate changes before mass extinction at the Cretaceous-Paleogene boundary: Proceedings of the National Academy of Sciences of the United States of America, v. 100, no. 2, p. 599–604, doi:10.1073/pnas.0234701100.

Witham, C.S., and Oppenheimer, C., 2005, Mortality in England during the 1783–1784 Laki Craters eruption: Bulletin of Volcanology, v. 67, p. 15–26, doi:10.1007/s00445-004-0357-7.

Woods, A.W., 1993, A model of the plumes above basaltic fissure eruptions: Geophysical Research Letters, v. 20, no. 12, p. 1115–1118, doi:10.1029/93GL01215.

Zhang, Y., Ren, Z.-Y., and Xu, Y.-G., 2013, Sulfur in olivine-hosted melt inclusions from the Emeishan picrites: Implications for S degassing and its impact on environment: Journal of Geophysical Research–Solid Earth, v. 118, no. 8, p. 4063–4070, doi:10.1002/jgrb.50324.

Manuscript Accepted by the Society 31 January 2014

Printed in the USA

The Geological Society of America
Special Paper 505
2014

Modeling the carbon-sulfate interplays in climate changes related to the emplacement of continental flood basalts

Mickaël Mussard
Guillaume Le Hir
Frédéric Fluteau
Équipe de Paléomagnétisme, Institut de Physique du Globe de Paris, Sorbonne Paris Cité, University Paris Diderot, UMR 7154 CNRS, 1 rue Jussieu, F-75005 Paris, France

Vincent Lefebvre
Laboratoire des Sciences du Climat et de l'Environnement, UMR 8212, CEA-CNRS-Université de Versailles Saint-Quentin-en-Yvelines, 91190 Gif-sur-Yvette, France

Yves Goddéris
Géosciences Environnement Toulouse, CNRS-Observatoire Midi-Pyrénées, 31400 Toulouse, France

ABSTRACT

Climatic and environmental changes are now widely recognized as the main cause of mass extinctions. Global warming that immediately preceded the Cretaceous-Tertiary boundary is regarded as a consequence of CO_2 released during the main phase of Deccan Trap emplacement. Modeling has shown that such global warming cannot be explained by the continuous release of volcanic carbon dioxide. In the present paper, we use a biogeochemical model, coupled to a climate model, to further our understanding of climate changes caused by continental flood basalts. The response of the global climate–carbon-cycle system to sulfur dioxide (SO_2) and carbon dioxide (CO_2) emissions is investigated, assuming a degassing history consisting of a series of evenly spaced pulses. We find that CO_2-related warming is enhanced when large-scale SO_2 injections are added. According to our model, we observe that the succession of drastic cooling events induced by sulfate aerosols decreases the efficiency of silicate weathering and destabilizes the carbon cycle during the full time span of trap emplacement. In the case of the Deccan Traps, these transient disequilibria lead to a 25% increase in pCO_2 and ensuing warming. The environmental consequences of emplacement of large igneous provinces appear to be even more complex: A SO_2-related climate feedback may have enhanced the long-term warming due to CO_2 emissions.

Mussard, M., Le Hir, G., Fluteau, F., Lefebvre, V., and Goddéris, Y., 2014, Modeling the carbon-sulfate interplays in climate changes related to the emplacement of continental flood basalts, *in* Keller, G., and Kerr, A.C., eds., Volcanism, Impacts, and Mass Extinctions: Causes and Effects: Geological Society of America Special Paper 505, p. 339–352, doi:10.1130/2014.2505(17). For permission to copy, contact editing@geosociety.org.

INTRODUCTION

The causes of the five major and 12 minor mass extinctions that occurred during the Phanerozoic have been a matter of debate for considerable time (e.g., Alvarez et al., 1980; Courtillot et al., 1994; Keller, 2003; Courtillot and Fluteau, 2010; Schulte et al., 2010). In the case of the Cretaceous-Tertiary mass extinction (ca. 66 Ma), two main theories, a bolide impact and large igneous province volcanism, have been proposed. The bolide impact theory relies on the discovery of an iridium anomaly in a thin clay layer coeval with the Cretaceous-Tertiary boundary (Alvarez et al., 1980). The presence of impact spherule ejecta in Haiti and NE Mexico of the same age (Izett et al., 1991) and an impact crater on Yucatan in Mexico (Hildebrand et al., 1991) have provided support for this hypothesis. However, the causal relationship between the bolide impact and the mass extinction has been highly disputed (for a synthesis, see Keller et al., 2012). Since the 1980s, an alternative theory has emerged. Huge volcanic eruptions leading to the formation of large igneous provinces (also named traps for continental flood basalts) have been proposed as a potential cause of global environmental perturbations (e.g., McLean, 1985; Courtillot et al., 1986, 1999). This theory relies on the following evidence: the emplacement of the Deccan Traps straddling the Cretaceous-Tertiary boundary, the duration of volcanism not exceeding 1 m.y., and the potential climatic effects of massive CO_2 released by some 2×10^6 km^3 of basalts (McLean, 1985; Courtillot et al., 1986). The potential role of large igneous provinces in mass extinction events has not been limited to the Cretaceous-Tertiary crisis. The synchronicity between large igneous provinces and most mass extinctions during the last 380 m.y. led Courtillot and Renne (2003) to conclude that each biotic crisis of that period can be linked with the emplacement of a large igneous province. Recently, Kravchinsky (2012) extended this conclusion to the whole Phanerozoic.

The Cretaceous-Tertiary mass extinction that occurred 66 m.y. ago, with a species loss estimated at 50% within less than 1 m.y. (Hallam and Wignall, 1997), is considered one of smallest mass extinctions of the big five (Late Ordovician, ca. 443 Ma; Franian-Famennian, ca. 372 Ma; Permian-Triassic, ca. 252.6 Ma; Triassic-Jurassic, ca. 201.3 Ma; Cretaceous-Tertiary boundary, ca. 66 Ma). This crisis led to the disappearance of dinosaurs as well as many groups of marine reptiles, the extinction of many marine invertebrates, and the decline of floral species. Most of the paleontological evidence for these events is observed in well-preserved Late Cretaceous–Early Paleocene sedimentary formations in North America (Nichols and Johnson, 2008), but on a global scale, a turnover in the marine ecosystem and a decline of primary producers are observed (for instance, Alegret et al., 2012). If large volcanic eruptions have played a crucial role in mass extinctions, the specific kill mechanisms that led to this loss of biodiversity are not clear.

In addition to the decline of species, a global warming of 3 °C is recorded during the late Maastrichtian (Li and Keller, 1998; Barrera and Savin, 1999; Abramovich and Keller, 2003). This warming would have been related to the massive release of CO_2 during the emplacement of the Deccan Traps. However, Dessert et al. (2001) estimated via a climate-geochemical model that the release of 70,000 Gt of CO_2 over 1 m.y. would only induce a small amount of global warming, not exceeding 2 °C. Dessert et al. (2001) concluded that this amount of CO_2 must be released over a period as short as 100,000 yr to cause a warming by 4 °C. However the amount of CO_2 emitted during the emplacement of the Deccan Traps has been estimated at ~28,000 Gt (Self et al., 2006) to 35,000 Gt (Chenet et al., 2009), and the 100-k.y.-long duration of injection is considered not to be realistic (Dessert et al., 2001). Using the inverse relationship between atmospheric CO_2 level and stomatal index, Beerling and Berner (2002) estimated that a peak of 2300 ppmv occurred at the Cretaceous-Tertiary boundary, significantly above the background level of 350–500 ppmv estimated 500 k.y. before and after the Cretaceous-Tertiary boundary. Beerling and Berner (2002) concluded that the bolide impact theory better explains the abrupt rise of pCO_2 close to the Cretaceous-Tertiary boundary compared to volcanism. Indeed, a long-duration extrusion of flood basalts over some 1–2 m.y. would lead to a small CO_2 flux. In this case, the CO_2 emissions do not appear to be large enough to cause significant climate changes and the mass extinction at the Cretaceous-Tertiary boundary (Caldeira and Rampino, 1990; Beerling and Berner, 2002).

However, large igneous provinces have not been built up at a steady-state rate over time. Analysis of the secular variation of Earth's magnetic field recorded by lava flows during their cooling reveals that large igneous provinces consist of a few tens of single eruptive events and single lava flows (Chenet et al., 2008, 2009; Courtillot and Fluteau, this volume, for a synthesis). Accordingly, the accumulated duration of volcanic activity would have been as short as a few tens of thousands of years distributed over a 1 or 2 m.y. period (Chenet et al., 2008, 2009; Courtillot and Fluteau, this volume). Moreover single eruptive events and single lava flows are not regularly distributed in time over this period, but are more probably concentrated in a few phases. In the case of the Deccan Traps, Chenet et al. (2009) suggested that the single eruptive events and lava flows were emplaced in three main phases: Deccan phase 1 at ca. 67.5 Ma, corresponding to ~20% of the volume of the whole Deccan Traps, Deccan phase 2, representing the main event, around the Cretaceous-Tertiary boundary, and associated with the outpouring of the major part of the Main Province (95% of the Main Province and 75% of the whole traps), and finally Deccan phase 3, emplaced during the Early Paleocene (Danian), corresponding to a small volume of the Deccan Traps (5%). Keller et al. (2011, p. 1) concluded that "Deccan phase-2 could have alone caused the Cretaceous-Tertiary boundary mass extinction."

Therefore, since large igneous provinces may have been emplaced in a few phases, with each phase consisting in single eruptive events and single lava flows, any attempt to estimate climate and environment changes must consider the tempo of emplacement. We propose hereafter to reevaluate the impact of the gases released by Deccan Traps volcanism using a biogeochemical model coupled to a three-dimensional (3-D) climate model GEOCLIM. The choice of the Deccan Traps is also driven by the fact that magmatic degassing is the main source of gases. This is not true for other traps. The absence of correlation between the volume of erupted basalt and the extinction rate for the Siberian Traps led Ganino and Arndt (2009) to address the role of thermogenic gases from the sediment pile beneath the traps. They concluded that the amount of gases released by contact metamorphism can be as large as or even larger than that released by magmas. The nature of gases (CO_2, SO_2, CH_4) depends on the type of sediments (evaporite, dolomite, organic-rich shale, coal) intruded by the large igneous province feeders (Ganino and Arndt, 2009; Svensen et al., 2010). In the case of the Deccan Traps, the substrate mainly consists of crystalline basement overlain by minor silt deposits, producing small amounts of thermogenic gases (Ganino and Arndt, 2009). More recently, Sobolev et al. (2011) proposed an additional source of massive degassing in the case of the Siberian Traps. Based on petrological evidence, they concluded that a massive degassing of CO_2 and HCl, mostly from the recycled oceanic crust in the plume head beneath the thick lithosphere of the Siberian craton, would have occurred prior to the magmatic degassing and could have resulted in the Permian-Triassic mass extinction. Such a mechanism requires specific conditions (notably an ancient recycled oceanic crust and a thick continental lithosphere), which are not present in the Deccan Traps.

Previous geochemical modeling of the impact of the Deccan Traps on global climate have focused on the effect of CO_2 (Caldeira and Rampino, 1990; Dessert et al., 2001; Beerling and Berner, 2002; Grard et al., 2005). However, traps volcanism also released large volumes of SO_2. Based on the observation of recent volcanic eruptions, it has been shown that volcanoes can emit large amounts of gases such as carbon and sulfur dioxides (as well as other gases; Textor et al., 2003). A single volcanic eruption emits a tiny amount of carbon dioxide compared to the amount of CO_2 in the atmospheric pool, and thus no climatic response is expected. Conversely, the injection of sulfur dioxide, almost absent in a nonperturbed atmosphere, can result in a global cooling of Earth's surface (Robock, 2000; Timmreck et al., 2010). Once injected into the atmosphere, the sulfur dioxide (SO_2) is oxidized and hydrated to form sulfate aerosols, thin hydrated sulfuric acid (H_2SO_4, $2H_2O$) particles. Sulfate aerosols alter Earth's radiative balance by scattering and backscattering the incoming solar radiation, leading to a global cooling of Earth's surface (Robock, 2000). The cooling impacts the vegetation, the hydrological cycle, sea-surface temperatures (SSTs), oceanic circulation, and biogeochemistry (Timmreck et al., 2012). A modeling study evaluating the climatic consequences of the Toba event (74 ka, Indonesia), the biggest eruption of the Quaternary period, predicts a surface cooling of 4 °C globally, assuming a release of 850 Mt of sulfur into the stratosphere (Timmreck et al., 2012).

In the case of traps emplacement, the climatic response appears to be more complex because both cooling and warming, not necessarily at the same time scale, are possible due to the release of SO_2 and CO_2 in the atmosphere. To estimate the climatic responses, one must consider several parameters: the residence time (or lifetime) of CO_2, SO_2, and sulfate aerosols, their respective effects on Earth's radiative budget, and a degassing scenario. The lifetime of sulfate aerosols does not exceed 1 yr, while atmospheric CO_2 remains over 100 yr at least (IPCC, 2007). Moreover, sulfate aerosols act on both short-wave and long-wave radiation, whereas CO_2 only absorbs long-wave radiation. To estimate the effects of traps volcanism, we account for these essential features and include the short-term processes (year) and estimate their effects over several hundred thousand years (duration of traps emplacement). To achieve this goal, we used a biogeochemical model coupled to a 3-D climate model GEOCLIM, including large-scale effects of SO_2 and CO_2 injections.

QUANTIFICATION OF GAS RELEASE

To quantify climate changes on short and long time scales, the amount of gas released and flux rates must be known. Studies on the Deccan Traps emplacement (Jay, 2005; Self et al., 2006; Chenet et al., 2007, 2008, 2009; Jay and Widdowson, 2008) provide useful constraints for modeling. To fix the durations of the single events of the Deccan flows, Chenet et al. (2009) used a relative chronometer based on the analysis of the secular geomagnetic variations recorded by lava flows. The absence of secular geomagnetic variations recorded in successive lava flows suggests that their emplacement (also named single eruptive events or volcanic pulses) may not have exceeded 100 yr. Thordarson and Self (1998) estimated on the basis of volcanological observations that the eruption of the 1300 km^3 of Roza flow field in the Columbia River flood basalt province occurred within a few months to a few years. Thus, we can conclude that the duration of emplacement of single eruptive events in the Deccan Traps is likely to be very similar to that of the Roza flow field. Chenet et al. (2009) proposed that the 3500-m-thick lava pile of the Main Province of the Deccan Traps erupted in ~30 volcanic pulses and some 41 individual lava flow units. The total duration of volcanic extrusions may not have exceeded 7000 yr (assuming a maximum emplacement time of 100 yr for each single eruptive event and each flow), and was possibly as short as 1000 yr (assuming a time of emplacement not exceeding a few years). The duration of volcanic activity does not account for the periods of quiescence between the emplacements of flows. The emplacement of the Main Province would have lasted ~400,000 yr (Chenet et al., 2009).

Assuming a simple geometry for lava flows and accounting for the estimations of SO_2 and CO_2 concentrations in basalts (Self et al., 2006), Chenet et al. (2009) estimated that a single volcanic event would have released up to 100 Gt of SO_2 and 200 Gt of CO_2 over a century or less. However, Chenet et al. (2009) estimated that the total volume of the 30 eruptive events and 41 single lava flows was ~253,000 km^3, 10% to 25% of the total original volume of the Deccan Traps. Consequently, either lava flows and/or volcanic pulses were not sampled and accounted for in the estimation of magma volume, leading to an underestimation of the duration of volcanic activity, or the volumes of lava flows and volcanic pulses are larger than Chenet et al. (2009) suggested, leading to an underestimation of the amounts of SO_2 and CO_2 injected within the atmosphere, or both. An underestimation of the number of volcanic pulses is more likely because the locations of volcanic centers feeding lava flows are not well known, and flows may not have been sampled. After scaling to the total volume of the main Deccan Traps province (~2 × 10^6 km^3), Chenet et al. (2009) estimated the amount of SO_2 released would have been between 6800 and 17,000 Gt and between 15,000 and 35,000 Gt for CO_2.

These constraints on time, volume, and thus fluxes of volatiles are used to build a scenario of injection of CO_2 and SO_2 within the atmosphere. We assume that:

1. The duration of Main Province emplacement was ~400,000 yr.
2. We scale the number of lava flows and single eruptive events to the total volume of the main Deccan Province, 280 volcanic events.
3. Given that each volcanic event lasts 10–100 yr, the total duration of volcanic activity ranges from 2800 to 28,000 yr.
4. The total time of volcanic quiescence is ~370,000–390,000 yr.
5. The time elapsed between each volcanic pulse is fixed at 1500 yr.
6. Flux rates vary between 0.1 Gt/yr and 10 Gt/yr for SO_2 and between 0.24 and 24 Gt/yr for CO_2 (depending on the duration of the volcanic events, 100 or 1 yr).

MODELING APPROACH

General Model Description

Changes in atmospheric CO_2 and SO_2 have a number of effects that can be quantified and modeled (for example, Segschneider et al., 2013). However, the consequences of the onset of Deccan Traps have not yet been investigated with a comprehensive climate-carbon model including SO_2. In this paper, we use GEOCLIM (Donnadieu et al., 2006a), which couples a general circulation model (GCM) called FOAM1.5 (Fast Ocean Atmospheric Model; Jacob, 1997) to a model of the main biogeochemical cycles called COMBINE (Goddéris and Joachimski, 2004) to simulate the consequences of the Deccan Traps volcanism over several hundred thousand years.

FOAM is forced with Late Maastrichtian boundary conditions (CO_2, paleogeography, solar constant). The atmospheric component of FOAM is a parallelized version of National Center for Atmospheric Research (NCAR) Community Climate Model 2 (CCM2) with the upgraded radiative and hydrologic physics incorporated in CCM3 v. 3.2 (Jacob, 1997). All simulations have been performed with an R15 spectral resolution (4.5° × 7.5°) and 18 vertical levels. FOAM is used in mixed-layer mode, meaning the atmospheric model is linked to a 50 m mixed-layer ocean with heat transport parameterized through diffusion. FOAM does not include an interactive chemistry module; all sulfate reactions and forcing are included in COMBINE (see section Incorporation of Sulfates into GEOCLIM).

COMBINE describes the biogeochemical cycles for carbon, phosphorus, alkalinity, and oxygen. The ocean is longitudinally divided into three parts, two polar oceans and a low-to-mid-latitude ocean. Polar oceans are located between −90° and −50° latitude for the Southern Hemisphere and +50° and +90° latitude for the Northern Hemisphere. Each polar ocean includes both surface (0–1000 m) and deep-ocean reservoirs (1000–3800 m). The low-to-mid-latitude ocean is divided into three layers: a photic zone (0–100 m), a thermocline (100–1000 m), and a deep-ocean reservoir (1000–3800 m). Two boxes representing epicontinental seas are also included, a first reservoir representing the photic zone (0–100 m) and a second representing the deep epicontinental reservoir (100–200 m). These two boxes are the exchange zones between lands and ocean. A twelfth box stands for the atmosphere. In absence of trap volcanism, we used a constant solid Earth degassing rate (6.8 × 10^{12} moles CO_2 per year), which is the degassing rate required to balance present-day CO_2 consumption through silicate weathering (Gaillardet et al., 1999). Field work indicates that silicate weathering is linearly dependent on mean annual continental runoff (hereafter R), associated with an exponential dependence on mean annual air temperature (hereafter T_{air}). In detail, the total CO_2 consumption through weathering is for granitic (F_{silw}) and basaltic (F_{basw}) lithologies, respectively (Dessert et al., 2001; Oliva et al., 2003).

$$F_{silw} = \sum_{i=1}^{n_{pixel}} k_{silw}.run(i).area(i).\exp\left[-\frac{E_a^{silw}}{R}\left(\frac{1}{T(i)}-\frac{1}{T_0}\right)\right] \quad (1)$$

$$F_{basw} = \sum_{i=1}^{n_{pixel}} k_{basw}.run(i).area(i).\exp\left[-\frac{E_a^{basw}}{R}\left(\frac{1}{T(i)}-\frac{1}{T_0}\right)\right] \quad (2)$$

$T(i)$ and run(i) are the mean annual air surface temperature and runoff for the grid cell i. R is the ideal gas constant; E_a^{silw} is the apparent activation energy (48,200 J/mol) for granite, and E_a^{basw} is the apparent activation energy for basalt (42,300 J/mol). *Area*(i) is the surface of the considered continental grid cell, and n_{pixel} is the number of continental grid cells. The terms k_{silw} and k_{basw} are calibration constants fixed so that, under present-day climatic conditions, basaltic dissolution consumes 30% of the total CO_2 consumption by continental silicate weathering (Dessert et al., 2003). The computation also considers a potential imbalance in the organic carbon cycle (burial vs. oxidation of sedimentary organic carbon), which may further modulate the atmospheric pCO_2 (Goddéris and Joachimski, 2004).

The coupling procedure between FOAM and COMBINE is asynchronous. FOAM is used to generate an offline catalogue of continental air temperature (T_{air}) and continental runoff (R) values for a range of pCO_2 values (see following section, Boundary Conditions and Simulation Design, for details). At each time step of the GEOCLIM calculation, and for each corresponding atmospheric pCO_2, T_{air} and R above each continental pixel are calculated through a linear interpolation procedure from the climatic catalogue. An exhaustive description of GEOCLIM and its components COMBINE and FOAM can be found in Goddéris and Joachimski (2004), Donnadieu et al. (2006a, 2006b), and Simon et al. (2007).

Incorporation of Sulfates into GEOCLIM

The computing time required to perform simulations with a GCM limits the length of simulations to a few thousand years, which means that a GCM cannot be used to simulate the impact of sulfate aerosols over the whole duration of the Deccan Traps emplacement (i.e., 400 k.y.). To solve this problem, we include the chemical reactions of sulfur dioxide into the atmospheric box of COMBINE. We use the parametric laws developed by Pierazzo et al. (2003) and Miles et al. (2004) to represent short-lived processes. We first assume that, for a given mass of SO_2 released by a volcanic event (F_{volc} in moles), the decay of SO_2 gas is governed by its oxidation:

$$\frac{dSO_2}{dt} = F_{volc}^{SO2} - k_a[SO_2], \qquad (3)$$

where $[SO_2]$ is in moles. Based on the Pinatubo eruption, SO_2 e-folding time (k_a) has been evaluated to 0.096 yr or 35 d (Rose et al., 2000). However, for larger eruptions, uncertainties exist for e-folding time values. According to Pinto et al. (1989), the SO_2 oxidation time increases with the mass of SO_2 injected into the atmosphere, notably because of the consumption by OH radicals, which limits the conversion. To account for this limiting effect, we tested two additional oxidation laws describing a limited oxidation rate (i.e., k_a varies as a function of the SO_2 content; see Pierazzo et al., 2003).

The rate of change in the concentration of gaseous H_2SO_4 is controlled by the SO_2 oxidation (source) and its hydration (sink):

$$\frac{dH_2SO_{4gas}}{dt} = k_a[SO_2] - k_b[H_2SO_{4gas}] \qquad (4)$$

where $[H_2SO_{4gas}]$ is in moles. According to Miles et al. (2004), the conversion of H_2SO_4 into aerosols (H_2SO_{4liq}) is supposed to be constant; the e-folding time (k_b) is fixed to 100 d (or 0.27 yr). Finally, Equation 5 describes the concentration of sulfate aerosols as a function of time, with the decay being due to gravitational sedimentation.

$$\frac{dH_2SO_{4liq}}{dt} = k_b[H_2SO_4] - k_c[H_2SO_{4liq}] \qquad (5)$$

where $[H_2SO_{4liq}]$ is in moles. The e-folding time (k_c), linked to the gravitational sedimentation, has been fixed to 1 yr (Miles et al., 2004). Recent studies, incorporating aerosol microphysics, obtained different gravitational sedimentation rates according to the size of aerosols (Timmreck et al., 2010). However, this process is largely beyond the scope of this study.

In the model, the incoming radiative flux reduction caused by atmospheric loads of aerosols is directly expressed in ΔT. ΔT is calculated as a function of the mass of sulfate aerosols (mH_2SO_{4liq}) and sea-ice thickness (H_{ice}).

$$T = Tbox(j) - 3.2mH_2SO_{4liq} \times (1 - H_{ice}) \qquad (6)$$

with ΔT in °C and mH_2SO_{4liq} in Gt. *Tbox(j)* is the mean annual temperature (°C) for box *j*. Only surface oceans and atmospheric boxes are affected by Equation 6. This equation takes into account the direct H_2SO_{4liq} effect and albedo feedback. H_{ice} (varying from 0 to 1) represents the theoretical sea-ice thickness (or the snow-layer over continents) formed when $Tbox(j) < 0$ °C. The term $(1 - H_{ice})$ simulates the lag time required to melt the H_{ice} when $Tbox(j) > 0$ °C. With continental runoff being dependent on the mean annual air temperature, the runoff is proportionally reduced. It is important to note that a large variety of ΔT-forcing relationships exists (Pollack et al., 1976; Jones et al., 1995; Oppenheimer, 2002; Robock et al., 2009), most of which have been calibrated on Pinatubo or estimated from Toba eruptions. Obviously, in the absence of direct measurements of optical depth caused by a Toba-like eruption, many cautions have to be used with temperature changes attributed to large eruptions (see sensitivity experiments in section Short-Term Response of a Single Volcanic Pulse).

To investigate the long-term climatic consequences of the Deccan Traps, we have also added long-lived processes. In keeping with Earth's oceanic/continental surface ratio, we assume that two-thirds of aerosols fall over oceans.

1. Over continents, H_2SO_{4liq} values are taken to instantaneously react with carbonates according to the following equation.

$$2CaCO_3 + H_2SO_{4liq} \rightarrow 2Ca^{2+} + 2HCO_3^- + SO_4^{2-} \qquad (7)$$

In the initial COMBINE version, the calculation of carbonate weathering is carried out assuming that calcium concentration $[Ca^{2+}]$ (unit mol/m^3) content of the runoff is at equilibrium with soil CO_2, and carbonate rocks. Here, we added $[SO_4^{2-}]$ in the charge balance calculation, where all $[SO_4^{2-}]$ comes from the dissociation of H_2SO_{4liq}. Since H_2SO_{4liq} decreases the soil pH, the phosphorus weathering is also affected (for details, see Guidry and MacKenzie, 2000; Donnadieu et al., 2006a).

2. For oceans, H_2SO_{4liq} mainly acts on the alkalinity budget and associated processes (e.g., carbonate saturation state). In a similar manner to the method described for continental runoff, we acidify the surface ocean reservoirs by adding [SO_4^{2-}] in the charge balance calculation. The H_2SO_{4liq} first affects the surface ocean, and then the deep ocean, according to the mixing ratio of the ocean (here fixed to the modern value). The riverine input flux of sulfate ions to the oceans is taken to be the product of runoff and ion concentrations.

Boundary Conditions and Simulation Design

To simulate the Late Cretaceous climate, FOAM simulations have been performed with a land-ocean distribution derived from a synthesis of paleomagnetic data, hotspot tracks, and geologic constraints (Blakey, 2007). To avoid the astronomic influence, orbital parameters are set at their present-day values in all experiments. The solar constant is fixed to its present-day value (Gough, 1981). FOAM is used to generate an offline catalogue of continental air temperature T_{air} and continental runoff (R) values for a range of pCO_2 values from 200 ppmv up to 4200 ppmv, with an increment of 200 ppmv. For COMBINE, as initial conditions for Cretaceous-Tertiary boundary pCO_2, we used the modern background-degassing rate (6.8 10^{12} mol C/yr), which corresponds to a steady-state pre-trap volcanism pCO_2 close to the preindustrial value (i.e., 280 ppmv). This atmospheric carbon dioxide value, also used in the Dessert et al. study (2001), is below other estimates for the Cretaceous-Tertiary boundary value (~400 ppmv—Caldeira and Rampino, 1990; ~500 ppmv—Beerling and Berner, 2002), but this does not have a significant impact on our results. Because the accumulation of neritic carbonates is dependent on the shelf surface, the shelf platform area has been modified to correspond to end-Cretaceous conditions (Walker et al., 2002).

Climate effects of continental flood basalts are tested using a variety of scenarios. To decipher short-term and long-term responses, we first focus on climatic consequences of a single volcanic pulse with duration of emission ranging from 1 to 100 yr. In the second step, we explore the climate effects of successive volcanic pulses representing Deccan Trap emplacement. We assume that the atmospheric CO_2, SO_2, H_2SO_4, and H_2SO_{4liq} are homogeneously distributed within the atmospheric reservoir. However, we have to keep in mind that the volcanic vent location and the height of injection are fundamental parameters for the dispersal of SO_2 and sulfate aerosols. We know that the dispersal of sulfate aerosols is global in the case of tropical volcanic eruptions (Robock, 2000). The location of the Deccan Traps at 66 Ma at subtropical latitudes may have favored such a global dispersal. In regard to the height of injection, it has been shown by Kaminski et al. (2011) that the heat flux released by huge lava flows alters the lower-atmospheric thermal structure, permitting the injection of SO_2 in the stratosphere. All factors other than CO_2 and/or SO_2 emissions were held constant in order to quantify the direct/indirect effects of these releases. Simulations are listed in Table 1.

RESULTS

We follow a two-step approach. First, the short-term response (decade-to-century time scales) enables investigation into the effects of a single large volcanic eruption. In a second step, we focus on the long-term response to estimate effects of the Deccan Traps volcanism on the Earth system. We focus on the temporal evolution of the surface temperature of the mid-to-low-latitudinal area (50°S–50°N). This will facilitate an easier comparison of the model output with geological data.

Short-Term Response of a Single Volcanic Pulse

The pCO_2 and SST perturbations in run 1 experiments are summarized in Table 2. In run 1a, 1b, and 1c, SST increases by 0.2 °C, while the pCO_2 increases by 7–10 ppmv (Table 2). All

TABLE 1. ALL SIMULATIONS REALIZED WITH COMBINE

Experiment names	Pulses (Nb)	Pulse duration (yr)	Total emitted (Gt)		CO_2 outgassing rate (Gt/yr)	SO_2 outgassing rate (Gt/yr)	SO_2 oxidation law used
			CO_2	SO_2			
RUN_1a	1	1	100	0	100	0	–
RUN_1b	1	10	100	0	10	0	
RUN_1c	1	100	100	0	1	0	
RUN_1d	1	1	10	0	10	0	
RUN_1e	1	10	10	0	1	0	
RUN_1f	1	100	10	0	0.1	0	
RUN_2a	1	10	0	24	0	2.4	Unlimited
RUN_2b	1	10	0	24	0	2.4	Asymptotic law
RUN_2c	1	10	0	24	0	2.4	Power law
RUN_2d	1	1	0	24	0	24	Asymptotic law
RUN_2e	1	100	0	24	0	0.24	Asymptotic law
RUN_3	280	10	28,000	0	10	0	–
RUN_4	280	10	0	6500	0	2.4	Asymptotic law
RUN_5	280	10	28,000	6500	10	2.4	Asymptotic law

Note: For runs 3, 4, and 5, the quiescent time between the pulses is 1500 yr. Outgassing rates concern flux emitted during one pulse. Two-thirds of SO_2 is deposited over the oceans and one-third over continents. The CO_2 degassing is 6.8 × 10^{12}.

TABLE 2. SHORT-RESPONSE EXPERIMENTS

Experiment names	Pulse duration (yr)	Total emitted (Gt)	CO_2 outgassing rate (Gt/yr)	pCO_{2max} (ppmv)	t_{max} (°C)	Perturbation duration (yr)
RUN_1a	1	100	100	10	0.2	100
RUN_1b	10	100	10	9	0.2	110
RUN_1c	100	100	1	7.5	0.2	160
RUN_1d	1	10	10	1.1	Undetectable	Undetectable
RUN_1e	10	10	1	1.1	Undetectable	Undetectable
RUN_1f	100	10	0.1	1.1	Undetectable	Undetectable

Note: The CO_2 degassing rate background is 6.8×10^{12} mol C/yr.

simulations performed with a release of CO_2 (run 1d, 1e, and 1f) not exceeding 10 Gt do not show any significant changes in the atmospheric carbon dioxide content. This suggests that the radiative effect of the carbon dioxide released by a single eruption is undetectable, a result already shown in previous studies (Caldeira and Rampino, 1990; Dessert et al., 2001). In the case of SO_2 (run 2), the largest SST cooling is simulated in response to the largest SO_2 mass injections, as expected (Fig. 1). However, the climatic response is modulated by the amplitude of the degassing rate and by the oxidation laws used.

To discriminate both factors, we performed two different sets of experiments. In the first set, constant SO_2 emissions are prescribed, and different oxidation laws are used: unlimited oxidation, asymptotic limitation, and exponential limitation. Figures 1A, 1B, and 1C indicate that the SO_2/H_2SO_{4liq} ratio largely depends on the oxidation phase, indicating this step as the most crucial. In run 2a, the SO_2 is instantaneously oxidized because the peak of SO_2 is not limited by the OH availability. The rapid chemical conversion of SO_2 favors high loads of H_2SO_{4liq} and an abrupt cooling. The SST drops by 9 °C during 10 yr and returns to its initial value rapidly after the volcanic event ceases. Conversely, when the SO_2 lifetime increases exponentially with increasing loads of SO_2 (run 2c), the H_2SO_{4liq} content remains very low, while the SO_2 accumulates in the atmosphere. The SST decreases by 1 °C, but it takes more than 100 yr before the temperature anomaly returns back to zero. This lengthening of the perturbation is the direct effect of the very slow chemical conversion of SO_2, with ~100 yr being required to return to pre-eruption levels.

To determine the most appropriate oxidation law to use, we compared COMBINE simulations with model outputs obtained from a more complex climate model including a fine interactive chemistry module (ESM for Earth System Model; for details, see Timmreck et al., 2010, 2012; Fig. 2). In case of Toba-like eruptions, the best fits are obtained for unlimited and asymptotic limitation laws. Temperature deviations from the pattern of the SST/T_{air} anomaly, predicted by GEOCLIM and ESM, respectively, are caused by the larger thermal inertia of the ocean compared to the atmosphere, 4200 and 1000 J kg^{-1} K^{-1}, respectively[1]. With regard to the SO_2 emitted during the Deccan pulses (Chenet et al., 2009), we found that the asymptotic law tends to be the best parameterization for describing SO_2 oxidation. Indeed, the unlimited oxidation law cannot be used to model large eruptions, because, for very high SO_2 emissions (i.e., Toba-like eruption), the available hydroxyl radical concentration (OH) should delay the $H_2SO_{4(gas)}$ conversion and thus the aerosol formation (Robock et al., 2009). For that reason, the following simulations were performed using the asymptotic oxidation law (Pierrazzo et al., 2003).

In the second set of simulations, we explored the climate response for various SO_2 emission rates with the total SO_2 released held constant. Volcanic pulses are based on the analyses of the secular variations of Earth's magnetic field (Chenet et al., 2008, 2009). Chenet et al. (2008, 2009) estimated that the volcanic pulse may have been as short as 100 yr, but this relative chronometer does not allow any better constraint of the duration. However, Thordarson and Self (1998) have shown that the emplacement of the 1300 km^3 Roza flow within the Columbia River Traps would have been as short as a few years. Thus, the duration of a single eruptive event may not have exceeded 1 yr.

We investigated the effect of different scenarios of SO_2 degassing rates (same amount of SO_2, but three durations of emplacement: 1 yr, 10 yr, and 100 yr). Figures 1B, 1D, and 1E show SST anomalies corresponding to these three cases. Looking at Figure 1B (i.e., run 2b, 10 yr of emissions), SST drops rapidly to a minimum value of 10 °C ($\Delta SST = -6$ °C) within the first 10 yr, followed by a relatively quick recovery over the next 20 yr. The maximum cooling is obtained with an emission rate ten times higher (run 2d, $\Delta SST = -9$ °C), while the SST drops by only 1 °C for an outgassing rate 10 times smaller (run 2e). These results show a pronounced nonlinearity between outgassing rates and climate changes, with stronger cooling being predicted when degassing rates overcome 2.4 Gt SO_2/yr. Consequently, a key question to quantify the effects of large volcanic eruptions is the volcanic dynamics. Up to now, this parameter could not be well constrained. For that reason, in the following experiments, we use a mean duration of 10 yr for each pulse, an assumption increasing the climate changes related to continental flood basalts events.

[1]This difference could be corrected in assuming a higher SST sensitivity to mH_2SO_{4liq} (see Eq. 6); however, this would imply that the Rosa flow event should trigger a snowball Earth (Thordarson and Self, 1996). Moreover, Oppenheimer (2002) claimed that the cooling caused by the Toba eruption was close to 1 °C in the tropics.

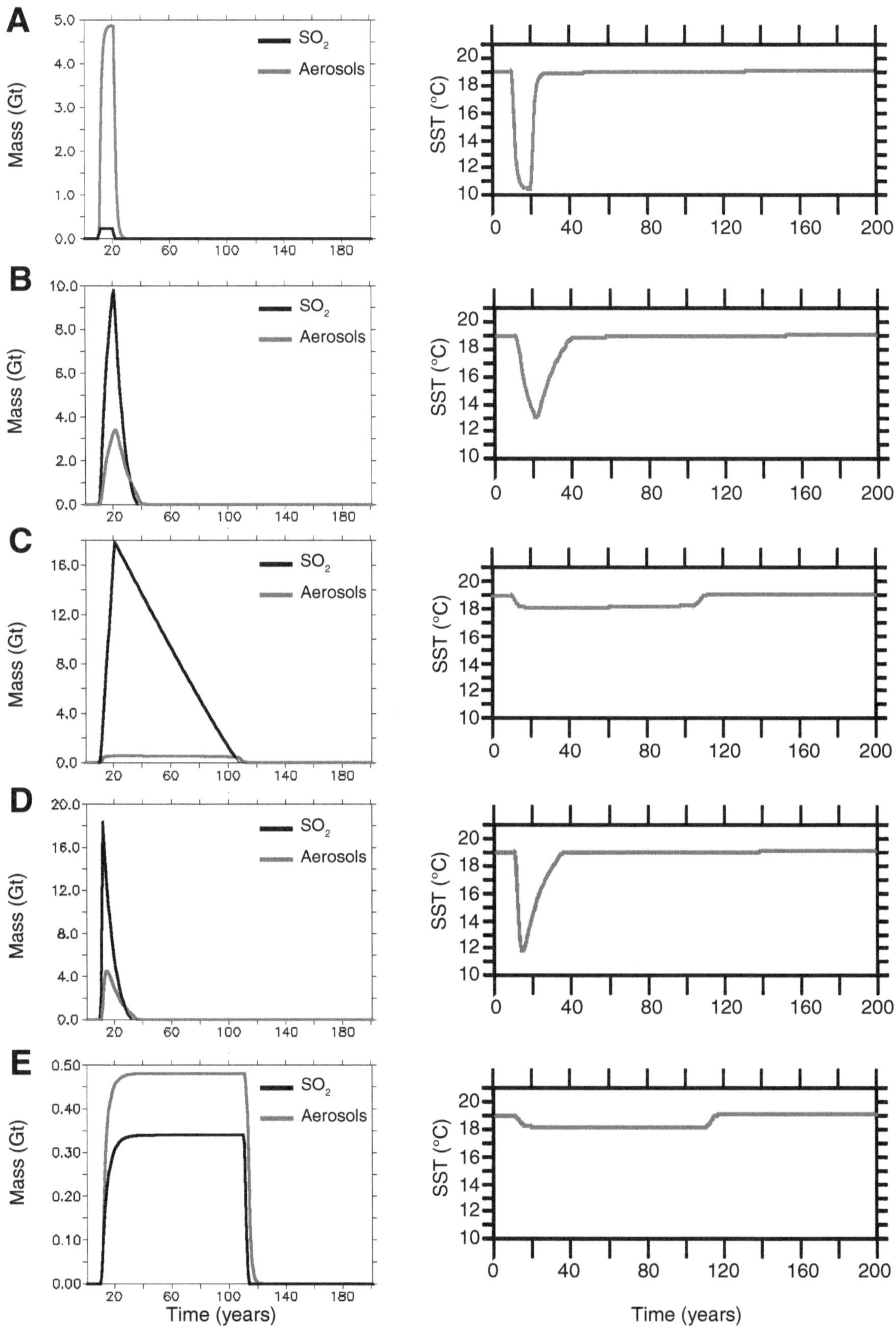

Figure 1. Evolution of the mass (left column) of the SO_2 (black line) and the aerosols (gray line) into the atmosphere, and of the sea-surface temperature (SST; right column) for run 2a, run 2b, run 2c, run 2d, and run 2e.

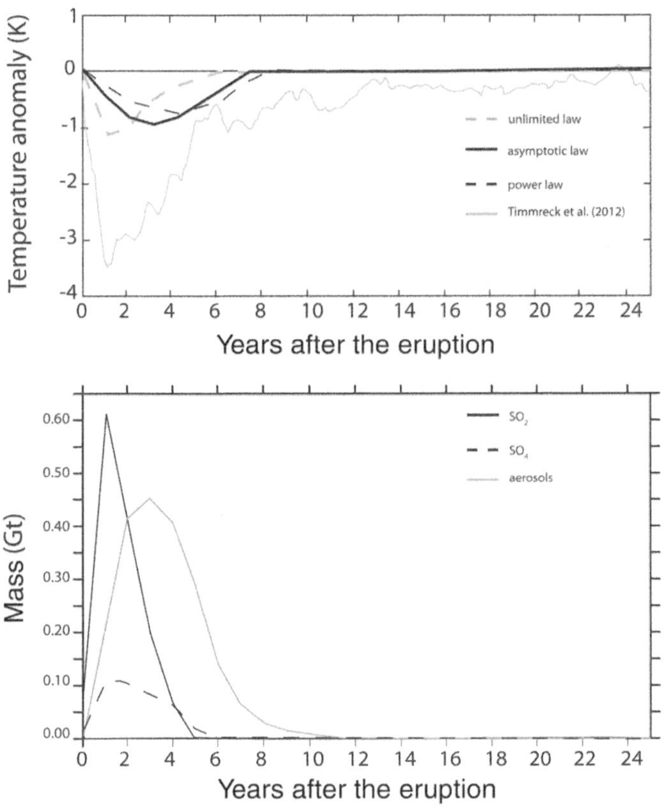

Figure 2. Comparison of the temperature anomaly (up) obtained by our model and by Timmreck's general circulation model (GCM; gray line) for a Toba-like eruption. We present here the sea-surface temperature (SST) anomaly, whereas Timmreck and collaborators worked on the air temperature. We considered different SO_2 oxidation laws: unlimited (dashed gray line), asymptotic (thick black line), and a power law (dashed black line). The zero-level is represented by the thin black line. Our evolution of sulfur species is shown in the figure below.

Long-Term Response of Successive Volcanic Pulses

The calculated evolutions of atmospheric pCO_2 (ppmv), SST (°C), and granite/basalt carbon consumption (10^{12} mol C/yr) of long-term runs are presented in the Figures 3–5. For the run 3 scenario, the variation of pCO_2 increases by 150 ppmv from its pre-Deccan value. Compared to the pioneer modeling study by Caldeira and Rampino (1990) for a total released three times larger (28,000 instead of 8800 Gt CO_2), our predicted pCO_2 is also three times larger than they obtained (150 ppmv instead of 40 ppmv). A similar behavior is observed for ΔSST, +0.4 °C for Caldeira and Rampino, instead of +1.5 °C in this study (Fig. 4). Hence, the mean temperature is not strikingly different for the same pCO_2. Our estimates also agree well with the results of Dessert et al. (2001). Their longest run (i.e., 1 m.y. for Deccan Traps emplacement for 78,000 Gt CO_2 outgassed) follows the same trend as our run after 400 k.y. of the simulation (200 ppmv/150 ppmv and 1.4 °C/1.5 °C). We conclude that the type of CO_2 emission, continuous or pulsed, over a 400 k.y. period induces a similar atmospheric pCO_2 increase.

This is not the case for SO_2 (run 4). A volcanic pulse releasing SO_2 during 10 yr every 1500 yr, leads to an abrupt cooling over ~30 yr. In the absence of coeval carbon dioxide emissions, the pCO_2 increases by 20 ppmv (Fig. 3). This CO_2 rise is directly linked to the increase of H_2SO_{4liq} loads. Indeed, sulfate aerosols change the radiative balance of the atmosphere by scattering solar radiation, which leads to cooling of the continental surfaces and decreased precipitation. Both effects reduce the efficiency of continental weathering as a CO_2 sink (Fig. 5). In detail (close-up of Fig. 5), the lowering of carbon consumption is followed by a slow recovery phase, lasting one century up to its pre-eruption level. This short-lived imbalance between carbon consumption and carbon release induces a rise in atmospheric CO_2 concentrations. The carbon dioxide added this way is not entirely consumed due to the periodicity of the cooling events. Hence, due

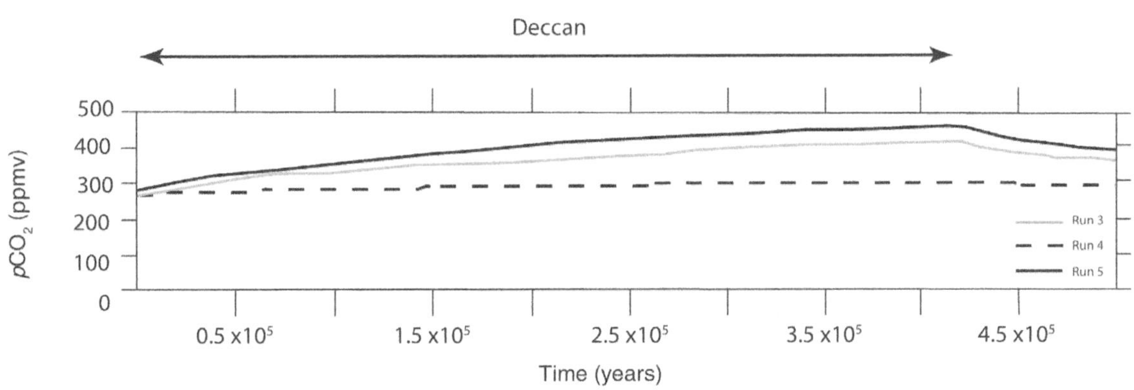

Figure 3. Evolution of the pCO_2 during Deccan Traps emplacement with a volcanic activity injecting CO_2 only (gray curve), SO_2 only (dashed black line), and CO_2 and SO_2 (black curve).

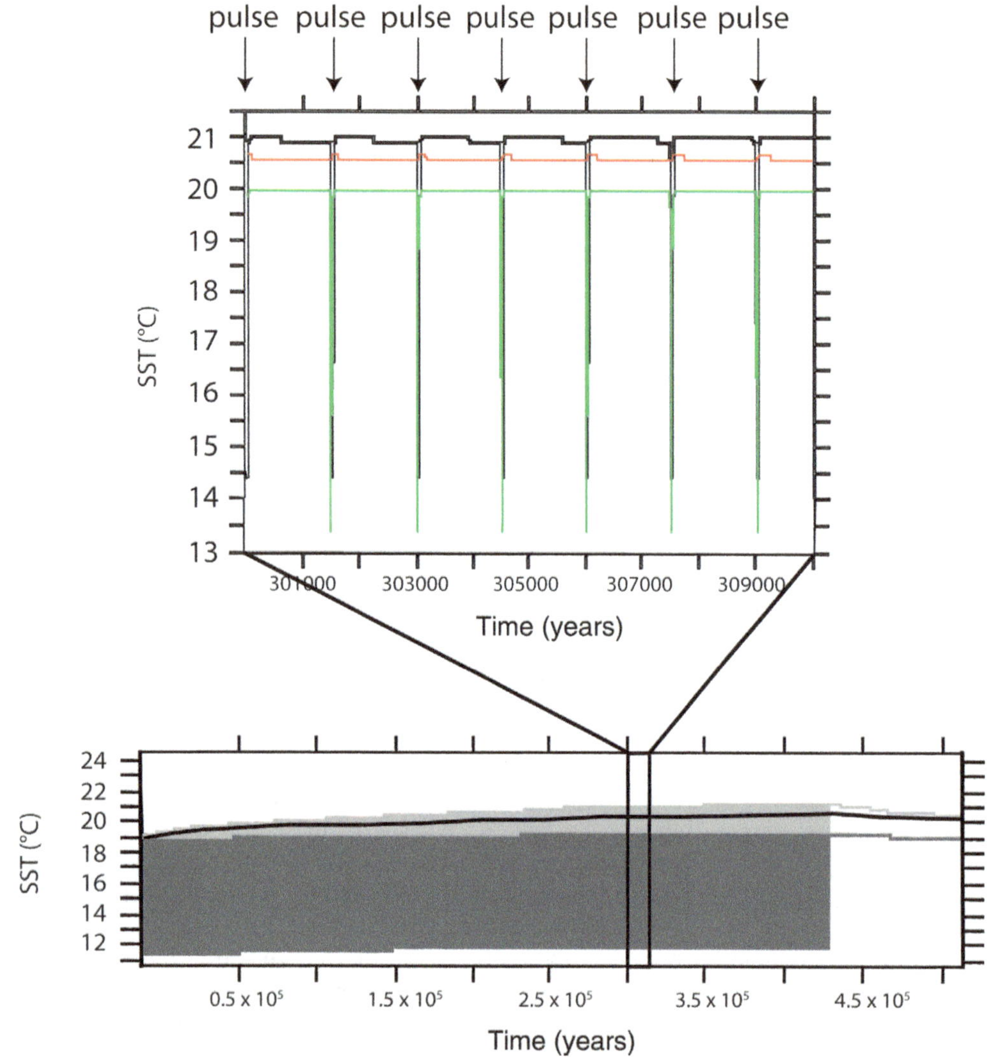

Figure 4. (Bottom) Evolution of the sea-surface temperature (SST) averaged on the low latitudes (−50°/50°) during Deccan Traps emplacement with a volcanic activity injecting CO_2 only (black curve), SO_2 only (dark gray), and CO_2 and SO_2 (light gray). (Top) To better distinguish the difference between the three runs, a close-up of 10,000 yr is shown. The lines represent the SST evolution with a volcanic activity injecting CO_2 only (red line), SO_2 only (green line) and CO_2 and SO_2 (black line).

to pulsed degassing of SO_2, the inorganic carbon cycle tends to be perturbed and its equilibrium altered. This important process cannot be observed in the case of a continuous degassing rate.

When CO_2 and SO_2 are injected simultaneously (run 5), the carbon dioxide content increases up to 480 ppmv (ΔpCO_2 = 200 ppmv; Fig. 3), leading to a 2.5 °C warming of the mid-to-low-latitude surface ocean (Fig. 4). For the same degassing rate, the pCO_2 perturbation is thus enhanced by 25% (200 ppmv instead of 150 ppmv, run 3 vs. run 5; Fig. 3) when SO_2 emissions are considered. This efficient carbon dioxide accumulation is due to the fact that during a pulse, the cooling inhibits the CO_2 consumption when the degassing rate is increasing. So, we can expect that brief cooling events induce greenhouse warming at longer time scales. Obviously, the value of 25% is a first-order estimation obtained with an idealized scenario. This feedback takes place during all trap province emplacements; however, its efficiency may vary as a function of the continental flood basalt province (e.g., Deccan, Siberia), with each trap having its own dynamics of emplacement.

DISCUSSION

Estimating the climate change caused by volcanic emissions is a challenging issue. Using an integrative carbon-sulfate model, we have provided evidence for a new climatic feedback mechanism. Instead of long-term greenhouse warming (run 3), we have demonstrated that continental flood basalts lead to abrupt climate changes (run 4) associated with a terminal warming greater than initially suggested (run 5; Caldeira and Rampino, 1990). This behavior, which we have called the cooling-induced warming paradox, is the result of the direct interplay between CO_2 and SO_2. During the trap emplacement, sulfate aerosols cool Earth's surface during short periods (a few years) but enhance the long-term warming due to an increase of pCO_2 by 25%. We used an amount of SO_2 that is slightly lower than the values proposed by Chenet et al. (2009) (6500 Gt vs. 6800 Gt to 17,000 Gt). A higher value would have more drastic consequences on the climate and environment and would enhance the final warming. The number of volcanic pulses, the duration of volcanic activity, and duration

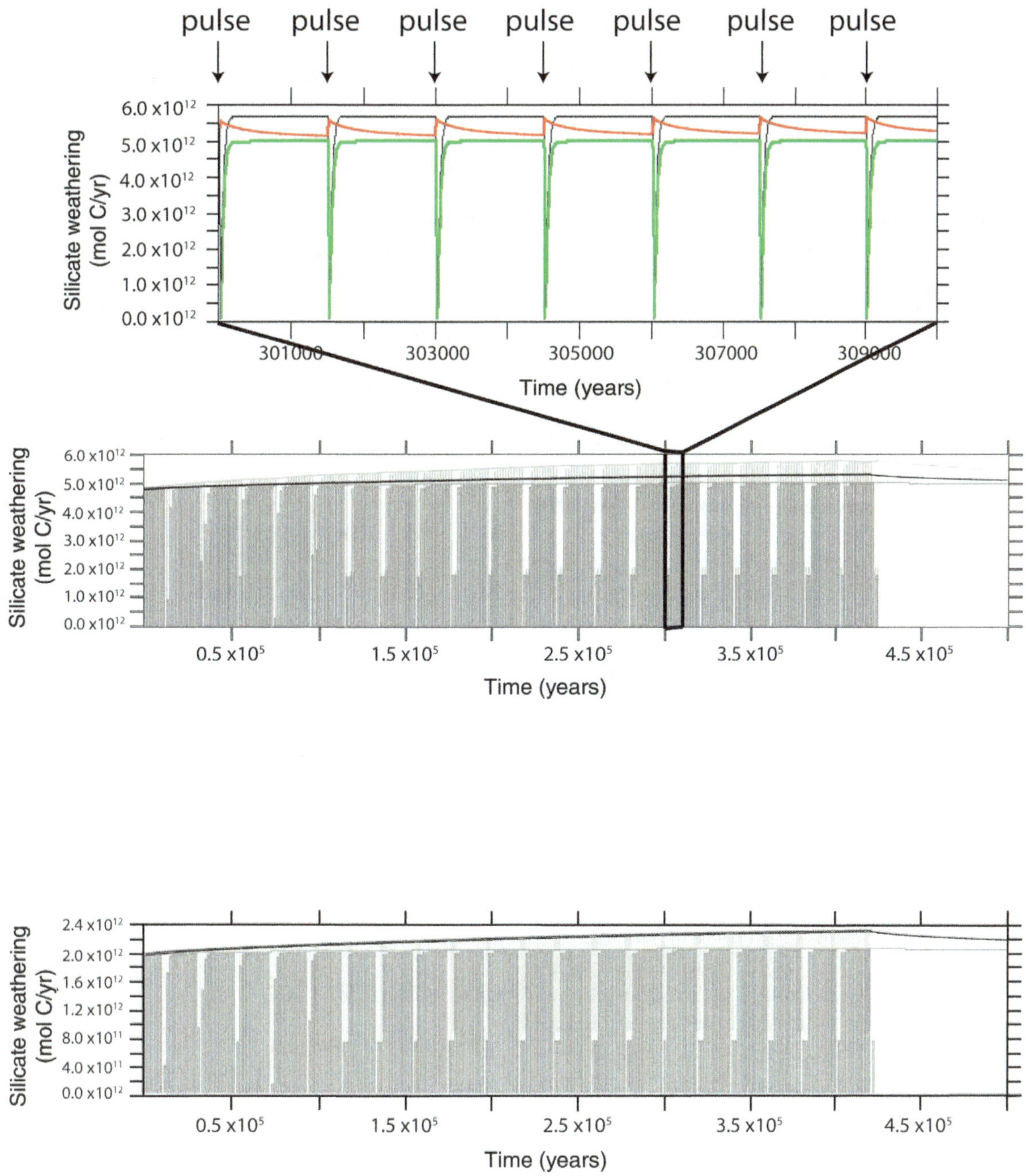

Figure 5. Evolution of the silicate (middle) and basalt (lower) weathering during Deccan Traps emplacement with a volcanic activity injecting CO_2 only (black curve), SO_2 only (dark gray), and CO_2 and SO_2 (light gray). (Top) To better distinguish the difference between the three runs, a close-up of 10,000 yr is shown for the evolution of the silicate weathering with injections of CO_2 only (red line), SO_2 only (green line), and CO_2 and SO_2 (black line).

of volcanic quiescence need to be analyzed for other traps in future work.

The complex climatic responses superimposed on short-term cooling events and a long-term warming remain to be observed in proxy data. Indeed, the short-term cooling events cannot be observed because of the low time resolution of proxy data. Based on the depositional resolution, we can assume that the level of achievable stratigraphic resolution is too crude to capture short-lived events, such as in the case of cooling (i.e., 10-yr-long pulse followed by 90 yr of recovery). In addition, short-term cooling may have favored a decrease in the sedimentation rate, reducing the record of those events. For deep time periods, only processes occurring over 10^5 yr can be seen (Erwin, 2006). Thus, long-term warming may appear as the only preserved climatic message. An abrupt increase of low-to-mid-latitude SSTs of 3–4 °C is clearly identified in the oxygen isotope records from the Atlantic, Pacific, and Indian Oceans (Li and Keller, 1998; Barrera and Savin, 1999; Abramovich and Keller, 2003), a warming also seen in high latitudes (Seymour Island, 64°S), along with higher intensity (i.e., 5–7 °C; Tobin et al., 2012).

Our attempt to simulate a pulse-like degassing provides significant results for our understanding of climate changes caused by continental flood basalts. Supported by an underlying coupling between SO_2 and CO_2, our results seem robust. Obviously, several aspects of our modeling have to be evaluated cautiously. Using a box model, we were able to mix short-term and long-term processes, producing a unified view of the impact of trap province eruption. However, the atmospheric component consists of only one box and assumes a homogeneous dispersal of the gases. These limitations can be solved using a GCM. A fruitful perspective would be to apply the methodology developed for CO_2 to SO_2 (i.e., an asynchronous coupling between COMBINE and GCM with an interactive chemistry for SO_2). This tool would be more appropriate to quantify OH depletion and to give more detailed information about the relationships between large igneous province emplacement and environmental changes.

Furthermore, other degassing processes must be considered as causal links between Phanerozoic mass extinctions and catastrophic geological events. Ganino and Arndt (2009) showed that sediments intruded by large igneous province feeders would have released large amounts of thermogenic gases (CH_4, CO_2, and SO_2). For the Siberian Traps, Iacano-Marziano et al. (2012) suspected that greenhouse gases (CO-CH_4 essentially) could be 7–20 times higher than initially assumed due to magma-sediment interactions. Finally, the massive degassing due to the interplay between an ancient oceanic crust and the Siberian Traps plume head proposed by Sobolev et al. (2011) would reinforce the global warming effect (CO_2). The resulting HCl release would be toxic for continental species and would favor the depletion of ozone in the stratosphere. This process must be investigated for other traps. In the case of the Deccan Traps, this process seems to be absent, with degassing driven primarily by erupted magmas. Concerning the Cretaceous-Tertiary mass extinction, the importance of the Chicxulub impact remains to be established along with that of the continental flood basalts.

CONCLUSIONS

In this study, we explored the geochemical and climatic impact of the eruption of the Deccan Traps with a coupled climate-geochemical model. Assuming a pulse-like degassing instead of a continuous release of volcanic gas, our study revealed some hitherto unaddressed aspects and has resulted in new interpretations for climatic consequences of continental flood basalts. Focusing on the Deccan Traps, our simulations show that each pulse triggers a substantial cooling forcing over multidecadal time scales. Successive cooling events driven by sulfate aerosols over a period of 400,000 yr are suspected to unbalance the long-term carbon cycle by decreasing the efficiency of silicate weathering.

We represent the Deccan Traps emplacement by a series of 280 volcanic pulses lasting 10 yr and occurring every 1500 yr. Using this scenario, we have demonstrated that the CO_2 consumption weakness is associated with an enhancement by 25% of pCO_2 and an increase in the SST by an additional 1 °C, compared to simulations ignoring the aerosol impact on climate. The SO_2 released by volcanic events might thus act as a positive feedback, enhancing the warming trend on long time scales. Nevertheless, this continental flood basalt scenario does not explain how climate changes caused mass extinctions. Even if climate changes are suspected to be associated with a reduced primary productivity and its potential shutdown, there are no straightforward links with loss of biodiversity. Paleontological investigations into the way species disappeared (step by step or linearly) could help in answering this question.

ACKNOWLEDGMENTS

We thank Gerta Keller and Andrew Kerr for inviting us to contribute to the Geological Society of America Special Paper *Volcanism, Impacts and Mass Extinctions: Causes and Effects*. Vincent Courtillot is greatly acknowledged for his comments that contribute to improve the original manuscript. We thank Clément Ganino and an anonymous reviewer for helpful comments on the original manuscript. This work was supported by Paris and its Emergence-program and was granted access to the High-Performance Computing resources of Centre de Calcul Recherche et Technologie under the allocation 2013-017013 made by GENCI (Grand Equipement National de Calcul Intensif). This is Institut de Physique du Globe de Paris contribution 3422.

REFERENCES CITED

Abramovich, S., and Keller, G., 2003, Planktonic foraminiferal response to the latest Maastrichtian abrupt warm event: A case study from South Atlantic DSDP Site 525a: Marine Micropaleontology, v. 48, no. 3, p. 225–249, doi:10.1016/S0377-8398(03)00021-5.

Alegret, L., Thomas, E., and Lohmann, K.C., 2012, End-Cretaceous marine mass extinction not caused by productivity collapse: Proceedings of the National Academy of Sciences of the United States of America, v. 109, no. 3, p. 728–732.

Alvarez, L., Alvarez, W., Asaro, F., and Michel, H., 1980, Extraterrestrial cause for the Cretaceous-Tertiary extinction—Experimental results and theoretical interpretation: Science, v. 208, no. 4448, p. 1095–1108, doi:10.1126/science.208.4448.1095.

Barrera, E., and Savin, S.M., 1999, Evolution of late Campanian–Maastrichtian marine climates and oceans, *in* Barrera, E., and Johnson, C.C., eds., Evolution of the Cretaceous Ocean Climate System: Geological Society of America Special Paper 332, p. 245–282.

Beerling, D., and Berner, R., 2002, Biogeochemical constraints on the Triassic-Jurassic boundary carbon cycle event: Global Biogeochemical Cycles, v. 16, no. 3, p. 10-1–10-13.

Blakey, R.C., 2007, Carboniferous-Permian paleogeography of the Assembly of Pangaea, *in* Wong, Th.E, ed., Proceedings of the XVth International Congress on Carboniferous and Permian Stratigraphy, Utrecht, 10–16 August 2003: Amsterdam, Royal Dutch Academy of Arts and Sciences, p. 443–456.

Caldeira, K., and Rampino, M., 1990, Carbon-dioxide emissions from Deccan volcanism and a K/T boundary greenhouse-effect: Geophysical Research Letters, v. 17, no. 9, p. 1299–1302, doi:10.1029/GL017i009p01299.

Chenet, A.-L., Quidelleur, X., Fluteau, F., Courtillot, V., and Bajpai, S., 2007, K-40-Ar-40 dating of the main Deccan large igneous province: Further evidence of KTB age and short duration: Earth and Planetary Science Letters, v. 263, no. 1–2, p. 1–15, doi:10.1016/j.epsl.2007.07.011.

Chenet, A.-L., Fluteau, F., Courtillot, V., Gérard, M., and Subbarao, K., 2008, Determination of rapid Deccan eruptions across the Cretaceous-Tertiary boundary using paleomagnetic secular variation: Results from a 1200-m-thick section in the Mahabaleshwar escarpment: Journal of Geophysical Research–Solid Earth, v. 113, no. B4, p. B04101, doi:10.1029/2006JB004635.

Chenet, A.-L., Courtillot, V., Fluteau, F., Gerard, M., Quidelleur, X., Khadri, S.F.R., Subbarao, K.V., and Thordarson, T., 2009, Determination of rapid Deccan eruptions across the Cretaceous-Tertiary boundary using paleomagnetic secular variation: 2. Constraints from analysis of eight new sections and synthesis for a 3500-m-thick composite section: Journal of Geophysical Research–Solid Earth, v. 114, no. B6, p. B06103, doi:10.1029/2008JB005644.

Courtillot, V., 1994, Mass extinctions in the last 300 million years: One impact and seven flood basalts: Israel Journal of Earth Sciences, v. 43, p. 255–266.

Courtillot, V., and Fluteau, F., 2010, Cretaceous extinctions: The volcanic hypothesis: Science, v. 328, no. 5981, p. 973–974, doi:10.1126/science.328.5981.973-b.

Courtillot, V., and Fluteau, F., 2014, this volume, A review of the embedded time scales of flood basalt volcanism with special emphasis on dramatically short magmatic pulses, *in* Keller, G., and Kerr, A., eds., Volcanism, Impacts, and Mass Extinctions: Causes and Effects: Geological Society of America Special Paper 505, doi:10.1130/2014.2505(15).

Courtillot, V., and Renne, P.R., 2003, On the ages of flood basalt events: Comptes Rendus Geoscience, v. 335, no. 1, p. 113–140, doi:10.1016/S1631-0713(03)00006-3.

Courtillot, V., Besse, J., Vandamme, D., Jaeger, J., and Montigny, R., 1986, Deccan Trap volcanism as a cause of biologic extinctions at the Cretaceous-Tertiary boundary: Comptes Rendus de l'Academie des Sciences, ser. II, v. 303, p. 863–868.

Courtillot, V., Jaupart, C., Manighetti, I., Tapponnier, P., and Besse, J., 1999, On causal links between flood basalts and continental breakup: Earth and Planetary Science Letters, v. 166, no. 3–4, p. 177–195, doi:10.1016/S0012-821X(98)00282-9.

Dessert, C., Dupre, B., Francois, L., Schott, J., Gaillardet, J., Chakrapani, G., and Bajpai, S., 2001, Erosion of Deccan Traps determined by river geochemistry: Impact on the global climate and the Sr-87/Sr-86 ratio of seawater: Earth and Planetary Science Letters, v. 188, no. 3–4, p. 459–474, doi:10.1016/S0012-821X(01)00317-X.

Dessert, C., Dupré, B., Gaillardet, J., François, L.M., and Allègre, C.J., 2003, Basalt weathering laws and the impact of basalt weathering on the global carbon cycle: Chemical Geology, v. 202, no. 3–4, p. 257–273, doi:10.1016/j.chemgeo.2002.10.001.

Donnadieu, Y., Goddéris, Y., Pierrehumbert, R., Dromart, G., Fluteau, F., and Jacob, R., 2006a, A GEOCLIM simulation of climatic and biogeochemical consequences of Pangea breakup: Geochemistry Geophysics Geosystems, v. 7, no. 11, p. Q11019, doi:10.1029/2006GC001278.

Donnadieu, Y., Pierrehumbert, R., Jacob, R., and Fluteau, F., 2006b, Modelling the primary control of paleogeography on Cretaceous climate: Earth and Planetary Science Letters, v. 248, no. 1–2, p. 426–437, doi:10.1016/j.epsl.2006.06.007.

Erwin, D.H., 2006, Dates and rates: Temporal resolution in the deep time stratigraphic record: Annual Review of Earth and Planetary Sciences, v. 34, p. 569–590, doi:10.1146/annurev.earth.34.031405.125141.

Gaillardet, J., Dupré, B., Louvat, P., and Allègre, C., 1999, Global silicate weathering and CO_2 consumption rates deduced from the chemistry of large rivers: Chemical Geology, v. 159, no. 1–4, p. 3–30, doi:10.1016/S0009-2541(99)00031-5.

Ganino, C., and Arndt, N.T., 2009, Climate changes caused by degassing of sediments during the emplacement of large igneous provinces: Geology, v. 37, no. 4, p. 323–326, doi:10.1130/G25325A.1.

Goddéris, Y., and Joachimski, M.M., 2004, Global change in the Late Devonian: Modelling the Frasnian-Famennian short-term carbon isotope excursions: Palaeogeography, Palaeoclimatology, Palaeoecology, v. 202, no. 3–4, p. 309–329, doi:10.1016/S0031-0182(03)00641-2.

Gough, D., 1981, Solar interior structure and luminosity variations: Solar Physics, v. 74, no. 1, p. 21–34, doi:10.1007/BF00151270.

Grard, A., Francois, L., Dessert, C., Dupre, B., and Godderis, Y., 2005, Basaltic volcanism and mass extinction at the Permo-Triassic boundary: Environmental impact and modeling of the global carbon cycle: Earth and Planetary Science Letters, v. 234, no. 1–2, p. 207–221, doi:10.1016/j.epsl.2005.02.027.

Guidry, M.W., and MacKenzie, F.T., 2000, Apatite weathering and the Phanerozoic phosphorus cycle: Geology, v. 28, no. 7, p. 631–634, doi:10.1130/0091-7613(2000)28<631:AWATPP>2.0.CO;2.

Hallam, A., and Wignall, P.B., 1997, Mass Extinctions and Their Aftermath: Oxford, UK, Oxford University Press, 309 p.

Hildebrand, A., Penfield, G., Kring, D., Pilkington, M., Camargo, A., Jacobsen, S., and Boynton, W., 1991, Chicxulub crater—A possible Cretaceous-Tertiary boundary impact crater on the Yucatan Peninsula, Mexico: Geology, v. 19, no. 9, p. 867–871, doi:10.1130/0091-7613(1991)019<0867:CCAPCT>2.3.CO;2.

Iacono-Marziano, G., Marecal, V., Pirre, M., Gaillard, F., Arteta, J., Scaillet, B., and Arndt, N.T., 2012, Gas emissions due to magma–sediment interactions during flood magmatism at the Siberian Traps: Gas dispersion and environmental consequences: Earth and Planetary Science Letters, v. 357, p. 308–318, doi:10.1016/j.epsl.2012.09.051.

Intergovernmental Panel on Climate Change (IPCC), 2007, Contribution of Working Groups I, II and III to the Fourth Assessment Report of the Intergovernmental Panel on Climate Change: Geneva, Switzerland, IPCC, 104 p.

Izett, G., Dalrymple, G., and Snee, L., 1991, Ar-40/Ar-39 age of Cretaceous-Tertiary boundary tektites from Haiti: Science, v. 252, no. 5012, p. 1539–1542, doi:10.1126/science.252.5012.1539.

Jacob, R.L., 1997, Low Frequency Variability in a Simulated Atmosphere Ocean System [Ph.D. dissertation]: Madison, Wisconsin, University of Wisconsin, 159 p. (http://ftp.mcs.anl.gov/pub/People/jacob/RLJdissertation.ps.gz).

Jay, A.E., 2005, Volcanic Architecture of the Deccan Traps, Western Maharashtra, India: An Integrated Chemostratigraphic and Paleomagnetic Study [Ph.D. dissertation]: Milton Keynes, UK, The Open University, 360 p.

Jay, A.E., and Widdowson, M., 2008, Stratigraphy, structure and volcanology of the SE Deccan continental flood basalt province: Implications for eruptive extent and volumes: Journal of the Geological Society, v. 165, no. 1, p. 177–188, doi:10.1144/0016-76492006-062.

Jones, P., Briffa, K., and Schweingruber, F., 1995, Tree-ring evidence of the widespread effects of explosive volcanic-eruptions: Geophysical Research Letters, v. 22, no. 11, p. 1333–1336, doi:10.1029/94GL03113.

Kaminski, E., Chenet, A.-L., Jaupart, C., and Courtillot, V., 2011, Rise of volcanic plumes to the stratosphere aided by penetrative convection above large lava flows: Earth and Planetary Science Letters, v. 301, no. 1–2, p. 171–178, doi:10.1016/j.epsl.2010.10.037.

Keller, G., 2003, Biotic effects of impacts and volcanism: Earth and Planetary Science Letters, v. 215, no. 1, p. 249–264, doi:10.1016/S0012-821X(03)00390-X.

Keller, G., Bhowmick, P., Upadhyay, H., Dave, A., Reddy, A., Jaiprakash, B., and Adatte, T., 2011, Deccan volcanism linked to the Cretaceous-Tertiary boundary mass extinction: New evidence from ONGC wells in the Krishna-Godavari Basin: Journal of the Geological Society of India, v. 78, no. 5, p. 399–428, doi:10.1007/s12594-011-0107-3.

Keller, G., Adatte, T., Bhowmick, P.K., Upadhyay, H., Dave, A., Reddy, A.N., and Jaiprakash, B.C., 2012, Nature and timing of extinctions in Cretaceous-Tertiary planktic foraminifera preserved in Deccan intertrappean sediments of the Krishna-Godavari Basin, India: Earth and Planetary Science Letters, v. 341, p. 211–221, doi:10.1016/j.epsl.2012.06.021.

Kravchinsky, V.A., 2012, Paleozoic large igneous provinces of northern Eurasia: Correlation with mass extinction events: Global and Planetary Change, v. 86–87, p. 31–36, doi:10.1016/j.gloplacha.2012.01.007.

Li, L., and Keller, G., 1998, Abrupt deep-sea warming at the end of the Cretaceous: Geology, v. 26, no. 11, p. 995–998, doi:10.1130/0091-7613(1998)026<0995:ADSWAT>2.3.CO;2.

McLean, D., 1985, Deccan Traps mantle degassing in the terminal Cretaceous marine extinctions: Cretaceous Research, v. 6, no. 3, p. 235–259, doi:10.1016/0195-6671(85)90048-5.

Miles, G., Grainger, R., and Highwood, E., 2004, The significance of volcanic eruption strength and frequency for climate: Quarterly Journal of the Royal Meteorological Society, v. 130(602, A), p. 2361–2376.

Nichols, D.J., and Johnson, K.R., 2008, Plants and the K-T boundary: Palynology, v. 32, p. 265–266.

Oliva, P., Viers, J., and Dupre, B., 2003, Chemical weathering in granitic environments: Chemical Geology, v. 202, no. 3–4, p. 225–256, doi:10.1016/j.chemgeo.2002.08.001.

Oppenheimer, C., 2002, Limited global change due to the largest known Quaternary eruption, Toba approximate to 74 kyr BP?: Quaternary Science Reviews, v. 21, no. 14–15, p. 1593–1609, doi:10.1016/S0277-3791(01)00154-8.

Pierazzo, E., Hahmann, A.N., and Sloan, L.C., 2003, Chicxulub and climate: Radiative perturbations of impact-produced S-bearing gases: Astrobiology, v. 3, no. 1, p. 99–118, doi:10.1089/153110703321632453.

Pinto, J., Turco, R., and Toon, O., 1989, Self-limiting physical and chemical effects in volcanic eruption clouds: Journal of Geophysical Research–Atmospheres, v. 94, no. D8, p. 11,165–11,174, doi:10.1029/JD094iD08p11165.

Pollack, J.B., Toon, O.B., Sagan, C., Summers, A., Baldwin, B., and Van Camp, W., 1976, Volcanic explosions and climatic change: A theoretical assessment: Journal of Geophysical Research, v. 81, no. 6, p. 1071–1083, doi:10.1029/JC081i006p01071.

Robock, A., 2000, Volcanic eruptions and climate: Reviews of Geophysics, v. 38, no. 2, p. 191–219, doi:10.1029/1998RG000054.

Robock, A., Ammann, C.M., Oman, L., Shindell, D., Levis, S., and Stenchikov, G., 2009, Did the Toba volcanic eruption of 74 ka BP produce widespread glaciation?: Journal of Geophysical Research–Atmospheres, v. 114, no. D10, p. D10107, doi:10.1029/2008JD011652.

Rose, W., Bluth, G., and Ernst, G., 2000, Integrating retrievals of volcanic cloud characteristics from satellite remote sensors: A summary: Philosophical Transactions of the Royal Society of London, ser. A, Mathematical Physical and Engineering Sciences, v. 358, p. 1585–1606, doi:10.1098/rsta.2000.0605.

Schulte, P., and 40 others, 2010, The Chicxulub asteroid impact and mass extinction at the Cretaceous-Paleogene boundary: Science, v. 327, p. 1214–1218, doi:10.1126/science.1177265.

Segschneider, J., Beitsch, A., Timmreck, C., Brovkin, V., Ilyina, T., Jungclaus, J., Lorenz, S.J., Six, K.D., and Zanchettin, D., 2013, Impact of an extremely large magnitude volcanic eruption on the global climate and carbon cycle estimated from ensemble Earth system model simulations: Biogeosciences, v. 10, no. 2, p. 669–687, doi:10.5194/bg-10-669-2013.

Self, S., Widdowson, M., Thordarson, T., and Jay, A.E., 2006, Volatile fluxes during flood basalt eruptions and potential effects on the global environment: A Deccan perspective: Earth and Planetary Science Letters, v. 248, no. 1–2, p. 518–532, doi:10.1016/j.epsl.2006.05.041.

Simon, L., Goddéris, Y., Buggisch, W., Strauss, H., and Joachimski, M.M., 2007, Modeling the carbon and sulfur isotope compositions of marine sediments: Climate evolution during the Devonian: Chemical Geology, v. 246, no. 1–2, p. 19–38, doi:10.1016/j.chemgeo.2007.08.014.

Sobolev, S.V., Sobolev, A.V., Kuzmin, D.V., Krivolutskaya, N.A., Petrunin, A.G., Arndt, N.T., Radko, V.A., and Vasiliev, Y.R., 2011, Linking mantle plumes, large igneous provinces and environmental catastrophes: Nature, v. 477, no. 7364, p. 312–316, doi:10.1038/nature10385.

Svensen, H., Planke, S., and Corfu, F., 2010, Zircon dating ties NE Atlantic sill emplacement to initial Eocene global warming: Journal of the Geological Society, v. 167, no. 3, p. 433–436, doi:10.1144/0016-76492009-125.

Textor, C., Graf, H., Herzog, M., and Oberhuber, J., 2003, Injection of gases into the stratosphere by explosive volcanic eruptions: Journal of Geophysical Research–Atmospheres, v. 108, no. D19, p. 4606, doi:10.1029/2002JD002987.

Thordarson, T., and Self, S., 1996, Sulfur, chlorine and fluorine degassing and atmospheric loading by the Roza eruption, Columbia River basalt group, Washington, USA: Journal of Volcanology and Geothermal Research, v. 74, no. 1–2, p. 49–73, doi:10.1016/S0377-0273(96)00054-6.

Thordarson, T., and Self, S., 1998, The Roza member, Columbia River basalt group: A gigantic pahoehoe lava flow field formed by endogenous processes?: Journal of Geophysical Research–Solid Earth, v. 103, no. B11, p. 27,411–27,445, doi:10.1029/98JB01355.

Timmreck, C., Graf, H.-F., Lorenz, S.J., Niemeier, U., Zanchettin, D., Matei, D., Jungclaus, J.H., and Crowley, T.J., 2010, Aerosol size confines climate response to volcanic supereruptions: Geophysical Research Letters, v. 37, p. L24705, doi:10.1029/2010GL045464.

Timmreck, C., Graf, H.-F., Zanchettin, D., Hagemann, S., Kleinen, T., and Krüger, K, 2012, Climate response to the Toba super eruption: Regional changes: Quaternary International, v. 258, p. 30–44, doi:10.1016/j.quaint.2011.10.008.

Tobin, T.S., Ward, P.D., Steig, E.J., Olivero, E.B., Hilburn, I.A., Mitchell, R.N., Diamond, M.R., Raub, T.D., and Kirschvink, J.L., 2012, Extinction patterns, $\delta^{18}O$ trends, and magnetostratigraphy from a southern high-latitude Cretaceous–Paleogene section: Links with Deccan volcanism: Palaeogeography, Palaeoclimatology, Palaeoecology, v. 350–352, p. 180–188, doi:10.1016/j.palaeo.2012.06.029.

Walker, L.J., Wilkinson, B.H., and Ivany, L.C., 2002, Continental drift and Phanerozoic carbonate accumulation in shallow-shelf and deep-marine settings: The Journal of Geology, v. 110, no. 1, p. 75–87, doi:10.1086/324318.

Manuscript Accepted by the Society 31 January 2014

Printed in the USA

The Geological Society of America
Special Paper 505
2014

GOLD OPEN ACCESS

Atmospheric halogen and acid rains during the main phase of Deccan eruptions: Magnetic and mineral evidence

Eric Font*
Instituto Dom Luís, Faculdade de Ciências, Universidade de Lisboa (IDL-FCUL), Edifício C8-8.3.22, Campo Grande, 1749-016, Portugal

Sébastien Fabre
Institut de Recherche en Astrophysique et Planétologie (IRAP), Observatoire Midi-Pyrénées (OMP), Université de Toulouse, 31028 Toulouse 24 Cedex 4, France

Anne Nédélec
Géosciences Environnement Toulouse (GET), UMR 5563, Université de Toulouse, 31400 Toulouse, France

Thierry Adatte
Earth Sciences Institute (ISTE), Université de Lausanne, Geopolis, CH-1015 Lausanne, Switzerland

Gerta Keller
Geosciences Department, Princeton University, Princeton, New Jersey 08544, USA

Cristina Veiga-Pires
Centro de Investigação Marinha e Ambiental (CIMA), Faculdade de Ciencias e Tecnologia, Universidade do Algarve, Campus de Gambelas, 8005-139 Faro, Portugal

Jorge Ponte
Instituto Dom Luís, Universidade de Lisboa (IDL-UL), Edifício C8-8.3.22, Campo Grande, 1749-016, Portugal

José Mirão
HERCULES Laboratory (HERança CULtural Estudos e Salvaguarda/Cultural Heritage Studies and Safeguard), University of Evora, 7000-809 Evora, Portugal

Hassan Khozyem
Earth Sciences Institute (ISTE), Université de Lausanne, Geopolis, CH-1015 Lausanne, Switzerland and *Geology Department, Faculty of Sciences, Aswan University, 81528-Aswan, Egypt*

Jorge E. Spangenberg
Institute of Earth Surface Dynamics (IDYST), University of Lausanne, Building Géopolis, CH-1015 Lausanne, Switzerland

*font_eric@hotmail.com

Font, E., Fabre, S., Nédélec, A., Adatte, T., Keller, G., Veiga-Pires, C., Ponte, J., Mirão, J., Khozyem, H., and Spangenberg, J., 2014, Atmospheric halogen and acid rains during the main phase of Deccan eruptions: Magnetic and mineral evidence, *in* Keller, G., and Kerr, A.C., eds., Volcanism, Impacts, and Mass Extinctions: Causes and Effects: Geological Society of America Special Paper 505, p. 353–368, doi:10.1130/2014.2505(18).

ABSTRACT

Environmental changes linked to Deccan volcanism are still poorly known. A major limitation resides in the paucity of direct Deccan volcanism markers and in the geologically short interval where both impact and volcanism occurred, making it hard to evaluate their contributions to the mass extinction. We investigated the low-magnetic-susceptibility interval just below the iridium-rich layer of the Bidart (France) section, which was recently hypothesized to be the result of paleoenvironmental perturbations linked to paroxysmal Deccan phase 2. Results show a drastic decrease of detrital magnetite and presence of scarce akaganeite, a hypothesized reaction product formed in the aerosols derived from reaction of a volcanic plume with water and oxygen in the high atmosphere. A weathering model of the consequences of acidic rains on a continental regolith reveals nearly complete magnetite dissolution after ~31,000 yr, which is consistent with our magnetic data and falls within the duration of the Deccan phase 2. These results highlight the nature and importance of the Deccan-related environmental changes leading up to the end-Cretaceous mass extinction.

INTRODUCTION

The history of Earth is punctuated by five severe global climatic and environmental catastrophes that led to nearly total mass extinctions of calcareous marine and many continental species. The most famous case is the Cretaceous-Paleogene boundary (KPB or Cretaceous-Tertiary boundary [KTB]) mass extinction, which is still the center of acrimonious debates between partisans of the bolide impact theory (Schulte et al., 2010) and those who favor a terrestrial origin linked to Deccan Traps volcanism (Courtillot, 2012; Keller et al., 2012). Others argue that each process, individually, is unlikely to have caused a global biological collapse, and a scenario of multiple causes is preferred (White and Saunders, 2005; Keller et al., 2009; Renne et al., 2013).

In recent years, Deccan volcanism studies have resulted in major advances that now facilitate recognition and evaluation of its global effects, such as attempted in this study. For example, three discrete Deccan volcanism phases with variable intensity have been dated based on magnetostratigraphy and ^{40}K-^{40}Ar ages (Chenet et al., 2007): Phase 1 in the early late Maastrichtian (base magnetozone C30n, ca. 67.4 Ma) accounts for ~6% of the total lava pile, which today forms mountain peaks of 3500 m (Self et al., 2006). Most of the lava pile (80%) erupted during phase 2 in the latest Maastrichtian paleomagnetic chron C29r. The final phase 3 erupted in the early Danian base of C29n and accounts for ~14% of total Deccan eruptions (Chenet et al., 2007, 2008; Keller et al., 2011, 2012). The original size prior to erosion is estimated to have been 1.5×10^6 km^2, and the volume of lava extruded is estimated at ~1×10^6 km^3 (Self et al., 2006; Chenet et al., 2007).

Most critical for the current study is Deccan volcanism phase 2, which has been directly linked to the Cretaceous-Paleogene boundary mass extinction in outcrops and deep wells of the Krishna-Godavari Basin in India (Keller et al., 2011, 2012). These studies revealed Earth's longest lava megaflows (Self et al., 2008a), 1500 km across India near the end of phase 2, with the final megaflow at the Cretaceous-Paleogene boundary mass extinction (Keller et al., 2008).

Current research thus narrows the main phase 2 of Deccan volcanism to the geologically short interval of the Cretaceous planktic foraminiferal zone CF1, which spans the last ~160,000 yr below to the Cretaceous-Paleogene boundary, whereas the Chicxulub impact is commonly placed at the Cretaceous-Paleogene boundary (Schulte et al., 2010) or within zone CF1 preceding the mass extinction (Keller et al., 2012). This short interval limits our ability to separate the main Deccan volcanic phase (i.e., phase 2) from the Chicxulub impact within resolvable radiometric ages and to evaluate their respective biotic effects and their contributions to the Cretaceous-Paleogene boundary mass extinction. Radiometric (K-Ar and Ar-Ar) methods are subject to intrinsic errors (~1%–2%), linked to interlaboratory calibration, imprecise K and Ar isotopic standards (Fish Canyon standard), and uncertainty of decay constants (Kuiper et al., 2008; Channell et al., 2010; Renne et al., 2010, 2013), preventing any precise radiometric age correlation between continental Deccan flood basalts and the mass extinction recorded in marine sections. In addition, because volcanic mineralogical markers (aerosols) are generally restricted to the vicinity of the volcanic source, the discrimination between Deccan phase 2 and the Chicxulub impact in distal sections is even harder within such a short temporal interval.

Are volcanic signals present but unrecognized outside India? A recent study of Cretaceous-Paleogene boundary sections at Bidart, France, and Gubbio, Italy, suggests so (Font et al., 2011). In these sections, a 50-cm-thick yellow-red interval of very low magnetic susceptibility (MS) directly underlies the Cretaceous-Paleogene boundary (Fig. 1), as also observed in Oman (Ellwood et al., 2003) and other marine sections. Preliminary data indicate that this low-MS interval contains an enigmatic Cl-rich iron oxyhydroxide that could have formed by interaction between HCl-rich volcanic gas and aerosols in the high atmosphere (Font et al., 2011). The plate-like morphology and well-preserved texture of this newly discovered mineral strongly contrast with the detrital aspect of titanomagnetite and suggest an eolian origin for its transport. Recent numerical modeling showed that penetrative convection driven by the cooling of the Deccan lavas may have generated buoyant atmospheric plumes (Kaminski et al., 2011). In such a situation, aerosols would have been able to rise to the stratosphere, providing an explanation for a rather long halogen residence time in the atmosphere and the eventual transport of their reaction products to Bidart and Gubbio sections.

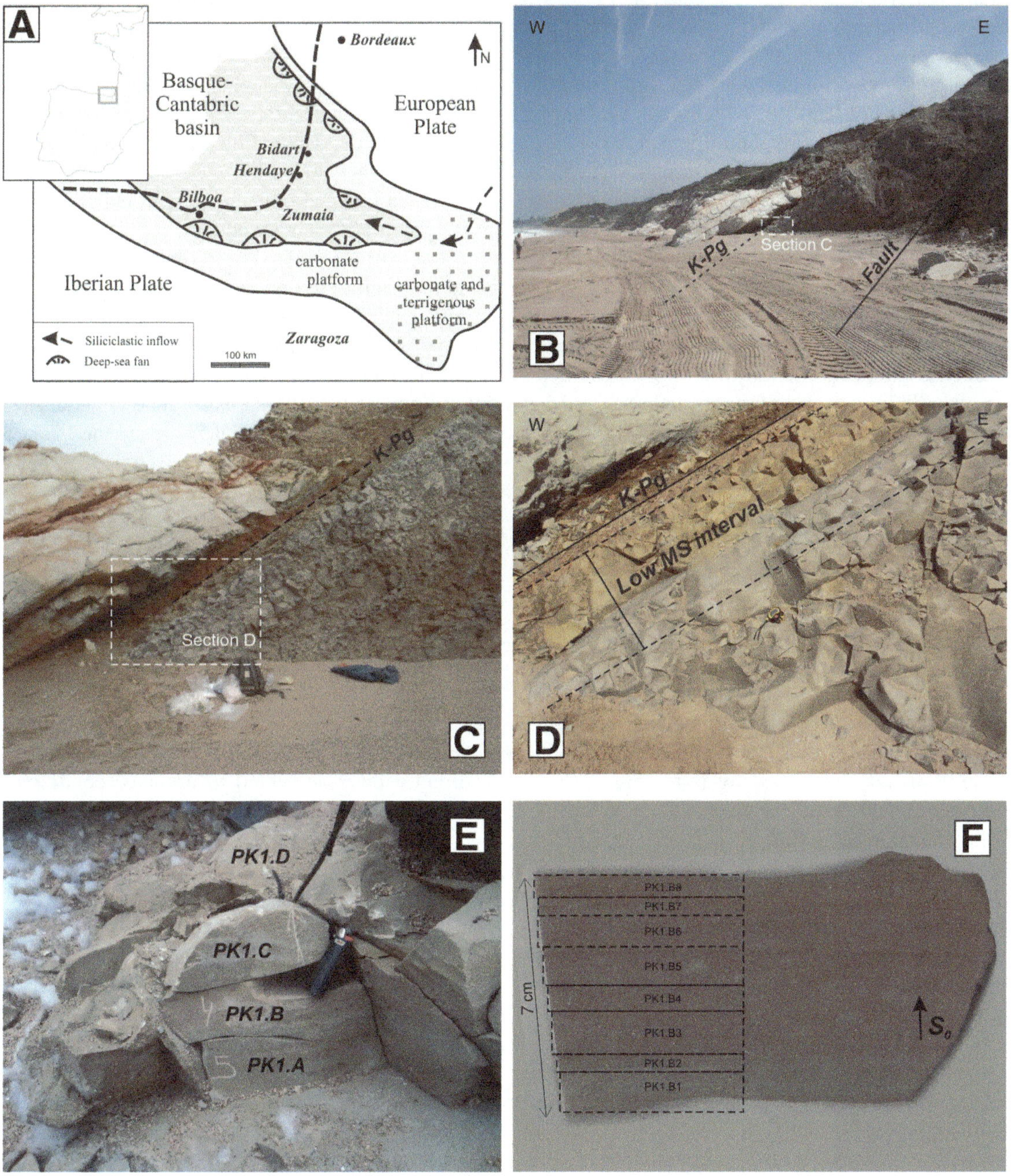

Figure 1. Geological settings and biostratigraphy of the Bidart section. (A) Paleoenvironmental context of the Biscay Bay. (B) The Erretegia Beach showing the white-pink Danian limestones and the Maastrichtian marls cut by a fault. (C) The Cretaceous-Paleogene (K-Pg) boundary showing (D) the typical dark clay layer and the yellow-red interval of low magnetic susceptibility (MS) values in the upper Maastrichtian. (E) Example of hand block collected at the outcrop. (F) Example of a block cut into small plates, by using a nonmagnetic Dremel saw, for subsequent magnetic analyses.

The main objective of this study was to evaluate the magnetic and mineralogical characteristics of the uppermost Maastrichtian at Bidart, France, in order to determine the potential signature of Deccan volcanism. To achieve this objective, the following evaluations were performed: (1) Determine the presence of zone CF1 (latest Maastrichtian correlative with Deccan phase 2) based on high-resolution biostratigraphy and chemostratigraphy (carbon and oxygen isotopes). (2) Perform high-resolution (centimeter-scale) magnetic analyses and microscope observations focused on the last meter of the Maastrichtian, including the low-MS interval, to determine its mineral content and origin. (3) Identify the enigmatic Cl-bearing iron oxyhydroxide and evaluate its possible formation and subsequent transport in relation with the Deccan volcanism. (4) Quantify the effects of acid rains on a continental regolith using a numerical weathering model.

METHODS

For biostratigraphic and faunal analyses, samples were processed using standard methods (Keller et al., 1995). Samples were washed through 38 µm and 63 µm sieves and oven dried at 50 °C. Species preservation is generally good, though shells are recrystallized and affected by dissolution. Dissolution effects are apparent below the Cretaceous-Paleogene boundary coincident with the low-MS interval and a major increase in benthic species. In the 25 cm below the Cretaceous-Paleogene boundary, almost all carbonate shells are dissolved due to intense dissolution. Faunal analysis was conducted on three size fractions (38–63 µm, 63–150 µm, >150 µm) with the quantitative analysis based on 38–63 µm and 63–150 µm size fractions.

We measured MS, saturation isothermal remanent magnetization (SIRM), and coercivity-dependent magnetic properties (e.g., $IRM_{-0.3T}/IRM_1$ = S-ratio, mean acquisition field [$B_{1/2}$]). Mass-specific MS is mainly controlled by ferromagnetic (i.e., iron oxide) and paramagnetic (i.e., clay) contents, whereas isothermal remanent magnetization (IRM) is mostly sensitive to ferro/ferrimagnetic and antiferromagnetic particles, even in very low amounts. MS was measured with a MFK1 (AGICO) apparatus at the Institute Dom Luís Laboratory (Portugal) and reported relative to mass-specific m^3/kg by dividing by density. Samples were cleaned by alternating fields (AF) treatment and submitted to IRM measurements using an impulse magnetizer (IM-10–30). IRM curves were then analyzed using a cumulative log-Gaussian function (Kruiver et al., 2001) in order to discriminate magnetic phases by their respective coercivity spectra. IRM data are normalized by the mass (unit in Am^2/kg) with a constant volume of 10 cm^3. Fresh rock fragments were also observed under a Hitachi S-3700N scanning electron microscope (SEM) coupled to a Bruker XFlash® 5010 energy dispersive spectra (EDS) detector at the Hercules Laboratory (Evora, Portugal). EDS provides semiquantitative compositional analysis.

We also measured the color by using diffuse reflectance spectrophotometry (DRS) analysis, which is based on the percentage reflectance of a sample relative to white light and provides qualitative information about mineralogical and grain-size variations. DRS data were converted into a color by using the Munsell Color System and are represented in L*a*b* coordinates following the Commission International de l'Éclairage (CIE, 1976), where L* is the lightness/darkness, a* is the redness/greenness, and b* is the yellowness/blueness. Data were obtained using a X-Rite® Colortron™ spectrophotometer from 86 rock powders, in order to avoid grain-size effects, and after three series of measurements.

Isotopic analyses were performed at the University of Lausanne on a Thermo Fisher Scientific (Bremen, Germany) GasBench II carbonate preparation device connected to a Thermo Fisher Scientific Delta Plus XL continuous-flow isotope ratio mass spectrometer (IRMS). CO_2 was extracted at 70 °C. Isotopic ratios of carbon and oxygen are reported in the delta (δ) notation as the per mil (‰) deviation relative to the Vienna Peedee belemnite standard (VPDB). The analytical reproducibility estimated from replicate analyses of the international calcite standard NBS-19 and the laboratory standards of Carrara marble was better than ±0.05‰ for $\delta^{13}C_{carb}$ and ±0.1‰ for $\delta^{18}O_{carb}$.

In order to discuss the origin of the loss in detrital magnetite just below the Cretaceous-Paleogene boundary, we processed a numerical weathering model. We used the PHREEQC program (Parkhurst and Appelo, 1999) developed by the U.S. Geological Survey for modeling water-rock interactions, coupled with the Lawrence Livermore National Laboratory database. The program determines the chemical speciation of aqueous solutes and the fugacities of gases, and it attributes each component of the solution to stable aqueous species. It also calculates the activity coefficients of water and the aqueous species, as well as the saturation index of the solution with respect to all the solid phases of the database. The weathered "rock" material is a monolithologic sandy regolith, assumed to have the same mineral composition as the mean continental upper crust (or typical granodiorite) and forming a layer subjected to drainage conditions and reacting with rainwater initially equilibrated with volcanic gases.

In our hypothesis, the initial magnetic particles accumulate from the parent rock during pedogenesis. The regolith alteration is assumed to proceed by a continuous batch process in water-saturated sediment, where the pore water is being periodically replaced by an equivalent volume of fresh rainwater at each rain event (Fabre et al., 2011; Fabre and Berger, 2012). The surface percentage of magnetite is taken equal to 1.8% (Table 1). The mineral reactive surface (*A*) is calculated according a simple parametric law assumed to be a correct approximation of the solid Brunauer-Emmet-Teller (BET) surface (i.e., method of adsorbing-desorbing gases at the mineral surface in order to measure the surface area; Brunauer et al., 1938; Sverdrup and Warfvinge, 1995):

$$A = (8.0x_{clay} + 2.2x_{silt} + 0.3x_{sand} + 0.0x_{coarse})\rho, \qquad (1)$$

where x_{clay}, x_{silt}, x_{sand}, and x_{coarse} represent fractions of clay, silt, sand, and coarse materials in the sandy regolith, and ρ is the

TABLE 1. RATE DISSOLUTION CONSTANTS OF SILICATE (FABRE ET AL., 2011) AND MAGNETITE (WHITE ET AL., 1994)

Granodiorite		Sandy regolith
Silicates	% of mineral surface	k_o (T = 20 °C) in mole Si m^{-2} s^{-1}
Quartz	29.9	3.16E–12
Albite	29.9	9.49E–12
Anorthite	22.8	6.32E–12
Orthose	15.6	9.49E–12
Oxides	% of mineral surface	k_o (T = 20 °C) in mole Fe m^{-2} s^{-1}
Magnetite (pH = 3.3)		5.01E–10
Magnetite (pH = 4.3)	1.8	2.39E–10
Magnetite (pH = 5.6)		1.30E–10

regolith density in g/m^3. Other relevant characteristics of the regolith are reported in Table 2.

The rain frequency is linked to the presumed Maastrichtian climate, namely, tropical hot and humid. The runoff is taken as equal to present-day values: 11 L s^{-1} km^2 (Gaillardet et al., 1999). Numerous studies of current conditions indicate the potential role of acidic rainwater in the mobilization of metals (Floor et al., 2011). In this study, we assumed a pH between 3.3 and 4.3, which is typical of rainwater resulting from the mixing between volcanic plumes and atmospheric aerosols (Chudaeva et al., 2006; Calabrese et al., 2011; Floor et al., 2011). These values were measured in the neighborhood of active volcanoes such as Mount Etna. After each rain event, magnetite and rainwater react to reach equilibrium. The overall dissolution rate of the minerals, R_k, considered as a first-order law, is equal to:

$$R_k = \left(\frac{dn}{dt}\right) = k_o A S_w \left(1 - \frac{Q}{K_k}\right) \quad (2)$$

where R_k stands for bulk dissolution rate of mineral k, k_o is the dissolution rate constant in mol m^{-2} s^{-1}, A is the reactive surface area of the solid (m^2/m^3), S_w is the soil moisture (fixed at 30% as a mean value; Sverdrup and Warfvinge, 1995), and Q/K is the saturation ratio. Thermodynamic constants are taken from Fabre et al. (2011) and White et al. (1994). No specific precipitation law was imposed for secondary phases, assuming that secondary phase formation is controlled by the dissolution of source minerals. Moreover, we assumed that the reaction is not controlled by the diffusion through the oxide layer as observed in closed reactors (White et al., 1994). The amount of dissolved magnetite in moles is calculated during each increment. The final value is the time required to dissolve a significant percentage of the initial magnetite mass (90%) corresponding to the observed MS decrease.

AGE AND CORRELATION WITH DECCAN PHASE 2

The Bidart section is generally considered as one of the most complete Cretaceous-Paleogene boundary sections worldwide because of a thin clay layer and Ir anomaly (Bonté et al., 1984; Ward and Kennedy, 1993; Vonhof and Smit, 1997; Galbrun and Gardin, 2004). However, in previous studies, the zone CF1 index species was not observed, leaving doubt as to the presence of a complete sediment record below the Cretaceous-Paleogene boundary. In this study, the zone CF1 index species (*Plummerita hantkeninoides*) was observed, confirming the presence of the correlative interval with Deccan phase 2.

A second independent test for the presence of the uppermost Maastrichtian can be obtained from oxygen and carbon isotopes, and more specifically the global climate warming, which begins in zone CF2 and continues into the lower part of CF1 and the subsequent cooling prior to the Cretaceous-Paleogene boundary (Fig. 2; Li and Keller, 1998; Stuben et al., 2003; Wilf et al., 2003; Abramovich et al., 2010). We measured carbon and oxygen isotope compositions for the entire section (~6 m) at 20 cm intervals, as well as at higher resolution (5 cm intervals) for the low-MS interval of the top 1 m, with comparable results (Figs. 2–3; Table 1). These data show the climate warming beginning in zone CF2 and proceeding into CF1, followed by cooling correlative with South Atlantic Site 525 (Fig. 2). Note that zones CF2 and CF1 span the last 160 and 120 k.y. at the end of the Maastrichtian, respectively. This study thus demonstrates the presence of the latest Maastrichtian at Bidart correlative with Deccan phase 2.

In this study, we hypothesize that Deccan volcanism caused acid rain and ocean acidification. If this is the case, then it should be reflected in the calcareous shells of planktic foraminifera and more specifically by the loss of dissolution-prone thin-walled species. We observed increasing dissolution effects in planktic foraminifera at the Bidart section in the 50 cm below the Cretaceous-Paleogene

TABLE 2. PHYSICAL PROPERTIES OF SANDY REGOLITH

Property	For 1 m^3 parcel
Density of regolith (kg m^{-3})	2.7
Thickness of soil (m)	0.32
Porosity (%)	20
Reactive surface (g m^{-3})	1,600,000
Runoff (L s^{-1} km^{-2})	11

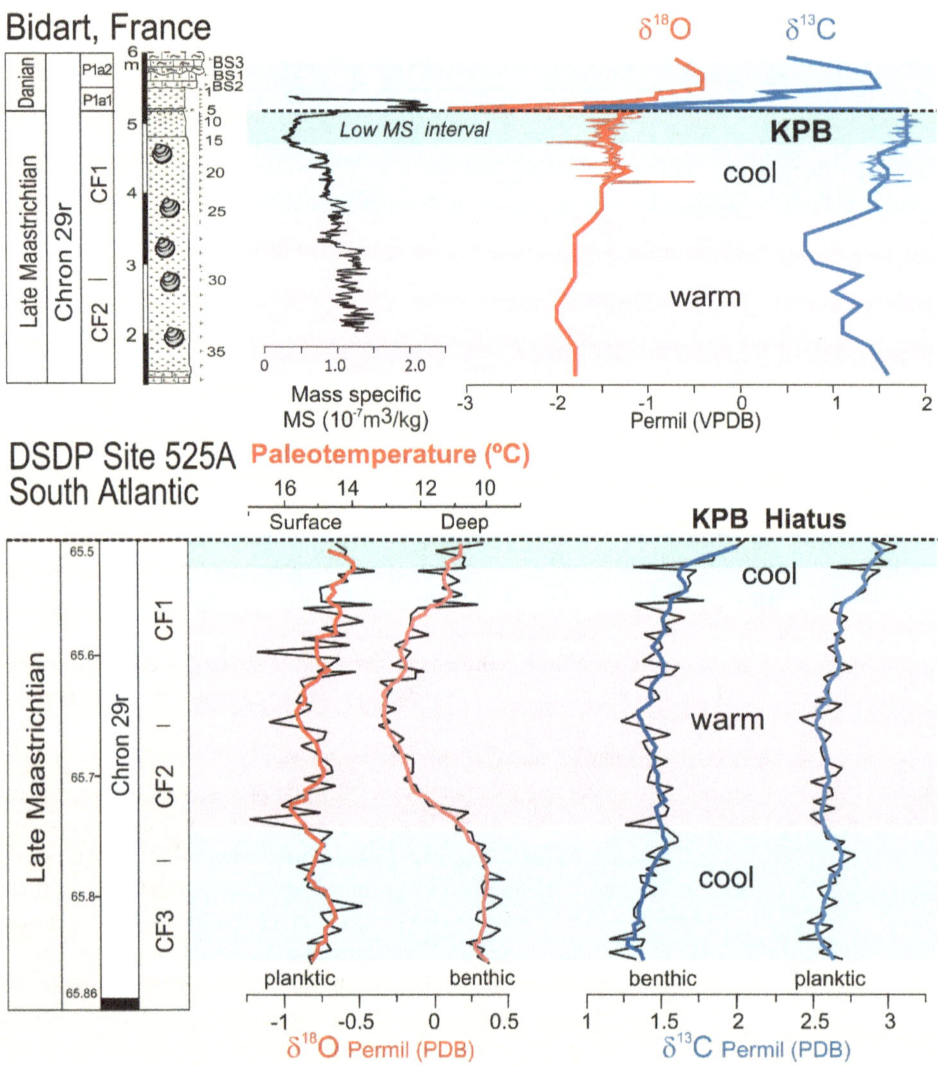

Figure 2. Carbon and oxygen isotopic compositions of the Bidart section and correlation with the South Atlantic (Deep Sea Drilling Project [DSDP] Site 525A; Li and Keller, 1998). The mass-specific magnetic susceptibility (MS) profiles compiled from Font et al. (2011) and the present study are also shown. KPB—Cretaceous-Paleogene boundary; VPDB—Vienna Peedee belemnite.

boundary and correlative with the low-MS interval, with maximum dissolution effects (nearly all species dissolved) in the 20 cm below the Cretaceous-Paleogene boundary (Fig. 2). Increased carbonate dissolution below the Cretaceous-Paleogene boundary has been observed in many sequences (e.g., Smit, 1990; Pardo et al., 1996), but no systematic recording is available to date because the cause has remained unknown. Only recently has Deccan volcanism been identified as a potential cause.

MAGNETIC DATA

The Bidart section was sampled at 1 cm intervals to determine the magnetic properties in the last meter of the Maastrichtian marls (Figs. 1E–1F). Our new high-resolution mass-specific MS analyses confirm the presence of the low-MS interval from −5 to −50 cm below the Cretaceous-Paleogene boundary (Fig. 3). The transition to the more characteristic MS values of the Maastrichtian marls at the base of the interval is gradual, suggesting a progressive change in paleoenvironmental conditions rather than the existence of a hiatus or a fault.

IRM curves show that the main magnetic carrier of the Danian limestones is a low-coercivity phase identified under SEM microscope as detrital titanomagnetite (Fig. 4A). The Danian dark clay layer and the Maastrichtian marls located below the low-MS interval are characterized by a bimodal association (low to high) of coercivity spectra (magnetite and hematite/goethite), whereas the high-coercivity phase alone characterizes the low-MS interval (Fig. 4A). After unmixing raw IRM curves by using the cumulative Log-Gaussian (CLG) function (Robertson and France, 1994; Kruiver et al., 2001), two to three magnetic phases (called components) were isolated (Fig. 4B). Components 1 and 2 correspond to low- to medium-coercivity phases with values of mean acquisition field ($B_{1/2}$) varying from 24 to 80 mT and 57 to 77 mT, respectively (Fig. 4B). CLG parameters of component 1 are in the range of magnetite and maghemite populations commonly found in sedimentary rocks (Kruiver et al.,

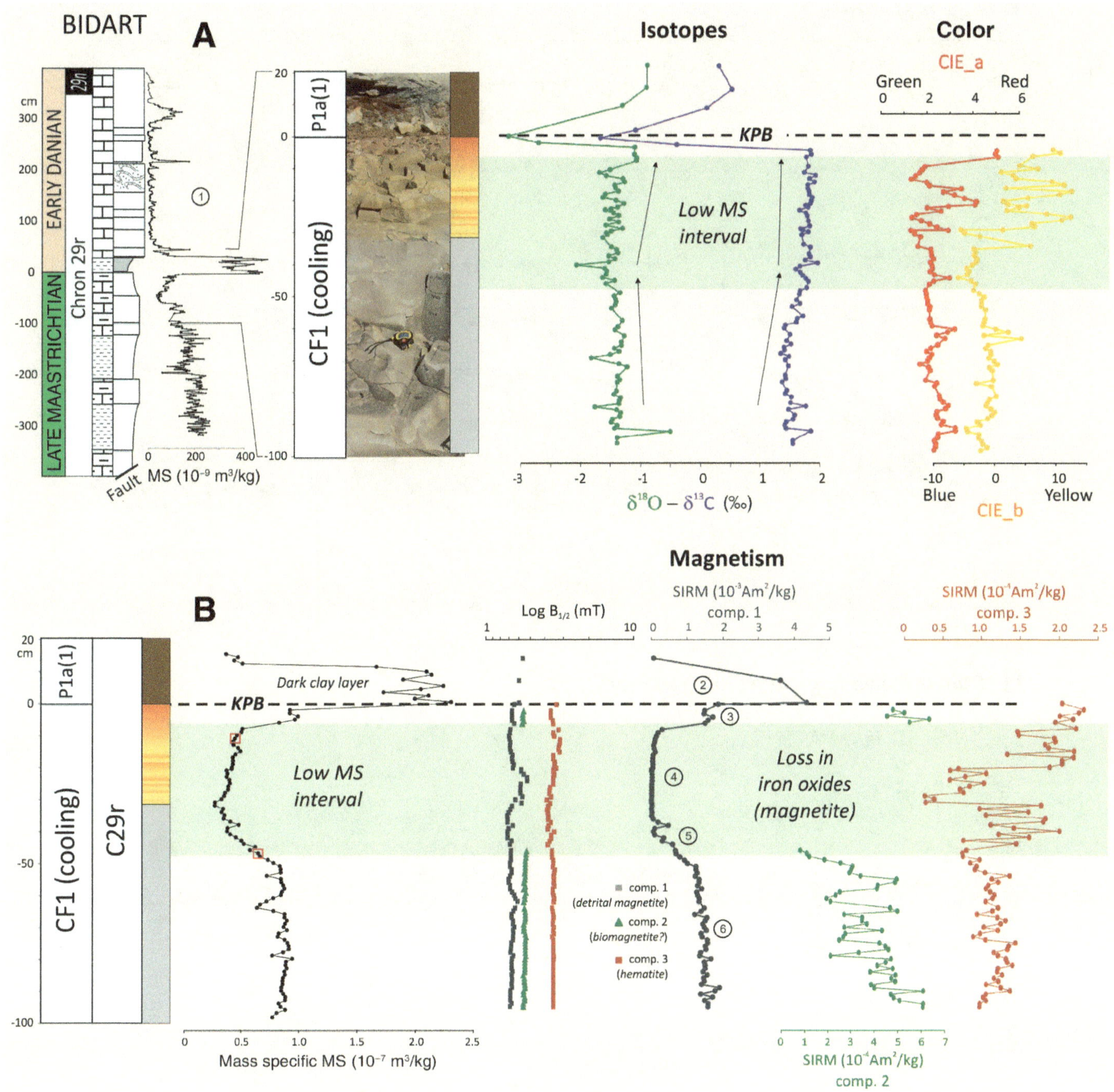

Figure 3. (Top) Magnetic susceptibility (Font et al., 2011; this study) and color profiles along with carbon and oxygen isotope data of the Maastrichtian outcrop at Bidart above the fault line. Carbon and oxygen isotopes mark the climate cooling of the uppermost zone CF1 (see Fig. 2). Diffuse reflectance spectrophotometry shows an increase in red-yellow–colored iron oxides (i.e., akaganeite) in the interval of low-magnetic-susceptibility (MS) values. (Bottom) High-resolution mass-specific MS and isothermal remanent magnetization (IRM) parameters (mean acquisition field [$B_{1/2}$], saturation isothermal remanent magnetization [SIRM]) of the last meter of Maastrichtian marls and the Cretaceous-Paleogene boundary (KPB). The Maastrichtian marls are marked by three magnetic phases (components 1, 2, and 3) assigned to detrital magnetite, biogenic magnetite, and hematite, respectively. The low-MS interval with akaganeite shows a loss in low- to medium-coercivity magnetic phases (components 1 and 2). Red squares on the MS profile show samples analyzed by scanning electron microscopy coupled to energy dispersive spectra (SEM-EDS) and illustrated in Figure 5. Numbered intervals (from 1 to 6) refer to IRM data (shown in Fig. 4).

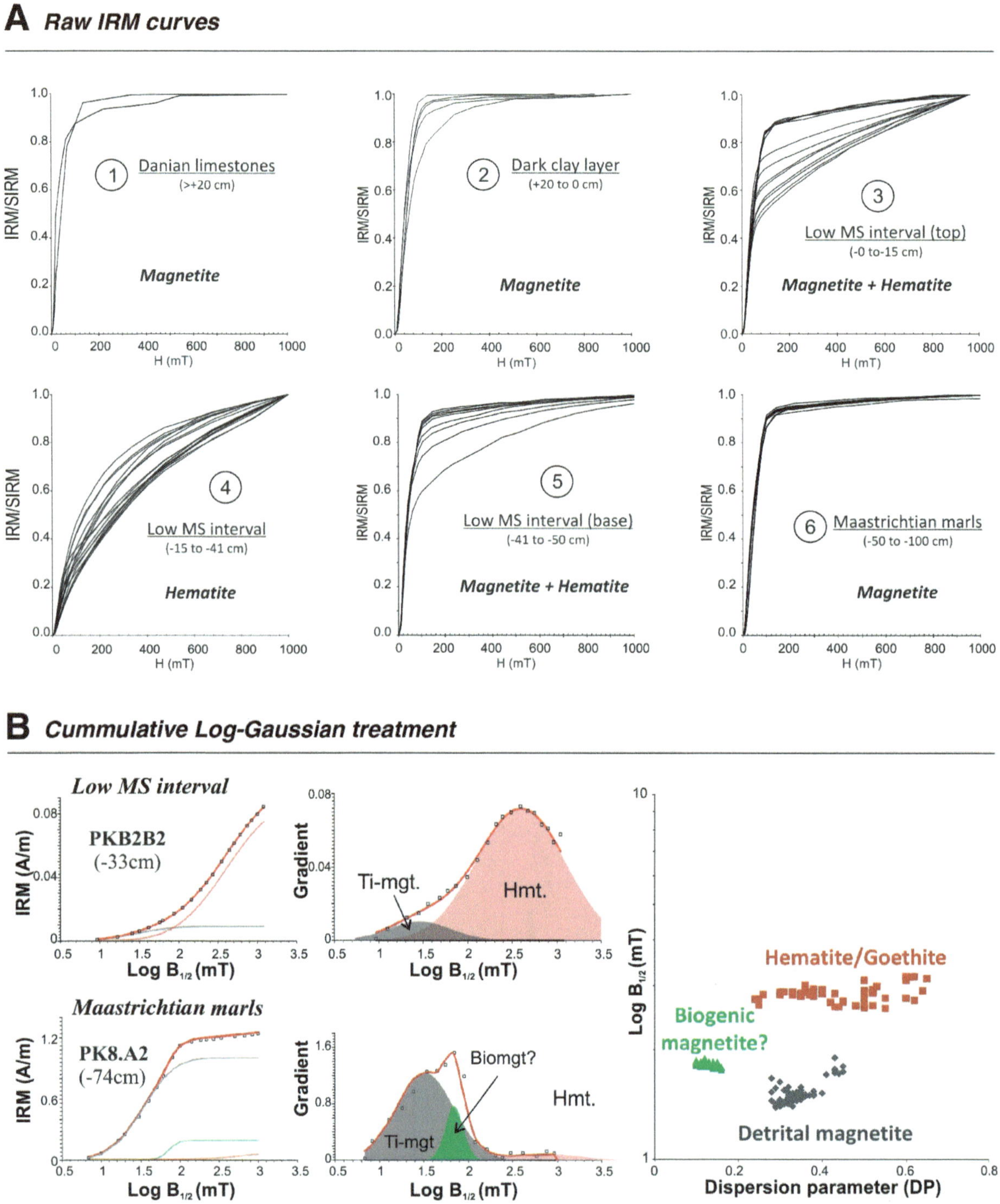

Figure 4. (A) Raw isothermal remanent magnetization (IRM; saturation isothermal remanent magnetization [SIRM]) curves of the Bidart section show that all sedimentary intervals are dominated by bimodal to trimodal distributions of remanent spectra, attributed to magnetite and hematite. (B) IRM data treated by the cumulative log-Gaussian function (Kruiver et al., 2001). In most Maastrichtian samples, component 1 (detrital magnetite) contributes to more than 90% of the magnetic signal, whereas component 2 (biomagnetite?) and component 3 (hematite) have minor contributions. The low-magnetic-susceptibility (MS) interval is characterized by very low magnetite content, disappearance of biogenic magnetite, and relative increase in hematite.

2001) and at Bidart (Galbrun and Gardin, 2004). Component 2 is observed in the whole section except in the low-MS interval, and it exhibits slightly higher coercivity (log $B_{1/2}$) and lower dispersion parameter (DP) values than component 1 (Fig. 4B). The very small dispersion of coercivity and DP parameters of component 2 (Fig. 4B) argues for the presence of a homogeneous population of the magnetic carrier in terms of grain size and composition (Egli, 2004). Such homogeneity and the very narrow distribution of component 2 (mean DP = 0.12) are usually indicative of the presence of magnetic grains (magnetosomes) produced by magnetotactic bacteria (Egli, 2004; Abrajevitch and Kodama, 2009, 2011). Values of remanent coercivity ($B_{1/2}$) of the hard fraction vary between 389 and 1479 mT and are typical of antiferromagnetic behavior, similar to hydroxides such as hematite or goethite commonly found in marine rocks (Kruiver et al., 2003). However, goethite has not been identified under thermomagnetic analysis (Font et al., 2011). Akaganeite is paramagnetic at ambient temperature (Néel temperature of 290 °K) and thus cannot be the carrier of the high-coercivity curve. Hematite is therefore likely the high-coercivity phase.

The decrease in MS values can be explained by reduced detrital titanomagnetite (and biogenic magnetite) content, the main magnetic carrier of the Maastrichtian marls at Bidart (Galbrun and Gardin, 2004; Font et al., 2011), as observed by a characteristic reduction in mass-normalized SIRM values of component 1 and 2 (Fig. 3–4). A maximum 90% reduction in component 1 + 2 remanence is observed in the middle part of the interval (Fig. 3).

SCANNING ELECTRON MICROSCOPY

We observed more than 10 samples, including the Danian and Maastrichtian sediments. SEM observations coupled to EDS show the occurrence of Cl-bearing iron hydroxides with a specular, plate-like morphology and grain sizes <10–40 µm (Fig. 5). This mineral was only observed in samples from the low-MS interval, whereas Fe-Ti iron oxides are the sole magnetic carriers in the rest of the section. EDS spectra show the presence of Fe, O, Cl, and Ca. Compositional mapping reveals only Fe and Cl together, whereas Ca is associated with the sediment matrix (Fig. 5). Ti-bearing iron oxides (Ti being an indicator of detrital origin) are rare in samples from the low-MS interval but exist either as small (<10 µm) cubic-like and eroded crystals of magnetite or as severely altered ilmenite (Fig. 6).

We infer that the Cl-rich iron hydroxide corresponds to akaganeite, in agreement with our SEM-EDS data. Concentration in Cl was estimated by semiquantitative analysis of EDS spectra (more than 20 analyses) and varies between 0.68 and 4.64 wt%, which we consider corresponds to a minimum value since other elements contained in the matrix also contribute to the calculation. The association Fe-Cl is explained by the hollandite-type structure of akaganeite (Post and Buchwald, 1991). In this arrangement, the octahedral sites are occupied by iron cations, whereas Cl^- ions partially fill the tunnels. The subeuhedral and plate-like morphologies of the crystals contrast with the eroded aspect of the detrital titanomagnetite and rather argue for an eolian transport mechanism (Fig. 6). Grain sizes from micrometer to tens of micrometers are in the range of Quaternary eolian dust (Maher, 2011). The presence of akaganeite, which strongly reflects in the yellow-brown region, is also consistent with DRS, which show significant trends in the low-MS interval in the red (CIE_a) and yellow (CIE_b) colors (Fig. 3). It also explains the characteristic yellow-brown color of the low-MS interval at Bidart (Figs. 1 and 3).

DISCUSSION

Akaganeite: A New Deccan Marker?

The discovery of akaganeite in the low-MS interval at Bidart (this study) and at Gubbio (Font et al., 2011) just below the Cretaceous-Paleogene boundary mass extinction and coeval with the main Deccan phase 2 argues for a volcanic origin. The subeuhedral and plate-like morphologies of the newly found akaganeite crystals contrast with the eroded aspect of detrital magnetite, and their scarcity (<0.1 wt%) in the sedimentary matrix is compatible with eolian transport from a distant source. Akaganeite (β-$Fe_2[OH]_3Cl$) is rare in natural conditions but a common product of corrosion of steel or archaeological items in marine environments because its formation requires large concentrations of both dissolved Fe(II) and chloride ions (Reguer et al., 2007; Remazeilles and Refait, 2007, 2008). Akaganeite is also described from weathered steel exposed to an acidic atmosphere (Li et al., 2008). Johnston (1977) described akaganeite co-occurring with jarosite near the steaming fumarole vents of the White Island volcano in New Zealand. Akaganeite also precipitates naturally as a pure mineral phase with an average Fe/Cl molar ratio of 6.7 from oxidation of iron sulfide minerals during occasional drying of sulfide sediments from Australian wetlands (Bibi et al., 2011). Naturally occurring akaganeite is also observed in weathered meteorites, especially from cold environments such as the Antarctic continent (Bland et al., 1997; Lee and Bland, 2004). Akaganeite from meteorites often contains a small amount of Ni.

Akaganeite crystals found in the Bidart and Gubbio sediments do not have the composition of meteorite-derived akaganeite. Their formation in periodically dried sulfidic coastal wetlands in the vicinity of the Deccan Traps is a possible scenario, but in this hypothesis their transportation to Gubbio (Italy) and Bidart (France) is highly unlikely (Font et al., 2011). However, their formation in relation with Deccan volcanic degassing followed by eolian transport is a possible scenario.

Metal transfer (notably Fe but also Na, K, Zn) from degassing lava to the atmosphere as simple chlorides has been observed and predicted by thermochemical modeling. These compounds are highly volatile, especially when the lava is at a high temperature. We infer that akaganeite results from a gaseous reaction between such iron chlorine compounds and volcanic/atmospheric gases, such as water and oxygen. The structural formula of akaganeite (oxyhydroxide and not oxide) pleads for a reaction at a rather

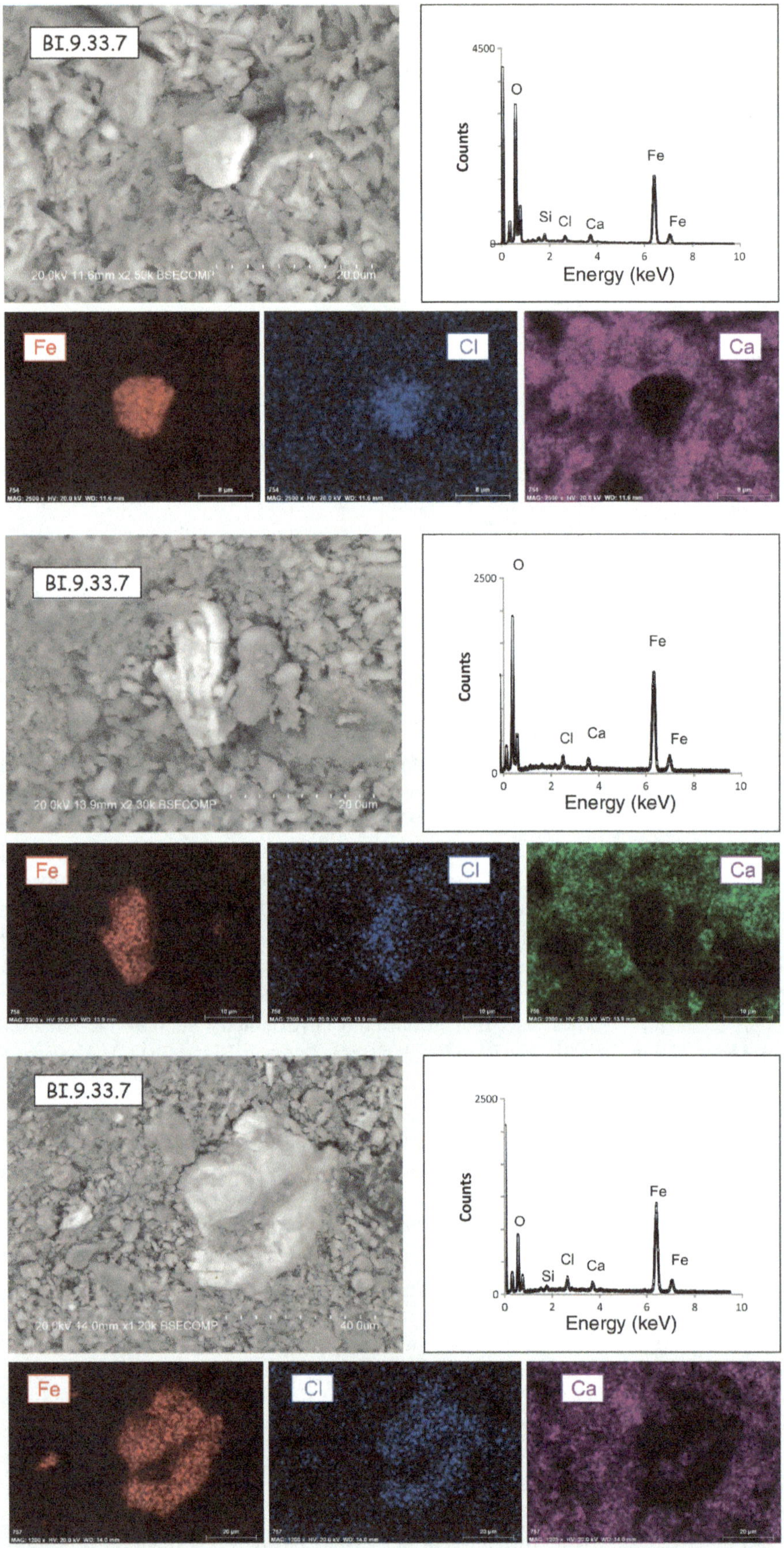

Figure 5. Scanning electron microscopy coupled to energy dispersive spectra (EDS) and composition mapping of the Cl-bearing iron hydroxide found in the interval of low-magnetic-susceptibility (MS) values at Bidart.

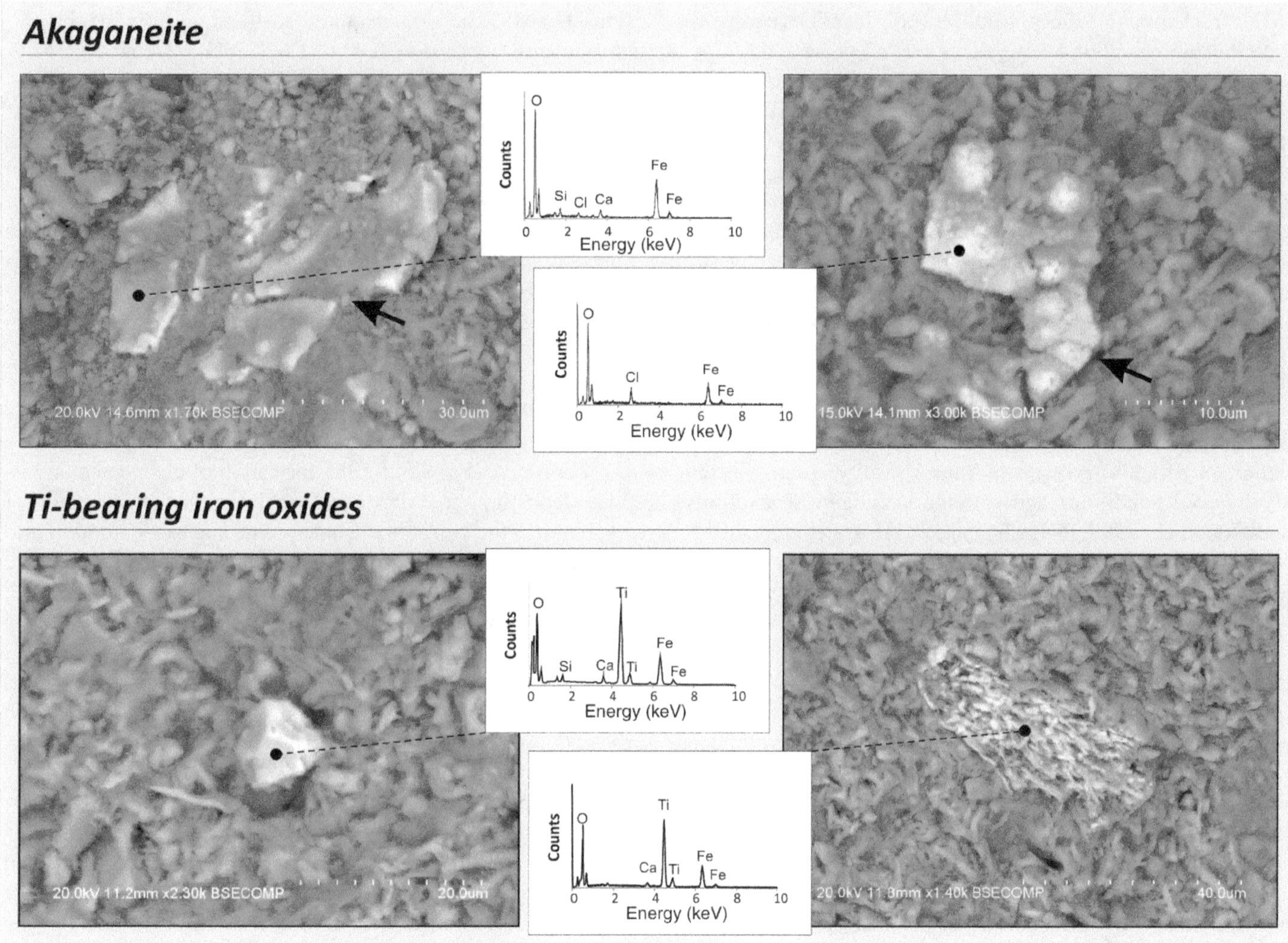

Figure 6. Additional scanning electron microscopy photographs and energy dispersive spectra (EDS) spectra of plate-like and semihexagonal akaganeite and eroded and severely altered Ti-bearing iron oxides.

low temperature. Indeed, at temperatures higher than 550 °C, akaganeite would react to form hematite. At these temperatures, $FeCl_2$ is the most abundant iron species, whereas at temperatures higher than 750 °C, it is replaced by $Fe(OH)_2$ (Symonds and Reed, 1993). $FeCl_2$ sublimates rapidly (64.4 mg/100 mL), allowing its dissolution in the aqueous and atmospheric medium. Fe^{2+} and Cl^- ions could then react according to the following equation to form akaganeite:

$$2Fe^{2+}_{(aq)} + Cl^-_{(aq)} + 3/2H_2O + 3/4O_2(g) = \beta\text{-}Fe_2(OH)_3Cl_{(s)}. \quad (3)$$

Considering a standard free energy of formation for β-$Fe_2(OH)_3Cl$ (Remazeilles and Refait, 2007) and enthalpies of aqueous Fe(II), Cl(II), and water (Wagman et al., 1982), the computed ΔG°_r is –280.74 kJ mol^{-1}. According to this reactive pathway, the akaganeite formation is thermodynamically possible.

Delivery of SO_2 and halogens to the atmosphere by active volcanoes is well known, especially for arc volcanism (Pyle and Mather, 2009), but it also occurs in continental flood basalt eruptions (Thordarson and Self, 1996; Self et al., 2008b). Glass analyses recently demonstrated that Deccan basalts were responsible for a huge HCl input into the atmosphere (4×10^{12} g SO_2 and 1×10^{12} g HCl released into the atmosphere per cubic kilometer of lava; Self et al., 2008b), confirming former scenarios (Glasby and Kunzendorf, 1996). The volume of lavas extruded during Deccan phase 2 is ~80% of the total lava volume, i.e., ~10^6 km^3 (Self et al., 2006). These figures are fairly consistent with the calculations of Black et al. (2012) for the Siberian Traps, which are suggested to have released 7×10^{12} g S and 3–9×10^{12} g Cl into the atmosphere. In the latter case, the variability of Cl degassing is due to the possible contribution of contact heating and metamorphism of the Tunguska sediments. The Kilauea volcano in Hawaii sometimes generates additional HCl when lavas flow into the sea (Edmonds and Gerlach, 2006). Using the conservative hypothesis of 1×10^{12} g HCl released by Deccan phase 2 for 200,000 yr, a mean value of 5 Mt Cl and 10 Mt S per year would have been released into the atmosphere (i.e., quantities equivalent to one Pinatubo eruption every year or every 2 yr for 200,000 yr). These calculations assume a rather unrealistic regular volcanic gas production and over a much longer time period than the currently estimated tens of thousands of years to at most 50,000–100,000 yr (Chenet et al., 2008). Despite these uncertainties, it is obvious that Deccan phase 2 corresponds to a severe atmospheric change and was responsible for major environmental hazards due to acid rains over tens of thousands of years, as recently suggested by Ward (2009).

The fate of the acidic gases depends on humidity at the emission site, because of their very high water solubility. At the time of Deccan phase 2 volcanism, India was a large island in the southern part of the Tethys Ocean located at ~21°S. Climate was relatively warm and humid, but arid close to the Deccan volcanic province as a result of "mock aridity" (e.g., volcanically induced conditions and extreme geochemical alkalinity; Khadkikar et al., 1999; Gertsch et al., 2011; Keller et al., 2012). Therefore, a large proportion of the released gases would have been transported into the high atmosphere (Kaminski et al., 2011) rather than scavenged locally by rains. Deccan volcanism likely occurred as large fissure eruptions with thermal plumes penetrating into the stratosphere due to the high lava volume and high temperatures (Self et al., 2006). A significant percentage of acid gases and halogens is therefore expected to have remained in the lower stratosphere for several years after each eruption (Ward, 2009). This is the likely explanation for environmental changes such as acid rains over a large portion of Earth, especially the intertropical and peri-Tethys regions.

Low-MS Interval: Consequence of Deccan Phase 2 Acid Rains?

Low-MS values just below the Cretaceous-Paleogene boundary may reflect a drastic decrease in the main magnetite component, that is, detrital titanomagnetite (component 1), originating from erosion of igneous rocks and/or their pedogenetic products and being transported into basin sediments. Absence of biomagnetite is also observed in the low-MS interval, but its origin still needs to be confirmed in the future. Variations in detrital titanomagnetite content at Bidart can be explained by several causes, such as (1) a rapid sea-level rise (e.g., dilution and remoteness to the source), (2) diagenetic reduction (Abrajevitch and Kodama, 2011), or (3) annihilation of iron oxides at or close to their continental source. The third hypothesis is tested here as a consequence of the huge Deccan volcanic gas emissions and resulting acid rains, which could have dissolved the continental magnetite grains before any riverine transfer to the oceanic basin. Rain in equilibrium with present-day atmospheric CO_2 is characterized by a pH of 5.6, whereas acidic volcanic rains have a much lower pH. For instance, the distribution of regional rain pH due to Mount Etna, a major gas emitter, has a pH of 3–4.5 (Calabrese et al., 2011).

Our weathering model indicates that equilibrium between magnetite and the weathering solution is reached between rain events (~1 d). The time required to dissolve 90% of the initial magnetite mass is between 31,000 (pH 3.3) and 68,000 yr (pH 4.3). These values are compatible with the temporal duration of the volcanic emissions based upon magneto-stratigraphic measurements and estimated at less than 100 k.y. (10–20 yr per flow; Chenet et al., 2007, 2008). Note that our calculations based on the pH of present-day rainwater (5.6) lead to a dissolution time of several million years, which is much longer than the expected landscape Davisian stability and consistent with previous work (White et al., 1994) giving very long residence time for sand-sized grains of magnetite subjected to present-day weathering conditions.

Conceptual Model

A conceptual model for the formation of akaganeite and its eolian deposition at Bidart and Gubbio infers the formation of akaganeite in the buoyant volcanic plume and transportation into the stratosphere by penetrative convection together with huge amounts of aerosols containing dissolved acid species (Fig. 7). It is well known that heat associated with widespread lava flows may generate convection in the atmosphere (Kaminski et al., 2011). These penetrative convections produce a thick well-mixed heated atmospheric layer where buoyancy ensures that volcanic gases get transported to the stratosphere, despite the low explosive nature of tholeiitic volcanism.

Subtropical eruptions such as the Deccan Traps are important contributors to global climate because stratospheric aerosols have a long lifetime in the tropical-subtropical stratosphere and can be transported into the midlatitudes of both hemispheres. For example, in the Soufrière Hills eruption (16.7°N) in 2006, the aerosol plume reached altitudes up to 20 km (Prata et al., 2007), which is well above the tropical tropopause layer. Two weeks after the eruption of the Nabro volcano (East Africa, 13.5°N) in 2011, the Cloud Aerosol Lidar and Infrared Pathfinder Satellite Observation (CALIPSO) detected enhanced aerosol backscatter in the lower stratosphere (~18 km; Vernier et al., 2013). During the 1991 Pinatubo eruption (15°N), the plume reached a maximum altitude of 39 km (Holasek et al., 1996), whereas the buoyant regions of the volcanic SO_2 and ash cloud are estimated at 25 km and 22 km, respectively (Guo et al., 2004b).

Once injected into the stratosphere, the large aerosol particles and small ones being formed by the volcanic gases are rapidly advected around the globe (Robock, 2000). The lifetime of sulfate aerosols is up to 3 yr in the stratosphere, but only 3 wk in the troposphere. Observations after the 1883 Krakatau eruption (6°S) showed that the aerosol cloud circled the globe in 2 wk (Robock, 2000). The El Chichón (17°N) eruption in 1982 and the Pinatubo eruption in 1991 are other examples of volcanic eruptions where aerosol clouds circled the globe within 3 wk and caused considerable cooling of the lower troposphere in the 2 yr after the eruptions (Robock, 2000; Guo et al., 2004a). The effect of the Laki eruption (1783) in Iceland, which extruded ~15 km^3 of lava over 8 mo (Thordarson and Self, 2003), is also well documented,

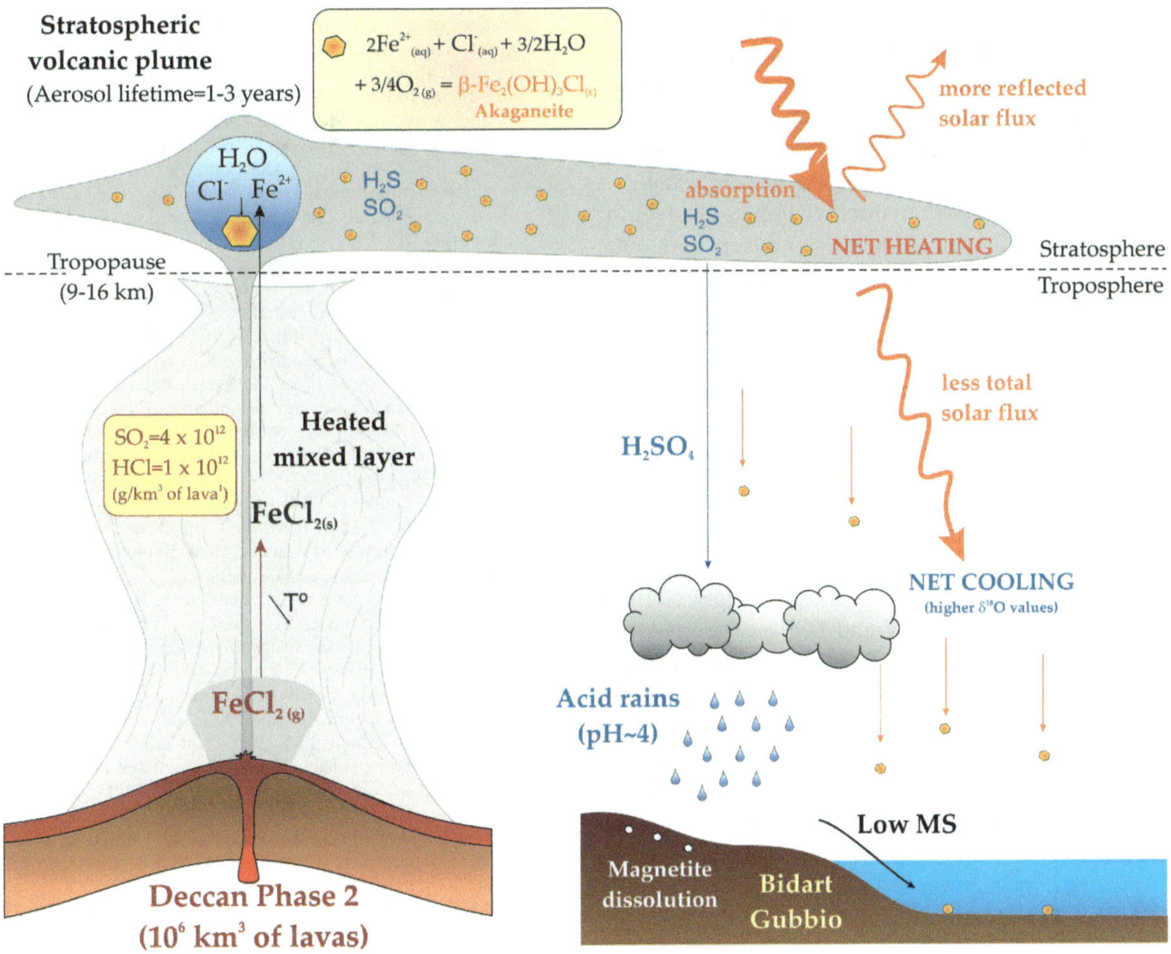

Figure 7. Conceptual model of the sedimentary marine record of continental magnetite dissolution caused by volcanic acidic rains (modified from Kaminski et al., 2011; Robock, 2000). MS—magnetic susceptibility.

including climate change and acid rains over most of Europe up to 3000 km from the volcano (Garnier, 2011). Numerical modeling of the Laki eruption confirmed that cooling was mostly due to stratospheric sulfate aerosols (Chenet et al., 2005).

In comparison, Deccan eruptions were much more intense, with an estimated 1 × 10^6 km^3 lava extruded (Self et al., 2006) over tens of thousands of years (Chenet et al., 2007). This likely caused acid rains through the intertropical peri-Tethys zone, reaching the North Atlantic and France (Bidart), in particular, ~7000 km from the erupting Deccan basalts. Our model assumes that Deccan volcanism caused acid rain at Bidart (France) and Gubbio (Italy), resulting in weathering and a dissolution effect on the continental surface. As noted earlier, these assumptions are well supported by historical comparisons, as well as our numerical weathering model. The subsequent cooling effect results from the huge input of stratospheric aerosols (Self et al., 2006, 2008b), as also supported by our carbon and oxygen isotopic data (Fig. 2).

We conclude that since volcanic plumes of relatively very small volcanic eruptions have reached the stratosphere and caused significant acid rain and climate changes over relatively very large areas (~3000 km for the Laki eruption, which extruded 15 km^3 of lava), then the giant pulsed Deccan eruptions near the end of the Maastrichtian, with an estimated 800,000 km^3 lava erupted (~53 times the Laki eruptions; Self et al., 2006), must have resulted in the most devastating global environmental changes. A comparison of stratospheric volcanic aerosol output from recent volcanos and its effect on Earth's radiation budget with the output from Deccan eruptions reveals further clues to the magnitude, nature, and importance of the enormous global Deccan-related environmental changes and their likely devastating contribution to the end-Cretaceous mass extinction.

CONCLUSION

The magnetic and mineral markers identified in the Bidart section are consistent with a major change in atmospheric conditions linked to Deccan-induced volcanic gas emissions. The low-MS signal is likely to be recognized elsewhere in the

Tethys sedimentary record, particularly in shallow and proximal marine environments where terrestrial input is appropriate and provided that wet local climatic conditions prevailed and that the continental surfaces contained significant amounts of magnetite. Presence of akaganeite is expected in the Tethys realm, but with a variable probability depending on the mechanisms of eolian transport. Moreover, both markers may help unravel the signature of other large igneous provinces emplaced in similar paleogeographic and paleoclimatic conditions, such as the Central Atlantic magmatic province. These newly discovered markers provide a promising tool with which to evaluate the environmental effects of large igneous provinces and their possible consequences for the biosphere.

ACKNOWLEDGMENTS

Funding was provided by Fundação para a Ciências e a Tecnologia (PTDC/CTE_GIX/110205/2010). We thank Celia Lee and Ana Sousa for technical and administrative supply, Bernard Peybernès for bibliographic supply and help in the field, and Dominique Serça and Céline Mari (Laboratoire d'Aérologie, Observatoire Midi-Pyrénées, Toulouse), Gilles Berger (IRAP, Institut de Recherche en Astrophysique et Planétologie, Toulouse), and Alexandra Abrajevitch (Institute of Tectonics and Geophysics, Russian Academy of Sciences, Khabarovsk) for helpful discussions. We thank anonymous reviewers for their constructive comments.

REFERENCES CITED

Abrajevitch, A., and Kodama, K., 2009, Biochemical vs. detrital mechanism of remanence acquisition in marine carbonates: A lesson from the K-T boundary interval: Earth and Planetary Science Letters, v. 286, p. 269–277, doi:10.1016/j.epsl.2009.06.035.

Abrajevitch, A., and Kodama, K., 2011, Diagenetic sensitivity of paleoenvironmental proxies: A rock magnetic study of Australian continental margin sediments: Geochemistry Geophysics Geosystems, v. 12, p. 1–18, doi:10.1029/2010GC003481.

Abramovich, S., Yovel-Corem, S., Almogi-Labin, A., and Benjamini, C., 2010, Global climate change and planktic foraminiferal response in the Maastrichtian: Paleoceanography, v. 25, p. PA2201, doi:10.1029/2009PA001843.

Bibi, I., Singh, B., and Silvester, E., 2011, Akaganeite (beta-FeOOH) precipitation in inland acid sulfate soils of south-western New South Wales (NSW), Australia: Geochimica et Cosmochimica Acta, v. 75, p. 6429–6438, doi:10.1016/j.gca.2011.08.019.

Black, B.A., Elkins-Tanton, L.T., Rowe, M.C., and Peate, I.U., 2012, Magnitude and consequences of volatile release from the Siberian Traps: Earth and Planetary Science Letters, v. 317, p. 363–373, doi:10.1016/j.epsl.2011.12.001.

Bland, P.A., Kelley, S.P., Berry, F.J., Cadogan, J.M., and Pillinger, C.T., 1997, Artificial weathering of the ordinary chondrite Allegan: Implications for the presence of Cl^- as a structural component in akaganeite: The American Mineralogist, v. 82, p. 1187–1197.

Bonté, P., Delacotte, O., Renard, M., Laj, C., Boclet, D., Jehanno, C., and Rocchia, R., 1984, An iridium-rich layer at the Cretaceous-Tertiary boundary in the Bidart section (southern France): Geophysical Research Letters, v. 11, p. 473–476, doi:10.1029/GL011i005p00473.

Brunauer, S., Emmett, P.H., and Teller, E., 1938, Adsorption of gases in multimolecular layers: Journal of the American Chemical Society, v. 60, p. 309–319, doi:10.1021/ja01269a023.

Calabrese, S., Aiuppa, A., Allard, P., Bagnato, E., Bellomo, S., Brusca, L., D'Alessandro, W., and Parello, F., 2011, Atmospheric sources and sinks of volcanogenic elements in a basaltic volcano (Etna, Italy): Geochimica et Cosmochimica Acta, v. 75, p. 7401–7425, doi:10.1016/j.gca.2011.09.040.

Channell, J.E.T., Hodell, D.A., Singer, B.S., and Xuan, C., 2010, Reconciling astrochronological and Ar-40/Ar-39 ages for the Matuyama-Brunhes boundary and late Matuyama chron: Geochemistry Geophysics Geosystems, v. 11, p. Q0AA12, doi:10.1029/2010GC003203.

Chenet, A.L., Fluteau, F., and Courtillot, V., 2005, Modelling massive sulphate aerosol pollution, following the large 1783 Laki basaltic eruption: Earth and Planetary Science Letters, v. 236, p. 721–731, doi:10.1016/j.epsl.2005.04.046.

Chenet, A.L., Quidelleur, X., Fluteau, F., Courtillot, V., and Bajpai, S., 2007, K-40-Ar-40 dating of the Main Deccan large igneous province: Further evidence of KTB age and short duration: Earth and Planetary Science Letters, v. 263, p. 1–15, doi:10.1016/j.epsl.2007.07.011.

Chenet, A.L., Fluteau, F., Courtillot, V., Gerard, M., and Subbarao, K.V., 2008, Determination of rapid Deccan eruptions across the Cretaceous-Tertiary boundary using paleomagnetic secular variation: Results from a 1200-m-thick section in the Mahabaleshwar escarpment: Journal of Geophysical Research–Solid Earth, v. 113, p. B04101, doi:10.1029/2006JB004635.

Chudaeva, V.A., Urchenko, S.G., Chudaev, O.V., Sugimory, K., Matsuo, M., and Kuno, A., 2006, Chemistry of rainwaters in the south Pacific area of Russia: Journal of Geochemical Exploration, v. 88, p. 101–105, doi:10.1016/j.gexplo.2005.08.020.

Commission Internationale de L'Eclairage (CIE), 1978, Recommendations on uniform color spaces, color difference equations, psychometric color terms: Supplement No 2 to Publication 15, Colorimetry: Paris, Bureau Central de la CIE.

Courtillot, V., 2012, Mass extinctions and massive volcanism: Journal of the Geological Society of India, v. 79, p. 107–108.

Edmonds, M., and Gerlach, T.M., 2006, The airborne lava-seawater interaction plume at Kilauea volcano, Hawai'i: Earth and Planetary Science Letters, v. 244, p. 83–96, doi:10.1016/j.epsl.2006.02.005.

Egli, R., 2004, Characterization of individual rock magnetic components by analysis of remanence curves: 1. Unmixing natural sediments: Studia Geophysica et Geodaetica, v. 48, p. 391–446, doi:10.1023/B:SGEG.0000020839.45304.6d.

Ellwood, B.B., MacDonald, W.D., Wheeler, C., and Benoist, S.L., 2003, The K-T boundary in Oman: Identified using magnetic susceptibility field measurements with geochemical confirmation: Earth and Planetary Science Letters, v. 206, p. 529–540, doi:10.1016/S0012-821X(02)01124-X.

Fabre, S., and Berger, G., 2012, How tillite weathering during the snowball Earth aftermath induced cap carbonate deposition: Geology, v. 40, p. 1027–1030, doi:10.1130/G33340.1.

Fabre, S., Berger, G., and Nedelec, A., 2011, Modeling of continental weathering under high-CO_2 atmospheres during Precambrian times: Geochemistry Geophysics Geosystems, v. 12, p. Q10001, doi:10.1029/2010GC003444.

Floor, G.H., Calabrese, S., Roman-Ross, G., D'Alessandro, W., and Aiuppa, A., 2011, Selenium mobilization in soils due to volcanic derived acid rain: An example from Mt. Etna volcano, Sicily: Chemical Geology, v. 289, p. 235–244, doi:10.1016/j.chemgeo.2011.08.004.

Font, E., Nedelec, A., Ellwood, B.B., Mirao, J., and Silva, P.F., 2011, A new sedimentary benchmark for the Deccan Traps volcanism?: Geophysical Research Letters, v. 38, p. L24309, doi:10.1029/2011GL049824.

Gaillardet, J., Dupre, B., Louvat, P., and Allegre, C.J., 1999, Global silicate weathering and CO_2 consumption rates deduced from the chemistry of large rivers: Chemical Geology, v. 159, p. 3–30, doi:10.1016/S0009-2541(99)00031-5.

Galbrun, B., and Gardin, S., 2004, New chronostratigraphy of the Cretaceous-Paleogene boundary interval at Bidart (France): Earth and Planetary Science Letters, v. 224, p. 19–32, doi:10.1016/j.epsl.2004.04.043.

Garnier, E., 2011, Les brouillards du Laki en 1783: Volcanisme et crise sanitaire en Europe: Bulletin de l'Académie Nationale de Médecine, v. 195, p. 1043–1055.

Gertsch, B., Keller, G., Adatte, T., Garg, R., Prasad, V., Berner, Z., and Fleitmann, D., 2011, Environmental effects of Deccan volcanism across the Cretaceous-Tertiary transition in Meghalaya, India: Earth and Planetary Science Letters, v. 310, p. 272–285, doi:10.1016/j.epsl.2011.08.015.

Glasby, G.P., and Kunzendorf, H., 1996, Multiple factors in the origin of the Cretaceous/Tertiary boundary: The role of environmental stress and Deccan Trap volcanism: Geologische Rundschau, v. 85, p. 191–210, doi:10.1007/BF02422228.

Guo, S., Bluth, G.J.S., Rose, W.I., Watson, I.M., and Prata, A.J., 2004a, Reevaluation of SO_2 release of the 15 June 1991 Pinatubo eruption using ultraviolet and infrared satellite sensors: Geochemistry Geophysics Geosystems, v. 5, p. Q04001, doi:10.1029/2003GC000654.

Guo, S., Rose, W.I., Bluth, G.J.S., and Watson, I.M., 2004b, Particles in the great Pinatubo volcanic cloud of June 1991: The role of ice: Geochemistry Geophysics Geosystems, v. 5, p. Q05003, doi:10.1029/2003GC000655.

Holasek, R.E., Self, S., and Woods, A.W., 1996, Satellite observations and interpretation of the 1991 Mount Pinatubo eruption plumes: Journal of Geophysical Research–Solid Earth, v. 101, p. 27,635–27,655, doi:10.1029/96JB01179.

Johnston, J.H., 1977, Jarosite and akaganeite from White Island volcano, New Zealand—An X-ray and Mossbauer study: Geochimica et Cosmochimica Acta, v. 41, p. 539–544, doi:10.1016/0016-7037(77)90291-5.

Kaminski, E., Chenet, A.L., Jaupart, C., and Courtillot, V., 2011, Rise of volcanic plumes to the stratosphere aided by penetrative convection above large lava flows: Earth and Planetary Science Letters, v. 301, p. 171–178, doi:10.1016/j.epsl.2010.10.037.

Keller, G., Li, L., and MacLeod, N., 1995, The Cretaceous-Tertiary boundary stratotype section at El Kef, Tunisia: How catastrophic was the mass extinction?: Palaeogeography, Palaeoclimatology, Palaeoecology, v. 119, p. 221–254, doi:10.1016/0031-0182(95)00009-7.

Keller, G., Adatte, T., Gardin, S., Bartolini, A., and Bajpai, S., 2008, Main Deccan volcanism phase ends at K-T mass extinction: Evidence from the Krishna-Godavari Basin, SE India: Earth and Planetary Science Letters, v. 268, p. 293–311.

Keller, G., Adatte, T., Juez, A.P., and Lopez-Oliva, J.G., 2009, New evidence concerning the age and biotic effects of the Chicxulub impact in NE Mexico: Journal of the Geological Society, v. 166, p. 393–411, doi:10.1144/0016-76492008-116.

Keller, G., Bhowmick, P.K., Upadhyay, H., Dave, A., Reddy, A.N., Jaiprakash, B.C., and Adatte, T., 2011, Deccan volcanism linked to the Cretaceous-Tertiary boundary mass extinction: New evidence from ONGC wells in the Krishna-Godavari Basin: Journal of the Geological Society of India, v. 78, p. 399–428, doi:10.1007/s12594-011-0107-3.

Keller, G., Adatte, T., Bhowmick, P.K., Upadhyay, H., Dave, A., Reddy, A.N., and Jaiprakash, B.C., 2012, Nature and timing of extinctions in Cretaceous-Tertiary planktic foraminifera preserved in Deccan intertrappean sediments of the Krishna–Godavari Basin, India: Earth and Planetary Science Letters, v. 341–344, p. 211–221, doi:10.1016/j.epsl.2012.06.021.

Khadkikar, A.S., Sant, D.A., Gogte, V., and Karanth, R.V., 1999, The influence of Deccan volcanism on climate: Insights from lacustrine intertrappean deposits, Anjar, western India: Palaeogeography, Palaeoclimatology, Palaeoecology, v. 147, p. 141–149, doi:10.1016/S0031-0182(98)00156-4.

Kruiver, P.P., Dekkers, M.J., and Heslop, D., 2001, Quantification of magnetic coercivity components by the analysis of acquisition curves of isothermal remanent magnetisation: Earth and Planetary Science Letters, v. 189, p. 269–276, doi:10.1016/S0012-821X(01)00367-3.

Kruiver, P.P., Langereis, C.G., Dekkers, M.J., and Krijgsman, W., 2003, Rock-magnetic properties of multicomponent natural remanent magnetization in alluvial red beds (NE Spain): Geophysical Journal International, v. 153, p. 317–332, doi:10.1046/j.1365-246X.2003.01880.x.

Kuiper, K.F., Deino, A., Hilgen, F.J., Krijgsman, W., Renne, P.R., and Wijbrans, J.R., 2008, Synchronizing rock clocks of Earth history: Science, v. 320, p. 500–504, doi:10.1126/science.1154339.

Lee, M.R., and Bland, P.A., 2004, Mechanisms of weathering of meteorites recovered from hot and cold deserts and the formation of phyllosilicates: Geochimica et Cosmochimica Acta, v. 68, p. 893–916, doi:10.1016/S0016-7037(03)00486-1.

Li, L.Q., and Keller, G., 1998, Abrupt deep-sea warming at the end of the Cretaceous: Geology, v. 26, p. 995–998, doi:10.1130/0091-7613(1998)026<0995:ADSWAT>2.3.CO;2.

Li, Q.X., Wang, Z.Y., Han, W., and Han, E.H., 2008, Characterization of the rust formed on weathering steel exposed to Qinghai salt lake atmosphere: Corrosion Science, v. 50, p. 365–371, doi:10.1016/j.corsci.2007.06.020.

Maher, B.A., 2011, The magnetic properties of Quaternary aeolian dusts and sediments, and their palaeoclimatic significance: Aeolian Research, v. 3, p. 87–144, doi:10.1016/j.aeolia.2011.01.005.

Pardo, A., Ortiz, M., and Keller, G., 1996, Latest Maastrichtian and Cretaceous-Tertiary boundary foraminiferal turnover and environmental changes at Agost, Spain, *in* MacLeod, N., and Keller, G., eds., Cretaceous-Tertiary Mass Extinction: Biotic and Environmental Changes: New York, W.W. Norton and Company, p. 139–171.

Parkhurst, D.L., and Appelo, C.A.J., 1999. User's Guide to PHREEQC Version 2—A Computer Program for Speciation, Batch-Reaction, One-Dimensional Transport, and Inverse Calculations: U.S. Geological Survey Water-Resources Investigations Report 99-4259, 312 p.

Post, J.E., and Buchwald, V.F., 1991, Crystal-structure refinement of akaganeite: The American Mineralogist, v. 76, p. 272–277.

Prata, A.J., Carn, S.A., Stohl, A., and Kerkmann, J., 2007, Long range transport and fate of a stratospheric volcanic cloud from Soufriere Hills volcano, Montserrat: Atmospheric Chemistry and Physics, v. 7, p. 5093–5103, doi:10.5194/acp-7-5093-2007.

Pyle, D.M., and Mather, T.A., 2009, Halogens in igneous processes and their fluxes to the atmosphere and oceans from volcanic activity: A review: Chemical Geology, v. 263, p. 110–121, doi:10.1016/j.chemgeo.2008.11.013.

Reguer, S., Dillmann, P., and Mirambet, F., 2007, Buried iron archaeological artefacts: Corrosion mechanisms related to the presence of Cl-containing phases: Corrosion Science, v. 49, p. 2726–2744, doi:10.1016/j.corsci.2006.11.009.

Remazeilles, C., and Refait, P., 2007, On the formation of beta-FeOOH (akaganeite) in chloride-containing environments: Corrosion Science, v. 49, p. 844–857, doi:10.1016/j.corsci.2006.06.003.

Remazeilles, C., and Refait, P., 2008, Formation, fast oxidation and thermodynamic data of Fe(II) hydroxychlorides: Corrosion Science, v. 50, p. 856–864, doi:10.1016/j.corsci.2007.08.017.

Renne, P.R., Mundil, R., Balco, G., Min, K.W., and Ludwig, K.R., 2010, Joint determination of K-40 decay constants and Ar-40*/K-40 for the Fish Canyon sanidine standard, and improved accuracy for Ar-40/Ar-39 geochronology: Geochimica et Cosmochimica Acta, v. 74, p. 5349–5367, doi:10.1016/j.gca.2010.06.017.

Renne, P.R., Deino, A.L., Hilgen, F.J., Kuiper, K.F., Mark, D.F., Mitchell, W.S., Morgan, L.E., Mundil, R., and Smit, J., 2013, Time scales of critical events around the Cretaceous-Paleogene boundary: Science, v. 339, p. 684–687, doi:10.1126/science.1230492.

Robertson, D.J., and France, D.E., 1994, Discrimination of remanence-carrying minerals in mixtures, using isothermal remanent magnetization acquisition curves: Physics of the Earth and Planetary Interiors, v. 82, p. 223–234, doi:10.1016/0031-9201(94)90074-4.

Robock, A., 2000, Volcanic eruptions and climate: Reviews of Geophysics, v. 38, p. 191–219, doi:10.1029/1998RG000054.

Schulte, P., Alegret, L., Arenillas, I., Arz, J.A., Barton, P.J., Bown, P.R., Bralower, T.J., Christeson, G.L., Claeys, P., Cockell, C.S., Collins, G.S., Deutsch, A., Goldin, T.J., Goto, K., Grajales-Nishimura, J.M., Grieve, R.A.F., Gulick, S.P.S., Johnson, K.R., Kiessling, W., Koeberl, C., Kring, D.A., MacLeod, K.G., Matsui, T., Melosh, J., Montanari, A., Morgan, J.V., Neal, C.R., Nichols, D.J., Norris, R.D., Pierazzo, E., Ravizza, G., Rebolledo-Vieyra, M., Reimold, W.U., Robin, E., Salge, T., Speijer, R.P., Sweet, A.R., Urrutia-Fucugauchi, J., Vajda, V., Whalen, M.T., and Willumsen, P.S., 2010, The Chicxulub asteroid impact and mass extinction at the Cretaceous-Paleogene boundary: Science, v. 327, p. 1214–1218, doi:10.1126/science.1177265.

Self, S., Widdowson, M., Thordarson, T., and Jay, A.E., 2006, Volatile fluxes during flood basalt eruptions and potential effects on the global environment: A Deccan perspective: Earth and Planetary Science Letters, v. 248, p. 518–532, doi:10.1016/j.epsl.2006.05.041.

Self, S., Jay, A.E., Widdowson, M., and Keszthelyi, L.P., 2008a, Correlation of the Deccan and Rajahmundry Trap lavas: Are these the longest and largest lava flows on Earth?: Journal of Volcanology and Geothermal Research, v. 172, p. 3–19, doi:10.1016/j.jvolgeores.2006.11.012.

Self, S., Blake, S., Sharma, K., Widdowson, M., and Sephton, S., 2008b, Sulfur and chlorine in Late Cretaceous Deccan magmas and eruptive gas release: Science, v. 319, p. 1654–1657, doi:10.1126/science.1152830.

Smit, J., 1990, Meteorite impact, extinctions and the Cretaceous-Tertiary boundary: Geologie en Mijnbouw, v. 69, p. 187–204.

Stuben, D., Kramar, U., Berner, Z.A., Meudt, M., Keller, G., Abramovich, S., Adatte, T., Hambach, U., and Stinnesbeck, W., 2003, Late Maastrichtian paleoclimatic and paleoceanographic changes inferred from Sr/Ca ratio and stable isotopes: Palaeogeography, Palaeoclimatology, Palaeoecology, v. 199, p. 107–127, doi:10.1016/S0031-0182(03)00499-1.

Sverdrup, H., and Warfvinge, P., 1995, Estimating field weathering rates using laboratory kinetics: Reviews in Mineralogy, v. 31, p. 485–541.

Symonds, R.B., and Reed, M.H., 1993, Calculation of multicomponent chemical-equilibria in gas-solid-liquid systems—Calculation methods, thermochemical data, and applications to studies of high-temperature volcanic gases with examples from Mount St-Helens: American Journal of Science, v. 293, p. 758–864, doi:10.2475/ajs.293.8.758.

Thordarson, T., and Self, S., 1996, Sulfur, chlorine and fluorine degassing and atmospheric loading by the Roza eruption, Columbia River Basalt Group, Washington, USA: Journal of Volcanology and Geothermal Research, v. 74, p. 49–73, doi:10.1016/S0377-0273(96)00054-6.

Thordarson, T., and Self, S., 2003, Atmospheric and environmental effects of the 1783–1784 Laki eruption: A review and reassessment: Journal of Geophysical Research–Atmospheres, v. 108, p. AAC 7-1–AAC 7-29.

Vernier, J.P., Thomason, L.W., Fairlie, T.D., Minnis, P., Palikonda, R., and Bedka, K.M., 2013, Comment on "Large volcanic aerosol load in the stratosphere linked to Asian monsoon transport": Science, v. 339, p. 347.

Vonhof, H.B., and Smit, J., 1997, High-resolution late Maastrichtian–early Danian oceanic Sr-87/Sr-86 record: Implications for Cretaceous-Tertiary boundary events: Geology, v. 25, p. 347–350, doi:10.1130/0091-7613(1997)025<0347:HRLMED>2.3.CO;2.

Wagman, D.D., Evans, W.H., Parker, V.B., Schumm, R.H., Halow, I., Bailey, S.M., Churney, K.L., and Nuttall, R.L., 1982, The NBS Tables of Chemical Thermodynamic Properties—Selected Values for Inorganic and C_1 and C_2 Organic Substances in SI Units: Journal of Physical and Chemical Reference Data, v. 11, p. 1.

Ward, P.D., and Kennedy, W.J., 1993, Maastrichtian ammonites from the Biscay region (France, Spain): Journal of Paleontology, v. 67, p. 1–58.

Ward, P.L., 2009, Sulfur dioxide initiates global climate change in four ways: Thin Solid Films, v. 517, p. 3188–3203, doi:10.1016/j.tsf.2009.01.005.

White, A.F., Peterson, M.L., and Hochella, M.F., 1994, Electrochemistry and dissolution kinetics of magnetite and ilmenite: Geochimica et Cosmochimica Acta, v. 58, p. 1859–1875, doi:10.1016/0016-7037(94)90420-0.

White, R.V., and Saunders, A.D., 2005, Volcanism, impact and mass extinctions: Incredible or credible coincidences?: Lithos, v. 79, p. 299–316, doi:10.1016/j.lithos.2004.09.016.

Wilf, P., Johnson, K.R., and Huber, B.T., 2003, Correlated terrestrial and marine evidence for global climate changes before mass extinction at the Cretaceous-Paleogene boundary: Proceedings of the National Academy of Sciences of the United States of America, v. 100, p. 599–604, doi:10.1073/pnas.0234701100.

Manuscript Accepted by the Society 31 January 2014

Printed in the USA

The Geological Society of America
Special Paper 505
2014

Impact volcanism and mass extinctions

Adrian P. Jones*
Earth Sciences, Gower Street, University College London, London WC1E 6BT, UK

ABSTRACT

For asteroid or comet impacts, the mass of the projectile or bolide and its velocity control the scale of damage and secondary catastrophes induced, and the impact flux can be used to determine whether such an impact was likely to occur at the time of interest. Impact cratering processes are still orders of magnitude more deadly than volcanism when considering the potential for atmospheric loading of deleterious particulate and gaseous materials, due to the extraordinarily rapid transfer of energy. Based on impact flux, there could have been sufficient large impactors to cause one or more of the "Big Five" mass extinctions in the last 300 m.y. The best contender so far is the Chicxulub event, but this did not trigger massive volcanism in situ, and the Deccan volcanism was not located correctly to be its antipodal pair. The combination of volcanism with impact cratering is a real possibility for the end-Cretaceous extinction, but there is no established connection. This contribution reviews the wider aspects of impact volcanism, including impact fluxes, impact melting, crater thermal anomalies, and secondary impact crises like antipodal volcanism in the context of Phanerozoic mass extinctions.

INTRODUCTION

A 10 km comet or asteroid hitting the ground at 25 km/s would deliver ~100 million megatons of energy, equivalent to a small hydrogen bomb on each square kilometer of Earth's surface, leaving a 100–200-km-diameter crater (French, 1998; Melosh, 1996; Osinski and Pierazzo, 2013). The ejection of dust cutting out sunlight, collapsing food chains, heat spread globally by incinerating rain of debris, possible volcanism, tsunamis, and many other effects, have naturally led to repeated suggestions that bolide impacts are intimately connected with mass extinctions, as reviewed elsewhere (Alvarez et al., 1984; Osinski and Pierazzo, 2013; Rampino and Stothers, 1984). The relative kill potentials of impacts, volcanism, and climate change are regularly considered as responsible for one or more of the top five mass extinctions in the Phanerozoic (Alvarez, 2003; Keller, 2005; Raup, 1991). This review of impact volcanism offers the chance to consider the following: Might have processes previously considered to be separate, or coincidentally overlapping (White and Saunders, 2005), in fact been triggered by a single or multiple bolide impact event?

Impact volcanism remains a contentious topic, and there has been no unequivocal demonstration that it has occurred on Earth. Arguments in support have appeared in several models (Abbott and Isley, 2002; Elkins Tanton and Hager, 2005; Jones, 2005; Jones et al., 2002). Arguments against impact volcanism have also come from hydrocode modeling (Ivanov and Melosh, 2003), from plume geochemical arguments (Tejada et al., 2004), and from statistical arguments of coincidence between impacts and volcanism (White and Saunders, 2005). However, there is no doubt that large volumes of impact melt have been produced by

*adrian.jones@ucl.ac.uk

Jones, A.P., 2014, Impact volcanism and mass extinctions, *in* Keller, G., and Kerr, A.C., eds., Volcanism, Impacts, and Mass Extinctions: Causes and Effects: Geological Society of America Special Paper 505, p. 369–381, doi:10.1130/2014.2505(19). For permission to copy, contact editing@geosociety.org.

large terrestrial impact craters reflecting a variety of crustal target rock compositions. The impact melt can resemble a wide variety of igneous melt compositions, from silicate melts that have differentiated in situ (Therriault et al., 2002), and immiscible sulfide-silicate melts (Keays and Lightfoot, 2004) to carbonate melts (Jones et al., 2000; Osinski and Spray, 2001), and complex carbonate-silicate and silicate-silicate immiscible melts (Graup, 1999; see also review by Osinski and Pierazzo, 2012).

In terrestrial impact craters up to ~250 km in diameter, the observed volume of impact melt increases nearly exponentially with crater diameter (Grieve and Cintala, 1992). Shock melting due to very high postshock temperatures far in excess of normal melting temperatures for most target rocks and minerals has been reviewed elsewhere (French, 1998). However, for this review, we are interested in impact craters the same or larger than ~200 km diameter, represented by the Chicxulub end-Cretaceous crater, formed by impact energy >~10^{23} J (Morgan et al., 1997), extending up to basin-forming craters, of more than ~200 km and up to perhaps ~1000 km diameter, which represents a size range of diminishing frequency in terms of impact flux (see later section). Geological evidence of this size range is missing from Earth, not because they did not happen, but likely due to the extensive and continual resurfacing driven by crustal plate tectonics. To understand large impact cratering processes and their effects of shock on rocks and minerals, we rely on computer simulations and software based on hydrocode models. The models are coupled with semi-empirical equations of state to predict the outcomes of impact cratering and associated melting (e.g., Jones, 2005; Melosh, 1996; Miljković et al., 2010; Osinski and Pierazzo, 2012). Although final impact craters are shallow-aspect structures, with typical diameter (D) to depth (d) ratios >10:1 or 20:1, during their hyper-rapid formation, they pass through a much deeper transient crater stage at maximum penetration, where D:d is closer to 1:1. Even larger impacts >1000 km, representing energies in the range of 10^{27}–10^{29} J, could be capable of modifying geophysical planetary processes, with deep embedded mantle melting and antipodal cratering both recently modeled (Marinova et al., 2011), and, although too large for Phanerozoic Earth, they provide additional insights into impact melting, and impact volcanic outcomes of the largest impacts during the Late Heavy Bombardment.

Large impact craters greater than ~200 km and up to 1000 km diameter capable of forming huge amounts of impact melt would involve melting of the subcrustal mantle, especially so for oceanic impacts with higher geothermal gradients and shallow peridotite mantle, which melts on decompression. These very large impact craters are often referred to as "basin forming," by comparison to the large maria-filled basins visible on the Moon (Bottke et al., 2012). Basin-forming impacts on Earth have been considered to trigger other terrestrial phenomena, including formation of large igneous provinces (LIPs; Glikson, 1999; Negi et al., 1993; Rampino, 1987), oceanic plateaus (Ingle and Coffin, 2004; Jones et al., 2002; Rogers, 1982), the breakup of crustal tectonic plates (Price, 2001; Seyfert and Sirkin, 1979), and komatiite-greenstone belts (Green, 1972). Empirical observations suggest that large impacts may also trigger mantle plume or hotspot activity (Abbott and Isley, 2002). Because Earth is a high-temperature planet with active internal convection, it has been argued (Jones et al., 2002) that the impact response to basin-forming impacts would be to generate additional volumes of melt by decompression mantle melting, perhaps reducing by half the diameter required to generate the benchmark volume of melt equivalent to a large igneous province (>10^6 km^3) or mantle plume (Coffin and Eldholm, 1992). The amount of decompression caused by a large impact crater is calculated from the mass of rock removed by cratering, and hence its reduced contribution to lithostatic load (Jones et al., 2002). Counter arguments have also been made that the size of this decompression effect will be substantially reduced because of prior downward movement during the early impact compression stage of cratering, followed by rebound (Melosh and Ivanov, 2004). There is some evidence in support of crater-scale dynamic two-way vertical components, recorded in the complex system of crustal penetrating faults and thrust faults, which characterize impact craters (Price, 2001), but this does not escape the fact that cratering removes mass. Part of the explanation for the discrepancy may be due to the challenging time scales and different methods developed to calculate mass removal and ejection by impact cratering (seconds) compared with geophysical modeling of mantle flow (millions of years).

On other planets, basin-scale impacts have been considered responsible for the formation of corona structures on Venus (Hansen, 2007) and impact-induced mantle plumes with long-lived volcanism on Mars (Reese et al., 2004). On the Moon, a timing delay between impact crater formation of basins and infilling by surface maria basalts on the lunar nearside is equivocal. Models suggest that impact magmatism could have persisted via decompression melting and mantle convection for up to 350 m.y., and that giant impacts account for a large proportion of the volume and longevity of mare basalt volcanism (Elkins Tanton et al., 2004), whether or not assisted by impact-induced mantle convection (Ghods and Arkani-Hamed, 2007). The enormous contribution of impact melting to the evolution of the surface of the Moon has recently been emphasized. The total volume of impact melt was estimated to be ~10^8 km^3, perhaps constituting as much as ~20% of the lunar crust and occurring mostly as slowly cooled and fractionated impact melt seas throughout ~15 km depths (e.g., Orientale; Vaughan et al., 2013). Some lunar impact-derived rocks with crystalline lithologies may therefore resemble pristine igneous rocks. On the Moon at least, it is important to differentiate between impact shock melts, and the voluminous mare basalts, which are mantle melts. The mare basalts may have been enabled by impacts (Elkins Tanton et al., 2004), but if they are younger, they may not have mixed with impact melts. Recent observations using petrographic textures and mineral chemistry of lunar basaltic breccia appear to indicate the process of mixing between large basin-forming impact melts and volcanic mantle melts (Snape et al., 2011).

So far, we have established that large impact craters on Earth can produce a lot of melt. There may be undiscovered craters that produced even more melt. The phenomenon of impact-induced melt production is widely acknowledged on the Moon, Venus, and Mars. In the context of mass extinctions, we would like to know (1) whether or not bolide impacts could have simultaneously triggered volcanism leading to global environmental catastrophe, and (2) whether or not statistically there could be suitably massive impactors that could have struck Earth at the times of known mass extinction events. This contribution examines what is known about both of these issues from the perspective of bolide impacts, although it does not examine research on the impact cratering process on Earth, which has advanced considerably in the last decade (Osinski and Pierazzo, 2012). The topic of impact volcanism is reviewed, and some new ideas are presented to test whether or not the phenomenon of impact volcanism might have caused mass extinction.

IMPACT FLUX

The impact flux record for the Earth-Moon system (Bland, 2005; Grieve, 1997) is essentially a size-frequency flux constructed from combining the well-dated but sparse and poorly preserved impact crater record on Earth with poorly dated but well-preserved abundant impact craters on the Moon. The impact flux can be used to predict the size and mass distribution of impactors (bolides, asteroids, comets), and it can be modeled or fitted to idealized curves of changing frequency with time for the age of the Earth-Moon system. Generally, the smoothed flux is visualized as consisting of two components: (1) a sharp influx of high mass bolides, which delivered an intense bombardment to the ancient lunar surface (the Late Heavy Bombardment) superimposed on (2) a background flux generally viewed as steadily declining since the formation of the solar system. However, some have observed that the actual impact flux record may be "lumpy" and highly irregular in detail (Fig. 1) with distinct episodes of high flux separated by long periods of low flux, both for the bulk of Earth history (Abbott and Isley, 2002) and a "sawtooth-like" flux for the early history of the Moon, including the Late Heavy Bombardment (Morbidelli et al., 2012). Dated impact spherules from one lunar site were used to determine lunar impact flux and suggest a sort of "hockey stick" curve, with a gradual decline from 3.5 Ga followed by an increase starting at 0.4 Ga back to the same level today as it was at 3.5 Ga (Muller, 2002). An appraisal of the way in which the impact flux translates into impact frequency, crater size, and energy also demonstrated an effect of target rock lithology on terrestrial crater morphology (Dence, 2006).

Late Heavy Bombardment Paradigm?

The Late Heavy Bombardment (Gomes et al., 2005) has been widely considered as a sudden increase in flux or a flux-spike, which started and ended abruptly between ca. 4.1 and ca. 3.9 Ga, respectively (called here the "spike model"), superimposed on an exponential background decay, but greater detail and potential mechanisms are emerging (Morbidelli et al., 2012). As can be seen from the widespread impact-damaged surface of the Moon, this was a catastrophic impact barrage, which caused massive surface modification and deep crustal-penetrating damage, as vividly revealed in the high-resolution gravity field (Zuber et al., 2012). However, recent interpretations, if correct, would dramatically change our view of the early additional impact flux and attribute the time scale of the Late Heavy Bombardment to a protracted event spanning >2 b.y., or nearly half the age of Earth. First, impactor size and velocities, back-calculated from a modeled fit to well-dated impact condensate spherule beds, support a gradual decline in impact flux after the Late Heavy Bombardment (Johnson and Melosh, 2012) for over >2 b.y. Second, astrophysical modeling of a possible asteroid impactor source could produce the sharp startup but was unable to produce an abrupt termination, supporting instead a gradual decline (Bottke et al., 2012). Clearly, such a shift in concept for the Late Heavy Bombardment could have far-reaching consequences for evolution,

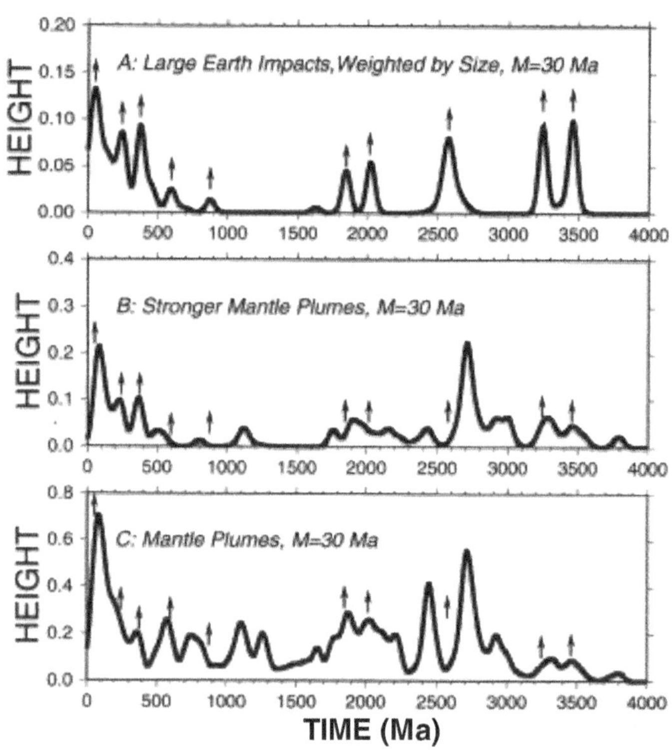

Figure 1. The bolide impact flux may not be smooth, but irregular and lumpy. Here the authors (Abbott and Isley, 2002) have drawn attention to apparent correlations between the irregular timing of large impacts, and mantle plume events inferred from large igneous provinces and other criteria in the geological record for the last 4 b.y. They contend that the correlation demonstrates a causative effect in that large impacts are connected to mantle plume activity. At another level, this implies that the flux of large impactors (and mantle plumes) has varied episodically through time.

not only of Earth and life (Abramov and Mozjsis, 2009), but also for other planets in the solar system, and it is receiving renewed attention (Fassett and Minton, 2013).

Further statistical comparisons between the impact fluxes for Earth and the Moon have also been made (Bland, 2005; Bottke et al., 2010). To compare the impact flux on the Moon with that on Earth requires an adjustment for size (cross-sectional area) and gravity, with the impact flux on Earth being correspondingly higher than that on the Moon. The new Late Heavy Bombardment impact flux (Bottke et al., 2012) curve may not have been corrected for surface area of Earth's oceans, since only continental impacts ($n = 2$ at 2 Ga) were calibrated, and it might be an underestimate for the background flux, perhaps by a factor of ~2. Figure 2, illustrating the Late Heavy Bombardment model flux curve (Bottke et al., 2012), shows that basin-scale impact craters would not be expected to persist into the Phanerozoic (ca. <0.5 Ga), but by the same token, Chicxulub-class impact craters are also not predicted, yet evidently at least one did occur. Unfortunately, for our purposes, the statistical uncertainty is highly skewed and is particularly poor for the low numbers of large-sized bolides, whereas the flux for more numerous small objects is much better constrained (Bland, 2005).

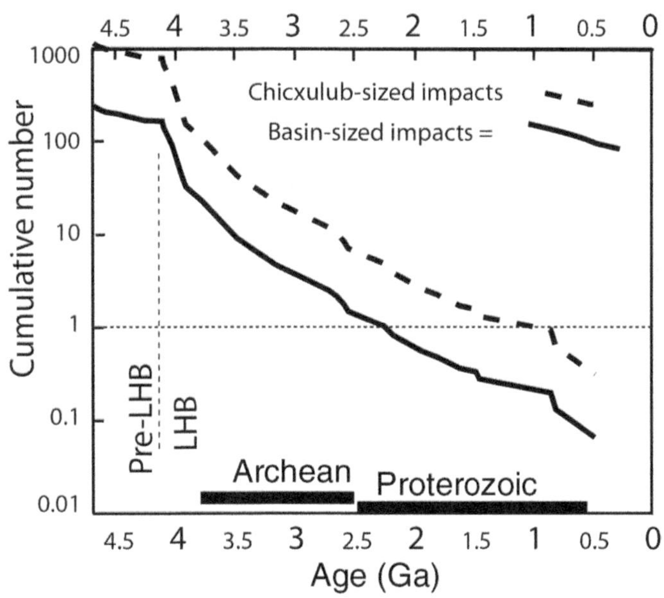

Figure 2. A revised Late Heavy Bombardment (LHB) impact flux model at Earth (Bottke et al., 2012) predicts a generally smooth decline after the hiatus at 4 Ga, and it is pegged against the known crater database at 2 Ga, where Sudbury and Vredefort provide $n = 2$ Chicxulub-class impact craters. Solid curve is for the largest, basin-scale impacts ($n > 1$ up to ca. 2.5 Ga); dashed curve is for Chicxulub-class events. Comment: This flux model may not adequately account for land surface by area, since oceanic impacts are not accounted for, in which case the impact curve is a minimum flux, too low by a factor of ~2. A similar correction of predicted low impact flux, also by a factor of ~2, has been noted on the Moon, for the "young " period of <1 Ga pegged by 2 "known" large impacts at 0.8 and 0.1 Ga (Kirchoff et al., 2013).

Comets versus Asteroids?

Asteroids are fairly well understood, both in terms of compositions, and their periodic delivery into the Earth system, mostly from the main asteroid belt (Morbidelli et al., 2001). Twenty years ago, the impact of asteroids was thought to dominate production of craters smaller than 30 km, whereas comet impacts were considered to probably form most craters >50 km (Shoemaker et al., 1990). The impact velocity of comets (~50 km^{-1}) versus that for asteroids (~18 km^{-1}), and the comet impactor flux for Earth as reviewed a few years ago (Weissman, 2007) highlighted the poor knowledge of the size distribution of their nuclei. Napier et al. (2004) calculated that comets may represent a much greater impact hazard than asteroids.

Giant comets in our solar system have been conjectured to lead to extended bombardment episodes (Napier, 2010). Giant comets (100–200 km across) might be disturbed from the Oort cloud when the solar system penetrates spiral arms of the galaxy. Two of the most damaging mass extinctions may have coincided with Earth's passage through spiral arms of the galaxy: the end-Permian to the Scutum arm (ca. 245 Ma), and the end-Cretaceous to the Sagittarius arm, 70–60 Ma (Wickramasinghe and Napier, 2008). It was proposed that following arrival of a supercomet in the solar system, dust to kilometer-scale fragments might have generated a particularly hazardous Earth environment for tens of thousands of years (Napier, 2010). Such a cosmic explanation for impact flux perturbations could be much more complicated than a single huge impact, with effects beyond the scope of this contribution, except as a potential provider of large bolides. In a similar vein, an interpreted ~32 m.y. periodicity in the flood basalt volcanism record has been postulated to have been caused by showers of impacting comets, which include those related to mass extinctions (Rampino and Stothers, 1984).

Flux of Large Impactors

Despite the uncertainty in bolide flux identified in the preceding sections, we can fall back on a pragmatic review of background impact flux in terms of binned crater diameters over geologically meaningful periods of time. Quantitative assessments of periodicity in the impact flux, whether or not taken as evidence for cometary bombardment episodes over the last 250 m.y., have also been made (Napier, 2006). The impact flux can be used as a first-order statistical assessment of the availability of classes of bolides and their maximum deliverable impact energies at important periods in the stratigraphic record, corrected for surface area of Earth (Dence, 2006; French, 1998). Throughout the Phanerozoic history of Earth, a large Chicxulub-sized, 200 km crater may have occurred every ~150 m.y., and a 500 km crater may have occurred every ~450 m.y. (Koulouris et al., 1999; Shoemaker et al., 1990). A review of the impact flux of comets and asteroids considered them to be similar (Grady et al., 1998) and estimated numbers of ~200 km class craters since 3 Ga to be ~25 and predicted 1–5 craters in the ~500 km class. These craters have not yet been identified.

Predicting craters from impact flux is only a guide, but why might they be "missing"? Large impact craters may not have been recognized, perhaps because thermal effects and melting overwhelmed the expected shock characteristics of the larger crater signatures (Jones et al., 2002), or they simply may not have occurred. Whether or not the revised Late Heavy Bombardment flux perturbed or in any way influenced the current day flux is an unknown question. Another assessment of Earth impact flux suggested that the numbers of known comets are rare (~1%) compared to asteroids, except in the large size range, where comet impact fluxes may be higher and capable of causing extinction events (Yeomans and Chamberlin, 2012). However, the flux of large comets (~1–10 km) was calculated to be large and proposed to have been radically underestimated from telescope observations due to their anomalously dark albedo (Napier et al., 2004).

LARGE-SCALE IMPACT MELTING AND LARGE IGNEOUS PROVINCES?

This section reviews the scope for large-scale melting, as a precursor to volcanism, from large bolide impacts in the context of a Phanerozoic Earth with life and plate tectonics.

Decompression Melting

Several processes might lead to decompression melting of the mantle beneath a basin-scale impact crater (Fig. 3): (1) Impact crater excavation reduces lithostatic load, possibly leading to rapid decompression melting in situ (Jones et al., 2002), (2) central uplift decompresses the mantle dynamically, possibly leading to rapid melting in a cylindrical conduit (Elkins Tanton et al., 2004), and (3) isostatic recovery causing upwarped isotherms across the lithosphere-asthenosphere boundary could induce mantle convection, leading to slow (> tens of Ma) decompression melting (Elkins Tanton et al., 2004). Very large volumes of mafic melts can be generated by sufficient decompression melting, assuming they migrate to the surface to match the required 10^6 km^3 melt volumes of a large igneous province (Jones et al., 2002). This is a much more likely process in oceanic lithosphere (Jones, 2005) due to the thin nature of oceanic crust (~10 km), and would be positively enhanced by high pre-impact geothermal gradients, such as proximity to a mid-oceanic ridge, proximity to young (~<10 Ma) oceanic crust, or coincidence or coupling with a preexisting mantle plume (Abbott and Isley, 2002). The same process, albeit in simpler form, was previously used as an explanation for impact formation of oceanic plateaus (Rogers, 1982).

Ivanov and Melosh (2003) recognized that at some size, their models for large impact craters could raise hot mantle rocks close to the surface, meaning this would initiate mantle melting by decompression, and they placed the lower size limit at a crater diameter of 400–500 km, considering this would be statistically a very low probability event after the Late Heavy Bombardment (see Fig. 1). However, the thermal state and melting characteristics of the target should have important effects on the volume of melt produced. In situ decompression melting in mafic mantle peridotite beneath a large impact crater could contribute large volumes of melt compared with crustal shock melting. Effective crater sizes on Earth should therefore be reduced due to additional decompression melting (Jones et al., 2002), and the threshold size to achieve $>10^6$ km^3 melt is estimated to require crater diameters in excess of ~200 km (Fig. 4). This is consistent with a lack of evidence for eruptive volcanism at the ~200 km Chicxulub crater in continental crust (Schulte et al., 2010). Additionally, energy-based arguments are persuasive in indicating that large-scale sustained eruptive volcanism from impact craters less than ~200 km is unlikely (Melosh and Ivanov, 2004). However, for larger, basin-scale impact craters, as on the Moon, significant decompression melting might be expected, provided that the Moon was impacted while the mantle was thermally active. Volumes and time scales for basin-scale impact-triggered adiabatic melting have been modeled (Elkins Tanton et al., 2004) to explain mare basaltic volcanism infilling lunar basins, and results suggest this may have continued for up to ~350 m.y. A similar process could have operated on Earth, extending the potential time frame between impact and melting. However, due to differences in gravity, geochemistry, and thermal states (and other parameters) between Earth and the Moon, direct comparison awaits the appropriate investigations.

IMPACT VOLCANISM

It is often implied that melting associated with meteorite impacts drives surface volcanism (Price, 2001; Rampino, 1987), but additional geophysical and geochemical conditions need to be satisfied relating to eruptibility, such as melt connectivity, volatile contents, and density of melts. Some assessment of the potential for a spectrum of eruptive volcanic processes, driven by terrestrial impact cratering (mostly <200 km diameter) by comparison with modern volcanic systems, has been debated (Glikson, 2004, 2005; Melosh and Ivanov, 2004). Volcanism characteristic of large igneous provinces entails eruption of high-volume lava flows, to generate the layer-cake topography of superimposed subhorizontal units so typical of "flood basalts." What kinds of volcanic events could be triggered by large impacts, and might any such impact volcanism have the potential to cause mass extinction, or to be preserved in the geological record? This section is exploratory and attempts to compare largely unknown impact volcanism with selected aspects of conventional volcanism.

In general, there are two classes of impact volcanism derived from impact-induced melting related to proximity to the impact point, or ground zero, which might be considered. Both would be impact-triggered, and here I use the word "melt" to distinguish them from nonimpact endogenic "magma," although impact melts might be expected to range from superheated liquids to conventional variable-degree partial melts (Jones et al., 2002).

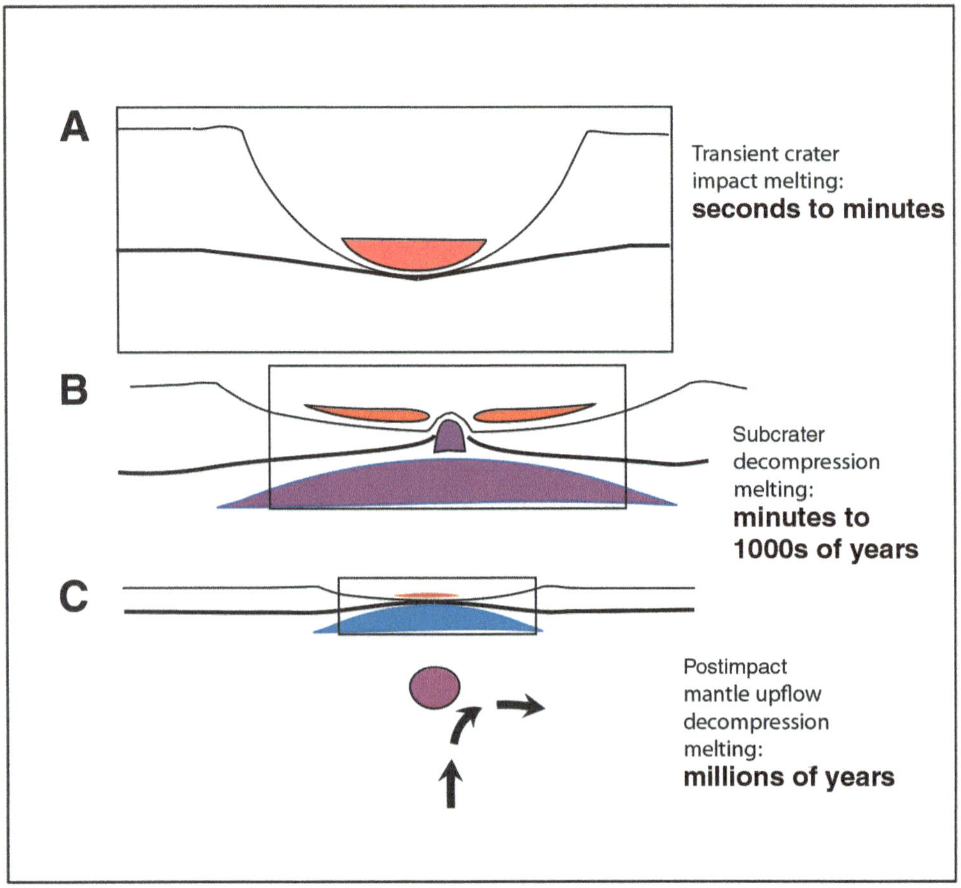

Figure 3. Schematic representation of decompression melting for different times (approximate ranges) after formation of a basin-scale impact crater (>200 km), from (A) transient crater impact melting (shock, rapid), (B) subcrater decompression melting (high volume, fast), including central uplift component, and (C) mantle convection–induced decompression melting (Elkins Tanton et al., 2004; Jones et al., 2003).

(1) Proximal melts could include shock-generated impact melt (Glass and Simonson, 2013), subcrater decompression melting, surface impact volcanism (Jones et al., 2002), and aftermath hydrovolcanism, as considered next. (2) Distal melts could include far-field volcanism triggered by secondary seismicity or faulting, tsunami-induced sector collapse of oceanic islands, or antipodal focusing on the opposite side of the globe, possibly related to oceanic large body impacts, volcanism, and megatsunamis (Hagstrum, 2005).

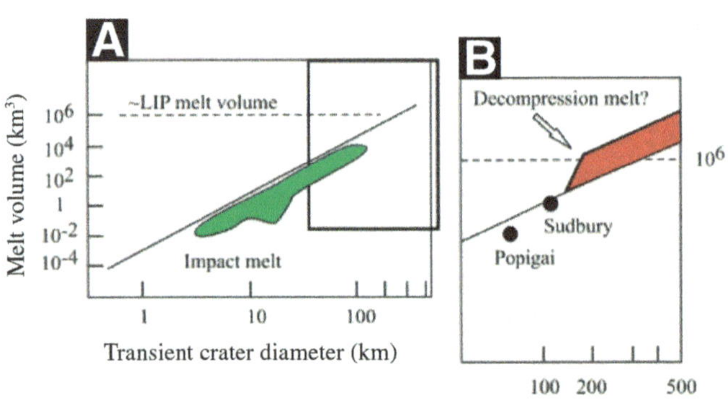

Figure 4. Volume of impact melts correlated with crater diameter for terrestrial impact craters (A) and the potential for large igneous provinces–scale melting (10^6 km^3) and (B) with additional contribution from impact-induced subcrater decompression melting (Jones et al., 2002, 2003).

Impact Melt Bodies, Magma, and Eruptions

Small impact melts are often characterized by fast cooling or quenching, producing a variety of distinctive textures often comprising elongated crystals and glass perhaps with immiscible features (French, 1998). Once formed, the cooling paths of all impact melts are subject to kinetics and thermodynamics (DeCarli et al., 2002), which for larger melt accumulations could lead them to evolve as magmas and to mimic igneous rocks. If large (greater than kilometer scale) impact melt bodies cool slowly due to thermal mass and latent heat released by crystallization, they may evolve to share characteristics of magma chambers in igneous petrology, although the initial impact conditions of superheating should be distinctive. An illustration of the likely cooling pathways and processes for superheated impact melt bodies is given in the petrologic model for the ~200 km impact crater at Sudbury by Keays and Lightfoot (2004). At Sudbury, the ~10^4 km^3 of accumulated within-crater silicate impact melt solidified slowly over >10^5 yr (Ivanov et al., 1997), resulting in crystalline rocks classified with conventional lithological names like norites or gabbro. In assessing the potential for fractional crystallization and gravity settling to segregate layers of crystals (cumulates), the importance of a melting to displacement ratio (m/d) and the cooling effect of unmelted clasts distributed through impact melts in smaller bodies are important (Warren et al., 1996). Sudbury was considered in an igneous petrology textbook as an example of a large layered intrusion (Best, 2003). There is no record of volcanism at Sudbury, but it may have been spectacular, although a 1400-m-thick sequence of impact breccias with evidence for extensive and prolonged hydrothermal activity was for some time considered to contain volcanic ash flows (Bohor and Betterton, 1992). Nearly 650 km to the west of Sudbury in the Gunflint Formation, mixed ash, lapilli, and glass ejecta are found in sedimentary beds ~7.6 m thick, considered to have been either "base surge" (a volcanological term) or impact tsunami deposits (Addison et al., 2010).

Shock melts formed within impact craters might be expected to retain very low contents of dissolved water due to very high temperatures, in which case their magmatic potential for violent eruption would be low. However, quite variable water contents are observed; for example, concentrations from 0.7 to 2.3 wt% H_2O were measured in fresh glassy impact melt rocks in the ~100 km Popigai crater (Vishnevsky and Montanari, 1999), and 1.3–3.8 wt% H_2O was measured in impact glass from the ~25 km Ries crater (Vennemann et al., 2001). Such water contents in silicate melts overlap with common igneous magmas like basalts, and target rock compositions likely exert a control on the water content inherited by the impact melt (Osinski et al., 2008), explaining much of the variability. In volcanic systems, the water content of magma exercises a strong control on eruption style and explosivity (Francis and Oppenheimer, 2003). By comparison, within-crater impact melts are variable but unremarkable, and few, so far identified, would resemble volatile-saturated magmas typical of explosive andesitic volcanism, which contains from ~2 to 10 wt% H_2O (Moore and Carmichael, 1998). Although water is the most common volatile component in volcanic magmas, melting within impact craters formed in unusual crustal target-rock lithologies may derive high contents of other volatiles, such as at Chicxulub, where high CO_2 and SO_2 were present in sedimentary target rocks and contributed to the impact melt (Jones et al., 2000; Osinski et al., 2008).

If there is little potential for driving volcanism from melt compositions within an impact crater, there may still be violent interaction between hot materials and water. The maximum length of time separating primary impact crater formation from secondary volcano-thermal events might be constrained from models of hydrothermal activity within large impact craters. For a Chicxulub-scale 200-km-class crater, a hydrothermal system is estimated to have operated for 1–3 m.y. after the initial impact (Osinski et al., 2013). Interaction of hot subsurface magma with subsurface water is a well-known process for driving modern volcanic eruptions and could generate a range of conventional volcanism from passive geysers to violent phreatomagmatic eruptions. Comparable eruptive mechanisms might have been triggered by impact melt bodies in the past and may not have been restricted simply to the largest impacts. Diverse magma compositions are capable of driving surface volcanic eruptions during interaction with water, whether from ocean water, freshwater lakes, melted ice, or subsurface aquifers. The distal reach of ash for some prehistoric volcanic eruptions includes 100–200 km for Lake Taupo in New Zealand, 150 km for ash and tsunami beds from the Santorini Minoan eruption, and historically >550 km in the South China Sea for tephra from the 1991 Pinatubo eruption (Fero et al., 2009). Phreatic to phreatoplinian eruptions are well documented in the Icelandic volcanic record, with frequent reach extending to Greenland and central Europe (2000 km), driven by interactions not only with liquid water but with ice, and in many Icelandic water-driven eruptions, the estimated magma volumes were very modest (Thordarson and Hoskuldsson, 2008).

Longevity of Impact Crater Thermal Anomalies

The potential scale for long-term impact disruption within the thermally active crust on Earth is only just beginning to be quantified and has potential both for destructive processes and for positive environments for life (Cockell, 2006). A large impact crater of ~200 km diameter, like Sudbury, Vredefort, or Chicxulub, with transcrustal faults, would be expected to fundamentally disturb regional geotherms and to have influenced the underlying mantle. At Vredefort, the eroded impact crater, possibly >250 km in diameter originally, still preserves substantial central uplift of ~12 km in the upper crust, and ~4 km uplift at the Moho or mantle (Henkel and Reimold, 2002); this brought lower-crustal temperatures near to the crater floor. Modeling (Henkel and Reimold, 2002) suggests that near-surface temperatures over hundreds of square kilometers within the Vredefort crater were elevated to in excess of 300 °C. If the model of Henkel and Reimold is correct, this would have transposed preexisting hot granulites to a concentric zone some ~50 km in diameter, independent of any additional impact heating. Furthermore, the central >20 km core of the central uplift would have potentially been close to adiabatic/isothermal melting (>900 °C from 0 to 30 km depth).

For impacts in Earth's continental crust in general, felsic crust would be expected to melt at relatively low temperatures of <800 °C, due to clustering around the "granite minimum" composition. Mobility of felsic melts is dependent on their viscosity, which in turn is strongly dependent on water content ("the water problem"), as discussed in a recent review (Aranovich et al., 2013), so that outcomes are difficult to predict. The middle and lower continental crust is typically more mafic, requiring higher temperatures to melt, with less dependence on water contents, but studies of the onset of melting ("anatexis") have focused on slow equilibrium metamorphism (Thompson, 2012) as opposed to rapid disequilibrium melting. This suggests that there may be a future need to consider a fundamentally different set of physical and mechanical processes for impact cratering into Earth's crust, where thermal gradients (geotherm) contrast with impact cratering into thermally dead or inactive planetary bodies in the solar system. This concept follows from the idea of decompression melting for hot oceanic mantle (Jones et al., 2002) to continental crust and revisits the concept of a critical catastrophe threshold to the outcome of impacts scaled to crater dimensions.

Unlike the predictable layered compositions of global oceanic crust, continental crust is much more complex, being compositionally heterogeneous and overall less mafic in composition. There is a strong compositional control on the likelihood of melting during quasi-isothermal decompression of Earth's crust. Simple phase diagrams familiar to students could be used in conjunction with the geothermal and isothermal contours of crater models (Henkel and Reimold, 2002) to predict dynamic melting regions. Erosion of perhaps 6 km at Vredefort has removed the original crater structure and evidence for crater impact melts, although impact melting contributed to subcrater pseudotachylyte dikes both at Vredefort and Sudbury (Lieger et al., 2009), possibly by drainage of superheated crater impact melt into tension fractures of up to 1 km width (Riller et al., 2010).

SECONDARY IMPACT CRISES?

The generation of melts at different depths within the crust, and underlying mantle, would have the potential for different classes of impact volcanism, but it is probably restricted to the largest known impacts. At the top end of the scale, the largest basin or maria-scale impacts have been considered sufficient to trigger large igneous provinces. One contender is the oceanic Ontong Java Plateau (Ingle and Coffin, 2004; Jones, 2005). Of course, this would require the availability of a suitable-sized bolide. It is true that unequivocal evidence for impact-triggered large igneous provinces has not yet been shown, but it is clear where to look. To test the hypothesis in the future, it would be important to search at the very base of any large igneous province, to identify any shock features when it started, and where there should be evidence for impact. This may be difficult if the critical zone lies buried beneath a thick accumulation of volcanic crust, as in the case of the Ontong Java Plateau. Impact evidence may be preserved far away from the impact site or ground zero, although global impact layers in the geological record are progressively more difficult to reconstruct as a function of age, but the end-Cretaceous-Paleogene layer acts as a fine example (Schulte et al., 2010).

A range of additional processes might also be considered as triggers for secondary volcano-thermal crises during or shortly after crater formation. Potentially significant factors for mass extinctions, but not treated here, are a number of significant possibly cascading secondary hazards like landslides, tsunamis, and slumping of continental shelf sediments, which could connect a wide variety of ground shock, earthquakes, and crustal-scale faulting. It has also been argued that paleomagnetic plate vectors act as passive recorders of lithospheric accelerations outside the bounds of mantle convection, the high coincidence of which with mass extinction events and major stratigraphic boundaries has been interpreted as due to major meteorite impacts (Price, 2001).

Antipodal Focusing

The process of antipodal focusing may have operated in the early Earth (Glikson, 2005), but at least in the Phanerozoic, it is unlikely for several reasons. At planetary scale, impacts even greater than "basin-scale," >1000 km events may be required to cause antipodal cratering and partial removal of crust (Marinova et al., 2011); this is unlikely during the Phanerozoic period in Earth history according to impact flux curves. Antipodal focusing of seismic waves due to a smaller than basin scale, ~200 km crater (Chicxulub-sized) oceanic impact has also been modeled in numerical simulations (Meschede et al., 2011), reinstating credibility, now better constrained, in the limited plausibility of this mechanism both for triggering antipodal volcanism and tsunamis. Although the impact energy for antipodal melting for the class of large impact craters is orders of magnitude insufficient (Ivanov and Melosh, 2003), the whole anomaly (antipodal volcanism) does not have to be created completely by the impact, and peak stresses at the surface are surprisingly close to those required for fracturing rock (Meschede et al., 2011). In this limited case scenario, this might perhaps enable volcanism that was about to occur anyway. However, in the specific case of the Chicxulub crater, the often-assumed antipodal distribution of the Deccan volcanism, even accounting for tectonic reconstruction, does not appear to be in the right place (Meschede et al., 2011).

IMPACTS WITHOUT MASS EXTINCTIONS

Not all large terrestrial impact craters are associated with mass extinctions. The anomalously thick oceanic crust of the Ontong Java Plateau has been the subject of many investigations (Tejada et al., 2004), and even if it were to be related to a large impact (Ingle and Coffin, 2004; Jones et al., 2005b; Rogers, 1982) and possibly associated with faunal extinctions and anoxic crisis, it is not dated to one of the major mass extinction events of the Phanerozoic. Due to the large volume of lava, geochemical signatures to discriminate between impact and plume origins are problematic, as an impact-triggered mantle plume (Elkins Tanton and Hager, 2005) cannot be excluded, so the mantle sources (oceanic-island basalt type) could be the same in both models. The answer will lie in the geology at the base of the Ontong Java Plateau lavas. Curiously, although this does not yet decide the issue of impact volcanism, it might support arguments against massive volcanism as a cause of mass extinctions. Perhaps in the case of the Ontong Java Plateau, serious modifiers to atmosphere degassing by active volcanism were imposed by the oceanic setting and the presumable dominance of deep submarine volcanism (Tejada et al., 2004).

The end-Eocene–Oligocene stratigraphic boundary has been interpreted as marking a climatic transition from greenhouse to icehouse conditions, and there were faunal extinctions (Pearson et al., 2007). Some have attributed the cause of progressive global cooling from the middle Eocene to plate tectonics and ocean circulation. There is, however, evidence of several bolide impact events close to this time, including a major tektite strewn field, at least five impact craters, including Chesapeake (85 km) and Popigai (100 km), and a whole host of mineralogical and geochemical impact indicators (Koeberl, 2009), including unambiguous Cr isotope evidence for extraterrestrial materials in impact spherules (Kyte et al., 2011). The end-Eocene fossil record does not appear to show a major mass extinction, but there may have been episodic "stepwise extinctions" (Keller, 1986). The effects of impact events in relation to rapid cooling of climate, with global temperature perturbations, and the onset of the Antarctic ice sheet are important to understand (Coccioni et al., 2009). The multiple Eocene bolide impacts may have played an important role in the deterioration of global climate. Recent refinements to age dates for some of these near-end-Eocene impact events still might change things, since new analyses (Fernandes et al., 2012) for the North American tektite swarm associated with the Chesapeake Bay impact give ages (33.14–34.76 Ma) slightly younger than the previously reported 35.3 Ma age, overlapping with the stratigraphic boundary age of between 33.5 and 34 Ma (Pearson et al., 2007).

A tentative conclusion from the Eocene impact record might be that solitary large impacts separated by long periods of time (perhaps like the Chicxulub event) may trigger larger lithospheric responses compared with the delivery of multiple impacts of equivalent energy spread relatively frequently across a longer period of geological time. This could be related to impact energy and size, wherein the total energy represented by the impact flux is dominated by scarce but more massive objects, and impacts much below 200 km in diameter lack kill potential (Poag, 1997; Raup, 1991).

CONCLUSIONS

The potential for impact volcanism to cause mass extinction remains a theoretical but distinct possibility, due primarily to the rapid time scales of energy transfer, global reach, and the arsenal of potential kill mechanisms. We have seen that the generation of decompression melting underneath large impact craters, both in situ, and by subsequent mantle convection, could feed and prolong the volcanic signal, and may have occurred on other planetary bodies like the Moon (Elkins Tanton et al., 2004; Vaughan et al., 2013). Although there is continued debate about impact fluxes, which may lead to future revisions (Abbott and Isley, 2002; Bottke et al., 2012; Johnson and Melosh, 2012), there is little if any effect on estimates for the current-day background flux. We might predict (French, 1998) that the background impact flux could have provided two or three sufficiently large objects to cause mass extinctions in the last 300 m.y. or more (Glikson, 2005), so this remains a distinct possibility.

In terms of specific mass extinction events, the clearest potentially causal impact signal yet discovered occurs in association with the Chicxulub crater (Schulte et al., 2010) and occurs as a recognizable global impact layer, so let us consider what we now know. This impact did not trigger volcanism in situ, although a coherent subsurface melt sheet similar to Sudbury has been identified (Barton et al., 2010). This suggests that the required threshold crater size for decompression melting was not reached, although this would perhaps have been more likely if it had occurred in a young oceanic target (Jones et al., 2005b). There may have been local to regional processes related to hydrothermal interaction, although there is no evidence for energetic secondary phreatic eruptions, and postimpact crater conditions may actually have promoted life (Osinski et al., 2013).

Could the Chicxulub impact have triggered Deccan volcanism through antipodal focusing? In the specific case of the end-Cretaceous Chicxulub impact, and although antipodal focusing remains a potential mechanism for triggering, the Deccan was seemingly not in the right place to be the antipode (Meschede et al., 2011). Since the main phase of Deccan volcanism was already under way in chron 29r and ended at the Cretaceous-Paleogene boundary, and since extinctions accelerated during the major eruptions (Keller et al., 2009, 2012), theoretically the Chicxulub impact may have caused a stronger pulse of volcanic activity in the Deccan, although such a trigger has not been identified. If such a connection could be confirmed, this would fit the concept of a secondary distal crisis related to impact, but it would not satisfy a definition of impact volcanism driven by (antipodal) impact melting.

Where does that leave the impact at Chicxulub in the context of the associated mass extinction event? If the kill mechanisms at the end of the Cretaceous match with those expected for bolide impacts, key factors are likely to hinge on the potential for atmospheric loading of deleterious particulate and gaseous materials due to the extraordinarily rapid transfer of energy from an impact. What about multiple impacts? From the perspective of this contribution, and although there is some evidence for multiple impacts at this time, such as Boltysh crater (Kelley and Gurov, 2002), a second event large enough to trigger a mass extinction in its own right has not been identified. Perhaps this is a dating issue, and we may be too concerned with the specific timing of individual impact events, yet we know surprisingly little about the broader aspects of Earth's encounter with, for example, long-term atmospheric loading from comets (e.g., Napier, 2006). Are any other large impact craters closely associated in time with the end-Cretaceous mass extinction? If we widen the time frame by just a few million years, several other contenders might deserve reconsideration, including the early Paleogene impact in west Greenland, where spherule beds occur very close to the base of a massive large igneous province volcanic formation (Jones et al., 2005a). The Eocene impact craters without mass extinctions tell us that 100 km is simply not big enough, and, at least in continental crust, a global killer crater needs to be at least ~200 km diameter.

Impacts clearly occur, some have global geological footprints, but the links with mass extinctions are complex and still need work. The preoccupation with coincidence dating is perceived as a major challenge—and indeed it is, because uncertainty of ages obscures confirmation of timing between geological events, despite ever-increasing levels of precision. Perhaps a bigger challenge is to understand the other impact processes we have yet to even consider, which might trigger environmental effects like global climate change leading to mass extinctions.

From a planetary perspective, the matching of bombardment episodes triggered by passages of the Sun through spiral arms of the galaxy was proposed as a mechanism for prolonged and complex mass extinctions (Napier and Clube, 1979), with some support (Leitch and Vasisht, 1998) showing, as predicted, the Sun was passing through a spiral arm of the galaxy during both the Permian-Triassic and Cretaceous-Tertiary extinctions around 250 and 65 m.y. ago, which also coincided with the Siberian Traps and Deccan Traps volcanism. The Permian-Triassic mass extinction was complex, with evidence for wildfires (Shen et al., 2011), and fewer well-preserved global geological signals to interrogate, although the Siberian Traps is the largest large igneous province by volume, and there is appeal to unusual hydrocarbon-rich volcanic degassing from the Tunguska Basin (Svensen et al., 2009). An important discovery has been that surprisingly diverse plant fossils are preserved in sediments within the Siberian Traps (Visscher et al., 2010), which could imply relatively benign environments near to active volcanism.

Whether or not a cometary bombardment (Napier et al., 2004) was involved in either of the two major mass extinction events (end-Permian, end-Cretaceous) is unclear. Perhaps the main point for this review is that impact volcanism, or more specifically impact-melt–driven volcanism, is not known to have been a factor in either the end-Permian or end-Cretaceous mass extinction. Conventional massive volcanism, we know, clearly occurred, yet both volcanic provinces (Siberian Traps, Deccan) preserve fossil evidence for diverse habitable environments, suggesting to some (Schulte et al., 2010) that large igneous province volcanism itself was not a mass killer. A large impact like the Chicxulub crater, and its associated enormous energy release, is thought by many to be the most significant disruption to the surface environment of Earth at the end of the Cretaceous (Schulte et al., 2010). However, the work by Schulte et al. (2010) was criticized for not fully assessing the potential role of volcanism in mass extinctions (Courtillot and Fluteau, 2010), so the debate continues.

Even in the "optimum" case where impact volcanism might have operated, such as the oceanic Ontong Java Plateau volcanism, there is global oceanic anoxia but no mass extinction (Tejada et al., 2009). Yet massive large igneous province volcanism is clearly associated in time during episodes of several mass extinction events, and ever-more-complex explanations have been proposed (Saunders and Reichow, 2009). Perhaps specific intense phases of volcanism will be identified, but that diminishes the arguments for greenhouse gases released when scaled to total large igneous province volume and magma flux. In this case, there might be a correlation between the rate of volcanic activity and the Cretaceous-Tertiary mass extinction. It has already been reported that one of three phases of the Deccan volcanism was exceptional (Chenet et al., 2009), with 80% of the eruptive volume immediately before the Cretaceous-Tertiary boundary, which has been identified as the most likely process responsible for the mass extinction (Keller et al., 2012). One might speculate that this individual volcanic episode was triggered by the Chicxulub impact, and both events contributed to the mass extinction.

It is probably too early to write off the role of impacts in other mass extinction events too, since we are still improving age correlations, learning about the geochemistry of comets, and learning about the secondary catastrophes that large impacts may precipitate.

ACKNOWLEDGMENTS

I would like to acknowledge the late Paul DeCarli for numerous discussions related to impact cratering, shock metamorphism, and melting. Norman MacLeod, Andrew Kerr, and Mike Widdowson (organizers of the meeting "Volcanism, impacts, and mass extinctions: causes and effects," at the Natural History Museum in London, 27–29 March, 2013) are thanked for providing the opportunity to discuss multidisciplinary aspects of mass extinctions. Lindy Elkins Tanton and two anonymous reviewers are thanked for their detailed and perceptive comments on the manuscript.

REFERENCES CITED

Abbott, D.H., and Isley, I.S., 2002, Extraterrestrial influences on mantle plume activity: Earth and Planetary Science Letters, v. 205, p. 53–62, doi:10.1016/S0012-821X(02)01013-0.

Abramov, O., and Mozjsis, S.J., 2009, Microbial habitability of the Hadean Earth during the Late Heavy Bombardment: Nature, v. 459, p. 419–422, doi:10.1038/nature08015.

Addison, W.D., Brumpton, G.R., Davis, D.W., Fralick, P.W., and Kissin, S.A., 2010, Debrisites from the Sudbury impact event in Ontario, north of Lake Superior, and a new age constraint: Are they base-surge deposits or tsunami deposits?, *in* Gibson, R.L., and Reimold, W.U., eds., Large Meteorite Impacts and Planetary Evolution IV: Geological Society of America Special Paper 465, p. 245–268, doi:10.1130/2010.2465(16).

Alvarez, W., 2003, Comparing evidence relevant to impact and flood basalt at times of major mass extinction: Astrobiology, v. 3, p. 153–161, doi:10.1089/153110703321632480.

Alvarez, W., Kauffman, E.G., Surlyk, F., Alvarez, L.W., Asaro, F., and Michel, H.V., 1984, Impact theory of mass extinctions and the invertebrate fossil record: Science, v. 223, p. 1135–1141, doi:10.1126/science.223.4641.1135.

Aranovich, L.Y., Newton, R.C., and Manning, C.E., 2013, Brine-assisted anatexis: Experimental melting in the system haplogranite-H_2O-NaCl-KCl at deep crustal conditions: Earth and Planetary Science Letters, v. 374, p. 111–120, doi:10.1016/j.epsl.2013.05.027.

Barton, P.J., Grieve, R.A.F., Morgan, J.V., Vermeesch, P.M., Christeson, G.L., Gulick, S.P.S., and Warner, M.R., 2010, Seismic images of Chicxulub impact melt sheet and comparison with the Sudbury impact structure, *in* Gibson, R.L., and Reimold, W.U., eds., Large Meteorite Impacts and Planetary Evolution IV: Geological Society of America Special Paper 465, p. 103–113, doi:10.1130/2010.2465(07).

Best, M.G., 2003, Igneous and Metamorphic Petrology: Wiley Blackwell, 752 p.

Bland, P.A., 2005, The impact rate of Earth: Philosophical Transactions of the Royal Society, v. 363, p. 2793–2810, doi:10.1098/rsta.2005.1674.

Bohor, B.F., and Betterton, W.J., 1992, Shocked zircons in the Onaping Formation: Further proof of impact origin, *in* Lunar and Planetary Science Conference, Volume XXIII: Houston, Texas, Lunar and Planetary Institute, p. 135–136.

Bottke, W.E., Walker, R.J., Day, J.M.D., Nesvorny, D., and Elkins-Tanton, L., 2010, Stochastic late accretion to Earth, the Moon, and Mars: Science, v. 330, p. 1527–1530, doi:10.1126/science.1196874.

Bottke, W.F., Vokrouhlicky, D., Minton, D., Nesvorny, D., Morbidelli, A., Brasser, R., Simonson, B., and Levinson, H.F., 2012, An Archaean heavy bombardment from a destabilized extension of the asteroid belt: Nature, v. 485, p. 78–81, doi:10.1038/nature10967.

Chenet, A.-L., Courtillot, V., Fluteau, F., Gerard, M., Quidelleur, X., Khadri, S., Subbarga, S.K., and Thordarson, T., 2009, Determination of rapid Deccan eruptions across the Cretaceous-Tertiary boundary using paleomagnetic secular variation: 2. Constraints from analysis of eight new sections and synthesis for a 3500-m-thick composite section: Journal of Geophysical Research, v. 114, p. B06103, doi:10.1029/2008JB005644.

Coccioni, R., Frontalini, F., and Spezzaferri, S., 2009, Late Eocene impact-induced climate and hydrological changes: Evidence from the Massignano global stratotype section and point (central Italy), *in* Koeberl, C., and Montanari, A., eds., The Late Eocene Earth—Hothouse, Icehouse and Impacts: Geological Society of America Special Paper 452, p. 97–118.

Cockell, C.S., 2006, The origin and emergence of life under impact bombardment: Philosophical Transactions of the Royal Society of London, ser. B, Biological Sciences, v. 361, p. 1845–1856, doi:10.1098/rstb.2006.1908.

Coffin, M.F., and Eldholm, O., 1992, Volcanism and continental break-up: A global compilation of large igneous provinces, *in* Storey, B.C., Alabaster, T., and Pankhurst, R.J., eds., Magmatism and the Causes of Continental Break-Up: Geological Society of London Special Publication 68, p. 17–30.

Courtillot, V., and Fluteau, F., 2010, Cretaceous extinctions: The volcanic hypothesis: Science, v. 328, p. 973–974, doi:10.1126/science.328.5981.973-b.

DeCarli, P.S., Bowden, E., Sharp, T.G., Jones, A.P., and Price, G.D., 2002, Evidence for kinetic effects on shock wave propagation in tectosilicates: American Institute of Physics Conference Proceedings, v. 620, p. 1381–1384, doi:10.1063/1.1483796.

Dence, M.R., 2006, Estimates of impact flux from terrestrial crater counts: The role of gravity and target properties, *in* 40th European Space Laboratory Conference Proceedings: First International Conference on Impact Cratering in the Solar System: Nordwijk, Netherlands, European Space Agency, p. 291326.

Elkins Tanton, L.T., and Hager, B.H., 2005, Giant meteoroid impacts can cause volcanism: Earth and Planetary Science Letters, v. 239, p. 219–232, doi:10.1016/j.epsl.2005.07.029.

Elkins Tanton, L.T., Hager, B.H., and Grove, T.L., 2004, Magmatic effects of the late lunar heavy bombardment: Earth and Planetary Science Letters, v. 222, p. 17–27, doi:10.1016/j.epsl.2004.02.017.

Fassett, C.I., and Minton, D.A., 2013, Impact bombardment of the terrestrial planets and the early history of the solar system: Nature Geoscience, v. 6, p. 520–524, doi:10.1038/ngeo1841.

Fernandes, V.A., Hopp, J., Schwarz, W., Trieloff, M., and Reimold, W.U., 2012, Re-evaluation of the Chesapeake Bay crater impact age: New ^{40}Ar-^{39}Ar step-heating results of North American tektites, *in* 43rd Lunar and Planetary Science Conference: Woodlands, Texas, Lunar and Planetary Institute.

Fero, J., Carey, S.N., and Merill, J.T., 2009, Simulating the dispersal of tephra from the 1991 Pinatubo eruption: Implications for the formation of widespread ash layers: Journal of Volcanology and Geothermal Research, v. 186, p. 120–131, doi:10.1016/j.jvolgeores.2009.03.011.

Francis, P., and Oppenheimer, C., 2003, Volcanoes: Oxford, UK, Oxford University Press, 536 p.

French, B.M., 1998, Traces of Catastrophe: Houston, Texas, Lunar and Planetary Institute, 120 p.

Ghods, A., and Arkani-Hamed, J., 2007, Impact-induced convection as the main mechanism for formation of lunar mare basalts: Journal of Geophysical Research, v. 112, p. E03005, doi:10.1029/2006JE002709.

Glass, B.P., and Simonson, B.M., 2013, Distal Impact Ejecta Layers: A Record of Large Impacts in Sediment Deposits: Berlin, Springer, 400 p.

Glikson, A.Y., 1999, Oceanic mega-impacts and crustal evolution: Geology, v. 27, p. 387–390, doi:10.1130/0091-7613(1999)027<0387:OMIACE>2.3.CO;2.

Glikson, A.Y., 2004, Impacts do not initiate volcanic eruptions: Eruptions close to the crater: Comment: Geology, v. 32, p. e48, doi:10.1130/0091-7613-32.1.e48.

Glikson, A.Y., 2005, Asteroid/comet impact clusters, flood basalts and mass extinctions: Significance of isotopic age overlaps: Earth and Planetary Science Letters, v. 236, p. 933–937, doi:10.1016/j.epsl.2005.05.007.

Gomes, R., Levison, H.F., Tsiganis, K., and Morbidelli, A., 2005, Origin of the cataclysmic Late Heavy Bombardment period of the terrestrial planets: Nature, v. 435, p. 466–469, doi:10.1038/nature03676.

Grady, M.M., Hutchinson, R., McCall, G.J., and Rothery, D.A., eds., 1998, Meteorites: Flux with Time and Impact Effects: Geological Society of London Special Publication 140, 278 p.

Graup, G., 1999, Carbonate-silicate liquid immiscibility upon impact melting: Ries crater, Germany: Meteoritics & Planetary Science, v. 34, p. 425–438, doi:10.1111/j.1945-5100.1999.tb01351.x.

Green, D.H., 1972, Archaean greenstone belts may include terrestrial equivalents of lunar mare?: Earth and Planetary Science Letters, v. 15, p. 263–270, doi:10.1016/0012-821X(72)90172-0.

Grieve, R.A.F., 1997, Target Earth: Evidence for large-scale impact events: Annals of the New York Academy of Sciences, v. 822, p. 319–352, doi:10.1111/j.1749-6632.1997.tb48350.x.

Grieve, R.A.F., and Cintala, M.J., 1992, An analysis of differential impact-melt crater-scaling and implications for the terrestrial impact record: Meteoritics, v. 27, p. 526–538, doi:10.1111/j.1945-5100.1992.tb01074.x.

Hagstrum, J.T., 2005, Antipodal hotspots and bipolar catastrophes: Were oceanic large-body impacts the cause?: Earth and Planetary Science Letters, v. 236, p. 13–27, doi:10.1016/j.epsl.2005.02.020.

Hansen, V.L., 2007, LIPs on Venus: Chemical Geology, v. 241, p. 354–374, doi:10.1016/j.chemgeo.2007.01.020.

Henkel, H., and Reimold, W.U., 2002, Magnetic model of the central uplift of the Vredefort impact structure, South Africa: Journal of Applied Geophysics, v. 51, p. 43–62, doi:10.1016/S0926-9851(02)00214-8.

Ingle, S., and Coffin, M.F., 2004, Impact origin of the Ontong Java Plateau?: Earth and Planetary Science Letters, v. 218, p. 123–134, doi:10.1016/S0012-821X(03)00629-0.

Ivanov, B.A., and Melosh, H.J., 2003, Impacts do not initiate volcanic eruptions: Eruptions close to the crater: Geology, v. 31, p. 869–872, doi:10.1130/G19669.1.

Ivanov, B.A., Deutsch, A., Ostermann, M., and Ariskin, A., 1997, Solidification of the Sudbury impact melt body and nature of the offset dikes—Thermal modeling, *in* Lunar and Planetary Science Conference, Volume XXVII: Houston, Texas, Lunar and Planetary Institute.

Johnson, B.B., and Melosh, H.J., 2012, Impact spherules as a record of an ancient heavy bombardment of Earth: Nature, v. 485, p. 75–77, doi:10.1038/nature10982.

Jones, A.P., 2005, Meteorite impacts as triggers to large igneous provinces: Elements, v. 1, p. 277–281, doi:10.2113/gselements.1.5.277.

Jones, A.P., Claeys, P., and Heuschkel, S., 2000, Impact melting of carbonates from the Chicxulub crater, *in* Gilmour, I., and Koeberl, C., eds., Impacts and the Early Earth: Lecture Notes in Earth Sciences: Berlin, Springer, v. 91, p. 343–361.

Jones, A.P., Price, G.D., Price, N.J., DeCarli, P.S., and Clegg, R.A., 2002, Impact induced melting and the development of large igneous provinces: Earth and Planetary Science Letters, v. 202, p. 551–561, doi:10.1016/S0012-821X(02)00824-5.

Jones, A.P., Price, G., DeCarli, P.S., Price, N., and Clegg, R., 2003, Impact decompression melting: A possible trigger for impact induced volcanism and mantle hotspots, *in* Koeberl, C., and Martínez Ruiz, F., eds., Impact Markers in the Stratigraphic Record: Berlin, Springer, p. 91–120.

Jones, A.P., Kearsley, A.T., Friend, C.R.L., Robin, E., Beard, A., Tamura, A., Trickett, S., and Claeys, P., 2005a, Are there signs of a large Paleocene impact, preserved around Disko Bay, West Greenland? Nuussuaq spherule beds origin by impact instead of volcanic eruption, *in* Kenkmann, T., Hörz, F., and Deutsch, A., eds., Large Meteorite Impacts III: Geological Society of America Special Paper 384, p. 281–298.

Jones, A.P., Wünemann, K., and Price, G.D., 2005b, Modeling impact volcanism as a possible origin for the Ontong Java Plateau, *in* Foulger, G.R., Natland, J.H., Presnall, D.C., and Anderson, D.L., eds., Plates, Plumes, and Paradigms: Geological Society of America Special Paper 388, p. 711–720.

Keays, R.R., and Lightfoot, P.C., 2004, Formation of Ni-Cu platinum group element sulfide mineralization in the Sudbury impact melt sheet: Mineralogy and Petrology, v. 82, p. 217–258, doi:10.1007/s00710-004-0050-8.

Keller, G., 1986, Stepwise mass extinctions and impact events: Late Eocene to early Oligocene: Marine Micropaleontology, v. 10, p. 267–293, doi:10.1016/0377-8398(86)90032-0.

Keller, G., 2005, Impacts, volcanism and mass extinction: Random coincidence or cause and effect?: Australian Journal of Earth Sciences, v. 52, p. 725–757, doi:10.1080/08120090500170393.

Keller, G., Sahni, A., and Bajpal, S., 2009, Deccan volcanism, the KT mass extinction and dinosaurs: Journal of Biosciences, v. 34, p. 709–728, doi:10.1007/s12038-009-0059-6.

Keller, G., Adatte, T., Bhowmick, P.K., Upadhyay, H., Dave, A., Reddy, A.N., and Jaiprakash, B.C., 2012, Nature and timing of extinctions in Cretaceous-Tertiary planktic foraminifera preserved in Deccan intertrappean sediments of the Krishna-Godavari Basin, India: Earth and Planetary Science Letters, v. 341–344, p. 211–221, doi:10.1016/j.epsl.2012.06.021.

Kelley, S.P., and Gurov, E., 2002, Boltysh, another impact at the KT boundary: Meteoritics & Planetary Science, v. 37, p. 1031–1043, doi:10.1111/j.1945-5100.2002.tb00875.x.

Kirchoff, M.R., Chapman, C.R., Marchi, S., and Bottke, W.F., 2013, Hints that the lunar impact flux has not been constant for large impacts during the last 3 Gyr, *in* Lunar and Planetary Science Conference, Volume 44: Houston, Texas, Lunar and Planetary Institute, p. 2689.

Koeberl, C., 2009, Late Eocene impact craters and impactoclastic layers—An overview, *in* Koeberl, C., and Montanari, A., eds., The Late Eocene Earth—Hothouse, Icehouse, and Impacts: Geological Society of America Special Paper 452, p. 17–26.

Koulouris, J., Janle, P., Milkereit, B., and Werner, S., 1999, Significance of large ocean impact craters, *in* Gersonde, R., and Deutsch, A., eds., Oceanic Impacts: Mechanisms and Environmental Perturbations: Berichte zur Polarforschung, v. 343, p. 48–50.

Kyte, F.T., Shukulyukov, A., Hildebrand, A.R., Lugmair, G.W., and Hanova, J., 2011, Chromium-isotopes in late Eocene impact spherules indicate a likely asteroid belt provenance: Earth and Planetary Science Letters, v. 302, p. 279–286, doi:10.1016/j.epsl.2010.12.006.

Leitch, E.M., and Vasisht, G., 1998, Mass extinctions and the Sun's encounters with spiral arms: New Astronomy, v. 3, p. 51–56, doi:10.1016/S1384-1076(97)00044-4.

Lieger, D., Riller, U., and Gibson, R.L., 2009, Generation of fragment-rich pseudotachylite bodies during central uplift formation in the Vredefort impact structure, South Africa: Earth and Planetary Science Letters, v. 279, p. 53–64, doi:10.1016/j.epsl.2008.12.031.

Marinova, M.M., Aharonson, O., and Asphaug, E., 2011, Geophysical consequences of planetary-scale impacts into a Mars-like planet: Icarus, v. 211, p. 960–985, doi:10.1016/j.icarus.2010.10.032.

Melosh, H.J., 1996, Impact Cratering, A Geologic Process: Oxford, UK, Oxford University Press, 254 p.

Melosh, H.J., and Ivanov, B.A., 2004, Impacts do not initiate volcanic eruptions: Eruptions close to the crater: Comment and reply: Geology, v. 32, p. e48–e49.

Meschede, M.A., Myhrvold, C.L., and Tromp, J., 2011, Antipodal focusing of seismic waves due to large meteorite impacts on Earth: Geophysical Journal International, v. 187, p. 529–537, doi:10.1111/j.1365-246X.2011.05170.x.

Miljković, K., Jones, A.P., and Price, M.C., 2010, High velocity impact modelling of ice into ice: Application to outer solar system landers and penetrators: European Planetary Science Congress Abstracts, v. 5, p. 321.

Moore, G., and Carmichael, I.S.E., 1998, The hydrous phase equilibria (to 3 kbar) of an andesite and basaltic andesite from western Mexico: Constraints on water content and conditions of phenocryst growth: Contributions to Mineralogy and Petrology, v. 130, p. 304–319, doi:10.1007/s004100050367.

Morbidelli, A., Bottke, W.F., Froeschle, C., and Michel, P., 2001, Origin and evolution of near Earth objects, *in* Bottke, W.E., Jr., Cellino, A., Paolicchi, P., and Binzel, R.P., eds., Asteroids III: Tucson, Arizona, University Arizona Press, p. 1–33.

Morbidelli, A., Marchi, S., Bottke, W.F., and Kring, D.A., 2012, A sawtooth-like timeline for the first billion years of lunar bombardment: Earth and Planetary Science Letters, v. 355–356, p. 144–151, doi:10.1016/j.epsl.2012.07.037.

Morgan, J., Warner, M., Brittan, J., Buffler, R., Camargo, A., Christeson, G., Denton, P., Hildebrand, A., Hobbs, R., Macintyre, H., Mackenzie, G., Maguire, P., Marin, L., Nakamura, Y., Pilkington, M., Sharpton, V., Snyder, D., Suarez, G., and Trejo, A., 1997, Size and morphology of the Chicxulub impact crater: Nature, v. 390, p. 472–476, doi:10.1038/37291.

Muller, R.A., 2002, Measurement of the lunar impact record for the past 3.5 b.y. and implications for the Nemesis theory, *in* Koeberl, C., and MacLeod, K.G., eds., Catastrophic Events and Mass Extinctions: Impacts and Beyond: Geological Society of America Special Paper 356, p. 659–665.

Napier, W.M., 2006, Evidence for cometary bombardment episodes: Monthly Notices of the Royal Astronomical Society, v. 366, p. 977–982.

Napier, W.M., 2010, Palaeolithic extinctions and the Taurid complex: Monthly Notices of the Royal Astronomical Society, v. 405, p. 1901–1906.

Napier, W.M., and Clube, S.V.M., 1979, A theory of terrestrial catastrophism: Nature, v. 282, p. 455–459, doi:10.1038/282455a0.

Napier, W.M., Wickramasinghe, J.T., and Wickmanrasinghe, N.C., 2004, Extreme albedo comets and the impact hazard: Monthly Notices of the Royal Astronomical Society, v. 355, p. 191–195, doi:10.1111/j.1365-2966.2004.08309.x.

Negi, J.G., Agrawal, P.K., Pandey, O.P., and Singh, A.P., 1993, A possible K-T boundary bolide impact site offshore Bombay and triggering of rapid Deccan volcanism: Earth and Planetary Science Letters, v. 76, p. 189–197, doi:10.1016/0031-9201(93)90011-W.

Osinski, G.R., and Pierazzo, E., 2012, Impact cratering: Processes and products, *in* Osinski, G.R., and Pierazzo, E., eds., Impact Cratering: Processes and Products: Hoboken, New Jersey, Wiley-Blackwell, p. 1–17.

Osinski, G.R., and Pierazzo, E., 2013, Impact Cratering: Processes and Products: Chichester, UK, Wiley, 330 p.

Osinski, G.R., and Spray, J.G., 2001, Impact-generated carbonate melts: Evidence from the Haughton impact structure, Canada: Earth and Planetary Science Letters, v. 194, p. 17–29, doi:10.1016/S0012-821X(01)00558-1.

Osinski, G.R., Grieve, R.A.F., Collins, G.S., Marion, C., and Sylvester, P., 2008, The effect of target lithology on the products of impact melting: Meteoritics & Planetary Science, v. 43, p. 1939–1954, doi:10.1111/j.1945-5100.2008.tb00654.x.

Osinski, G.R., Tornabene, L.L., Banerjee, N.R., Cockell, C.S., Flemming, R., Izawa, M.R.M., McCutcheon, J., Parnell, J., Preston, L.J., Pickersgill, A.E., Pontefract, A.E., Sapers, H.M., and Southam, G., 2013, Impact-generated hydrothermal systems on Earth and Mars: Icarus, v. 224, p. 347–363, doi:10.1016/j.icarus.2012.08.030.

Pearson, P.N., McMillan, I.K., Wade, B.S., Dunkley-Jones, T., Coxall, H.K., Brown, P.R., and Lear, C.H., 2007, Extinction and environmental change across the Eocene-Oligocene boundary in Tanzania: Geology, v. 36, p. 179–182, doi:10.1130/G24308A.1.

Poag, C.W., 1997, Roadblocks on the kill curve: Testing the Raup hypothesis: Palaios, v. 12, p. 582–590, doi:10.2307/3515413.

Price, N.J., 2001, Major Impacts and Plate Tectonics: A Model for the Phanerozoic Evolution of the Earth's Lithosphere: London, Routledge, 368 p.

Rampino, M.R., 1987, Impact cratering and flood basalt volcanism: Nature, v. 327, p. 468–469, doi:10.1038/327468a0.

Rampino, M.R., and Stothers, R.B., 1984, Terrestrial mass extinctions, cometary impacts and the Sun's motion perpendicular to the galactic plane: Nature, v. 308, p. 709–712, doi:10.1038/308709a0.

Raup, D.M., 1991, Extinction: Bad Genes or Bad Luck?: New York, W.W. Norton, 228 p.

Reese, C.C., Solomatov, V.S., Baumgardner, J.R., Stegman, D.R., and Vezolainen, A.V., 2004, Magmatic evolution of impact induced Martian mantle plumes and the origin of Tharsis: Journal of Geophysical Research, v. 109, p. E08009, doi:10.1029/2003JE002222.

Riller, U., Lieger, D., Gibson, R.L., Grieve, R.A.F., and Stoffler, D., 2010, Origin of large-volume pseudotachylite in terrestrial impact structures: Geology, v. 38, p. 619–622, doi:10.1130/G30806.1.

Rogers, G.C., 1982, Oceanic plateaus as meteorite impact signatures: Nature, v. 299, p. 341–342, doi:10.1038/299341a0.

Saunders, A., and Reichow, M., 2009, The Siberian Traps and the end-Permian mass extinction: A critical review: Chinese Science Bulletin, v. 54, p. 20–37, doi:10.1007/s11434-008-0543-7.

Schulte, P., Alegret, L., Arenillas, I., Arz, J.A., Barton, P.J., Bown, P.R., Bralower, T.J., Christeson, G.L., Claeys, P., Cockell, C., Collins, G.S., Deutsch, A., Goldin, T.J., Goto, K., Grajales-Nishimura, J.M., Grieve, R.A.F., Gulick, S.P.S., Johnson, K.R., Kiessling, W., Koeberl, C., Kring, D.A., MacLeod, K., Matsui, T., Melosh, H.J., Montanari, A., Morgan, J.V., Neal, C.R., Nichols, D.J., Norris, R.D., Pierazzo, E., Ravizza, G., Rebolledo-Vieyra, M., Reimold, W.U., Robin, E., Salge, T., Speijer, R.P., Sweet, A.R., Urrutia-Fucugauchi, J., Vajda, V., Whalen, M.T., and Willumsen, P.S., 2010, The Chicxulub asteroid impact and mass extinction at the Cretaceous-Paleogene boundary: Science, v. 327, p. 1214–1218, doi:10.1126/science.1177265.

Seyfert, C.K., and Sirkin, L.A., 1979, Earth History and Plate Tectonics: New York, Harper and Row, 96 p.

Shen, W., Sun, Y., Lin, Y., Liu, D., and Chai, P., 2011, Evidence for wildfire in the Meishan section and implications for Permian-Triassic events: Geochimica et Cosmochimica Acta, v. 75, p. 1992–2006, doi:10.1016/j.gca.2011.01.027.

Shoemaker, E.M., Wolfe, R.F., and Shoemaker, C.S., 1990, Asteroid and comet flux in the neighborhood of Earth, *in* Sharpton, V.L., and Ward, P.D., eds.,

Global Catastrophes in Earth History: An Interdisciplinary Conference on Impacts, Volcanism, and Mass Mortality: Geological Society of America Special Paper 247, p. 155–170, doi:10.1130/SPE247-p155.

Snape, J.F., Joy, K.H., and Crawford, I.A., 2011, Characterization of multiple lithologies within lunar feldspathic regolith breccia meteorite Northeast Africa 001: Meteoritics & Planetary Science, v. 46, p. 1288–1312, doi:10.1111/j.1945-5100.2011.01230.x.

Svensen, H., Planke, S., Polozov, A.G., Schmidbauer, N., Corfu, F., Podladchikov, Y.Y., and Jamtveit, B., 2009, Siberian gas venting and the end-Permian environmental crisis: Earth and Planetary Science Letters, v. 277, p. 490–500, doi:10.1016/j.epsl.2008.11.015.

Tejada, M.L.G., Mahoney, J.J., Castillo, P.R., Ingle, S., Sheth, H.C., and Weis, D., 2004, Pin-pricking the elephant: Evidence on the origin of the Ontong Java Plateau from Pb-Sr-Hf-Nd isotopic characteristics of ODP Leg 192 basalts, *in* Fitton, J.G., Mahoney, J.J., Wallace, P.J., and Saunders, A.D., eds., Origin and Evolution of the Ontong Java Plateau: Geological Society of London Special Publication 229, p. 133–150.

Tejada, M.L.G., Suzuki, K., Kuroda, J., Coccioni, R., Mahoney, J.J., Ohkouchi, N., Sakamoto, T., and Tatsumi, Y., 2009, Ontong Java Plateau eruption as a trigger for the early Aptian oceanic anoxic event: Geology, v. 37, p. 855–858, doi:10.1130/G25763A.1.

Therriault, A.M., Fowler, A.D., and Grieve, R.A.F., 2002, The Sudbury igneous complex: A differentiated impact melt sheet: Economic Geology and the Bulletin of the Society of Economic Geologists, v. 97, p. 1521–1540, doi:10.2113/gsecongeo.97.7.1521.

Thompson, A.B., 1995, Melting of the continental crust: Some thermal and petrological constraints on anatexis in continental collision zones and other tectonic settings: Journal of Geophysical Research–Solid Earth, v. 100, p. 15,565–15,579, doi:10.1029/95JB00191.

Thordarson, T., and Hoskuldsson, A., 2008, Postglacial volcanism in Iceland: Jokull, v. 58, p. 197–228.

Vaughan, W.M., Head, J.W., Wilson, L., and Hess, P.C., 2013, Geology and petrology of enormous volumes of impact melt on the Moon: A case study of the Orientale basin impact melt sea: Icarus, v. 223, p. 749–765, doi:10.1016/j.icarus.2013.01.017.

Vennemann, T.W., Morlok, A., Engelhardt, W.V., and Kyser, K., 2001, Stable isotope composition of impact glasses from the Nordlinger Ries impact crater, Germany: Geochimica et Cosmochimica Acta, v. 65, p. 1325–1336, doi:10.1016/S0016-7037(00)00600-1.

Vishnevsky, S., and Montanari, A., 1999, The Popigai impact crater (Arctic Siberia, Russia): Geology, petrology, geochemistry, and geochronology of glass-bearing impactites, *in* Dressler, B.O., and Sharpton, V.L., eds., Large Meteorite Impacts and Planetary Evolution II: Geological Society of America Special Paper 339, p. 19–60.

Visscher, H., Svensen, H., Looy, C., Fristad, K., Polozov, A., and Planke, S., 2010, Palynological constraints on timing and duration of Siberian Traps volcanic events, *in* EGU General Assembly 2010: Vienna, Austria, European Geophysical Union, p. 11100.

Warren, P.H., Claeys, P., and Cedillo-Pardo, E., 1996, Mega-impact melt petrology (Chicxulub, Sudbury, and the Moon): Effects of scale and other factors on potential for fractional crystallization and development of cumulates, *in* Ryder, G., Fastovsky, D., and Gartner, S., eds., The Cretaceous-Tertiary Event and Other Catastrophes in Earth History: Geological Society of America Special Paper 307, p. 105–124.

Weissman, P.R., 2007, The cometary impact flux at the Earth, *in* Milani, A., Valseechi, G.B., and Vokrouhlicky, D., eds., Near Earth Objects, our Celestial Neighbours: Opportunity and Risk, Volume 236: International Astronomy Union, p. 441–450.

White, R.V., and Saunders, A.D., 2005, Volcanism, impact and mass extinctions: Incredible or credible coincidences?: Lithos, v. 79, p. 299–316, doi:10.1016/j.lithos.2004.09.016.

Wickramasinghe, J.T., and Napier, W.M., 2008, Impact cratering and the Oort cloud: Monthly Notices of the Royal Astronomical Society, v. 387, p. 153–157, doi:10.1111/j.1365-2966.2008.13098.x.

Yeomans, D.K., and Chamberlin, A.B., 2012, Comparing the Earth impact flux from comets and near-Earth asteroids: Acta Astronautica, v. 90, p. 3–5, doi:10.1016/j.actaastro.2012.03.006.

Zuber, M.T., Smith, D.E., Watkins, M.M., Asmar, S.W., Konopliv, A.S., Lemoine, F.G., Melosh, H.J., Neumann, G.A., Phillips, R.J., Solomon, S.C., Wieczorek, M.A., Williams, J.G., Goossens, S.J., Kruizinga, G., Mazarico, E., Park, R.S., and Yuan, D.-N., 2012, Gravity field of the Moon from the gravity recovery and interior laboratory (GRAIL) mission: Science, v. 339, p. 668–671, doi:10.1126/science.1231507.

Manuscript Accepted by the Society 31 January 2014

Printed in the USA

The Geological Society of America
Special Paper 505
2014

The role of giant comets in mass extinctions

W.M. Napier*
Astrobiology Centre, University of Buckingham, Buckingham MK18 1EG, UK

ABSTRACT

Dynamical studies of the asteroid belt reveal it to be an inadequate source of terrestrial impactors of more than a few kilometers in diameter. A more promising source for large impactors is an unstable reservoir of comets orbiting between Jupiter and Neptune. Comets 100–300 km across leak from this reservoir into potentially hazardous orbits on relatively short time scales. With a mass typically 10^3–10^4 times that of a Chicxulub-sized impactor, the fragmentation of a giant comet yields a highly enhanced impact hazard at all scales, with a prodigious dust influx into the stratosphere over the duration of its breakup, which could be anywhere from a few thousand to a few hundred thousand years. Repeated fireball storms of a few hours' duration, occurring while the comet is fragmenting, may destroy stratospheric ozone and enhance incident ultraviolet light. These storms, as much as large impacts, may be major contributors to biological trauma. Thus, the debris from such comets has the potential to create mass extinctions by way of prolonged stress. Large impact craters are expected to occur in episodes rather than at random, and this is seen in the record of well-dated impact craters of the past 500 m.y. There is a strong correlation between these bombardment episodes and mass extinctions of marine genera.

INTRODUCTION

The idea that geological disturbances on Earth may be occasioned by impacts goes back at least to the Babylonian Talmud, when the "Flood" was ascribed to two falling stars (tractate Berakoth, folio 59a, ca. A.D. 200). In Western Civilization, Whiston (1696) proposed that the Great Flood was caused by passage through the tail of a comet. That the Cretaceous-Paleocene collapse in particular was caused by a giant impact has been suggested from time to time (De Laubenfels, 1956; Urey, 1973), but with scores of ad hoc hypotheses for these extinctions (Benton, 1990), there was no obvious reason to give it weight.

Beginning in the 1960s, however, space probes were bringing back images of heavily cratered planets and satellites in the solar system. Lunar soil samples, retrieved from 1969 onward, settled the volcano versus impact debate with regard to lunar craters. With the help of satellite imagery, impact craters began to be found in increasing numbers on Earth too, and early attempts were made to establish their size-frequency distribution (Grieve and Robertson, 1979). Commensurate with these findings, wide-angle telescope surveys were beginning to detect increasing numbers of small near-Earth objects (NEOs), in orbits which made them a contemporary hazard (Helin and Shoemaker, 1979), and by the end of the 1970s, it was possible to provide a quantitative estimate of the rate of bombardment of Earth by impactors of various sizes. Napier and Clube (1979) pointed out that this rate was high enough for mass extinctions by impact to move from speculation to expectation. Thus, 10 km planetesimals were

*bill.napier@buckingham.ac.uk

Napier, W.M., 2014, The role of giant comets in mass extinctions, *in* Keller, G., and Kerr, A.C., eds., Volcanism, Impacts, and Mass Extinctions: Causes and Effects: Geological Society of America Special Paper 505, p. 383–395, doi:10.1130/2014.2505(20). For permission to copy, contact editing@geosociety.org.

expected to have struck the Earth every ~60 m.y., yielding impact energies of 160 million megatons and a crater, in the event of a land impact, ~200 km across. Modern estimates are within a factor of two of these early values (Bland, 2005). The extinction mechanisms proposed were collapse of food chains due to the obscuration of sunlight by injection of dust into the stratosphere, destruction of the ozone layer due to NO_x production with a resulting ultraviolet (UV) deluge, and straightforward blast.

In 1980, Alvarez et al. (1980) published their seminal paper on the discovery of an excess of iridium at the end-Cretaceous boundary at Gubbio, Italy. They interpreted this as debris from the impact of a 10 km asteroid and likewise proposed that the Cretaceous-Paleogene extinctions were caused by the collapse of food chains due to the obscuration of sunlight by injection of dust into the stratosphere. The "Alvarez hypothesis" has become widely accepted in both popular and professional literature, particularly following the discovery of the 170 km Chicxulub impact crater at, or close to, the Cretaceous-Paleogene boundary (Hildebrand et al., 1991). Sudden death is an expectation of this hypothesis; whether there actually was such has been a contentious issue for 30 yr (cf. Kring, 2007; Schulte et al., 2010; Keller et al., 2009; Gertsch et al., 2011).

Clearly, the ground for the Alvarez hypothesis had been prepared by the astronomical findings of the previous decade. However, our understanding of this environment has significantly evolved since 1980. In particular, it now appears that the near-Earth environment is, on time scales of geological relevance, dominated by the disintegration products of exceptionally large comets. These may cause prolonged environmental stress through fireball swarms, multiple impacts, and dustings. A one-off stray asteroid yields a "sudden death" set of expectations; an episode of cometary bombardment, which involves prolonged multiple impacts and atmospheric dusting, may yield another. Such episodes are expected to happen quite frequently, raising the question of whether physical changes in the astronomical environment affect biodiversity on a more continuous scale, with the "Big Five" extinctions lying at the extreme end of a spectrum of extinction events.

COMETARY AND ASTEROIDAL RESERVOIRS

The orbits of asteroids in the main belt are affected by collisions, planetary perturbations, and long-acting radiation forces. Because of these disturbances, asteroids may drift into orbits that resonate with the cyclic perturbing forces exerted by the major planets, in particular, Jupiter and Saturn. Once an asteroid enters such a resonance, its orbit becomes chaotic, and it may be rapidly expelled from the main belt, either falling into the Sun, hitting a planet, or being ejected from the solar system altogether. This leakage from the main asteroid belt appears to be an adequate source for the bulk of kilometer-sized terrestrial impactors (Bottke et al., 2006).

As one goes to larger bodies, however, the main asteroid belt becomes increasingly inadequate as a replenishing source (Menichella et al., 1996). Bodies more than ~5 km in diameter need an exceptionally energetic collision to be injected into an "escape hatch." Minton and Malhotra (2010) found that weak chaotic evolution in the main asteroid belt causes bodies over 10 km across to escape on billion-year time scales, with only one in 300 of such escapers striking Earth. They found an expectation of one such impact over the last billion years, as against 6–18 inferred from lunar impact cratering statistics. Thus, the present-day flux of asteroids from the main belt is inadequate by an order of magnitude to supply the largest impactors. This can be seen qualitatively from the fact that there is a brisk traffic of comets over 10 km across with perihelia $q < 1$ au, as for example, comets Hale-Bopp ($q = 0.91$ au, $D = 30–60$ km, perihelion passage in 1997), Swift-Tuttle ($q = 0.96$ au, $D = 26$ km, perihelion passage in 1992), and Halley ($q = 0.59$ au, $D = 15 \times 8 \times 8$ km, passage in 1986), whereas there are no asteroidal Earth-crossers at all reaching these sizes. The mean encounter velocity with such cometary bodies is ~56 km/s, generally ranging 40–70 km/s, and they would carry an impact energy in the billions of megatons. Probably ~10 active long-period comets with $D > 1$ km pass within the orbit of Earth each year. A significant unknown in this reckoning is the number of dark comets, going unseen, that enter Earth-crossing orbits (Bailey and Emel'yanenko, 1998).

The Chicxulub impactor may have had the chemistry of a carbonaceous chondrite (Shukolyukov and Lugmair, 1998), and that which made the Morokweng crater in South Africa had the chemistry of an ordinary chondrite (McDonald et al., 2001). However, the identification of a parent body as an asteroid can probably not be inferred simply from its chondritic composition. Samples of material recovered from Comet Wild 2 by the *Stardust* spacecraft in 2006 reveal that it had elemental abundance equal to that of ordinary C1 meteorites, although with a relative overabundance of moderately volatile elements (Flynn et al., 2006). Chondrule-like objects in the comet were also found, with oxygen isotope ratios similar to those found in ordinary carbonaceous chondrites derived from the asteroid belt (Nakamura et al., 2008). It appears that refractory solids were mixed throughout the solar system while it was still a protoplanetary disk; such transport has been observed in a young eruptive star with a protoplanetary disk and is probably a common process during the formation of planetary systems (Juhasz et al., 2012; Liffman and Brown, 1995). The important difference between asteroids and comets may be primarily one of volatile abundance and hence tensile strength. In this respect, the more distant the prime object is from the Sun, the more likely it is to have a volatile composition and to be vulnerable to multiple disintegration.

There are several major dynamical groups of comets:

1. Long-period comets have orbital periods (P) in the approximate range 200 yr < P < 4 m.y. They derive from an inferred cloud of comets, the Oort cloud, distributed isotropically around the Sun, with a radius of at least 40,000 astronomical units (au) and a population of ~10^{11} comets. There may be an inner cloud of ~10^{12}

comets, but this does not contribute to the observed cometary population.

2. Halley-type comets have orbital periods in the range $20 < P < 200$ yr and likewise form a spheroidal system, with a similar distribution of impact speeds, i.e., ~40–70 km/s, with mean 56 km/s.
3. Jupiter family comets have $P < 20$ yr and orbital inclinations $i < 20°–30°$ and so move in direct orbits relatively close to the ecliptic. Their orbital evolution is dominated by Jupiter.
4. Encke-type comets have orbital periods of 4 yr or less, low orbital inclinations, and Earth-crossing orbits. They lie outside the Jovian sphere of influence, giving them longer dynamical lifetimes and shorter orbital periods, and so making them a hazard out of proportion to their number. Currently, there is only one active member of this family, but it appears to belong to a group of dormant bodies derived from a recent fragmentation (Napier and Asher, 2009; Napier, 2010).

Uncertainties attend estimates of impact probabilities from comets in all these orbital classes, in part because the known, active members of each class may be outnumbered by dormant comets, which, being inactive, tend to go undiscovered.

The dynamical lifetimes of Jupiter-family and Encke-type comets are short (~0.3 m.y.): they may fall into the Sun, be returned to the outer solar system from whence they came, be thrown into interstellar space, or simply disintegrate into meteoritic dust. These families must therefore be replenished from other sources. The Oort cloud is likely to be the prime replenishing source of the Halley-type system. From population balance considerations, it has long been known that the Jupiter-family population cannot be accounted for by direct capture of the Oort cloud comets. However, deep telescopic surveys of the last couple of decades have revealed the existence of significant populations of large bodies on the fringes of the planetary system. There is a dynamically cold disk, in the range 39–48 au, consisting of objects in relatively stable orbits that probably do not contribute much to the other populations. Beyond that, there is the scattered disk population, with high eccentricities and inclinations, and perihelia q generally within the orbit of Neptune. Long-acting perturbations of Neptune perturb objects in the scattered disk, with orbital control cascading down through Uranus and Saturn into Jupiter-family orbits. It is expected that some proportions of comets arriving from the Oort cloud eventually find their way into the scattered disk and hence, indirectly, contribute to the Jupiter family. These trans-Neptunian bodies are thought, largely on cosmological grounds, to be icy in nature. Thus, both the trans-Neptunian and Oort cloud reservoirs are expected to feed comets into the planetary system, although their relative contributions remain uncertain, and there are other secondary sources (Fig. 1). Emel'yanenko et al. (2013) considered that virtually all Halley-type comets and nearly half of the observed Jupiter-family comets come from the Oort cloud.

Objects in transition from the scattered disk to the Jupiter-family population have been detected and constitute the system of centaurs. The inhabitants of this migrant population, buffeted by the gravity of the giant planets from Jupiter to Neptune, are in highly unstable orbits, and their mean lifetimes are a few million years, declining strongly with decreasing perihelion distance. Most of them, like the Jupiter-family comets, are thrown back to their source region or are ejected from the solar system; others may collide with Jupiter; and a few are destined to become terrestrial collision hazards.

CENTAUR EVOLUTION

Because their orbits are chaotic, the evolution of centaurs can only be followed statistically. The standard procedure is to construct "clones," that is, fictitious bodies with orbital elements closely similar to those of a real centaur, and follow their evolution by numerical integrations. The hazard represented by the arrival of centaurs is determined by their rate of arrival in the near-Earth environment, and this rate is dominated by that of the most dynamically unstable ones at any given time.

We may take the orbit of Chiron, the first centaur to be discovered, as archetypal for the centaurs. It is in a chaotic orbit

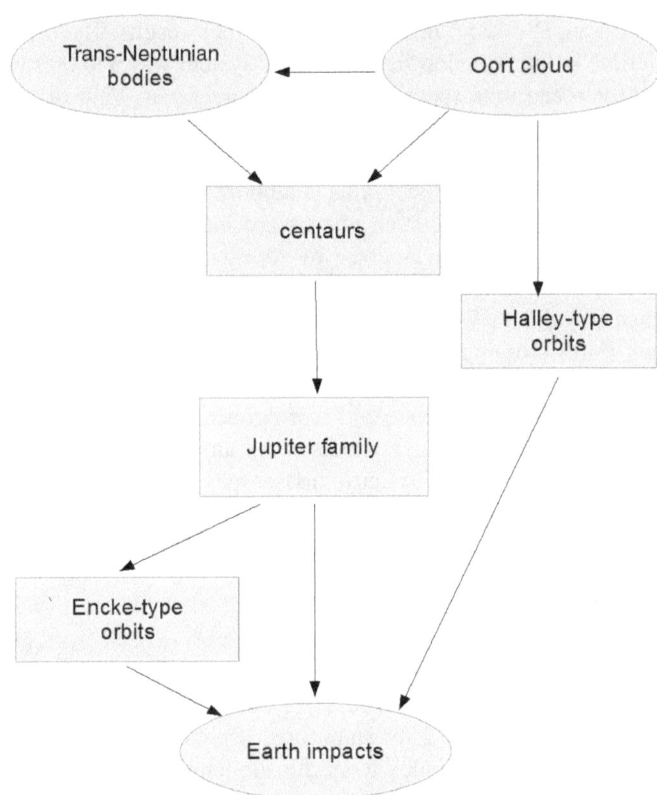

Figure 1. The main cometary populations, and the major dynamical highways tracing their Earthward migrations. Minor reservoirs are excluded, and there may be a two-way traffic between the Oort cloud and the trans-Neptunians.

between Saturn and Uranus, with semimajor axis $a \sim 13.7$ au and a modest eccentricity. At present, it never approaches the Sun closer than $q = 8.5$ au. From observations with the Spitzer Space Telescope, its diameter is estimated to be 233 ± 14 km (Stansberry et al., 2007). The distances of the centaurs and trans-Neptunian comets are such that discovery is seriously incomplete, and only the largest bodies can be detected. The number of centaurs larger than Chiron has been estimated as $360 < N < 650$, of which there are expected to be $4 < N < 7$ inside 18 au. This is consistent with the fact that four centaurs larger than Chiron have been discovered, three of them within 18 au.

Chiron may have entered an Earth-crossing configuration several times in the past (Hahn and Bailey, 1990). Horner et al. (2004), simulating the evolution of Chiron with 729 clones for up to 3 m.y., found that more than half of the clones became short-period comets at some point, and ~1 in 8 became Earth-crossers. When they entered this latter state, they did so repeatedly, typically 10 times per m.y. Similar rates were arrived at from other centaurs with similar orbits.

Once a comet is thrown into an orbit with perihelion $q <$ 2.5 au, it crosses the water snowline. It begins to outgas, growing one or more tails, and becomes liable to disintegration. Numerical runs of 100 Chiron clones over 1 m.y., using the same input conditions and program as Horner et al. (2004), revealed that about half of them temporarily entered orbits with $q < 2.5$ au, $P < 12$ yr, in agreement with their results. The mean lifetime before ejection from the solar system was 350,000 yr, and the mean time spent as an Earth-crosser was 0.2% of this lifetime. If there are say five centaurs with $D > 250$ km in such orbits at a given time, then over a 500,000 yr interval, Earth will have a Chiron-sized object in its neighborhood for a mean time of 2000 yr. In reality, such epochs are highly sporadic, with long quiet periods interspersed by repeated encounters: Only nine of the 100 clones became temporary Earth-crossers, but when they did, they did so repeatedly, with the mean duration of their Earth-crossing epochs being ~20,000 yr and varying from 3000 yr to over 80,000 yr. Figure 2 illustrates the dynamical evolution of a Chiron clone with a prolonged short-period phase that includes a 40,000 yr residence as an Earth-crosser. The transition from centaur to Earth-crosser may occur rapidly, on a time scale of ~1000 yr.

The cumulative size distribution of centaurs and trans-Neptunians is

$$N(> D) \propto D^{-s}, \tag{1}$$

with $s = 1.7 \pm 0.3$ (Bauer et al., 2013), consistent with the theoretical expectation $s = 2$ for an evolved population (Kenyon and Bromley, 2012). The index found for the long-period comets in the diameter range $D = 20$–200 km is $s = 2.15 \pm 0.75$ (Fernandez and Sosa, 2012). Adopting $D = 240$ km for Chiron (Stansberry et al., 2007), there is then a 6% probability that a centaur >100 km across will be in an Earth-crossing state at any given time. If it is, its dynamical lifetime in that configuration, along with that of its fragments, will likely be 3–100 millennia, adequate time to flood the inner planetary system with debris.

There is supporting evidence for such past events in the form of the fossil remains of two such bodies. One of them was the progenitor of the current Kreutz sungrazing comets. About 1000 fragments have been detected with the *Solar and Heliospheric Observatory* (*SOHO*) heliocentric orbiting observatory, out of a total probably running to many thousands, and it appears that disintegration of the progenitor has proceeded through cascades of fragmentation (Sekanina and Chodas, 2007). Both provenance and size of the original are, however, uncertain. It was on a high-inclination retrograde orbit at breakup, and it may have come from the Halley system, although evolution from a low-inclination prograde orbit to this state is also possible (Bailey et al., 1992). Sekanina and Chodas (2004) considered that the Kreutz progenitor may have been 100 km across and broken

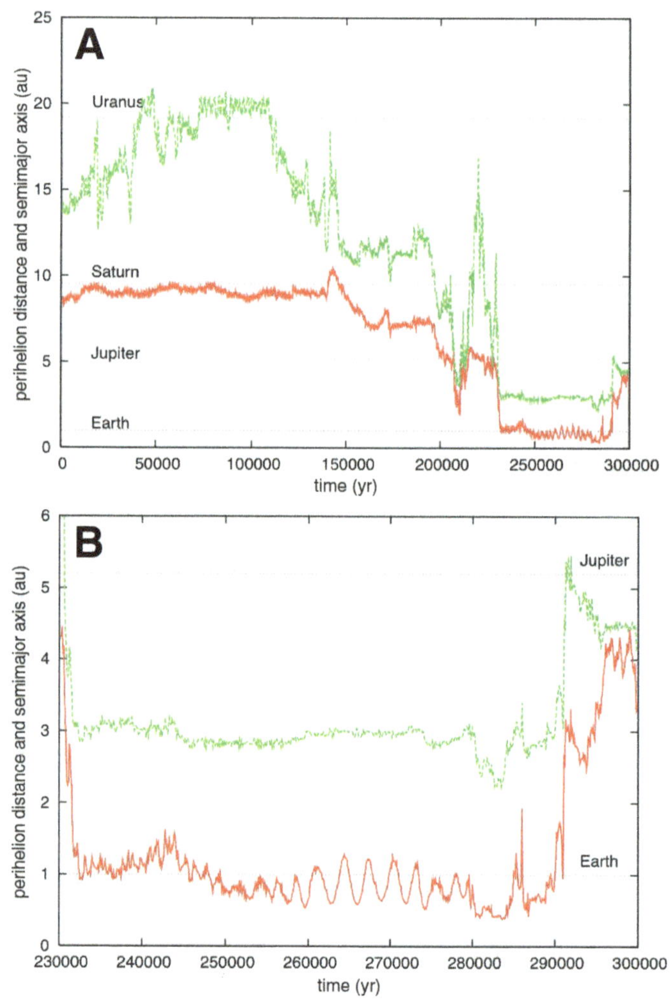

Figure 2. Top: Evolution over 1 m.y. of a Chiron clone that becomes an Earth-crosser. Bottom: The Jupiter-family phase in more detail. Substantial disintegration is likely to take place over the ~60,000 yr of its existence in this state. The green line shows the course of the semimajor axis; the red line shows that of the perihelion.

up 1700 yr ago, with an orbital period of 500–1000 yr immediately before fragmentation. The other recent large comet in our neighborhood was the progenitor of the Taurid complex, which may have been of similar size (Napier, 2010) but greater age (≳20,000 yr; Steel and Asher, 1996). The debris is low inclination and short period ($P \sim 3.3$ yr); it includes several large near-Earth "asteroids" (probably dormant comets), and it dominates the contemporary meteoroid environment (Wiegert et al., 2009), although significant uncertainty remains about its initial mass and the membership of the complex.

If the size distribution represented by Equation 1 can be extrapolated upward, then one expects at least one centaur over 600 km in diameter to have disintegrated in an Earth-crossing orbit at some time during the Phanerozoic. However, the distribution must eventually turn down, and the observational data are too sparse to say where the turndown occurs. A 200 km comet striking Earth would carry ~10,000 times the energy involved in creating the 180 km Chicxulub crater, and it would presumably remove the biosphere. Impact crater statistics show this to have been an unlikely event over the 600 m.y. time scale of the Phanerozoic. However, the prime hazard appears to come from their disintegration products. Destruction through entry into a sungrazing orbit is quite common for comets with initial perihelia $q < 2$ au or so, arising from correlated changes in the orbital elements due to the long-term secular perturbations by giant planets (Bailey et al., 1992). Most comets a few kilometers across would not survive passage within 0.1 au of the Sun, but those in direct orbits (e.g., the centaurs) that avoid close encounters with Jupiter can reach small perihelion distances within a few dozen perihelion passages.

DISRUPTION OF NEAR-EARTH GIANT COMETS

Most comets are porous, virtually strengthless bodies, typically with mean densities about half that of water. A cohesive strength of only 0.001 bar would have prevented the tidal destruction of both Shoemaker-Levy 9 and the Kreutz parent body. They may be loose aggregates of "building blocks," perhaps up to 1 km in diameter, perhaps remnants of previous disruptions, barely held together by their own gravity (A'Hearn, 2011). Modeling of the collisional evolution of bodies in the trans-Neptunian region indicates that perhaps all bodies between 10 and 100 km diameter are so constituted. Irrespective of size, a comet will disintegrate when its self-gravity is exceeded by the ambient tide. This happens when it passes within Roche's limit, given by

$$L = 2.44\, R_0\, (\rho_0/\rho)^{1/3}, \qquad (2)$$

where R_0 is the radius of the large body, and ρ_0 is its density; ρ is the density of the comet. For the Sun, r_0 = 1.41 g/cm^3, while for a comet ρ = 0.4 g/ cm^3. The Roche limit for a sungrazing comet is then 3.7 solar radii.

Bailey et al. (1992) considered that 5%–15% of comets that attain $q < 2$ au become sungrazers. Of those that enter Encke-type orbits, the figure is ~40%. If we adopt, overall, a 5% probability that an Earth-crossing comet in a short-period orbit will become a sungrazer, then we expect a >100 km object to disrupt completely in the terrestrial neighborhood through tidal disintegration every 100 m.y. or so. The disintegration is expected to be hierarchical, with short-lived fragments cascading down to dust.

Knight et al. (2010) found that the *SOHO* sungrazers has a cumulative size distribution

$$n(>R) \propto R^{-s}, \qquad (3)$$

with $s \sim 2.2$, which translates to a differential mass distribution

$$n\,(m)dm \propto m^{-\alpha}, \qquad (4)$$

with $\alpha \sim 1.73$. Such mass distributions are top heavy, with the largest single fragment typically containing almost half the total mass of the system. Even so, this fragment size distribution completely fails to predict the great sungrazing comets from historical times that belong to this system and that may have been up to 60 km in diameter (A'Hearn, 2011). This must involve splitting rather than surface fragmentation, and in fact records of comet splitting go back ~2000 yr, and with the advent of telescopes, it has become clear that the phenomenon is a major route of comet demise. The short-period comet 73P/Schwassmann-Wachmann 3, having been seen as a single object in its 1990 apparition, was observed to split into three fragments when it returned in late 1995; many fresh fragments were seen on subsequent returns, and by 2006, upward of 150 fragments had been detected (Fig. 3). Ishiguro et al. (2009) found that the >150 fragments from the

Figure 3. Fragments of a fragment: 2006 *Hubble* image showing fragment B of periodic comet Schwassmann-Wachmann 3, itself undergoing disintegration. This small Jupiter-family comet began to break up in 1995. Scale is about 3 arcsec, or 3200 km. North is to the top, East to the left. Credit: NASA and STScI.

split comet 73P/Schwassmann-Wachmann 3B had a cumulative size distribution with index 2.34 ± 0.05, in which case almost half the total mass is contained in the largest fragment. Comet Hyakutake split into seven subnuclei in 1996; the long-period comet Linear S4 disintegrated into thousands of fragments the year after its discovery in 1999; and so on. Di Sisto et al. (2009) considered that, on average, a single Jupiter-family comet will undergo a major splitting every ~77 revolutions. There is a weak tendency for such disruptions to happen after perihelion passage. The splitting mechanism is not understood, however, and whether Chiron-sized objects undergo this mode of disruption is unknown, although the existence of large historical comets from the Kreutz progenitor, and multikilometer fragments of the Taurid complex, suggest that they do.

The progress of a comet from dirty snowball to dust may also be interrupted by a dormant phase. A comet is usually considered to be an icy matrix in which grains are embedded. Once thrown into an orbit with perihelion $q < 2.5$ au, it begins to outgas, growing a coma and one or more tails. Bodies some tens of meters in diameter may detach from the nucleus, fading rapidly from sight. As the ice sublimates, grains larger than a critical radius fall back on the nucleus and eventually create a stable mantle that chokes off the escaping gas (e.g., Rickman et al., 1990); the comet is then asteroidal in appearance. For a water-ice composition, the time scale for mantle growth is of an order of a year at 2.5 au, and ~1000 yr at 5 au, these time scales being virtually independent of size (Jewitt, 2004). The mantle is only centimeters thick, however, and the dormancy phase may only be temporary. If the perihelion evolves inward under the influence of planetary perturbations, the comet may be reactivated.

Given that the largest observed centaurs have masses 10^{21}–10^{22} g, as opposed to $<10^{19}$ g for the entire current near-Earth asteroid system, there is the potential for a substantially enhanced hazard during the breakup of such a body in the inner planetary system. Such a breakup might involve multiple bombardments and substantial dust input (Clube and Napier, 1984), leading perhaps to a qualitatively different astronomical scenario for mass extinctions from that of a single, 10 km asteroid impact (Alvarez et al., 1980).

BOMBARDMENT EPISODES

Known impact craters tend to concentrate in stable regions such as the Canadian and Baltic Shields, and the Australian desert. Throughout India, Asia, and the Far East, only one small impact crater has been discovered (the Lonar crater in central India, about 1 km across and perhaps 50,000 yr old). The Australasian strewn tektite field is dated at ca. 0.77 Ma; estimates of the diameter of the associated impact crater are in the range ~30–120 km, but no crater has been found (Lee and Wei, 2000). This patchy geographical distribution indicates that the evidence of past impacts is rapidly removed. This can be seen also when the ages of well-dated impact craters are plotted as a function of time (Fig. 4).

The mean interval δt (m.y.) between impacts by bodies that could yield craters of diameter $D > 20$ km has been estimated as

$$\delta t = 2.8\ (D/20)^2 \tag{5}$$

(Grieve, 1984), where D is measured in kilometers. Because of subduction, erosion, and so on, this estimate is heavily biased toward structures <120 m.y. old (Fig. 4). The sample of large impact craters on which this rate is based is very small, in part

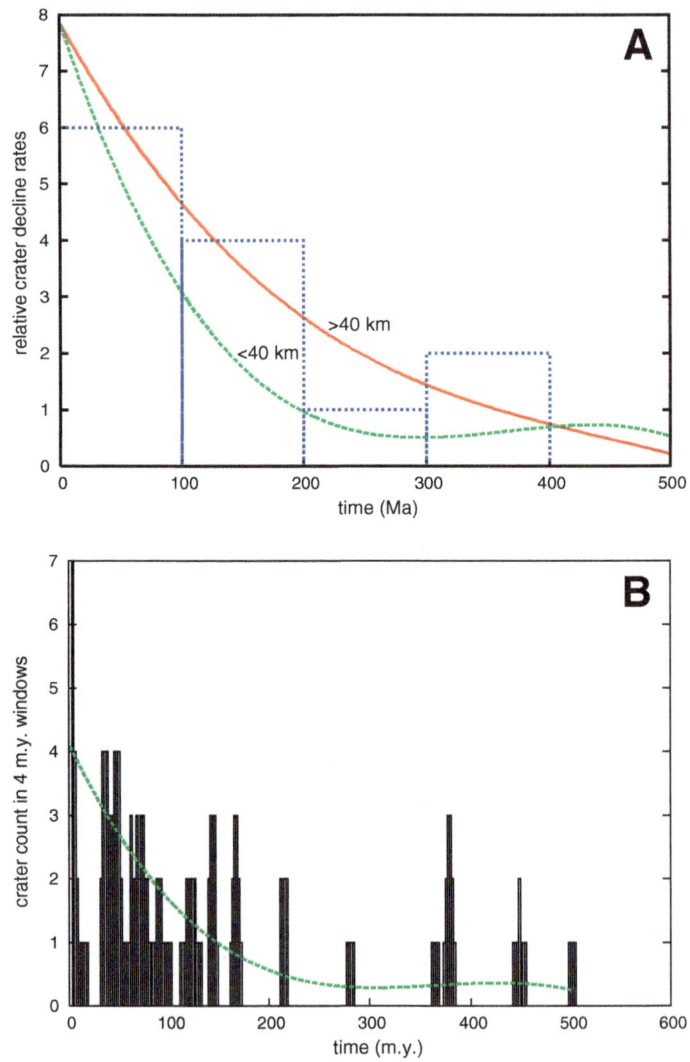

Figure 4. Top: Decline of known impact craters with increasing age. Discovery completeness declines to 50% within 120 m.y. for impact craters with $D > 40$ km, and ~70 m.y. for smaller craters, assuming 100% completeness of discovery of Miocene craters. The data employed are 47 craters with $D > 3$ km and radiometric ages dated to $2\sigma <$ 10 m.y. (35 with $2\sigma \leq 5$ m.y.). Best-fit cubics are shown. The histogram is of the 13 craters with $D > 40$ km. Bottom: The crater age distribution smoothed by a rectangular window of width 4 m.y., stepping in 1 m.y. intervals. The impression of clustering in time is verified statistically to a high confidence level (updated from Napier, 2006).

because of their rarity, and in part because of the limited area of ancient cratons available for recording large structures.

Crater data were taken from the 184 confirmed craters listed in the Earth Impact Database (2005, accessed 18 May 2013). Of these, 47 were culled with the criteria $D > 3$ km, dating accuracy $2\sigma \leq 10$ m.y., age ≤ 500 m.y. These 47 constitute the data set analyzed here for patterns. There are 24 craters in this list with $D > 20$ km, against an expectation of 180 over the past 500 m.y. Thus, one in seven impacts of this size have been detected. Detection of craters over 10 km across is ~5% complete over the Phanerozoic.

Table 1 lists groups of craters that are indistinguishable in age within their errors. There are 36 craters within these groups, which are tightly bunched in time and are evidence of discrete bombardment episodes. Monte Carlo simulations indicate that this degree of clustering arises by chance about once in 1000 trials, or a few times in 1000 trials if group 1 is excluded as being due to the pull of the recent. There are likely to be interlopers in the list, but the probability that the groupings as a whole are due to chance is negligible. Thus, the rate of creation of impact craters is not random but rather occurs in sharp episodes, with characteristic durations of at most a few million years and separated by tens of millions of years. Taking account of the incompleteness of discovery, it appears that, at any rate for the large impact craters, the intensity of these episodes is at least an order of magnitude greater than the background continuum.

TABLE 1. POSSIBLE BOMBARDMENT EPISODES

Group	Crater	*D* (km)	*t* (Ma)	2σ (m.y.)	Δ*t* (m.y.)
1	Zhamanshin	14.0	0.90	0.10	5.0
	Bosumtwi	10.5	1.07	----	
	New Quebec	3.4	1.40	0.10	
	Kara-Kul	52.0	<5.0	----	
	El'gygytgyn	18.0	3.5	0.50	
	Karla	10.0	5.0	1.0	
2	Chesapeake	40.0	35.3	0.10	1.9
	Popigai	90.0	35.7	0.20	
	Mistastin	28.0	36.4	0.40	
	Wanapitei	7.5	37.2	1.20	
3	Ragozinka	9.0	46.0	3.00	4.5
	Chiyli	5.5	46.0	0.70	
	Kamensk	25.0	49.0	0.20	
	Montagnais	45.0	50.50	0.76	
4	Chicxulub	150.0	64.98	0.05	0.0
	Boltysh	24.0	65.17	0.64	
5	Kara	65.0	70.3	2.20	3.8
	Lappajarvi	23.0	73.3	5.30	
	Manson	35.0	74.1	0.10	
6	Dellen	19.00	89.0	2.70	2.0
	Steen	25.00	91.0	7.00	
7	Mien	9.00	121.0	2.30	2.0
	Vargeao	12.00	123.0	1.40	
8	Mjolnir	40.00	142.0	2.60	3.0
	Gosses Bluff	22.00	142.5	0.80	
	Morokweng	70.00	145.0	0.80	
9	Zapadnay	3.20	165.0	5.00	4.0
	Puchezh-Katunki	40.00	167.0	3.00	
	Obolon'	20.00	169.0	7.00	
10	Manicouagan	85.00	214.0	1.00	0.0
	Rochechouart	23.00	214.0?	8.00	
11	Siljan	52.00	376.8	1.70	3.3
	Ilyinets	8.50	378.0	5.00	
	Kaluga	15.00	380.0	5.00	
12	Pilot	6.00	445.0	2.00	5.0
	Calvin	8.50	450.0	10.0	

Some comments on these groups:

Group 1

It might be argued that the apparent grouping of Miocene craters is an illusion caused by their relative completeness of discovery, and that the impactors are unrelated. However, this group is followed by a 30 m.y. gap, empty except for one impact structure, Nördlinger Ries at 15 Ma. Further, the Australasian tektites, at 0.77 Ma, which cover 10% of the surface area of Earth, imply that another large impactor should be added to this episode. On this interpretation, we are in a bombardment episode now.

Group 2

The 23 km Haughton crater, dated at 39 Ma, might belong to this group but has been excluded because no probable error has been attached to its radiometric age.

Group 4

The Chicxulub and Boltysh craters certainly appear to have been formed by a single astronomical event, the Ukrainian Boltysh crater having preceded the Mexican Chicxulub one by only a few thousand years (Jolley et al., 2010); hence $\Delta t = 0.0$ m.y. in the table. The only known astronomical mechanism that could plausibly account for this coincidence is the disintegration of a giant comet in the terrestrial neighborhood. Moore and Sharma (2013) have pointed out that there is a mismatch between the size of the Chicxulub crater and the iridium abundance at the boundary, in that an asteroid large enough to generate the crater would yield an overabundance of iridium. They argued that the impactor was probably a high velocity comet or, somewhat less plausibly, a short-period comet. However, there is a substantial variation in the ice-to-dust ratio from comet to comet, and one cannot reliably identify the dynamical population from which the comet arrived.

Group 10

The Rochechouart impact crater has been controversially redated to 201 ± 2 (2σ) Ma, removing its temporal connection with Manicouagan. The proposed change has not made it to the Earth Impact Database and so is not incorporated here. If the new date is correct, then the crater was formed at the Triassic-Jurassic boundary within the errors (Schmieder et al., 2010). Simms (2003, 2007) has identified a seismite layer extending over 250,000 km^2 at the late Triassic, and proposed that this may have been caused by a bolide impact at least 600 km W or NW of central Britain. If so, multiple impacts at the Triassic-Jurassic boundary are indicated.

Groups 7–12

There are 16 craters in the range 140–500 Ma, three of which do not belong to any group yet discovered. The other 13 do belong to identifiable groups, if we accept the Earth Impact Database age for the Rochechouart impact structure. Thus, the impact episodes remain unambiguous, even when, delving further into the past, the data become sparse.

TABLE 2. GROUP CONCENTRATIONS

D (km)	In	Out
>40	11	2
<40	19	15

The fact that, of the 47 craters analyzed, 36 can be assigned to groups, implies that episodes of bombardment are a prime means of delivery of large impactors. Of the 13 craters in the selected sample with $D \geq 40$ km, 11 belong to a group, with an additional "possible" member. Table 2, analyzed via the Fisher test, reveals that the smallest craters have about the same tendency to group membership as the largest ones. Thus the background drizzle of small asteroids is not a major contributor to impact history.

OTHER SOURCES OF BOMBARDMENT EPISODES

Other possible sources of multiple bombardment that have been discussed in the literature are the breakup of a main belt asteroid, passage of a star through the Oort cloud, and a disturbance of the Oort cloud by a passing massive nebula. None of them seems adequate to account for the cratering data.

Asteroid Breakup

A conceivable source of impact episodes is disintegration of main belt asteroids through mutual collisions, with feeding of debris into a resonance and their transfer to the terrestrial environment. If disintegration occurs close to a resonance, rapid transfer of fragments to the terrestrial environment is possible. Such collisions leave identifiable families of asteroids. It has been proposed that the late Eocene episode (group 2), with its accompaniment of extraterrestrial dust laid down over 2.5 m.y., was due to asteroid breakup (Kyte et al., 2011). This is based on chromium isotope ratios of impact spherules from the late Eocene, which exclude an (assumed cometary) carbonaceous chondrite source. However, the rise and fall time of the enhanced extraterrestrial dust input do not match the expectations of an asteroid breakup (Farley, 2009), and the cause of this episode remains uncertain. The majority of asteroid showers expected from the known asteroid families fail by about an order of magnitude to reproduce the sharpness, amplitude, and frequency of the observed bombardment episodes (Zappala et al., 1998; Napier, 2006; Napier and Asher, 2009).

Stellar Penetration of the Oort Cloud

Another possibility is the passage of a star through a hypothetical dense Oort cloud, disturbing its comets and sending a shower of them into the planetary system. The mean interval between passages of a star within 10,000 au of the Sun is

~30 m.y., and the duration of such showers is of the order of 1 m.y., the orbital period of in-falling comets. The inner regions of the Oort cloud (3000–10,000 au) are inaccessible to observations, and until recently, the concentrations of comets in this region, and hence the strength of any shower generated by a stellar penetration, were unknown. However, Kaib and Quinn (2009) have shown that substantial fractions of long-period comets arriving from the outer Oort cloud have leaked from the inner one, so permitting their rate of arrival to constrain the density of the unseen inner cloud. They find that probably only one comet shower has reached the likely intensity of the Eocene bombardment episode over the Phanerozoic. Close stellar encounters are therefore an unlikely cause for the bulk of the episodes documented in Table 2.

Encounters with Massive Nebulae

In the course of its orbit around the Galaxy, the Sun occasionally passes through spiral arms, within which are concentrations of giant molecular clouds, with masses up to 5×10^5 times that of the Sun and radii of order 20 parsecs (~65 light years). An encounter with such a massive nebula has a devastating effect on the Oort cloud, one consequence of which is to flood the planetary system with comets over a period of some millions of years (Fig. 5). The impact rate on Earth is correspondingly increased, both directly and through feeding long-period comets into the Halley-type and Jupiter-family reservoirs. The solar system has probably penetrated such clouds about half a dozen times during the course of its history, and a close encounter of this sort has probably taken place only about once since the Precambrian explosion. Encounters with less massive nebulae are more frequent but have correspondingly lesser effect and are not capable of reproducing the observed bombardment episodes.

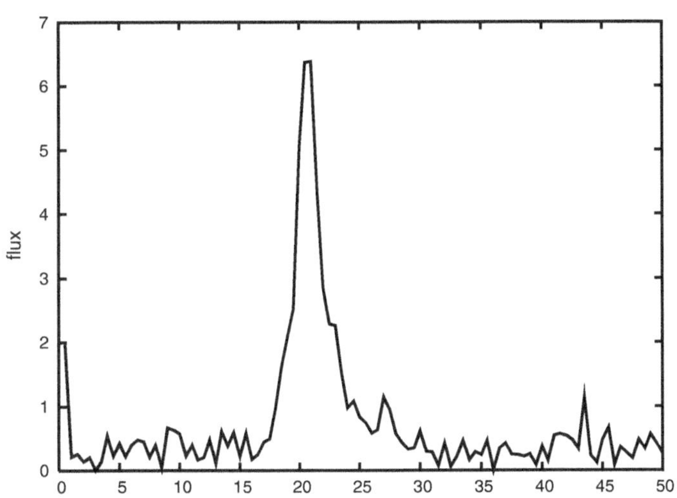

Figure 5. Flux of long-period comets into the planetary system in the course of a grazing encounter with a giant molecular cloud of mass 5×10^5 solar masses and radius 20 parsec (~65 light years). An order of magnitude increase in the background flux occurs over a period of 5 m.y.

Question of Periodicity

There is an extensive literature both supporting and disputing the existence of periodicities in the geological, biological, and impact cratering records, with periodicities in the range 26–35 m.y. often being proposed. An ~30 m.y. cycle was discussed by Holmes (1927) in connection with sea-level variations and orogenies, by Dorman (1968) in connection with global climatic variations, by Tiwari and Rao (1998) in connection with CO_2 variations, by Steiner and Grillmair (1973) in connection with glaciations, by Ager (1981) again in connection with sea-level variations, by Fischer and Arthur (1977) in relation to marine extinctions; and by many others. A link to the galactic environment has often been suggested, with periodicities arising from tidal stress exerted on the Oort cloud by spiral arm penetrations, or by the Sun's vertical oscillations through the galactic plane. These claims have been reviewed by Feng and Bailer-Jones (2013) and Melott and Bambach (2013) and are not considered here.

TERRESTRIAL CONSEQUENCES

A large impact on Earth carries energy equivalent to the explosion of a small hydrogen bomb on each square kilometer of Earth's surface. Much of this energy is quickly spread around the globe, yielding a red-hot sky that incinerates everything on the ground. It loads the atmosphere with nitric acid at toxic levels, blackens the sky for months, and, when the sky eventually clears, exposes the surface to UV radiation for years afterwards. It is not surprising that much research effort has gone into trying to connect the great mass extinctions with these violent events. Impacts of this order occur at a calculable rate, and from the astronomer's perspective, it is simply unsound to regard cosmic bombardment as a "deus ex machina" (Hallam, 2004), an arbitrary device to be invoked as a last resort when internal hypotheses fail.

Nevertheless, the simple equation "large impact = mass extinction" now seems incomplete. A new feature to emerge from the increased understanding of the astronomical environment is the importance of large comets in this process, and their tendency to break up in the inner planetary system, generating dust, icy boulders, and larger fragments up to Chicxulub class and beyond. This has several consequences. First, the trauma experienced is likely to be prolonged, with multiple large impacts and dustings occurring over 10^5 yr, often in repeated bursts (Fig. 2). Second, recurring intense fireball swarms may act to damage ecosystems starting at the base of the food chain.

The most damage that can be done to the biosphere with the least impact energy is through a modest ocean impact (Birks et al., 2007, Pierazzo et al., 2010). Halogens in saltwater reaching the upper stratosphere will deplete atmospheric ozone, the threshold for global environmental effects being crossed for a 1 km impactor. In the present scenario, bombardment episodes may enhance the frequency of such impacts by two or more powers of 10 during the disintegration time of a giant short-period comet.

A third possible effect, complex (Weart, 2013) and virtually unexplored, is climatic perturbation due to stratospheric dusting. The dust generated in the course of a giant comet breakup is subject to mutual collisions and radiation effects and will generally clear from the inner planetary system within ~20,000 yr. At present, the terrestrial mass accretion rate of cosmic dust is ~40,000 tons/yr (Love and Brownlee, 1993). However, detailed modeling (Napier, 2001) reveals that intermittent flux enhancements in excess of a million tons per year, recurring for some millennia, are expected in the course of a bombardment episode. If this incoming material is ablated to meteoric smoke (micron-sized particles) upon atmospheric entry (Klekociuk et al., 2005), then τ may approach unity for some millennia; there is then the prospect of prolonged climatic effects (Hoyle and Wickramasinghe, 1978; Clube and Napier, 1984). These phenomena raise the prospect of a more-continuous control on biological phenomena, even in the absence of huge impacts. This does not, of course, exclude multiple-cause scenarios (McLeod, 2005).

Unlike (say) large igneous provinces, the impact cratering record is highly incomplete, with many impact episodes represented by only two or three large craters. A test was carried out to see whether, nevertheless, there is a correlation between the impact episodes of Table 1 and marine mass extinctions. The latter were obtained from the relative extinction intensities (genera going extinct divided by genera extant) derived by Melott and Bambach (2010) from the Sepkoski data, using events meeting the criteria for mass extinctions given by Bambach (2006). There are 20 such events, 18 of which could be used since they span the range of the cratering data.

The statistic employed was a count of the Bambach mass extinctions found within 4 m.y. of the midpoint of an impact episode. The outcome of 50,000 Monte Carlo trials, in which synthetic mass extinction data sets were constructed by randomizing the real one, is shown in Figure 6 and Table 3. It turns out that 8 of the 18 extinction events lie within 4 m.y. of the known bombardment episodes, a result that has chance probability $p \sim 0.004$. A tighter constraint ($dt = 2.5$ m.y.) yields fewer coincidences (three) but a stronger signal ($p \sim 0.003$). If the Rochechouart redating to 201 Ma is accepted, along with the claim by Simms (2003, 2007) for at least one other bolide impact at that period, then the end-Triassic also joins this list (Fig. 7).

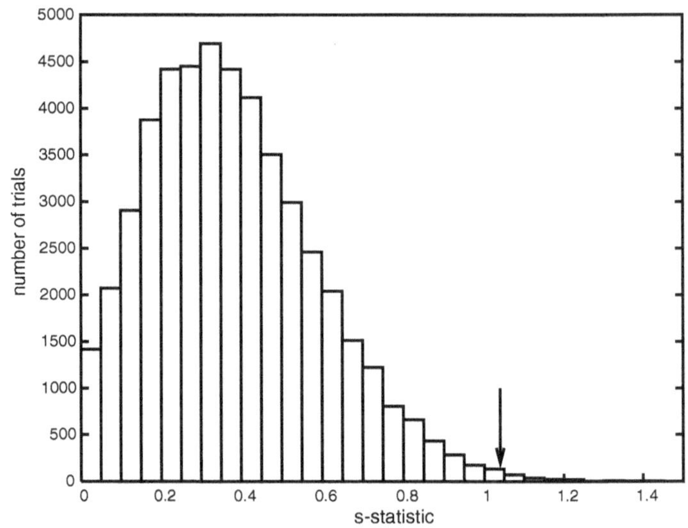

Figure 6. The arrow shows a correlation statistic between bombardment episodes and peak genera extinction rates from Bambach. The histogram shows the distribution of this statistic for 50,000 Monte Carlo trials in which the extinction peaks are randomized in time, as discussed in the text.

TABLE 3. COMPARISON OF MASS EXTINCTION AND IMPACT EPISODE SERIES

No.	Extinction event (m.y.)	Episode (m.y.)	Δt (m.y.)
1	1.81	2.50	0.69
2	33.90	36.15	2.25
3	65.50	65.00	0.50
4	93.50	90.00	3.50
5	145.50	143.17	2.33
6	183.00	166.00	17.00
7	199.60	214.00	14.40
8	251.00	214.00	37.00
9	260.40	214.00	46.40
10	322.25	378.26	56.01
11	359.20	378.26	19.06
12	374.50	378.26	3.76
13	385.30	378.26	7.04
14	443.70	447.50	3.80
15	445.47	447.50	2.03
16	488.30	447.50	40.80
17	494.65	447.50	47.15
18	498.90	447.50	51.40

Note: Mass extinction events as defined by Bambach (2006) versus known nearest neighbor bombardment episodes.

CONCLUSIONS

The impact of 20 fragments of Shoemaker-Levy 9 on Jupiter in July 1994, following a close encounter with the planet 2 yr earlier, was a vivid example of the process described here. Its occurrence within a human lifetime illustrates the fact that this is a common, continuing process. The parent comet was probably 2 km or less in diameter, and the fragments were probably <1 km. Each fragment had impact energy of ~100,000 Mt, and any one of these Jovian impacts, if it had taken place on Earth, would have collapsed civilization. There is a brisk traffic between bodies in the environments of Jupiter and Earth.

However, the mass distribution of incoming comets is top heavy, and on time scales of geological interest, the occasional arrival of bodies 100–600 km diameter will dominate the inner interplanetary environment. One expects a major disintegration (say >250 km) every 100 m.y., with a spectrum of lesser events, where the disintegration products flood the terrestrial environment with dust, icy boulders, and fragments of Chicxulub-creating dimensions and more. The dynamical clearing time is of order 0.3 m.y., but fragments may become dormant only to reactivate later, and repeated transits of fragments in and out of the terrestrial environment may extend this period. There is, however, a spectrum of comet masses, and if their disintegration products do indeed have an effect on UV exposure at ground level, then a more-continuous control of terrestrial phenomena is possible. A progressive "toughening" of marine genera with time is noted by Bambach (2006) and might at least speculatively be understood in terms of adaptation to a varying UV hazard.

The terrestrial impact record provides some support for this scenario. Impact craters are found to be clustered in episodes, within which impact rates are enhanced over the background by at least an order of magnitude. There is no one-to-one correlation between mass extinctions and specific large impacts (with the controversial exception of the Cretaceous-Paleogene extinctions), but there is a strong statistical correlation between marine extinction peaks and these episodes. It is possible that prolonged and repeated exposure to damaging UV radiation during such episodes is the prime mode of extinction. Since some near-surface aquatic species may be better adapted to endure these conditions than others, it may be possible to put this mechanism to the test.

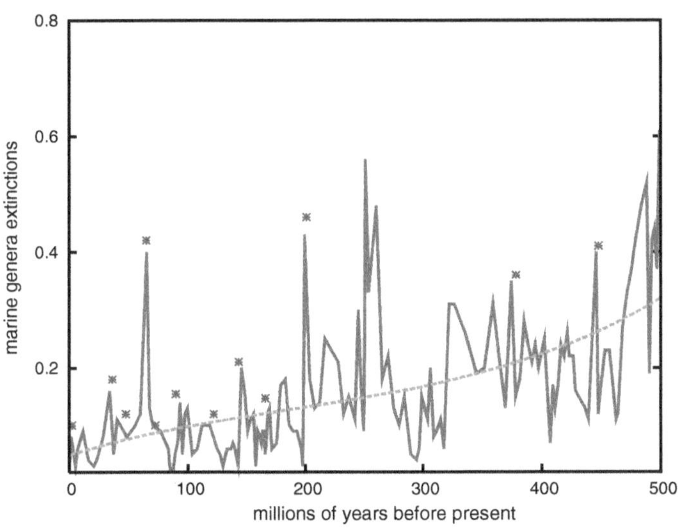

Figure 7. Bombardment episodes (asterisks) and relative extinction intensities of marine genera from Sepkoski as revised by Bambach. As discussed in the text, the revised date of the Rochechouart impact crater is used here. The dashed line is a cubic fit to the extinction data.

ACKNOWLEDGMENTS

I wish to thank Julio Fernandez and Hans Rickman, whose critical and constructive reviews of an early draft of the manuscript led to its material improvement. It is also a pleasure to acknowledge detailed discussions I have had with colleagues over the years on these issues, in particular, David Asher, Mark Bailey, Victor Clube, and Duncan Steel.

REFERENCES CITED

Ager, D., 1981, Major marine cycles in the Mesozoic: Journal of the Geological Society, v. 138, p. 159–166, doi:10.1144/gsjgs.138.2.0159.

A'Hearn, M.F., 2011, Comets as building blocks: Annual Review of Astronomy and Astrophysics, v. 49, p. 281–299, doi:10.1146/annurev-astro-081710-102506.

Alvarez, W., Alvarez, L.W., Asaro, F., and Michel, H.V., 1980, Extraterrestrial cause for the Cretaceous-Tertiary extinction: Science, v. 208, p. 1095–1108, doi:10.1126/science.208.4448.1095.

Bailey, M.E., and Emel'yanenko, V.V., 1998, Capture of Halley-type comet from the near-parabolic flux: Monthly Notices of the Royal Astronomical Society, v. 298, p. 212–222, doi:10.1046/j.1365-8711.1998.01628.x.

Bailey, M.E., Chambers, J.E., and Hahn, G., 1992, Origin of sungrazers: A frequent cometary end-state: Astronomy and Astrophysics, v. 257, p. 315–322.

Bambach, R.K., 2006, Phanerozoic biodiversity mass extinctions, *in* Jeanloz, R., Albee, A.L., Burke, K.C., and Freeman, K.H., eds., Annual Review of Earth and Planetary Sciences, v. 34, p. 127–155, doi:10.1146/annurev.earth.33.092203.122654.

Bauer, J.M., Grav, T., Blauvelt, E., Mainzer, A.K., Masiero, J.R., Stevenson, R., Kramer, E., Fernandez, Y.R., Lisse, C.M., Cutri, R.M., Weissman, P.R., Dailey, J.W., Frank, M., Russel, W., Waszczak, A., Nugent, C., Meech, K., Lucas, A., Pearman, G., Wilkins, A., Watkins, J., Kulkarni, S., and Wright, E.L., 2013, Centaurs and scattered disc objects in the thermal infrared: Analysis of WISE/NEOWISE observations: The Astrophysical Journal, v. 773, p. 12.

Benton, M.J., 1990, Scientific methodologies in collision: The history of the study of the extinction of the dinosaurs: Evolutionary Biology, v. 24, p. 371–400.

Birks, J.W., Crutzen, P.J., and Roble, R.G., 2007, Frequent ozone depletion resulting from impacts of asteroids and comets, *in* Bobrowsky, P., and Rickman, H., eds., Comet/Asteroid Impacts and Human Society: Berlin, Heidelberg, Springer, p. 225—245.

Bland, P.A., 2005, The impact rate on Earth: Philosophical Transactions of the Royal Society, ser. A, v. 363, p. 2793–2810, doi:10.1098/rsta.2005.1674.

Bottke, W.F., Vokrouhlicky, D., Rubincam, D.P., and Nesvorny, D., 2006, The Yarkovsky and YORP effects: Implications for asteroid dynamics: Annual Review of Earth and Planetary Sciences, v. 34, p. 157–191, doi:10.1146/annurev.earth.34.031405.125154.

Clube, S.V.M., and Napier, W.M., 1984, The microstructure of terrestrial catastrophism: Monthly Notices of the Royal Astronomical Society, v. 211, p. 953–968.

De Laubenfels, M.W., 1956, Dinosaur extinction: One more hypothesis: Journal of Paleontology, v. 30, p. 207–218.

Di Sisto, R.P., Fernandez, J.A., and Brunini, A., 2009, On the population, physical decay and orbital distribution of Jupiter family comets:

Numerical simulations: Icarus, v. 203, p. 140–154, doi:10.1016/j.icarus.2009.05.002.

Dorman, F.M., 1968, Some Australian oxygen isotope temperatures and a theory for a 30 m.y. world temperature cycle: The Journal of Geology, v. 76, p. 297–313, doi:10.1086/627330.

Earth Impact Database, 2005, Earth Impact Database: http://www.unb.ca/passc/ImpactDatabase/ (accessed 18 May 2013).

Emel'yanenko, V.V., Asher, D.J., and Bailey, M.E., 2013, A model for the common origin of Jupiter family and Halley type comets: Earth, Moon, and Planets, v. 110, p. 105–130, doi:10.1007/s11038-012-9413-z.

Farley, K.A., 2009, Late Eocene and late Miocene cosmic dust events: Comet showers, asteroid collisions, or lunar impacts, *in* Koeberl, C., and Montanari, A., eds., The Late Eocene Earth—Hothouse, Icehouse and Impacts: Geological Society of America Special Paper 452, p. 27–35.

Feng, C., and Bailer-Jones, C.A.L., 2013, Assessing the influence of the solar orbit on terrestrial biodiversity: The Astrophysical Journal, v. 768, p. 152, doi:10.1088/0004-637X/768/2/152.

Fernandez, J.D., and Sosa, A., 2012, Magnitude and size distribution of long-period comets in Earth-crossing or approaching orbits: Monthly Notices of the Royal Astronomical Society, v. 423, p. 1674–1690, doi:10.1111/j.1365-2966.2012.20989.x.

Fischer, A.G., and Arthur, M.A., 1977, Secular variations in the pelagic realm, *in* Cook, H.E., and Enos, P., eds., Deep-Water Carbonate Environments: Society of Economic Palaeontologists and Mineralogists Special Publication 25, p. 19–50.

Flynn, G.J., and 79 others, 2006, Elemental composition of Comet 81 P/Wild 2 samples collected by *Stardust*: Science, v. 314, p. 1731–1735, doi:10.1126/science.1136141.

Gertsch, B., Keller, G., Adatte, T., Prasad, V., Berner, Z., and Fleitmann, D., 2011, Environmental effects of the Deccan volcanism across the Cretaceous-Tertiary transition in Meghalaya, India: Earth and Planetary Science Letters, v. 310, p. 272–285, doi:10.1016/j.epsl.2011.08.015.

Grieve, R.A.F., 1984, The impact cratering rate in recent time: Journal of Geophysical Research–Solid Earth, v. 89, no. S02, p. 403–408, doi:10.1029/JB089iS02p0B403.

Grieve, R.A.F., and Robertson, P.B., 1979, The terrestrial cratering record: Icarus, v. 38, p. 212–229, doi:10.1016/0019-1035(79)90179-9.

Hahn, G., and Bailey, M.E., 1990, Rapid dynamical evolution of giant comet Chiron: Nature, v. 348, p. 132–136, doi:10.1038/348132a0.

Hallam, T., 2004, Catastrophes and Lesser Calamities: Oxford, UK, Oxford University Press, 240 p.

Helin, E.F., and Shoemaker, E.M., 1979, Palomar planet-crossing asteroid survey, 1973–1978: Icarus, v. 40, p. 321–328, doi:10.1016/0019-1035(79)90021-6.

Hildebrand, A.R., Penfield, G.T., Kring, D.A., Pilkington, M., Camargo, Z.A., Jacobson, S.B., and Boynton, W.V., 1991, A possible Cretaceous-Tertiary boundary impact crater on the Yucatán Peninsula, Mexico: Geology, v. 19, p. 867–871, doi:10.1130/0091-7613(1991)019<0867:CCAPCT>2.3.CO;2.

Holmes, A., 1927, The Age of the Earth: London, Benn, 222 p.

Horner, J., Evans, N.W., and Bailey, M.E., 2004, Simulations of the population of centaurs: I. Te bulk statistics: Monthly Notices of the Royal Astronomical Society, v. 354, p. 798–810, doi:10.1111/j.1365-2966.2004.08240.x.

Hoyle, F., and Wickramasinghe, N.C., 1978, Comets, ice ages, and ecological catastrophes: Astrophysics and Space Science, v. 53, p. 523–526, doi:10.1007/BF00645040.

Ishiguro M., Usui, F., Sarugaku Y., and Ueno, M., 2009, 2006 fragmentation of Comet 73P/Schwassmann-Wachmann 3B observed with Subaru/Suprime-Cam: Icarus, v. 203, p. 560–570. doi.org/10.1016/j.icarus.2009.04.030.

Jewitt, D.C., 2004, From cradle to grave: The rise and demise of the comets, *in* Festou, M., Weaver, H., and Keller, U., eds., Comets II: Tucson, Arizona, University of Arizona, p. 659–676.

Jolley, D., Gilmour, I., Gurov, E., Kelley, S., and Watson, J., 2010, Two large meteorite impacts at the Cretaceous-Paleogene boundary: Geology, v. 38, p. 835–838, doi:10.1130/G31034.1.

Juhasz, A., Dullemond, C.P., van Boekel, R., Bouwman, J., Acosta-Pulido, J.A., Henning, Th., Kospal, A., Sicilia-Aguilar, A., Jones, A., Moor, A., Mosoni, L., Regaly, Z.S., Szokoly, G.Y., and Sipos, N., 2012, The 2008 outburst of EX Lup—Silicate crystals in motion: The Astrophysical Journal, v. 744, p. 118, doi:10.1088/0004-637X/744/2/118.

Kaib, N.A., and Quinn, T., 2009, Reassessing the source of the long-period comets: Science, v. 325, p. 1234–1236, doi:10.1126/science.1172676.

Keller, G., Sahni, A., and Bajpai, S., 2009, Deccan volcanism, the KT mass extinction and dinosaurs: Journal of Biosciences, v. 34, p. 709–728, doi:10.1007/s12038-009-0059-6.

Kenyon, S.J., and Bromley, B.C., 2012, Coagulation calculations of icy planet formation at 15–150 au: A correlation between the maximum radius and the slope of the size distribution for trans-Neptunian objects: The Astronomical Journal, v. 143, p. 63, doi:10.1088/0004-6256/143/3/63.

Klekociuk, A.R., Brown, P.G., Pack, D.W., ReVelle, D.O., Edwards, W.N., Spalding, R.E., Tagliaferri, E., Yoo, B.B., and Zagari, J., 2005, Meteoritic dust from the atmospheric disintegration of a large meteoroid: Nature, v. 436, p. 1132–1135, doi:10.1038/nature03881.

Knight, M.M., A'Hearn, M.F., Biesecker, D.A., Faury, G., Hamilton, D.P., Lamy, P., and Llebaria, A., 2010, Photometric study of the Kreutz comets observed by *SOHO* from 1996 to 2005: The Astronomical Journal, v. 139, p. 926–949, doi:10.1088/0004-6256/139/3/926.

Kring, D.A., 2007, The Chicxulub impact event and its environmental consequences at the Cretaceous-Tertiary boundary: Palaeogeography, Palaeoclimatology, Palaeoecology, v. 255, p. 4–21, doi:10.1016/j.palaeo.2007.02.037.

Kyte, F.T., Shukolyukov, A., Hildebrand, A.R., Lugmair, G.W., and Hanova, J., 2011, Chromium-isotopes in late Eocene impact spherules indicate a likely asteroid belt provenance: Earth and Planetary Science Letters, v. 302, p. 279–286, doi:10.1016/j.epsl.2010.12.006.

Lee, M.Y., and Wei, K.Y., 2000, Australasian microtektites in the South China Sea and West Philippine Sea: Implications for age, size and location of the impact crater: Meteoritics & Planetary Science, v. 35, p. 1151–1155, doi:10.1111/j.1945-5100.2000.tb01504.x.

Liffman, K., and Brown, M., 1995, The motion and sorting of particles ejected from protostellar accretion disc: Icarus, v. 116, p. 275–290, doi:10.1006/icar.1995.1126.

Love, E.G., and Brownlee, D.E., 1993, A direct measurement of the terrestrial mass accretion rate of cosmic dust: Science, v. 262, p. 550–553, doi:10.1126/science.262.5133.550.

McDonald, I., Andreoli, M.A.G., Hart, R.J., and Tredoux, M., 2001, Platinum-group elements in the Morokweng impact structure, South Africa: Evidence for the impact of a large ordinary chondrite projectile at the Jurassic-Cretaceous boundary: Geochimica et Cosmochimica Acta, v. 65, p. 299–309, doi:10.1016/S0016-7037(00)00527-5.

McLeod, N., 2005, Mass extinction causality: Statistical assessment of multiple-cause scenarios: Russian Geology and Geophysics, v. 46, p. 993–1001.

Melott, A.L., and Bambach, R.K., 2010, Nemesis reconsidered: Monthly Notices of the Royal Astronomical Society, v. 407, p. L99–L102, doi:10.1111/j.1745-3933.2010.00913.x.

Melott, A.L., and Bambach, R.K., 2013, Do periodicities in extinction—with possible astronomical connections—survive a revision of the geological timescale?: The Astrophysical Journal, v. 773, p. 6, doi:10.1088/0004-637X/773/1/6.

Menichella, M., Paolicchi, P., and Farinella, P., 1996, The main belt as a source of near-Earth asteroids: Earth, Moon, and Planets, v. 72, p. 133–149, doi:10.1007/BF00117512.

Minton, D.A., and Malhotra, R., 2010, Dynamical erosion of the asteroid belt and implications for large impacts in the inner solar system: Icarus, v. 207, p. 744–757, doi:10.1016/j.icarus.2009.12.008.

Moore, J.R. and Sharma, M., 2013, The K-Pg impactor was likely a high velocity comet: Lunar and Planetary Institute Contributions, v. 1719, abstract 2431.

Nakamura, T., Noguchi, T., Tsuchiyama, A., Ushikubo, T., Kita, N.T., Valley, J.W., Zolensky, M.E., Kakazu, Y., Mashio, E., Uesugi, K., and Nakano, T., 2008, Chondrule-like objects in short-period comet 81P/Wild 2: Science, v. 321, p. 1664–1667, doi:10.1126/science.1160995.

Napier, W.M., 2001, Temporal variation of the zodiacal dust cloud: Monthly Notices of the Royal Astronomical Society, v. 321, p. 463–470, doi:10.1046/j.1365-8711.2001.04020.x.

Napier, W.M., 2006, Evidence for cometary bombardment episodes: Monthly Notices of the Royal Astronomical Society, v. 366, p. 977–982.

Napier, W.M., 2010, Palaeolithic extinctions and the Taurid Complex: Monthly Notices of the Royal Astronomical Society, v. 405, p. 1901–1906.

Napier, W.M., and Asher, D.J., 2009, The Tunguska impact event and beyond: Astronomy & Geophysics, v. 50, p. 18–26, doi:10.1111/j.1468-4004.2009.50118.x.

Napier, W.M., and Clube, S.V.M., 1979, A theory of terrestrial catastrophism: Nature, v. 282, p. 455–459, doi:10.1038/282455a0.

Pierazzo, E., Garcia, R.R., Kinnison, D.E., Marsh, D.R., Lee-Taylor, J., and Crutzen, P.J., 2010, Ozone perturbation from medium-size asteroid impacts in the ocean: Earth and Planetary Science Letters, v. 299, p. 263–272. doi:10.1016/j.epsl.2010.08.036.

Rickman, H., Fernandez, J.A., and Gustafson, B.A.S., 1990, Formation of stable dust mantles on short-period comet nuclei: Astronomy and Astrophysics, v. 237, p. 524–535.

Schmieder, M., Buchner, E., Schwarz, W., Trieloff, M., and Lambert, P., 2010, A Rhaetian ^{40}Ar/^{39}Ar age for the Rochechouart impact structure (France) and implications for the latest Triassic sedimentary record: Meteoritics & Planetary Science, v. 45, p. 1225–1242.

Schulte, P., Alegret, L., Arenillas, I., Arz, J.A., Barton, P.J., Bown, P.R., Bralower, T.J., Christeson, G.L., Claeys, P., Cockell, C.S., Collins, G.S., Deutsch, A., Goldin, T.J., Goto, K., Grajales-Nishimura, J.M., Grieve, R.A.F., Gulick, S.P.S., Johnson, K.R., Kiessling, W., Koeberl, C., Kring, D.A., MacLeod, K.G., Matsui, T., Melosh, J., Montanari, A., Morgan, J.V., Neal, C.R., Nichols, D.J., Norris, R.D., Pierazzo, E., Ravizza, G., Rebolledo-Vieyra, M., Reimold, W.U., Robin, E., Salge, T., Speijer, R.P., Sweet, A.R., Urrutia-Fucugauchi, J., Vajda, V., Whalen, M.T., and Willumsen, P.S., 2010, The Chicxulub asteroid impact and mass extinction at the Cretaceous-Paleogene boundary: Science, v. 327, p. 1214–1218, doi:10.1126/science.1177265.

Sekanina, Z., and Chodas, P.W., 2004, Fragmentation hierarchy of bright sungrazing comets and the birth and orbital evolution of the Kreutz system: I. Two-superfragment model: The Astrophysical Journal, v. 607, p. 620–639, doi:10.1086/383466.

Sekanina, Z., and Chodas, P.W., 2007, Fragmentation hierarchy of bright sungrazing comets and the birth and orbital evolution of the Kreutz system: II. The case for cascading fragmentation: The Astrophysical Journal, v. 663, p. 657–676, doi:10.1086/517490.

Shukolyukov, A., and Lugmair, G.W., 1998, Isotopic evidence for the Cretaceous-Tertiary impactor and its type: Science, v. 282, p. 927–930, doi:10.1126/science.282.5390.927.

Simms, M.E., 2003, Uniquely extensive seismite from the latest Triassic of the United Kingdom: Evidence for bolide impact?: Geology, v. 31, p. 557–560, doi:10.1130/0091-7613(2003)031<0557:UESFTL>2.0.CO;2.

Simms, M.E., 2007, Uniquely extensive soft-sediment deformation in the Rhaetian of the UK: Evidence for earthquake or impact?: Palaeogeography, Palaeoclimatology, Palaeoecology, v. 244, p. 407–423, doi:10.1016/j.palaeo.2006.06.037.

Stansberry, J., Grundy, W., Brown, M., Cruikshank, D., Spencer, J., Trilling, D., and Margot, J.-L., 2008, Physical properties of Kuiper belt and centaur objects: Constraints from Spitzer space telescope, *in* Barucci, M.A., Boehnhardy, H., Cruikshank, D.P., and Morbidelli, A., eds., The Solar System beyond Neptune: University of Arizona Press, p. 161–179.

Steel, D.I., and Asher, D.J., 1996, The orbital dispersion of the macroscopic Taurid objects: Monthly Notices of the Royal Astronomical Society, v. 280, v. 806–822.

Steiner, A., and Grillmair, E., 1973, Possible Galactic causes for periodic and episodic glaciations: Geological Society of America Bulletin, v. 84, p. 1003–1018.

Tiwari, R.K., and Rao, K.N.N., 1998, Correlated variations and periodicity of global CO_2, biological mass extinctions and extra-terrestrial bolide impacts over the past 250 million years and possible geodynamical implications: Geofizika, v. 15, p. 103–117.

Urey, H.C., 1973, Cometary collisions and geological periods: Nature, v. 242, p. 32–33, doi:10.1038/242032a0.

Weart, S., 2013, Aerosols: Volcanoes, Dust, Clouds and Climate: http://www.aip.org/history/climate/aerosol.htm (accessed 14 February 2014).

Whiston, W., 1696, A New Theory of the Earth from its Original to its Consummation of All Things: London, Benjamin Tooke, 388 p.

Wiegert, P.A., Vaubaillon, J., and Campbell-Brown, M.D., 2009, A dynamical model of the sporadic meteoroid complex: Icarus, v. 201, p. 295–310, doi:10.1016/j.icarus.2008.12.030.

Zappala, V., Cellino, A., Gladman, B.J., Manley, S., and Migliorini, F., 1998, Asteroid showers on Earth after family breakup events: Icarus, v. 134, p. 176–179, doi:10.1006/icar.1998.5946.

Manuscript Accepted by the Society 31 January 2014

Printed in the USA

The Geological Society of America
Special Paper 505
2014

Implications of the centaurs, Neptune-crossers, and Edgeworth-Kuiper belt for terrestrial catastrophism

Duncan Steel*
Centre for Astrobiology, University of Buckingham, and *Armagh Observatory, UK*

ABSTRACT

The discovery of many substantial objects in the outer solar system demands a reassessment of extraterrestrial factors putatively implicated in mass extinction events. These bodies, despite their formal classification as minor (or dwarf) planets, actually are physically similar to comets observed passing through the inner solar system. By dint of their sizes (typically 50–100 km and upward), these objects should be considered to be giant comets.

Here, I complement an accompanying paper by Napier, who describes how giant comets should be expected to cause major perturbations of the interplanetary environment as they disintegrate, leading to fireball storms, atmospheric dustings, and bursts of impacts by Tunguska- and Chelyabinsk-class bodies into the atmosphere, along with less-frequent arrivals of large (>10 km) objects. I calculate the terrestrial impact probability for all known asteroids and discuss why the old concept of single, random asteroid impacts causing mass extinctions is deficient, in view of what we now know of the inventory of small bodies in the solar system. Also investigated is how often giant comets might be thrown directly into Earth-crossing orbits, with implications for models of terrestrial catastrophism.

A theme of this paper is an emphasis on the wide disparity of ideas amongst planetary and space scientists regarding how such objects might affect the terrestrial environment, from a purely astronomical perspective. That is, geoscientists and paleontologists should be aware that there is no uniformity of thought in this regard amongst the astronomical community.

PREAMBLE

Papers appearing in this volume result from the international conference on "Volcanism, Impacts and Mass Extinctions." While that meeting had "impacts" (meaning hypervelocity arrivals on Earth of extraterrestrial objects) in its title, comparatively little expertise in that particular area was represented amongst the attendees. That is not, in itself, surprising, and this is also reflected in the fact that the Geological Society of America (rather than some astronomical society) is publishing these papers.

What I found more surprising was the apparently limited knowledge of the huge advances made in astronomical knowledge

*Contact address: P.O. Box 9461, Marion Square, Wellington 6141, New Zealand; e-mail: TMA1@duncansteel.com.

Steel, D., 2014, Implications of the centaurs, Neptune-crossers, and Edgeworth-Kuiper belt for terrestrial catastrophism, *in* Keller, G., and Kerr, A.C., eds., Volcanism, Impacts, and Mass Extinctions: Causes and Effects: Geological Society of America Special Paper 505, p. 397–410, doi:10.1130/2014.2505(21). For permission to copy, contact editing@geosociety.org.

and understanding of the smaller bodies in the solar system over the past two decades, with the discovery of myriad asteroids and comets in distinct populations that had hitherto only been suspected to exist. That is, in terms of the cosmic environment in which Earth (and therefore, obviously, the biosphere) orbits the Sun, there are now recognized by space scientists to be:

> ...more things in heaven and earth, Horatio,
> Than are dreamt of in your philosophy.
>
> —William Shakespeare, *Hamlet*, Act 1, Scene 5.

In view of this, I have prepared this report in a deliberately chatty manner, contrary to the practice of most researchers, in the hope that it will be more accessible and therefore more useful than most dry, abstruse scientific papers.

This report is not intended as a complete review of the field, however, and I have taken very much a personal view. Others in the worldwide astronomical–planetary science community have quite different opinions on substantive matters discussed herein; and so, to the reader: *caveat emptor*.

Further, as I explain herein, I have benefited from the fact that there is also a paper in this volume written by Bill Napier, who has covered much material that otherwise I might have needed to discuss here; thus my paper should be read in conjunction with that of Napier (this volume).

To bring this preamble to a conclusion, I explain that my comment about there being "limited knowledge" of astronomical topics represented at the conference is intended solely as an observation, and not in any way a criticism of either the attendees or the organizers, who come predominantly from the geological and paleontological fraternities and sororities. For myself, I found the conference extremely interesting and useful specifically because in those geoscience areas I must confess to profound ignorance and a lack of solid background. I sat and listened enthralled for three days, trying to understand as best I could what various experts in the fields of geology and paleontology were describing in terms of their research outcomes. Very often I found myself thinking: "Yes, that is just what one might expect to result on Earth from a major upset of the inner solar system environment as a big comet breaks apart, spreading its decay products around, rather than the simple (and rather passé) notion of a random single large asteroid strike."

What I have tried to outline here, then, is a picture of what we now know the outer solar system to contain in terms of giant, comet-like bodies; how and why they are on unstable orbits (on time scales of millions of years); and what this must mean for the inner solar system and therefore the terrestrial environment.

INTRODUCTION

The outer solar system contains a cosmic zoo of large, cold, solid objects variously classified as being centaurs, Edgeworth-Kuiper belt objects (EKBOs), Neptune-crossers, dwarf planets, minor planets[1], and (giant) comets, although other terminology, classifications, and subdivisions are also in use amongst the planetary science community (e.g., trans-Neptunian objects [TNOs], cis-Neptunian objects, plutinos, cubewanos, scattered disk objects [SDOs], distant detached objects [DDOs], inner Oort [or Oort-Öpik] cloud objects). Considering that the first centaur[2] was only discovered in 1977, and the first trans-Neptunian object[3] was discovered in 1992, this proliferation of terms is bewildering to the ingénue.

It is lamentable that the reclassification in 2006 of some of the "minor planets" (and one "planet": Pluto) as "dwarf planets" by the International Astronomical Union (IAU) proceeded as it did, because the opportunity was missed to rectify the anomalous situation whereby many objects that appear stellar to the eye in a small telescope (i.e., literally "asteroidal") might have been better defined by dint of their physical nature as being "giant comets." That is, many centaurs, Neptune-crossers, and Edgeworth-Kuiper belt objects are actually icy bodies that would certainly be counted as being comets should they have been discovered on paths bringing them sufficiently close to the Sun (e.g., the water-ice initiation-of-sublimation heliocentric distance of around 2.9 astronomical units [au]) such that cometary emissions of gas and dust would soon have been observed. That is, rather than appearing asteroidal, they would be seen to be "hairy stars" (the etymology of the word "comet"). The revised classification of comet-like bodies suggested by Horner et al. (2003) has, in my opinion, much to recommend it.

A particular example is the centaur (2060) Chiron = 95P/Chiron. The first designation there signifies Chiron to be a minor planet; the second designation signifies a periodic comet, number

[1]The term "minor planet" is that which the International Astronomical Union applies to those objects that most people call "asteroids"; that is, these are broadly synonyms. For this reason, I will use the term "asteroid" in this paper to refer to subplanetary rocky/metallic bodies in inner solar system (cis-Jovian) orbits.

[2]Here I have classed (2060) Chiron as the first centaur, in terms of the recognition of such objects constituting a new class of outer solar system body, although (944) Hidalgo was discovered over half a century earlier, in 1920. "Centaurs" are "small" (compared to major planets) outer solar system bodies that are observed mainly in the region between Jupiter and Neptune and cross the orbit of at least one of the Jovian planets: A technical definition of centaurs has been discussed recently by Emel'yanenko et al. (2013), and their definition would exclude Hidalgo on the basis of its relatively small perihelion distance. The largest-known centaur is (10199) Chariklo, which is about 260 km across; (2060) Chiron is around 200 km in size. Several of the known centaurs display comet-like outgassing and have reflectance spectra similar to volatile-rich carbonaceous (C-type) asteroids and cometary nuclei, and so it is not clear whether they should be classed as minor planets (as they generally are), or comets. Because their physical nature appears to be basically cometary, I argue in this paper that they should be classified as being "comets."

[3]The first trans-Neptunian object to be discovered was 1992 QB1, hence the peculiar class name "cubewano," although the existence of such a belt of large bodies had been predicted far earlier, resulting in the term "Edgeworth-Kuiper belt," containing the names of those who had suggested its unseen existence in the 1940s and 1950s, respectively.

95 in the IAU list of such bodies. This double-designation occurs because it was classified as a minor planet on discovery in 1977 but was later recognized to exhibit comet-like outgassing (and one might argue that it should actually be called "95P/Kowal," for the discoverer). Many such crossover objects[4] are now known to exist, confusing the IAU's illogical classification scheme.

This mixed bag of distant objects contains bodies that must occasionally be forced to misbehave through long-term gravitational perturbations, collisions with each other, and/or close encounters with the gas giant planets. In a small fraction of such contingent events, the intact objects (or perhaps their decay products) will be thrown into orbits with small perihelia. Then chaos will reign in the inner solar system as they disintegrate due to thermal stresses and other factors yet to be fully comprehended.

I am indebted to an accompanying paper by my long-time colleague Bill Napier in which he discussed a range of considerations with regard to the feasible consequences of a giant comet entering the inner solar system and becoming trapped in a cis-Jovian orbit, especially in terms of what this could mean for the biosphere. Those are largely matters, then, that I do not need to repeat herein, with an adequate prefatory statement regarding the present paper being contained in the next paragraph.

Consider what might happen should a 100 km object (i.e., a giant comet) be diverted first into an orbit with perihelion distance $q < 1$ au, and then into a cis-Jovian orbit similar to that of 2P/Encke. Such an orbit is dynamically stable for a period of order 10^5 yr, over which time it would be expected (based on the disparate behavior of observed comets) to undergo a hierarchical disintegration into a vast complex of smaller bodies (which we might classify as minor planets, near-Earth asteroids, comets, meteoroids, and dust). The currently observed Earth-crossing interplanetary dust and meteoroid complex has a total mass equivalent to a comet or asteroid only ~20 km in diameter, so that the influx of such small debris to Earth would be hugely enhanced, with substantial environmental effects, in this giant-comet-breakup scenario. A 100 km comet, if subdivided into neat 1 km fragments, could also result in catastrophic terrestrial impacts on a once-per-century time scale for a million years.

This might be considered to be wildly speculative, but our expanding knowledge and understanding of the outer solar system should be leading us to an expectation that events such as this—long-lived (10^5–10^6 yr) severe perturbations of the inner solar system interplanetary environment with concomitant consequences for the biosphere—must have occurred many times over the course of our planet's history.

In this paper, I examine the likelihood of such events occurring. I will start with some calculations and comments regarding the frequency of asteroid impacts on Earth, largely because it is important that a distinction be drawn between the simple idea of stray cosmic rocks upsetting the terrestrial environment for a short interval (1–100 yr?) and the rather more complex scenario involving large comets breaking asunder and bombarding our planet over prolonged (10^5–10^6 yr) episodes. I then continue with an estimation of the probability per unit time of a Neptune-crosser or a centaur being diverted directly into an Earth-crossing orbit, from there deducing how often episodes like those envisioned earlier might have occurred across the Phanerozoic.

The results have implications for the interpretation of Earth's geological and biological history, and also for our understanding of the present and future hazards posed by asteroids, comets, meteoroids, and dust to our contemporary civilizations.

ASTEROIDS VERSUS COMETS

A relatively small contingent of astronomers and space scientists attended the conference on "Volcanism, Impacts and Mass Extinctions" held at the Natural History Museum in London in late March of 2013, compared to the number of attendees from the geological and paleontological communities. Just over 2 weeks later, I also attended the Planetary Defense Conference ("Protecting Earth from Asteroids") in Flagstaff, Arizona, organized by the International Academy of Astronautics; there, the lack of equilibrium between the scientific disciplines was reversed.

One thing that these two disparate international conferences did have in common, however, was that the discussions focused squarely on "asteroids" as being the cosmic culprits for extinction events and/or civilization-interrupting catastrophes, rather than "comets." I believe this is misguided, for both meetings, but for different reasons.

In terms of planetary defense—defending ourselves from the deleterious effects of terrestrial collisions by extraterrestrial bodies of substantial size—for every single mention of comets at the Flagstaff conference, I heard 99 mentions of asteroids. In some ways, this is because comets represented the elephant in the room: Tackling an earthbound asteroid would be difficult, but tackling an earthbound comet would be impossible, at least in the foreseeable future, plus we know how to predict asteroid impacts decades ahead of time, but we might expect only a couple of years' warning for a new, long-period comet found to be heading our way (Marsden and Steel, 1994). For planetary

[4]These are not all in the outer solar system. The Earth-crossing object 107P/Wilson-Harrington (i.e., a periodic comet designation) = (4015) Wilson-Harrington (i.e., a minor planet/asteroid designation) was discovered in 1949, classed then as being a comet due to its nebulous appearance and dust tail, but quickly lost; it was rediscovered in 1979 as an asteroid, and it was not until 1992 that the two discoveries were tied together as being a single object that displayed comet-like activity in 1949 but not in subsequent detections. Based on its absolute magnitude, it is about 2–4 km in size. Whether it is actually an extinct, moribund, or dormant comet is a question that should be answered by future observations, including perhaps by space probes; it is also feasible that Wilson-Harrington might truly be an asteroid (i.e., a rocky/metallic body), which appeared slightly fuzzy in 1949 due to dust and meteoroids having recently been lost from its surface regolith in some way, such as a collision with a smaller but substantial object. There are several other examples: Three crossover asteroid-comet objects were studied in terms of both dynamics and physical properties by Di Martino et al. (1998). In addition to the above objects in high- and moderate-eccentricity (i.e., comet-like) orbits, there are also now several objects known in the (low-eccentricity) main asteroid belt that exhibit comet-like physical activity, and these have been termed the "Lazarus comets" (Ferrin et al., 2013).

defense purposes, however, this neglect of comets is likely to be justifiable: the next significant (e.g., 50–200 m) impactors to strike Earth will most probably be small asteroids, and indeed their arrivals are, in principle, predictable, enabling ameliorative action to be taken (even if that only consists of evacuation of the target area or other civil defense actions).

From the perspective of understanding past mass extinction events, though, a mindset that is fixated on asteroids is most likely *not* justifiable. Yes, comets generally have substantially higher impact speeds (and therefore impact energies per unit mass that are around an order of magnitude greater than for asteroids), but this ignores the *qualitative* differences between the two broad types of interplanetary object in terms of how they might affect the terrestrial environment. The frequencies with which comets are observed to fragment on-orbit, their far higher masses, and the plausible consequences of flooding the inner solar system with small meteoroids and dust, have all been discussed by Napier (this volume) in his accompanying paper, and so they are not repeated here.

At the Volcanism, Impacts and Mass Extinctions Conference, it was apparent that most attendees regarded cosmic impacts onto Earth as being largely the consequence of single, monolithic, stray asteroids. From a knowledgeable astronomical perspective, that might be regarded as being simple to the point of being oversimplistic. On the other hand, with the benefit of hindsight, it would appear that much of the research published by astronomers on this topic in the 1980s and 1990s (cometary waves; Planet X; the Nemesis star; and so on) may have been sophisticated, but also exhibited sophistry. Simply put, I do not believe that we knew enough about giant comets back then to be able to say much definitively about their relevance (or not) with regard to mass extinction events; I have already pointed out that the first trans-Neptunian object was not discovered until late 1992, and only three centaurs were known by then, so how could we say anything definitive about the ways and frequencies with which giant comets might affect the biosphere? That is a rhetorical question anticipating a negative response, but it does have a partial positive answer, which is: Study the papers by Napier and Clube (1979) and Clube and Napier (1984).

Previously, I wrote that at the Planetary Defense Conference it seemed to me that the subject of comets represented the elephant in the room. Now I will use another pachydermic metaphor, the well-known tale of the blind men and the elephant. The relevant *Wikipedia* page concerning this story begins:

> "The story of the blind men and an elephant originated in the Indian subcontinent from where it has widely diffused. It has been used to illustrate a range of truths and fallacies. At various times it has provided insight into the relativism, opaqueness or inexpressible nature of truth, the behavior of experts in fields where there is a deficit or inaccessibility of information, the need for communication, and respect for different perspectives."

In trying to understand the biosphere and its evolution, we blind men and women need to be sure we are feeling around the correct end of the elephant. My experience was that at both the Planetary Defense and the Volcanism, Impacts and Mass Extinctions conferences, most participants were thinking of cosmic impacts as being like the currant buns that a miscreant ungulate might occasionally throw our way, using its prehensile proboscis, whereas in fact the things we need to regard as being of most significance in terms of unraveling the history of mass extinctions is the less-cohesive, bigger lumps of stuff that come winging at us from under the elephant's tail. Metaphorically speaking, of course.

PRESENT-EPOCH ASTEROID INFLUX TO EARTH

In this section, I consider the frequency of impacts on Earth by asteroids, based on the presently observed population, and discuss briefly what this implies with regard to the understanding of how asteroid arrivals may (or may not) contribute to mass extinction events or other biospheric upsets.

The atmospheric entry on 15 February 2013 of a small (~17 m) asteroid just to the south of the city of Chelyabinsk in eastern Russia has led to conflicting reports on the population of Earth-crossing asteroids of such a size, and therefore the frequency with which such minor collisions might be anticipated. Obviously, the total population of Earth-crossing asteroids at some size limit multiplied by the mean impact probability of the ensemble should render the frequency of collisions with Earth. A population of about 10 million Earth-crossing asteroids down to this (15–20 m) size has been argued, and a frequency of impacts of about one per century has been proposed. Those two figures may be combined to provide a mean impact probability of 1.0×10^{-9} per year, and on realizing this, I recognized that something must be wrong, because my earlier calculations of mean impact probabilities (e.g., Steel, 2002) indicated values significantly larger than this, by a factor of at least three or four. Thus, either the population of small Earth-crossing asteroids is numerically lower than stated above, or the frequency of such minor Earth-crossing asteroid arrivals is higher than once per century, or else something else is affecting the sums (e.g., some resonance effect is protecting Earth from being struck as often as one might anticipate).

Whatever the cause of the apparent anomaly, it was clear to me that an updated value for the mean Earth-crossing asteroid collision probability with Earth as a function of asteroid size would be of utility, and so I accessed all the Earth-crossing asteroid orbits available from the Minor Planet Center[5] on 11 April 2013 and set about calculating the individual—and thence average—impact probabilities, and also speeds (v). The cutoff date has no meaning other than the fact that I was about to depart the National Aeronautics and Space Administration (NASA) Ames Research Center in California for the Planetary Defense Conference (Flagstaff, Arizona, 15–19 April 2013), and I thought it might be useful to prepare an impromptu poster on this topic.

Previously I have used the technique of Kessler (1981) to calculate the terrestrial impact probabilities and speeds of

[5]http://www.minorplanetcenter.net/iau/lists/MPLists.html

various planet-crossing objects based on the known populations of asteroids and comets (e.g., see Steel and Baggaley, 1985; Olsson-Steel, 1987a; Marsden and Steel, 1994; Steel, 1998, 2002). Because the discovered population of Earth-crossing asteroids has grown markedly over the past 11 yr, a revisit of such calculations seems timely.

Kessler's method is based on the concept of spatial densities, which might be regarded as likelihoods of particular orbiting objects (e.g., Earth and an asteroid) being present in the same volume of space at the same time, whereupon a collision is possible; the probability of such an event occurring depends also on the intersection velocities and the (speed-dependent) collision cross section of the planet. This method allows for both the objects to be in noncircular orbits, and also for both to have nonzero inclinations (although in the present instance, all Earth-crossing asteroid orbits are referred to the ecliptic, and so regions of space at nonzero ecliptic latitudes are not relevant; that is, any impact on Earth must occur in the ecliptic plane). Kessler's method is superior to that of Öpik (1976, and many earlier papers) in that it is not limited to circular planetary orbits, and Öpik's geometrical method also strictly requires the impactors to be deep-crossers of the planet's orbit (cf. Bottke et al., 1994). This is particularly important because shallow-crossers of Earth's orbit have intrinsically elevated impact probabilities, meaning that they should not be neglected in collision rate/probability assessments. A comparison between various statistical-type collision probability techniques has been given by Manley et al. (1998).

A fundamental assumption intrinsic in Kessler's technique is that the longitude of the node and the argument of periapsis of each orbiting object are taken to be random. In essence, this requires the precession time scales of those angular elements to be briefer than the collisional and dynamical lifetimes. It is feasible, however, that some orbiting objects cannot collide, due to mean-motion resonance effects (e.g., the Trojan asteroids in a 1:1 resonance with Jupiter; or Pluto in a 3:2 resonance with Neptune).

Another possible deficiency of Kessler's technique (and similar statistical methods, such as that of Öpik) is that the precession of the argument of perihelion (ω) does not occur at a uniform rate, but rather is dependent upon the value of ω, whereas collisions can only occur with a planet for specific values of ω given a potential impactor's values of semimajor axis and eccentricity, as discussed by Shoemaker et al. (1979). For example, 2P/Encke's rate of apsidal precession is higher when ω is close to 90° and 270° than when it is near 0° or 180°, and both the eccentricity and inclination also vary with the phase in the apsidal precession cycle (i.e., the value of ω): See figure A2 in Asher and Clube (1993).

In total, 5633 Earth-crossing asteroids were available for use in the calculations presented here, their orbits being downloaded from the Minor Planet Center Web site as noted above. Of these, 762 are Atens and 4871 are Apollos (differentiated by the former having semimajor axes smaller than 1 au, and the latter having larger semimajor axes and thus orbital periods longer than one year), based on them crossing the osculating terrestrial orbit (defined by a semimajor axis $a = 1$ au and eccentricity $e = 0.0167$). There are also 12 known asteroids with orbits entirely interior to Earth's present orbit (i.e., their aphelion distances are less than the current terrestrial perihelion distance of 0.9833 au), and these were not included in the calculations because they cannot collide with Earth in the present epoch.

Of these 5633 Earth-crossing asteroids, there are 13 with absolute magnitude $H > 30$, implying sizes (equivalent diameters) of below 2.6–5.9 m (for albedos of 0.25 and 0.05 respectively[6]), and these I have neglected. This leaves 5620 Earth-crossing asteroids with $H \leq 30$ that were used in the calculations discussed in the following.

All "impact speeds" discussed here are the speeds with which an asteroid would strike the terrestrial surface in the absence of any atmospheric deceleration. That is, these impact speeds are the orbital intersection velocities added in quadrature with Earth's escape speed (~11.2 km/s).

For each of the 5620 Earth-crossing asteroids, the collision probability with Earth was calculated and the individual results stored for later analysis. In addition, the software used was adjusted such that a tally was kept, in working on each Earth-crossing asteroid, of the probability distribution of different impact speeds; the impact speeds for individual Earth-crossing asteroids vary, for example, dependent upon whether the putative collision occurs near Earth's perihelion, or aphelion. This means that values of the ranges of impact speeds were accumulated in the analysis with allowance for weighting by the impact probabilities, so that a properly weighted mean impact speed could be derived based on all Earth-crossing asteroids brighter than whichever limiting absolute magnitude H_{lim} was selected. The resulting mean values are shown in Table 1; they are also plotted in Figure 1.

It is also possible to weight the collision probabilities by some other power of the impact speed than unity: For example, if a weighting according to values of v^2 were used, then the derived collision probability distribution might be interpreted in terms of the likelihoods of different impact kinetic energies. Such weightings (for powers on v from 0.1 to 4.0) were used for both asteroidal and parabolic comet orbits that might impact Earth or Mars by Steel (1998).

It is apparent from Table 1 and Figure 1 that the calculated mean terrestrial impact probability, based on the presently known Earth-crossing asteroid population, grows markedly as one admits ever-fainter/smaller asteroids into the sample, while the weighted mean impact speed reduces. Going out from (say) $H_{lim} = 15$ to $H_{lim} = 30$ results in the calculated mean impact probability increasing by more than an order of magnitude, while the derived average impact speed tumbles by 40%.

These trends should not be thought to reflect reality precisely, but rather are most likely affected by discovery selection effects. For the larger Earth-crossing asteroids (size > 1 km, $H < 18$), the discovery is believed to be almost complete, with over

[6]http://www.minorplanetcenter.net/iau/lists/Sizes.html

TABLE 1. MEAN TERRESTRIAL IMPACT PROBABILITIES AND SPEEDS

Limiting absolute magnitude H_{lim}	Typical size (km)	Number of known Earth-crossing asteroids	Mean impact probability per 10^9 yr	Weighted mean impact speed (km/s)
13	5	1	0.97	27.7
14	4	5	1.19	27.7
15	2.6	26	2.14	21.1
16	1.7	101	2.57	20.6
17	1.0	244	3.01	19.4
18	0.67	549	3.00	19.5
19	0.42	1077	3.48	19.0
20	0.26	1777	4.43	18.3
22	0.11	2997	5.67	17.5
24	0.04	3919	8.42	16.0
26	0.017	4941	18.64	14.0
28	0.007	5503	27.92	13.4
30	0.003	5620	36.01	13.0

90% of Earth-crossing asteroids in cis-Jovian orbits now being known. For smaller/fainter Earth-crossing asteroids, however, a far smaller fraction of the entire population has so far been identified, with that fraction diminishing as the asteroids in question get fainter/smaller. The small Earth-crossing asteroids that *have* been discovered will tend to be those that pass relatively close by Earth, and are therefore preferentially discovered by the telescopic search techniques because they have orbits making more frequent near-Earth passages. Broadly, these will have low-inclination (i), low-eccentricity (e) orbits, and thus exhibit higher-than-normal collision probabilities, and the collisions would occur at lower-than-normal speeds. This presumed preferential discovery of small Earth-crossing asteroids in such near-Earth orbits is most evident in the abrupt upturn at $H_{lim} = 24$ in the impact probability curve in Figure 1.

There is an alternative, additional, consideration that might be worthy of investigation. Earlier, I noted that Earth may be protected from collisions by some Earth-crossing asteroids due to them being in some form of orbital resonance (e.g., Trojan orbits, or horseshoe orbits such as those occupied by [3753] Cruithne and 2010 SO16), but this would seem likely to be a minor *reducer* of the overall terrestrial collision rate. A physical effect that could significantly *enhance* the collision rate by smaller Earth-crossing asteroids might be this: Any large but weak (i.e., "rubble pile") Earth-crossing asteroid that migrates into a low-e, low-i orbit near 1 au will generally make many close approaches past Earth before any collision occurs. During these close approaches, the tidal force, especially on a spinning or tumbling Earth-crossing asteroid of low tensile strength, may disrupt that object into several/many smaller bodies, each of which would then follow its own trajectory in terms of its future orbital evolution (Richardson et al., 1998; Michel and Holsapple, 2007). Thus, it is feasible that at least part of the small Earth-crossing asteroid population might derive from larger Earth-crossing asteroids that have split apart *after* reaching low-e, low-i orbits; this would imply that we should not expect the orbital distribution of small Earth-crossing asteroids (sub–100 m, say) to mimic precisely the large Earth-crossing asteroid orbital distribution. There are also reasons to

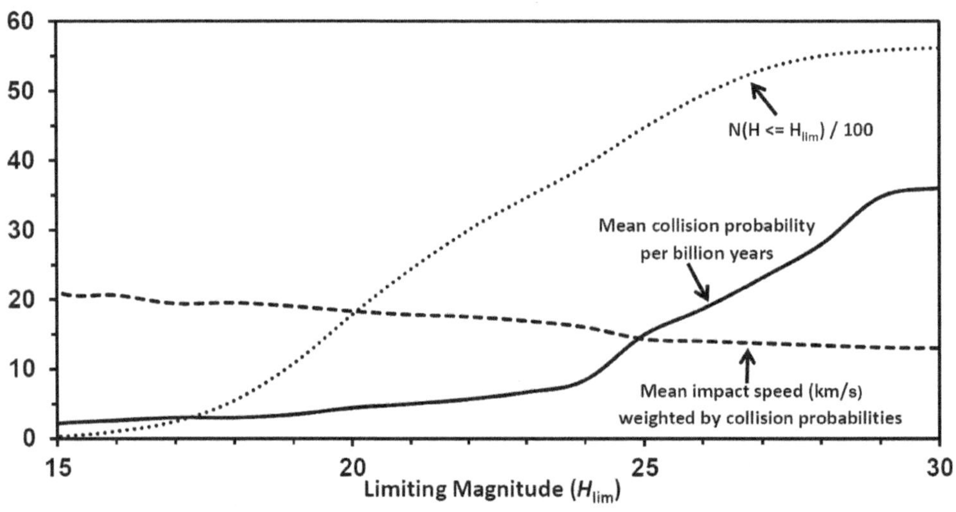

Figure 1. The solid curve indicates (on the ordinate) the mean collision probability per billion years for all discovered Earth-crossing asteroids (ECAs) that are brighter than some limiting magnitude H_{lim} (abscissa); the dashed line shows the mean impact speeds (km/s), weighted by the individual collision probabilities; and the dotted line indicates the number (divided by 100) of Earth-crossing asteroids contributing to the calculations at each limiting magnitude (cf. Table 1).

believe that radiative effects, and in particular the Yarkovsky pseudo-force, might deliver smaller asteroids more efficiently than larger ones from the main asteroid belt into Earth-crossing trajectories (Michel et al., 2005).

Bearing in mind the near-completeness of discovery of cis-Jovian Earth-crossing asteroids at $H < 18$, the recommended outcome from this analysis is that a mean terrestrial impact probability of 3×10^{-9} per year should be used for large Earth-crossing asteroids, with a weighted mean impact speed of ~20 km/s. These values might also be applicable to the population of smaller Earth-crossing asteroids, but as suggested in the preceding paragraph, it could be that the larger mean impact probabilities and smaller impact speeds calculated herein for the discovered smaller/fainter Earth-crossing asteroids, especially at $H > 23$–24, actually reflect not only discovery selection effects but perhaps also some real differences in the orbital distributions of large-versus-small Earth-crossing asteroids, with a greater fraction of small Earth-crossing asteroids occupying low-e, low-i orbits near Earth, these having intrinsically higher collision probabilities and lower impact speeds.

If there are indeed 10 million small asteroids of the type that arrived near Chelyabinsk on 15 February 2013 (size ~17 m, hence $H \approx 26$), and a mean collision probability of $\sim 3 \times 10^{-9}$ per year applies to them, then such events would occur somewhere on Earth about once every 30–35 yr on average. However, the calculations presented herein (cf. Table 1) indicate a mean terrestrial impact probability of almost 2×10^{-8} per year for the discovered Earth-crossing asteroids with $H \le 26$. If, after allowing for discovery probabilities, it were found that at $H = 26$ a mean impact probability of as much as 1×10^{-8} per year were applicable, then if there are 10 million Earth-crossing asteroids similar in size to the Chelyabinsk object, then the impact rate on Earth would be about once per decade; this does not appear to be in accord with what is known of impact events within the past century, but on the other hand, global surveillance (i.e., covering unpopulated regions) using space-based infrared systems has only been conducted for the past 20 yr or so, so this is a field in which comparatively little definitive knowledge exists. This remains an open question deserving further investigation.

VERY LARGE ASTEROID IMPACTS IN OTHER EPOCHS?

Might the influx of asteroids to Earth have been different in other epochs during the Phanerozoic? The answer to that question would appear to be "obviously, yes," if one assumes that at least some of the larger terrestrial craters formed during the last 600 m.y. were due to asteroid (rather than comet) impacts, because there are no cis-Jovian Earth-crossing asteroids as large as 10 km in size at present, and so no asteroids large enough to form craters substantially larger than ~100 km diameters.

The previous calculations and analysis were based on the presently observed population of near-Earth asteroids, and specifically those that have osculating orbits crossing that of our planet, and that population might not reflect a long-term average situation, then. The largest current Earth-crossing asteroid is (1866) Sisyphus, which has an absolute magnitude $H = 12.4$ (i.e., it is the sole asteroid at $H_{lim} < 13$ in Table 1), which is indicative of a size of 8–9 km given its moderate albedo of 0.16.

Larger near-Earth asteroids are known, with perihelia just outside Earth's orbit. Amor-type asteroid (1036) Ganymed has $q =$ 1.24 au, and its magnitude $H = 9.45$ and albedo 0.29 indicate a size/diameter just above 30 km; it may become an Earth-crosser on a time scale of 10 m.y. (Michel et al., 1999). The prime target of the *NEAR-Shoemaker* spacecraft mission was (433) Eros, which has a measured size of around 34 × 11 × 11 km and $q =$ 1.13 au, bringing it closer by Earth than Ganymed at present.

How such larger asteroids may arrive in Earth-crossing orbits, and so become potential impactors, has been extensively studied (e.g., Michel et al., 1996, 1998, 2005; Gladman et al., 1997, 2000; Migliorini et al., 1998). The time scales involved (for evolution into potential Earth-impacting paths) are brief, compared to geological time frames, typically being measured in millions of years. It follows that over the Phanerozoic, it should be anticipated that the population of potential Earth-impacting asteroids has varied substantially.

Quite apart from the population of Earth-crossing asteroids being different in other epochs, also the eccentricity (e) of Earth's heliocentric orbit oscillates with a period of only ~10^5 yr, at times reducing to $e \approx 0.0$ (i.e., a circular orbit), while at others a value as high as $e = 0.0579$ may be attained, i.e., more than three times the current $e = 0.0167$. This alters the individual impact probabilities of asteroids (or comets) with Earth, but more importantly when e is near its upper limit, there would be larger populations of asteroids with perihelion distances (q) near 1.02–1.06, which then become potential impactors despite the fact that, assuming (invalidly) unchanging asteroid orbits, they do not cross Earth's path when our planet has a lower value of e. The way in which the overall and mean impact probability by asteroids varies with Earth's orbital eccentricity was discussed by Steel (2002).

In view of these issues, one should not imagine that the currently observed Earth-crossing asteroid population is diagnostic of the long-term population of potential Earth-impacting asteroids. There are reservoirs of asteroids from which large impactors might be delivered into Earth-approaching orbits on time scales less than 10 m.y. (see Fig. 2 for the observed asteroid population larger than 50 km in size). The populations of impactors in the inner solar system have been reviewed by Michel and Morbidelli (2007), although as I will indicate later, I believe that any analysis that neglects the distant solar system as a source of Earth-impacting bodies is short-sighted, given what we now know of the host of large objects amongst and beyond the Uranus and Neptune region.

ENTER THE COMETS

In the preceding section I have outlined how, while in the present epoch no Earth-crossing asteroids exist that have

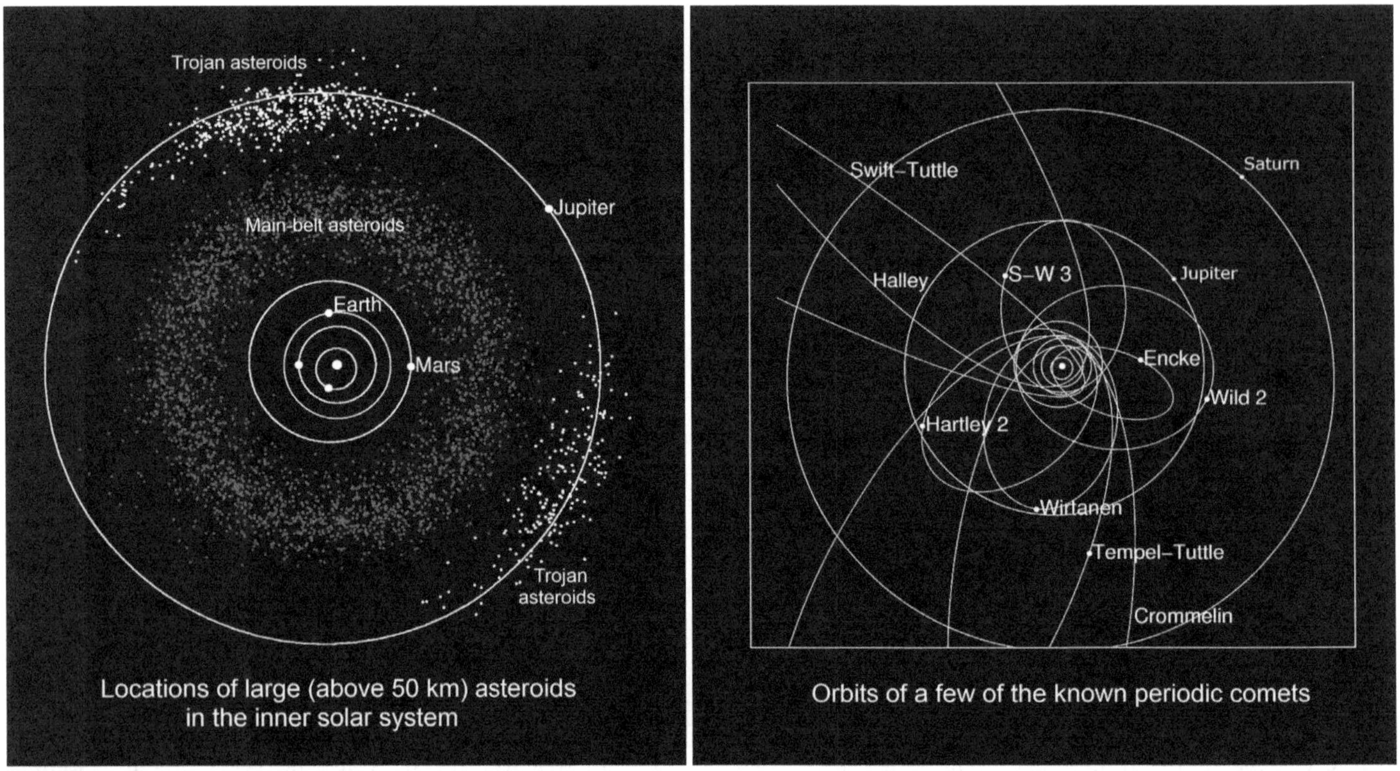

Figure 2. (Left) The presently observed population of asteroids in the inner solar system (Jupiter's orbit sunward) with sizes above ~50 km. There are no cis-Jovian Earth-approaching asteroids of such size in the current epoch, although over time scales above 1 m.y. it is feasible that such objects might be delivered from the main belt between Mars and Jupiter. For an up-to-date graphic of all known asteroids and comets in the inner solar system, see http://www.minorplanetcenter.net/iau/lists/InnerPlot.html. (Right) The orbits of a handful of the known short-period (orbital period < 20 yr) and Halley-type (20 yr < orbital period < 200 yr) comets that pass close by Earth in the present epoch. Many of these would, if they collided with our planet, deliver energy releases of the order of those associated with the formation of the Chicxulub crater, or larger. For example, 1P/Halley has an elongated shape around 15 × 8 × 8 km, a mass estimated to be above 2×10^{14} kg, and it would impact Earth at ~66 km/s based on its present orbit.

diameters as large as 10 km, in the longer term (1–10 m.y. time scales), it is feasible, indeed likely, that asteroids of such sizes will arrive in Earth-approaching orbits, having been perturbed in one way or another out of the main belt, or indeed currently having orbits with perihelia just outside Earth's distance from the Sun (like Eros). Such asteroids might then represent the sorts of energetic impactors visualized by some as being the root cause of mass extinction events.

There is, though, no reason to wait (proverbially speaking) such an extended period for such large impactors to appear close to our planet. For example, Comet Hale-Bopp, which passed perihelion inside Earth's orbit in 1997, had a solid nucleus ~60 km in size, and it will be back in a few thousand years. Before that, we can expect many more giant comets to come winging by.

The notion that comets can impact Earth with calamitous consequences is not a new one, dating back far before the discovery of the first main-belt asteroid in 1801, the first Earth-approaching asteroid in 1898, and the first Earth-crossing asteroids in the 1930s. Napier (this volume) has mentioned some of the early suggestions regarding mass extinctions being caused by comet or asteroid impacts. To the early writers that he lists, I might add that Edmond Halley, in 1694, discussed how comet impacts must have occurred in the past and that this was a "strange Catastrophe we may be sure has at least once happened to the earth."

Halley's ideas in this regard were picked up by many other intellectuals of the era, not always appreciatively, because they tended to be regarded as contrary to Christian beliefs. It was possibly at the suggestion of Isaac Newton that this cometary impact concept was altered by Halley so as to suggest that this was how a former world was reduced to chaos, with the Biblical Creation story then being regarded as an explanation of how order was produced.

Other authors also incorporated such notions into their writing. In *Gulliver's Travels* (1726), Jonathan Swift described "people under continual disquietudes" because they believed that the next return of Halley's Comet "one and thirty years hence, will probably destroy us."

Still on the theme of Halley's Comet, I turn now to the great American statesman and scientist, Benjamin Franklin. In his *Poor Richard's Almanac* for 1757, anticipating the earnestly awaited return of that comet as foreseen by Halley (and Swift), Franklin wrote that:

> "Should a Comet in its Course strike the Earth, it might instantly beat it to Pieces, or carry it off out of the Planetary System. The great Conflagration may also, by Means of a Comet, be easily brought about. All the Disputes between the Powers of Europe would be settled in a Moment; the World, to such a Fire, being no more than a Wasp's Nest thrown into an Oven. But our Comfort is, the same great Power that made the Universe, governs it by his Providence. And such terrible Catastrophes will not happen till 'tis best they should."

Franklin did not, though, say when he thought this might be.

To select one more example of the opinions of well-educated men prior to the era of modern science, in 1822 we have the poet Byron, fully conversant with the concept of extinction events and terrestrial upheavals in the light of Georges Cuvier's excavations in Paris, stating the following:

> "Who knows whether, when a comet shall approach this globe to destroy it, as it often has been and will be destroyed, men will not tear rocks from their foundations by means of steam, and hurl mountains, as the giants are said to have done, against the flaming mass?—and then we shall have traditions of Titans again, and of wars with Heaven."
>
> —Medwin's *Conversations of Lord Byron* (1824).

Due to my mention of "modern science" above, the reader may have imagined that during and after the nineteenth century, we must have entered a more-sophisticated and rational era, in which professional scientists made only solid, supportable statements based on sound knowledge of the phenomena involved; however, the reality is far from that. After its perihelion passages in 1759 and 1835, the expected return of Halley's Comet in 1910 was again the cause of widespread concern. As a consequence, the following appeared as part of a calming note to the public from someone who should have been an expert in the subject (and therefore should have known what he did *not* know):

> "For a hundred million years life has been continuous on this earth, though we have been visited by at least five comets every year. If comets could ever have done the earth any harm they would have done it long ago and you and I would not be discussing comets or anything else."
>
> —Robert Stawell Ball, Lowndean Professor of Astronomy and Geometry, University of Cambridge, in a letter to *The Times*, published on 10 February 1910.

In my personal experience, things have not changed much in the intervening century, with astrophysicists in the main making false witness to the public (and governments) with regard to the hazard posed to us by comets and asteroids, based on their (the astrophysicists') ignorance of the reality and also the fear that funding might be diverted from their own esoteric research interests.

POPULATIONS OF COMETS

From the preceding section, it should be clear that the thought that comets might impinge on Earth with significant consequences for the terrestrial environment and therefore the biosphere is not at all new. Edmond Halley realized this when he determined the orbit of his eponymous comet, late in the seventeenth century.

Over the decades and centuries, growing numbers of periodic (i.e., routinely returning) comets have been discovered and their orbits determined. Limited numbers (for clarity) have their paths shown in Figure 2.

While certain fractions of comets are periodic, like these, greater numbers of observed comets are nonperiodic (at least compared to historical time scales), or follow near-parabolic orbits around the Sun; an example is the aforementioned Comet Hale-Bopp. That is, such comets are seen once only and then depart again into the region far beyond Neptune and Pluto. The source of these is broadly believed to be a spherical distribution or cloud that stretches out to a substantial proportion of the distance to the nearest stars. This cometary reservoir is known as the Oort-Öpik cloud, and it has been discussed in more detail in the accompanying paper by Napier (this volume); see also, for example, Bailey et al. (1990) and Emel'yanenko et al. (2013). Simply to provide a time scale for readers not familiar with the history, this cloud was hypothesized by Ernst Öpik in 1932, and independently by Jan Oort in 1950.

For present purposes, I want to concentrate on cometary reservoirs/populations much closer to (and indeed, to some extent, amongst) the outer planetary region: This is the Edgeworth-Kuiper belt and the centaur objects, as foreshadowed in the introduction to this paper. My reason for concentrating on these is that they represent source regions for massive comets. For a more detailed description of how and why these various locations should be considered to be reservoirs from which comet-like (i.e., icy, primordial) objects occasionally entering the inner solar system are derived, see Horner et al. (2003) and Emel'yanenko et al. (2013).

While Comet Hale-Bopp was a behemoth amongst comets, at a size estimated at 60 ± 20 km, and other very large near-parabolic comets have been witnessed during recorded history, from the perspective of wreaking havoc in the inner solar system (in the future, and in the geological past), the centaurs and the Edgeworth-Kuiper belt objects must surely be amongst the prime suspects, based on what little we know about them so far.

Centaurs are known to range up to 260 km in size, and so (with some central compression to be anticipated) have masses a hundred times as high as Hale-Bopp. At the periphery of the planetary system, close to Neptune's orbit, far larger objects are known, all discovered in the past two decades. See Figure 3 to gain an appreciation of just how many big objects are out there, each figuratively representing a Sword of Damocles hanging over life on Earth.

Napier (this volume) has provided some discussion of orbital evolution studies of centaurs, following on the work of Horner et al. (2004b). Since (2060) Chiron was discovered in 1977, there have been several such orbital evolution studies done (e.g., Hahn and Bailey, 1990), involving clones (i.e., slight initial orbital variants) of this object, and such studies have been reviewed by Horner et al. (2004a, 2004b). Other centaurs have also been investigated in this way.

All such work has resulted in indications of the computational clones having nonzero probabilities of evolving into Earth-crossing orbits. A variety of fates awaits the clones, depending on contingent events including (in particular) close passages by a large planet, Jupiter being most effective in this regard. The clones may end their virtual lives by colliding with a planet; by being subject to temporary satellite capture by a major planet; by becoming sungrazers (including dipping below the solar surface and so being lost); or, more likely, being ejected from the solar system on hyperbolic heliocentric orbits.

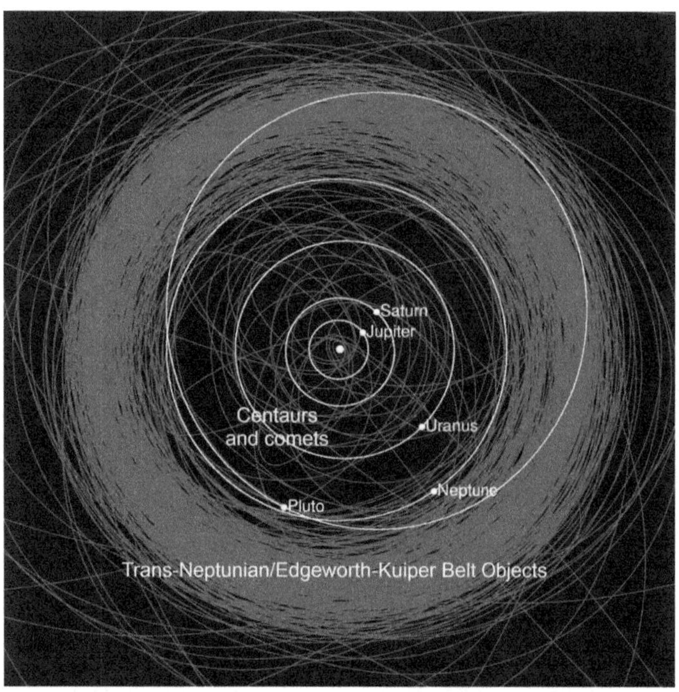

Figure 3. The ghost in the machine: This diagram shows the orbits of the planets Jupiter, Saturn, Uranus, Neptune, and Pluto, with their positions as at the start of the year 2000. Also shown are the orbits of the discovered centaurs and trans-Neptunian objects (plus a few of the known short-period and Halley-type comets for comparison with Fig. 2), less than a decade after the first trans-Neptunian object was found. For up-to-date maps of the positions of large outer solar system objects, visit the Minor Planet Center Web site at http://www.minorplanetcenter.net/iau/lists/OuterPlot.html. The obvious vast ring of trans-Neptunian objects in Figure 3 represents a reservoir from which giant comets are occasionally perturbed into Neptune-crossing orbits, a few of those then will likely be diverted into the inner solar system, where they must wreak havoc amongst the terrestrial planets.

A shortcoming of such research, of necessity, is that only the dynamical evolution of the clones is followed. Any comet-like body[7] may also be expected to undergo physical evolution, for example, the sublimation of different volatile constituents occurring at distinct heliocentric distances; thermal-stress–induced splitting; spontaneous or tidally induced fragmentation events; the loss of silicates as meteoroid trails and dust tails; and so on. At our present state of understanding of the physical nature of comets, such combined dynamical and physical evolution studies seem somewhat intractable: One could build a model, and run it, but would the outcomes be useful? In orbital evolution studies of comets (contra asteroid dynamical evolution), it is often necessary to include the jetting effect ("nongravitational force") imposed by the asymmetric outgassing of sublimating volatiles in order to understand specific cometary trajectories, but that is often a matter of fitting input parameters to observed deviations by a comet from the path indicated by a pure gravitational solution. This can render useful information about the comets (e.g., the observed nongravitational accelerations can give an indication of their masses), but almost the only thing predictable about the physical evolution of any particular comet is the fact that the physical evolution will be unpredictable: Comets are often seen to fall apart in interplanetary space for no apparent reason.

Returning to the purely dynamical orbital evolution studies, I note the following. These involve extensive numerical integrations of orbits in order to delineate dynamical avenues, and they represent a special mathematical art. Also, they require extensive computational power; see Horner et al. (2004a, 2004b), for example. Earlier in this paper, I cited publications involving authors including Farinella, Froeschlé, Gladman, Michel, Morbidelli, and others, and their work similarly involves such numerical integrations of orbits.

SINGLE-IMPULSE DEFLECTION INTO THE INNER SOLAR SYSTEM

All such research work is useful in developing an understanding of how comet and asteroid orbits evolve. However, for present purposes, I would like to be able to deduce, using limited computing power, and short execution times, the relative likelihoods of centaurs or other planet-crossers being deflected by singular close encounters with planets into wildly different orbits. For example, for a centaur with a low-eccentricity orbit crossing only the orbit of Uranus, what is the probability that a near miss of that planet will divert it (the centaur) into an orbit with a much-smaller perihelion distance? What is the chance of it attaining a path crossing Saturn? Or Jupiter? Or Mars? Or Earth? To answer such a question would require a vast amount of computing power

[7]Asteroidal bodies should also be expected to undergo physical evolution during the course of their dynamical evolution; for example, a close passage by a planet may cause the disruption of a rubble-pile asteroid into discrete aggregations or individual monoliths, or at least rearrange the shape of the overall object.

should the precise-numerical-integration-of-orbits technique be employed, as previously discussed.

Some years ago, when (944) Hidalgo and (2060) Chiron were the only known large objects in short-period orbits crossing the Jovian planets, I developed an alternative technique for estimating such gross orbital change probabilities and applied it to these two specific objects (Olsson-Steel, 1987b).

The technique in question is an extension of the collision probability method introduced by Kessler (1981) and also described by Steel and Baggaley (1985), except that instead of a collision cross section being used, instead I employ a close encounter cross section, rendering the probability (per year, or per orbit) of a minor body on some arbitrary planet-crossing orbit passing through the sphere of influence of that planet. Then, for each possible miss distance of the planet and each possible azimuthal angle around the planet (both suitably subdivided into bins of impact parameter and position angle), the new heliocentric orbit that would result from passage through that particular planetocentric position is computed on the basis of a deflection derived using the well-known Rutherford scattering formulae. That is, a two, two-body approximation is used, with the pre-encounter orbit of the minor body (i.e., the centaur, here) being a heliocentric ellipse; the orbit during the encounter being a planetocentric hyperbola; and the postencounter orbit being heliocentric but maybe either elliptical (i.e., a new orbit gravitationally bound to the Sun) or hyperbolic (i.e., the body is ejected from the solar system because it has attained a heliocentric speed greater than the escape speed at this solar distance). The definition of the dimension of the sphere of influence is not greatly important here, because the most-interesting alterations in the minor body's orbit generally occur as a result of close misses of the planet. More details of the model are given by Olsson-Steel (1987b).

As aforesaid, I developed this technique over a quarter-century ago, largely in order to investigate how likely it might be that small bodies/giant comets in the outer solar system would be diverted directly into orbits entering the inner solar system, a matter connected with terrestrial catastrophism that was also examined using numerical integration techniques by other researchers such as Hahn and Bailey (1990) and in particular Bailey et al. (1994). I have also used the technique, for example, to derive probabilities for near-parabolic comets being deflected into short-period orbits in single planetary encounters (Olsson-Steel et al., 1989; Olsson-Steel, 1990).

What I have done in the present instance is to access over 370 orbits of centaurs and scattered disk objects (SDOs) from the listing maintained by the Minor Planet Center.[8] This list begins[9] with (2060) Chiron, (5145) Pholus, and (7066) Nessus, discovered in 1977, 1992, and 1993, respectively, and then continues with the hundreds of such objects that have been discovered over the past two decades. The list[10] of trans-Neptunian objects (TNOs) is substantially longer, emphasizing to the reader that there are myriad large comet-like objects orbiting at the periphery of the planetary system, of which just a small subset has been spotted since the first was identified 21 yr ago.

From the 370-odd centaurs and scattered disk objects, I selected four example objects: two that have Neptune-crossing orbits with perihelia near that planet, one object with a low-eccentricity orbit that crosses just Uranus, and one with a high-eccentricity long-period orbit that crosses both Neptune and Uranus. Their relevant orbital parameters and estimated sizes are shown in Table 2.

For each of these four objects, I executed my single-impulse new-orbit probability code, and derived results as also shown in Table 2. Close encounters (i.e., passages through a planet's sphere of influence: for definitions, see Olsson-Steel, 1987b) with the outermost two Jovian planets are seen to occur typically once every 100–600 orbits of the objects in question.

It is the outcomes from such close encounters that are of most interest here. The two Neptune-crossers 2008 LP17 and 2011 GM27 are found to have nonzero probabilities of being diverted into Saturn-crossing orbits (and, of course, Uranus-crossing orbits) in single-impulse events, whereupon it would be expected that Saturn would dominate their further orbital evolution with a high probability of being passed inward to orbits with yet smaller perihelion distances. The probability figures for these two Neptune-crossers are quite different (by three orders of magnitude), however, and it was for this reason that I have presented results for both of them, to highlight that contrasting results can eventuate for initial orbits that appear broadly similar. Object 2011 GM27 is much more likely to reach a Saturn-crossing orbit quickly. If there were a thousand such objects (note its size is around 250 km) in similar orbits, then single-impulse deflections by Neptune would deliver one per 10,000 yr into Saturn-crossing trajectories.

Uranus-crosser 2012 CE17 was selected so as to demonstrate that low-eccentricity orbits near Uranus and Neptune are unlikely to be transformed directly into orbits with substantially smaller perihelion distances: Note the low probability figures pertaining to this object quickly entering a substantially reduced perihelion distance orbit, as given in the final two columns in Table 2.

The Uranus- and Neptune-crosser (65489) Ceto is more interesting. Uranus can divert it directly into Jupiter (and indeed Mars) crossing orbits, but Neptune is able to deflect it directly into an Earth-crossing trajectory, in fact an orbit with perihelion as small as 0.2 au, so that it would in that case be a crosser of all the major planets. The probability of becoming an Earth-crosser may seem small (~8×10^{-13} per year), but with a semimajor axis of just above 100 au, it has an orbital period of over 1000 yr, and

[8]http://www.minorplanetcenter.net/iau/lists/Centaurs.html

[9]Again, it is not clear whether (944) Hidalgo should be classed as a centaur; according to Emel'yanenko et al. (2103), it is not.

[10]http://www.minorplanetcenter.net/iau/lists/TNOs.html

TABLE 2. EXAMPLE CENTAUR AND SCATTERED DISK OBJECTS AND THEIR PROBABILITIES OF BEING THROWN INWARD BY NEPTUNE OR URANUS

Object	Estimated size (km)	Perihelion distance (q, au)	Eccentricity (e)	Inclination (i, degrees)	Planetary encounter rates	Minimum perihelion distance attained (au)	Resultant planet-crossing probabilities per year	Resultant planet-crossing probabilities per orbit
Neptune-crosser 2008 LP17	120	29.97	0.6564	14.12	Neptune: once per 100 orbits	6	Saturn: 2×10^{-10}	1.6×10^{-7}
Neptune-crosser 2011 GM27	250	29.16	0.4366	13.10	Neptune: once per 200 orbits	8	Saturn: 1.3×10^{-7}	5×10^{-5}
Uranus-crosser 2012 CE17	45	18.95	0.1767	5.90	Uranus: once per 200 orbits	10	Saturn: 2.5×10^{-11}	2.7×10^{-9}
Uranus- and Neptune-crosser (65489) Ceto	140	17.76	0.8241	22.30	Uranus: once per 350 orbits	1.1	Mars: 7×10^{-13} Jupiter: 1×10^{-11}	7×10^{-10} 1×10^{-8}
(65489) Ceto					Neptune: once per 600 orbits	0.2	Earth: 8×10^{-13}	8×10^{-10}

so most such objects spend the vast majority of their lifetimes at aphelion far beyond the reach of our telescopes. If there were a million objects in similar orbits, which would seem to be feasible, then the rate of delivery into Earth-crossing orbits would be about one per million years.

The previous calculations pertain only to objects being thrown directly into smaller-perihelion orbits, and they would remain (at least for some time) on paths with aphelia in the outer planetary region. However, the results of these computations indicate that we should be anticipating that radical alterations in the inner solar system environment will occur on time scales which are at most millions of years, with frequent arrivals of giant comets of sizes 100 km and upward, which would then (based on observed comet behavior) undergo hierarchical disintegrations as described by Napier (this volume). Apart from being of interest in themselves, the results for particular centaur and scattered disk object orbits may be useful in making preliminary identifications of those objects that are most likely to undergo rapid orbital evolution, thus informing the choice of sample orbits in numerical experiments such as those undertaken by Horner et al. (2004a, 2004b), Emel'yanenko et al. (2103), and others.

A final point with regard to my own calculations as presented here: One might anticipate that radical orbital changes through close planetary encounters should be evidenced in the results of such centaur orbit numerical integrations as those mentioned previously. Horner et al. (2004a) followed 23,328 test particles for 3 m.y. both into the past, and into the future. The number of particle-years involved in their experiment was therefore $\sim 2 \times 10^4$ particles $\times$ 6×10^6 yr $\approx 10^{11}$ particle-years. My calculations of the individual probabilities of being plunged directly into Earth-crossing orbits from the outer solar system through a close planetary encounter are of order 10^{-12} per year. Therefore, it is not surprising that no such events were identified in the study by Horner et al. (2004a): They would need to increase the extent of their experiment (the number of particle-years) by a factor of 10 before one might expect to find such an event occurring[11]. This might be interpreted as evidence of the utility of the probabilistic method I have employed here, which may be regarded as an adjunct to numerical integrations of orbits, rather than a replacement.

CONCLUSIONS

Our understanding of the inventory of the smaller bodies in the solar system is currently undergoing a revolution, and the

[11]Emel'yanenko et al. (2013) recently followed test orbits for approximately 5×10^{14} particle-years (8925 objects with 200 clones of each, for 300 m.y.), but because these were selected on the basis of initial perihelion distances q_0 below 36 au, many were not Neptune-crossers; however, many test particles did migrate into small q orbits, including falling into the Sun. A search of their (and others') integrations for planetary close encounters causing immediate gross changes of heliocentric orbits would be of interest in the present context, and so is to be encouraged.

discovery of many hundreds of large bodies in and around the outer planetary region over the past 20 yr has important repercussions for interpretations of Earth's collisional history.

The old idea of occasional singular asteroid impacts by large rogue rocks escaping from the main belt between Mars and Jupiter should be reconsidered in the light of discoveries such as those mentioned herein. Models for terrestrial catastrophism should encompass what we see happening on a yearly basis—comets decaying and disintegrating into smaller debris that persist in the inner solar system for time scales of ~10^5–10^6 yr—with researchers' imaginations being engaged so as to visualize and understand what must happen should a giant (larger than ~100 km) comet enter the inner solar system and become trapped in a cis-Jovian orbit. Such prolonged episodes of interplanetary upset, and concomitant terrestrial chaos, must have occurred many times over geological history.

Earlier in this paper, I stated that I would be giving a personal view, and that other researchers in the planetary and space science community would not necessarily agree with me with regard to the significance of the large outer solar system objects in terms of the supply of Earth impactors and therefore the importance (or otherwise) of cosmic impacts in mass extinction events. In order to illustrate this, I describe here the following.

Michel and Morbidelli (2007) supposedly have reviewed knowledge of "the population of impactors and the impact cratering rate in the inner solar system," but in doing so, they have made almost no mention of the observed outer solar system population of minor bodies and the implications of these, despite there having been more than two decades of history of investigations of the dynamics of centaurs using numerical integration techniques essentially the same as Michel and Morbidelli themselves used. These investigations of centaurs have shown that they must occasionally enter short-period orbits and indeed Earth-crossing trajectories, and must therefore have contributed to the cratering record of the solid planetary surfaces from which we obtain ground-truth in terms of cratering rates. The only paper cited by Michel and Morbidelli (2007) that is relevant in this specific regard is that by Koeberl (2003), a useful paper in itself, but only one from substantial numbers of publications that have dealt with the question of the contribution of presently observed outer solar system minor bodies (which are claimed here to be, in essence, gigantic comets) to the Earth-cratering population and terrestrial catastrophism over the history of the solar system.

In view of this situation, which I have identified here simply as an example of how single-minded researchers can be in the astronomical community (and, one might therefore presume, also in the geoscience community), I must repeat my earlier warning of *caveat emptor*, with regard to this paper of mine and also all others. It would certainly be possible to be marvelously mistook if one were not to adopt a critical stance. We yet have much to learn about the cosmic context of our planet, and the ways in which it may affect the biosphere.

ACKNOWLEDGMENTS

I am grateful to David Asher, Mark Bailey, Victor Clube, and Bill Napier for various discussions over the decades; I also thank Richard Firestone and an anonymous reviewer for their helpful comments. This paper was partly written while I was a visiting space scientist at National Aeronautics and Space Administration Ames Research Center, Moffett Field, California. This work was not supported by any funding agency, but was personally financed.

REFERENCES CITED

Asher, D.J., and Clube, S.V.M., 1993, An extraterrestrial influence during the current glacial-interglacial: Quarterly Journal of the Royal Astronomical Society, v. 34, p. 481–511.

Bailey, M.E., Clube, S.V.M., and Napier, W.M., 1990, The Origin of Comets: Oxford, Pergamon Press, 604 p.

Bailey, M.E., Clube, S.V.M., Hahn, G., Napier, W.M., and Valsecchi, G.B., 1994, Hazards due to giant comets: Climate and short-term catastrophism, *in* Gehrels, T., Matthews, M.S., and Schumann, A., eds., Hazards Due to Comets and Asteroids: Tucson, Arizona, University of Arizona Press, p. 479–533.

Bottke, W.F., Nolan, M.C., Greenberg, R., and Kolvoord, R.A., 1994, Collisional lifetimes and impact statistics of near-Earth asteroids, *in* Gehrels, T., Matthews, M.S., and Schumann, A., eds., Hazards Due to Comets and Asteroids: Tucson, Arizona, University of Arizona Press, p. 337–357.

Clube, S.V.M., and Napier, W.M., 1984, The microstructure of terrestrial catastrophism: Monthly Notices of the Royal Astronomical Society, v. 211, p. 953–968.

Di Martino, M., Carusi, A., Dotto, E., Lazzarin, M., Marzari, F., and Migliorini, F., 1998, Dynamical and physical properties of comet–asteroid transition objects: Astronomy and Astrophysics, v. 329, p. 1145–1151.

Emel'yanenko, V.V., Asher, D.J., and Bailey, M.E., 2013, A model for the origin of Jupiter family and Halley type comets: Earth, Moon, and Planets, v. 110, p. 105–130, doi:10.1007/s11038-012-9413-z.

Ferrin, I., Zuluaga, J., and Cuartas, P., 2013, The location of asteroidal belt comets (ABCs) in a comets' evolutionary diagram: The Lazarus comets: Monthly Notices of the Royal Astronomical Society, v. 434, p. 1821–1837, doi:10.1093/mnras/stt839.

Gladman, B.J., Migliorini, F., Morbidelli, A., Zappalá, V., Michel, P., Cellino, A., Froeschlé, Ch., Levison, H.F., Bailey, M., and Duncan, M., 1997, Dynamical lifetimes of objects injected into asteroid belt resonances: Science, v. 277, p. 197–201, doi:10.1126/science.277.5323.197.

Gladman, B.J., Michel, P., and Froeschlé, Ch., 2000, The near-Earth object population: Icarus, v. 146, p. 176–189, doi:10.1006/icar.2000.6391.

Hahn, G., and Bailey, M.E., 1990, Rapid dynamical evolution of giant comet Chiron: Nature, v. 348, p. 132–136, doi:10.1038/348132a0.

Horner, J., Evans, N.W., Bailey, M.E., and Asher, D.J., 2003, The populations of comet-like bodies in the solar system: Monthly Notices of the Royal Astronomical Society, v. 343, p. 1057–1066, doi:10.1046/j.1365-8711.2003.06714.x.

Horner, J., Evans, N.W., and Bailey, M.E., 2004a, Simulation of the populations of centaurs: I. The bulk statistics: Monthly Notices of the Royal Astronomical Society, v. 354, p. 798–810, doi:10.1111/j.1365-2966.2004.08240.x.

Horner, J., Evans, N.W., and Bailey, M.E., 2004b, Simulation of the populations of centaurs: II. Individual objects: Monthly Notices of the Royal Astronomical Society, v. 355, p. 321–329, doi:10.1111/j.1365-2966.2004.08342.x.

Kessler, D.J., 1981, Derivation of the collision probability between orbiting objects: The lifetimes of Jupiter's outer moons: Icarus, v. 48, p. 39–48, doi:10.1016/0019-1035(81)90151-2.

Koeberl, C., 2003, The Late Heavy Bombardment in the inner solar system: Is there any connection to Kuiper belt objects?: Earth, Moon, and Planets, v. 92, p. 79–87, doi:10.1023/B:MOON.0000031927.85937.9e.

Manley, S.P., Migliorini, F., and Bailey, M.E., 1998, An algorithm for determining collision probabilities between small solar system bodies: Astronomy and Astrophysics, v. 133, supplement, p. 437–444.

Marsden, B.G., and Steel, D.I., 1994, Warning times and impact probabilities for long-period comets, *in* Gehrels, T., Matthews, M.S., and Schumann,

A., eds., Hazards Due to Comets and Asteroids: Tucson, Arizona, University of Arizona Press, p. 221–239.

Michel, P., and Holsapple, K.A., 2007, Tidal disturbances of small cohesionless bodies: Limits on planetary close approach distances, *in* Milani, A., Valsecchi, G.B., and Vokrouhlický, D., eds., Proceedings IAU Symposium 236, Near Earth Objects, Our Celestial Neighbors: Opportunity and Risk: Cambridge, UK, Cambridge University Press, p. 201–210.

Michel, P., and Morbidelli, A., 2007, Review of the population of impactors and the impact cratering rate in the inner solar system: Meteoritics & Planetary Science, v. 42, p. 1861–1869, doi:10.1111/j.1945-5100.2007.tb00545.x.

Michel, P., Farinella, P., and Froeschlé, Ch., 1996, The orbital evolution of the asteroid Eros and implications for collision with the Earth: Nature, v. 380, p. 689–691, doi:10.1038/380689a0.

Michel, P., Farinella, P., and Froeschlé, Ch., 1998, Dynamics of Eros: The Astronomical Journal, v. 116, p. 2023–2031, doi:10.1086/300562.

Michel, P., Gonczi, R., Farinella, P., and Froeschlé, Ch., 1999, Dynamical evolution of 1036 Ganymed, the largest near-Earth asteroid: Astronomy and Astrophysics, v. 347, p. 711–719.

Michel, P., Morbidelli, A. and Bottke, W.F., 2005, Origin and dynamics of near Earth objects: Compte Rendus–Physique, v. 6, no. 3, p. 291–301.

Migliorini, F., Michel, P., Morbidelli, A., Nesvorný, D., and Zappalá, V., 1998, Origin of multikilometer Earth- and Mars-crossing asteroids: A quantitative simulation: Science, v. 281, p. 2022–2024, doi:10.1126/science.281.5385.2022.

Napier, W.M., 2014, this volume, The role of giant comets in mass extinctions, *in* Keller, G., and Kerr, A.C., eds., Volcanism, Impacts, and Mass Extinctions: Causes and Effects: Geological Society of America Special Paper 505, doi:10.1130/2014.2505(20).

Napier, W.M., and Clube, S.V.M., 1979, A theory of terrestrial catastrophism: Nature, v. 282, p. 455–459, doi:10.1038/282455a0.

Olsson-Steel, D., 1987a, Collisions in the solar system: IV. Cometary impacts upon the planets: Monthly Notices of the Royal Astronomical Society, v. 227, p. 501–524.

Olsson-Steel, D., 1987b, Planetary close-encounters: Probability distributions of resultant orbital elements and application to Hidalgo and Chiron: Icarus, v. 69, p. 51–69, doi:10.1016/0019-1035(87)90006-6.

Olsson-Steel, D., 1990, The long-period to short-period comet transition: A new method for estimating capture probabilities, *in* Lagerkvist, C.-I., Rickman, H., Lindblad, B.A., and Lindgren, M., eds., Asteroids, Comets, Meteors III: Uppsala, Sweden, University of Uppsala Press, p. 401–404.

Olsson-Steel, D., Taaffe, L., and Williams, S., 1989, Short-period comet capture in close encounters with Jupiter: Proceedings of the Astronomical Society of Australia, v. 8, p. 108–118.

Öpik, E.J., 1976, Interplanetary Encounters: Amsterdam, Elsevier, 164 p.

Richardson, D.C., Bottke, W.F., and Love, S.G., 1998, Tidal distortion and disruption of Earth-crossing asteroids: Icarus, v. 134, p. 47–76, doi:10.1006/icar.1998.5954.

Shoemaker, E.M., Williams, J.G., Helin, E.F., and Wolfe, R.F., 1979, Earth-crossing asteroids: Orbital classes, collision rates with Earth, and origin, *in* Gehrels, T., ed., Asteroids: Tucson, Arizona, University of Arizona Press, p. 253–282.

Steel, D., 2002, Terrestrial impact rates for the known population of Earth-crossing asteroids, *in* Warmbein, B., ed., Proceedings of Asteroids, Comets, Meteors: European Space Agency Special Publication SP-500, p. 145–147.

Steel, D.I., 1998, Distributions and moments of asteroid and comet impact speeds upon the Earth and Mars: Planetary and Space Science, v. 46, p. 473–478, doi:10.1016/S0032-0633(97)00232-8.

Steel, D.I., and Baggaley, W.J., 1985, Collisions in the solar system: I. Impacts of the Apollo-Amor-Aten asteroids upon the terrestrial planets: Monthly Notices of the Royal Astronomical Society, v. 212, p. 817–836.

Manuscript Accepted by the Society 31 January 2014

Printed in the USA

The Geological Society of America
Special Paper 505
2014

Calendar-year dating of the Greenland Ice Sheet Project 2 (GISP2) ice core from the early sixth century using historical, ion, and particulate data

Dallas H. Abbott
Dee Breger[†]
Pierre E. Biscaye
Lamont-Doherty Earth Observatory of Columbia University, Palisades, New York 10964, USA

Robert A. Juhl
Independent researcher, 1-4-1 Rokko Heights 906, Shinkawa, Chuo-ku, Tokyo 104-0033, Japan

ABSTRACT

We use the occurrence of unusual or out-of-season dust storms and dissolved ion data as proxies for dust to propose a calendar-year chronology for a portion of the Greenland Ice Sheet Project 2 (GISP2) ice core during the early sixth century A.D. Our new time scale moves a small sulfate peak to early 537 A.D., which is more consistent with recent findings of a 6 mo to 18 mo time lag between volcanic eruptions and atmospheric fallout of their sulfate aerosols. Our new time scale is consistent with a small volcanic input to the A.D. 536–537 climate downturn. We use the time range of Ni-rich fragments and cosmic spherules to provide an independent test of the chronology. The time range of Ni-rich fragments and cosmic spherules matches historical observations of "dancing stars" starting in the summer of A.D. 533 and lasting until A.D. 539 or 540. These dancing stars have been previously attributed to cosmogenic dust loading of Earth's atmosphere. The time scale cannot be shifted to be either younger or older by 1 yr without destroying the match to historical accounts of dancing stars.

INTRODUCTION

The Greenland Ice Sheet Project 2 (GISP2) ice cores from central Greenland are much better dated than those from Antarctica. Errors in dating of the portion of the GISP2 ice core covering the sixth century A.D. are on the order of ±3 yr. In contrast, typical errors in dating of Antarctic ice cores are about ±10 yr. Because precise chronology is essential to the solution of many geological problems, more accurate dating of ice cores could produce new insights.

In this paper, we discuss a section of the GISP2 ice core that was purposely selected for its potentially high content of

[†]Also at Micrographic Arts, Saratoga Springs, New York 12866, USA.

Abbott, D.H., Breger, D., Biscaye, P.E., and Juhl, R.A., 2014, Calendar-year dating of the Greenland Ice Sheet Project 2 (GISP2) ice core from the early sixth century using historical, ion, and particulate data, *in* Keller, G., and Kerr, A.C., eds., Volcanism, Impacts, and Mass Extinctions: Causes and Effects: Geological Society of America Special Paper 505, p. 411–420, doi:10.1130/2014.2505(22). For permission to copy, contact editing@geosociety.org.

extraterrestrial material (Baillie, 1999). The time period from A.D. 534 to 541 has been the subject of an ongoing controversy about the exact dating of these sections within ice cores from Greenland (Baillie, 2008; Larsen et al., 2008). If we can resolve this controversy for the GISP2 core, it will provide important constraints on the relative importance of volcanic and extraterrestrial drivers for climate downturns during this time period (Rigby et al., 2004; Baillie, 2007).

CHRONOLOGY OF GREENLAND ICE CORES—HOW WELL CAN WE DATE THEM?

We start with the Meese-Sowers time scale for the GISP2 core (Sowers et al., 1993; Meese et al., 1994a, 1994b). They estimated errors of ±3 yr during the middle of the sixth century. We checked this inference using our ion data from the GISP2 ice core. Larsen et al., (2008) used layer counts and other methods to cross-date three Greenland ice cores: Greenland Ice Core Project (GRIP), North Greenland Ice Core Project (NGRIP), and DYE-3. Within those cores, they found a Tambora-sized sulfate peak for which they estimated an age of 533–534 ± 2 yr. They also found a larger sulfate peak for which they estimated an age of A.D. 529 ± 2. This chronologically earlier sulfate peak is lacking in our time series (Fig. 1), meaning our time series starts at A.D. 530 ± 2 or later. Our time series of ion data (Tables 1 and 2) has two moderately large sulfate peaks, one at A.D. 539.2 ± 3 and a second peak at A.D. 543.8 ± 3. The older of these sulfate peaks corresponds to the A.D. 533–534 ± 2 sulfate peak of Larsen et al. (2008). Within their estimated errors, our older sulfate peak could be as young as A.D. 536, but no younger if Larsen et al. (2008) are correct. Although Larsen et al. (2008) estimated that their absolute ages had errors of only 2 yr, another recent paper suggests that their errors are too small by 4–5 yr (Baillie, 2008). We used historical data to support Baillie's inferences, but we find a more modest error of 3 yr in Larson et al.'s ages. This latter age error is consistent with the Meese-Sowers time scale and its error estimates.

Our first historical data set (Table 3) chronicles intense, unusual dust storms recorded in Chinese annals. Dust storms are common in north China in spring, as seen in recently studied Greenland snow at NGRIP (Bory et al., 2003a, 2003b), but much less common in south China. During the period A.D. 500–560, Chinese annals recorded four dust-related events, all of which were unusual in some sense. Table 3 shows the record and notable features associated with each event.

The first three events were considered unusual for the season, having taken place late in the year. And although the 537 event took place "in season," it was considered unusual at the time, in that "yellow ashes rained from a cloudless sky," which is presumably why it was mentioned in the annals. The A.D. 535 and 536 events were also notable for being heavy falls, and the last four events were unusual in having been recorded in southern annals.

High deserts in China are the major source of dust in high-altitude Greenland ice cores (GISP2, DYE3, NGRIP, and GRIP; Biscaye et al., 1997; Bory et al., 2003a, 2003b). The Mg content

TABLE 1. ION DATA FROM THE GISP 2 ICE CORE

Sample number	Na (μg/L)	K (μg/L)	Mg (μg/L)	Ca (μg/L)	Cl (μg/L)	SO_4 (μg/L)
1A	124.9	127.1	20.8	202.8	178.5	141.3
1B	14.2	11	3.3	31.7	35.7	30.5
2-S1	21.3	13.8	7.9	63.7	38.3	24.3
2-S2	21.9	14.4	8.4	65.8	39.2	26.5
3-S1	23	19.9	5.3	54.9	49	26.1
3-S2	22.5	19.9	5.4	56.7	48.7	26.1
4A-S1	280.7	144.4	10.8	105.7	485.8	122.6
4A-S2	307.8	168.5	12.8	126.4	413.5	107.5
4B	24.2	26.9	6.6	80.6	102.3	81
4C	40.5	41.2	2.8	49.5	78.4	34.1
5A	25.7	24.1	12.2	144.9	56.8	52
5B	38.5	22.6	14.9	236.2	63.3	86.3
5C	269	240.1	22.8	220.8	404.3	143.3
6	63.4	61.3	4.4	40.6	113.8	21.9
7	33.5	31.7	20.4	574.8	77.6	92.9
8	24.4	28.1	4.7	51.2	65.7	46.5
9	108.2	88.9	12.7	105.1	197.8	91
10A	16.2	10.6	2.9	51.5	47	32.2
10B	19.4	13	2.4	10.1	52.8	21.8
11-S1	24.4	30	3.6	42.1	53.8	31.5
11-S2	24.2	29.6	3.8	43	52.2	29.8
12A	36.5	38.2	3.6	59.8	70.8	36.1
12B	11.8	11.5	2.7	53.7	37.3	29.1
Blank 1	0.4	0.3	0	0.9	45.5	0
Blank 2	0.3	0.3	0	0.8	45.6	0.7

Note: Cl values are used to correct for the effects of sea salt on the concentrations of K, Mg, and Ca (Table 2). GISP2—Greenland Ice Sheet Project 2.

TABLE 2. ION DATA FROM GISP2 ICE CORE

Mean depth (m)	Mg_{nss}	Ca_{nss}	K_{nss}	SO_4 (µg/L)	Age* (yr A.D., Meese-Sowers)
360.05	5.9	198.1	122.5	141.3	543.8 ± 3.26
360.15	1.6	31.2	10.5	30.5	543.3 ± 3.26
360.29	5.4	62.8	13	24.3	542.6 ± 3.46
360.29	5.8	65	13.6	26.5	542.6 ± 3.46
360.445	2.5	54	19	26.1	541.8 ± 3.33
360.445	2.7	55.9	19	26.1	541.8 ± 3.33
360.575	–22.7	95	134	122.6	541.1 ± 3.33
360.575	–23.9	114.6	157.1	107.5	541.1 ± 3.33
360.67	3.7	79.7	26	81	540.6 ± 3.15
360.735	–2	47.9	39.7	34.1	540.3 ± 3.18
360.805	9.1	143.9	23.1	52	540 ± 3.18
360.88	10.3	234.7	21.2	86.3	539.6 ± 3.2
360.96	–9.3	210.5	230.2	143.3	539.2 ± 3.2
361.04	–3.2	38.1	59	21.9	538.8 ± 3.2
361.16	16.4	573.5	30.5	92.9	538.1 ± 3.41
361.315	1.8	50.3	27.2	46.5	537.4 ± 3.38
361.42	–0.2	101	84.9	91	536.8 ± 3.15
361.5	0.9	50.9	10	32.2	536.4 ± 3.26
361.595	0.1	9.4	12.2	21.8	535.9 ± 3.23
361.72	0.7	41.2	29.1	31.5	535.3 ± 3.41
361.72	0.9	42.1	28.7	29.8	535.3 ± 3.41
361.85	–0.8	58.4	36.8	36.1	534.6 ± 3.26
361.95	1.3	53.2	11.1	29.1	534.1 ± 3.26

Note: Mg, Ca, and K are corrected for addition of ions from seawater (nss). GISP2—Greenland Ice Sheet Project 2. Units for all data are micrograms/liter.

*Age according to time scale of Sowers et al. (1993) and Meese et al. (1994a, 1994b).

of ice-core meltwater, corrected for the Mg contribution from seawater salt, is an accepted proxy for continental dust (Kang et al., 2003). Although it is less reliable due to oceanic carbonate sources, the Ca content, corrected for the Ca contribution from seawater salt, is also a proxy for continental dust (O'Brien et al., 1995; Zdanowicz et al., 2000). Our ion data show two peaks in Mg and Ca that are separated by about a year. If the Meese-Sowers time scale is shifted backward by 2 yr, the Ca and Mg peaks line up with the ages of the recorded dust storms in south China (Fig. 1).

LABORATORY METHODS

Our ice-core samples were processed under a laminar-flow hood. The hood has a filter and positive pressure that prevent contamination from local dust. The outside surface of each ice-core sample was rinsed with quartz-distilled water to remove contamination from drilling and atmospheric dust. The remaining ice was melted under the laminar-flow hood. We centrifuged the meltwater, and the supernatant was analyzed for ion concentrations using ion chromatography. A large drop of the remaining particulate-rich water was placed inside cleaned Teflon washers glued to cleaned glass slides and left to evaporate in the laminar-flow hood. All final cleaning and rinsing was done with quartz-distilled water. Each slide was then covered and subsequently imaged and analyzed using a Zeiss Supra 50 scanning electron microscope (SEM) with attached EDAX energy-dispersive X-ray spectrometer (EDS). Micrographs were collected in secondary electron (SE), backscattered electron (BSE), or mixed SE and BSE modes. Imaging and analyses were conducted in high-vacuum mode at 15 kV accelerating voltage. In between SEM sessions, each slide was covered and stored inside the laminar-flow hood to prevent contamination from atmospheric dust. Covers were removed immediately prior to SEM/EDS examination and replaced immediately after. Marine-sediment cores were stored in a different building.

TABLE 3. DUST-RELATED EVENTS IN CHINA A.D. 500–560

Date/time frame	Sources*	Record
Approximately Sep to mid-Nov 522	Wei shu 105, Tianxiang Zhi: 2341	N. China: "Since the 8th lunar month (6 Sep–5 Oct 522), yellow dust has hidden the sun…." [Quote is from a document addressed to the emperor from a minister. Document is dated 14 Nov 522.]
11 Nov–9 Dec 535	Nan shi 7, Liang Benji: 211	S. China: Winter, 10th month (11 Nov–9 Dec 535). Yellow dust fell like snow.
29 Nov–27 Dec 536	Nan shi 7, Liang Benji: 212	S. China: 11th month (29 Nov–27 Dec 536), yellow dust fell like snow. It filled scoops when picked up.
1 Feb 537	Liang shu 3, Wudixia: 81	S. China: On 1 Feb 537, the sky was cloudless, but it rained ashes of yellow color.
8 Feb 550	Nan shi 8, Liang Benji: 230	S. China: On 8 Feb 550, Yellow sand rained from the sky.

*Volume number appears before comma, page number appears after colon.

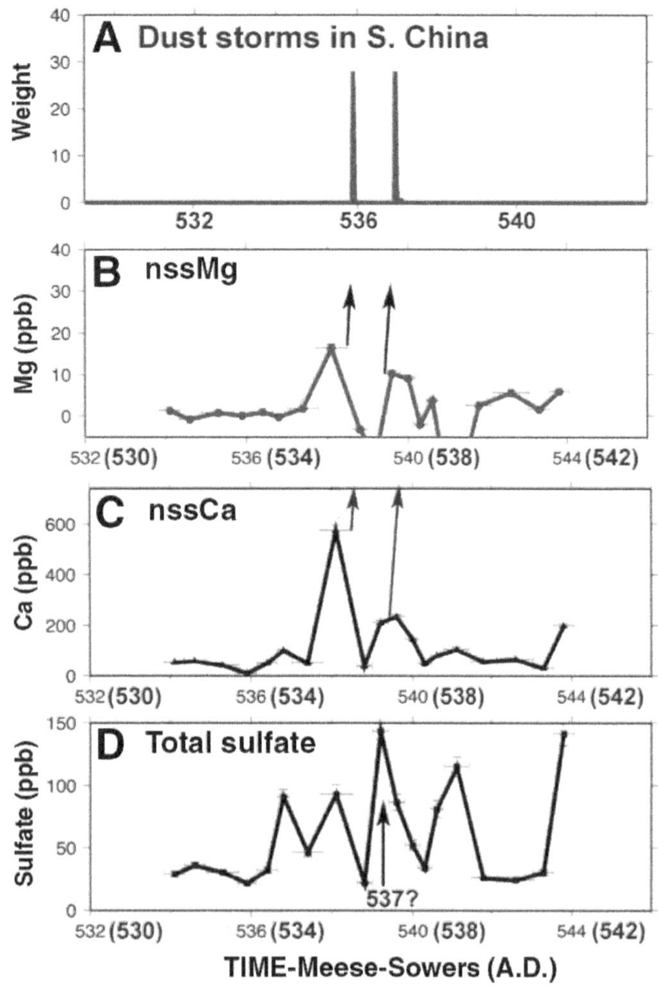

Figure 1. Time series of (A) dust storms in south China and (B–D) ion data from ice-core meltwater in ppb. Absolute ages and relative ages are shown in different fonts and sizes. Smaller numbers—relative ages from the Meese-Sowers time scale. Larger bold numbers—calendar-year age. Bold numbers in parentheses—inferred calendar-year age. (A) Historical time series of dust storms. Each event is plotted as a Gaussian distribution. (B) Ion data, where nssMg is Mg corrected for the effect of salt from the ocean. (C) Ion data, where nssCa is Ca corrected for the effect of salt from the ocean. (D) Total sulfate. The sulfate peak labeled 537? corresponds to the A.D. 533–534 ± 2 sulfate peak of Larsen et al. (2008).

CHRONOLOGY, ION DATA, AND VOLCANIC HYPOTHESIS FOR THE A.D. 536–537 CLIMATE CATASTROPHE

Our ion data show two relatively large sulfate peaks, one ca. A.D. 537 and the second ca. A.D. 541–542 (Fig. 1). Both peaks are approximately the same size. This is important, because historical sources indicate that the climate downturn in A.D. 536–537 was far more pronounced than in A.D. 541–542. Ancient histories have provided at least four contemporary accounts of a dramatic decrease in the Sun's brightness during A.D. 536 and 537 (Stothers, 1984). In Mesopotamia (30–37°N), a source later quoted by Michael the Syrian (ca. 1166–1199 A.D.) and Bar-Hebraeus (ca. 1246–1286) stated "In the year 848 of the Greeks [536 A.D.]...the Sun was dark and its darkness lasted for eighteen months, each day it shone for about 4 hours, and still this light was only a feeble shadow... the fruits did not ripen and wine tasted like sour grapes" (translated to English from Chabot [in French], 1901, p. 220–221). According to an anonymous writer in Constantinople (41°N) (Hamilton and Brooks, 1899, p. 267), associated by tradition with Zacharius of Mytilene, "In the year 14 [536 A.D.]...the Sun began to be darkened by day and the Moon by night... from the 24th of March in this year until the 24th of June in the following year 15." Because the dimming was less severe and shorter in duration (15 mo instead of 18) at higher latitudes, it is generally agreed that the atmosphere was filled with fine dust from an unknown, low-latitude source. Volcanic ash and impact debris are both promoted as the primary sources of that dust (Rigby et al., 2004; Larsen et al., 2008).

These observations are important because the Tambora eruption produced a dim Sun but no reduction in the length of the day (Rampino et al., 1988). In contrast, during the A.D. 536–537 event, the length of the day was reduced to 4 h. Thus, if the sole driver of the A.D. 536–537 climate downturn was sulfate from volcanic sources, the average size of the sulfate peak should be significantly larger than the peak from the Tambora eruption. As observed in Greenland, the A.D. 534 ± 2 sulfate peak in three ice cores has the same amplitude as the sulfate peak from the Tambora eruption (Fig. 2).

Our proposed time scale places the same sulfate peak at ca. A.D. 537.2. This is consistent with recent research on the time lag between the deposition of volcanic ash (nearly simultaneous with the eruption) and later deposition of volcanic sulfate (the driver of climate cooling from tropical eruptions). The time lag is between 6 and 18 mo, depending on the height of the eruptive column (Koffman et al., 2011). Because ice cores show pronounced seasonal changes in dust deposition and in acidity (Wei et al., 2008), it is relatively easy to estimate the time range covered by individual ice core samples. Based on the depth range of our samples, the ca. A.D. 537 sulfate peak has a proposed age range of ca. A.D. 537–537.4. Using a time lag of 6–18 mo, the source volcanic eruption occurred between A.D. 535.5 and 536.9. Because the A.D. 536–537 climate downturn started in March of A.D. 536, the new time scale is consistent with a small volcanic contribution to the climate cooling.

Although our new time scale is consistent with a volcanic contribution to the A.D. 536 event, the relatively small size of the sulfate peak—similar to that from the Tambora eruption—suggests that the story is more complicated. As we will show, the ice core contains particles of extraterrestrial provenance. These particles suggest that significant amounts of extraterrestrial dust were falling through Earth's atmosphere during this time.

EXTRATERRESTRIAL DUST IN THE ICE CORE

We found two types of debris of possible extraterrestrial origin in the particles from the GISP2 ice core. The first type of material is rich in Ni and is typically not spherical, although some particles have locally melted surfaces or ablation rinds (Fig. 3). The second type contains Fe-oxide–rich spherules (I-type cosmic spherules; Fig. 4). I-type cosmic spherules contain mainly FeO with minor amounts of other oxides. Ni-rich iron metal can occur but is usually inside the spherule and would not be detectable using our methodology (Genge et al., 2008). Larger I-type cosmic spherules have visible quench textures on their surfaces. The two types of material have slightly different origins.

Intact fragments of Ni-rich material resemble unmelted or partially melted extraterrestrial debris. Cosmic populations with diameters ranging from 53 to 106 μm contain only ~3% of the expected number of completely melted cosmic spherules (Taylor et al., 2000). The remainder of the material is partially melted and nonspherical. Consequently, individual fragments of extraterrestrial material less than ~50 μm in diameter are unlikely to completely melt to form cosmic spherules.

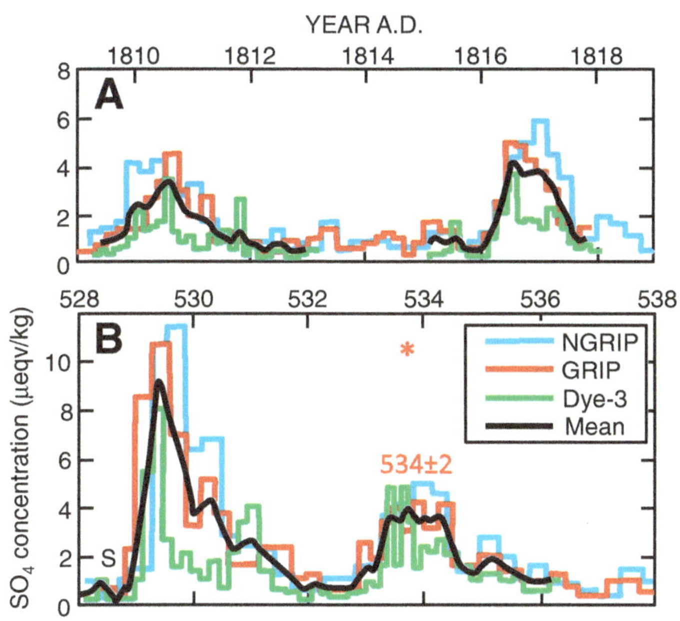

Figure 2. Sulfate (SO_4) concentration in microequivalents per kilogram (μeqv/kg) in three Greenland ice cores: Greenland Ice Core Project (GRIP), North Greenland Ice Core Project (NGRIP), and DYE-3 (after Larson et al., 2008). (A) A.D. 1809–1819. The Tambora eruption was in April A.D. 1815. The sulfate peak lags the eruption by about a year. The time during which sulfate remains in the atmosphere should match the duration of increased atmospheric dimming and cooling. (B) A.D. 528–538. The sulfate peak is smaller than that from A.D. 529 ± 2. In A.D. 530, there is a modest growth reduction recorded by some tree-ring records that is interpreted as a climate downturn (Baillie, 2008). There is no frost ring in bristlecone pines at A.D. 530 (Salzer and Hughes, 2007). The red letters show the suggested dating of the sulfate peak by Larson et al. (2008). The height of the red star is a minimum estimate of the amount of sulfate needed to produce a prominent climate downturn like that during A.D. 536–537.

Although unmelted Ni-rich particles can arise from a Chelyabinsk-type explosion (Galimov et al., 2013) when a relatively small impactor explodes high in the atmosphere, the maximum diameter of our Ni-rich debris is 24 μm. These small, sharp-edged fragments do not require a Chelyabinsk-type event to descend to the surface of Earth. Such small fragments often experience some local melting and ablation, as is evidenced by the presence of small spherules on the upper left of the fragment in Figure 3A. Nevertheless, despite a relatively high velocity of entry into Earth's atmosphere (typically 17 km/s for asteroidal debris and 51 km/s for cometary debris; Collins et al., 2005), these tiny extraterrestrial fragments float down to Earth's surface relatively unaltered.

The second type of particle is cosmic spherules dominantly composed of Fe oxide. The larger cosmic spherules exhibit quench textures on their surfaces (Figs. 4A and 4C). Fe-rich cosmic spherules are derived from ablation of larger Fe-rich bodies of extraterrestrial origin. When a large extraterrestrial body enters Earth's atmosphere, it is heated by atmospheric friction. As a result, its surface simultaneously melts and ablates. Cosmic spherules spall off the surface of the ablating body. Because they are secondary in origin, these I-type cosmic spherules can be much smaller than sharp-edged Ni-rich fragments. The quench textures on the surfaces of the Fe-rich cosmic spherules require that the spherule was first molten and then rapidly cooled.

Both the cosmic spherules and Ni-rich fragments must have been derived from larger bodies. Because Ni tends to concentrate in nuggets during melting (Borisov and Palme, 1997), the presence of Ni-rich material within a Fe-rich matrix implies a source body that did not experience large-scale melting during atmospheric transit. In contrast, the Fe-oxide–rich cosmic spherules require extensive melting and ablation of their source body. The Fe-oxide–rich cosmic spherules are probably derived from melting and ablation of the edges of larger pieces of extraterrestrial material or from particles with greater momentum for their size. Their apparent lack of Ni may reflect the low levels of Ni (1%–2%) in the fusion crust of meteorites (Fechtig and Utech, 1964; Genge and Grady, 1999), or it may reflect their derivation from an extraterrestrial source body or mineral with only a few percent Ni (Taylor et al., 2000). Ni contents below ~1%–2% cannot be detected reliably using the EDS X-ray microanalyzer on the SEM.

Both the cosmic spherules and the Ni-rich particles could represent dust derived from recent fragmentation of an extraterrestrial body. Because sharp edges would be smoothed out by intergranular collisions and the solar wind over time, Ni-rich particles with sharp edges are ephemeral features. Our Ni-rich particles are in the group of smaller fragments (less than ~50 μm in diameter) that would survive their transit into Earth's atmosphere with no ablation or only modest ablation of their surfaces. In contrast, larger fragments (>50 μm in diameter) would ablate extensively or completely on transit through Earth's atmosphere.

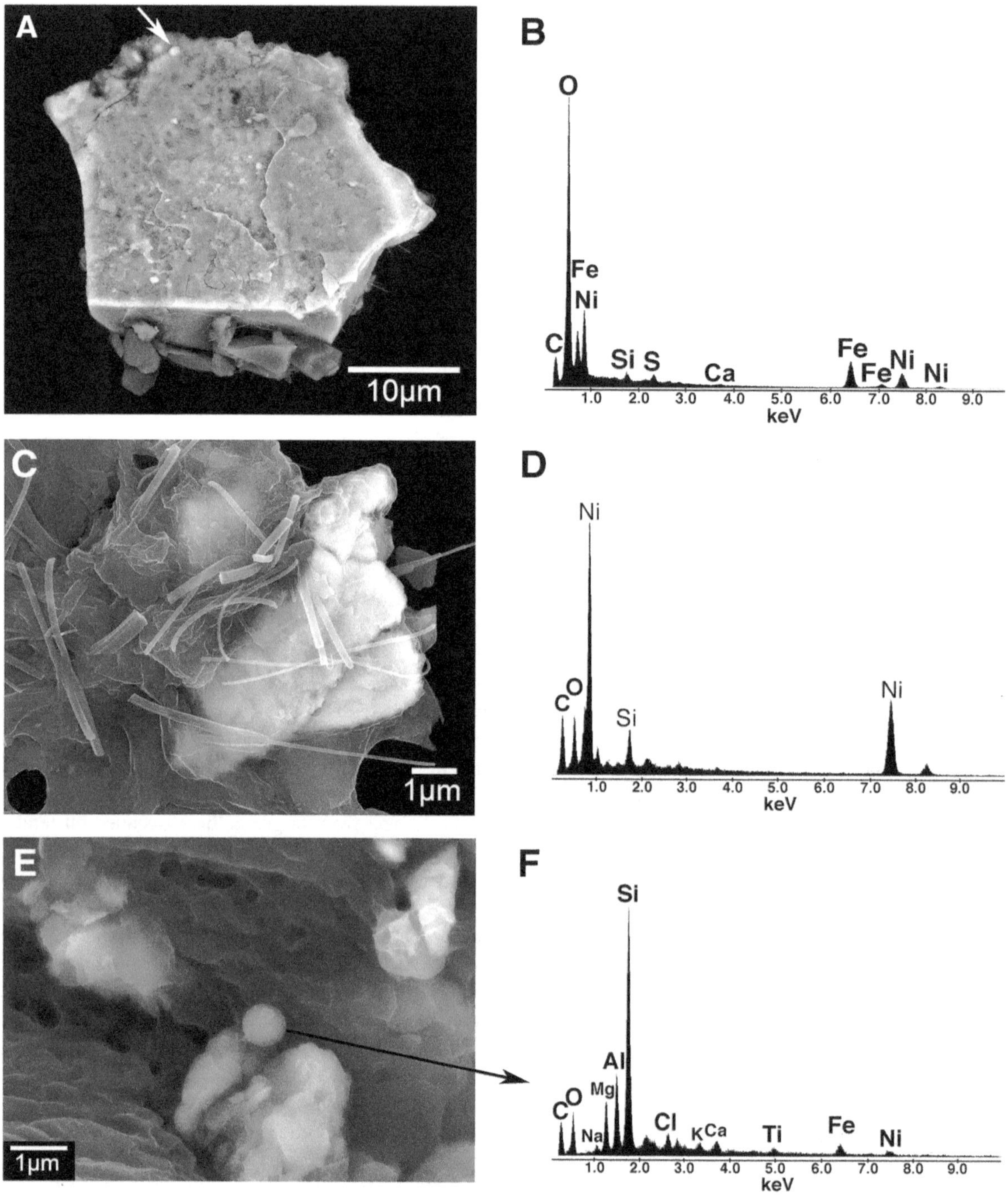

Figure 3. Representative Ni-rich particles from different stratigraphic levels of the Greenland Ice Sheet Project 2 (GISP2) ice core. Left side: Scanning electron microscope (SEM) micrograph. Right side: Qualitative energy-dispersive X-ray spectrometer (EDS) analysis of particle on the left side. (A–B) Arrow points to nucleating ablation spherule on top of Ni-rich particle. This particle contains significant Fe, C, Ca, S, O, and Si in addition to Ni. (C–D) Bright particle in micrograph contains Ni with some Si, O, and C. The surface fibers are common in uncontaminated, primary samples. Particles interpreted as being surface contamination from brief contacts with the laboratory air during Au-Pd coating and transfer into the SEM do not contain fibers on their surfaces. In some cases, fibers are observed lying underneath the contaminants. (E–F) Arrow points from bright spherule in E to its chemical composition in F. The spherule is mostly Si but contains Na, Mg, Al, Cl, K, Ca, Ti, Fe and Ni. The Ni peak suggests an extraterrestrial component. The K peak suggests a terrestrial component. Micrograph imaging conditions: (A) 85% BSE (backscattered electrons), 15% ILSE (in-lens, secondary electrons); (C) 50% BSE, 50% SE2 (secondary electrons); (E) 50% BSE, 50% ILSE. (Note that the peak heights are not absolute so no scale is given for peak heights. This means that the relative heights of nearby spectral peaks provide a good estimate of relative abundances of these elements. In contrast, the relative heights of peaks with different energies are only loosely correlated to their relative abundance.)

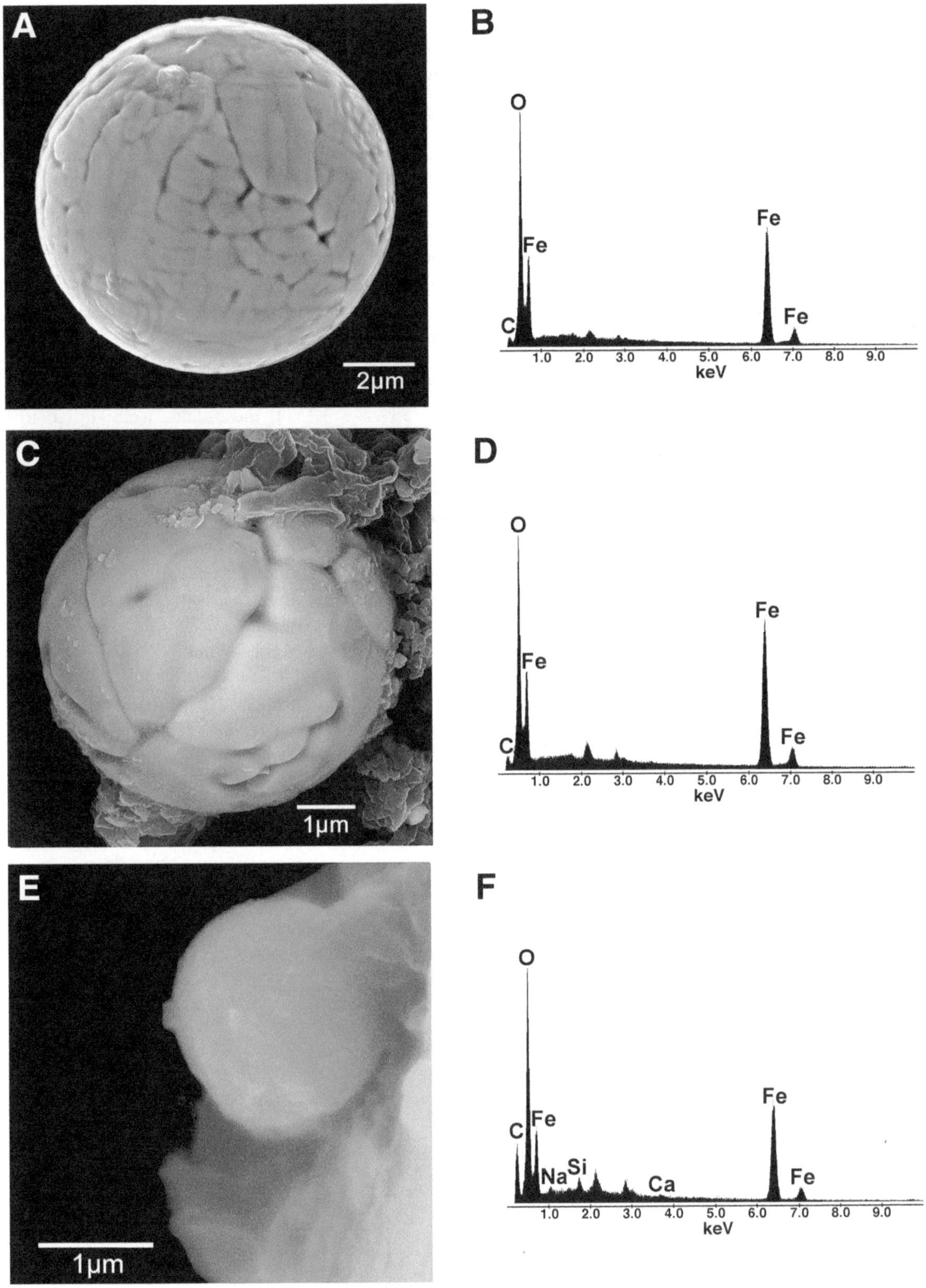

Figure 4. Representative Fe-oxide–rich spherules from different stratigraphic levels of the Greenland Ice Sheet Project 2 (GISP2) ice core. Left: Scanning electron microscope (SEM) micrograph. Right: Qualitative energy-dispersive X-ray spectrometer (EDS) spectrum. Spherules are largely iron oxide with minor C, Mg, Na, and Si in some cases. Peaks around 2.2 and 2.9 keV are from the Au-Pd coating used to improve the image quality of the SEM micrographs. Micrograph imaging conditions: (A) 80% ILSE (in-lens, secondary electrons), 20% BSE (backscattered electrons); (C) 50% BSE, 50% SE2 (secondary electrons); (E) 50% BSE, 50% ILSE. (Note that the peak heights are not absolute so no scale is given for peak heights. This means that the relative heights of nearby spectral peaks provide a good estimate of relative abundances of these elements. In contrast, the relative heights of peaks with different energies are only loosely correlated to their relative abundance.)

The melt droplets that spalled off these larger fragments would solidify as Fe-oxide–rich droplets (I-type cosmic spherules.) Unless they were quite large (meters in diameter), these larger fragments would not survive intact to the surface of Earth. Only their cosmic spherules would remain to be incorporated into the geological record. Thus, both the cosmic spherules and the Ni-rich fragments could have been derived from the same extraterrestrial body.

Independent Test of Chronology from Historical Data

According to Zacharius of Mytilene "And the stars in the sky had appeared dancing in a strange manner and it was the summer of the year eleven" (A.D. 533; Hamilton and Brooks, 1899, p. n240); "And it lasted about 6 or 7 years, until the year three" (A.D. 540; Hamilton and Brooks, 1899, p. n240). This observation has been interpreted by Baillie as a sign of increased cosmic dust loading of Earth's atmosphere between A.D. 533 and 540 (Baillie, 1999).

Two A.D. 533 celestial events recorded in a north Chinese historical record may reinforce Zacharias' observation. Both suggest a soupy or hazy atmosphere.

(1) "On 15 Oct 533, Mars and Jupiter were in conjunction about one inch apart near alpha Crateris. Their rays covered one another" (Ershisishi: dianjiaoben [all further references to Ershisishi: dianjiaoben] Weishu 105-1 Tianxiangzhi 4: 2445). In other words, the night sky was so hazy that bright objects appeared to project long rays (554).

(2) "On 1 Nov 533, Venus and Mars were in conjunction, 7 inches apart, near gamma Corvi; their rays reached to one another" (Weishu 105-1 Tianxiangzhi 4: 2445).

In addition, there is the account of Cassiodorus in Italy, which was grouped with the A.D. 536 event by Stothers (1984). "The sun…seems to have lost its wonted light, and appears to be of a bluish color. We marvel to see no shadows of our bodies at noon, to feel the mighty vigor of the sun's heat wasted into feebleness, and the phenomena that accompany an eclipse prolonged through almost a whole year. We have had …a summer without heat … the crops have been chilled by north winds… the rain is denied" (Barnish, 1992, p. 179). According to Ruggini as quoted by Barnish, this account dates to the late spring of A.D. 534.

If increased cosmic dust produced atmospheric dimming, our observed fluxes of extraterrestrial material should be high compared to the average flux during the Holocene. The average rate of deposition of meteoritic smoke at high latitudes during the Holocene was 78 ± 30 metric tons/yr extrapolated over the entire Earth's surface (Gabrielli et al., 2004). (The average Holocene flux at low latitudes extrapolated over the entire Earth's surface is lower; 15 ± 4 metric tons/yr.) When the high-latitude flux is converted to volume per unit area using an average particle density of 3.0 g/cm^3, this translates into a flux range of 3.1E–9 to 7.1E–9 cm^3/cm^2/yr. Each of our ice-core samples represents the flux over about 1 cm^2 of the surface area of Earth. If we take the stratigraphic level with the largest extraterrestrial particles (one example is in Fig. 3A), it has a total volume of 2.73E–8 cm^3. Corrected for the time range covered by this level (0.82 yr), this represents between 5 and 11 times the average Holocene flux of meteoritic smoke per square centimeter at high latitudes of Earth. This is a minimum calculation, as we cannot include extraterrestrial particles below a few microns in diameter. Thus, the flux of extraterrestrial material into one level of the ice core appears to have been significantly higher than the average flux at high latitudes.

Using our new time scale, our observations of Ni-rich fragments and cosmic spherules in the ice core also extend over this time period (Fig. 5). The oldest Ni-rich fragments are dated as A.D. 533.3 ± 0.41. The youngest cosmic spherule is dated as A.D. 539.1 ± 0.33. If these dates were shifted to be older by 1 yr, the time range of cosmic input would be A.D. 532–538. This is inconsistent with the historical data. If these dates were shifted to be younger by 1 yr, the time range with errors would encompass the end of A.D. 533 (i.e., A.D. 533.89) to the start of 540 A.D. Because the dancing of the stars started in the summer of A.D. 533, we cannot shift the time scale to be 1 yr younger. Our calendar-year time scale is the only allowable match.

Suggestions for Further Work

Our dating can be further tested by examining the GISP2 core for the chemical record of unusual, or out-of-season dust storms, as expressed in Mg and Ca ion data corrected for the signal from seawater. There were unusual dust storms in A.D. 522 and in A.D. 550. Other than the dust storms in A.D. 535, 536, and 537, these are the only unusual dust storms documented between

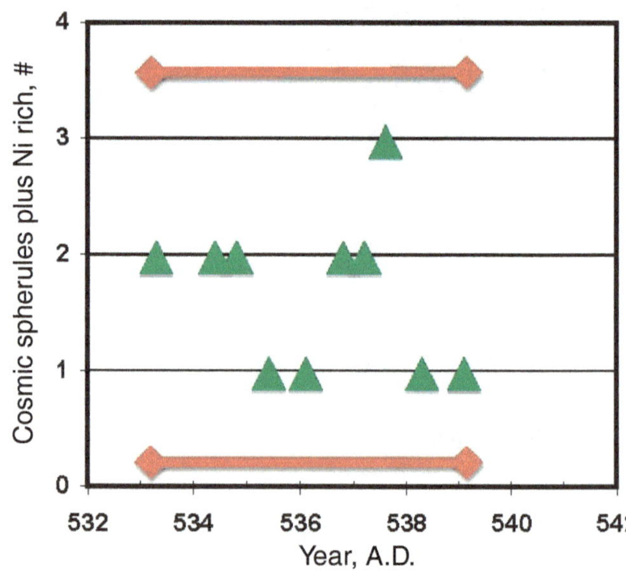

Figure 5. Proposed calendar-year time scale and the occurrence of dancing stars. Green triangles—numbers of Ni-rich fragments and cosmic spherules versus time. Red lines with diamonds on end—time range of historical observations of dancing stars.

A.D. 500 and 560. Although a complete scan is not available to us at present, it is possible that a decomposition of color scans of archived ice from the GISP2 core might produce peaks in yellow. If so, color scans coupled with historical data could provide a nondestructive method of dating and cross correlation of high-altitude ice cores from Greenland.

Although the Mg and Ca signal from dust storms in central Asia is not present in Antarctica, cosmic dust loading should be planetwide. Thus, Ni-rich particles and cosmic spherules are likely to be present in high-altitude ice cores from Antarctica. These particles could be searched for directly using an SEM and EDS analyzer, or they could be assessed indirectly by measuring the flux of meteoritic smoke (Gabrielli et al., 2004; Lanci and Kent, 2006).

CONCLUSIONS

Historical accounts of unusual dust storms in China, coupled with ion data, offer a method for calendar-year dating the GISP2 ice. This time scale meets a test based on an independent assessment of the load of Ni-rich particulates and cosmic spherules during this time. This time scale could be further tested by looking at longer time series of ion and particulate data and by color decompositions of color scans of the corresponding sections of ice cores. Our methods offer a potentially powerful tool for refining the dating of both Greenland and Antarctic ice cores.

ACKNOWLEDGMENTS

We thank Jihong Cole-Dai for the ion data on our supernatant water samples. We are grateful to the Lamont-Doherty Earth Observatory Climate Center for their support of our laboratory and analytical costs. This work would have been impossible without their generous support and encouragement. We thank Norm Sleep and Michael Ram for helpful reviews. We thank the Electron Microscopy Center at The City College of New York for access to their facilities.

REFERENCES CITED

Baillie, M.G., 1999, Exodus to Arthur: Catastrophic Encounters with Comets: London, UK, Batsford, 256 p.

Baillie, M.G.L., 2007, Tree-rings indicate global environmental downturns that could have been caused by comet debris, *in* Bobrowsky, P.T., and Rickman, H., eds., Comet/Asteroid Impacts and Human Society: Berlin, Springer, p. 105–122.

Baillie, M.G.L., 2008, Proposed re-dating of the European ice core chronology by seven years: Geophysical Research Letters, v. 35, p. L15813, doi:10.1029/12008GL034755.

Bar-Hebraeus, [ca. 1246–1286], 1932, Chronography (translated by Budge, E.A.W.): London, Oxford University Press.

Barnish, S.J.B., 1992, The Variae of Magnus Aurelius Cassiodorus Senator: Liverpool, UK, Liverpool University Press, 254 p.

Biscaye, P., Grousset, E.E., Revel, M., Gaast, S.V., Zielinski, G.A., Vaars, A., and Kukla, G., 1997, Asian provenance of glacial dust (stage 2) in the Greenland Ice Sheet Project 2 core, Summit, Greenland: Journal of Geophysical Research, v. 102, p. 26,765–26,781.

Borisov, A., and Palme, H., 1997, Experimental determination of the solubility of platinum in silicate melts: Geochimica et Cosmochimica Acta, v. 61, p. 4349–4357, doi:10.1016/S0016-7037(97)00268-8.

Bory, A.J.-M., Biscaye, P.E., and Grousset, F.E., 2003a, Two distinct seasonal Asian source regions for mineral dust deposited in Greenland (NorthGRIP): Geophysical Research Letters, v. 30, p. 1167, doi:10.1029/2002GL016446.

Bory, A.J.-M., Biscaye, P.E., Piotrowski, A.M., and Steffensen, J.P., 2003b, Regional variability of ice core dust composition and provenance in Greenland: Geochemistry Geophysics Geosystems, v. 4, no. 12, p. 1107, doi:1110.1029/2003GC000627.

Collins, G.S., Melosh, H.J., and Marcus, R.A., 2005, Earth Impact Effects Program: A Web-based computer program for calculating the regional environmental consequences of a meteoroid impact on Earth: Meteoritics & Planetary Science, v. 40, p. 817–840, doi:10.1111/j.1945-5100.2005.tb00157.x.

Fechtig, H., and Utech, K., 1964, On the presence or absence of nickel in dark magnetic cosmic spherules and their mechanics of origin: Annals of the New York Academy of Sciences, v. 119, p. 243–249, doi:10.1111/j.1749-6632.1965.tb47437.x.

Gabrielli, P., Barbante, C., Plane, J.M., Varga, A., Hong, S., Cozzi, G., Gaspari, V., Planchon, F.M., Cairns, W., Ferrari, C., Crutzen, P., Cescon, P., and Boutron, C.F., 2004, Meteoric smoke fallout over the Holocene Epoch revealed by iridium and platinum in Greenland ice: Nature, v. 432, p. 1011–1014, doi:10.1038/nature03137.

Galimov, E.M., Kolotov, V.P., Nazarov, M.A., Kostitsyn, Yu. A., Kubrakova, I.V., Kononkova, N.N., Roshchina, I.A., Alexeev, V.A., Kashkarov, L.L., Badyukov, D.D., and Sevast'yanov, V.S., 2013, Analytical results for the material of the Chelyabinsk meteorite: Geochemistry International, v. 51, p. 522–539.

Genge, M.J., and Grady, M.M., 1999, The fusion crusts of stony meteorites: Implications for the atmospheric reprocessing of extraterrestrial materials: Meteoritics & Planetary Science, v. 34, p. 341–356, doi:10.1111/j.1945-5100.1999.tb01344.x.

Genge, M.J., Engrand, C., Gounelle, M., and Taylor, S., 2008, The classification of micrometeorites: Meteoritics & Planetary Science, v. 43, p. 497–515, doi:10.1111/j.1945-5100.2008.tb00668.x.

Hamilton, F.J., and Brooks, E.W., 1899, The Syriac Chronicle Known as that of Zachariah of Mitylene: London, UK, Methuen and Co., Byzantine Texts, p. 24–37.

Kang, S., Mayewski, P.A., Yan, Y., Qin, D., Yao, T., and Ren, J., 2003, Dust records from three ice cores: Relationships to spring atmospheric circulation over the Northern Hemisphere: Atmospheric Environment, v. 37, p. 4823–4835, doi:10.1016/j.atmosenv.2003.08.010.

Koffman, B.G., Kreutz, K.J., Breton, D.J., Dunbar, N.W., and Kurbatov, A., 2011, Depositional phasing of volcanic aerosols in the WAIS Divide ice core over the past 2400 years: Washington, DC, American Geophysical Union, Fall Meeting, abstract V11D-2539.

Lanci, L., and Kent, D.V., 2006, Meteoric smoke fallout revealed by superparamagnetism in Greenland ice: Geophysical Research Letters, v. 33, doi:10.1029/2006GL026480.

Larsen, L.B., Vinther B.M., Briffa, K.R., Melvin, T.M., Clausen, H.B., Jones, P.D., Siggard-Anderson, M.-L., Hammer, C.U., Eronen, M., Grudd, H., Gunnarson, B.E., Hantemirov, R.M., Naurzbaev, M.N., and Nicolussi, K., 2008, New ice core evidence for a volcanic cause of the A.D. 536 dust veil: Geophysical Research Letters, v. 35, p. L04708, doi:10.1029/02007GL032450.

Li Yanshou, editor, 650, Nan shi, Liangbenji (History of the Southern Dynasties, Records of the Liang Dynasty): Ershisishi, dianjiaoben, 1962–75 [The 24 standard dynastic histories, punctuated and collated; 241 volume sets]: Beijing, Zhonghua Book Company (Zhonghua Shuju) [in Chinese].

Meese, D.A., Gow, A.J., Grootes, P.M., Mayewski, P.A., Ram, M., Stuiver, M., Taylor, K.C., Waddington, E.D., and Zielinski, G.A., 1994, The accumulation record from the GISP2 core as an indicator of climate change throughout the Holocene: Science, v. 266, p. 1680–1682, doi:10.1126/science.266.5191.1680.

Meese, D.A., Gow, A.J., Alley, R.B., Zielinski, G.A., Grootes, P.M., Ram, M., Taylor, K.C., Mayewski, P.A., and Bolzan, J.F., 1997, The Greenland Ice Sheet Project 2 depth-age scale: Methods and results: Journal of Geophysical Research, v. 102, no. C12, p. 26,411–26,423, doi:10.1029/97JC00269.

Michael the Syrian, [ca. 1166–1199 A.D.], 1901, Michael the Syrian Chronicle (translated to French by Chabot, J.–B.): Paris, Belles-Lettres.

O'Brien, S.R., Mayewski, P.A., Meeker, L.D., Meese, D.A., Twickler, M.S., and Wicklow, S.I., 1995, Complexity of Holocene climate as reconstructed from a Greenland ice core: Science, v. 270, p. 1962–1964, doi:10.1126/science.270.5244.1962.

Rampino, M.R., Self, S., and Stothers, R.B., 1988, Volcanic winters: Annual Review of Earth and Planetary Sciences, v. 16, p. 73–99, doi:10.1146/annurev.ea.16.050188.000445.

Rigby, E., Symonds, M., and Ward-Thompson, D., 2004, A comet impact in AD536?: Astronomy and Geophysics, v. 45, p. 1.23–21.26.

Salzer, M.W., and Hughes, M.K., 2007, Bristlecone pine tree rings and volcanic eruptions over the last 5000 yr: Quaternary Research, v. 67, p. 57–68, doi:10.1016/j.yqres.2006.07.004.

Sowers, T., Bender, M., Labeyrie, L., Martinson, D., Jouzel, J., Raynaud, D., Pichon, J.J., and Korotkevich, Y.S., 1993, A 135,000 year Vostok-SPEC-MAP common temporal framework: Paleoceanography, v. 8, p. 737–766, doi:10.1029/93PA02328.

Stothers, R.B., 1984, Mystery cloud of AD536: Nature, v. 307, p. 344–345, doi:10.1038/307344a0.

Taylor, S., Lever, J.H., and Harvey, R.P., 2000, Numbers, types, and compositions of an unbiased collection of cosmic spherules: Meteoritics & Planetary Science, v. 35, p. 651–666, doi:10.1111/j.1945-5100.2000.tb01450.x.

Wei, L., Mosley-Thompson, E., Gabrielli, P., Thompson, L.G., and Barbante, C., 2008, Synchronous deposition of volcanic ash and sulfate aerosols over Greenland in 1783 from the Laki eruption (Iceland): Geophysical Research Letters, v. 35, p. L16501, doi:10.1029/2008GL035117.

Wei Shu, Tianxiangzhi, 554, Wei Shou, editor, tianxiangzhi (History of the Wei Dynasty, Celestial Phenomena): Ershisishi, dianjiaoben, 1962–75 [The 24 standard dynastic histories, punctuated and collated; 241 volume sets]: Beijing, Zhonghua Book Company (Zhonghua Shuju) [in Chinese] (all Chinese references in this chapter are to Ershisishi: dianjiaoben).

Yao Cha and Yao Silian, editors, 635, Liang shu, wudiji (History of the Liang Dynasty, Records of Emperor Wudi): Ershisishi, dianjiaoben, 1962–75 [The 24 standard dynastic histories, punctuated and collated; 241 volume sets]: Beijing, Zhonghua Book Company (Zhonghua Shuju) [in Chinese].

Zdanowicz, C.M., Zielinski, G.A, Wake, C.P., Fisher, D.A., and Koerner, R.M., 2000, A Holocene record of atmospheric dust deposition on the Penny Ice Cap, Baffin Island, Canada: Quaternary Research, v. 53, p. 62–69, doi:10.1006/qres.1999.2091.

Manuscript Accepted by the Society 31 January 2014

Printed in the USA

The Geological Society of America
Special Paper 505
2014

What caused terrestrial dust loading and climate downturns between A.D. 533 and 540?

Dallas H. Abbott
Dee Breger[†]
Pierre E. Biscaye
Lamont-Doherty Earth Observatory of Columbia University, Palisades, New York 10964, USA

John A. Barron
U.S. Geological Survey, 345 Middlefield Road, Menlo Park, California 94025, USA

Robert A. Juhl
Independent researcher, 1-4-1 Rokko Heights 906, Shinkawa, Chuo-ku, Tokyo 104-0033, Japan

Patrick McCafferty
Independent researcher, 7 Windsor Avenue North, Belfast BT9 6EL, Northern Ireland

ABSTRACT

Sn-rich particles, Ni-rich particles, and cosmic spherules are found together at four discrete stratigraphic levels within the 362–360 m depth interval of the Greenland Ice Sheet Project 2 (GISP2) ice core (72.6°N, 38.5°W, elevation: 3203 m). Using a previously derived calendar-year time scale, these particles span a time of increased dust loading of Earth's atmosphere between A.D. 533 and 540. The Sn-rich and Ni-rich particles contain an average of 10–11 wt% C. Their high C contents coupled with local enrichments in the volatile elements I, Zn, Cu, and Xe suggest a cometary source for the dust. The late spring timing of extraterrestrial input best matches the Eta Aquarid meteor shower associated with comet 1P/Halley. An increased flux of cometary dust might explain a modest climate downturn in A.D. 533. Both cometary dust and volcanic sulfate probably contributed to the profound global dimming during A.D. 536 and 537 but may be insufficient sources of fine aerosols. We found tropical marine microfossils and aerosol-sized $CaCO_3$ particles at the end A.D. 535–start A.D. 536 level that we attribute to a low-latitude explosion in the ocean. This additional source of dust is probably needed to explain the solar dimming during A.D. 536 and 537. Although there has been no extinction documented at A.D. 536, our results are relevant because mass extinctions may also have multiple drivers. Detailed examinations of fine particles at and near extinction horizons can help to determine the relative contributions of cosmic and volcanic drivers to mass extinctions.

[†]Also at Micrographic Arts, P.O. Box 3088, Saratoga Springs, New York 12866, USA.

Abbott, D.H., Breger, D., Biscaye, P.E., Barron, J.A., Juhl, R.A., and McCafferty, P., 2014, What caused terrestrial dust loading and climate downturns between A.D. 533 and 540?, *in* Keller, G., and Kerr, A.C., eds., Volcanism, Impacts, and Mass Extinctions: Causes and Effects: Geological Society of America Special Paper 505, p. 421–437, doi:10.1130/2014.2505(23). For permission to copy, contact editing@geosociety.org.

INTRODUCTION

We have previously used historical, particle, and ion data to propose a calendar-year time scale for the Greenland Ice Sheet Project 2 (GISP2) ice core from central Greenland at depths between 362 m and 360 m during part of the early sixth century A.D. Because previous work suggested a significant extraterrestrial influence on climate and tree rings between ca. A.D 533 and 541 (Baillie, 1999, 2007), we selected ice-core samples that covered this time. We found Ni-rich particles and I-type (Fe-oxide) cosmic spherules from 361.8 to 360.51 m depth, dated by us to between 533 and 540 A.D., the same time range as observations of increased atmospheric dust i.e., "dancing stars," in historical records (Abbott et al., this volume, Chapter 22). There are at least five possible sources for the dust in our samples: asteroidal debris, cometary dust, volcanic sulfates, windblown particles, and terrestrial impact ejecta. In this paper, we address the multiple origins of this dust and its possible effects on climate from 533 to 540 A.D.

Previous Work on Extraterrestrial Debris within Polar Ice

There are many accepted tracers for extraterrestrial debris in polar ice: noble gas abundances and isotopic compositions (Brook et al., 2000; Lal and Jull, 2005; McGee and Mukhopadhyay, 2013), Ir, Pt, and Ni abundances (Brocas and Picciotto, 1967; Karner et al., 2003; Gabrielli et al., 2004), nonvolcanic Fe- and Ni-rich spherules (Hodge and Wright, 1964; Hodge et al., 1964; Taylor et al., 1998, 2000), meteoritic fragments (Yada et al., 2004; Misawa et al., 2010), and magnetic properties (Lanci and Kent, 2006). A major drawback to most of these tracers is that they require relatively large volumes and stratigraphic thicknesses of ice. For example, a study of extraterrestrial He^3 abundance in replicate samples found that around 1000 g of ice were needed to achieve reproducible results with 2σ errors on the order of ±30% (Brook et al., 2009). Another study achieved lower errors, but individual samples represented over 100 yr of ice accumulation (Winckler and Fischer, 2006). In addition, noble gases are lost from larger or higher-velocity particles as they heat up during atmospheric entry (Farley et al., 1997; Mukhopadhyay and Farley, 2006). Thus, noble gases can only provide a lower limit to the deposition rate of extraterrestrial debris (Marcantonio et al., 1996, 1998, 1999).

One study of Ir and Pt abundances required 1000–2000 g samples to achieve reproducible results (Karner et al., 2003). Magnetic studies use smaller 360 g samples but constrain only the background levels of meteoritic smoke recorded in particles 7–17 nanometers in diameter (Lanci and Kent, 2006). In our study, we assessed the composition and origin of particles from a relatively small amount of ice meltwater, between 27 and 102 g (mL) per sample (Appendix Table 1A). Individual samples of meltwater represent between 0.3 and 0.92 yr of time. Based on previous ice-core research, such brief intervals of time and masses of ice would be expected to yield no micrometer-sized extraterrestrial particles. As we have shown (Abbott et al., this volume, Chapter 22) and will further demonstrate in this paper, we instead found surprisingly large abundances and sizes of inferred extraterrestrial particles within a restricted stratigraphic interval. These results could mean that the time between A.D. 533 and 540 had an unusually high deposition rate of extraterrestrial particles. In this paper, we make a case for a cometary origin of many of these particles.

COMETS OR ASTEROIDS AS SOURCES OF EXTRATERRESTRIAL DUST IN THE GISP2 ICE CORE

Our previous findings of Ni-rich particles and cosmic spherules do not require a cometary source for the historically documented dust loading of Earth's atmosphere, i.e., "the dancing stars," between A.D. 533 and 540 (Abbott et al., this volume, Chapter 22). Ni-rich particles and cosmic spherules could come from either asteroids or comets. They are both sources of micrometer-sized cosmic dust particles and interplanetary dust particles (IDPs).

Asteroids are composed of material that was differentiated at the same temperatures as igneous rocks, that is, at minimum primitive liquidus temperatures of ~1100 °C (McKenzie and Bickle, 1988). Because comets represent material that condensed from the primordial solar nebula at lower temperatures than asteroids, comets should contain higher concentrations of material with low-equilibrium nebular condensation temperatures, in particular C (Table 1; Lodders, 2003). Some ultravolatile materials, for example, water ice and noble gases, are likely to be very rare or absent in asteroids and relatively abundant in comets.

Some cosmic dust contains concentrations of volatile elements such as Sn, Zn, and Bi (Table 2). During the 1990s, all or most of the volatile elements in cosmic dust particles were considered to be volcanogenic contaminants that condensed onto the particle surfaces after they entered Earth's atmosphere (Jessberger et al., 1992; Rietmeijer, 1993, 1995). The most recent research suggests that these volatile elements are integral parts of the grains (e.g., cometary) rather than volcanogenic surface precipitates (Flynn et al., 2006). In the late 1990s, ultra-carbon-rich micrometeorites were found in Antarctica (Engrand and Maurette, 1998). As a result of this and other discoveries, the percentage of cosmic dust consisting of carbonaceous chondrite-like material has been steadily revised upward. At present, it is estimated that comets (e.g., carbonaceous chondrite-like dust) probably represent the ultimate source of over 80% of the cosmic dust entering Earth's atmosphere (Engrand, 2011).

Observational Evidence for Excess Concentrations of Volatile Elements in Comets

Several types of observations constrain the concentrations of volatile elements in comets relative to ordinary chondritic meteorites. The first are direct observations of the coma of comets. The

TABLE 1. ELEMENTS WITH LOW NEBULAR CONDENSATION TEMPERATURES (DATA FROM LODDERS, 2003)

Element	50% condensation temperature (°K)	Phase
C	40	$CH_4 \cdot 7H_2O$ plus CH_4 (ice)
Xe	68	Water
I	535	Cl apatite
Sn	704	Fe alloy
Zn	726	Forsterite plus enstatite
Pb	727	Fe alloy
Bi	746	Fe alloy
K	1006	Feldspar
Cu	1037	Fe alloy

spectra of comet Ikeyi Seki indicate excess Cu (Rietmeijer, 1988); the spectra from Comet 1P/Halley indicate excess Zn (Krueger and Kissel, 1987). The second are observations of the volatile content of chondritic, porous, cosmic dust, the type of cosmic dust that comes from comets (Bradley and Brownlee, 1986; Brownlee, 1987; Bradley et al., 1999). Chondritic, porous, cosmic dust includes particles with excess Bi, Sn, and Zn relative to chondritic meteorites (Rietmeijer, 1989; Schulze et al., 1997; Ciucci et al., 2010). The third are observations of high concentrations of volatiles in the lunar soil. Excess Zn and Pb are attributed to input from comets (Dikov et al., 1998). When these concentrations are normalized relative to the concentrations of the same element in chondritic meteorites (e.g., asteroids), they provide an index that can be used to distinguish cometary from asteroidal material.

The volatile elements most diagnostic of a cometary origin are those for which concentrations are highest relative to ordinary chondritic meteorites. The excess concentrations of volatiles can be very high, typically several times chondritic for Sn, but possibly up to 350,000 times chondritic for the same element in cometary dust in an ice core (Table 1; La Violette, 1987). Zn also has very high excess concentrations: 3–4 times chondritic to possibly 610 times chondritic in material from the 1908 explosion over Tunguska in Siberia. Thus, Sn and Zn are good proxies to use to distinguish between cometary and asteroidal material.

LABORATORY METHODS

Our ice-core samples were processed under a laminar-flow hood. The hood has a filter and positive pressure that prevent contamination from local dust. The outside surface of each ice-core sample was rinsed with quartz-distilled water to remove contamination from drilling and atmospheric dust. The remaining ice was melted under the laminar-flow hood and centrifuged. The supernatant was analyzed for ion concentrations using ion chromatography. All of the remaining particle-rich water was placed inside cleaned Teflon washers, glued to cleaned glass slides, and left to evaporate in the laminar-flow hood. All final cleaning and rinsing was done with quartz-distilled water. Each slide was then covered until immediately before insertion into a Zeiss Supra 50 scanning electron microscope (SEM) with attached EDAX energy-dispersive X-ray spectrometer (EDS). Micrographs were collected in secondary electron (SE), backscattered electron (BSE), or mixed SE and BSE modes. Imaging and analyses were conducted in high-vacuum mode at 15 kV accelerating voltage. In between SEM sessions, each slide was re-covered immediately upon removal from the SEM and stored inside the laminar-flow hood to prevent contamination from atmospheric dust. Marine-sediment cores are stored in a different building, thus precluding them as a source of contamination.

Our standard method of searching for interesting particles on the SEM combines 50% backscattered and 50% secondary electrons to make a composite image in order to maximize both structural and chemical data. We survey for interesting particles at a magnification of 200×, increasing the magnification when we find an interesting particle or particle group. We select particles that have one or more of the following attributes: distinct sharp edges, a spherical morphology, a biological morphology, or a bright appearance. Each slide of particles from a given

TABLE 2. VOLATILE ELEMENTS AND THEIR ENRICHMENTS IN COMETS AND COMETARY MATERIAL

Metal	Where observed	Amount	Reference
Bi	Chondritic porous IDP	4–35 times chondritic (best 20 times)	Rietmeijer (1989)
Cu	Comet Ikeya Seki	high	Rietmeijer (1988)
Pb	Lunar soil	0.1–0.2 at%	Dikov et al. (1998)
Sn	Chondritic aggregate IDP	3 grains of Sn oxide	Ciucci et al. (2010)
Sn	Chondritic porous IDP	3–8 times chondritic (best 6.3 times), includes Sn-rich particles	Rietmeijer (1989)
Sn	Cometary dust in ice core	350,000 times chondritic?	La Violette (1987)
Sn	Tunguska (comet)	11.6 times chondritic?	La Violette (1987)
Zn	Lunar soil	4 at%	Dikov et al. (1998)
Zn	Dust from comet 1P/Halley	3–4 times chondritic	Schulze et al. (1997)
Zn	Comet 1P/Halley	high	Krueger and Kissel (1987)
Zn	Tunguska (comet)	610 times chondritic	La Violette (1987)

Note: IDP—interplanetary dust particle.

stratigraphic level contains hundreds to thousands of grains; we cannot analyze every grain. The most carbon-rich or fluffy grains are usually not selected. This means that most of our particles are in one or more of the following classes: metal-rich, minerals, biological skeletons, spherules, or glass.

A COMETARY SOURCE FOR Ni-RICH AND Sn-RICH PARTICLES

Particles of cometary origin typically contain at least a few percent carbon (C). The most primitive asteroids are those most like comets, i.e., carbonaceous chondrites. Although the amount of carbon is quite variable, the most primitive carbonaceous chondrites (CI chondrites) contain ~4% carbon (Alexander et al., 2012; Marty, 2012). The more primitive chondritic IDPs contain ~12% carbon (Thomas et al., 1993). Dust from comet 1P/Halley contains ~18% carbon (Delsemme, 1991; Marty et al., 2013).

Because we coat our samples with Au/Pd, our semiquantitative X-ray microanalyses are a rough measure of the carbon content of our samples. We use the measured C content of terrestrial SiO_2 grains in our samples to estimate background C levels (the maximum level of C potentially attributable to terrestrial sources) of ~2–3 wt% (Appendix Table 1B). On average, we find ~11 wt% C (range 5–19 wt%) in our samples of Ni-rich particles (Table 3). This C content is in the right range for cometary materials but is somewhat low. (We attribute this to our criteria for sample selection, which systematically avoids the most C-rich particles.) We also find local enrichments of the volatile elements Zn and Cu.

Because the brightness in backscatter mode of C is directly proportional to its average atomic weight, a particle of nearly pure C appears dark gray. In contrast, a pure heavy metal oxide or native heavy metal appears light gray or bright white, respectively. With the exception of a grain of native Ni, our Ni-rich samples did not appear as bright in backscatter mode (fig. 3 in Abbott et al., this volume, Chapter 22) as is typical of pure Fe-Ni oxides or pure Fe-Ni metal. The lower relative brightness of our Ni-rich particles in backscatter mode is consistent with the presence of significant amounts of carbon (10% level) with the metal.

Sn-Rich Particles in the GISP2 Ice Core

We found small Sn-rich particles at four different horizons in the GISP2 ice core (Figs. 1 and 2; Table 4). Because Sn is a heavy metal, small, Sn-rich particles appear bright in backscatter mode on the SEM. Their relative brightness enhances the visibility of small, Sn-rich particles.

The Sn-rich particles have several characteristics. The first is their small size (Fig. 1). The largest particle is ~3 µm in diameter. The particles are platy in appearance. The brightness and compositions of the particles are heterogeneous (Fig. 1; Table 4). The Sn-rich particles all contain substantial C, on average ~10 wt%. The Sn-rich particles contain Fe with the Sn or in neighboring grains. Two spectra of Sn-rich particles are best fit with the addition of about 1 wt% of the noble gas Xe. To achieve a best fit (comparison of actual spectrum with theoretical model using same kV and element list), a third spectrum required the addition of 3.2 wt% I in addition to 3.5 wt% Xe (Fig. 2). A few of the spectra show significant peaks (>1%) in either Cu or Zn.

The Sn-rich particles show several similarities to other particles of inferred cometary origin. The first is their platy appearance, which resembles that of two Sn-rich particles previously found in cometary-type IDPs (Rietmeijer, 1989; Ciucci et al., 2010). The second similarity is their average 10 wt% C content (Delsemme, 1991; Marty et al., 2013), which is closer to that of cometary IDPs (12 wt% C) than to carbonaceous chondrite meteorites (4 wt% C). The Fe associated with the Sn is consistent with models of primordial condensation from the solar nebula (Lodders, 2003), which predict that Sn (and Cu) should initially condense within an iron alloy (Table 1). The last and perhaps most compelling factor is the need to frequently add Xe and in one case I to fit the spectra of the Sn-rich particles (Fig. 2). (We have carefully checked for sum and escape peaks of Sn, O, C, and Si and found that these are not present

TABLE 3. SEMIQUANTITATIVE EDS ANALYSES OF Ni-RICH PARTICLES

Sample no.	C (wt%)	Si (wt%)	Ca (wt%)	Cr (wt%)	Fe (wt%)	Ni (wt%)	Zn (wt%)	Cu (wt%)
11-8a	8.8	1.1	0.3	0	24.9	25.9	0	0
11-15c	10.9	1.7	0.8	0.2	34	34.9	0	0
8-8a	5.1	15.1	1.8	0.4	0.4	3	14.9	36.6
8-8b	5.4	12.8	1.4	0.4	0.5	5.5	15	37.6
8-8c	6.4	17.6	2.3	0.4	0.4	3.1	12.2	33.4
6-20a	18.4	3.6	0.1	0	0	65.4	0	0
6-20b	10.9	1.2	0.4	0.1	0.2	80	0.4	0
5C-9	19.2	29.2	2.7	0.2	8.6	7.3	0	0

Note: Sample no.—stratigraphic level is the number before the dash; the number after the dash is the grain number within a given stratigraphic level. Suffixes of a, b, and c represent multiple analyses of the same grain. Differences of a few percent between analyses of the same grain primarily reflect local heterogeneity rather than analytical error. None of these samples contained more than 1 wt% Ti, which would be present in terrestrial material (Hodge and Wright, 1964). Sample 11-8 in Figures 3A and 3B (Abbott et al., this volume, Chapter 22), micrograph (85% backscattered electrons [BSE]), 15% in-lens secondary electrons (ILSE). Sample 8-8 in Figure 1I micrograph (50% BSE, 50% ILSE). This might be a refractory grain like that found by Wark (1986). Sample 6-20 in Figures 3C and 3D (Abbott et al., this volume, Chapter 22), micrograph 50% BSE, 50% SE2 (secondary electron mode). Sample 5C-9 in Figures 3E and 3F (Abbott et al., this volume, Chapter 22) micrograph 50% BSE, 50% ILSE.

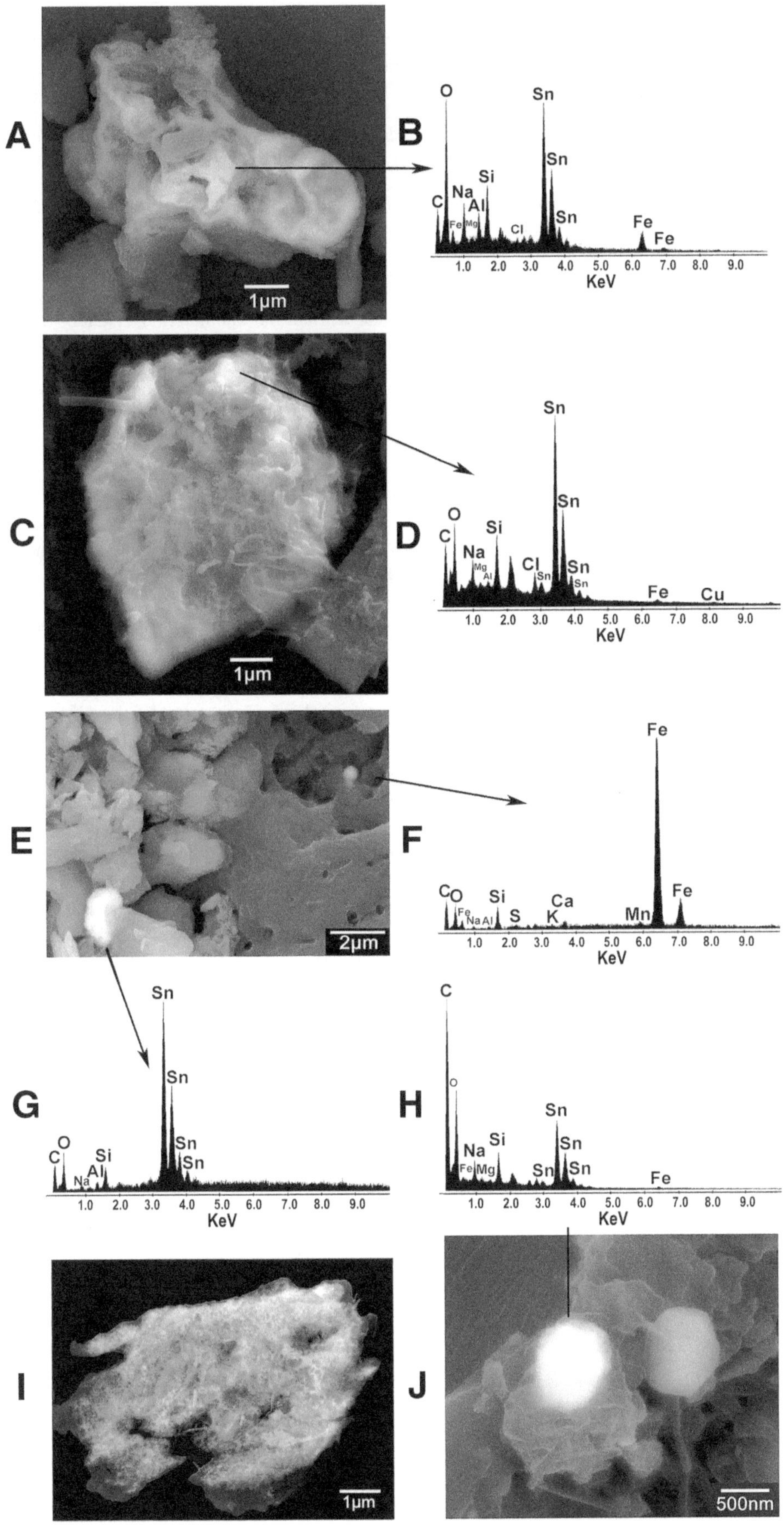

Figure 1. Scanning electron microscope (SEM) micrographs of particles from the ice core. (A, C, E, J) Sn-rich particles with one Fe-rich spherule in E. (B, D, F, G, H) Qualitative energy-dispersive X-ray microanalyses of compositions. (Note that the peak heights are not absolute so no scale is given for peak heights. This means that the relative heights of nearby spectral peaks provide a good estimate of relative abundances of these elements. In contrast, the relative heights of peaks with different energies are only loosely correlated to their relative abundance.) Four out of five analyses are of Sn-rich particles. Analysis D shows a small but distinct Cu peak in addition to Sn peaks. The remaining analysis is of a Fe-rich spherule (F). (I) Ni-rich particle-analysis in Table 3. Ratios of backscattered and secondary electrons were adjusted to maximum the image quality: (A) 100% BSE (backscattered electron mode); (C, I) 50% BSE, 50% ILSE (in-lens secondary electron mode); (E) 90% BSE, 10% ILSE, (J) 50% BSE, 50% SE2 (standard secondary electron mode). Black arrows or lines point from particles to their spectra.

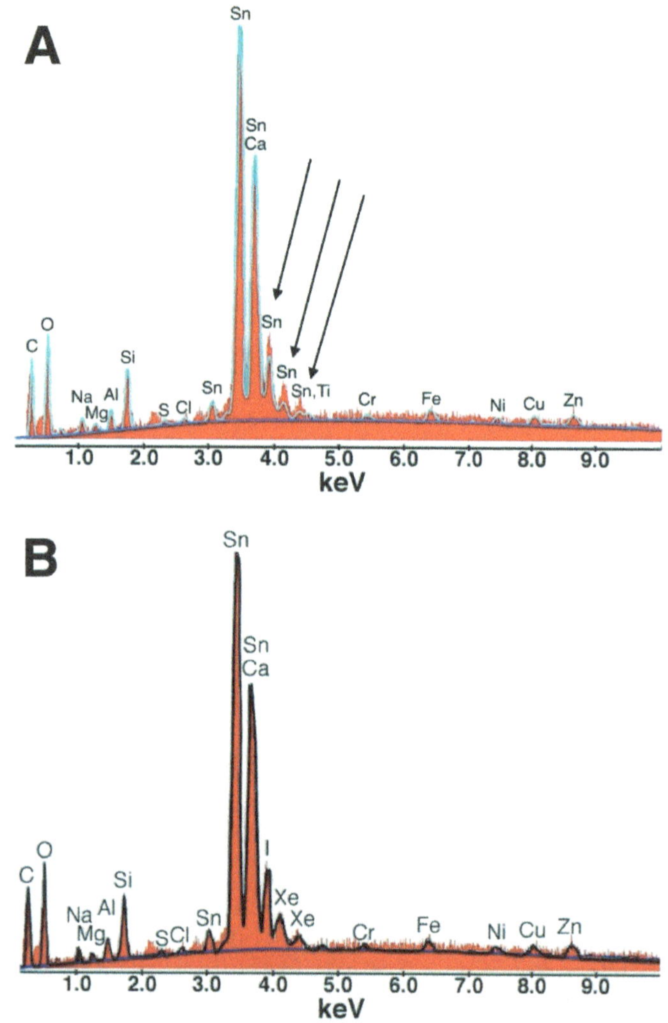

Figure 2. Comparison of two models for the composition of Sn-rich particle. (Note that the peak heights are not absolute so no scale is given for peak heights. This means that the relative heights of nearby spectral peaks provide a good estimate of relative abundances of these elements. In contrast, the relative heights of peaks with different energies are only loosely correlated to their relative abundance.) Spectral peaks from the Au/Pd coating are not included in either model. (Micrograph of particle is given in Figure 1E; spectrum is in Figure 1G.) Red area—raw spectrum. Dark blue line—background to the spectrum. (A) Light-blue line—model of the spectral peaks based on selected element peak locations and the accelerating voltage used for the analysis. Red area extends above the light-blue line to the right of the peak labeled Sn, Ca. The data are poorly fit here at the three peaks designated by arrows. (B) Black line—model of spectral peaks. The spectrum is best fit by adding both Xe and I to the Sn and Ca peaks between 3.8 and 4.8 keV. Cu, Zn, Ni, and Fe are also present at modest levels.

at the right energy levels to fit the spectra without the addition of Xe and I. No rare earth elements fit the spectra.) Xe and I are both highly volatile elements with nebular condensation temperatures of 68 K and 535 K, respectively (Table 1). Xe is thought to condense with water in the solar nebula (Lodders, 2003). In addition, some Xe isotopes can be produced during bombardment of iodine by subatomic particles in the solar wind (Pepin et al., 1995). The Sn-rich particles sometimes contain significant Zn and Cu, i.e., two other volatile elements that are enriched in comets relative to asteroids. Thus, the overall patterns of morphology and elemental abundance of the Sn-rich particles are most consistent with a cometary source.

The final argument for a cometary origin of the Sn particles is their co-occurrence with the Ni-rich particles (Fig. 3). The Ni-rich particles, more generally accepted as extraterrestrial, also have high C contents that resemble the C contents of cometary dust (Table 3). Thus, if the high C contents of the Ni-rich particles are accepted as being indicative of cometary origin, it follows that the four co-occurrences of high-C, Sn-rich particles are also likely to have a cometary origin.

POTENTIAL SOURCES OF COMETARY DEBRIS

Now that we have made a case for a cometary source of the Sn-rich and Ni-rich dust in our samples, we seek to determine whether or not a specific comet or comets can be linked to this dust. We use two data sets to provide constraints: the changes in volume of extraterrestrial debris over time, and the seasonality of input of extraterrestrial debris. Finally, we use modeling of cometary orbits to determine those comets that could have shed debris that would affect Earth during its rotation about the Sun.

In the 10 yr period between A.D. 532 and 542 covered by our ice-core samples, there were seven observed comets (Hasegawa,

TABLE 4. SEMIQUANTITATIVE MICROANALYSES OF Sn-RICH PARTICLES

Sample no.	C (wt%)	O (wt%)	Na (wt%)	Al (wt%)	Si (wt%)	Sn (wt%)	Ca (wt%)	Cr (wt%)	Fe (wt%)	Ni (wt%)	Zn (wt%)	Cu (wt%)	Other elements
11-57a	5.1	24.4	4.1	1.7	3.7	47.2	1.9	0.2	5.7	0.6	2.2	0.7	0.8% Xe
11-57b	5.3	38	6.4	2.2	11.4	22.7	1.7	0	9	0	0.4	0	0.7% Xe
11-57c	5.9	23.9	4.2	1.7	3.8	45.9	2.1	0	8.3	0.3	1.7	0.4	0.7% Xe
6-1b	32.3	23.8	2.8	0.2	3.1	33.1	1.6	0	1.2	0.1	0.7	0.4	No Xe
5C-8a	4	6.2	0.7	0.7	1.8	59	3.5	0.8	2.48	2.1	7	4.3	3.2% I, 3.5% Xe
4A-30a	6.8	13.1	2.9	0.3	2.1	68.9	1.3	0	0.7	0.4	0.4	0.4	1.2% Xe
4A-30b	7.4	14.8	3.4	0.5	4.1	63.9	1.4	0	0.9	0.4	0	0.9	0.8% Xe
4A-30c	11	20.4	5.2	0.5	6.7	52	1.4	0	0.3	0	0.2	0	0.7% Xe

Note: Differences above the percent level reflect sample heterogeneity. Sample 11-57 is Figures 1A and 1B; sample 5C-8 is Figures 1E and 1G; sample 4A-30 is Figures 1C and 1D; sample 6-1b is Figures 1H and 1J.

1980; Kronk, 1984). In comparison, there were seven comet passes observed between A.D. 500 and 532 (a 32 yr period) and five comet passes between A.D. 542 and 560 (a 28 yr period). It is possible that one or more of the debris fields left behind by comets (i.e., the sources of meteor showers) might have been strengthened prior to A.D. 533. If more than one meteor shower was strengthened, the volumes of extraterrestrial debris would not decrease uniformly over time but would show multiple peaks and lows. If only one meteor shower was strengthened, the volume of extraterrestrial debris should decrease relatively uniformly over time.

The change in volume of extraterrestrial material over time (Fig. 4) is derived from summing the estimated volumes of Ni-rich particles, Sn-rich particles, and I-type cosmic spherules in each sample that contains them. The upper envelope of volumes (four samples) shows a linear power of 2 decline from A.D. 533 to 540.

Discussion

The volume of extraterrestrial dust from A.D. 533 to 540 measured in our GISP2 samples declines over time. The first and simplest interpretation is that dust from a single comet pass produced declining volumes of cosmic dust starting in A.D. 533 and ending in A.D. 540. The data also permit a second interpretation involving multiple injections of cosmic dust. The volume of cosmic dust is greatest in A.D. 533 and declines until A.D. 536. Later in A.D. 536, there is a new injection of cosmic dust that declines until A.D. 537. There is a final injection of cosmic dust in A.D. 538 that declines until A.D. 539–540. With our data set, we cannot distinguish among these interpretations.

There are other caveats. Our ice samples are small, far smaller than is necessary to obtain good constraints on the size distribution of extraterrestrial particles (Peucker-Ehrenbrink and Ravizza, 2000). Our methods do not allow us to constrain the abundance of nanometer-sized extraterrestrial particles. The latter are fine aerosol sized, the most likely particles to remain in the atmosphere for months at a time and to reduce the transparency of the atmosphere during their residence. Thus, we do not know if a reduced or increased volume of fine-aerosol-sized particles accompanied the apparently reduced volume of larger extraterrestrial particles during A.D. 536 and 537.

Seasonality of Cosmic Input

Although ice-core records usually have some uncertainty as to their absolute age, the seasons within annual ice-core records are relatively well defined. The GISP2 ice core shows seasonal changes in bubble and near-surface grain structure, $\delta^{18}O$, and conductivity that allow identification of annual layers (Alley et al., 1997; Meese et al., 1997). Within the Holocene, the counting of annual layers is reproducible at the 1% level. This means that it is possible to estimate the seasonality of input of extraterrestrial debris into the GISP2 ice core.

If the debris can be related to a specific season and meteor shower, we have a stronger basis for relating it to a specific comet. Many of our ice-core samples cover nearly an entire year, so they contribute less to an evaluation of the seasonality of extraterrestrial input. Some of our samples cover only part of a year, and these permit a more precise evaluation of the seasonality of extraterrestrial input.

To evaluate the seasonality of input of extraterrestrial debris, we constructed a time series from the estimated time range of each of the 10 samples that contained some cosmic debris. (There are intercalated samples that contain no cosmic debris.) We interpolated at daily intervals to produce a time series with

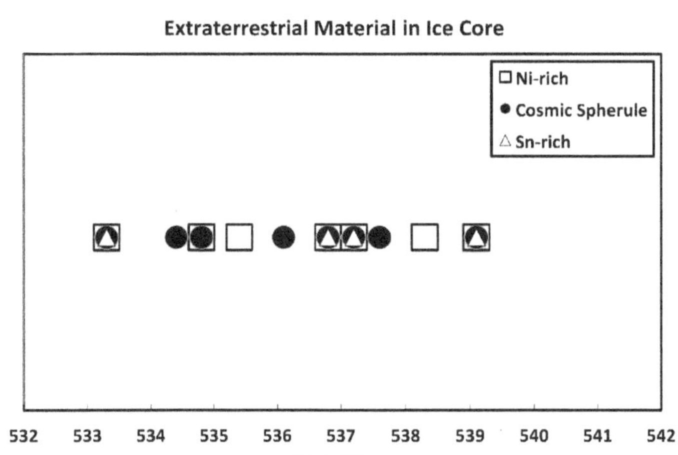

Figure 3. Extraterrestrial material versus calendar-year age in the Greenland Ice Sheet Project 2 (GISP2) ice core. Squares with black rim—Ni-rich particles; solid black circles—I-type cosmic spherules; triangles with black rim—Sn-rich particles.

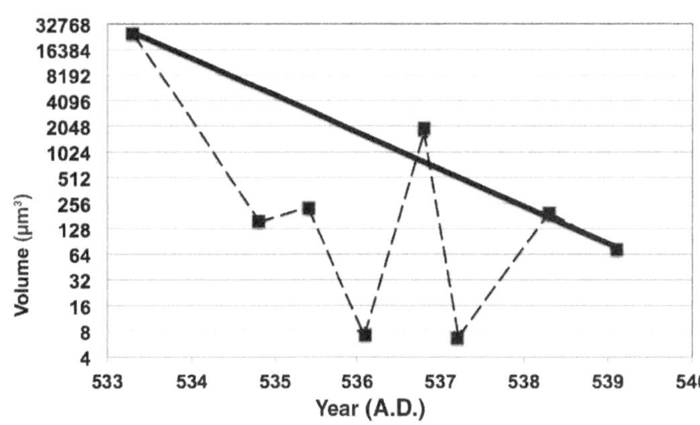

Figure 4. Change in volume of extraterrestrial material versus time in our ice-core samples. *Y*-axis is the log base 2 of the total volume in cubic micrometers of extraterrestrial material. Solid black line—estimate of the decline in volume of the largest samples over time if there was a single source of extraterrestrial material. Dashed lines—estimate of the change in volume over time if there were multiple sources of extraterrestrial material.

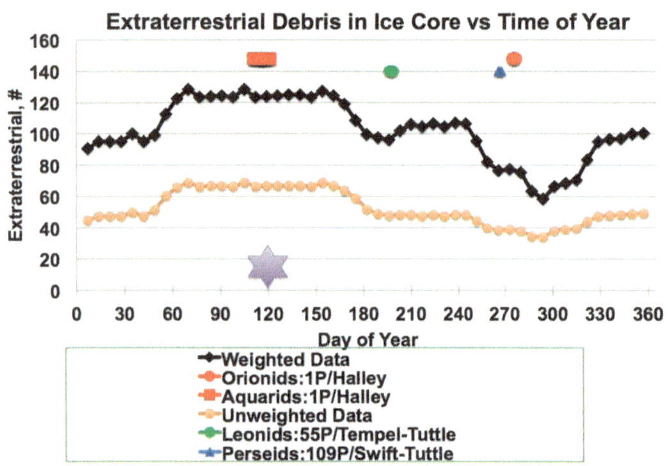

Figure 5. Seasonality of input of extraterrestrial dust particles. Black line with diamonds—total number of days in 2 wk intervals with extraterrestrial particles as weighted by total number of particles per sample. Brown line with circles—total number of days in 2 wk intervals with extraterrestrial particles. Red symbols—meteor showers derived from Comet 1P/Halley. Blue triangles—meteor shower derived from Comet 109P/Swift-Tuttle. Dark green circles—meteor shower derived from Comet 55P/Tempel-Tuttle. Purple star—most prominent peak in the number of meteor showers observed by Chinese astronomers prior to A.D. 918 (Yang et al., 2005).

one data point per day. To produce the weighted time series, we multiplied each time series of days of the year by the number of particles of cosmic debris in that particular sample. We then stacked both the weighted and unweighted time series to search for peaks in the seasonal input of cosmic debris. Finally, we reduced noise by summing and binning each time series over 2 wk intervals (Fig. 5).

Not every comet is a potential source of fine cosmic dust that might cloud Earth's atmosphere. Only a few comets pass close enough to Earth's orbit for Earth to cross their fields of abandoned debris. These latter comets are the main sources of meteor showers and cosmic dust on Earth.

During the first millennium A.D., ancient astronomers documented four meteor showers with known cometary associations: the Eta Aquarid, Orionid, Perseid, and Leonid meteor showers (Imoto and Hasegawa, 1958). The dates of these meteor showers were somewhat different during the sixth century A.D. (Table 5). With the exception of the Orionids, the adjusted dates for each shower came from models of three or more observations of each shower. The adjusted dates for the Orionids are from the two extant observations of this shower.

Implications of Seasonality

The seasonality of particle input suggests that comet 1P/Halley was the main source of the extraterrestrial dust in the GISP2 core. Comet 1P/Halley passed near Earth in A.D. 530. The Chinese observed comet 1P/Halley between 29 August 530 A.D. and 27 September 530 A.D. Why was this particular apparition of comet 1P/Halley a significant source of extraterrestrial dust to Earth's atmosphere? One important variable is the amount of dust and vapor that degasses from the comet. Because degassing removes the black outer rind of a comet, it increases the brightness of the comet. Therefore, we use the absolute magnitude of a comet as a proxy for the intensity of cometary outgassing. Out of 22 apparitions of comet 1P/Halley between 239 B.C. and A.D. 1682, the comets of A.D. 530 and A.D. 1222 share the highest absolute magnitude of 5.5 (Kronk, 1984). We infer that comet 1P/Halley was ejecting unusually large amounts of dust and vapor during its A.D. 530 transit through the inner solar system.

Modeling of Comet Intersections with Earth's Orbit

Over time, the intersections of cometary debris trails with Earth's orbit slowly change. At present, Earth's orbit is most likely to intersect those of larger comet fragments when it crosses the path of prior passes of comet 2P/Encke during late June and early July. This is the time of the Taurid meteor showers. Two recent events are suspected of being caused by the Taurid meteor stream. The first was the Tunguska explosion on 30 June 1908. The second was a meteor storm on the Moon in June 1975 (Hartung, 1993). During the sixth century A.D., the most prominent intersection of Earth's orbit with a meteor shower was with the Eta Aquarid meteor shower of Comet 1P/Halley.

Orbital Simulations—Methods and Implications

Orbital simulations were carried out using a software program for N-body integrations (Chambers, 1999). This technique considers the positions and velocities of all bodies in a system and predicts where each body will end up a short time later based on the combined gravitational pull caused by all other bodies. By repeating the calculations, the positions and

TABLE 5. DATES OF PROMINENT METEOR SHOWERS DURING THE SIXTH CENTURY A.D.

Adjusted date	Day range	Meteors	Parent comet
27 Sep–9 Oct	270–282	Leonids	55P/Tempel-Tuttle
15 Jul–18 Jul	196–199	Perseids	109P/Swift-Tuttle
13 Apr–22 Apr	103–112	Aquarids	1P/Halley
~23 Sep–25 Sep	~266–268	Orionids	1P/Halley

velocities of the bodies are established for a series of time increments. Similar integrations have been performed for comets over 400,000 yr (Valsecchi et al., 1995) and 40,000 yr periods (Harris and Bailey, 1998).

For this paper, simulations were performed over the past 6000 yr for short-period comets. In general, comets are usually on orbits that do not intersect Earth's orbit. With precession and perturbations of their orbits, there may be occasions when the orbits intersect. During such occasions, the node of the comet's orbit (where it intersects the ecliptic plane) will be equal to 1.0 ± 0.05 astronomical units (au). From this simulation, Comet 1P/Halley emerged as significant during the sixth century.

The orbital simulations have several important implications (Fig. 6). They suggest that the Eta Aquarid meteor shower, caused by the descending node of comet 1P/Halley's orbit, was an important source of meteors during the sixth century A.D. During the period from ca. A.D. 528 to 537, the descending node was closest to the plane of Earth's orbit. After the comet passed close to the sun in 530 A.D. (Fig. 6A), there could have been a renewed source of cometary debris to the Eta Aquarid meteor shower. As we will show, this debris may be related to subsequent climate downturns.

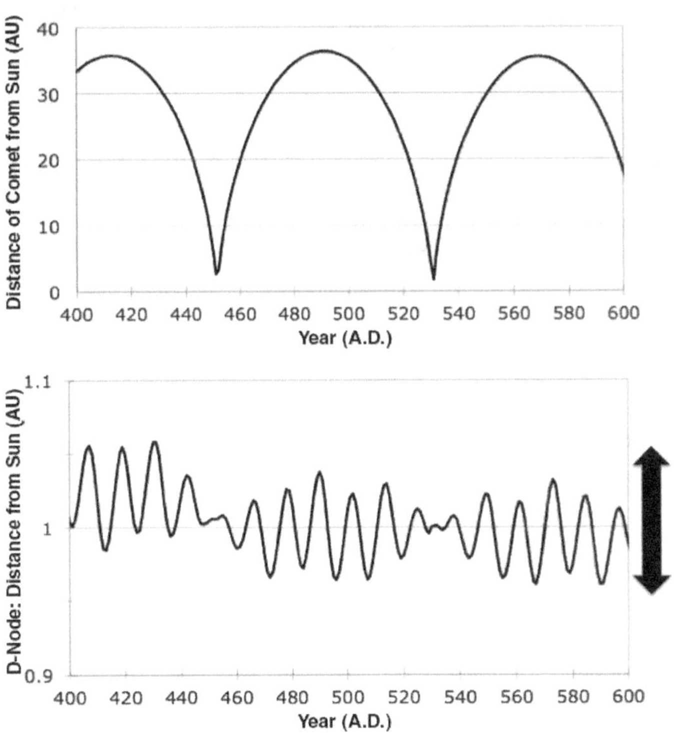

Figure 6. Comet 1P/Halley between A.D. 400 and 600. Distances are in au (astronomical units). (Top) Distance of 1P/Halley from the Sun. (Bottom) Distance of the descending node of Haley's orbit (source of Eta Aquarid meteor shower) from the Sun. Arrow on the right shows the range of distances of the orbital node (0.95–1.05 au) at which 1P/Halley could be a source of debris for a meteor shower that would affect Earth.

SEVERE CLIMATE DOWNTURN OF A.D 536–537

Ancient histories have provided us with at least four contemporary accounts of a dramatic decrease in the Sun's brightness during A.D. 536 and 537 (Stothers, 1984). In Mesopotamia (30–37°N), a source later quoted by Michael the Syrian (ca. 1166–1199 A.D.) and Bar-Hebraeus (ca. 1246–1286 A.D.) stated "in the year 848 of the Greeks [536 A.D.] … the Sun was dark and its darkness lasted for eighteen months, each day it shone for about 4 hours, and still this light was only a feeble shadow… the fruits did not ripen and wine tasted like sour grapes" (translated to English from Chabot [in French], 1901, p. 220–221). According to an anonymous writer, associated by tradition with Zacharius of Mytilene in Constantinople (41°N) (Hamilton and Brooks, 1899, p. 267), "in the year 14 [536 A.D.]… the Sun began to be darkened by day and the Moon by night… from the 24th of March in this year until the 24th of June in the following year 15." Because the dimming was less severe and shorter in duration (15 mo instead of 18) at higher latitudes, it is generally agreed that the atmosphere was filled with fine dust from an unknown, low-latitude source. Volcanic ash and impact debris as the source of that dust are vigorously promoted by different writers (Rigby et al., 2004; Larsen et al., 2008).

What Are the Forcing Factors for the Climate Downturn of A.D. 536–537?

An increased input of extraterrestrial dust from A.D. 533 to 540 could have been one factor that contributed to the prominent climate downturn between A.D. 536 and 537. We know that there was also a modest volcanic eruption during this time that probably injected sulfate aerosols into the stratosphere early in A.D. 536. The question is whether the sulfate associated with this volcanic eruption was enough to have caused what is arguably the most significant solar dimming in the past 2000 yr. Our data suggest that the maximum volume of extraterrestrial dust loading was in A.D. 533. This is intriguing, as some tree rings record reduced growth, implying a modest climate cooling in A.D. 533 (Baillie, 2007). There is no record of a volcanic eruption at this time. Thus, high volumes of extraterrestrial dust could have dimmed Earth's atmosphere and cooled the climate in 533 A.D.

The problem with explaining the prominent climate downturn between A.D. 536 and 537 with a combination of extraterrestrial dust loading and a relatively small volcanic eruption is that the maximum volume of extraterrestrial dust was apparently around 16 times smaller in A.D. 536 than it was in A.D. 533 (Fig. 4). This amount of dust seems insufficient to force an unprecedented climate downturn. It is possible that there was an injection of fine-aerosol-size (<~0.5 μm in diameter) extraterrestrial dust early in A.D. 536 that we did not observe due to our small sample sizes and relatively low magnifications. However, we should also ascertain if our data contain any evidence for another source of aerosol-sized dust. Next, we make a case for an additional source of dust beyond volcanic sulfate and extraterrestrial aerosols.

Was There a Low-Latitude, Submarine Explosion Early in A.D. 536?

Our ice-core data in two samples spanning the end of 535 and the start of A.D. 536 contain an unusual assemblage of marine and terrestrial microfossils. Based on our calendar-year time scale, we found and identified low-latitude diatoms and silicoflagellates in two samples centered at A.D. 535.4 ± 0.4 and 536.1 ± 0.41 (Appendix Table 1A). All but one of the 23 diatoms are marine. Based on the presence of *Thalassiosira oestrupii* var. *oestrupii* and *Dictyocha stapedia*, the A.D. 535.4 ± 0.4–536.1 ± 0.41 microfossil assemblage is most likely from a low latitude (Fig. 7).

Diatoms in general, and marine diatoms in particular, are rare in ice cores (Gayley and Ram, 1984; Burckle et al., 1988). So far as we know, no one had ever found diatoms in the GISP2 ice core before this research. Kellogg and Kellogg (2005) performed a search of the GISP2 core and found no diatoms. In addition, the concentration of diatoms at the A.D. 536 ± 0.41 level (667 diatoms/L) is the highest ever recorded in any ice core (Kellogg and Kellogg, 2005; Kellogg, 2010, written commun.).

The marine source of the diatoms and silicoflagellates is also unusual, given the elevation of 3203 m at the GISP2 site. Of the ice cores taken at elevations above 3000 m where diatoms have been found, nonmarine diatoms dominate the assemblages. Ninety-eight percent of the diatoms at Dome C (3240 m) in Antarctica are nonmarine (Burckle et al., 1988). In the Crete ice core from Greenland (elevation 3174 m), 100% of the diatoms are nonmarine (Gayley and Ram, 1984). In contrast, a comprehensive study of 136 samples from the South Pole water well (elevation 2835 m) found 29% nonmarine diatoms (Kellogg and Kellogg, 1996; Kellogg, 2010, written commun.).

The characterization of the diatoms in meltwater from the South Pole water well is the most detailed we know of (Kellogg and Kellogg, 1996, 2005). Tom Kellogg (2010, written commun.) sent us his data on the number of diatoms versus depth and the volume of water filtered per sample. The maximum number of diatoms per liter in the South Pole water well is 455 diatoms/L. Our water samples are much smaller (27–102 mL; Appendix Table 1A) than theirs (modal volumes of 800–900 mL). If our diatom count is spread over non-diatom-bearing samples from the stratigraphic levels immediately above and below them (328 mL of water total), the diatom count averages 79 diatoms/L of water. Of the 136 samples from the South Pole water well, five contained more than 79 diatoms/L of water. Thus, the diatom concentrations that we found are high but not unprecedented.

Kellogg and Kellogg (1996, 2005) identified 1179 diatoms from the South Pole water well. They identified 839 specimens of marine diatoms. All but two marine diatoms are high latitude or cosmopolitan. A single low-latitude diatom and a single Miocene-age diatom were found in the two samples with the highest concentrations of diatoms and over 100 diatoms total. Their detailed study shows that low-latitude diatoms are sometimes moved to high latitudes, but such transport is rare. Other workers found evidence that diatoms are ablated from Antarctic ice and subsequently blown elsewhere by the wind (McKay et al., 2008). Although some marine diatoms are present, the wind-blown assemblages are dominated by terrestrial species. The diatom surfaces appear ablated and smoothed, as would be expected for reworked material (see figures in McKay et al., 2008). In contrast, our diatoms exhibit clear, sharp surface textures (Fig. 7). Thus, our finding of a dominantly marine, well-preserved, low-latitude assemblage of diatoms in a high-altitude ice core is not matched in previous studies.

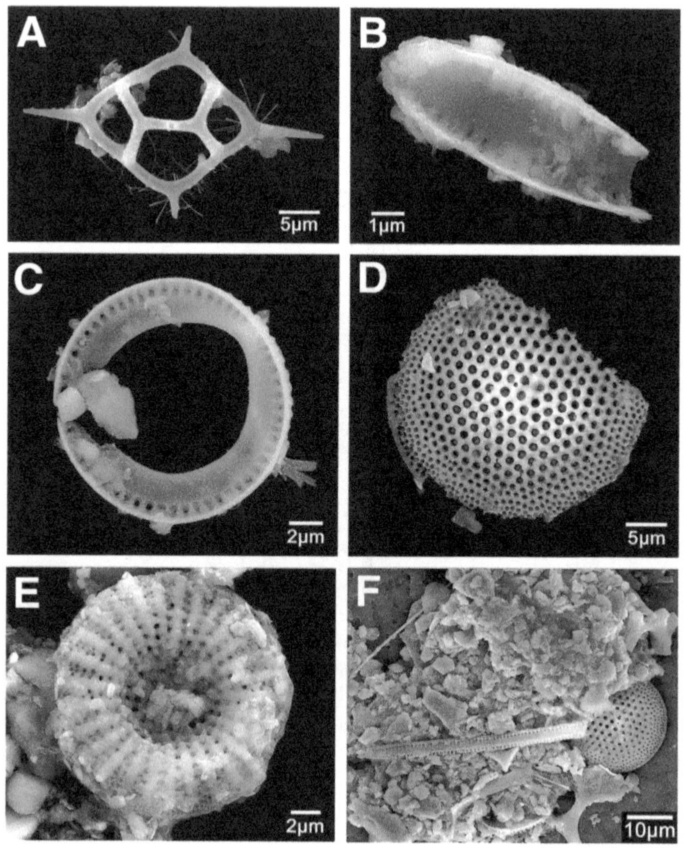

Figure 7. Scanning electron microscope (SEM) micrographs of marine microfossils from the Greenland Ice Sheet Project 2 (GISP2) core: (A–E) A.D. 536 ± 0.41 and (F) A.D. 535 ± 0.4. The total number of occurrences follows each species name. (A) *Dictyocha stapedia* (1), an equatorial to subequatorial marine silicoflagellate; (B) *Opephora* (1), a benthic marine diatom; (C) girdle band from marine diatom *Thalassiosira oestrupii* (4); (D) *Thalassiosira oestrupii* var. *oestrupii* (7), a marine equatorial to subequatorial planktonic diatom; (E) *Cyclostephanos* sp. (1), a nonmarine planktonic diatom; (F) marine planktonic diatoms *Thalassiosira oestrupii* var. *oestrupii* (right) and *Thalassiothrix* (left middle) (5). Total diatoms: A.D. 536 ± 0.41 layer (23); A.D. 535 ± 0.4 layer (3). Ratios of backscattered and secondary electron imaging modes were chosen to maximize the display of the images: A, D, F: 50% ILSE (in-lens secondary electron mode), 50% BSE (backscattered electron mode); B: 100% SE2 (standard secondary electron mode); C, E: 50% SE2, 50% BSE.

Provenance of Marine Microfossils

The marine silicoflagellate is from equatorial to subtropical latitudes (Barron and Bukry, 2007). *Thalassiosira oestrupii* is identified as the variety *oestrupii* on the basis of its diameter (Lloyd Burckle, 2010, oral commun.). *Thalassiosira oestrupii*, a tropical to subtropical diatom, can occur at higher latitudes but is abundant (>10%) only in low-latitude assemblages. Our assemblage is >20% *Thalassiosira oestrupii* (Burckle, 1977). *Opephora* is a benthic marine diatom. Because they are photosynthetic, benthic diatoms live within the photic zone at water depths shallower than ~100 m, so their presence in oceanic sediments normally indicates proximity to land. The other marine diatoms are planktonic. The occurrence of the terrestrial diatom *Cyclostephanos* is suggestive of a continental margin environment (Abrantes, 2007). If all of the diatoms and the silicoflagellate are from the same location, they are from a continental margin at water depths of less than 100 m in the tropics or subtropics.

Marine Dust as an Additional Forcer of the A.D. 536–537 Climate Downturn

Some of the diatoms are found among piles of particles (Fig. 7E). We suggest that the marine sediment containing these diatoms and associated fine particles may have provided an additional source of dust during the severe climate downturn between A.D. 536 and 537. The dominant component of the dust may be constrained by the presence of high Ca (574.8 µg/L) at the A.D. 536 level. This is the highest Ca value in our ion data (Table 1, sample 7; Abbott et al., this volume, Chapter 22). In general, high Ca values in ion data are considered to be a proxy for $CaCO_3$ dust (Zdanowicz et al., 2000; Bory et al., 2003; Kang et al., 2003; Fischer et al., 2007), which is usually interpreted as being terrestrial dust. The finding of low-latitude marine microfossils suggests another interpretation—the high Ca is derived from marine carbonate dust. Because sulfate aerosols from Icelandic volcanic eruptions acidify the snow at the GISP2 site, calcium carbonate is usually not preserved. Instead, the calcium carbonate is dissolved and contributes Ca to meltwater (ion data) derived from the ice.

The concentration of Ca at the A.D. 536 level is one of the highest ever observed. For example, the average Ca concentration in the detailed ion data from 96 to 593 m depth (GISP2-D) is 7.2 µg/L (Mayewski et al., 1997). The maximum Ca concentration is 87 µg/L. This depth interval represents over 2000 yr of time (259 yr B.P. to 2589 yr B.P.). If the Ca value of 575 µg/L is a proxy for dust and by implication dust loading of the atmosphere, then the dust loading is ~80 times the background level. We do not know how much fine dust/aerosol was needed to produce the A.D. 536 solar dimming, but 80 times the background level seems promising.

We found calcium carbonate crystals in the same samples as those containing tropical diatoms. The calcium carbonate is in the form of single crystals or crystal aggregates (Fig. 8). Calcium carbonate often precipitates within sediments during diagenesis. However, diagenetic $CaCO_3$ precipitates have many morphologies, and only a few are equant single crystals (Morse et al., 2007; Dupraz et al., 2009; Moore and Wade, 2013). It is common for marine microfossils and nanofossils to become more coarsely crystalline during early diagenesis (Reid and Macintyre, 1998; Crudeli et al., 2004). If broken, these coarsely crystalline carbonate aggregates may appear hollow. The presence of hollow $CaCO_3$ aggregates could mean that the fine $CaCO_3$ crystals in our ice-core particles are primary rather than secondary. The single crystals in our samples may represent either broken fragments of the hollow aggregates or diagenetic carbonate crystals derived directly from low-latitude marine sediment. Either of these origins would mean that the fine calcium carbonate crystals were suspended in the atmosphere before they and the diatoms were deposited in Greenland.

The individual $CaCO_3$ crystals are typically less than 1 µm in diameter (Fig. 8A). This puts these crystals into the aerosol size range. Aerosols can remain in the stratosphere for over a year. Calcium carbonate is typically white or light gray in color. Black aerosols like black carbon act as black bodies that absorb heat (Jacobson, 2001; Roberts and Jones, 2004; Ramanathan and Carmichael, 2008). White or light-gray aerosols reflect sunlight and nucleate clouds, helping to cool the climate. Thus, if significant percentages of the fine carbonate crystals were primary, they could be the reason it became cold and dark during A.D. 536 and 537. We will not know until we can examine less acid ice from Antarctica covering the same time period.

In addition to calcium carbonate aggregates, we found one distinctive plagioclase feldspar grain at the same stratigraphic level. The grain has micro-offsets in directions that define a staircase-like pattern (Figs. 8B and 8C). It is possible that this is the ladder structure found in impact ejecta (French, 1998), but we cannot be sure without further study, which is impossible (for us) with such small grains.

The source of the explosion that transported diatoms, feldspar, and possibly carbonate dust from low latitudes to Greenland is still in question. Recent work has shown that terrestrial diatoms can be lofted high into the atmosphere during explosive volcanic eruptions (Van Eaton et al., 2013); thus, the occurrence of diatoms is not by itself a marker for any particular type of explosion. Because transport from low latitudes to the poles is most probable in the stratosphere, a large explosion is required. Regardless of whether the dimming was primarily produced by mineral aerosols or sulfate aerosols, the aerosols must have resided in the stratosphere for 15–18 mo. The primary source of the explosion can be determined by a combination of microprobe work on possible volcanic glass particles in the GISP2 core and an assessment of the types of particles within the portions of Antarctic ice cores dating to ca. A.D. 536–537.

MASS EXTINCTIONS—RELEVANCE AND APPLICATIONS

Our work on the GISP2 ice core shows the complications that are present in a 10 yr period involving one or more extraterrestrially driven perturbations of the climate with some additional forcing from volcanic sulfate aerosols. Although only the Cretaceous-Tertiary mass extinction has unequivocal evidence of extraterrestrial input, some studies suggest that coupling of both volcanism and impact are needed to produce extinctions. If the terrestrial system is stressed by large-scale volcanism, a large impact event could provide the coup de grâce. Therefore, we favor the pulse-press model of mass extinctions with coupled volcanic and extraterrestrial (or other) forcing (Arens and West, 2008).

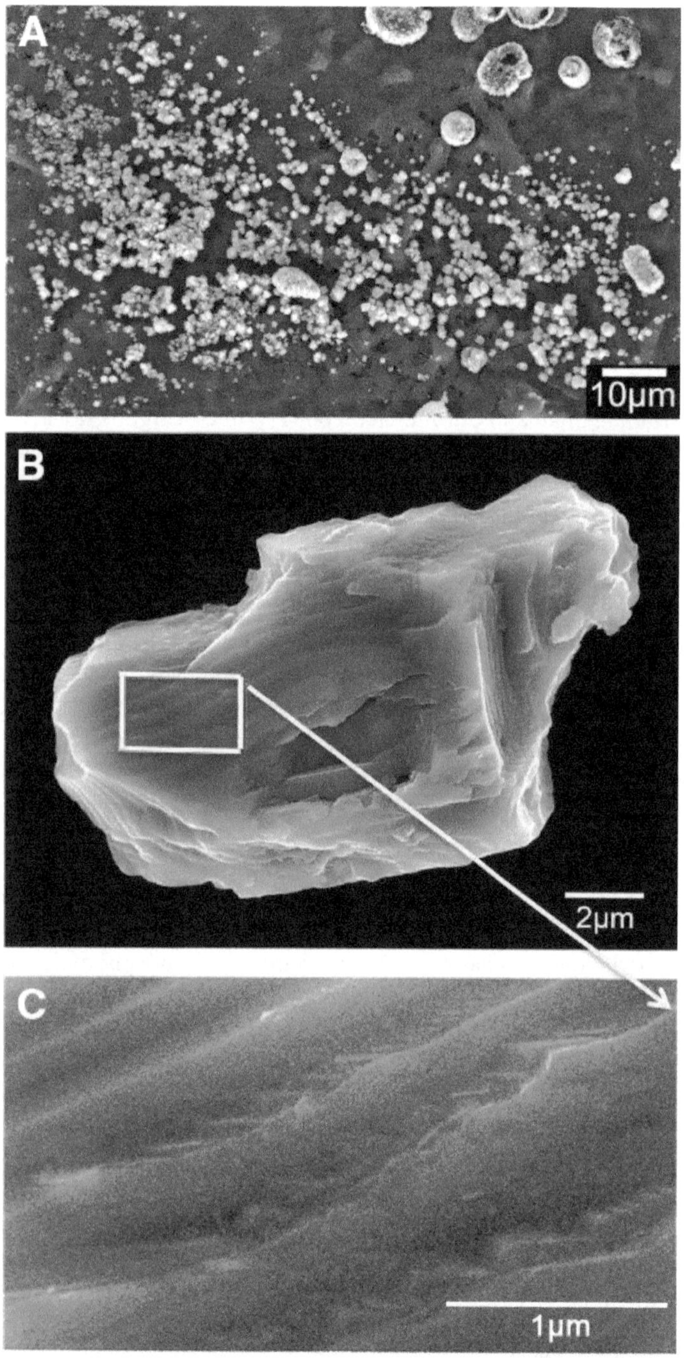

The largest extraterrestrial bodies are comets, not asteroids. Ninety-five percent of the extraterrestrial bodies big enough to produce a large impact crater (20–100 km in diameter) are comets (Hughes, 1998). For impact craters over 100 km in diameter, the proportion of comets will be even higher. The problem is how to detect their signal in the geological record. This is not easy. There are still arguments about the cometary versus asteroidal nature of the 65 Ma Cretaceous-Tertiary impactor (Hut et al., 1987; Bottke et al., 2007). Our study suggests another novel marker for extraterrestrial input that may help to distinguish cometary from asteroidal sources.

Improved Strategies for Finding Extraterrestrial Markers

Another problem in detecting extraterrestrial inputs to mass extinctions is that the signal of extraterrestrial forcing is confined to a narrow stratigraphic interval and may be obscured by diagenesis, erosion, or tiny grain size. We show two examples in the following section.

In the past, bulk chemical analyses for Ir, Pt, and Ni have been the method of choice, along with a search for shocked quartz. The Cretaceous-Tertiary impact event was relatively "easy" to prove because it was partially on land and occurred at a time when the Atlantic Ocean was narrow. In the case of a distal oceanic impact in the middle of a paleo–Pacific Ocean, both bulk chemical analyses and searches for shocked quartz are less likely to yield results. Because the shock lamellae in the largest quartz grains are annealed out with increasing distance from an impact site (Croskell et al., 2002), the preserved, still-shocked grains are very small. For example, the average size of distal ejecta grains deposited 5000 km away from an abyssal crater is ~10–32 μm (Collins et al., 2005) for impacts with recurrence intervals up to the length of the Phanerozoic (Fig. 9). This is far below the minimum grain size of ~125 μm that is needed to make a good thin section of shocked quartz. Thus, any plausible size of impactors hitting near the center of a paleo–Pacific Ocean at an average abyssal water depth of 3800 m (Parsons and Sclater, 1977) would be unlikely to produce a detectable, verifiable layer of shocked quartz on land.

Figure 8. Scanning electron microscope (SEM) micrographs of unusual grains from stratigraphic level containing marine diatoms. (A) $CaCO_3$ crystals. Micrograph image is 50% BSE (backscattered electron mode), 50% SE2 (standard secondary electron mode). Aggregates of crystal on upper right may represent recrystallized nanofossils. (B) Feldspar grain with unusual cleavage patterns. Micrograph 50% BSE, 50% ILSE (in-lens secondary electron mode). (C) Enlargement of area inside of white rectangle in B.

Bulk chemical analyses for Ir, Pt, and Ni may be affected by the development of water-soluble chlorides during a major oceanic impact. We know that these elements exhibit similar chemical behavior and are often associated with Fe. Although the carrier phases for Ir and Pt are uncertain, both Fe and Ni weather to form water-soluble salts, in particular, Fe and Ni chlorides. In one study of polar snow, ~35% of the Ni was estimated to reside in a water-soluble Ni salt (Brocas and Picciotto, 1967). Because most of the Cl within polar snow and ice comes from sea salt (O'Brien et al., 1995), we suggest that an abyssal oceanic impact might produce water-soluble salts of Ni, Ir, and Pt within the impact ejecta layer. If so, these soluble salts would make it difficult to preserve a local peak in Ir, Pt, or Ni over geological time. (Interestingly, only one out of five of the known tektite strewn fields has a significant Ir anomaly [Glass, 1990; Koeberl and Shirey, 1993].)

Recent research has found a new marker for extraterrestrial inputs by examining small particles: Fe-Ni silicides (formation temperature >2000 °C) within impact spherules (Wu et al., 2013). This particular spherule-bearing layer does not produce a detectable Ir anomaly in analyses of the bulk sediment. Our ice-core results further suggest that examination of individual small particles can complement bulk analyses. The finer particles should be examined for shocked ilmenite (Sclar et al., 1973; Harris et al., 2005) and cometary markers such as excess Sn and noble gases. Although noble gases are removed by heating the sediment to temperatures above 70 °C (Farley et al., 2005), the other markers can survive low-grade metamorphism.

Because atmospheric deposition of fine particles is often enhanced at high latitudes (Megner et al., 2008), extinction-level stratigraphic sections with high paleolatitudes are preferred if they exist. At Cenozoic-age high-latitude sites, out-of-place equatorial diatoms or foraminifera might provide marker horizons. In addition, many of the stratigraphic markers for impact either are intrinsically magnetic (e.g., shocked ilmenite, magnetic spherules, Fe-Ni silicides) or are typically associated with Fe oxide (Sn-rich particles). Magnetic separation of disaggregated sediment followed by examination of the magnetic fine particles may produce more reliable results.

CONCLUSIONS

Cometary dust is the probable source of micrometer-sized C-rich, Sn-rich particles found at four stratigraphic levels in the GISP2 ice core. The input of extraterrestrial dust is concentrated in late spring, suggesting that dust from Comet 1P/Halley contributed to increased opacity of Earth's atmosphere between A.D. 533 and 540. High volumes of cometary dust in A.D. 533 may have caused the minor climate downturn recorded by tree rings and historical sources. Later on, in A.D. 536 and 537, the volumes of cometary dust were possibly much smaller but with an uncertain contribution from fine-aerosol-sized particles. Atmospheric aerosols derived from multiple sources mostly likely produced the profound solar dimming in A.D. 536 and 537. By analogy, synergistic catastrophic events may also cause mass extinctions. Detailed examinations of fine particles may complement previous methods of testing models of synergistic impact and volcanic drivers of mass extinctions.

ACKNOWLEDGMENTS

We thank Jihong Cole-Dai for the ion data on our supernatant water samples. We are grateful to the Lamont-Doherty Earth Observatory Climate Center for their support of our laboratory and analytical costs. This work would have been impossible without their generous support and encouragement. We thank anonymous reviewers for helpful reviews. We thank the Electron Microscopy Center at The City College of New York for access to their facilities.

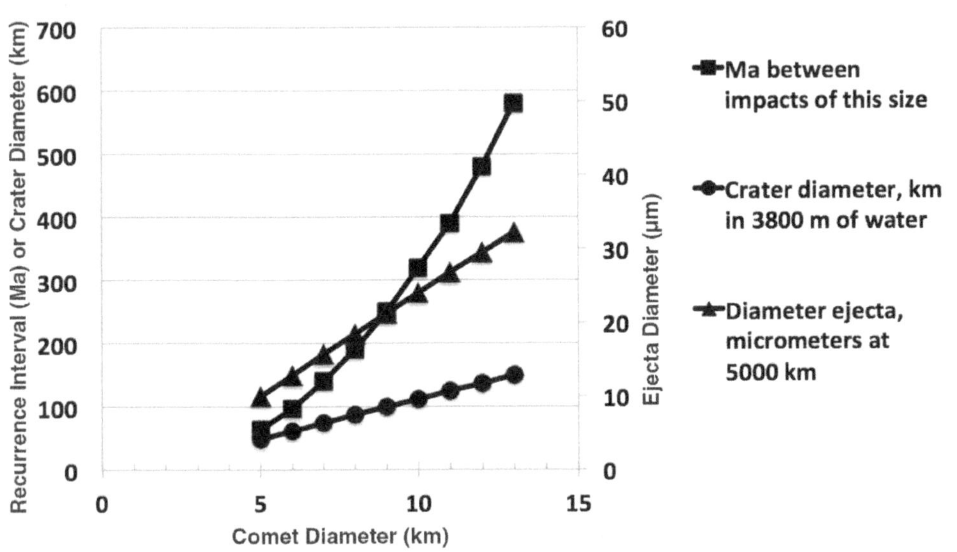

Figure 9. Models of ejecta diameter for comets hitting in the middle of a paleo–Pacific Ocean, 5000 km from any site on land. The models are for comets between 5 and 13 km in diameter hitting at the average oceanic water depth of 3800 m. Black squares—impactor diameter in kilometers. Black circles—average recurrence interval of the impactor in millions of years. Black triangles—average ejecta diameter in micrometers.

APPENDIX 1

Appendix 1 (Appendix Table 1A) contains the metadata and data for all of the samples analyzed for ions and particles. The data in this appendix will provide a framework for comparison of our data to other data sets from the GISP2 core.

APPENDIX 2

Appendix 2 (Appendix Table 1B) contains representative EDS analyses of clean quartz grains. This data shows the background levels of carbon measured on clean quartz that was not carbon coated.

APPENDIX TABLE 1A. TIME SCALE AND VOLUMES OF WATER CENTRIFUGED TO DERIVE PARTICLES

Sample no.	Meltwater (mL)	Depth range (m)	Age (yr A.D., Meese-Sowers)	Age (yr A.D., adjusted)
1A	33	360–360.1	543.8 ± 3.26	541.8 ± 0.26
1B	42	360.1–360.2	543.3 ± 3.26	541.3 ± 0.26
2-S1	51	360.2–360.38	542.6 ± 3.46	540.6 ± 0.46
2-S2	51	360.2–360.38	542.6 ± 3.46	540.6 ± 0.46
3-S1	45	360.38–360.51	541.8 ± 3.33	539.8 ± 0.33
3-S2	45	360.38–360.51	541.8 ± 3.33	539.8 ± 0.33
4A-S1	39	360.51–360.64	541.1 ± 3.33	539.1 ± 0.33
4A-S2	32.5	360.51–360.64	541.1 ± 3.33	539.1 ± 0.33
4B	38.5	360.64–360.7	540.6 ± 3.15	538.6 ± 0.15
4C	40	360.7–360.77	540.3 ± 3.18	538.3 ± 0.18
5A	30	360.77–360.84	540 ± 3.18	538 ± 0.18
5B	34	360.84–360.92	539.6 ± 3.2	537.6 ± 0.2
5C	52	360.92–361	539.2 ± 3.2	537.2 ± 0.2
6	42	361–361.08	538.8 ± 3.2	536.8 ± 0.2
7	33	361.08–361.24	538.1 ± 3.41	536.1 ± 0.41
8	37	361.24–361.39	537.4 ± 3.38	535.4 ± 0.38
9	27	361.39–361.45	536.8 ± 3.15	534.8 ± 0.15
10A	32.5	361.45–361.55	536.4 ± 3.26	534.4 ± 0.26
10B	32.5	361.55–361.64	535.9 ± 3.23	533.9 ± 0.23
11-S1	46	361.64–361.8	535.3 ± 3.41	533.3 ± 0.41
11-S2	47	361.64–361.8	535.3 ± 3.41	533.3 ± 0.41
12A	46	361.8–361.9	534.6 ± 3.26	532.6 ± 0.26
12B	45	361.9–362	534.1 ± 3.26	532.1 ± 0.26

Note: The Meese-Sowers time scale (Meese et al., 1997) is primarily derived from annual layer counts that rely on the formation of coarse-grained hoar layers during the Arctic summer (Alley et al, 1997). When the annual signal is poorly resolved in visual stratigraphy, it is supplemented by measurements of electrical conductivity, oxygen isotopes, laser light scattering of dust, and major ion chemistry.

APPENDIX TABLE 1B. SEMIQUANTITATIVE ENERGY-DISPERSIVE SPECTROMETRY ANALYSES OF CLEAN TERRESTRIAL QUARTZ GRAINS IN LEVEL 11

Sample no.	C (wt%)	O (wt%)	Si (wt%)	Total
11-20	1.7	42	54.4	98.1
11-19	2.9	39.3	55.5	97.7
11-31	2.8	43	52.4	98.2
11-39	1.7	41.6	54.2	97.5
Average	2.3	41.5	54.1	

REFERENCES CITED

Abbott, D.H., Breger, D., Biscaye, P.E., and Juhl, R.A., 2014, this volume, Chapter 22, Calendar-year dating of the Greenland Ice Sheet Project 2 (GISP2) ice core from the early sixth century using historical, ion, and particulate data, *in* Keller, G., and Kerr, A.C., eds., Volcanism, Impacts, and Mass Extinctions: Causes and Effects: Geological Society of America Special Paper 505, doi:10.1130/2014.2505(22).

Abrantes, F., 2007, Marine diatoms, *in* Elias, S.A., ed., Encyclopedia of Quaternary Sciences: Amsterdam, Elsevier, p. 1668–1678.

Alexander, C.O.D., Bowden, R., Fogel, M.L., Howard, K.T., Herd, C.D.K., and Nittler, L.R., 2012, The provenances of asteroids, and their contributions to the volatile inventories of the terrestrial planets: Science, v. 337, p. 721–723, doi:10.1126/science.1223474.

Alley, R.B., Shuman, C.A., Meese, D.A., Gow, A.J., Taylor, K.C., Cuffey, K.M., and Elder, B., 1997, Visual-stratigraphic dating of the GISP2 ice core: Basis, reproducibility, and application: Journal of Geophysical Research, v. 102, p. 26,367–26,381, doi:10.1029/96JC03837.

Arens, N.C., and West, I.D., 2008, Press-pulse: A general theory of mass extinction?: Paleobiology, v. 34, p. 456–471, doi:10.1666/07034.1.

Baillie, M.G., 1999, Exodus to Arthur: Catastrophic Encounters with Comets: London, UK, Batsford, 256 p.

Baillie, M.G.L., 2007, Tree-rings indicate global environmental downturns that could have been caused by comet debris, *in* Bobrowsky, P.T., and Rickman, H., eds., Comet/Asteroid Impacts and Human Society: Berlin, Springer, p. 105–122.

Bar-Hebraeus, [ca. 1246–1286], 1932, Chronography (translated by Budge, E.A.W.): London, Oxford University Press.

Barron, J.A., and Bukry, D., 2007, Development of the California Current during the past 12,000 years based on diatoms and silicoflagellates: Palaeogeography, Palaeoclimatology, Palaeoecology, v. 248, p. 313–338, doi:10.1016/j.palaeo.2006.12.009.

Bory, A.J.-M., Biscaye, P.E., Piotrowski, A.M., and Steffensen, J.P., 2003, Regional variability of ice core dust composition and provenance in Greenland: Geochemistry Geophysics Geosystems, v. 4, no. 12, p. 1107, doi:10.1029/2003GC000627.

Bottke, W.F., Vokrouhlický, D., and Nesvorný, D., 2007, An asteroid breakup 160 Myr ago as the probable source of the K/T impactor: Nature, v. 449, p. 48–53, doi:10.1038/nature06070.

Bradley, J.P., and Brownlee, D.E., 1986, Cometary particles: Thin sectioning and electron beam analysis: Science, v. 231, p. 1542–1544, doi:10.1126/science.231.4745.1542.

Bradley, J.P., Keller, L.P., Snow, T.P., Hanner, M.S., Flynn, G.J., Gezo, J.C., and Bowey, J.E., 1999, An infrared spectral match between GEMS and interstellar grains: Science, v. 285, p. 1716–1718, doi:10.1126/science.285.5434.1716.

Brocas, J., and Picciotto, E., 1967, Nickel content of Antarctic snow: Implications of the influx rate of extraterrestrial dust: Journal of Geophysical Research, v. 72, p. 2229–2236, doi:10.1029/JZ072i008p02229.

Brook, E., Kurz, M., Curtice, J., and Cowburn, S., 2000, Accretion of interplanetary dust in polar ice: Geophysical Research Letters, v. 27, p. 3145–3148, doi:10.1029/2000GL011813.

Brook, E., Kurz, M.D., and Curtice, J., 2009, Flux and size fractionation of He-3 in interplanetary dust from Antarctic ice core samples: Earth and Planetary Science Letters, v. 286, p. 565–569, doi:10.1016/j.epsl.2009.07.024.

Brownlee, D.E., 1987, Morphological, chemical and mineralogical studies of cosmic dust: Philosophical Transactions of the Royal Society of London, ser. A, Mathematical and Physical Sciences, v. 323, p. 305–311, doi:10.1098/rsta.1987.0087.

Burckle, L.H., 1977, Pliocene and Pleistocene diatom levels from the equatorial Pacific: Quaternary Research, v. 7, p. 330–340, doi:10.1016/0033-5894(77)90025-4.

Burckle, L.H., Gayley, R.I., Ram, M., and Petit, J.R., 1988, Biogenic particles in Antarctic ice cores and the source of Antarctic dust: Antarctic Journal of the United States, v. 23, p. 71–72.

Chambers, J.E., 1999, A hybrid symplectic integrator that permits close encounters between massive bodies: Monthly Notices of the Royal Astronomical Society, v. 304, p. 793–799, doi:10.1046/j.1365-8711.1999.02379.x.

Ciucci, A., Rietmeijer, F.J.M., Della Corte, V., Brunetto, R., de Angelis, S., Palumbo, P., and Rotundi, A., 2010, Collecting dust in the upper stratosphere at high latitudes: Constraints on the compositions of cosmic dust particles, *in* Proceedings European Planetary Science Congress, Volume 1: Rome, Italy, European Planetary Science Congress, p. 175.

Collins, G.S., Melosh, H.J., and Marcus, R.A., 2005, Earth impact effects program: A Web-based computer program for calculating the regional environmental consequences of a meteoroid impact on Earth: Meteoritics & Planetary Science, v. 40, p. 817–840, doi:10.1111/j.1945-5100.2005.tb00157.x.

Croskell, M., Warner, M., and Morgan, J., 2002, Annealing of shocked quartz during atmospheric re-entry: Geophysical Research Letters, v. 29, p. 1-1–1-4.

Crudeli, D., Young, J.R., Erba, E., de Lange, G.J., Henriksen, K., Kinkel, H., and Ziveri, P., 2004, Abnormal carbonate diagenesis in Holocene–late Pleistocene sapropel-associated sediments from the Eastern Mediterranean; evidence from *Emiliania huxleyi* coccolith morphology: Marine Micropaleontology, v. 52, p. 217–240, doi:10.1016/j.marmicro.2004.04.010.

Delsemme, A.H., 1991, Cometary origin of carbon, nitrogen and water on the Earth: Origins of Life and Evolution of the Biosphere, v. 21, p. 279–298, doi:10.1007/BF01808303.

Dikov, Y.P., Ivanov, A.V., Wlotzka, F., Galimov, E.M., and Wänke, H., 1998, High enrichment of carbon and volatile elements in the surface layers of Luna 16 soil sample 1635: Result of comet or meteorite impact?: Earth and Planetary Science Letters, v. 155, p. 197–204, doi:10.1016/S0012-821X(97)00212-4.

Dupraz, C., Reid, R.P., Braissant, O., Decho, A.W., Norman, R.S., and Visscher, P.T., 2009, Processes of carbonate precipitation in modern microbial mats: Earth-Science Reviews, v. 96, p. 141–162.

Engrand, C., 2011, Meteorites and cosmic dust: Interstellar heritage and nebular processes in the early solar system: European Physical Journal Web of Conferences, v. 18, p. 5001–5026, doi:10.1051/epjconf/20111805001.

Engrand, C., and Maurette, M., 1998, Carbonaceous micrometeorites from Antarctica: Meteoritics & Planetary Science, v. 33, p. 565–580, doi:10.1111/j.1945-5100.1998.tb01665.x.

Farley, K.A., Love, S.G., and Patterson, D.B., 1997, Atmospheric entry heating and helium retentivity of interplanetary dust particles: Geochimica et Cosmochimica Acta, v. 61, p. 2309–2316, doi:10.1016/S0016-7037(97)00068-9.

Farley, K.A., Ward, P., Garrison, G., and Mukhopadhyay, S., 2005, Absence of extraterrestrial helium 3 in Permian–Triassic age sedimentary rocks: Earth and Planetary Science Letters, v. 240, p. 265–275, doi:10.1016/j.epsl.2005.09.054.

Fischer, H., Siggaard-Andersen, M.L., Ruth, U., Röthlisberger, R., and Wolff, E., 2007, Glacial/interglacial changes in mineral dust and sea salt records in polar ice cores: Sources, transport, and deposition: Reviews of Geophysics, v. 45, p. 1–26, doi:10.1029/2005RG000192.

Flynn, G.J., Keller, L.P., and Sutton, S.R., 2006, Sub-micrometer scale minor element mapping in interplanetary dust particles: A test for stratospheric contamination, *in* 35th Lunar and Planetary Science Conference: Houston, Texas, Lunar and Planetary Institute, abstract 1334.

French, B.M., 1998, Traces of Catastrophe: A Handbook of Shock-Metamorphic Effects in Terrestrial Materials: Houston, Texas, Lunar and Planetary Institute, 120 p.

Gabrielli, P., Barbante, C., Plane, J.M., Varga, A., Hong, S., Cozzi, G., and Boutron, C.F., 2004, Meteoric smoke fallout over the Holocene Epoch revealed by iridium and platinum in Greenland ice: Nature, v. 432, p. 1011–1014.

Gayley, R.I., and Ram, M., 1984, Observations of diatoms in Greenland ice: Art Journal, v. 23, p. 71–72.

Glass, B.P., 1990, Tektites and microtektites: Key facts and inferences: Tectonophysics, v. 171, p. 393–404, doi:10.1016/0040-1951(90)90112-L.

Harris, N.W., and Bailey, M.E., 1998, Dynamical evolution of cometary asteroids: Monthly Notices of the Royal Astronomical Society, v. 297, p. 1227–1236, doi:10.1046/j.1365-8711.1998.01683.x.

Harris, R.S., Schultz, P.H., and Bunch, T.E., 2005, Accessory phases in Argentine impact breccias: Implications for shock history, emplacement dynamics, vapor composition and target lithologies, *in* 36th Lunar and Planetary Science Conference: Houston, Texas, Lunar and Planetary Institute, p. 1952.

Hartung, J.B., 1993, Giordano Bruno, the June 1975 meteoroid storm, Encke, and other Taurid Complex objects: Icarus, v. 104, p. 280–290, doi:10.1006/icar.1993.1101.

Hasegawa, I., 1980, Catalogue of ancient and naked-eye comets: Vistas in Astronomy, v. 24, p. 59–102, doi:10.1016/0083-6656(80)90005-7.

Hodge, P.W., and Wright, F.W., 1964, Studies of particles for extraterrestrial origin: 2. A comparison of microscopic spherules of meteoritic and volcanic origin: Journal of Geophysical Research, v. 69, p. 2449–2454, doi:10.1029/JZ069i012p02449.

Hodge, P.W., Wright, F.W., and Langway, C.C., Jr., 1964, Studies of particles for extraterrestrial origin: 3. Analyses of dust particles from polar ice deposits: Journal of Geophysical Research, v. 69, p. 2919–2931, doi:10.1029/JZ069i014p02919.

Hughes, D.W., 1998, The mass distribution of crater-producing bodies, *in* Grady, M.M., Hutchison, R., McCall, G.J.H., and Rothery, D.A., eds., Meteorites: Flux with Time and Impact Effects: Geological Society of London Special Publication 140, p. 31–42.

Hut, P., Alvarez, W., Elder, W.P., Hansen, T., Kauffman, E.G., Keller, G., and Weissman, P.R., 1987, Comet showers as a cause of mass extinctions: Nature, v. 329, p. 118–126.

Imoto, S., and Hasegawa, I., 1958, Historical records of meteor showers in China, Korea, and Japan: Smithsonian Contributions to Astrophysics, v. 2, p. 131–144.

Jacobson, M.Z., 2001, Strong radiative heating due to the mixing state of black carbon in atmospheric aerosols: Nature, v. 409, p. 695–697, doi:10.1038/35055518.

Jessberger, E.K., Bohsung, J., Chakaveh, S., and Traxel, K., 1992, Chondritic interplanetary dust particles and their volatile element enrichment: Earth and Planetary Science Letters, v. 112, p. 91–99, doi:10.1016/0012-821X(92)90009-K.

Kang, S., Mayewski, P.A., Yan, Y., Qin, D., Yao, T., and Ren, J., 2003, Dust records from three ice cores: Relationships to spring atmospheric circulation over the Northern Hemisphere: Atmospheric Environment, v. 37, p. 4823–4835, doi:10.1016/j.atmosenv.2003.08.010.

Karner, D.B., Levine, J., Muller, R.A., Asaro, F., Ram, M., and Stolz, M.R., 2003, Extraterrestrial accretion from the GISP2 ice core: Geochimica et Cosmochimica Acta, v. 67, p. 751–763, doi:10.1016/S0016-7037(02)01145-6.

Kellogg, D.E., and Kellogg, T.E., 1996, Diatoms in South Pole ice: Implications for eolian contamination of Sirius Group deposits: Geology, v. 24, p. 115–118, doi:10.1130/0091-7613(1996)024<0115:DISPII>2.3.CO;2.

Kellogg, D.E., and Kellogg, T.E., 2005, Frozen in time: The diatom record in ice cores from remote drilling sites on the Antarctic ice sheets, *in* Rogers, S.O., and Castello, J.D., eds., Life in Ancient Ice: Princeton, New Jersey, Princeton University Press, p. 69–93.

Koeberl, C., and Shirey, S.B., 1993, Detection of a meteoritic component in Ivory Coast tektites with rhenium-osmium isotopes: Science, v. 261, p. 595–598, doi:10.1126/science.261.5121.595.

Kronk, G.W., 1984, Comets: A Descriptive Catalog: Hillside, New Jersey, Enslow Publishers, 344 p.

Krueger, F.R., and Kissel, J., 1987, The chemical composition of the dust of comet P/Halley as measured by "PUMA" on board VEGA-1: Naturwissenschaften, v. 74, p. 312–316, doi:10.1007/BF00367925.

Lal, D., and Jull, A., 2005, On the fluxes and fates of He-3 accreted by the Earth with extraterrestrial particles: Earth and Planetary Science Letters, v. 235, p. 375–390, doi:10.1016/j.epsl.2005.04.011.

Lanci, L., and Kent, D.V., 2006, Meteoric smoke fallout revealed by superparamagnetism in Greenland ice: Geophysical Research Letters, v. 33, p. L13308, doi:10.1029/2006GL026480.

Larsen, L.B., Vinther, B.M., Briffa, K.R., Melvin, T.M., Clausen, H.B., Jones, P.D., Siggard-Anderson, M.-L., Hammer, C.U., Eronen, M., Grudd, H., Gunnarson, B.E., Hantemirov, R.M., Naurzbaev, M.N., and Nicolussi, K., 2008, New ice core evidence for a volcanic cause of the A.D. 536 dust veil: Geophysical Research Letters, v. 35, p. L04708, doi:10.1029/02007GL032450.

La Violette, P.A., 1987, The cometary breakup hypothesis re-examined: Monthly Notices of the Royal Astronomical Society, v. 224, p. 945–951.

Lodders, K., 2003, Solar system abundances and condensation temperatures of the elements: The Astrophysical Journal, v. 591, p. 1220–1247, doi:10.1086/375492.

Marcantonio, F., Anderson, R., Stute, M., Kumar, N., Schlosser, P., and Mix, A., 1996, Extraterrestrial He-3 as a tracer of marine sediment transport and accumulation: Nature, v. 383, p. 705–707, doi:10.1038/383705a0.

Marcantonio, F., Higgins, S., Anderson, R., Stute, M., Schlosser, P., and Rasbury, E., 1998, Terrigenous helium in deep-sea sediments: Geochimica et Cosmochimica Acta, v. 62, p. 1535–1543, doi:10.1016/S0016-7037(98)00091-X.

Marcantonio, F., Turekian, K.K., Higgins, S., Anderson, R.F., Stute, M., and Schlosser, P., 1999, The accretion rate of extraterrestrial ^{3}He based on oceanic ^{230}Th flux and the relation to Os isotope variation over the past 200,000 years in an Indian Ocean core: Earth and Planetary Science Letters, v. 170, p. 157–168, doi:10.1016/S0012-821X(99)00100-4.

Marty, B., 2012, The origins and concentrations of water, carbon, nitrogen and noble gases on Earth: Earth and Planetary Science Letters, v. 313, p. 56–66, doi:10.1016/j.epsl.2011.10.040.

Marty, B., Alexander, C.M.D., and Raymond, S.N., 2013, Primordial origins of Earth's carbon: Reviews in Mineralogy and Geochemistry, v. 75, p. 149–181, doi:10.2138/rmg.2013.75.6.

Mayewski, P.A., Meeker, L.D., Twickler, M.S., Whitlow, S.I., Yang, Q., Lyons, W.B., and Prentice, M., 1997, Major features and forcing of high-latitude Northern Hemisphere atmospheric circulation using a 110,000-year-long glaciochemical series: Journal of Geophysical Research, v. 102, p. 26,345–26,366, doi:10.1029/96JC03365.

McGee, D., and Mukhopadhyay, S., 2013, Extraterrestrial He in sediments: From recorder of asteroid collisions to timekeeper of global environmental changes, *in* Burnard, P., ed., The Noble Gases as Geochemical Tracers: Berlin, Springer-Verlag, p. 155–176.

McKay, R.M., Barrett, P.J., Harper, M.A., and Hannah, M.J., 2008, Atmospheric transport and concentration of diatoms in surficial and glacial sediments of the Allan Hills, Transantarctic Mountains: Palaeogeography, Palaeoclimatology, Palaeoecology, v. 260, p. 168–183, doi:10.1016/j.palaeo.2007.08.014.

McKenzie, D., and Bickle, M.J., 1988, The volume and composition of melt generated by extension of the lithosphere: Journal of Petrology, v. 29, p. 625–679, doi:10.1093/petrology/29.3.625.

Meese, D.A., Gow, A.J., Alley, R.B., Zielinski, G.A., Grootes, P.M., Ram, M., and Bolzan, J.F., 1997, The Greenland Ice Sheet Project 2 depth-age scale: Methods and results: Journal of Geophysical Research, v. 102, p. 26,411–26,423, doi:10.1029/97JC00269.

Megner, L., Siskind, D.E., Rapp, M., and Gumbel, J., 2008, Global and temporal distribution of meteoric smoke: A two-dimensional simulation study: Journal of Geophysical Research, v. 113, p. D03202, doi:10.1029/2007JD009054.

Michael the Syrian, [ca. 1166–1199 A.D.], 1901, Michael the Syrian Chronicle (translated to French by Chabot, J.–B.): Paris, Belles-Lettres.

Misawa, K., Kohno, M., Tomiyama, T., Noguchi, T., Nakamura, T., Nagao, K., and Nishiizumi, K., 2010, Two extraterrestrial dust horizons found in the Dome Fuji ice core, East Antarctica: Earth and Planetary Science Letters, v. 289, p. 287–297, doi:10.1016/j.epsl.2009.11.016.

Moore, C.H., and Wade, W.J., 2013, Carbonate diagenesis: Introduction and tools, *in* Carbonate Reservoirs—Porosity and Diagenesis in a Sequence Stratigraphic Framework: Amsterdam, Elsevier, Developments in Sedimentology, v. 67, p. 67–89.

Morse, J.W., Arvidson, R.S., and Lüttge, A., 2007, Calcium carbonate formation and dissolution: Chemical Reviews, v. 107, p. 342–381, doi:10.1021/cr050358j.

Mukhopadhyay, S., and Farley, K.A., 2006, New insights into the carrier phase(s) of extraterrestrial ^{3}He in geologically old sediments: Geochimica et Cosmochimica Acta, v. 70, p. 5061–5073, doi:10.1016/j.gca.2006.06.1566.

O'Brien, S.R., Mayewski, P.A., Meeker, L.D., Meese, D.A., Twickler, M.S., and Wicklow, S.I., 1995, Complexity of Holocene climate as reconstructed from a Greenland ice core: Science, v. 270, p. 1962–1964, doi:10.1126/science.270.5244.1962.

Parsons, B., and Sclater, J.G., 1977, An analysis of the variation of ocean floor bathymetry and heat flow with age: Journal of Geophysical Research, v. 82, p. 803–827, doi:10.1029/JB082i005p00803.

Pepin, R.O., Becker, R.H., and Rider, P.E., 1995, Xenon and krypton isotopes in extraterrestrial regolith soils and in the solar wind: Geochimica et Cosmochimica Acta, v. 59, p. 4997–5022, doi:10.1016/0016-7037(96)80916-1.

Peucker-Ehrenbrink, B., and Ravizza, G., 2000, The effects of sampling artifacts on cosmic dust flux estimates: A reevaluation of nonvolatile tracers (Os, Ir): Geochimica et Cosmochimica Acta, v. 64, p. 1965–1970, doi:10.1016/S0016-7037(99)00429-9.

Ramanathan, V., and Carmichael, G., 2008, Global and regional climate changes due to black carbon: Nature Geoscience, v. 1, p. 221–227, doi:10.1038/ngeo156.

Reid, R.P., and Macintyre, I.G., 1998, Carbonate recrystallization in shallow marine environments: A widespread diagenetic process forming micritized grains: Journal of Sedimentary Research, v. 68, p. 928–946, doi:10.2110/jsr.68.928.

Rietmeijer, F.J., 1988, Sulfides and oxides in comets: The Astrophysical Journal, v. 331, p. L137, doi:10.1086/185253.

Rietmeijer, F.J., 1989, Tin in a chondritic interplanetary dust particle: Meteoritics, v. 24, p. 43–47, doi:10.1111/j.1945-5100.1989.tb00940.x.

Rietmeijer, F.J., 1993, The bromine content of micrometeorites: Arguments for stratospheric contamination: Journal of Geophysical Research–Planets, v. 98, p. 7409–7414, doi:10.1029/93JE00118.

Rietmeijer, F.J., 1995, Post-entry and volcanic contaminant abundances of zinc, copper, selenium, germanium and gallium in stratospheric micrometeorites: Meteoritics, v. 30, p. 33–41, doi:10.1111/j.1945-5100.1995.tb01209.x.

Rigby, E., Symonds, M., and Ward-Thompson, D., 2004, A comet impact in AD536?: Astronomy and Geophysics, v. 45, p. 1.23–21.26.

Roberts, D.L., and Jones, A., 2004, Climate sensitivity to black carbon aerosol from fossil fuel combustion: Journal of Geophysical Research–Atmospheres (1984–2012), v. 109, D16202, doi:10.1029/2004JD004676.

Schulze, H., Kissel, J., and Jessberger, E.K., 1997, Chemistry and mineralogy of Comet Halley's dust, *in* Pendleton, Y.J., and Tielens, R.G.G.M., eds., From Stardust to Planetesimals, Volume 122: Santa Clara, California, Astronomical Society of the Pacific, p. 397–414.

Sclar, C.B., Bauer, J.F., Pickart, S.J., and Alperin, H.A., 1973, Shock effects in experimentally shocked terrestrial ilmenite, lunar ilmenite of rock fragments in 1–10 mm fines (10085, 19) and lunar rock 60015, 127: Geochimica et Cosmochimca Acta, v. 1, supplement 4, p. 841–859.

Stothers, R.B., 1984, Mystery cloud of AD536: Nature, v. 307, p. 344–345, doi:10.1038/307344a0.

Taylor, S., Lever, J.H., and Harvey, R.P., 1998, Accretion rate of cosmic spherules measured at the South Pole: Nature, v. 392, p. 899–903, doi:10.1038/31894.

Taylor, S., Lever, J.H., and Harvey, R.P., 2000, Numbers, types, and compositions of an unbiased collection of cosmic spherules: Meteoritics & Planetary Science, v. 35, p. 651–666, doi:10.1111/j.1945-5100.2000.tb01450.x.

Thomas, K.L., Blanford, G.E., Keller, L.P., Klöck, W., and McKay, D.S., 1993, Carbon abundance and silicate mineralogy of anhydrous interplanetary dust particles: Geochimica et Cosmochimica Acta, v. 57, p. 1551–1566, doi:10.1016/0016-7037(93)90012-L.

Valsecchi, G.B., Morbidelli, A., Gonczi, R., Farinella, P., Froeschle, C., and Froeschle, C., 1995, The dynamics of objects in orbits resembling that of P/Encke: Icarus, v. 118, p. 169–180, doi:10.1006/icar.1995.1183.

Van Eaton, A.R., Harper, M.A., and Wilson, C.J., 2013, High-flying diatoms: Widespread dispersal of microorganisms in an explosive volcanic eruption: Geology, v. 41, p. 1187–1190, doi:10.1130/G34829.1.

Wark, D.A., 1986, Evidence for successive episodes of condensation at high temperature in a part of the solar nebula: Earth and Planetary Science Letters, v. 77, p. 129–148, doi:10.1016/0012-821X(86)90155-X.

Winckler, G., and Fischer, H., 2006, 30,000 years of cosmic dust in Antarctic ice: Science, v. 313, p. 491, doi:10.1126/science.1127469.

Wu, Y., Sharma, M., LeCompte, M.A., Demitroff, M.N., and Landis, J.D., 2013, Origin and provenance of spherules and magnetic grains at the Younger Dryas boundary: Proceedings of the National Academy of Sciences of the United States of America, v. 110, p. E3557–E3566, doi:10.1073/pnas.1304059110.

Yada, T., Nakamura, T., Takaoka, N., Noguchi, T., Terada, K., Yano, H., and Kojima, H., 2004, The global accretion rate of extraterrestrial materials in the last glacial period estimated from the abundance of micrometeorites in Antarctic glacier ice: Earth, Planets, and Space, v. 56, p. 67–80.

Yang, H.J., Park, C., and Park, M.G., 2005, Analysis of historical meteor and meteor shower records: Korea, China, and Japan: Icarus, v. 175, p. 215–225, doi:10.1016/j.icarus.2004.10.007.

Zdanowicz, C.M., Zielinski, G.A., Wake, C.P., Fisher, D.A., and Koerner, R.M., 2000, A Holocene record of atmospheric dust deposition on the Penny Ice Cap, Baffin Island, Canada: Quaternary Research, v. 53, p. 62–69, doi:10.1006/qres.1999.2091.

Manuscript Accepted by the Society 31 January 2014

Printed in the USA

The Geological Society of America
Special Paper 505
2014

The public impact of impacts: How the media play in the mass extinction debates

Steve Miller
Department of Science and Technology Studies, University College London, Gower Street, London WC1E 6BT, UK

ABSTRACT

"Mass media" presentations of the dinosaurs and their co-inhabitants have been around for some 200 years. The question of what exterminated the dinosaurs and allowed mammals to take their leading place on Earth has a similarly lengthy history in the scientific arena and in public. However, there are amazingly few communication studies of the debates around mass extinctions and impacts. Those that do exist have picked up on the fact that these debates involve scientists from several disciplines, scientists who are often unused to reading each other's research. Under these circumstances, more public or leading journals play a key role, not only in getting ideas out into the public arena, but in informing scientists across disciplinary boundaries. "Normal" communication processes, in which articles in peer-reviewed journals inform the scientific community and "simplified" versions may trickle out to the public via the mass media, become more complex.

The dramatic impact answer to the question of the death of the dinosaurs seems to have attracted limited media attention at the time, confined to the "elite" newspapers. This paper analyzes the newspaper coverage of the death of the dinosaurs during the period from 1980 to 2008. I find that the period from 1991 to 1995 was critical in terms of changing public perceptions, insofar as they are determined/reflected in articles in general newspapers. I argue that the "Great Crash of 1994," when Comet Shoemaker-Levy 9 collided with the giant planet Jupiter, played an important role in propelling the impact scenario for the death of the dinosaurs into the (mass) public eye, and that the news value co-option was important in this process.

INTRODUCTION

"Two proposals that involved catastrophes that enveloped the Earth in sun-shading debris have been advanced to account for past episodes of widespread extinction and climate change. One was a direct hit on the Earth by an asteroid, causing an explosion whose debris cut off all or most sunlight for several years. This could explain the extinction of the dinosaurs and numerous other creatures some 65 million years ago" (Sullivan, 1980, p. C3).

"Just about every astronomer in the world will be looking at only one place this coming July—the planet Jupiter. It's going to be the most important astronomical event for centuries, and the most eagerly awaited. In July, a massive speeding comet will smash into Jupiter, unleashing the most powerful forces seen since our local star, the Sun, was born" (Turner, 1994).

Miller, S., 2014, The public impact of impacts: How the media play in the mass extinction debates, *in* Keller, G., and Kerr, A.C., eds., Volcanism, Impacts, and Mass Extinctions: Causes and Effects: Geological Society of America Special Paper 505, p. 439–455, doi:10.1130/2014.2505(24). For permission to copy, contact editing@geosociety.org.

Separated by almost 14 years, these two quotations deal with the first widely accepted (if strongly contested) theory to account for the demise of the dinosaurs and their fellow inhabitants of Earth some 65 m.y. ago at the end of the Cretaceous Period, and an unparalleled astronomical event that did much to lend credence to that theory amongst the wider scientific community and the general public at large. On 6 June 1980, Luis and Walter Alvarez and their coworkers Frank Asaro and Helen Michel published an article in the prestigious, interdisciplinary journal *Science*, claiming that an iridium-rich layer of clay found at the Cretaceous-Tertiary boundary was the result of a 10-km-diameter asteroid hitting Earth and creating climatic conditions that caused the mass extinction observed at that time (Alvarez et al., 1980). The first quotation above is science journalist Walter Sullivan's report about this paper in the *New York Times* a few days later. (Note that this had been foreshadowed by coverage of the 1979 meeting of the American Geophysical Union, when this theory was presented there.)

On 26 March 1993, comet-hunters Caroline and Eugene Shoemaker and their colleague David Levy sent out an International Astronomical Union telegram announcing that, 3 days earlier, they had discovered a "squashed comet" that appeared to be orbiting Jupiter (International Astronomical Union, 1993). Over the next couple of months, further observations produced an orbit reliable enough to demonstrate that (1) Comet Shoemaker-Levy 9 (SL9) was indeed orbiting Jupiter, (2) that in July 1992, it had undergone a close encounter with Jupiter that had caused it to fragment in ~20 pieces (giving the "squashed" appearance first noted in the low-resolution images taken by the Shoemakers and Levy), and (3) that in July 1994, the comet would actually crash into the giant planet.

The second quote above is from British Broadcasting Corporation (BBC) television presenter Anthea Turner giving a heads-up to British 10 years olds that Jupiter was the place to watch that coming summer. Her report came direct from the Barringer Crater in Arizona and went out on the children's program *Blue Peter*, one of the UK's most watched television shows by both preteens and their parents. Toward the end of her piece, Turner opined: "Some scientists believe that it was a comet that hit the Earth, causing a climatic change that resulted in the extinction of the dinosaurs." SL9 and the death of the dinosaurs were linked. This paper looks at original data from general newspaper articles to see how this worked out in practice.

INTERDISCIPLINARY SCIENCE AND THE PUBLIC SPHERE

Dinosaurs have fascinated the public ever since scientists first attempted to put flesh on their fossilized bones in the nineteenth century. Victorian promenaders could gaze with wonder at Benjamin Waterhouse Hawkins' reconstructions of them in the Crystal Palace Park in South London from the mid-1850s onwards. These life-sized statues—despite their anatomical imperfections that reflected the state of knowledge of the time and the disputes between paleontological giants of the stature of Richard Owen and Gideon Mantell—remain treasured features of the park today. (Visitors to New York might also have had the same pleasure had Waterhouse Hawkins not run afoul of William Boss Tweed in the 1870s, when he was working on models for Central Park [see Bramwell and Peck, 2008].)

Understanding the dinosaurs and their fellow creatures always was an interdisciplinary issue, although their study dates back to a time when the boundaries between individual scientific disciplines—and, indeed, between science and general culture/knowledge—were still in the process of crystallizing (Young, 1985; Yeo, 1993). Tensions abounded—geologists demanding "deep time" to account for measured rock formations to accumulate and erode according to Charles Lyell's (1830) principle of uniformitarianism were faced with Lord Kelvin's estimate of the age of Earth of a few tens to a hundred million years, based on principles of (astro)-physics (Burchell, 1975). Kelvin was sufficiently scary that Darwin withdrew his estimate of 300 m.y. for the time taken for the "denudation of the Weald" (Darwin and Costa, 1859, 2009).

Scientific disputes were fought out in the public sphere: in meetings of the British Association for the Advancement of Science and (later) the American Association for the Advancement of Science or in the pages of *The Athenaeum* or *Macmillan's Magazine* (e.g., Broks, 2006; Fyfe and Lightman, 2007; Lightman, 2007; Rudwick, 1985), and from the dinner parties of the Victorian upper classes to the meetings of the working class Mechanics Institutes (Secord, 2000), until such time as the sheer volume of science and its increasing specialization meant that such disputes were no longer comprehensible to the wider public (Young, 1985). At that point, they tended to disappear into "specialist" journals such as *Nature* (an 1869 split off from *Macmillan's Magazine*) and the burgeoning scientific societies. In the case of the dinosaurs, their public display came just prior to this period of change. So controversy over the posture of iguanodon—bird-like according to Owen; lizard-like according to Mantell—was (and is still) publicly visible in the two statues made by Waterhouse Hawkins for Crystal Palace Park. (Neither statue has the thumb-spike right, however; it is firmly attached to both noses.)

By the time the Alvarez team were at work, scientific disciplines were much more firmly entrenched than a century earlier. Their 1980 paper cut across several of these. This is evident from the summary alone, which clearly covers matters geological, astronomical, and paleontological (see Fig. 1). The themes outlined in the summary are expanded in the main article, with evidence deduced from all three disciplines as well as modern biological studies. The paper was "politically" timely: Many scientists were looking into the possible effects of nuclear war, in particular, the climatic effect known as a "nuclear winter" that would arise from the dust and gas generated by massive fusion explosions; the asteroid impact theory relied on mechanisms akin to "nuclear winter" for its killing power (Glasstone, 1983; Turco et al., 1983).

The three science disciplines already mentioned each come with their own prevailing (but not necessarily exclusive) "ideologies." Geology, as already mentioned, has a strong allegiance to the uniformitarianism proposed by Lyell to break the "Neptunist-Plutonist" rivalries of the previous century (Lyell, 1830; Davy, 1805/1980). Paleontology draws heavily on the gradualism of Darwin's *Origin of Species* (Darwin and Costa, 1859/2009). Catastrophism and extreme events are to the fore amongst the astronomical community—What could be more "catastrophic" than the Big Bang origin of the universe? Standard theories of the formation of the solar system, such as the "Nice model," involve multiple and massive impacts, including such epochs as the Late Heavy Bombardment some 3.95 b.y. ago (Gomes et al., 2005).

Standard accounts of how scientific results are popularized—a process that the French call *vulgarisation* and the Spanish *divulgacion*, with slightly different nuances—tend to have a somewhat linear structure. Upstream in the process, there are the peer-reviewed journals, in which well-behaved scientists test out their ideas on their colleagues and await their approval. If they are sufficiently interesting, these ideas may be channeled downriver via the media, eventually making it into the muddy delta of the public sphere. This journey from scientific purity to public imperfection and oversimplification is one that allows the scientific community to exercise maximum control over what gets "out" and how (Hilgartner, 1990).

However, communication scholar Bruce Lewenstein (1995) has demonstrated that where the science crosses disciplinary boundaries, and particularly where big claims are made with important implications, this neat, linear process just does not work. In its place, Lewenstein proposes a "web" of communication channels, not least of which are the mass media, and through which the wider scientific community, as well as the general public, are informed of the latest developments. Lewenstein's web model arose out of his study of the "cold fusion" debates of the late 1980s, in which Stanley Pons and Martin Fleischmann claimed to have generated nuclear fusion in a simple electrochemical cell. Lewenstein's model was independently supported by Frank Close, one of a number of physicists who debunked cold fusion. Close claimed that—since Pons and Fleischmann refused to publish their data in peer-reviewed journals—other scientists were reduced to photographing purported plots of neutron emission on their television screens as they watched the evening news (Close, 1990). Lewenstein's web model (1995) has also been applied to other areas, such as "life from space" (Gregory, 2003) and general physics (Mellor, 2003).

One scholar who has looked seriously at the public impact of the impact hypothesis for the Cretaceous-Tertiary extinctions is Elisabeth S. Clemens (1986, 1994). She points out that those involved in the various debates have been brought together as a result of their support for or opposition to (or even their wish to modify) a particular theory, rather than because they all belong to the same scientific discipline, adding:

> Broad-ranging interdisciplinary science strains the ability of personal and professional ties to promote consensus over the definition of problems and the criteria for proof. Under such conditions, considerable light may be shed on procedures and criteria that usually go unexamined (Clemens, 1986, p. 425).

Clemens demonstrated the importance of the general science journals such as *Science* and *Nature* in advancing the discussion. These two general journals accounted for 117 (nearly a quarter) of the 498 citations to the Alvarez et al. paper (1980) that occurred during the 1980s in the 130-plus journals that she looked at. Moreover, nearly 140 articles discussing impacts and extinctions were published in the American popular (science) press, including publications such as *Scientific American* and *The New Yorker*, over the same period. Clearly, the impact hypothesis was making its mark well outside of disciplinary scientific journals, and involving a wide range of scientists and

Summary. Platinum metals are depleted in the Earth's crust relative to their cosmic abundance; concentrations of these elements in deep-sea sediments may thus indicate influxes of extraterrestrial material. Deep-sea limestones exposed in Italy, Denmark, and New Zealand show iridium increases of about 30, 160, and 20 times, respectively, above the background level at precisely the time of the Cretaceous-Tertiary extinctions, 65 million years ago. Reasons are given to indicate that this iridium is of extraterrestrial origin, but did not come from a nearby supernova. A hypothesis is suggested which accounts for the extinctions and the iridium observations. Impact of a large earth-crossing asteroid would inject about 60 times the object's mass into the atmosphere as pulverized rock; a fraction of this dust would stay in the stratosphere for several years and be distributed worldwide. The resulting darkness would suppress photosynthesis, and the expected biological consequences match quite closely the extinctions observed in the paleontological record. One prediction of this hypothesis has been verified: the chemical composition of the boundary clay, which is thought to come from the stratospheric dust, is markedly different from that of clay mixed with the Cretaceous and Tertiary limestones, which are chemically similar to each other. Four different independent estimates of the diameter of the asteroid give values that lie in the range 10 ± 4 kilometers.

Figure 1. Summary of the 1980 *Science* article by Alvarez, Alvarez, Asaro, and Michel. The section shaded in blue is mainly geology, the section shaded in red is mainly astronomy, and the section shaded in green is mainly paleontology/geology.

individuals with a more amateur interest in science (Clemens, 1994). In interdisciplinary debates, the wider public sphere has to be addressed by those arguing for and/or against particular points of view. To extend the work of Clemens on the high-profile general science journals, I have chosen to look at *Nature* for the three decades from 1979 onwards: There is a plausible—though undocumented—argument that Luis Alvarez's associations with *Science* may have influenced that journal's early stance on the impact theory; I have had private and convincing correspondence with the editorial staff of *Nature* indicating that, at the time, no such influence could have been brought to bear there.

Figures 2 and 3, respectively, show coverage for the term "dinosaur" in the "Research" section (including "Articles," "Letters to Nature," and "Scientific Correspondence") and "News" section (including "News and Views") of *Nature* for the 30 years covered by this study. This is wider than Clemens' original search, but it is intended to give some background to those articles that deal either mainly or in passing with extinction issues. Overall, this search of *Nature* produced 804 articles, of which 208 came under the "Research" heading and 276 under "News," after some "spuriously selected" articles from the automated search were manually rejected. Thus, over half of the total *Nature* coverage of "dinosaur" issues was analyzed. For both sections, articles were characterized as "pro" impact theory, "anti" impact theory, "balanced," or "other" if they mentioned dinosaurs but made no comment about extinction. Interest in dinosaurs is clearly much greater in both sections analyzed in the second half of the period analyzed than in the first.

For the "Research" section of *Nature*, interest in the extinction of the dinosaurs is almost totally confined to the first half of the sample, however. Up to 1994, there were 12 articles that were clearly in favor of the Alvarez et al. (1980) impact theory (or variations on it), and nine against. There were five research articles characterized as "balanced" up to 1994. After 1994, there is one balanced in 2001 and one in 2007, but no pro– or anti–impact theory articles at all. Overall, dinosaur articles that concerned their extinction made up nearly 55% of the 66 articles in the "Research" section up to 1994. In 1990, strong evidence was produced that an undersea crater off the coast of Chicxulub in Mexico marked the site of an impact that occurred between Earth and an asteroid ~10 km in diameter 65 m.y. ago (Hildebrand et al., 1991; Sigurdsson et al., 1991), and, at least as far as "Research" in *Nature* was concerned, the heat seems to have largely gone out of the debate, even though an "anti" article was published in 1993. We conclude that, for the "Research" section, this issue had been settled somewhat in favor of the "pros" by the mid-1990s. The dinosaur-related articles in the second half of our sample are far more concerned with characterizing these animals and their fellow creatures—feathers, flight, and the relationship to birds being one strong interest theme. The enormous numbers of new (Chinese, in particular) fossil specimens becoming available also boosted this aspect of paleontology.

The picture for the "News" section is somewhat different. Here, 69 articles appeared prior to the start of 1994, and 207 afterward. Of these, prior to 1994, nine were "pro," three were "anti," and 11 were "balanced," together making one third of the total. So "News" was proportionately less extinction-oriented in its discussion of the dinosaurs than "Research," although the total numbers of dinosaur articles (66 "Research," 69 "News") were very similar. After 1994, the proportion of "News" articles dealing with extinction issues is still relatively high: ~20% of 207, with the majority (22/41) "pro," as against just three "anti" and 16 "balanced." Given that the majority of extinction-orientated or extinction-mentioning "News" articles occur in the second half of our sample, and that they are far more "pro" than "anti," one might conclude that the issue, as far as *Nature* overall is concerned, is settled, and, insofar as the issue is newsworthy, the position of this leading scientific journal is pretty much "pro-impact." The picture, of course, is more complicated; in September 2003, *Nature* published an article that highlighted "squabbles" over access to core samples from the Chicxulub crater, which indicated that there were several in the scientific community who were far from convinced that the "dinosaur killer" had indeed been identified, either in particular or in general (Dalton, 2003). That said, if Clemens is correct in her assessment of the role of *Nature* (and *Science*) in informing the wider scientific community, one might deduce that—outside of particular disciplinary niches—the impact theory was accepted as being correct throughout the 1990s, and as the leading explanation beyond then.

NARRATIVES, NEWS VALUES, AND THE ROLE OF THE MASS MEDIA

One of the most compelling mass media presentations of the demise of the dinosaurs is to be found in Walt Disney's *Fantasia* (1940). To the strains of Igor Stravinsky's *Rite of Spring*, colossal beasts tramp wearily across a sunbaked desert, scrapping over a wilted leaf or pool of not-quite-solid mud, crashing to the ground, carnivore next to herbivore, lesser on top of greater. In the end, there is nothing but a trail of footprints leading to the inevitable bone pile. The animation ends with floods and earthquakes, burying and re-exposing the dinosaurs' remains. Something has caused changes that make the dinosaurs lives impossible. That something is not explained, but their final trudge looks painful and prolonged; there is no coup de grâce. Had Disney's animators been aware of the impact theory, given their natural tendency to the dramatic and Stravinsky's stirring ballet, they would undoubtedly have used it. In contrast, the final cartoon in Anthea Turner's *Blue Peter* presentation (1994) has the dinosaurs being literally bowled over by the blast caused by the meteor impact. Nice and quick and "they never knew what hit 'em." Framing a story for the media—including the "News" section of journals such as *Nature*—has to follow certain conventions: For the print media, news values (Nelkin, 1987; Gregory and Miller, 1998) are key; for television, storytelling devices and rhetorics may be employed, as will be discussed later (Silverstone, 1987).

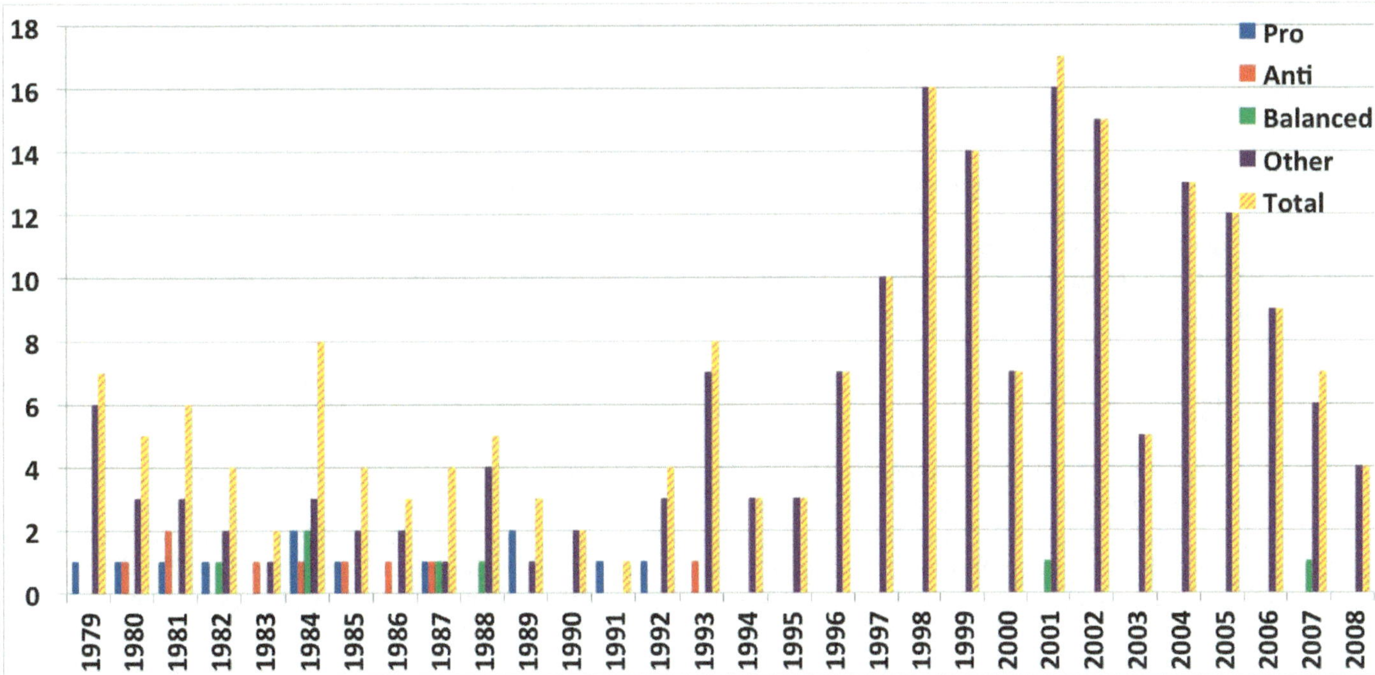

Figure 2. "Research" section coverage of the impact theory of the extinction of the dinosaurs by the leading general science journal *Nature* from 1979 to 2008.

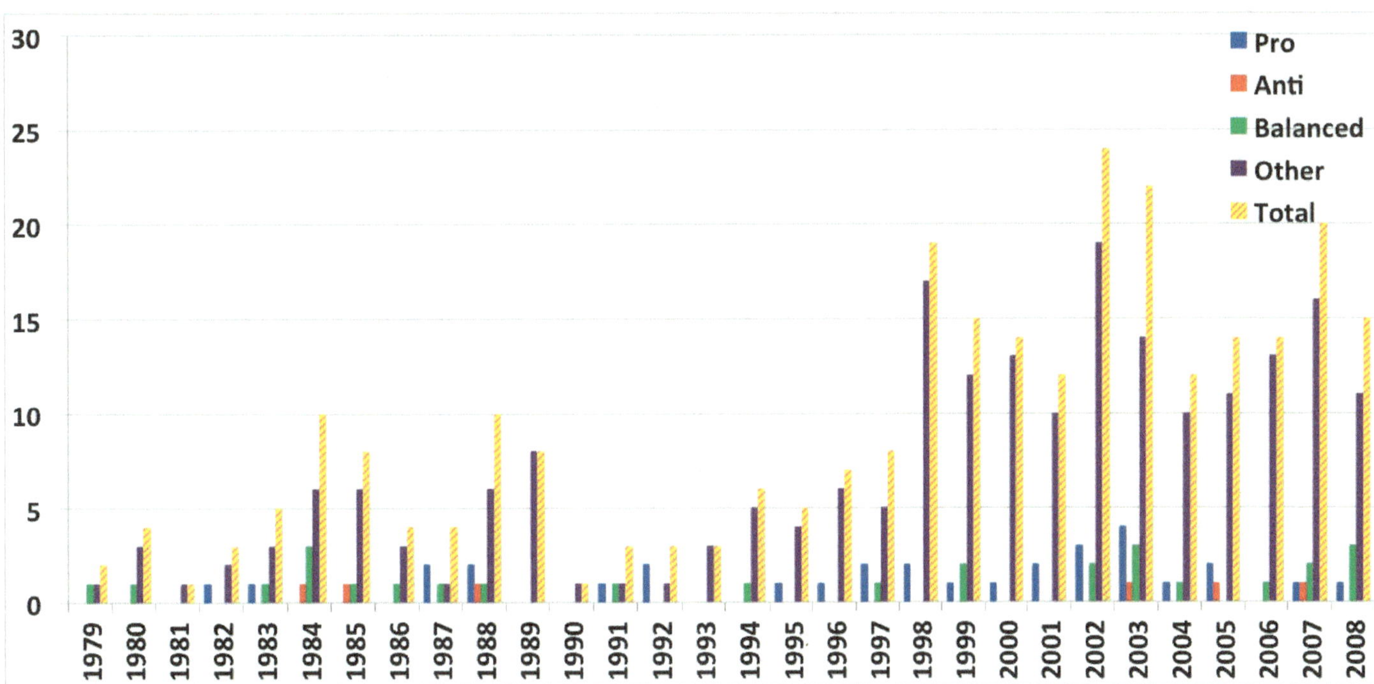

Figure 3. "News" section coverage of the impact theory of the extinction of the dinosaurs by the leading general science journal *Nature* from 1979 to 2008.

Dorothy Nelkin's (1987, p. 111–112) groundbreaking study of science in the press emphasized the importance of "breaking news" and the urgency with which journalists had to work as factors that greatly influenced what could be covered, and in how much detail. Gregory and Miller (1998) built on the work of several media scholars to set out eight categories of news values. Table 1 lists these and picks out those that came into play in making Sullivan's (1980) article "newsworthy." Table 1 shows that the article hit many, if not all, of the key news value groups that journalists are looking for in a story. Most of these values continued to play a part in the popularization of research into the Cretaceous-Tertiary mass extinctions as time went on.

SHOEMAKER-LEVY 9

It was a comet with attitude and a problem. Captured by Jupiter sometime in the 1920s, skimming too close to the surface of the giant planet in July 1992, Shoemaker-Levy 9 (SL9) was in fragments and on its final orbit when it was discovered by Carolyn and Gene Shoemaker and David Levy on 23 March 1993 (Shoemaker and Shoemaker, 1995). Once the orbital parameters of what Carolyn Shoemaker called the "squashed comet" were determined a few weeks after its discovery, it was clear that the 20-plus fragments would collide with Jupiter during the week of 16–23 July 1994, creating a series of multi-megaton explosions (Marsden, 1995), although there was public and scientific controversy as to whether these would be real "bangs" or just "whimpers" (Radford, 1994; Weissman, 1994; Zahnle, 1995).

With a year to prepare, the international planetary science community was able to put together a major campaign to make use of every available telescope to watch Jupiter during that critical week (A'Hearn, 1995). Impact Week itself put to rest any doubts as to whether or not the collision would produce bangs or whimpers: There was the odd "whimper" as a fragment that was not very large or made of dust rather than rock-ice fizzled out, but for the most part it was "bangs" (Spencer, 1995). Jupiter was seen to be ringed with dark impact sites, some large enough to stretch from Washington to the mid-Atlantic, had they happened on Earth (Spencer, 1995; Chapman, 1995). The impacts had superheated and shocked the atmosphere of Jupiter for thousands of kilometers around the impact sites, creating chemical species new to the giant planet, throwing dust and debris high into the atmosphere, and even firing some of it off into space (Spencer, 1995; Dinelli et al., 1997). The immediate results filled much of the research section of one issue of *Science* (1995, v. 267), and there are now well over 1000 scientific papers that derive wholly or in part from the comet and its impacts, according to the National Aeronautics and Space Administration (NASA) Astrophysics Data System that logs them.

Prior to and during the run-up to the SL9 collisions, the scientific community was itself assessing the extent to which our own planet might be at risk from impacts. There were four key workshops between 1991, after the discovery of the Chicxulub crater, and 1993, just after SL9 had been discovered, that gave rise to a considerable volume of scientific work (Gehrels, 1994). In January 1994, *Nature* published a key review article by Clarke Chapman and David Morrison that estimated the annual likelihood of the average American being killed by a comet or asteroid impact to be somewhere between 1/3000 and 1/250,000, with a mean of 1/20,000, making it comparable to dying in an airplane crash. They commented:

> The impact hazard must be considered in parallel with, and balanced against, debates over society's priorities in dealing with other potential ecological disasters and hazards in general... Thus, by choosing whether or not to do something about this threat from the skies, society may establish a standard against which its responses to other hazards are measured. (Chapman and Morrison, 1994)

During 1993–1994, newspaper articles on SL9 ranged from features preparing for the impacts, such as the *Independent*'s "By Jupiter, What a Bang" (Miller, 1993) and the *New York Times*' "Comet to Hit Jupiter with a Texas-Sized Bang" (Wilford, 1993). John Noble Wilford (1993, p. C1) told New Yorkers:

> The astronomers are eagerly anticipating their first opportunity next July to observe explosive impacts on a planet that could be comparable in strength to the collision of a large asteroid or comet with Earth 65 million years ago, the catastrophe implicated in the mass extinctions of dinosaurs and many other species of life.

Assessing the likely scale of the impacts at somewhere between 1 and 100 million megatons of trinitrotoluene (TNT), Steve Miller (1993, p. 10) explained in the *Independent*:

> To put the collisions in perspective, in 1908, our Earth ran into an asteroid. Exploding high above the ground, the mere 10 megatons of energy this collision produced were enough to flatten thousands of square kilometres of forest in the wilds of Siberia. At the other end of the scale, the comet that crashed into the Earth some 65 million years ago—and so changed our climate that the dinosaurs were wiped out—is reckoned to have produced a 100 million megaton detonation. Whatever the precise outcome, no one doubts that we will end up with a better understanding of the catastrophes which befall planets from time to time. How nice to be able to watch it happen on someone else's world. But comets are among the most unpredictable inhabitants of our solar system. One day it could be us.

As the comet fragments struck the giant planet (Fig. 4), all the major newspapers carried blow-by-blow reports. The *Sydney Morning Herald* told readers "Jupiter Under Bombardment" and had SL9 codiscoverer Eugene Shoemaker declaring: "For us it's *Star Wars*, only it's for real" (McCarthy, 1994, p. 1). A day later, Shoemaker was telling *USA Today* that "a fragment called G flared as bright as Jupiter itself on heat-measuring telescopes" (Hoversten, 1994, p. 3A), which he estimated to be equivalent to 6 million megatons of TNT for the Canadian *Globe and Mail* (Immen, 1994).

As far as the implications of the impacts were concerned for the safety of humanity from asteroid or cometary impacts (Fig.

TABLE 1. NEWS VALUES AFTER GREGORY AND MILLER (1998) APPLIED TO WALTER SULLIVAN'S (1980) ARTICLE

News value	Brief description	Application to Sullivan (1980) article
Threshold	A story has to be sufficiently important to make it into the press. Local newspapers may require a "lower" threshold, involving proximity, than national or international press outlets.	*Yes:* Wiping out a large part of creation—even if it happened 65 m.y. ago—is a pretty big deal. And (some) dinosaurs were very big.
Meaningfulness, relevance, and consonance	A story has to have some meaning to its readership and be as relevant as possible. It often helps if the story chimes with what readers are already thinking.	*Yes(-ish): Nova* (WGBH, 1981) and BBC *Horizon* (BBC, 1981) both made programs relevant with claims humans would (probably) not have evolved had the dinosaurs still been around. To what extent Sullivan's readers would already have been aware of that theory, however, may be questionable, and it is doubtful that, until they opened the newspaper, many were thinking "now, I wonder what happened to the dinosaurs" as they drank their morning coffee.
Co-option and composition	Many stories that may not quite make it through the threshold barrier on their own get into print nonetheless because they can tie into other, ongoing, "big" stories. At present, anything to do with climate change gets coverage. Newspapers also have their own composition and sections: so a science story may not be "big" enough for the main pages, but it may make it into the science section.	*Not really:* At the time of Sullivan's article, there did not seem to be much debate in public about the death of the dinosaurs, although the story could play into discussions around "nuclear winter" that were around in the 1980s. In later stories, however, co-option was an important news value. Sullivan's article was not in a special science section, so not really a question of composition, although two medical articles accompanied it on the page.
Frequency, unexpectedness, and continuity	Although the press is used to dealing with irregular items, some—like weekly football results—have a regular frequency. Journalists covering the science beat have to invest time to research a story properly and like the idea of one that may continue for some time. That said, there is nothing like taking the public by surprise to get readers.	*Not really, for the first two:* But the controversy implied in Sullivan highlighting two rival theories meant that he might expect to be writing more than once about the subject, as more evidence—one way or another—came to light. Sullivan himself had written a popular geology book (Sullivan, 1974), so had a personal "continuity" factor.
Competition	Journalists like to get exclusive stories and hate to miss "big" ones.	*No and yes:* The Alvarez et al. (1980) paper was published openly in *Science*, and any journalist could have used the story had they so wished. However, Sullivan seems to have been one of very few journalists to take up this story.
Unambiguity and negativity	Journalists prefer straight "yes" or "no" answers. Press coverage of the "no" answers—the government has NOT done what it should—predominate over "yes" articles.	*Yes(-ish) to both:* The Alvarez et al. paper was very definite that it had *the* answer to the Cretaceous-Tertiary extinctions, ruling out other scenarios for the iridium-rich clays found at the boundary. Sullivan's article, however, offers its readers two rival theories (including lunar volcanism). Again, wiping out a large percentage of creation is pretty negative.
Facts, sources, and their reliability	Authenticity is very important for journalists, and being able to cite facts, or to quote reliable people giving facts.	*Yes:* The fact that Alvarez et al.'s article was published in the prestige peer-reviewed journal *Science* gave it great credibility, and it was full of facts, including numbers and calculations.
Elitism and personalization	Humans like to read about other humans, especially if they are "famous."	*Up to a point:* Luis Alvarez enjoyed some popular recognition in the 1980s as a Nobel prize–winning physicist who had worked on the Manhattan Project and so would have been fairly well known to older readers. Other team members were not in the public limelight at the time.

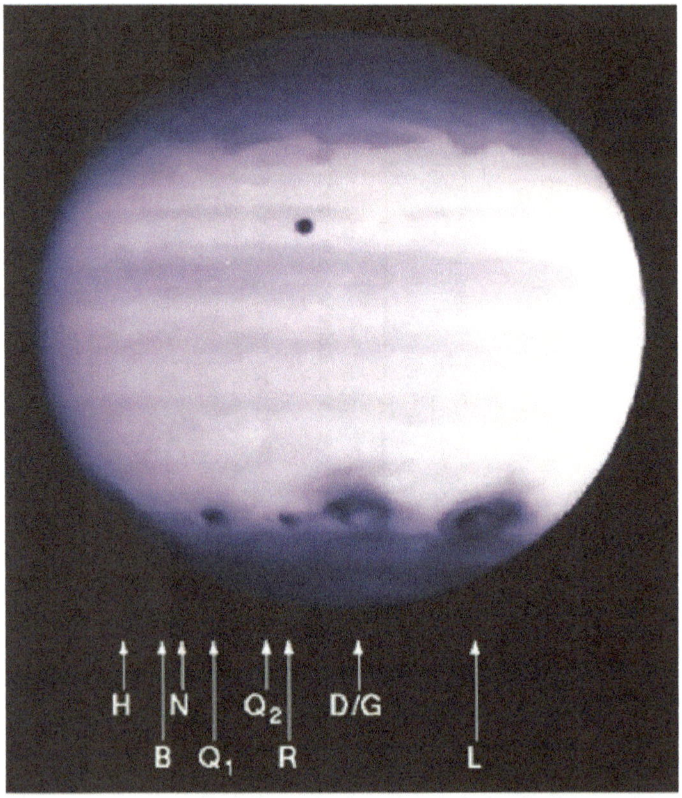

Figure 4. Hubble space telescope image of Jupiter immediately after Impact Week, with the impact sites denoted by letters. Credit: NASA.

Figure 5. The impact site of Fragment G transposed onto an image of Earth and centered on Washington DC. Credit: NASA.

5), Tim Radford (1994, p. 10) was typical of many: Writing in the *Guardian*'s "Serial Killers from Heaven," he revealed:

> Two scientists from Arizona and California this year calculated that there was a one in 10,000 chance of a 2 km diameter comet or asteroid colliding with the Earth in the next century, killing a very large section of humanity. A third calculated that the chances of death from an asteroid was about the same as death in an airline accident. This school of thought is called catastrophism. It argues that the planet, and with it the solar system, are not as they are because of the slow accretion of processes over time: they are also subject to periodic and catastrophic bumps.

In the middle of Impact Week, *Nature* opined:

> The greater than expected impact of Comet Shoemaker-Levy 9 as it crashed into Jupiter this week is set to intensify the debate over what steps, if any, should be taken to deal with the prospects of such an object striking the Earth. (MacIlwain and Verrell, 1994)

Nature noted that the proposal by the Spaceguard group, to build six 2.5 m telescopes dedicated to watching out for rogue asteroids or comets potentially on a collision course with our home planet, would receive a new impetus, although some leading scientists such as Carl Sagan were concerned about the use of nuclear weapons as part of a planetary defense strategy. So SL9 played into several preexisting scientific debates that had already been covered by several newspaper articles.

LONG-TERM NEWSPAPER COVERAGE OF THE DEATH OF THE DINOSAURS

While Clemens (1986) looked at a variety of popular publications, many of which clearly labeled themselves as "science" (e.g., *Scientific American*), another indicator of the popularization of theories concerned with the death of the dinosaurs can be obtained from looking at the general media. It has long been understood that, although newspapers may play a role in educating their readers, they operate under constraints that limit their ability to do so (for extensive and classic discussions of these issues, see Nelkin, 1987; Friedman et al., 1986). Privately owned newspapers are, fundamentally, commercial enterprises (witness the recent sale of the *Washington Post*). They may also be seen as cultural indicators, for our purposes giving an indication of the extent to which ideas from science have "diffused" into the wider citizenry, and can be referenced in such a way that all understand the allusion. So this article makes use of the Nexis database of "Major World Newspapers" (using the UK version of Nexis, to which the author has access) as a proxy for this popularization process, sampling newspapers for 30 years from the period 1 January 1979 to 31 December 2008. Thus, this general newspaper sample goes more than 20 years beyond Clemens' original (1986) work.

In this section, I set out to quantify the coverage of Alvarez' theory of the extinction of the dinosaurs, in terms of the number of articles published each year as a way of tracing both the "educational" and "cultural indicator" roles of general newspapers. In addition, I wish also to measure the extent to which the articles linked this "story" to events surrounding the planet Jupiter. Initially, therefore, two searches were carried out using the terms "dinosaur" and "impact/collision" and "term," where the final term was "asteroid/comet" (search 1), covering the key elements of the Alvarez impact theory, and "Jupiter" (search 2) in order, particularly, to uncover articles that linked the demise of the dinosaurs to the SL9 story.

For search 1, Nexis delivered 1770 returns. This sample was then refined by using only the 20 newspapers that returned 25 articles or more, so that each newspaper had the ability to generate a reasonable time series. The selected newspapers and the numbers of articles retrieved from Nexis for the two searches are shown in Table 2. Of the selected newspapers, five are from the United States, and six are from Australia (including one from Tasmania, the *Hobart Mercury*). Four have UK-wide coverage, and two more British papers are more restricted to Scotland. There are two Canadian papers, and the *Irish Times* is also included. So our sample is representative of the English-speaking world, but for some reason, the UK version of Nexis does not pick up major U.S. West Coast newspapers such as the *Los Angeles Times* or the *San Francisco Chronicle*. (These newspapers appear to have carried the first coverage of the Alvarez impact theory, in an article on 25 June 1979, reporting on Alvarez' discovery of the link between the iridium layer at Gubbio and the Cretaceous-Tertiary [K-T] boundary.)

Since we are interested in widespread public acceptance of the impact theory of the death of the dinosaurs, the selected articles are not limited to those that address this question directly. Some do, as in the case of Sullivan's (1980) original article reporting on the Alvarez et al. (1980) paper or the Australian *Advertiser*'s "Victorian Fossil May Smash Theory on Dinosaurs' Demise" (Australian Associated Press, 1986). However, others simply make reference to this issue as part of another topic that is their main focus, as in the case of the articles dealing with SL9 cited earlier or Dick Ahlstrom (1995) writing in the *Irish Times*, "Rare comet coming to the sky near you," alerting readers to the bright comet Hale Bopp. Articles in our sample ranged from short news items or announcements, some under 200 words, such as the *Daily Mail*'s warning that asteroids heading toward Earth might be "indestructible" (Associated Press, 1998), or *USA Today*'s report on efforts at Tel Aviv University to characterize asteroids (Weise, 2008). Others were well over 1000 words in length: in the *New York Times*, John Noble Wilford (1983) had nearly 2000 words to report in detail on the work of William Clemens at Hell Hollow, and how it contradicted the impact theory of the death of the dinosaurs; Stephen Mcginty (2002) used nearly 1800 words on page 2 of the *Scotsman* to try to persuade his fellow citizens that the danger of an impact ought to be of more importance to them than Mick Jagger's knighthood. Articles that had search terms completely unassociated with one another were manually rejected, but all others were retained, irrespective of their main topic, their position in the newspaper, or their overall length. Moreover, I placed no lower limit on the number of words or the percentage of the article that had to be devoted to the death of the dinosaurs: An article that simply contained a phrase such as "… wiped out the dinosaurs …" in a story dealing with the threat to Earth from asteroid impacts (e.g., Mcginty, 2002) was taken as relevant, "culturally," if not "educationally," as well as implicitly endorsing the impact theory of their demise (see following).

Table 2 shows that the *New York Times* (USA) produced the most articles for both searches, followed by the *Guardian* (UK) for search 1, and *The Times* (UK) for search 2. The *New York Times* produced nearly eight times as many articles as the least productive newspaper in our sample—the *Philadelphia Enquirer*—for search 1, and over 14 times as many as the least

TABLE 2. NEWSPAPERS USED IN ANALYSIS OF THE MEDIA COVERAGE OF THE DEATH OF THE DINOSAURS

Newspaper (Country)	Publication	Search 1	Search 2
New York Times (USA)	Continuous	198	43
Guardian/Observer (UK)	Continuous	136	23
Washington Post (USA)	Continuous	127	26
The Times (including Sunday) (UK)	Continuous	119	35
Independent (UK)	From 1986	91	23
Globe and Mail (Canada)	Continuous	88	16
Courier Mail (Australia)	Continuous	73	24
Mail (UK)	Continuous	57	13
Toronto Star (Canada)	Continuous	55	18
Sydney Morning Herald (Australia)	Continuous	53	19
Washington Times (USA)	From 1982	48	3
USA Today (USA)	From 1982	44	19
Advertiser (Australia)	Continuous	44	8
Irish Times (Ireland)	Continuous	42	13
Hobart Mercury (Tasmania)	Continuous	40	11
Herald Sun (Australia)	From 1990	39	6
The Age (Australia)	Continuous	38	11
Scotsman (Scotland)	Continuous	37	12
Glasgow Herald (Scotland)	Continuous	37	6
Philadelphia Enquirer (USA)	Continuous	25	8
TOTAL		1391	337

productive—the *Washington Times*—for search 2. The *New York Times* published at least one article on the demise of the dinosaurs every year from 1980 to 2008, with its coverage peaking in 1996 with 22 articles. In terms of consistency, the *Washington Post* (USA) missed just 2 years (1982 and 2008) in our sample, followed by the *Globe and Mail* of Canada, which returned in all but 1982, 1991, and 1999, and the *Guardian*, which returned articles starting in 1984, and from then on failed to publish an article only in 1987 and 1990. The consistency of the "elite" newspapers contrasted with the record of some of the more "down market" in our sample: *USA Today* published nothing prior to 1989, and then missed 1996 and 1999, and the *Mail* (UK) started in 1992, and missed 1995 and 1996.

Figure 6 shows the total number of articles published for our two searches (search 1, red; search 2, green), along with the articles that overlapped in the two searchers (purple). Allowing for this overlap, Figure 6 also shows the total numbers returned from Nexis for the total search on the term asteroid/comet/Jupiter. Of the 20 chosen newspapers, 16 have continuous publication throughout the 30 years from 1979, two commenced publication in 1982 (*USA Today* and *Washington Times*), the UK *Independent* began life in 1986, as a breakaway from *The Times*, and the Australian *Herald Sun* started in 1990. I have allowed for this by weighting the numbers of articles by 20 divided by the number of sample newspapers that were publishing. (In practice, this changes overall numbers by a maximum of two in any one year.)

It is possible to divide the period covered by the sample into three: 1979–1988, 1989–1998, and 1999–2008. In the first period, numbers of articles selected by search 1 rise fairly steadily to reach 20+ per year by 1985, and then level off. During this period, only eight of the selected newspapers (44%) published search 1 articles, with only the *New York Times* publishing at least one article every year. During the next decade, article numbers increase steeply, with the percentage of newspapers having published at least one article rising from 58% in 1989 to 100% by the end of 1994. As a result, total numbers increase by (nearly) a factor of 5 from 24 in 1989 to 119 in 1998. From 1999 to the end of the sample period, the general trend is for the number of articles to go down to the high forties in 2007 and 2008. As Table 3 shows, overall numbers of articles in each of the three decades went from 118 (1979–1988) to 658 (1989–1998) and then dropped slightly to 615 (1999–2008).

Several factors from within the relevant scientific disciplines can be linked to the considerable increase in articles published during the middle decade of the sample. (Note that I discuss some of the features within science communication and journalism later in this paper.) As noted already, 1991 marked publication of the detection of the Chicxulub crater (Hildebrand et al., 1991; Sigurdsson et al., 1991). In March of 1993, Comet Shoemaker-Levy 9 was discovered on collision course for Jupiter the following July: Article numbers more than trebled from 25 in 1989 to 82 in 1994, and leaping from 40 in 1991 to 70 in 1992 and 1993.

Although individual years in the final decade never reached the high of nearly 120 in 1998, at an average of over 60 articles a year in the sample, it is clear that considerable interest in the demise of the dinosaurs, triggered by the events of the 1990s,

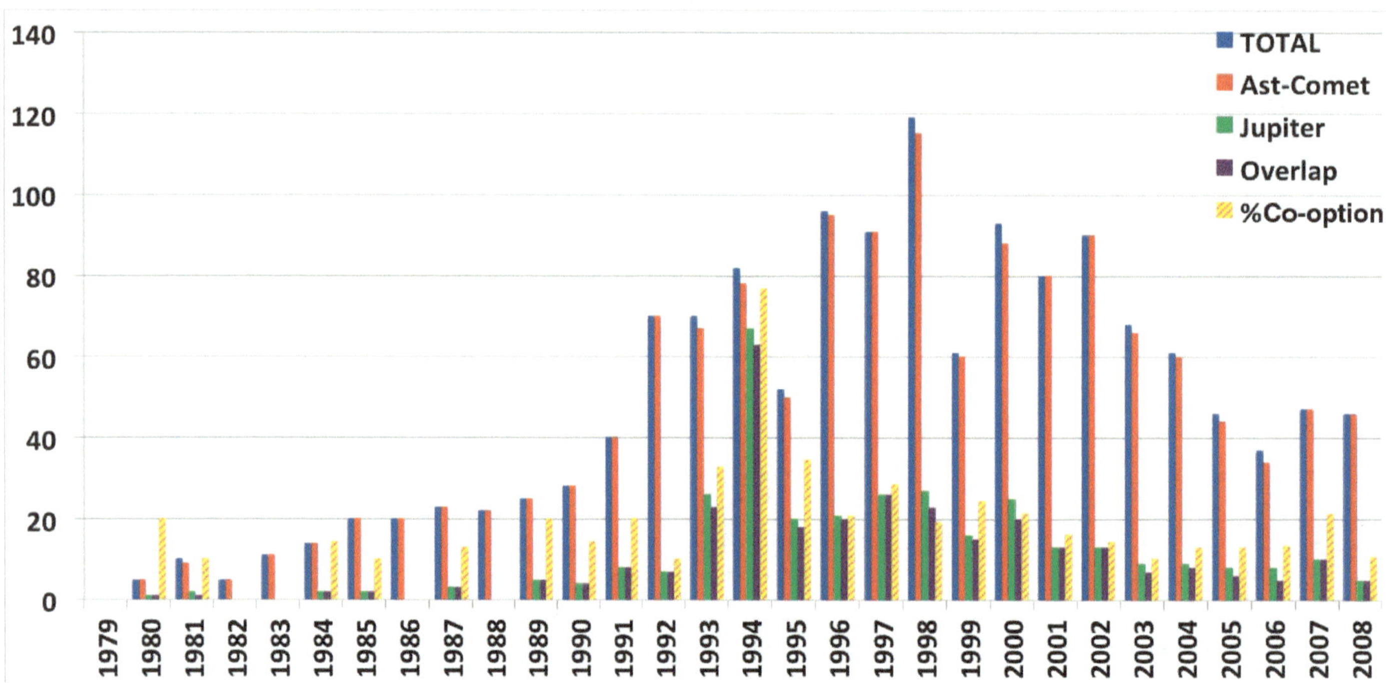

Figure 6. Numbers of newspaper articles in Nexis UK Major World Newspapers database from searches 1 and 2 (given in text). Numbers prior to 1982 are weighted by 1.25, between 1982 and 1986 by 1.11, and between 1986 and 1990 by 1.05, to allow for the number of publishing newspapers in the sample. Also shown is the overlap between the two searches and the percentage cooption, N_{coop}, defined in the text.

TABLE 3. NEWSPAPER COVERAGE BY DECADE AS SHOWN BY SEARCH 1

Newspaper (Country)	1979–1988	1989–1998	1999–2008
New York Times (USA)	51	93	54
Guardian/Observer (UK)	10	63	63
Washington Post (USA)	25	58	44
The Times (including Sunday) (UK)	4	55	60
Independent (UK)	0	46	45
Globe and Mail (Canada)	17	42	29
Courier Mail (Australia)	5	41	27
Mail (UK)	0	16	41
Toronto Star (Canada)	0	35	20
Sydney Morning Herald (Australia)	4	24	25
Washington Times (USA)	0	15	33
USA Today (USA)	0	28	16
Advertiser (Australia)	2	26	16
Irish Times (Ireland)	0	16	26
Hobart Mercury (Tasmania)	0	24	16
Herald Sun (Australia)	0	15	24
The Age (Australia)	0	24	14
Scotsman (Scotland)	0	9	28
Glasgow Herald (Scotland)	0	16	21
Philadelphia Enquirer (USA)	0	12	13
TOTAL	118	658	615

persisted. Within the sample, different newspapers showed varying behaviors. After the excitement of the middle decade, the *New York Times* settled back to its earlier coverage levels, but the UK's *Daily Mail*—always on the lookout for something to worry about—increased its reports two-and-a-half-fold in 1999–2008.

CO-OPTION AS A NEWS VALUE IN POPULARIZING THE DEATH OF THE DINOSAURS

One of the news values listed in Table 1 is that of "co-option," a value that (usually) involves adding one story to another, longer-running, narrative to get over the required "threshold" for publication and create more public interest. In this article, I am investigating the link between the media interest generated by the impact of Comet SL9 on Jupiter in July 1994 and the process of popularizing the Alvarez theory. As well as the numbers of articles returned by searches 1 and 2, and the total number of individual articles the combined searches produced, Figure 6 also shows the overlap between the two searches (purple), and a "percentage co-option" (orange and yellow stripe), defined as:

$$N_{coop} = [(\text{No. of Overlap Articles})/(\text{Total No. of Articles})] \times 100,$$

where the total number of articles is given by the sum of searches 1 and 2 less the number of Overlap Articles. For the first dozen years of our sample, N_{coop} remains at 20% or below, dropping to just 10% in 1992, as the total number of articles exceeds 50 for the first time. However, it increases to 33% in 1993, as SL9 is discovered, and goes to 77% in 1994, when it is clear that the majority of articles dealing with the death of the dinosaurs are linking the impacts on Jupiter with the impact(s) on Earth 65 m.y. previously; in 1995, where there is a postimpact lull in the total number of articles, N_{coop} is around 35%. Figure 6 shows that the percentage co-option remains at around 20% or above until the start of the new millennium, before falling back to around 10%; with the general increase in coverage, however, the number of articles showing considerable overlap is always higher in the last decade of our sample than it was in the first. It would appear therefore that the co-option wrought in 1993–1995 around the time of the SL9-Jupiter collisions had an effect that lasted for at least 5 years, and probably longer.

Another way of thinking about co-option, however, is that a shorter-lived but highly newsworthy "event" may provide a "peg" on which to hang articles about a longer running story: the UK royal birth of Prince George (22 July 2013) gave newspapers an opportunity to run large numbers of articles giving the proud couple (and their readers) parenting advice, a topic many newspapers cover on a regular basis. To assess this effect, we modified search 1 so that the first term was "Jupiter" rather than "dinosaur" coupled with "impact/collision" and "asteroid/comet" (search 3). The results are shown in Figure 7 on a year-by-year basis. The figure shows the numbers returned from search 3 across the sample of 20 newspapers, together with the overlap between those stories and those also containing the word "dinosaur." The figure shows that there was a very large peak in articles returned by search 3 in 1994, at the time of the SL9-Jupiter collision: 299 articles were found, i.e., four times as many as the next highest year, 73 in 1997. Once more, in Figure 7, I have shown the percentage co-option, N_{coop}. In this case, however, this is defined by:

$$N_{coop} = [(\text{No. of Overlap Articles})/(\text{No. of Search 3 Articles})] \times 100.$$

This results in some high values of N_{coop} in the years leading up to 1992, when the total number of search 3 articles is low (only

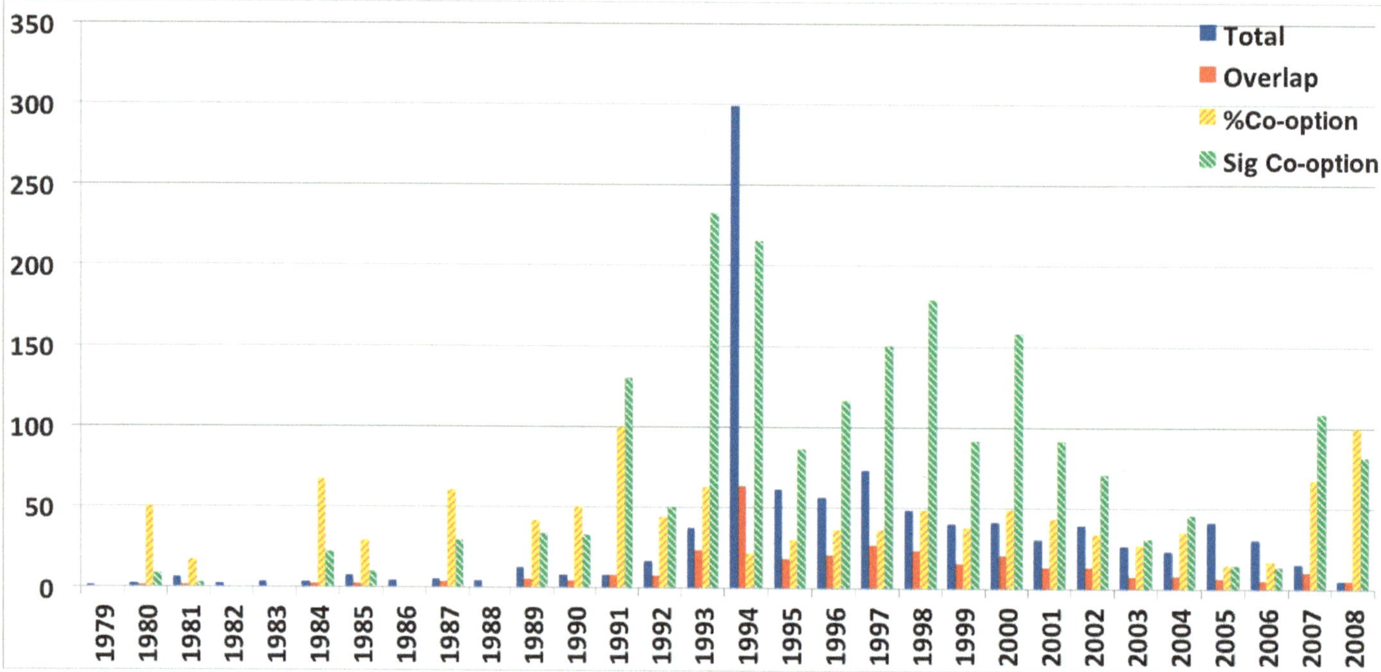

Figure 7. Numbers of newspaper articles in Nexis UK Major World Newspapers database from search 3 (given in text). Numbers prior to 1982 are weighted as Figure 8. Also shown is the overlap between the two searches and the percentage cooption, N_{coop}. The figure also shows "significant co-option," defined in the text. (Note that this has been scaled by 50, so as to be easily seen on this figure.)

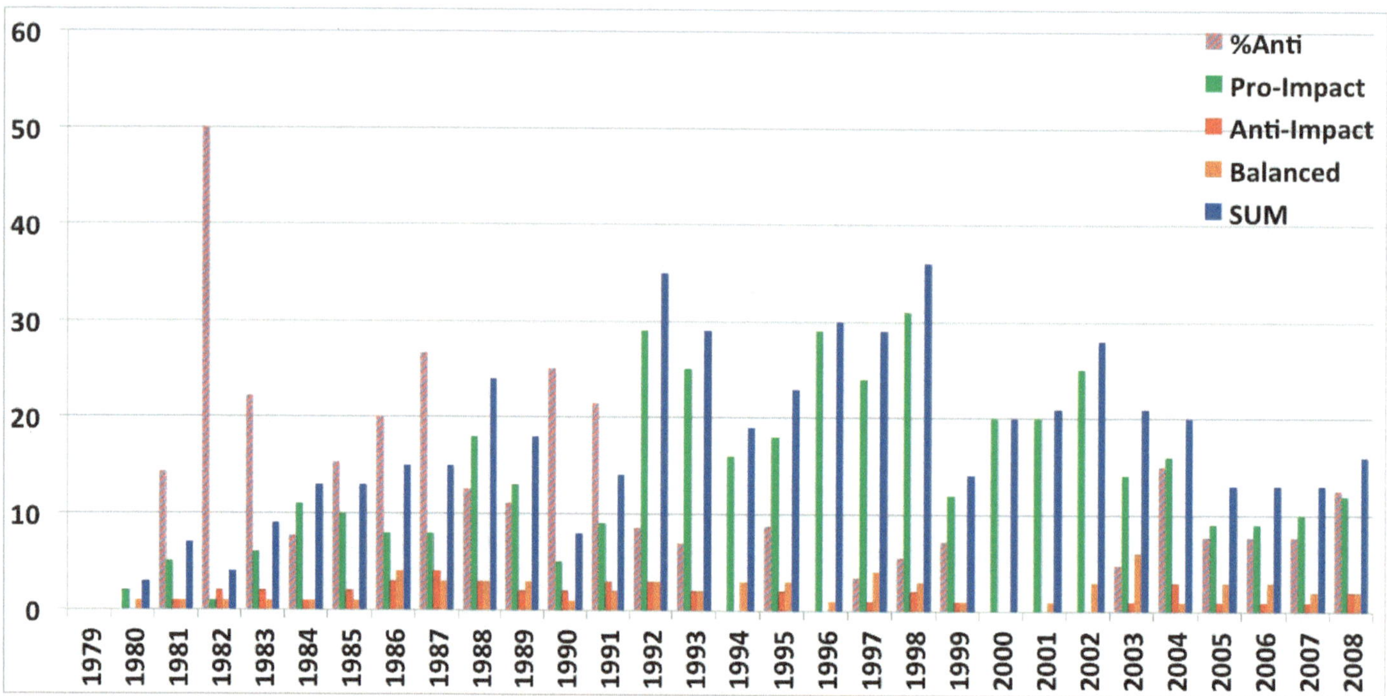

Figure 8. Coverage of impact theories of the demise of the dinosaurs in the *New York Times* (USA), the *Washington Post* (USA), the *Guardian* (UK) and the *Globe and Mirror* (Canada). For each year, the numbers giving coverage judged to be in favor of the impact theory (green), against (red), and balanced (orange) are shown, together with the percentage of articles against the impact theory (%Anti, blue and red diagonal stripe, obtained by dividing the number of "anti" articles by the total number published that year [pro, anti, and balanced], expressed as a percentage). The total number (SUM, blue) for each year is also shown.

after 1989 does it get above 10 in the sample), and relatively low values, when the overall numbers are large. However, the number of overlapping articles—i.e., those showing the co-option of the demise of the dinosaurs onto the stories about Jupiter and impacts (particularly that of SL9)—is highest in 1994, and I have attempted to illustrate the *significance* of this co-option by taking the number of overlapping articles into account, defining a "significant overlap" parameter:

$$\sigma_{\text{Coop}} = [N_{\text{Coop}} \times (\text{No. of Overlap Articles})/(\Sigma_{\text{All\_Years}} \text{ No. of Overlap Articles})],$$

where the denominator is the number of overlap articles in total, summed from 1979 to 2008. Here, σ_{Coop} peaks at over 4 in the SL9 years of 1993 and 1994, with a secondary peak of 3.5 in 1998 (the year of two Hollywood blockbusters discussed later). (Note that in Figure 7, σ_{Coop} has been scaled by 50 to make it clearly visible on the plot.) Between them, Figures 6 and 7 show that whether the co-option is defined as enhancing the news value of the short-lived SL9 events through its link with the long-running demise of the dinosaurs or vice versa, there is a significant effect. Clearly, this is most significant in the SL9 years (1993 and 1994), but the effect can be seen continuing for the next decade.

NEWSPAPER ATTITUDES TOWARD THE IMPACT THEORY

To look more closely at the message that readers were getting about the demise of the dinosaurs, we have analyzed the articles in the *New York Times* (U.S.), the *Washington Post* (U.S.), the *Guardian* (UK), and the *Globe and Mail* (Canada), four newspapers that had the most consistent coverage of the demise of the dinosaurs from 1980 until 2008. In order not to miss articles that left out any reference to "impacts" or "collisions," the search was widened to include a search on the terms "dinosaur" and "death/demise/extinction/wiped out" plus "volcano/volcanism," since this is the main rival explanation to impacts as the cause of the Cretaceous-Tertiary mass extinction. The results of this additional search were added to those found in search 1, with any overlapping articles counted just once, producing a combined search, search 4. In this widened search, the *New York Times* had at least one article in each of the years covered, and the *Washington Post* now missed just 1982. The *Globe and Mail* missed 3 years (1982, 1991, and 1999), while the *Guardian* started its coverage in 1984 and then missed just 1987 and 1990, according to the sample. Overall, 523 articles resulted from these four newspapers covering all of the search items outlined earlier.

The results of search 4 are shown in Figure 8, where I have plotted the total number of articles in our four newspapers on a yearly basis (SUM, blue). Each article was read and analyzed to be put into one of three categories, "pro-impact" (green), "anti-impact" (red), and "balanced" (orange). Pro-impact articles argued for, reported positively on others' arguments, or mentioned the impact explanation for the death of the dinosaurs as if it were a matter of fact. Anti-impact articles argued for alternative explanations to the impact explanation, criticized it (with or without reporting evidence against impacts), or reported others' criticisms of it. Balanced articles reported impact theories alongside alternative explanations, reported on impact theories while indicating that there were considerable doubts, or simply left unanswered what was the cause of the demise of the dinosaurs. Figure 8 shows the numbers of the articles by category on a yearly basis.

In all years, "pro" articles outnumbered "anti" articles, with the exception of 1982, which had the second lowest number of articles in the search 3 sample (4; lowest is 1980 with 3). Over the whole period from 1980 to 2008, I categorized 415 articles as pro-impact, 45 as anti-impact, and 63 as balanced. The year 1998 saw the highest number of articles in the sample, 36, with the highest number of "pro" articles, 31. The highest number of "anti" articles was published in 1987, with 4 out of 15 against the impact theory (still outnumbered by 8 in favor). To allow for the differing numbers of articles in each year, Figure 8 also gives the percentage of those articles (%Anti, blue and red diagonal stripe in Fig. 8) that were characterized as "anti." This is obtained by dividing the number of "anti" articles by the total number published that year (pro, anti, and balanced), expressed as a percentage. The years with the highest percentages of "anti" articles were 1982 (50%, 2 out of 4) and 1987 (27%, 4 out of 15). The average percentage of "anti" articles is 8.6% over the three decades.

Figure 8 shows that the time period can be split into two halves, centered on 1994, the first year after the original 1980 Alvarez et al. paper when *no* articles against the impact theory were published in any of the four newspapers. Up to 1993, the sample generated 207 articles, and after 1994, it generated 297, with 19 occurring in 1994 itself. In the 14 years from 1980 to 1993, only the first year has no "anti" articles; in the 14 years following 1994, there are four such years. (As pointed out already, 1994—the year of the SL9-Jupiter impacts—also had no "anti" articles.) In the first 14 years, half have percentage "anti" articles greater than the overall average of 8.6%, although numbers decline in 1992 (8.6%) and 1993 (6.9%) from higher values in the two preceding years (25% and 21%, respectively). In the second 14 years, there are just three years, 1995 (8.7%), 2004 (15%, the year that Keller et al. [2004a, 2004b] published evidence that the Chicxulub impact predated the Cretaceous-Tertiary mass extinction), and 2008 (12.5%, the year that Keller et al. [2008] published evidence linking the Cretaceous-Tertiary mass extinctions with Deccan Traps volcanism), when this happens. Consequently, the pre-1994 average of "anti" articles is 17.3%—twice the overall average—compared with just 5.7% afterward. After 1994, post-SL9, the landscape for journalism about the demise of the dinosaurs had clearly changed strongly in favor of the impact explanation.

CHANGING NEWSPAPER ENVIRONMENT FOR SCIENCE

Alongside the numbers of newspaper articles and their characterization outlined herein, it is also important to take account of changes in the newspaper environment for science that may affect levels and styles of reporting. The period from the mid-1980s to the mid-1990s saw the numbers of newspapers with science sections and the numbers of journalists on the "science beat" grow, although not evenly across the English-speaking world: in the U.S., 95 newspapers had their own science section by 1989 (Brumfiel, 2009), while in the UK, the number of science journalists doubled from this date to reach over 80 by 2005 (Williams and Clifford, 2009). This coincided with greater efforts to increase "scientific literacy" (U.S.) and "public understanding of science" (UK): The American Association for the Advancement of Science of published its "Project 2061: Benchmarks for Scientific Literacy" in 1993 (AAAS, 1993) following an intense 4-year long effort to work out what the average American should know about science by the time Comet Halley next returned (in 2061); in the UK, the Royal Society's "Report on the Public Understanding of Science" (Bodmer, 1985) had, amongst other things, recommended an increase in the amount of science that British newspapers should carry, and that scientists should "consider it their duty" to tell their fellow citizens what they were doing.

By the middle of the 2000s, however, commercial pressures and those from "new media" had led to a sharp decline in science sections in U.S. newspapers to 34 in 2004 (Brumfiel, 2009), while Williams and Clifford (2009) noted that the numbers of UK science journalists have remained static since around that time, with some downward pressure. Looking at Figure 6 and Table 3, one can see evidence for these trends at work in the case of the demise of the dinosaurs, although individual newspapers—such as the *Washington Times* (USA), the *Daily Mail* (UK), the *Herald Sun* (Australia), and the three newspapers from the Celtic fringe—go against this trend, at least until 2008.

It is far more difficult to explain away the change in attitude toward the impact explanation for the death of the dinosaurs (Fig. 8) solely by a changing media environment, unless one is going to argue that along with an expansion in science coverage, there was dumbing down to the extent that only one explanation was "allowed." The figures shown herein give little evidence for this, however: While the number of "anti" articles halved from 30 to 15 in the post-1994 period, the number of "balanced" articles increased from 27 to 36 in the four, somewhat "elite," newspapers I used in this part of the study, all of which have good coverage prior to and subsequent to 1994.

DOCUMENTARIES AND FILMS

The fact that the story sparked by Alvarez et al.'s paper was newsworthy and noteworthy was *sine qua non* for it to be turned into a television documentary or a film. Television as a medium is much more demanding in terms of time and effort than print; to justify such investment thus requires any story to be a proportionately "bigger deal." The first television depiction of the impact theory came quite soon after the Alvarez et al. (1980) paper was published: In 1981, the WGBH channel put out a documentary in its *Nova* series called "The Asteroid and the Dinosaur" (WGBH, 1981). Its publicity material announced: "For 150 million years, dinosaurs dominated the Earth. Then, 65 million years ago, they suddenly vanished, along with a great deal of the planet's animal and plant life. NOVA examines a remarkable theory about the cause of the catastrophe—in which the first clue to the solution was a piece of clay." Later that year, the BBC's flagship science documentary program *Horizon* broadcast *The Death of the Dinosaurs*, a 50-min-long program, much of it shot on location and making use of the latest graphics and modeling, in which the Alvarez team featured heavily, although not, ultimately, as the heroes of the hour (BBC, 1981). That honor went to Italian paleontologist Cesare Emiliani, who explained the "kill mechanism" in now-familiar impact-provokes-devastating-climate-change terms.

This program formed the basis of a detailed examination of television science by media researcher Roger Silverstone (1987). His analysis of television science in general centers on the use of "narrative strategies" and their accompanying "rhetorics." The "mythic strategy" is very much about telling stories about heroes and villains, facing challenges and trials, and their eventual triumph. In this case, "detectives" Alvarez, Alvarez, Asaro, and Michel face the mystery of the missing dinosaurs, with only the "smoking gun" of iridium-rich clays to mark where they came to grief. Undaunted, they use their forensic skills to solve the puzzle. However, from an initial high, the story turns downward into despair as every attempt to find the clinching evidence fails. Enter the new hero, Emiliani, who comes up with the solution, winning the day for science-kind against the alien invader.

Silverstone's "mimetic strategy" plots the way—using rhetorics of image, lighting and framing, and voice—to mimic the scientific process and the way scientists actually work. So the puzzle is set: Why and how did the dinosaurs die out, and what difference would it have made if they did not? Scientists are seen in the field, the office, and the analysis laboratory, and the evidence is analyzed and conclusions reached. However, Silverstone (1987, p. 314) warns:

> The narrative of a television programme, even of a documentary programme, has no *necessary* relationship to the world to which it refers ... The narrative is constructed on the basis of conventions and rules which are generally applicable, recognisable and effective...Television's texts are therefore not true but plausible (my emphasis).

Television's products can thus only give an indication of how the science is done and to what effect; the product seen is always a "dramatization." Nonetheless, the applicability of television to the debate around the death of the dinosaurs means it is a subject to which broadcasters have returned on many occasions.

During the 2000s, surveys have shown that adults get much of their general information and knowledge about science from the broadcast media (e.g., Commission of the European Communities, 2005), although the internet was fast increasing as a source of information about specific items (particularly health information). This article focuses on general newspapers as sources of information and as cultural indicators, and a detailed analysis of the way in which television, radio, and film industries have dealt with issues surrounding the demise of the dinosaurs is outside of its scope. However, a few comments are in order.

In the run-up to the SL9 impacts, during, and after them, television and radio news programs around the world broadcast items about the science of the event, the biter and the bit, many of which made at least passing reference to impact causes for the Cretaceous-Tertiary mass extinctions. These are so many and—by their nature—so ephemeral, and often embedded in newsroom conversations rather than actual items, as to be impossible to quantify and analyze. However, television documentaries do give a picture of changing media—and by implication, public—perceptions.

Prior to the impact of SL9 with Jupiter, only the 1981 *Nova* (WGBH, 1981) and *Horizon* (BBC, 1981) documentaries had dealt with the death of the dinosaurs, both of them publicizing the work of Alvarez et al. (1980), although not uncritically. After a decade during which the program-makers ignored the subject, SL9 brought on a flourish of impact-related documentaries—eight in the following decade—all of which made mention of the death of the dinosaurs, if not the Alvarez team explicitly. The majority of these—explicitly or implicitly—endorsed the impact theory, particularly if the program centered on threats to Earth (e.g., "Hunt for the Doomsday Asteroid," BBC, 1994). Additionally, in 1998, Hollywood released two blockbusters—*Deep Impact* and *Armageddon*—both of which had a comet en route to hit Earth as their central plotline. Scripted into those films were allusions to the Cretaceous-Tertiary extinctions, pretty much uncritically supporting impacts as the killer blow.

So what of alternative explanations? In 2000, the BBC's *Horizon* series had screened "Supervolcanoes" (BBC, 2000), which centered on the mass killing of rhinoceroses in Nebraska some 10 m.y. ago as the result of a supervolcanic eruption 1600 km away, but this did not mention the Cretaceous-Tertiary extinctions, despite 25 years of scientific papers that linked the Deccan Traps eruptions with them (McLean, 1985; Courtillot et al., 1986). Not until 2004, 10 years after SL9, when "What Really Killed the Dinosaurs?" (BBC, 2004) was broadcast, did a major counter-theory to the original Alvarez et al. (1980) asteroid impact explanation get significant air time. In that, U.S. researcher Gerta Keller was able to demonstrate some of the inconsistencies of the single impact killer theory, ideas she had been championing for well over a decade (e.g., Ward et al., 1995). That said, in "Asteroids—the Good, the Bad, and the Ugly," *Horizon* once more cast impacts by these bodies as the killers responsible for the Cretaceous-Tertiary extinctions (BBC, 2010). In the broadcast media, even more so than in print, impacts really do have impact.

CONCLUSIONS

A fringe often builds up around major news events. SL9, once it became publicly well known, gave rise to several. One self-styled "astronomer and plain-clothes nun," who called herself Sister Marie Gabriel, decided that what was going to happen to Jupiter was a wake-up call to the world, its religious and secular leadership in particular (Gabriel, 1994). After the impacts, pigeon racers (known as fanciers) across the UK blamed the loss of their favorite birds to what was happening to Jupiter (Wainwright, 1994), and the cartoon world had a field day. A tongue-in-cheek analysis of the reason for the extensive media coverage of SL9 came in the form of a *Guardian* cartoon (see Fig. 9), according to which journalists were left with nothing to fill their pages during the summer months when various legislative bodies were in recess, and the soccer World Cup had finished.

As noted at the time in *Nature* (MacIlwain and Verrell, 1994), SL9 also played into the policy discussions around planet

Figure 9. Cartoon in the *Guardian* newspaper explaining the reason for the extensive coverage of the collision of Comet Shoemaker-Levy 9 (SL9) with Jupiter. Reproduced by kind permission of Nick Newman and Ben Woolley.

protection from impact hazards, with the dinosaurs being cited as the way *not* to go. Some of this discussion may have been stimulated by astronomers keen to boost their own funding, if the work of Mellor (2010) is correct. That said, post-SL9 policy documents have taken on board the potential threats, and NASA and the European Space Agency have a joint program to track and—potentially—deflect or destroy potential impactors (e.g., Atkinson et al., 2000; European Space Agency Web site, 2014).

For the "general public"—and that includes many in the scientific community who are not intimately involved in debates around the numerous mass extinctions that have occurred during the history of life on Earth, as well as politicians—understandings about what was responsible for the major extinctions at the time of the Cretaceous-Tertiary boundary (whenever that is exactly dated) are formed by the sources of information available to them. In the wider scientific community, journals such as *Nature* and *Science* can be hugely influential in the case of cross-disciplinary debates (Clemens 1986, 1994), as well as the more mass-circulation media (Lewenstein, 1995).

Looking at the "Research" section of *Nature*, we conclude that the Alvarez et al. (1980) impact theory had been very much settled by the early 1990s, after the discovery of the Chicxulub crater and its accompanying tektites (Hildebrand et al., 1991; Sigurdsson et al., 1991; Smit, 1991). The general newspaper coverage examined in this article gives a somewhat different picture, however (as does the "News" section of *Nature*). Although it is clear that the impact theory was gaining public ground in the early 1990s as scientists found their "smoking gun" (Hildebrand et al., 1991; Sigurdsson et al., 1991; Smit, 1991), the watershed year in terms of the numbers of articles in the general newspapers in favor and against seems to be 1994, the year of SL9: In that year, for the first time since Alvarez et al. (1980) published their *Science* paper, there are no articles against the impact theory in the *New York Times*, the *Washington Post*, the *Guardian*, or the *Globe and Mail*, and the degree of co-option—SL9 onto the death of the dinosaurs and vice versa—measured across all 25 newspapers studied pivots on this year. The impact theory thus received considerable exposure throughout the general public as it was carried along with the mass media coverage of the impacts of Comet SL9 with Jupiter, and it is clear that this explanation for the death of the dinosaurs remains the main "public perception," insofar as nonspecialist citizens have a view on the matter and those views are reflected in the mass media, even if the SL9 effect has now faded.

ACKNOWLEDGMENTS

I would like to acknowledge the work of several students during the University College London course "Popularisation of the Physical Sciences," run either as HPSC C317 or HPSC 3010, by the Department of Science and Technology Studies, which augmented my own researches. In particular, thanks go to Anna Lewcock, Stephanie McGovern, and Becky Whitham, for their work on analyzing press coverage of the death of the dinosaurs and SL9, which prompted my own, wider, searches, to Nimish Lad and Nathaniel Hass, for work on the coverage by *Nature* and *Science*, which again I have extended, and to Hunain Dosani, Lucy Harmer, and Alison Somerville, for work on television documentaries. Linda Mao pointed out to me the role of Boss Tweed in disrupting the building of dinosaur statues in Central Park.

REFERENCES CITED

A'Hearn, M.F., 1995, The international observing campaign, *in* Spencer, J.R., and Mitton, J., eds., The Great Comet Crash: The Collision of Comet Shoemaker-Levy 9 and Jupiter: Cambridge, UK, University Press, p. 39–44.

Ahlstrom, D., 1995, Rare comet coming to the sky near you: Irish Times, 18 March 1995, p. 3.

Alvarez, L.W., Alvarez, W., Asaro, F., and Michel, H.V., 1980, Extraterrestrial cause for the Cretaceous-Tertiary extinction: Science, v. 208, p. 1095–1108, doi:10.1126/science.208.4448.1095.

American Association for the Advancement of Science, 1993, Benchmarks for Science Literacy: Oxford, UK, Oxford University Press, 448 p.

Associated Press, 1998, The Indestructible Asteroids: Daily Mail, 4 June 1998, p. 28.

Atkinson, H., Tickell, C., and Williams, D.A., 2000, Report of the Task Force on Potentially Hazardous Near Earth Objects: London, Her Majesty's Stationary Office, 59 p.

Australian Associated Press, 1986, Victorian fossil may smash theory on dinosaurs' demise: The Advertiser, 28 April 1986, p. 7.

BBC, 1994, Hunt for the Doomsday Asteroid: *Horizon*, v. 30.2, February 28.

Bodmer, W., Chair, 1985, The Public Understanding of Science: Report of a Royal Society ad hoc Group endorsed by the Council of the Royal Society: London, The Royal Society, 46 p.

Bramwell, V., and Peck, R.M., 2008, All in the Bones: A Biography of Benjamin Waterhouse Hawkins: Philadelphia, Pennsylvania, Academy of Natural Sciences, 128 p.

British Broadcasting Corporation (BBC), 1981, Death of the dinosaurs: Horizon, v. 18.8, 16 November 1981.

British Broadcasting Corporation (BBC), 2000, Supervolcanos: Horizon, v. 36.8, 3 February 2000.

British Broadcasting Corporation (BBC), 2004, What really killed the dinosaurs: Horizon, v. 41.4, 7 October 2004.

British Broadcasting Corporation (BBC), 2010, Asteroids—The good, the bad and the ugly: Horizon, v. 47.6, 3 November 2010.

Broks, P., 2006, Understanding Popular Science: Maidenhead, UK, Open University Press, 200 p.

Brumfiel, G., 2009, Supplanting the old media: Nature, v. 458, p. 274–277, doi:10.1038/458274a.

Burchell, J.D., 1975, Lord Kelvin and the Age of the Earth: London, Macmillan Press, 278 p.

Chapman, C.R., 1995, What if...?, *in* Spencer, J.R., and Mitton, J., eds., The Great Comet Crash: The Collision of Comet Shoemaker-Levy 9 and Jupiter: Cambridge, UK, Cambridge University Press, p. 103–108.

Chapman, C.R., and Morrison, D., 1994, Impacts on the Earth by asteroids and comets: Assessing the hazard: Nature, v. 367, p. 33–40, doi:10.1038/367033a0.

Clemens, E.S., 1986, Of asteroids and dinosaurs: The role of the press in the shaping of scientific debate: Social Studies of Science, v. 16, p. 421–456, doi:10.1177/030631286016003002.

Clemens, E.S., 1994, The impact hypothesis and popular science, *in* Glenn, W., ed., The Mass Extinction Debates: How Science Works in a Crisis: Stanford, California, Stanford University Press, p. 92–120.

Close, F., 1990, Too Hot to Handle: The Race for Cold Fusion: London, W.H. Allen, 167 p.

Commission of the European Communities, 2005, Eurobarometer 2005: Brussels, Commission of the European Communities, 336 p.

Courtillot, V., Besse, J., Vandamme, D., Montigny, R., Jaeger, J.-J., and Capetta, H., 1986, Deccan flood basalts at the Cretaceous/Tertiary boundary?: Earth and Planetary Science Letters, v. 80, p. 361–374, doi:10.1016/0012-821X(86)90118-4.

Dalton, R., 2003, Hot tempers, hard core: Nature, v. 425, p. 13–14, doi:10.1038/425013a.

Darwin, C.R., and Costa, J.T., 1859/2009, The Annotated Origin: A Facsimile of the First Edition of *On the Origin of Species*: Cambridge, Massachusetts, Harvard College, 286 p.

Davy, H., 1805/1980, Humphry Davy on Geology: The 1805 Lectures for the General Audience: Madison, Wisconsin, University of Wisconsin Press Ltd., p. 57–70.

Dinelli, B.M., Miller, S., Achilleos, N., Lam, H.A., Cahilll, M., Tennyson, J., Jagod, M.-F., Oka, T., Helico, J.-C., and Geballe, T.R., 1997, UKIRT observations of the impact and consequences of Comet Shoemaker-Levy 9 on Jupiter: Icarus, v. 126, p. 107–125, doi:10.1006/icar.1996.5630.

Fantasia, 1940, Disney, W., director, Burbank, California, Walt Disney Studios.

European Space Agency, Near-Earth Objects—NEO Segment, http://www.esa.int/Our_Activities/Operations/Space_Situational_Awareness/Near-Earth_Objects_-_NEO_Segment (accessed 25 October 2013).

European Space Agency, 2014, http://www.esa.int/ESA (accessed 25 October 2013).

Friedman, S.M., Dunwoody, S., and Rogers, C.L., eds., 1986, Scientists and Journalists: Reporting Science as News: New York, The Free Press, Macmillan, Inc., 333 p.

Fyfe, A., and Lightman, B.V., 2007, Science in the Marketplace: Chicago, Illinois, University of Chicago Press, 432 p.

Gabriel, Sister M., 1994, World news flash: An SOS! to all world leaders: Various UK newspapers, paid advertisement during February 1994.

Gehrels, T., ed., 1994, Hazards due to Comets and Asteroids: Tucson, Arizona, University of Arizona Press, 1300 p.

Glasstone, S., 1983, The Effects of Nuclear Weapons: Washington, D.C., Government Printing Office, 69 p.

Gomes, R., Levison, H.F., Tsiganis, K., and Morbidelli, A., 2005, Origin of the cataclysmic Late Heavy Bombardment period of the terrestrial planets: Nature, v. 435, p. 466–469, doi:10.1038/nature03676.

Gregory, J., 2003, The popularisation and excommunication of Fred Hoyle's "life from space" theory: Public Understanding of Science (Bristol, England), v. 12, p. 25–46, doi:10.1177/0963662503012001245.

Gregory, J., and Miller, S., 1998, Science in Public: Communication, Culture and Credibility: New York, Plenum, p. 110–114.

Hildebrand, A.R., Penfield, G.T., Kring, D.A., Pilkington, M., Camargo, A.Z., Jacobsen, S.B., and Boynton, W.V., 1991, Chicxulub crater: A possible Cretaceous/Tertiary boundary impact crater on the Yucatán Peninsula, Mexico: Geology, v. 19, p. 867–871, doi:10.1130/0091-7613(1991)019<0867:CCAPCT>2.3.CO;2.

Hilgartner, S., 1990, The dominant view of popularization: Conceptual problems, political uses: Social Studies of Science, v. 20, p. 519–539, doi:10.1177/030631290020003006.

Hoversten, P., 1994, Jupiter hit with a 'big wallop': USA Today, 19 July 1994, p. 3A.

Immen, W., 1994, Comet collision blinds scopes: The Globe and Mail, 19 July 1994, p. 1.

International Astronomical Union, 1993, Comet Shoemaker-Levy (1993e): Central Bureau for Astronomical Telegrams Circular 5725, March, 26 p.

Keller, G., Adatte, T., Stinnesbeck, W., Rebolledo-Vieyra, M., Urrutia Fucugauchi, J., Kramar, U., and Stueben, D., 2004a, Chicxulub crater predates K-T boundary mass extinction: Proceedings of the National Academy of Sciences of the United States of America, v. 101, no. 11, p. 3753–3758, doi:10.1073/pnas.0400396101.

Keller, G., Adatte, T., and Stinnesbeck, W., 2004b, More evidence that Chicxulub predates KT boundary: Meteoritics & Planetary Science, v. 39, p. 1127–1144, doi:10.1111/j.1945-5100.2004.tb01133.x.

Keller, G., Adatte, T., Gardin, S., Bartolini, A., and Bajpai, S., 2008, Main Deccan volcanism phase ends at K-T mass extinction: Evidence from the Krishna-Godavari Basin, SE India: Earth and Planetary Science Letters, v. 268, p. 293–311, doi:10.1016/j.epsl.2008.01.015.

Lewenstein, B.V., 1995, From fax to facts: Communication in the cold fusion: Social Studies of Science, v. 25, p. 403–436, doi:10.1177/030631295025003001.

Lightman, B.V., 2007, Victorian Popularisers of Science: Designing Nature for New Audiences: Chicago, Illinois, University of Chicago Press, 568 p.

Lyell, C., 1830, Principles of Geology: London, John Murray, 532 p.

MacIlwain, C., and Verrell, M., 1994, Comet collision boosts controversy over global protection strategy: Nature, v. 370, p. 165, doi:10.1038/370165a0.

Marsden, B.G., 1995, The path to destruction, *in* Spencer, J.R., and Mitton, J., eds., The Great Comet Crash: The Collision of Comet Shoemaker-Levy 9 and Jupiter: Cambridge, UK, Cambridge University Press, p. 8–18.

McCarthy, P., 1994, Jupiter under bombardment: Sydney Morning Herald, 18 July 1994, p. 1.

Mcginty, S, 2002, In the line of fire: The Scotsman, 28 June 2002, p. 2.

McLean, D.M., 1985, Deccan Traps mantle degassing in the terminal Cretaceous marine extinctions: Cretaceous Research, v. 6, p. 235–259, doi:10.1016/0195-6671(85)90048-5.

Mellor, F., 2003, Between "fact" and fiction: Demarcating science from nonscience in popular physics books: Social Studies of Science, v. 33, p. 509–538, doi:10.1177/0306312703334002.

Mellor, F., 2010, Negotiating uncertainty: Asteroids, risk and the media: Public Understanding of Science (Bristol, England), v. 19, p. 16–33, doi:10.1177/0963662507087307.

Miller, S., 1993, By Jupiter, what a bang: The Independent, 25 October 1993, p. 10.

Nelkin, D., 1987, Selling Science: How the Press Covers Science and Technology: New York, W.H. Freeman and Co., 217 p.

Noble, J.W., 1993, Comet to hit Jupiter with a Texas-sized bang: New York Times, 19 October 1993, p. C1.

Radford, T., 1994, Serial killers from space: The Guardian, 7 July 1994, p. 10.

Rudwick, M.J.S., 1985, The Great Devonian Controversy: The Shaping of Scientific Knowledge among Gentleman Specialists: Chicago, Illinois, University of Chicago Press, 528 p.

Secord, J.A., 2000, Victorian Sensation: The Extraordinary Publication, Reception, and Secret Authorship of Vestiges of the Natural History of Creation: Chicago, Illinois, University Of Chicago Press, 624 p.

Shoemaker, C.S., and Shoemaker, E.M., 1995, A comet like no other, *in* Spencer, J.R., and Mitton, J., eds. The Great Comet Crash: The Collision of Comet Shoemaker-Levy 9 and Jupiter: Cambridge, UK, Cambridge University Press, p. 7–12.

Sigurdsson, H., D'Hondt, S., Arthur, M.A., Bralower, T.J., Zachos, J.C., van Fossen, M., and Channell, J.E.T., 1991, Glass from the Cretaceous/Tertiary boundary in Haiti: Nature, v. 349, p. 482–487, doi:10.1038/349482a0.

Silverstone, R., 1987, Narrative strategies in television science, *in* Curran, J., Smith, A., and Wingate, P., eds., Impacts and Influences: Essays on Media Power in the Twentieth Century: London, Methuen and Co., Ltd., p. 291–330.

Smit, J., 1991, Where did it happen: Nature, v. 349, p. 461–462, doi:10.1038/349461a0.

Spencer, J.R., 1995, The Great Crash, *in* Spencer, J.R., and Mitton, J., eds., The Great Comet Crash: The Collision of Comet Shoemaker-Levy 9 and Jupiter: Cambridge, UK, Cambridge University Press, p. 55–95.

Sullivan, W., 1974, Continents in Motion: The New Earth Debate: New York, McGraw-Hill, 399 p.

Sullivan, W., 1980, Two New Theories Offered on Mass Extinctions in Earth's Past: New York Times, 10 June 1980, p. C3.

Turco, R.P., Toon, O.B., Ackerman, T.P., Pollack, J.B., and Sagan, C., 1983, Nuclear winter: Global consequences of multiple nuclear explosions: Science, v. 222, p. 1283–1292, doi:10.1126/science.222.4630.1283.

Turner, A., 1994, BBC-1 Blue Peter, 19 May 1994.

Wainwright, M., 1994, Pigeon fanciers blame Jupiter's big bang as birds go off course: The Guardian, 25 July 1994, p. 3.

Ward, W.C., Keller, G., Stinnesbeck, W., and Adatte, T., 1995, Yucatán subsurface stratigraphy: Implications and constraints for the Chicxulub impact: Geology, v. 23, p. 873–876, doi:10.1130/0091-7613(1995)023<0873:YNSSIA>2.3.CO;2.

Weise, E., 2008, Crash course in obliterating a danger: USA Today, 8 December 2008, p. 5D.

Weissman, P., 1994, The big fizzle is coming: Nature, v. 370, p. 94, doi:10.1038/370094a0.

WGBH, 1981, The asteroid and the dinosaur: Nova, 10 March 1981.

Wilford, J.N., 1983, Dinosaurs: Catastrophic theory is contested at Hell Hollow: New York Times, 16 August 1983, p. C1.

Wilford, J.N., 1993, Comet to hit Jupiter with a Texas-sized bang: New York Times, 19 October 1993, p. C1.

Williams, A., and Clifford, S., 2009, Mapping the Field: Specialist Science News Journalism in the UK National Media: Cardiff, UK, Cardiff University Press, 82 p.

Yeo, R., 1993, Defining Science: William Whewell, Natural Knowledge and Public Debate in Early Victorian Britain: Cambridge, UK, Cambridge University Press, 298 p.

Young, R.M., 1985, Darwin's Metaphor: Nature's Place in Victorian Culture: Cambridge, UK, Cambridge University Press, p. 126–165.

Zahnle, K., 1995, Bangs or whimpers?, *in* Spencer, J.R., and Mitton, J., eds., The Great Comet Crash: The Collision of Comet Shoemaker-Levy 9 and Jupiter: Cambridge, UK, Cambridge University Press, p. 29–37.

Manuscript Accepted by the Society 31 January 2014

Printed in the USA